Lecture Notes in Computer Science 12577

More information about this subseries at http://www.springer.com/series/7407

Weili Wu · Zhongnan Zhang (Eds.)

Combinatorial Optimization and Applications

14th International Conference, COCOA 2020
Dallas, TX, USA, December 11–13, 2020
Proceedings

Editors
Weili Wu
The University of Texas at Dallas
Richardson, TX, USA

Zhongnan Zhang (iD)
Xiamen University
Xiamen, China

ISSN 0302-9743   ISSN 1611-3349   (electronic)
Lecture Notes in Computer Science
ISBN 978-3-030-64842-8   ISBN 978-3-030-64843-5   (eBook)
https://doi.org/10.1007/978-3-030-64843-5

LNCS Sublibrary: SL1 – Theoretical Computer Science and General Issues

This Springer imprint is published by the registered company Springer Nature Switzerland AG
The registered company address is: Gewerbestrasse 11, 6330 Cham, Switzerland

Preface

The 14th Annual International Conference on Combinatorial Optimization and Application (COCOA 2020) was planned to take place in Dallas, Texas, USA, December 11–13, 2020. As a result of extenuating circumstances due to COVID-19, COCOA 2020 was held as a fully virtual conference. COCOA 2020 provided a forum for researchers working in the area of combinatorial optimization and its applications with all aspects, including algorithm design, theoretical and experimental analysis, and applied research of general algorithmic interest.

The Program Committee received a total of 104 submissions from 17 countries and regions, among which 55 were accepted for presentations in the conference. Each contributed paper was rigorously peer-reviewed by reviewers who were drawn from a large pool of Technical Committee members.

We wish to thank all authors for submitting their papers to COCOA 2020 for their contributions. We also wish to express our deepest appreciation to the Program Committee members and all subreviewers for their hard work within demanding constraints, so that each paper received two to three review reports. Especially, we wish to thank the Organization Committee members and conference sponsor from The University of Texas at Dallas, USA, for various assistance.

October 2020

Weili Wu
Zhongnan Zhang

Organization

Program Chairs

Weili Wu   The University of Texas at Dallas, USA
Zhongnan Zhang   Xiamen University, China

Program Committee

Smita Ghosh   Santa Clara University, USA
Xiao Li   The University of Texas at Dallas, USA
Hsu-Chun Yen   National Taiwan University, Taiwan
Bhadrachalam Chitturi   The University of Texas at Dallas, USA
Stanley Fung   University of Leicester, UK
Zhi-Zhong Chen   Tokyo Denki University, Japan
Michael Khachay   Krasovsky Institute of Mathematics and Mechanics,
                         Russia
Xiaoming Sun   Institute of Computing Technology, Chinese Academy
                         of Sciences, China
Vladimir Boginski   University of Central Florida, USA
Haipeng Dai   Nanjing University, China
Annalisa De Bonis   Università di Salerno, Italy
Weili Wu   The University of Texas at Dallas, USA
Tsan-sheng Hsu   Academia Sinica, Taiwan
Lidong Wu   The University of Texas at Tyler, USA
Yingshu Li   Georgia State University, USA
Weitian Tong   Eastern Michigan University, USA
Zhao Zhang   Zhejiang Normal University, China
Sun-Yuan Hsieh   National Cheng Kung University, Taiwan
Kun-Mao Chao   National Taiwan University, Taiwan
Yi Li   The University of Texas at Tyler, USA
Huaming Zhang   University of Alabama in Huntsville, USA
Pavel Skums   Georgia State University, USA
Zhiyi Tan   Zhejiang University, China
Viet Hung Nguyen   University of Auvergne, France
Ovidiu Daescu   The University of Texas at Dallas, USA
Yan Shi   University of Wisconsin-Platteville, USA
Chuangyin Dang   City University of Hong Kong, Hong Kong
Xianyue Li   Lanzhou University, China
Guohui Lin   University of Alberta, Canada
Suneeta Ramaswami   Rutgers University, USA
Jie Wang   University of Massachusetts Lowell, USA
Maggie Cheng   Illinois Institute of Technology, USA

Contents

Search, Facility and Graphs

Geometric Problem

Approximation Algorithms

Approximate Ridesharing of Personal Vehicles Problem

Qian-Ping Gu[1], Jiajian Leo Liang[1(✉)], and Guochuan Zhang[2]

[1] School of Computing Science, Simon Fraser University, Burnaby, Canada
{qgu,leo_liang}@sfu.ca
[2] College of Computer Science and Technology, Zhejiang University,
Hangzhou, China
zgc@zju.edu.cn

Abstract. The ridesharing problem is that given a set of trips, each trip consists of an individual, a vehicle of the individual and some requirements, select a subset of trips and use the vehicles of selected trips to deliver all individuals to their destinations satisfying the requirements. Requirements of trips are specified by parameters including source, destination, vehicle capacity, preferred paths of a driver, detour distance and number of stops a driver is willing to make, and time constraints. We analyze the relations between time complexity and parameters for two optimization problems: minimizing the number of selected vehicles and minimizing total travel distance of the vehicles. We consider the following conditions: (1) all trips have the same source or same destination, (2) no detour is allowed, (3) each participant has one preferred path, (4) no limit on the number of stops, and (5) all trips have the same earliest departure and same latest arrival time. It is known that both minimization problems are NP-hard if one of Conditions (1), (2) and (3) is not satisfied. We prove that both problems are NP-hard and further show that it is NP-hard to approximate both problems within a constant factor if Conditions (4) or (5) is not satisfied. We give $\frac{K+2}{2}$-approximation algorithms for minimizing the number of selected vehicles when condition (4) is not satisfied, where K is the largest capacity of all vehicles.

Keywords: Ridesharing problem · Optimization problems · Approximation algorithms · Algorithmic analysis

1 Introduction

As the population grows in urban areas, the number of cars on the road also increases. As shown in [20], personal vehicles are still the main transportation mode in 218 European cities between 2001 and 2011. In the United States, the estimated cost of congestion is around \$121 billion per year [3]. Based on reports in 2011 [6,18], occupancy rate of personal vehicles in the United States is 1.6 persons per vehicle, which can be a cause for congestion. Shared mobility (carpooling or ridesharing) can be an effective way to increase occupancy rate [5].

© Springer Nature Switzerland AG 2020
W. Wu and Z. Zhang (Eds.): COCOA 2020, LNCS 12577, pp. 3–18, 2020.
https://doi.org/10.1007/978-3-030-64843-5_1

Caulfield [4] estimated that ridesharing to work in Dublin, Ireland can reduce 12,674 tons of CO_2 emissions annually. Ma et al. [15] showed taxi-ridesharing in Beijing has the potential to save 120 million liter of gasoline per year. Systems that provide ridesharing services are known as mobility-on-demand (MoD) systems, such as Uber, Lyft and DiDi. These systems are cable of supporting dynamic ridesharing, meaning the ridesharing requests enter and leave the system in real-time. Although the analysis and algorithms discussed in this paper are for static ridesharing, one can view a dynamic ridesharing instance as a sequence of static ridesharing instances (computing a solution for a fixed interval, e.g. [19]).

We consider the following ridesharing problem: given a set of trips (requests) in a road network, where each trip consists of an individual, a vehicle of the individual and some requirements, select a subset of trips and use the vehicles of the selected trips to deliver the individuals of all trips to their destinations satisfying the requirements. An individual of a selected trip is called a *driver* and an individual other than a driver is called a *passenger*. The requirements of a trip are specified by parameters including: the source and destination of the trip, the vehicle capacity (number of seats to serve passengers), the preferred paths of the individual when selected as a driver, the detour distance and number of stops the driver is willing to make to serve passengers, and time constraints (e.g., departure/arrival time). There are different benefits of shared mobility [5], but the main goal considered in this paper is to reduce the number of cars on the roads. This can be achieved by the following optimization goals: minimize the number of vehicles (or equivalently drivers) and minimize the total travel distance of vehicles (or equivalently drivers) to serve all trips.

In general, the ridesharing problem is NP-hard as it is a generalization of the vehicle routing problem (VRP) and Dial-A-Ride problem (DARP) [17]. A common approach for solving the ridesharing problem is to use a Mixed Integer Programming (MIP) formulation and solve it by an exact method or heuristics [3,11,12]. MIP based exact algorithms are time consuming and not practical for large-scale ridesharing, so most previous studies rely on (meat)heuristics or focus on simplified ridesharing for large instances [1,13,19]. The optimization goals of the ridesharing problem are typically classified into two categories: *operational objectives* and *quality-related objectives* [17]. Operational objectives are system-wide optimizing goals, whereas quality-related objectives focus on the performance from the individual (driver/passenger) perspective. Some variants and mathematical formulation of ridesharing problem come from the well studied DARP. A literature review on DARP can be found in [16]. For the ridesharing problem, we refer readers to literature surveys and reviews [2,5,17].

Most previous works focus on computational studies of the ridesharing problems and do not have a clear model for analyzing the relations between the time complexity of the ridesharing problem and its parameters. A recent model was introduced in [7] for analyzing the computational complexity of simplified ridesharing problem with parameters of source, destination, vehicle capacity, detour distance limit, and preferred paths only. The work in [7] gives an algorithmic analysis on the simplified ridesharing problems based on the following conditions: (1) all trips have the same destination or all trips have the same source; (2) no detour is allowed; (3) each trip has a unique preferred path. It was shown in [7] that if any one of the three conditions is not satisfied, both minimization problems (minimizing the number of drivers and travel distance of vehicles) are NP-hard. When all three conditions are satisfied, Gu et al. [8] developed a dynamic programming algorithm that finds an exact solution for each minimization problem in $O(M + l^3)$ time, where M is the size of the road network and l is the number of trips. In [8], a greedy algorithm was also proposed that finds a solution with minimum number of drivers in $O(M + l \cdot \log l)$ time.

A closely related problem was studied by Kutiel and Rawitz [14], called the maximum carpool matching problem (MCMP). Given a set V of trips, the goal of MCMP is to find a set of drivers $S \subseteq V$ to serve all trips of V such that the number of passengers is maximized. It was shown that MCMP is NP-hard [10]. Algorithms are proposed in [14] with $\frac{1}{2}$-approximation ratio, that is, the number of passengers found by the algorithms is at least half of that for the optimal solution. The algorithms in [14] can be modified to $\frac{K+2}{2}$-approximation algorithms for the ridesharing problem with minimizing the number of drivers.

In this paper, we extend the time complexity analysis of simplified ridesharing problems in [7] to more generalized ridesharing problems with three additional parameters considered: number of stops a driver willing to make to serve passengers, arrival time and departure time of each trip. We introduce two more conditions: (4) Each driver is willing to stop to pick-up passengers as many times as its vehicle capacity. (5) All trips have the same earliest departure and same latest arrival time. We call Condition (4) the *stop constraint condition* and (5) the *time constraint condition*. Our results in this paper are:

1. We prove that both ridesharing minimization problems are NP-hard and further show that it is NP-hard to approximate both problems within a constant factor if stop constraint or time constraint condition is not satisfied.
2. We present two $\frac{K+2}{2}$-approximation algorithms for minimizing the number of drivers when the input instances satisfy all conditions except stop constraint condition, where K is the largest capacity of all vehicles. For a ridesharing instance containing a road network of size M and l trips, our first algorithm, which is a modification of a $\frac{1}{2}$-approximation algorithm (StarImprove) in [14], runs in $O(M + K \cdot l^3)$ time. Our second algorithm is more practical and runs in $O(M + K \cdot l^2)$ time.

An instance satisfying Conditions (1–3) and (5) may reflect a ridesharing in school commute: in the morning many staffs and students go to the school around the same time (Conditions (1) and (5)), each driver has a fixed path from home to the school (Condition (3)) and does not want to detour (Condition (2)); in the afternoon, staffs and students leave from the school (Conditions (1) and (5)). The rest of the paper is organized as follows. Section 2 gives the preliminaries of the paper. In Sect. 3, we show the NP-hardness results for stop constraint condition and time constraint condition. Section 4 presents the approximation algorithms for minimizing the number of drivers for stop constraint condition. Section 5 concludes the paper.

2   Preliminaries

A (undirected) graph G consists of a set $V(G)$ of vertices and a set $E(G)$ of edges, where each edge $\{u, v\}$ of $E(G)$ is a (unordered) pair of vertices in $V(G)$. A digraph H consists of a set $V(H)$ of vertices and a set $E(H)$ of arcs, where each arc (u, v) of $E(H)$ is an ordered pair of vertices in $V(H)$. A graph G (digraph H) is weighted if every edge of G (arc of H) is assigned a real number as the edge length. A *path* between vertex v_0 and vertex v_k in graph G is a sequence $e_1, .., e_k$ of edges, where $e_i = \{v_{i-1}, v_i\} \in E(G)$ for $1 \leq i \leq k$ and $v_i \neq v_j$ for $i \neq j$ and $0 \leq i, j \leq k$. A path from vertex v_0 to vertex v_k in a digraph H is defined similarly with each $e_i = (v_{i-1}, v_i)$ an arc in H. The *length* of a path P is the sum of the lengths of edges (arcs) in P. For simplicity, we express a road network by a weighted undirected graph $G(V, E)$ with non-negative edge length: $V(G)$ is the set of locations in the network, an edge $\{u, v\}$ represents the road segment between u and v.

In the ridesharing problem, we assume that the individual of every trip can be assigned as a driver or passenger. In general, in addition to a vehicle and individual, each trip has a source, a destination, a capacity of the vehicle, a set of preferred (optional) paths (e.g., shortest paths) to reach the destination, a limit (optional) on the detour distance/time from the preferred path to serve other individuals, a limit (optional) on the number of stops a driver wants to make to pick-up passengers, an earliest departure time, and a latest arrival time. Each trip in the ridesharing problem is expressed by an integer label i and specified by parameters $(s_i, t_i, n_i, d_i, \mathcal{P}_i, \delta_i, \alpha_i, \beta_i)$, which are defined in Table 1.

Table 1. Parameters for every trip i.

Parameter	Definition
s_i	The source (start location) of i (a vertex in G)
t_i	The destination of i (a vertex in G)
n_i	The number of seats (capacity) of i available for passengers
d_i	The detour distance limit i willing to make for offering services
$\mathcal{P}_i$	The set of preferred paths of i from s_i to t_i in G
δ_i	The maximum number of stops i willing to make to pick-up passengers
α_i	The earliest departure time of i
β_i	The latest arrival time of i

When the individual of trip i delivers (using i's vehicle) the individual of a trip j, we say trip i *serves* trip j and call i a *driver* and j a *passenger*. The *serve relation* between a driver i and a passenger j is defined as follows. A trip i can serve i itself and can serve a trip $j \neq i$ if i and j can arrive at their destinations by time β_i and β_j respectively such that j is a passenger of i, the detour of i is at most d_i, and the number of stops i has to make to serve j is at most δ_i. When a trip i *can serve* another trip j, it means that i-j is a feasible assignment of a driver-passenger pair. We extend this notion to a set $\sigma(i)$ of passenger trips that can be served by a driver i ($i \in \sigma(i)$). A driver i can serve all trips of $\sigma(i)$ if the total detour of i is at most d_i, the number of stops i have to make to pick-up $\sigma(i)$ is at most δ_i, and every $j \in \sigma(i)$ arrives at t_i before β_j. At any specific time point, a trip i can serve at most $n_i + 1$ trips. If trip i serves some trips after serving some other trips (known as *re-take passengers* in previous studies), trip i may serve more than $n_i + 1$ trips. In this paper, we study the ridesharing problem in which no re-taking passenger is allowed. A serve relation is *transitive* if i can serve j and j can serve k imply i can serve k. Let (G, R) be an instance of the ridesharing problem, where G is a road network (weighted graph) and $R = \{1, .., l\}$ is a set of trips. (S, σ), where $S \subseteq R$ is a set of trips assigned as drivers and σ is a mapping $S \to 2^R$, is a partial solution to (G, R) if

- for each $i \in S$, i can serve $\sigma(i)$,
- for each pair $i, j \in S$ with $i \neq j$, $\sigma(i) \cap \sigma(j) = \emptyset$, and
- $\sigma(S) = \cup_{i \in S} \sigma(i) \subseteq R$.

When $\sigma(S) = R$, (S, σ) is called a solution of (G, R). For a (partial) solution (S, σ), we sometimes call S a (partial) solution when σ is clear from the context.

We consider the problem of minimizing $|S|$ (the number of drivers) and the problem of minimizing the total travel distance of the drivers in S. To investigate the relations between the computational complexity and problem parameters, Gu et al. [7] introduced the simplified minimization (ridesharing) problems with parameters $(s_i, t_i, n_i, d_i, \mathcal{P}_i)$ only and the following conditions:

(1) All trips have the same destination or all trips have the same source, that is, $t_i = D$ for every $i \in R$ or $s_i = \chi$ for every $i \in R$.
(2) Zero detour: each trip can only serve others on his/her preferred path, that is, $d_i = 0$ for every $i \in R$.
(3) Fixed path: $\mathcal{P}_i$ has a unique preferred path P_i.

It is shown in [7] that if any one of Conditions (1), (2) and (3) is not satisfied, both minimization problems are NP-hard. Polynomial-time exact algorithms are given in [8] for the simplified minimization problems if all of Conditions (1–3) and transitive serve relation are satisfied. In this paper, we study more generalized minimization problems with all parameters in Table 1 considered. To analyze the computational complexity of the more generalized minimization problems, we introduce two more conditions:

(4) The number of stops each driver is willing to make to pick-up passengers is at least its capacity, that is, $\delta_i \geq n_i$ for every $i \in R$.
(5) All trips have the same earliest departure and same latest arrival time, that is, for every $i \in R$, $\alpha_i = \alpha$ and $\beta_i = \beta$ for some $\alpha < \beta$.

The polynomial-time exact algorithms in [8] can still apply to any ridesharing instance when all of Conditions (1–5) and transitive serve relation are satisfied.

3   NP-Hardness Results

We first show the NP-hardness results for the stop constraint condition, that is, when Conditions (1)–(3) and (5) are satisfied but Condition (4) is not. When Condition (1) is satisfied, we assume all trips have the same destination (since it is symmetric to prove the case that all trips have the same source). If all trips have distinct sources, one can solve both minimization problems by using the polynomial-time exact algorithms in [8]: when Conditions (1–3) are satisfied and each trip has a distinct source s_i, each trip is represented by a distinct vertex i in the serve relation graph in [8]. Each time a driver i serves a trip j, i must stop at $s_j \neq s_i$ to pick-up j. When Condition (4) is not satisfied ($\delta_i < n_i$), i can serve at most δ_i passengers. Therefore, we can set the capacity n_i to $\min\{n_i, \delta_i\}$ and apply the exact algorithms to solve the minimization problems. In what follows, we assume trips have arbitrary sources (multiple trips may have a same source).

3.1   Both Minimization Problems Are NP-Hard

We prove both minimization problems are NP-hard. The proof is a reduction from the 3-partition problem. The decision problem of 3-partition is that given a set $A = \{a_1, a_2, ..., a_{3r}\}$ of $3r$ positive integers, where $r \geq 2$, $\sum_{i=1}^{3r} a_i = rM$ and $M/4 < a_i < M/2$, whether A can be partitioned into r disjoint subsets $A_1, A_2, ..., A_r$ such that each subset has three elements of A and the sum of integers in each subset is M. Given a 3-partition instance $A = \{a_1, ..., a_{3r}\}$, construct a ridesharing problem instance (G, R_A) as follows (see Fig. 1).

- G is a graph with $V(G) = \{D, u_1, ..., u_{3r}, v_1, ..., v_r\}$ and $E(G)$ having edges $\{u_i, v_1\}$ for $1 \leq i \leq 3r$, edges $\{v_i, v_{i+1}\}$ for $1 \leq i \leq r - 1$ and $\{v_r, D\}$. Each edge $\{u, v\}$ has weight of 1, representing the travel distance from u to v. It takes $r + 1$ units of distance traveling from u_i to D for $1 \leq i \leq 3r$.
- $R_A = \{1, ..., 3r + rM\}$ has $3r + rM$ trips. Let α and β be valid constants representing time.
  - Each trip i, $1 \leq i \leq 3r$, has source $s_i = u_i$, destination $t_i = D, n_i = a_i, d_i = 0, \delta_i = 1, \alpha_i = \alpha$ and $\beta_i = \beta$. Each trip i has a preferred path $\{u_i, v_1\}, \{v_1, v_2\}, ..., \{v_r, D\}$ in G.
  - Each trip i, $3r + 1 \leq i \leq 3r + rM$, has source $s_i = v_j$, $j = \lceil (i - 3r)/M \rceil$, destination $t_i = D$, $n_i = 0$, $\delta_i = 0$, $d_i = 0$, $\alpha_i = \alpha$, $\beta_i = \beta$ and a unique preferred path $\{v_j, v_{j+1}\}, \{v_{j+1}, v_{j+2}\}, ..., \{v_r, D\}$ in G.

Lemma 1. *Any solution for the instance (G, R_A) has every trip i, $1 \leq i \leq 3r$, as a driver and total travel distance at least $3r \cdot (r + 1)$.*

Proof. Since condition (2) is satisfied (detour is not allowed), every trip $i, 1 \leq i \leq 3r$, must be a driver in any solution. A solution with exactly $3r$ drivers has total travel distance $3r \cdot (r + 1)$, and any solution with a trip i, $3r + 1 \leq i \leq 3r + rM$, as a driver has total travel distance greater than $3r \cdot (r + 1)$. □

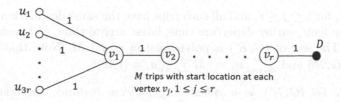

Fig. 1. Ridesharing instance based on a given 3-partition problem instance.

Theorem 1. *Minimizing the number of drivers in the ridesharing problem is NP-hard when Conditions (1–3) and (5) are satisfied, but Condition (4) is not.*

Proof. We prove the theorem by showing that an instance $A = \{a_1, ..., a_{3r}\}$ of the 3-partition problem has a solution if and only if the ridesharing problem instance (G, R_A) has a solution of $3r$ drivers.

Assume that instance A has a solution $A_1, ..., A_r$ where the sum of elements in each A_j is M. For each $A_j = \{a_{j_1}, a_{j_2}, a_{j_3}\}$, $1 \leq j \leq r$, assign the three trips whose $n_{j_1} = a_{j_1}$, $n_{j_2} = a_{j_2}$ and $n_{j_3} = a_{j_3}$ as drivers to serve the M trips with sources at vertex v_j. Hence, we have a solution of $3r$ drivers for (G, R_A).

Assume that (G, R_A) has a solution of $3r$ drivers. By Lemma 1, every trip i, $1 \leq i \leq 3r$, is a driver in the solution. Then, each trip j for $3r + 1 \leq j \leq 3r + rM$ must be a passenger in the solution, total of rM passengers. Since $\sum_{1 \leq i \leq 3r} a_i = rM$, each driver i, $1 \leq i \leq 3r$, serves exactly $n_i = a_i$ passengers. Since $a_i < M/2$ for every $a_i \in A$, at least three drivers are required to serve the M passengers with sources at each vertex v_j, $1 \leq j \leq 3r$. Due to $\delta_i = 1$, each driver i, $1 \leq i \leq 3r$, can only serve passengers with the same source. Therefore, the solution of $3r$ drivers has exactly three drivers j_1, j_2, j_3 to serve the M passengers with sources at vertex v_j, implying $a_{j_1} + a_{j_2} + a_{j_3} = M$. Let $A_j = \{a_{j_1}, a_{j_2}, a_{j_3}\}$, $1 \leq j \leq r$, we get a solution for the 3-partition instance.

The size of (G, R_A) is polynomial in r. It takes a polynomial time to convert a solution of (G, R_A) to a solution of the 3-partition instance and vice versa. □

The proof of Theorem 2 can be found in Sect. 3 of [9].

Theorem 2. *Minimizing the total travel distance of drivers in the ridesharing problem is NP-hard when Conditions (1–3) and (5) are satisfied but Condition (4) is not.*

3.2 Inapproximability Results

Based on the results in Sect. 3.1, we extent our reduction to further show that it is NP-hard to approximate both minimization problems within a constant factor if Condition (4) is not satisfied. Let (G, R_A) be the ridesharing problem instance constructed based on a given 3-partition instance A as described above for Theorem 1. We modify (G, R_A) to get a new ridesharing instance (G, R') as follows. For every trip i, $1 \leq i \leq 3r$, we multiply n_i with rM, that is, $n_i = a_i \cdot rM$, where r and M are given in instance A. There are now rM^2 trips with sources at vertex v_j for $1 \leq j \leq r$, and all such trips have the same destination, capacity, detour, stop limit, earlier departure time, latest arrival time, and preferred path as before. The size of (G, R') is polynomial in r and M. Note that Lemma 1 holds for (G, R') and $\sum_{i=1}^{3r} n_i = rM \sum_{i=1}^{3r} a_i = (rM)^2$.

Lemma 2. *Let (G, R') be a ridesharing problem instance constructed above from a 3-partition problem instance $A = \{a_1, \ldots, a_{3r}\}$. The 3-partition problem instance A has a solution if and only if the ridesharing problem instance (G, R') has a solution (σ, S) s.t. $3r \leq |S| < 3r + rM$, where S is the set of drivers.*

Proof. Assume that instance A has a solution $A_1, \ldots, A_r$ where the sum of elements in each A_j is rM^2. For each $A_j = \{a_{j_1}, a_{j_2}, a_{j_3}\}$, $1 \leq j \leq r$, we assign the three trips whose $n_{j_1} = a_{j_1} \cdot rM$, $n_{j_2} = a_{j_2} \cdot rM$ and $n_{j_3} = a_{j_3} \cdot rM$ as drivers to serve the rM^2 trips with sources at vertex v_j. Hence, we have a solution of $3r$ drivers for (G, R').

Assume that (G, R') has a solution with $3r \leq |S| < 3r + rM$ drivers. Let $R'(1, 3r)$ be the set of trips in R' with labels from 1 to $3r$. By Lemma 1, every trip $i \in R'(1, 3r)$ is a driver in S. Since $a_i < M/2$ for every $a_i \in A$, $n_i < rM \cdot M/2$ for every trip $i \in R'(1, 3r)$. From this, it requires at least three drivers in $R'(1, 3r)$ to serve the rM^2 trips with sources at each vertex v_j, $1 \leq j \leq r$. For every trip $i \in R'(1, 3r)$, i can only serve passengers with the same source due to $\delta_i = 1$. There are two cases: (1) $|S| = 3r$ and (2) $3r < |S| < 3r + rM$.

(1) It follows from the proof of Theorem 1 that every three drivers j_1, j_2, j_3 of the $3r$ drivers serve exactly rM^2 passengers with sources at vertex v_j. Then similar, let $A_j = \{a_{j_1}, a_{j_2}, a_{j_3}\}$, $1 \leq j \leq r$, we get a solution for the 3-partition problem instance.

(2) For every vertex v_j, let X_j be the set of trips with source v_j not served by drivers in $R'(1, 3r)$. Then $0 \leq |X_j| < rM$ due to $|S| < 3r + rM$. For every trip $i \in R'(1, 3r)$, $n_i = a_i \cdot rM$ is a multiple of rM. Hence, the sum of capacity for any trips in $R'(1, 3r)$ is also a multiple of rM, and further, $n_{i_1} + n_{i_2} = (a_{i_1} + a_{i_2}) \cdot rM < rM \cdot (M-1)$ for every $i_1, i_2 \in R'(1, 3r)$ because $a_{i_1} < M/2$ and $a_{i_2} < M/2$. From these and $|X_j| < rM$, there are 3 drivers $j_1, j_2, j_3 \in R'(1, 3r)$ to serve trips with source v_j and $n_{j_1} + n_{j_2} + n_{j_3} \geq rM^2$. Because $n_{j_1} + n_{j_2} + n_{j_3} \geq rM^2$ for every $1 \leq j \leq r$ and $\sum_{1 \leq i < 3r} n_i = (rM)^2$, $n_{j_1} + n_{j_2} + n_{j_3} = rM^2$ for every j. Thus, we get a solution with $A_j = \{a_{j_1}, a_{j_2}, a_{j_3}\}$, $1 \leq j \leq r$, for the 3-partition problem.

It takes a polynomial time to convert a solution of (G, R') to a solution of the 3-partition instance and vice versa. $\square$

Theorem 3. *Let (G, R') be the ridesharing instance stated above based on a 3-partition instance. Approximating the minimum number of drivers for (G, R') within a constant factor is NP-hard. This implies that it is NP-hard to approximate the minimum number of drivers within a constant factor for a ridesharing instance when Conditions (1–3) and (5) are satisfied and Condition (4) is not.*

Proof. Assume that there is a polynomial time c-approximation algorithm C for instance (G, R') for any constant $c > 1$. This means that C will output a solution (σ_C, S_C) for (G, R') such that $OPT(R') \leq |S_C| \leq c \cdot OPT(R')$, where $OPT(R')$ is the minimum number of drivers for (G, R'). When the 3-partition instance is a "No" instance, the optimal value for (G, R') is $OPT(R') \geq 3r + rM$ by Lemma 2. Hence, algorithm C must output a value $|S_C| \geq 3r + rM$. When the 3-partition instance is a "Yes" instance, the optimal value for (G, R') is $OPT(R') = 3r$. For any constant $c > 1$, taking M such that $c < M/3 + 1$. The output $|S_C|$ from algorithm C on (G, R') is $3r \leq |S_C| \leq 3rc < 3r + rM$ for a 3-partition "Yes" instance. Therefore, by running the c-approximation algorithm C on any ridesharing instance (G, R') and checking the output value $|S_C|$ of C, we can answer the 3-partition problem in polynomial time, which contradicts that the 3-partition problem is NP-hard unless $P = NP$. ☐

The proof of Theorem 4 can be found in Sect. 3 of [9].

Theorem 4. *It is NP-hard to approximate the total travel distance of drivers within any constant factor for a ridesharing instance when Conditions (1–3) and (5) are satisfied and Condition (4) is not.*

3.3 NP-Hardness Result for Time Constraint Condition

Assume that Conditions (1–4) are satisfied but Condition (5) is not, that is, trips can have arbitrary departure time and arrival time. Then we have the following results (detailed proofs for the results are given in Sect. 4 of [9]).

Theorem 5. *Minimizing the number of drivers in the ridesharing problem is NP-hard when Conditions (1–4) are satisfied but Condition (5) is not.*

Theorem 6. *Minimizing the total travel distance of drivers in the ridesharing problem is NP-hard when Conditions (1–4) are satisfied but Condition (5) is not.*

Theorem 7. *It is NP-hard to approximate the minimum number of drivers within any constant factor for a ridesharing instance satisfying Conditions (1–4) but not Condition (5).*

Theorem 8. *It is NP-hard to approximate the minimum total travel distance of drivers within any constant factor for a ridesharing instance satisfying Conditions (1–4) but not Condition (5).*

4 Approximation Algorithms for Stop Constraint Condition

For short, we call the ridesharing problem with all conditions satisfied except Condition (4) as *ridesharing problem with stop constraint*. Let $K = \max_{i \in R} n_i$ be the largest capacity of all vehicles. Kutiel and Rawitz [14] proposed two $\frac{1}{2}$-approximation algorithms for the maximum carpool matching problem. We first show that the algorithms in [14] can be modified to $\frac{K+2}{2}$-approximation algorithms for minimizing the number of drivers in the ridesharing problem with stop constraint. Then we propose a more practical $\frac{K+2}{2}$-approximation algorithm for the minimization problem.

4.1 Approximation Algorithms Based on MCMP

An instance of the maximum carpool matching problem (MCMP) consists of a directed graph $H(V, E)$, a capacity function $c : V \rightarrow \mathbb{N}$, and a weight function $w : E \rightarrow \mathbb{R}^{+}$, where the vertices of V represent the individuals and there is an arc $(u, v) \in E$ if v can serve u. We are only interested in the unweighted case, that is, $w(u, v) = 1$ for every $(u, v) \in E$. Every $v \in V$ can be assigned as a driver or passenger. The goal of MCMP is to find a set of drivers $S \subseteq V$ to serve all V such that the number of passengers is maximized. A solution to MCMP is a set $\mathcal{S}$ of vertex-disjoint stars in H. Let S_v be a star in $\mathcal{S}$ rooted at center vertex v, and leaves of S_v is denoted by $P_v = V(S_v) \setminus \{v\}$. For each star $S_v \in \mathcal{S}$, vertex v has out-degree of 0 and every leaf in P_v has only one out-edge towards v. The center vertex of each star S_v is assigned as a driver and the leaves are assigned as passengers. The set of edges in $\mathcal{S}$ is called a matching M. An edge in M is called a *matched* edge. Notice that $|M|$ equals to the number of passengers. For an arc $e = (u, v)$ in H, vertices u and v are said to be *incident to* e. For a matching M and a set $V' \subseteq V$ of vertices, let $M(V')$ be the set of edges in M incident to V'. Table 2 lists the basic notation and definition for this section.

Table 2. Common notation and definition used in this section.

Notation	Definition
$\mathcal{S}$	A set of vertex-disjoint stars in H (solution to MCMP)
S_v and P_v	Star S_v rooted at center vertex v with leaves $P_v = V(S_v) \setminus \{v\}$
$c(v)$	Capacity of vertex v (equivalent to n_v in Table 1)
Matching M	The set of edges in $\mathcal{S}$
$M(V')$	The set of edges in M incident to a set V' of vertices
$N^{in}(v)$	The set of *in-neighbors* of v, $N^{in}(v) = \{u \mid (u, v) \in E\}$
$E^{in}(v)$	The set of arcs entering v, *in-arcs* $E^{in}(v) = \{(u, v) \mid (u, v) \in E\}$
δP_v	The number of stops required for v to pick-up P_v

Two approximation algorithms (StarImprove and EdgeSwap) are presented in [14]; both can achieve $\frac{1}{2}$-approximation ratio, that is, the number of passengers found by the algorithm is at least half of that for the optimal solution.

EdgeSwap. The EdgeSwap algorithm requires the input instance to have a bounded degree graph (or the largest capacity K is bound by a constant) to have a polynomial running time. The idea of EdgeSwap is to swap i matched edges in M with $i + 1$ edges in $E \setminus M$ for $1 \leq i \leq k$ and k is a constant integer. The running time of EdgeSwap is in the order of $O(|E|^{2k+1})$. EdgeSwap can directly apply to the minimization problem to achieve $\frac{K+2}{2}$-approximation ratio in $O(l^{2K})$ time, which may not be practical even if K is a small constant.

StarImprove. Let $(H(V, E), c, w)$ be an instance of MCMP. Let $\mathcal{S}$ be the current set of stars found by StarImprove and M be the set of matched edges. The idea of the StarImprove algorithm is to iteratively check in a *for-loop* for every vertex $v \in V(G)$:

– check if there exists a star S_v with $E(S_v) \cap M = \emptyset$ s.t. the resulting set of stars $\mathcal{S} \backslash M(V(S_v)) \cup S_v$ gives a larger matching.

Such a star S_v is called an *improvement* and $|P_v| \leq c(v)$. Given a ridesharing instance (G, R) satisfying all conditions, except Condition (4). The StarImprove algorithm cannot apply to (G, R) directly because the algorithm assumes a driver v can serve any combination of passengers corresponding to vertices adjacent to v up to $c(v)$. This is not the case for (G, R) in general. For example, suppose v can serve u_1 and u_2 with $n_v = 2$ and $\delta_v = 1$. The StarImprove assigns v as a driver to serve both u_1 and u_2. However, if u_1 and u_2 have different sources ($s_v \neq s_{u_1} \neq s_{u_2}$), this assignment is not valid for (G, R). Hence, we need to modify StarImprove for computing a star. For a vertex v and star S_v, let $N_{-M}^{in}(v) = \{i \mid i \in N^{in} \setminus V(M)\}$ and δP_v be the number of stops required for v to pick-up P_v. Suppose the in-neighbors $N_{-M}^{in}(v)$ are partitioned into $g_1(v), \ldots, g_m(v)$ groups such that trips with same source are grouped together. When stop constraint is considered, finding a star S_v with maximum $|P_v|$ is similar to solving a fractional knapsack instance using a greedy approach as shown in Fig. 2.

Algorithm 1 Greedy algorithm

1: $P_v = \emptyset$; $c = c(v)$; $\delta P_v = 0$;
2: **if** $\exists$ a group $g_j(v)$ s.t. $s_u = s_v$ for any $u \in g_j(v)$ **then**
3:     **if** $|g_j(v)| \leq c$ **then** $P_v = P_v \cup g_j(v)$; $c = c - |g_j(v)|$;
4:     **else** $P_v = P_v \cup g_j'(v)$, where $g_j'(v) \subseteq g_j(v)$ with $|g_j'(v)| = c$. $c = 0$;
5: **end if**
6: **while** $c > 0$ and $\delta P_v < \delta_v$ **do**
7:     Select $g_i(v) = \max_{1 \leq i \leq m}\{|g_i(v) \setminus P_v|\}$;
8:     **if** $|g_i(v)| \leq c$ **then** $P_v = P_v \cup g_i(v)$; $c = c - |g_i(v)|$; $\delta P_v = \delta P_v + 1$;
9:     **else** $P_v = P_v \cup g_i'(v)$, where $g_i'(v) \subseteq g_i(v)$ with $|g_i'(v)| = c$. $c = 0$; $\delta P_v = \delta P_v + 1$;
10: **end while**
11: **return** the star S_v induced by $P_v \cup \{v\}$;

Fig. 2. Greedy algorithm for computing S_v.

Lemma 3. *Let v be the trip being processed and S_v be the star found by Algorithm 1 w.r.t. current matching M. Then $|P_v| \geq |P_v'|$ for any star S_v' s.t. $P_v' \cap M = \emptyset$.*

The proof of Lemma 3 is in Sect. 5 of [9].

Definition 1. *A star S_v rooted at v is an improvement with respect to matching M if $|P_v| \leq c(v), \delta P_v \leq \delta_v$ and $|S_v| - \sum_{(u,v)\in E(S_v)} |M(u)| > |M(v)|$.*

Definition 1 is equivalent to the original definition in [14], except the former is for the unweighted case and stop constraint. When an improvement is found, the current matching M is increased by exactly $|S_v| - \sum_{(u,v) \in E(S_v)} |M(u)|$ edges. For a vertex v and a subset $S \subseteq E^{in}(v)$, let $N_S^{in}(v) = \{u \mid (u,v) \in S\}$.

Lemma 4. *Let M be the current matching and v be a vertex with no improvement. Let $S_v \subseteq E^{in}(v)$ s.t. $|S_v| \le c(v)$ and $\delta P_v \le \delta_v$, then $|S_v| \le |M(v)| + |M(N_{S_v}^{in}(v))|$. Further, if the star S_v found by Algorithm 1 w.r.t. M is not an improvement, then no other S_v' is an improvement.*

The proof of Lemma 4 is in Sect. 5 of [9]. By Lemma 4 and the same argument of Lemma 6 in [14], we have the following lemma.

Lemma 5. *The modified StarImprove algorithm computes a solution to an instance of ridesharing problem with stop constraint with $\frac{1}{2}$-approximation.*

By Lemma 5 and a straight forward implementation of the modified StarImprove algorithm, we have the following theorem (proof in Sect. 5 of [9]).

Theorem 9. *Let (G, R) be a ridesharing instance satisfying all conditions, except condition (4). Let $|S^*|$ be the minimum number of drivers for (G, R), $l = |R|$ and $K = \max_{i \in R} n_i$. Then,*

- *The EdgeSwap algorithm computes a solution (σ, S) for (G, R) s.t. $|S^*| \le |S| \le \frac{K+2}{2}|S^*|$ with running time $O(M + l^{2K})$.*
- *The modified StarImprove algorithm computes a solution (σ, S) for (G, R) s.t. $|S^*| \le |S| \le \frac{K+2}{2}|S^*|$ with running time $O(M + K \cdot l^3)$, where M is the size of a ridesharing instance which contains a road network and l trips.*

4.2 A More Practical New Approximation Algorithm

For our proposed algorithm, we assume the serve relation is transitive, that is, trip i can serve trip j and j can serve trip k imply i can serve k. In general, if each trip has a unique preferred path and trip i can serve trip j implies j's preferred path is a subpath of i's preferred path, then the serve relation is transitive. Given a ridesharing instance (G, R), we construct a directed meta graph $\Gamma(V, E)$ to express the serve relation, where $V(\Gamma)$ represents the start locations of all trips in (G, R). Each node μ of $V(\Gamma)$ contains all trips with the same start location μ. There is an arc (μ, ν) in $E(\Gamma)$ if a trip in μ can serve a trip in ν. Since Conditions (1–3) are satisfied, if one trip in μ can serve a trip in ν, any trip in μ can serve any trip in ν. We say node μ can serve node ν. An arc (μ, ν) in Γ is called a *short cut* if after removing (μ, ν) from Γ, there is a path from μ to ν in Γ. We simplify Γ by removing all short cuts from Γ. In what follows, we use Γ for the simplified meta graph. Notice that Γ is an inverse tree and for every pair of nodes μ and ν in Γ, if there is a path from μ to ν then μ can serve ν. We label the nodes of Γ as $V(\Gamma) = \{\mu_p, \mu_{p-1}, ..., \mu_1\}$, where $p = |V(\Gamma)|$, in such a way that for every arc (μ_b, μ_a) of Γ, $b > a$, and we say μ_b has a larger label than μ_a. The labeling is done by the procedure in [8] (see Appendix of [9]). Figure 3 shows an example of a graph $\Gamma(V, E)$. Each node in Γ without an incoming arc is called an *origin*, and μ_1 is the unique sink. Table 3 contains the basic notation and definition for this section.

Table 3. Basic notation and definition used in this section.

Notation	Definition
$\Gamma(V, E)$	A directed graph expressing the serve relation and $p = \|V(\Gamma)\|$
μ is an ancestor of ν	If $\exists$ a nonempty path from μ to ν in Γ (ν is a descendant of μ)
A_μ and A_μ^*	Set of ancestors of μ and $A_\mu^* = A_\mu \cup \{\mu\}$ respectively
D_μ and D_μ^*	Set of descendants of μ and $D_\mu^* = D_\mu \cup \{\mu\}$ respectively
$R(\mu)$ and $R(U)$	Set of trips in a node μ and in a set U of nodes respectively
$S(\mu)$ and $S(U)$	Set of drivers in a node μ and in nodes U respectively
$node(i)$	The node that contains trip i (if $i \in R(\mu)$ then $node(i) = \mu$)

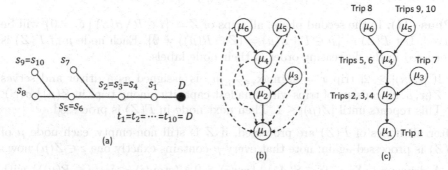

Fig. 3. (a) a set R of 10 trips with same destination D in the road network graph G. (b) The directed meta graph expressing the serve relation of these trips with shortcuts in dashed arcs. (c) The simplified meta graph, which is an inverse tree.

We divide all trips of R into two sets W and X as follows. Let $W = \{i \in R \mid n_i = 0\} \cup \{i \in R(\mu) \mid \delta_i = 0 \text{ and } |R(\mu) = 1| \text{ for every node } \mu \in V(\Gamma)\}$ and $X = R \setminus W$. For a node μ in Γ, let $X(\mu) = X \cap R(\mu)$ and $W(\mu) = W \cap R(\mu)$. Our algorithm tries to minimize the number of drivers that only serve itself. There are three phases in the algorithm. In Phase-I, it serves all trips of W and tries to minimize the number of trips in W that are assigned as drivers since each trip of W can serve only itself. Let Z be the set of unserved trips after Phase-I such that for every $i \in Z$, $\delta_i = 0$. In Phase-II, it serves all trips of Z and tries to minimize the number of such drivers of Z, each only serves itself. In Phase III, it serves all remaining trips. Let (S, σ) be the current partial solution and $i \in R$ be a driver. Denoted by $free(i) = n_i - |\sigma(i)| + 1$ is the remaining capacity of i w.r.t. solution (S, σ). Denoted by $stop(i)$ is the number of stops i has to made in order to serve all of $\sigma(i)$ w.r.t. (S, σ). For a driver i and node μ, we define $R(i, \mu, S)$ as the set of min$\{free(i), |R(\mu) \setminus \sigma(S)|\}$ trips in $R(\mu) \setminus \sigma(S)$ and $W(i, \mu, S)$ as the set of min$\{free(i), |W(\mu) \setminus \sigma(S)|\}$ trips in $W(\mu) \setminus \sigma(S)$, and similarly for $Z(i, \mu, S)$. We start with initial solution $(S, \sigma) = (\emptyset, \emptyset)$, $free(i) = n_i$

and stop$(i) = 0$ for all $i \in R$. The three phases of the approximation algorithm are described in following, and the pseudo code is given in Sect. 6 of [9].

(Phase-I). In this phase, the algorithm assigns a set of drivers to serve all trips of W. Let $\Gamma(W) = \{\mu \in V(\Gamma) \mid W(\mu) \setminus \sigma(S) \neq \emptyset\}$, and each node of $\Gamma(W)$ is processed. In each iteration, the node $\mu = \mathrm{argmax}_{\mu \in \Gamma(W)} |W(\mu) \setminus \sigma(S)|$ is selected and a subset of trips in $W(\mu) \setminus \sigma(S)$ is served by a driver as follows:

- Let $\hat{X}_1 = \{i \in S(A_\mu) \mid \mathrm{free}(i) > 0 \wedge \mathrm{stop}(i) < \delta_i\}$ and $\bar{X} = \{i \in X \cap R(A_\mu^*) \setminus \sigma(S) \mid \mathrm{stop}(i) < \delta_i \vee i \in R(\mu)\}$. The algorithm finds and assigns a trip x as a driver to serve $W(x, \mu, S)$ s.t. $x = \mathrm{argmin}_{x \in \hat{X}_1 \cup \bar{X} \,:\, n_x \geq |W(\mu) \setminus \sigma(S)|} \delta_x - \mathrm{stop}(x)$.
  - If such a trip x does not exist, it means that $n_x < |W(\mu) \setminus \sigma(S)|$ for every $x \in \hat{X}_1 \cup \bar{X}$ assuming $\hat{X}_1 \cup \bar{X} \neq \emptyset$. Then, $x = \mathrm{argmax}_{x \in \hat{X}_1 \cup \bar{X}} \mathrm{free}(x)$ is assigned as a driver to serve $W(x, \mu, S)$. If there is more than one x with same $\mathrm{free}(x)$, the trip with smallest $\delta_x - \mathrm{stop}(x)$ is selected.
- When $\hat{X}_1 \cup \bar{X} = \emptyset$, assign every $w \in W(\mu) \setminus \sigma(S)$ as a driver to serve itself.

(Phase-II). In the second phase, all trips of $Z = \{i \in R \setminus \sigma(S) \mid \delta_i = 0\}$ will be served. Let $\Gamma(Z) = \{\mu \in \Gamma \mid Z(\mu) = (Z \cap R(\mu)) \neq \emptyset\}$. Each node μ of $\Gamma(Z)$ is processed in the decreasing order of their node labels.

- If $|Z(\mu)| \geq 2$, trip $x = \mathrm{argmax}_{x \in Z(\mu)} n_x$ is assigned as a driver and serves $Z(x, \mu, S)$ consists of trips with smallest capacity among trips in $Z(\mu) \setminus \sigma(S)$.
- This repeats until $|Z(\mu)| \leq 1$. Then next node in $\Gamma(Z)$ is processed.

After all nodes of $\Gamma(Z)$ are processed, if Z is still non-empty, each node μ of $\Gamma(Z)$ is processed again; note that every μ contains exactly one $z \in Z(\mu)$ now.

- A driver $x \in \hat{X}_2 = \{i \in S(A_\mu^*) \mid \mathrm{free}(i) > 0 \wedge (\mathrm{stop}(i) < \delta_i \vee i \in R(\mu))\}$ with largest $\mathrm{free}(x)$ is selected to serve $z = Z(\mu)$ if $\hat{X}_2 \neq \emptyset$.
- If $\hat{X}_2 = \emptyset$, a trip $x \in \bar{X} = \{i \in X \cap R(A_\mu^*) \setminus \sigma(S) \mid \mathrm{stop}(i) < \delta_i \vee i \in R(\mu)\}$ with largest δ_x is selected to serve $z = Z(\mu)$.

(Phase-III). To serve all remaining trips, the algorithm processes each node of Γ in decreasing order of node labels from μ_p to μ_1. Let μ_j be the node being processed by the algorithm. Suppose there are trips in $R(\mu_j)$ that have not be served, that is, $R(\mu_j) \not\subseteq \sigma(S)$.

- A driver $x \in \hat{X}_2 = \{i \in S(A_{\mu_j}^*) \mid \mathrm{free}(i) > 0 \wedge (\mathrm{stop}(i) < \delta_i \vee i \in R(\mu_j))\}$ with largest $\mathrm{free}(x)$ is selected if $\hat{X}_2 \neq \emptyset$.
- If $\hat{X}_2 = \emptyset$, a trip $x = \mathrm{argmax}_{x \in X(\mu_j) \setminus \sigma(S)} n_x$ is assigned as a driver. If the largest n_x is not unique, the trip with the smallest δ_x is selected.
- In either case, x is assigned to serve $R(x, \mu_j, S)$. This repeats until all of $R(\mu_j)$ are served. Then, next node μ_{j-1} is processed.

Theorem 10. *Given a ridesharing instance (G, R) of size M and l trips satisfying Conditions (1–3) and (5). Algorithm 1 computes a solution (S, σ) for (G, R) such that $|S^*| \leq |S| \leq \frac{K+2}{2} |S^*|$, where (S^*, σ^*) is any optimal solution and $K = \max_{i \in R} n_i$, with running time $O(M + l^2)$.*

The complete proof of the theorem is in Sect. 6 of [9].

5   Conclusion

We proved that the problems of minimizing the number of vehicles and minimizing the total distance of vehicles are NP-hard, and further it is NP-hard to approximate these two problems within a constant factor if neither Condition (4) nor (5) is satisfied. Combining these with the results of [7,8], both minimization problems are NP-hard if one of Conditions (1)–(5) is not satisfied. We also presented $\frac{K+2}{2}$-approximation algorithms for minimizing number of drivers for problem instances satisfying all conditions except Condition (4), where K is the largest capacity of all vehicles. It is worth developing approximation algorithms for other NP-hard cases; for example, two or more of the five conditions are not satisfied. It is interesting to study applications of the approximation algorithms for other related problems, such as multimodal transportation with ridesharing (integrating public and private transportation).

Acknowledgments. The authors thank the anonymous reviewers for their constructive comments. This work was partially supported by Canada NSERC Discovery Grant 253500 and China NSFC Grant 11531014.

References

1. Agatz, N., Erera, A., Savelsbergh, M., Wang, X.: Dynamic ride-sharing: a simulation study in metro Atlanta. Trans. Res. Part B **45**(9), 1450–1464 (2011)
2. Agatz, N., Erera, A., Savelsbergh, M., Wang, X.: Optimization for dynamic ride-sharing: a review. Eur. J. Oper. Res. **223**, 295–303 (2012)
3. Alonso-Mora, J., Samaranayake, S., Wallar, A., Frazzoli, E., Rus, D.: On-demand high-capacity ride-sharing via dynamic trip-vehicle assignment. In: Proceedings of the National Academy of Sciences (PNAS), vol. 114, no. 3, pp. 462–467 (2017)
4. Caulfield, B.: Estimating the environmental benefits of ride-sharing: a case study of Dublin. Transp. Res. Part D **14**(7), 527–531 (2009)
5. Furuhata, M., Dessouky, M., Ordóñez, F., Brunet, M., Wang, X., Koenig, S.: Ridesharing: the state-of-the-art and future directions. Transp. Res. Part B **57**, 28–46 (2013)
6. Ghoseiri, K., Haghani, A., Hamed, M.: Real-time rideshare matching problem. Final Report of UMD-2009-05, U.S. Department of Transportation (2011)
7. Gu, Q.-P., Liang, J.L., Zhang, G.: Algorithmic analysis for ridesharing of personal vehicles. In: Chan, T.-H.H., Li, M., Wang, L. (eds.) COCOA 2016. LNCS, vol. 10043, pp. 438–452. Springer, Cham (2016). https://doi.org/10.1007/978-3-319-48749-6_32
8. Gu, Q., Liang, J.L., Zhang, G.: Efficient algorithms for ridesharing of personal vehicles. Theor. Comput. Sci. **788**, 79–94 (2019)
9. Gu, Q.P., Liang, J.L., Zhang, G.: Approximate Ridesharing of Personal Vehicles Problem. eprint arXiv:2007.15154 [cs.DS], 2020
10. Hartman, I.B.-A., et al.: Theory and practice in large carpooling problems. In: Proceedings of the 5th International Conference on ANT, pp. 339–347 (2014)
11. Herbawi, W., Weber, M.: The ridematching problem with time windows in dynamic ridesharing: a model and a genetic algorithm. In: Proceedings of ACM Genetic and Evolutionary Computation Conference (GECCO), pp. 1–8 (2012)

12. Huang, Y., Bastani, F., Jin, R., Wang, X.S.: Large scale real-time ridesharing with service guarantee on road networks. Proc. VLDB Endow. **7**(14), 2017–2028 (2014)
13. Jung, J., Jayakrishnan, R., Park, J.Y.: Dynamic shared-taxi dispatch algorithm with hybrid-simulated annealing. Comput.-Aided Civ. Inf. **31**(4), 275–291 (2016)
14. Kutiel, G., Rawitz, D.: Local search algorithms for maximum carpool matching. In: Proceedings of 25th Annual European Symposium on Algorithms, pp. 55:1–55:14 (2017)
15. Ma, S., Zheng, Y., Wolfson, O.: Real-time city-scale taxi ridesharing. IEEE Trans. Knowl. Data Eng. **27**(7), 1782–1795 (2015)
16. Molenbruch, Y., Braekers, K., Caris, A.: Typology and literature review for dial-a-ride problems. Ann. Oper. Res. **259**(1), 295–325 (2017). https://doi.org/10.1007/s10479-017-2525-0
17. Mourad, A., Puchinger, J., Chu, C.: A survey of models and algorithms for optimizing shared mobility. Transp. Res. Part B **123**, 323–346 (2019)
18. Santos, A., McGuckin, N., Nakamoto, H.Y., Gray, D., Liss, S.: Summary of travel trends: 2009 national household travel survey. Technical report, US Department of Transportation Federal Highway Administration (2011)
19. Santos, D.O., Xavier, E.C.: Taxi and ride sharing: a dynamic dial-a-ride problem with money as an incentive. Exp. Syst. Appl. **42**(19), 6728–6737 (2015)
20. Sierpiński, G.: Changes of the modal split of traffic in Europe. Arch. Transp. Syst. Telemat. **6**(1), 45–48 (2013)

A Sub-linear Time Algorithm for Approximating k-Nearest-Neighbor with Full Quality Guarantee

Hengzhao Ma and Jianzhong Li[✉]

Harbin Institute of Technology, Harbin 150001, Heilongjiang, China
hzma@stu.hit.edu.cn, lijzh@hit.edu.cn

Abstract. In this paper we propose an algorithm for the approximate k-Nearest-Neighbors problem. According to the existing researches, there are two kinds of approximation criteria. One is the distance criterion, and the other is the recall criterion. All former algorithms suffer the problem that there are no theoretical guarantees for the two approximation criteria. The algorithm proposed in this paper unifies the two kinds of approximation criteria, and has full theoretical guarantees. Furthermore, the query time of the algorithm is sub-linear. As far as we know, it is the first algorithm that achieves both sub-linear query time and full theoretical approximation guarantees.

Keywords: Computation geometry · Approximate k-nearest-neighbors

1 Introduction

The k-Nearest-Neighbor (kNN) problem is a well-known problem in theoretical computer science and applications. Let (U, D) be a metric space, then for the input set $P \subseteq U$ of elements and a query element $q \in U$, the kNN problem is to find the k elements with smallest distance to q. Since the exact results are expensive to compute when the size of the input is large [19], and approximate results serve as good as the exact ones in many applications [30], the approximate kNN, kANN for short, draws more research efforts in recent years. There are two kinds of approximation criteria for the kANN problem, namely, the distance criterion and the recall criterion. The distance criterion requires that the ratio between the distance from the approximate results to the query and the distance from the exact results to the query is no more than a given threshold. The recall criterion requires that the size of the intersection of the approximate result set and the exact result set is no less than a given threshold. The formal description will be given in detail in Sect. 2. Next we brief the existing algorithms for the kANN problem to see how these two criteria are considered by former researchers.

This work was supported by the National Natural Science Foundation of China under grant 61732003, 61832003, 61972110 and U1811461.

The algorithms for the kANN problem can be categorized into four classes. The first class is the tree-based methods. The main idea of this method is to recursively partition the metric space into sub-spaces, and organize them into a tree structure. The K-D tree [6] is the representative idea in this category. It is efficient in low dimensional spaces, but the performance drops rapidly when the number of dimension grows up. Vantage point tree (VP-tree) [31] is another data structure with a better partition strategy and better performance. The FLANN [25] method is a recent work with improved performance in high dimensional spaces, but it is reported that this method would achieve in sub-optimal results [20]. To the best of our knowledge, the tree based methods can satisfy neither the distance nor the recall criterion theoretically.

The second class is the permutation based methods. The idea is to choose a set of pivot points, and represent each data element with a permutation of the pivots sorted by the distance to it. In such a representation, close objects will have similar permutations. Methods using the permutation idea include the MI-File [2] and PP-Index [13]. Unfortunately, the permutation based method can not satisfy either of the distance or the recall criterion theoretically, as far as we know.

The third class is the Locality Sensitive Hashing (LSH) based methods. LSH was first introduced by Indyk et al. [19] for the kANN problem where $k = 1$. Soon after, Datar et al. [11] proposed the first practical LSH function, and since then there came a burst in the theoretical and applicational researches on the LSH framework. For example, Andoni et al. proved the lower bound of the time-space complexities of the LSH based algorithms [3], and devised the optimal LSH function which meets the lower bound [4]. On the other hand, Gao et al. [15] proposed an algorithm that aimed to close the gap between the LSH theory and kANN search applications. See [29] for a survey. The basic LSH based method can satisfy only the distance criterion when $k = 1$ [19]. Some existing algorithms made some progress. The C2LSH algorithm [14] solved the kANN problem with the distance criterion, but it has a constraint that the approximation factor must be a square of an integer. The SRS algorithm [28] is another one aimed at the distance criterion. However, it only has partial guarantee, that is, the results satisfy the distance criterion only when the algorithm terminates on a specific condition.

The forth class is graph based methods. The specific kind of graphs used in this method is the proximity graphs, where the edges in this kind of graph are defined by the geometric relationship of the points. See [23] for a survey. The graph based kANN algorithms usually conduct a navigating process on the proximity graphs. This process selects an vertex in the graph as the start point, and move to the destination point following some specific navigating strategy. For example, Paredes et al. [27] used the kNN graph, Ocsa et al. [26] used the Relative Neighborhood Graph (RNG), and Malkov et al. [22] used the Navigable Small World Graph (NSW) [22]. None of these algorithms have theoretical guarantee on the two approximation criteria.

In summary, most of the existing algorithms do not have theoretical guarantee on either of the two approximation criteria. The recall criterion is only used as a measurement of the experimental results, and the distance criterion is only partially satisfied by only a few algorithms [14,28]. In this paper, we propose a sub-linear time algorithm for kANN problem that unifies the two kinds of approximation criteria, which overcomes the disadvantages of the existing algorithms. The contributions of this paper are listed below.

1. We propose an algorithm that unifies the distance criterion and the recall criterion for the approximate k-Nearest-Neighbor problem. The result returned by the algorithm can satisfy at least one criterion in any situation. This is a major progress compared to the existing algorithms.
2. Assuming the input point set follows the Poisson Point Process, the algorithm takes $O(n \log n)$ time of preprocessing, $O(n \log n)$ space, and answers a query in $O(dn^{1/d} \log n + kn^\rho \log n)$ time, where $\rho < 1$ is a constant.
3. The algorithm is the first algorithm for kANN that provides theoretical guarantee on both of the approximation criteria, and it is also the first algorithm that achieves sub-linear query time while providing theoretical guarantees. The former works [14,28] with partial guarantee both need linear query time.

The rest of this paper is organized as follows. Section 2 introduces the definition of the problem and some prerequisite knowledge. The detailed algorithm are presented in Sect. 3. Then the time and space complexities are analyzed in Sect. 4. Finally the conclusion is given in Sect. 5.

2 Preliminaries

2.1 Problem Definitions

The problem studied in this paper is the approximate k-Nearest-Neighbor problem, which is denoted as kANN for short. In this paper the problem is constrained to the Euclidean space. The input is a set P of points where each $p \in P$ is a d-dimensional vector $(p^{(1)}, p^{(2)}, \cdots, p^{(n)})$. The distance between two points p and p' is defined by $D(p, p') = \sqrt{\sum_{i=1}^{d} (p^{(i)} - p'^{(i)})^2}$, which is the well known Euclidean distance. Before giving the definition of the kANN problem, we first introduce the exact kNN problem.

Definition 2.1 (kNN). *Given the input point set $P \subset R^d$ and a query point $q \in R^d$, define $kNN(q, P)$ to be the set of k points in P that are nearest to q. Formally,*

1. *$kNN(q, P) \subseteq P$, and $|kNN(q, P)| = k$;*
2. *$D(p, q) \leq D(p', q)$ for $\forall p \in kNN(q, P)$ and $\forall p' \in P \setminus kNN(q, P)$.*

Next we will give the definition of the approximate kNN. There are two kinds of definitions based on different approximation criteria.

Definition 2.2 ($kANN_c$). *Given the input point set $P \subset R^d$, a query point $q \in R^d$, and a approximation factor $c > 1$, find a point set $kANN_c(q, P)$ which satisfies:*

1. $kANN_c(q, P) \subseteq P$, and $|kANN_c(q, P)| = k$;
2. *let* $T_k(q, P) = \max\limits_{p \in kNN(q, P)} D(p, q)$, *then* $D(p', q) \leq c \cdot T_k(q, P)$ *holds for* $\forall p' \in$
    $kANN_c(q, P)$.

Remark 2.1. The second requirement in Definition 2.2 is called the distance criterion.

Definition 2.3 ($kANN_\delta$). *Given the input point set $P \subset R^d$, a query point $q \in R^d$, and a approximation factor $\delta < 1$, find a point set $kANN_\delta(q, P) \subseteq P$ which satisfies:*

1. $kANN_\delta(q, P) \subseteq P$, and $|kANN_\delta(q, P)| = k$;
2. $|kANN_\delta(q, P) \cap kNN(q, P)| \geq \delta \cdot k$.

Remark 2.2. If a kANN algorithm returned a set S, the value $\frac{|S \cap kNN(q, P)|}{|kNN(q, P)|}$ is usually called the recall of the set S. This is widely used in many works to evaluate the quality of the kANN algorithm. Thus we call the second statement in Definition 2.3 as the recall criterion.

Next we give the definition of the problem studied in this paper, which unifies the two different criteria.

Definition 2.4. *Given the input point set $P \subset R^d$, a query point $q \in R^d$, and approximation factors $c > 1$ and $\delta < 1$, find a point set $kNN_{c,\delta}(q, P)$ which satisfies:*

1. $kANN_{c,\delta}(q, P) \subseteq P$, and $|kANN_{c,\delta}(q, P)| = k$;
2. $kANN_{c,\delta}(q, P)$ *satisfies at least one of the distance criterion and the recall criterion. Formally, either* $D(p', q) \leq c \cdot T_k(q, P)$ *holds for* $\forall p' \in kANN_{c,\delta}(q, P)$, *or* $|kANN_{c,\delta}(q, P) \cap kNN(q, P)| \geq \delta \cdot k$.

According to Definition 2.4, the output of the algorithm is required to satisfy one of the two criteria, but not both. It will be our future work to devise an algorithm to satisfy both of the criteria.

In the rest of this section we will introduce some concepts and algorithms that will be used in our proposed algorithm.

2.2   Minimum Enclosing Spheres

The d-dimensional spheres is the generalization of the circles in the 2-dimensional case. Let c be the center and r be the radius. A d-dimensional sphere, denoted as $S(c, r)$, is the set $S(c, r) = \{x \in R^d \mid D(x, c) \leq r\}$. Note that the boundary is included. If $q \in S(c, r)$ we say that q falls inside sphere $S(c, r)$, or the sphere encloses point p. A sphere $S(c, r)$ is said to pass through point p iff $D(c, p) = r$.

Given a set P of points, the minimum enclosing sphere (MES) of P, is the d-dimensional sphere enclosing all points in P and has the smallest possible radius. It is known that the MES of a given finite point set in R^d is unique, and can be calculated by a quadratic programming algorithm [32]. Next we introduce the approximate minimum enclosing spheres.

Definition 2.5 (AMES). *Given a set of points $P \subset R^d$ and an approximation factor $\epsilon < 1$, the approximate minimum enclosing sphere of P, denoted as $AMES(P, \epsilon)$, is a d-dimensional sphere $S(c, r)$ satisfies:*

1. $p \in S(c, r)$ for $\forall p \in P$;
2. $r < (1 + \epsilon)r^*$, where r^* is the radius of the exact MES of P.

The following algorithm can calculate the AMES in $O(n/\epsilon^2)$ time, which is given in [5].

Algorithm 1: Compute $AMES$

Input: a point set P, and an approximation factor ϵ.
Output: $AMES(P, \epsilon)$

1   $c_0 \leftarrow$ an arbitrary point in P;
2   **for** $i = 1$ *to* $1/\epsilon^2$ **do**
3      $p_i \leftarrow$ the point in P farthest away from c_{i-1};
4      $c_i \leftarrow c_{i-1} + \frac{1}{i}(p_i - c_{i-1})$;
5   **end**

The following Lemma gives the complexity of Algorithm 1 .

Lemma 2.1 [5]. *For given ϵ and P where $|P| = n$, Algorithm 1 can calculate $AMES(P, \epsilon)$ in $O(n/\epsilon^2)$ time.*

2.3   Delaunay Triangulation

The Delaunay Triangulation (DT) is a fundamental data structure in computation geometry. The definition is given below.

Definition 2.6 (DT). *Given a set of points $P \subset R^d$, the Delaunay Triangulation is a graph $DT(P) = (V, E)$ which satisfies:*

1. $V = P$;
2. *for $\forall p, p' \in P$, $(p, p') \in E$ iff there exists a d-dimensional sphere passing through p and p', and no other $p'' \in P$ is inside it.*

The Delaunay Triangulation is a natural dual of the Voronoi diagram. We omit the details about their relationship since it is not the focus of this paper.

There are extensive research works about the Delaunay triangulation. An important problem is to find the expected properties of DT built on random point sets. Here we focus on the point sets that follow the Poisson Point Process in d-dimensional Euclidean space. In this model, for any region $\mathcal{R} \subset R^d$, the probability that $\mathcal{R}$ contains k points follows a Poisson-like distribution. See [1] for more details. We cite one important property of the Poisson Point Process in the following lemma.

Lemma 2.2 [1]. *Let $S \subset R^d$ be a point set following the Poisson Point Process. Suppose there are two regions $B \subseteq A \subset R^d$. For any point $p \in S$, if p falls inside A then the probability that p falls inside B is the ratio between the volume of B and A. Formally, we have*

$$\Pr[p \in B \mid p \in A] = \frac{volume(B)}{volume(A)}.$$

Further, we cite some important properties of the Delaunay triangulation built on point sets which follow the Poisson Point Process.

Lemma 2.3 [7]. *Let $S \subset R^d$ be a point set following the Poisson Point Process, and $\Delta(G) = \max\limits_{p \in V(G)} |\{(p,q) \in E(G)\}|$ be the maximum degree of G. Then the expected maximum degree of $DT(S)$ is $O(\log n / \log \log n)$.*

Lemma 2.4 [9]. *Let $S \subset R^d$ be a point set following the Poisson Point Process. The expected time to construct $DT(S)$ is $O(n \log n)$.*

2.4 Walking in Delaunay Triangulation

Given a Delaunay Triangulation DT, the points and edges of DT form a set of simplices. Given a query point q, there is a problem to find which simplex of DT that q falls in. There is a class of algorithms to tackle this problem which is called Walking. The Walking algorithm starts at some simplex, and *walks* to the destination by moving to adjacent simplices step by step. There are several kinds of walking strategy, including Jump&Walk [24], Straight Walk [8] and Stochastic Walk [12], etc. Some of these strategies are only applicable to 2 or 3 dimensions, while Straight Walk can generalize to higher dimension. As Fig. 1 shows, the Straight Walk strategy only considers the simplices that intersect the line segment from the start point to the destination. The following lemma gives the complexity of this walking strategy.

Lemma 2.5 [10]. *Given a Delaunay Triangulation DT of a point set $P \subset R^d$, and two points p and p' in R^d as the start point and destination point, the walking from p to p' using Straight Walk takes $O(n^{1/d})$ expected time.*

2.5 (c, r)-NN

The Approximate Near Neighbor problem is introduced in [19] for solving the $kANN_c$ problem with $k = 1$. Usually the Approximate Near Neighbor problem is denoted as (c, r)-NN since there are two input parameters c and r. The definition is given below. The idea to use (c, r)-NN to solve $1ANN_c$ is via Turing reduction, that is, use (c, r)-NN as an oracle or sub-procedure. The details can be found in [16, 17, 19, 21].

Definition 2.7. *Given a point set P, a query point q, and two query parameters $c > 1, r > 0$, the output of the (c, r)-NN problem should satisfy:*

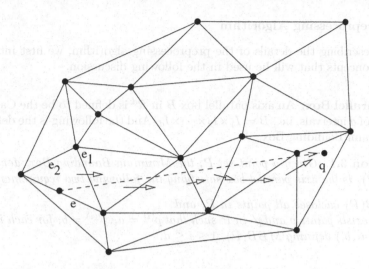

Fig. 1. Illustration of the straight walk

1. *if $\exists p^* \in S(q, r) \cap P$, then output a point $p' \in S(q, c \cdot r) \cap P$;*
2. *if $D(p, q) > c \cdot r$ for $\forall p \in P$, then output No;*

Since we aim to solve kANN problem in this paper, we need the following definition of (c, r)-kNN.

Definition 2.8. *Given a point set P, a query point q, and two query parameters c, r, the output of the (c, r)-kNN problem is a set $kNN_{(c,r)}(q, P)$, which satisfies:*

1. *if $|P \cap S(q, r)| \geq k$, then output a set $Q \subseteq P \cap S(q, c \cdot r)$, where $|Q| = k$;*
2. *if $|P \cap S(q, c \cdot r)| < k$, then output $\emptyset$;*

It can be easily seen that the (c, r)-kNN problem is a natural generalization of the (c, r)-NN problem. Recently, there are several algorithms proposed to solve this problem. The following Lemma 2.6 gives the complexity of the (c, r)-kNN algorithm. The proof can be found on the online version of this paper [18].

Lemma 2.6. *There is an algorithm that solves (c, r)-kNN problem in $O(kn^\rho)$ of time, requiring $O(kn^{1+\rho} \log n)$ time of preprocessing and $O(kn^{1+\rho})$ of space. The parameter ρ is a constant depending on the LSH function used in the algorithm, and $\rho < 1$ always holds.*

3   Algorithm

The proposed algorithm consists of two phases, i.e., the preprocessing phase and the query phase. The preprocessing phase is to built a data structure, which will be used to guide the search in the query phase. Next we will describe the algorithm of the two phases in detail.

3.1   Preprocessing Algorithm

Before describing the details of the preprocessing algorithm, we first introduce several concepts that will be used in the following discussion.

Axis Parallel Box. An axis parallel box B in R^d is defined to be the Cartesian product of d intervals, i.e., $B = I_1 \times I_2 \times \cdots \times I_d$. And the following is the definition of Minimum Bounding Box.

Definition 3.1. *Given a point set P, the Minimum Bounding Box, denoted as $MBB(P)$, is the axis parallel box satisfying the following two requirements:*

1. *$MBB(P)$ encloses all points in P, and*
2. *there exists points p and p' in P such that $p^{(i)} = a_i, p'^{(i)} = b_i$ for each interval $I_i = (a_i, b_i)$ defining $MBB(P)$, $1 \leq i \leq d$.*

Median Split. Given a point set P and its minimum bounding box $MBB(P)$, we introduce an operation on P that splits P into two subsets, which is called median split. This operation first finds the longest interval I_i from the intervals defining $MBB(P)$. Then, the operation finds the median of the set $\{p^{(i)} \mid p \in P\}$, which is the median of the i-th coordinates of the points in P. This median is denoted as $med_i(P)$. Finally P is split into two subsets, i.e., $P_1 = \{p \in P \mid p^{(i)} \leq med_i(P)\}$ and $P_2 = \{p \in P \mid p^{(i)} > med_i(P)\}$. Here we assume that no two points share the same coordinate in any dimension. This assumption can be assured by adding some random small shift on the original coordinates.

Median Split Tree. By recursively conducting the median split operation, a point set P can be organized into a tree structure, which is called the Median Split Tree (MST). The definition of MST is given below.

Definition 3.2. *Given the input point set P, a Median Split Tree (MST) based on P, denoted as $MST(P)$, is a tree structure satisfying the following requirements:*

1. *the root of $MST(P)$ is P, and the other nodes in $MST(P)$ are subsets of P;*
2. *there are two child nodes for each interior node $N \in MST(P)$, which are generated by conducting a median split on N;*
3. *each leaf node contains only one point.*

Balanced Median Split Tree. The depth of a node N in a tree T, denoted as $dep_T(N)$, is defined to be the number of edges in the path from N to the root of T. It can be noticed that the leaf nodes in the MST may have different depths. So we introduce the Balanced Median Split Tree (BMST), where all leaf nodes have the same depth.

Let $L_T(i) = \{N \in T \mid dep_T(N) = i\}$, which is the nodes in the i-th layer in tree T, and $|N|$ be the number of points included in node N. For a median split tree $MST(P)$, it can be easily proved that either $|N| = \lceil n/2^i \rceil$ or $|N| = \lfloor n/2^i \rfloor$ for $\forall N \in L_{MST(P)}(i)$. Given $MST(P)$, the $BMST(P)$ is constructed as follows. Find the smallest i such that $\lfloor n/2^i \rfloor \leq 3$, then for each node $N \in L_{MST(P)}(i)$, connect all the nodes in the sub-tree rooted at N directly to N.

Hierarchical Delaunay Graph. Given a point set P, we introduce the most important concept for the preprocessing algorithm in this paper, which is the Hierarchical Delaunay Graph (HDG). This structure is constructed by adding edges between nodes in the same layer of $BMST(P)$. The additional edges are called the graph edges, in contrast with the tree edges in $BMST(P)$. The definition of the HDG is given below. Here $Cen(N)$ denotes the center of $AMES(N)$.

Definition 3.3. *Given a point set P and the balanced median split tree $BMST(P)$, a Hierarchical Delaunay Graph HDG is a layered graph based on $BMST(P)$, where each layer is a Delaunay triangulation. Formally, for each $N, N' \in HDG(P)$, there is an graph edge between N, N' iff*

1. *$dep_{BMST(P)}(N) = dep_{BMST(P)}(N')$, and*
2. *there exists a d-dimensional sphere S passing through $Cen(N), Cen(N')$, and there is no $N'' \in HDG(P)$ such that $Cen(N'')$ falls in S, where N'' is in the same layer with N and N'. That is, the graph edges connecting nodes in the same layer forms the Delaunay Triangulation.*

The Preprocessing Algorithm. Next we describe the preprocessing algorithm which aims to build the HDG. The algorithm can be divided into three steps.

Step 1, split and build tree. The first step is to recursively split P into smaller sets using the median split operation, and the median split tree is built. Finally the nodes near the leaf layer is adjusted to satisfy the definition of the balanced median split tree.

Step 2, compute spheres. In this step, the algorithm will go over the tree and compute the AMES for each node using Algorithm 1.

Step 3, construct the HDG. In this step, an algorithm given in [9] which satisfies Lemma 2.4 is invoked to compute the Delaunay triangulation for each layer.

The pseudo codes of the preprocessing algorithm is given in Algorithm 2.

3.2 Query Algorithm

The query algorithm takes the HDG built by the preprocessing algorithm, and executes the following three steps.

The first is the descending step. The algorithm goes down the tree and stops at level i such that $k \leq n/2^i < 2k$. At each level, the child node with smallest distance to the query is chosen to be visited in next level.

Algorithm 2: Preprocessing Algorithm

Input: a point set P
Output: a hierarchical Delaynay graph $HDG(P)$
1 $T \leftarrow$ SplitTree(P);
2 Modify T into a BMST;
3 ComputeSpheres(T);
4 HierarchicalDelaunay(T);
5 **Procedure** SplitTree(N):
6 | Conduct median split on N and generate two sets N_1 and N_2;
7 | $T_1 \leftarrow$ SplitTree(N_1);
8 | $T_2 \leftarrow$ SplitTree(N_2);
9 | Let T_1 be the left sub-tree of N, and T_2 be the right sub-tree of N;
10 **end**
11 **Procedure** ComputeSpheres(T):
12 | **foreach** $N \in T$ **do**
13 | | Call AMES($N, 0.1$) (Algorithm 1);
14 | **end**
15 **end**
16 s **Procedure** HierarchicalDelaunay(T):
17 | Let dl be the depth of the leaf node in T;
18 | **for** $i = 0$ *to* dl **do**
19 | | Delaunay($L_T(i)$) (Lemma 2.4);
20 | **end**
21 **end**

The second is the navigating step. The algorithm marches towards the local nearest AMES center by moving on the edges of the HDG.

The third step is the answering step. The algorithm finds the answer of $kANN_{c,\delta}(q, P)$ by invoking the (c, r)-kNN query. The answer can satisfy the distance criterion or the recall criterion according to the different return result of the (c, r)-kNN query.

Algorithm 3 describes the above process in pseudo codes, where $Cen(N)$ and $Rad(N)$ are the center and radius of the $AMES$ of node N, respectively.

4   Analysis

The analysis in this section will assume that the input point set P follows the Poisson Point Process. The proofs can be found in the online version of this paper [18], and are all omitted here due to space limitation.

4.1   Correctness

Lemma 4.1. *If Algorithm 3 terminates when $i = 0$, then the returned point set Res is a δ-kNN of q in P with at least $1 - e^{-\frac{n-k}{n^d}}$ probability.*

Algorithm 3: Query

Input: a query points q, a point set P, approximation factors $c > 1, \delta < 1$, and $HDG(P)$

Output: $kANN_{c,\delta}(q, P)$

1   $N \leftarrow$ the root of $HDG(P)$;

2   **while** $|N| > 2k$ **do**

3      $Lc \leftarrow$ the left child of N, $Rc \leftarrow$ the right child of N;

4   **if** $D(q, Cen(Lc)) < D(q, Cen(Rc)))$ **then**

5         $N \leftarrow Lc$;

6   **else**

7         $N \leftarrow Rc$;

8   **end**

9   **end**

10   **while** $\exists N' \in Nbr(N)$ s.t. $D(q, Cen(N')) < D(q, Cen(N)))$ **do**

11      $N \leftarrow \arg \min\limits_{N' \in Nbr(N)} \{D(q, Cen(N'))\}$;

12   **end**

13   **for** $i = 0$ *to* $\log_c n$ **do**

14   Invoke (c, r)-kNN query where $r = \frac{D(q, Cen(N)) + Rad(N)}{n} c^i$;

15   **if** *the query returned a set Res* **then**

16   **return** *Res* as the final result;

17   **end**

18   **end**

Lemma 4.2. *If Algorithm 3 terminates when $i > 0$, then the returned point set Res is a c-kNN of q in P.*

Theorem 4.1. *The result of Algorithm 3 satisfies the requirement of $kNN_{c,\delta}(q, P)$ with at least $1 - e^{-\frac{n-k}{n^d}}$ probability.*

4.2 Complexities

Theorem 4.2. *The expected time complexity of Algorithm 2, which is the pre-processing time complexity, is $O(n \log n)$.*

Theorem 4.3. *The space complexity of Algorithm 2 is $O(n \log n)$.*

Theorem 4.4. *The time complexity of Algorithm 3, which is the query complexity, is $O(dn^{1/d} \log n + kn^\rho \log n)$, where $\rho < 1$ is a constant.*

5 Conclusion

In this paper we proposed an algorithm for the approximate k-Nearest-Neighbors problem. We observed that there are two kinds of approximation criteria in the history of this research area, which is called the distance criterion and the recall criterion in this paper. But we also observed that all existing works do not have theoretical guarantees on the two criteria. We raised a new definition for the approximate k-Nearest-Neighbor problem which unifies the distance criterion and the recall criterion, and proposed an algorithm that solves the new problem. The result of the algorithm can satisfy at least one of the two criteria. In our future work, we will try to devise new algorithms that can satisfy both of the criteria.

References

1. Poisson Point Process. https://wikimili.com/en/Poisson_point_process
2. Amato, G., Gennaro, C., Savino, P.: MI-file: using inverted files for scalable approximate similarity search. Multimed. Tools Appl. **71**(3), 1333–1362 (2012). https://doi.org/10.1007/s11042-012-1271-1
3. Andoni, A., Laarhoven, T., Razenshteyn, I., Waingarten, E.: Optimal hashing-based time-space trade-offs for approximate near neighbors. In: Proceedings of the Twenty-Eighth Annual ACM-SIAM Symposium on Discrete Algorithms, pp. 47–66, January 2017
4. Andoni, A., Razenshteyn, I.: Optimal data-dependent hashing for approximate near neighbors. In: Proceedings of the Forty-Seventh Annual ACM on Symposium on Theory of Computing - STOC 2015, pp. 793–801 (2015)
5. Bâdoiu, M., Bâdoiu, M., Clarkson, K.L., Clarkson, K.L.: Smaller core-sets for balls. In: Proceedings of the Fourteenth Annual ACM-SIAM Symposium on Discrete Algorithms, pp. 801–802 (2003)
6. Bentley, J.L.: Multidimensional binary search trees used for associative searching. Commun. ACM **18**(9), 509–517 (1975)
7. Bern, M., Eppstein, D., Yao, F.: The expected extremes in a delaunay triangulation. Int. J. Comput. Geom. Appl. **01**(01), 79–91 (1991)
8. Bose, P., Devroye, L.: On the stabbing number of a random Delaunay triangulation. Comput. Geom.: Theory Appl. **36**(2), 89–105 (2007)
9. Buchin, K., Mulzer, W.: Delaunay triangulations in O(sort(n)) time and more. In: Proceedings - Annual IEEE Symposium on Foundations of Computer Science, FOCS , vol. 5, pp. 139–148 (2009)
10. de Castro, P.M.M., Devillers, O.: Simple and efficient distribution-sensitive point location in triangulations. In: 2011 Proceedings of the Thirteenth Workshop on Algorithm Engineering and Experiments (ALENEX), pp. 127–138, January 2011
11. Datar, M., Immorlica, N., Indyk, P., Mirrokni, V.S.: Locality-sensitive hashing scheme based on p-stable distributions. In: Proceedings of the Twentieth Annual Symposium on Computational Geometry - SCG 2004, pp. 253–262 (2004)
12. Devillers, O., Pion, S., Teillaud, M.: Walking in a triangulation. Int. J. Found. Comput. Sci. **13**(02), 181–199 (2002)
13. Esuli, A.: Use of permutation prefixes for efficient and scalable approximate similarity search. Inf. Process. Manage. **48**(5), 889–902 (2012)
14. Gan, J., Feng, J., Fang, Q., Ng, W.: Locality-sensitive hashing scheme based on dynamic collision counting. In: Proceedings of the 2012 International Conference on Management of Data - SIGMOD 2012, pp. 541–552 (2012)
15. Gao, J., Jagadish, H., Ooi, B.C., Wang, S.: Selective hashing: closing the gap between radius search and k-NN Search. In: Proceedings of the 21th ACM SIGKDD International Conference on Knowledge Discovery and Data Mining - KDD 2015, pp. 349–358 (2015)

16. Har-Peled, S.: A replacement for Voronoi diagrams of near linear size. In: Proceedings 42nd IEEE Symposium on Foundations of Computer Science, pp. 94–103. IEEE (2001)
17. Har-Peled, S., Indyk, P., Motwani, R.: Approximate nearest neighbor: towards removing the curse of dimensionality. Theory Comput. 8(1), 321–350 (2012)
18. Hengzhao Ma, J.L.: A sub-linear time algorithm for approximating k-nearest-neighbor with full quality guarantee (2020). https://arxiv.org/abs/2008.02924
19. Indyk, P., Motwani, R.: Approximate nearest neighbors: towards removing the curse of dimensionality. In: Proceedings of the Thirtieth Annual ACM Symposium on Theory of Computing - STOC 1998, pp. 604–613 (1998)
20. Lin, P.C., Zhao, W.L.: Graph based Nearest Neighbor Search: Promises and Failures, pp. 1–8 (2019)
21. Ma, H., Li, J.: An algorithm for reducing approximate nearest neighbor to approximate near neighbor with O(log n) query time. In: 12th International Conference on Combinatorial Optimization and Applications - COCOA 2018, pp. 465–479 (2018)
22. Malkov, Y., Ponomarenko, A., Logvinov, A., Krylov, V.: Approximate nearest neighbor algorithm based on navigable small world graphs. Inf. Syst. 45, 61–68 (2014)
23. Mitchell, J.S., Mulzer, W.: Proximity algorithms. In: Handbook of Discrete and Computational Geometry, Third Edition, pp. 849–874 (2017)
24. Mücke, E.P., Saias, I., Zhu, B.: Fast randomized point location without preprocessing in two- and three-dimensional Delaunay triangulations. Comput. Geom. 12(1–2), 63–83 (1999)
25. Muja, M., Lowe, D.G.: Scalable nearest neighbor algorithms for high dimensional data. IEEE Trans. Pattern Anal. Mach. Intell. 36(11), 2227–2240 (2014)
26. Ocsa, A., Bedregal, C., Cuadros-vargas, E., Society, P.C.: A new approach for similarity queries using proximity graphs, pp. 131–142. Simpósio Brasileiro de Banco de Dados (2007)
27. Paredes, R., Chávez, E.: Using the k-nearest neighbor graph for proximity searching in metric spaces. In: International Symposium on String Processing and Information Retrieval, pp. 127–138 (2005)
28. Sun, Y., Wang, W., Qin, J., Zhang, Y., Lin, X.: SRS: solving c-approximate nearest neighbor queries in high dimensional Euclidean space with a tiny index. Proc. VLDB Endow. 8, 1–12 (2014)
29. Wang, J., Shen, H.T., Song, J., Ji, J.: Hashing for similarity search: a survey. In: ArXiv:1408.2927 (2014)
30. Weber, R., Schek, H.J., Blott, S.: A quantitative analysis and performance study for similarity-search methods in high-dimensional spaces. In: Proceedings of 24rd International Conference on Very Large Data Bases, pp. 194–205 (1998)
31. Yianilos, P.N.: Data structures and algorithms for nearest neighbor search in general metric spaces. In: Proceedings of the Fourth Annual ACM-SIAM Symposium on Discrete Algorithms, pp. 311–321 (1993)
32. Yildirim, E.A.: Two algorithms for the minimum enclosing ball problem. SIAM J. Optim. 19(3), 1368–1391 (2008)

Sampling-Based Approximate Skyline Calculation on Big Data

Xingxing Xiao and Jianzhong Li$^{(\boxtimes)}$

Harbin Institute of Technology, Harbin 150001, Heilongjiang, China
{xiaoxx,lijzh}@hit.edu.cn

Abstract. The existing algorithms for processing skyline queries cannot adapt to big data. This paper proposes two approximate skyline algorithms based on sampling. The first algorithm obtains a fixed size sample and computes the approximate skyline on the sample. The error of the first algorithm is relatively small in most cases, and is almost independent of the input relation size. The second algorithm returns an (ϵ, δ)-approximation for the exact skyline. The size of sample required by the second algorithm can be regarded as a constant relative to the input relation size, so is the running time.

Keywords: Sampling · Skyline · Approximation · Big data

1 Introduction

Skyline queries are important in many applications involving multi-criteria decision making. Given a relation $T(A_1, A_2, ..., A_d)$ and a set of skyline criteria $C \subseteq \{A_1, ..., A_d\}$, a skyline query on T is to find a subset of T such that each tuple t in the subset is not dominated by any tuple in T, where t' dominates t, written as $t' \prec t$, means that $t'.A_i \leq t.A_i$ for all $A_i \in C$ and there is an attribute $A_j \in C$ such that $t'.A_j < t.A_j$. $t.A_i$ is the value of tuple $t \in T$ on attribute A_i. Skyline queries can also be defined using $\geq$ and $>$. Without loss of generality, this paper only considers the skyline queries defined by $\leq$ and $<$. The answers to a skyline query are all the potentially best tuples to users, and skyline queries provide good mechanisms for merging user's preferences into queries.

Studies on skyline queries originated in theoretical computer science area in the last century. Skyline was called as the set of maximals or the pareto set in that time. Many algorithms for finding the maximals were proposed [3,4,16]. The lowest time complexity of these algorithms is $O(n \log^{d-2} n)$ in the worst case, and $O(n)$ in the average case. However, all the algorithms are based on Divide&Conquer strategy and assume that their input tuples are stored in main memory.

This work was supported by the National Natural Science Foundation of China under grant 61732003, 61832003, 61972110 and U1811461.

© Springer Nature Switzerland AG 2020
W. Wu and Z. Zhang (Eds.): COCOA 2020, LNCS 12577, pp. 32–46, 2020.
https://doi.org/10.1007/978-3-030-64843-5_3

Borzsony first introduced skyline queries to the database field [5]. It attracted considerable attention to design efficient algorithms for processing skyline queries on relations stored in external storage. Many algorithms have been proposed [2, 5, 8, 12]. The lowest time complexity of the algorithms is $O(n^2)$ in the worst case, and $O(n)$ in the average case.

Nowadays, big data is coming to the force in a lot of applications [10]. Processing a skyline query on big data in more than linear time is by far too expensive and often even linear time may be too slow. Thus, designing a subliner time algorithm for processing skyline queries becomes a highly concerned research subject. Many index-based algorithms for processing skyline queries have been proposed to achieve the sublinear running time in the average case [5, 13, 15, 17, 24, 26]. However, all the algorithms have serious limitations. Firstly, the algorithms require much time for pre-computation, which is at least $\Omega(n)$. Secondly, they need expensive extra space overhead for indexes. Thirdly, there is much overhead to maintain indexes while the input relations are updated.

Approximation computation [7, 20, 21, 23] of the skyline is the only way to break trough the three limitations. Fortunately, approximate skyline results are enough in many applications. An example of skyline queries is to find restaurants near the workplace that provide delicious foods and excellent services. To get the answer quickly, users can accept approximate skyline results that are the good restaurants but not the best ones. Actually, users prefer to get approximate results in seconds rather than exact results in hours or more in many applications.

There have been many researches on approximate algorithms for skyline queries [14, 18, 19, 25, 27], but their goal is to reduce the skyline size and approximate the best subset of k input tuples to represent the skyline under various measures. Moreover, they have higher running time than the precise algorithms for processing skyline queries.

In this paper, we propose two approximate algorithms based on sampling [22], for processing skyline queries on big data. The proposed algorithms don't need any extra space or pre-computation overhead. Viewing the skyline as a covering, the error of a approximate algorithm is defined as $|\frac{\mathcal{DN}(Sky)-\mathcal{DN}(\widetilde{Sky})}{\mathcal{DN}(Sky)}|$, where $\mathcal{DN}(\widetilde{Sky})$ is the number of tuples dominated by the approximate result $\widetilde{Sky}$, and $\mathcal{DN}(Sky)$ is the number of tuples dominated by the exact result Sky. If $|\frac{\mathcal{DN}(Sky)-\mathcal{DN}(\widetilde{Sky})}{\mathcal{DN}(Sky)}| \leq \epsilon$, then ϵ is called as the error bound of the approximate algorithm.

The first algorithm draws a random sample from the input relation at the beginning, and then computes the approximate skyline on the sample. The algorithm has two advantages. First, the expected error of the algorithm is almost independent of the input relation size. Second, the standard deviation of the error is relatively small.

The second algorithm, DOUBLE, is a random algorithm and returns an (ϵ, δ)-approximation for the exact skyline efficiently. The size of sample required by DOUBLE is almost a constant relative to the input relation size. DOUBLE first draws an initial sample, and then computes the approximate skyline on the sample. Afterwards, it judges whether the current result meets the requirement by *Monte Carlo method*. If not, it doubles the sample size and repeats the above process. Otherwise it terminates.

The main contributions of the paper are listed below.

(1) A baseline approximate algorithm for processing skyline queries is proposed, which is based on a sample of size m. The running time of the algorithm is $O(m \log^{d-2} m)$ in the worst case and $O(m)$ in the average case. If m is equal to $n^{\frac{1}{k}}$ ($k > 1$), the baseline algorithm is in sublinear time. If all skyline criteria are independent of each other, the expected error of the algorithm is

$$\overline{\varepsilon} \le \frac{n-m}{n} \sum_{i=0}^{d-1} \frac{(\log(m+1))^i}{i!(m+1)}.$$

And the standard deviation of the error is $o(\overline{\varepsilon})$.

(2) An approximate algorithm, DOUBLE, is proposed to return an (ϵ, δ)-approximation for the exact skyline efficiently. The expected sample size required by DOUBLE is $O(\mathcal{M}_{\frac{\epsilon}{3}, \delta})$, and the expected time complexity of DOUBLE is $O(\mathcal{M}_{\frac{\epsilon}{3}, \delta} \log^{d-1} \mathcal{M}_{\frac{\epsilon}{3}, \delta})$, where $\mathcal{M}_{\frac{\epsilon}{3}, \delta}$ is the size of sample required by the baseline algorithm to return an $(\frac{\epsilon}{3}, \delta)$-approximation. $\mathcal{M}_{\frac{\epsilon}{3}, \delta}$ is almost unaffected by the relation size.

The remainder of the paper is organized as follows. Section 2 provides problem definitions. Section 3 describes the baseline algorithm and its analysis. Section 4 presents DOUBLE and its analysis. Section 5 concludes the paper.

2   Problem Definition

2.1   Skyline Definition

Let $T(A_1, A_2, ..., A_d)$ be a relation with n tuples and d attributes, abbreviated as T. In the following, we assume that all attributes are skyline criteria. First, we formally define the dominance relationship between tuples in T.

Definition 2.1 (Dominance between Tuples). *Let t and t' be tuples in the relation $T(A_1, A_2, ..., A_d)$. t dominates t' with respect to the d attributes of T, denoted by $t \prec t'$, if $t.A_i \le t'.A_i$ for all $A_i \in \{A_1, ..., A_d\}$, and $\exists A_j \in \{A_1, ..., A_d\}$ such that $t.A_j < t'.A_j$.*

Based on the dominance relationship between tuples, we can define the dominance relationship between sets. In the following, $t \preceq t'$ denotes $t \prec t'$ or $t = t'$ with respect to d attributes of T.

Definition 2.2 (Dominance between Sets). *A tuple set Q dominates another set Q', denoted by $Q \preceq Q'$, if for each tuple t' in Q', there is a tuple t in Q such that $t \prec t'$ or $t = t'$, i.e. $t \preceq t'$. $Q \preceq \{t\}$ can be abbreviated as $Q \preceq t$.*

Now, we define the skyline of a relation.

Definition 2.3 (Skyline). *Given a relation* $T(A_1, A_2, ..., A_d)$*, the skyline of* T *is* $Sky(T) = \{t \in T | \forall t' \in T, t' \nprec t\}$.

Definition 2.4 (Skyline Problem). *The skyline problem is defined as follows.*
Input: a relation $T(A_1, A_2, ..., A_d)$*.*
Output: $Sky(T)$*.*

The skyline problem can be equivalently defined as following optimization problem.

Definition 2.5 (OP-Sky Problem). *OP-Sky problem is defined as follows.*
Input: a relation $T(A_1, A_2, ..., A_d)$*.*
Output: $Q \subseteq T$ *such that* $|\{t \in T | Q \preceq t\}|$ *is maximized and* $\forall t_1, t_2 \in Q, t_1 \nprec t_2$.

The following Theorem 2.1 shows that the *Skyline Problem* is equivalent to the *OP-Sky*.

Theorem 2.1. *The skyline of* T *is one of the optimal solutions of the problem* OP_1*. If there is no duplicate tuples in* T*,* $Sky(T)$ *is the unique optimal solution.*

This paper focus on approximate algorithms for solving the *OP-Sky* problem. The error of an approximate algorithm for an input relation T is defined as $|\frac{\mathcal{DN}(Sky) - \mathcal{DN}(\widetilde{Sky})}{\mathcal{DN}(Sky)}|$, where $\mathcal{DN}(\widetilde{Sky})$ is the number of tuples in T dominated by the approximate solution $\widetilde{Sky}$, and $\mathcal{DN}(Sky)$ is the number of tuples in T dominated by the exact solution Sky. If $|\frac{\mathcal{DN}(Sky) - \mathcal{DN}(\widetilde{Sky})}{\mathcal{DN}(Sky)}| \le \epsilon$, then ϵ is called as the error bound of the approximate algorithm.

In the following sections, we will present two approximate algorithms for solving the *OP-Sky* problem.

3 The Baseline Algorithm and Analysis

3.1 The Algorithm

The baseline algorithm first obtains a sample S of size m from the input relation T, and then computes the approximate skyline result on S. Any existing skyline algorithm can be invoked to compute the skyline of S.

Algorithm 1: The Baseline Algorithm

    Input: The relation $T(A_1, A_2, ..., A_d)$ with n tuples, and the sample size m;
    Output: $\widetilde{Sky}$, i.e. the approximate skyline of T
  1 S is the sample of m tuples from T;
  2 **return** getSkyline(S). /* getSkyline can be any exact skyline algorithm */

3.2   Error Analysis of the Baseline Algorithm

To facilitate the error analysis of the baseline algorithm, we assume that the baseline algorithm is based on sampling without replacement. Let ε be the error of the algorithm, $\overline{\varepsilon}$ be the expected error of the algorithm, and σ^2 be the variance of the error.

The Expected Error. We first analyze the expected error $\overline{\varepsilon}$ of the baseline algorithm. Assume each tuple in T is a d-dimensional *i.i.d.* (independent and identically distributed) random variable.

If the n random variables are continuous, we assume that they have the joint probability distribution function $F(v_1, v_2, ..., v_d) = F(\overline{V})$, where $\overline{V} = (v_1, v_2, ...v_d)$. Let $f(v_1, v_2, ..., v_d) = f(\overline{V})$ be the joint probability density function of the random variables. Without loss of generality, the range of variables on each attribute is $[0, 1]$, since the domain of any attribute of T can be transformed to $[0, 1]$.

Theorem 3.1. *If all the n tuples in T are d-dimensional i.i.d. continuous random variables with the distribution function $F(\overline{V})$, then the expected error of the baseline algorithm is*

$$\overline{\varepsilon} = \frac{n-m}{n} \int_{[0,1]^d} f(\overline{V})(1 - F(\overline{V}))^m d\overline{V}$$

where m is the sample size, $f(\overline{V})$ is the density function of the variables, and the range of variables on each attribute is $[0, 1]$.

Proof. Due to $\mathcal{DN}(Sky(S)) = \mathcal{DN}(S) \leq \mathcal{DN}(Sky(T)) = n$, where n is the size of the relation T, we have

$$\varepsilon = \left| \frac{\mathcal{DN}(Sky(T)) - \mathcal{DN}(Sky(S))}{\mathcal{DN}(Sky(T))} \right| = \frac{n - \mathcal{DN}(S)}{n}$$

Let X_i be a random variable for $1 \leq i \leq n$, and t_i be the ith tuple in T. $X_i = 0$ if t_i in T is dominated by the sample S, otherwise $X_i = 1$. Thus, we have $\mathcal{DN}(S) = n - \sum_{i=1}^{n} X_i$ and $\varepsilon = \frac{\sum_{i=1}^{n} X_i}{n}$. By the linearity of expectations, the expected error of the baseline algorithm is $\overline{\varepsilon} = \frac{\sum_{i=1}^{n} EX_i}{n} = \frac{\sum_{i=1}^{n} Pr(X_i=1)}{n} = Pr(X_i = 1)$, where $Pr(X_i = 1)$ is the probability that t_i in T is not dominated by S.

Let Y_i be a random variable for $1 \leq i \leq n$. $Y_i = 0$ if t_i in T is picked up into the sample S, otherwise $Y_i = 1$. According to the conditional probability formula, we have

$$Pr(X_i = 1) = Pr(Y_i = 0)Pr(X_i = 1|Y_i = 0) + Pr(Y_i = 1)Pr(X_i = 1|Y_i = 1)$$

If t_i is selected into in S, then it is dominated by S. Therefore, we have $Pr(X_i = 1|Y_i = 0)$ is equal to 0. Due to sampling with replacement, $Pr(Y_i = 1)$ is equal to $\frac{n-m}{n}$. In short, we have

$$Pr(X_i = 1) = \frac{n - m}{n} Pr(X_i = 1 | Y_i = 1)$$

Assume t_i is not selected into S. Let t_i have the value $\overline{V} = (v_1, v_2, ..., v_d)$. Subsequently, for the j_{th} tuple t'_j in S, t'_j satisfies the distribution F and is independent of t_i. It is almost impossible that t_i has a value equal to t'_j on an attribute. The probability of $t'_j \prec t_i$ is $F(\overline{V})$. In turn, we have $Pr(t'_j \not\prec t_i | Y_i = 1) = 1 - F(\overline{V})$.

Because S is a random sample without replacement, all tuples in S are distinct tuples from T. All the tuples in T are independently distributed, so are the tuples in S. Therefore, the probability that S doesn't dominate $\{t_i\}$ is

$$Pr(S \not\prec \{t_i\} | Y_i = 1) = \prod_{j=1}^{m} Pr(t'_j \not\prec t_i | Y_i = 1) = (1 - F(\overline{V}))^m$$

In the analysis above, $\overline{V}$ is regarded as a constant vector. Since $\overline{V}$ is a variable vector and has the density function $f(\overline{V})$, we have

$$Pr(X_i = 1 | Y_i = 1) = \int_{[0,1]^d} f(\overline{V})(1 - F(\overline{V}))^m d\overline{V}$$

Thus the probability that t_i is not dominated by S is

$$Pr(X_i = 1) = \frac{n - m}{n} \int_{[0,1]^d} f(\overline{V})(1 - F(\overline{V}))^m d\overline{V}$$

$\square$

Corollary 3.1. *If all the n tuples in T are d-dimensional i.i.d. continuous random variables, then the expected error of the baseline algorithm is*

$$\overline{\varepsilon} = \frac{n - m}{n} \frac{\mu_{m+1,d}}{m + 1}$$

where m is the sample size and $\mu_{m+1,d}$ is the expected skyline size of a set of $m + 1$ d-dimensional i.i.d. random variables with the same distribution.

Proof. Let Q be a set of $m + 1$ d-dimensional i.i.d. random variables with the distribution function F, then the expected skyline size of Q is $\mu_{m+1,d} = (m + 1) \int_{[0,1]^d} f(\overline{V})(1 - F(\overline{V}))^m d\overline{V}$. Based on Theorem 3.1, we get the corollary. $\square$

If the n random variables are discrete, we assume that they have the joint probability mass function as follows

$$g(v_1, v_2, ..., v_d) = Pr(A_1 = v_1, A_2 = v_2, ..., A_d = v_d)$$

Let $G(v_1, v_2, ..., v_d) = G(\overline{V})$ be the probability distribution function of the variables. Assume that $\mathcal{V}$ is the set of all tuples in T, i.e. all value vectors of the d-dimensional variables.

Theorem 3.2. *If all the n tuples in T are d-dimensional i.i.d. discrete random variables with the distribution function $G(\overline{V})$, then the expected error of the baseline algorithm is*

$$\overline{\varepsilon} = \frac{n-m}{n} \sum_{\overline{V} \in \mathcal{V}} g(\overline{V})(1 - G(\overline{V}))^m$$

where m is the sample size, $\mathcal{V}$ is the set of all value vectors of the d-dimensional variables and g is the mass function.

The proof is basically the same as Theorem 3.1, except that duplicate tuples need to be considered.

Based on Theorem 3.1 and 3.2, the relation size has almost no effect on the expected error of the baseline algorithm. Indeed, m is equal to $o(n)$, and $\frac{n-m}{n}$ approaches to 1 in most cases.

Corollary 3.2. *If all the n tuples in T are d-dimensional i.i.d. discrete random variables, then the expected error of the baseline algorithm is*

$$\overline{\varepsilon} \leq \frac{n-m}{n} \frac{\mu_{m+1,d}}{m+1} \tag{1}$$

where m is the sample size and $\mu_{m+1,d}$ is the expected skyline size of a set of $m+1$ d-dimensional i.i.d. random variables with the same distribution. If there is no duplicate tuples in T, then the equality of (1) holds.

Proof. Let Q be a set of $m+1$ d-dimensional i.i.d. random variables with the distribution function G, then the expected skyline size of Q is

$$\mu_{m+1,d} = (m+1) \sum_{\overline{V} \in \mathcal{V}} g(\overline{V})(1 - G(\overline{V}) + g(\overline{V}))^m$$

$$\geq (m+1) \sum_{\overline{V} \in \mathcal{V}} g(\overline{V})(1 - G(\overline{V}))^m \tag{2}$$

the equality of (2) holds if and only if there is no duplicate tuples in Q. Based on Theorem 3.2, the corollary is proved. $\square$

By the analysis of the expected skyline size under stronger assumptions in [11], we further analyze the expected error of the baseline algorithm.

Definition 3.1 (Component independence). $T(A_1, A_2, ..., A_d)$ *satisfies component independence (CI), if all n tuples in T follow the conditions below.*

1. *(Attribute Independence) the values of tuples in T on a single attribute are statically independent of the values on any other attribute;*
2. *(Distinct Values) T is sparse, i.e. any two tuples in T have different values on each attribute.*

Theorem 3.3. *Under CI, the error of the baseline algorithm is unaffected by the specific distribution of T.*

Proof. If T satisfies component independence, it can be converted into an uniformly and independently distributed set. After conversion, the error of the basline algorithm remains unchanged. The specific conversion process is as follows. Consider each attribute in turn. For the attribute A_i, sort tuples in ascending order by values on A_i. Then rank 0 is allocated the lowest value 0 on A_i, and so forth. Rank j is allocated the value j/n on A_i.   □

From [11], we have the following lemma.

Lemma 3.1. *Under CI, the expected skyline size of T is equal to the $(d-1)^{th}$ order harmonic of n, denoted by $H_{d-1,n}$.*

For integers $k > 0$ and integers $n > 0$, $H_{d,n} = \sum_{i=1}^{n} \frac{H_{d-1,i}}{i}$. From [6] and [9], we have

$$H_{d,n} = \frac{(\log n)^d}{d!} + \gamma \frac{(\log n)^{d-1}}{(d-1)!} + O((\log n)^{d-2}) \le \sum_{i=0}^{d} \frac{(\log n)^i}{i!},$$

where $\gamma = 0.577...$ is Euler's constant.

From Definition 3.1, there is no duplicate tuples in T under CI. Thus, based on Corollary 3.1 and 3.2, we have the following corollary.

Corollary 3.3. *If the relation $T(A_1, A_2, ..., A_d)$ with n tuples satisfies CI, then the expected error of the baseline algorithm is*

$$\overline{\varepsilon} = \frac{n-m}{n(m+1)} H_{d-1,n} \le \frac{n-m}{n(m+1)} \sum_{i=0}^{d-1} \frac{(\log(m+1))^i}{i!}$$

where m is the sample size.

If there are tuples in T with the same values on an attribute and Attribute Independence in Definition 3.1 holds, we have the following corollary.

Corollary 3.4. *If all attributes in $T(A_1, A_2, ..., A_d)$ are independent of each other, then the expected error of the baseline algorithm is*

$$\overline{\varepsilon} \le \frac{n-m}{n(m+1)} H_{d-1,n} \le \frac{n-m}{n(m+1)} \sum_{i=0}^{d-1} \frac{(\log(m+1))^i}{i!}$$

where m is the sample size.

Corollary 3.5. *If all attributes in $T(A_1, A_2, ..., A_d)$ are independent of each other, with sample size m equal to $n^{\frac{1}{k}} - 1 (k > 1)$, then the expected error of the baseline algorithm is*

$$\overline{\varepsilon} \le \frac{n - n^{\frac{1}{k}} + 1}{n} \sum_{i=0}^{d-1} \frac{(\log n)^i}{k^i n^{\frac{1}{k}} i!}$$

where m is the sample size.

Variance of the Error. We assume that each tuple in T is a d-dimensional *i.i.d* random variable and T satisfies component independence (CI). Without losing generality, all random variables are uniformly distributed over $[0,1]^d$.

Theorem 3.4. *If the relation $T(A_1, A_2, ..., A_d)$ with n tuples satisfies CI, then $\sigma^2 = O(\frac{\mu_{m,d}}{m^2})$, and $\sigma = O(\frac{\sqrt{\mu_{m,d}}}{m}) = o(\bar{\varepsilon})$.*

Proof. Let X_i be a random variable for $1 \leq i \leq n$. $X_i = 0$ if t_i in T is dominated by the sample S, otherwise $X_i = 1$. From the proof in Theorem 3.1, we have

$$\sigma^2 = D(\varepsilon) = D(\frac{\sum_{i=1}^n X_i}{n}) = D(\sum_{i=1}^n X_i)/n^2$$

$$= (E(\sum_{i=1}^n X_i^2) + E(\sum_{i \neq j} X_i X_j) - E^2(\sum_{i=1}^n X_i))/n^2$$

$$= \frac{1}{n}Pr(X_i = 1) + \frac{n-1}{n}Pr_{i \neq j}(X_i = X_j = 1) - Pr^2(X_i = 1).$$

Assume the t_i in T has the value $\overline{U} = (u_1, u_2, ..., u_d)$ and the j_{th} tuple t_j has the value $\overline{V} = (v_1, v_2, ..., v_d)$. Let (η) be

$$\{(\overline{U}, \overline{V}) | \overline{U} \in [0,1]^d, \overline{V} \in [0,1]^d, u_1 \leq v_1, ..., u_\eta \leq v_\eta, v_{\eta+1} < u_{\eta+1}, ..., v_d < u_d\}.$$

(η) represents the set of all possible $(\overline{U}, \overline{V})$, in which $\overline{U}$ has values no more than $\overline{V}$ on the first η attributes and has higher values on the subsequent attributes. Then we have

$$Pr_{i \neq j}(X_i = X_j = 1)$$
$$= \frac{(n-m)(n-1-m)}{n(n-1)} \sum_{\eta=0}^d \binom{d}{\eta} \int_{(\eta)} (1 - \prod_{i=1}^d u_i - \prod_{i=1}^d v_i + \prod_{i=1}^\eta u_i \prod_{i=\eta+1}^d v_i)^m d\overline{U} d\overline{V}.$$

In the above equation, $\frac{(n-m)(n-1-m)}{n(n-1)}$ is the probability that two distinct tuples both are not selected into the sample. Based on [1], we have

$$\sum_{\eta=1}^{d-1} \binom{d}{\eta} \int_{(\eta)} (1 - \prod_{i=1}^d u_i - \prod_{i=1}^d v_i + \prod_{i=1}^\eta u_i \prod_{i=\eta+1}^d v_i)^m d\overline{U} d\overline{V} = \frac{\mu_{m+2,d}^2 + O(\mu_{m+2,d})}{(m+1)(m+2)}.$$

Thus,

$$Pr_{i \neq j}(X_i = X_j = 1)$$
$$= \frac{(n-m)(n-1-m)}{n(n-1)} (2 \int_{[0,1]^d} (1 - \prod_{i=1}^d v_i)^m \prod_{i=1}^d v_i d\overline{V} + \frac{\mu_{m+2,d}^2 + O(\mu_{m+2,d})}{(m+1)(m+2)}) \tag{3}$$

$$= \frac{(n-m)(n-1-m)}{n(n-1)} \frac{2\mu_{m+2,d}(2) + \mu_{m+2,d}^2 + O(\mu_{m+2,d})}{(m+1)(m+2)}. \tag{4}$$

Equation (3) is based on variable substitution. In (4), $\mu_{n,d}(r)$ denotes the expected size of the r_{th} layer skyline of T, where the rth layer skyline of T is the set of tuples in T that are dominated by exactly $r - 1$ tuples in T, and its expected size is equal to

$$\mu_{n,d}(r) = n\binom{n-1}{r-1} \int_{[0,1]^d} (1 - \prod_{i=1}^{d} v_i)^{n-r} (\prod_{i=1}^{d} v_i)^{r-1} d\overline{V}.$$

Due to $Pr(X_i = 1) = \frac{n-m}{n} \frac{\mu_{m+1,d}}{m+1}$, we have

$$D(\sum_{i=1}^{n} X_i)$$

$$= (n - m)\frac{\mu_{m+1,d}}{m+1} - \frac{(n-m)(n+1)}{(m+1)^2(m+2)}\mu_{m+1,d}^2 + \frac{(n-m)(n-1-m)}{(m+1)(m+2)}(\mu_{m+2,d}+$$

$$\mu_{m+1,d})(\mu_{m+2,d} - \mu_{m+1,d}) + \frac{(n-m)(n-1-m)}{(m+1)(m+2)}(2\mu_{m+2,d}(2) + O(\mu_{m+2,d}))$$

$$= (n - m)\frac{\mu_{m+1,d}}{m+1} - \frac{(n-m)(n+1)}{(m+1)^2(m+2)}\mu_{m+1,d}^2 + \frac{(n-m)(n-1-m)}{(m+1)(m+2)}O(\mu_{m+2,d}).$$

$$(5)$$

Equation (5) holds because $\mu_{m+2,d} - \mu_{m+1,d} \leq 1$ and $\mu_{m+2,d}(2) \leq \mu_{m+2,d}$. With $\mu_{m+1,d} \leq m + 1$, it is true that $D(\sum_{i=1}^{n} X_i) = O(\frac{n^2}{m^2}\mu_{m,d})$. $\qquad \square$

3.3   Analysis of the Time Complexity

Theorem 3.5. *If getSkyline in step 2 is based on FLET [3], then the time complexity of the baseline algorithm is $O(m \log^{d-2} m)$ in the worst case, and $O(m)$ in the average case.*

Proof. Since the time complexity of FLET [3] is $O(n \log^{d-2} n)$ in the worst case, and $O(n)$ in the average case, step 2 of the algorithm needs $O(m \log^{d-2} m)$ time. Thus, the time complexity of the algorithm is $O(m \log^{d-2} m)$ because that step 1 of the algorithm needs $O(m)$ time. $\qquad \square$

Corollary 3.6. *If sample size m equal to $n^{\frac{1}{k}}(k > 1)$, then the running time of the baseline algorithm is $O(n^{\frac{1}{k}} \log^{d-2} n^{\frac{1}{k}})$ in the worst case, and $O(n^{\frac{1}{k}})$ in the average case.*

Corollary 3.6 tells that the baseline algorithm is in sublinear time if the sample size m is equal to $n^{\frac{1}{k}}(k > 1)$.

4   DOUBLE and Analysis

In this section, we devise a sampling-based algorithm, DOUBLE, to return an (ϵ, δ)-approximation efficiently for the exact skyline of the given relation T. It

---

Algorithm 2: DOUBLE

Input:
  T: the input relation with n tuples and d attributes;
  ϵ: the error bound;
  δ: the error probability;
Output:
  an (ϵ, δ)-approximation $\widetilde{Sky}$ of the skyline of T;

1 $m = s_I$, and $S[1, ..., m]$ is the sample of m tuples;
2 $\widetilde{Sky} = \text{getSkyline}(S[1, ..., m])$;
3 $\hat{\varepsilon} = \text{verifyError}(\widetilde{Sky})$;
4   **While** $(\hat{\varepsilon} > \frac{2\epsilon}{3})$ **Do**
5 $m = 2m$, and $S[\frac{m}{2} + 1, ..., m]$ is the sample of $m/2$ tuples;
6 $\widetilde{Sky} = \text{mergeSkyline}(\widetilde{Sky}, \text{getSkyline}(S[\frac{m}{2} + 1, ..., m]))$;
7 $\hat{\varepsilon} = \text{verifyError}(\widetilde{Sky})$;
8   **return** $\widetilde{Sky}$;

---

first draws an initial sample of size s_I (line 1). The value of s_I can be set to any positive integer. Afterwards, DOUBLE computes the approximate skyline result on the sample (line 2), and then verifies the error ε of the current result $\widetilde{Sky}$ (lines 3–4). If it is guaranteed that $Pr(\varepsilon \leq \epsilon)$ is at least $1 - \delta$, then DOUBLE terminates (line 8). Otherwise, it doubles the sample size and repeats the above process (lines 5–7).

DOUBLE judges whether ε meets the requirement by *Monte Carlo method*. In the subroutine verifyError, s_v is the sample size for each verification, and is equal to $\lceil \frac{18(\ln \log_2 n + \ln(\frac{1}{\delta}))}{\epsilon} \rceil$ (line 2). DOUBLE first obtains a random sample V of size s_v (line 3). Then it counts and returns the proportion of tuples in V not dominated by the approximate result $\widetilde{Sky}$ (lines 4–7), which is denoted by $\hat{\varepsilon}$. If $\hat{\varepsilon} \leq \frac{2\epsilon}{3}$, it is guaranteed that the error of $\widetilde{Sky}$ is not higher than the error bound ϵ with a probability no less than $1 - \delta$. In the following, we prove the above in detail.

Subroutines 1

1 **verifyError** $(\widetilde{Sky})$
2 $count = 0$, and $s_v = \lceil \frac{18(\ln \log_2 n + \ln(\frac{1}{\delta}))}{\epsilon} \rceil$;
3 V is the sample of s_v tuples;
4   **for** *each tuple t in V* **do**
5   **if** *t is not dominated by $\widetilde{Sky}$* **then**
6   count+=1;
7   **return** $count/s_v$;

4.1 Error Analysis of DOUBLE

Let q be the total number of times to invoke verifyError. For $1 \leq j \leq q$, m_j (respectively, ε_j) denotes the value of m (respectively, ε) when verfyError is being invoked for the jth time. $\widetilde{Sky_j}$ is defined in a similar way. $\hat{\varepsilon}_j$ is the value returned by the jth invocation of verifyError. Then we have the following theorem.

Theorem 4.1. *For the jth invocation of verifyError, if $\varepsilon_j > \epsilon$, then $Pr(\hat{\varepsilon}_j \leq \frac{2}{3}\epsilon) < \frac{\delta}{\log_2 n}$.*

Proof. Let X_i be a random variable for $1 \leq i \leq s_v$. For the jth invocation of verifyError, $X_i = 0$ if tuple t_i in V is dominated by the approximate result $\widetilde{Sky_j}$, otherwise $X_i = 1$. Obviously, $\hat{\varepsilon}_j$ is equal to $\frac{1}{s_v}\sum_{i=1}^{s_v} X_i$. According to the definition of ε, $E(\hat{\varepsilon}_j) = Pr(X_i = 1) = \varepsilon_j$.

By the Chernoff bound, we have

$$Pr(\hat{\varepsilon}_j \leq \frac{2}{3}\varepsilon_j) \leq \frac{1}{e^{s_v \varepsilon_j / 18}}.$$

With $\varepsilon_j > \epsilon$ and $s_v \geq \frac{18(\ln \log_2 n + \ln(\frac{1}{\delta}))}{\epsilon}$, we get the theorem. □

$\widetilde{Sky_q}$ is the final result returned by DOUBLE. Next, we show that $\widetilde{Sky_q}$ is an (ϵ, δ)-approximation of the exact skyline.

Corollary 4.1. *If DOUBLE terminates normally, it returns an (ϵ, δ)-approximation $\widetilde{Sky_q}$, i.e. the error ε_q of $\widetilde{Sky_q}$ satisfies*

$$Pr(\varepsilon_q \leq \epsilon) = Pr(|\frac{\mathcal{DN}(\widetilde{Sky_q}) - \mathcal{DN}(Sky)}{\mathcal{DN}(Sky)}| \leq \epsilon) \geq 1 - \delta.$$

Proof. DOUBLE finally returns an (ϵ, δ)-approximation $\widetilde{Sky_q}$, if and only if, for any positive integer $j < q$, the jth invocation of verifyError with the error $\varepsilon_j > \epsilon$ must return an estimated value $\hat{\varepsilon}_j > \frac{2\epsilon}{3}$. The number of times to invoke verifyError is at most $\log_2 n$. Based on Theorem 4.1, the probability in this corollary is at least $(1 - \frac{\delta}{\log_2 n})^{\log_2 n} > 1 - \delta$. □

4.2 Analysis of Sample Size and Time Complexity

m_q is the final value of m. Assume $\mathcal{M}_{\epsilon,\delta}$ is the size of sample required by the baseline algorithm running on T to return an (ϵ, δ)-approximation. Based on analysis in Sect. 3, $\mathcal{M}_{\epsilon,\delta}$ and m_q are almost unaffected by the relation size n. Here we analyze the relationship between $\mathcal{M}_{\epsilon,\delta}$ and m_q.

Theorem 4.2. *If $\delta \leq 1/8$, the expected value of m_q is $O(\mathcal{M}_{\frac{\epsilon}{3},\delta})$.*

Proof. If m_q is less than $\mathcal{M}_{\frac{\epsilon}{3},\delta}$, then the theorem holds. Otherwise, for the jth invocation of verifyError with $\mathcal{M}_{\frac{\epsilon}{3},\delta} \leq m_j < 2\mathcal{M}_{\frac{\epsilon}{3},\delta}$, the error ε_j of $\widetilde{Sky}_j$ satisfies $Pr(\varepsilon_j \leq \frac{\epsilon}{3}) \geq 1 - \delta$. Under the condition $\varepsilon_j \leq \frac{\epsilon}{3}$, the probability of $\hat{\varepsilon}_j \leq \frac{2\epsilon}{3}$ is at least $1 - (\frac{\delta}{\log_2 n})^2$. Thus the probability of $q \leq j$ is at least $(1 - \delta)(1 - (\frac{\delta}{\log_2 n})^2) > 1 - 2\delta$. Similarly, for any $k > j$, the probability of $q > k$ is less than 2δ. Thus the expected value of m_q is at most

$$2\mathcal{M}_{\frac{\epsilon}{3},\delta} + 2^2\mathcal{M}_{\frac{\epsilon}{3},\delta} \times 2\delta + \dots + 2^i\mathcal{M}_{\frac{\epsilon}{3},\delta} \times (2\delta)^{i-1} + \dots$$

With $\delta \leq 1/8$, it is $O(\mathcal{M}_{\frac{\epsilon}{3},\delta})$.   □

Based on analysis in Sect. 3, $\mathcal{M}_{\frac{\epsilon}{3},\delta}$ is up-bounded by $O(\mathcal{M}_{\epsilon,\delta})$ in most cases. Thus, the sample used by DOUBLE has the same order of magnitude as the baseline algorithm. Hereafter, we analyze the time complexity of DOUBLE on m_q.

Theorem 4.3. *If getSkyine and mergeSkyline are based on FLET [3], then the time complexity of DOUBLE is $O(m_q \log^{d-1} m_q + (\ln \log n + \ln \frac{1}{\delta})m_q \log m_q)$, where m_q is the final value of m in DOUBLE.*

Proof. Except for verifyError, the algorithm process is completely equivalent to SD&C [4]. Therefore, the total running time of getSkyline and mergeSkyline is $T'(m_q, d) = 2T'(m_q/2, d) + M(m_q, d)$, where $M(m_q, d)$ is $O(m_q \log^{d-2} m_q)$. Thus we have $T'(m_q, d) = O(m_q \log^{d-1} m_q)$. The number of verifications is $O(\log m_q)$. Based on Corollary 3.1 and 3.2, the size of $\widetilde{Sky}$ is $O(\epsilon m_q)$. Therefore, due to $\epsilon s_v = O(\ln \log n + \ln \frac{1}{\delta})$, the total running time of verifyError is $O(\epsilon m_q s_v \log m_q)$ $= O((\ln \log n + \ln \frac{1}{\delta})m_q \log m_q)$. Finally, the time complexity of the algorithm is $O(m_q \log^{d-1} m_q + (\ln \log n + \ln \frac{1}{\delta})m_q \log m_q)$.   □

Even if n is up to 2^{70}, $\ln \log_2 n$ is less than 5. Without loss of generality, the time complexity of DOUBLE is $O(m_q \log^{d-1} m_q)$. Through a proof similar to Theorem 4.2, we get the following corollary.

Corollary 4.2. *For $\mathcal{M}_{\frac{\epsilon}{3},\delta} \geq \frac{1}{2^{\frac{1}{d-1}}-1}$ and $\delta \leq 1/16$, if getSkyine and mergeSky-line are based on FLET [3], then the expected time complexity of DOUBLE is $O(\mathcal{M}_{\frac{\epsilon}{3},\delta} \log^{d-1} \mathcal{M}_{\frac{\epsilon}{3},\delta})$.*

5   Conclusion

In this paper, we proposed two sampling-based approximate algorithms for processing skyline queries on big data. The first algorithm draws a random sample of size m and computes the approximate skyline on the sample. The expected error of the algorithm is almost independent of the input relation size and the standard deviation of the error is relatively small. The running time of the algorithm is $O(m \log^{d-2} m)$ in the worst case and $O(m)$ in the average case. With a moderate size sample, the algorithm has a low enough error. Given ϵ and δ, the second algorithm returns an (ϵ, δ)-approximation of the exact skyline. The expected time complexity of the algorithm is $O(\mathcal{M}_{\frac{\epsilon}{3}, \delta} \log^{d-1} \mathcal{M}_{\frac{\epsilon}{3}, \delta})$, where $\mathcal{M}_{\frac{\epsilon}{3}, \delta}$ is the size of sample required by the first algorithm to return an $(\frac{\epsilon}{3}, \delta)$-approximation. $\mathcal{M}_{\frac{\epsilon}{3}, \delta}$ is up-bounded by $O(\mathcal{M}_{\epsilon, \delta})$ in most cases, and is almost unaffected by the relation size.

References

1. Bai, Z.D., Chao, C.C., Hwang, H.K., Liang, W.Q.: On the variance of the number of maxima in random vectors and its applications. In: Advances in Statistics, pp. 164–173. World Scientific (2008)
2. Bartolini, I., Ciaccia, P., Patella, M.: Efficient sort-based skyline evaluation. ACM Trans. Database Syst. (TODS) 33(4), 1–49 (2008)
3. Bentley, J.L., Clarkson, K.L., Levine, D.B.: Fast linear expected-time algorithms for computing maxima and convex hulls. Algorithmica 9(2), 168–183 (1993)
4. Bentley, J.L., Kung, H.T., Schkolnick, M., Thompson, C.D.: On the average number of maxima in a set of vectors and applications. J. ACM (JACM) 25(4), 536–543 (1978)
5. Borzsony, S., Kossmann, D., Stocker, K.: The skyline operator. In: Proceedings 17th International Conference on Data Engineering, pp. 421–430. IEEE (2001)
6. Buchta, C.: On the average number of maxima in a set of vectors. Inf. Process. Lett. 33(2), 63–65 (1989)
7. Cai, Z., Miao, D., Li, Y.: Deletion propagation for multiple key preserving conjunctive queries: approximations and complexity. In: 2019 IEEE 35th International Conference on Data Engineering (ICDE), pp. 506–517. IEEE (2019)
8. Chomicki, J., Godfrey, P., Gryz, J., Liang, D.: Skyline with presorting. In: ICDE, vol. 3, pp. 717–719 (2003)
9. Devroye, L.: A note on finding convex hulls via maximal vectors. Inf. Process. Lett. 11(1), 53–56 (1980)
10. Gao, X., Li, J., Miao, D., Liu, X.: Recognizing the tractability in big data computing. Theor. Comput. Sci. 838, 195–207 (2020)
11. Godfrey, P.: Skyline cardinality for relational processing. In: Seipel, D., Turull-Torres, J.M. (eds.) FoIKS 2004. LNCS, vol. 2942, pp. 78–97. Springer, Heidelberg (2004). https://doi.org/10.1007/978-3-540-24627-5_7
12. Godfrey, P., Shipley, R., Gryz, J., et al.: Maximal vector computation in large data sets. VLDB 5, 229–240 (2005)
13. Han, X., Li, J., Yang, D., Wang, J.: Efficient skyline computation on big data. IEEE Trans. Knowl. Data Eng. 25(11), 2521–2535 (2012)
14. Koltun, V., Papadimitriou, C.H.: Approximately dominating representatives. In: Eiter, T., Libkin, L. (eds.) ICDT 2005. LNCS, vol. 3363, pp. 204–214. Springer, Heidelberg (2004). https://doi.org/10.1007/978-3-540-30570-5_14
15. Kossmann, D., Ramsak, F., Rost, S.: Shooting stars in the sky: an online algorithm for skyline queries. In: VLDB 2002: Proceedings of the 28th International Conference on Very Large Databases, pp. 275–286. Elsevier (2002)
16. Kung, H.T., Luccio, F., Preparata, F.P.: On finding the maxima of a set of vectors. J. ACM (JACM) 22(4), 469–476 (1975)
17. Lee, K.C., Lee, W.C., Zheng, B., Li, H., Tian, Y.: Z-sky: an efficient skyline query processing framework based on z-order. VLDB J. 19(3), 333–362 (2010)

18. Lin, X., Yuan, Y., Zhang, Q., Zhang, Y.: Selecting stars: the k most representative skyline operator. In: 2007 IEEE 23rd International Conference on Data Engineering, pp. 86–95. IEEE (2007)
19. Magnani, M., Assent, I., Mortensen, M.L.: Taking the big picture: representative skylines based on significance and diversity. VLDB J. **23**(5), 795–815 (2014)
20. Miao, D., Cai, Z., Li, J.: On the complexity of bounded view propagation for conjunctive queries. IEEE Trans. Knowl. Data Eng. **30**(1), 115–127 (2017)
21. Miao, D., Cai, Z., Li, J., Gao, X., Liu, X.: The computation of optimal subset repairs. Proc. VLDB Endow. **13**(12), 2061–2074 (2020)
22. Miao, D., Liu, X., Li, J.: On the complexity of sampling query feedback restricted database repair of functional dependency violations. Theor. Comput. Sci. **609**, 594–605 (2016)
23. Miao, D., Yu, J., Cai, Z.: The hardness of resilience for nested aggregation query. Theor. Comput. Sci. **803**, 152–159 (2020)
24. Papadias, D., Tao, Y., Fu, G., Seeger, B.: An optimal and progressive algorithm for skyline queries. In: Proceedings of the 2003 ACM SIGMOD International Conference on Management of Data, pp. 467–478 (2003)
25. Søholm, M., Chester, S., Assent, I.: Maximum coverage representative skyline. In: EDBT, pp. 702–703 (2016)
26. Tan, K.L., Eng, P.K., Ooi, B.C., et al.: Efficient progressive skyline computation. VLDB **1**, 301–310 (2001)
27. Tao, Y., Ding, L., Lin, X., Pei, J.: Distance-based representative skyline. In: 2009 IEEE 25th International Conference on Data Engineering, pp. 892–903. IEEE (2009)
28. Xingxing Xiao, J.L.: Sampling based approximate skyline calculation on big data (2020). https://arxiv.org/abs/2008.05103

Approximating k-Orthogonal Line Center

Barunabha Chakraborty[1], Arun Kumar Das[2(✉)], Sandip Das[2],
and Joydeep Mukherjee[1]

[1] Ramakrishna Mission Vivekananda Educational and Research Institute,
Howrah, India
chakrabortybarunabha@yahoo.com, joydeep.m1981@gmail.com
[2] Indian Statistical Institute, Kolkata, India
arund426@gmail.com, sandipdas@isical.ac.in

Abstract. k-orthogonal line center problem computes a set of k axis-parallel lines for a given set of n points such that the maximum among the distances between each point and its nearest line is minimized. In this paper, we design a deterministic bi-criteria approximation algorithm that runs in $O(k^2 n \log n)$ time and returns at most $\frac{3}{2}k$ lines such that the minimized distance is within $16r$. Here r is the minimized distance in the optimal solution with k line centers for the given input.

Keywords: Line centers · Approximation algorithm · Computational geometry · Bi-criteria approximation

1 Introduction

The problem of computing k-*orthogonal line centers* for a given set of points is to find k axis-parallel lines such that the maximum among the distances from each point to its nearest line is minimized. These lines are called *centers* and the minimized distance is called *radius*. We define them formally in the preliminaries section. A direct application of this problem is in the designing of transport networks where the roads or any other transport tracks are orthogonal to each other and the designer has to plan them such that the maximum among all the distances between each user and its nearest track is minimized. A similar scenario happens while designing a circuit board. The designer needs to optimize the distances between wires and components. This problem also finds application in facility location [19,20], machine learning [9] etc.

A closely related problem to the orthogonal line center problem is *stabbing or hitting problem*. In this problem, a set of geometric objects are given as input and one has to find a set of predetermined objects with minimum cardinality such that each member of the given set is intersected by at least one member of the output set. This problem also has wide application in facility location [11,15], medical researches [18], statistical analysis [7] etc. The decision version of the k-orthogonal line centers can be viewed as *square stabbing* as well. Here the set of squares have side lengths $2r$, where r is the radius in the decision version of k-line center problem. The classical problem of *orthogonal stabbing of unit squares*

W. Wu and Z. Zhang (Eds.): COCOA 2020, LNCS 12577, pp. 47–60, 2020.
https://doi.org/10.1007/978-3-030-64843-5_4

computes a minimum set of axis-parallel lines such that each member of the given set of axis-parallel unit squares is stabbed by at least one line. This problem is known to be **W[1]**-hard [12]. In the problem of k-orthogonal line centers for unit squares, one has to find a set of axis-parallel lines such that the maximum among the *distances* between a square and its nearest line is minimized. Here the distance between a square and a horizontal line is the distance between the line and the horizontal side of the square which is nearest to the line. The distance between a vertical line and a square is defined similarly. Orthogonal stabbing of unit squares is **NP**-hard. Thus we can not expect to find a finite factor approximation in terms of the radius for the problem of k-orthogonal line centers for unit squares with running time $c \times poly(n, k)$, where c is independent of n and k. We can only expect to get a (α, β) bi-criteria approximation for this problem. In this approximate solution, we use at most αk ($\alpha > 1$) axis-parallel lines such that every square is within βr ($\beta > 1$) distance from some line. Here r is the optimal radius for the k-orthogonal line center problem for unit squares. Although we design our algorithm for planar point set, it is not hard to see that the algorithm can be seamlessly applied to get a constant factor bi-criteria approximation in the case of squares as well. This gives a clear motivation behind designing the bi-criteria approximation algorithm that we present in this paper.

The k-orthogonal line center problem is a special case of *projective clustering*. Projective clustering problem takes a point set with n points in $\mathbb{R}^d$ and two integers $k < n$ and $q \le d$ as input and asks for k q-dimensional flats such that the point set is partitioned into k subsets (clusters) so that when a cluster is projected on the flats, the maximum distance between each point to its projection is minimized. In our problem, the q-dimensional flats are lines, and the input point set is in $\mathbb{R}^2$. Megiddo and Tamir [16] showed that finding k lines which cover a given set of points, as well as the projective clustering, is **NP**-hard.

1.1   Previous Results

A significant amount of work has been done in the domain of clustering and stabbing problems due to their wide application. Stabbing of compact sets with lines was studied by Hassin and Megiddo [13]. In this paper, they devised a polynomial time algorithm for stabbing a set of planar points with axis-parallel lines. They also showed it is **NP**-hard if the lines are in arbitrary orientation. Agarwal and Procopiuc [1] studied the problem of covering a point set in $\mathbb{R}^d$ with k cylinders. They presented the following results. For $d = 2$, they gave a randomized algorithm, which runs in $O(nk^2 \log^4 n)$ expected time, when $k^2 \log k < n$ to compute $O(k \log k)$ strips of width at most w^* that cover the given set of points. Here w^* is the width in the optimal solution. But for higher values of k the expected running time is $O(n^{\frac{2}{3}} k^{\frac{8}{3}} \log^4 n)$. For $d = 3$, the expected time is $O(n^{\frac{3}{2}} k^{\frac{9}{4}} polylog(n))$. They also compute a set of $O(dk \log k)$ d-cylinders of diameter at most $8w^*$ in expected time $O(dnk^3 \log^4 n)$ to cover a set of n points in $\mathbb{R}^d$. Aggarwal et al. [3] designed a randomized algorithm that computes k cylinders of radius $(1 + \varepsilon)r^*$ that cover a set of n given points in $\mathbb{R}^d$. Here r^* is the optimal radius. The expected running time of the algorithm is $O(n \log n)$, but the constant of proportionality depends on k, d, ε. Besides the theoretical results on projective clustering, the practical implementation and heuristics for the problems are also studied [4–6,17]. Special case like 2 line center problem was studied by Jaromczyk and Kowaluk [14] designed a $O(n^2 \log^2 n)$ time algorithm. Agarwal et al. [2] provided an $(1 + \varepsilon)$ approximation scheme for the problem which runs in $O(n(\log n + \varepsilon^{-2} \log \frac{1}{\varepsilon}) + \varepsilon^{-\frac{7}{2}} \log \frac{1}{\varepsilon})$ time. Feldman et al. [9] gave a randomized linear time bi-criteria approximation for generalized k-center, mean and median problems. Gaur et al. [11] gave a 2-factor approximation for the orthogonal rectangle stabbing problem. Dom et al. [8] showed that stabbing squares of same size with axis-parallel lines is $\mathbf{W}[1]$-hard and designed a fixed-parameter tractable algorithm for the problem with disjoint squares of same size, which runs in $(4k + 1)^k n^{O(1)}$ time.

1.2   Our Result

We devise a $(\frac{3}{2}, 16)$ deterministic bi-criteria approximation algorithm for k-orthogonal line center problem which runs in $O(k^2 n \log n)$ time. To the best of our knowledge, this result is a first deterministic constant factor bi-criteria approximation for this problem.

1.3   Organisation of the Paper

We give a brief overview of the algorithm in Subsect. 1.4. In Sect. 2, we describe a set of problems formally which are related to each other in terms of determining a constant factor approximation. We present an algorithm to these problems and discuss the running time of the algorithm in Sect. 3. Finally, we conclude the paper by stating some further direction of the work.

1.4   Overview of the Algorithm

Our algorithm consists of three phases. In the beginning, we choose a subset of the given point set S. We call these chosen points as *pivots*. We denote the set of pivots as $\mathcal{P}$. We choose the pivots greedily in such a way so that any two pivots are *well separated* both horizontally and vertically. The notion of well separateness will be explained when we describe the algorithm formally below. Next, we employ a local search technique to update the set of pivots so that at the end of each update step we increase the number of pivots by *one* yet maintaining the property of well separateness among the updated set of pivots. We maintain the condition of well separateness at every step to guarantee that the cardinality of the set of pivots never exceeds k. Finally, we show that we can choose at most $\frac{3}{2}k$ lines passing through the members of $\mathcal{P}$ as line centers such that radius is within 16 times the optimal radius.

2   Preliminaries

> *Problem 1.* A set S of n points in 2-dimensional plane is given, along with an integer k such that $k < n$. A set C of axis-parallel lines are to be returned as output such that $|C| = k$ and the maximum among the distances between each point and its nearest line is minimized.

The distance between two points of S is defined as $Min\{|x_1 - x_2|, |y_1 - y_2|\}$, where (x_1, y_1) and (x_2, y_2) are the coordinates of the two points respectively. The distance between a point and a line is the perpendicular distance between them.

The distance between a point and the set C is defined as the distance between the point and the member of C which is nearest to the point. We denote this as $d(s, C)$ The radius r of a set C is defined as $max_{s \in S} d(s, C)$.

> *Problem 2.* A set S of n points in 2-dimensional plane is given, along with an integer k such that $k < n$. A set C of axis-parallel lines are to be returned as output such that the lines pass through at least one member of S with $|C| = k$ and the maximum among the distances between each point and its nearest line is minimized.

Lemma 1. *An optimal solution to Problem 2 is a $(1, 2)$ bi-criteria approximate solution to Problem 1.*

Proof. Let C^* be a set of line centers with a radius r^* that is an optimal solution to Problem 1 for a given set of points S. Let $l \in C^*$ be a line center. Let $C(l) = \{s \in S | d(s, l) \leq d(s, l'), \forall l' \in C^* \setminus \{l\}\}$. So, $d(s, l) \leq r^*$ for all $s \in C(l)$. Let l^x be the line of same orientation as l, that passes through some member x of $C(l)$. Then for any member $y \in C(l)$ we have $d(y, l^x) \leq d(y, l) + d(x, l) \leq 2r^*$. Thus we can construct a set C_x that consists of l^x, for all $l \in C^*$. Here x is a point nearest to l and on the below of l, if l is horizontal. Otherwise x is a point nearest to l and on the right of l if l is vertical. Existence of such x is guaranteed by the optimality of C^*. Thus $d(s, C_x) \leq 2r^*$ for all $s \in S$. Thus C_x is a $(1, 2)$ bi-criteria approximation for Problem 1.

Furthermore, C_x is a solution to Problem 2 which may not be optimal. The radius of an optimal solution to Problem 2 must be less than or equal to the radius of C_x. Hence the lemma follows.   □

Note that the radius of C is defined by a pair consisting of a point and a vertical line or a point and a horizontal line. So in Problem 2, the possible values for the radius of C are the distances between a member of S and a line passing through another member of S. Since there are $O(n^2)$ such choices, we have the following observation.

Observation 1. *There are $O(n^2)$ candidates for optimal radius in Problem 2.*

Problem 3. A set $\mathcal{S}$ of n points in 2-dimensional plane is given, along with k and r. A set $\mathcal{C}$ of k axis-parallel lines are to be returned as output such that the lines pass through at least one member of $\mathcal{S}$ and the maximum among the distances between each point and its nearest line is within r.

Lemma 2. *Problem 2 can be solved in $O((T+n)\log n)$ time if Problem 3 can be solved in $O(T)$ time.*

Proof. There are $O(n^2)$ candidates for radius in Problem 2 as noted in Observation 1. So we can solve Problem 2 by solving 3 with these radii as input. We return the solution of Problem 3 as a solution to Problem 2 with the minimum r among these candidates. We can do a binary search on these $O(n^2)$ candidates to achieve the solution to Problem 2. The technique by Frederickson and Johnson [10] helps to determine the middle element in the binary search in linear time. Thus the lemma holds.   □

We design an algorithm for Problem 3 in the following. We denote the horizontal and vertical lines passing through a member s of $\mathcal{S}$ by h^s and v^s respectively. Let $\mathcal{C}$ be a set of line centers with radius ξ for a set of points $\mathcal{S}$. If a line $h^s \in \mathcal{C}$ is the nearest line to a point $t \in \mathcal{S}$, then t must be within a distance of ξ from this line. We observe this scenario in a different way. Let (x_s, y_s) be the coordinates of s. A *horizontal strip* generated from s, is the planar region bounded by two horizontal lines $y = y_s + \xi$ and $y = y_s - \xi$. We denote it by h_ξ^s. Then t lies in the interior of this horizontal strip. Similarly we define a *vertical strip*, v_ξ^s, generated from s, as the planar region bounded by two vertical lines $x = x_s + \xi$ and $x = x_s - \xi$. s is called the *generator* of the strips h_ξ^s and v_ξ^s.

The *upper shadow zone* of a horizontal strip h_ξ^s is defined as the planar region bounded by the two horizontal lines $y = y_s + \xi$ and $y = y_s + 2\xi$. The *lower shadow zone* of a horizontal strip h_ξ^s is defined as the planar region bounded by the two horizontal lines $y = y_s - \xi$ and $y = y_s - 2\xi$. Union of these two regions are called *shadow zone* of a horizontal strip. Similarly we define the *left shadow zone, right shadow zone* and *shadow zone* of a vertical strip. (Fig. 1)

In our algorithm, each strip will correspond to a unique pivot of $\mathcal{P}$ and each pivot will correspond to exactly one horizontal and exactly one vertical strip. Let $H(p)$ and $V(p)$ denotes the horizontal and vertical strip corresponding to $p \in \mathcal{P}$ respectively. The horizontal and vertical lines passing through the generator of $H(p)$ and $V(p)$ respectively are called *generating lines* of the strips.

Now we define a *private neighbour* of a pivot $p \in \mathcal{P}$. Horizontal private neighbours of p (depicted in Fig. 1) are those points $s \in \mathcal{S}$, which satisfy the following properties:

- s lies in the interior of $H(p)$,
- s does not lie in the interior of $V(q)$ and their shadow zones, for all $q \in \mathcal{P}$.

Similarly, we define *vertical private neighbour* of a pivot. Note that if we consider all strips of same width then a point $s \in \mathcal{S}$ can be a private neighbour

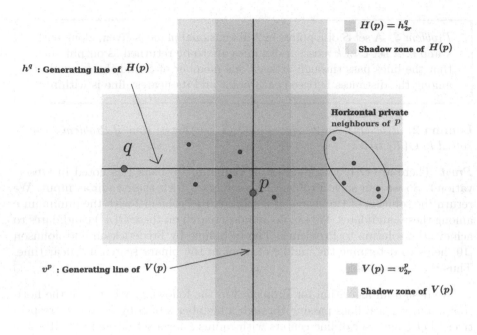

Fig. 1. Strips and private neighbours of a pivot p

of two pivots in same orientation. But it can not be a private neighbour of two pivots in different orientations.

We say that two pivots $p, q \in \mathcal{P}$ are *well separated* if they are at a distance more than $2r$ from each other. We say that a pivot $p \in \mathcal{P}$ is *conflicting horizontally* (depicted in Fig. 2) with another pivot $q \in \mathcal{P}$, if one of them has a horizontal private neighbour which is within a distance of $2r$ from the other. Similarly, we say they are *conflicting vertically* if there exists such a vertical private neighbour. We say p is *conflicting* with q if they are either conflicting horizontally or vertically or both. The corresponding private neighbour is called a *conflict witness*.

Observation 2. *A pivot $p \in \mathcal{P}$ can conflict with at most two members of $\mathcal{P} \setminus p$ in one orientation.*

Proof. Since the members of $\mathcal{P}$ are well separated from each other, p can conflict horizontally with at most one pivot in $\mathcal{P} \setminus \{p\}$, which lies above p and at most one pivot in $\mathcal{P} \setminus \{p\}$, which lies below p. Thus the observation holds.   □

3   Description of the Algorithm

3.1   Phase I of the Algorithm

In the first phase, we choose a set of pivots maintaining the well separateness among all of them. We maintain this well separateness throughout the algorithm.

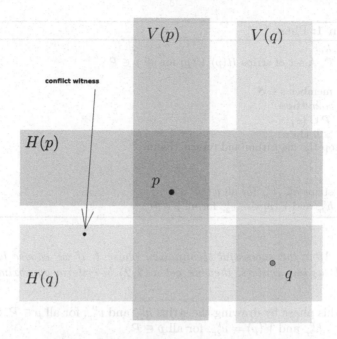

Fig. 2. Horizontal Conflict of two pivots p and q

We begin with any point s_1 of the given set S. We take this point in the set of pivots $\mathcal{P}$. Now we choose the next point s_2 which is at a distance greater than $2r$ from s_1. We choose a point $s_i \in S$ to be included in $\mathcal{P}$ if the point is at a distance greater than $2r$ from all the existing members of $\mathcal{P}$. We stop the iteration when we can not find a point to be included in $\mathcal{P}$. We terminate the algorithm if $|\mathcal{P}|$ exceeds k and report "failure" as we can not find a set of line centers with cardinality k for this given r.

Lemma 3. *Cardinality of $\mathcal{P}$ is at most k, if all the members of $\mathcal{P}$ are well separated from each other.*

Proof. Let p and q are two points such that $d(p,q) > 2r$. Then there can not be a single line l such that $d(p,l) \leq r$ and $d(q,l) \leq r$. We need two different lines for that. If r is the radius of the optimal set of k line centers for S, then we can not find more than k points which are well separated. $\quad\square$

Lemma 4. *All point of S are covered with h_{2r}^p or v_{2r}^p for some $p \in \mathcal{P}$.*

Proof. If any point $s \in S$ is not covered with h_{2r}^p or v_{2r}^p for some $p \in \mathcal{P}$, then s is well separated from all the members of $\mathcal{P}$. So we can increase the cardinality of $\mathcal{P}$ by including s in $\mathcal{P}$. So when we stop the above mentioned iteration, the lemma holds. $\quad\square$

As a consequence of Lemma 3 and Lemma 4 we can claim the following lemma. But note that this is not an output of the algorithm.

Algorithm 1: Phase I

Input: $\mathcal{S}, k, r$
Output: $\mathcal{P}$, A set of strips $H(p), V(p)$ for all $p \in \mathcal{P}$
$\mathcal{P} \longleftarrow \phi$;
For every member $s \in \mathcal{S}$
if $d(s, \mathcal{S}) > 2r$ **then**
$\quad \mid \quad \mathcal{P} \longleftarrow \mathcal{P} \cup \{s\}$;
$\quad \mid \quad$ **if** $|\mathcal{P}| > k$ **then**
$\quad \mid \quad \mid \quad$ Stop the algorithm and return "Failure".
$\quad \mid \quad$ **end**
end
Draw the strips h_{2r}^p, v_{2r}^p for all $p \in \mathcal{P}$;
$H(p) \longleftarrow h_{2r}^p$ and $V(p) \longleftarrow v_{2r}^p$ for all $p \in \mathcal{P}$;

Lemma 5. *After the successful termination phase I, if we choose h^p and v^p for all $p \in \mathcal{P}$ as line centers, then we get a $(2,2)$ bi-criteria approximation for Problem 3.*

We end this phase by drawing the strips h_{2r}^p and v_{2r}^p, for all $p \in \mathcal{P}$. So in this phase $H(p) = h_{2r}^p$ and $V(p) = v_{2r}^p$, for all $p \in \mathcal{P}$.

3.2 Phase II of the Algorithm

Algorithm 2: Phase II

Input: $\mathcal{S}, k, r, \mathcal{P}$, A set of strips corresponding to the members of $\mathcal{P}$
Output: $\mathcal{P}$, A set of strips corresponding to the members of $\mathcal{P}$
For every member $p \in \mathcal{P}$
if *p has horizontal and vertical private neighbours t_h and t_v respectively, such that $d(t_h, (\mathcal{P} \setminus \{p\})) > 2r$ and $d(t_h, (\mathcal{P} \setminus \{p\})) > 2r$* **then**
$\quad \mid \quad \mathcal{P} \longleftarrow \mathcal{P} \setminus \{p\} \cup \{t_h, t_v\}.$;
$\quad \mid \quad$ Draw $v_{2r}^{t_h}$ and $h_{2r}^{t_v}$;
$\quad \mid \quad H(t_h) \leftarrow h_{2r}^p$;
$\quad \mid \quad V(t_h) \leftarrow v_{2r}^{t_h}$;
$\quad \mid \quad H(t_v) \leftarrow h_{2r}^{t_v}$;
$\quad \mid \quad V(t_v) \leftarrow v_{2r}^p$;
$\quad \mid \quad$ **if** $|\mathcal{P}| > k$ **then**
$\quad \mid \quad \mid \quad$ Return "failure" and terminate algorithm.
$\quad \mid \quad$ **end**
end

In this phase, we try to increase the cardinality of $\mathcal{P}$ by choosing two members of $\mathcal{S}$ as pivots in place of one existing pivot. Thus in each step of this phase, we increase the cardinality of $\mathcal{P}$ by one. The following lemma helps us to identify the pivots which we can replace by two other members of $\mathcal{S}$.

Lemma 6. *Let $p \in \mathcal{P}$ be a pivot having private neighbours in its strips of both orientations such that they are well separated from all members of $\mathcal{P} \setminus \{p\}$, then we can increase the cardinality of $\mathcal{P}$ by one without violating the well separateness of the members in $\mathcal{P}$.*

Proof. Let $p \in \mathcal{P}$ have a horizontal private neighbour t_h and a vertical private neighbour t_v such that t_h and t_v are well separated from all the members of $\mathcal{P} \setminus \{p\}$. It follows from the definition of private neighbours of a pivot that t_h and t_v are well separated from each other as well. So we can update $\mathcal{P} \leftarrow (\mathcal{P} \setminus \{p\}) \cup \{t_h, t_v\}$. Thus the lemma holds. $\square$

In the following, we describe the pivot replacement and other associated steps performed in phase II. We implement this phase in an iterative fashion.

If for some $p \in \mathcal{P}$ has two private neighbours t_h and t_v in $H(p)$ and $V(p)$ respectively, such that t_h and t_v are well separated from all the members of $\mathcal{P} \setminus \{p\}$, then by Lemma 6, we update $\mathcal{P} \leftarrow (\mathcal{P} \setminus \{p\}) \cup \{t_h, t_v\}$. We also draw two new strips. They are $v_{2r}^{t_h}$ and $h_{2r}^{t_v}$. Note that $H(t_h) = h_{2r}^p, V(t_h) = v_{2r}^{t_h}$ and $H(t_v) = h_{2r}^{t_v}, V(t_v) = v_{2r}^p$. At any point if $|\mathcal{P}|$ exceeds k, we terminate the algorithm and return "failure".

We terminate phase II when we can not increase the cardinality of $\mathcal{P}$ in this manner. Note that we have not removed any strip that is drawn at phase I.

Observation 3. *Let t is a member of $\mathcal{P}$ that is chosen in phase II, by replacing an old member p of $\mathcal{P}$. Then the private neighbours of t is a subset of the private neighbours of p.*

3.3 Phase III of the Algorithm

After termination of phase II, if a pivot $p \in \mathcal{P}$ has private neighbours in both orientations, then the following observation holds.

Observation 4. *All the private neighbours of p in at least one orientation are conflict witnesses.*

Proof. If p has at least one private neighbours in both orientations which are well separated from all other pivots in $\mathcal{P}$, then we can increase the cardinality of $\mathcal{P}$ by virtue of Lemma 6. So, after completion of phase II this observation holds. $\square$

We can classify all the members $p \in \mathcal{P}$ in two classes.

- **Case 1:** p has private neighbours in both orientations. We call the set of these pivots as $\mathcal{P}_{con}$.
- **Case 2:** p has private neighbours only in one orientation. In such a case we call p as a *loner*. We denote the set of such pivots as $\mathcal{P}_{lon}$.

Now we assign line centers for the points covered by the strips corresponding to the members of $\mathcal{P}_{con}$ and $\mathcal{P}_{lon}$ in the following two steps.

Step 1: Assigning Line Centers Corresponding to the Members of $\mathcal{P}_{con}$
Consider a member $p \in \mathcal{P}_{con}$. As mentioned earlier that after the termination of phase II, p has to satisfy any one of the following.

- All the private neighbours of p in $H(p)$ are conflict witnesses. Then $p \in \mathcal{P}_{con}^{H}$.
- All the private neighbours in $V(p)$ are conflict witnesses. Then $p \in \mathcal{P}_{con}^{V}$.

Note that as we defined $\mathcal{P}_{con}^{H}$ and $\mathcal{P}_{con}^{V}$, their intersection may not be empty. If we put the members in their intersection only in $\mathcal{P}_{con}^{H}$, we can partition them in two disjoint sets $\mathcal{P}_{con}^{H}$ and $\mathcal{P}_{con}^{V}$ such that $\mathcal{P}_{con} = \mathcal{P}_{con}^{H} \cup \mathcal{P}_{con}^{V}$. Let $|\mathcal{P}_{con}^{H}| = c_1$ and $|\mathcal{P}_{con}^{V}| = c_2$.

Let $p \in \mathcal{P}_{con}^{H}$. Now p can conflict horizontally with at most two members of $\mathcal{P}$ as noted in Observation 2. Let p horizontally conflicts with $q \in \mathcal{P}$ such that q lies below p.

Lemma 7. *We can choose 3 lines such that all points, which are covered with $H(p), V(p), H(q), V(q)$, and their shadow zones, are within $8r$ distance from these lines.*

Proof. We choose the generating lines of $V(p), V(q)$ and $H(q)$. Any point covered with these two vertical strips or their shadow zones is within a distance of $4r$ from these vertical lines. Let $H(q) = h_{2r}^{u}$, for some $u \in \mathcal{S}$. Any point covered with $H(q)$ or its shadow zone is also within $4r$ distance from h^{u}. Now consider a point s which is covered with $H(p)$. Since there is a conflict witness t which is a horizontal private neighbour of p or q and lies within a distance of $2r$ from both p and q, then the distance between h^{p} and h^{q} is at most $4r$. So, if s lies in $H(p)$, then s lies within $6r$ distance from h^{u}. If s lies in the shadow zone of $H(p)$, s is within a distance of $8r$ from h^{u}. Thus the lemma holds. □

We call these chosen lines as the line centers *assigned* to the pivots p and q. We say that all the points which are within $8r$ distance from the assigned line centers are covered with these centers. Using Lemma 7, we assign line centers to the members of $\mathcal{P}_{con}^{H}$. We choose the members of $\mathcal{P}_{con}^{H}$ with respect to the decreasing order of their y-coordinates. Once we assign the line centers corresponding to a strip, we "mark" the pivots also. Initially, no member of $\mathcal{P}$ is marked. We traverse $\mathcal{P}_{con}^{H}$ from top to bottom and use the following rules to assign line centers for a pivot $p \in \mathcal{P}_{con}^{H}$, which is not already marked.

- **R1:** If p is conflicting horizontally with q such that q lies below p, then we assign the generating lines of $H(q), V(q)$ and $V(p)$ as line centers. We mark both p and q.
- **R2:** If p is conflicting horizontally only with q such that q lies above p, then we assign line centers based on the following two scenarios.
    - If q is already marked while we assigned line centers from top to bottom, the generating lines of $H(q)$ and $V(q)$ are already chosen. We assign the generating line of $V(p)$ as a center and mark p.

– If q is not already marked we assign the generating lines of $V(q), V(p)$, and $H(p)$ as centers and mark p and q.

Lemma 8. *Consider $p \in \mathcal{P}_{con}^{H}$ is horizontally conflicting with a marked pivot, q, then all the points covered by $H(p)$ and its shadow zone are also within a distance of $8r$ from the generating line of $H(q)$.*

Proof. p is horizontally conflicting only with q. So the distance between h^p and h^q are at most $4r$. So this lemma follows from a similar argument as in Lemma 7. □

Thus we assign line centers to the members of $\mathcal{P}_{con}^{H}$ according to the above-mentioned rules. We can argue similarly for all the members of $\mathcal{P}_{con}^{V}$. We assign the line centers corresponding to them from left to right, that is by the increasing order of their x-coordinate. Using similar argument as in Lemma 7 and Lemma 8, we can conclude that, the points those are covered with $H(p), V(p)$ and their shadow zones, for all $p \in \mathcal{P}_{con}^{V}$, are within a distance of $8r$ from the assigned line centers. Note that some pivots that are not members of $\mathcal{P}_{con}$ are also marked while assigning line centers to the members of $\mathcal{P}_{con}$. We denote the set of these pivots as $\mathcal{P}_{sup}$. Let $|\mathcal{P}_{sup}| = c_3$. We denote the set of these assigned line centers as $\mathcal{C}_{con}$.

Lemma 9. $|\mathcal{C}_{con}| \leq \frac{3}{2}(c_1 + c_2 + c_3)$.

Proof. We have assigned at most 3 line centers for every two members of $\mathcal{P}_{con}^{H}$, $\mathcal{P}_{con}^{V}$ and $\mathcal{P}_{sup}$. Furthermore in the rules **R1** and **R2** we have not assigned line centers for any pivot more than once. Hence, $|\mathcal{C}_{con}| \leq \frac{3}{2}(c_1 + c_2 + c_3)$. □

Note that in this fashion all pivots of $\mathcal{P}_{con}$ will be marked. All the points which are covered with the strips $H(p), V(p)$, and their shadow zones, for all $p \in \mathcal{P}_{con} \cup \mathcal{P}_{sup}$ are covered by the members of $\mathcal{C}_{con}$ by virtue of Lemma 7 and Lemma 8.

Step 2: Assigning Line Centers with $\mathcal{P}_{lon}$

Now we are left with some unmarked members of $\mathcal{P}_{lon}$. Such a member $p \in \mathcal{P}_{lon}$ can be one of the following types.

– **Type 1:** p has only vertical private neighbours. We denote the set of such pivots as $\mathcal{P}_{lon}^{V}$. Let $|\mathcal{P}_{lon}^{V}| = c_4$.
– **Type 2:** p has only horizontal private neighbours. We denote the set of such pivots as $\mathcal{P}_{lon}^{H}$. Let $|\mathcal{P}_{lon}^{H}| = c_5$.
– **Type 3:** p does not have a private neighbour. We denote the set of such pivots as $\mathcal{P}_{lon}^{N}$. Let $|\mathcal{P}_{lon}^{N}| = c_6$.

Note that $|\mathcal{P}_{lon}^{H} \cup \mathcal{P}_{lon}^{V}| = c_4 + c_5$, since $\mathcal{P}_{lon}^{H} \cap \mathcal{P}_{lon}^{V} = \phi$.

Lemma 10. *We can choose at most $\frac{3}{2}(c_4 + c_5 + c_6)$ lines such that all the points covered by the strips $H(p)$ and $V(p)$ for all $p \in \mathcal{P}_{lon}^{V} \cup \mathcal{P}_{lon}^{H} \cup \mathcal{P}_{lon}^{N}$, which are not covered by any assigned line centers in step 1, are within $4r$ distance from these new lines.*

Proof. With out loss of generality let us assume $c_5 \leq \frac{1}{2}(c_4 + c_5)$. We choose the generating lines of the following strips.

- $V(p)$, for all $p \in \mathcal{P}_{lon}^{N}$
- $V(p)$, for all $p \in \mathcal{P}_{lon}^{V}$
- $H(p)$ and $V(p)$, for all $p \in \mathcal{P}_{lon}^{H}$

We denote the set of these lines as $\mathcal{C}_{lon}$. So, $|\mathcal{C}_{lon}| = (c_6 + c_4 + 2c_5) \leq c_6 + \frac{3}{2}(c_4 + c_5) < \frac{3}{2}(c_4 + c_5 + c_6)$.

Now we show that $d(s, \mathcal{C}_{lon}) \leq 4r$ for all $s \in \mathcal{S}$ such that s is not already covered with a line center assigned in step 1. Such an s must lie in some $H(p)$ or $V(p)$ for some $p \in \mathcal{P}_{lon}^{V} \cup \mathcal{P}_{lon}^{H} \cup \mathcal{P}_{lon}^{N}$.

From the construction of $\mathcal{C}_{lon}$ all the private neighbours of the pivots in $\mathcal{P}_{lon}^{V} \cup \mathcal{P}_{lon}^{H} \cup \mathcal{P}_{lon}^{N}$ are within a distance of $2r$ from $\mathcal{C}_{lon}$. If a point which is not a horizontal private neighbour of any pivot and lies only in a shadow zone of a horizontal strip and not in any other vertical strip or its shadow zone, then that point is well separated from all the members of $\mathcal{P}$. But this implies we could have selected this point in phase I, which contradicts the termination of phase I. So the remaining points must be in some vertical strip or in the shadow zone of some vertical strip corresponding to some member of $\mathcal{P}_{lon}^{V} \cup \mathcal{P}_{lon}^{H} \cup \mathcal{P}_{lon}^{N}$. We have chosen the generating lines of all the vertical strips corresponding to the pivots belonging to $\mathcal{P}_{lon}^{V} \cup \mathcal{P}_{lon}^{H} \cup \mathcal{P}_{lon}^{N}$. Hence, they are within $4r$ distance from $\mathcal{C}$. Thus the lemma holds.   □

We assign the members of $\mathcal{C}_{lon}$ as line centers corresponding to the members of $\mathcal{P}_{lon}^{V} \cup \mathcal{P}_{lon}^{H} \cup \mathcal{P}_{lon}^{N}$ and mark them. At this step we have all the members of $\mathcal{P}$ marked. We return the set $\mathcal{C} = \mathcal{C}_{con} \cup \mathcal{C}_{lon}$ as the output to the algorithm.

Theorem 1. $\mathcal{C}$ *is a* $(\frac{3}{2}, 8)$ *bi-criteria approximation for Problem 3.*

Proof. The proof follows from Lemma 9 and Lemma 10.   □

3.4   Running Time of the Algorithm

We are constructing the set $\mathcal{P}$ by choosing the members from $\mathcal{S}$ which constitutes phase I. In this phase, we check each member of $\mathcal{S}$ if they can be included in $\mathcal{P}$. This checking for each member requires $O(k)$ time. Since each member of $\mathcal{S}$ is checked only once, phase I finishes in $O(kn)$ time. Then we check for the private neighbours of each pivot in phase II. This checking requires $O(kn)$ time for each pivot. Any update of $\mathcal{P}$ requires $O(kn)$ time. We have to check this for at most k times. So phase II takes $O(k^2 n)$ time. In phase III we label the members of $\mathcal{P}_{lon}$ and $\mathcal{P}_{con}$. This takes $O(k^2 n)$ time. Then we sort the members of $\mathcal{P}$ which takes $O(k \log k)$ time. We can return the line centers from the sorted list. So in total, our algorithm runs in $O(nk^2)$ amount of time.

Theorem 2. *A* $(\frac{3}{2}, 16)$ *bi-criteria approximation for k-orthogonal line centers can be found in* $O(k^2 n \log n)$ *time.*

Proof. The proof follows from the above running time analysis, Theorem 1, Lemma 1, and Lemma 2.   □

4   Conclusion

We have presented a bi-criteria approximation algorithm for k-orthogonal line center problem for planar point sets. An interesting direction would be to investigate whether our algorithm can be extended to a bi-criteria approximation scheme for the k-orthogonal line center problem. Another direction would be to design bi-criteria approximation for k-orthogonal median and mean with a similar approach.

References

1. Agarwal, P.K., Procopiuc, C.M.: Approximation algorithms for projective clustering. J. Algorithms **46**(2), 115–139 (2003)
2. Agarwal, P.K., Procopiuc, C.M., Varadarajan, K.R.: A $(1+\varepsilon)$-approximation algorithm for 2-line-center. Comput. Geom. **26**(2), 119–128 (2003). https://doi.org/10.1007/s10878-012-9579-3
3. Agarwal, P.K., Procopiuc, C.M., Varadarajan, K.R.: Approximation algorithms for a k-line center. Algorithmica **42**(3), 221–230 (2005). https://doi.org/10.1007/s00453-005-1166-x
4. Aggarwal, C.C., Wolf, J.L., Yu, P.S., Procopiuc, C., Park, J.S.: Fast algorithms for projected clustering. SIGMOD Rec. **28**, 61–72 (1999)
5. Aggarwal, C.C., Yu, P.S.: Finding generalized projected clusters in high dimensional spaces. In: Proceedings of the 2000 ACM SIGMOD International Conference on Management of Data, SIGMOD 2000, pp. 70–81. Association for Computing Machinery, New York (2000)
6. Agrawal, R., Gehrke, J., Gunopulos, D., Raghavan, P.: Automatic subspace clustering of high dimensional data for data mining applications. ACM SIGMoD Rec. **27**(2), 94–105 (1998)
7. Calinescu, G., Dumitrescu, A., Karloff, H., Wan, P.J.: Separating points by axis-parallel lines. Int. J. Comput. Geom. Appl. **15**, 575–590 (2005)
8. Dom, M., Fellows, M.R., Rosamond, F.A.: Parameterized complexity of stabbing rectangles and squares in the plane. In: Das, S., Uehara, R. (eds.) WALCOM 2009. LNCS, vol. 5431, pp. 298–309. Springer, Heidelberg (2009). https://doi.org/10.1007/978-3-642-00202-1_26
9. Feldman, D., Fiat, A., Sharir, M., Segev, D.: Bi-criteria linear-time approximations for generalized k-mean/median/center. In: Proceedings of the Twenty-Third Annual Symposium on Computational Geometry, SCG 2007, pp. 19–26. Association for Computing Machinery, New York (2007)
10. Frederickson, G.N., Johnson, D.B.: Generalized selection and ranking: sorted matrices. SIAM J. Comput. **13**(1), 14–30 (1984)
11. Gaur, D.R., Ibaraki, T., Krishnamurti, R.: Constant ratio approximation algorithms for the rectangle stabbing problem and the rectilinear partitioning problem. J. Algorithms **43**(1), 138–152 (2002)
12. Giannopoulos, P., Knauer, C., Rote, G., Werner, D.: Fixed-parameter tractability and lower bounds for stabbing problems. Comput. Geom. **46**(7), 839–860 (2013). euroCG 2009
13. Hassin, R., Megiddo, N.: Approximation algorithms for hitting objects with straight lines. Discret. Appl. Math. **30**(1), 29–42 (1991)

14. Jaromczyk, J.W., Kowaluk, M.: The two-line center problem from a polar view: a new algorithm and data structure. In: Akl, S.G., Dehne, F., Sack, J.-R., Santoro, N. (eds.) WADS 1995. LNCS, vol. 955, pp. 13–25. Springer, Heidelberg (1995). https://doi.org/10.1007/3-540-60220-8_47
15. Kovaleva, S., Spieksma, F.C.R.: Approximation algorithms for rectangle stabbing and interval stabbing problems. SIAM J. Discret. Math. **20**(3), 748–768 (2006)
16. Megiddo, N., Tamir, A.: Finding least-distances lines. SIAM J. Algebraic Discret. Methods **4**, 207–211 (1983)
17. Procopiuc, C.M., Jones, M., Agarwal, P.K., Murali, T.M.: A Monte Carlo algorithm for fast projective clustering. In: Proceedings of the 2002 ACM SIGMOD International Conference on Management of Data, SIGMOD 2002, pp. 418–427. Association for Computing Machinery, New York (2002)
18. Renner, W.D., Pugh, N.O., Ross, D.B., Berg, R.E., Hall, D.C.: An algorithm for planning stereotactic brain implants. Int. J. Radiat. Oncol. Biol. Phys. **13**(4), 631–637 (1987)
19. Tansel, B.C., Francis, R.L., Lowe, T.J.: State of the art-location on networks: a survey. Part I: the p-center and p-median problems. Manag. Sci. **29**(4), 482–497 (1983)
20. Zanjirani Farahani, R., Hekmatfar, M.: Facility Location: Concepts, Models, Algorithms and Case Studies. Springer, Heidelberg (2009). https://doi.org/10.1007/978-3-7908-2151-2

Selecting Sources for Query Approximation with Bounded Resources

Hongjie Guo, Jianzhong Li$^{(\boxtimes)}$, and Hong Gao

Harbin Institute of Technology, Harbin 150001, Heilongjiang, China
hjguo@stu.hit.edu.cn, {lijzh,honggao}@hit.edu.cn

Abstract. In big data era, the Web contains a big amount of data, which is extracted from various sources. Exact query answering on large amounts of data sources is challenging for two main reasons. First, querying on big data sources is costly and even impossible. Second, due to the uneven data quality and overlaps of data sources, querying low-quality sources may return unexpected errors. Thus, it is critical to study approximate query problems on big data by accessing a bounded amount of the data sources. In this paper, we present an efficient method to select sources on big data for approximate querying. Our approach proposes a gain model for source selection by considering sources overlaps and data quality. Under the proposed model, we formalize the source selection problem into two optimization problems and prove their hardness. Due to the NP-hardness of problems, we present two approximate algorithms to solve the problems and devise a bitwise operation strategy to improve efficiency, along with rigorous theoretical guarantees on their performance. Experimental results on both real-world and synthetic data show high efficiency and scalability of our algorithms.

Keywords: Big data · Data quality · Source selection · Query approximation

1 Introduction

Traditional query processing mainly focuses on efficient computation of exact answers $Q(D)$ to a query Q in a dataset D. Nowadays, with the dramatic growth of useful information, dataset can be collected from various sources, i.e., websites, data markets, and enterprises. In applications with huge amounts of heterogeneous and autonomous data sources, exact query answering on big data is not only infeasible but also unnecessary due to the following reasons.

(1) Query answering is *costly*, even simple queries that require a single scan over the entire dataset cannot be answered within an acceptable time bound. Indeed, a linear-time query processing algorithm may take days on a dataset D of PB size [1].

(2) *Overlaps* among data sources are significant. Due to the autonomy of data sources, data sources are likely to contain overlap information, querying redundant sources may bring some gain, but with a higher extra cost.

© Springer Nature Switzerland AG 2020
W. Wu and Z. Zhang (Eds.): COCOA 2020, LNCS 12577, pp. 61–75, 2020.
https://doi.org/10.1007/978-3-030-64843-5_5

(3) Data sources are often *low-quality*. Even for domains such as flight and stock, which people consider to be highly reliable, a large amount of inconsistency has been observed in data collected from multiple sources [2], querying low-quality data sources can even deteriorate the quality of query results.

We next use a real-world example to illustrate these.

We consider the dataset D of all Computer Science books from an online bookstore aggregator. There are 894 bookstore sources, together providing 1265 Computer Science books. Each of these sources can be viewed as a relation: *Book (ISBN, Name, Author)*, describing the ISBN, name and author of Book, and each source identifies a book by its ISBN. Consider the following.

(1) A query Q_1 is to find the total number of books in D:

SELECT * FROM Bookstores

It is costly to answer Q_1, for not only the number of data sources is large, but also a data source contains a large volume of data. If we process the sources in decreasing order of their provide books, and query the total number of books after adding each new source. We can find that the largest source already provides 1096 books (86%), and we can obtain all 1265 books after querying 537 sources. In other words, after querying the first 537 sources, the rest of the sources do not bring any new gain [3].

(2) As another example, consider a query Q_2 to find the author lists of the book with ISBN = 201361205, query Q_2 is written as:

SELECT *Name* FROM Bookstores

WHERE *ISBN* = 201361205

We observed that D returns 5 different answers for Q_2, it shows that sources may have quality problems and can provide quite different answers for a query. In fact, in this dataset, each book has 1 to 23 different provided author lists.

These examples show that it is inappropriate to accessing the whole dataset when querying data from a large number of sources, which motivates us to select proper data sources before querying. Due to the importance, data source selection draws attention recently. [3] selects a set of sources for data fusion, and maximizes the profit by optimizing the gain and cost of integrating sources. However, it does not consider data self-conflicting and incomplete, and could hardly scale to big data sources for high time complexity of algorithm. [12] proposes a probabilistic coverage model to evaluate the quality of data and consider overlaps information. However, this method requires some prior statistics truth probability of each value, furthermore, this paper adopts a simple greedy method to solve the problem which is not optimal.

Based on the above discussion, in this paper, given an upper bound on the amount of data that can be used, we propose efficient source selection methods for approximate query answering. We propose a gain model for source selection by measuring the sources according to their coverage, overlaps, quality and cost. Considering these measures, we formulate the data source selection problem as two optimization problems from different aspects, both of which are proven to be NP-hard. Due to the hardness, we present two approximate algorithms to solve the optimization problems. These algorithms are proved can obtain the best possible approximate factor, furthermore, the running time of these methods are linear or polynomial in source number for the two propose problems, respectively. To improve the efficiency and scalability, we prestore the coverage information of each source offline and eliminate overlaps between sources online using bitwise operation, in this way, we can greatly accelerate the source selection algorithms.

In this paper, we make the following contributions.

First, we propose the gain model for data source selection from big autonomous data sources. This model take into consideration data quality and sources overlaps.

Second, under the proposed gain model, We formalize two optimization goals for source selection, called BNMG and BCMG, respectively. We show these two problems both are NP-hard.

Third, We present a greedy approximation algorithm for BNMG. We show that the greedy heuristic algorithm for BCMG has an unbounded approximation factor, and present a modified greedy algorithm to achieve a constant approximation factor for BCMG. A bitwise operation strategy for our algorithms is proposed to achieve better performance.

Finally, we conduct experiments on real-life and synthetic data to verify the effectiveness, efficiency and scalability of our proposed algorithms.

2 Problem Definition

This section first formally defines the basic notions used in the paper, then formulates the source selection problems, and then analyzes the complexity of these problems.

2.1 Basic Notions and Quality Metric

Definition 1 *(Data source). A dataset D is specified by a relational schema $\mathcal{R}$, which consists of a collection of relation schemas $(R_1, \cdots, R_m)$. Each relation schema R_i is defined over a set of attributes. A dataset consists of a set of data sources $\mathbb{S} = \{S_i | 1 \leq i \leq m\}$, where each source S_i includes a schema R_i. We assume that the schemas of sources have been mapped by existing schema mapping techniques.*

Definition 2 *(Data item). Consider a data item domain $\mathcal{DI}$. A data item $DI \in \mathcal{DI}$ represents a particular aspect of a real-world entity in a domain, such as the name of a book, and each data item should be identified by a key. For each data item $DI \in \mathcal{DI}$, a source $S_i \in \mathbb{S}$ can (but not necessarily) provide a value, and $\mathcal{DI}_i$ denotes that a set of data items provided by S_i.*

Definition 3 *(Functional dependency (FD)). An FD φ: $[A_1, \cdots, A_l] \rightarrow [B]$, where A_i and B are attributes in relational table. The semantic of φ is that any two tuples are equal on the left-hand side attribute of φ, they should also be equal on the right-hand side, otherwise, we say such tuples violate the FD [4].*

Definition 4 *(Claim). Let S be a set of sources and C a set of claims, each of which is triple $< S, k, v >$, meaning that the source, S, claims a value, v, on a data item with key, k. Sources often provide claims in the form of tuples.*

For example, in Table 1 there are two sources, s_1 and s_2, providing 3 claim-tuples. Note that a source can provide conflicting claims for a data item, for instance, consider data item: $ISDN = 02013.Name$, s_1 claims that the name of $ISDN = 02013$ are *Java* and *C++*, respectively. However, only one of the conflicting values is true. s_2 also miss the value of the author of the book with $ISDN = 02014$.

Table 1. Claims tuple of books.

$Source_{id}$	$Tuple_{id}$	ISDN	Name	Author
s_1	t_1	02013	Java	Robert
s_1	t_2	02013	C++	Robert
s_2	t_3	02014	Analysis of Algorithms	

Next, we consider quality metrics. Selecting sources should consider three sides. First, we wish to select a source that has a high *coverage* and a low *overlap*: such a source would return more answers and contribute more new answers than another source with the same coverage. Second, the answers returned by a high-quality source are high *reliability*. Third, a source with low *cost* will yield better performance. We next formally define these measures considered in source selection.

Definition 5 *(Coverage). We define the* coverage *of source S as the number of its provided data items, denoted by $V(S)$. Formally,*

$$V(S_i) = |\mathcal{DI}_i|, 1 \leq i \leq m \tag{1}$$

For source set $\mathcal{S}$ (a subset of source $\mathbb{S}$), we have

$$V(\mathcal{S}) = \cup_{S_i \in \mathcal{S}} V(S_i) \tag{2}$$

coverage of the source S reflects the expected number of answers returned by S, and *coverage* of the source set $\mathcal{S}$ represents the total distinct data items provided by sources, which has already eliminated the *overlap* information.

A source may provide self-conflicting or incomplete data, which means that the source has low reliability, querying low reliability may lead to an unexpectedly bad result. Thus, it is non-trivial to select sources according to their reliability, there can be many different ways to measure source reliability, in this paper, we measure it as the maximum correct number of claims provided by source, called the *reliability* of the source. The *reliability* of source S is denoted by $R(S)$.

In Table 1, s_1 claims two different values for the name of book, and no more than one of these can be correct. The upper bound of correct claims number provided by source S over key k is

$$u_{s,k} = \max_v(N_{s,k,v}) \tag{3}$$

where $N_{s,k,v}$ is the number of claims provided by S for k with a value v.

Definition 6 *(Reliability). The reliability of source S according to the upper bounds as*

$$R(S) = \sum_k u_{s,k} \tag{4}$$

For source set $\mathcal{S}$, we have

$$R(\mathcal{S}) = \sum_{S_i \in \mathcal{S}} R(S_i) \tag{5}$$

Collecting sources for querying come with a *cost*. First, many data sources charge for their data. Second, collecting and integrating them requires resources and time.

Definition 7 *(Cost). We define the cost of source S as the total number of its provided claims, denoted by $C(S)$.*

$$C(S) = \sum_{k,v}(N_{s,k,v}) \tag{6}$$

Similarly, for source set $\mathcal{S}$, we have

$$C(\mathcal{S}) = \sum_{S_i \in \mathcal{S}} C(S_i) \tag{7}$$

Definition 8 *(Gain). Now we define the gain model of source selection.*

$$G(\mathcal{S}) = \alpha V(\mathcal{S}) + (1 - \alpha)R(\mathcal{S}) \tag{8}$$

where $\alpha \in [0, 1]$ is a parameter controlling how much coverage and reliability have to be taken into account.

2.2 Problems

It is impractical to maximize the *gain* while minimizing the *cost*. Thus, we define the following two constrained optimization problems to select sources.

In some scenarios, a query system gives an upper **B**ound on the **N**umber of data sources can be accessed, and the optimization goal is to **M**aximize the **G**ain, we call this problem as BNMG.

Definition 9 *(BNMG). Given a source set* $\mathbb{S}$ *and a positive integer* K, *the BNMG problem is to find a subset* S *of* $\mathbb{S}$, *such that* $|S| \leq K$, *and* $G(S)$ *is maximized.*

In some scenarios, a query system gives an upper **B**ound on the **C**ost as the constraint, and wishes to obtain the **M**aximum **G**ain, we define this problem as BCMG.

Definition 10 *(BCMG). Given a source set* $\mathbb{S}$ *and* τ_c *be a budget on cost, the BCMG problem is to find a subset* S *of* $\mathbb{S}$, *such that maximizes* $G(S)$ *under* $C(S) \leq \tau_c$.

2.3 Complexity Results

Theorem 1. *The gain function of formulation (8) is non-negative, monotone and submodular.*

Proof. **non-negative**. *Obviously.*
monotone. *For* $x \in \mathbb{S} - S$, *if* $G(S \cup x) \geq G(S)$, *the gain model is monotone.*

$$\alpha V(S \cup x) \geq \alpha V(S), obviously \tag{9}$$

$$(1 - \alpha)R(S \cup x) = (1 - \alpha)R(S) + (1 - \alpha)R(x) \geq (1 - \alpha)R(S) \tag{10}$$

According to Eqs. (9) and (10), we get

$$G(S \cup x) = \alpha V(S \cup x) + (1 - \alpha)R(S \cup x) \geq \alpha V(S) + (1 - \alpha)R(S) = G(S) \tag{11}$$

Submodular. *For* $\mathcal{R} \subset S$ *and* $x \in \mathbb{S} - S$, *if* $G(S \cup x) - G(S) \leq G(\mathcal{R} \cup x) - G(\mathcal{R})$, *the gain model is submodular.*

$$V(S \cup x) - V(S) = V(S) + V(x) - V(S \cap x) - V(S)$$
$$= V(x) - V(S \cap x) \tag{12}$$

Similarly,

$$V(\mathcal{R} \cup x) - V(\mathcal{R}) = V(x) - V(\mathcal{R} \cap x) \tag{13}$$

Since, $\mathcal{R} \subset S$, *then* $V(S \cap x) \geq V(\mathcal{R} \cap x)$. *Hence*

$$\alpha V(S \cup x) - \alpha V(S) \leq \alpha V(\mathcal{R} \cup x) - \alpha V(\mathcal{R}) \tag{14}$$

And

$$(1 - \alpha)R(S \cup x) - (1 - \alpha)R(S) = (1 - \alpha)R(S) + (1 - \alpha)R(x) - (1 - \alpha)R(S)$$
$$= (1 - \alpha)R(x) \tag{15}$$

Similarly, $(1 - \alpha)R(\mathcal{R} \cup x) - (1 - \alpha)R(\mathcal{R}) = (1 - \alpha)R(x)$, *we have*

$$(1 - \alpha)R(S \cup x) - (1 - \alpha)R(S) = (1 - \alpha)R(\mathcal{R} \cup x) - (1 - \alpha)R(\mathcal{R}) \tag{16}$$

Combining Eqs. (14) and (16), we get

$$G(S \cup x) - G(S) \leq G(\mathcal{R} \cup x) - G(\mathcal{R}) \tag{17}$$

Theorem 2. *Both BNMG and BCMG are NP-hard problems.*

Proof. For a submodular function f, if f only takes non-negative value, and is monotone. Finding a K-element set S for which $f(S)$ is maximized is an NP-hard optimization problem [5,6]. For BNMG problem, function f is the *gain* model, thus, BNMG is NP-hard.

The BCMG problem is an instance of the Budgeted Maximum Coverage Problem (BMC) that is proven to be NP-hard [7]. Given an instance of BMC: A collection of sets $\mathbb{S} = \{S_1, S_2, \cdots, S_m\}$ with associated costs $\{C_i\}_{i=1}^m$ is defined over a domain of elements $X = \{x_1, x_2, \cdots, x_n\}$ with associated equivalent-weights. The goal is to find a collection of sets $S \subseteq \mathbb{S}$, such that the total cost of elements in S does not exceed a given budget L, and the total weight of elements covered by S is maximized. BCMG can be captured by the BMC problem in the following way: 1) the sets in BMC represent the sources in $\mathbb{S}$ of BCMG; 2) the elements in BMC represent the data items in BCMG; 3) the parameter α of the *gain* model in BCMG is equal to 1. Since the reduction could be accomplished in polynomial time, BCMG is an NP-hard problem.

3 Algorithm for Source Selection

Due to the NP-hardness of BNMG and BCMG, in this section, firstly, we devise a greedy approximation algorithm for BNMG and analyze the complexity and approximation ratio (Sect. 3.1). Then, we show that a greedy strategy is insufficient for solving the BCMG problem. Indeed, it can get arbitrary bad results. Thus, we generate a modified greedy algorithm using the enumeration technique for BCMG, and demonstrate that such algorithm has the best possible approximation factor (Sect. 3.2). We devise a bitwise operation strategy to accelerate the running of proposed algorithms (Sect. 3.3).

3.1 Algorithm for BNMG

For a submodular and nondecreasing function f, f satisfies a natural "diminishing returns" property: The marginal gain from adding a source to a set of sources S is at least as high as the marginal gain from adding the same source to a superset of S. Here, the marginal gain $(G(S \cup S_i) - G(S)$ in this algorithm) is the difference between the gain after and before selecting the new source. Such problem is well-solved by a simple greedy algorithm, denoted by Greedy (shown in Algorithm 1), that selects K sources by iteratively picking the source that provides the largest marginal gain (line 6).

Algorithm 1. Greedy

Input: $\mathbb{S}$, K
Output: a subset $\mathcal{S}$ of $\mathbb{S}$ with $|\mathcal{S}| \leq K$
1: Initialize $\mathcal{S} \leftarrow \emptyset$
2: **while** $|\mathcal{S}| < K$ **do**
3:     **for all** $S_i \in \mathbb{S}$ **do**
4:        $G(\mathcal{S} \cup S_i) \leftarrow \text{CompGain}(\mathcal{S} \cup S_i)$;
5:     **end for**
6:     $S_{opt} \leftarrow \arg\max_{S_i \in \mathbb{S}} G(\mathcal{S} \cup S_i) - G(\mathcal{S})$;
7:     $\mathcal{S} \leftarrow \mathcal{S} \cup S_{opt}$;
8:     $\mathbb{S} \leftarrow \mathbb{S} \backslash S_{opt}$;
9: **end while**

Time Complexity Analysis. The time complexity of Algorithm 1 is determined by the complexity to compute the *gain* of $(\mathcal{S} \cup S_i)$, this complexity is $O(n)$, n is the maximal number of data items in S_i. Clearly, the complexity of Algorithm 1 is $O(K * n * m)$, where K is the number of selected sources, and m is the number of sources in $\mathbb{S}$.

Theorem 3. *Algorithm 1 is a* $(1 - 1/e) - approximation$ *algorithm, where e is the base of the natural logarithm.*

Proof. The greedy algorithm has $(1 - 1/e)$ approximation ratio for a submodular and monotone function with a cardinality constraint [6].

3.2   Algorithm for BCMG

The greedy heuristic algorithm that picks at each step a source maximizing the ratio $\frac{G(\mathcal{S} \cup S_i) - G(\mathcal{S})}{C(S_i)}$ has an unbounded approximation factor. Namely, the worst case behavior of this algorithm might be very far from the optimal solution. In Table 2 for example, two sources S_1 and S_2 are subjected to an FD: *key $\rightarrow$ value*. According to our problem definition, S_1 has $V(S_1) = 1$, $R(S_1) = 1$, $C(S_1) = 1$; S_2 has $V(S_2) = p$, $R(S_2) = p$, $C(S_2) = p+1$. Let $\mathbb{S} = \{S_1, S_2\}$, $\alpha = 0.5$, and the budget of cost $\tau_c = p + 1$. The optimal solution contains the source S_2 and has *gain* p, while the solution picked by the greedy heuristic contains the source S_1 and has *gain* 1. The approximation factor of this instance is p, and is therefore unbounded (since p is not a constant).

Table 2. Two sources for an example

(a) Source S_1

id	key	value
t_1	1	1

(b) Source S_2

id	key	value
t_1	1	1
...	...	...
t_p	p	p
t_{p+1}	p	m

We modify the greedy heuristic using the enumeration technique, so as to achieve a constant approximation factor for the BCMG problem. The main idea is to apply the partial enumeration technique [8] before calling greedy algorithm, denoted by EnumGreedy (shown in Algorithm 2). Let l be a fixed integer. Firstly, we enumerate all subsets of $\mathbb{S}$ of cardinality less than l which have cost at most τ_c, and select the subset that has the maximal *gain* as the candidate solution (line 2). Then, we consider all subsets of $\mathbb{S}$ of cardinality l which have cost at most τ_c, and we complete each subset to a candidate solution using the greedy heuristic (line 3–17). The algorithm outputs the candidate solution having the greatest *gain* (line 18–22). *Time Complexity Analysis.* The running time of Algorithm 2 is

---

Algorithm 2. EnumGreedy

Input: $\mathbb{S}, \tau_c, l$
Output: a subset $\mathcal{S}$ of $\mathbb{S}$ with $C(\mathcal{S}) \leq \tau_c$
1: Initialize $\mathcal{S} \leftarrow \emptyset, \overline{\mathcal{S}'} \leftarrow \emptyset, \overline{\mathcal{S}''} \leftarrow \emptyset$
2: $\overline{\mathcal{S}'} \leftarrow \arg\max_{\overline{\mathcal{S}'} \subseteq \mathbb{S}} \{G(\overline{\mathcal{S}'}) | \ C(\overline{\mathcal{S}'}) \leq \tau_c, |\overline{\mathcal{S}'}| < l\}$
3: **for all** $\mathcal{S}'' \subseteq \mathbb{S}, |\mathcal{S}''| = l, C(\mathcal{S}'') \leq \tau_c$ **do**
4:    $\mathbb{S} \leftarrow \mathbb{S} \backslash \mathcal{S}''$
5:    **for all** $S_i \in \mathbb{S}$ **do**
6:       $G(\mathcal{S}'' \cup S_i) \leftarrow \text{CompGain}(\mathcal{S}'' \cup S_i);$
7:       $C(S_i) \leftarrow \text{CompCost}(S_i);$
8:    **end for**
9:    $S_{opt} \leftarrow \arg\max_{S_i} \frac{G(\mathcal{S}'' \cup S_i) - G(\mathcal{S}'')}{C(S_i)};$
10:    **if** $C(\mathcal{S}'') + C(S_i) \leq \tau_c$ **then**
11:       $\mathcal{S}'' \leftarrow \mathcal{S}'' \cup S_{opt};$
12:       $\mathbb{S} \leftarrow \mathbb{S} \backslash S_{opt};$
13:    **end if**
14:    **if** $G(\mathcal{S}'') > G(\overline{\mathcal{S}''})$ **then**
15:       $\overline{\mathcal{S}''} \leftarrow \mathcal{S}'';$
16:    **end if**
17: **end for**
18: **if** $G(\overline{\mathcal{S}'}) > G(\overline{\mathcal{S}''})$ **then**
19:    $\mathcal{S} \leftarrow \overline{\mathcal{S}'};$
20: **else**
21:    $\mathcal{S} \leftarrow \overline{\mathcal{S}''};$
22: **end if**

$O((n \cdot m)^{(l-1)})$ executions of enumeration and $O((n \cdot m)^{2l})$ executions of greedy, where m is the number of sources, n is the maximal number of data items in S_i. Therefore, for every fixed l, the running time is polynomial in $n \cdot m$.

Discussion. When $l = 1$, Algorithm 2 is actually a simple greedy method which has an unbounded approximation factor as previously mentioned. When $l = 2$, Algorithm 2 finds a collection of sources according to the greedy heuristic as the first candidate for solution. The second candidate is a single set in $\mathbb{S}$ for which *gain* is maximized. The algorithm outputs the candidate solution having the maximum *gain*.

Theorem 4. *For $l = 2$, Algorithm 2 achieves an approximation factor of $\frac{1}{2}(1 - \frac{1}{e})$ for the BCMG problem. For $l \geq 3$, Algorithm 2 achieves a $(1 - \frac{1}{e})$ approximation ratio for the BCMG, and this approximation factor is the best possible.*

Proof. The proof is by generalized the proof of approximation factor for the BMC problem, presented in [7]. The detail is omitted due to space limitation.

3.3   Improvement for Algorithms

The time complexities of proposed algorithms are determined by the complexity to compute the *gain*. In fact, the time complexities are dominated by the computing of *coverage* since *reliability* and *cost* can be computed in constant time. To reduce the computation time, we transform the process of computing *coverage* into building bit vectors and conducting bitwise *or* operation between them so that we can compute *coverage* without accessing original data. We build a bit vector $\boldsymbol{B}(S_i)$ for each S_i offline and use them to compute the coverage of sources online.

Constructing a bit vector for each source in an offline phase. Given $\mathbb{S}$, we consider all data items of $\mathbb{S}$ as bit vector space. For a source S_i, we maintain a bit vector $\boldsymbol{B}(S_i) = \{(b_1, b_2, \cdots, b_M) | b_i \in \{0, 1\}\}$, where M is the total number of data items provided by $\mathbb{S}$, b_j equals 1 if S_i contains j-th data item of $\mathbb{S}$. The bit vector building algorithm for each source S_i is described in Algorithm 3.

Algorithm 3. Bit Vector Building

Input: $\mathbb{S}$, Σ (FDs set)
Output: $\{\boldsymbol{B}(S_i) | S_i \in \mathbb{S}, 0 \leq i \leq m\}$
 1: Initialize $\boldsymbol{B}(S_i) \leftarrow \emptyset$
 2: **for all** $S_i \in \mathbb{S}$ **do**
 3:     **for all** $\varphi \in \Sigma$ **do**
 4:         **for all** Left-hand side A_j of φ **do**
 5:             **if** S_i contains A_j **then**
 6:                 $b_j = 1$;
 7:             **else**
 8:                 $b_j = 0$;
 9:             **end if**
10:             Add b_j to $\boldsymbol{B}(S_i)$;
11:         **end for**
12:     **end for**
13: **end for**

Then, the *coverage* of S_i is the number of 1 in $\boldsymbol{B}(S_i)$, denoted as $\boldsymbol{B}(S_i).cardinality$.

$$V(S_i) = \boldsymbol{B}(S_i).cardinality = \sum_1^M b_j \qquad (18)$$

Computing source *coverage* online. Given a source set $\mathcal{S}$, the *coverage* of $\mathcal{S}$ can be easily computed by bitwise *or* operation:

$$V(\mathcal{S}) = \cup_{S_i \in \mathcal{S}} V(S_i) = (\vee_{S_i \in \mathcal{S}} B(S_i)).cardinality \qquad (19)$$

where $\vee$ is the bitwise *or* operation.

Time Complexity Analysis. The expected time of Algorithm 3 is $O(m * |\Sigma| * n_{avg})$, where $n_{avg} = avg_{0 \leq i \leq m} |S_i|$. Although the complexity is quite high, such bit vectors are computed offline. Hence it will not affect the performance of the algorithm.

We denote the improvement greedy algorithm for BNMG problem which combines the bit vector operation as BitGreedy, and BitEnumGreedy for BCMG problem similarly.

4   Experimental Results

In this section, we study the proposed algorithms experimentally. The goals are to investigate (1) the comparison of performance between Greedy and Bit-Greedy for BNMG problem, as well as EnumGreedy and BitEnumGreedy for BCMG problem, and (2) how our algorithms perform in terms of efficiency and scalability.

4.1   Experiment Setup

We conducted our comparison experiments over a real-world dataset: *Book* [3]. In addition, to investigate the efficiency and scalability of our algorithm, we evaluated the performance of BitGreedy and BitEnumGreedy on synthetic datasets that yielded more sources and more tuples.

The *Book* dataset contains 894 data sources. Information on Computer Science books was collected from online bookstore aggregator AbeBooks.com, there are two FDs between the attributes: $ISDN \rightarrow Name$ and $ISDN \rightarrow Author$.

The *Synthetic Data* is synthetic data sets with various data source number and data size. We used 10 attributes $A1 - A10$ and 8 FDs: $A1 \rightarrow A8$, $A1 \rightarrow A9$, $A1 \rightarrow A10$, $A2 \rightarrow A6$, $A2 \rightarrow A7$, $A3 \rightarrow A6$, $A3 \rightarrow A7$, $[A4, A5] \rightarrow A8$. Each data source randomly chose an attribute with 20% probability, and each source contains at least one of the FDs, and the size of each data source is a random number in the range of $[2000, 10000]$.

In practical situations, due to the enormous number and volume of sources, Algorithm 2 needs to consume considerable time even when $l = 3$ to guarantee the best approximation ratio. Thus, in this paper, we set $l = 2$.

In this paper, we focus more on selecting data sources with high coverage, thus, we set $\alpha = 0.9$. Users can set different values for α according to their preferences.

All experiments are implemented in Java and executed on a PC with Windows 7, a 16 GB of RAM and a 3.6 GHz Intel i7-7700U CPU.

4.2   Comparison

For BNMG problem, we firstly compare the effectiveness of Greedy and Bit-Greedy with #Bounded Source (K in problem definition) varying from 1 to 10 on *Book* dataset, shown in Fig. 1(a) and Fig. 1(b), then we vary #Bounded Source from 10 to 100 on *Synthetic Data* with #Source = 100, #Tuple = 2000, shown in Fig. 1(c) and Fig. 1(d).

For BCMG problem, we compare EnumGreedy and BitEnumGreedy with Bounded cost varying from 5k to 50k on *Book* dataset and Bounded cost varying from $5*10^4$ to $5*10^5$ on *Synthetic Data*, The results are shown in Fig. 1(e)–1(h).

We have the following observations. (1) For BNMG problem, both on real-world data set and Synthetic data. Greedy and BitGreedy achieve the same Gain, and with the increase of #Bounded sources, the runtime of Greedy is linear to #Bounded sources, while the runtime of BitGreedy grows much slowly with #Bounded and outperforms Greedy significantly. (2) BCMG problem get a similar result. Whether the data set is *Book* or *Synthetic Data*, compare to EunmGreedy, BitEumGreedy achieves hundreds of speed up while not sacrifice the Gain. (3) Due to the higher time complexity, the algorithm for BCMG problem requires much runtime than that of BNMG, it also signifies that algorithm EnumGreedy, which without improvement strategy, for BCMG problem can not apply to real-time query systems.

4.3   Efficiency and Scalability

To further test how the number of sources and source size affects efficiency and scalability, we conduct experiments on synthetic datasets. (1) Fig. 1(i) and Fig. 1(j) report the runtimes of both algorithms with varying the data size. We observe that the runtimes of BitGreedy and EnumBitGreedy are very stable, it shows that the high efficiency and stability of our method when the volume of data source grows. Figure 1(k) and Fig. 1(l) plot the running time of BitGreedy and EnumBitGreedy respectively, as we vary the number of data sources from 100 to 1000. These show that the runtimes of both BitGreedy and EnumBit-Greedy increase nearly linearly. BitGreedy costs 47ms when the source number reaches 1000, and EnumBitGreedy finishes in 20316ms when the source number is 1000, showing the great scalability of our methods.

Summary. (1) The sources selected by BitGreedy and BitEnumGreedy are same as the selections of Greedy and EnumGreedy. (2) The algorithms using bitwise operation outperform original methods both on efficiency and scalability significantly. (3) The effectiveness of BitGreedy and BitEnumGreedy are insensitive to the source size. (4) Our algorithms scale well on both the data size and the number of sources.

5   Related Work

Source selection [3, 9–13] has been recently studied. [3] selects a set of sources for data fusion, it efficiently estimates fusion accuracy, and maximizes the profit by

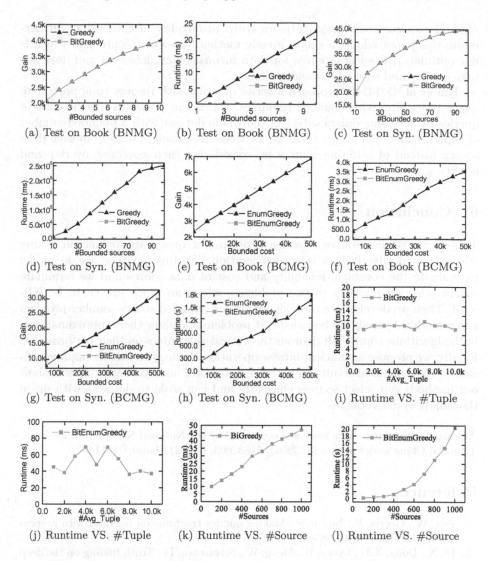

Fig. 1. Experimental results

optimizing the gain and cost of integrating sources. However, it does not consider data self-conflicting and incomplete, and could hardly scale to big data sources. [9] studies online ordering sources by estimating overlaps of sources, but this method requires some prior statistics and neglects data quality. [10] only takes freshness as quality metric without a comprehensive consideration for the quality of data sources such as functional dependency, completeness and the required resources. [11] focuses on finding sources relevant to a given query and does not take overlap into consideration. [12] proposes a probabilistic coverage model to evaluate the quality of data and consider overlaps information. However, this method requires some prior statistics truth probability of each value, furthermore, this paper adopts a simple greedy method to solve the problem which is not optimal. [13] selects sources for data inconsistency detection and does not take quality and cost into consideration.

Fan et al. [1,14–18] proposed a series of work with respect to approximate query processing with bounded resources, which can answer a specific class of queries by accessing only a subset of the whole dataset with a bounded number of tuples. Unlike our study, these works access a portion of data from each data source instead of selecting sources by considering their coverage, overlap and quality.

6 Conclusion

This paper studies source selection problem for query approximation taking efficiency and effectiveness into consideration. We first propose a gain model to evaluate the coverage, reliability and cost of data source and we formulate source selection problem to two problems, which are both proven to be NP-hard. Then we develop a greedy algorithm for bounded source number problem and a modified greedy for bounded cost problem and show their approximations, both algorithms come with rigorous theoretical guarantees on their performance. Finally, we propose an efficient bitwise operation strategy to further improve efficiency. Experimental results on both the real-world and synthetic datasets show our methods can select sources efficiently and can scale to datasets with up to thousands of data sources.

Acknowledgments. This work was supported by the National Natural Science Foundation of China under grants 61732003, 61832003, 61972110 and U1811461.

References

1. Fan, W., Geerts, F., Neven, F.: Making queries tractable on big data with preprocessing. PVLDB **6**(9), 685–696 (2013)
2. Li, X., Dong, X.L., Lyons, K., Meng, W., Srivastava, D.: Truth finding on the deep web: is the problem solved? VLDB **6**(2), 97–108 (2012)
3. Dong, X.L., Saha, B., Srivastava, D.: Less is more: selecting sources wisely for integration. In: VLDB, vol. 6, pp. 37–48. VLDB Endowment (2012)
4. Codd, E.F.: Relational completeness of data base sublanguages. In: Courant Computer Science Symposia, vol. 6, pp. 65–98. Data Base Systems (1972)
5. Cornuejols, G., Fisher, M.L., Nemhauser, G.L.: Location of bank accounts to optimize float: an analytic study of exact and approximate algorithms. Manag. Sci. **23**(8), 789–810 (1977)
6. Nemhauser, G.L., Wolsey, L.A., Fisher, M.L.: An analysis of approximations for maximizing submodular set functions-I. Math. Program. **14**(1), 265–294 (1978). https://doi.org/10.1007/BF01588971
7. Khuller, S., Moss, A., Naor, J.: The budgeted maximum coverage problem. Inf. Process. Lett. **70**(1), 39–45 (1999)

8. Shachnai, H., Tamir, T.: Polynomial time approximation schemes. In: Handbook of Approximation Algorithms and Metaheuristics, pp. 9.1–9.21. Chapman & Hall/CRC Computer and Information Science Series (2007)

9. Salloum, M., Dong, X.L., Srivastava, D., Tsotras, V.J.: Online ordering of overlapping data sources. VLDB **7**(3), 133–144 (2013)

10. Rekatsinas, T., Dong, X.L., Srivastava, D.: Characterizing and selecting fresh data sources. In: SIGMOD, pp. 919–930. ACM (2014)

11. Lin, Y., Wang, H., Zhang, S., Li, J., Gao, H.: Efficient quality-driven source selection from massive data sources. J. Syst. Softw. **118**(1), 221–233 (2016)

12. Lin, Y., Wang, H., Li, J., Gao, H.: Data source selection for information integration in big data era. Inf. Sci. **479**(1), 197–213 (2019)

13. Li, L., Feng, X., Shao, H., Li, J.: Source selection for inconsistency detection. In: Pei, J., Manolopoulos, Y., Sadiq, S., Li, J. (eds.) DASFAA 2018. LNCS, vol. 10828, pp. 370–385. Springer, Cham (2018). https://doi.org/10.1007/978-3-319-91458-9_22

14. Cao, Y., Fan, W., Wo, T., Yu, W.: Bounded conjunctive queries. PVLDB **7**(12), 1231–1242 (2014)

15. Fan, W., Geerts, F., Libkin, L.: On scale independence for querying big data. In: PODS, pp. 51–62. ACM (2014)

16. Fan, W., Geerts, F., Cao, Y., Deng, T., Lu, P.: Querying big data by accessing small data. In: PODS, pp 173–184. ACM (2015)

17. Cao, Y., Fan, W.: An effective syntax for bounded relational queries. In: SIGMOD, pp 599–614. ACM (2016)

18. Cao, Y., Fan, W.: Data driven approximation with bounded resources. PVLDB **10**(9), 973–984 (2017)

Parameterized Complexity of Satisfactory Partition Problem

Ajinkya Gaikwad, Soumen Maity$^{(\boxtimes)}$, and Shuvam Kant Tripathi

Indian Institute of Science Education and Research, Pune 411008, India
{ajinkya.gaikwad,tripathi.shuvamkant}@students.iiserpune.ac.in,
soumen@iiserpune.ac.in

Abstract. The SATISFACTORY PARTITION problem consists in deciding if the set of vertices of a given undirected graph can be partitioned into two nonempty parts such that each vertex has at least as many neighbours in its part as in the other part. The Balanced Satisfactory Partition problem is a variant of the above problem where the two partite sets are required to have the same cardinality. Both problems are known to be NP-complete. This problem was introduced by Gerber and Kobler [European J. Oper. Res. 125 (2000) 283-291] and further studied by other authors, but its parameterized complexity remains open until now. We enhance our understanding of the problem from the viewpoint of parameterized complexity. The three main results of the paper are the following: (1) The Satisfactory Partition problem is polynomial-time solvable for block graphs, (2) The Satisfactory Partition problem and its balanced version can be solved in polynomial time for graphs of bounded clicque-width, and (3) A generalized version of the Satisfactory Partition problem is W[1]-hard when parametrized by treewidth.

Keywords: Parameterized complexity · FPT · W[1]-hard · Treewidth · Clique-width

1 Introduction

Gerber and Kobler [8] introduced the problem of deciding if a given graph has a vertex partition into two non-empty parts such that each vertex has at least as many neighbours in its part as the other part. A graph satisfying this property is called *partitionable*. For example, complete graphs, star graphs, complete bipartite graphs with at least one part having odd size are not partitionable, where as some graphs are easily partitionable: cycles of length at least 4, trees that are not star graphs [5].

Given a graph $G = (V, E)$ and a subset $S \subseteq V(G)$, we denote by $d_S(v)$ the degree of a vertex $v \in V$ in $G[S]$, the subgraph of G induced by S. For $S = V$, the subscript is omitted, hence $d(v)$ stands for the degree of v in G. In

---

S. Maity—The author's research was supported in part by the Science and Engineering Research Board (SERB), Govt. of India, under Sanction Order No. MTR/2018/001025.

W. Wu and Z. Zhang (Eds.): COCOA 2020, LNCS 12577, pp. 76–90, 2020.
https://doi.org/10.1007/978-3-030-64843-5_6

this paper, we study the parameterized complexity of SATISFACTORY PARTITION and BALANCED SATISFACTORY PARTITION problems. We define these problems as follows:

SATISFACTORY PARTITION

Input: A graph $G = (V, E)$.

Question: Is there a nontrivial partition (V_1, V_2) of V such that for every $v \in V$, if $v \in V_i$ then $d_{V_i}(v) \geq d_{V_{3-i}}(v)$?

A variant of this problem where the two parts have equal size is:

BALANCED SATISFACTORY PARTITION

Input: A graph $G = (V, E)$ on an even number of vertices.

Question: Is there a nontrivial partition (V_1, V_2) of V such that $|V_1| = |V_2|$ and for every $v \in V$, if $v \in V_i$ then $d_{V_i}(v) \geq d_{V_{3-i}}(v)$?

Given a partition (V_1, V_2), we say that a vertex $v \in V_i$ is *satisfied* if $d_{V_i}(v) \geq d_{V_{3-i}}(v)$, or equivalently if $d_{V_i}(v) \geq \lceil \frac{d(v)}{2} \rceil$. A graph admitting a non-trivial partition where all vertices are satisfied is called *satisfactory partitionable*, and such a partition is called *satisfactory partition*. For the standard concepts in parameterized complexity, see the recent textbook by Cygan et al. [6].

Our Results: Our main results are the following:

- The SATISFACTORY PARTITION problem is polynomial-time solvable for block graphs,
- The SATISFACTORY PARTITION and BALANCED SATISFACTORY PARTITION problems can be solved in polynomial time for graphs of bounded clique-width.
- A generalized version of the SATISFACTORY PARTITION problem is W[1]-hard when parameterized by treewidth.

Related Work: In the first paper on this topic, Gerber and Kobler [8] considered a generalized version of this problem by introducing weights for the vertices and edges and showed that a general version of the problem is strongly NP-complete. For the unweighted version, they presented some sufficient conditions for the existence of a solution. This problem was further studied in [2,7,9]. The SATISFACTORY PARTITION problem is NP-complete and this implies that BALANCED SATISFACTORY PARTITION problem is also NP-complete via a simple reduction in which we add new dummy vertices and dummy edges to the graph [3,5]. Both problems are solvable in polynomial time for graphs with maximum degree at most 4 [5]. They also studied generalizations and variants of this problem when a partition into $k \geq 3$ nonempty parts is required. Bazgan, Tuza, and Vanderpooten [2,4] studied an "unweighted" generalization of SATISFACTORY PARTITION, where each vertex v is required to have at least $s(v)$ neighbours in its own part, for a given function s representing the degree of satisfiability. Obviously, when $s = \lceil \frac{d}{2} \rceil$, where d is the degree function, we obtain satisfactory partition. They gave a polynomial-time algorithm for graphs of bounded treewidth which decides if a graph admits a satisfactory partition, and gives such a partition if it exists.

2 Polynomial Time Algorithm on Block Graphs

A single vertex whose removal disconnects the graph is called a cut-vertex. A maximal connected subgraph without a cut-vertex is called a block. Thus, every block of a graph G is either a maximal 2-connected subgraph, or a cut-edge with its end points. By their maximality, different blocks of G overlap in at most one vertex, which is then a cut-vertex of G. Hence, every edge of G lies in a unique block, and G is the union of its blocks. A block graph is a graph whose blocks are cliques. An end block of a block graph is a block that contains exactly one cut-vertex of G. A block graph that is not complete graph has at least two end blocks. A block graph admits a nice tree structure, called *cut-tree*. Let $\{B_1, B_2, \ldots, B_r\}$ and $\{c_1, c_2, \ldots, c_s\}$ be the set of blocks and the set of cut-vertices of a block graph G, respectively. A cut-tree of G is a tree $T_G = (V', E')$, where $V' = \{B_1, B_2, \ldots, B_r, c_1, c_2, \ldots, c_s\}$ and $E' = \{B_i c_j \mid c_j \in B_i, i \in [r] \text{ and } j \in [s]\}$. Note that the cut-tree T_G is a tree in which the leaves are the end blocks of G. The computation of blocks in a graph G and construction of T_G can be done in $O(|V| + |E|)$ time, using Depth First Search (DFS) [1].

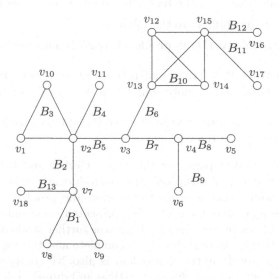

Fig. 1. A block graph G

Lemma 1. If a cut vertex v is adjacent to an end block $B' = \{v, v'\}$ of size two then v and v' must lie in the same part in any satisfactory partition.

Proof. Suppose v and v' are in different parts. As B' is an end-block of size 2, v' has no neighbour in its own part and one neighbour v in the other part. Hence v' is not satisfied. This proves the lemma.

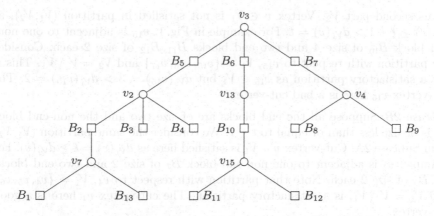

Fig. 2. The cut tree T_G of the block graph G shown in Fig. 1

Let G be a block graph and T_G be the cut-tree of G. Every cut-vertex is adjacent to at least two blocks. For simplicity, suppose cut-vertex v is adjacent to two blocks B_1 and B_2. Then $T_G - v$ has two components T_{G,B_1} and T_{G,B_2}, where T_{G,B_i} is the component that contains block B_i, $i = 1, 2$. There are two possible partitions with respect to v: $\left(V(T_{G,B_1}) \setminus \{v\}, V(T_{G,B_2})\right)$ and $\left(V(T_{G,B_1}), V(T_{G,B_2}) \setminus \{v\}\right)$. A cut-vertex $v \in V(T_G)$ is said to be a *good cut-vertex* if there is a satisfactory partition of G with respect to v; otherwise v is called a *bad cut-vertex*. In general, suppose cut-vertex $v \in V(T_G)$ is adjacent to k non-end blocks $B_1, B_2, \ldots, B_k$, and ℓ end blocks $B'_1, B'_2, \ldots, B'_\ell$. As each cut-vertex is adjacent to at least two blocks, we have $k + \ell \geq 2$. We consider the following cases and decide if a cut-vertex v is *good* or *bad* in each case (Fig. 2).

Case 1: Let $k \geq 2$ and $\ell \geq 0$. That is, v is adjacent to at least two non-end blocks and $\ell \geq 0$ end blocks. Let B_1 be a smallest non-end block adjacent to v. Consider the partition $V_1 = V(T_{G,B_1}) \setminus \{v\}$ and $V_2 = V \setminus V_1$. Note that $v \in V_2$ and $d_{V_2}(v) \geq d_{V_1}(v)$, hence v is satisfied and, clearly all other vertices are also satisfied. Thus (V_1, V_2) forms a satisfactory partition. For example in Fig. 1, v_2 is adjacent to two non-end blocks B_2, B_5 and two end block B_3, B_4. Here B_2 is a smallest non-end block, thus $V_1 = V(T_{G,B_2}) \setminus \{v_2\} = \{v_7, v_8, v_9, v_{18}\}$ and $V_2 = V \setminus V_1$ form a satisfactory partition. The cut-vertex v_2 here is a good cut-vertex.

Case 2: Let $k = 1$ and $\ell \geq 1$. That is, v is adjacent to exactly one non-end block B_1 and at least one end block.

Subcase 2A: Suppose all the end blocks are of size two and the non-end block B_1 is of size greater than or equal to $\ell + 2$. As end blocks $B'_1, B'_2, \ldots, B'_\ell$ are of size two each, using Lemma 1, we know the vertices of blocks $B'_1, B'_2, \ldots, B'_\ell$ are in one part along with v. Thus $\bigcup_{i=1}^{\ell} B'_i$ forms first part V_1 and $V(T_{G,B_1}) \setminus \{v\}$

forms second part V_2. Vertex $v \in V_1$ is not satisfied in partition (V_1, V_2), as $d_{V_2}(v) \geq \ell + 1 > d_{V_1}(v) = \ell$. For example in Fig. 1, v_{15} is adjacent to one non-end block B_{10} of size 4 and two end blocks B_{11}, B_{12} of size 2 each. Consider the partition with respect to v_{15}, $V_1 = \{v_{15}, v_{16}, v_{17}\}$ and $V_2 = V \setminus V_1$. This is not a satisfactory partition as $v_{15} \in V_1$ but $d_{V_2}(v_{15}) = 3 > d_{V_1}(v_{15}) = 2$. The cut-vertex v_{15} here is a bad cut-vertex.

Subcase 2B: Suppose all the end blocks are of size two and the non-end block B_1 is of size less than or equal to $\ell + 1$. We consider the same partition (V_1, V_2) as in Subcase 2A. Cut-vertex $v \in V_1$ is satisfied here as $d_{V_1}(v) = \ell \geq d_{V_2}(v)$. For example, v_4 is adjacent to one non-end block B_7 of size 2 and two end blocks B_8, B_9 of size 2 each. Note that partition with respect to v_4, $V_1 = \{v_4, v_5, v_6\}$ and $V_2 = V \setminus V_1$, is a satisfactory partition. The cut-vertex v_4 here is a good cut-vertex.

Subcase 2C: At least one end block is of size greater than 2. Without loss of generality suppose $|B_1'| > 2$. If $|B_1| \geq |B_1'|$, then $V_1 = V(T_{G,B_1} \cup \bigcup_{i=2}^{\ell} B_i'$ and $V_2 = B_1' \setminus \{v\}$ form a satisfactory partition. If $|B_1'| \geq |B_1|$, then $V_1 = \bigcup_{i=1}^{\ell} B_i'$ and $V_2 = B_1 \setminus \{v\}$ form a satisfactory partition. For example, v_7 is adjacent to one non-end block B_2 and two end blocks B_1 and B_{13}; B_1 is of size 3 and B_{13} is of size 2. Note that $V_1 = \{v_7, v_8, v_9, v_{18}\}$ and $V_2 = V \setminus V_1$ form a satisfactory partition. The cut-vertex v_7 is a good cut-vertex.

This suggests the following theorem.

Theorem 1. Let G be a block graph. If T_G has a good cut-vertex then G is satisfactory partitionable.

Note that although the condition of this theorem is sufficient to assure that a block graph is satisfactory partitionable, this certainly is not a necessary condition. For example, the block graph shown in Fig. 3 is satisfactory partitionable but does not have any good cut-vertices; clearly such block graphs always have at least two bad cut-vertices. If a block graph has exactly one cut-vertex and that too is a bad cut-vertex, then the graph is not satisfactory partitionable. Now, we consider block graphs G having no good cut-vertices but the number m of bad cut-vertices is at least two; such graphs satisfy following two conditions:

1. There is exactly one non-end block in G and every cut-vertex is adjacent to it.
2. All the end blocks of G are of size exactly equal to 2.

Suppose B is the only non-end block in G and B' is obtained from B by removing its cut vertices. Let D_i represent the union of all end-blocks of size two that contain v_i. For graph G in Fig. 3, we have $D_3 = \{v_3, v_7, v_6, v_8\}$, $D_4 = \{v_4, v_9, v_{10}, v_{11}\}$, $D_5 = \{v_5, v_{12}, v_{13}\}$; and $B = \{v_1, v_2, v_3, v_4, v_5\}$, $B' = \{v_1, v_2\}$. By Lemma 1, all the vertices of D_i must lie in one part in any satisfactory partition. Let $(D_1, D_2, \ldots, D_m)$ be a decreasing ordering of D_i's according to cardinalities, that is, $|D_1| \geq |D_2| \geq \ldots \geq |D_m|$.

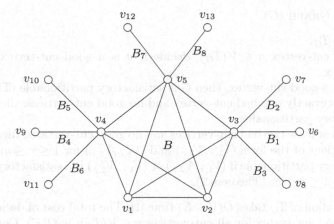

Fig. 3. A satisfactory partitionable block graph $G = (V, E)$ with no good cut-vertices. Note that G has three cut-vertices v_3, v_4, v_5 and all of them are bad cut-vertices. $V_1 = \{v_3, v_4, v_6, v_7, v_8, v_9, v_{10}, v_{11}\}$ and $V_2 = V \backslash V_1$ form a satisfactory partition of G.

Theorem 2. Let G be a block graph satisfying Conditions 1 and 2 above. Then G is satisfactory partitionable if and only if G has a satisfactory partition of the form either

$$V_{1,r} = \bigcup_{i=1}^{r} D_i, \quad V_{2,r} = B' \cup \bigcup_{i=r+1}^{m} D_i \quad \text{or}$$

$$V'_{1,r} = B' \cup \bigcup_{i=1}^{r} D_i, \quad V'_{2,r} = \bigcup_{i=r+1}^{m} D_i,$$

for some $1 \le r \le m$.

Proof. Suppose G is satisfactory partitionable, and $V_1 = \bigcup_{i=1}^{j-1} D_i \cup \bigcup_{i=j+1}^{r+1} D_i$, $V_2 = B' \cup D_j \cup \bigcup_{i=r+2}^{m} D_i$ form a satisfactory partition of G. It is easy to see that $V'_1 = \bigcup_{i=1}^{r} D_i$ and $V'_2 = B' \cup \bigcup_{i=r+1}^{m} D_i$, obtained from (V_1, V_2) by swapping two sets D_j and D_{r+1}, also form a satisfactory partition in the required form. On the other hand, if there is a satisfactory partition of the form $(V_{1,r}, V_{2,r})$ or $(V'_{1,r}, V'_{2,r})$ for some r, then G is satisfactory partitionable. This proves the theorem.

The following algorithm determines if a given block graph G is satisfactory partitionable.

SP-BLOCK GRAPH (G)

1. compute T_G.
2. for each cut-vertex $v \in V(T_G)$, decide if v is a good cut-vertex or a bad cut-vertex.
3. if T_G has a good cut-vertex, then G is satisfactory partitionable (Theorem 1).
4. if T_G has exactly one bad cut-vertex and no good cut-vertices, then G is not satisfactory partitionable.
5. if T_G has at least two bad cut-vertices and no good cut-vertices, then compute all partitions of the form $(V_{1,r}, V_{2,r})$ and $(V'_{1,r}, V'_{2,r})$ for $1 \le r \le m$, and G is satisfactory partitionable if $(V_{1,r}, V_{2,r})$ or $(V'_{1,r}, V'_{2,r})$ is a satisfactory partition for some $1 \le r \le m$ (Theorem 2).

Computation of T_G takes $O(V + E)$ time [1]. The total cost of deciding if v is a good or bad cut-vertex for all cut-vertices $v \in V(T_G)$, is $O(E)$. Computation of all partitions of the form $(V_{1,r}, V_{2,r})$ and $(V'_{1,r}, V'_{2,r})$ for $1 \le r \le m$, requires a decreasing ordering $(D_1, D_2, \ldots, D_m)$ of D_i's according to their cardinalities. This takes $O(V \log V)$ time as m can be at most $|V|$. The running time of SP-BLOCK GRAPH is therefore $O(V \log V + E)$.

3   Graphs of Bounded Clique-Width

This section presents a polynomial time algorithm for the SATISFACTORY PARTITION and BALANCED SATISFACTORY PARTITION problems for graphs of bounded clique-width. The clique-width of a graph G is a parameter that describes the structural complexity of the graph; it is closely related to treewidth, but unlike treewidth it can be bounded even for dense graphs. In a vertex-labeled graph, an i-vertex is a vertex of label i.

A c-expression is a rooted binary tree T such that

- each leaf has label o_i for some $i \in \{1, \ldots, c\}$,
- each non-leaf node with two children has label $\cup$, and
- each non-leaf node with only one child has label $\rho_{i,j}$ or $\eta_{i,j}$ $(i, j \in \{1, \ldots, c\}, i \ne j)$.

Each node in a c-expression represents a vertex-labeled graph as follows:

- a o_i node represents a graph with one i-vertex;
- a $\cup$-node represents the disjoint union of the labeled graphs represented by its children;
- a ρ_{ij}-node represents the labeled graph obtained from the one represented by its child by replacing the labels of the i-vertices with j;
- a η_{ij}-node represents the labeled graph obtained from the one represented by its child by adding all possible edges between i-vertices and j-vertices.

A c-expression represents the graph represented by its root. A c-expression of a n-vertex graph G has $O(n)$ vertices. The *clique-width* of a graph G, denoted by $\text{cw}(G)$, is the minimum c for which there exists a c-expression T representing a graph isomorphic to G.

A c-expression of a graph is *irredundant* if for each edge $\{u, v\}$, there is exactly one node $\eta_{i,j}$ that adds the edge between u and v. It is known that a c-expression of a graph can be transformed into an irredundant one with $O(n)$ nodes in linear time. Here we use irredundant c-expression only.

Computing the clique-width and a corresponding c-expression of a graph is NP-hard. For $c \leq 3$, we can compute a c-expression of a graph of clique-width at most c in $O(n^2 m)$ time, where n and m are the number of vertices and edges, respectively. For fixed $c \geq 4$, it is not known whether one can compute the clique-width and a corresponding c-expression of a graph in polynomial time. On the other hand, it is known that for any fixed c, one can compute a $(2^{c+1} - 1)$-expression of a graph of clique-width c in $O(n^3)$ time. For more details see [10]. We now have the following result:

Theorem 3. Given an n-vertex graph G and an irredundant c-expression T of G, the SATISFACTORY PARTITION and BALANCED SATISFACTORY PARTITION problems are solvable in $O(n^{8c})$ time.

For each node t in a c-expression T, let G_t be the vertex-labeled graph represented by t. We denote by V_t the vertex set of G_t. For each i, we denote the set of i-vertices in G_t by V_t^i. For each node t in T, we construct a table $dp_t(\mathbf{r}, \bar{\mathbf{r}}, \mathbf{s}, \bar{\mathbf{s}}) \in \{\text{true, false}\}$ with indices $\mathbf{r} : \{1, \ldots, c\} \to \{0, \ldots, n\}$, $\bar{\mathbf{r}} : \{1, \ldots, c\} \to \{0, \ldots, n\}$, $\mathbf{s} : \{1, \ldots, c\} \to \{-n+1, \ldots, n-1\} \cup \{\infty\}$, and $\bar{\mathbf{s}} : \{1, \ldots, c\} \to \{-n+1, \ldots, n-1\} \cup \{\infty\}$ as follows. We set $dp_t(\mathbf{r}, \bar{\mathbf{r}}, \mathbf{s}, \bar{\mathbf{s}}) = \text{true}$ if and only if there exists a partition $(S, \bar{S})$ of V_t such that for all $i \in \{1, 2, \ldots, c\}$

- $\mathbf{r}(i) = |S \cap V_t^i|$;
- $\bar{\mathbf{r}}(i) = |\bar{S} \cap V_t^i|$;
- if $S \cap V_t^i \neq \emptyset$, then $\mathbf{s}(i) = \min_{v \in S \cap V_t^i} \{|N_{G_t}(v) \cap S| - |N_{G_t}(v) \setminus S|\}$, otherwise $\mathbf{s}(i) = \infty$;
- if $\bar{S} \cap V_t^i \neq \emptyset$, then $\bar{\mathbf{s}}(i) = \min_{v \in \bar{S} \cap V_t^i} \{|N_{G_t}(v) \cap \bar{S}| - |N_{G_t}(v) \setminus \bar{S}|\}$, otherwise $\bar{\mathbf{s}}(i) = \infty$.

That is, $\mathbf{r}(i)$ denotes the number of the i-vertices in S; $\bar{\mathbf{r}}(i)$ denotes the number of the i-vertices in $\bar{S}$; $\mathbf{s}(i)$ is the "surplus" at the weakest i-vertex in S and $\bar{\mathbf{s}}(i)$ is the "surplus" at the weakest i-vertex in $\bar{S}$.

Let τ be the root of the c-expression T of G. Then G has a satisfactory partition if there exist $\mathbf{r}, \bar{\mathbf{r}}, \mathbf{s}, \bar{\mathbf{s}}$ satisfying

1. $dp_\tau(\mathbf{r}, \bar{\mathbf{r}}, \mathbf{s}, \bar{\mathbf{s}}) = \text{true}$;
2. $\min\{\mathbf{s}(i), \bar{\mathbf{s}}(i)\} \geq 0$.

For the BALANCED SATISFACTORY PARTITION problem, we additionally ask that $\sum_{i=1}^c \mathbf{r}(i) = \sum_{i=1}^c \bar{\mathbf{r}}(i)$. If all entries $dp_\tau(\mathbf{r}, \bar{\mathbf{r}}, \mathbf{s}, \bar{\mathbf{s}})$ are computed in advance, then we can verify above conditions by spending $O(1)$ time for each tuple $(\mathbf{r}, \bar{\mathbf{r}}, \mathbf{s}, \bar{\mathbf{s}})$.

In the following, we compute all entries $dp_t(\mathbf{r}, \bar{\mathbf{r}}, \mathbf{s}, \bar{\mathbf{s}})$ in a bottom-up manner. There are $(n+1)^c \cdot (n+1)^c \cdot (2n)^c \cdot (2n)^c = O(n^{4c})$ possible tuples $(\mathbf{r}, \bar{\mathbf{r}}, \mathbf{s}, \bar{\mathbf{s}})$. Thus, to prove Theorem 3, it is enough to prove that each entry $dp_t(\mathbf{r}, \bar{\mathbf{r}}, \mathbf{s}, \bar{\mathbf{s}})$ can be computed in time $O(n^{4c})$ assuming that the entries for the children of t are already computed.

Lemma 2. For a leaf node t with label o_i, $dp_t(\mathbf{r}, \bar{\mathbf{r}}, \mathbf{s}, \bar{\mathbf{s}})$ can be computed in $O(1)$ time.

Proof. Observe that $dp_t(\mathbf{r}, \bar{\mathbf{r}}, \mathbf{s}, \bar{\mathbf{s}}) = $ **true** if and only if $\mathbf{r}(j) = 0$, $\bar{\mathbf{r}}(j) = 0$, $\mathbf{s}(j) = 0$, and $\bar{\mathbf{s}}(j) = 0$ for all $j \neq i$ and either

- $\mathbf{r}(i) = 0$, $\bar{\mathbf{r}}(i) = 1$, $\mathbf{s}(i) = \infty$, $\bar{\mathbf{s}}(i) = 0$, or
- $\mathbf{r}(i) = 1$, $\bar{\mathbf{r}}(i) = 0$, $\mathbf{s}(i) = 0$, $\bar{\mathbf{s}}(i) = \infty$.

The first case corresponds to $S = \emptyset$, $\bar{S} = V_t^i$, and the second case corresponds to $S = V_t^i$, $\bar{S} = \emptyset$. These conditions can be checked in $O(1)$ time.

Lemma 3. For a $\cup$-node t, $dp_t(\mathbf{r}, \bar{\mathbf{r}}, \mathbf{s}, \bar{\mathbf{s}})$ can be computed in $O(n^{4c})$ time.

Proof. Let t_1 and t_2 be the children of t in T. Then $dp_t(\mathbf{r}, \bar{\mathbf{r}}, \mathbf{s}, \bar{\mathbf{s}}) = $ **true** if and only if there exist $\mathbf{r}_1, \bar{\mathbf{r}}_1, \mathbf{s}_1, \bar{\mathbf{s}}_1$ and $\mathbf{r}_2, \bar{\mathbf{r}}_2, \mathbf{s}_2, \bar{\mathbf{s}}_2$ such that $dp_t(\mathbf{r}_1, \bar{\mathbf{r}}_1, \mathbf{s}_1, \bar{\mathbf{s}}_1) = $ **true**, $dp_t(\mathbf{r}_2, \bar{\mathbf{r}}_2, \mathbf{s}_2, \bar{\mathbf{s}}_2) = $ **true**, $\mathbf{r}(i) = \mathbf{r}_1(i) + \mathbf{r}_2(i)$, $\bar{\mathbf{r}}(i) = \bar{\mathbf{r}}_1(i) + \bar{\mathbf{r}}_2(i)$, $\mathbf{s}(i) = \min\{\mathbf{s}_1(i), \mathbf{s}_2(i)\}$ and $\bar{\mathbf{s}}(i) = \min\{\bar{\mathbf{s}}_1(i), \bar{\mathbf{s}}_2(i)\}$ for all i. The number of possible pairs for $(\mathbf{r}_1, \mathbf{r}_2)$ is at most $(n+1)^c$ as $\mathbf{r}_2$ is uniquely determined by $\mathbf{r}_1$; the number of possible pairs for $(\bar{\mathbf{r}}_1, \bar{\mathbf{r}}_2)$ is at most $(n+1)^c$ as $\bar{\mathbf{r}}_2$ is uniquely determined by $\bar{\mathbf{r}}_1$. There are at most $2^c(2n)^c$ possible pairs for $(\mathbf{s}_1, \mathbf{s}_2)$ and for $(\bar{\mathbf{s}}_1, \bar{\mathbf{s}}_2)$ each. In total, there are $O(n^{4c})$ candidates. Each candidate can be checked in $O(1)$ time, thus the lemma holds.

Lemma 4. For a η_{ij}-node t, $dp_t(\mathbf{r}, \bar{\mathbf{r}}, \mathbf{s}, \bar{\mathbf{s}})$ can be computed in $O(1)$ time.

Proof. Let t' be the child of t in T. Then, $dp_t(\mathbf{r}, \bar{\mathbf{r}}, \mathbf{s}, \bar{\mathbf{s}}) = $ **true** if and only if $dp_t(\mathbf{r}, \bar{\mathbf{r}}, \mathbf{s}', \bar{\mathbf{s}}') = $ **true** for some $\mathbf{s}', \bar{\mathbf{s}}'$ with the following conditions:

- $\mathbf{s}(h) = \mathbf{s}'(h)$ and $\bar{\mathbf{s}}(h) = \bar{\mathbf{s}}'(h)$ hold for all $h \notin \{i, j\}$;
- $\mathbf{s}(i) = \mathbf{s}'(i) + 2\mathbf{r}(j) - |V_t^j|$ and $\mathbf{s}(j) = \mathbf{s}'(j) + 2\mathbf{r}(i) - |V_t^i|$;
- $\bar{\mathbf{s}}(i) = \bar{\mathbf{s}}'(i) + 2\bar{\mathbf{r}}(j) - |V_t^j|$ and $\bar{\mathbf{s}}(j) = \bar{\mathbf{s}}'(j) + 2\bar{\mathbf{r}}(i) - |V_t^i|$.

We now explain the condition for $\mathbf{s}(i)$. Recall that T is irredundant. That is, the graph $G_{t'}$ does not have any edge between the i-vertices and the j-vertices. In G_t, an i-vertex has exactly $\mathbf{r}(j)$ more neighbours in S and exactly $|V_t^j| - \mathbf{r}(j)$ more neighbours in $\bar{S}$. Thus we have $\mathbf{s}(i) = \mathbf{s}'(i) + 2\mathbf{r}(j) - |V_t^j|$. The lemma holds as there is only one candidate for each $\mathbf{s}'(i)$, $\mathbf{s}'(j)$, $\bar{\mathbf{s}}'(i)$ and $\bar{\mathbf{s}}'(j)$.

Lemma 5. For a ρ_{ij}-node t, $dp_t(\mathbf{r}, \bar{\mathbf{r}}, \mathbf{s}, \bar{\mathbf{s}})$ can be computed in $O(n^4)$ time.

Proof. Let t' be the child of t in T. Then, $dp_t(\mathbf{r}, \bar{\mathbf{r}}, \mathbf{s}, \bar{\mathbf{s}}) = $ **true** if and only if there exist $\mathbf{r}', \bar{\mathbf{r}}', \mathbf{s}', \bar{\mathbf{s}}'$ such that $dp_{t'}(\mathbf{r}', \bar{\mathbf{r}}', \mathbf{s}', \bar{\mathbf{s}}') = $ **true**, where :

- $\mathbf{r}(i) = 0$, $\mathbf{r}(j) = \mathbf{r}'(i) + \mathbf{r}'(j)$, and $\mathbf{r}(h) = \mathbf{r}'(h)$ if $h \notin \{i, j\}$;
- $\bar{\mathbf{r}}(i) = 0$, $\bar{\mathbf{r}}(j) = \bar{\mathbf{r}}'(i) + \bar{\mathbf{r}}'(j)$, and $\bar{\mathbf{r}}(h) = \bar{\mathbf{r}}'(h)$ if $h \notin \{i, j\}$;
- $\mathbf{s}(i) = \infty$, $\mathbf{s}(j) = \min\{\mathbf{s}'(i), \mathbf{s}'(j)\}$, and $\mathbf{s}(h) = \mathbf{s}'(h)$ if $h \notin \{i, j\}$;
- $\bar{\mathbf{s}}(i) = \infty$, $\bar{\mathbf{s}}(j) = \min\{\bar{\mathbf{s}}'(i), \bar{\mathbf{s}}'(j)\}$, and $\bar{\mathbf{s}}(h) = \bar{\mathbf{s}}'(h)$ if $h \notin \{i, j\}$.

The number of possible pairs for $(\mathbf{r}'(i), \mathbf{r}'(j))$ is $O(n)$ as $\mathbf{r}'(j)$ is uniquely determined by $\mathbf{r}'(i)$; similarly the number of possible pairs for $(\bar{\mathbf{r}}'(i), \bar{\mathbf{r}}'(j))$ is $O(n)$ as $\bar{\mathbf{r}}'(j)$ is uniquely determined by $\bar{\mathbf{r}}'(i)$. There are at most $O(n)$ possible pairs for $(\mathbf{s}'(i), \mathbf{s}'(j))$ and for $(\bar{\mathbf{s}}'(i), \bar{\mathbf{s}}'(j))$. In total, there are $O(n^4)$ candidates. Each candidate can be checked in $O(1)$ time, thus the lemma holds.

4 W[1]-Hardness Parameterized by Treewidth

In this section we show that a generalization of SATISFACTORY PARTITION is W[1]-hard when parameterized by treewidth. We consider the following generalization of SATISFACTORY PARTITION, where some vertices are forced to be in the first part V_1 and some other vertices are forced to be in the second part V_2.

SATISFACTORY PARTITION^{FS}
Input: A graph $G = (V, E)$, a set $V_\triangle \subseteq V(G)$, and a set $V_\square \subseteq V(G)$.

Question: Is there a satisfactory partition (V_1, V_2) of V such that (i) $V_\triangle \subseteq V_1$ (ii) $V_\square \subseteq V_2$.

In this section, we prove the following theorem:

Theorem 4. *The* SATISFACTORY PARTITION^{FS} *is W[1]-hard when parameterized by the treewidth of the graph.*

Let $G = (V, E)$ be an undirected and edge weighted graph, where V, E, and w denote the set of nodes, the set of edges and a positive integral weight function $w : E \to Z^+$, respectively. An orientation Λ of G is an assignment of a direction to each edge $\{u, v\} \in E(G)$, that is, either (u, v) or (v, u) is contained in Λ. The weighted outdegree of u on Λ is $w_{\text{out}}^u = \sum_{(u,v) \in \Lambda} w(\{u, v\})$. We define MINIMUM MAXIMUM OUTDEGREE problem as follows:

MINIMUM MAXIMUM OUTDEGREE
Input: A graph G, an edge weighting w of G given in unary, and a positive integer r.
Question: Is there an orientation Λ of G such that $w_{\text{out}}^u \leq r$ for each $u \in V(G)$?

It is known that MINIMUM MAXIMUM OUTDEGREE is W[1]-hard when parameterized by the treewidth of the input graph [11]. We reduce this problem to the following generalization of SATISFACTORY PARTITION problem:

SATISFACTORY PARTITION^{FSC}
Input: A graph $G = (V, E)$, a set $V_\triangle \subseteq V(G)$, a set $V_\square \subseteq V(G)$, and a set $C \subseteq V(G) \times V(G)$.
Question: Is there a satisfactory partition (V_1, V_2) of V such that (i) $V_\triangle \subseteq V_1$ (ii) $V_\square \subseteq V_2$, and (iii) for all $(a, b) \in C$, V_1 contains either a or b but not both?

To prove Theorem 4, we give a 2-step reduction. In the first step of the reduction, we reduce MINIMUM MAXIMUM OUTDEGREE to SATISFACTORY PARTITION^{FSC}. In the second step of the reduction we reduce the SATISFACTORY PARTITION^{FSC} to SATISFACTORY PARTITION^{FS}. To measure the treewidth of a SATISFACTORY PARTITION^{FSC} instance, we use the following definition. Let $I = (G, V_\triangle, V_\square, C)$ be a SATISFACTORY PARTITION^{FSC} instance. The *primal graph* G' of I is defined as follows: $V(G') = V(G)$ and $E(G') = E(G) \cup C$.

Lemma 6. *The* SATISFACTORY PARTITION^{FSC} *is W[1]-hard when parameterized by the treewidth of the primal graph.*

Proof. Let $G = (V, E, w)$ and a positive integer r be an instance of MINIMUM MAXIMUM OUTDEGREE. We construct an instance of SATISFACTORY PARTITION^{FSC} as follows. An example is given in Fig. 4. For each vertex $v \in V(G)$, we introduce a set of new vertices $H_v = \{h_1^{v\triangle}, \ldots, h_{2r}^{v\triangle}\}$. For each edge $(u, v) \in E(G)$, we introduce the set of new vertices $V_{uv} = \{u_1^v, \ldots, u_{w(u,v)}^v\}$, $V_{uv}' = \{u_1'^v, \ldots, u_{w(u,v)}'^v\}$, $V_{vu} = \{v_1^u, \ldots, v_{w(u,v)}^u\}$, $V_{vu}' = \{v_1'^u, \ldots, v_{w(u,v)}'^u\}$, $V_{uv}^\square = \{u_1^{v\square}, \ldots, u_{w(u,v)}^{v\square}\}$, $V_{uv}'^\square = \{u_1'^{v\square}, \ldots, u_{w(u,v)}'^{v\square}\}$, $V_{vu}^\square = \{v_1^{u\square}, \ldots, v_{w(u,v)}^{u\square}\}$, $V_{vu}'^\square = \{v_1'^{u\square}, \ldots, v_{w(u,v)}'^{u\square}\}$. We now define the graph G' with

$$V(G') = V(G) \bigcup_{v \in V(G)} H_v \bigcup_{(u,v) \in E(G)} (V_{uv} \cup V_{uv}^\square \cup V_{vu} \cup V_{vu}^\square)$$

$$\bigcup_{(u,v) \in E(G)} (V_{uv}' \cup V_{uv}'^\square \cup V_{vu}' \cup V_{vu}'^\square)$$

and

$$E(G') = \left\{(v, h) \mid v \in V(G), h \in H_v\right\} \bigcup \left\{(u, x) \mid (u, v) \in E(G), x \in V_{uv} \cup V_{uv}^\square\right\}$$

$$\bigcup \left\{(x, v) \mid (u, v) \in E(G), x \in V_{vu} \cup V_{vu}^\square\right\}$$

$$\bigcup \left\{(u_i^v, u_i'^v), (u_i^{v\square}, u_i'^{v\square}), (v_i^u, v_i'^u), (v_i^{u\square}, v_i'^{u\square}) \mid (u, v) \in E(G), 1 \le i \le w(u, v)\right\}.$$

We define the complementary vertex pairs

$$C = \left\{(u_i'^v, v_i'^u), (u_{i+1}'^v, v_i'^u), (u_i^v, v_i'^u), (u_i'^v, v_i^u) \mid (u, v) \in E(G), 1 \le i \le w(u, v)\right\}$$

Complementary vertex pairs are shown in dashed lines in Fig. 4. Finally we define $V_\triangle = V(G) \bigcup_{v \in V(G)} H_v$ and $V_\square = \bigcup_{(u,v) \in E(G)} (V_{uv}^\square \cup V_{uv}'^\square \cup V_{vu}^\square \cup V_{vu}'^\square)$. We use I to denote $(G', V_\triangle, V_\square, C)$ which is an instance of SATISFACTORY PARTITION^{FSC}.

Clearly, it takes polynomial time to compute I. We now prove that the treewidth of the primal graph G' of I is bounded by a function of the treewidth of G. We do so by modifying an optimal tree decomposition τ of G as follows:

– For every edge (u, v) of G, there is a node in τ whose bag B contains both u and v; add to this node a chain of nodes $1, 2, \ldots, w(u, v) - 1$ where the bag of node i is $B \cup \{u_i^v, u_i'^v, v_i^u, v_i'^u, u_{i+1}^v, u_{i+1}'^v, v_{i+1}^u, v_{i+1}'^u\}$.
– For every edge (u, v) of G, there is a node in τ whose bag B contains u; add to this node a chain of nodes $1, 2, \ldots, w(u, v)$ where the bag of node i is $B \cup \{u_i^{v\square}, u_i'^{v\square}\}$.
– For every edge (u, v) of G, there is a node in τ whose bag B contains v and add to this node a chain of nodes $1, 2, \ldots, w(u, v)$ where the bag of node i is $B \cup \{v_i^{u\square}, v_i'^{u\square}\}$.

– For every vertex v of G, there is a node in τ whose bag B contains v and add to this node a chain of nodes $1, 2, \ldots, 2r$ where the bag of node i is $B \cup \{h_i^{v\triangle}\}$.

Clearly, the modified tree decomposition is a valid tree decomposition of the primal graph of I and its width is at most the treewidth of G plus eight.

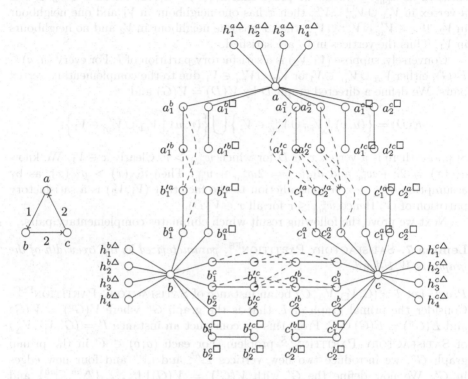

Fig. 4. Result of our reduction on a MINIMUM MAXIMUM OUTDEGREE instance G with $r = 2$. The graph G is shown at the left; and G' is shown at the right. Complementary vertex pairs are shown using dashed lines. The vertices in the first part of satisfactory partition (V_1, V_2) of G' are shown in red for the given orientation of G. (Color figure online)

Let D be the directed graph obtained by an orientation of the edges of G such that for each vertex the sum of the weights of outgoing edges is at most r. Consider the partition

$$V_1 = V_\triangle \bigcup_{(u,v) \in E(D)} (V_{vu} \cup V'_{vu}) = V(G) \bigcup_{v \in V(G)} H_v \bigcup_{(u,v) \in E(D)} (V_{vu} \cup V'_{vu})$$

and

$$V_2 = \bigcup_{(u,v) \in E(D)} (V_{uv} \cup V'_{uv} \cup V_{uv}^\square \cup V_{uv}'^\square) \bigcup_{(u,v) \in E(D)} (V_{vu}^\square \cup V_{vu}'^\square).$$

To prove that (V_1, V_2) is a satisfactory partition, first we prove that $d_{V_1}(x) \geq d_{V_2}(x)$ for all $x \in V_1$. If x is a vertex in H_v or $V_{vu} \cup V'_{vu}$, then clearly all neighbours of x are in V_1, hence x is satisfied. Suppose $x \in V(G)$. Let w^x_{out} and w^x_{in} denote the sum of the weights of outgoing and incoming edges of vertex x, respectively. Hence $d_{V_1}(x) = 2r + w^x_{in}$ and $d_{V_2}(x) = 2w^x_{out} + w^x_{in}$ in G'. This shows that x is satisfied as $w^x_{out} \leq r$. Now we prove that $d_{V_2}(x) \geq d_{V_1}(x)$ for all $x \in V_2$. If x is a vertex in $V_{uv} \cup V^{\square}_{uv} \cup V^{\square}_{vu}$ then x has one neighbour in V_1 and one neighbour in V_2. If $x \in V'_{uv} \cup V'^{\square}_{uv} \cup V'^{\square}_{vu}$ then x has one neighbour in V_2 and no neighbours in V_1. Thus the vertices in V_2 are satisfied.

Conversely, suppose (V_1, V_2) is a satisfactory partition of I. For every $(u, v) \in E(G)$, either $V_{uv} \cup V'_{uv} \in V_1$ or $V_{vu} \cup V'_{vu} \in V_1$ due to the complementary vertex pairs. We define a directed graph D by $V(D) = V(G)$ and

$$E(D) = \Big\{ (u, v) \mid V_{vu} \cup V'_{vu} \in V_1 \Big\} \bigcup \Big\{ (v, u) \mid V_{uv} \cup V'_{uv} \in V_1 \Big\}.$$

Suppose there is a vertex x in D for which $w^x_{out} > r$. Clearly $x \in V_1$. We know $d_{V_1}(x) = 2r + w^x_{in}$ and $d_{V_2}(x) = 2w^x_{out} + w^x_{in}$. Then $d_{V_2}(x) > d_{V_1}(x)$, as by assumption $w^x_{out} > r$, a contradiction to the fact that (V_1, V_2) is a satisfactory partition of G'. Hence $w^x_{out} \leq r$ for all $x \in V(D)$.

Next we prove the following result which eliminates complementary pairs.

Lemma 7. SATISFACTORY PARTITION^{FS}, *parameterized by the treewidth of the graph, is W[1]-hard.*

Proof. Let $I = (G, V_{\square}, V_{\triangle}, C)$ be an instance of SATISFACTORY PARTITION^{FSC}. Consider the primal graph of I, that is the graph G^p where $V(G^p) = V(G)$ and $E(G^p) = E(G) \cup C$. From this we construct an instance $I' = (G', V'_{\square}, V'_{\triangle})$ of SATISFACTORY PARTITION^{FS} problem. For each $(a, b) \in C$ in the primal graph G^p, we introduce two new vertices $\triangle^{ab}$ and $\square^{ab}$ and four new edges in G'. We now define the G' with $V(G') = V(G) \bigcup_{(a,b) \in C} \{\triangle^{ab}, \square^{ab}\}$ and

$$E(G') = E(G) \bigcup_{(a,b) \in C} \Big\{ (a, \triangle^{ab}), (a, \square^{ab}), (b, \triangle^{ab}), (b, \square^{ab}) \Big\}.$$ Finally, we define the sets $V'_{\triangle} = V_{\triangle} \bigcup_{(a,b) \in C} \{\triangle^{ab}\}$ and $V'_{\square} = V_{\square} \bigcup_{(a,b) \in C} \{\square^{ab}\}$. We illustrate our construction in Fig. 5. It is easy to see that we can compute I' in polynomial time and its treewidth is linear in the treewidth of I.

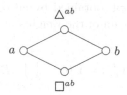

Fig. 5. Gadget for a pair of complementary vertices (a, b) in the reduction from SATISFACTORY PARTITION^{FSC} to SATISFACTORY PARTITION^{FS}.

The following holds for every solution (V_1', V_2') of I': V_1' contains $\triangle^{ab}$ for every $(a, b) \in C$, so it must also contain a or b. It cannot contain both a and b for any $(a, b) \in C$, because $\square^{ab} \in V_2'$. Restricting (V_1', V_2') to the original vertices thus is a solution to I. Conversely, for every solution (V_1, V_2) of I, the partition (V_1', V_2') where $V_1' = V_1 \bigcup_{(a,b) \in C} \{\triangle^{ab}\}$ and $V_2' = V_2 \bigcup_{(a,b) \in C} \{\square^{ab}\}$, is a solution of I'.

This proves Theorem 4.

5   Conclusion

In this work we proved that the SATISFACTORY PARTITION problem is polynomial time solvable for block graphs; the SATISFACTORY PARTITION and BALANCED SATISFACTORY PARTITION problems are polynomial time solvable for graphs of bounded clique width; and a generalized version of the SATISFACTORY PARTITION problem is W[1]-hard when parameterized by treewidth. The parameterized complexity of the SATISFACTORY PARTITION problem remains unsettle when parameterized by other important structural graph parameters like cliquewidth and modular width.

Acknowledgments. The first author gratefully acknowledges support from the Ministry of Human Resource Development, Government of India, under Prime Minister's Research Fellowship Scheme (No. MRF-192002-211).

References

1. Aho, A.V., Hopcroft, J.E.: The Design and Analysis of Computer Algorithms, 1st edn. Addison-Wesley Longman Publishing Co., Inc., New York (1974)
2. Bazgan, C., Tuza, Z., Vanderpooten, D.: On the existence and determination of satisfactory partitions in a graph. In: Ibaraki, T., Katoh, N., Ono, H. (eds.) ISAAC 2003. LNCS, vol. 2906, pp. 444–453. Springer, Heidelberg (2003). https://doi.org/10.1007/978-3-540-24587-2_46
3. Bazgan, C., Tuza, Z., Vanderpooten, D.: Complexity and approximation of satisfactory partition problems. In: Wang, L. (ed.) COCOON 2005. LNCS, vol. 3595, pp. 829–838. Springer, Heidelberg (2005). https://doi.org/10.1007/11533719_84
4. Bazgan, C., Tuza, Z., Vanderpooten, D.: Degree-constrained decompositions of graphs: bounded treewidth and planarity. Theoret. Comput. Sci. **355**(3), 389 395 (2006)
5. Bazgan, C., Tuza, Z., Vanderpooten, D.: The satisfactory partition problem. Discret. Appl. Math. **154**(8), 1236–1245 (2006)
6. Cygan, M., et al.: Parameterized Algorithms. Springer, Heidelberg (2015). https://doi.org/10.1007/978-3-319-21275-3
7. Gerber, M., Kobler, D.: Classes of graphs that can be partitioned to satisfy all their vertices. Aust. J. Comb. **29**, 201–214 (2004)
8. Gerber, M.U., Kobler, D.: Algorithmic approach to the satisfactory graph partitioning problem. Eur. J. Oper. Res. **125**(2), 283–291 (2000)

9. Gerber, M.U., Kobler, D.: Algorithms for vertex-partitioning problems on graphs with fixed clique-width. Theoret. Comput. Sci. **299**(1), 719–734 (2003)
10. Kiyomi, M., Otachi, Y.: Alliances in graphs of bounded clique-width. Discret. Appl. Math. **223**, 91–97 (2017)
11. Szeider, S.: Not so easy problems for tree decomposable graphs. CoRR, abs/1107.1177 (2011)

An Approximation of the Zero Error Capacity by a Greedy Algorithm

Marcin Jurkiewicz$^{(\boxtimes)}$ [ID]

Faculty of Electronics, Telecommunications and Informatics,
Gdańsk University of Technology, Narutowicza 11/12, 80-233 Gdańsk, Poland
marjurki@pg.edu.pl

Abstract. We present a greedy algorithm that determines a lower bound on the zero error capacity. The algorithm has many new advantages, e.g., it does not store a whole product graph in a computer memory and it uses the so-called distributions in all dimensions to get a better approximation of the zero error capacity. We also show an additional application of our algorithm.

Keywords: Shannon capacity · Greedy algorithm · Strong product · Independence number.

1 Introduction

Let $G = (V, E)$ be a graph[1]. If $u, v \in V(G)$ and $\{u, v\} \in E(G)$, then we say that u is *adjacent* to v and we write $u \sim v$. The *open neighborhood* of a vertex $v \in V(G)$ is $N_G(v) = \{u \in V(G) \colon \{u, v\} \in E(G)\}$, and its *closed neighborhood* is the set $N_G[v] = N_G(v) \cup \{v\}$. The *degree* of a vertex v, denoted by $d_G(v)$, is the cardinality of its open neighborhood. The *minimum* and *maximum degree* of G is the minimum and maximum degree among the vertices of G and is denoted by $\delta(G)$ and $\Delta(G)$, respectively. A graph G is *regular* if $\delta(G) = \Delta(G)$. By the *complement* of G, denoted by $\overline{G}$, we mean a graph which has the same vertices as G, and two vertices of $\overline{G}$ are adjacent if and only if they are not adjacent in G. If U is a subset of vertices of G, we write $G[U]$ and $G - U$ for $(U, E(G) \cap [U]^2)$ and $G[V(G) \setminus U]$, respectively. Furthermore, if $U = \{v\}$, then we write $G - v$ rather than $G - \{v\}$.

Given two graphs $G_1 = (V(G_1), E(G_1))$ and $G_2 = (V(G_2), E(G_2))$, the *strong product* $G_1 \boxtimes G_2$ is defined as follows. The vertices of $G_1 \boxtimes G_2$ are all pairs of the Cartesian product $V(G_1) \times V(G_2)$. There is an edge between (v_1, v_2) and (u_1, u_2) if and only if $\{v_1, u_1\} \in E(G_1)$ and $\{v_2, u_2\} \in E(G_2)$, or $v_1 = u_1$

---

[1] The number of vertices and edges of G we often denote by n and m, respectively, thus $|V(G)| = n$ and $|E(G)| = m$.

This article was partially supported by the Narodowe Centrum Nauki under grant DEC-2011/02/A/ST6/00201.

W. Wu and Z. Zhang (Eds.): COCOA 2020, LNCS 12577, pp. 91–104, 2020.
https://doi.org/10.1007/978-3-030-64843-5_7

and $\{v_2, u_2\} \in E(G_2)$, or $v_2 = u_2$ and $\{v_1, u_1\} \in E(G_1)$. The *union* $G_1 \cup G_2$ is defined as $(V(G_1) \cup V(G_2), E(G_1) \cup E(G_2))$. In addition, if $\circ$ is a binary graph operation, then we write $G^{\circ r}$ to denote the *rth power* of G, i.e., $G \circ G \circ \ldots \circ G$, where G occurs r-times.

A *clique* (*independent vertex set*, resp.) in a graph $G = (V, E)$ is a subset $V' \subseteq V$ such that all (no, resp.) two vertices of V' are adjacent. The size of a largest clique (independent vertex set, resp.) in a graph G is called the *clique* (*independence*, resp.) *number* of G and is denoted by $\omega(G)$ ($\alpha(G)$, resp.). A *split graph* is one whose vertex set can be partitioned as the disjoint union of an independent set and a clique. A *legal coloring* of a graph G is an assignment of colors to the vertices of G ($C \colon V(G) \to \mathbb{N}$) such that any two adjacent vertices are colored differently.

A *discrete channel* $W \colon \mathcal{X} \to \mathcal{Y}$ (or simply W) is defined as a stochastic matrix[2] whose rows are indexed by the elements of a finite *input set* $\mathcal{X}$ while the columns are indexed by a finite *output set* $\mathcal{Y}$. The (x, y)th entry is the probability $W(y|x)$ that y is received when x is transmitted. A sequence of channels $\{W^n \colon \mathcal{X}^n \to \mathcal{Y}^n\}_{n=1}^{\infty}$, where $W^n \colon \mathcal{X}^n \to \mathcal{Y}^n$ is the nth *direct power* of W, i.e.,

$$W^n(y_1 y_2 \ldots y_n | x_1 x_2 \ldots x_n) = \prod_{i=1}^{n} W(y_i | x_i)$$

and $\mathcal{X}^n$ is the nth Cartesian power of $\mathcal{X}$, is called a *discrete memoryless channel* (DMC) with stochastic matrix W and is denoted by $\{W \colon \mathcal{X} \to \mathcal{Y}\}$ or simply $\{W\}$. See [7,8,24,29,32] for more details.

Let $W \colon \mathcal{X} \to \mathcal{Y}$ be a discrete channel. We define the ω-*characteristic graph* G of W as follows. Its vertex set is $V(G) = \mathcal{X}$ and its set of edges $E(G)$ consists of input pairs that cannot result in the same output, namely, pairs of orthogonal rows of the matrix W. We define α-*characteristic graph* $G(W)$ (we call it *characteristic graph* for short) of W as the complement of the ω-characteristic graph of W. Let $\{W \colon \mathcal{X} \to \mathcal{Y}\}$ be a DMC and so $W \colon \mathcal{X} \to \mathcal{Y}$ is the corresponding discrete channel. We define the *characteristic graph* $G(\{W\})$ of the discrete memoryless channel $\{W\}$ as $\{G(W^n)\}_{n=1}^{\infty}$. The *Shannon* (*zero-error*) *capacity* $C_0(W)$ of the DMC $\{W \colon \mathcal{X} \to \mathcal{Y}\}$ is defined as $C(G(W))$, where

$$C(G) = \sup_{n \in \mathbb{N}} \frac{\log \alpha(G^{\boxtimes n})}{n} = \lim_{n \to \infty} \frac{\log \alpha(G^{\boxtimes n})}{n}.$$

See [7,8,24,29] for more details. Let G be the characteristic graph of W and $\Theta(G) = \sup_{n \in \mathbb{N}} \sqrt[n]{\alpha(G^{\boxtimes n})}$. Then $\Theta(G)$ uniquely determines $C_0(W)$.

---

[2] We assume that W is non-empty.

2   Fractional Independence Number

Computing the independence number of a graph $G = (V, E)$ can be formulated by the following integer program.

$$\text{Maximize} \quad \sum_{v \in V} x_v$$

$$\text{subject to} \quad \underset{\{u,v\} \in E}{\forall} x_u + x_v \leqslant 1 \text{ and } \underset{v \in V}{\forall} x_v \in S,$$

$$(1)$$

where $S = \{0, 1\}$. Now let $S = [0, 1]$. Given a graph G, by $\alpha_2^*(G)$ we denote the optimum of the objective function in the integer program (1). However, for a graph G and a set of not necessarily all its cliques[3] C by $\alpha_C^*(G)$ we denote the optimum of the objective function in the following integer program.

$$\text{Maximize} \quad \sum_{v \in V} x_v$$

$$\text{subject to} \quad \underset{C \in \mathcal{C}}{\forall} \sum_{v \in C} x_v \leqslant 1 \text{ and } \underset{v \in V}{\forall} x_v \in [0, 1].$$

$$(2)$$

If C is the set of all maximal cliques of size at most r in G, then we denote $\alpha_C^*(G)$ by $\alpha_r^*(G)$. If C contains the set of all cliques (or equivalently all maximal cliques) of G, then we denote $\alpha_C^*(G)$ by $\alpha^*(G)$ and it is called *the fractional independence number* of G. It is worth to note that α^* is multiplicative with respect to the strong product [31].

The following results present some properties of the linear program (2). In particular, the first observation establishes an order between the above-mentioned measures.

Observation 1. *Let G be a graph and C, C' be sets of its cliques. If $C' \subseteq C$, then $\alpha_C^*(G) \leqslant \alpha_{C'}^*(G)$.*

Observation 2. *Let G be a graph. If $\omega(G) = r$, then $\alpha^*(G) = \alpha_r^*(G)$.*

Lemma 1. *For every graph G and a non-empty set of its cliques C we have*

$$\frac{|V(G)|}{\omega(G)} \leqslant \alpha_C^*(G) \leqslant \frac{|C|}{\varsigma(C)} + R_C(G), \qquad (3)$$

where $\varsigma(C) = \min\{\sum_{C \in \mathcal{C}} |\{v\} \cap C| : v \in \bigcup_{C \in \mathcal{C}} C\}$ and $R_C(G) = |V(G)| - |\bigcup_{C \in \mathcal{C}} C|$. Furthermore, the equalities hold in the inequality chain (3) if G is vertex-transitive[4] and C is the set of all largest cliques in G.

---

[3] It is worth to note that the number of maximal cliques in G is at most exponential with respect to $|V(G)|$ [11], while the number of edges in G is at most quadratic with respect to $|V(G)|$.

[4] A graph is *vertex transitive* if for any two vertices u and v of this graph, there is an automorphism such that the image of u is v.

Proof. It is well known [14] that for every graph G we have

$$\alpha^*(G) \geqslant \frac{|V(G)|}{\omega(G)}. \tag{4}$$

From (4) and Observation 1, the left inequality holds in (3).

Given a linear program (2) and its optimum $\alpha_{\mathcal{C}}^*(G)$. Since $\sum_{C \in \mathcal{C}} \sum_{v \in C} x_v \leqslant |\mathcal{C}|$, so

$$\sum_{v \in V(G)} x_v \leqslant \frac{|\mathcal{C}|}{\varsigma(\mathcal{C})} + R_{\mathcal{C}}(G).$$

Hence $\alpha_{\mathcal{C}}^*(G) \leqslant |\mathcal{C}|/\varsigma(\mathcal{C}) + R_{\mathcal{C}}(G)$.

If G is vertex-transitive and $\mathcal{C}$ is the set of all largest cliques in G, then $\mathcal{C}$ covers the whole vertex set, i.e., $V(G) = \bigcup_{C \in \mathcal{C}} C$. Hence $R_{\mathcal{C}}(G) = 0$. Furthermore, every vertex is contained in the same number of largest cliques. Hence $\varsigma(\mathcal{C})|V(G)| = \omega(G)|\mathcal{C}|$.   □

It is interesting that the measure α^* has a particular interpretation in information theory [24, 32].

3   Capacity Approximation

It is well known [32] that[5]

$$\alpha(G) \leqslant \sqrt[i]{\alpha(G^{\boxtimes i})} \leqslant \Theta(G) \leqslant \alpha^*(G) \tag{5}$$

for each positive integer i. A graph G is of *type* I if $\Theta(G) = \alpha(G)$, otherwise is of *type* II. Furthermore, Hales [16] showed that for arbitrary graphs G and H we have

$$\alpha(G \boxtimes H) \leqslant \min\{\alpha(G)\alpha^*(H), \alpha(H)\alpha^*(G)\}.$$

In contrast to the above results, in the next section we use the fractional independence number to calculate lower bounds on the Shannon capacity and the independence number of strong products.

A function $\beta \colon \mathcal{G} \to \mathbb{R}$ is *supermultiplicative* (resp. *submultiplicative*) on $\mathcal{G}$ with respect to the operation $\circ$, if for any two graphs $G_1, G_2 \in \mathcal{G}$ we have $\beta(G_1 \circ G_2) \geqslant \beta(G_1) \cdot \beta(G_2)$ (resp. $\beta(G_1 \circ G_2) \leqslant \beta(G_1) \cdot \beta(G_2)$). A supermultiplicative and submultiplicative function is called *multiplicative*. The independence number α is supermultiplicative on the set of all graphs with respect to the strong product, i.e., $\alpha(G \boxtimes H) \geqslant \alpha(G) \cdot \alpha(H)$ for any graphs G and H. Let B be a lower bound on the independence number α, i.e., $\alpha(G) \geqslant B(G)$. If $B(G^{\boxtimes i}) > (\alpha(G))^i$ $(i \geqslant 2)$, then G is of type II and is more interesting from an information theory point of view [32]. It is possible if $B(G^{\boxtimes i}) > (B(G))^i$. Thus we require that B has the

---

[5] There is a better upper bound on $\Theta(G)$, the so-called Lovász theta function [28].

last two properties for at least one graph, i.e., B *recognizes* some graphs of type II.

The *residue* R of a graph G of degree sequence $S : d_1 \geqslant d_2 \geqslant d_3 \ldots \geqslant d_n$ is the number of zeros obtained by the iterative process consisting of deleting the first term d_1 of S, subtracting 1 from the d_1 following ones, and re-sorting the new sequence in non-increasing order [10]. It is well known that $\alpha(G) \geqslant R(G)$ [9]. Unfortunately, the following negative result holds.

Proposition 1. *Let G and H be regular*[6] *or split graphs. Then $R(G \boxtimes H) \leqslant R(G) \cdot R(H)$.*

Proof. Let G and H be regular graphs. For a regular graph G, from [9], we have $R(G) = \lceil \sum_{i=1}^{n}(1/(1 + d_i)) \rceil = \lceil (n/(1 + d(G))) \rceil$, where $d(G)$ is the degree of each vertex of G. From [21] we know that the ceiling function is submultiplicative on non-negative real numbers with respect to the multiplication. Hence $R(G \boxtimes H) = \lceil |V(G)||V(H)|/(1 + (d(G)d(H) + d(G) + d(H))) \rceil \leqslant \lceil |V(G)|/(1 + d(G)) \rceil \lceil |V(H)|/(1 + d(H)) \rceil = R(G) \cdot R(H)$, since a strong product of regular graphs is regular.

Let G and H be split graphs. From [2] and [17], we have $\alpha(G) = R(G)$ and $\alpha(G \boxtimes H) = \alpha(G) \cdot \alpha(H)$, respectively. Finally, we have $R(G \boxtimes H) \leqslant \alpha(G \boxtimes H) = \alpha(G) \cdot \alpha(H) = R(G) \cdot R(H)$. $\square$

We conjecture that the residue is submultiplicative on the set of all graphs with respect to the strong product. This probably means that the residue does not recognize any graphs of type II. There are more such bounds, e.g., the average distance [21], the radius [20], the Caro–Wei bound and the Wilf bound [23]. On the other hand, it is hard to find bounds that recognize at least one graph of type II.

4   Greedy Algorithms

In this section we present a new greedy algorithm that determines a lower bound on the independence number of a strong product. Furhtermore, this value, from (5), determines a lower bound on the Shannon capacity. An advantage of a greedy algorithm is that it produces an existing solution of an optimization problem. In the considered problem, it produces independent sets that correspond to DMCs codes [24,32].

We modify Algorithm 4.1, the so-called greedy algorithm MIN [18], which works for an arbitrary graph. Our goal is to get larger independent sets for strong products by a modification[7] of the mentioned algorithm. The algorithm MIN has complexity $O(n^2)$.

A greedy algorithm always makes the choice that looks best at the moment. That is, it makes a locally optimal choice in the hope that this choice will lead

---

[6] This part of the proposition was found by my colleague [30].

[7] Our algorithm works for an arbitrary graph product, as well as for an arbitrary single graph.

Algorithm 4.1. Greedy Algorithm MIN

1: **function** MIN(G)
2:      $G_1 \leftarrow G, \ j \leftarrow 1, \ I \leftarrow \emptyset$
3:      **while** $V(G_j) \neq \emptyset$ **do**
4:          choose $i_j \in V(G_j)$ with $d_{G_j}(i_j) = \delta(G_j)$
5:          $G_{j+1} \leftarrow G_j - N_{G_j}[i_j]$
6:          $I \leftarrow I \cup \{i_j\}$
7:          $j \leftarrow j + 1$
8:      **return** I

---

to a globally optimal solution [6]. Vertices chosen (in such a way) by MIN often strongly block an eventual choice of vertices in a further stage of the algorithm, making generated independent sets are small, especially for strong products of graphs of type II. In Table 1, we summarize results produced by MIN for these graphs. We try to improve MIN, since from our research it follows that it does not work well, i.e., it does not recognize graphs of type II. We begin by introducing definitions required in the rest of the paper.

Table 1. For each graph $G \in \mathcal{G}_{n,2}^+ = \{H : \alpha(H^{\boxtimes 2}) > \alpha^2(H) \wedge |V(H)| = n\}$ we determine an independent set I of the graph $G^{\boxtimes 2}$ using the algorithm MIN.

Greedy Algorithm MIN (results)				
n	$|\mathcal{G}_{n,2}^+|$	$|I| \leqslant \alpha^2(G)$	$|I| > \alpha^2(G)$	$|I| = \alpha(G^{\boxtimes 2})$
5	1	1	0	0
6	4	4	0	0
7	36	36	0	0
8	513	513	0	0
9	16015	16015	0	0
10	908794	908794	0	0

A *semigroup* is a set S with an associative binary operation on S. A *semiring* is defined as an algebra $(S, +, \cdot)$ such that $(S, +)$ and $(S, \cdot)$ are semigroups and for any $a, b, c \in S$ we have $a \cdot (b + c) = a \cdot b + a \cdot c$, $(b + c) \cdot a = b \cdot a + c \cdot a$ [1]. Note that $(\mathcal{G}, \cup, \boxtimes)$ is a semiring, where $\mathcal{G}$ is the set of all finite graphs. In addition, $\cup$ and $\boxtimes$ are commutative operations with neutral elements $(\emptyset, \emptyset)$ and K_1, respectively.

Lemma 2. *Let p, r be positive integers and $G_1, G_2, \ldots, G_r$ be graphs. Then*

$$\left(\bigcup_{i \in [r]} G_i \right)^{\boxtimes p} = \bigcup_{p_1 + p_2 + \ldots + p_r = p} \left[\binom{p}{p_1, p_2, \ldots, p_r} \boxtimes_{i \in [r]} G_i^{\boxtimes p_i} \right]$$

and

$$\alpha\left(\left(\bigcup_{i\in[r]} G_i\right)^{\boxtimes p}\right) = \sum_{p_1+p_2+\ldots+p_r=p}\left[\binom{p}{p_1,p_2,\ldots,p_r}\alpha\left(\boxtimes_{i\in[r]} G_i^{\boxtimes p_i}\right)\right],$$

where summations extend over all ordered sequences $(p_1, p_2, \ldots, p_r)$ of nonnegative integers that sum to p.

Proof. The first part of the theorem can be proved in analogous way to the one in [27, Theorem 2.12] for rings (we only need the above mentioned properties of the semiring $(\mathcal{G}, \cup, \boxtimes)$).

The second part of the theorem follows from the fact that the independence number is multiplicative with respect to the disjoint union $\cup$ for all graphs. □

The considered modification of the greedy algorithm takes as input arbitrary graphs $G_1, G_2, \ldots, G_r$ and produces as output an independent set of $G^{\boxtimes} = G_1 \boxtimes G_2 \boxtimes \ldots \boxtimes G_r$. From Lemma 2, we can find connected components of $G^{\boxtimes}$. Hence, our greedy algorithm can be applied to each connected component of $G^{\boxtimes}$ separately, or to the entire graph $G^{\boxtimes}$ at once. We prefer the first method.

The next step of our modification is a reduction of factors of a strong product. For each $i \in \{1, 2, \ldots, r\}$ and any $u, v \in V(G_i)$, if $N_{G_i}[u] \subseteq N_{G_i}[v]$, then $\alpha(G_1 \boxtimes G_2 \boxtimes \ldots \boxtimes G_i \boxtimes \ldots \boxtimes G_r) = \alpha(G_1 \boxtimes G_2 \boxtimes \ldots \boxtimes (G_i - v) \boxtimes \ldots \boxtimes G_r)$ [19]. Let G be a factor of a strong product $G^{\boxtimes}$, for example $G = G_i$. Let $>$ be a strict total order on $V(G)$. We reduce the factor G by Algorithm 4.2 (Reduction GR), which has complexity $O(\Delta^2 m)$. This algorithm is correct (in the considered context) since we remove vertices from the strong product, and hence we can only decrease or leave its unchanged independence number.

Algorithm 4.2. Reduction GR

1: **function** GR(G)
2:    $R \leftarrow \emptyset$
3:    **for all** $\{u, v\} \in E(G)$ **do**
4:       **if** $u > v$ **then**
5:          **if** $N_G[u] = N_G[v]$ **then**
6:             $R \leftarrow R \cup \{u\}$
7:          **else if** $N_G[u] \subseteq N_G[v]$ **then**
8:             $R \leftarrow R \cup \{v\}$
9:          **else if** $N_G[v] \subseteq N_G[u]$ **then**
10:            $R \leftarrow R \cup \{u\}$
11:   **return** $G - R$

For some graphs, which we take as input, e.g., for a path on $n \geqslant 6$ vertices, we need to recursively repeat (at most n times) the algorithm GR to get a smaller graph. Sometimes, the algorithm GR produces vertices with degree zero. Such vertices should be removed from a graph, but taken into account in the outcome.

Let G be a graph and k be a positive integer. Let A be a k-tuple of subsets of $V(G)$. By $B_G(A)$ we denote a sequence containing upper bounds on $\alpha(G[A_i])$ for $i \in \{1, 2, \ldots, k\}$. Let $G = 2K_3$. Then, for example, $V(G) = \{1, 2, \ldots, 6\}$, $E(G) = \{\{1, 2\}, \{2, 3\}, \{3, 1\}, \{4, 5\}, \{5, 6\}, \{6, 4\}\}$, and $B_G((\{1, 2, 3\}, \{4, 5, 6\})) = (1, 1)$. Let A' be a k'-tuple of subsets of $V(G)$. A *distribution* $D_G(A')$ is a k'-tuple of non-negative integers, and is our prediction about an arrangement of independent vertices of G in sets from A'. Let $G^{\boxtimes} = G_1 \boxtimes G_2 \boxtimes \ldots \boxtimes G_r$. Let $i \in \{1, 2, \ldots, r\}$, $S \subseteq V(G_i)$ and $V_{G_i}(S) = V(G_1) \times V(G_2) \times \ldots \times V(G_{i-1}) \times S \times V(G_{i+1}) \times \ldots \times V(G_r)$. From [17], for each clique Q of G_i we have

$$\alpha \left(\underset{j \in [r] \setminus \{i\}}{\boxtimes} G_j \right) = \alpha(G_1 \boxtimes G_2 \boxtimes \ldots \boxtimes G_{i-1} \boxtimes G_i[Q] \boxtimes G_{i+1} \boxtimes \ldots \boxtimes G_r).$$

Thus, if $\mathcal{Q} = \{Q_1, Q_2, \ldots, Q_k\}$ ($k \in \mathbb{N}_+$) is a set of cliques of G_i and

$$\alpha_i \geqslant \alpha \left(\underset{j \in [r] \setminus \{i\}}{\boxtimes} G_j \right), \tag{6}$$

then we can choose $B_{G^{\boxtimes}}((V_{G_i}(Q_1), V_{G_i}(Q_2), \ldots, V_{G_i}(Q_k))) = (\alpha_i, \alpha_i, \ldots, \alpha_i)$, where α_i occurs k times. Let

$$\alpha_i = \left\lfloor \prod_{j \in [r] \setminus \{i\}} \alpha^*_{E(G_j)}(G_j) \right\rfloor.$$

The function α^* is multiplicative with respect to the strong product for all graphs [31]. Thus, from Observation 1 and (5) we get

$$\alpha_i \geqslant \left\lfloor \prod_{j \in [r] \setminus \{i\}} \alpha^*(G_j) \right\rfloor = \left\lfloor \alpha^* \left(\underset{j \in [r] \setminus \{i\}}{\boxtimes} G_j \right) \right\rfloor \geqslant \alpha \left(\underset{j \in [r] \setminus \{i\}}{\boxtimes} G_j \right) \tag{7}$$

and finally α_i holds the condition (6). Furthermore, from (7) and (3), for graphs without vertices with degree zero, also the following substitution

$$\alpha_i = \left\lfloor \prod_{j \in [r] \setminus \{i\}} \left(|E(G_j)| \varsigma^{-1}(E(G_j)) + R_{E(G_j)}(G) \right) \right\rfloor = \left\lfloor \prod_{j \in [r] \setminus \{i\}} \frac{|E(G_j)|}{\delta(G_j)} \right\rfloor$$

holds the condition (6).

Let $i \in \{1, 2, \ldots, r\}$. Algorithm 4.4 (Distribution DISTR), which takes as input a graph $G = G_i$ and an upper bound $\alpha_b = \alpha_i$, determines a distribution for a graph $G^{\boxtimes}$. The algorithm DISTR, whose running time is $O(n^2)$, uses Algorithm 4.3 (Greedy Coloring GC), which has complexity $O(n + m)$ [25]. The algorithm GC takes as input a graph G and an arbitrary permutation P of the vertex set of G. GC in DISTR legally colors the complement of G and hence produces a partition $\mathcal{Q}$ of the vertex set of G into cliques (the so-called *clique cover* of G). Subsequently, DISTR distributes $\alpha_b = \alpha_i$ potential elements of an independent

Algorithm 4.3. Greedy Coloring GC

1: **function** GC(G, P)
   comment: In all algorithms, loops contained the keyword **in** are performed in a given order.
2:    **for each** v **in** P **do**
3:        assign to v the smallest possible legal color $C(v)$ in G
4:    **return** C

Algorithm 4.4. Distribution DISTR

1: **function** DISTR(G, α_b)
   comment: $V(G) = \{v_1, \ldots, v_n\}$
2:    **for each** $v \in V(G)$ **do**
3:        $distr_v \leftarrow 0$
4:    assign to P vertices of G arranged in non-increasing order according to their degrees
5:    $C \leftarrow GC(\overline{G}, P)$
6:    create the clique cover CC of G from the coloring C
7:    sort cliques from CC in non-increasing order according to their sizes
8:    sort vertices in cliques from CC in non-decreasing order according to their degrees
9:    **for each** Q **in** CC **do**
10:        $q \leftarrow \lfloor \alpha_b / |Q| \rfloor$
11:        $r \leftarrow \alpha_b \bmod |Q|$
12:        $i \leftarrow 0$
13:        **for each** v **in** Q **do**
14:            $K \leftarrow q$
15:            **if** $i < r$ **then**
16:                $K \leftarrow K + 1$
17:            $m \leftarrow \alpha_b$
18:            **for each** Q' **in** CC **do**
19:                $M \leftarrow \max \left\{ k \in \{0, \ldots, K\} : \sum_{v' \in N(v) \cap Q'} distr_{v'} + k \leqslant \alpha_b \right\}$
20:                **if** $m > M$ **then**
21:                    $m \leftarrow M$
22:            $distr_v \leftarrow m$
23:            $i \leftarrow i + 1$
24:    **return** $(distr_{v_1}, \ldots, distr_{v_n})$

---

set roughly evenly (about $\alpha_b / |Q|$ elements or less depending on the sum from line 19) among all vertices of Q (as well as among all subgraphs of $G_1 \boxtimes G_2 \boxtimes \ldots \boxtimes G_{i-1} \boxtimes G_i[Q] \boxtimes G_{i+1} \boxtimes \ldots \boxtimes G_r$) for all $Q \in \mathcal{Q}$.

As we mentioned before, vertices chosen by MIN strongly block an eventual choice of vertices in a further stage of the algorithm. Our greedy algorithm, i.e., Algorithm 4.5 (Greedy Algorithm MIN-SP), significantly diminishes the mentioned effect by the use of generated distributions. The vertex set of $G_1 \boxtimes G_2 \boxtimes \ldots \boxtimes G_r$ can be interpreted as the r-dimensional cuboid of the size $|V(G_1)| \cdot$

$|V(G_2)| \cdot \ldots \cdot |V(G_r)|$. MIN-SP uses distributions in all r dimensions. Earlier [3, 5, 22], only one distribution was used at one time in algorithms for the maximum independent set problem in subclasses of the strong product of graphs to reduce a search space. The important point to note here is that in cases that are more interesting from an information theory point of view, i.e., if $G_1 = G_2 = \ldots = G_r$, some parts of MIN-SP are much simpler, e.g., we can determine one distribution and then we use it in all dimensions.

Algorithm 4.5. Greedy Algorithm MIN-SP

1: **function** MIN-SP$((G_1, G_2, \ldots, G_r))$
   comment: $\bar{v} = (v_1, v_2, \ldots, v_r)$, $\bar{v}^* = (v_1^*, v_2^*, \ldots, v_r^*)$
2:     $I \leftarrow \emptyset$
3:     **for** $i \leftarrow 1$ **to** r **do**
4:         $distr^{(i)} \leftarrow \text{DISTR}(G_i, \alpha_i)$
5:     $V \leftarrow V(G_1) \times V(G_2) \times \ldots \times V(G_r)$
6:     **for each** $\bar{v} \in V$ **do**
7:         $d(\bar{v}) \leftarrow |N_{G_1}[v_1]| \cdot |N_{G_2}[v_2]| \cdot \ldots \cdot |N_{G_r}[v_r]| - 1$
8:     **while** $V \neq \emptyset$ **do**
9:         assign to $\bar{v}^*$ an element $\bar{v} \in V$ with the smallest $d(\bar{v})$
10:        $N \leftarrow N_{G_1}[v_1^*] \times N_{G_2}[v_2^*] \times \ldots \times N_{G_r}[v_r^*]$
11:        $N \leftarrow N \cap V$
12:        $F \leftarrow \emptyset$
13:        **for** $i \leftarrow 1$ **to** r **do**
14:            $distr_{v_i^*}^{(i)} \leftarrow distr_{v_i^*}^{(i)} - 1$
15:            **if** $distr_{v_i^*}^{(i)} = 0$ **then**
16:                append to F elements $\bar{v} \in V$ with $v_i = v_i^*$
17:        $V \leftarrow V \setminus (N \cup F)$
18:        **for each** $\bar{v} \in V$ **do**
19:            **for each** $\bar{v}' \in N$ **do**
20:                **if** $\bar{v} \sim \bar{v}'$ **then**
21:                    $d(\bar{v}) \leftarrow d(\bar{v}) - 1$
22:        $I \leftarrow I \cup \{\bar{v}^*\}$
23:     **return** I

MIN-SP defines four sets N, V, F and I, where N is the closed neighborhood of a chosen vertex $\bar{v}^*$ (line 10), V is a set of vertices that are available for the next iterations, F is a set of forbidden vertices that are not available for the next iterations and I is an actual solution (an actual independent set). In lines 13–14 and lines 18–21, MIN-SP updates distributions and degrees of all vertices from V, respectively. In line 17, elements of N and F are removed from V, but only degrees of vertices from N are updated.

An advantage of MIN-SP is that we do not need to store edges of $G^{\boxtimes} = G_1 \boxtimes G_2 \boxtimes \ldots \boxtimes G_r$ in a computer memory. This is important since $|E(G^{\boxtimes})|$ almost always fastly increases with r. In the memory, we only keep factors of

$G^{\boxtimes}$, and the adjacency relation $\sim$ is directly checked from the conditions specified in the definition of the strong product (line 20).

Sometimes MIN-SP produces I such that $V(G^{\boxtimes}) - N_{G^{\boxtimes}}[I] \neq \emptyset$, where $N_G[I] = \bigcup_{v \in I} N_G[v]$ for $I \in V(G)$ and a graph G. Thus, finally, it is possible to get a larger independent set of $G^{\boxtimes}$, i.e., $I' = I \cup \text{MIN}(G^{\boxtimes} - N_{G^{\boxtimes}}[I])$. We prefer such a method in our computations. It turns out that we also do not need to store edges of $G^{\boxtimes}$ if we want to execute $\text{MIN}(G^{\boxtimes} - N_{G^{\boxtimes}}[I])$. It can be done by a modification of MIN similar to that we performed, when we constructed MIN-SP.

In Table 2, we summarize results produced by MIN-SP. The algorithm has a running time of $O(|V|^3)$.

Table 2. For each graph $G \in \mathcal{G}_{n,2}^+ = \{H : \alpha(H^{\boxtimes 2}) > \alpha^2(H) \wedge |V(H)| = n\}$ we determine an independent set I' of the graph $G^{\boxtimes 2}$ using the algorithm MIN-SP. It is worth to note that the gap between $\alpha^2(G)$ and $\alpha(G^{\boxtimes 2})$ is small for the n that are small [15].

Greedy Algorithm MIN-SP (results)				
n	$|\mathcal{G}_{n,2}^+|$	$|I'| \leqslant \alpha^2(G)$	$|I'| > \alpha^2(G)$	$|I'| = \alpha(G^{\boxtimes 2})$
5	1	0	1	1
6	4	0	4	4
7	36	4	32	32
8	513	127	386	386
9	16015	6306	9709	9652
10	908794	505089	403705	403469

We can approximate the Shannon capacity using (5) and the algorithm MIN-SP. We show it by the following example.

Example 1. We consider strong products of some fullerenes[8], since they are regular, symmetrical [12] and hence are not so easy for solvers and programs that calculate the independence number. Furthermore, fullerenes are often of type II. The algorithm MIN-SP produced the following upper bounds: $\alpha(C_{20} \boxtimes C_{20}) \geqslant 56$, $\alpha(C_{24} \boxtimes C_{24}) \geqslant 85$ and $\alpha(C_{28} \boxtimes C_{28}) \geqslant 123$, where symbols C_{20}, C_{24} and C_{28} mean 20-fullerene (dodecahedral graph), 24-fullerene and 28-fullerene ($\alpha(C_{20}) = 8$, $\alpha(C_{24}) = 9$ and $\alpha(C_{28}) = 11$ [13]), respectively. Therefore, from (5), the Shannon capacity $\Theta(C_{24}) \geqslant \sqrt[2]{85} = 9.21954.. > \alpha(C_{24}) = 9$ and $\Theta(C_{28}) \geqslant \sqrt[2]{123} = 11.09053.. > \alpha(C_{28}) = 11$, but $\Theta(C_{20}) \geqslant \alpha(C_{20}) = 8$ (we conjecture that $\alpha(C_{20} \boxtimes C_{20}) = 64$).

---

[8] A *fullerene* graph is the graph formed from the vertices and edges of a convex polyhedron, whose faces are all pentagons or hexagons and all vertices have degree equal to three.

5 Community Detection Problems

Chalupa et al. [4] investigated the growth of large independent sets in the Barabási–Albert model of scale-free complex networks. They formulated recurrent relations describing the cardinality of typical large independent sets and showed that this cardinality seems to scale linearly with network size. Independent sets in social networks represent groups of people, who do not know anybody else within the group. Hence, an independent set of a network plays a crucial role in community detection problems, since vertices of this set are naturally unlikely to belong to the same community [4,33]. These facts imply that the number of communities in scale-free networks seems to be bounded from below by a linear function of network size [4].

Leskovec et al. [26] introduced the Kronecker graph network model that naturally obeys common real network properties. In particular, the model assumes that graphs have loops and corresponds to the strong product [17]. Let $i \geqslant 1$ and $G' = G^{\boxtimes i}$. As mentioned earlier, the function α is supermultiplicative and α^* is multiplicative with respect to the strong product for all graphs. Thus $(\alpha^*(G))^i = \alpha^*(G') \geqslant \alpha(G') \geqslant (\alpha(G))^i$ and hence $|V(G')|^{c'} \geqslant \alpha(G') \geqslant |V(G')|^c$, where[9] $c' = \log(\alpha^*(G))/\log(V(G))$ and $c = \log(\alpha(G))/\log(V(G))$. We have just showed that the cardinality of maximum independent sets, in the mentioned model, scale sublinearly with network[10] size. Furthermore, if G is of type I, then $\alpha(G') = |V(G')|^c$. These considerations show that the number of communities in scale-free networks seems to be bounded from below by a sublinear (rather than a linear) function of network size. It is worth to note that we can approximate (resp. predict) the number of communities, in the mentioned model (resp. real complex network), using Algorithm 4.5 (Greedy Algorithm MIN-SP).

References

1. Adhikari, M.R., Adhikari, A.: Basic Modern Algebra with Applications. Springer, New Delhi (2014). https://doi.org/10.1007/978-81-322-1599-8
2. Barrus, M.D.: Havel-Hakimi residues of unigraphs. Inform. Process. Lett. **112**(1–2), 44–48 (2012). https://doi.org/10.1016/j.ipl.2011.10.011
3. Baumert, L.D., McEliece, R.J., Rodemich, E., Rumsey, Jr., H.C., Stanley, R., Taylor, H.: A combinatorial packing problem. In: Computers in algebra and number theory (Proceedings of SIAM-AMS Symposium on Applied Mathematics, New York, 1970), SIAM-AMS Proceedings, vol. IV, pp. 97–108. American Mathematical Society, Providence (1971)
4. Chalupa, D., Pospíchal, J.: On the growth of large independent sets in scale-free networks. In: Zelinka, I., Suganthan, P.N., Chen, G., Snasel, V., Abraham, A., Rössler, O. (eds.) Nostradamus 2014: Prediction, Modeling and Analysis of Complex Systems. AISC, vol. 289, pp. 251–260. Springer, Cham (2014). https://doi.org/10.1007/978-3-319-07401-6_24

---

[9] We can use the so-called Lovász theta function instead of α^*, since it is also multiplicative with respect to the strong product for all graphs [28].

[10] We assume that a network (graph) G is non-empty, i.e., $|E(G)| \neq 0$.

5. Codenotti, B., Gerace, I., Resta, G.: Some remarks on the Shannon capacity of odd cycles. Ars Combin. **66**, 243–257 (2003)
6. Cormen, T.H., Leiserson, C.E., Rivest, R.L., Stein, C.: Introduction to Algorithms, 3rd edn. MIT Press, Cambridge (2009)
7. Cover, T.M., Thomas, J.A.: Elements of Information Theory. Wiley-Interscience, 2nd edn. Wiley, Hoboken (2006)
8. Csiszár, I., Körner, J.: Information Theory, 2nd edn. Cambridge University Press, Cambridge (2011). https://doi.org/10.1017/CBO9780511921889. Coding theorems for discrete memoryless systems
9. Favaron, O., Mahéo, M., Saclé, J.F.: On the residue of a graph. J. Graph Theory **15**(1), 39–64 (1991). https://doi.org/10.1002/jgt.3190150107
10. Favaron, O., Mahéo, M., Saclé, J.F.: Some eigenvalue properties in graphs (conjectures of Graffiti. II). Discrete Math. **111**(1–3), 197–220 (1993). https://doi.org/10. 1016/0012-365X(93)90156-N. Graph theory and combinatorics (Marseille-Luminy, 1990)
11. Fomin, F.V., Kratsch, D.: Exact Exponential Algorithms. Texts in Theoretical Computer Science. An EATCS Series. Springer, Heidelberg (2010). https://doi. org/10.1007/978-3-642-16533-7
12. Fowler, P.W., Manolopoulos, D.: An Atlas of Fullerenes. Courier Corporation (2007)
13. Fowler, P., Daugherty, S., Myrvold, W.: Independence number and fullerene stability. Chem. Phys. Lett. **448**(1), 75–82 (2007). https://doi.org/10.1016/j.cplett.2007. 09.054. http://www.sciencedirect.com/science/article/pii/S0009261407012948
14. Gross, J.L., Yellen, J., Zhang, P. (eds.): Handbook of Graph Theory. Discrete Mathematics and its Applications, 2nd edn. CRC Press, Boca Raton (2014)
15. Gyárfás, A., Sebő, A., Trotignon, N.: The chromatic gap and its extremes. J. Combin. Theory Ser. B **102**(5), 1155–1178 (2012). https://doi.org/10.1016/j.jctb. 2012.06.001
16. Hales, R.S.: Numerical invariants and the strong product of graphs. J. Combin. Theory Ser. B **15**, 146–155 (1973)
17. Hammack, R., Imrich, W., Klavžar, S.: Handbook of Product Graphs. Discrete Mathematics and its Applications. CRC Press, Boca Raton (2011). With a foreword by Peter Winkler
18. Harant, J., Schiermeyer, I.: On the independence number of a graph in terms of order and size. Discrete Math. **232**(1–3), 131–138 (2001). https://doi.org/10.1016/ S0012-365X(00)00298-3
19. Jurkiewicz, M.: A generalization of the Shannon's theorem and its application to complex networks. preprint
20. Jurkiewicz, M.: Relevant measures of product networks. preprint
21. Jurkiewicz, M.: Average distance is submultiplicative and subadditive with respect to the strong product of graphs. Appl. Math. Comput. **315**, 278–285 (2017). https://doi.org/10.1016/j.amc.2017.06.025
22. Jurkiewicz, M., Kubale, M., Ocetkiewicz, K.: On the independence number of some strong products of cycle-powers. Found. Comput. Decis. Sci. **40**(2), 133–141 (2015). https://doi.org/10.1515/fcds-2015-0009
23. Jurkiewicz, M., Pikies, T.: Selected topics in modern mathematics. Chap. Some classical lower bounds on the independence number and their behavior on the strong product of graphs. Publishing House AKAPIT, Kraków, inst. Politechnika Krakowska (2015)

24. Körner, J., Orlitsky, A.: Zero-error information theory. IEEE Trans. Inform. Theory **44**(6), 2207–2229 (1998). https://doi.org/10.1109/18.720537. Information theory: 1948-1998

25. Kubale, M. (ed.): Graph Colorings, Contemporary Mathematics, vol. 352. American Mathematical Society, Providence (2004). https://doi.org/10.1090/conm/352

26. Leskovec, J., Chakrabarti, D., Kleinberg, J., Faloutsos, C., Ghahramani, Z.: Kronecker graphs: an approach to modeling networks. J. Mach. Learn. Res. **11**, 985–1042 (2010)

27. Loehr, N.A.: Bijective Combinatorics. Discrete Mathematics and its Applications. CRC Press, Boca Raton (2011)

28. Lovász, L.: On the Shannon capacity of a graph. IEEE Trans. Inform. Theory **25**(1), 1–7 (1979). https://doi.org/10.1109/TIT.1979.1055985. http://dx.doi.org.mathematical-reviews.han.bg.pg.edu.pl/10.1109/TIT.1979. 1055985

29. McEliece, R.J.: The Theory of Information and Coding, Encyclopedia of Mathematics and Its Applications, vol. 86. Cambridge University Press, Cambridge (2004). http://dx.doi.org.mathematical-reviews.han.bg.pg.edu.pl/10.1017/CBO9780511819896. With a foreword by Mark Kac

30. Pikies, T.: Personal communication (2015)

31. Scheinerman, E.R., Ullman, D.H.: Fractional Graph Theory. Dover Publications Inc., Mineola (2011). A rational approach to the theory of graphs, With a foreword by Claude Berge, Reprint of the 1997 original

32. Shannon, C.E.: The zero error capacity of a noisy channel. Inst. Radio Eng. Trans. Inf. Theory **IT-2**(September), 8–19 (1956)

33. Whang, J.J., Gleich, D.F., Dhillon, I.S.: Overlapping community detection using seed set expansion. In: Proceedings of the 22nd ACM International Conference on Information & Knowledge Management, CIKM 2013, pp. 2099–2108. Association for Computing Machinery, New York (2013). https://doi.org/10.1145/2505515. 2505535

Scheduling

On the Complexity of a Periodic Scheduling Problem with Precedence Relations

Richard Hladík[1,2], Anna Minaeva[1]([✉]), and Zdeněk Hanzálek[1]

[1] Czech Institute of Informatics, Robotics, and Cybernetics,
Czech Technical University in Prague, Prague, Czech Republic
minaevaana@gmail.com
[2] Charles University, Prague, Czech Republic

Abstract. Periodic scheduling problems (PSP) are frequently found in a wide range of applications. In these problems, we schedule a set of tasks on a set of machines in time, where each task is to be executed repeatedly with a given period. The tasks are assigned to machines, and at any moment, at most one task can be processed by a given machine. Since no existing works address the complexity of PSPs with precedence relations, we consider the most basic PSP with chains and end-to-end latency constraints given in the number of periods. We define a degeneracy of a chain as the number of broken precedence relations within the time window of one period. We address the general problem of finding a schedule with the minimum total degeneracy of all chains. We prove that this PSP is strongly NP-hard even when restricted to unit processing times, a common period, and 16 machines, by a reduction from the job shop scheduling problem. Finally, we propose a local search heuristic to solve the general PSP and present its experimental evaluation.

1 Introduction

Periodic scheduling problems (PSPs) are frequently found in a wide range of applications, including communications [16], maintenance [17], production [1], avionics [7], and automotive [5]. A control loop is a typical example of an application that requires periodic data transmission from sensors over control units and gateways to actuators. The result depends not only on a logically correct computation but also on the end-to-end latency measured from the moment when the sensor acquires a physical value to the moment when the actuator performs its action. Due to the periodic nature of the problem, the end-to-end latency is typically expressed in a number of periods [15].

In a PSP, we are given a set of tasks and a set of machines. Each task has a *processing time* p and is to be executed repeatedly with a given *period* T on a (given) machine. The goal is to schedule the tasks in time so that at any given moment, at most one task is processed by each machine, and the periodical nature of the tasks is satisfied. A PSP can be either preemptive or non-preemptive, when an execution of a task can or cannot be preempted by

© Springer Nature Switzerland AG 2020
W. Wu and Z. Zhang (Eds.): COCOA 2020, LNCS 12577, pp. 107–124, 2020.
https://doi.org/10.1007/978-3-030-64843-5_8

the execution of another task, respectively. In this work, we deal with the *non-preemptive* version of the PSP.

Many works have addressed the complexity of non-preemptive PSPs. Jeffay et al. in [10] address a PSP on a single machine with arbitrary release dates and deadlines equal to one period from the corresponding release dates, where the release date is the earliest time a task can start in its period and the deadline is the latest time it must complete. In the three-field Graham notation $\alpha|\beta|\gamma$ introduced in [8], where the α field characterizes the resources, the β field reflects properties of tasks, and the γ field contains the criterion, this problem is denoted as $1|T_i, r_i, d_i = r_i + T_i|-$. The authors prove that this problem is strongly NP-hard by a reduction from the 3-Partition problem. However, the proof relies on the different release dates of the tasks. Furthermore, the case of the harmonic period set, when larger periods are divisible by smaller periods, seems to be an easier problem, since there are efficient heuristics to solve it (e.g., in [4]). However, Cai et al. in [3] strengthened the result of [10] by proving strong NP-hardness of the PSP $1|T_i^{harm}, d_i = T_i|-$ with zero release dates and harmonic periods. Later, Nawrocki et al. in [14] prove that this complexity result holds even if the ratio between the periods is a power of 2, i.e., for $1|T_i^{pow2}, d_i = T_i|-$.

A PSP with a zero jitter requirement (also known as strictly or perfectly periodic, where the position of a task within a period is the same in all periods) is widely assumed [4,6] since a non-zero jitter represents a disturbance to control systems. Korst et al. in [12] show that the PSP with zero jitter requirements on a single machine, $1|T_i, jit_i = 0|-$, is strongly NP-hard. Moreover, the same problem with unit processing times, $1|T_i, jit_i = 0, p_i = 1|-$, is shown to be NP-hard by Bar-Noy et al. [1] by the reduction from the graph coloring problem. Jacobs et al. in [9] strengthen this result by proving that this PSP is strongly NP-hard. As a matter of fact, even deciding whether a single task can be added to the set of already scheduled tasks for this PSP is NP-complete, since it is the problem of computing simultaneous incongruences.

There are no results on the complexity of non-preemptive PSPs with precedence relations. Therefore, in this paper, we focus on a PSP with chains of precedence relations, i.e., a task can only be scheduled after the completion of its predecessor unless it is the first task in the chain. End-to-end latency of a chain is the time from the start time of its first task to the completion time of its last task. We define the degeneracy of a chain as its end-to-end latency divided by its period. Alternatively, it is the number of broken precedence relations within the time window of one period.

We consider a general PSP$_{gen}$, PD$|T_i^{harm}, jit_i = 0, \text{chains}| \sum \delta$, where tasks with harmonic periods and zero jitter requirements are scheduled on multiple dedicated machines (i.e., assignment of tasks to machines is given) so that the total degeneracy of all chains is minimized. Furthermore, we address the complexity of a special case of PSP$_{gen}$ called PSP$_{com}$, PD16$|T_i = T, jit_i = 0, p_i = 1, \text{chains}, \delta_l = 0|-$, where tasks with a *common* period and unit processing times are scheduled on 16 machines, and chains are 0-degenerated (i.e., all precedence relations are satisfied within one period).

The main three contributions of this paper are: 1) We propose a novel formulation of a PSP with chains of precedence constraints called PSP$_{gen}$ based on chains degeneracy. The degeneracy offers a coarser alternative to the widely used end-to-end latency and may be a more suitable metric for some real-world problems. 2) We establish that PSP$_{gen}$ is strongly NP-hard even when restricted to unit execution times, common period, and 16 machines by a reduction from the job shop scheduling problem $J3 \mid p_i = 1 \mid C_{\max}$. This problem is called PSP$_{com}$ and denoted as $PD16|T_i = T, jit_i = 0, p_i = 1, \text{chains}, \delta_s = 0|-$. 3) We provide a local search heuristic algorithm that solves PSP$_{gen}$. Moreover, we experimentally demonstrate the soundness of our algorithm and show that it can solve 92 % of our instances (with up to 9 000 of tasks) in a few minutes on a desktop computer, and the provably optimal solution is found for more than 75% instances.

2   Problem Description

In this section, we present a general problem PSP$_{gen}$ considered in this work. We first introduce non-collision constraints and then the optimization criterion based on how well the precedence constraints are satisfied within one period. Finally, we constructively prove that the existence of a solution satisfying the non-collision constraints is equivalent to the existence of the solution satisfying both non-collision and precedence constraints.

2.1   Problem Statement

We are given a set of *tasks* $\mathcal{T} = \{\tau_1, \ldots, \tau_n\}$ and a set of *machines* $\mathcal{M} = \{\mu_1, \ldots, \mu_{|\mathcal{M}|}\}$. Each task τ_i has a *processing time* $p(\tau_i) \in \mathbb{N}^*$ and executes repeatedly with a given *period* $T(\tau_i) \in \mathbb{N}^*$. Here, $\mathbb{N}^*$ is a set of natural numbers without zero and $\mathbb{N}_0$ is a set of natural numbers with zero. Each task τ_i is also *assigned* to machine $m(\tau_i) \in \mathcal{M}$, on which it must be executed.

Our goal is to find a *schedule*, which is a function $s : \mathcal{T} \rightarrow \mathbb{N}_0$ that assigns a *start time* $s(\tau_i)$ to each task τ_i. The task τ_i is then executed every $T(\tau_i)$ units of time, i.e., with zero jitter; its k-th execution (for $k \in \mathbb{N}_0$) spans the interval $[s(\tau_i) + k \cdot T(\tau_i), s(\tau_i) + p(\tau_i) + k \cdot T(\tau_i))$. Let $\mathcal{R}(\tau_i)$ denote the union of all such intervals for task τ_i.

A schedule s *has no collisions* if there is at most one task executed on each machine at any given moment, that is:

$$\mathcal{R}(\tau_i) \cap \mathcal{R}(\tau_j) = \emptyset, \quad \forall i \neq j : m(\tau_i) = m(\tau_j) \tag{1}$$

Korst et al. [11] have shown that for zero-jitter case, Equation (1) is equivalent to

$$p(\tau_i) \leq (s(\tau_j) - s(\tau_i)) \bmod g_{i,j} \leq g_{i,j} - p(\tau_j), \tag{2}$$

where $g_{i,j} = \gcd(T(\tau_i), T(\tau_j))$.

There are precedence constrains in the form of task chains. Let $\mathcal{C} = \{C_1, \ldots, C_k\}$ be a partition of $\mathcal{T}$ into pairwise disjoint ordered sets $C_1, \ldots, C_k$ such that in each set, all tasks have the same period.[1] Each of these sets is called a *(precedence) chain*. The r-th task of the c-th chain is denoted by C_c^r (for $r \in \{1, \ldots, |C_c|\}$). We call C_c^{r-1} and C_c^{r+1} (if they exist) *predecessor* and *successor* of C_c^r. Tasks without a predecessor are called *root tasks*. Note that formally, $C_c^r \in \mathcal{T}$.

A schedule s *satisfies precedence relations*, if

$$s(C_c^r) \geq s(C_c^{r-1}) + p(C_c^{r-1}), \ \forall C_c \in \mathcal{C}, \ r = 2, \ldots, |C_c| \tag{3}$$

that is, each task starts only after its predecessor finishes execution. Given that all tasks in a chain have the same period, all further executions of the chain are also ordered correctly.

The end-to-end latency $L(C_c)$ of a chain C_c is the distance from the start time of the first task to the completion time of the last task in the chain as given by Eq. (4). Then, the *degeneracy* $\delta_s(C_c)$ with respect to schedule s is defined in Eq. (5).

$$L(C_c) = s(C_c^{|C_c|}) + p(C_c^{|C_c|}) - s(C_c^1), \tag{4}$$

$$\delta_s(C_c) = \left\lceil \frac{L(C_c)}{T(C_c^1)} \right\rceil - 1. \tag{5}$$

In other words, a chain degeneracy is the number of crossed relative period boundaries, with the first period starting at the start time of the first task in the chain. For the example in Fig. 1, the degeneracy of chain C_1 is 2, since its first task $C_1^1 = \tau_1$ starts at 0 and crosses its relative period boundary (in this case coinciding with its absolute period boundary) 2 times. On the other hand, although C_4 crosses its absolute period boundary at time 28, its degeneracy equals 0, since its relative period boundary is at time $10 + 28 = 38$.

The *degeneracy* of a schedule $\delta(s)$ is then defined as:

$$\delta(s) = \begin{cases} \sum_{C_c \in \mathcal{C}} \delta_s(C_c) & \text{if } s \text{ is feasible,} \\ +\infty & \text{otherwise.} \end{cases}$$

We also say a schedule with degeneracy k is *k-degenerated*. Note that by this definition, the minimum possible degeneracy is zero. In this case, its end-to-end latency does not exceed the length of chain's period.

With the definitions provided, PSP_{gen} is: given a description of tasks and precedence chains, find a schedule with minimal degeneracy. Formally, find

$$\underset{s \,:\, \mathcal{T} \to \mathbb{N}_0}{\arg \min} \ \delta(s), \tag{6}$$

---

[1] On the other hand, two tasks with equal period are not necessarily in the same C_c.

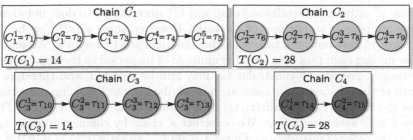

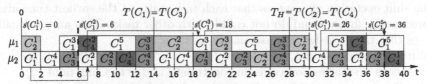

(a) Chains of precedence relations C_1, C_2, C_3, and C_4.

(b) An example schedule with degeneracies $\delta_s(C_1) = 2$, $\delta_s(C_2) = 0$, $\delta_s(C_3) = 1$, and $\delta_s(C_4) = 0$. Solid vertical lines mark absolute period boundaries, whereas bold dashed lines depict relative period boundaries for chain C_4.

Fig. 1. A periodic scheduling problem with an example solution. There are four chains: $C_1 = (\tau_1, \tau_2, \tau_3, \tau_4, \tau_5)$, $C_2 = (\tau_6, \tau_7, \tau_8, \tau_9)$, $C_3 = (\tau_{10}, \tau_{11}, \tau_{12}, \tau_{13})$, and $C_4 = (\tau_{14}, \tau_{15})$ with periods 14, 28, 14, and 28 time units, respectively. The task assignments are $m(\tau_3) = m(\tau_5) = m(\tau_6) = m(\tau_7) = m(\tau_8) = m(\tau_{12}) = m(\tau_{15}) = \mu_1$ and $m(\tau_1) = m(\tau_2) = m(\tau_4) = m(\tau_9) = m(\tau_{10}) = m(\tau_{11}) = m(\tau_{13}) = m(\tau_{14}) = \mu_2$, the processing times are 2 except for $p(\tau_5) = p(\tau_7) = 4$.

such that non-collision (1) and precedence (3) constraints hold. In the three-field notation, PSP_{gen} is denoted as $\mathrm{PD}|T_i^{harm}, jit_i = 0, \mathrm{chains}|\sum \delta$.

2.2 Equivalence Proof

We show that if there is a schedule satisfying non-collision constraints for PSP_{gen}, it is possible to modify it to also satisfy precedence constraints. We actually formulate this in a stronger form, which we use later in Sect. 3:

Lemma 1. *Let s be a schedule satisfying* (1), *and let $\tau^\star$ be a fixed root task. Then there exists a schedule s' satisfying both* (1) *and* (3) *such that $s'(\tau^\star) = 0$ and $s'(\tau) \leq T(\tau)$ for all other root tasks τ. Moreover, if s already satisfies* (3), *then $\delta(s') \leq \delta(s)$.*

Proof. We define s'' as s moved in time so that the start time of the chosen root task $\tau^\star$ is zero:

$$s''(\tau_i) = s(\tau_i) - s(\tau^\star).$$

Schedule s'' satisfies non-collision constraint (2) since the move does not change relative positions of start times of the tasks. However, some $s''(\tau_i)$ may be negative since $\tau^\star$ may not be a task with the minimum start time in s. We may fix that by moving each task by a suitable multiple of its period to the right in time. Nevertheless, precedence constraint (3) may still be violated, and therefore we fix both of these issues simultaneously as described in the following paragraph.

For $\tau_i \in \mathcal{T}$, $t_0 \in \mathbb{N}_0$, let $\mathrm{shift}(\tau_i, t_0)$ be a minimum value $t = s''(\tau_i) + k \cdot T(\tau_i)$ (where $k \in \mathbb{Z}$) such that $t \geq t_0$. We construct s' chain by chain, traversing each chain in the order of precedences. Given a chain C_c, we set $s'(C_c^1) = \mathrm{shift}(C_c^1, 0)$. For each subsequent task C_c^r, we set $s'(C_c^r) = \mathrm{shift}(C_c^r, s'(C_c^{r-1}) + p(C_c^{r-1}))$. That is, the shift operation guarantees that each task starts at the earliest time after its predecessor finishes and do not collide with other tasks. This automatically ensures that the precedence constraint holds for s'.

With this construction, $s'(\tau_i) \bmod T(\tau_i) = s''(\tau_i) \bmod T(\tau_i)$ for each τ_i, which means s' satisfies non-collision constraint (1) due to Constraint (2).

Finally, since $\tau^\star$ is a root task, $s'(\tau^\star) = \mathrm{shift}(\tau^\star, 0) = 0$. For all other root tasks τ, $s'(\tau) < T(\tau)$ by the definition of shift. Assume s satisfies precedence constraints (3) and observe that $\delta(s'') = \delta(s)$. For each C_c, $\delta_{s'}(C_c) \leq \delta_{s''}(C_c)$, since the tasks are scheduled as close as possible. Thus, $\delta(s') \leq \delta(s'') = \delta(s)$. $\square$

An example of the result of this constructive proof can be seen in Fig. 1(b), where for chain C_1, an original schedule s can be $s(\tau_1) = 0, s(\tau_2) = 6, s(\tau_3) = 4, s(\tau_4) = 2, s(\tau_5) = 8$, and a constructed schedule s' is $s'(\tau_1) = 0, s'(\tau_2) = 6, s'(\tau_3) = 18, s'(\tau_4) = 30$, and $s'(\tau_5) = 36$.

3   Problem Complexity

In this section, we prove that even a less general version of PSP_{gen} is strongly NP-hard by a polynomial transformation from a special version of the job shop scheduling problem.

3.1 PSP_{com}

The proof of NP-hardness will be carried out on a restricted variant of PSP_{gen} called PSP_{com}. This is a decision problem based on PSP_{gen} with the following modifications: the number of machines is at most 16, all tasks have unit processing time and a common period T_H, and the problem is to decide whether there exists a 0-degenerated schedule. Thus, it is denoted as $\mathrm{PD16}|T_i = T, jit_i = 0, p_i = 1, \mathrm{chains}, \delta_l = 0|-$ in the three-field notation. Note that (strong) NP-hardness of PSP_{com} implies (strong) NP-hardness of PSP_{gen} since the latter is more general.

Definition 1. *PSP_{com} is defined by a 4-tuple $(\mathcal{T}, \mathcal{M}, \mathcal{C}, T_H)$ consisting of the task set, $\mathcal{T}$, machine set, $\mathcal{M}$, chain set, $\mathcal{C}$, and the common period T_H. The* *problem is to decide if there exists a feasible schedule s such that precedence constraint (3), non-collision constraint (7), and 0-degeneracy constraint (8) hold.*

$$s(\tau_i) \bmod T_H \neq s(\tau_j) \bmod T_H, \ \forall i \neq j : m(\tau_i) = m(\tau_j) \tag{7}$$

$$s(C_c^k) - s(C_c^1) < T_H, \qquad \forall C_c \in \mathcal{C}. \tag{8}$$

Due to unit processing times, non-collision constraint (1) resulted in Constraint (7) and 0-degeneracy constraint (8) is simply a constraint on chains' end-to-end latency (4).

Finally, a schedule of PSP_{com} *is feasible if it satisfies precedence constraint (3), non-collision constraint (7), and 0-degeneracy constraint (8).*

3.2 Job Shop Scheduling Problem JS3

We prove NP-hardness of PSP_{com} by reduction from a specific variant of the *job shop scheduling problem* JS3 denoted in the three-field notation as $J3 \mid p_i = 1 \mid C_{\max}$. Thus, the tasks with unit processing times are scheduled on three machines so that the makespan (i.e., the completion time of the latest task) is minimal. We formulate it briefly in the following paragraphs.

Definition 2. *JS3 is defined by a 4-tuple $(\hat{\mathcal{T}}, \hat{\mathcal{M}}, \hat{\mathcal{C}}, L)$ consisting of the task set, $\hat{\mathcal{T}} = \{\hat{\tau}_1, \ldots, \hat{\tau}_{\hat{n}}\}$, machine set, $\hat{\mathcal{M}} = \{\hat{\mu}_1, \hat{\mu}_2, \hat{\mu}_3\}$, chain set, $\hat{\mathcal{C}} = \{\hat{C}_1, \ldots, \hat{C}_{\hat{k}}\}$, and the maximum makespan L. The problem is to decide if there exist schedule s such that non-collision constraint (1) and precedence constraint (3) (with s substituted for $\hat{s}$, τ_i for $\hat{\tau}_i$, and similarly for other variables) such that makespan constraint (9) holds.*

$$\max_{\hat{\tau}_i \in \hat{\mathcal{T}}} \{\hat{s}(\hat{\tau}_i) + \hat{p}(\hat{\tau}_i)\} - \min_{\hat{\tau}_i \in \hat{\mathcal{T}}} \{\hat{s}(\hat{\tau}_i)\} \leq L, \tag{9}$$

The definitions of the task set, machine set, and chain set are the same as those of PSP_{com}. However, the tasks are not periodic. Therefore, 0-degeneracy constraint states that the time elapsed between the first task starting and the last task finishing among all tasks must be at most L.

Lenstra et al. have shown [13] that this problem is strongly NP-hard.

3.3 Naive Incomplete Reduction

We show that PSP_{com} (and, therefore, PSP_{gen}) is strongly NP-hard by constructing a polynomial reduction from JS3 to PSP_{com}.

An obvious, but incorrect attempt at the reduction is: given an arbitrary JS3 instance of $\hat{\mathcal{I}} = (\hat{\mathcal{C}}, \hat{\mathcal{M}}, \hat{\mathcal{C}}, L)$, we create one PSP_{com} task for each JS3 task: $\mathcal{T} = \{\tau_i \mid \hat{\tau}_i \in \hat{\mathcal{T}}\}$, and similarly $\mathcal{M} = \{\mu_i \mid \hat{\mu}_i \in \hat{\mathcal{M}}\}$. We also define $\mathcal{C} = \{C_1, \ldots, C_k\}$, where $k = \hat{k}$ and $C_i = \hat{C}_i$. At last, for each $\tau_i \in \mathcal{T}$ we define $p(\tau_i) = 1$ and $T(\tau_i) = T_H = L$.

Unlike PSP_{com}, which imposes the time limit T_H on time elapsed between the first task starting and the last task finishing *for each chain separately*, JS3 requires L to be a global limit *for tasks among all chains*. Thus, some solutions feasible for PSP_{com} are infeasible for JS3. For the example in Fig. 1(b), the schedule for chains C_2 and C_4 with period $T_H = 28$ is feasible for PSP_{com}, but not feasible for JS3 due to makespan constraint (9). Although both chains individually span over less than 28 time units (C_2 from 0 to 26 and C_4 from 10 to 36), together, they run from time 0 to time 36, which is more than the corresponding makespan value of $L = 28$. Thus, *the two conditions are equivalent only if we ensure that all chains start at the same time*. We focus on that in the following subsections.

3.4   Anchoring Chains

We allocate new machines, tasks, and chains to enforce a particular configuration of the schedule. By introducing several "dense" chains (i.e., chains with the number of tasks equal to the hyper-period), we make all chains start at the same time. To simplify the analysis, we limit ourselves to the following subclass of schedules.

Definition 3. *Let τ be a root task. A schedule s is τ-initial, if $s(\tau) = 0$, and $s(\tau') < T_H$ for all other root tasks τ'. A schedule s is* initial *if it is τ-initial for some task τ.*

Without loss of generality, we may consider only τ-initial schedules: Lemma 1 guarantees that if the instance has a feasible schedule, it also has a τ-initial feasible schedule. The converse is true since τ-initial feasible schedule is a feasible schedule by definition.

To make the reduction in Sect. 3.3 complete, we formulate and prove Lemma 2. It states that for any PSP_{com} instance, except for special cases, we can create another instance that is feasible if and only if there exists a schedule for the initial instance satisfying the makespan constraint in job shop scheduling problem.

Lemma 2. *Given a PSP_{com} instance $\mathcal{I} = (\mathcal{T}, \mathcal{M}, \mathcal{C}, T_H)$ with $\mathcal{M} = \{\mu_1, \mu_2, \mu_3\}$, $|\mathcal{C}| > 1$ and $T_H > 2$, it is possible to create a PSP_{com} instance $\mathcal{I}'$ such that $\mathcal{I}'$ is feasible if and only if*

$$\exists s, \text{ a feasible schedule of } \mathcal{I}, \text{ such that} \\ \forall \tau_i \in \mathcal{T} : s(\tau_i) + p(\tau_i) \leq T_H. \tag{10}$$

To prove Lemma 2, we formulate Lemma 3. It states that the space of solutions (schedules) for a PSP_{com} problem instance with 16 machines and no additional restrictions is equivalent to the space of schedules for a PSP_{com} instance with 4 machines and the enforced configuration shown in Fig. 2. In this configuration, tasks mapped to one machine may be executed in a time interval $[0, x]$, whereas all other tasks may be executed in a time interval $[x, T_H'']$ for a fixed (of our choosing) $x \in \{2, \dots, T_H - 3\}$. Therefore, this lemma allows for working in this constrained space of schedules.

Lemma 3. *Given a PSP$_{com}$ instance $\mathcal{I}'' = (\mathcal{T}'', \mathcal{M}'', \mathcal{C}'', T_H'')$ and a parameter $x > 1$, where $\mathcal{M}'' = \{\mu_1, \mu_2, \mu_3, \mu_\star\}$, and $T_H'' > x + 2$, it is possible to create an instance $\mathcal{I}'$ such that $\mathcal{I}'$ is feasible if and only if*

$$\exists s, \text{ a feasible schedule of } \mathcal{I}'', \text{ such that:}$$
$$\forall \tau_i \in \mathcal{T}'' : m(\tau_i) = \mu_\star \implies s(\tau_i) \bmod T_H'' \in [0, x), \tag{11}$$
$$\forall \tau_i \in \mathcal{T}'' : m(\tau_i) \neq \mu_\star \implies s(\tau_i) \bmod T_H'' \in [x, T_H'').$$

Equation (11) states that the tasks from $\mathcal{T}''$ are forbidden to execute in the gray solid areas of Fig. 2 (and in any of their congruent copies).

We first prove Lemma 2 using the result of Lemma 3 and then we prove Lemma 3.

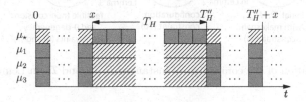

Fig. 2. Illustration of the complexity proof. Gray solid area stand for the tasks added in Lemma 3. In the lower right hatched area, JS3 tasks are located, whereas the upper left hatched area contains the anchoring tasks added in Lemma 2.

Proof (Lemma 2). The main idea is to use Lemma 3 to work with schedules with the configuration displayed in Fig. 2. We "anchor" each chain by prepending to it a special task assigned to $\mu_\star$. This guarantees that each chain starts before time x. Since tasks of $\mathcal{I}$ (i.e., JS3 tasks) must be executed in the time interval $[0, T_H'' + x)$ due to the end-to-end latency constraint, and at the same time cannot be executed in time intervals $[0, x)$ and $[T_H'', T_H'' + x)$, they must be executed in interval $[x, T_H'']$. Therefore, the makespan of the resulting JS3 chains is not more than $T_H = T_H'' - x$. The details follow.

For each chain, we create a new anchor task τ_{a_c} and prepend it to the chain: $C_c'' = (\tau_{a_c}, C_c^1, \dots, C_c^{|C_c|})$. We create an instance $\mathcal{I}'' = (\mathcal{T}'', \mathcal{M}'', \mathcal{C}'', T_H'')$ with $\mathcal{T}'' = \mathcal{T} \cup \{\tau_{a_1}, \dots, \tau_{a_k}\}$, $\mathcal{M}'' = \mathcal{M} \cup \{\mu_\star\}$, $\mathcal{C}'' = \{C_1'', \dots, C_k''\}$, and $T_H'' = T_H + k$. We assign the anchor tasks to the auxiliary machine: $m(\tau_{a_c}) = \mu_\star$ for all c.

We use Lemma 3 on $\mathcal{I}''$ with $x = k$ to obtain $\mathcal{I}' = (\mathcal{T}', \mathcal{M}', \mathcal{C}', T_H')$. We want to prove that $\mathcal{I}'$ simulates the JS3 makespan constraint (10) in $\mathcal{I}$. By Lemma 3, $\mathcal{I}'$ enforces the configuration depicted in Fig. 2 in $\mathcal{I}''$. We only have to prove that $\mathcal{I}''$ with the configuration constraints (11) simulates the JS3 makespan constraint (10) in $\mathcal{I}$. Formally, proving the lemma is now equivalent to proving that $\mathcal{I}''$ satisfies configuration Constraints (11) if and only if $\mathcal{I}$ satisfies configuration Constraint (10) as depicted in Fig. 3. Next, we prove both implications.

"$\mathcal{I}$ satisfies Constraint (10) $\Rightarrow$ $\mathcal{I}''$ satisfies Constraint (11)": Let s be the feasible schedule of $\mathcal{I}$ satisfying Constraint (10). We create s'' as follows:

$$s''(\tau_i) = \begin{cases} c - 1 & \text{If } \tau_i = \tau_{a_c} \text{ for some } \tau_{a_c}, \\ s(\tau_i) + k & \text{otherwise.} \end{cases}$$

The definition is valid, since if $\tau_i \neq \tau_{a_c}$, then $\tau_I \in \mathcal{T}$, and $s(\tau_i)$ is defined. Both the satisfaction of Constraints (11) of s'' and its 0-degeneracy with $T_H'' = T_H + k$ are guaranteed by the construction.

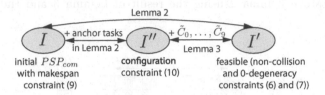

Fig. 3. Illustration of the connection of instances $\mathcal{I}$, $\mathcal{I}'$, and $\mathcal{I}''$ in Lemmas 2 and 3

"$\mathcal{I}''$ satisfies Constraint (11) $\Rightarrow$ $\mathcal{I}$ satisfies Constraint (10)": Let s'' be the feasible schedule of $\mathcal{I}''$ satisfying Constraints (11). Since there are k anchor tasks assigned to machine $\mu_\star$ and k time moments where the tasks may be scheduled, there must exist τ_{a_c} such that $s''(\tau_{a_c}) \bmod T_H'' = 0$. We may assume s'' is τ_{a_c}-initial, otherwise we invoke Lemma 1 with τ_{a_c} and make it such. Observe that the shifted schedule still satisfies Constraints (11). Since s'' is initial, $s''(\tau_{a_c}) \leq T_H''$ and therefore, due to configuration constraints (11), $s''(\tau_{a_c}) \leq k$ for all τ_{a_c}. Then, $s(\tau)'' < T_H'' + k$ for all $\tau \in \mathcal{T}$ because of 0-degeneracy, and, finally, $s''(\tau) < T_H'' = T_H + k$ since Constraints (11) hold for s''.

We may thus set $s(\tau_i) = s''(\tau_i) - k$ for all $\tau_i \in \mathcal{T}$ and observe the resulting schedule is feasible and satisfies Constraint (10).

$\square$

Now we proceed by proving Lemma 3.

Proof (Lemma 3). We create 12 new machines and $10 \times T_H$ new tasks. We create a new instance $\mathcal{I}' = (\mathcal{T}', \mathcal{M}', \mathcal{C}', T_H')$ with $\mathcal{M}' = \mathcal{M}'' \cup \{\mu_{t_1}, \ldots, \mu_{t_9}, \mu_{s_1}, \mu_{s_2}, \mu_{s_3}\}$, $T_H' = T_H''$, $\mathcal{C}' = \mathcal{C}'' \cup \{\tilde{C}_0, \ldots, \tilde{C}_9\}$, and $\mathcal{T}' = \mathcal{T}'' \cup \tilde{C}_0 \cup \cdots \cup \tilde{C}_9$. Each chain consists of T_H'' new tasks with unit processing times and we shall prove that the start times of all auxiliary tasks are predetermined across all initial feasible schedules.

We prove the lemma in two steps: first, we describe the assignment of the tasks in $\tilde{C}_0, \ldots, \tilde{C}_5$, and show that they enforce a special configuration on machines μ_{s_1}, μ_{s_2}, and μ_{s_3}. Then, we describe the assignment of the remaining chains and conclude the proof.

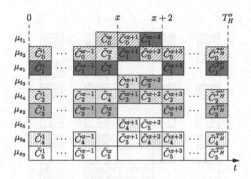

Fig. 4. Assignment of tasks in $\tilde{C}_0, \ldots, \tilde{C}_5$ to machines $\mu_{t_1}, \ldots, \mu_{t_6}, \mu_{s_1}, \ldots, \mu_{s_3}$ and the only possible schedule of these tasks. Tasks in the same chain have the same background. The purpose of this configuration is to make free only intervals $[x, x + 2)$ on machines μ_{s_1}, μ_{s_2} and μ_{s_3}.

The assignment of tasks in $\tilde{C}_0, \ldots, \tilde{C}_5$ to machines is shown in Fig. 4. All tasks in each chain except for two tasks are assigned to the same machine, whereas the remaining two tasks are assigned to two other mutually distinct machines. This assignment ensures that the shown schedule s is the only possible for any $\tilde{C}_0^1$-initial feasible schedule. Since each added chain contains exactly T_H tasks, fixing the start time of one task in a chain results in uniquely determined start times of all other tasks in this chain due to non-collision and 0-degeneracy constraints (7) and (8), respectively. Thus, the start times of the tasks in $\tilde{C}_0$ are uniquely determined because of $\tilde{C}_0^1$. Then, $\tilde{C}_1^{x+1}$ can only start at x since for any other choice, $\tilde{C}_0$ and $\tilde{C}_1$ would collide. Therefore, $s(\tilde{C}_1^r) = r - 1$ for all r. We proceed with the same reasoning, and conclude that $s(\tilde{C}_c^r) = r - 1$ for all $c \in \{0, \ldots, 5\}, r \in \{1, \ldots, T_H\}$.

The assignment of the tasks in chains $\tilde{C}_6, \ldots, \tilde{C}_9$, is shown in Fig. 5. The first x tasks in each chain are assigned to the same machine μ_1, μ_2, μ_3 and μ_{t_8}, the next two tasks in each chain are assigned to the same machine $\mu_{s_1}, \mu_{s_2}, \mu_{s_3}$ and $\mu_\star$, and the rest of the tasks in each chain are assigned to the same machine $\mu_{t_7}, \mu_{t_8}, \mu_\star$, and μ_{t_9}, respectively. Note that machines $\mu_1, \mu_2, \mu_3, \mu_\star$ are free exactly at times indicated by Constraint (11) (compare with Fig. 2).

Using the same reasoning as in the previous part of the proof, we can prove that the configuration shown in Fig. 5 is the only possible for any $\tilde{C}_0^1$-initial feasible schedule. We shall now verify that $\mathcal{I}'$ satisfies the requirements of the lemma, i. e., that it simulates configuration constraints (11) in $\mathcal{I}''$. We prove both implications:

"$\mathcal{I}''$ satisfies configuration constraints (11) $\implies$ $\mathcal{I}'$ is feasible": Let s'' be the schedule of $\mathcal{I}''$ satisfying Constraints (11). We define $s'(\tau_i) = s''(\tau_i)$ for $\tau_i \in \mathcal{T}$ and use the idea from Figs. 4 and 5 for the remaining auxiliary tasks. By the construction, s' is 0-degenerated and has no collisions.

"$\mathcal{I}'$ is feasible $\implies$ $\mathcal{I}''$ satisfies configuration constraints (11)": Let s' be a feasible schedule of $\mathcal{I}'$. We assume without loss of generality that s' is

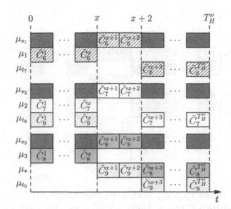

Fig. 5. Assignment of tasks in $\tilde{C}_6, \ldots, \tilde{C}_9$ to machines $\mu_{s_1}, \mu_{s_2}, \mu_{s_3}, \mu_{t_7}, \mu_{t_8}, \mu_{t_9}, \mu_1,$ $\mu_2, \mu_3, \mu_\star$ and the only possible schedule of these tasks. Dark gray rectangles are the tasks fixed in Fig. 4. The purpose of this configuration is to occupy interval $[0, x]$ on machines $\mu_1, \mu_2,$ and μ_3 and $[x, T_H'']$ on $\mu_\star$.

$\tilde{C}_0^1$-initial, otherwise we make it such using Lemma 1. We define $s''(\tau_i) = s'(\tau)$ for all $\tau \in \mathcal{T}$. Due to s' being $\tilde{C}_0^1$-initial, we know that it must schedule tasks from the auxiliary chain as displayed in Fig. 5. Then Constraints (11) become just the no-collision constraints on s' between tasks from $\mathcal{T}$ and tasks from $\tilde{C}_6,$ $\tilde{C}_7,$ $\tilde{C}_8$ and $\tilde{C}_9$. Since these hold by construction, we are done.

$\square$

Observation 1. *If Lemma 2 is used on $\mathcal{I} = (\mathcal{T}, \mathcal{M}, \mathcal{C}, T_H)$, the resulting instance has $\mathcal{O}(|\mathcal{T}| + T_H)$ tasks, and the time complexity of constructing it is also $\mathcal{O}(|\mathcal{T}| + T_H)$.*

Theorem 1. *There exists a polynomial reduction from JS3 to PSP_{com}.*

Proof. Let $\mathcal{J} = (\hat{\mathcal{T}}, \hat{\mathcal{M}}, \hat{c}, L)$ be a JS3 instance. Let $\mathcal{I} = (\mathcal{T}, \mathcal{M}, \mathcal{C}, T_H)$ be the instance obtained from $\mathcal{J}$ by the naive reduction described in Sect. 3.3. We consider two cases:

- The conditions of Lemma 2 do not hold. Then either $|\mathcal{C}| = 1$, or $T_H = L \leq 2$. In both cases, we may decide the feasibility of the instance in polynomial time.
- If $T_H > |\mathcal{T}|$, the instance is always feasible.
- Otherwise, we use Lemma 2 to obtain in time $\mathcal{O}(|\mathcal{T}| + T_H) = \mathcal{O}(|\mathcal{T}|)$ an instance $\mathcal{I}'$ that is feasible if and only if there is a feasible schedule s of $\mathcal{I}$ satisfying Constraint (10), which is equivalent to $\mathcal{J}$ having a schedule satisfying Constraint (9).

Corollary 1. *PSP_{com} and PSP_{gen} are strongly NP-hard.*

4   Heuristic Approach

We propose a local search heuristic to solve PSP_{gen} described in Sect. 2. The algorithm first generates a (possibly infeasible) schedule and then, for a fixed number of iterations, creates a new schedule based on the current one. The old schedule is swapped with the new one if the quality of the latter is not worse in terms of degeneracy.

Schedule Representation. In PSPs, schedules are usually represented directly by task start times. However, inspired by the disjunctive graphs in the job shop scheduling problem [2], we represent a schedule as a *queue* of tasks. We reconstruct the schedule whenever we want to compute its degeneracy. A queue Q is a totally ordered list of tasks: $Q = (\tau_{\pi(1)}, \dots, \tau_{\pi(n)})$, where π is a permutation of $\{1, \dots, n\}$. Let $Q(i) = \tau_{\pi(i)}$.

Reconstruction. A *reconstruction function* $f : Q \to s$ is a function that takes Q as an argument and returns a schedule (i.e., task start times). An important property of reconstruction functions is that a small change in a queue results in a small change in the reconstructed schedule. This property is the main motivation for the design decisions of this function. Moreover, two schedules represented by start times might look very different, but in fact they might be nearly identical from the "search space" viewpoint.

The reconstruction starts with an empty schedule (i.e., $s(\tau_i) = \emptyset$ for all τ_i) and it schedules task $Q(\ell)$ in the ℓ-th iteration such that non-collision (1) and precedence (3) (if the predecessor of $Q(\ell)$ is already scheduled) constraints are respected. Once a task is scheduled, it remains fixed until the end of the reconstruction. Since tasks may not be in their precedence order in a queue Q, missing precedence constraints are handled at the end of the reconstruction by the shifting procedure from Lemma 1 in Sect. 2.2. Allowing broken precedence constraints in Q gives the heuristic the freedom to decide that a particular precedence relation should be broken. Note that the partial schedule may not be extendable to a feasible schedule. Then, we return $\emptyset$ instead of the schedule.

We next describe the strategy to assign the start time for task $Q(\ell) = \tau_i$ in details. For each machine μ_q, we maintain the time $head(\mu_q) = \max\{s(\tau_i) + p(\tau_i) \mid s(\tau_i) \neq \emptyset, m(\tau_i) = \mu_q\}$ at which the last task assigned to this machine (scheduled so far) finishes executing. We schedule task τ_i at the earliest time t such that: 1. it is not sooner than the last task on the corresponding machine, i.e., $t \geq head(m(\tau_i))$; 2. non-collision constraints are satisfied, i.e., no tasks are already scheduled in intervals $[t + k \cdot T(\tau_i), t + k \cdot T(\tau_i) + p(\tau_i))$; and 3. $t \geq s(\tau_j) + p(\tau_j)$, where τ_j is the predecessor of τ_i (if $\tau_j = \emptyset$ or $s(\tau_j) = \emptyset$, we set $s(\tau_j) = -\infty$). If no such t exists, we return $\emptyset$. Due to the efficient schedule representation, finding the smallest t is done quickly. However, in this work we do not elaborate on it due to the space limit.

Neighbor Function. A neighbor function takes the current Q and modifies it. The resulting queue is a neighbor. Multiple functions have been tested, out of which the most notable are presented in Table 1. Note that in this work, all random choices are assumed to be in a form of the uniform distribution. The idea used by the most successful functions, as shown later in the experiments, is to rearrange the tasks of a randomly chosen chain such that they go in the order of precedence relations (chain_sort function). For example, queue $Q = (\cdots, C_c^2, \cdots, C_c^3, \cdots, C_c^1)$ becomes $Q' = (\cdots, C_c^1, \cdots, C_c^2, \cdots, C_c^3)$ for a chosen chain C_c.

Table 1. List of neighbor functions

Name	Description
swap_uniform	Swap two randomly chosen tasks
chain_uniform	Perform swap_uniform for two tasks of a random chain
chain_adjacent	Swap two adjacent tasks in a random chain
swap_ran	Perform one of the previous three randomly
chain_sort	Rearrange tasks of a random chain to satisfy precedence relations
chain_sort_patient	Perform chain_sort on a chain with broken precedences
chain_sort_loc	Perform chain_sort until the neighbor degeneracy decreases
combined_local	Perform chain_sort_loc; if all chains are sorted, do swap_ran
combined_random	Perform swap_ran or chain_sort_patient randomly
switch	In the initial iterations, perform chain_sort_loc, later perform combined_random

In chain_sort_loc, the change is discarded if the degeneracy of a queue with the rearranged tasks is worse, and the change is kept if the degeneracy has not changed. In one iteration, we apply this function until the neighbor degeneracy becomes less than the initial degeneracy or until the queue is sorted. The intuition behind is that although sorting the current chain may not change the degeneracy, it may nevertheless improve the solution. Finally, switch is the only function that change its strategy with iterations. Here, chain_sort_loc is used until all chains are sorted. From the first iteration when there is no unsorted chain onward, we always perform combined_random. The idea is that in the initial burn-in phase, we want to sort as many chains as possible. Once we achieve that, we want to avoid getting stuck in a local minimum. Thus, we switch to combined_random, allowing more unpredictable changes.

Initialization. To find an initial Q, we sort all tasks in the ascending order using a custom comparison $\prec$ defined as follows ("$<$" denoting the lexicographic comparison on ordered pairs): $\tau_i \prec \tau_j \iff (T(\tau_i), -p(\tau_i)) < (T(\tau_j), -p(\tau_j))$. In other words, we place tasks with smaller periods first, and in case of a tie, we place longer-executing tasks first.

```
   Input: N, N+∞, time_limit
 1 Q.initialize();
 2 while elapsed time < time_limit do
 3     s = Q.reconstruct(), δ₁ = δ(s);
 4     if δ₁ = +∞ then  Q' = N+∞(Q)  else  Q' = N(Q) ;
 5     s' = Q'.reconstruct(), δ₂ = δ(s');
 6     if δ₂ ≤ δ₁ then
 7         Q = Q';
 8         if δ(Q)' = 0 then
             | Output: s
 9         end
10     end
11 end
   Output: s
```

Algorithm 1: Local search heuristic

4.1   Algorithm Overview

The proposed local search heuristic is presented in Algorithm 1. It is parameterized by the choice of two neighbor functions, $\mathcal{N}$ and $\mathcal{N}_{+\infty}$, and the running time *time_limit*. We use function $\mathcal{N}$ when the current schedule is feasible (i.e., the reconstruction managed to find a non-collision schedule for a queue in the corresponding iteration), whereas we use $\mathcal{N}_{+\infty}$ when the schedule is not feasible. The parameter *time_limit* can be chosen based on the time that is acceptable by the system designer. The algorithm starts by generating an initial queue on Line 1. Then, the heuristic repeats the following procedure for a fixed number of iterations (finishing early if $\delta(Q) = 0$ (Line 8)). First, the schedule is reconstructed, and its degeneracy is computed (Line 3). If the reconstructed schedule is infeasible or the reconstruction fails, we set $\delta(Q) = +\infty$. In this case, as mentioned earlier, a neighbor is generated by calling $\mathcal{N}_{+\infty}(Q)$. Otherwise, it is done using $\mathcal{N}(Q)$. If the new schedule is not worse than the old one (Line 6), the old schedule is swapped with the new one (Line 7), and the next iteration continues working with this updated schedule. The idea behind using non-strict inequality on Line 6 is that we want to explore as much solution space as possible and not get stuck in a local optimum.

4.2   Experimental Results

Experimental Setup. We randomly generated 7 sets of problem instances differing in the following parameters presented in Table 2: minimum and maximum number of tasks in each chain, $l(C)_{\min}$ and $l(C)_{\max}$, respectively, and the upper limit on the utilization of each machine, $\sigma_{\max} = \max_{\mu \in \mathcal{M}} \sigma_\mu$. The utilization is defined as $\sigma_\mu = \sum_{\tau:\, m(\tau)=\mu} \frac{p(\tau)}{T(\tau)}$, which is the fraction of time this machine is occupied. In Sets 1–6, the utilization of 95% of the machines lies in the interval $[\sigma_{\max} - 0.2, \sigma_{\max}]$. In Set 7, the utilization of more than 90% of the

122   R. Hladík et al.

Table 2. Parameters of the generated sets

Name	σ_{max}	$l(C)_{min}$	$l(C)_{max}$
Set 1	0.8	5	15
Set 2	0.9	2	5
($\star$) Set 3	0.9	5	15
Set 4	0.9	15	25
Set 5	0.93	5	15
Set 6	0.96	5	15
Set 7	1	5	15

Table 3. Results on Set 3 with various functions $\mathcal{N}$

Neighbor function	average degeneracy
chain_sort	1.49
chain_sort_patient	1.49
combined_local	0.13
combined_random	1.93
switch	0.12

Table 4. Results of the heuristic with $\mathcal{N} = \texttt{combined\_local}$ on different sets

Set	% of solved instances	Average degeneracy
Set 1, $\sigma_{max} = 0.8$, medium chains	99.56	0
Set 5, $\sigma_{max} = 0.93$, medium chains	97.98	0.49
Set 6, $\sigma_{max} = 0.96$, medium chains	96.65	1.34
Set 7, $\sigma_{max} = 1$, medium chains	54.50	58.52
Set 2, $\sigma_{max} = 0.9$, short chains	98.56	0.01
Set 3, $\sigma_{max} = 0.9$, medium chains	98.98	0.13
Set 4, $\sigma_{max} = 0.9$, long chains	98.90	0.96

machines equals 1, and the utilization of the rest of the machines lies in the interval $[0.96, 1)$. The number of tasks in all sets varies from 100 to 9 000, with more than half of the instances having more than 1 000 tasks in each set. Each of these seven test sets consists of 4 groups of 200 problem instances, with the following parameters ($\mathcal{P}$ is the set of periods, $|\mathcal{M}|$ is the number of machines): 1. $\mathcal{P} = \{2^0, 2^1, \ldots, 2^{10}\}$, $|\mathcal{M}| = 5$, 2. $\mathcal{P} = \{2^0, 2^1, \ldots, 2^{10}\}$, $|\mathcal{M}| = 10$, 3. $\mathcal{P} = \{2, 10, 20, 100, 200, 1000, 2000, 4000\}$, $|\mathcal{M}| = 5$, and 4. $\mathcal{P} = \{8, 16, 64, 256, 1024, 2048\}$, $|\mathcal{M}| = 5$. The generation procedure ensures that each of the 5 400 generated problem instances has a 0-degenerated schedule.

We ran the heuristic 6 times on each instance with $\mathcal{N}_{+\infty} = \texttt{chain\_uniform}$ with a different random seed. For each run of the heuristic, we set *time_limit* to 3 min. We choose the best of these runs as a result. Finally, we performed the experiments on a machine equipped with four Intel(R) Xeon(R) Silver 4110 CPUs, all clocked at 2.10 GHz, each having 8 cores and 16 threads. Similar results were achieved on a middle-end laptop.

Results. Table 3 presents the average degeneracy of the heuristic algorithm with the most promising neighbor functions on problem instances of Set 3, which we consider moderately difficult. The percentage of problem instances for which the heuristic found a feasible solution is 98.98. The degeneracy of the heuristic with all neighbor functions is relatively small, however `combined_local` and `switch` show the best results with statistically insignificant difference. Since the former is conceptually easier, we use it in the second experiment.

Table 4 shows the percentage of problem instances for which a feasible solution was found and the average degeneracy for each test set. Note that Sets 1, 3, 5, 6, and 7 have equal parameters except for σ_{max}. As expected, increased machine utilization leads to larger degeneracy with the significant difference for Set 7 with $\sigma_{max} = 1$. On the other hand, Sets 2, 3, and 4 differ in the chain length only. Instances with longer chains have larger average degeneracy, but the results do not suggest such a dramatic increase as in Set 7.

5 Conclusions and Future Work

This paper addresses the periodic scheduling problem PSP_{gen} with chains of precedence relations. In this problem, periodic tasks are scheduled in time on dedicated machines so that at any moment, at most one task is executed on each machine. We define a degeneracy of a chain as the number of broken precedence relations within the time window of one period. The problem is to find a schedule with the minimum degeneracy.

We prove that this problem is strongly NP-hard even when restricted to unit processing times, a single period, and 16 machines (called PSP_{com}), by a reduction from a variant of a job shop scheduling problem. In this reduction, by introducing auxiliary tasks, machines, and chains, we prove that the entire space of solutions of PSP_{com} is equivalent to the space of solutions respecting the job shop constraint on the total length of the schedule (missing in PSP_{com}). Furthermore, we present a local search heuristic algorithm that solves PSP_{gen}. We generated 5 400 problem instances allowing 0-degenerated solutions. On them, we experimentally demonstrate the soundness of our algorithm and show that it can successfully solve 92% of the instances with average resulting degeneracy of 8.78, each in a few minutes.

As future work, it would be interesting to explore the complexity of PSP_{com} on a smaller number of machines. Whereas the PSP_{com} with a common period on one machine is polynomially solvable (by placing one task after another of the first chain in precedence order, then the second chain, etc., in an arbitrary order), scheduling on two and more resources (up to 15) is an open question.

Acknowledgments. Research leading to these results has received funding from the EU ECSEL Joint Undertaking and the Ministry of Education of the Czech Republic under grant agreement 826452 (project Arrowhead Tools).

References

1. Bar-Noy, A., Bhatia, R., Naor, J.S., Schieber, B.: Minimizing service and operation costs of periodic scheduling. Math. Oper. Res. **27**(3), 518–544 (2002)

2. Błażewicz, J., Pesch, E., Sterna, M.: The disjunctive graph machine representation of the job shop scheduling problem. Eur. J. Oper. Res. **127**(2), 317–331 (2000)
3. Cai, Y., Kong, M.: Nonpreemptive scheduling of periodic tasks in uni-and multi-processor systems. Algorithmica **15**, 572–599 (1996)
4. Dvorak, J., Hanzalek, Z.: Multi-variant time constrained FlexRay static segment scheduling. In: 2014 10th IEEE Workshop on Factory Communication Systems (WFCS 2014), pp. 1–8 (2014)
5. Dvořák, J., Hanzálek, Z.: Multi-variant scheduling of critical time-triggered communication in incremental development process: application to flexray. IEEE Trans. Veh. Technol. **68**(1), 155–169 (2018)
6. Eisenbrand, F., Hähnle, N., Niemeier, M., Skutella, M., Verschae, J., Wiese, A.: Scheduling periodic tasks in a hard real-time environment. In: Abramsky, S., Gavoille, C., Kirchner, C., Meyer auf der Heide, F., Spirakis, P.G. (eds.) ICALP 2010. LNCS, vol. 6198, pp. 299–311. Springer, Heidelberg (2010). https://doi.org/10.1007/978-3-642-14165-2_26
7. Eisenbrand, F., et al.: Solving an avionics real-time scheduling problem by advanced IP-methods. In: de Berg, M., Meyer, U. (eds.) ESA 2010. LNCS, vol. 6346, pp. 11–22. Springer, Heidelberg (2010). https://doi.org/10.1007/978-3-642-15775-2_2
8. Graham, R.L., Lawler, E.L., Lenstra, J.K., Kan, A.R.: Optimization and approximation in deterministic sequencing and scheduling: a survey. Ann. Discret. Math. **5**, 287–326 (1979)
9. Jacobs, T., Longo, S.: A new perspective on the windows scheduling problem. arXiv preprint arXiv:1410.7237 (2014)
10. Jeffay, K., Stanat, D.F., Martel, C.U.: On non-preemptive scheduling of periodic and sporadic tasks. In: IEEE Real-time Systems Symposium, pp. 129–139. IEEE, USA (1991)
11. Korst, J., Aarts, E., Lenstra, J.K., Wessels, J.: Periodic multiprocessor scheduling. In: Aarts, E.H.L., van Leeuwen, J., Rem, M. (eds.) Parle 1991 Parallel Architectures and Languages Europe. LNCS, pp. 166–178. Springer, Heidelberg (1991). https://doi.org/10.1007/978-3-662-25209-3_12
12. Korst, J., Aarts, E., Lenstra, J.K.: Scheduling periodic tasks. INFORMS J. Comput. **8**(4), 428–435 (1996)
13. Lenstra, J.K., Kan, A.R.: Computational complexity of discrete optimization problems. In: Annals of Discrete Mathematics, vol. 4, pp. 121–140. Elsevier (1979)
14. Nawrocki, J.R., Czajka, A., Complak, W.: Scheduling cyclic tasks with binary periods. Inf. Process. Lett. **65**(4), 173–178 (1998)
15. Ogata, K.: Discrete-Time Control Systems, vol. 2. Prentice Hall, Englewood Cliffs (1995)
16. Oliver, R.S., Craciunas, S.S., Steiner, W.: IEEE 802.1 Qbv gate control list synthesis using array theory encoding. In: 2018 IEEE Real-Time and Embedded Technology and Applications Symposium (RTAS), pp. 13–24. IEEE (2018)
17. Wei, W., Liu, C.: On a periodic maintenance problem. Oper. Res. Lett. **2**(2), 90–93 (1983)

Energy-Constrained Drone Delivery Scheduling

Rafael Papa, Ionut Cardei, and Mihaela Cardei$^{(\boxtimes)}$

Department of Computer and Electrical Engineering and Computer Science,
Florida Atlantic University, Boca Raton, FL 33431, USA
{rpapa2013,icardei,mcardei}@fau.edu

Abstract. In recent years drones have been used in many applications such as surveillance, geographic mapping, search and rescue, and weather forecast. Motivated by the increased use of drones in shipping and delivery, in this paper we tackle the problem of parcel delivery by taking into consideration important aspects such as serving on-demand requests, flight duration limitation due to energy constraints, maintaining the safety distance to avoid collisions, and using warehouses as starting and ending points in parcel delivery. In this paper we define the UAS Energy-constrained Delivery Scheduling problem and propose a scheduling mechanism using a multi-source A* algorithm variant. Simulation results show that our algorithm is efficient and scalable with the number of requests and network size.

Keywords: UAS scheduling · Traffic management system · Energy constraints · Multi-source A* · Graph search

1 Introduction

With the recent advances in drone technologies, a new delivery method is emerging as new trend among the e-commerce companies, *the drone delivery*. This method uses a vehicle that can be operated without the cost of a human pilot and provides a fast and low cost alternative to the regular truck delivery. Amazon and few other high-profile drone delivery companies such as Wing, UPS Flight Forward, and Zipline have presented their solutions recently [4]. This exciting market is growing, but it is also facing challenges.

Flight restrictions and flight duration can make it difficult to use drones in a large operation area. For instance, the Federal Aviation Administration (FAA) guidelines prevent drones to fly over (or near) airports, stadiums and sporting events, Washington-DC territory or special use airspace [5]. In addition to that, a commercial drone can fly only for about half an hour in good conditions (i.e. no wind or rain) and it is able to carry only one small parcel per time.

There are several research works that tackle the problem of delivery task[1] using drones. Some of them are using a combination of trucks and drones. We

---

[1] The action of delivering goods requested by clients within a limited time is called delivery task.

© Springer Nature Switzerland AG 2020
W. Wu and Z. Zhang (Eds.): COCOA 2020, LNCS 12577, pp. 125–139, 2020.
https://doi.org/10.1007/978-3-030-64843-5_9

observed few limitations and drawbacks in the related works. The solutions are not scalable with the number of request and network size. Many formulations present results for a small number of clients. Another issue is that the models assume that all requests are known upfront. This is not the case in practical applications, where requests can arrive at any time. Another limitation is to assume that drones can fly to the destination along the euclidean distance. As mentioned previously, there are no-fly areas such as airports that must be taken into account.

In this paper we address these limitations and focus on drone delivery task without using trucks. 86% of all orders shipped by Amazon weight less than 5 pounds and can potentially be delivered in small packages by drones. Our objective is to increase the drone flight time by allowing drones to land on charging stations to replace batteries along the delivery path. The main contribution of this paper is a UAS path scheduler that computes safe collision-free trajectories constrained by vehicle battery capacity and application deadline demands. Our algorithm is a variant of A* with multiple sources and branching factor of 1. Its running time complexity is polynomial and good enough to scale to metropolitan size networks with tens of thousands of vertices and edges supporting admission rates of 4 requests per second (i.e. 8000 requests in 2000 s).

The rest of the paper is organized as follows. Related works are presented in Sect. 2. Section 3 discusses the motivation and introduces the UAS-ECDS problem definition. In Sect. 4 we propose a scheduling mechanism using a variant of multi-source A* algorithm. The performance of our algorithm is illustrated in Sect. 5. The conclusions are stated in Sect. 6.

2   Related Works

Paper [9] proposes a hybrid truck-drone delivery system for the last-mile delivery in logistics operations. In the proposed Flying Sidekick Traveling Salesman Problem (FSTSP), a single drone carried by the truck is launched to make deliveries to few selected customers and then joins the truck while on route. The objective is to find a route that minimizes the time to complete the delivery task using both vehicles. The paper solves FSTSP using mixed integer linear programming (MILP) and uses a heuristic to calculate an upper bound on the time required to visit all customers and return to the depot. The FSTSP is NP-hard and the proposed MILP took several hours to complete for 10 clients.

There are few other formulations for TSP with drone (TSP-D) problem in literature. Article [1] presents an integer programming (IP) formulation and several fast route first-cluster second heuristics based on local search and dynamic programming. It took about 2 h to achieve a near-optimal solution for 12 customers. The recent work [2] uses the Bellman-Held-Karp dynamic programming algorithm for the TSP-D problem. This method was able to find the optimal solution much faster for an input with 10 vertices. The authors were able to solve problems with a higher complexity with a single drone and simple operations.

Few recent works address the same problem using multiple drones and multiple trucks. The recent paper [10] proposes the Multiple Flying Sidekicks Traveling Salesman Problem (mFSTSP) extending the previous FSTSP to multiple drones. A new three-phase heuristic solution approach for the MILP formulation was proposed. In Phase I the customers are divided into two sets (served by truck and served by UAVs), in Phase II the routes are computed and in the Phase III MILP is used to determine the exact time of launch, recovery and service activities for the truck and the UAVs. About 66% of the problems were optimally solved within the time limit of 1 hour for scenarios with 8 customers.

Article [8] solves the Multiple Traveling Salesman Problem with Drones (mTSPD) by extending the previous TSP-D to multiple drones and multiple trucks. Authors propose a MIP formulation to minimize the trucks' and drones' arrival time at the depot after delivery. The MIP model was used to solve scenarios with 10 customers. For larger problem sizes, they use a heuristic approach based on the greedy node(s)-insertion strategy. All truck-drone delivery paths are computed along Euclidean distances.

The hybrid truck-drone delivery systems must account for the drone energy consumption. High speeds, different parcels weight, successive battery replacements and frequent recharging may limit the use of drones in such applications. For example, a faster speed or heavier weight parcel will demand a larger power consumption by the device. Using charging stations is a promising strategy for drone delivery. Paper [7] extended the FSTSP model by taking into consideration the parcel weight and the no-fly zone areas for energy consumption purposes.

Article [6] proposes a model where drones can charge to increase the flight time. The work proposes a scheduling algorithm based on game theory where drones compete for charging slots on limited charging stations along the path. Using these charging capabilities enable drones to fly longer (i.e. for hours) for surveillance, delivery tasks and rescue operations. The proposed drone charging model is based on Hedera Hashgraph, an asynchronous Byzantine Fault Tolerance (ABFT) consensus algorithm capable of securing the platform against attacks. This algorithm is used to compute the consensus time-stamp for scheduling the charging requests. The game-theoretic approach is used to allocate drones to the charging slots in a cost-optimal manner.

3   Motivation and Problem Definition

Many applications are using drone technology recently such as surveillance, aerial photography, shipping and delivery, geographic mapping, search and rescue, and weather forecast. In this paper we tackle the problem of parcel delivery by taking into consideration important aspects such as on-demand (or real-time) requests, limitation of drone flight duration due to energy constraints, maintaining a safety distance to avoid collisions, and using warehouses as starting and ending points in parcel delivery. We assume that an aerial highway system is predefined and it is represented by a directed graph where each edge represents an aerial lane. The set of vertices include intersections, client locations, warehouse locations, and charging station locations.

Next, we define the **UAS Energy-constrained Delivery Scheduling (UAS-ECDS)** problem: We are given an aerial highway system modeled as a directed graph $G(V, E)$. The set of vertices $C = \{c_1, ..., c_k\}$ is the set of charging stations, $C \subset V$. Client requests can arrive at any time. A request is characterized by the following fields: client location (a vertex in V), the set of start warehouses $W_s \subset V$, the set of final warehouses $W_f \subset V$, where $W_s \subseteq W_f$, the request start time, the deadline to reach the client, and the deadline to reach a final warehouse. The flight time is limited by the battery capacity. The objective is to design an efficient, on-demand UAS delivery scheduling which accounts for the battery constraints and is collision-free. We assume that all UAS's have the same maximum speed v_{max} and they must be separated by a distance greater than or equal to the safety distance d when in flight, in order to be collision-free.

4   UAS Energy-Constrained Scheduling Algorithm

In this section we describe our solution for the UAS-ECDS problem. We make the following assumptions:

- UAS's move along the edges of the aerial highway traffic system $G^{init}(V, E^{init})$. This is a weighted graph, where the weight of an edge $dist(u, v)$ is the Euclidean distance from u to v.
- the minimum speed of each UAS is $0\,\mathrm{m/s}$, such as for VTOL (vertical takeoff and landing) UAS. More specifically we assume that at each time instance, the speed of each UAS is between $0\,\mathrm{m/s}$ and v_{max}.
- we assume that the UAS can charge at most once on the path from the start warehouse (a vertex in W_s) to the client, and at most once on the path from the client to the final warehouse (a vertex in W_f). A delivery path looks as follows: start warehouse $\rightarrow$ charging station (optional) $\rightarrow$ client $\rightarrow$ charging station (optional) $\rightarrow$ final warehouse.

Our solution has the following steps: (i) clients send requests to a Base Station (BS) or traffic management system for drones, (ii) based on the directed graph G and the trajectory of the UAS's already in transit, the BS computes the trajectory of the UAS serving the new request, and (iii) if the delivery path can be computed such that to meet the deadline for reaching the client and the deadline for reaching a final warehouse, then a success message is sent to the client and a message with the computed trajectory is sent to the corresponding start warehouse. If the request cannot be satisfied, then an error message is sent to the client.

For a *collision-free system*, the BS has to ensure that the distance between moving UAS's is at least the safety distance d when scheduling UAS trajectories. To model the safety distance requirement, we use the mechanism from [12] to transform the graph G^{init} to a *unit-graph* $G_u^{init} = (V_u, E_u^{init})$ where the weight of each edge is d. Similar to [12], we transform the graph to a unit-graph and use a time-space graph to avoid collisions. We divide each edge into segments of length d and a shorter segment with the remainder.

Each UAS takes time δ to traverse an edge (u, v) in the unit-graph, where $\delta = d/v_{max}$. The time is discretized with 1 time unit equal to δ. Let us assume that uas_j is in transit, and it is located at the vertex u at time t. The following options are possible during the next time-interval $[t, t+1)$: (i) uas_j pauses (e.g. hovers), thus uas_j is in u at time $t+1$, (ii) uas_j traverses the edge (u, v), thus uas_j is in v at time $t+1$, and (iii) uas_j is charging at the charging station u. Based on our modeling of the problem, a UAS traverses an edge of the unit-graph in 1 time unit.

To model the charging of a UAS, we assume that charging takes α time units, where α is a configuration parameter. Charging takes a relatively small time if the only operation performed is to replace the battery.

Another important aspect in computing the schedule of an UAS is designing a rule to determine which unit-edges are available at a certain time t, that means during the time interval $[t, t+1)$. Note that based on the way that we model conflict prevention, an edge (u, v) can be traversed by a single UAS during a time interval $[t, t+1)$. In addition, we need to ensure that the safety distance d is maintained between any two UAS's in transit.

To model the availability of edges in time, we define a *space-time* graph denoted $G_u(t) = (V_u, E_u(t))$. The set of edges $E_u(t)$ represents the edges which are available during the time interval $[t, t+1)$. Initially, $E_u(t) = E_u^{init}$ for each $t = 0, 1, 2, ...$ but when UAS trajectories are scheduled, the set of edges is pruned accordingly.

We use the **PRUNE-RULE** algorithm from [12] to prune the available set of edges. If a UAS is scheduled to use the edge (u, v) at time t (i.e. during the time interval $[t, t+1)$), then $E_u(t)$ is pruned as follows:

1. delete (u, v) from $E_u(t)$
2. delete edges (a, b) with distance$\{(u, v), (a, b)\} < d$ from $E_u(t)$, with the following exceptions:
   (a) an edge (a, u) with vertices a, u, and v collinear and $dist(a, u) = dist(u, v)$ $= d$ is not deleted from $E_u(t)$
   (b) an edge (v, b) with vertices u, v, and b collinear and $dist(u, v) = dist(v, b)$ $= d$ is not deleted from $E_u(t)$.
   (c) the reverse edge (v, u) is not deleted from $E_u(t)$.

The same edge (u, v) cannot be used by more than one UAS in the same time interval. If an adjacent edge lies on the same straight line as (u, v) and both of these edges have distance d, then they can be active at the same time, therefore the adjacent edge will not be removed in this case. Lastly, edges (a, b) which are located at a distance smaller than d to the edge (u, v) are deleted since they may conflict with maintaining the safety distance. The reverse direction edges are not removed.

There are few ways in which we can model the battery-constraint uas delivery. Since in our framework uas travel along unit-edges and each uas takes δ time to traverse an edge, we assume that a fully charged battery lasts $H \times \delta$ time, thus the uas can traverse at most H hops before recharging. An uas can replenish its battery at a charging station and we assume that the charging takes α time-units.

A request has the following fields: $req = (clientVertex, W_s, W_f, C, t_s,$ $maxTimeClient, maxTimeWf)$, where $clientVertex$ is the client location, W_s is the set of start warehouses, W_f is the set of final warehouses, C is the set of charging stations, t_s is the request start time, $maxTimeClient$ is the deadline to reach the client, and $maxTimeWf$ is the deadline to reach a final warehouse. Our proposed algorithm COMPUTE-PATH is a multi-source A* algorithm variant with the following specific features: (I) there are multiple sources (or roots) in the search tree, and (II) the branching factor is one.

An **Initialization phase** is performed by the BS before processing client requests. The Breath-First-Search (BFS) algorithm [3] runs on G_u^{init} starting from each charging station vertex as a source vertex and it runs only for distances less than or equal to H. Each vertex in the graph stores the cost (i.e. the number of edges) and the predecessor node for each charging station.

Next we describe our algorithm COMPUTE-PATH which is used to process a client request. The input parameters are: (i) $G_u[t_s..maxTimeWf]$ which is the unit space-time graph from t_s to $maxTimeWf$, and (ii) the client request req. We start in line 1 by running BFS [3] from the client vertex, with a span (or radius) of at most $2H$ edges. Each node in the graph (within distance $2H$) has the shortest path (number of edges) to the client vertex.

In presenting our algorithm r, p, and q are nodes in the search tree, while u, v, and w are nodes in the unit space-time graph G_u. A node p in the search tree has the following fields:

- *level* - the level in the search tree represents the time when the uas has reached the node $p.crtVertex$.
- *crtVertex* - the current vertex in the unit space-time graph G_u.
- *state* - takes one of the values {ST_TO_CH1, ST_TO_CLIENT, ST_TO_CH2, ST_TO_WF} depending on the next objective vertex on the path.
- *timeAtClient* - initially has the value infinity. It records the time when the uas reaches the client and it remains unchanged thereafter.
- *parent* - stores the parent node in the search tree.
- *score* - the priority for the minimum priority queue PQ. The score is a tuple $(score_1, score_2)$ where $score_1$ is 0 if the uas is traveling between the client and the final warehouse and 1 if the uas is traveling between the the start warehouse and the client. If $score_1$ equals 0 then $score_2$ is the expected time to reach the final warehouse. If $score_1$ equals 1 then $score_2$ is the expected time to reach the client.
- *batteryCharging* - boolean value used to indicate whether the uas is charging or not in the current search tree node.
- *batteryLevel* - takes a value between 0 and H. After the battery is charged, the value is reset to H. The value is decremented for each new time unit.
- *ch1* - stores the charging station on the path from the start warehouse to the client, if any. Otherwise it is NIL.
- *ch2* - stores the charging station on the path from the client to the final warehouse, if any. Otherwise it is NIL.
- *wf* - stores the final warehouse vertex.

ALGORITHM 1: COMPUTE-PATH (G_u[req.t_s .. req.maxTimeWf], req)

```
 1:  BFS($G_u$, req.clientVertex)
 2:  PQ = ∅
 3:  for each warehouse w ∈ $W_s$ do
 4:      r = INITIALIZE-ROOT($G_u$, req, w)
 5:      if r ≠ NIL then insert(PQ, r)
 6:  end for
 7:  while PQ ≠ ∅ do
 8:      p = remove(PQ) // dequeue based on the score field
 9:      v = NEXT-NODE($G_u$, p)
10:      create a new node q in the search tree
11:      INITIALIZE(p, q) /* initializes the fields of q with default values*/
12:      if p.batteryCharging == true then q.level = p.level + 1 + α
13:      /* if the edge is available then the path advances, otherwise the uas waits */
14:      if (p.crtVertex, v) ∈ $E_u$(q.level-1) then q.crtVertex = v
15:      if (p.state == ST_TO_CH1) AND (q.crtVertex == q.ch1) then
16:          q.batteryCharging = true
17:          q.batteryLevel = H  //reset after battery charging
18:          q.state = ST_TO_CLIENT
19:      end if
20:      if (p.state == ST_TO_CLIENT)AND(q.crtVertex == req.clientVertex) then
21:          q.timeAtClient = q.level
22:          if q.ch2 ≠ NIL then q.state = ST_TO_CH2 else q.state = ST_TO_WF
23:      end if
24:      if (p.state == ST_TO_CH2) AND (q.crtVertex == q.ch2) then
25:          q.batteryCharging = true
26:          q.batteryLevel = H  //reset after battery charging
27:          q.state = ST_TO_WF
28:      end if
29:      if (p.state == ST_TO_WF) AND (q.crtVertex == q.wf) AND FEASIBLE-NODE(q) then
30:          /* feasible solution found */
31:          PRUNE-GRAPH($G_u$[$t_s$..req.maxTimeWf], q)
32:          return ASSIGN-PATH(q)
33:      end if
34:      if (q.state == ST_TO_CH1) OR (q.state == ST_TO_CLIENT) then
35:          q.score = (1, q.level + cost to client)
36:      else
37:          /* state is ST_TO_CH2 or ST_TO_WF */
38:          q.score = (0, q.level + cost to wf)
39:      end if
40:      if FEASIBLE-NODE(q) then insert(PQ, q)
41:  end while
```

We are using a minimum priority queue PQ where the priority is decided by the field *score*. There are at most $|W_s|$ source nodes (or root nodes) in the search tree that are added to the priority queue PQ, see lines 3 to 6. The root node initialization is done in the procedure INITIALIZE-ROOT. We explain next the lines 8–15 of this procedure. For each warehouse w, we are seeking to compute a shortest path to a final warehouse, in the following order:

– **Group1.** Paths are in the form: w → client → wf, where wf ∈ W_f. Paths in this group do not contain charging stations. The shortest subpath w → client is obtained directly from the BFS tree. The least-cost subpath client → wf is selected among all final warehouses wf in the BFS tree rooted in the client. A path is feasible if: (i) the cost from w to wf is at most the battery level

---

ALGORITHM 2: INITIALIZE-ROOT(G_u, req, w)

1: create the root node r of a new search tree
2: r.parent = NIL
3: r.crtVertex = w
4: r.level = req.t_s // the field *level* stores the time
5: r.batteryCharging = false
6: r.batteryLevel = H
7: r.timeAtClient = ∞ // time when customer is reached
8: **if** feasible path from w to the client to a final warehouse exists **then**
9:    select the best path
10:   set r.state, r.$ch1$, r.$ch2$, $r.wf$ based on the path
11:   r.score = (1, r.level + cost to client)
12:   return r
13: **else**
14:   return NIL
15: **end if**

ALGORITHM 3: INITIALIZE(p, q)

1: /* initializes the fields of the node q*/
2: q.parent = p
3: q.level = p.level + 1
4: q.crtVertex = p.crtVertex
5: q.batteryLevel = p.batteryLevel - 1
6: q.batteryCharging = false
7: q.timeAtClient = p.timeAtClient
8: q.state = p.state
9: q.ch1 = p.ch1; q.ch2 = p.ch2; q.wf = p.wf

H, (ii) t_s + cost from w to the client $\leq$ maxTimeClient, and (iii) t_s + cost from w to wf $\leq$ maxTimeWf. Once the path has been established we set up the search node: r.state = ST_TO_CLIENT; r.$ch1$ = r.$ch2$ = NIL; r.wf is set up to the final warehouse; r.score = (1, cost from w to the client). If there are no feasible paths in this group, then move to Group2.

- **Group2.** Paths are in the form: w $\rightarrow$ client $\rightarrow$ ch2 $\rightarrow$ wf, where ch2 is a charging station. The subpath w $\rightarrow$ client is available from the client's BFS tree. The client $\rightarrow$ ch2 $\rightarrow$ wf subpath is selected as the least-cost among all chargers ch2 in the client's BFS tree and warehouses wf in ch2's BFS tree. A path is feasible if (i) the cost from w to ch2 is at most H, (ii) the cost from ch2 to wf is at most H, (iii) the time to reach the client does not exceed maxTimeClient, and (iv) the time to reach wf does not exceed maxTimeWf. If a path has been selected, set up the fields r.state = ST_TO_CLIENT; r.$ch1$ = NIL; r.$ch2$ is set to the charging station; r.wf is set to the final warehouse; r.score = (1, cost from w to the client). If there are no feasible paths in this group, then move to Group3.

- **Group3.** Paths are in the form: w → ch1 → client → wf, where ch1 is a charging station in the client's BFS tree. The shortest subpath w → ch1 → client is found by checking all charging stations and selecting the least cost one for which w is in ch1's BFS tree. The shortest path client → wf is directly available from the client's BFS tree. A path is feasible if the battery and path length constraints are met similar to the previous cases. If a path has been selected, set-up the fields r.state = ST_TO_CH1; r.ch1 is set up to the charging station; r.ch2 = NIL; r.Wf is set up to the final warehouse; r.score = (1, cost from w to the client). If there are no feasible paths in this group, then move to Group4.
- **Group4.** Paths are in the form: w → ch1 → client → ch2 → wf. The process is similar to the prior groups. We check that the battery constraint is met from w to ch1, from ch1 to ch2, and from ch2 to wf. The fields of r are initialized similarly to Group 3, except r.ch2 is set up to the corresponding ch2. If there are no feasible paths in this group, then return NIL (line 14).

Lines 7–41 in the COMPUTE-PATH algorithm extract one node in each iteration, based on the score field, until a path is found. The path of the node dequeued is examined. Note that the paths follow the precomputed shortest path, but in each iteration a path is either advanced or the uas has to pause if the edge is not available (e.g. pruned edge). We note that the branching factor of our algorithm is 1, and this is our strategy to limit the algorithm complexity and achieve scalability with the number of requests. A branching factor larger than 1 results in an exponential complexity to process each request.

Let p be the node dequeued in line 8. Vertex v (line 9) is the next vertex on the shortest path precomputed in the INITIALIZE-ROOT procedure. In lines 10–11, we create a new node q which is the descendant of p in the search tree, and initialize the fields with default values. Line 12 sets-up the *level* for the case when p is a charging station. In line 14, the crtVertex is set to v if the edge is available, otherwise if the edge is pruned then the uas pauses a time interval.

Lines 15–19 address the case when the uas reaches the charging station before the client. In this case the battery related fields are updated and the status changes to ST_TO_CLIENT, indicating that the uas is now traveling toward the client vertex. Lines 20–23 consider the case when the uas is reaching the client and lines 24–28 address the case when the uas has reached the charging station after the client was visited. Lines 29–33 show the case when the uas has reached a final warehouse. We check that the node is feasible: the uas has enough energy to reach the final warehouse vertex and the time (or q.level) does not exceed $maxTimeWf$. If the node is feasible, then a feasible solution was found. In this case the graph is pruned using the PRUNE-RULE described previously and the path is returned. ASSIGN-PATH starts from the node q and follows the *parent* field up to the root node. The path is given by the *crtVertex* field printed in reverse order.

Lines 34–39 compute the score of the node q. The score is a tuple. Since the paths that have already reached the client have higher priority (i.e. chances to be completed sooner), the first element of the tuple is 0, while the paths that have not reached the client get assign the value 1. The second element of the tuple is the expected time to reach the final warehouse or the expected time to reach the client, respectively. A shorter time is preferred thus it has a smaller cost and therefore a higher priority.

After all the fields have been set-up, line 40 checks whether the node q is feasible. For this we check: (i) energy resources, (ii) the time to reach the client does not exceed $maxTimeClient$, and (iii) the time to reach the final warehouse does not exceed $maxTimeWf$. If the node q is feasible, then it is inserted in the minimum priority queue PQ.

For the complexity analysis, $|C|$ is the number of charging stations and $|W|$ = $|Wf|$ is the number of warehouses with shortest path (number of edges) at most $2H$ from the client. Let $T = req.maxTimeWf - req.t_s$ is the maximum flight time. The unit graph has E_u edges and V_u vertices, and since the graph is connected $V_u = O(E_u)$. The Initialization phase takes $O(C(V_u + E_u)) = O(CE_u)$ and this is computed only once, before requests are processed.

Line 1 of COMPUTE-PATH runs the BFS algorithm with complexity $O(E_u)$. We note that the algorithm is run only on a subgraph of the unit graph, for vertices with cost at most $2H$ from the client. Lines 3–6 take $O(C^2W^2)$ mainly due to the INITIALIZE-ROOT procedure which has to considers all feasible paths traversing the sets of charging stations and warehouses.

The *while* loop in line 7 has at most $O(TW)$ iterations. The priority queue insert and remove operations take $O(lgW)$ time. If a solution is found, then the graph is pruned in $O(TE_u)$ and the path is assigned in $O(T)$. Lines 7–41 take $O(TWlgW + T + TE_u)$. The complexity to compute the path for a request is polynomial and upper-bounded asymptotically by $O(E_u + C^2W^2 + TWlgW + T + TE_u)$ which is simplified to $O(C^2W^2 + TWlgW + TE_u)$.

5   Performance Evaluation

We analyzed the performance of our algorithm with simulation programs written in Python 3.8. We ran simulations on a Linux PC with 24 GB RAM and an Intel Core i5 4-core CPU running at a sustained clock frequency of 2 GHz. Results for a particular simulation scenario configuration were averaged over 10 random iterations and we reported the average.

The simulation scenarios use a graph (see Fig. 1) extracted from the Open-StreetMap [11] Miami road map that had tertiary roads removed. The map covers downtown Miami (3.6 km × 2.5 km) and the initial graph has 402 vertices and 1176 edges. The corresponding discretized unit graph has 1176 vertices and 14264 edges for conflict distance $d = 10$ m. Table 1 shows the simulation parameters.

The independent variables considered for experiments are: the request rate (incoming requests/s), the safety distance d, the number of charging stations, the number of start (W_s) and final (W_f) warehouses. All scenarios used the same UAS maximum speed $v_{max} = 10$ m/s. The performance metrics are the algorithm runtime, the trajectory duration from request start time to when the

Table 1. Simulation parameters

Miami map	*Approx* 3.6 km × 2.5 km
Number of UAS	100, 200, 1000, 2000, 3000, 4000, 6000, 8000
Request rate	0.05/s, 0.1/s, 0.5/s, 1/s, 1.5/s, 2/s, 3/s, 4/s
Safety distance d	5 m, 10 m, 20 m
δ	0.5 s, 1 s, 2 s
Max flight time	250 s, 280 s, 300 s
Max route time to client	520 s, 600 s, 620 s
Max time back at warehouse	1000 s
Total simulation time	3000 s

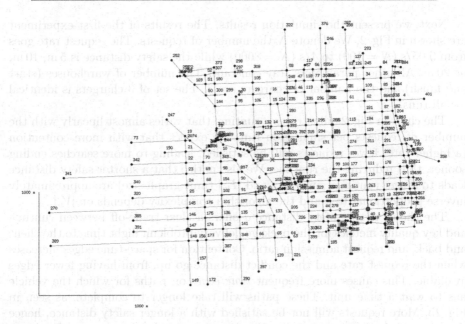

Fig. 1. The Miami graph with a sample route. The green circle shows the start warehouse, the red X marks the client, the red circle is the finish warehouse and the light blue circle is a charging station. (Color figure online)

client is reached, the total path delay, and the request satisfaction ratio. The charger locations for all runs in particular scenario are selected from unit graph vertices closest located to the k-means of unit graph vertices in order to have a balanced placement on the map.

The graph pruning algorithm employs a grid partitioning scheme and a caching mechanism [12] for segment-to-segment distances that achieves a 99.99% hit ratio in our simulations. This high ratio is caused by vehicles moving mostly on BFS paths to chargers and warehouses, which tend to stay the same for many scenarios.

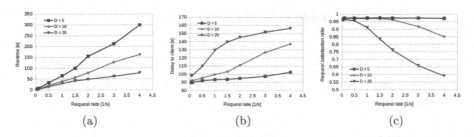

Fig. 2. Results when the request rate varies from 0.05/s ($N = 100$) to 5/s ($N = 8000$) and the safety distance is 5 m, 10 m, or 20 m. (a) Running time. (b) Path delay to reach the client. (c) Request satisfaction ratio.

Next, we present the simulation results. The results of the first experiment are shown in Fig. 2. We denote N the number of requests. The request rate goes from 0.05/s ($N = 100$) to 4/s ($N = 8000$) while the safety distance is 5 m, 10 m, or 20 m. All other parameters stay constant. The number of warehouses (start and finish) and the number of chargers are 3. The set of 3 chargers is identical for all runs.

The chart in Fig. 2a illustrates a runtime that varies almost linearly with the number of requests per simulation run. We expect that with more contention (a higher rate) fewer requests will be satisfied, leading to more searches ending sooner, thus reducing the runtime. We also notice that a shorter safety distance leads to longer runtimes since $|V_u|$ and $|E_u|$ (unit graph size) are approximately inverse proportional to d and the algorithm complexity depends on $|V_u|$.

The choice of the safety distance d offers a clear trade-off between runtime and key quality metrics for the package delivery problem: flight time to the client and back, and request admission ratio. Contention for space-time edges increases when the request rate and the conflict distance go up, from having fewer edges available. This causes more frequent *time edges* on paths for which the vehicle has to wait a time unit. These paths will take longer to complete, as seen in Fig. 2b. More requests will not be satisfied with a longer safety distance, hence the request admission rate seen in Fig. 2c.

The admission ratio is also far more sensitive to higher contention (from a higher request rate) when the safety distance is shorter. When $d = 5$ m the admission ratio stays above 96.6% for all tested request rates. When $d = 10$ m the admission ratio drops below 90% at about 3.5 requests/s, while with $d = 20$ m the 90% threshold is hit just above 1 request/s. A study like this could be used in practice to determine the ideal conflict distance, guided by safety regulations, engineering constraints, and available computation power.

For the second experiment (Fig. 3) we keep the safety distance constant at 10 m and the request rate at 0.1 requests/s ($N = 200$), and we vary the number of warehouses $|W_f|$ from 1 to 10. The set of start warehouses W_s is either equal to the set of finish warehouses ($W_s = W_f$) or it has just one warehouse ($|W_s| = 1$) picked randomly from W_f. The set of chargers has 4 vertices selected once for

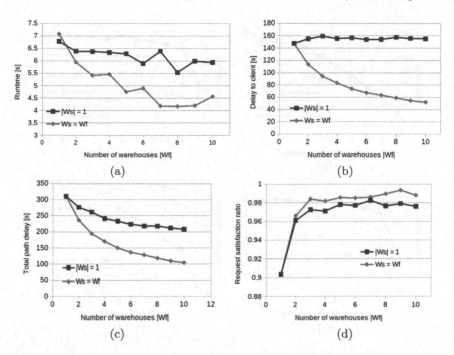

Fig. 3. Performance results when the number of warehouses $|W_f|$ varies from 1 to 10 and the number of start warehouses $|W_s| = 1$ and picked randomly from W_f or the two sets are the same, $W_s = W_f$. (a) Running time. (b) Path delay to reach the client. (c) Total path delay. (d) Request satisfaction ratio.

all scenarios, with vertices placed near the 4-means locations of all unit graph vertices.

Figure 3a shows the algorithm runtime. We notice the runtime dropping with more warehouses and its slope steeper for scenarios where $W_s = W_f$ compared with scenarios where there is only one starting warehouse. While the algorithm runtime for seeding the priority queue with search nodes depends on $|W_s|$ and $|W_f|$, higher $|W_s|$ and $|W_f|$ also increase the probability of finding shorter paths to the client (for $W_s = W_f$), and then, to a final warehouse, as evidenced in Fig. 3b and 3c, respectively. The reduction in client path and total path delays for going from one warehouse to two is about 30% for both, which could provide an economic benefit for more spread out warehouses instead of centralized package distribution. For the $|W_s| = 1$ case the client path delay stays fairly constant regardless of the number of final warehouses since it does not depend on it.

The third experiment looks at the impact of the number of charging stations. We selected a low battery capacity of only 250 s in order to stress the system into a lower admission ratio at the beginning. Indeed, Fig. 4d shows the admission ratio starting from 33% for $|W_s| = 1$ and 54% for $|W_s| = 2$ with no charging stations. With the addition of just one charger the admission ratio jumps to 57% for $|W_s| = 1$ and to 70% for $|W_s| = 2$. The proportional improvement in

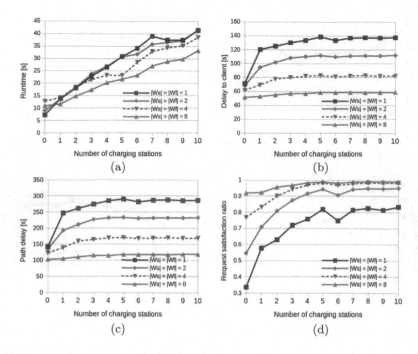

Fig. 4. Performance results when the number of charging stations varies from 0 to 10 and the number of warehouses varies in set $\{1, 2, 4, 8\}$ and $W_s = W_f$. (a) Running time. (b) Path delay to reach the client. (c) Total path delay. (d) Request satisfaction ratio.

admission ratio is lower for $|W_s| = |W_f| = 4$ or 8. The growth in admission ratio seems to stabilize with 5 chargers or more. The outlier case with 6 chargers that yields results similar to 4 chargers can be explained by the k-means algorithm generating an aberrant set of means where 3 of the 6 are too close to each other to make a sufficient difference in results.

The runtime chart (Fig. 4a) indicates a general almost-linear growth trend depending on the number of chargers. With more warehouses the paths are shorter from more choices becoming available. Thus, the runtimes will be shorter, too. The higher admission ratio for scenarios with more chargers implies that more paths will actually be computed and not rejected outright. Therefore, the runtime will grow with the number of chargers, with a growth slope depending on $|W_f|$.

The client and full path delay charts in Fig. 4b and 4c show growth with an increased number of chargers that may seem surprising. The considerable jump in delay from none to one charger for cases $|W_f| \in \{1, 2, 4\}$ mirrors the jump in admission ratio for those scenarios. Having one or more chargers increases the probability of finding a path, but at the cost of longer paths on average since they now include chargers. The reduction in delays in scenarios with more warehouses is consistent with the results from Fig. 3.

6   Conclusions

This paper proposes a scheduling algorithm for drone delivery that processes on-demand requests and accounts for energy constraints. We use a multi-source A* algorithm variant to compute drone scheduling. Simulation results show that our algorithm is efficient asymptotically and it is scalable with the number of requests and network size.

References

1. Agatz, N., Bouman, P., Schmidt, M.: Optimization approaches for the traveling salesman problem with drone. Transp. Sci. **52**, 1–17 (2018)
2. Bouman, P., Agatz, N., Schmidt, M.: Dynamic programming approaches for the traveling salesman problem with drone. Networks **72**, 528–542 (2018). https://doi.org/10.1002/net.21864
3. Cormen, T.H., Leiserson, C.E., Rivest, R.L., Stein, C.: Introduction to Algorithms, 3rd edn. The MIT Press, Cambridge (2009)
4. DRONEII: The Drone Delivery Market Map. https://dronelife.com/2019/11/07/droneii-the-drone-delivery-market-map/. Accessed August 2020
5. FAA - Federal Aviation Administration, Airspace Restrictions. https://www.faa.gov/uas/recreational_fliers/where_can_i_fly/airspace_restrictions/.   Accessed August 2020
6. Hassija, V., Saxena, V., Chamola, V.: Scheduling drone charging for multi-drone network based on consensus time-stamp and game theory. Comput. Commun. **149**, 51–61 (2019). https://doi.org/10.1016/j.comcom.2019.09.021
7. Jeong, H., Lee, S., Song, B.: Truck-drone hybrid delivery routing: payload-energy dependency and no-fly zones. Int. J. Prod. Econ. **214**, 220–233 (2019). https://doi.org/10.1016/j.ijpe.2019.01.010
8. Kitjacharoenchai, P., Ventresca, M., Moshref-Javadi, M., Lee, S., Tanchoco, J., Brunese, P.: Multiple traveling salesman problem with drones: mathematical model and heuristic approach. Comput. Ind. Eng. **129**, 14–30 (2019). https://doi.org/10.1016/j.cie.2019.01.020
9. Murray, C.C., Chu, A.G.: The flying sidekick traveling salesman problem: optimization of drone-assisted parcel delivery. Transp. Res. Part C **54**, 86–109 (2015). https://doi.org/10.1016/j.trc.2015.03.005
10. Murray, C.C., Raj, R.: The multiple flying sidekicks traveling salesman problem: parcel delivery with multiple drones. Transp. Res. Part C **110**, 368–398 (2020). https://doi.org/10.1016/j.trc.2019.11.003
11. Query Features—OpenStreetMap, OpenStreetMap (2020). https://www.openstreetmap.org/query?lat=26.3678&lon=-80.0780#map=14/26.3497/-80.0777. Accessed August 2020
12. Steinberg, A., Cardei, M., Cardei, I.: UAS path planning using a space-time graph. In: IEEE SysCon, August 2020

Scheduling Jobs with Precedence Constraints to Minimize Peak Demand

Elliott Pryor, Brendan Mumey, and Sean Yaw[(✉)]

School of Computing, Montana State University, Bozeman, MT 59717, USA
{brendan.mumey,sean.yaw}@montana.edu

Abstract. Job scheduling to minimize peak demand occurs in the context of smart electric power grids. Some jobs (e.g. certain household appliances) may have flexibility in their start times and so can be shifted in order to lower the peak power demand of the schedule. In this work, we consider a version of peak-demand scheduling where jobs are non-preemptible and have precedence constraints (e.g. job j cannot begin until job i has finished). This problem occurs in the setting of industrial processes, where resource-consuming tasks may have completion dependencies. Our main contribution is the first polynomial time approximation algorithm for this problem. The algorithm is randomized and finds a $O(\Delta \frac{\log n}{\log \log n})$-approximation with probability at least $1 - O(1/n)$, where n is the number of jobs to be scheduled and Δ is the length of the input's longest precedence chain. We demonstrate that the algorithm is practical on realistic inputs, finds solutions that are close to optimal, and improves over existing algorithms on the data sets tested.

Keywords: Precedence job scheduling · Smart grid · Approximation algorithm · Randomized algorithm

1 Introduction

Job scheduling is a well-studied problem with many variations (e.g. maximize throughput, minimize makespan) and application areas. A variation that is important to smart power grid management is to find a schedule that has the lowest *peak demand*, as this minimizes the maximum instantaneous resource consumption. Power consumption has historically been dictated by the consumer, with the electricity supplier modifying their output to meet instantaneous demand. The introduction of two-way communication between electricity consumers and suppliers has begun to allow coordination of electricity utilization, generically called *demand response*. One key goal of demand response is to reduce the peak demand placed on an electrical power grid. Peaks in power demand are proportionally more expensive, compared to constant demand, to supply and provision the distribution network for. It is thus advantageous to both consumers and suppliers to have schedules where peak power demand is minimized.

© Springer Nature Switzerland AG 2020
W. Wu and Z. Zhang (Eds.): COCOA 2020, LNCS 12577, pp. 140–150, 2020.
https://doi.org/10.1007/978-3-030-64843-5_10

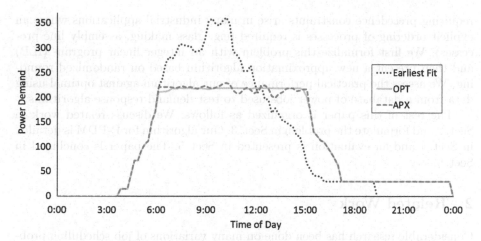

Fig. 1. Electricity demand over a 24-h period for jobs scheduled to start at their earliest valid start times (Earliest Fit), an optimal schedule (OPT), and the schedule created by the algorithm presented in this paper (APX) showing a reduction in peak demand when jobs are actively scheduled.

Individual job scheduling is one mechanism by which peaks in power demand can be reduced. Jobs that have flexibility in their execution timelines (e.g. dishwasher, hot water heater) can be scheduled to start later than they are initiated, if doing so moves execution of the job to an off-peak time. The basic problem of scheduling power jobs to minimize peak demand is called the Peak Demand Minimization (PDM) problem. An instance of PDM consists of a set of non-preemptive jobs, where each job j has an arrival time a_j, deadline d_j, width (i.e. duration) w_j, and height (i.e. instantaneous power demand) h_j. Each job needs to be assigned a start time s_j such that $[s_j, s_j + w_j) \subseteq [a_j, d_j)$. The demand of the schedule at time t is the sum of the job heights that are scheduled to run during t:

$$H(t) = \sum_{j : t \in [s_j, s_j + w_j)} h_j$$

Then, the peak demand $H^{\max}$ of the schedule is the maximum demand over all times t.

$$H^{\max} = \max_t H(t)$$

The objective of the PDM problem is to determine a job schedule, $\mathcal{S} = \langle s_j \rangle$, that minimizes $H^{\max}$. Figure 1 shows the result of scheduling jobs to reduce the peak demand of a schedule. Details of the simulation are discussed in Sect. 5.

In this research, we consider a generalization of the PDM problem, called Precedence-PDM (P-PDM), where explicit orderings of jobs (i.e. precedence constraints) need to be enforced. In addition to the four parameters described above, each job j also has a list of other jobs P_j that must be completed before j begins. In other words, $s_i + w_i \leq s_j$ for each job i in P_j. Electrical power jobs requiring precedence constraints arise in many industrial applications where an explicit ordering of processes is required (e.g. glass making, assembly line processes). We first formalize this problem with an integer linear program (ILP) and then present a new approximation algorithm based on randomized rounding. We assess the practical performance of our algorithm against optimal using data from a database of power jobs used to test demand response algorithms.

The rest of this paper is organized as follows: We discuss related work in Sect. 2 and formulate the problem in Sect. 3. Our algorithm for P-PDM is detailed in Sect. 4 and an evaluation is presented in Sect. 5. The paper is concluded in Sect. 6.

2   Related Work

Considerable research has been done on many variations of job scheduling problems. Scheduling non-preemptible jobs to minimize peak demand is similar to the machine minimization problem. Approximation algorithms have been developed for the machine minimization problem; a seminal work was [4], which presented a randomized algorithm based on randomized rounding. More recently, the online variant has drawn interest [3]. The machine minimization problem, however, assumes uniform height jobs and does not consider precedence constraints.

In the context of smart power grid management, scheduling non-preemtible jobs to minimize peak demand is called the peak demand minimization (PDM) problem. The PDM problem without precedence constraints has been extensively studied for some time [11]. An optimal fixed-parameter tractable algorithm based on dynamic programming has been introduced for the PDM problem [14]. Heuristics that lack performance guarantees have also been proposed based on decomposing integer linear programs [5] and simulated annealing [6]. Geometric strip packing techniques have been used to construct approximation algorithms for key special cases of the PDM problem (e.g. identical arrival times and deadlines, preemptible jobs) [9,12,13]. An approximation algorithm for the general PDM problem was given in [15] that was based on the approach of [4]. It was shown to provide a schedule with height within $O\left(\frac{\log n}{\log \log n}\right)$ of optimal, with probability at least $1 - O(1/n)$. Our new precedence-aware PDM scheduling algorithm described in Sect. 4 makes use of this algorithm as a subroutine.

In contrast to the basic PDM problem, the version with precedence constraints (P-PDM) has not yet been studied in great detail. Optimal algorithms have been developed for generalizations of the P-PDM problem based on Mixed Integer Linear Program implementations [1,10]. A heuristic has been introduced that introduces new precedence constraints to force jobs off the schedule's peak thereby reducing peak demand [2]. A meta-heuristic based on the greedy randomized adaptive search procedure has also been introduced that was slightly modified to apply to the P-PDM problem [2,8]. As far as we are aware, our work is the first approximation algorithm for the P-PDM problem.

3   Problem Formulation

In the general version of the PDM problem, each job j is assumed to have an arrival time a_j, deadline d_j, duration w_j, and instantaneous demand h_j. This model supports instances with either discrete (i.e. finite number of timeslots and possible start times) or continuous timescales. In this research, we assume a discrete timescale which allows us to replace the arrival time, deadline, and duration for each job with a set of possible intervals. For job j, each of its intervals is of the form $[l, r)$, where $l \geq a_j$, $r \leq d_j$, $0 \leq l < r$, and $r = l + w_j$. Since the timescale is discrete, there is a finite number of such intervals starting with $l = a_j$ and going to $l = d_j - w_j$. We use the following notation in the problem formulation and algorithm description:

J	Set of n jobs
h_j	Height of job j
I_j	Set of intervals for job j indexed by k of the form $[l_j^k, r_j^k)$
$I_j[\geq t_1; \leq t_2]$	Intervals for job j that start on or after t_1 (i.e. $l_j^k \geq t_1$) and end on or before t_2 (i.e. $r_j^k \leq t_2$)
E	Precedence rules: (i, j) means job i must precede job j
P_j	Set of jobs that must precede j
$G = (J, E)$	Graph formed by jobs J and precedence edges E
Δ	Length (number of vertices) of longest directed path in G
$\Delta(j)$	Length of longest directed path in G ending at j

Definition 1. *Given a set of jobs J with precedence constraints, the* ***Precedence-PDM (P-PDM)*** *problem seeks an interval for each job j in J, $[l_j, r_j) \in I_j$, such that:*

1. *All precedence constraints are met (i.e. if $(i, j) \in E$, then $r_i \leq l_j$).*
2. $\max\limits_{t} height(t)$ *is minimized, where* $height(t) = \sum\limits_{j : t \in [l_j, r_j)} h_j$.

The P-PDM problem can be formulated as an integer linear program (ILP) with the following decision variables:

$$x_{jk} \in \{0, 1\} \quad \text{Indicates if } [l_j^k, r_j^k) \text{ was selected to be scheduled}$$
$$H \in \mathbb{R} \qquad \text{Height of the schedule}$$

The ILP is driven by the objective function:

$$\min H$$

Subject to the following constraints:

$$\sum_k x_{jk} = 1, \qquad\qquad\qquad \forall j \qquad (1)$$

$$\sum_{k \in I_j [<t; \,]} x_{jk} + \sum_{k \in I_i [\,;\geq t]} x_{ik} \leq 1, \qquad\qquad \forall t \geq 0, \forall j, \forall i \in P_j \qquad (2)$$

$$\sum_{j,k: t \in [l_j^k, r_j^k)} h_j x_{jk} \leq H. \qquad\qquad\qquad \forall t \geq 0 \qquad (3)$$

Constraint (1) ensures that exactly one interval will be selected for each job, and constraint (2) enforces all precedence constraints. Constraint (3) ensures the H reflects the maximum height of the schedule. We note that there are only $O(\sum_j |I_j|)$ distinguishable constraints from (2) and (3).

In [14], the PDM problem was shown to be NP-hard to approximate within a ratio of 2 by reducing it from the *Scheduling with Release Times and Deadlines on a Minimum Number of Machines* (SRDM) problem. Since the PDM problem is a special case of the P-PDM problem, P-PDM is also NP-hard to approximate within a ratio of 2.

4   Approximation Algorithm

In this section we present the first approximation algorithm for the P-PDM problem. The approximation algorithm begins by topologically sorting the jobs according to the precedence constraints. Then, it solves a relaxed version of the ILP described in Sect. 1. Since the x_{jk} variables sum to one for job j, they can be treated as the probability that interval k is selected for j. Using the relaxed solution, it computes left and right boundaries (L_j, R_j) for each job j that guarantee all precedence relations are met, provided the interval $[l_j, r_j)$ chosen falls inside these new boundaries, i.e. $L_j \leq l_j < r_j \leq R_j$. Finally, the *RoundLP* algorithm described in [15] is used to schedule the jobs using only intervals that fall within these new boundaries.

Theorem 1. *If there is a feasible solution, Algorithm 1 produces an* $O(\Delta \frac{\log n}{\log \log n})$*-approximation with probability at least* $1 - O(1/n)$.

We will prove this theorem in two parts: (1) If there is a feasible solution, the algorithm is always able to find R_j in Step 4(b). (2) The schedule produced in Step 5. has the desired approximation ratio.

Lemma 1. *If there is a feasible solution, For any job j, we have*

$$\sum_{k \in I_j [\geq L_j; \geq R_j]} x_{jk} \geq 1 - \frac{\Delta(j)}{\Delta}.$$

Algorithm 1. P-PDM APX

Step 1 Topologically sort J by precedence constraints so that $(i,j) \in E \Rightarrow i < j$.

Step 2 Relax all x_{jk} variables in the ILP formulated in Sect. 3 to be real-valued in $[0,1]$.

Step 3 Solve this LP to find an optimal real-valued solution $\langle x_{jk}, H \rangle$. If the LP is infeasible, report that no solution exists.

Step 4 **for** $j = 1 \ldots n$:
   a. Let $L_j = \max\{\{R_i : i \in P_j\} \cup \{0\}\}$. (left boundary for job j)
   b. Compute:

$$R_j = \mathrm{argmin}_{t \geq L_j} \sum_{k \in I_j[\geq L_j; \leq t]} x_{jk} \geq \frac{1}{\Delta}.$$

   (right boundary for job j)
   endfor

Step 5 Apply the *RoundLP* algorithm [15] to schedule jobs where each job j is restricted to use an interval $I_j[\geq L_j; \leq R_j]$.

Proof. By induction on j.
Basis $(j = 1)$: Observe $L_1 = 0$, so

$$R_1 = \mathrm{argmin}_{t \geq 0} \sum_{k \in I_1[\geq 0, \leq t]} x_{1k} \geq \frac{1}{\Delta}.$$

By the choice of R_1 it follows that

$$\sum_{k \in I_1[\geq 0; < R_1]} x_{1k} < \frac{1}{\Delta}.$$

But then,

$$\sum_{k \in I_1[\geq L_1; \geq R_1]} x_{1k} \geq 1 - \frac{1}{\Delta}.$$

Since $\Delta(1) = 1$, the basis holds.

Inductive step $(j > 1)$: There are two cases: if $\Delta(j) = 1$, the same argument as the basis can be applied. If $\Delta(j) > 1$, then $L_j = R_i$ for some $i \in P_j$. By induction,

$$\sum_{k \in I_i[\ ; \geq R_i]} x_{ik} \geq \sum_{k \in I_i[\geq L_i; \geq R_i]} x_{ik} \geq 1 - \frac{\Delta(i)}{\Delta}.$$

By the LP constraints and the fact that $L_j = R_i$,

$$\sum_{k \in I_j[<L_j;\,]} x_{jk} + \sum_{k \in I_i[\,;\geq R_i]} x_{ik} \leq 1.$$

Combining the above two inequalities, and the fact that $i \in P_j$ we have,

$$\sum_{k \in I_j[<L_j;\,]} x_{jk} \leq \frac{\Delta(i)}{\Delta} \leq \frac{\Delta(j) - 1}{\Delta}.$$

Since

$$\sum_{k \in I_j[<L_j;\,]} x_{jk} + \sum_{k \in I_j[\geq L_j;\,]} x_{jk} = 1,$$

it follows that

$$\sum_{k \in I_j[\geq L_j;\,]} x_{jk} \geq 1 - \frac{\Delta(j) - 1}{\Delta} = \frac{\Delta + 1 - \Delta(j)}{\Delta}.$$

Since $\Delta(j) \leq \Delta$, Step 4(b) successfully finds an R_j such that $\sum_{k \in I_j[\geq L_j;\leq R_j]} x_{jk} \geq \frac{1}{\Delta}$. Furthermore, by the choice of R_j it follows that

$$\sum_{k \in I_j[\geq L_j;<R_j]} x_{jk} < \frac{1}{\Delta}.$$

But then,

$$\sum_{k \in I_j[\geq L_j;\geq R_j]} x_{jk} \geq \frac{\Delta + 1 - \Delta(j)}{\Delta} - \frac{1}{\Delta}$$

$$= \frac{\Delta - \Delta(j)}{\Delta}$$

$$= 1 - \frac{\Delta(j)}{\Delta},$$

as desired.

Corollary 1. *As noted in the proof, the algorithm is able to identify L_j and R_j for each job j such that $\sum_{k \in I_j[\geq L_j;\leq R_j]} x_{jk} \geq \frac{1}{\Delta}$. Also, if $(i, j) \in E$, then $R_i \leq L_j$, so all precedence constraints will be satisfied.*

We can now prove Theorem 1:

Proof. Let $\langle x_{jk}^*, H^{\mathrm{opt}} \rangle$ be an optimal solution to the ILP Formulation and let $\langle x_{jk}, H^{\mathrm{lp}} \rangle$ be the optimal solution of the LP Relaxation as found by Step 1. Clearly, $H^{\mathrm{lp}} \leq H^{\mathrm{opt}}$, as $\langle x_{jk}^*, H^{\mathrm{opt}} \rangle$ is a valid but not necessarily optimal LP solution.

In Step 5 of the algorithm we consider a restricted scheduling problem in which each job j must be scheduled with an interval from $I_j[L_j, R_j]$. By Corollary 1, any solution to this restricted problem will satisfy all precedence constraints. Let $\langle x_{jk}^r, H^r \rangle$ be a solution to the LP relaxation of the restricted ILP. Let $m_j = (\sum_{k \in I_j[\geq L_j; \leq R_j]} x_{jk})^{-1}$ and consider $x'_{jk} = m_j x_{jk}$. Clearly, the $\{x'_{jk}\}$ provide a solution to the LP relaxation of the restricted ILP with height $H' \leq \Delta H^{\mathrm{lp}}$, since $m_j \leq \Delta$, by Corollary 1. It follows that

$$H^r \leq H' \leq \Delta H^{\mathrm{lp}} \leq \Delta H^{\mathrm{opt}}.$$

The scheduling algorithm from [15] produces a job schedule $\langle x_{jk}^a, H^a \rangle$ with $H^a \leq \frac{4 \log n}{\log \log n} H^{opt}$. It is shown in [15] (using a Chernoff-bound based approach) that

$$\Pr[H^a > O\left(\frac{\log n}{\log \log n}\right) H^r] = O(1/n).$$

So we have,

$$\Pr[H^a > O\left(\frac{\log n}{\log \log n}\right) \Delta H^{\mathrm{opt}}] \leq \Pr[H^a > O\left(\frac{\log n}{\log \log n}\right) H^r] = O(1/n).$$

Thus, Step 5 produces a feasible solution that is an $O(\Delta \frac{\log n}{\log \log n})$-approximation with probability at least $1 - O(1/n)$.

5   Evaluation

In this section, we evaluate the performance of the algorithm presented in Sect. 4 (denoted APX) against optimal (denoted OPT) and the SWAG algorithm introduced in [2]. All trials were run on a PC with an Intel i5-4690K processor and 32GB memory. Linear programs were solved using IBM's CPLEX optimization tool version 12.10.

Synthetic job instances were generated based on the benchmark data set for industrial jobs introduced in [7]. In that work, the authors extracted job flexibility characteristics from a real-world data set of a small industrial facility. They used the job characteristics to construct a set of recurring motifs that can then be used to build job instances for benchmark testing of demand response algorithms. To generate job instances, we first randomly selected a motif with probability based on its prevalence amongst the other motifs. A job was then created based on this motif with the duration, and instantaneous power demand calculated as a randomly weighted average of three random occurrences of that motif. The initial release time was also calculated as the same randomly weighted average of the three random occurrences. The initial deadline was set to be the initial release time plus the duration. This initial execution window of the job (i.e. release time to deadline) does not leave any flexibility for scheduling, so we stretched out the execution window of the job by moving the release time earlier

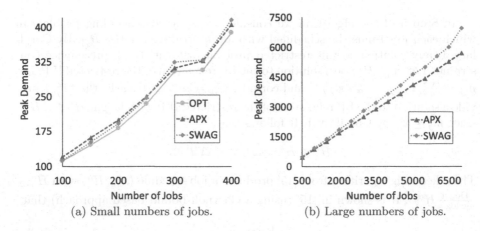

(a) Small numbers of jobs.   (b) Large numbers of jobs.

Fig. 2. Peak demand found by an optimal schedule (OPT), the schedule created by the algorithm presented in Sect. 4 (APX), and the schedule created by the heuristic introduced in [2] (SWAG) on a varying number of jobs.

and deadline later. The amount we stretched the execution window is calculated as 0.15 times the difference between the latest time any occurrence of the selected motif ends and the earliest any occurrence starts. Half of this quantity is subtracted from the release time and half is added to the deadline to increase the execution window and provide opportunity for scheduling. Finally, precedence constraints were made by randomly adding precedence relations between jobs with overlapping execution windows.

To evaluate the effectiveness of the algorithms in reducing the peak demand of a schedule, they were used to schedule scenarios of between 100 and 400 jobs. 20 instances were run for each number of jobs and the peak demands of the resulting schedules were averaged. Δ was fixed at 4 for all instances. SWAG was run for 10 s on each instance, consistent with the evaluation in [2]. Figure 2a shows the average peak demand of the schedules produced by OPT, APX, and SWAG. Both APX and SWAG are very close to optimal, with peak demand at most 10% and on average about 5% over optimal.

Scenarios with between 500 and 7000 jobs were solved to evaluate the performance of the algorithms on larger instances. The optimal algorithms used (ILP formulated in Sect. 3 and [1]) could not efficiently solve instances over 400 jobs, so only APX and SWAG were used for these larger instances. SWAG did not have good performance for larger instances with only 10 s of run time, so it was given 60 s. 20 instances were again run for each number of jobs and the peak demands of the resulting schedules were averaged. Δ was also fixed at 4 for all instances. Figure 2b shows the average peak demand of the schedules produced by APX and SWAG. The average peak demand found by APX ranges from 3% to 22% less than the peak found by SWAG.

A limited test was also done on a single instance with 7000 jobs where SWAG was allowed to run for one hour. After one hour, the peak found by SWAG was almost 2% less than the peak found by SWAG at 60 s, but was still about 11% larger than the peak found by APX. Since APX took an average of 16.4 s to solve instances of size 7000, this suggests that APX finds higher quality solutions than SWAG in much less time, while also benefiting from the performance guarantee detailed in Sect. 4.

These results indicate that APX is a practical algorithm for solving P-PDM instances as it can produce very high quality results with very little computation time.

6   Conclusions

In this work we introduce the first approximation algorithm for the P-PDM problem. This problem is relevant to industrial applications where an explicit ordering of processes must be enforced. Simulations with realistic data suggest that this algorithm is not only the first algorithm with a performance guarantee, but it also performs better than existing algorithms. Interesting future work includes considering not just precedence constraints, but constraints that dictate a job must execute immediately after another is complete. This would enable modeling jobs whose profiles are not simple rectangles (i.e. non-constant demand throughout the life of the job). It also remains to be seen if using output from the algorithm presented to seed the heuristic from [2] could lead to an improved solution. Finally, it could be interesting to study the online version of the P-PDM problem.

References

1. Barth, L., Ludwig, N., Mengelkamp, E., Staudt, P.: A comprehensive modelling framework for demand side flexibility in smart grids. Comput. Sci. - Res. Dev. **33**, 13–23 (2017). https://doi.org/10.1007/s00450-017-0343-x
2. Barth, L., Wagner, D.: Shaving peaks by augmenting the dependency graph. In: Proceedings of the Tenth ACM International Conference on Future Energy Systems (e-Energy), pp. 181–191 (2019)
3. Chen, L., Megow, N., Schewior, K.: An $\mathcal{O}(\log m)$-competitive algorithm for online machine minimization. SIAM J. Comput. **47**(6), 2057–2077 (2018)
4. Chuzhoy, J., Guha, S., Khanna, S., Naor, J.S.: Machine minimization for scheduling jobs with interval constraints. In: 45th Annual IEEE Symposium on Foundations of Computer Science, pp. 81–90 (2004)
5. Hong, Y., Wang, S., Huang, Z.: Efficient energy consumption scheduling: towards effective load leveling. Energies **10**(1), 105 (2017)
6. Jewell, N., Bai, L., Naber, J., McIntyre, M.L.: Analysis of electric vehicle charge scheduling and effects on electricity demand costs. Energy Syst. **5**(4), 767–786 (2013). https://doi.org/10.1007/s12667-013-0114-0
7. Ludwig, N., Barth, L., Wagner, D., Hagenmeyer, V.: Industrial demand-side flexibility: a benchmark data set. In: Proceedings of the Tenth ACM International Conference on Future Energy Systems (e-Energy), pp. 460–473 (2019)

8. Petersen, M.K., Hansen, L.H., Bendtsen, J., Edlund, K., Stoustrup, J.: Heuristic optimization for the discrete virtual power plant dispatch problem. IEEE Trans. Smart Grid 5(6), 2910–2918 (2014)
9. Ranjan, A., Khargonekar, P., Sahni, S.: Offline preemptive scheduling of power demands to minimize peak power in smart grids. In: 2014 IEEE Symposium on Computers and Communications (ISCC), pp. 1–6. IEEE (2014)
10. Sou, K.C., Weimer, J., Sandberg, H., Johansson, K.H.: Scheduling smart home appliances using mixed integer linear programming. In: 50th IEEE Conference on Decision and Control and European Control Conference, pp. 5144–5149 (2011)
11. Tang, S., Huang, Q., Li, X., Wu, D.: Smoothing the energy consumption: peak demand reduction in smart grid. In: 2013 Proceedings IEEE INFOCOM, pp. 1133–1141 (2013)
12. Tang, S., Yuan, J., Zhang, Z., Du, D.Z.: iGreen: green scheduling for peak demand minimization. J. Glob. Optim. 69(1), 45–67 (2017)
13. Yaw, S., Mumey, B., McDonald, E., Lemke, J.: Peak demand scheduling in the smart grid. In: 2014 IEEE International Conference on Smart Grid Communications (SmartGridComm), pp. 770–775 (2014)
14. Yaw, S., Mumey, B.: An exact algorithm for non-preemptive peak demand job scheduling. In: Zhang, Z., Wu, L., Xu, W., Du, D.-Z. (eds.) COCOA 2014. LNCS, vol. 8881, pp. 3–12. Springer, Cham (2014). https://doi.org/10.1007/978-3-319-12691-3_1
15. Yaw, S., Mumey, B.: Scheduling non-preemptible jobs to minimize peak demand. Algorithms 10(4), 122 (2017)

Reachability Games for Optimal Multi-agent Scheduling of Tasks with Variable Durations

Dhananjay Raju$^{(\boxtimes)}$, Niklas Lauffer, and Ufuk Topcu

The University of Texas at Austin, Austin, USA
{draju,nlauffer,utopcu}@utexas.edu

Abstract. Scheduling tasks with variable durations across multiple agents is an NP-hard problem for even two agents. Typically, the runtime of any exact algorithm is dominated by the number of tasks because of an exponential dependence. We shift this exponential dependency from the number of tasks to a new parameter, which we call *window length*. This novel parameterization enables to reduce the problem of finding an optimal schedule to one of searching for winning strategies in a two-player reachability game on graphs of *size polynomial in the number of tasks*. As such, the complexity of finding an optimal schedule is polynomial in the number of tasks but exponential in the window length. We demonstrate that, in practice our algorithm runs faster than the worst-case complexity. The approach we present is applicable for most common optimization criteria, such as minimization of *makespan* and *total load*. We demonstrate the practical value of this technique by finding optimal schedules for astronauts aboard the International Space Station. Finally, experiments on randomly generated instances show that, on average, this technique is at least two orders of magnitude faster than an integer program formulation.

Keywords: Multi-agent scheduling · Graph games · Variable durations · Linear optimization criteria · Schedulability

1 Introduction

We aim to understand the role of task structure in the complexity of finding an optimal schedule for the *agent resource-constrained project scheduling problem* (ARCPSP) with variable task durations [14]. The ARCPSP is an extension of the *resource-constrained project scheduling problem* (RCPSP) with a notion of agents that execute tasks in parallel. The problem includes lower and upper limits for each task's duration. Therefore, scheduling also involves assigning execution durations from the interval for each task.

This work has been supported in part by the grants NASA NNX17AD04G, NSF 1652113 and NSF 1646522.

© Springer Nature Switzerland AG 2020
W. Wu and Z. Zhang (Eds.): COCOA 2020, LNCS 12577, pp. 151–167, 2020.
https://doi.org/10.1007/978-3-030-64843-5_11

The ARCPSP is a combinatorial optimization problem and is NP-hard for even two agents [17]. It is possible to solve the problem efficiently if one can divide the planning horizon into smaller time windows so that tasks start and end within individual windows. However, this condition is often impractical and tasks do spillover. We relax this restrictive condition. Specifically, we allow such spillovers but limit them to the next window, but not the windows beyond. This relaxed condition is natural in many scheduling scenarios. We describe two of them here. Consider software development teams (a group of agents) that aim to create software through *sprints*. A sprint is a short time period when a team works to complete a number of tasks. Ideally, tasks that start in a sprint (window) should end in the same sprint. However, it is not always possible to satisfy this condition. In practice, task spillovers are pushed to the next sprint and special efforts are made to avoid additional spillovers [13]. Consider a second scenario where tasks correspond to goods being delivered to customers by a fixed number of agents. After a customer chooses her delivery slot, the company has to ensure that the deliveries are performed with minimal latencies. In other words, the task spillovers are restricted.

In Sect. 2, we introduce a new parameter which we call *window length* (Δ). The window length is chosen such that task spillovers are restricted to adjacent windows. More specifically, the window length is the smallest integer such that tasks that start in a window only spillover to the next window but not the windows after. This parameterization enables to encode all feasible schedules as paths in a graph of *size polynomial in the number of tasks*. An alternate notion of *windows* has been used to restrict the difference between the start times of various tasks, when the task durations are fixed [19]. However, we use windows to restrict the length of individual tasks.

Uncertainty is prevalent in scheduling due to a lack of accurate process models and variability on the process and environmental data [11]. As such, it is impossible to estimate the durations for tasks without uncertainty. Following [15], we model the uncertainty in the task durations by allowing the tasks to have variable durations. In this paper, we compensate for such uncertainty by allocating each task the maximum permissible duration while simultaneously ensuring that the resulting schedule does not violate the resource constraints.

We study optimization criteria defined as functions on *non-idling* durations of individual agents. Due to the high complexity of the problem, such objectives are seldom studied even if they have a wide range of practical applications. For example, such optimization criteria enable to assign weights to agents to give preference to schedules that maximize the use of agents with higher weights. In Sect. 5, we find optimal schedules for astronauts aboard the International Space Station (ISS). We model a scenario where some of the astronauts can be ill or injured and try to minimize the assignment of tasks to such astronauts while simultaneously finding a schedule that maximizes the execution time for the tasks.

We cast the problem of finding an optimal schedule as a two-player reachability game on graphs of *size polynomial in the number of tasks*. In the context of scheduling, reachability games have previously been used for finding feasible schedules for sporadic tasks [7]. On the other hand, we use these games to find optimal schedules. In Table 1, we list the complexities of finding optimal schedules for different optimization criteria when the window length Δ and the number of agents k are fixed constants. The definition of Δ implies that the *planning horizon* $\mathcal{H}$ is $\mathcal{O}(2n\Delta)$, where n is the number of tasks (we assume that there are no empty windows). Even though the worst-case complexities are exponential in k and Δ, the experiments in Sect. 5 show that we can find optimal schedules for five agents that have to complete 100 tasks within a day, in under a minute, when a) each time step is ten minutes long b) the maximum duration of every task is at most five hours and c) the difference between the earliest start time and latest start time for every task is at most five hours.

Table 1. The complexity of finding optimal schedules. $\mathcal{H}$ is the length of the planning horizon, k is the number of agents and Δ is the window length.

Type of optimization	Complexity
None (feasibility)	$\mathcal{O}_{k,\Delta}(\mathcal{H})$
Linear function	$\mathcal{O}_{k,\Delta}(\mathcal{H}^3)$
Min total load	$\mathcal{O}_\Delta(\mathcal{H}^{k+2})$
Min makespan	$\mathcal{O}_{k,\Delta}(\mathcal{H}^3)$

In Sect. 5, we validate the technique for different objectives on randomly generated instances. The experiments show that the technique works well, even for a large number of tasks and long planning horizons. Lastly, we compare our technique against an integer programming encoding of the problem that we run in Gurobi [10]. Experiments show that the technique is at least two orders of magnitude faster.

2 The Agent Resource-Constrained Project Scheduling Problem and Windows

An instance I of the *agent resource-constrained project scheduling problem* is a tuple $(\mathcal{A}, \mathcal{T}, \mathcal{D}, \mathcal{S}, \mathcal{C}, B, R)$ defined as follows.

- $\mathcal{A} = \{1, 2, \ldots, k\}$ is the set of agents.
- $\mathcal{T} = \{1, 2, \ldots, n\}$ is the set of tasks.
- $\mathcal{D} = \{(d_t^{min}, d_t^{max}) \mid t \in \mathcal{T} \text{ and } d_t^{min}, d_t^{max} \in \mathbb{N}\}$ is a set of pairs of *minimum and maximum duration* for every task. Each task has to be scheduled for at least the minimum duration and at most the maximum duration.
- $\mathcal{S} = \{(s_t^e, s_t^\ell) \mid t \in \mathcal{T} \text{ and } s_t^e, s_t^\ell \in \mathbb{N}\}$ is a set of pairs of *earliest start time* and *latest start time* for every task. Every task has to be scheduled at or after the earliest start time and before or at the latest start time.

- For any t in $\mathcal{T}$, $\mathcal{C}_t \subseteq \mathcal{A}$ is the set of agents that can perform task t. Let $\mathcal{C} = \{\mathcal{C}_1, \ldots, \mathcal{C}_n\}$.
- There are m types of renewable resources. The maximum quantities of the resources are encoded in a vector B in $\mathbb{N}^m$. B_i gives the maximum quantity of resource i.
- R in $\mathbb{N}^{n \times m}$ is a matrix that encodes the resource requirements of the tasks. R_{tj} gives the quantity of resource j required by task t.

The *planning horizon* $\mathcal{H}$ of an instance I is defined as $\max\{s_t^\ell + d_t^{max} \mid t \in \mathcal{T}\}$. In the rest of the paper, we use x to denote an arbitrary non-negative integer.

Associated with every ARCPSP instance is a parameter, *window length* denoted by Δ, that is intrinsic to the instance. Formally, Δ is the smallest positive integer such that, for all tasks t, if $x\Delta < s_t^e \leqslant (x+1)\Delta$, then $s_t^\ell + d_t^{max} \leqslant (x+2)\Delta$. The definition of Δ implies that, if the entire planning horizon is partitioned into intervals of size Δ, then the tasks that start in an interval can only spill over to the next interval but not the interval after. We refer to the interval $(x\Delta, (x+1)\Delta]$ as *window x*. The definition of window length implies that $0 < \Delta \leqslant d_{max}$, where $d_{max} = \max\{(s_t^\ell + d_t^{max}) - s_t^e + 1 \mid t \in \mathcal{T}\}$. Therefore, given a scheduling instance I, the window length can be determined in time polynomial in the number n of tasks and d_{max}. Henceforth, we assume that Δ is given.

Remark 1. The window length restricts the maximum duration for the tasks depending on the starting time inside a window. For example, consider a task t that has a maximum duration of 2Δ and can start in the interval $(x\Delta, (x+1)\Delta]$. If it starts at $x\Delta + 1$, then its maximum duration is 2Δ. If it starts at $x\Delta + r + 1$ (where, $r < 2\Delta$ and $r \in \mathbb{N}$), then the maximum duration is at most $2\Delta - r$. The scheduling of tasks with starting-time-dependent execution times has been extensively studied [5,6]. More specifically, it is known to be NP-hard [12].

A set $\mathcal{T}' \subseteq \mathcal{T}$ of tasks is *permissible* if the sum of the quantities of each type of resource required by the tasks in $\mathcal{T}'$ is less than or equal to its maximum quantity. Formally, $\mathcal{T}'$ is permissible, if for every resource $j \in [m]$, $\sum_{t \in \mathcal{T}'} R_{tj} \leqslant B_j$. The set of all permissible sets is the complement of the set of all *forbidden sets of tasks* (a set of tasks that cannot be scheduled together) [20]. In general, the enumeration of forbidden sets (consequently of permissible sets) for any instance is computationally expensive [21]. Intuitively, if the earliest starting time of a task is after another task ends, then the two tasks can never interfere with each other. By the definition of window length, only tasks with the earliest start time in the intervals $((x-1)\Delta, x\Delta]$, $(x\Delta, (x+1)\Delta]$ or $((x+1)\Delta, (x+2)\Delta]$ can interfere with a task with the earliest start time in the interval $(x\Delta, (x+1)\Delta]$. Permissible sets over such tasks are said to be *relevant*. We denote the set of all relevant permissible sets by $\mathcal{P}$. Algorithm 1 computes all relevant permissible sets in time $\mathcal{O}((k\Delta)^k \mathcal{H})$. In the rest of the paper, we assume that $\mathcal{P}$ is given.

A *schedule* is a set of tuples of the form (s_t, a_t, d_t) for every task t, where s_t is the actual start time, a_t is the agent assigned to the task and d_t is the exact

---

Algorithm 1. Finding all the relevant permissible sets.

Input an instance I of the ARCPSP.
Output the relevant permissible sets in I.
1: $\mathcal{P} \leftarrow \varnothing$
2: **for** $i := 0$ **to** floor($\mathcal{H}/\Delta$) **do**
3:     $\mathcal{X} \leftarrow \{t : s_t^e \in (i\Delta, (i+1)\Delta]\}$
4:     **for all** $U \subset \mathcal{X}$ s.t. $|U| \leqslant k$ **do**
5:         **if** $\sum_{k \in U} R_{kj} < B_j$ for all $j \in [1, m]$ **then**
6:             $\mathcal{P} \leftarrow \mathcal{P} \cup \{U\}$
7:         **end if**
8:     **end for**
9: **end for**
10: **return** $\mathcal{P}$

---

duration allocated to the task. A schedule is *valid* if the following conditions are satisfied.

(a) At any time, every agent is assigned at most one task.
(b) At any time, the set of tasks scheduled together is permissible.
(c) Tasks that have been scheduled are not preempted.
(d) For every task t in $\mathcal{T}$, $s_t^e \leqslant s_t \leqslant s_t^\ell$, a_t in $\mathcal{C}_t$ and $d_t^{min} \leqslant d_t \leqslant d_t^{max}$.

One reason for the nonexistence of a valid schedule may be the lack of sufficiently many agents. In any window, k agents can complete at most $k\Delta$ tasks. Therefore, if there is a window $(x\Delta, (x+1)\Delta]$ such that the number of tasks that have to be completed in the window is more than $k\Delta$, then no valid schedule exists. In this case, we say that the number of agents is *insufficient*. In the rest of the paper, we assume that the number of agents is sufficient. For clarity, we restate the assumptions.

(A1) The number k of agents is sufficient.
(A2) The window length Δ is given.
(A3) The set $\mathcal{P}$ of all relevant permissible sets is given.

3   Encoding Valid Schedules as Paths in a Graph

With every instance $I = (\mathcal{A}, \mathcal{T}, \mathcal{D}, \mathcal{S}, \mathcal{C}, B, R)$, we associate a graph $G_I = (V_I, E_I)$. Intuitively, each vertex in V_I corresponds to a configuration of the agents at some time in the planning horizon of I. There is an edge (u, v) in E_I if and only if it is possible for the configuration of the agents corresponding to vertex u to progress to the configuration corresponding to vertex v in the following time step without violating the constraints of I. If a vertex has more than one out-edge, it means that the corresponding configuration of the agents can progress in different ways depending on the scheduling decision.

Every path in the graph G_I corresponds to a valid sequence of configurations of the agents. By designating an initial vertex v_{init} and a final vertex v_f corresponding, respectively, to the initial configuration of agents before execution and the final configuration of agents after completing all the tasks, a path from v_{init} to v_f corresponds to a sequence of scheduling decisions constituting a valid schedule for I.

Every vertex in V_I has four components. The first component is a vector that holds the task assignment for every agent along with the duration left to complete the task. An *idling* agent, i.e., an agent which is not assigned any task from $\mathcal{T}$, is assigned a dummy task 0. The second component is the current time. The third component has a set of tasks that have to be completed by the end of the window corresponding to the current time. The fourth component records the set of tasks completed so far in the window. Formally,

$$v = \left(((1, t_1, \ell_1), \dots, (k, t_k, \ell_k)), \tau, F, C\right) \in ([k] \times \mathcal{T} \times [2\Delta])^k \times [\mathcal{H} + 1] \times 2^{\mathcal{T}} \times 2^{\mathcal{T}}$$

belongs to V_I if the following conditions are satisfied.

1. There exists $P \in \mathcal{P}$ such that $\{t_1, \dots, t_k\} \subseteq P \cup \{0\}$, i.e., the set of tasks scheduled at any time is permissible.
2. For every currently assigned task $t_a \in \{t_1, \dots, t_k\}$, $a \in \mathcal{C}_{t_a}$, i.e., the agent assigned to the task can perform it.
3. For all pairs t, t' of tasks in $\{t_1, \dots, t_k\} \setminus \{0\}$, $a_t \neq a_{t'}$, i.e., the same agent cannot be assigned multiple tasks (not idling).
4. For every agent a in $\mathcal{A}$, $0 \leqslant \ell_a \leqslant d_{t_a}^{max}$, i.e., the duration left for the task allocated to agent a is shorter than or equal to the maximum duration.
5. $0 \leqslant \tau \leqslant \mathcal{H} + 1$.
6. $F \subseteq \mathcal{T}$ and if $x\Delta < \tau \leqslant (x + 1)\Delta$, then, for every task t in F, $x\Delta < s_t^\ell + d_t^{max} \leqslant (x + 1)\Delta$.
7. $C \subseteq \mathcal{T}$ and if $x\Delta < \tau \leqslant (x + 1)\Delta$, then, for every task t in C, $(x - 1)\Delta < s_t^e + d_t^{min} \leqslant s_t^\ell + d_t^{max} \leqslant (x + 1)\Delta$.
8. For every task t in $\{t_1, \dots, t_k\}$, if $t \neq 0$, then $t \in F$ and $t \notin C$.

Let $v = \left(((1, t_1, \ell_1), \dots, (k, t_k, \ell_k)), \tau, F, C\right)$ and $v' = \left(((1, t_1', \ell_1'), \dots, (k, t_k', \ell_k')), \tau', F', C'\right)$ be two vertices in G_I. The vertex v is said to be *in the window* x if $\tau \in (x\Delta, (x + 1)\Delta]$. The initial vertex v_{init} is $\left(((1, 0, 1), (2, 0, 1), \dots, (k, 0, 1)), 0, \varnothing, \varnothing\right)$ and the final vertex v_f is $\left(((1, 0, 1), (2, 0, 1), \dots, (k, 0, 1)), H + 1, \varnothing, \varnothing\right)$.

There are three types of edges in G_I.

(E1) *The edges between two vertices in the same window.*
(E2) *The edges from vertices in a window to vertices in the next window.*
(E3) *The edges to the final vertex.*

We formally define the three types of edges in the graph G_I in Table 2. The edges of type **(E1)** and **(E1)** correspond to the assignment of new tasks; some of the agents may be assigned new tasks, while others continue their previously assigned task. The edges of type **(E1)** are from vertices corresponding to the completion of all the tasks to the final vertex v_f.

The following lemma provides a necessary and sufficient condition for the existence of a valid schedule.

Table 2. The three types of edges in G_I.

Edge type	Conditions for edge between v and v'
(E1)	$\forall x \in \mathbb{N}_{\geqslant 0} : \tau \neq x\Delta$ and $\tau \neq \mathcal{H}$.
	$\forall a \in [k] :$ if $t_a = t'_a \neq 0$, then $\ell'_a = \ell_a - 1$. $\forall a \in [k] :$ if $t_a = t'_a = 0$, then $\ell'_a = 1$. $\forall a \in [k] :$ if $t_a \neq t'_a$, then $\ell_a \leqslant d^{max}_{t_a} - d^{min}_{t_a}$, $\qquad \ell'_a = d^{max}_{t'_a}$, $\qquad s^e_{t'_a} \leqslant \tau + 1 \leqslant s^\ell_{t'_a}$ and $\qquad$ if $t'_a = 0$, then $\ell'_a = 1$.
	$(\tau', F', C') = (\tau + 1, F \backslash \{t_a \mid t_a \neq t'_a\}, C \cup \{t_a \mid t_a \neq t'_a\})$.
(E2)	$\exists x \in \mathbb{N}_{\geqslant 0} : x < \lfloor \mathcal{H}/\Delta \rfloor$ and $\tau = x\Delta$.
	$\forall a \in [k] :$ if $t_a = t'_a \neq 0$, then $\ell'_a = \ell_a - 1$. $\forall a \in [k] :$ if $t_a = t'_a = 0$, then $\ell'_a = 1$. $\forall a \in [k] :$ if $t_a \neq t'_a$, then $\ell_a \leqslant d^{max}_{t_a} - d^{min}_{t_a}$, $\qquad \ell'_a = d^{max}_{t'_a}$, $\qquad s^e_{t'_a} \leqslant \tau + 1 \leqslant s^\ell_{t'_a}$ and $\qquad$ if $t'_a = 0$, then $\ell'_a = 1$.
	$\tau' = \tau + 1$. $F' = (F \cup \{t \in \mathcal{T} : x\Delta < s^\ell_t + d^{max}_t \leqslant (x+1)\Delta\}) \backslash (\{t_a \mid t_a \neq t'_a\} \cup C)$. $C' = \varnothing$.
(E3)	$\tau = \mathcal{H}$ and $\tau' = \mathcal{H} + 1$. $F = \{t \in \mathcal{T} : \exists\, a \in [k],\ t = t_a \text{ and } \ell_a \leqslant d^{max}_{t_a} - d^{min}_{t_a}\}$.
	$\forall a \in [k] :$ if $t_a \neq 0$, then $\ell_a \leqslant d^{max}_{t_a} - d^{min}_{t_a}$, $\qquad \ell'_a = 1$ and $t'_a = 0$.
	$(F', C') = (\varnothing, \varnothing)$.

Lemma 1. *There is a valid schedule for the instance I if and only if there is a path from v_{init} to v_f in G_I.*

Proof. ($\Rightarrow$) In any path ρ from v_{init} to v_f, task t has been assigned to an agent a if there is a vertex of the form $((\ldots, (a, t, d^{max}_t), \ldots), \tau, F, C)$. If $x\Delta \leqslant s^\ell_t + d^{max}_t \leqslant (x+1)\Delta$ and the task is not completed by either time $x\Delta$ or $x + 1\Delta$, then there will be no out-edge from the vertex with time $\tau + 1$ in the path. Additionally, by the definition of edge types **(E1)** and **(E2)**, when task t is completed, it is removed from F, hence it cannot be reassigned. Furthermore, no task can be assigned simultaneously to multiple agents. The start time for task t is $s_t = \min \{\tau \mid ((\ldots, (a, t, d^{max}_t), \ldots), \tau, F, C) \in \rho\}$ and the end time is $e_t = \max \{\tau \mid ((\ldots, (a, t, d^{max}_t), \ldots), \tau, F, C) \in \rho\}$. The duration is the difference between the end time and the start time, i.e., $d_t = e_t - s_t$.

($\Leftarrow$) Every valid schedule induces a path from v_{init} to v_f. $\qquad\qquad\square$

Given an ARCPSP instance I, we compute a valid schedule by constructing the corresponding graph G_I and searching for a path from v_{init} to v_f in G_I. Given a path from v_{init} to v_f, Algorithm 2 presents the procedure to extract a valid schedule corresponding to a path from v_{init} to v_f.

Lemma 2. *A valid path can be computed in time* $\mathcal{O}\big(\mathcal{H}\big((2k\Delta^2)^k \cdot 2^{4k\Delta}\big)^2\big)$.

Proof. The number of vertices in the graph G_I is $\mathcal{O}\big((2k\Delta^2)^k \cdot \mathcal{H} \cdot 2^{4k\Delta}\big)$. The first component of any vertex in V_I has a task assignment for each agent and the duration left for the task. Since the number of agents is sufficient (Assumption (**A1**)), any agent may start at most $2k\Delta$ tasks over the course of the schedule. Moreover, the duration left for the task is at most 2Δ. The size of the first component is $\mathcal{O}(2k\Delta^2)^k$. The size of the second component is $\mathcal{H}$. The third component F encodes the set of tasks to be completed by the end of the current window. If the number of agents are sufficient, then the maximum number of tasks that can finish in the window is $2k\Delta$. In the worst case, the size of F (third component) is $2^{2k\Delta}$. The size of the third component is the same as the size of the fourth component. The number of edges in G_I is $\mathcal{O}\big(\mathcal{H}\big((2k\Delta^2)^k \cdot 2^{4k\Delta}\big)^2\big)$. In any graph, a path between any two vertices can be computed in $\mathcal{O}(|V| + |E|)$. $\qquad\square$

Algorithm 2. Schedule corresponding to a path from v_{init} to v_f in G_I.

Input path $\rho = (v_{init}, v_1, ..., v_k, v_f)$ in G_I.
Output valid schedule S.
1: $S \leftarrow \{0, ..., 0\}$ {empty schedule of length n}
2: **for** $\Big(\big((1, i, \ell_i), ..., (k, j, \ell_j)\big), \tau, F, C\Big) \in \rho$ **do**
3:    **for** $(a, t, \ell) \in \big((1, i, \ell_i), ..., (k, j, \ell_j)\big)$ **do**
4:       **if** $t \neq 0$ **then**
5:          $d \leftarrow d_t^{max} - \ell$ {duration that t has run for so far}
6:          $s \leftarrow \tau - d$ {time that t was started}
7:          $S_t \leftarrow (s, a, d)$
8:       **end if**
9:    **end for**
10: **end for**
11: **return** S

Remark 2. The complexity of finding a valid schedule depends on the number of tasks that have to be completed in each window. If we fix $\mathcal{H}$ and Δ and increase n, then the number of tasks per window increases. The worst-case complexity for particular values of $\mathcal{H}$, k and Δ occurs when we have to schedule close to $k\Delta$ tasks in every window.

4   Reachability Games for Optimal Scheduling

In this section, we define optimal schedules and provide a technique to compute optimal schedules by building upon the graph construction presented in Sect. 3.

Let S be a schedule for an instance $I = (\mathcal{A}, \mathcal{T}, \mathcal{D}, \mathcal{S}, \mathcal{C}, B, R)$. For every agent a, let $w_a \in \mathbb{Z}$ be the *weight* of the agent. The *value of the agent* in the schedule S denoted by $values_S(a)$ is defined as

$$values_S(a) = \sum_{(s_t, a, d_t) \in S} d_t,$$

i.e., the value of the agent a in the schedule S is defined as the sum of the duration of the tasks assigned to the agent a in schedule S. The value of the schedule S denoted by $val(S)$ is defined as

$$val(S) = \sum_{a \in \mathcal{A}} w_a \cdot values_S(a).$$

A schedule S for instance I is *optimal* if for every other schedule S', $val(S') \leqslant val(S)$. Let $\tau \in [0, \mathcal{H} + 1]$, where $\mathcal{H}$ is the planning horizon of I, then the *value of a schedule up to a time* τ denoted by $val(\tau; S)$ is defined as

$$val(\tau; S) = \sum_{a \in \mathcal{A}} w_a \sum_{(s_t, a, d_t): s_t \leqslant \tau} \min\{d_t, \tau - s_t + 1\}.$$

Since the value of a schedule is a linear function on the value of each agent in the schedule, we observe that

$$val(\tau + 1; S) = val(\tau; S) + \sum_{\substack{a \in \mathcal{A}: \\ a \text{ is not idle at } \tau+1}} w_a. \tag{1}$$

We incentivize the assignment of tasks to a particular agent by giving it a greater weight compared to the weights of the other agents.

4.1   Two-Player Reachability Games

For each problem instance, we construct a *two-player reachability game* such that we can obtain an optimal schedule from any *memoryless winning strategy* for one of the players, which we call the *reachability player*.

A two-player *reachability game* [16] is played on a graph $G = (V, E)$ between a reachability player and a safety player. Initially, a token is placed on a designated *initial vertex* v_η in V. The two players take turns moving the token along the edges of the graph. The objective of the reachability player is to move the token to a vertex in $\mathcal{F} \subseteq V$. The safety player tries to prevent the token from reaching a vertex in $\mathcal{F}$. A play ρ is a (possibly infinite) sequence $v_\eta v_1 \ldots$ of vertices such that $(v_i, v_{i+1}) \in E$ for all $0 \leqslant i$. The play ρ is *winning* for the reachability player if, for some v_i in ρ, $v_i \in \mathcal{F}$. A *memoryless strategy* $\sigma : V \to V$ maps every vertex to one of its successors. The play ρ is said to *agree* with the strategy σ for player P if $v_{i+1} = \sigma(v_i)$ whenever player P has to play from v_i. A memoryless strategy is said to be *winning* for player P if all plays that agree with it are winning for player P. Such games are *determined*, i.e., one of the players has a memoryless winning strategy [16]. Additionally, a memoryless winning strategy for the winning player can be found in $\mathcal{O}(|V| + |E|)$ [4,18]. An extensive survey of reachability games on graphs can be found in [2,9].

Two-player reachability games on graphs have been used for *online scheduling* of *sporadic tasks* [7,8]. In online scheduling, the reachability player tries to create tasks that will miss the deadline and the safety player is the scheduler who tries to ensure that none of the tasks miss their deadlines.

4.2 Optimal Scheduling

In the setting of this paper, the reachability player tries to find an optimal schedule, whereas the safety player tries to produce an alternate schedule with a value greater than the one produced by the reachability player.

With every instance $I = (\mathcal{A}, \mathcal{T}, \mathcal{D}, \mathcal{S}, \mathcal{C}, B, R)$, we associate a graph $G_I^{linear} = (V, E)$. The vertices of the graph have four components. We use the first and second components to record the schedules for the reachability player and the safety player, respectively. The third component, which we call the *counter*, holds the difference between the values of the two schedules. The difference between the values of any two valid schedules is at most $c\mathcal{H}$ and at least $-c\mathcal{H}$, where $c = k \cdot \max\{|w_a| \mid a \in \mathcal{A}\}$. The last component records the *turn*, i.e., the player that has to move the token next. Let $G_I = (V_I, E_I)$ be the graph corresponding to instance I as defined in Sect. 3.

The set of vertices of the graph G_I^{linear} is $V = V_I \times V_I \times [-c\mathcal{H}, c\mathcal{H}] \times \{r, s\}$. The initial vertex v_η is the tuple $(v_{init}, v_{init}, 0, r)$ and the set of winning vertices for the reachability player is $\mathcal{F} = \{(v_f, v_f, \delta, \alpha) \mid \delta \geqslant 0, \ \alpha = s \vee \alpha = r\}$. There are three types of edges in G_I^{linear}.

(E4) *The edges starting from vertices with turn r.*
(E5) *The edges starting from vertices with turn s.*
(E6) *The direct edges to $(v_f, v_f, 0, r)$.*

Table 3. The three types of edge in G_I^{linear}.

Edge type	Conditions for edge between χ and χ'
(E4)	$\alpha = r, \ \alpha' = s, \ v = v', \ (u, u') \in E_I$ and $\delta' = \delta + \sum_{t'_a \neq 0 \ \in u'} w_a.$
(E5)	$\alpha = s, \ \alpha' = r, \ u = u', \ (v, v') \in E_I$ and $\delta' = \delta - \sum_{t'_a \neq 0 \ \in v'} w_a.$
(E6)	There is no out going edge from v in G_I, $\alpha = s, \ \alpha' = r, \ u' = v' = v_f$ and $\delta' = 0.$

The edges of type **(E4)** and **(E5)** record the scheduling choices of the reachability player and the safety player, respectively. The edges of type **(E6)** ensure that the reachability player wins the game when the safety player has no action to extend its schedule. Let $\chi = (u, v, \delta, \alpha)$ and $\chi' = (u', v', \delta', \alpha')$ be two vertices in V. We formally define the three types of edges in Table 3.

By Lemma 1, every valid schedule for I induces a valid path from v_{init} to v_f in the graph G_I. Since the first component of the vertex set V is V_I, every valid schedule induces a valid path on this component. Thus, reachabilty player can use any valid schedule as a memoryless strategy. Moreover, any play ρ that agrees with such a strategy, induces a path from v_ι to a vertex of the form (v_f, v, δ, s), where $v \in V_I$ and $\delta \in [-c\mathcal{H}, c\mathcal{H}]$.

In the reachability game on the graph G_I^{linear}, the two players take turns to assign tasks to agents for each time step in the scheduling horizon such that the constructed schedules are valid. If the reachability and safety players follow schedules S_1 and S_2 respectively, then Eq. (1) implies that

$$val(\tau + 1; S_1) - val(\tau + 1; S_2) = (val(\tau; S_1) - val(\tau; S_2))$$

$$+ \sum_{\substack{a \text{ is not idle at } \tau \\ \text{in } S_1}} w_a - \sum_{\substack{a \text{ is not idle at } \tau \\ \text{in } S_2}} w_a. \quad (2)$$

For finding the optimal schedule, we require that the value of the schedule chosen by the reachability player is greater than or equal to the value of the schedule chosen by the safety player, i.e., the difference between these values is non-negative. We maintain and update this difference according to Eq. (2), by recording the difference in the counter. If the difference in the value of the two chosen schedules is non-negative at the end of the scheduling horizon, then the reachability player reaches a vertex in $\mathcal{F}$.

Lemma 3. *If the reachability player follows an optimal schedule, then it wins the game.*

Proof. In the reachability game on G_I^{linear}, the two players take turns in constructing their respective schedules. We maintain the difference between the values of the two schedules constructed so far in the counter. This difference is always contained in the closed interval $[-c\mathcal{H}, c\mathcal{H}]$. If the safety player follows a valid schedule and the reachability player follows an optimal schedule, by construction of G_I^{linear}, the difference is non-negative and a vertex in $\mathcal{F}$ is reached. However, if the safety player does not follow a valid schedule, then the reachability player wins by using an edge of type **(E6)**. □

Corollary 1. *If the reachability player does not follow an optimal schedule and the safety player follows one, then the safety player wins the game.*

Theorem 1. *A memoryless winning strategy for the reachability player can be computed in time $\mathcal{O}\big(\mathcal{H}^3\big((2k\Delta^2)^k \cdot 2^{4k\Delta}\big)^2\big)$.*

Proof. The number of vertices in G_I^{linear} is $\mathcal{O}(|V_I|^2 \cdot 2k\mathcal{H})$. Suppose it is the turn of the reachability player, upon fixing the first component of the vertex, the other three components are directly determined. Therefore, the number of edges in G_I^{max} is $|E_I|^2$.   □

4.3   Extracting the Optimal Schedule

We present a technique to extract the optimal schedule corresponding to any memoryless winning strategy σ for the reachability player. Consider the scenario where the safety player uses the same memoryless winning strategy σ. Let ρ denote the play corresponding to this scenario. Algorithm 3 presents the procedure to construct this play ρ and extract the optimal schedule.

Algorithm 3. Extracting the optimal schedule.

Input memoryless winning strategy σ.
Output optimal schedule S.
 1: $v \leftarrow (v_{init}, v_{init}, 0, r)$
 2: $\rho \leftarrow \varnothing$
 3: **while** $v \neq (v_f, *, 0, s)$ **do**
 4:     **if** fourth component of $v = r$ **then**
 5:         $\rho \leftarrow \rho \cup \{v\}$
 6:     **end if**
 7:     $v \leftarrow \sigma(v)$
 8: **end while**
 9: apply Algorithm 2 to ρ to retrieve S
10: **return** S

4.4   Minimizing Total Load and Makespan

The *completion time* of an agent is the time when the agent finishes all of its assigned tasks. *Total load* is defined as the sum of the completion times of all the agents [5]. To minimize the total load, we modify the counter in construction from Sect. 4; it now has k components, one for each agent. In the counter corresponding to agent a, we maintain the difference between the latest time when agent a is not idle across the two schedules (corresponding to the moves of the reachability player and the safety player). Since the smallest value of completion time is zero and the greatest value of completion time is $\mathcal{H}$, this difference is contained in the closed interval $[-\mathcal{H}, \mathcal{H}]$. Thus, the counter takes values from this interval.

Makespan is the total length of the schedule, i.e., the maximum value among the completion times of the agents. Makespan minimization is another common optimization criterion in the literature [1]. For minimizing the makespan, we modify the counter in the construction from Sect. 4. The counter takes values from the interval $[-1, \mathcal{H}]$. In the counter, we maintain the difference between the latest time when all the agents are idle across the two schedules (corresponding to the moves of the reachability player and the safety player). In Table 4, we present the counters for each type of optimization criteria.

Table 4. The number of components in the counter and the range of each component corresponding to the optimization criterion.

Optimization type	#components	Range of each component
Minimize makespan	1	$[-1, \mathcal{H}]$
Linear function	1	$[-c\mathcal{H}, c\mathcal{H}]$
Minimize total load	k	$[-\mathcal{H}, \mathcal{H}]$

5 Experimental Evaluation

In this section, we validate the reachability game approach for finding optimal schedules using 1) a case study for scheduling tasks for astronauts aboard the International Space Station and 2) randomized experiments for different optimization criteria.

5.1 Qualitative Evaluation

We solve a scheduling problem for six astronauts aboard the International Space Station (ISS). The astronauts have to perform a set $\mathcal{T}$ of lab tasks with variable durations. Due to power requirements, the astronauts can perform only three lab tasks at any time. Additionally, to stay healthy, the astronauts have to a) eat, b) use a treadmill and c) lift weights. An astronaut cannot exercise after she eats. Additionally, we also model a scenario where some of the astronauts are unhealthy (they may be injured or ill). We penalize schedules that use unhealthy astronauts. Thus, we assign an unhealthy astronaut a weight of -1 and assign all others a weight of 1.

Table 5. Time(s) for computing a valid schedule and an optimal schedule.

Number (n) of tasks	Valid	Optimal
10	0.07	0.179
20	0.08	0.191
30	0.18	0.592
40	0.20	0.608
50	3.56	4.450
60	3.62	4.515

Let $\mathcal{A}_h$ denote the set of healthy astronauts. We compute schedules that are optimal with respect to the linear function $\sum_{a \in \mathcal{A}_h} value_S(a) - \sum_{a \notin \mathcal{A}_h} value_S(a)$. For the experiments, we fix the window length Δ as 20 and the planning horizon $\mathcal{H}$ as 300. The time in seconds for computing both a valid schedule and an optimal schedule versus the number n of lab tasks is presented in Table 5.

Since only three lab tasks can be performed at any time, the worst-case complexity for finding an optimal schedule is reached when around 3Δ tasks ($n = 50$) have to be scheduled.

5.2 Randomized Evaluation and Comparison with an Integer Programming Formulation

For each optimization criterion, we generate random problem instances and record the time for synthesizing an optimal schedule for these instances. We fix the number of agents as five and assume that all the agents can perform all

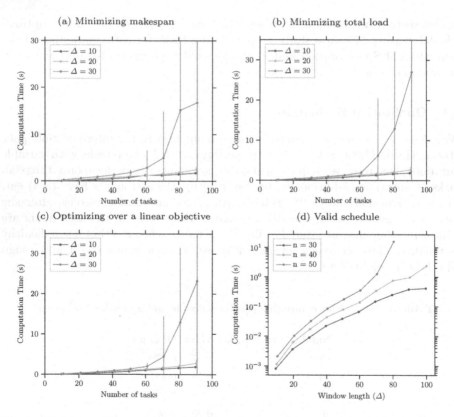

Fig. 1. (a)–(c) provide average running time for computing an optimal schedule for the corresponding optimization criterion. (d) provides average running time for computing a valid schedule as a function of window length.

the tasks. We fix the planning horizon $\mathcal{H} = 200$. We vary the window length Δ and the number n of tasks and observe its impact on the running time.

For each value of Δ and n, we generate 100 random instances and record the running time as the average over the running times of these 100 instances. In total, we run 10800 experiments to generate the graphs presented in Fig. 1. We performed the experiments on an Ubuntu 18.04 system with an Intel i7-8550U (1.80 GHz) processor and 16 GB memory.

The experiments show that we can compute optimal schedules in less than a minute for up to 100 tasks when $\Delta = 30$. After we fix the values of Δ and $\mathcal{H}$, if the number n of tasks is small compared to $\mathcal{H}$, the tasks are distributed sparsely across the windows. In this case, the $\mathcal{H}$ term dominates the complexity of finding an optimal schedule. However, as we increase n (until its upper bound $\mathcal{O}(k\mathcal{H})$), the density of tasks in each window increases. As a result, the $2^{4k\Delta}$ term dominates the complexity of finding an optimal schedule. This observation is consistent with Remark 2.

Finally, we compare the reachability game technique against an integer programming encoding for the ARCPSP problem. We use Gurobi [10], a state-of-the-art mixed-integer linear programming (MILP) solver for solving the integer programming formulation. Figure 2 contains the results of this comparison. The experiments show that the reachability game technique is at least two orders of magnitude faster than the integer program that we run on Gurobi.

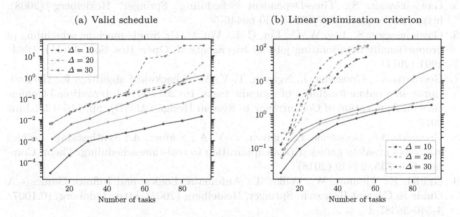

Fig. 2. Average running times for computing (a) a valid schedule and (b) an optimal schedule with respect to a linear optimization criterion using the reachability game formulation versus an IP encoding run in Gurobi. The solid lines correspond to run-times obtained by using the reachability game technique and the dashed lines correspond to the Gurobi implementation.

6 Conclusion

We identified a new parameter called *window length* for the agent resource-constrained project scheduling problem (ARCPSP). Using this parameter, we provide a novel algorithm for finding optimal schedules that scales polynomially in the number of tasks as long as the window length is a fixed constant. We illustrate the applicability of this method by solving a scheduling problem for astronauts aboard the International Space Station (ISS). Furthermore, a direct comparison with an integer program formulation that we run in Gurobi shows that this technique is at least two orders of magnitude faster.

References

1. Artigues, C., Demassey, S., Neron, E.: Resource-Constrained Project Scheduling: Models, Algorithms. Extensions and Applications, ISTE (2007)
2. Chatterjee, K.: Graph games with reachability objectives. In: Delzanno, G., Potapov, I. (eds.) RP 2011. LNCS, vol. 6945, p. 1. Springer, Heidelberg (2011). https://doi.org/10.1007/978-3-642-24288-5_1
3. Cheng, T., Ding, Q., Lin, B.: A concise survey of scheduling with time-dependent processing times. Eur. J. Oper. Res. **152**(1), 1–13 (2004)
4. Emerson, E.A., Jutla, C.S.: Tree automata, mu-calculus and determinacy. IEEE, October 1991
5. Gawiejnowicz, S.: Time-Dependent Scheduling. Springer, Heidelberg (2008). https://doi.org/10.1007/978-3-540-69446-5
6. Gawiejnowicz, S., Lee, W.-C., Lin, C.-L., Wu, C.-C.: Single-machine scheduling of proportionally deteriorating jobs by two agents. J. Oper. Res. Soc. **62**(11), 1983–1991 (2011)
7. Geeraerts, G., Goossens, J., Nguyen, T.-V.-A.: A backward algorithm for the multiprocessor online feasibility of sporadic tasks. In: 2017 17th International Conference on Application of Concurrency to System Design (ACSD), pp. 116–125, June 2017
8. Geeraerts, G., Goossens, J., Nguyen, T.-V.-A., Stainer, A.: Synthesising succinct strategies in safety games with an application to real-time scheduling. Theor. Comput. Sci. **735**, 24–49 (2018)
9. Grädel, E., Thomas, W., Wilke, T.: Automata, Logics, and Infinite Games - A Guide to Current Research. Springer, Heidelberg (2002). https://doi.org/10.1007/3-540-36387-4
10. Gurobi Optimization: L. Gurobi optimizer reference manual (2020)
11. Hughes, M.: Why projects fail: the effect of ignoring the obvious. Ind. Eng. **18**, 14–18 (1986)
12. Kononov, A.: Scheduling problems with linear increasing processing times. Operations Research Proceedings 1996, pp. 208–212. Springer, Heidelberg (1997). https://doi.org/10.1007/978-3-642-60744-8_38
13. Larman, C.: Agile and Iterative Development. Addison-Wesley, Boston (2004)
14. Lauffer, N.T., Topcu, U.: Human-understandable explanations of infeasibility for resource-constrained scheduling problems. In: Workshop on Explainable Planning (XAIP 2019) (2019)

15. Lombardi, M., Milano, M.: A precedence constraint posting approach for the RCPSP with time lags and variable durations. In: Gent, I.P. (ed.) CP 2009. LNCS, vol. 5732, pp. 569–583. Springer, Heidelberg (2009). https://doi.org/10.1007/978-3-642-04244-7_45

16. McNaughton, R.: Infinite games played on finite graphs. Ann. Pure Appl. Logic **65**(2), 149–184 (1993)

17. Mosheiov, G.: Multi-machine scheduling with linear deterioration. INFOR: Inf. Syst. Oper. Res. **36**(4), 205–214 (1998)

18. Mostowski, A.W.: Games with Forbidden Positions. UG (1991)

19. Neumann, K., Schwindt, C., Zimmermann, J.: Resource-constrained project scheduling with time windows. In: Józefowska, J., Weglarz, J. (eds.) Perspectives in Modern Project Scheduling. ISOR, pp. 375–407. Springer, Heidelberg (2006). https://doi.org/10.1007/978-0-387-33768-5_15

20. Radermacher, F.J.: Scheduling of project networks. Ann. Oper. Res. **4**(1), 227–252 (1985)

21. Stork, F., Uetz, M.: Enumeration of circuits and minimal forbidden sets. Electron. Notes Discret. Math. **13**, 108–111 (2003)

Improved Scheduling with a Shared Resource via Structural Insights

Christoph Damerius[1]([✉]), Peter Kling[1], Minming Li[2], Florian Schneider[1], and Ruilong Zhang[2]

[1] Universität Hamburg, Hamburg, Germany
christoph.damerius@uni-hamburg.de
[2] City University Hong Kong, Kowloon, Hong Kong SAR, China

Abstract. We consider a scheduling problem with resource-dependent processing speeds in which n jobs have to be scheduled on m machines that share a common resource. The resource may be distributed arbitrarily among the machines. This distribution is under the control of the scheduler and can be changed over time. Each job j has a processing volume $p_j \in \mathbb{N}$ and a resource requirement $r_j \in (0, 1]$. The latter indicates what fraction of the resource a job requires to run at full speed. Providing it with a larger share is not beneficial, but lowering its share results in a proportionally lowered processing speed. The goal is to schedule all jobs non-preemptively while minimizing the latest completion time.

This problem was introduced by Kling et al. [SPAA'17], who proved NP-hardness and gave an efficient algorithm with approximation ratio $2 + 1/(m - 2)$. The (asymptotic) tightness of that bound was left as an open question. We focus on the case of two machines and derive a strong, structural lower bound. This lower bound is based on a relaxed version and allows us to design an asymptotic 3/2-approximation that runs in time $O(n \cdot \log n)$. As an immediate consequence we also get an improved 9/4-approximation for the case of three machines.

Keywords: Approximation algorithm · Multiprocessor scheduling · Relaxation · Resource constraints · Shared resource · Makespan

1 Introduction

Resource allocation is probably among the oldest and most well-studied optimization problems. In the context of computing systems, the resource typically corresponds to computational power, often in the form of a number of machines that must process a set of incoming jobs while optimizing a suitable quality of service measure. Even for this restricted scenario, there is a huge variety of models, differing in both machine and job properties as well as in the considered quality of service measures (see [11] for a detailed overview).

However, computational power is not the only contended resource in computing systems. In fact, in modern HPC environments computational power is

© Springer Nature Switzerland AG 2020
W. Wu and Z. Zhang (Eds.): COCOA 2020, LNCS 12577, pp. 168–182, 2020.
https://doi.org/10.1007/978-3-030-64843-5_12

rarely the performance bottleneck. Instead, often other shared resources, like the bandwidth or an I/O bus, constitute the performance bottleneck of such systems. Thus, the distribution of these additional shared resources can severely impact the system performance [14]. In this work, we study systems with such an additional shared resource.

Both standard *resource constrained scheduling* [4,8,12] (in which jobs require a certain amount of the resource to be able to run) and scheduling models with *resource dependent processing times* [6–9] (where a job's processing time depends on the amount of resource it receives) found considerable interest in the research community. The model we study falls into the category of resource dependent processing times. However, a particular feature is that we not only allow the scheduler to assign a resource share to a job *once* (when it is started). Instead, the scheduler may readjust the resource distribution adaptively at integral time points (processor cycles).

As an example, consider a multiprocessor system with a shared communication bus of limited bandwidth. The processed jobs may have different communication requirements, depending on their data-processing (generation or consumption) rate. Assigning a job a bandwidth that saturates its data-processing rate yields optimal performance, while throttling its bandwidth typically results in an immediate, proportional efficiency drop. On the other hand, increasing a job's bandwidth above its data-processing rate has no beneficial effect. As jobs enter and leave the system, the scheduler should adjust the resource distribution to the new situation. Note that while the linear efficiency drop is natural in the described setting, the resource dependency may be more complex (e.g., concave), as in the case of a shared power supply or cooling system [13].

We obviously adopt an idealized perspective by disregarding aspects like how CPU-intensive a given job is and by assuming that the shared resource is the performance bottleneck. Nevertheless, we aim at understanding exactly this aspect of resource allocation in modern data driven computing centers, in which computational power is often available in abundance.

1.1 Basic System Model

The following scheduling problem, originally proposed by Kling et al. [10], models the scenario described above. There are $m \in \mathbb{N}$ *machines* from the set $M := [m] = \{1, 2, \ldots, m\}$ and $n \in \mathbb{N}$ *jobs* from the set $J := [n]$. Time is partitioned into integral (time) *slots* $t \in \mathbb{N}_0$, representing the time interval $[t, t + 1)$. The machines share a common, finite *resource*. During any slot t, each machine $i \in M$ is assigned a fraction $R_i(t) \in [0, 1]$ of the resource. The resource may not be overused, so we require $\sum_{i \in M} R_i(t) \leq 1$. At any time, each machine can be assigned at most one job (no machine sharing) and each job can be assigned to at most one machine (no parallelism). A job $j \in J$ is defined via two parameters, its *processing volume* $p_j > 0$ and its *resource requirement* $r_j \in (0, 1]$.[1] If j is assigned

---

[1] In [10], $r_j > 1$ is allowed. Our restriction is without loss of generality, as we can assign such jobs a resource requirement of 1 and increase their processing volume by a factor r_j to get an equivalent instance.

to machine i during slot t, it is processed at speed min $\{1, R_i(t)/r_j\}$, which is also the amount of the job's processing volume that finishes during this slot. Job j *finishes* in the first slot t after which all p_j units of its processing volume are finished. Preemption of jobs is not allowed, so once a job j is assigned to machine i, no other job can be assigned to i until j is finished. The objective is to find a *schedule* (a resource and job assignment adhering to the above constraints) that has minimum *makespan* (the first time when all jobs are finished).

This problem is known as SHARED RESOURCE JOB-SCHEDULING *(SRJS)* [10]. In the bandwidth example from before, the resource requirement r_j models how communication intensive a job is. For example, $r_j = 0.5$ means that the process can utilize up to half of the available bandwidth. Assigning it more than this does not help (it cannot use the excess bandwidth), while decreasing its bandwidth results in a linear drop of the job's processing speed.

Simplifying Assumptions. To simplify the exposition, throughout this paper we assume $r_j \neq 1$ for all $j \in J$. This simplifies a few notions and arguments (e.g., we avoid slots in which formally two jobs are scheduled but only one job is processed at non-zero speed) but is not essential for our results. We also assume $p_j \in \mathbb{N}$, which is not required in [10] but does not change the hardness or difficult instances and is a natural restriction given the integral time slots.

1.2 Related Work

In the following we give an overview of the most relevant related work. In particular, we survey known results on SRJS and other related resource constrained scheduling problems. We also briefly explain an interesting connection to bin packing.

Known Results for SRJS. The SRJS problem was introduced by Kling et al. [10], who proved strong NP-hardness even for two machines and unit size jobs. Their main result was a polynomial time algorithm with an approximation ratio of $2 + 1/(m - 2)$. If jobs have unit size, a simple modification of their algorithm yields an asymptotic $(1 + 1/(m - 1))$-approximation.

Althaus et al. [1] considered the SRJS problem for *unit size* jobs in a slightly different setting, in which the jobs' assignment to machines and their orders on each machine are fixed. They prove that this variant is NP-hard if the number of machines is part of the input and show how to efficiently compute a $(2 - 1/m)$-approximation. Furthermore, they provide a dynamic program that computes an optimal schedule in (a rather high) polynomial time, when m is not part of the input. For the special case $m = 2$, a more efficient algorithm with running time $O(n^2)$ is given.

Resource Constrained Scheduling. One of the earliest results for scheduling with resource constraints is due to Garey and Graham [4]. They considered n jobs that must be processed by m machines sharing s different resources. While processed, each job requires a certain amount of each of these resources. Garey and

Graham [4] considered list-scheduling algorithms for this problem and proved an upper bound on the approximation ratio of $s + 2 - (2s + 1)/m$. Garey and Johnson [5] proved that this problem is already NP-complete for a single resource, for which the above approximation ratio becomes $3 - 3/m$. The best known absolute approximation ratio for this case is due to Niemeier and Wiese [12], who gave a $(2 + \epsilon)$-approximation. Using a simple reduction from the PARTITION problem, one can see that no efficient algorithm can give an approximation ratio better than $3/2$, unless $P = NP$. While this implies that there cannot be a PTAS (polynomial-time approximation scheme), Jansen et al. [8] recently gave an APTAS (asymptotic PTAS).

Resource Dependent Processing Times. A common generalization of resource constrained scheduling assumes that the processing time of a job depends directly on the amount of resource it receives. Among the first to consider this model was Grigoriev et al. [6], who achieved a 3.75-approximation for unrelated machines. For identical machines, Kellerer [9] improved the ratio to $3.5 + \epsilon$. In a variant in which jobs are already preassigned to machines and for succinctly encoded processing time functions, Grigoriev and Uetz [7] achieved a $(3 + \epsilon)$-approximation. Recently, Jansen et al. [8] gave an asymptotic PTAS for scheduling with resource dependent processing times.

Note that in resource dependent processing times, the resource a job gets assigned is fixed and cannot change over time. In contrast, the model we consider gives the scheduler the option to adjust the resource assignment over time, which may be used to prioritize a short but resource intensive job during the processing of a long but less resource hungry job.

Connection to Bin Packing. Resource constrained scheduling problems are often generalizations of bin packing problems. For example, for a single resource, unit processing times and k machines, resource constrained scheduling is equivalent to *bin packing with cardinality constraints* [2,8] (where no bin may contain more than k items). Similarly, the SRJS problem is a generalization of *bin packing with cardinality constraints and splittable items*. Here, items may be larger than the bin capacity but can be split, and no bin may contain more than k item parts. This problem is (up to preemptiveness) equivalent to SRJS for k machines if all resource requirements are 1 and processing volumes correspond to item sizes.[2] In this case, each time slot can be seen as one bin.

Since we consider arbitrary resource requirements, we refer to [3] for the state of the art in bin packing with cardinality constraints and splittable items.

1.3 Our Contribution

We construct an efficient algorithm for SRJS on two machines that improves upon the previously best known approximation factor 2. Specifically, our main

---

[2] Alternatively, one can allow resource requirements > 1 and use these as item sizes while setting all processing volumes to 1, as described in [10].

result is the following theorem: (Due to space constraint, some proofs are omitted in this conference version.)

Theorem 1. *There is an asymptotic 1.5-approximation algorithm for SRJS with $m = 2$ machines that has running time $O(n \log n)$.*

This is the first algorithm reaching an approximation ratio below 2. As a simple consequence, we also get an improved asymptotic 9/4-approximation for $m = 3$ machines (compared to the bound $2 + 1/(m - 2) = 3$ from [10] for this case).

Our approach is quite different from Kling et al. [10]. They order jobs by increasing resource requirement and repeatedly seek roughly m jobs that saturate the full resource. By reserving one machine to maintain a suitable invariant, they can guarantee that, during the first phase of the algorithm, either always the full resource is used or always (almost) m jobs are given their full resource requirement. In the second phase, when there are less than m jobs left and these cannot saturate the resource, those jobs are always given their full resource requirement. Both phases are easily bounded by the optimal makespan, which is where the factor 2 comes from. While the bound of $2 + 1/(m - 2)$ becomes unbounded for $m = 2$ machines, the algorithm in this case basically ignores one machine and yields a trivial 2-approximation.

The analysis of [10] relies on two simple lower bounds: the optimal makespan OPT is at least $\sum_{j \in J} r_j \cdot p_j$ (job j must be assigned a total of $r_j \cdot p_j$ resource share over time) and at least $\sum_{j \in J} \lceil p_j \rceil / m$ (job j occupies at least p_j/m time slots on some machine). Our improvement uses a more complex, structural lower bound based on a relaxed version of SRJS as a blueprint to find a good, non-relaxed schedule. The relaxed version allows (a) preemption and (b) that the resource and job assignments is changed at arbitrary (non-integral) times. More exactly, we show that there is an *optimal* relaxed *structured schedule* S_R in which, except for a single *disruptive* job j_D, one machine schedules the jobs of large resource requirement in descending order of resource requirement and the other machine schedules the jobs of small resource requirement in ascending order of resource requirement. We further simplify such a structured schedule S_R by assigning jobs j with small r_j their full resource requirement, yielding an *elevated schedule* $\hat{S}_R$. This elevated schedule is no longer necessarily optimal, but we can show that it becomes not much more expensive. This elevated schedule $\hat{S}_R$ yields the aforementioned structural lower bound, which we use to guide our algorithm when constructing a valid, non-relaxed schedule. The following theorem states a slightly simplified version of the guarantees provided by our structural lower bound. See Theorem 3 for the full formal statement.

Theorem 2. *There is an optimal structured relaxed schedule S_R and an elevated structured relaxed schedule $\hat{S}_R$ with a distinguished disruptive job j_D such that*

1. *if $r_{j_D} \leq 1/2$, then $|\hat{S}_R| \leq |S_R|$ and*
2. *if $r_{j_D} > 1/2$, then $|\hat{S}_R| \leq c_{j_D} \cdot |S_R| + 0.042 \cdot A_{j_D}$.*

Here, A_{j_D} denotes the total time for which j_D is scheduled in S_R and the value c_{j_D} depends on r_{j_D} but lies in $[1, 1.18)$.

While our lower bound does not immediately extend to the case of more machines, we believe that this is an important first step towards designing an improved approximation algorithm for arbitrary number of machines.

Note that we also show that our bound from Theorem 1 is tight in the sense that there are instances for which an optimal relaxed schedule is by a factor of 3/2 shorter than an optimal non-relaxed schedule. Thus, improving upon the asymptotic 3/2-approximation would require new, stronger lower bounds.

2 Preliminaries

Before we derive the structured lower bound in Sect. 3 and use it to derive and analyze our algorithm in Sect. 4, we introduce some notions and notation which are used in the remainder of this paper.

Schedules. We model a schedule as a finite sequence S of *stripes*. A stripe $s \in S$ represents the schedule in a maximal time interval $I(s)$ with integral endpoints in which the job and resource assignments remain the same. The order of these stripes in the sequence S corresponds to the temporal order of these intervals. We let $J(s) \subseteq J$ denote the jobs scheduled during $I(s)$. For $j \in J(s)$ we let $R_j(s)$ denote the resource share that job j receives during $I(s)$ (i.e., the resource assigned to the machine that runs j). To ease the exposition, we sometimes identify a stripe s with the time interval $I(s)$. This allows us, e.g., to speak of a job $j \in J(s)$ scheduled during stripe s, to use $|s| := |I(s)|$ to refer to the *length* of a stripe s, or to write $s \subseteq T$ if a stripe's interval $I(s)$ is contained in another time interval T.

Any finite sequence S of stripes can be seen as a – possibly invalid – schedule for SRJS. To ensure a valid schedule, we need additional constraints: The resource may not be overused and we may never schedule more than m jobs, so we must have $\sum_{j \in J(s)} R_j(s) \leq 1$ and $|J(s)| \leq m$ for all $s \in S$. Since j is processed at speed $\min\{1, R_j(s)/r_j\}$ during s, we can assume (w.l.o.g.) that $R_j(s) \leq r_j$ for all $s \in S$ and $j \in J(s)$. With this assumption, the requirement that a schedule finishes all jobs can be expressed as $\sum_{s \in S: \, j \in J(s)} |s| \cdot R_j(s)/r_j \geq p_j$ for all $j \in J$. Not allowing preemption implies that the *active time* $A_j := \bigcup_{s \in S: \, j \in J(s)} I(s)$ of each job j must form itself an interval. While the sequence S does not give a specific assignment of jobs to machines, we can derive such an assignment easily via a greedy round robin approach.

W.l.o.g., we assume that for all $s \in S$ we have $J(s) \neq \emptyset$, since otherwise we can delete s and move subsequent stripes by $|s|$ to the left. Thus, we can define the *makespan* of S as $|S| := \sum_{s \in S} |s|$. When dealing with multiple schedules, we sometimes use superscript notation (e.g., A_j^S) to emphasize the schedule to which a given quantity refers.

Relaxed Schedules. We also consider a relaxed version of the SRJS problem (r-SRJS), in which the resource and job assignments may change at non-integral times and in which jobs can be preempted (and migrated) finitely often. This

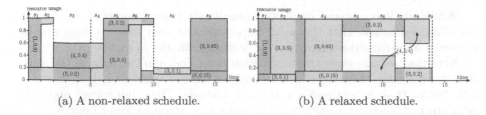

(a) A non-relaxed schedule.   (b) A relaxed schedule.

Fig. 1. A non-relaxed and a relaxed schedule for $m = 2$ machines and $n = 8$ jobs, with their parameters indicated in forms of tuples (p_j, r_j). Stripes are indicated by color grading. Note that in the non-relaxed schedule, all stripes start/end at integral time points and no job is preempted. (Color figure online)

gives rise to *relaxed schedules* S, which are finite sequences of stripes adhering to the same constraints as schedules except that the time intervals $I(s)$ for $s \in S$ may now have *non-integral* endpoints and the jobs' active times A_j are not necessarily intervals. Figure 1 illustrates how relaxed schedules differ from non-relaxed schedules. Relaxed schedules can be considerably shorter.

Subschedules, Substripes, and Volume. For a schedule S we define a *subschedule* S' of S as an arbitrary, not necessarily consecutive subsequence of S. Similarly, a *relaxed subschedule* S'_R of a relaxed schedule S_R is a subsequence of S_R. A *substripe* s' of a stripe s is a restriction of s to a subinterval $I(s') \subseteq I(s)$ (we denote this by $s' \subseteq s$). In particular, s' has the same job set $J(s') = J(s)$ and the same resource assignments $R_j(s') = R_j(s)$ for all $j \in J(s')$. For a (sub-)stripe s we define the *volume* of job $j \in J(s)$ in s as $V_j(s) := R_j(s) \cdot |s|$. The *volume* of a (sub-)stripe $s \in S$ is defined as $V(s) := \sum_{j \in J(s)} V_j(s)$ and the *volume* of a subschedule S' as $V(S') := \sum_{s \in S'} V(s)$.

Big and Small Jobs. Using the resource requirements, we partition the job set $J = J_B \uplus J_S$ into *big jobs* $J_B := \{ j \in J \mid r_j > 1/2 \}$ and *small jobs* $J_S := \{ j \in J \mid r_j \leq 1/2 \}$. Given a (relaxed or non-relaxed) schedule S, a *region* is a maximal subschedule of consecutive stripes during which the number of big jobs and the number of small jobs that are simultaneously scheduled remain the same. The *type* $\mathcal{T} \in \{ B, S, (B, B), (S, S), (B, S) \}$ of an interval, (sub-)stripe, region indicates whether during the corresponding times exactly one big/small job, exactly two big/small jobs, or exactly one big job and one small job are scheduled. We call a stripe of type $\mathcal{T}$ a $\mathcal{T}$-stripe and use a similar designation for (sub-)stripes, regions, and intervals. If there exists exactly one $\mathcal{T}$-region for schedule S, then we denote that region by $S^\mathcal{T}$.

3   A Lower Bound for SRJS

To derive a lower bound, we aim to "normalize" a given optimal relaxed schedule such that it gets a favorable structure exploitable by our algorithm. While this normalization may increase the makespan, Theorem 3 will bound this increase by a small factor. In the following we first provide the high level idea of our approach and major notions. Afterwards we give the full formal definitions and results.

High Level Idea. We can assume that, at each time point, the optimal relaxed schedule either uses the full resource or both two jobs reach their full resource requirement (see Definition 1). A relaxed schedule that satisfies this property is called *reasonable*. Any unreasonable schedule can be easily transformed into a reasonable schedule.

Next, we balance the jobs by ordering them. We show that it is possible to transform a relaxed optimal schedule such that one machine schedules the jobs of large resource requirement in decreasing order of resource requirement and, similarly, the other machine schedules the jobs of small resource requirement in ascending order of resource requirement. However, the load of machines may not be equal in the end. Therefore, we may need a *disruptive* job j_D to equalize the loads. The aforementioned order then only holds up to some point where j_D starts. Intuitively, if this job would be scheduled later, the load of both machines would be further imbalanced and the makespan would increase. A relaxed schedule that satisfies the above ordering constraint is called *ordered* (see Definition 3 for the formal definition). If an ordered relaxed schedule satisfies an additional property about the scheduling of j_D, we call that schedule *structured* (see Definition 4). We prove that we can always transform an optimal relaxed schedule into an ordered one without increasing the makespan (see Lemma 1).

As a further simplification, we increase the resource of all small jobs to their full resource requirement. This process is called *elevating* (see Definition 5). Intuitively, elevating makes the small jobs be processed with a higher speed, but this may come at the price of processing big jobs with a lower speed and thus increase the makespan. To be more precise, we show that the makespan may only increase when small jobs scheduled together with a disruptive big job are elevated. In Theorem 3, we analyze the makespan increase incurred by elevating in detail.

Formal Definitions and Results

Definition 1. For a job set J, we call a (sub-)stripe s with $J(s) \subseteq J$ *reasonable* if $\sum_{j \in J} R_j(s) = 1$ or $R_j(s) = r_j$ for all jobs $j \in J(s)$. A subschedule is called *reasonable* if all of its (sub-)stripes are reasonable and *unreasonable* otherwise.

Definition 2. Define a strict total order $\prec$ on $j, j' \in J$ where $r_j \neq r_{j'}$ as $j \prec j' :\Leftrightarrow r_j < r_{j'}$. Otherwise order j, j' arbitrarily but consistently by $\prec$.

The following Definition 3 formalizes the *ordered*-property of relaxed schedules S_R, which is central to our lower bound. Intuitively, it requires that S_R can be split into a left part and a right part (separated by a stripe index l). The left part schedules the smallest jobs $J_{\prec}^{S_R}$ in ascending order (according to $\prec$) and the biggest jobs $J_{\succ}^{S_R}$ in descending order. It is separated from the right part

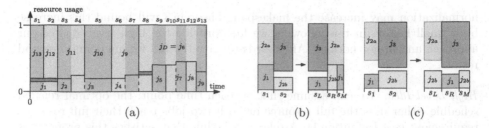

(a)   (b)   (c)

Fig. 2. (a) An ordered relaxed schedule $S_R = (s_i)_{i=1,...,13}$ with the disruptive job $j_D = j_6$, where $j_1 \prec \cdots \prec j_{13}$. In this example, $l = 8$, $J_R^{S_R} = \{j_4, ..., j_9\}$, $j_D \in J_B$ and as such the other jobs in $s_8, ..., s_{13}$ are sorted ascendingly after $\prec$. (b+c) Exchange argument of Observation 1.

throughout which the disruptive job j_D and the remaining jobs are scheduled. j_D could be any of the jobs from $J \setminus (J_\prec^{S_R} \cup J_\succ^{S_R})$ and adheres to no further ordering constraint, hence its disruptive nature. The ordering of the remaining jobs which are scheduled together with j_D is either ascending (if j_D is big) or otherwise descending. (See Fig. 2a for an example.) The definition also includes a special case where all stripes schedule j_D (then S_R only comprises out of its right part). We start by giving the main definitions for *ordered*, *structured* and *elevated* relaxed schedules.

Definition 3. For a job set J, we call a reasonable relaxed schedule $S_R = (s_1, ..., s_k)$ *ordered* for the *disruptive* job $j_D \in J$, if

1. If s is a stripe with $|J(s)| = 1$, then $s = s_k$ and $A_{j_D} = [0, |S_R|)$.
2. If $\forall i = 1, ..., k : |J(s_i)| = \{j_{i,1}, j_{i,2}\}$ with $j_{i,1} \prec j_{i,2}$, then there exists an $l \in [k]$ such that $j_D \in J(s_i)$ iff $i \geq l$ and for $J_R^{S_R} := \{j \mid R_j(s_i) > 0, i \in \{l, ..., k\}\}$, we have $j_{1,1} \preceq \cdots \preceq j_{l-1,1} \preceq J_R^{S_R} \preceq j_{l-1,2} \cdots \preceq j_{1,2}$. Further, define $J_\prec^{S_R} = (j_{i,1})_{i=1,...,l-1}$ and $J_\succ^{S_R} = (j_{i,2})_{i=1,...,l-1}$.
3. For all $i \leq k - 1$ where $J(s_i) = \{j_D, j\}$, $J(s_{i+1}) = \{j_D, j'\}$ with $j \neq j'$, we have $j \prec j'$ iff $j_D \in J_B$.

Remark 1. In the context of Definition 3, we can assume that j_D is always chosen such that there exists a stripe s with $j_D \in J(s)$ such that $j \preceq j_D$ for all $j \in J(s)$. Otherwise, choose s with $J(s) = \{j_D', j_D\}$ with j_D' being minimal according to $\prec$. We assign j_D' to be the new disruptive job, and reverse the order of all stripes that schedule j_D if necessary, to reobtain property 3 of Definition 3.

Remark 2. The orderedness immediately implies the existence of up to three regions in the order $(B, S), (B, B), B$ or $(B, S), (S, S), S$. For example, in Fig. 2a, $j_D \in J_B$ and therefore the jobs $j_6 = j_D, ..., j_{13}$ are big jobs, while $j_1, ..., j_5$ might all be small jobs. Then the stripes $s_1, ..., s_9$ form a (B, S) region, while $s_{10}, ..., s_{13}$ form a (B, B) region. No B region exists in this case.

We will first show that any optimal relaxed schedule can be transformed into an ordered schedule without losing its optimality (Lemma 1). The proof mainly relies on an exchange argument (Observation 1) to deal with unreasonable relaxed schedules.

Observation 1. Let s_1, s_2 be stripes with $J(s_1) = \{j_1, j_{2a}\}$, $J(s_2) = \{j_{2b}, j_3\}$, where $j_1 \prec j_{2x} \prec j_3 \ \forall x \in \{a, b\}$ and $j_{2a} \neq j_{2b}$. We can transform them into stripes s_L, s_R and possibly s_M, such that $J(s_L) = \{j_1, j_3\}$, $J(s_R) = \{j_{2a}, j_{2b}\}$ and $J(s_M)$ is either $J(s_1)$ or $J(s_2)$, such that the volume of jobs scheduled and total length of stripes is unchanged.

Lemma 1. *For any job set J there exists an ordered, optimal relaxed schedule S_R.*

Definition 4. We call an ordered relaxed schedule S_R for J and for $j_D \in J$ *structured*, if no stripes s with $J(s) = \{j_D, j'\}$ with $j_D \prec j'$ exist or $R_{j_D}(A_{j_D} \setminus S_R^{(B,S)}) = r_{j_D}$.

Definition 5. We call a subschedule $S_R' \subseteq S_R$ *elevated* in a relaxed schedule S_R, if $R_j(A_j^{S_R'}) = r_j$ for all $j \in J_S$.

The following two lemmas show the existence of structured, optimal relaxed schedules. Lemma 2 essentially tells us that an optimal relaxed schedule can either be fully elevated or structured and at least partially elevated. Lemma 3 then gives rise to structured optimal elevated relaxed schedules if the full elevation in Lemma 2 was possible. Unfortunately, the full elevation step can not always be pursued while staying optimal. Theorem 3 gives details about the makespan increase.

Lemma 2. *For every job set J there exists an optimal ordered (for job j_D) relaxed schedule S_R which is either elevated, or is elevated in $S_R \setminus A_{j_D}^{S_R}$, $R_{j_D}(A_{j_D}^{S_R} \setminus S_R^{(B,S)}) = r_{j_D}$ and $j_D \in J_B$.*

Lemma 3. *For any optimal elevated ordered relaxed schedule S_R for a job set J there exists a structured optimal elevated relaxed schedule $\hat{S}_R$ for J.*

Theorem 3. *For every job set J there exists a structured (for $j_D \in J$) optimal relaxed schedule S_R and a structured (for j_D), elevated relaxed schedule $\hat{S}_R$ such that one of the following holds:*

1. *$j_D \in J_S$ and $|\hat{S}_R| \leq |S_R|$.*
2. *$j_D \in J_B$ and $\hat{S}_R^B = \varnothing$. Let $a_X := |X^{(B,S)} \setminus A_{j_D}|$ and $b_X := |X^{(B,S)} \cap A_{j_D}|$ for $X \in \{S_R, \hat{S}_R\}$. Then $|\hat{S}_R| \leq |S_R| + \lambda b_{S_R}$, $a_{\hat{S}_R} \leq a_{S_R}$ and $b_{\hat{S}_R} \leq b_{S_R}$, where λ is the smallest positive root of $(\lambda + 1)^3 - 27\lambda$.*
3. *$j_D \in J_B$, $\hat{S}_R^B \neq \varnothing$ and $|\hat{S}_R| \leq (4 - 2r_{j_D} - \frac{1}{r_{j_D}})|S_R| \leq (4 - \sqrt{8})|S_R|$.*

4   Approximation Algorithm and Analysis

Our approximation algorithm ALG for SRJS constructs the schedule by using the structure derived in Theorem 3 as a starting point. To accomplish this, ALG is designed to first gain information about the relaxed schedule $\hat{S}_R$ given by Theorem 3 by essentially replicating the order given by the orderedness property. Based on this information, either ALG_{big} or $\text{ALG}_{\text{small}}$ is executed. Essentially, ALG determines whether j_D, as given by Theorem 3, is in J_S or in J_B and branches into $\text{ALG}_{\text{small}}$ or ALG_{big} accordingly.

$\text{ALG}_{\text{small}}$ processes the jobs by first scheduling the small jobs in descending order of their processing volumes (using ASSIGN), and scheduling the big jobs in arbitrary order afterwards. For this case it can be easily shown that our bound is satisfied (see the proof of Theorem 4).

For the case of $j_D \in J_B$, a more sophisticated approach is needed. ALG_{big} roughly schedules the jobs in the (B, S) region, mimicking the order as in $\hat{S}_R$. Afterwards, ALG_{big} schedules all remaining big jobs using a slightly modified longest-first approach. Care has to be taken on the transition between both parts of ALG_{big}. For that reason, we calculate two schedules S and S', one of which can be shown to match the desired bound. Their job scheduling order mainly differs in whether the longest (S) or most resource intensive remaining job (S') is scheduled first. The remaining jobs are then scheduled in a longest-first order. Lastly, the machine loads are balanced by adjusting the resources given to the big jobs in the second part (BALANCELENGTH).

We will first give the pseudocode of the algorithm, then describe the subroutines involved and then give some analysis.

ALG(job set J)
1   $S'_R :=$ empty relaxed schedule
2   $J_S := \{j \in J \mid r_j \le 0.5\}$
3   $J_B := \{j \in J \mid r_j > 0.5\}$
4   Sort J_S ascendingly after $\prec$
5   Sort J_B descendingly after $\prec$
6   for $j \in J_S$ do ASSIGN$(S'_R, j, 1, 0)$
7   $V := \sum_{j \in J_B} p_j r_j$
8   $H := J_B$
9   for $j \in J_B$ do
10      $t = load^{S'_R}(2)$
11   if GETENDPOINT2$(S'_R, j) \ge load^{S'_R}(1)$:
12   return $\text{ALG}_{\text{big}}(J_S, H, J_B \setminus H, t)$
13   if $|H| = 1$:
14   break
15      $j_l := \text{argmax}_{j' \in H \setminus \{j\}} p_{j'}$
16   ASSIGN$(S'_R, j, 2, 0)$
17      $V := V - p_j r_j$
18      $u := \text{GETENDPOINT2}(S'_R, j_l)$
19   if $(u - load^{S'_R}(1))(1 - r_{j_l}) > V - r_{j_l} p_{j_l}$:
20   return $\text{ALG}_{\text{big}}(J_S, H, J_B \setminus H, t)$
21      $H := H \setminus \{j\}$
22   return $\text{ALG}_{\text{small}}(J_S, J_B)$

ALG_{big}(job set J_S, job set H, job set J'_B, t)
1   $S := \varnothing$
2   $t' := \lceil \lfloor t \rfloor / 2 \rceil + 2$
3   Sort J'_B descendingly after $\prec$
4   Sort H descendingly after $\prec$
5   for $j \in J_S$ do ASSIGN$(S, j, 1, t')$
6   for $j \in J'_B$ do ASSIGN$(S, j, 2, 0)$
7   $j_f := \text{POP}(H)$
8   Sort $j \in H$ descendingly after p_j
9   if $H = \varnothing$:
10   ASSIGN$(S, j_f, 1, t')$ and return S
11   $j_l := \text{POP}(H)$
12   $S' := S$
13   ASSIGN$(S, j_l, 2, t')$
14   ASSIGN$(S', j_f, 2, t')$
15   if $i_{\min}(S) = 2$: ASSIGN$(S, j_f, 2, t')$
16   else : $H = H \cup \{j_f\}$
17   ASSIGN$(S', j_l, i_{\min}(S'), t')$
18   for $j \in H$ do ASSIGN$(S, j, i_{\min}(S), 1, t')$
19   for $j \in H \setminus \{j_f\}$ do
20   ASSIGN$(S', j, i_{\min}(S'), 1, t')$
21   for $\hat{S} \in \{S, S'\}$ do BALANCELENGTH$(\hat{S})$
22   return $\text{argmin}_{\hat{S} \in \{S, S'\}} |\hat{S}|$

$\text{ALG}_{\text{small}}$(job set J_S, job set J_B)
1   $S :=$ empty schedule
2   Sort the $j \in J_S$ descendingly after p_j
3   for $j \in J_S$ do ASSIGN$(S, j, i_{\min}(S), 0)$
4   $i := i_{\min}(S)$
5   for $j \in J_B$ do ASSIGN$(S, j, i, 0)$
6   return S

Description of Subroutines. ASSIGN(S, j, i, t) schedules a job j into a (relaxed or unrelaxed) schedule S on machine i, starting at the earliest time point possible, but not before t. In contrast to the lower bound, where we did not state on which machine we schedule, here we always give a specific machine to schedule on, as it simplifies further analysis. The resource is assigned to j by consecutively considering the slots $s = t, t + 1, \ldots$ and in each of these slots giving j the maximum possible resource. The given resource is only restricted by r_j, other already scheduled jobs (their given resources remain unchanged) and the remaining processing volume that j needs to schedule in s.

For a (relaxed) schedule S, we denote by $load^S(i)$ the earliest time point after which machine i remains idle for the rest of S. Furthermore, define $i_{min}(S) \in \{1, 2\}$ to be a machine that has the lowest load in S. We assume that job sets will retain their ordering when passed as arguments or when elements are removed.

GETENDPOINT2(S, j) simulates scheduling j into S as ASSIGN$(S, j, 2, 0)$ does, but instead returns $load^S(2)$ without altering S. This is useful so that we do not have to revert the scheduling of jobs.

BALANCELENGTH processes a given schedule S as follows: It only changes S if $\Delta := |load^S(2) - load^S(1)| \geq 2$. If so, it checks if the job j, scheduled in the last slot of S, is given r_j resource during $A_j^S \cup S^{(B,B)}$. As we have scheduled longest-first, there can be at most one stripe s with $J(s) = \{j, j'\}$ in $S^{(B,B)}$ during which j does not get r_j resource. The algorithm then gives j' less resource during s, pushing all other jobs scheduled with j after s to the right, shortening Δ. Simultaneously, BALANCELENGTH gives j more resource during s, shortening Δ even further. All jobs but j scheduled after s are given at most $1 - r_j$ resource and as such their resource does not change when moved. Hence, it is straightforward to calculate how much volume has to be redistributed to shorten Δ until either $\Delta \leq 1$ or j is given r_j resource in s.

Outline of the Analysis. We will first analyze how ALG branches into ALG$_{small}$ or ALG$_{big}$ and with which arguments. In the case that ALG branches into ALG$_{small}$, we can easily derive the 1.5-approximation factor in Theorem 4.

ALG$_{big}$ basically consists of a first part where roughly the jobs from the (B, S)-region in $\hat{S}_R$ are scheduled and a second part where the remaining jobs are scheduled longest-first. To take care of the transition between both parts, we define the notion of a *bridge* job j_β:

Definition 6. For a structured relaxed schedule S_R, the *bridge* job j_β is the smallest job scheduled in $S_R^{(B,S)}$ that is not scheduled together with jobs $j \succ j_\beta$ in S_R (if it exists).

Using this definition, we are now able to give Lemma 4, which shows which sub-algorithm is executed with precisely which arguments by ALG.

Lemma 4. *Let J be a job set and S_R, $\hat{S}_R$ be the relaxed schedules given by Theorem 3 for J, with j_D being the disruptive job of $\hat{S}_R$. If property 1 of Theorem 3 holds, then calling ALG(J) will execute ALG$_{small}(J_S, J_B)$ with*

J_S (J_B) being ascendingly (descendingly) sorted after $\prec$, respectively. Otherwise, $\text{ALG}_{big}(J_S, H, J_B \setminus H, t)$ will be executed with arguments $t = \inf(A_{j_\beta}^{\hat{S}_R})$ if there exists a bridge job j_β for $\hat{S}_R$, or $t = 0$ otherwise. Furthermore, $H = \{ j \in J_B \mid \inf(A_j^{\hat{S}_R}) \geq t \} \neq \varnothing$.

H is the set of all big jobs scheduled not before the bridge job. For the first part, ALG_{big} schedules J_S and $J_B \setminus H$, mimicking the order as in $\hat{S}_R$ into the interval $[0, t' + \lfloor t \rfloor)$. The following observation guarantees that ALG can fit these jobs into said interval.

Observation 2. Let S_R be a relaxed schedule for a job set J as obtained by Theorem 3, with $S_R^{(B,S)} \neq \varnothing$. Then ASSIGN in line 6 of ALG_{big} does not schedule jobs beyond $t' + \lfloor t \rfloor$.

For the second part, only big jobs remain to be scheduled. We can show they need at most $\approx 4/3 \cdot V$ slots, where V is their total volume, or the number of slots they require is dominated by one long job (where all other big jobs can fit onto the other machine). We can then guarantee the bound for one of the schedules procured by ALG_{big}, which then helps us to prove the overall bound.

In summary, we show the following theorem.

Theorem 4. *For any job set J we have $|\text{ALG}(J)| \leq 1.5 OPT + O(1)$, and this bound is tight for ALG.*

Proof. We only show the asymptotic lower bound here. Construct a job set with $k \in \mathbb{Z}_{2n}$ unit-size jobs j with $r_j = 1/3$ and $k - 1$ unit-size jobs j with $r_j = 2/3$. We can obviously construct a schedule of makespan k. The corresponding relaxed schedule obtained by Theorem 3 will have $j_D \in J_S$, so $\text{ALG}_{\text{small}}$ will be executed. It will first schedule all small jobs in $k/2$ slots. Afterwards, all big jobs will be scheduled on the same machine, using $k - 1$ slots. This gives the asymptotic lower bound of $3/2$.

5   Additional Results

Note that ALG, as stated in Sect. 4, does not necessarily have a running time of $O(n \log n)$. However, we prove this running time using a slightly modified algorithm in the following lemma.

Lemma 5. *ALG can be implemented to run in $O(n \log n)$ time for a job set J with $|J| = n$.*

Our results for the two-machine case also imply an improved bound for the three-machine case, which is based on ignoring one of the three machines.

Corollary 1. *For $m = 3$ and any job set J with $|J| = n$, we have $\text{ALG} \leq 9/4 \cdot OPT + O(1)$, where ALG runs in $\mathcal{O}(n \log n)$ time and OPT is the optimal solution for 3 machines.*

Lemma 6. *For any $P \in \mathbb{N}$ there exists a job set J with $\sum_{j \in J} p_j \geq P$ such that $3/2 \cdot |S_R| \leq |S|$, where S_R and S are optimal relaxed (unrelaxed) schedules for J, respectively. Furthermore, S_R can be chosen not to use preemption.*

In Lemma 6, S_R did not even use preemption. Thus, the slotted time is the reason for the 3/2-gap between relaxed and unrelaxed schedules. To beat the approximation ratio of 3/2, we would have to improve our understanding of how slotted time affects optimal schedules.

6   Conclusion and Open Problems

Using structural insights about the SRJS problem, we were able to improve approximation results from [10] for the cases of $m \in \{2, 3\}$ machines in the SRJS problem. As mentioned in Sect. 5, our (asymptotic) 3/2-approximation for $m = 2$ is the best possible result that can be achieved with the lower bound based on our definition of relaxed schedules and can be computed in time $O(n \log n)$. This leaves two natural research questions.

First, can a similar approach improve further improve the competitive ratio of [10] for larger values of m? While the lower bound we constructed in Sect. 3 is tailored towards $m = 2$, the underlying principle may be used to design improved algorithms for $m > 2$ machines. Indeed, a key insight behind our improvement stems from a worst case instance for the algorithm of [10]: Consider an instance with a job j that has small resource requirement r_j and processing volume $p_j \approx OPT$. An optimal schedule must begin to process p_j early, in parallel to the rest of the instance. However, the algorithm from [10] is *resource-focused*, in the sense that it orders jobs by resource requirement and basically ignores the processing volume when selecting jobs to be processed. This might result in j being processed at the very end, after all other jobs have been finished, possibly yielding an approximation ratio of roughly 2 (for large m). One could fix this problem using an algorithm focused on processing-volume, but that would run into similar issues caused by different resource requirements. Our algorithm for $m = 2$ basically identifies jobs like j (the disruptive job) and uses it to balance between these two extremes. A key question when considering such an approach for larger m is how to identify a suitable (set of) disruptive job(s).

A second possible research direction is to beat the lower bound limit of our structural approach. Given its relation to other resource constrained scheduling problems and to bin packing variants, it seems possible that one can even find a PTAS for SRJS. While a PTAS has typically a worse runtime compared to more direct, combinatorial algorithms, it would yield solutions that are arbitrarily close to optimal schedules. A difficulty in constructing a PTAS seems to stem from the partition of time into *discrete* slots. An incautious approach might yield cases where all machines are stuck with finishing an ε-portion of work, forcing them to waste most of the available resource in such a time slot. If the average job processing time is small, this might have a comparatively large influence on the approximation factor. Previous work [10] reserved one of the m machines to deal with such problems (which is wasteful for small m). Also augmenting the available resource in each slot by, e.g., a $1 + \varepsilon$ factor should help to circumvent such difficulties.

Acknowledgement. Peter Kling and Christoph Damerius were partially supported by the DAAD PPP with Project-ID 57447553. Minming Li is also from City University of Hong Kong Shenzhen Research Institute, Shenzhen, P.R. China. The work described in this paper was partially supported by Project 11771365 supported by NSFC.

References

1. Althaus, E., et al.: Scheduling shared continuous resources on many-cores. J. Sched. **21**(1), 77–92 (2017). https://doi.org/10.1007/s10951-017-0518-0
2. Epstein, L., Levin, A.: AFPTAS results for common variants of bin packing: a new method for handling the small items. SIAM J. Optim. **20**(6), 3121–3145 (2010)
3. Epstein, L., Levin, A., van Stee, R.: Approximation schemes for packing splittable items with cardinality constraints. Algorithmica **62**(1–2), 102–129 (2012)
4. Garey, M.R., Graham, R.L.: Bounds for multiprocessor scheduling with resource constraints. SIAM J. Comput. **4**(2), 187–200 (1975)
5. Garey, M.R., Johnson, D.S.: Complexity results for multiprocessor scheduling under resource constraints. SIAM J. Comput. **4**(4), 397–411 (1975)
6. Grigoriev, A., Sviridenko, M., Uetz, M.: Machine scheduling with resource dependent processing times. Math. Program. **110**(1), 209–228 (2007)
7. Grigoriev, A., Uetz, M.: Scheduling jobs with time-resource tradeoff via nonlinear programming. Discret. Optim. **6**(4), 414–419 (2009)
8. Jansen, K., Maack, M., Rau, M.: Approximation schemes for machine scheduling with resource (in-)dependent processing times. ACM Trans. Algorithms **15**(3), 31:1–31:28 (2019)
9. Kellerer, H.: An approximation algorithm for identical parallel machine scheduling with resource dependent processing times. Oper. Res. Lett. **36**(2), 157–159 (2008)
10. Kling, P., Mäcker, A., Riechers, S., Skopalik, A.: Sharing is caring: multiprocessor scheduling with a sharable resource. In: Scheideler, C., Hajiaghayi, M.T. (eds.) Proceedings of the 29th ACM Symposium on Parallelism in Algorithms and Architectures, SPAA 2017, Washington DC, USA, 24–26 July 2017, pp. 123–132. ACM (2017)
11. Leung, J.Y. (ed.): Handbook of Scheduling - Algorithms, Models, and Performance Analysis. Chapman and Hall/CRC (2004)
12. Niemeier, M., Wiese, A.: Scheduling with an orthogonal resource constraint. Algorithmica **71**(4), 837–858 (2015)
13. Rózycki, R., Weglarz, J.: Improving the efficiency of scheduling jobs driven by a common limited energy source. In: 23rd International Conference on Methods & Models in Automation & Robotics, MMAR 2018, Międzyzdroje, Poland, 27–30 August 2018, pp. 932–936. IEEE (2018)
14. Trinitis, C., Weidendorfer, J., Brinkmann, A.: Co-scheduling: prospects and challenges. In: Trinitis, C., Weidendorfer, J. (eds.) Co-Scheduling of HPC Applications [Extended Versions of All Papers from COSH@HiPEAC 2016, Prague, Czech Republic, 19 January 2016]. Advances in Parallel Computing, vol. 28, pp. 1–11. IOS Press (2016)

Network Optimization

Two-Stage Pricing Strategy with Price Discount in Online Social Networks

He Yuan, Ziwei Liang, and Hongwei Du[(⊠)]

Department of Computer Science and Technology, Key Laboratory of Internet
Information Collaboration, Harbin Institute of Technology (Shenzhen),
Shenzhen, China
heyuan@micc.hitsz.edu.cn, 20B951013@stu.hit.edu.cn, hongwei.du@ieee.org

Abstract. With the rapid development of online social networks
(OSNs), more and more product companies are focusing on viral marketing of products through the word-of-mouth effect. For product companies, designing effective marketing strategies is important for obtaining
profit. However, most existing research focuses on effective influence maximization analysis to disseminate information widely, rather than explicitly incorporating pricing factors into the design of intelligent marketing
strategies. In this paper, we have studied the product's marketing strategy and pricing model. We assume that the monopolistic seller divides
product marketing into two stages, the regular price stage and the discount price stage. All users have their own expected price of the product.
Only when the product price is not higher than the user's expected price,
the user will adopt the product. Therefore, we propose a pricing model
named Two-stage Pricing with Discount Model (TPDM). We propose
that companies use two marketing methods: Advertisement Marketing
(AM) and Word-of-mouth Marketing (WM). To achieve the goal of maximizing the profit of product companies, we propose a Two-stage with
Discount Greedy Algorithm (TSDG) to determine product price and
discount rate. In order to study the impact of advertising and word-of-
mouth marketing on product pricing on online social networks, we use
several real social network data sets for experiments. The experimental results show that advertising marketing can significantly increase the
profit of product companies.

Keywords: Pricing marketing · Profit maximization · Social networks

1 Introduction

With the rapid development of science and technology, online social networks
(OSNs) such as Facebook, Twitter and Google+ have become important platforms for people to communicate and share information. These social network
platforms make large-scale real-time communication become possible, which also
provides the potential for viral marketing innovation and a surge in opportu-
nities [13]. However, most of the existing research focuses on the problem of

© Springer Nature Switzerland AG 2020
W. Wu and Z. Zhang (Eds.): COCOA 2020, LNCS 12577, pp. 185–197, 2020.
https://doi.org/10.1007/978-3-030-64843-5_13

maximizing influence, and does not explicitly consider product pricing factors in product influence marketing strategies. Due to the fast information dissemination in OSNs, product companies want to place product advertisements on social networks to expand product visibility and profits. We assume that product companies want to take advantage of the rapid spread of information in OSNs, and place product advertisements in it to expand product visibility and profit. This method is to conduct product marketing through OSNs, which is also the following research problem that we need to solve.

Viral marketing does not really carry out marketing by spreading viruses, but by users' spontaneous word-of-mouth promotion of a product or service in OSNs, because this marketing method can deliver information to thousands of users. Viral marketing is often used for company product promotion. This marketing method that uses word of mouth between users is word-of-mouth marketing (WM). In addition to the WM, advertisement marketing (AM) can also increase product visibility and influence. For most companies, AM can greatly increase product attention and exposure, and gain greater influence [18].

There are three ways for companies to improve the competitiveness and profit of their products: reduce products cost; increase market share; adjust prices to adapt to market conditions [5]. Among them, adjusting prices is called a pricing strategy. The goal of the pricing strategy is to determine the optimal price under the circumstances of maximizing the current profit and maximizing the quantity of products sales. Historically, in production management, the goal is to reduce production costs and expand market share. The company has made tremendous efforts to reduce expenses (high cost). At now, more and more researchers are now trying to solve the problem of price adjustments. It should be noted that changing prices is obviously easier and faster than developing a process to reduce production costs or increase market share. In addition, the price parameter directly and strongly affects profit and market share. It has been shown that changing the price by 1% will cause daily consumption to change by at least 10%. Therefore, price as the adjustment parameter of profit is the simplest and fastest way to improve competitiveness.

For product companies, price is the main parameter that affects the companies' profit. In this paper, we study the market environment of single product and single oligarch. Companies use both advertising marketing and word-of-mouth marketing to attract consumers. We propose the company divides product promotion into two time stages, the regular price stage and the discount price stage, and maximizes profit through the two-stage pricing decision of the product. When a user is affected by a product advertisement in a social network, the user will have an expected price for the product. Only when the product price is not higher than the expected price will the user purchase the product. Therefore, in this paper, we consider the role of product price in the spread of product influence. And our contribution are as follows:

1) We divide the company's product marketing into two parts, namely Advertising Marketing (AM) and Word-of-mouth Marketing (WM). These two marketing methods greatly increase the influence of the product.

2) We divide product promotion into two stages, considering the effect of product prices on users' adoption. And propose a Two-stage Pricing with Discount Model (TPDM), which is closer to the situation of products selling in real life.

3) In order to solve the problem of product pricing, we propose a Two-stage with Discount Greedy algorithm (TSDG), which effectively improves the company's profits.

4) We test our algorithm on several real OSN datasets and use TPDM to compare different pricing strategies. The experimental results confirmed the effectiveness and the efficiency of our algorithm.

The rest of the paper is organized as follows. Section 2 reviews some related work. Section 3 proposes the TPDM and formulates the PROFIT problem. Section 4 introduces the pricing strategy and Two-stage with Discount Greedy algorithm (TSDG). Experiments are discussed in Section Sect. 5. Section 6 provides the conclusions.

2   Related Work

Our work is related to both Propagation Modelling and Pricing Strategy. We briefly discuss both of them.

2.1   Propagation Model

There has been a large amount of research [2, 4, 10, 12] on maximizing the influence in OSNs. Domingos and Richardson [6] first studied the influence diffusion as an algorithmic problem for optimizing marketing strategies. They treated the market as a social network and modeled it as a Markov random field, and then devised a heuristic algorithm to solve the problem. Kempe et al. [12] proposed independent cascade (IC) and linear threshold (LT) models. These two models are the basic models for the subsequent research [3, 9] on the diffusion model. In management science, people adopt new products with two steps, they are awareness and adoption respectively [11]. Lu et al. [16] proposed that users have their own valuations of product price, and users will only purchase the product if the product price is lower than the users' valuation. They raised a Linear Threshold Model with User Valuation (LT-V) model and defined a profit maximization problem which combined the product price and user's valuation. Zhu et al. [21] considered the relationship between influence and profit, and proposed price related PR frame and expanded the IC and LT models respectively. Zhang et al. [20] considered the multi-product influence in OSNs, and these products were not purely competitive. They presented a Multiple Thresholds (MT) model which extended from LT model.

2.2 Pricing Strategy

Many researchers start to study the problem of profit maximization on the basis of influence maximization, but they rarely consider the impact of product pricing on profits. Hartline et al. [8] and Dismitris et al [7]. researched the optimal marketing of digital goods in OSNs, and proposed an "Influence and Exploit" (IE) strategy. In IE strategy, product prices are different for seed nodes and non-seed nodes, and product prices are free for seed nodes. However, although price discrimination helps maximize profit, it may cause negative reactions from buyers. For example, Shor et al. [19] suggested that price discrimination can reduce the likelihood of purchase. Moreover, it is impractical for sellers to adjust prices too frequently. In order to solve price discrimination, Akhlaghpour et al. [1] based on the assumption that prices are open to all users, a descending pricing sequence is proposed. Similarly, the order of pricing in [17] is also falling. Niu et al. [17] proposed a multi-state diffusion scheme, and designed the best sequential pricing strategy based on dynamic programming. Based on the multi-state diffusion scheme, Li et al. [15] designed a pricing strategy with limited promotional time to study the problem of maximizing profit.

3   Preliminaries

In the section, we describe our proposed TPDM model in details and formulate the PROMAX problem. The relevant basic notations are shown in Table 1.

3.1   Propagation Model

In a social network $G(V, E)$, where all nodes $u_i \in V$ and all edges $(u_i, u_{j(j \neq i)}) \in E$, $i, j = 1, 2, ..., n$. User nodes have different judgements on the value of product, thus, all users in OSNs have an expected price $e(i)$ for the product. The u_i's expected price $e(i)$ indicates the highest price of the product that the u_i can accept. There is a gap price $g(i)$ between the actual product price and the expected price, e.g., $g(i) = e(i) - P$. If $g(i)$ is large, it means that the product price P is far lower than the user's expected price $e(i)$, and the u_i is more willing to recommend the product to its friends. According to the expected price of all nodes, we obtain the all users' gap prices $g = \{g(1), g(2), ..., g(n)\}$. For node u_i, the diffusion probability $p(i) = \frac{g(i)}{P}$. If the $p(i) < 0$, the diffusion probability $p(i) = 0$, if the $p(i) > 1$, the diffusion probability $p(i) = 1$. For all nodes, the probability $p(i) \in [0, 1]$. Since product advertisements also have a certain impact on users, we use q denotes the probability of users viewing advertisements. Based on the IC model in the [12], we propose a Two-stage Pricing with Discount Model (TPDM). Figure 1 shows the influence spread procession under the TPDM model. There are three states of nodes in social networks, they are INACTIVE, INFLUENCED and ACTIVE respectively. In the INACTIVE state, nodes not aware the product (Users not view the ads of the product). In the INFLUENCED state, nodes know the product but the product price exceed the expected price

Table 1. Important notations in the article.

Notation	Description
$G(V, E)$	The social network G with nodes set V and edges set E
n, m	The number of nodes and edges in G are n and m respectively
$e(i)$	The expected price of user i for company product
$g(i)$	The gap price between the actual product price and the expected price
$p(i)$	The probability of node i being influenced
q	The probability that node i views the ads
c	The cost of a single product
$t(i)$	The price tag of user i
$\tau(S)$	The total costs of selecting seed set S
$\rho(S, P, d)$	The profit of company

(Users view the ads of the product but not adopt). In the ACTIVE state, nodes spread the influence of the product (Users adopt the product and recommend it to friends). Our marketing strategy is divided into two time stages, regular price sales stage and discount sales stage. In Fig. 2, we can see the influence of product prices on whether users adopt products, Those two stages are showing as follows.

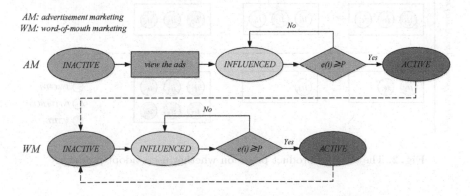

Fig. 1. The influence spread procession under the TPDM model.

(1) Regular Price Stage: The product company sets a price P for the product, and then targets product advertisements on social networks and selects some influential users to promote the product. For all users in social network, they view the ads of product with q probability. If the user i views the ads of product, their state will change to INFLUENCED. And if the product price P is not higher than their expected price $e(i)$, the user change the state to ACTIVE. The user i will adopt the product and recommends the product to friends with a $p(i)$ probability. For user j, the friends which state are ACTIVE are $1, 2, ..., n$. The probability that the state becomes INFLUENCED is $1 - (1 - q) \prod_{k=1}^{n} (1 - p(i))$, and the probability that the state becomes ACTIVE when $P < e(j)$. This diffusion process continues until no user can change its state.

(2) Discount Price Stage: In order to promote more users to adopt the product and spread the product influence, the product company sets a price discount d for the product promotion. The final product price becomes $P * d$, some users who did not adopt the product because of the price higher than their expected price become ACTIVE. For the user i that state is INFLUENCED, if the discount price $P * d \leq e(i)$, the state change to ACTIVE and begin to adopt the product. And if the discount price $P * d > e(i)$, the user's state maintain INFLUENCED. Those ACTIVE users can spread the product influence to the INACTIVE friends. This diffusion process continues until no user can change its state.

Following [11], we set that only ACTIVE users can propagate product influence and the state will not change after the user becomes ACTIVE. In order to facilitate the calculation of the number of user nodes affected by the company product, we use A_1, A_2 to represent the number of users that state is ACTIVE in the two sales stages respectively.

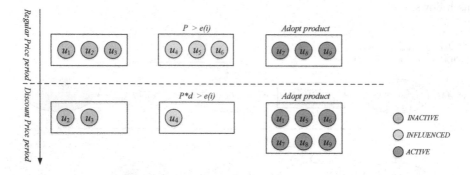

Fig. 2. The effect of product prices on whether users adopt products

3.2   Problem Formulation

The company needs to consume some materials when producing products, we use $c \in [0, 1]$ to represent the cost of a single product. Generally speaking, product sales are divided into two stages. The goal of product company is to expand the demand for products by setting reasonable price P and discount rate d. In OSNs, each user node has its own price tag selected as the seed node. And the more friend nodes a user has, the higher the price tag. The $t(i)$ denotes the price tag of user node u_i and $\tau(S)$ denotes the total costs of selecting seed set S.

$$\tau(S) = \sum_{u \in S_s} t(u) \tag{1}$$

Company makes profits by selling products, and the company's profits increase as the number of users who purchase products increases. We use A_1P and A_2P*d to represent the profits of the two sale stages respectively. We use $\rho(S, P, d)$ to denote the entire profit of company.

$$\rho(S, P, d) = A_1P + A_2P * d - c(A_1 + A_2)) - \tau(S) \tag{2}$$

Definition 1. *(PROMAX problem). In a social network $G(V, E)$, product company sets a optimal product price P and price discount d under TPDM model for product promotion. And select a optimal seed set S that maximizing the finally profit $\rho(S, P, d)$.*

In addition to advertisement marketing, we need to consider word-of-mouth marketing. In order to increase word-of-mouth marketing, we initially activate some seed nodes to make them stay ACTIVE state. Choosing the reasonable seed users can increase the product's scope of influence and increase the company's profit. Obviously, $\rho(\emptyset, P, d) < \rho(u_i, P, d)$, and $\rho(u_i, P, d) > \rho(V, P, d)$. Therefore, the profit $\rho(S, P, d)$ is non-monotone. We need to choose a suitable seed node set to optimize profit and the problem is NP-hard.

4 Solution

In this section, we present the Construction of Nodes' State Algorithm to obtain the number of active nodes. For Maximizing the profit of the product company, we propose a pricing strategy for setting product price **P** and a Two-stage with Discount Greedy (TSDG) Algorithm to maximizing the product profits.

4.1 Pricing Strategies

In our TDPM model, product companies can design incentive pricing strategies, such as providing rewards to user nodes that help diffuse the product influence. There are two types of rewards, one is free product samples and the other is advertising costs (The cost is proportional to the number of users' fans). In this paper, we design an incentive strategies. The product company first selects some influential user nodes, provides them with rewards, and makes them active. These user nodes will actively promote the product's influence in product marketing.

Definition 2. *(IS(Incentive Strategies)). The reward provided to the seed user corresponding their price tag (proportional to the number of neighbor users).*

Algorithm 1. Construction of Nodes' State Algorithm

Input: $G(V, E)$, $u'_i s$ initial state, product price P
Output: $u'_i s$ state
1: **while** True **do**
2:   **if** $u'_i s$ state is INACTIVE **then**
3:     **if** u_i views the product ads **then**
4:       $u'_i s$ state becomes INFLUENCED with probability q
5:       **if** $P >$ expected price $e(i)$ **then**
6:         $u'_i s$ state becomes ACTIVE
7:         $u'_i s$ INACTIVE friend users' state becomes INFLUENCED
8:       **else**
9:         $u'_i s$ state still INFLUENCED
10:   **else if** $u'_i s$ state is INFLUENCED **then**
11:     **if** $P >$ expected price $e(i)$ **then**
12:       $u'_i s$ state becomes ACTIVE
13:     **else**
14:       $u'_i s$ state still INFLUENCED
15:   **else**
16:     $u'_i s$ INACTIVE friend users' states become INFLUENCED
17: **return** $u'_i s$ state

The seed users' state are ACTIVE, and the users will affect his/her neighbor users. In our strategy, users will evaluate the product price and the expected price when they are INFLUENCED. When the product price is higher than the expected price, the user will only be INFLUENCED and will not become active and purchase the product. If the price of the product is lower than the price expected by the user, the user will decide to purchase the product and turn the state to be ACTIVE. Algorithm 1 shows the state changing of nodes under our pricing strategy (Nodes' State Changing Algorithm).

The product company divides the promotion of the product into two stages, the regular price stage and the discount stage. Some users are influenced in the regular price stage because the product price is higher than the expected price. In the discount stage, because the product price is affected by the discount rate and the product price is lower than the expected price, these users purchase products from the affected state to become active status. Based on the optimal myopic price (OMP) [8], we add the factor of product cost to calculate the optimal price **P**.

$$p_i = argmax(p - c) * (1 - F_i(p)) \tag{3}$$

The regular price $\mathbf{P} = \{p_1, p_2, ...p_{|V|}\}$, and the F_i is the distribution function of expected price $e(i)$. In the Discount Price Sales Stage, company will reduce the price of the product to attract many users to buy the product. We combine INACTIVE users and INFLUENCED users to a new set I, and then determine the optimal discount rate through this new set I.

$$d = argmax(p_i - c) * (1 - F_i(p_i)) \tag{4}$$

Algorithm 2. Two-stage with Discount Greedy (TSDG) Algorithm

Input: $G(V, E)$, $T = 1$, $S = \emptyset$, $P = 0$, $d = 1$
Output: S, P, d

1: Calculate the Optimal price $P = \{p_1, p_2, ...p_{|V|}\}$
2: **for** each user $u_i \in V$ **do**
3:    Remove all users whose maximum marginal profit is negative, obtain a new user set T;
4: **for** each user $u_i \in T$ **do**
5:    Select the users whose maximum marginal profit is positive, add them into seed set S. The remaining users are in candidates set C.
6: Obtain the number of ACTIVE nodes A_1, and obtain the profit ρ_1 of the first stage of the product
7: Calculate the Optimal discount d, product price becomes $p_i * d$
8: Construction of Nodes' State
9: **for** each user $u_i \in C$ **do**
10:    Obtain the number of users who have newly become ACTIVE A_2 and the ρ_2 of the first stage of the product
11: **return** $\rho(S, P, d)$

The F_i is the distribution function of expected price $e(i)$ of the user node of the set I.

4.2 Two-Stage with Discount Greedy (TSDG) Algorithm

In word-of-mouth marketing, we increase the spread of product influence by selecting seed nodes. Since the number of nodes in OSNs is very large, we propose the TSDG algorithm that can improve the efficiency of selection. Our profit function has sub-modularity, through which we can reduce the search space of nodes. Our algorithm is divided into three steps. First, in regular price stage, calculating the optimal P of the products. Then, selecting the optimal seed set S. Finally, in discount price stage, calculating the optimal discount d of the products Before selecting the seed nodes, we remove some nodes that are not impossible to select as seed nodes, which can improve the efficiency of selection. We obtain the maximum margin profit of all nodes, and remove all nodes whose maximum profit margin is negative. Then calculate the minimum profit margin of all nodes, and add nodes with a positive minimum profit margin to the seed set S.

The purpose of the company's product marketing strategy to select the seed set node set S is to activate a part of influential user nodes in advance, so that the number of people buying products increases, and ultimately maximize the company's profits. Here, we use $A_1(S)$ and $A_2(S)$ to denote the node set activated by the seed node set S in two stages respectively. Our TSDG algorithm is divided into two parts, first determine the product price P and select the appropriate seed set S, and then determine the discount d.

Our TSDG algorithm greatly reduces the search space of nodes and improves the efficiency of seed node selection. By activating the word-of-mouth marketing of some nodes in advance, the spread of product influence can be accelerated and the company's profits can be improved.

5   Evaluations

5.1   Experimental Setup

Datasets: We select several real OSNs datasets from [14], they are Amazon, Epinions, HepPh and P2P respectively. These graphs contain various social relationships, for example, follower and trust relationships. Table 2 shows the important statistics of these graphs:

Table 2. Important information for networks.

Dataset	Nodes	Edges	Average Degree
Amazon	262K	1235K	4.7
Epinions	76K	509K	6.7
HepPh	15K	59K	3.9
P2P	9K	32K	3.6

Algorithms. We compare the following algorithms.

- All-OMP: The pricing strategy of this algorithm is providing optimal myopic price (OMP) to all nodes regardless of whether a node is a seed or how influential it is. Then use a simple greedy algorithm to select the seed node set. The simple greedy algorithm is showed in Algorithm 1.
- Free-For-Seeds (FFS): The algorithm gives different prices to seed nodes and non-seed nodes. it provides optimal myopic price (OMP) to all non-seed nodes and the price of seeds is free. Then use a simple greedy algorithm to select the seed node set.
- TSDG (Two-stage with Discount Greedy): The algorithm is showed in Algorithm 2.

Expected prices Distribution: Until now, we have not obtained the expected price of the product from users. Since product manufacturing requires a certain cost, in general, the user's expected price of the product is higher than the product's cost, and the user's expected price of the product is related to the product's cost. For example, the expected price of a piece of paper is different from the expected price of a book. Therefore, in the experiment, we assume that users expect that the normal distribution of prices is related to cost.

5.2   Profit Results of Different Algorithms

In this section we show the experimental results of four data sets. We used TPDM model in the experiment. We assume that only those who adopt the product have influence and that throughout the propagation process, users' expected price of the product unchanged.

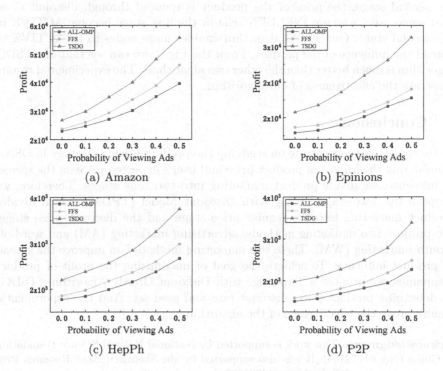

Fig. 3. The influence of advertising propagation probability on profit of different data sets

It can be seen from Fig. 3, as the probability q of viewing advertisements continues to increase, the profit of the product company continues to increase. This is because more users are affected by product advertisements to purchase products, and these users indirectly become free seed nodes to spread the influence of the product to neighboring nodes. The increase in the probability of browsing product advertisements causes more user nodes to be ACTIVE to purchase the product, and the profit of the product company increases accordingly. We can conclude that, regardless of the cost of advertising, increasing the number of ads will cause more users to be affected, and advertising marketing can increase the profits of product companies.

When the advertising propagation probability is 0, the company's profit is lower, which illustrates the importance of advertising marketing. The product price in the All-OMP algorithm is the same for seed users and non-seed users, which makes some seed users inactive because they expected prices are lower than the product price, and cannot spread the influence of the product. However, for FFS algorithm, the product price for the seed node is free, which makes the seed node must be ACTIVE and improves the spread of influence. Therefore, the profit obtained by the FFS algorithm is higher than that of the All-OMP algorithm. Our TSDG algorithm divides product marketing into two stages. In the second stage, the price of the product is reduced through discount d, so that some users who are INFLUENCED in the first stage become ACTIVE in the second stage. Our TSDG algorithm enables more nodes to be ACTIVE to spread the influence of the product. From the Fig. 3, we can see that our TSDG algorithm is much better than the other two algorithms. The experimental results illustrate the effectiveness of our algorithm.

6   Conclusion

In this paper, we concentrate on studying the product pricing strategy in OSNs. Considering the impact of product price and user's expected price on the spread of influence, we divide product marketing into two time stages. Therefore, we propose the Two-stage Pricing with Discount Model (TPDM), which divides product marketing into the regular price stage and the discount price stage. We propose two marketing methods: advertising marketing (AM) and word-of-mouth marketing (WM). These two marketing methods can improve the spread of product influence. To achieve the goal of maximizing the profit of product companies, we propose a Two-stage with Discount Greedy Algorithm (TSDG) to determine product price, discount rate and seed set. And the experimental results prove the effectiveness of the algorithm.

Acknowledgment. This work is supported by National Natural Science Foundation of China (No. 61772154). It was also supported by the Shenzhen Basic Research Program (Project No. JCYJ20190806143011274).

References

1. Akhlaghpour, H., Ghodsi, M., Haghpanah, N., Mirrokni, V.S., Mahini, H., Nikzad, A.: Optimal iterative pricing over social networks (extended abstract). In: Saberi, A. (ed.) WINE 2010. LNCS, vol. 6484, pp. 415–423. Springer, Heidelberg (2010). https://doi.org/10.1007/978-3-642-17572-5_34
2. Borgs, C., Brautbar, M., Chayes, J., Lucier, B.: Maximizing social influence in nearly optimal time. In: Proceedings of the Twenty-Fifth Annual ACM-SIAM Symposium on Discrete Algorithms, pp. 946–957. SIAM (2014)
3. Chen, W., et al.: Influence maximization in social networks when negative opinions may emerge and propagate. In: Proceedings of the 2011 SIAM International Conference on Data Mining, pp. 379–390. SIAM (2011)
4. Chen, W., Wang, C., Wang, Y.: Scalable influence maximization for prevalent viral marketing in large-scale social networks. In: Proceedings of the 16th ACM SIGKDD International Conference on Knowledge Discovery and Data Mining, pp. 1029–1038 (2010)

5. Dolgui, A., Proth, J.M.: Pricing strategies and models. Ann. Rev. Control **34**(1), 101–110 (2010)
6. Domingos, P., Richardson, M.: Mining the network value of customers. In: Proceedings of the Seventh ACM SIGKDD International Conference on Knowledge Discovery and Data Mining, pp. 57–66 (2001)
7. Fotakis, D., Siminelakis, P.: On the efficiency of influence-and-exploit strategies for revenue maximization under positive externalities. In: Goldberg, P.W. (ed.) WINE 2012. LNCS, vol. 7695, pp. 270–283. Springer, Heidelberg (2012). https://doi.org/10.1007/978-3-642-35311-6_20
8. Hartline, J., Mirrokni, V., Sundararajan, M.: Optimal marketing strategies over social networks. In: Proceedings of the 17th International Conference on World Wide Web, pp. 189–198 (2008)
9. He, X., Song, G., Chen, W., Jiang, Q.: Influence blocking maximization in social networks under the competitive linear threshold model. In: Proceedings of the 2012 SIAM International Conference on Data Mining, pp. 463–474. SIAM (2012)
10. Jung, K., Heo, W., Chen, W.: Irie: scalable and robust influence maximization in social networks. In: 2012 IEEE 12th International Conference on Data Mining, pp. 918–923. IEEE (2012)
11. Kalish, S.: A new product adoption model with price, advertising, and uncertainty. Manage. Sci. **31**(12), 1569–1585 (1985)
12. Kempe, D., Kleinberg, J., Tardos, É.: Maximizing the spread of influence through a social network. In: Proceedings of the Ninth ACM SIGKDD International Conference on Knowledge Discovery and Data Mining, pp. 137–146 (2003)
13. Leskovec, J., Adamic, L.A., Huberman, B.A.: The dynamics of viral marketing. ACM Trans. Web **1**(1), 5 (2007)
14. Leskovec, J., Krevl, A.: Snap datasets: stanford large network dataset collection (2014)
15. Li, Y., Li, V.O.K.: Pricing strategies with promotion time limitation in online social networks, pp. 254–261 (2018)
16. Lu, W., Lakshmanan, L.V.: Profit maximization over social networks. In: 2012 IEEE 12th International Conference on Data Mining, pp. 479–488. IEEE (2012)
17. Niu, G., Li, V.O.K., Long, Y.: Sequential pricing for social networks with multi-state diffusion, pp. 3176–3181 (2013)
18. Shareef, M.A., Mukerji, B., Dwivedi, Y.K., Rana, N.P., Islam, R.: Social media marketing: comparative effect of advertisement sources. J. Retail. Consum. Serv. **46**, 58–69 (2019)
19. Shor, M., Oliver, R.L.: Price discrimination through online couponing: impact on likelihood of purchase and profitability. J. Econ. Psychol. **27**(3), 423–440 (2006)
20. Zhang, H., Zhang, H., Kuhnle, A., Thai, M.T.: Profit maximization for multiple products in online social networks. In: IEEE INFOCOM 2016-The 35th Annual IEEE International Conference on Computer Communications, pp. 1–9. IEEE (2016)
21. Zhu, Y., Li, D., Yan, R., Wu, W., Bi, Y.: Maximizing the influence and profit in social networks. IEEE Trans. Comput. Soc. Syst. **4**(3), 54–64 (2017)

Almost Linear Time Algorithms
for Minsum k-Sink Problems
on Dynamic Flow Path Networks

Yuya Higashikawa[1]($\boxtimes$), Naoki Katoh[1], Junichi Teruyama[1], and Koji Watase[2]

[1] University of Hyogo, Kobe, Japan
{higashikawa,naoki.katoh,junichi.teruyama}@sis.u-hyogo.ac.jp
[2] Kwansei Gakuin University, Sanda, Japan
fnt43517@kwansei.ac.jp

Abstract. We address the facility location problems on dynamic flow path networks. A *dynamic flow path network* consists of an undirected path with positive edge lengths, positive edge capacities, and positive vertex weights. A path can be considered as a road, an edge length as the distance along the road and a vertex weight as the number of people at the site. An edge capacity limits the number of people that can enter the edge per unit time. In the dynamic flow network, given particular points on edges or vertices, called *sinks*, all the people evacuate from the vertices to the sinks as quickly as possible. The problem is to find the location of sinks on a dynamic flow path network in such a way that the aggregate evacuation time (i.e., the sum of evacuation times for all the people) to sinks is minimized. We consider two models of the problem: the *confluent flow model* and the *non-confluent flow model*. In the former model, the way of evacuation is restricted so that all the people at a vertex have to evacuate to the same sink, and in the latter model, there is no such restriction. In this paper, for both the models, we develop algorithms which run in almost linear time regardless of the number of sinks. It should be stressed that for the confluent flow model, our algorithm improves upon the previous result by Benkoczi et al. [Theoretical Computer Science, 2020], and one for the non-confluent flow model is the first polynomial time algorithm.

Keywords: Dynamic flow networks · Facility location problems · Minimum k-link path problem · Persistent data structures

1 Introduction

Recently, many disasters, such as earthquakes, nuclear plant accidents, volcanic eruptions and flooding, have struck in many parts of the world, and it has been recognized that orderly evacuation planning is urgently needed. A powerful tool

---

A full version of the paper is available at [14]; https://arxiv.org/abs/2010.05729.

© Springer Nature Switzerland AG 2020
W. Wu and Z. Zhang (Eds.): COCOA 2020, LNCS 12577, pp. 198–213, 2020.
https://doi.org/10.1007/978-3-030-64843-5_14

for evacuation planning is the *dynamic flow model* introduced by Ford and Fulk-erson [10], which represents movement of commodities over time in a network. In this model, we are given a graph with *source* vertices and *sink* vertices. Each source vertex is associated with a positive weight, called a *supply*, each sink vertex is associated with a positive weight, called a *demand*, and each edge is associated with positive length and capacity. An edge capacity limits the amount of supply that can enter the edge per unit time. One variant of the dynamic flow problem is the *quickest transshipment problem*, of which the objective is to send exactly the right amount of supply out of sources into sinks with satisfying the demand constraints in the minimum overall time. Hoppe and Tardos [15] pro-vided a polynomial time algorithm for this problem in the case where the transit times are integral. However, the complexity of their algorithm is very high. Find-ing a practical polynomial time solution to this problem is still open. A reader is referred to a recent survey by Skutella [18] on dynamic flows.

This paper discusses a related problem, called the *k-sink problem* [2,4,5,7–9, 12,13,16], of which the objective is to find a location of k sinks in a given dynamic flow network so that all the supply is sent to the sinks as quickly as possible. For the optimality of location, the following two criteria can be naturally considered: the minimization of *evacuation completion time* and *aggregate evacuation time* (i.e., *average evacuation time*). We call the k-sink problem that requires finding a location of k sinks that minimizes the evacuation completion time (resp. the aggregate evacuation time) the *minmax* (resp. *minsum*) *k-sink problem*. Several papers have studied the minmax k-sink problems on dynamic flow networks [2,7–9,12,13,16]. On the other hand, the minsum k-sink problems on dynamic flow networks have not been studied except for the case of path networks [4,5,13].

Moreover, there are two models on the way of evacuation. Under the *confluent flow model*, all the supply leaving a vertex must evacuate to the same sink through the same edges, and under the *non-confluent flow model*, there is no such restriction. To our knowledge, all the papers which deal with the k-sink problems [2,4,5,7–9,13] adopt the confluent flow model.

In order to model the evacuation behavior of people, it might be natural to treat each supply as a discrete quantity as in [15,16]. Nevertheless, almost all the previous papers on sink problems [2,7–9,12,13] treat each supply as a continuous quantity since it is easier for mathematically handling the problems and the effect is small enough to ignore when the number of people is large. Throughout the paper, we also adopt the model with continuous supplies.

In this paper, we study the minsum k-sink problems on dynamic flow path networks under both the confluent flow model and the non-confluent flow model. A path network can model a coastal area surrounded by the sea and a hilly area, an airplane aisle, a hall way in a building, a street, a highway, etc., to name a few. For the confluent flow model, the previous best results are an $O(kn \log^3 n)$ time algorithm for the case with uniform edge capacity in [4], and $O(kn \log^4 n)$ time algorithm for the case with general edge capacities in [5], where n is the number of vertices on path networks. We develop algorithms which run in time $\min\{O(kn \log^2 n), n2^{O(\sqrt{\log k \log \log n})} \log^2 n\}$ for the case with uniform edge capacity, and in time $\min\{O(kn\log^3 n), n2^{O(\sqrt{\log k \log\log n})}\log^3 n\}$ for the case with general edge capacities, respectively. Thus, our algorithms improve upon the complexities by [4,5] for any value of k. Especially, for the non-confluent flow model, this paper provides the first polynomial time algorithms.

Since the number of sinks k is at most n, we confirm $2^{O(\sqrt{\log k \log\log n})} = n^{O(\sqrt{\log\log n/\log n})} = n^{o(1)}$, which means that our algorithms are the first ones which run in almost linear time (i.e., $n^{1+o(1)}$ time) regardless of k. The reason why we could achieve almost linear time algorithms for the minsum k-sink problems is that we newly discover a convex property from a novel point of view. In all the previous papers on the k-sink problems, the evacuation completion. time and the aggregate evacuation time (called CT and AT, respectively) are basically determined as functions in "distance": Let us consider the case with a 1-sink. The values $CT(x)$ or $AT(x)$ may change as a sink location x moves along edges in the network. In contrast, we introduce a new metric for CT and AT as follows: assuming that a sink is fixed and all the supply in the network flows to the sink, for a positive real z, $CT(z)$ is the time at which the first z of supply completes its evacuation to the sink and then $AT(z)$ is the integral of $CT(z)$, i.e., $AT(z) = \int_0^z CT(t)dt$. We can observe that $AT(z)$ is convex in z since $CT(z)$ is increasing in z. Based on the convexity of $AT(z)$, we develop efficient algorithms.

The rest of the paper is organized as follows. In Sect. 2, we introduce the terms that are used throughout the paper and explain our models. In Sect. 3, we show that our problem can be reduced to the *minimum k-link path problem with links satisfying the concave Monge condition*. This immediately implies by Schieber [17] that the optimal solutions for our problems can be obtained by solving $\min\{O(kn), n2^{O(\sqrt{\log k \log\log n})}\}$ subproblems of computing the optimal aggregate evacuation time for subpaths, in each of which two sinks are located on its endpoints. Section 3 subsequently shows an overview of the algorithm that solves the above subproblems. In Sect. 4, we introduce novel data structures, which enable to solve each of the above subproblems in $O(\text{poly}\log n)$ time. Section 5 concludes the paper.

2   Preliminaries

2.1   Notations

For two real values a, b with $a < b$, let $[a, b] = \{t \in \mathbb{R} \mid a \le t \le b\}$, $[a, b) = \{t \in \mathbb{R} \mid a \le t < b\}$, $(a, b] = \{t \in \mathbb{R} \mid a < t \le b\}$, and $(a, b) = \{t \in \mathbb{R} \mid a < t < b\}$, where $\mathbb{R}$ is the set of real values. For two integers i, j with $i \le j$, let $[i..j] = \{h \in \mathbb{Z} \mid i \le h \le j\}$, where $\mathbb{Z}$ is the set of integers. A dynamic flow path network $\mathcal{P}$ is given as a 5-tuple $(P, \mathbf{w}, \mathbf{c}, \mathbf{l}, \tau)$, where P is a path with vertex set $V = \{v_i \mid i \in [1..n]\}$ and edge set $E = \{e_i = (v_i, v_{i+1}) \mid i \in [1..n - 1]\}$, $\mathbf{w}$ is a vector $\langle w_1, \ldots, w_n \rangle$ of which a component w_i is the *weight* of vertex v_i representing the amount of supply (e.g., the number of evacuees, cars) located at v_i, $\mathbf{c}$ is a vector $\langle c_1, \ldots, c_{n-1} \rangle$ of which a component c_i is the *capacity* of edge e_i representing the upper bound on the flow rate through e_i per unit time, $\mathbf{l}$ is a vector $\langle \ell_1, \ldots, \ell_{n-1} \rangle$ of which a component ℓ_i is the *length* of edge e_i, and τ is the time which unit supply takes to move unit distance on any edge.

We say a point p lies on path $P = (V, E)$, denoted by $p \in P$, if p lies on a vertex $v \in V$ or an edge $e \in E$. For two points $p, q \in P$, $p \prec q$ means that p lies to the left side of q. For two points $p, q \in P$, $p \preceq q$ means that $p \prec q$ or p and q lie on the same place. Let us consider two integers $i, j \in [1..n]$ with $i < j$. We denote by $P_{i,j}$ a *subpath* of P from v_i to v_j. Let $L_{i,j}$ be the distance between v_i and v_j, i.e., $L_{i,j} = \sum_{h=i}^{j-1} \ell_h$, and let $C_{i,j}$ be the minimum capacity for all the edges between v_i and v_j, i.e., $C_{i,j} = \min\{c_h \mid h \in [i..j-1]\}$. For $i \in [1..n]$, we denote the sum of weights from v_1 to v_i by $W_i = \sum_{j=1}^{i} w_j$. Note that, given a dynamic flow path network $\mathcal{P}$, if we construct two lists of W_i and $L_{1,i}$ for all $i \in [1..n]$ in $O(n)$ preprocessing time, we can obtain W_i for any $i \in [1..n]$ and $L_{i,j} = L_{1,j} - L_{1,i}$ for any $i, j \in [1..n]$ with $i < j$ in $O(1)$ time. In addition, $C_{i,j}$ for any $i, j \in [1..n]$ with $i < j$ can be obtained in $O(1)$ time with $O(n)$ preprocessing time, which is known as the *range minimum query* [1,3].

A k-*sink* $\mathbf{x}$ is k-tuple $(x_1, \ldots, x_k)$ of points on P, where $x_i \prec x_j$ for $i < j$. We define the function Id for point $p \in P$ as follows: the value $\mathrm{Id}(p)$ is an integer such that $v_{\mathrm{Id}(p)} \preceq p \prec v_{\mathrm{Id}(p)+1}$ holds. For a k-sink $\mathbf{x}$ for $\mathcal{P}$, a *divider* $\mathbf{d}$ is $(k-1)$-tuple $(d_1, \ldots, d_{k-1})$ of real values such that $d_i < d_j$ for $i < j$ and $W_{\mathrm{Id}(x_i)} \leq d_i \leq W_{\mathrm{Id}(x_{i+1})}$. Given a k-sink $\mathbf{x}$ and a divider $\mathbf{d}$ for $\mathcal{P}$, the portion $W_{\mathrm{Id}(x_i)} - d_{i-1}$ supply that originates from the left side of x_i flows to sink x_i, and the portion $d_i - W_{\mathrm{Id}(x_i)}$ supply that originates from the right side of x_i also flows to sink x_i. For instance, under the non-confluent flow model, if $W_{h-1} < d_i < W_h$ where $h \in [1..n]$, $d_i - W_{h-1}$ of w_h supply at v_h flows to sink x_i and the rest of $W_h - d_i$ supply to do sink x_{i+1}. The difference between the confluent flow model and the non-confluent flow model is that the confluent flow model requires that each value d_i of a divider $\mathbf{d}$ must take a value in $\{W_1, \ldots, W_n\}$, but the non-confluent flow model does not. For the notation, we set $d_0 = 0$ and $d_k = W_n$.

For a dynamic flow path network $\mathcal{P}$, a k-sink $\mathbf{x}$ and a divider $\mathbf{d}$, the *evacuation completion time* $\mathrm{CT}(\mathcal{P}, \mathbf{x}, \mathbf{d})$ is the time at which all the supply completes the evacuation. The *aggregate evacuation time* $\mathrm{AT}(\mathcal{P}, \mathbf{x}, \mathbf{d})$ is that the sum of the evacuation completion time for all the supply. Their explicit definitions are given later. In this paper, our task is, given a dynamic flow path network $\mathcal{P}$, to find a k-sink $\mathbf{x}$ and a divider $\mathbf{d}$ that minimize the aggregate evacuation time $\mathrm{AT}(\mathcal{P}, \mathbf{x}, \mathbf{d})$ in each evacuation model.

2.2   Aggregate Evacuation Time on a Path

For the confluent flow model, it is shown in [5,13] that for the minsum k-sink problems, there exists an optimal k-sink such that all the k sinks are at vertices. This fact also holds for the non-confluent flow model. Indeed, if a divider $\mathbf{d}$ is fixed, then we have k subproblems for a 1-sink and the optimal sink location for each subproblem is at a vertex. Thus, we have the following lemma.

Lemma 1 ([13]). *For the minsum k-sink problem in a dynamic flow path network, there exists an optimal k-sink such that all the k sinks are at vertices under the confluent/non-confluent flow model.*

Lemma 1 implies that it is enough to consider only the case that every sink is at a vertex. Thus, we suppose $\mathbf{x} = (x_1, \ldots, x_k) \in V^k$, where $x_i \prec x_j$ for $i < j$.

A Simple Example with a 1-sink. In order to give explicit definitions for the evacuation completion time and the aggregate evacuation time, let us consider a simple example for a 1-sink. We are given a dynamic flow path network $\mathcal{P} = (P, \mathbf{w}, \mathbf{c}, \mathbf{l}, \tau)$ with n vertices and set a unique sink on a vertex v_i, that is, $\mathbf{x} = (v_i)$ and $\mathbf{d} = ()$ which is the 0-tuple. In this case, all the supply on the left side of v_i (i.e., at $v_1, \ldots, v_{i-1}$) will flow right to sink v_i, and all the supply on the right side of v_i (i.e., at $v_{i+1}, \ldots, v_n$) will flow left to sink v_i. Note that in our models all the supply at v_i immediately completes its evacuation at time 0.

To deal with this case, we introduce some new notations. Let the function $\theta^{i,+}(z)$ denote the time at which the first $z - W_i$ of supply on the right side of v_i completes its evacuation to sink v_i (where $\theta^{i,+}(z) = 0$ for $z \in [0, W_i]$). Higashikawa [11] shows that the value $\theta^{i,+}(W_n)$, the evacuation completion time for all the supply on the right side of v_i, is given by the following formula:

$$\theta^{i,+}(W_n) = \max \left\{ \frac{W_n - W_{j-1}}{C_{i,j}} + \tau \cdot L_{i,j} \mid j \in [i+1..n] \right\}. \tag{1}$$

Recall that $C_{i,j} = \min\{c_h \mid h \in [i..j-1]\}$. We can generalize formula (1) to the case with any $z \in [0, W_n]$ as follows:

$$\theta^{i,+}(z) = \max\{\theta^{i,+,j}(z) \mid j \in [i+1..n]\}, \tag{2}$$

where $\theta^{i,+,j}(z)$ for $j \in [i+1..n]$ is defined as

$$\theta^{i,+,j}(z) = \begin{cases} 0 & \text{if } z \leq W_{j-1}, \\ \frac{z - W_{j-1}}{C_{i,j}} + \tau \cdot L_{i,j} & \text{if } z > W_{j-1}. \end{cases} \tag{3}$$

Similarly, let $\theta^{i,-}(z)$ denote the time at which the first $W_{i-1} - z$ of supply on the left side of v_i completes its evacuation to sink v_i (where $\theta^{i,-}(z) = 0$ for $z \in [W_{i-1}, W_n]$). Then,

$$\theta^{i,-}(z) = \max\{\theta^{i,-,j}(z) \mid j \in [1..i-1]\}, \tag{4}$$

where $\theta^{i,-,j}(z)$ is defined as

$$\theta^{i,-,j}(z) = \begin{cases} \frac{W_j - z}{C_{j,i}} + \tau \cdot L_{j,i} & \text{if } z < W_j, \\ 0 & \text{if } z \geq W_j. \end{cases} \tag{5}$$

The aggregate evacuation times for the supply on the right side and the left side of v_i are

$$\int_{W_i}^{W_n} \theta^{i,+}(z)dz = \int_0^{W_n} \theta^{i,+}(z)dz \text{ and } \int_0^{W_{i-1}} \theta^{i,-}(z)dz = \int_0^{W_n} \theta^{i,-}(z)dz,$$

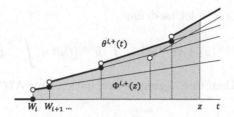

Fig. 1. The thick half-open segments indicate function $\theta^{i,+}(t)$ and the gray area indicates $\Phi^{i,+}(z)$ for some $z > W_i$.

respectively. Thus, the aggregate evacuation time $\mathsf{AT}(\mathcal{P}, (v_i), ())$ is given as

$$\mathsf{AT}(\mathcal{P}, (v_i), ()) = \int_0^{W_n} \left\{ \theta^{i,+}(z) + \theta^{i,-}(z) \right\} dz.$$

Aggregate Evacuation Time with a k-Sink. Suppose that we are given a k-sink $\mathbf{x} = (x_1, \ldots, x_k) \in V^k$ and a divider $\mathbf{d} = (d_1, \ldots, d_{k-1})$. Recalling the definition of $\mathrm{Id}(p)$ for $p \in P$, we have $x_i = v_{\mathrm{Id}(x_i)}$ for all $i \in [1..k]$. In this situation, for each $i \in [1..k]$, the first $d_i - W_{\mathrm{Id}(x_i)}$ of supply on the right side of x_i and the first $W_{\mathrm{Id}(x_i)-1} - d_{i-1}$ of supply on the left side of x_i move to sink x_i.

By the argument of the previous section, the aggregate evacuation times for the supply on the right side and the left side of x_i are

$$\int_{W_{\mathrm{Id}(x_i)}}^{d_i} \theta^{\mathrm{Id}(x_i),+}(z)dz = \int_0^{d_i} \theta^{\mathrm{Id}(x_i),+}(z)dz \quad \text{and}$$

$$\int_{d_{i-1}}^{W_{\mathrm{Id}(x_i)-1}} \theta^{\mathrm{Id}(x_i),-}(z)dz = \int_{d_{i-1}}^{W_n} \theta^{\mathrm{Id}(x_i),-}(z)dz,$$

respectively. In order to give the general form for the above values, let us denote by $\Phi^{i,+}(z)$ the aggregate evacuation time when the first $z - W_i$ of supply on the right side of v_i flows to sink v_i. Similarly, we denote by $\Phi^{i,-}(z)$ the aggregate evacuation time when the first $W_{i-1} - z$ of supply on the left side of v_i flows to sink v_i. Therefore, we have

$$\Phi^{i,+}(z) = \int_0^z \theta^{i,+}(t)dt \quad \text{and} \quad \Phi^{i,-}(z) = \int_z^{W_n} \theta^{i,-}(t)dt = \int_{W_n}^z -\theta^{i,-}(t)dt \quad (6)$$

(see Fig. 1). Let us consider a subpath $P_{\mathrm{Id}(x_i),\mathrm{Id}(x_{i+1})}$ which is a subpath between sinks x_i and x_{i+1}. The aggregate evacuation time for the supply on $P_{\mathrm{Id}(x_i),\mathrm{Id}(x_{i+1})}$ is given by

$$\int_0^{d_i} \theta^{\mathrm{Id}(x_i),+}(z)dz + \int_{d_i}^{W_n} \theta^{\mathrm{Id}(x_{i+1}),-}(z)dz = \Phi^{\mathrm{Id}(x_i),+}(d_i) + \Phi^{\mathrm{Id}(x_{i+1}),-}(d_i).$$

For $i, j \in [1..n]$ with $i < j$, let us define

$$\Phi^{i,j}(z) = \Phi^{i,+}(z) + \Phi^{j,-}(z) = \int_0^z \theta^{i,+}(t)dt + \int_z^{W_n} \theta^{j,-}(t)dt \tag{7}$$

for $z \in [W_i, W_{j-1}]$. Then, the aggregate evacuation time $\mathsf{AT}(\mathcal{P}, \mathbf{x}, \mathbf{d})$ is given as

$$\mathsf{AT}(\mathcal{P}, \mathbf{x}, \mathbf{d}) = \Phi^{\mathrm{Id}(x_1),-}(0) + \sum_{i=1}^{k-1} \Phi^{\mathrm{Id}(x_i),\mathrm{Id}(x_{i+1})}(d_i) + \Phi^{\mathrm{Id}(x_k),+}(W_n). \tag{8}$$

In the rest of this section, we show the important properties of $\Phi^{i,j}(z)$. Let us first confirm that by Eq. (6), both $\Phi^{i,+}(z)$ and $\Phi^{j,-}(z)$ are convex in z since $\theta^{i,+}(z)$ and $-\theta^{j,-}(z)$ are non-decreasing in z, therefore $\Phi^{i,j}(z)$ is convex in z. On the condition of the minimizer for $\Phi^{i,j}(z)$, we have a more useful lemma.

Lemma 2. *For any $i, j \in [1..n]$ with $i < j$, there uniquely exists*

$$z^* \in \underset{z \in [W_i, W_{j-1}]}{\arg\min} \ \max\{\theta^{i,+}(z), \theta^{j,-}(z)\}.$$

Furthermore, $\Phi^{i,j}(z)$ is minimized on $[W_i, W_{j-1}]$ when $z = z^*$.

See Lemma 2 in [14] for the proof. In the following sections, such z^* is called the *pseudo-intersection point*[1] of $\theta^{i,+}(z)$ and $\theta^{j,-}(z)$, and we say that $\theta^{i,+}(z)$ and $\theta^{j,-}(z)$ *pseudo-intersect* on $[W_i, W_{j-1}]$ at z^*.

3   Algorithms

In order to solve our problems, we reduce them to *minimum k-link path problems*. In the minimum k-link path problems, we are given a weighted complete directed acyclic graph (DAG) $G = (V', E', w')$ with $V' = \{v'_i \mid i \in [1..n]\}$ and $E' = \{(v'_i, v'_j) \mid i, j \in [1..n], i < j\}$. Each edge (v'_i, v'_j) is associated with weight $w'(i, j)$. We call a path in G a *k-link* path if the path contains exactly k edges. The task is to find a k-link path $(v'_{a_0} = v'_1, v'_{a_1}, v'_{a_2}, \ldots, v'_{a_{k-1}}, v'_{a_k} = v'_n)$ from v'_1 to v'_n that minimizes the sum of weights of k edges, $\sum_{i=1}^k w'(a_{i-1}, a_i)$. If the weight function w' satisfies the *concave Monge property*, then we can solve the minimum k-link path problems in almost linear time regardless of k.

Definition 1 (Concave Monge property). *We say function $f : \mathbb{Z} \times \mathbb{Z} \to \mathbb{R}$ satisfies the concave Monge property if for any integers i, j with $i + 1 < j$, $f(i, j) + f(i+1, j+1) \leq f(i+1, j) + f(i, j+1)$ holds.*

Lemma 3 ([17]). *Given a weighted complete DAG with n vertices, if the weight function satisfies the concave Monge property, then there exists an algorithm that solves the minimum k-link path problem in time $\min\{O(kn), n2^{O(\sqrt{\log k \log \log n})}\}$.*

---

[1] The reason why we adopt a term "pseudo-intersection" is that two functions $\theta^{i,+}(z)$ and $\theta^{j,-}(z)$ are not continuous in general while "intersection" is usually defined for continuous functions.

We describe how to reduce the k-sink problem on a dynamic flow path network $\mathcal{P} = (P = (V, E), \mathbf{w}, \mathbf{c}, \mathbf{l}, \tau)$ with n vertices to the minimum $(k + 1)$-link path problem on a weighted complete DAG $G = (V', E', w')$. We prepare a weighted complete DAG $G = (V', E', w')$ with $n + 2$ vertices, where $V' = \{v'_i \mid i \in [0..n + 1]\}$ and $E' = \{(v'_i, v'_j) \mid i, j \in [0..n + 1], i < j\}$. We set the weight function w' as

$$
w'(i, j) = \begin{cases} \mathsf{OPT}(i, j) & i, j \in [1..n], i < j, \\ \varPhi^{i,+}(W_n) & i \in [1..n] \text{ and } j = n + 1, \\ \varPhi^{j,-}(0) & i = 0 \text{ and } j \in [1..n], \\ \infty & i = 0 \text{ and } j = n + 1, \end{cases} \tag{9}
$$

where $\mathsf{OPT}(i, j) = \min_{z \in [W_i, W_{j-1}]} \varPhi^{i,j}(z)$.

Now, on a weighted complete DAG G made as above, let us consider a $(k+1)$-link path $(v'_{a_0} = v'_0, v'_{a_1}, \ldots, v'_{a_k}, v'_{a_{k+1}} = v'_{n+1})$ from v'_0 to v'_{n+1}, where $a_1, \ldots, a_k$ are integers satisfying $0 < a_1 < a_2 < \cdots < a_k < n + 1$. The sum of weights of this $(k + 1)$-link path is

$$
\sum_{i=0}^{k} w'(a_i, a_{i+1}) = \varPhi^{a_1, -}(0) + \sum_{i=1}^{k-1} \mathsf{OPT}(a_i, a_{i+1}) + \varPhi^{a_k, +}(W_n).
$$

This value is equivalent to $\min_{\mathbf{d}} \mathsf{AT}(\mathcal{P}, \mathbf{x}, \mathbf{d})$ for a k-sink $\mathbf{x} = (v_{a_1}, v_{a_2}, \ldots, v_{a_k})$ (recall Eq. (8)), which implies that a minimum $(k+1)$-link path on G corresponds to an optimal k-sink location for a dynamic flow path network $\mathcal{P}$.

We show in the following lemma that the function w' defined as formula (9) satisfies the concave Monge property under both of evacuation models. See Lemma 4 in [14] for the proof.

Lemma 4. *The weight function w' defined as formula (9) satisfies the concave Monge property under the confluent/non-confluent flow model.*

Lemmas 3 and 4 imply that if we can evaluate $w'(i, j)$ in time at most t for any $i, j \in [0..n + 1]$ with $i < j$, then we can solve the k-sink problem in time $\min\{O(knt), n2^{O(\sqrt{\log k \log \log n})}t\}$.

In order to obtain $w'(i, j)$ for any $i, j \in [0..n + 1]$ with $i < j$ in $O(\operatorname{poly} \log n)$ time, we introduce novel data structures and some modules using them. Basically, we construct a *segment tree* [6] $\mathcal{T}$ with root ρ such that its leaves correspond to indices of vertices of P arranged from left to right and its height is $O(\log n)$. For a node $u \in \mathcal{T}$, let $\mathcal{T}_u$ denote the subtree rooted at u, and let l_u (resp. r_u) denote the index of the vertex that corresponds to the leftmost (resp. rightmost) leaf of $\mathcal{T}_u$. Let p_u denote the parent of u if $u \neq \rho$. We say a node $u \in \mathcal{T}$ *spans* subpath P_{l_u, r_u}. If $P_{l_u, r_u} \subseteq P'$ and $P_{l_{p_u}, r_{p_u}} \not\subseteq P'$, node u is called a *maximal subpath node* for P'. For each node $u \in \mathcal{T}$, let m_u be the number of edges in subpath P_{l_u, r_u}, i.e., $m_u = r_u - l_u$. As with a standard segment tree, $\mathcal{T}$ has the following properties.

Property 1. For $i, j \in [1..n]$ with $i < j$, the number of maximal subpath nodes for $P_{i,j}$ is $O(\log n)$. Moreover, we can find all the maximal subpath nodes for $P_{i,j}$ by walking on $\mathcal{T}$ from leaf i to leaf j in $O(\log n)$ time.

Property 2. If one can construct data structures for each node u of a segment tree $\mathcal{T}$ in $O(f(m_u))$ time, where $f : \mathbb{N} \rightarrow \mathbb{R}$ is some function independent of n and bounded below by a linear function asymptotically, i.e., $f(m) = \Omega(m)$, then the running time for construction of data structures for every node in $\mathcal{T}$ is $O(f(n) \log n)$ time in total.

At each node $u \in \mathcal{T}$, we store four types of the information that depend on the indices of the vertices spanned by u, i.e., $l_u, \ldots, r_u$. We will introduce each type in Sect. 4. As will be shown there, the four types of the information at $u \in \mathcal{T}$ can be constructed in $O(m_u \log m_u)$ time. Therefore, we can construct $\mathcal{T}$ in $O(n \log^2 n)$ time by Property 2.

Recall that for $i, j \in [1..n]$ with $i < j$, it holds $w'(i,j) = \mathsf{OPT}(i,j)$. We give an outline of the algorithm that computes $\mathsf{OPT}(i,j)$ only for the non-confluent flow model since a similar argument holds even for the confluent flow model with minor modification. The main task is to find a value z^* that minimizes $\Phi^{i,j}(z)$, i.e., $\mathsf{OPT}(i,j) = \Phi^{i,j}(z^*)$. By Lemma 2, such the value z^* is the pseudo-intersection point of $\theta^{i,+}(z)$ and $\theta^{j,-}(z)$ on $[W_i, W_{j-1}]$.

Before explaining our algorithms, we need introduce the following definition:

Definition 2. *For integers* $i, \ell, r \in [1..n]$ *with* $i < \ell \leq r$, *we denote by* $\theta^{i,+,[\ell..r]}(z)$ *the upper envelope of functions* $\{\theta^{i,+,h}(z) \mid h \in [\ell..r]\}$, *that is,*

$$\theta^{i,+,[\ell..r]}(z) = \max\{\theta^{i,+,h}(z) \mid h \in [\ell..r]\}.$$

For integers $i, \ell, r \in [1..n]$ *with* $\ell \leq r < i$, *we denote by* $\theta^{i,-,[\ell..r]}(z)$ *the upper envelope of functions* $\{\theta^{i,-,h}(z) \mid h \in [\ell..r]\}$, *that is,*

$$\theta^{i,-,[\ell..r]}(z) = \max\{\theta^{i,-,h}(z) \mid h \in [\ell..r]\}.$$

Algorithm for computing $\mathsf{OPT}(i,j)$ for given $i, j \in [1..n]$ with $i < j$

Phase 1: Find a set U of the maximal subpath nodes for $P_{i+1,j-1}$ by walking on segment tree $\mathcal{T}$ from leaf $i+1$ to leaf $j-1$.

Phase 2: For each $u \in U$, compute a real interval $\mathcal{I}_u^+$ such that $\theta^{i,+}(z) = \theta^{i,+,[\ell_u..r_u]}(z)$ holds on any $z \in \mathcal{I}_u^+$, and a real interval $\mathcal{I}_u^-$ such that $\theta^{j,-}(z) = \theta^{j,-,[\ell_u..r_u]}(z)$ holds on any $z \in \mathcal{I}_u^-$, both of which are obtained by using information stored at node u. See Sect. 5.1 in [14] for the details.

Phase 3: Compute the pseudo-intersection point z^* of $\theta^{i,+}(z)$ and $\theta^{j,-}(z)$ on $[W_i, W_{j-1}]$ by using real intervals obtained in Phase 2. See Sect. 5.2 in [14] for the details.

Phase 4: Compute $\mathsf{OPT}(i,j) = \Phi^{i,j}(z^*)$ as follows: By formula (7), we have

$$\Phi^{i,j}(z^*) = \int_0^{z^*} \theta^{i,+}(t)dt + \int_{z^*}^{W_n} \theta^{j,-}(t)dt$$

$$= \sum_{u \in U} \left\{ \int_{\mathcal{I}_u^+ \cap [0, z^*]} \theta^{i,+,[\ell_u..r_u]}(t)dt + \int_{\mathcal{I}_u^- \cap [z^*, W_n]} \theta^{j,-,[\ell_u..r_u]}(t)dt \right\}.$$

For each $u \in U$, we compute integrals $\int \theta^{i,+,[\ell_u..r_u]}(t)dt$ and $\int \theta^{j,-,[\ell_u..r_u]}(t)dt$ by using the information stored at u. See Sect. 5.3 in [14] for the details.

For the cases of $i = 0$ or $j = n + 1$, we can also compute $w'(0,j) = \Phi^{j,-}(0)$ and $w'(i, n + 1) = \Phi^{i,+}(W_n)$ by the same operations except for Phase 3.

We give the following lemma about the running time of the above algorithm for the case of general edge capacities. See Lemma 8 in [14] for the proof.

Lemma 5 (Key lemma for general capacity). *Let us suppose that a segment tree $\mathcal{T}$ is available. Given two integers $i, j \in [0..n+1]$ with $i < j$, one can compute a value $w'(i,j)$ in $O(\log^3 n)$ time for the confluent/non-confluent flow model.*

Recalling that the running time for construction of data structure $\mathcal{T}$ is $O(n \log^2 n)$, Lemmas 3, 4 and 5 imply the following main theorem.

Theorem 1 (Main theorem for general capacity). *Given a dynamic flow path network $\mathcal{P}$, there exists an algorithm that finds an optimal k-sink under the confluent/non-confluent flow model in time $\min\{O(kn \log^3 n), n2^{O(\sqrt{\log k \log \log n})} \log^3 n\}$.*

When the capacities of $\mathcal{P}$ are uniform, we can improve the running time for computing $w'(i,j)$ to $O(\log^2 n)$ time with minor modification. See Lemma 9 in [14] for the proof.

Lemma 6 (Key lemma for uniform capacity). *Let us suppose that a segment tree $\mathcal{T}$ is available. Given two integers $i, j \in [1..n]$ with $i < j$, one can compute a value $w'(i,j)$ in $O(\log^2 n)$ time for the confluent/non-confluent flow model when the capacities are uniform.*

Theorem 2 (Main theorem for uniform capacity). *Given a dynamic flow path network $\mathcal{P}$ with a uniform capacity, there exists an algorithm that finds an optimal k-sink under the confluent/non-confluent flow model in time $\min\{O(kn \log^2 n), n2^{O(\sqrt{\log k \log \log n})} \log^2 n\}$.*

4   Data Structures Associated with Nodes of $\mathcal{T}$

In the rest of the paper, we introduce novel data structures associated with each node u of segment tree $\mathcal{T}$, which are used to compute $\mathsf{OPT}(i,j)$ in $O(\text{poly} \log n)$ time. Note that our data structures generalize the *capacities and upper envelopes tree (CUE tree)* provided by Bhattacharya et al. [7].

Recall the algorithm for computing $\mathsf{OPT}(i,j)$ shown in Sect. 3. To explain the data structures, let us see more precisely how the algorithm performs in Phase 2. Confirm that for $z \in [W_i, W_{j-1}]$, it holds $\theta^{i,+}(z) = \max\{\theta^{i,+,[\ell_u..r_u]}(z) \mid u \in U\}$, where U is a set of the maximal subpath nodes for $P_{i+1,j-1}$. Let us focus on function $\theta^{i,+,[\ell_u..r_u]}(z)$ for a node $u \in U$ only on interval $(W_{\ell_u-1}, W_n]$ since it holds $\theta^{i,+,[\ell_u..r_u]}(z) = 0$ if $z \leq W_{\ell_u-1}$. Interval $(W_{\ell_u-1}, W_n]$ consists of three

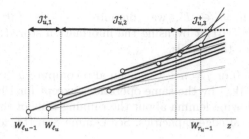

Fig. 2. Illustration of $\mathcal{J}_{u,1}^+$, $\mathcal{J}_{u,2}^+$ and $\mathcal{J}_{u,3}^+$. The thick half lines have the same slope of $1/C_{i,\ell_u}$, the gray half lines have slopes $\leq 1/C_{i,\ell_u}$, and the regular half lines have slopes $> 1/C_{i,\ell_u}$. The upper envelope of all the thick half lines and the regular half lines is function $\theta^{i,+,[\ell_u..r_u]}(z)$.

left-open-right-closed intervals $\mathcal{J}_{u,1}^+$, $\mathcal{J}_{u,2}^+$ and $\mathcal{J}_{u,3}^+$ that satisfy the following conditions: (i) For $z \in \mathcal{J}_{u,1}^+$, $\theta^{i,+,[\ell_u..r_u]}(z) = \theta^{i,+,\ell_u}(z)$. (ii) For $z \in \mathcal{J}_{u,2}^+$, $\theta^{i,+,[\ell_u..r_u]}(z) = \theta^{i,+,[\ell_u+1..r_u]}(z)$ and its slope is $1/C_{i,\ell_u}$. (iii) For $z \in \mathcal{J}_{u,3}^+$, $\theta^{i,+,[\ell_u..r_u]}(z) = \theta^{i,+,[\ell_u+1..r_u]}(z)$ and its slope is greater than $1/C_{i,\ell_u}$. See also Fig. 2. Thus in Phase 2, the algorithm computes $\mathcal{J}_{u,1}^+$, $\mathcal{J}_{u,2}^+$ and $\mathcal{J}_{u,3}^+$ for all $u \in U$, and combines them one by one to obtain intervals $\mathcal{I}_u^+$ for all $u \in U$. To implement these operations efficiently, we construct some data structures at each node u of $\mathcal{T}$. To explain the data structures stored at u, we introduce the following definition:

Definition 3. *For integers* $i, \ell, r \in [1..n]$ *with* $i < \ell \leq r$ *and a positive real* c, *let* $\bar\theta^{i,+,[\ell..r]}(c,z) = \max\{\bar\theta^{i,+,h}(c,z) \mid h \in [\ell..r]\}$, *where*

$$\bar\theta^{i,+,j}(c,z) = \begin{cases} 0 & \text{if } z \leq W_{j-1}, \\ \frac{z-W_{j-1}}{c} + \tau \cdot L_{i,j} & \text{if } z > W_{j-1}. \end{cases} \tag{10}$$

For integers $i, \ell, r \in [1..n]$ *with* $\ell \leq r < i$ *and a positive real* c, *let* $\bar\theta^{i,-,[\ell..r]}(c,z) = \max\{\bar\theta^{i,-,h}(c,z) \mid h \in [\ell..r]\}$, *where*

$$\bar\theta^{i,-,j}(c,z) = \begin{cases} \frac{W_j-z}{c} + \tau \cdot L_{j,i} & \text{if } z < W_j, \\ 0 & \text{if } z \geq W_j. \end{cases} \tag{11}$$

We can see that for $z \in \mathcal{J}_{u,2}^+$, $\theta^{i,+,[\ell_u+1..r_u]}(z) = \bar\theta^{\ell_u,+,[\ell_u+1..r_u]}(C_{i,\ell_u}, z) + \tau \cdot L_{i,\ell_u}$, and for $z \in \mathcal{J}_{u,3}^+$, $\theta^{i,+,[\ell_u+1..r_u]}(z) = \theta^{\ell_u,+,[\ell_u+1..r_u]}(z) + \tau \cdot L_{i,\ell_u}$. We then store at u of $\mathcal{T}$ the information for computing in $O(\text{poly}\log n)$ time $\theta^{\ell_u,+,[\ell_u+1..r_u]}(z)$ for any $z \in [0, W_n]$ as TYPE I, and also one for computing in $O(\text{poly}\log n)$ time $\bar\theta^{\ell_u,+,[\ell_u+1..r_u]}(c,z)$ for any $c > 0$ and any $z \in [0, W_n]$ as TYPE III.

In Phase 4, the algorithm requires computing integrals $\int_0^z \theta^{\ell_u,+,[\ell_u+1..r_u]}(t)dt$ for any $z \in [0, W_n]$, and $\int_0^z \bar\theta^{\ell_u,+,[\ell_u+1..r_u]}(c,t)dt$ for any $c > 0$ and any $z \in [0, W_n]$, for which the information is stored at each $u \in \mathcal{T}$ as TYPEs II and IV, respectively.

In a symmetric manner, we also store at each $u \in \mathcal{T}$ the information for computing $\theta^{r_u,-,[\ell_u..r_u-1]}(z)$, $\int_z^{W_n} \theta^{r_u,-,[\ell_u..r_u-1]}(t)dt$, $\bar{\theta}^{r_u,-,[\ell_u..r_u-1]}(c,z)$, and $\int_z^{W_n} \bar{\theta}^{r_u,-,[\ell_u..r_u-1]}(c,t)dt$ as TYPEs I, II, III, and IV, respectively.

Let us introduce what information is stored as TYPEs I–IV at $u \in \mathcal{T}$. See Sect. 4 in [14] for the detail.

TYPE I. We give the information only for computing $\theta^{\ell_u,+,[\ell_u+1..r_u]}(z)$ stored at $u \in \mathcal{T}$ as TYPE I since the case for $\theta^{r_u,-,[\ell_u..r_u-1]}(z)$ is symmetric. By Definition 2, the function $\theta^{\ell_u,+,[\ell_u+1..r_u]}(z)$ is the upper envelope of m_u functions $\theta^{\ell_u,+,h}(z)$ for $h \in [\ell_u+1..r_u]$. Let $\mathcal{B}^{u,+} = (b_1^{u,+} = 0, b_2^{u,+}, \ldots, b_{N^{u,+}}^{u,+} = W_n)$ denote a sequence of breakpoints of $\theta^{\ell_u,+,[\ell_u+1..r_u]}(z)$, where $N^{u,+}$ is the number of breakpoints. For each $p \in [1..N^{u,+} - 1]$, let $H_p^{u,+} \in [\ell_u+1..r_u]$ such that $\theta^{\ell_u,+,[\ell_u+1..r_u]}(z) = \theta^{\ell_u,+,H_p^{u,+}}(z)$ holds for any $z \in (b_p^{u,+}, b_{p+1}^{u,+}]$. As TYPE I, each node $u \in \mathcal{T}$ is associated with following two lists:

1. Pairs of breakpoint $b_p^{u,+}$ and value $\theta^{\ell_u,+,[\ell_u+1..r_u]}(b_p^{u,+})$, and
2. Pairs of range $(b_p^{u,+}, b_{p+1}^{u,+}]$ and index $H_p^{u,+}$.

Note that the above lists can be constructed in $O(m_u \log m_u)$ time for each $u \in \mathcal{T}$. By Property 2, we can obtain the whole information of TYPE I of $\mathcal{T}$ in time $O(n \log^2 n)$. We now give the application of TYPE I. See Lemma 11 in [14] for the proof.

Lemma 7 (Query with TYPE I). *Suppose that TYPE I of $\mathcal{T}$ is available. Given a node $u \in \mathcal{T}$ and a real value $z \in [0, W_n]$, we can obtain*

(i) index $H \in [\ell_u+1..r_u]$ such that $\theta^{\ell_u,+,H}(z) = \theta^{\ell_u,+,[\ell_u+1..r_u]}(z)$, and
(ii) index $H \in [\ell_u..r_u - 1]$ such that $\theta^{r_u,-,H}(z) = \theta^{r_u,-,[\ell_u..r_u-1]}(z)$

in time $O(\log n)$ respectively. Furthermore, if the capacities of $\mathcal{P}$ are uniform and $z \notin [W_{\ell_u}, W_{r_u-1}]$, we can obtain the above indices in time $O(1)$.

TYPE II. We give the information only for computing $\int_0^z \theta^{\ell_u,+,[\ell_u+1..r_u]}(t)dt$ stored at $u \in \mathcal{T}$ as TYPE II since the case for $\int_z^{W_n} \theta^{r_u,-,[\ell_u..r_u-1]}(t)dt$ is symmetric. Each node $u \in \mathcal{T}$ contains a list of all pairs of breakpoint $b_p^{u,+}$ and value $\int_0^{b_p^{u,+}} \theta^{\ell_u,+,[\ell_u+1..r_u]}(t)dt$. This list can be constructed in $O(m_u)$ time for each $u \in \mathcal{T}$ by using TYPE I of u. By Property 2, we obtain the whole information of TYPE II of $\mathcal{T}$ in time $O(n \log n)$. We give the application of TYPEs I and II. See Lemma 12 in [14] for the proof.

Lemma 8 (Query with TYPEs I and II). *Suppose that TYPEs I and II of $\mathcal{T}$ is available. Given a node $u \in \mathcal{T}$ and a real value $z \in [0, W_n]$, we can obtain (i) value $\int_0^z \theta^{\ell_u,+,[\ell_u+1..r_u]}(t)dt$, and (ii) value $\int_z^{W_n} \theta^{r_u,-,[\ell_u..r_u-1]}(t)dt$ in time $O(\log n)$ respectively. Furthermore, if the capacities of $\mathcal{P}$ are uniform and $z \notin [W_{\ell_u}, W_{r_u-1}]$, we can obtain the above values in time $O(1)$.*

TYPE III. We give the information only for computing $\bar{\theta}^{\ell_u,+,[\ell_u+1..r_u]}(c,z)$ stored at $u \in \mathcal{T}$ as TYPE III since the case for $\bar{\theta}^{r_u,-,[\ell_u..r_u-1]}(c,z)$ is symmetric. Note that it is enough to prepare for the case of $z \in (W_{\ell_u}, W_{r_u}]$ since it holds that $\bar{\theta}^{\ell_u,+,[\ell_u+1..r_u]}(c,z) = 0$ for $z \in [0, W_{\ell_u}]$ and

$$\bar{\theta}^{\ell_u,+,[\ell_u+1..r_u]}(c,z) = \bar{\theta}^{\ell_u,+,[\ell_u+1..r_u]}(c, W_{r_u}) + \frac{z - W_{r_u}}{c}$$

for $z \in (W_{r_u}, W_n]$, of which the first term is obtained by prepared information with $z = W_{r_u}$ and the second term is obtained by elementally calculation.

For each $u \in \mathcal{T}$, we construct a *persistent segment tree* as TYPE III. Referring to formula (10), each function $\bar{\theta}^{\ell_u,+,j}(c,z)$ for $j \in [l_u + 1..r_u]$ is linear in $z \in (W_{j-1}, W_n]$ with the same slope $1/c$. Let us make parameter c decrease from ∞ to 0, then all the slopes $1/c$ increase from 0 to ∞. As c decreases, the number of subfunctions that consist of $\bar{\theta}^{\ell_u,+,[\ell_u+1..r_u]}(c,z)$ also decreases one by one from m_u to 1. Let $c_h^{u,+}$ be a value c at which the number of subfunctions of $\bar{\theta}^{\ell_u,+,[\ell_u+1..r_u]}(c,z)$ becomes $m_u - h$ while c decreases. Note that we have $\infty = c_0^{u,+} > c_1^{u,+} > \cdots > c_{m_u-1}^{u,+} > 0$. Let us define indices $j_1^h, \ldots, j_{m_u-h}^h$ with $l_u + 1 = j_1^h < \cdots < j_{m_u-h}^h \leq r_u$ corresponding to the subfunctions of $\bar{\theta}^{\ell_u,+,[\ell_u+1..r_u]}(c_h^{u,+}, z)$, that is, for any integer $p \in [1..m_u - h]$, we have

$$\bar{\theta}^{\ell_u,+,[\ell_u+1..r_u]}(c_h^{u,+}, z) = \bar{\theta}^{\ell_u,+,j_p^h}(c_h^{u,+}, z) \quad \text{if } z \in (W_{j_p^h-1}, W_{j_{p+1}^h-1}], \qquad (12)$$

where $j_{m_u-h+1}^h - 1 = r_u$. We give the following lemma about the property of $c_h^{u,+}$. See Lemma 13 in [14] for the proof.

Lemma 9. *For each node $u \in \mathcal{T}$, all values $c_1^{u,+}, \ldots, c_{m_u-1}^{u,+}$ can be computed in $O(m_u \log m_u)$ time.*

By the above argument, while $c \in (c_h^{u,+}, c_{h-1}^{u,+}]$ with some $h \in [1..m_u]$ (where $c_{m_u}^{u,+} = 0$), the representation of $\bar{\theta}^{\ell_u,+,[\ell_u+1..r_u]}(c,z)$ (with $m_u - h + 1$ subfunctions) remains the same. Our fundamental idea is to consider segment trees corresponding to each interval $(c_h^{u,+}, c_{h-1}^{u,+}]$ with $h \in [1..m_u]$, and construct a persistent data structure for such the segment trees.

First of all, we introduce a segment tree T_h with root ρ_h to compute $\bar{\theta}^{\ell_u,+,[\ell_u+1..r_u]}(c,z)$ for $c \in (c_h^{u,+}, c_{h-1}^{u,+}]$ with $h \in [1..m_u]$. Tree T_h contains m_u leaves labeled as $l_u + 1, \ldots, r_u$. Each leaf j corresponds to interval $(W_{j-1}, W_j]$. For a node $\nu \in T_h$, let ℓ_ν (resp. r_ν) denote the label of the leftmost (resp. rightmost) leaf of the subtree rooted at ν. Let p_ν denote the parent of ν if $\nu \neq \rho_h$. We say a node $\nu \in T_h$ *spans* an interval $(W_{\ell_\nu-1}, W_{r_\nu}]$. For some two integers $i, j \in [l_u + 1..r_u]$ with $i < j$, if $(W_{\ell_\nu-1}, W_{r_\nu}] \subseteq (W_{i-1}, W_j]$ and $(W_{\ell_{p_\nu}-1}, W_{r_{p_\nu}}] \nsubseteq (W_{i-1}, W_j]$, then ν is called a *maximal subinterval node* for $(W_{i-1}, W_j]$. A segment tree T_h satisfies the following property similar to Property 1: For any two integers $i, j \in [l_u + 1..r_u]$ with $i < j$, the number of maximal subinterval nodes in T_h for $(W_{i-1}, W_j]$ is $O(\log m_u)$. For each $p \in [1..m_u - h + 1]$, we store function $\bar{\theta}^{\ell_u,+,j_p^{h-1}}(c,z)$ at all the maximal subinterval nodes for interval $(W_{j_p^{h-1}-1}, W_{j_{p+1}^{h-1}-1}]$, which takes $O(m_u \log m_u)$ time by the property. The other nodes in T_h contains NULL.

If we have T_h, for given $z \in (W_{\ell_u}, W_{r_u}]$ and $c \in (c_h^{u,+}, c_{h-1}^{u,+}]$, we can compute value $\bar{\theta}^{\ell_u,+,[\ell_u+1..r_u]}(c,z)$ in time $O(\log m_u)$ as follows: Starting from root ρ_h, go down to a child such that its spanned interval contains z until we achieve a node that contains some function $\bar{\theta}^{\ell_u,+,j}(c,z)$ (not NULL). Now, we know $\bar{\theta}^{\ell_u,+,[\ell_u+1..r_u]}(c,z) = \bar{\theta}^{\ell_u,+,j}(c,z)$, which can be computed by elementally calculation.

If we explicitly construct T_h for all $h \in [1..m_u]$, it takes $O(m_u^2 \log m_u)$ time for each node $u \in T$, which implies by Property 2 that $O(n^2 \log^2 n)$ time is required in total. However, using the fact that T_h and T_{h+1} are almost same except for at most $O(\log m_u)$ nodes, we can construct a persistent segment tree in $O(m_u \log m_u)$ time, in which we can search as if all of T_h are maintained. Thus, we obtain the whole information of TYPE III of T in time $O(n \log^2 n)$ by Property 2.

Using this persistent segment tree, we can compute $\bar{\theta}^{\ell_u,+,[\ell_u+1..r_u]}(c,z)$ for any $z \in [0, W_n]$ and any $c > 0$ in $O(\log m_u)$ time as follows: Find integer h over $[1..m_u]$ such that $c \in (c_h^{u,+}, c_{h-1}^{u,+}]$ in $O(\log m_u)$ time by binary search, and then search in the persistent segment tree as T_h in time $O(\log m_u)$.

Lemma 10 (Query with TYPE III). *Suppose that TYPE III of T is available. Given a node $u \in T$, real values $z \in [0, W_n]$ and $c > 0$, we can obtain*

(i) index $H \in [\ell_u + 1..r_u]$ such that $\bar{\theta}^{\ell_u,+,H}(c,z) = \bar{\theta}^{\ell_u,+,[\ell_u+1..r_u]}(c,z)$, and
(ii) index $H \in [\ell_u..r_u - 1]$ such that $\bar{\theta}^{r_u,-,H}(c,z) = \bar{\theta}^{r_u,-,[\ell_u..r_u-1]}(c,z)$ in time $O(\log n)$ respectively.

TYPE IV. We give the information only for computing $\int_0^z \bar{\theta}^{\ell_u,+,[\ell_u+1..r_u]}(c,t)dt$ stored at $u \in T$ as TYPE IV since the case for $\int_z^{W_n} \bar{\theta}^{\ell_u,-,[\ell_u..r_u-1]}(c,t)dt$ is symmetric. Similar to TYPE III, we prepare only for the case of $z \in (W_{\ell_u}, W_{r_u}]$ since it holds that $\int_0^z \bar{\theta}^{\ell_u,+,[\ell_u+1..r_u]}(c,t)dt = 0$ for $z \in [0, W_{\ell_u}]$ and

$$\int_0^z \bar{\theta}^{\ell_u,+,[\ell_u+1..r_u]}(c,t)dt = \int_0^{W_{r_u}} \bar{\theta}^{\ell_u,+,[\ell_u+1..r_u]}(c,t)dt + \frac{(z - W_{r_u})^2}{2c}$$

for $z \in (W_{r_u}, W_n]$, of which the first term can be obtained by prepared information with $z = W_{r_u}$ and the second term by elementally calculation.

For each $u \in T$, we construct a persistent segment tree again, which is similar to one shown in the previous section. To begin with, consider the case of $c \in (c_h^{u,+}, c_{h-1}^{u,+}]$ with some $h \in [1..m_u]$ (where recall that $c_0^{u,+} = \infty$ and $c_{m_u}^{u,+} = 0$), and indices $j_1^{h-1}, \cdots, j_{m_u-h+1}^{h-1}$ that satisfy (12). In this case, for $z \in (W_{j_p^{h-1}-1}, W_{j_{p+1}^{h-1}-1}]$ with $p \in [1..m_u - h + 1]$, we have

$$\int_0^z \bar{\theta}^{\ell_u,+,[\ell_u+1..r_u]}(c,t)dt = \sum_{q=1}^{p-1}\left\{\int_{W_{j_q^{h-1}-1}}^{W_{j_{q+1}^{h-1}-1}} \bar{\theta}^{\ell_u,+,j_q^{h-1}}(c,t)dt\right\} + \int_{W_{j_p^{h-1}-1}}^{z} \bar{\theta}^{\ell_u,+,j_p^{h-1}}(c,t)dt.$$

$$(13)$$

For ease of reference, we use $F^{h,p}(c, z)$ instead of the right hand side of (13).

Similarly to the explanation for TYPE III, let T_h be a segment tree with root ρ_h and m_u leaves labeled as $l_u + 1, \ldots, r_u$, and each leaf j of T_h corresponds to interval $(W_{j-1}, W_j]$. In the same manner as for TYPE III, for each $p \in [1..m_u - h + 1]$, we store function $F^{h,p}(c, z)$ at all the maximal subinterval nodes in T_h for interval $(W_{j_p^{h-1}-1}, W_{j_{p+1}^{h-1}-1}]$. Using T_h, for any $z \in (W_{\ell_u}, W_{r_u}]$ and any $c \in (c_h^{u,+}, c_{h-1}^{u,+}]$, we can compute value $\int_0^z \bar{\theta}^{\ell_u,+,[\ell_u+1..r_u]}(c, t)dt$ in time $O(\log m_u)$ by summing up all functions of nodes on a path from root ρ_h to leaf with an interval that contains z. Actually, we store functions in a more complicated way in order to maintain them as a persistent data structure. We construct a persistent segment tree at $u \in T$ in $O(m_u \log m_u)$ time, in which we can search as if all of T_h are maintained. Using this persistent segment tree, we can compute $\int_0^z \bar{\theta}^{\ell_u,+,[\ell_u+1..r_u]}(c, t)dt$ for any $z \in [0, W_n]$ and any $c > 0$ in $O(\log m_u)$ time in the same manner as for TYPE III.

Lemma 11 (Query with TYPE IV). *Suppose that TYPE IV of T is available. Given a node $u \in T$, real values $z \in [0, W_n]$ and $c > 0$, we can obtain (i) value $\int_0^z \bar{\theta}^{\ell_u,+,[\ell_u+1..r_u]}(c, t)dt$, and (ii) value $\int_z^{W_n} \bar{\theta}^{r_u,-,[\ell_u..r_u-1]}(c, t)dt$ in time $O(\log n)$ respectively.*

5 Conclusion

We remark here that our algorithms can be extended to the minsum k-sink problem in a dynamic flow path network, in which each vertex v_i has the cost λ_i for locating a sink at v_i, and we minimize $\mathsf{AT}(\mathcal{P}, \mathbf{x}, \mathbf{d}) + \sum_i \{\lambda_i \mid \mathbf{x} \text{ consists of } v_i\}$. Then, the same reduction works with link costs $w''(i, j) = w'(i, j) + \lambda_i$, which still satisfy the concave Monge property. This implies that our approach immediately gives algorithms of the same running time.

Acknowledgement. Yuya Higashikawa: Supported by JSPS Kakenhi Grant-in-Aid for Young Scientists (20K19746), JSPS Kakenhi Grant-in-Aid for Scientific Research (B) (19H04068), and JST CREST (JPMJCR1402).
Naoki Katoh: Supported by JSPS Kakenhi Grant-in-Aid for Scientific Research (B) (19H04068), and JST CREST (JPMJCR1402).
Junichi Teruyama: Supported by JSPS Kakenhi Grant-in-Aid for Scientific Research (B) (19H04068), and JST CREST (JPMJCR1402).

References

1. Alstrup, S., Gavoille, C., Kaplan, H., Rauhe, T.: Nearest common ancestors: a survey and a new distributed algorithm. In: Proceedings of the the 14th Annual ACM Symposium on Parallel Algorithms and Architectures, pp. 258–264 (2002)
2. Belmonte, R., Higashikawa, Y., Katoh, N., Okamoto, Y.: Polynomial-time approximability of the k-sink location problem. CoRR abs/1503.02835 (2015)

3. Bender, M.A., Farach-Colton, M.: The LCA problem revisited. In: Gonnet, G.H., Viola, A. (eds.) LATIN 2000. LNCS, vol. 1776, pp. 88–94. Springer, Heidelberg (2000). https://doi.org/10.1007/10719839_9

4. Benkoczi, R., Bhattacharya, B., Higashikawa, Y., Kameda, T., Katoh, N.: Minsum k-Sink problem on dynamic flow path networks. In: Iliopoulos, C., Leong, H.W., Sung, W.-K. (eds.) IWOCA 2018. LNCS, vol. 10979, pp. 78–89. Springer, Cham (2018). https://doi.org/10.1007/978-3-319-94667-2_7

5. Benkoczi, R., Bhattacharya, B., Higashikawa, Y., Kameda, T., Katoh, N.: Minsum k-sink problem on path networks. Theor. Comput. Sci. **806**, 388–401 (2020)

6. de Berg, M., Cheong, O., van Kreveld, M., Overmars, M.: Computational Geometry: Algorithms and Applications, 3rd edn. Springer, Heidelberg (2010)

7. Bhattacharya, B., Golin, M.J., Higashikawa, Y., Kameda, T., Katoh, N.: Improved algorithms for computing k-Sink on dynamic flow path networks. In: Ellen, F., Kolokolova, A., Sack, J.R. (eds.) WADS 2017. LNCS, vol. 10389, pp. 133–144. Springer, Cham (2017). https://doi.org/10.1007/978-3-319-62127-2_12

8. Chen, D., Golin, M.J.: Sink evacuation on trees with dynamic confluent flows. In: 27th International Symposium on Algorithms and Computation (ISAAC 2016). Schloss Dagstuhl-Leibniz-Zentrum fuer Informatik (2016)

9. Chen, D., Golin, M.J.: Minmax centered k-partitioning of trees and applications to sink evacuation with dynamic confluent flows. CoRR abs/1803.09289 (2018)

10. Ford, L.R., Fulkerson, D.R.: Constructing maximal dynamic flows from static flows. Oper. Res. **6**(3), 419–433 (1958)

11. Higashikawa, Y.: Studies on the space exploration and the sink location under incomplete information towards applications to evacuation planning. Ph.D. thesis, Kyoto University, Japan (2014)

12. Higashikawa, Y., Golin, M.J., Katoh, N.: Minimax regret sink location problem in dynamic tree networks with uniform capacity. J. Graph Algorithms Appl. **18**(4), 539–555 (2014)

13. Higashikawa, Y., Golin, M.J., Katoh, N.: Multiple sink location problems in dynamic path networks. Theor. Comput. Sci. **607**, 2–15 (2015)

14. Higashikawa, Y., Katoh, N., Teruyama, J., Watase, K.: Almost linear time algorithms for minsum k-sink problems on dynamic flow path networks (a full version of the paper). CoRR abs/2010.05729 (2020)

15. Hoppe, B., Tardos, E.: The quickest transshipment problem. Math. Oper. Res. **25**(1), 36–62 (2000)

16. Mamada, S., Uno, T., Makino, K., Fujishige, S.: An $O(n \log^2 n)$ algorithm for a sink location problem in dynamic tree networks. Discret. Appl. Math. **154**, 2387–2401 (2006)

17. Schieber, B.: Computing a minimum weight k-link path in graphs with the concave monge property. J. Algorithms **29**(2), 204–222 (1998)

18. Skutella, M.: An introduction to network flows over time. In: Cook, W., Lovász, L., Vygen, J. (eds.) Research Trends in Combinatorial Optimization, pp. 451–482. Springer, Heidelberg (2009). https://doi.org/10.1007/978-3-540-76796-1_21

Matched Participants Maximization
Based on Social Spread

Guoyao Rao[1], Yongcai Wang[1], Wenping Chen[1], Deying Li[1(⊠)], and Weili Wu[2]

[1] School of Information, Renmin University of China, Beijing 100872, China
{gyr,ycw,wenpingchen,deyingli}@ruc.edu.cn
[2] Department of Computer Science, University of Texas at Dallas,
Richardson, TX 75080, USA
weiliwu@utdallas.edu

Abstract. With the great advantage of information spread in the social network, more and more team activities would like to be organized through the social platforms. These activities require users to match partners and then participate in groups, e.g., group-buying and blind-date. In this paper, we consider to organize such activities with matching constraints in social networks to attract as many matched participants as possible through a limited seed set. An Interest-based Forwarding model which is similar to Independent Cascading is used to model information propagation. We investigate two matching strategies to forming groups: (1) neighbor matching (NM), i.e., only direct neighbors can match and (2) global matching (GM), i.e., matching is organized by an external organizer. We prove the *matched participants maximization (MPM)* problem to optimize the seed set selection to maximize the expected number of final participants is NP-hard and the computation of the target function is #P-hard, under both the NM and GM strategies. To solve MPM-NM efficiently, we propose a Matching Reachable Set method and a $(1 - 1/e - \epsilon)$-approximation algorithm. Sandwich method is used for solving MPM-GM by using the result of MPM-NM as a lower-bound and constructing an upper bound in an extended graph. A $\beta(1 - 1/e - \epsilon)$-approximation algorithm is proposed for MPM-GM. At last, experiments on the real-world databases verifies the effectiveness and efficiency of the proposed algorithms.

Keywords: Social influence · Group-buying · Matching

1 Introduction

The online social network has a profound impact on our daily life. People usually publish their opinions and receive others' information on social platforms such as Facebook, Twitter and Wechat. Nowadays, many businesses would like to

This work is supported by the National Natural Science Foundation of China (Grant NO. 12071478, 11671400, 61972404, 61672524), and partially by NSF 1907472.

W. Wu and Z. Zhang (Eds.): COCOA 2020, LNCS 12577, pp. 214–229, 2020.
https://doi.org/10.1007/978-3-030-64843-5_15

promote their products or activities through social network platforms. They generally choose a few customers to experience the product firstly, and then exploit word-of-mouth to spread positive information about their products to attract more users. Since Kemp's original work in [10], different scenes of information propagation, different purposes of optimization, and different approaching algorithms have been studied [2,4,6,7,11,13,14,16–21].

However, to the best of our knowledge, we haven't seen research works considering the influence maximization problem with matching constraint. Businesses often would like to organize team activities through social network platforms, and some kinds of businesses allow only group-type participation. Users can participate only if they can form a group, and cannot participate individually. This is usual seen in online group-buying markets [1,9] such as Pinduoduo, a famous company gives discounts to group buyers, while an individual buyer cannot purchase. Users who are interested in the event and want to participate are called *latent participants*. Whether some latent participants can form a participant group is based on some *matching index* of the users, e.g. similar shipping addresses for reducing mailing cost, or common interest to joint the same event etc. Only the latent participants who form groups finally participate, which are called the *final participants*. The business's target is to attract as many final participants as possible through optimizing the selection of seed set using limited budget to initialize the information propagation.

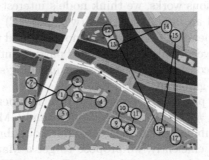

Fig. 1. An example of group-buying marketed through social network.

An simple example of matched participation is shown in Fig. 1. We use connected yellow blocks in the map to represent common addresses. The black line represents the friendships among people in a social network. Without loss of generality, we suppose that once a user gets the activity information from neighbors in the social network, he/she will be interested and then forwards the information to other friends.

If we choose only one node as seed to spread the group-buying activity and require the matching condition to be the common shipping address, then any node in $\{v_1, v_2, v_3, v_4, v_5, v_6, v_7\}$ is the best choice to make the information spread with 7 latent participants. But note that none of them can finally join the activity as each node can't find a partner with common address. If we choose node v_9 instead, it spreads less widely with four latent participants $\{v_8, v_9, v_{10}, v_{11}\}$. But all these nodes can finally join the activity to be final participants by grouping as $\{(v_8, v_9), (v_{10}, v_{11})\}$. So the traditional strategy based on widest spreading may not generate proper seed set for *matched participate maximization (MPM)*.

Moreover, let's further consider the matching conditions. In some events, people won't like to group with strangers who are not neighbors in social network. Such events include watching movies or having a meal together, which require only friends on the social network can match. On the other hand, when we conduct group-buy or play online game, we don't mind to group with strangers recommended by the system. Corresponding to these two kinds of matching scenarios, we summarize two matching strategies commonly seen in reality. (1) *Neighbor matching (NM)*, in which, each latent participants seek partners among his/her friends on the social network. (2) *Global matching (GM)*, in which an administrator or back-end system will seek partner for each latent participant globally. Specifically in this example, any node in $\{v_{12}, v_{13}, v_{14}, v_{15}, v_{16}, v_{17}\}$ is the best choice in global matching, and any node in $\{v_8, v_9, v_{10}, v_{11}\}$ is the best choice in neighbor matching. So different matching strategies need to be considered.

More specifically, this paper considers MPM problem in a given social network $G(V, E, p, m)$, where V is the user set; E is the edge set; p models users' interests to the event, and m models the matching indices of users. The key contributions are as following

1. Different from previous works, we think node's interest plays the key role in node activation and an Interest-based Forwarding (IF) diffusion model similar to classical independent Cascade (IC) model is used to model information propagation. Then we formulate the matched participants maximization problems(MPM) which considers two usual seen matching strategies as neighbor matching(NM) and global matching (GM).
2. We prove the MPM problem is NP-hard and the number computation for the final participants is #P-hard for both the NM and GM cases. We prove the target function is submodular in NM case, but not submodular in GM case.
3. To solve MPM-NM efficiently, we propose to get a Matching Reachable Set (MRS) to approximately estimate the target function. Then an approximation algorithm is proposed for the MPM-NM problem with the $(1 - 1/e - \epsilon)$ ratio. For the MPM-GM, we use the idea of Sandwich [14] by seeking an upper bound with extending the graph since MPM-NM provides a lower bound. Hence the sandwich algorithm guarantees the $\beta(1 - 1/e - \epsilon)$ approximation ratio.
4. At last, we experience our methods based on the real-word labeled databases. The results of experiment demonstrate that the proposed methods outperform existing comparing methods.

2   Related Work

Soical Influence Problems: Kemp et al. [10] firstly formulate the basic problem of influence maximization (IM) in social influence and propose two basic spread models as IC and LT. There are already many extended researches to investigate IM problem from different perspectives., such as time-constrained [13], topic-aware [2], competition [4], rumor-control [7], location-target [12], companies-balance [3]. Specially, in recent work [24], the authors proposed the group influence maximization problem in social networks, in which some scenes they considered are similar to our paper but not the same. In their model, they supposed there are already definite groups and the key problems they need to solve is how to activate these groups as more as possible, in which a group is said to be activated if a certain ratio of nodes in this group are activated under IC model. Group-fairness in influence maximization was considered in [22]. They also considered the groups exist in advance and pursued fair information propagation among the groups. However, we usually don't know the group in advance and so in our paper, we consider the groups are dynamically online formed by common matching index of the latent participants.

Influence Maximization Algorithms: Besides, there are many works in improving algorithms to solve the basic IM problem. Usually, these algorithms can also be applied to solve the extension problem mentioned above. Kempe [10] proved that the basic IM problem is NP-hard and the influence computation is #P-hard. With the good property of submodularity for the target function, the basic greedy method can provide a $(1 - 1/e - \epsilon)$-approximation solution where ϵ is the loss caused by influence estimation. but costs too much time using the heavy Monte Carlo simulations to estimate the marginal gain of node's influence, although there are many improvements such as [11,17]. These algorithms are not efficient for the large scale network. Tang et al. [21] and Borgs et al. [5] proposed the reverse influence set (RIS) sampling method to estimate the influence. The RIS-based methods are efficient for the large scale network, but they have a key problem that how to sample the RIS sets as less as possible to reduce the time cost and guarantee the $(1 - 1/e - \epsilon)$-approximation with high confidence. So next there are many RIS-based extensions and improvements such as the Influence Maximization via Martingales (IMM) [20], Stop-and-Stare (SSA) and Dynamic Stop-and-Stare (D-SSA) [8,16]. Recently, as far as we have known, the Online Processing Information Maximization (OPIM) [19] has superior empirical effectiveness.

3 Problem Model and Analysis

We study how to choose seed set with limited size to maximize the number of final participants considering the matching constraint. The social network(SN) we consider is modeled as $G(V, E, p, m)$. Each vertex in $i \in V$ represents a user. Each edge $(i, j) \in E$ denotes a friendship between two users in the social network. The vertex weight p_i (in the range [0.1]) represents the user i's interest probability to the activity; and the vertex weight $m_i \in \mathbb{N}$ is the matching index of the user i. As we analyze above, we need to solve two stage problems: (1) information spread from a seed set S to activate latent participants, which are denoted by set V^l and (2) matching process to generate final participants, which are denoted by set V^p.

3.1   Interest-Based Forwarding Model (IF)

There are many models for the information spreading such as the widely used IC and LT models [10]. However in participating problems, whether a user turns to be a latent participant is mainly determined by his/her own interest when he/she firstly knows the information. It is less matter with whom they get the information. So we adopt a information propagation model, called Interest-based Forwarding (IF) which is very similar to the classical IC model except that a node's activation is determined by the node's interest probability instead of the activating edge's probability.

Starting from the seed set, when a node v knows the information for the first time, it will decide to be a latent participant by its interest probability p_v. If it turns to be a latent participant, it will forward the information to its neighbors. Otherwise, it stays inactive and cannot be activated by later received information. The process terminates when there is no any new latent participant.

We show an IF instance based on the network in Fig. 2(a). The node colors indicate their matching indices and the values on the nodes are the interest probabilities. And as shown in Fig. 2(b), when we choose node h as a seed, in first round, h gets interested and forwards the information to its neighbor a. In next round, a gets interested and forwards it to its neighbors$\{b, e, f\}$. But only $\{b, f\}$ get interested and b continues to forward the information to its neighbors $\{c, d, g\}$. At last, uninterested node g won't forward it to its neighbor j and the spread terminates. The latent participants in this example is $V^l = \{h, a, b, c, d, f\}$ as shown in Fig. 2(b).

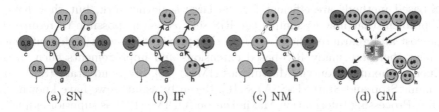

(a) SN	(b) IF	(c) NM	(d) GM

Fig. 2. An example of matching participating based on social spread.

The Equivalent Generating Model. For each non-seed node during the IF, the following steps are checked to decide whether the node is a latent participant; (i) Check whether the node gets the information, i.e. whether the node has a neighbor who is a latent participant. (ii) Decide node to be interested or not by the interest probability. From such view, we can decide all nodes to be interested or not without really propagating the information. The IF propagation process is equivalent as the following process:

1. Firstly remove each G's node v by probability $1 - p_v$ and denote all remain nodes as set T;
2. Then get G's vertex-induced subgraph $g = G(T)$ and we mark $Pr_{[g]}$ as the probability of g being induced in step 1.

3. Each node u in g will be a latent participant iff there exists a route in g connecting u and any seed in S. So the set of all latent participants is the set of all nodes in g that S can reach to and we mark it as g_S.

According to the equivalent generating model, the probability of V^l triggered by S is the sum of the probabilities of all the vertex-induced subgraphs where the set of nodes that S can reach to equals to V^l.

3.2 Matching Strategies

After knowing the latent participants, they are matched to generate final participants. We consider two matching strategies.

Neighbor Matching (NM): Some activities, such as watching movies require friends on the social network to participate. In such case, each latent participant seeks common matching index friends to be his/her partners. As shown in Fig. 2(c), the latent participant b can group with d who has the same color. a can group with h. c and f cannot find matched friends. So under neighbor matching strategy, $\{b, d, a, h\}$ will be the final participants. In NM, each latent participant v can be a final participant if it has common matching-index neighbors. Let $V_\triangle^p$ be the set of final participants through the neighbor matching process $\mathcal{M}_\triangle$. We denote $V_\triangle^p = \mathcal{M}_\triangle(V^l)$. Then

$$v \in V_\triangle^p \Leftrightarrow \exists u \in N_v \cap V^l \text{ s.t. } m_u = m_v \tag{1}$$

where N_v is v's one-hop neighbor set in G.

Global Matching (GM): In some activities, such as group buying, matching is determined by a back-end system, which may match any latent participants with common matching index into one group. As shown in Fig. 2(d), through the back-end system, two strangers c and f with common matching-index is formed into one group to become final participants. Under GM strategy, each latent participant v turns to be a final participant if there is at least one common matching index latent participant. Let $V_\square^p$ be the set of final participants through the global matching process $\mathcal{M}_\square$, i.e., $V_\square^p = \mathcal{M}_\square(V^l)$. Then,

$$v \in V_\square^p \Leftrightarrow \exists u \in V^l \text{ s.t. } m_u = m_v \tag{2}$$

3.3 The MPM Problems

Based the above models, we formulate our matching participants maximization problem as following.

Definition 1 (Matching Participants Maximization Problem (MPM)).
Given the graph $G(V, E, p, m)$, let $\sigma(S)$ be the expected number of final participants V^p through the stochastic IF process from the seeds S in G. The MPM problem is to find a seed set S^* with size k such that

$$S^* := \underset{S \subseteq V, |S| = k}{argmax} \ \sigma(S). \tag{3}$$

More specifically, according to the generating model, for MPM problem using neighbor matching strategy, denoted by **MPM-NM** problem, we have

$$\sigma_\triangle(S) = E(|V_\triangle^p|) = E(|\mathcal{M}_\triangle(V^l)|) = \sum_{g \subseteq G} Pr_{[g]} |\mathcal{M}_\triangle(g_S)| \tag{4}$$

For MPM using global matching, denoted by **MPM-GM**, we have

$$\sigma_\square(S) = E(|V_\square^p|) = E(|\mathcal{M}_\square(V^l)|) = \sum_{g \subseteq G} Pr_{[g]} |\mathcal{M}_\square(g_S)| \tag{5}$$

Note that $\mathcal{M}_\triangle(g_S) \subseteq \mathcal{M}_\square(g_S)$ and then from the definition, we can easily have the following Lemma.

Lemma 1. The number of expected final participants in **MPM-GM** is not less that in **MPM-NM**, i.e., $\sigma_\square(S) \geq \sigma_\triangle(S)$.

Problem Hardness. Note that our spread model is different from the traditional spread model in previous IM problems. We can't see traditional IM problem as a special case of our MPM problem to show the hardness.

Theorem 1. The MPM problem is NP-hard.

Proof 1. Consider any instance of the NP-complete Set Cover problem with a collection **S** of sets $S_1, S_2, \ldots, S_m$, whose union is the node set $U = \{u_1, u_2, \ldots, u_\triangle\}$; Supposing that $k < n < m$, we wish to know whether there exist k sets in **S**, $S_{q_1}, S_{q_2}, \ldots, S_{q_k}$ whose union also is U. Actually, it's equivalent to a special MPM problem with the $G(V, E, p, m)$ constructed as following. There is a node s_i corresponding to each set $S_i (1 \leq i \leq m)$, and there are two nodes v_j, t_j corresponding to each node $u_j (1 \leq j \leq n)$. There is an edge $<s_i, v_j>$ whenever $u_j \in S_i$, and an edge $< v_j, t_j >$ for each v_j and t_j. Let all $p_{s_i} = 1$, $p_{v_j} = 0.5$ and $p_{t_j} = 1$. Let each $m_{s_i} = i$ and $m_{v_j} = m_{t_j} = m + j$. We don't show later details of the proof here because of the limitation of space.

Calculating $\sigma(S)$ is #P-hard. In fact, the computation of expected number of final-participants σ is also complex and we have following theorem.

Theorem 2. Given a seed set S, computing the expected number for final participants $\sigma(S)$ is #P-hard.

Proof 2. Consider any instance of the #P-complete s-t connectedness counting problem with $G(V, E)$ and two vertix s and t. We wish to count the number of G's subgraphs in which s is connected to t, and we denote all of such subgraphs as a set $\mathcal{G}$. We show this problem is equivalent to the following special computing problem of σ with one seed on $G'(V', E', p, m)$. Let $V' = V \cap \{t'\}$, $E' = E \cup \{<t, t'>\}$, and $p_{v_i} = 0.5$, $m_{v_i} = i$ for each node $v_i \in V$. Let $p_\triangle = p_t = p_{t'} = 1$, and $m_\triangle = |V| + 1$, $m_t = m_{t'} = |V| + 2$. Given a seed set $S = \{s\}$, we can easily have that $\sigma_\square(S) = \sigma_\triangle(S) = 2Pr\{\{s\} \vdash t\} = 2\sum_{g \in \mathcal{G}} 0.5^{|V|} = 0.5^{|V|-1}|\mathcal{G}|$. Thus we can get the size of $\mathcal{G}$ by $2^{|V|-1}\sigma(S)$. So the computation problem of σ is #P-hard since the s-t connectedness counting problem is #P-complete.

Next we design algorithms to solve the MPM under the NM and GM strategies respectively.

4 The Algorithms Designs

As the NP-hard property of the problem we proved above, we hope to find efficient algorithms with good approximation-guaranteed. Nemhauser, Wolsey, and Fisher [15] show that the general greedy hill-climbing algorithm approximates the optimum to within a factor of $(1-1/e)$, when the target set function $f : 2^V \rightarrow R$ is non-negative, non-decreasing and submodular. In our MPM problem, it's easy to get that the target function σ is non-negative and non-decreasing. Next, we design the algorithms by considering the submodularity.

4.1 MRS-Based Algorithm for MPM-NM

Before analyzing the submodularity of $\sigma_\triangle$, we first consider following estimation problem. Note that the computation problem of $\sigma_\triangle$ is #P-hard, which means it's hard to compute the expectation exactly. And it's heavy to estimate the value by using Monte Carlo to directly simulate the process of firstly IF spreading and then matching. To solve such problem, we need other efficient method to approximately estimate the target function. Firstly, we introduce the concept of a random set as following.

Definition 2. (Matching Reachable Set(MRS)) *Given a Graph $G(V, E, p, m)$ and a node v randomly choose from $\mathcal{M}_\triangle(V)$ with node probability $\frac{p_v}{\tau}$, where τ is the sum of all nodes' probabilities in $\mathcal{M}_\triangle(V)$. Let g be a vertex-induced subgraph of $G/\{v\}$ by removing each node u with probability $1 - p_u$. The Matching Reachable Set(MRS) of v is the set of nodes that can be reached by any v's neighbor in g if there is a v's matched neighbor in g, otherwise it is an empty set.*

According to the previous problem definition, we can infer as following,

$$\sigma_\triangle(S) = \sum_{g \subseteq G} Pr_{[g]}|\mathcal{M}_\triangle(g_S)| \tag{6}$$

$$= \sum_{g \subseteq G} Pr_{[g]} \sum_{v \in \mathcal{M}_\triangle(V)} \mathbb{1}_{[v \in \mathcal{M}_\triangle(g_S)]} \tag{7}$$

$$= \sum_{v \in \mathcal{M}_\triangle(V)} \sum_{g \subseteq G} Pr_{[g]} \mathbb{1}_{[v \in \mathcal{M}_\triangle(g_S)]} \tag{8}$$

$$= \sum_{v \in \mathcal{M}_\triangle(V)} \sum_{g \subseteq G} Pr_{[g]} \left(\bigvee_{u \in g \cap N_v} \mathbb{1}_{[m_u = m_v]} \right) \mathbb{1}_{[g_v \cap S \neq \emptyset]} \tag{9}$$

$$= \sum_{v \in \mathcal{M}_\triangle(V)} p_v \sum_{g \subseteq G/\{v\}} Pr_{[g]}\left(\bigvee_{u \in g \cap N_v} \mathbb{1}_{[m_u = m_v]}\right)\mathbb{1}_{[g_{N_v} \cap S \neq \emptyset]} \tag{10}$$

$$= \tau \sum_{v \in \mathcal{M}_\triangle(V)} \sum_{g \subseteq G/\{v\}} \frac{p_v}{\tau} Pr_{[g]}\left(\bigvee_{u \in g \cap N_v} \mathbb{1}_{[m_u = m_v]}\right)\mathbb{1}_{[g_{N_v} \cap S \neq \emptyset]} \tag{11}$$

$$= \tau E(\mathbb{1}_{[S \cap S \neq \emptyset]}) \tag{12}$$

where $\mathbb{S}$ is the random MRS set. It's easy to have that $E(\mathbb{1}_{[\mathbb{S} \cap S \neq \emptyset]})$ is submodular, and hence we have following theorem.

Theorem 3. $\sigma_\triangle$ *is submodular for MPM-NM.*

As we analyse the equivalence between IF and the generating model, we can get a MRS set $\mathbb{S}$ by a random Breadth-First-Search as shown in Algorithm 1 and avoid to heavily sample large numbers of vertex-induced subgraphs. Firstly, choose a random node v by the probability $\frac{p_v}{\tau}$ from $\mathcal{M}_\triangle(V)$, then do the random Breadth-First-Search in G starting from v by the node probability in rounds. If there is no any v's matched neighbor in first search round, then terminate the process ahead of time and let $\mathbb{S}$ be the $\emptyset$, else continue the process and let $\mathbb{S}$ be all nodes searched.

Algorithm 1: MRS($G(V, E, p, m)$)

1 Initialize a queue $Q \leftarrow \emptyset$, a set $S \leftarrow \emptyset$, the search flag $F \leftarrow False$;
2 Choice a node v by the probability $\frac{p_v}{\tau}$ from $M_\triangle(V)$;
3 Mark v be *visited*;
4 **for** $u \in N_v$ **do**
5   Mark u be *visited*;
6   Push u into Q by the probability of p_u;
7   **if** u *is pushed into* Q *and* $m_u = m_v$ **then**
8         $F \leftarrow True$;

9 **if** F *is True* **then**
10   **while** Q *is not empty* **do**
11         $q \leftarrow Q.pop()$ and add q to S;
12   **for** $o \in N_q$ *not visited* **do**
13   Mark o be *visited*;
14   Push o into Q by the probability of p_o;

15 **return** S;

Let θ be the number of MRS sets. When we get θ outcomes $X_\theta = \{x_1, x_2, ..., x_\theta\}$ from random MRS sampling, we have $F_{X_\theta}(S) := \frac{1}{\theta} \sum_{i=1}^\theta \mathbb{1}_{[S \cap x_i \neq \emptyset]}$ which is the proportion of sets in X_θ covered[1] by S. So $F_{X_\theta}(S)$ is a statistical estimation for $E(\mathbb{1}_{[\mathbb{S} \cap S \neq \emptyset]})$, that is we have $\sigma_\triangle \approx |\tau|F_{X_\theta}$. Then the

---

[1] We say that a set A covers a set B that is $A \cap B \neq \emptyset$, and a node a covers a set B that is $a \in B$.

greedy Max-Coverage(Algorithm 2) provides a $(1-1/e)$-approximation solution for the maximum coverage problem [23]to choose k nodes that cover the maximum number of sets in X_θ. So at the same time we may get same approximation solution for our neighbor matching maximization problem ignoring the estimation error. We call such idea as MRS-based.

Algorithm 2: Max-Coverage(X_θ, k)

1 $S_k \leftarrow \emptyset$;
2 **for** $i = 1$ *to* k **do**
3   Get the node s_i which *covers* most sets in X_θ;
4   Remove all sets from X_θ, which is *covered* by s_i;
5   Let $S_k \leftarrow S_k \cup \{s_i\}$;
6 **return** S_k *as the selected seeds*

Algorithm 3: OP-MPM($G(V, E, p, m)), k, \epsilon, \delta$)

1 Get the samples number θ by the OPIM-C[19] algorithm with the estimation error ϵ and confidence value δ ;
2 Get the MRS sets X_θ with the numer of θ;
3 $S_k \leftarrow$ Max-Coverage(X_θ, k) ;
4 **return** *The selected seeds* S_k

However, as we know, the more samples lead to less error between the statistical estimation and the truth in Monte Carlo, but more sampling cost. Then our MRS-based problem is similar with the RIS-based IM problem [21] and both of them face such sampling cost problem that how to sample less sets under the premise of ensuring certain accuracy of the solution, that is how to balance the accuracy and efficiency. Then to solve our sampling problem, we adapt the Algorithm 3 from the OPIM [19] which provides an algorithm framework to solve the above set sampling problem when the target function can be estimated by the statistical method of covering a random set. According to the property of OPIM and without loss of generality by replacing the set cover estimation of $I(S) = nPr(RIS \cap S)$ in IM with $\sigma_\triangle(S) = \tau Pr(MRS \cap S)$ in our MPM-NM, the Algorithm 3 has following formulations.

Theorem 4. *The adapted OP-MPM can guarantee: The output S_k is an $(1-1/e-\epsilon)$-approximation solution with probability at least $1-\delta(0 < \delta < 1)$; When $\delta \leq 1/2$, the expected sampling number of MRS sets is $O\Big((k\ln n - \ln(\delta))\tau\epsilon^{-2}/\sigma_\triangle(S_k^*)\Big)$, where S_k^* is the unknown optimum solution.*

Hence, by this theorem, we have the expected time cost in sampling(the main cost of the algorithm) is $O\Big(\frac{\text{ESMRS}\big(k\ln n - \ln(\delta)\big)\tau\epsilon^{-2}}{\sigma_\triangle(S_k^*)}\Big)$, where ESMRS is the expected number of the nodes searched in sampling a MRS set.

4.2   Sandwich Algorithm for MPM-GM

Actually, $\sigma_\square$ may not be submodular in MPM-GM.

Theorem 5. $\sigma_\square$ *can't be guaranteed to be submodular for MPM-GM.*

Fig. 3. A special case

Proof 3. *We prove it by a special case shown in the Fig. 3. We can have* $\sigma_\square(\{a\}) = \sigma_\square(\{c\}) = 2p_a p_b p_c = 0.01$, $\sigma_\square(\{a,c\}) = 2p_a p_c = 0.05$, *and then* $\sigma_\square(\{a,c\}) - \sigma_\square(\{a\}) \geq \sigma_\square(\{c\})$. *So in this sample, $\sigma_\square$ is not submodular.*

For the unsubmodularity of $\sigma_\square$, we use the idea of Sandwich [14] which firstly gets corresponding solutions in two maximization problems with the lower bound and the upper bound of the target function, then choose the best one between them as the solution for the maximization problem with the target function. It can provide approximation analysis for this solution to the target maximization problem, specially when the solutions of each bound can guarantee good approximations for their own problem. So we consider to find such upper and lower bounds for our target function $\sigma_\square$. As we know that $\sigma_\triangle$ is a lower bound of $\sigma_\square$, we will construct an upper bound as following.

A upper bound of $\sigma_\square$. We extend a graph $G'(V, E', p, m)$ from $G(V, E, p, m)$ by adding edges as following: for any two nodes u and v without edge connecting each other in G, add an edge $<u.v>$ to G if $m_u = m_v$. Then G is a subgraph of G' and we mark the target function of MPM-NM on G' as $\sigma'_\triangle$. Nextly we will prove that $\sigma'_\triangle$ is an upper bound of $\sigma_\square$.

Theorem 6. $\sigma'_\triangle$ *is an upper bound of $\sigma_\square$.*

Proof 4. *By the generating model, we have*

$$\sigma_\square(S) = \sum_{T \in 2^V} \prod_{v \in T} \prod_{u \notin T} p_v (1 - p_u) |M_\square(G(T)_S)| \tag{13}$$

$$\sigma'_\triangle(S) = \sum_{T \in 2^V} \prod_{v \in T} \prod_{u \notin T} p_v (1 - p_u) |M_\triangle(G'(T)_S)| \tag{14}$$

Actually, for any node $u \in M_\square(G(T)_S)$, we must have $u \in G(T)_S$ and u has a matched node v in $G(T)_S$. It's obviously that any node in $G(T)_S$ must also exist in $G'(T)_S$ as $G(T)$ is a subgraph of $G'(T)$. So both u and v are in $G'(T)_S$ and as any matched nodes are neighbors in G', we have $u \in M_\triangle(G'(T)_S)$. Hence $M_\square(G(T)_S) \subseteq M_\triangle(G'(T)_S)$. So $\sigma_\square(S) \leq \sigma'_\triangle(S)$ since $|M_\square(G(T)_S)| \leq |M_\triangle(G'(T)_S)|$.

Sandwich Algorithm (SA-MPM). We first run the OP-MPM algorithm to give MPM-NM's solutions S_l^k, S_u^k corresponding to the target function $\sigma_\triangle$ and $\sigma'_\triangle$, then return our MPM-GM's solution $S_{sa}^k := argmax_{S\in\{S_l^k, S_u^k\}}\sigma_\square(S)$, which provides following approximation guarantee.

Theorem 7. *The solution given by SA-MPM satisfies*

$$\sigma_\square(S_{sa}^k) \geq \beta(1 - 1/e - \epsilon).\sigma_\square(S_\square^k) \tag{15}$$

with probability at least $1 - 2\delta$, where $S_\square^k$ is the optimum solution for the MPM-GM problem, and $\beta = max\{\frac{\sigma_\square(S_u^k)}{\sigma'_\triangle(S_u^k)}, \frac{\sigma_\triangle(S_\square^k)}{\sigma_\square(S_\square^k)}\}$.

5 Experiments

In this section, our experiments aim to analysis our proposed model and evaluate the performance of the proposed methods, based on 4 open real-world labeled datasets[2](BlogcCatalog, Flickr, DBLP, Twitter) as shown in Table 1. All codes of the experiments are written in c++ with parallel optimization, and all experiments run in a linux machine with 12 cores, 24 threads, 2.4 G hz, CPU and 64G RAM.

Table 1. Statistics of the datasets

Dataset	#Node	#Edge	#Label
BlogCatalog	10K	333K	2k
Flickr	80K	5.9M	8K
DBLP	203K	382K	10K
Twitter	580K	717K	20K

5.1 Experiment Setup

We set the interest probability of each node randomly and uniformly from $[0, 1]$. We model the matching index from node label, i.e., $m_i := j$ where j is the index of the node v_i's label. We compare our algorithms with some baseline algorithms as following.

- **OP-MPM**: The algorithm we proposed to solve the MPM in neighbor matching.
- **SA-MPM**: The algorithm we proposed to solve the MPM in global matching.
- **MCGreedy**: Add the seed with maximum marginal gain to current seeds set, which is computed by the value estimation of target function with Monte Carlo simulations. We set the simulation number to be 10000.
- **IM-IF**: Get the seed set with the maximum spread influence based on IF process and we can get such seed set by using OP-MPM to solve a special case of MPM in which all nodes' matching indices are same.
- **Random**: The basic baseline algorithms by choosing seeds randomly.

---

[2] http://networkrepository.com.

5.2 Experiments Result

For all OP-MPM and the OP-MPM used in SA-MPM and IM-IF as a subalgorithm, we set same $\epsilon = 0.1$ and $\delta = 0.01$. We chose different seed size k increasing from 0 to 100 by a step of 10, and evaluate the expected number of final participants with the average number in 10000 simulations.

MPM-NM: For the MPM problems in neighbor matching, we compared methods OP-MPM, MCGreedy, IM-IF and Random on each network. As shown in Fig. 4, almost all the curves except Random is increasing and gradually stabilize, and the expected number satisfy following comparisons: Random<IM-IF<MCGreedy≈OP-MPM. Of course, given a large and enough simulations number in MCGreedy, it can also provide good accuracy equivalent to OP-MPM, but it will cost too much more time than OP-MPM.

MPM-GM: For the MPM problems in global matching, we compare SA-MPM, MCGreedy, IM-IF and Random. As shown in Fig. 5, it's similar to MPM-NM that the expected number satisfies following comparisons: Random<IM-IF<MCGreedy<SA-MPM. Note that the target function in MPM-GM may not be submodular and the greedy method has no approximation guarantee, so the accuracy of MCGreedy is not as good as SA-MPM. Actually once again, both in our experiments in MPM-NM and MPM-GM, the less expected number of final participants by IM-IF than MCGreedy and OP-MPM tells that widely spread strategy may not be reliable.

Running Time: We run each algorithm 10 times and compute the average of running time. As the experiments shown in Fig. 6, on all datasets, the MCGreedy is much heavier than the MRS-based method because of the heavy Monte Carlo simulations. The running time of SA-MPM is nearly two times more than OP-MPM because two similar algorithms of OP-MPM need to run in SA-MPM. The experiments results show that IM-IF is heavier than OP-MPM and sometimes even heavier than SA-MPM (e.g., on Flickr), because IM-IF may cost more time than others to sample a MTR set.

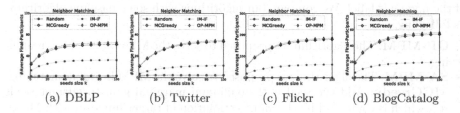

(a) DBLP   (b) Twitter   (c) Flickr   (d) BlogCatalog

Fig. 4. Expected final participants comparison in neighbor matching.

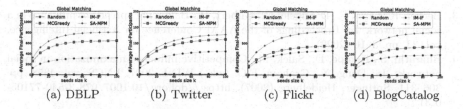

(a) DBLP   (b) Twitter   (c) Flickr   (d) BlogCatalog

Fig. 5. Expected final participants comparison in global matching.

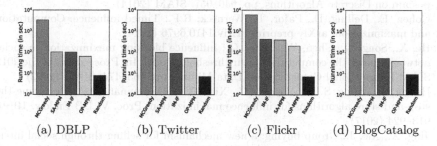

(a) DBLP   (b) Twitter   (c) Flickr   (d) BlogCatalog

Fig. 6. Average running time comparisons with k = 20.

6 Conclusions

In this paper, we are the first one to study the matched participating problem based on the result of social spread. We propose the information spread model IF which is based on the node probability. Considering the matching behavior in reality, we model two matching means as neighbor matching (NM) and global matching (GM). Combining the matching model and the information spread model, we formulate the MPM problem in which the goal is to find a small size seed set such that the expected number of final participants is maximized. We show the MPM is NP-hard and the computation of target function is #P-hard. We design the MRS-based method and propose the OP-MPM algorithm which has a $(1 - 1/e - \epsilon)$ approximation ratio for the MPM-NM. and the sandwich algorithm SA-MPM which has a $\beta(1 - 1/e - \epsilon)$ approximation ratio for the MPM-GM. At last, a lot of experiments have been conducted on real-world datasets show that the method we proposed outperforms other comparison methods.

References

1. Anand, K.S., Aron, R.: Group buying on the web: a comparison of price-discovery mechanisms. Manag. Sci. **49**(11), 1546–1562 (2003)
2. Barbieri, N., Bonchi, F., Manco, G.: Topic-aware social influence propagation models. Knowl. Inf. Syst. **37**(3), 555–584 (2013). https://doi.org/10.1007/s10115-013-0646-6

3. Becker, R., Coro, F., D'Angelo, G., Gilbert, H.: Balancing spreads of influence in a social network. In: Proceedings of the AAAI Conference on Artificial Intelligence, vol. 34, pp. 3–10 (2020)
4. Bharathi, S., Kempe, D., Salek, M.: Competitive influence maximization in social networks. In: Deng, X., Graham, F.C. (eds.) WINE 2007. LNCS, vol. 4858, pp. 306–311. Springer, Heidelberg (2007). https://doi.org/10.1007/978-3-540-77105-0_31
5. Borgs, C., Brautbar, M., Chayes, J., Lucier, B.: Maximizing social influence in nearly optimal time. In: Proceedings of the Twenty-Fifth Annual ACM-SIAM Symposium on Discrete Algorithms, pp. 946–957. SIAM (2014)
6. Cohen, E., Delling, D., Pajor, T., Werneck, R.F.: Timed influence: Computation and maximization. arXiv preprint arXiv:1410.6976 (2014)
7. He, X., Song, G., Chen, W., Jiang, Q.: Influence blocking maximization in social networks under the competitive linear threshold model. In: Proceedings of the 2012 SIAM International Conference on Data Mining, pp. 463–474. SIAM (2012)
8. Huang, K., Wang, S., Bevilacqua, G., Xiao, X., Lakshmanan, L.V.: Revisiting the stop-and-stare algorithms for influence maximization. Proc. VLDB Endow. **10**(9), 913–924 (2017)
9. Jing, X., Xie, J.: Group buying: a new mechanism for selling through social interactions. Manag. Sci. **57**(8), 1354–1372 (2011)
10. Kempe, D., Kleinberg, J., Tardos, É.: Maximizing the spread of influence through a social network. In: Proceedings of the Ninth ACM SIGKDD International Conference on Knowledge Discovery and Data Mining, pp. 137–146. ACM (2003)
11. Leskovec, J., Krause, A., Guestrin, C., Faloutsos, C., VanBriesen, J., Glance, N.: Cost-effective outbreak detection in networks. In: Proceedings of the 13th ACM SIGKDD International Conference on Knowledge Discovery and Data Mining, pp. 420–429. ACM (2007)
12. Li, G., Chen, S., Feng, J., Tan, K.l., Li, W.S.: Efficient location-aware influence maximization. In: Proceedings of the 2014 ACM SIGMOD International Conference on Management of Data, pp. 87–98. ACM (2014)
13. Liu, B., Cong, G., Xu, D., Zeng, Y.: Time constrained influence maximization in social networks. In: 2012 IEEE 12th International Conference on Data Mining, pp. 439–448. IEEE (2012)
14. Lu, W., Chen, W., Lakshmanan, L.V.: From competition to complementarity: comparative influence diffusion and maximization. Proc. VLDB Endow. **9**(2), 60–71 (2015)
15. Nemhauser, G.L., Wolsey, L.A., Fisher, M.L.: An analysis of approximations for maximizing submodular set functions–i. Math. Program. **14**(1), 265–294 (1978)
16. Nguyen, H.T., Thai, M.T., Dinh, T.N.: Stop-and-stare: optimal sampling algorithms for viral marketing in billion-scale networks. In: Proceedings of the 2016 International Conference on Management of Data, pp. 695–710. ACM (2016)
17. Ohsaka, N., Akiba, T., Yoshida, Y., Kawarabayashi, K.I.: Fast and accurate influence maximization on large networks with pruned Monte-Carlo simulations. In: Twenty-Eighth AAAI Conference on Artificial Intelligence (2014)
18. Song, C., Hsu, W., Lee, M.L.: Targeted influence maximization in social networks. In: Proceedings of the 25th ACM International on Conference on Information and Knowledge Management, pp. 1683–1692. ACM (2016)
19. Tang, J., Tang, X., Xiao, X., Yuan, J.: Online processing algorithms for influence maximization. In: Proceedings of the 2018 International Conference on Management of Data, pp. 991–1005 (2018)

20. Tang, Y., Shi, Y., Xiao, X.: Influence maximization in near-linear time: a Martingale approach. In: Proceedings of the 2015 ACM SIGMOD International Conference on Management of Data, pp. 1539–1554. ACM (2015)
21. Tang, Y., Xiao, X., Shi, Y.: Influence maximization: near-optimal time complexity meets practical efficiency. In: Proceedings of the 2014 ACM SIGMOD International Conference on Management of Data, pp. 75–86. ACM (2014)
22. Tsang, A., Wilder, B., Rice, E., Tambe, M., Zick, Y.: Group-fairness in influence maximization. In: Proceedings of the 28th International Joint Conference on Artificial Intelligence, pp. 5997–6005. AAAI Press (2019)
23. Vazirani, V.V.: Approximation Algorithms. Springer, Heidelberg (2013)
24. Zhu, J., Ghosh, S., Wu, W.: Group influence maximization problem in social networks. IEEE Trans. Comput. Soc. Syst. **6**(6), 1156–1164 (2019)

Mixed-Case Community Detection
Problem in Social Networks

Yapu Zhang[1], Jianxiong Guo[2], and Wenguo Yang[1(✉)]

[1] School of Mathematical Sciences, University of Chinese Academy of Sciences,
Beijing 100049, China
zhangyapu16@mails.ucas.ac.cn, yangwg@ucas.edu.cn
[2] Department of Computer Science, University of Texas at Dallas,
Richardson, TX 75080, USA
jianxiong.guo@utdallas.edu

Abstract. The problem of detecting communities is one of the essential problems in the study of social networks. To devise the algorithms of community detection, one should first define high-quality communities. In fact, there are no agreed methods to measure the quality of the community. In this paper, we consider a novel objective function of this problem. Our goal is to maximize not only the average of the sum of edge weights within communities (i.e., average-case) but also the sum of edge weights within the minimum community (i.e., worst-case). To balance both the average-case and worst-case problems, we introduce a parameter into our objective function and call it the mixed-cased community detection problem. We devise several approximation algorithms for the worst-case, such as the Greedy, Semi-Sandwich Approximation, and Local Search algorithms. For the average-case, an efficient Terminal-based algorithm is proposed. We prove that the best solution between the average-case and worst-case problems still can provide an approximate guarantee for any mixed-case community detection problem.

Keywords: Social networks · Community detection · Approximation algorithms.

1 Introduction

Nowadays, the social network plays a significant role in our daily life. It can serve platforms that allow people to share information, make friends, and sell products. Therefore, it has attracted widespread attention in sociology, economics, marketing, etc. One of the essential problems of social networks is the community detection problem. Generally, a social network is modeled as a graph, where vertices represent individuals, and edges represent the relationships among the

This work is supported by National Science Foundation under Grant No. 1907472 and by the National Natural Science Foundation of China under Grants No. 11991022 and No. 12071459.

individuals. The community detection problem aims to organize individuals into groups based on the graph structure.

Weiss and Jacobson [21] are among the first researchers who studied the problem of community detection. Following them, there are extensive works regarding this problem. However, the community detection problem is still an ill-defined problem. That is, there is no universal definition of the objective function that measures the quality of a partition. In fact, it is a hot issue of the study on identifying the benchmark of the community. Generally, one expects that the links within each community are denser than those between different communities [16]. Based on this concept, a popular modularity-based objective function [16] is proposed. Also, there are the density-based [15] and spectrum-based objective functions [11, 19].

The above objective functions are too complicated to solve. A simple objective function is to minimize the number of cut-edges. However, it may obtain a bad solution from a simple cut-based view [10, 18]. For example, as shown in Fig. 1, our goal is to cluster the nodes into 2 communities. It is easy to know that v_1 will separate from other nodes considering the minimum cut-edges. According to the graph topology, it is better to let $v_1, \ldots, v_5$ be into a community and $v_6, \ldots v_{10}$ be into a community (Fig. 1).

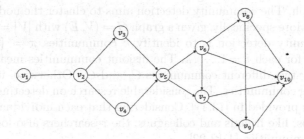

Fig. 1. Example illustrating the weakness of the cut-based method.

Observe that it performs poorly due to ignoring to guarantee the quality of the worst community. Motivated by this idea, one may also expect to make the size of links in each community as much close as possible. In this paper, we propose a novel objective function of community detection. On the one hand, the objective is to maximize the average of the sum of edge weights in a community partition. Notice that maximizing the sum of edge weights within communities means minimizing the edge weights between different communities. On the other hand, our goal is to maximize the sum of edge weights in the worst community. We denote them as the average-case and worst-case community detention problems, respectively. Combined with the average-case and the worst-case, we define a mixed community detection problem. In summary, our contributions are as follows.

First, we consider a new objective function of the community detection problem, which takes into account the average-case and the worst-case problems. Then, some properties, such as the supermodularity and monotonicity of the objective function, are studied. For the worst-case objective function, we devise a greedy algorithm. Based on the Sandwich Approximation strategy [14], a Semi-Sandwich algorithm with a data-dependent ratio is proposed. Moreover, we further improve the approximate ratio through iteratively finding an optimal local solution. For the average-case problem, different from using the cut-based methods directly, we derive a more efficient heuristic algorithm. Based on the results of the average-case and worst-case problems, it is proved that the better solution can provide the appropriation guarantee for the general case.

The following paper is arranged as below. We review the related work in Sect. 2. Section 3 presents the problem definition. Meanwhile, the properties of the objective function are studied. We discuss the algorithms for the worst-case, average-case, and general-case problems in Sects. 4, 5, and 6, respectively. Finally, our work is concluded in Sect. 7.

2   Related Works

The community detection problem attracts much attention due to its important role in social networks. There are a huge body of the related works [1,7,16,23,24]. The work of the community detection problem can be traced back to Weiss et al. [21]. A community is a subgraph in which the nodes are denser than the rest of the graph. The community detection aims to cluster the nodes into some communities. More specifically, given a graph $G = (V, E)$ with $|V| = n$ and $|E| = m$, the community detection is to identify k communities $\pi = \{C_1, \ldots, C_k\}$, where $C_i \subseteq V$ for each $i = 1, \ldots, k$. The disjoint communities means that $C_i \cap C_j = \emptyset$ for any two different communities C_i and C_j. Otherwise, it is called as the overlapping communities. The considerable research on detecting the disjoint communities is provided in [13,17]. Considering that each individual can be into different groups, like friends and colleagues, the researchers also focused on the overlapping communities [4,12,22].

In fact, the definition of community is ambiguous. To measure the quality of the community, some metrics are proposed. For instance, given a partition $\pi = \{C_1, \ldots, C_k\}$, a classic strategy is to measure it from the cut-based view. In [10], the ratio cut is written as $RatioCut(C_1, \ldots, C_k) = \sum_{i=1}^{k} \frac{cut(C_i, V \setminus C_i)}{|C_i|}$, where $cut(C_i, V \setminus C_i)$ is the sum of the edge weights between C_i and $V \setminus C_i$. The density-based method is designed based on the internal degree $d^{int}(C)$ and external degree $d^{out}(C)$ of the community C. More specifically, authors [8] denote by $\frac{d^{int}(C)}{|C|(|C|-1)}$ and $\frac{d^{out}(C)}{|C|(n-|C|)}$ the internal edge density and eternal edge density of the community C, respectively. Newman et al. [16] defined the modularity. The modularity of the partition π can be defined as $Q(\pi) = \frac{1}{2m} \sum_{i=1}^{k} \sum_{u,v \in C_i} [A_{uv} - \frac{d_u d_v}{2m}]$, where A_{uv} are the adjacency matrix elements, d_u is the degree of node u. Different metrics can lead to different algorithms.

In this paper, we study a novel objective function, in which both the average-case and the worst-case problems are combined. To the best of our knowledge, Wei et al. [20] first considered this pattern. Several algorithms including the greedy, minorization-maximization are proposed. Different from our work, their objective function is monotone non-decreasing submodular. Here, a set function

$f : 2^V \rightarrow R^+$ is monotone, if $f(S) \leq f(T)$ whenever $S \subseteq T \subseteq V$. f is submodular if $f(S \cup \{v\}) - f(S) \geq f(T \cup \{v\}) - f(T)$, for any $S \subseteq T \subseteq V$ and $v \in V \backslash T$. f is supermodular if and only if $-f$ is submodular. And our objective function is monotone non-decreasing supermodular.

3 Problem Definition

Let a social network be modeled by an undirected graph $G = (V, E)$, in which the nodes denote the users and the edges represent the links among users. Suppose that there exists a weight $w_{uv} \in [0, 1]$ for each edge $(u, v) \in E$. The weight w_{uv} describes the social influence between nodes u and v. Note that, in general, an undirected graph G can also be viewed as a directed graph, where $w_{uv} = w_{vu}$ for each $(u, v) \in E$ and $w_{uv} = 0$ for each $(u, v) \notin E$.

In this paper, we study the Community Detection problem. Given a social network $G = (V, E)$, one expects to organize the users into groups. We denote these groups as communities. Let k be the size of communities. Our goal is to divide a network into k disjoint communities. More specifically, we ask for a partition $\pi = \{C_1, \ldots, C_k\}$ such that $C_i \neq \emptyset$, $\bigcup_{i=1}^{k} C_i = V$ and $C_i \cap C_j = \emptyset$ for any $i \neq j$.

Denote a set function $f : 2^V \rightarrow R^+$ such that $f(C) = \sum_{u,v \in C} w_{uv}$ for any subset $C \subseteq V$. Then, we can use $f(C)$ to measure the internal connectedness [8] in the community C.

Lemma 1. *As defined above, f is a monotone non-decreasing supermodular function.*

Proof. It is trivial to prove that f is monotone non-decreasing. Next, it suffices to show its supermodularity. Let $C_1 \subseteq C_2 \subseteq V$ and $c \in V \setminus C_2$. According to definitions, we have $f(C_1 \cup \{c\}) - f(C_1) = \sum_{u,v \in C_1 \cup \{c\}} w_{uv} - \sum_{u,v \in C_1} w_{uv} = \sum_{v \in C_1}(w_{cv} + w_{vc})$. Similarly, we have $f(C_2 \cup \{c\}) = \sum_{v \in C_2}(w_{cv} + w_{vc})$. Then, the inequality $\sum_{v \in C_1}(w_{cv} + w_{vc}) \leq \sum_{v \in C_2}(w_{cv} + w_{vc})$ holds since $C_1 \subseteq C_2$. Thus, $f(C_1 \cup \{c\}) - f(C_1) \leq f(C_2 \cup \{c\}) - f(C_2)$ and the lemma follows.

Based on the set function f, we consider a following problem.

Problem 1. (Mixed-case Community Detection problem) Given a social network G and an integer $k \geq 2$, a Mixed-case Community Detection problem is:

$$\max_{\pi \in \Pi}[(1 - \lambda) \min_i f(C_i^\pi) + \frac{\lambda}{k} \sum_{i=1}^{k} f(C_i^\pi)], \tag{1}$$

where $\lambda \in [0, 1]$, a partition $\pi = \{C_1^\pi, \ldots, C_k^\pi\}$, and Π is a set containing all partitions of V into k communities.

According to definitions, our goal is to maximize not only the average of the sum of the function f but also the worst value of f of a community. We define this problem as the worst-case, average-case and general-case problem when $\lambda = 0$, $\lambda = 1$, and $\lambda \in (0, 1)$, respectively. In the following, we discuss the above cases one by one.

4 Worst-Case Community Detection Problem

First, we consider solving the worst-case community detection problem. A simple but efficient way is to use a greedy strategy. As shown in Algorithm 1, we initialize $C_1, \ldots, C_k$ with the empty sets. This algorithm aims to iteratively select a node with the maximum marginal increment to the community whose value is minimum until sets $C_1, \ldots, C_k$ form a partition.

Algorithm 1. Greedy

Input: $G = (V, E), k$
Output: A partition π
1: Initialize $C_1, = \ldots, = C_k = \emptyset$ and $S = V$
2: **while** $S \neq \emptyset$ **do**
3:     $j \leftarrow \arg\min_{i=1,\ldots,k} f(C_i)$
4:     $s \leftarrow \arg\max_{s \in S}\{f(C_j \cup \{s\}) - f(C_j)\}$
5:     $C_j \leftarrow C_j \cup \{s\}$
6:     $S \leftarrow S \setminus \{s\}$
7: **end while**
8: **return** $\pi = \{C_1, \ldots, C_k\}$

Algorithm 2. Semi-Sandwich

Input: $G = (V, E), k$
Output: A partition π
1: $\pi_1 \leftarrow$ a partition returned by Algorithm 1
2: $\pi_2 \leftarrow$ a partition returned by Algorithm 1 where f is replaced by $\overline{f}$
3: **return** $\pi \leftarrow \arg\max_{\pi = \pi_1, \pi_2} \min_i f(C_i^\pi)$

Generally, the solution of the Greedy is good. However, it is hard to derive an approximation guarantee. To address this issue, we define $\overline{f}(C) = \sum_{u \in C, v \in V} w_{uv} + \sum_{u \in V, v \in C} w_{uv}$. It is easy to prove that $\overline{f}$ is a modular upper bound of f. More specifically, for any $C \subseteq V$, we have that $\overline{f}(C) \geq f(C)$ and $\overline{f}(C) = \sum_{c \in C} \overline{f}(\{c\})$. Using Algorithm 1 , we compute the solutions for the original function and upper bound, respectively. That is, we obtain the solutions for $\max_{\pi \in \Pi} \min_i f(C_i^\pi)$, $\max_{\pi \in \Pi} \min_i \overline{f}(C_i^\pi)$. Similar to the Sandwich Approximation framework [14], our final result will be the partition with the largest value for the original function. This process is concluded in Algorithm 2 and this solution returned has a data-dependent approximation ratio.

Theorem 1. *Algorithm 2 returns a $\frac{f(A_i^{\pi_2})}{k\overline{f}(A_i^{\pi_2})}$-approximate solution.*

Proof. It is proved that the Greedy returns a $1/k$-approximate solution if f is submodular [20]. The modular function $\overline{f}$ can also satisfy this property. That is, we have $\min_{i=1,...,k} \overline{f}(C_i^{\pi_2}) \geq \frac{1}{k} \max_{\pi \in \Pi} \min_{i=1,...,k} \overline{f}(C_i)$. Suppose that $\overline{\pi}^*$ and π^* are the optimal solution for the upper bound and the original function, respectively. Then, we have

$$\min_i f(C_i^{\pi_2}) = \min_i \frac{f(C_i^{\pi_2})}{\overline{f}(C_i^{\pi_2})} \overline{f}(C_i^{\pi_2})$$

$$\geq \min_i \frac{f(C_i^{\pi_2})}{\overline{f}(C_i^{\pi_2})} \frac{1}{k} \overline{f}(C_i^{\overline{\pi}^*})$$

$$\geq \min_i \frac{f(C_i^{\pi_2})}{\overline{f}(C_i^{\pi_2})} \frac{1}{k} \overline{f}(C_i^{\pi^*})$$

$$\geq \min_i \frac{f(C_i^{\pi_2})}{\overline{f}(C_i^{\pi_2})} \frac{1}{k} f(C_i^{\pi^*})$$

Thus, the theorem is proved.

Algorithm 3. Local Search

Input: $G = (V, E), k, d, \lambda$
Output: A partition π

1: Select a partition $\hat{\pi}$ as the initial partition
2: **repeat**
3:     $\pi \leftarrow \hat{\pi}$
4:     $i^* \leftarrow \arg\min_i f(C_i^\pi)$
5:     $u \leftarrow \arg\max_{v \in V \setminus C_{i^*}^\pi} \{f(C_{i^*}^\pi \cup \{v\}) - f(C_{i^*}^\pi)\}$
6:     $j^* \leftarrow$ the community which contains node u
7:     update $\hat{\pi}$ such that $C_{i^*}^{\hat{\pi}} = C_{i^*}^\pi \cup \{u\}$ and $C_{j^*}^{\hat{\pi}} = C_{j^*}^\pi \setminus \{u\}$
8: **until** $\min_i f(C_i^\pi) \geq \min_i f(C_i^{\hat{\pi}})$
9: **return** π

Next, we propose another method to solve the worst-case problem. We define a given partition as the optimal local solution as follows.

Definition 1. *Given a partition* $\pi = \{C_1^\pi, \ldots, C_k^\pi\}$, *we call it the local optimal solution if changing the community of any node cannot improve the value of the objective function.*

Our method is to obtain an optimal local solution iteratively. As shown in Algorithm 3, we select an initial partition, which is any solution to the worst-case problem. The key idea of this strategy is to iteratively improve the solution by taking an item from the current community to another community. At each iteration, the algorithm first finds out the minimum community, i.e.,

$i^* = \arg\min_i f(C_i^\pi)$. Then, it finds out an element that can maximize the value of this maximum community and its corresponding community. That is, this element $u = \arg\max_{v \in V \setminus C_{i^*}^\pi} \{f(C_{i^*}^\pi \cup \{v\}) - f(C_{i^*}^\pi)\}$. A new partition is obtained by removing the element u from its current community j^* and then adding it into the minimum community i^*. If the objective function of the new community partition is no better than before, i.e., $\min_i f(C_i^\pi) \geq \min_i f(C_i^{\hat{\pi}})$, our algorithms terminates.

In fact, we can utilize Algorithm 3 to improve the solution returned from Algorithm 2. This method is to consider the result obtained from Algorithm 2 as our initial partition in Algorithm 3.

Notice that the number of iterations is ambiguous. A simple way to accelerate our algorithm is to give a number of iterations. That is, the algorithm terminates when the number of iterations is equal to a given integer.

5   Average-Case Community Detection

In this section, we consider the average-case community detection problem. In fact, maximizing the average of the sum of the edge weights within communities is equivalent to minimizing the edge weights crossing different communities. That is, our problem $\max_{\pi \in \Pi} \frac{1}{k} \sum_{i=1}^{k} f(C_i^\pi)$ can be viewed as:

$$\min_{\pi \in \Pi} \sum_{i=1}^{k} g(C_i^\pi), \tag{2}$$

where $g(C_i^\pi) = \sum_{u \in C_i^\pi, v \in V \setminus C_i^\pi} (w_{uv} + w_{vu})$.

Lemma 2. *As defined above, g is a submdoular function.*

Proof. Let $C_1 \subseteq C_2 \subseteq V$ and $c \in V \setminus C_2$. Similar to Lemma 1, we have $g(C_1 \cup \{c\}) - g(C_1) = \sum_{v \in V \setminus (C_1 \cup \{c\})} (w_{cv} + w_{vc})$ and $g(C_2 \cup \{c\}) - g(C_2) = \sum_{v \in V \setminus (C_2 \cup \{c\})} (w_{cv} + w_{vc})$. Then, $g(C_1 \cup \{c\}) - g(C_1) \leq g(C_2 \cup \{c\}) - g(C_2)$ holds since $C_1 \subseteq C_2$. Thus, the lemma is proved.

The above problem can be viewed as the Minimum k-Cut problem. Observe that this problem is also equivalent to the minimum $s - t$ cut problem when $k = 2$. It can be solved in polynomial time using a maximum flow algorithm. Moreover, it is polynomial solvable for any fixed k [9]. Unfortunately, the time complexity of this polynomial algorithm is $n^{O(k)}$. Moreover, the authors proved that the time of any improved algorithm will still grow exponentially with k. Thus, this kind of algorithms cannot apply to large networks. Also, as shown the example in Sect. 1, the partition obtained by solving the Minimum k-cut can perform badly. Thus, a new strategy should be proposed.

The Minimum k-cut problem is closely related to the Multiway Partition problem. Next, we propose a heuristic algorithm based on the Multiway Partition problem.

Definition 2. *(Multiway Partition problem) Let $g : 2^V \leftarrow R^+$ be a submodular set function and $S = \{s_1, \ldots, s_k\}$, where $s_i \in V$ is called as a terminal for any $i = 1, \ldots, k$. The Multiway Cut problem asks for a partition $\pi = \{C_1^\pi, \ldots, C_k^\pi\}$ of V such that $s_i \in C_i^\pi$ and $\sum_{i=1}^k g(C_i^\pi)$ is minimized.*

Multiway Partition problem is also equivalent to the minimum $s - t$ cut problem when $k = 2$, and it can be solved in polynomial time. However, the Multiway Partition problem is NP-hard when $k \geq 3$. Fortunately, the considerable approximation algorithms are proposed [2,3,5,6]. For instance, Dahlhaus et al. [6] devised a $2(1 - \frac{1}{k})$-approximate solution. Their stagey is based on the minimum cut. For each terminal s_i, we compute a minimum cut M_i. Among these cuts, the algorithm discards the largest one. After removing these cuts, the left graph will form a partition with k parts.

Next, we study the relationship between the Minimum k–cut problem and the Multiway Partition problem. Let $\pi^* = \{C_1^{\pi^*}, \ldots, C_k^{\pi^*}\}$ be the optimal solution to the Minimum k–cut problem and $S^* = \{s_1^*, \ldots, s_k^*\}$ be a corresponding terminal set, where $s_i^* \in C_i^{\pi^*}$ for any $i = 1, \ldots, k$. Accordingly, there is an optimal solution $\hat{\pi}^* = \{C_1^{\hat{\pi}^*}, \ldots, C_k^{\hat{\pi}^*}\}$ to the Multiway Partition problem. We can show that the quality of these two partition are equal. Then, the following result holds.

Lemma 3. *Suppose that $\hat{\pi}$ is a β–approximate solution to the Multiway Partition problem with the terminal set S^*. This partition $\hat{\pi}$ is also a β–approximate solution to the Minimum k–cut problem.*

Proof. We have that $\sum_{i=1}^k g(C_i^{\hat{\pi}^*}) \geq \sum_{i=1}^k g(C_i^{\pi^*})$ since π^* is a feasible solution to the Multiway Partition problem. Furthermore, we have that $\sum_{i=1}^k g(C_i^{\hat{\pi}^*}) \leq \sum_{i=1}^k g(C_i^{\pi^*})$ since π^* is the optimal solution, which separates the graph into k parts. Then, $\sum_{i=1}^k g(C_i^{\hat{\pi}^*}) = \sum_{i=1}^k g(C_i^{\pi^*})$ holds. Thus, the partition $\hat{\pi}$ is also a β–approximate solution to the Minimum k–cut problem.

The key point is to guess the terminal set S^*. However, there are $\binom{n}{k}$ possibilities for such a subset S^*, and then it is unavailable for large networks. In the meantime, a good subset S^* can lead to a highly-quality community partition. To address this problem, we propose the Algorithm 4.

Algorithm 4. Terminal-based

Input: $G = (V, E), k, d$
Output: A partition π

1: Initialize set S_d containing d nodes with the highest degrees
2: Initialize $\Pi \leftarrow \emptyset$
3: **for each** $S = \{s_1, \ldots, s_k\} \subseteq S_d$ **do**
4:     $\pi \leftarrow$ a partition using the Dahlhaus's method
5:     $\Pi \leftarrow \Pi \cup \pi$
6: **end for**
7: **return** $\pi \leftarrow \arg\max_{\pi \in \Pi} \sum_{i=1}^k f(C_i^\pi)$

Initially, we control the selection of set S^* in a small range. That is, we choose a set S_d containing d nodes with the highest degrees. For each subset S which includes k nodes in S_d, we compute its corresponding partition π using the method proposed by Dahlhaus et al. [6]. Notice that $\sum_{i=1}^{k} f(C_i^\pi) = f(V) - \sum_{i=1}^{k} g(C_i^\pi)$ and g is nonnegative function. Thus, among all these partitions, the one with the minimum value of $\sum_{i=1}^{k} g(C_i^\pi)$ becomes our final result. Furthermore, we can conclude that $f(V) \geq \sum_{i=1}^{k} f(C_i^\pi)$ holds for any partition. The approximation ratio for the average-case problem is at least equal to $\frac{f(V) - \sum_{i=1}^{k} g(C_i^\pi)}{f(V)}$.

6 General-Case Community Detection

In this section, we start to consider the general-case problem with $0 < \lambda < 1$. A natural idea is to combine the result of the worst-case (i.e., $\lambda = 0$) with the result of the average-case (i.e., $\lambda = 1$). Denote WCD algorithm as the approximation algorithm for solving the worst-case problem and ACD algorithm as the approximation algorithm for solving the average-case problem. The key scheme is to obtain a partition $\hat{\pi}_1$ by running WCD algorithm, and then obtain a partition $\hat{\pi}_2$ by running ACD algorithm. Between partitions $\hat{\pi}_1$ and $\hat{\pi}_2$, the one with a higher value for the objective function is selected as our final result. That is, our solution is $\hat{\pi} = \arg\max_{\pi \in \hat{\pi}_1, \hat{\pi}_2}[(1 - \lambda)\min_i f(C_i^\pi) + \frac{\lambda}{k}\sum_{i=1}^{k} f(C_i^\pi)]$. We denote such an idea as the Combination algorithm. For convenience, we let $F_1(\pi) = \min_i f(C_i^\pi)$ and $F_2(\pi) = \frac{1}{m}\sum_{i=1}^{k} f(C_i^\pi)$ in the following. Suppose that the WCD algorithm and ACD algorithm return the approximate solution with factor $\alpha \geq 0$ and $\beta \geq 0$. We can conclude the following theorem.

Theorem 2. *The Combination algorithm solves the General-case Community Detection problem with a factor* $\max\{\frac{(1-\lambda)\beta}{(1-\lambda)\beta+\alpha}, \lambda\beta\}$.

Proof. Let $\hat{\pi}_1$ and $\hat{\pi}_2$ be the solution of the WCD algorithm and ACD algorithm, respectively. Suppose that π^* is the optimal solution to the general-case community detection problem. That is, $\pi^* \in \arg\max_{\pi \in \Pi}(1 - \lambda)F_1(\pi) + \lambda F_2(\pi)$.

According to definitions, we have $F_1(\hat{\pi}_1) \geq \alpha F_1(\pi)$ and $F_2(\hat{\pi}_2) \geq \beta F_2(\pi)$. Moreover, for any partition π, $F_1(\pi) \leq F_2(\pi)$. On the one hand, we have

$$
\begin{aligned}
(1 - \lambda)F_1(\hat{\pi}_1) + \lambda F_2(\hat{\pi}_1) &= \mu[(1 - \lambda)F_1(\hat{\pi}_1) + \lambda F_2(\hat{\pi}_2)] \\
&\quad + (1 - \mu)[(1 - \lambda)F_1(\hat{\pi}_1) + \lambda F_2(\hat{\pi}_2)] \\
&\geq \mu[(1 - \lambda)\alpha F_1(\hat{\pi}^*) + \lambda\alpha F_1(\hat{\pi}^*)] + (1 - \mu)[0 + \lambda\beta F_2(\hat{\pi}^*)] \\
&\geq \frac{\mu\alpha}{1 - \lambda}(1 - \lambda)F_1(\pi^*) + (1 - \mu)\beta\lambda F_2(\pi^*) \\
&\geq \min\{\frac{\mu\alpha}{1 - \lambda}, (1 - \mu)\beta\}[(1 - \lambda)F_1(\pi^*) + \lambda F_2(\pi^*)].
\end{aligned}
$$

If u satisfies $\frac{\mu\alpha}{1-\lambda} = (1 - \mu)\beta$, then u is the optimal solution to maximize $\min\{\frac{\mu\alpha}{1-\lambda}, (1 - \mu)\beta\}$. At this time, $u = \frac{(1-\lambda)\beta}{(1-\lambda)\beta+\alpha}$. Then, the following inequality holds:

$$(1 - \lambda)F_1(\hat{\pi}_1) + \lambda F_2(\hat{\pi}_2) \geq \frac{(1 - \lambda)\beta}{(1 - \lambda)\beta + \alpha}[(1 - \lambda)F_1(\pi^*) + \lambda F_2(\pi^*)]. \quad (3)$$

On the other hand, we have

$$\begin{aligned}
(1 - \lambda)F_1(\hat{\pi}_2) + \lambda F_2(\hat{\pi}_2) \\
\geq \lambda F_2(\hat{\pi}_2) \\
\geq \lambda \beta F_2(\pi^*) \\
\geq \lambda \beta[(1 - \lambda)F_1(\pi^*) + \lambda F_2(\pi^*)].
\end{aligned} \quad (4)$$

Combining Eq. (3) with Eq. (4), we have

$$\begin{aligned}
\max\{(1 - \lambda)F_1(\hat{\pi}_1) + \lambda F_2(\hat{\pi}_2), (1 - \lambda)F_1(\hat{\pi}_2) + \lambda F_2(\hat{\pi}_2)\} \\
\geq \max\{\frac{(1 - \lambda)\beta}{(1 - \lambda)\beta + \alpha}, \lambda \beta\}[(1 - \lambda)F_1(\pi^*) + \lambda F_2(\pi^*)].
\end{aligned} \quad (5)$$

7 Conclusions

In this paper, we focus on the community detection problem. To measure the quality of the communities, we consider a novel objective function. The objective aims to maximize not only the average of the sum of edge weights within communities (i.e., average case) but also the sum of edge weights within the minimum community (i.e., worst case). Taking into account the average-case and the worst-case, we introduce a parameter to balance them. Then, the mixed-cased community detection problem is defined. According to the value of this parameter, we define the average-case, worst-case, and general-case problems when the parameter equals one, zero, and between zero and one, respectively. The Greedy, Semi-Sandwich Approximation and Local Search algorithms are designed to solve the worst-case problem. A Terminal-based algorithm is proposed with respect to the average-case. We prove that the best solution between the worst-case and average-case problems still can provide an approximate guarantee.

References

1. Abbe, E.: Community detection and stochastic block models: recent developments. J. Mach. Learn. Res. **18**(1), 6446–6531 (2017)
2. Buchbinder, N., Schwartz, R., Weizman, B.: Simplex transformations and the multiway cut problem. In: Proceedings of the Twenty-Eighth Annual ACM-SIAM Symposium on Discrete Algorithms, pp. 2400–2410. SIAM (2017)
3. Călinescu, G., Karloff, H., Rabani, Y.: An improved approximation algorithm for multiway cut. In: Proceedings of the Thirtieth Annual ACM Symposium on Theory of Computing, pp. 48–52 (1998)
4. Chakraborty, T., Ghosh, S., Park, N.: Ensemble-based overlapping community detection using disjoint community structures. Knowl.-Based Syst. **163**, 241–251 (2019)

5. Chekuri, C., Ene, A.: Approximation algorithms for submodular multiway partition. In: 2011 IEEE 52nd Annual Symposium on Foundations of Computer Science, pp. 807–816. IEEE (2011)
6. Dahlhaus, E., Johnson, D.S., Papadimitriou, C.H., Seymour, P.D., Yannakakis, M.: The complexity of multiway cuts. In: Proceedings of the Twenty-Fourth Annual ACM Symposium on Theory of Computing, pp. 241–251 (1992)
7. De Bacco, C., Power, E.A., Larremore, D.B., Moore, C.: Community detection, link prediction, and layer interdependence in multilayer networks. Phys. Rev. E **95**(4), 042317 (2017)
8. Fortunato, S., Hric, D.: Community detection in networks: a user guide. Phys. Rep. **659**, 1–44 (2016)
9. Goldschmidt, O., Hochbaum, D.S.: A polynomial algorithm for the k-cut problem for fixed k. Math. Oper. Res. **19**(1), 24–37 (1994)
10. Hagen, L., Kahng, A.B.: New spectral methods for ratio cut partitioning and clustering. IEEE Trans. Comput.-aided Des. Integr. Circ. Syst. **11**(9), 1074–1085 (1992)
11. Kannan, R., Vempala, S., Vetta, A.: On clusterings: good, bad and spectral. J. ACM (JACM) **51**(3), 497–515 (2004)
12. Kelley, S., Goldberg, M., Magdon-Ismail, M., Mertsalov, K., Wallace, A.: Defining and discovering communities in social networks. In: Thai, M., Pardalos, P. (eds.) Handbook of Optimization in Complex Networks, pp. 139–168. Springer, Heidelberg (2012). https://doi.org/10.1007/978-1-4614-0754-6_6
13. Leskovec, J., Lang, K.J., Mahoney, M.: Empirical comparison of algorithms for network community detection. In: Proceedings of the 19th International Conference on World Wide Web, pp. 631–640 (2010)
14. Lu, W., Chen, W., Lakshmanan, L.V.: From competition to complementarity: comparative influence diffusion and maximization. Proc. VLDB Endow. **9**(2), 60–71 (2015)
15. Mancoridis, S., Mitchell, B.S., Rorres, C., Chen, Y., Gansner, E.R.: Using automatic clustering to produce high-level system organizations of source code. In: Proceedings of 6th International Workshop on Program Comprehension. IWPC 1998 (Cat. No. 98TB100242), pp. 45–52. IEEE (1998)
16. Newman, M.E., Girvan, M.: Finding and evaluating community structure in networks. Phys. Rev. E **69**(2), 026113 (2004)
17. Taha, K.: Disjoint community detection in networks based on the relative association of members. IEEE Trans. Comput. Soc. Syst. **5**(2), 493–507 (2018)
18. Tong, G., Cui, L., Wu, W., Liu, C., Du, D.Z.: Terminal-set-enhanced community detection in social networks. In: IEEE INFOCOM 2016-The 35th Annual IEEE International Conference on Computer Communications, pp. 1–9. IEEE (2016)
19. Von Luxburg, U.: A tutorial on spectral clustering. Stat. Comput. **17**(4), 395–416 (2007)
20. Wei, K., Iyer, R.K., Wang, S., Bai, W., Bilmes, J.A.: Mixed robust/average submodular partitioning: fast algorithms, guarantees, and applications. In: Advances in Neural Information Processing Systems, pp. 2233–2241 (2015)
21. Weiss, R.S., Jacobson, E.: A method for the analysis of the structure of complex organizations. Am. Sociol. Rev. **20**(6), 661–668 (1955)
22. Xie, J., Kelley, S., Szymanski, B.K.: Overlapping community detection in networks: the state-of-the-art and comparative study. ACM Comput. Surv. (CSUR) **45**(4), 1–35 (2013)

23. Zeng, X., Wang, W., Chen, C., Yen, G.G.: A consensus community-based particle swarm optimization for dynamic community detection. IEEE Trans. Cybern. **50**(6), 2502–2513 (2019)
24. Zhe, C., Sun, A., Xiao, X.: Community detection on large complex attribute network. In: Proceedings of the 25th ACM SIGKDD International Conference on Knowledge Discovery & Data Mining, pp. 2041–2049 (2019)

How to Get a Degree-Anonymous Graph Using Minimum Number of Edge Rotations

Cristina Bazgan[1], Pierre Cazals[1](✉), and Janka Chlebíková[2]

[1] Université Paris-Dauphine, Université PSL, CNRS, LAMSADE,
75016 Paris, France
{cristina.bazgan,pierre.cazals}@dauphine.eu
[2] School of Computing, University of Portsmouth, Portsmouth, UK
janka.chlebikova@port.ac.uk

Abstract. A graph is k-degree-anonymous if for each vertex there are at least $k - 1$ other vertices of the same degree in the graph. MIN ANONYMOUS-EDGE-ROTATION asks for a given graph G and a positive integer k to find a minimum number of edge rotations that transform G into a k-degree-anonymous graph. In this paper, we establish sufficient conditions for an input graph and k ensuring that a solution for the problem exists. We also prove that the MIN ANONYMOUS-EDGE-ROTATION problem is NP-hard even for $k = n/3$, where n is the order of a graph. On the positive side, we argue that under some constraints on the number of edges in a graph and k, MIN ANONYMOUS-EDGE-ROTATION is polynomial-time 2-approximable. Moreover, we show that the problem is solvable in polynomial time for any graph when $k = n$ and for trees when $k = \theta(n)$.

1 Introduction

Huge amounts of data has been aggregated on social networks in recent years. To assure the privacy of network's users is one of the key research task in this field. One possible study model was introduced by Liu and Terzi [15] who transferred the k-degree-anonymity concept from tabular data in databases [9] to graphs which are often used as a representation of networks. Therefore, a graph is called k-degree-anonymous if for each vertex there are at least $k - 1$ other vertices with the same degree. The parameter k represents the number of vertices that are mixed together and thus the increasing value of k increases the level of anonymity. In [18], Wu et al. presented a survey of different anonymization models and some of their weaknesses.

In this paper we consider the k-degree-anonymous concept of Liu and Terzi [15]. Different graph operations of transforming a graph into a k-degree-anonymous one are considered in several papers where the operations maybe the following: delete vertex/edge, add vertex/edge, or add/delete of an edge (see the references later). One advantage in that approaches of vertex/edge deletion/adding is that a solution always exists since in the worst case scenario one can consider the empty or the complete graph that is k-degree-anonymous for any k (at most the number of vertices of the graph). However, the basic graph parameters as the number of vertices and edges could be modified.

In this paper we consider the version of transforming a graph into a k-degree-anonymous one using edge rotations which don't modify the number of vertices/edges, however a solution may not always exists, as we show later.

Vertex/edge modification versions associated to k-degree-anonymity have been relatively well studied. Hartung et al. [12, 13] studied the edge adding modification as proposed by Liu and Terzi [15]. For this version Chester et al. [6] established a polynomial time algorithm for bipartite graphs

The variant of adding vertices instead of edges was studied by Chester et al. and in [5] they presented an approximation algorithm with an additive error. Bredereck et al. [2] investigated the parameterized complexity of several variants of vertex adding which differ in the way the inserted vertices can be adjacent to existing vertices. Concerning the vertex deletion variant, Bazgan et al. [1] showed the NP-hardness even on very restricted graph classes such as trees, split graphs, or trivially perfect graphs. Moreover, in [1] the vertex and edge deletion variants are proved intractable from the approximability and parameterized complexity point of view.

Several papers study the basic properties of edge rotations, including some bounds for the minimum number of edge rotations between two graphs [3, 4, 8, 14]. To the best of our knowledge the problem of transforming a graph to a k-degree anonymous graph using the edge rotations has not been fully explored. In some particular cases some research has been done, e.g. in [16] they study an edge rotation distance and various other metric between the degree sequences to find a "closest" regular graph.

Our Results. In this paper we study the various aspects of the MIN ANONYMOUS-EDGE-ROTATION problem. An input to the problem is an undirected graph $G = (V, E)$ with n vertices and m edges and an integer $k \leq n$. The goal is to find a shortest sequence of edge rotations that transforms G into a k-degree-anonymous graph, if such a sequence exists. We first show that when $\frac{n}{2} \leq m \leq \frac{n(n-3)}{2}$ and $k \leq \frac{n}{4}$ a solution always exists. Moreover for trees a solution exists if and only if $\frac{2m}{n}$ is an integer. We prove that MIN ANONYMOUS-EDGE-ROTATION is NP-hard even when $k = \frac{n}{3}$ and provide a polynomial-time 2-approximable algorithm under some constraints. Finally, we demonstrate that MIN ANONYMOUS-EDGE-ROTATION is solvable in polynomial time for trees when $k = \theta(n)$ and for any graph when $k = n$.

Our paper is organized as follows. Some preliminaries about edge rotations and our formal definitions are given in Sect. 2. The study of feasibility is established in Sect. 3. Section 4 presents the NP-hardness proof. In Sect. 5 we establish a lower bound that is used in Sect. 6 to present a polynomial-time 2-approximation algorithm and in Sect. 7 to demonstrate the polynomial time algorithm for trees. Moreover in Sect. 7 we consider the case $k = n$ in general graphs. Some conclusions are given at the end of the paper. The omitted proofs can be found in the full version of the paper.

2   Preliminaries

In this paper we assume that all graphs are undirected, without loops and multiple edges, and not necessary connected graphs.

Let $G = (V, E)$ be a graph. For a vertex $v \in V$, let $deg_G(v)$ be the degree of v in G, and Δ_G be the maximum degree of G. A vertex v with degree $deg_G(v) = |V| - 1$ is called a universal vertex. The neighborhood of v in G is denoted by $\mathcal{N}_G(v) = \{u \in V : uv \in E\}$ and $Inc_G(v)$ is the set of all edges incident to v, $Inc_G(v) = \{e \in E : v \in e\}$. If the underlying graph G is clear from the context, we omit the subscript G.

Definition 1. *Given a graph $G = (V, E)$ of order n, the degree sequence S_G of G is the non-increasing sequence of its vertex degrees, $S_G = (deg(v_1), \ldots, deg(v_n))$, $deg(v_1) \geq deg(v_2) \geq \cdots \geq deg(v_n)$. A sequence D of non-negative integers $D = (d_1, d_2, \ldots, d_n)$ is **graphic** if there exists a graph G such that its degree sequence coincides with D.*

As follows from Erdős-Gallai theorem (see e.g. [7]) the necessary and sufficient conditions for a non-increasing sequence $D = (d_1, d_2, \ldots, d_n)$ to be graphic are:

$$\sum_{i=1}^{n} d_i \text{ is even} \tag{1}$$

$$\sum_{i=1}^{\ell} d_i \leq \ell(\ell - 1) + \sum_{i=\ell+1}^{n} \min(d_i, \ell) \text{ holds for any } 1 \leq \ell \leq n. \tag{2}$$

Furthermore, it is an easy exercise to prove that a sequence of integers $D = (d_1, d_2, \ldots, d_n)$ corresponds to a degree sequence of a tree on n vertices if and only if each $d_i \geq 1$ and $\sum_{i=1}^{n} d_i = 2(n-1)$.

Let $\mathbf{G}(n, m)$ be the set of all graphs with n vertices and m edges.

Definition 2. *Let $G, G' \in \mathbf{G}(n, m)$. We say that G' can be obtained from G by an **edge rotation** (uv, uw) if $V(G) = V(G')$ and there exist three distinct vertices u, v and w in G such that $uv \in E(G)$, $uw \notin E(G)$, and $E(G') = (E(G) \setminus \{uv\}) \cup \{uw\}$.*

Remark 1. Let G be a graph. For the vertices u, v, w in G the edge rotation (uv, uw) modifies G into the graph G' such that $deg_{G'}(v) = deg_G(v) - 1$, $deg_{G'}(w) = deg_G(w) + 1$, and the degree of the other vertices is not changed. Let define a $(+1, -1)$-degree modification of the degree sequence $D = (d_1, \ldots, d_n)$ in such a way that $d_i := d_i + 1$, $d_j := d_j - 1$ for any two indices i, j such that $i, j \in \{1, \ldots, n\}$. Note that each edge rotation corresponds to a $(+1, -1)$-degree modification, but not opposite.

Definition 3. *A sequence of integers $D = (d_1, d_2, \ldots, d_n)$ is called k-**anonymous** where $k \in \{1, \ldots, n\}$, if for each element d_i from D there are at least $k - 1$ other elements in D with the same value. A graph G is called k-**degree-anonymous** if its degree sequence is k-**anonymous**. The vertices of the same degree correspond to a **degree class**.*

In this paper we study the following anonymization problem:

MIN ANONYMOUS-EDGE-ROTATION
Input: (G, k) where $G = (V, E)$ is an undirected graph and k a positive integer, $k \in \{1, \ldots, |V|\}$.
Output: If there is a solution, find a sequence of a minimum number $\ell + 1$ of graphs $G_0 = G, G_1, G_2, \ldots, G_\ell$ such that G_{i+1} can be obtained from G_i by one edge rotation, and G_ℓ is k-degree-anonymous.

Note that a solution to the MIN ANONYMOUS-EDGE-ROTATION problem may not exist for all instances, e.g. there is no solution if G is a graph obtained from the complete graph K_n, $n \geq 6$, removing an edge together with $k = 3$. Therefore, we are only interested in studying of **feasible instances** (G, k) defined as an instance for which there exists a solution to MIN ANONYMOUS-EDGE-ROTATION. Our initial study of sufficient conditions for feasibility is covered in Sect. 3.

Obviously, since all graphs are 1-degree-anonymous, we are only interested in cases where $k \geq 2$.

The decision version associated to MIN ANONYMOUS-EDGE-ROTATION is defined as follows for a feasible instance (G, k):

ANONYMOUS-EDGE-ROTATION
Input: (G, k, r) where $G = (V, E)$ is an undirected graph, $k \in \{1, \ldots, |V|\}$, and r be a positive integer.
Question: Is there a sequence of $\ell + 1$ graphs $G_0 = G, G_1, G_2, \ldots, G_\ell$ such that $\ell \leq r$, G_{i+1} can be obtained from G_i by one edge rotation, and G_ℓ is k-degree-anonymous?

We also consider the MIN ANONYMOUS-EDGE-ROTATION problem in restricted graph classes, e.g. trees. In that case we require that all graphs in the sequence $G_0, \ldots, G_\ell$ must be from the same graph class. Note that the problem can also be studied without this requirement, but the results may be different.

The following theorem highlights important properties of the edge rotations, the proof can be found e.g. in [4].

Theorem 1. *For any two graphs $G, G' \in \mathbf{G}(n, m)$ there exists a sequence of edge rotations transforming G into G'.*

Corollary 1. *For any two graphs $G, G' \in \mathbf{G}(n, m)$, the edge distance between G and G' is bounded by $2m$.*

3   Feasibility

As it was discussed in Sect. 2, the MIN ANONYMOUS-EDGE-ROTATION problem does not have a solution for every input instance. It is not difficult to see that if a graph is 'almost' complete or 'almost' empty, then there are only restricted options on the number of different degree classes.

First we present some sufficient conditions for an instance to be feasible showing that if a graph is not 'almost' complete or an empty graph, then a solution of the problem exists for all $k \leq \frac{n}{4}$, where n is the order of the graph.

Theorem 2. *Let* $G \in \mathbf{G}(n, m)$ *such that* $\frac{n}{2} \leq m \leq \frac{n(n-3)}{2}$ *and* $n \geq 4$. *Then there exists a feasible solution for the* MIN ANONYMOUS-EDGE-ROTATION *problem, hence a k-degree-anonymous graph* $G' \in \mathbf{G}(n, m)$, *for any* $k \leq \frac{n}{4}$.

Proof. Let m, n, k be fixed. Any graph $G \in \mathbf{G}(n, m)$ is a 1-degree-anonymous graph, hence we can suppose $k \geq 2$.

In the first part of the proof we describe a construction of a k-anonymous sequence $D = (d_1, d_2, \ldots, d_n)$ with property $\sum_{i=1}^{n} d_i = 2m$ for any m, n, k satisfying the restriction of the theorem. In the second part we show that the sequence D is graphic, hence that the sequence satisfies the conditions (1) and (2) from Sect. 2. As $\sum_{i=1}^{n} d_i = 2m$ is the condition for a constructed sequence, the property (1) trivially holds.

Now we construct three distinct k-anonymous sequences Type 1, 2, 3 of integers based on the values of k and $s \equiv 2m \bmod n$. Denote by d the average degree of the graph G defined as $d = \lfloor \frac{2m}{n} \rfloor$.

Type 1: $k \leq s \leq n - k$
Let $D_1 = (d_1^1, d_2^1, \ldots, d_s^1, d_1^2, d_2^2, \ldots, d_{n-s}^2)$ be a sequence of positive integers where for all i, $1 \leq i \leq s$, $d_i^1 = d + 1$ and for all j, $1 \leq j \leq n - s$, $d_j^2 = d$ (see Fig. 1). The sequence contains n elements and it is easy to see that $\sum_{i=1}^{s} (d+1) + \sum_{j=1}^{n-s} d = 2m$.
Following the assumptions $s \geq k$ and $n - s \geq k$, therefore D_1 is a k-anonymous sequence.

Type 2 : $s < k$
Let $D_2 = (d_1^1, d_2^1, \ldots, d_{s+k}^1, d_1^2, d_2^2, \ldots, d_{n-s-2k}^2, d_1^3, d_2^3, \ldots, d_k^3)$ be a sequence of positive integers where for all i, $1 \leq i \leq s + k$, $d_i^1 = d + 1$; for all r, $1 \leq r \leq n - s - 2k$, $d_r^2 = d$; for all j, $1 \leq j \leq k$, $d_j^3 = d - 1$ (see Fig. 1). The sequence contains n elements and $\sum_{i=1}^{s+k} (d+1) + \sum_{j=1}^{k} (d-1) + \sum_{\ell=1}^{n-s-2k} d = 2m$
Since $n \geq 4k$ and $s < k$, $n - s - 2k \geq k$, D_2 is a k-anonymous sequence.

Type 3: $s > n - k$
Let $D_3 = (d_1^1, d_2^1, \ldots, d_k^1, d_1^2, d_2^2, \ldots, d_{s-2k}^2, d_1^3, d_2^3, \ldots, d_{k+n-s}^3)$ be a sequence of positive integers where for all i, $1 \leq i \leq k$, $d_i^1 = d + 2$; for all r, $1 \leq r \leq s - 2k$, $d_r^2 = d + 1$; for all j, $1 \leq j \leq k + n - s$, $d_j^3 = d$ (see Fig. 1). The sequence has n elements and $\sum_{i=1}^{k} (d+2) + \sum_{j=1}^{k+n-s} d + \sum_{\ell=1}^{s-2k} (d+1) = 2m$.

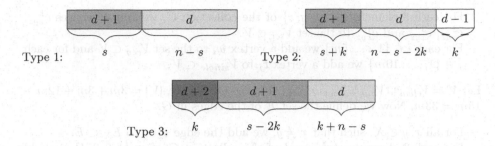

Fig. 1. The sequences of Type 1, 2, 3

Because $n > s$, the number d appears more than k-times in D_3. Due to the assumptions $n \geq 4k$ and $s > n - k$, also $s - 2k \geq k$. Hence D_3 is a k-anonymous sequence.

The proof that all three sequences are graphic can be found in the full version of the paper. □

Now we extend the feasibility study to the case $k = n$ for which we get necessary and sufficient conditions.

Theorem 3. *Let $G \in G(n, m)$ for some positive integers n and m. Then (G, n) is a feasible instance of* MIN ANONYMOUS-EDGE-ROTATION *if and only if $\frac{2m}{n}$ is an integer.*

4 NP-Hardness

In this section we show that the decision version of MIN ANONYMOUS-EDGE-ROTATION, the problem ANONYMOUS-EDGE-ROTATION, is NP-complete. The proof is based on a reduction from the restricted version of a cover set problem, EXACT COVER BY 3-SETS, which is known to be NP-complete [10].

EXACT COVER BY 3-SETS (X3C)
Input: A set X of elements with $|X| = 3m$ and a collection C of 3-elements subsets of X where each element appears in exactly 3 sets.
Question: Does C contain an exact cover for X, i.e. a subcollection $C' \subseteq C$ such that every element occurs in exactly one member set of C' ?

We define a polynomial-time reduction and then prove the NP-completeness of ANONYMOUS-EDGE-ROTATION.

Reduction. Let $I = (X, C)$ be an instance of $X3C$ with $|X| = |C| = 3m$ and m even. We describe the construction σ transforming an instance I into the graph $G := \sigma(I)$ where $G = (V, E)$ is defined as follows:

- For each element $x \in X$, we add a vertex v_x to the set $V_{elem} \subset V$ and a vertex u_x to the set $V_{hub} \subset V$.

- For each 3-element set $\{x, y, z\}$ of the collection C, we add 4 vertices c^1_{xyz}, c^2_{xyz}, c^3_{xyz} and c^4_{xyz} to the set $V_{set} \subset V$.
- For each $i \in \{1, \ldots, 5m\}$ we add a vertex w_i to the set $V_{reg} \subset V$ and for each $j \in \{1, \ldots, 10m\}$ we add a vertex t_j to $V_{single} \subset V$.

Let $V = V_{elem} \cup V_{hub} \cup V_{set} \cup V_{reg} \cup V_{single}$. Obviously, $|V| = 3m + 3m + 12m + 15m = 33m$. Now we define the set E of the edges in G.

- For all $x, y \in X$, such that $x \neq y$, we add the edge $v_x u_y$ to $E_X \subset E$.
- For each 3-element set $\{x, y, z\}$ of the collection C, $\forall i \in \{1, 2, 3, 4\}$, we add the edges $c^i_{xyz} u_x$, $c^i_{xyz} u_y$ and $c^i_{xyz} u_z$ to the set $E_C \subset E$.
- We add the set of edges $E' \subset E$ to the vertex set V_{elem} such that (V_{elem}, E') is a 11-regular graph. Since the number of vertices in the set $|V_{elem}| = 3m$ is even (m is even) and $11 < 3m$ such a regular graph exists [17]. Furthermore, such a graph can be constructed in polynomial time using Havel-Hakimi algorithm [11].
- We add the set of the edges $E'' \subset E$ to the vertex set V_{reg} such that (V_{reg}, E'') is a $(3m + 11)$-regular graph. Since the number of vertices of V_{reg} is even and $3m + 11 < 5m$,

  similarly to the previous case such a regular graph exists and can be constructed in polynomial time.

Finally, let $E = E_X \cup E_C \cup E' \cup E''$.

Obviously, the graph $G = (V, E)$ has the following properties: (i) $10m$ vertices of degree 0 (the vertices of the set V_{single}), (ii) $12m$ vertices of degree 3 (the vertices of the set V_{set}), (iii) $8m$ vertices of degree $3m + 11$ (the vertices of the set V_{reg} and V_{hub}), (iv) $3m$ vertices of degree $3m + 10$ (the vertices of the set V_{elem}).

Theorem 4. ANONYMOUS-EDGE-ROTATION *is NP-complete even in case* $k = \frac{n}{3}$ *where* n *is the order of the graph* G *for an input instance* (G, k, r).

Proof. Let $I = (X, C)$ be an instance of X3C with $|X| = |C| = 3m$ and consider the instance $I' = (G, k, r)$ of ANONYMOUS-EDGE-ROTATION where $G = \sigma(I)$, $k = 11m$ and $r = 3m$. We claim that I is a yes-instance if and only if I' is a yes-instance.

Let $C' \subseteq C$ be an exact cover for X of size m. Now we define $3m$ rotations which are independent from each other: for every 3-element set $\{x, y, z\} \in C'$, we replace the edge $u_x c^1_{xyz}$ by the edge $u_x v_x$, and similarly $u_y c^1_{xyz}$ by $u_y v_y$ and $u_z c^1_{xyz}$ by $u_z v_z$. Since C' is of size m, we define exactly $3m$ rotations. Let G' be the graph obtained from G after applying all $3m$ rotations. Since C' is an exact cover of size m: (i) there are m vertices of type c^1_{xyz} that lost all 3 neighbours and become of degree 0 in G', (ii) all $3m$ vertices of type v_x are attached to a new neighbour, so they become of degree $3m + 11$ in G'.

Then the graph G' has $10m + m = 11m$ vertices of degree 0, $12m - m = 11m$ of degree 3 vertices, $8m + 3m = 11m$ of degree $3m + 11$ vertices, hence we conclude that G' is the $11m$-anonymous graph.

Let I' be a yes-instance of ANONYMOUS-EDGE-ROTATION. Then there exists a sequence of $3m$ rotations such that the graph $G' = (V, E')$ obtained after applying the rotations to G is a $11m$-anonymous graph. Since $|V| = 33m$, there must be only three different degrees classes in G'. Note that with one rotation, we can change the degree of two vertices, therefore the degree at most $6m$ vertices can be changed by $3m$ rotations. Since the graph G has more than $6m$ vertices of the degrees $3m+11$, 3, and 0, all these degree classes must be in G'. Furthermore, due to the number of vertices of G, these are the only degree classes in G'. This means that in G' the number of vertices of degree $3m + 11$ must be increased by $3m$, the number of vertices of degree 0 must be increased by m, the number of vertices of degree 3 must be decreased by m and there are no vertices of degrees $3m + 10$ in G'. A single rotation can increase or decrease the degree of a vertex by 1 therefore using $3m$ rotations no vertex of degree $3m + 10$ in G can have degree 0 in G' and similarly, no vertex of degree 3 in G can have degree $3m + 11$ in G'. Therefore the $3m$ new vertices of degree $3m + 11$ in G' must have degree $3m + 10$ in G. This is only possible if the degree of each vertex v_x from the set V_{elem} is increased by 1. Similarly, the m new vertices of degree 0 in G' must have degree 3 in G, let $C_{G'}$ be the set of such vertices. Obviously, $C_{G'}$ must be a subset of V_{set}, in which the vertices have the form c_{xyz}^{ℓ} with $x, y, z \in X$, for any set $\{x, y, z\} \in C$, and $\ell \in \{1, 2, 3, 4\}$.

To reach the requested degree configuration in G' with exactly $3m$ edge rotations, in each rotation the degree of each vertex from V_{elem} must be increased by 1 and the degree of each vertex from the set $C_{G'}$ must be decrease by 1. To achieve that, for each vertex v_x from V_{elem}, the only possible rotation is to add the edge $u_x v_x$ where $u_x \in V_{hub}$ and remove the edge $u_x c_{xyz}^{\ell}$ where $c_{xyz} \in C_{G'}$. To fulfil the condition about the degree classes and the number of the rotations, the only way to achieve that is that $C'' = \{\{x, y, z\} \mid c_{xyz}^{\ell} \in C_{G'}\}$ is an exact cover of X.   □

5   Lower Bound for a k-Degree-Anonymous Graph

In this section we suppose that (G, k) is a feasible instance. For any such instance we define a k-anonymous degree sequence S_{bound} that can be computed in polynomial time if $k = \theta(n)$. We show that with the $(+1, -1)$-degree modifications (Remark 1) the graph G can be transformed into a k-degree-anonymous graph G' with degree sequence S_{bound} using at most double of edge rotations as in an optimal solution of MIN ANONYMOUS-EDGE-ROTATION for (G, k).

Note that in general a $(+1, -1)$-degree modification doesn't correspond to an edge rotation, but as we show later in Sect. 7.1, it is true for trees.

Now in the following steps we show how to define the degree sequence S_{bound}.

Step 1: Compute Every Available Target Sequence
Let $S = (s_1, \ldots, s_n)$ be a non-increasing sequence of non-negative integers, $r \in \{1, \ldots, n\}$. Any partition of S into r contiguous subsequences (i.e. if $S[a]$ and $S[b]$ are in one part, then all $S[i]$, $a \leq i \leq b$ must be in the same part) is called a contiguous r-partition. The number of contiguous r-partitions of S is $\binom{n-1}{r-1}$,

therefore bounded by $(n-1)^{r-1}$. Then the number of contiguous partitions of S with at most r parts can be bounded by $\sum_{i=0}^{r-1}(n-1)^i \leq 2n^{r-1}$.

For each contiguous ℓ-partition p, $1 \leq \ell \leq r$, we use notation $p = [p_1, \ldots, p_\ell]$, where p_i denotes the number of elements in part i, $1 \leq i \leq \ell$. Note that at this stage important is the number of elements in each part, not which elements from S are in it.

Let G be a graph of order n and k an integer, $k \geq 2$. If G is a k-degree-anonymous graph, then the vertices of G can be partitioned into at most $c = \lfloor \frac{n}{k} \rfloor$ parts where the vertices in each part have the same degree. Let P be the set of all such contiguous partitions with at most c parts. As it follows from the initial discussion, the number of such partitions is bounded by $2n^{c-1}$.

Now for each contiguous partition $p = [p_1, p_2, \ldots, p_\ell] \in P$, $\ell \in \{1, \ldots, c\}$, we compute all non-increasing sequences $(d_1, d_2, \ldots, d_\ell)$ of ℓ integers d_i such that $0 \leq d_i < |V|$. Let $\hat{P}_p$ be the set of all feasible k-anonymous degree sequences for p, i.e.

$$S = (\underbrace{d_1, \ldots, d_1}_{p_1-\text{times}}, \underbrace{d_2, \ldots, d_2}_{p_2-\text{times}}, \ldots, \underbrace{d_\ell, \ldots, d_\ell}_{p_\ell-\text{times}}) = (d_1^{p_1}, d_2^{p_2}, \ldots, d_\ell^{p_\ell}) \in \hat{P}_p$$

if and only if $\sum_{i=1}^{\ell} p_i d_i = 2|E|$, S is graphic and k-anonymous.

For each contiguous partition p with ℓ parts, $1 \leq \ell \leq c$, there are at most n possibilities for a degree on each position. The test whether the generated sequence is graphic and k-anonymous can be done in $O(n)$ operations. Since $|P| = O(n^{c-1})$, there are at most $O(n^{c-1} \times n^\ell \times n) \leq O(n^{2c})$ operations to compute all feasible degree sequences of every partition, where $c = \lfloor \frac{n}{k} \rfloor$. Obviously, if c is a constant, such number of operations is polynomial.

Step 2: Find the Best One

Now based on the previous analysis we can define the degree sequence S_{bound} and prove some basic properties.

Definition 4. *Let G be a graph with the degree sequence S_G. Then define S_{bound} for G as a degree sequence for which the sum $\sum_{i=1}^{n} |S_G[i] - S[i]|$ achieves the minimum for all elements $S \in \hat{P}_p$ and $p \in P$.*

Remark 2. Similarly to a k-anonymous sequence S_{bound} defined in Definition 4 for a graph, we can define a k-anonymous sequence S_{Tbound} for a tree. The only difference is that in the set $\hat{P}_p$, every feasible solution must have $d_i \geq 1$, which would be a subset of $\hat{P}_p$. Also for the testing, we don't need to check whether S is graphic, the condition $\sum_{i=1}^{\ell} p_i d_i = 2|E|$, is enough for the degree sequence of a tree.

The following lemmas describes the basic properties of sequences.

Lemma 1. *Let S be a n-sequence of non-negative integers and denote by S' the sequence S sorted in non-increasing order. Let S_s be another n-sequence of non-negative integers sorted in non-increasing order. Then*

$$|\sum_{i=1}^{n} |S_s[i] - S'[i]|| \le \sum_{i=1}^{n} |S_s[i] - S[i]| \tag{3}$$

Lemma 2. *Let (G, k) be a feasible instance for the* MIN ANONYMOUS-EDGE-ROTATION *problem. Let OPT be an optimum solution that is a minimum set of rotations that transform G to a k-degree-anonymous graph G'. Then $\sum_{i=1}^{n} |S_G[i] - S_{bound}[i]| \le 2|OPT|$, where the degree sequence S_{bound} is defined in Definition 4.*

Proof. Let $S_{G'}$ be the degree sequence of G' sorted in the same order as S_G (i.e. for every $v \in V$, if $deg_G(v)$ is in the position i in S_G then $deg_{G'}(v)$ is in the position i in $S_{G'}$). Let $S'_{G'}$ be the degree sequence $S_{G'}$ sorted in non-increasing order. As in the definition of S_{bound} we considered all the options, there must exist $p \in P$ and $S \in \hat{P}_p$ such that $S = S'_{G'}$, and $\sum_{i=1}^{n} |S_G[i] - S_{bound}[i]| \le \sum_{i=1}^{n} |S_G[i] - S'_{G'}[i]|$.

Since the degree sequence $S'_{G'}$ is sorted in non-increasing order, then $\sum_{i=1}^{n} |S_G[i] - S'_{G'}[i]| \le \sum_{i=1}^{n} |S_G[i] - S_{G'}[i]|$ by Lemma 1. One rotation from the graph G_j to G_{j+1} in the sequence of the graphs from G to G' can only decrease the degree of a vertex by one and increase the degree of another one by one, hence $\sum_{i=1}^{n} |S_{G_j}[i] - S_{G'}[i]| \le \sum_{i=1}^{n} |S_{G_{j+1}}[i] - S_{G'}[i]| + 2$. This means by one rotation the value $\sum_{i=1}^{n} |S_G[i] - S_{G'}[i]|$ decreases by at most 2. After $|OPT|$ rotations, the last graph G_{j+1} in the sequence is G', therefore $\sum_{i=1}^{n} |S_G[i] - S_{G'}[i]| \le 2|OPT|$ and the lemma follows. $\square$

6 Approximation

In this section we show that under some constraints on the number of edges and k, there exists a polynomial time 2-approximation algorithm for the MIN ANONYMOUS-EDGE-ROTATION problem for all feasible inputs (G, k).

Remark 3. Let $S = (x_1, x_2, \ldots, x_n)$ be a non-increasing sequence of n non-negative integers. Denote by $R = x_1 - x_n$, $A_0 = \frac{x_1 + x_n}{2}$, and let $A = \frac{\sum_{i=1}^{n} x_i}{n}$.

The standard deviation of S is defined as $\sigma(S) = \sqrt{\frac{\sum(x_i - A)^2}{n}}$. It can be shown that

$$\sum_{i=1}^{n}(x_i - A)^2 \le \sum_{i=1}^{n}(x_i - A_0)^2 \le \frac{nR^2}{4},$$

hence $\sigma(S) \le \frac{R}{2}$.

The mean absolute derivation of S is defined as $MAD[S] = \frac{1}{n}\sum_{i=1}^{n}|x_i - A|$. It is well known (e.g. applying Jensen's inequality) that $MAD[S] \leq \sigma(S)$.

Based on the correlation mentioned in Remark 3, we calculate an upper bound on the values in the degree sequence S_{bound} in the following lemma.

Lemma 3. *Let (G,k) be an instance of the* MIN ANONYMOUS-EDGE-ROTATION *problem where G is the graph with n vertices and m edges. Suppose that $\frac{n}{2} \leq m \leq \frac{n(n-3)}{2}$, $k \leq \frac{n}{4}$, and let the constant c be defined as $c = \lfloor \frac{n}{k} \rfloor$, hence $k = \theta(n)$. Let S_{bound} be the k-anonymous degree sequence associated with G defined following Definition 4. Then for every i, $S_{bound}[i] \leq \min\{(1 + \frac{n}{4k} + \frac{n}{k\Delta})\Delta, n-1\}$, $1 \leq i \leq n$.*

In the following two lemmas we prove that if a graph has 'sufficiently' many edges than edge rotations with the specific properties exist in a graph.

Lemma 4. *Let $G = (V,E)$ be a graph with $|E| > \Delta^2$, let $uv \in E$. Then there exists an edge $ab \in E$ such that both vertices a and b are different from u and v and at most one of the following edges $\{av, au, bv, bu\}$ is in E.*

Lemma 5. *Let $G = (V,E)$ be a graph and suppose $|E| > \Delta^2$. Let $v^+, v^- \in V$ such that $1 \leq d_G(v^-) \leq \Delta$ and $0 \leq d_G(v^+) \leq \Delta < |V| - 1$. Then there exists a sequence of at most two edge rotations that transform G to G' such that $d_{G'}(v^+) = d_G(v^+) + 1$, $d_{G'}(v^-) = d_G(v^-) - 1$ and degrees of other vertices in G are not changed. These rotations can be found in $O(|E|^2)$ steps.*

Theorem 5. *The* MIN ANONYMOUS-EDGE-ROTATION *problem is polynomial time 2-approximable for all instances (G,k), $k \leq \frac{n}{4}$ where $k = \theta(n)$ and G is the graph with n vertices and m edges, where $\max\{\frac{n}{2}, (1 + \frac{n}{4k} + \frac{n}{k\Delta})^2 \Delta^2\} \leq m \leq \frac{n(n-3)}{2}$, and the constant c is defined as $c = \lfloor \frac{n}{k} \rfloor$.*

Proof. Let $(G = (V,E), k)$ be an instance of MIN ANONYMOUS-EDGE-ROTATION and S_G be the degree sequence of G. Let the constant c be defined as $c = \lfloor \frac{n}{k} \rfloor$. Due to our assumptions about the number of edges and k, all such instances are feasible as follows from Sect. 3. First we compute a k-anonymous degree sequence S_{bound} following Definition 4 in $O(n^{2c})$ steps. Due to the assumption $k = \theta(n)$ and consequently c being a constant, such number of steps is polynomial. Furthermore, the condition on the number of edges ensures that we can always apply Lemma 5 and find suitable edge rotations.

If there exist two vertices v^+, $v^- \in V$ such that $0 \leq S_G[v^+] < S_{bound}[v^+] \leq (1 + \frac{n}{4k} + \frac{n}{k\Delta})\Delta < |V| - 1$ and $S_G[v^-] > S_{bound}[v^-]$ we apply Lemma 5 to transform G to a graph G_1 with at most two rotations such that $d_{G_1}(v^+) = d_G(v^+) + 1$ and $d_{G_1}(v^-) = d_G(v^-) - 1$.

We'll be executing the above transformations while there are two vertices v^+, $v^- \in V$ with the required properties. In each such transformation we decrease the degree of one vertex by 1 and increase the degree of another one by 1 with at most two rotations. Hence we transform G to a final graph G' with degree sequence S_{bound} by at most $\sum_{i=1}^{n} |S_G[i] - S_{bound}[i]|$ rotations. By Lemma 2 we know that $\sum_{i=1}^{n} |S_G[i] - S_{bound}[i]| \leq 2|OPT|$, hence we use at most 2 times the numbers of rotations of an optimal solution. In each transformation loop searching for the vertices v^+ and v^- can be done in time $O(n)$ and searching for an edge ab in time $O(m^2)$ (Lemma 4). Due to the modifications in each transformation loop, there can be at most $O(n^2)$ loops. Therefore the time complexity is bounded by $O(n^{2c} + n^2 \times m^2 \times n)$. Since $c \geq 4$, $O(n^{2c} + m^2 \times n^3) \leq O(n^{2c})$.

Finally, since S_{bound} is k-anonymous, G' is a k-degree-anonymous graph. $\square$

7   Polynomial Cases

As follows from Sect. 4, the MIN ANONYMOUS-EDGE-ROTATION problem is NP-hard even for $k = \frac{n}{3}$, where n is the order of an input graph. In this section we show that the problem can be solved in polynomial time on trees when $k = \theta(n)$ or in case of any graph when $k = n$.

7.1   Trees

For a tree $T = (V, E)$ rooted in a vertex r, for any $v \in V$, $v \neq r$, $child(v)$ is a vertex that is a neighbor of v not on the path from r to v.

Lemma 6. *Let $T = (V, E)$ be a tree and v^-, v^+ vertices from V such that v^- is not a leaf and v^+ is not a universal vertex. Then using one rotation we can transform T into a tree T' such that $d_{T'}(v^-) = d_T(v^-) - 1$ and $d_{T'}(v^+) = d_T(v^+) + 1$.*

Theorem 6. *The MIN ANONYMOUS-EDGE-ROTATION problem is polynomial-time solvable for any instance (T, k) where T is a tree of the order n, $k \leq \frac{n}{4}$ and such that $c = \lfloor \frac{n}{k} \rfloor$ is a constant, hence $k = \theta(n)$.*

Proof. Let T be a tree and $S_T = (d_1, d_2, \ldots, d_n)$ its degree sequence sorted in non-increasing order. As it was mentioned in Sect. 2, for a degree sequence of a tree only the following conditions must hold $\sum_{i=1}^{n} d_i = 2(n-1)$ and $d_i \geq 1$ for all i, $1 \leq i \leq n$. Now based on S_T define a k-anonymous sequence S_{Tbound} as discussed in Sect. 5.

Let x and y be integers such that $S_T[x] > S_{Tbound}[x]$ and $S_T[y] < S_{Tbound}[y]$. Since S_{Tbound} correspond to a tree, $S_{Tbound}[x] \geq 1$ then $S_T[x] > 1$ then v_x is not a leave in T. Moreover since $S_{Tbound}[y] \leq n - 1$, v_y is not a universal vertex in T.

By Lemma 6 there exists a tree T_1 such that $S_{T_1}[x] = S_T[x] - 1$ and $S_{T_1}[y] = S_T[y] + 1$. Repeat this operation until reaching a tree T' with the degree sequence S_{Tbound}. The cost of one operation is $O(n)$ and we repeat it $\frac{\sum_{i=1}^{n} |S_T[i] - S_{Tbound}[i]|}{2} \leq n^2$ times. Since S_{Tbound} is k-anonymous, T' is a k-degree-anonymous tree. Since we use $\frac{\sum_{i=1}^{n} |S_T[i] - S_{Tbound}[i]|}{2} \leq |OPT|$ rotations (Lemma 2), the algorithm is optimal. The total cost of the algorithm is bounded by $O(n^{2c} + n^2) = O(n^{2c})$, where $c = \lfloor \frac{n}{k} \rfloor$ is a constant.   □

7.2   One Degree Class, $k = n$

In this part we show that MIN ANONYMOUS-EDGE-ROTATION is polynomial-time solvable for instances where k coincides with the number of vertices of the graph, that means all vertices must be in the same degree class.

Lemma 7. *Let $G = (V, E)$ be a graph and $u, v \in V$. If $\mathcal{N}_G(u) \nsubseteq \mathcal{N}_G(v)$, then there is an edge rotation that leads to a graph G' such that $d_{G'}(u) = d_G(u) - 1$ and $d_{G'}(v) = d_G(v) + 1$.*

Remark 4. Let $G = (V, E)$ be a graph, $\forall u, v \in V$, if $d_G(u) > d_G(v)$, then there is an edge rotation that leads to a graph G' such that $d_{G'}(u) = d_G(u) - 1$ and $d_{G'}(v) = d_G(v) + 1$.

Lemma 8. *Let (G, n) be an instance of MIN ANONYMOUS-EDGE-ROTATION where $G \in \mathbf{G}(n, m)$ for some positive integers m, n, and $\frac{2m}{n}$ is an integer. Then the optimum value of MIN ANONYMOUS-EDGE-ROTATION on (G, n) is $\frac{\sum_{w \in V} |d_G(w) - 2m/n|}{2}$,*

Theorem 7. *The MIN ANONYMOUS-EDGE-ROTATION problem is polynomial-time solvable for instances (G, k) when $k = n$, where n is the order of the graph G.*

Proof. In case $k = n$, we are looking for a n-degree-anonymous graph with only one degree class, hence for a regular graph. Due to Theorem 3, we can easily decide whether (G, n) is a feasible instance of MIN ANONYMOUS-EDGE-ROTATION: if for $G \in \mathbf{G}(n, m)$ the fraction $\frac{2m}{n}$ is not an integer, (G, n) is not a feasible input.

For a feasible input (G, n), the result is based on Algorithm 1 and its correctness follows from Lemmas 7 and 8.

Obviously, the algorithm runs in polynomial time.   □

Input   : A graph $G = (V, E)$
Output: A sequence S of edge rotations if $\frac{2|E|}{|V|}$ is an integer
NO otherwise

$S = \emptyset$;
$d = \frac{2|E|}{|V|}$;
if *if d is not integer* **then**
| return NO ;
else
| **while** $\exists u, v \in V$ *such that* $d_G(u) < d$ *and* $d_G(v) > d$ **do**
| | Let $w \in \mathcal{N}(v) \setminus \mathcal{N}(u)$;
| | $E = E \setminus \{vw\}$;
| | $E = E \cup \{uw\}$;
| | $S = S \cup \{(wv, wu)\}$;
| **end**
end

Algorithm 1: Algorithm for $k = |V|$

8   Conclusion

In this paper we initiate the study of the complexity of MIN ANONYMOUS-EDGE-ROTATION problem in which the task is to transform a given graph to a k-degree anonymous graph using the minimum number of edge rotations. The problem doesn't have a solution for all graphs and all possible values of k, but our initial feasibility study already covers the majority of instances rotations. The problem doesn't have a solution for all graphs and all possible values of k, but our initial feasibility study already covers the majority of instances. The extensions of these results are still possible, mainly to find necessary and sufficient conditions for feasibility.

As we were able to prove NP-hardness in case where the number of vertices k in each degree class is $\theta(n)$, further research could explore stronger hardness results or cases when k is a constant. Our next research step includes relaxation of the condition on the number of the edges in the presented 2-approximation algorithm as well as extension of the graph classes in which the MIN ANONYMOUS-EDGE-ROTATION problem can be solved in polynomial time.

References

1. Bazgan, C., Bredereck, R., Hartung, S., Nichterlein, A., Woeginger, G.J.: Finding large degree-anonymous subgraphs is hard. Theor. Comput. Sci. **622**, 90–110 (2016)
2. Bredereck, Robert., Froese, Vincent., Hartung, Sepp., Nichterlein, André., Niedermeier, Rolf, Talmon, Nimrod: The complexity of degree anonymization by vertex addition. In: Gu, Qianping, Hell, Pavol, Yang, Boting (eds.) AAIM 2014. LNCS, vol. 8546, pp. 44–55. Springer, Cham (2014). https://doi.org/10.1007/978-3-319-07956-1_5

3. Chartrand, G., Gavlas, H., Johnson, H.H.M.A.: Rotation and jump distances between graphs. Discussiones Mathematicae Graph Theory **17**, 285–300 (1997)
4. Chartrand, G., Saba, F., Zou, H.B.: Edge rotations and distance between graphs. Časopis pro pěstování matematiky **110**(1), 87–91 (1985)
5. Chester, S., Kapron, B.M., Ramesh, G., Srivastava, G., Thomo, A., Venkatesh, S.: Why waldo befriended the dummy? k-anonymization of social networks with pseudo-nodes. Social Netw. Analys. Mining **3**(3), 381–399 (2013)
6. Chester, S., Kapron, B.M., Srivastava, G., Venkatesh, S.: Complexity of social network anonymization. Social Netw. Analys. Mining **3**(2), 151–166 (2013)
7. Erdos, P., Gallai, T.: Gráfok eloírt fokú pontokkal (graphs with points of prescribed degrees, in Hungarian). Mat. Lapok **11**, 264–274 (1961)
8. Faudree, R.J., Schelp, R.H., Lesniak, L., Gyárfás, A., Lehel, J.: On the rotation distance of graphs. Discrete Math. **126**(1–3), 121–135 (1994)
9. Fung, B., Wang, K., Chen, R., Yu, P.: Privacy-preserving data publishing: a survey of recent developments. ACM Comput. Surv. **42**(4), 14:1–14:53 (2010)
10. Garey, M.R., Johnson, D.S.: Computers and Intractability, vol. 174. Freeman, San Francisco (1979)
11. Hakimi, S.L.: On realizability of a set of integers as degrees of the vertices of a linear graph. i. J. Soc. Ind. Appl. Math. **10**(3), 496–506 (1962)
12. Hartung, S., Hoffmann, C., Nichterlein, A.: Improved upper and lower bound heuristics for degree anonymization in social networks. In: Gudmundsson, J., Katajainen, J. (eds.) SEA 2014. LNCS, vol. 8504, pp. 376–387. Springer, Cham (2014). https://doi.org/10.1007/978-3-319-07959-2_32
13. Hartung, S., Nichterlein, A., Niedermeier, R., Suchý, O.: A refined complexity analysis of degree anonymization in graphs. Inf. Comput. **243**, 249–262 (2015)
14. Jarrett, E.B.: Edge rotation and edge slide distance graphs. Comput. Math. Appl. **34**(11), 81–87 (1997)
15. Liu, K., Terzi, E.: Towards identity anonymization on graphs. In: Proceedings of the ACM SIGMOD International Conference on Management of Data, SIGMOD, pp. 93–106. ACM (2008)
16. Salas, J., Torra, V.: Graphic sequences, distances and k-degree anonymity. Discrete Appl. Math. **188**, 25–31 (2015)
17. Tomescu, I.: Problems in combinatorics and graph theory. Wiley-Interscience Series Discrete Math. 212–213 (1961)
18. Wu, X., Ying, X., Liu, K., Chen, L.: A survey of privacy-preservation of graphs and social networks. In: Aggarwal, C., Wang, H. (eds.) Managing and Mining Graph Data. Advances in Database Systems, vol. 40, pp. 421–453. Springer, Heidelberg (2010). https://doi.org/10.1007/978-1-4419-6045-0_14

The Small Set Vertex Expansion Problem

Soumen Maity[✉]

Indian Institute of Science Education and Research, Pune 411008, India
soumen@iiserpune.ac.in

Abstract. Given a graph $G = (V, E)$, the vertex expansion of a set $S \subset V$ is defined as $\Phi^V(S) = \frac{|N(S)|}{|S|}$. In the SMALL SET VERTEX EXPANSION (SSVE) problem, we are given a graph $G = (V, E)$ and a positive integer $k \leq \frac{|V(G)|}{2}$, the goal is to return a set $S \subset V(G)$ of k nodes minimizing the vertex expansion $\Phi^V(S) = \frac{|N(S)|}{k}$; equivalently minimizing $|N(S)|$. SSVE has not been as well studied as its edge-based counterpart SMALL SET EXPANSION (SSE). SSE, and SSVE to a less extend, have been studied due to their connection to other hard problems including the Unique Games Conjecture and Graph Colouring. Using the hardness of MINIMUM k-UNION problem, we prove that SMALL SET VERTEX EXPANSION problem is NP-complete. We enhance our understanding of the problem from the viewpoint of parameterized complexity by showing that (1) the problem is W[1]-hard when parameterized by k, (2) the problem is fixed-parameter tractable (FPT) when parameterized by the neighbourhood diversity $\mathtt{nd}$, and (3) it is fixed-parameter tractable (FPT) when parameterized by treewidth $\mathtt{tw}$ of the input graph.

Keywords: Parameterized complexity · FPT · W[1]-hard · Treewidth · Neighbourhood diversity

1 Introduction

Covering problems are very well-studied in theoretical computer science. Given a set of elements $\{1, 2, ..., n\}$ (called the universe) and a collection $\mathcal{S}$ of m sets whose union equals the universe, the SET COVER problem is to identify the smallest sub-collection of $\mathcal{S}$ whose union equals the universe, and MAX k-COVER is the problem of selecting k sets from $\mathcal{S}$ such that their union has maximum cardinality. MAX k-COVER is known to admit a $(1 - \frac{1}{e})$-approximation algorithm (which is also known to be tight) [5]. A natural variation of MAX k-COVER problem is instead of covering maximum number of elements, the problem is to cover minimum number of elements of the universe by the union of k sets. MINIMUM k-UNION [2,16] is one of such problems, where we are given a family of sets within a finite universe and an integer k and we are asked to choose k sets from this family in order to minimise the number of elements of universe

The author's research was supported in part by the Science and Engineering Research Board (SERB), Govt. of India, under Sanction Order No. MTR/2018/001025.

W. Wu and Z. Zhang (Eds.): COCOA 2020, LNCS 12577, pp. 257–269, 2020.
https://doi.org/10.1007/978-3-030-64843-5_18

that are covered. MINIMUM k-UNION has not been studied until recently, when an $O(\sqrt{m})$-approximation algorithm is given by Eden Chlamtác et al. [3], where m is the number of sets in $\mathcal{S}$. Given an instance of MINIMUM k-UNION, we can construct the obvious bipartite graph in which the left side represents sets and the right side represents elements and there is an edge between a set node and an element node if the set contains the element. Then MINIMUM k-UNION is clearly equivalent to the problem of choosing k left nodes in order to minimize the size of their neighbours. This is known as the SMALL SET BIPARTITE VERTEX EXPANSION (SSBVE) problem [2]. This is the bipartite version of the SMALL SET VERTEX EXPANSION, in which we are given an arbitrary graph and are asked to choose a set S of k nodes minimizing the vertex expansion $\Phi^V(S) = |N(S)|$. SMALL SET VERTEX EXPANSION problem is vertex version of the SMALL SET EXPANSION (SSE) problem, in which we are asked to choose a set of k nodes to minimize the number of edges with exactly one endpoint in the set. SSVE has not been as well studied as SSE, but has recently received significant attention [12]. SSE, and SSVE to a less extend, have been studied due to their connection to other hard problems including the Unique Games Conjecture [8]. These problems recently gained interest due to their connection to obtain sub-exponential time, constant factor approximation algorithm for may combinatorial problems like Sparsest Set and Graph Colouring [1].

A problem with input size n and parameter k is said to be 'fixed-parameter tractable (FPT)' if it has an algorithm that runs in time $\mathcal{O}(f(k)n^c)$, where f is some (usually computable) function, and c is a constant that does not depend on k or n. What makes the theory more interesting is a hierarchy of intractable parameterized problem classes above FPT which helps in distinguishing those problems that are not fixed parameter tractable. For the standard concepts in parameterized complexity, see the recent textbook by Cygan et al. [4].

Our main results are the following:

- The SMALL SET VERTEX EXPANSION (SSVE) problem is NP-complete.
- SSVE is W[1]-hard when parameterized by k.
- SSVE is fixed-parameter tractable (FPT) when parameterized by neighbourhood diversity of the input graph.
- SSVE is FPT when parameterized by treewidth of the input graph.

Related Results: Despite being a very natural problem, MINIMUM k-UNION/ SSBVE has received surprisingly little attention. Chlamtác et al. [3] gave an $O(\sqrt{n})$-approximation algorithm for SSBVE and equivalently $O(\sqrt{m})$-approximation algorithm for MINIMUM k-UNION problem. Louis and Makarychev [12] studied approximation algorithms for hypergraph small set expansion and small set vertex expansion problem. They provided a polylogarithmic approximation when k is very close to n, namely, $k \geq \frac{n}{\text{ploylog}(n)}$. To the best of our knowledge, the parameterized complexity of SSVE and SSE problems have not been studied before. Raghavendra and Steurer [13] have investigated the connection between Graph Expansion and the UNIQUE GAMES CONJECTURES. They proved that a simple decision version of the problem of approximately small set expansion reduces to UNIQUE GAMES.

2   Preliminaries

The vertex and edge expansion in graphs have been a subject of intense study with applications in almost all branches of theoretical computer science. From an algorithmic standpoint SSVE and SSE are fundamental optimization problems with numerous applications. The computational complexity of computing and approximating expansion is still not very well understood. Throughout this article, $G = (V, E)$ denotes a finite, simple and undirected graph of order $|V(G)| = n$. For a vertex $v \in V$, we use $N(v) = \{u \ : \ (u, v) \in E(G)\}$ to denote the (open) neighbourhood of vertex v in G, and $N[v] = N_G(v) \cup \{v\}$ to denote the closed neighbourhood of v. The degree $d(v)$ of a vertex $v \in V(G)$ is $|N(v)|$. For a subset $S \subseteq V(G)$, we define its closed neighbourhood as $N[S] = \bigcup_{v \in S} N[v]$ and its open neighbourhood as $N(S) = N[S] \setminus S$. Given a graph $G = (V, E)$, the vertex expansion of a set $S \subset V$ is defined as

$$\Phi^V(S) = \frac{|N(S)|}{|S|}.$$

Definition 1. [2] In the SMALL SET VERTEX EXPANSION (SSVE) problem, we are given a graph $G = (V, E)$ and an integer $k \leq \frac{|V|}{2}$. The goal is to return a subset $S \subset V$ with $|S| = k$ minimizing the vertex expansion $\Phi^V(S) = \frac{|N(S)|}{k}$; equivalently minimizing $|N(S)|$.

The edge expansion of a subset of vertices $S \subset V$ in a graph G measures the fraction of edges that leaves S. For simplicity we consider regular graphs in the definition of SMALL SET EXPANSION (SSE). In a d-regular graph, the edge expansion $\Phi(S)$ of a subset $S \subset V$ is defined as

$$\Phi(S) = \frac{|E(S, V \setminus S)|}{d|S|}$$

where $E(S, V \setminus S)$ denotes the set of edges with one endpoint in S and other endpoint in $V \setminus S$.

Definition 2. [2] In the SMALL SET EXPANSION (SSE) problem, we are given a d-regular graph $G = (V, E)$ and an integer $k \leq \frac{|V|}{2}$. The goal is to return a subset $S \subset V$ with $|S| = k$ minimizing the edge expansion $\Phi(S) = \frac{|E(S,V \setminus S)|}{kd}$; equivalently minimizing $|E(S, V \setminus S)|$.

Among the two notions of expansion, this work will concern with vertex expansion. The decision version of the problem studied in this paper is formalized as follows:

SMALL SET VERTEX EXPANSION

Input: An undirected graph $G = (V, E)$ and two positive integers $k \leq \frac{|V|}{2}$, $\ell \leq |V(G)|$.

Question: Is there a set $S \subset V(G)$ with $|S| = k$ such that the vertex expansion $\Phi^V(S) = |N(S)| \leq \ell$?

We now recall some graph parameters used in this paper. The graph parameters we explicitly use in this paper are neighbourhood diversity $\mathbf{nd}$ and treewidth $\mathbf{tw}$. We now review the concept of a tree decomposition, introduced by Robertson and Seymour in [14].

Definition 3. A *tree decomposition* of a graph G is a pair $(T, \{X_t\}_{t \in V(T)})$, where T is a tree and each node t of the tree T is assigned a vertex subset $X_t \subseteq V(G)$, called a bag, such that the following conditions are satisfied:

1. Every vertex of G is in at least one bag.
2. For every edge $uv \in E(G)$, there exists a node $t \in T$ such that bag X_t contains both u and v.
3. For every $u \in V(G)$, the set $\{t \in V(T) \mid u \in X_t\}$ induces a connected subtree of T.

Definition 4. The *width* of a tree decomposition is defined as $width(T) = max_{t \in V(T)} |X_t| - 1$ and the treewidth $tw(G)$ of a graph G is the minimum width among all possible tree decomposition of G.

A special type of tree decomposition, known as a *nice tree decomposition* was introduced by Kloks [9]. The nodes in such a decomposition can be partitioned into four types.

Definition 5. A tree decomposition $(T, \{X_t\}_{t \in V(T)})$ is said to be *nice tree decomposition* if the following conditions are satisfied:

1. All bags correspond to leaves are empty. One of the leaves is considered as root node r. Thus $X_r = \emptyset$ and $X_l = \emptyset$ for each leaf l.
2. There are three types of non-leaf nodes:
   - **Introduce node:** a node t with exactly one child t' such that $X_t = X_{t'} \cup \{v\}$ for some $v \notin X_{t'}$; we say that v is *introduced* at t.
   - **Forget node:** a node t with exactly one child t' such that $X_t = X_{t'} \setminus \{w\}$ for some $w \in X_{t'}$; we say that w is *forgotten* at t.
   - **Join node:** a node with two children t_1 and t_2 such that $X_t = X_{t_1} = X_{t_2}$.

Note that, by the third property of tree decomposition, a vertex $v \in V(G)$ may be introduced several time, but each vertex is forgotten only once. To control introduction of edges, sometimes one more type of node is considered in nice tree decomposition called introduce edge node. An *introduce edge node* is a node t, labeled with edge $uv \in E(G)$, such that $u, v \in X_t$ and $X_t = X_{t'}$, where t' is the only child of t. We say that node t introduces edge uv. It is known that if a graph G admits a tree decomposition of width at most $\mathbf{tw}$, then it also admits a nice tree decomposition of width at most $\mathbf{tw}$, that has at most $O(n \cdot \mathbf{tw})$ nodes [4].

3   Proving SMALL SET VERTEX EXPANSION is NP-complete

Using the hardness of MINIMUM k-UNION problem, we prove that SMALL SET VERTEX EXPANSION problem is NP-complete. We state the decision version of MINIMUM k-UNION problem.

Definition 6. [2] In MINIMUM k-UNION problem, we are given an universe $U = \{1, 2, \ldots, n\}$ of n elements and a collection of m sets $\mathcal{S} \subseteq 2^U$, as well as two integers $k \leq m$ and $\ell \leq n$. Does there exist a collection $T \subseteq \mathcal{S}$ with $|T| = k$ such that $|\cup_{S \in T} S| \leq \ell$?

It is known that MINIMUM k-UNION problem is NP-complete [16]. Now we prove the following hardness result.

Theorem 1. *The* SMALL SET VERTEX EXPANSION *problem is NP-complete.*

Proof. We first show that SMALL SET VERTEX EXPANSION problem is in NP. Given a graph $G = (V, E)$ with n vertices and two integers $k \leq \frac{n}{2}$ and $\ell \leq n$, a certificate could be a set $S \subset V$ of size k. We could then check, in polynomial time, there are k vertices in S, and the vertex expansion $\Phi^V(S) = |N(S)|$ is less than or equal to ℓ. We prove the SMALL SET VERTEX EXPANSION problem is NP-hard by showing that MINIMUM k-UNION $\leq_P$ SMALL SET VERTEX EXPAN-SION. Given an instance $(U, \mathcal{S}, k, \ell)$ of MINIMUM k-UNION problem, we construct a graph H with vertex sets X and Y. The vertices in $X = \{s_1, s_2, \ldots, s_m\}$ correspond to sets in $\mathcal{S} = \{S_1, S_2, \ldots, S_m\}$; the vertices in $Y = \{u_1, u_2, \ldots, u_n\}$ are the elements in U. We make $s_j \in X$ adjacent to $u_i \in Y$ if and only if $u_i \in S_j$. Additionally, for each vertex u_i, we add a clique of size $n + 1$, K^i_{n+1} and we make u_i adjacent to each vertex in K^i_{n+1}.

We show that there is a collection of k sets $\left\{S_{i_1}, S_{i_2}, \ldots, S_{i_k}\right\} \subseteq \mathcal{S}$ such that $|\cup_{j=1}^{k} S_{i_j}| \leq \ell$, for MINIMUM k-UNION problem if and only if there is a set $S \subset V(H)$ of $k \leq \frac{|V(H)|}{2}$ vertices such that $|N_H(S)| \leq \ell$, for SMALL SET VERTEX EXPANSION problem. Suppose there is a collection of k sets $\left\{S_{i_1}, S_{i_2}, \ldots, S_{i_k}\right\} \subseteq \mathcal{S}$ such that $|\cup_{j=1}^{k} S_{i_j}| \leq \ell$. We choose the vertices $\{s_{i_1}, s_{i_2}, \ldots, s_{i_k}\} \subseteq X$ correspond to sets $S_{i_1}, S_{i_2}, \ldots, S_{i_k}$. As the size of the union of these k sets $S_{i_1}, S_{i_2}, \ldots, S_{i_k}$ is less or equal to ℓ, the vertex expansion of $\{s_{i_1}, s_{i_2}, \ldots, s_{i_k}\}$ is also at most ℓ. If $k > \frac{|V(H)|}{2}$, then S is any size $\frac{|V(H)|}{2}$ subset of $\{s_{i_1}, s_{i_2}, \ldots, s_{i_k}\}$ and it has vertex expansion at most ℓ.

Conversely, suppose there is a subset $S \subseteq V(H)$ of k vertices such that $\Phi^V(S) \leq \ell$ where $\ell \leq n$. Note that S cannot contain any vertex from Y as each vertex in Y has at least $n + 1$ neighbours in H. Similarly S cannot contain any vertex from K^i_{n+1}, as each vertex in K^i_{n+1} has at least $n + 1$ neighbours in H. Thus $S \subseteq X$ and let $S = \{s_{j_1}, s_{j_2}, \ldots, s_{j_k}\}$. We consider the k sets $S_{j_1}, S_{j_2}, \ldots, S_{j_k}$ correspond to these k vertices. As $\Phi^V(S) = |N(S)| \leq \ell$, we have $|\cup_{i=1}^{k} S_{j_i}| \leq \ell$. This completes the proof.

4   W[1]-Hardness Parameterized by k

The input to the decision version of SSVE is a graph G with two integers $k \leq \frac{n}{2}$ and $\ell \leq n$, and (G, k, ℓ) is a yes-instance if G has a set S of k vertices such that the vertex expansion $\Phi^V(S) = |N(S)| \leq \ell$. In this section we show that SSVE is $W[1]$-hard when parameterized by k, via a reduction from CLIQUE.

Theorem 2. *The* SMALL SET VERTEX EXPANSION *problem is* $W[1]$*-hard when parameterized by* k.

Proof. Let (G, k) be an instance of CLIQUE. We construct an instance $(G', \frac{k(k-1)}{2}, k)$ of SMALL SET VERTEX EXPANSION problem as follows. We construct a graph G' with vertex sets X and Y, where $X = V(G) = \{v_1, v_2, \ldots, v_n\}$ and $Y = E(G) = \{e_1, e_2, \ldots, e_m\}$, the edge set of G. We make v_i adjacent to e_j if and only if v_i is an endpoint of e_j. We further add a set $P = \{p_1, p_2, \ldots, p_{k^2}\}$ of k^2 vertices; the vertices in P are adjacent to every element of X and all vertices in P are pairwise adjacent.

We claim that there is a set S of $\frac{k(k-1)}{2}$ vertices in G' with vertex expansion $\Phi^V(S) = |N(S)| \leq k$ if and only if G contains a clique on k vertices. Suppose first that G contains a clique on k vertices $\{v_1, v_2, \ldots, v_k\}$; we set S to be the set of edges belonging to this clique, and notice that in G all endpoints of edges in S belong to the set $\{v_1, v_2, \ldots, v_k\}$. Thus the vertex expansion of S in G' is exactly $\{v_1, v_2, \ldots, v_k\}$ and $\Phi^V(S) = |N(S)| = k$, so we have a yes-instance for $(G', \frac{k(k-1)}{2}, k)$.

Conversely, suppose that G' contains a set S of $\frac{k(k-1)}{2}$ vertices such that $\Phi^V(S) = |N(S)| \leq k$. As $d(v) \geq k+1$ for every vertex $v \in X \cup P$, we cannot include any vertex of X or P in the set S. So we conclude that $S \subseteq Y$ is a set of edges in G. All edges in S belong to the subgraph of G induced by $N(S)$, which by assumption has at most k vertices. Since $|S| = \frac{k(k-1)}{2}$, this is only possible if $|N(S)| = k$ and $N(S)$ in fact induces a clique in G, as required.

5   FPT Algorithm Parameterized by Neighbourhood Diversity

In this section, we present an FPT algorithm for the SMALL SET VERTEX EXPANSION problem parameterized by neighbourhood diversity. We say two vertices u and v have the same type if and only if $N(u) \setminus \{v\} = N(v) \setminus \{u\}$. The relation of having the same type is an equivalence relation. The idea of neighbourhood diversity is based on this type structure.

Definition 7. [10] *The neighbourhood diversity of a graph* $G = (V, E)$*, denoted by* $\mathbf{nd}(G)$*, is the least integer* w *for which we can partition the set* V *of vertices into* w *classes, such that all vertices in each class have the same type.*

If neighbourhood diversity of a graph is bounded by an integer w, then there exists a partition $\{C_1, C_2, \ldots, C_w\}$ of $V(G)$ into w type classes. It is known that such a minimum partition can be found in linear time using fast modular decomposition algorithms [15]. Notice that each type class could either be a clique or an independent set by definition. For algorithmic purpose it is often useful to consider a *type graph* H of graph G, where each vertex of H is a type class in G, and two vertices C_i and C_j are adjacent iff there is complete bipartite clique between these type classes in G. It is not difficult to see that there will be either a complete bipartite clique or no edges between any two type classes. The key property of graphs of bounded neighbourhood diversity is that their type graphs have bounded size. In this section, we prove the following theorem:

Theorem 3. *The* SMALL SET VERTEX EXPANSION *problem is fixed-parameter tractable when parameterized by the neighbourhood diversity.*

Given a graph $G = (V, E)$ with neighbourhood diversity $\mathbf{nd}(G) \leq w$, we first find a partition of the vertices into at most w type classes $\{C_1, \ldots, C_w\}$. Next we guess a set of type classes C_i for which $C_i \cap S \neq \emptyset$, where S is a set with k vertices such that the vertex expansion $\Phi^V(S) = |N(S)|$ is minimum. Let $\mathcal{P} \subseteq \{C_1, \ldots, C_w\}$ be a collection of type classes for which $C_i \cap S \neq \emptyset$. There are at most 2^w candidates for $\mathcal{P}$. Finally we reduce the problem of finding a set S that minimizes the vertex expansion $\Phi^V(S)$ to 2^w integer linear programming (ILP) optimizations with at most w variables in each ILP optimization. Since ILP optimization is fixed parameter tractable when parameterized by the number of variables [6], we conclude that our problem is fixed parameter tractable when parameterized by the neighbourhood diversity w.

ILP Formulation: For each C_i, we associate a variable x_i that indicates $|S \cap C_i| = x_i$. As the vertices in C_i have the same neighbourhood, the variables x_i determine S uniquely, up to isomorphism. We define

$$r(C_i) = \begin{cases} 1 & \text{if } C_i \text{ is adjacent to some } C_j \in \mathcal{P}; i \neq j \\ 0 & \text{otherwise} \end{cases}$$

Let $\mathcal{C}$ be a subset of $\mathcal{P}$ consisting of all type classes which are cliques; $\mathcal{I} = \mathcal{P} \setminus \mathcal{C}$ and $\mathcal{R} = \{C_1, \ldots, C_w\} \setminus \mathcal{P}$. Given a $\mathcal{P} \subseteq \{C_1, \ldots, C_w\}$, our goal is to minimize

$$\Phi^V(S) = |N(S)| = \sum_{C_i \in \mathcal{R}} r(C_i)|C_i| + \sum_{C_i \in \mathcal{C}} (|C_i| - x_i) + \sum_{C_i \in \mathcal{I}} r(C_i)(|C_i| - x_i) \quad (1)$$

under the condition $x_i \in \{1, \ldots, |C_i|\}$ for all i : $C_i \in \mathcal{P}$ and $x_i = 0$ for all i : $C_i \in \mathcal{R}$ and the additional conditions described below. Note that if $C_i \in \mathcal{R}$ and it is adjacent to some type class in $\mathcal{P}$, then C_i is contained in $N(S)$; if $C_i \in \mathcal{C}$ then $|C_i| - x_i$ vertices of C_i are in $N(S)$; finally if $C_i \in \mathcal{I}$ and it is adjacent to some type class in $\mathcal{P}$, then $|C_i| - x_i$ vertices of C_i are in $N(S)$. It is easy to see that minimizing the expansion $\Phi^V(S)$ in Eq. 1 is equivalent to maximizing $\sum_{C_i \in \mathcal{C}} x_i + \sum_{C_i \in \mathcal{I}} r(C_i)x_i$. Given a $\mathcal{P} \subseteq \{C_1, \ldots, C_w\}$, we present ILP formulation of SSVE problem as follows:

$$\text{Maximize} \sum_{C_i \in \mathcal{C}} x_i + \sum_{C_i \in \mathcal{I}} r(C_i) x_i$$

Subject to

$$\sum x_i = k$$

$$x_i \in \{1, \dots, |C_i|\} \text{ for all } i \ : \ C_i \in \mathcal{C} \cup \mathcal{I}$$

Solving the ILP: Lenstra [11] showed that the feasibility version of p-ILP is FPT with running time doubly exponential in p, where p is the number of variables. Later, Kannan [7] designed an algorithm for p-ILP running in time $p^{O(p)}$. In our algorithm, we need the optimization version of p-ILP rather than the feasibility version. We state the minimization version of p-ILP as presented by Fellows et al. [6].

p-VARIABLE INTEGER LINEAR PROGRAMMING OPTIMIZATION (p-OPT-ILP): Let matrices $A \in Z^{m \times p}$, $b \in Z^{p \times 1}$ and $c \in Z^{1 \times p}$ be given. We want to find a vector $x \in Z^{p \times 1}$ that minimizes the objective function $c \cdot x$ and satisfies the m inequalities, that is, $A \cdot x \geq b$. The number of variables p is the parameter. Then they showed the following:

Lemma 1. [6] p-OPT-ILP can be solved using $O(p^{2.5p+o(p)} \cdot L \cdot log(MN))$ arithmetic operations and space polynomial in L. Here L is the number of bits in the input, N is the maximum absolute value any variable can take, and M is an upper bound on the absolute value of the minimum taken by the objective function.

In the formulation for SSVE problem, we have at most w variables. The value of objective function is bounded by n and the value of any variable in the integer linear programming is also bounded by n. The constraints can be represented using $O(w^2 \log n)$ bits. Lemma 1 implies that we can solve the problem with the guess $\mathcal{P}$ in FPT time. There are at most 2^w choices for $\mathcal{P}$, and the ILP formula for a guess can be solved in FPT time. Thus Theorem 3 holds.

6   FPT Algorithm Parameterized by Treewidth

This section presents an FPT algorithm using dynamic programming for the SMALL SET VERTEX EXPANSION problem parameterized by treewidth. Given a graph $G = (V, E)$, an integer $k \leq \frac{n}{2}$ and its nice tree decomposition $(T, X_t : t \in V(T))$ of width at most $\mathtt{tw}$, subproblems will be defined on $G_t = (V_t, E_t)$ where V_t is the union of all bags present in subtree of T rooted at t, including X_t and E_t is the set of edges e introduced in the subtree rooted at t. We define a colour function $f : X_t \mapsto \{0, 1, \hat{1}\}$ that assigns three different colours to the vertices of X_t. The meanings of three different colours are given below:

1 (black vertices): all black vertices are contained in set S whose vertex expansion $\Phi^V(S)$ we wish to calculate in G_t.

0 (white vertices): white vertices are adjacent to black vertices, these vertices are in the expansion $N(S)$ in G_t.

$\hat{1}$ (gray vertices): gray vertices are neither in S nor in $N(S)$.

Now we introduce some notations. Let $X \subseteq V$ and consider a colouring $f : X \mapsto \{1, 0, \hat{1}\}$. For $\alpha \in \{1, 0, \hat{1}\}$ and $v \in V(G)$ a new colouring $f_{v \mapsto \alpha} : X \cup \{v\} \mapsto \{1, 0, \hat{1}\}$ is defined as follows:

$$f_{v \mapsto \alpha}(x) = \begin{cases} f(x) & \text{when } x \neq v \\ \alpha & \text{when } x = v \end{cases}$$

Let f be a colouring of X, then the notation $f_{|Y}$ is used to denote the restriction of f to Y, where $Y \subseteq X$.

For a colouring f of X_t and an integer i, a set $S \subseteq V_t$ is said to be compatible for tuple (t, f, i) if

1. $|S| = i$,
2. $S \cap X_t = f^{-1}\{1\}$ which is the set of vertices of X_t coloured black, and
3. $N(S) \cap X_t = f^{-1}\{0\}$, which is the set of vertices of X_t coloured white.

We call a set S a *minimum compatible set* for (t, f, i) if its vertex expansion $\Phi^V(S) = |N_{V_t}(S)|$ is minimum. We denote by $c[t, f, i]$ the minimum vertex expansion for (t, f, i), that is, $c[t, f, i]$ equals to $|N_{V_t}(S)|$, where S is a minimum compatible set for (t, f, i). If no such S exists, then we put $c[t, f, i] = \infty$ also $c[t, f, i < 0] = \infty$. Since each vertex in X_t can be coloured with 3 colours $(1, 0, \hat{1})$, the number of possible colouring f of X_t is $3^{|X_t|}$ and for each colouring f we vary i from 0 to k. The smallest value of vertex expansion $\Phi^V(S) = |N(S)|$ for a set S with k nodes will be $c[r, \phi, k]$, where r is the root node of tree decomposition of G as $G = G_r$ and $X_r = \emptyset$. We only show that $\Phi^V(S)$ can be computed in the claimed running time in Theorem 4. Corresponding set S can be easily computed in the same running time by remembering a corresponding set for each tuple (t, f, i) in the dynamic programming above. Now we present the recursive formulae for the values of $c[t, f, i]$.

Leaf Node: If t is a leaf node, then the corresponding bag X_t is empty. Hence the colour function f on X_t is an empty colouring; the number i of vertices coloured black cannot be greater than zero. Thus we have

$$c[t, \emptyset, i] = \begin{cases} 0 & \text{if } i = 0 \\ \infty & \text{otherwise} \end{cases}$$

Introduce Node: Suppose t is an introduce node with child t' such that $X_t = X_{t'} \cup \{v\}$ for some $v \notin X_{t'}$. The introduce node introduces the vertex v but does not introduce the edges incident to v in G_t. So when v is introduced by node t it is an isolated vertex in G_t. Vertex v cannot be coloured white 0; as it is isolated and it cannot be neighbour of any black vertex. Hence if $f(v) = 0$, then $c[t, f, i] = \infty$. When $f(v) = 1$, v is contained in S. As v is an isolated vertex, it does not contribute towards the size of $N_{V_t}(S)$, hence $c[t, f, i] = c[t', f_{|_{X_{t'}}}, i-1]$. When $f(v) = \hat{1}$, v does not contribute towards the size of $N_{V_t}(S)$. Here the sets compatible for $(t', f_{|_{X_{t'}}}, i)$ are also compatible for (t, f, i). So, $c[t, f, i] = c[t', f_{|_{X_{t'}}}, i]$. Combining all the cases together, we get

$$c[t, f, i] = \begin{cases} \infty & \text{if } f(v) = 0 \\ c[t', f_{|_{X_{t'}}}, i-1] & \text{if } f(v) = 1 \\ c[t', f_{|_{X_{t'}}}, i] & \text{if } f(v) = \hat{1} \end{cases}$$

Introduce Edge Node: Let t be an introduce edge node that introduces the edge (u, v), let t' be the child of t. Thus $X_t = X_{t'}$; the edge (u, v) is not there in $G_{t'}$, but it is there in G_t. Let f be a colouring of X_t. We consider the following cases:

- Suppose $f(u) = 1$ and $f(v) = \hat{1}$. This means $u \in S$ and v is non-adjacent to black vertices in G_t. But u and v are adjacent in G_t. Thus $c[t, f, i] = \infty$. The same conclusion can be drawn when v is coloured black and u is coloured gray.
- Suppose $f(u) = 1$ and $f(v) = 0$. This means $u \in S$ and $v \in N(S)$ in G_t. In order to get a solution for (t, f, i), we consider two cases.
  Case 1: While considering precomputed solution for t' we can relax the colour of v from white to gray. Then the minimum vertex expansion for (t, f, i) is one more than the minimum vertex expansion for $(t', f_{v \to \hat{1}}, i)$, that is, $c[t, f, i] = 1 + c[t', f_{v \to \hat{1}}, i]$.
  Case 2: While considering precomputed solution for t' we keep the colour of v be white. Then the minimum vertex expansion for (t, f, i) is equal to the minimum vertex expansion for (t', f, i), that is, $c[t, f, i] = c[t', f, i]$.
  Combining above two cases we get

$$c[t, f, i] = \min\left\{ c[t', f, i], 1 + c[t', f_{v \to \hat{1}}, i] \right\}$$

  The same conclusion can be drawn when v is coloured black and u is coloured white.
- Other colour combinations of u and v do not affect the size of $N(S)$ or do not contradict the definition of campatability. So the compatible sets for (t', f, i) are also compatible for (t, f, i) and hence $c[t, f, i] = c[t', f, i]$.

Combining all the cases together, we get

$$c[t, f, i] = \begin{cases} \infty & \text{if } (f(u), f(v)) = (\hat{1}, 1) \\ \infty & \text{if } (f(u), f(v)) = (1, \hat{1}) \\ \min\{c[t', f, i], 1 + c[t', f_{v \to \hat{1}}, i]\} & \text{if } (f(u), f(v)) = (1, 0) \\ \min\{c[t', f, i], 1 + c[t', f_{u \to \hat{1}}, i]\} & \text{if } (f(u), f(v)) = (0, 1) \\ c[t', f, i] & \text{otherwise} \end{cases}$$

Forget Node: Let t be a forget node with the child t' such that $X_t = X_{t'} \setminus \{w\}$ for some vertex $w \in X_{t'}$. Here the bag X_t forgets the vertex w. At this stage we decides the final colour of the vertex w. We observe that $G_{t'} = G_t$. The compatible sets for $(t', f_{w \mapsto 1}, i)$, $(t', f_{w \mapsto 0}, i)$, $(t', f_{w \mapsto \hat{1}}, i)$ are also compatible for (t, f, i). On the other hand compatible sets for (t, f, i) are also compatible for $(t', f_{w \mapsto 1}, i)$ if $w \in S$, for $(t', f_{w \mapsto 0}, i)$ if $w \in N(S)$ or for $(t', f_{w \mapsto \hat{1}}, i)$ if $w \notin N[S]$. Hence

$$c[t, f, i] = \min \left\{ c[t', f_{w \mapsto 1}, i], c[t', f_{w \mapsto 0}, i], c[t', f_{w \mapsto \hat{1}}, i] \right\}$$

Join Node: Let t be a join node with children t_1 and t_2, such that $X_t = X_{t_1} = X_{t_2}$. Let f be a colouring of X_t. We say that colouring f_1 of X_{t_1} and f_2 of X_{t_2} are consistent for colouring f of X_t, if the following conditions are true for each $v \in X_t$:

1. $f(v) = 1$ if and only if $f_1(v) = f_2(v) = 1$,
2. $f(v) = \hat{1}$ if and only if $f_1(v) = f_2(v) = \hat{1}$,
3. $f(v) = 0$ if and only if $(f_1(v), f_2(v)) \in \{(0, \hat{1}), (\hat{1}, 0), (0, 0)\}$.

Let f be a colouring of X_t; f_1 and f_2 be two colouring of X_{t_1} and X_{t_2} respectively consistent with f. Suppose S_1 is a compatible set for (t_1, f_1, i_1) and S_2 is a compatible set for (t_2, f_2, i_2), where $|S_1| = i_1$ and $|S_2| = i_2$. Set $S = S_1 \cup S_2$, clearly $|S| = |S_1| + |S_2| - |f^{-1}\{1\}|$. It is easy to see that S is a compatible set for (t, f, i), where $i = i_1 + i_2 - |f^{-1}\{1\}|$. According to Condition 3 in the definition of consistent function, each $v \in X_t$ that is white in f, we make it white either in f_1, f_2 or in both f_1 and f_2. Consequently, we have the following recursive formula:

$$c[t, f, i] = \min_{f_1, f_2} \left\{ \min_{i_1, i_2 \,:\, i = i_1 + i_2 - |f^{-1}\{1\}|} \left\{ c[t_1, f_1, i_1] + c[t_2, f_2, i_2] - \alpha_{f_1, f_2} \right\} \right\},$$

where $\alpha_{f_1, f_2} = |\{v \in X_t \mid f_1(v) = f_2(v) = 0\}|$.

We now analyse the running time of the algorithm. We compute all entries $c[t, f, i]$ in a bottom-up manner. Clearly, the time needed to process each leaf node, introduce vertex node, introduce edge node or forget node is $3^{tw+1} \cdot k^{O(1)}$ assuming that the entries for the children of t are already computed. The computation of $c[t, f, i]$ for join node takes more time and it can be done as follows. If a pair (f_1, f_2) is consistent with f, then for every $v \in X_t$, we have $(f(v), f_1(v), f_1(v)) \in \{(1, 1, 1), (\hat{1}, \hat{1}, \hat{1}), (0, 0, 0), (0, 0, \hat{1}), (0, \hat{1}, 0)\}$. Hence there are exactly $5^{|X_t|}$ triples of colouring (f, f_1, f_2) such that f_1 and f_2 are consistent with f, since for every vertex v, we have 5 possibilities for $(f(v), f_1(v), f_2(v))$. In order to compute $c(t, f, i)$, we iterate through all consistent pairs (f_1, f_2); then for each considered triple (f, f_1, f_2) we vary i_1 and i_2 from 0 to k such that $i = i_1 + i_2 - |f^{-1}\{1\}|$. As $|X_t| \leq tw + 1$, the time needed to process each join node is $5^{tw+1} k^{O(1)}$. Since we assume that the number of nodes in a nice tree decomposition is $O(n \cdot tw)$, we have the following theorem.

Theorem 4. *Given an n-vertex graph G and its nice tree decomposition of width at most* tw, *the* SMALL SET VERTEX EXPANSION *problem can be solved in* $O(5^{tw}n)$ *time.*

7 Conclusion

In this work we proved that the SMALL SET VERTEX EXPANSION problem is W[1]-hard when parameterized by k, the number of vertices in S; it is FPT when parameterized neighbourhood diversity; and the problem is FPT when parameterized by treewidth of the input graph. The parameterized complexity of the SMALL SET VERTEX EXPANSION problem remains unsettle when parameterized by $k + \ell$, and when parameterized by other important structural graph parameters like clique-width, modular width and treedepth.

Acknowledgement. We are grateful to Dr. Kitty Meeks, University of Glasgow, for useful discussions and her comments on the proof of Theorem 2.

References

1. Arora, S., Ge, R.: New tools for graph coloring. In: Goldberg, L.A., Jansen, K., Ravi, R., Rolim, J.D.P. (eds.) APPROX/RANDOM -2011. LNCS, vol. 6845, pp. 1–12. Springer, Heidelberg (2011). https://doi.org/10.1007/978-3-642-22935-0_1
2. Chlamtác, E., Dinitz, M., Makarychev, Y.: Minimizing the union: Tight approximations for small set bipartite vertex expansion. ArXiv, abs/1611.07866 (2017)
3. Chlamtáč, E., Dinitz, M., Konrad, C., Kortsarz, G., Rabanca, G.: The densest k-subhypergraph problem. SIAM J. Discrete Math. **32**(2), 1458–1477 (2018)
4. Cygan, M., et al.: Parameterized Algorithms. Springer, Cham (2015). https://doi.org/10.1007/978-3-319-21275-3_15
5. Feige, U.: A threshold of ln n for approximating set cover. J. ACM **45**(4), 634–652 (1998)
6. Fellows, M.R., Lokshtanov, D., Misra, N., Rosamond, F.A., Saurabh, S.: Graph layout problems parameterized by vertex cover. In: Hong, S.-H., Nagamochi, H., Fukunaga, T. (eds.) ISAAC 2008. LNCS, vol. 5369, pp. 294–305. Springer, Heidelberg (2008). https://doi.org/10.1007/978-3-540-92182-0_28
7. Kannan, R.: Minkowski's convex body theorem and integer programming. Math. Oper. Res. **12**(3), 415–440 (1987)
8. Khot, S.A., Vishnoi, N.K.: The unique games conjecture, integrality gap for cut problems and embeddability of negative type metrics into l/sub 1/. In: 46th Annual IEEE Symposium on Foundations of Computer Science (FOCS 2005), pp. 53–62 (2005)
9. Kloks, T. (ed.): Treewidth. LNCS, vol. 842. Springer, Heidelberg (1994). https://doi.org/10.1007/BFb0045375
10. Lampis, M.: Algorithmic meta-theorems for restrictions of treewidth. Algorithmica **64**, 19–37 (2012)
11. Lenstra, H.W.: Integer programming with a fixed number of variables. Math. Oper. Res. **8**(4), 538–548 (1983)
12. Louis, A., Makarychev, Y.: Approximation algorithms for hypergraph small-set expansion and small-set vertex expansion. Theory Comput. **12**(17), 1–25 (2016)

13. Raghavendra, P., Steurer, D.: Graph expansion and the unique games conjecture. In: Proceedings of the Forty-Second ACM Symposium on Theory of Computing, STOC 2010, pp. 755–764. Association for Computing Machinery, New York (2010)
14. Robertson, N., Seymour, P.: Graph minors. iii. planar tree-width. J. Comb. Theory Ser. B **36**(1), 49–64 (1984)
15. Tedder, M., Corneil, D., Habib, M., Paul, C.: Simpler linear-time modular decomposition via recursive factorizing permutations. In: Aceto, L., Damgård, I., Goldberg, L.A., Halldórsson, M.M., Ingólfsdóttir, A., Walukiewicz, I. (eds.) ICALP 2008. LNCS, vol. 5125, pp. 634–645. Springer, Heidelberg (2008). https://doi.org/10.1007/978-3-540-70575-8_52
16. Vinterbo, S.A.: A Note on the Hardness of the k-Ambiguity Problem. Technical report, Harvard Medical School, Boston, MA, USA, 06 2002

Complexity and Logic

On Unit Read-Once Resolutions and Copy Complexity

P. Wojciechowski and K. Subramani[✉]

LDCSEE, West Virginia University, Morgantown, WV, USA
pwojciec@mix.wvu.edu, k.subramani@mail.wvu.edu

Abstract. In this paper, we discuss the copy complexity of unit resolution with respect to Horn formulas. A Horn formula is a boolean formula in conjunctive normal form (CNF) with at most one positive literal per clause. Horn formulas find applications in a number of domains such as program verification and logic programming. Resolution as a proof system for boolean formulas is both sound and complete. However, resolution is considered an inefficient proof system when compared to other stronger proof systems for boolean formulas. Despite this inefficiency, the simple nature of resolution makes it an integral part of several theorem provers. *Unit resolution* is a restricted form of resolution in which each resolution step needs to use a clause with only one literal (unit literal clause). While not complete for general CNF formulas, unit resolution is complete for Horn formulas. A *read-once resolution* (ROR) refutation is a refutation in which each clause (input or derived) may be used at most once in the derivation of a refutation. As with unit resolution, ROR refutation is incomplete in general and complete for Horn clauses. This paper focuses on a combination of unit resolution and read-once resolution called *Unit read-once resolution* (UROR). UROR is **incomplete** for Horn clauses. In this paper, we study the *copy complexity* problem in Horn formulas under UROR. Briefly, the copy complexity of a formula under UROR is the smallest number k such that replicating each clause k times guarantees the existence of a UROR refutation. This paper focuses on two problems related to the copy complexity of unit resolution. We first relate the copy complexity of unit resolution for Horn formulas to the copy complexity of the addition rule in the corresponding Horn constraint system. We also examine a form of copy complexity where we permit replication of derived clauses, in addition to the input clauses. Finally, we provide a polynomial time algorithm for the problem of checking if a 2-CNF formula has a UROR refutation.

1 Introduction

In this paper, we discuss the copy complexity of unit resolution with respect to Horn formulas. A Horn formula is a boolean formula in conjunctive normal

---

K. Subramani—This research was supported in part by the Air-Force Office of Scientific Research through Grant FA9550-19-1-0177 and in part by the Air-Force Research Laboratory, Rome through Contract FA8750-17-S-7007.

W. Wu and Z. Zhang (Eds.): COCOA 2020, LNCS 12577, pp. 273–288, 2020.
https://doi.org/10.1007/978-3-030-64843-5_19

form (CNF) with at most one positive literal per clause [2]. Unit resolution is a restricted form of resolution in which each resolution step needs to use a clause with only one literal. While not complete for general CNF, unit resolution is complete for Horn formulas. We focus on copy complexity as it relates to Horn formulas and also present a polynomial time algorithm for the unit read-once resolution (UROR) problem in 2-CNF formulas.

The primary focus of this paper is copy complexity. It is closely related to the concept of clause duplication described in [5]. A CNF formula Φ has copy complexity k if the formula Φ' formed by taking k copies of each clause $\phi_i \in \Phi$ has a read-once refutation. In this paper, we focus on Horn formulas. Note that Horn formulas always have read-once resolution refutations [2], but don't always have unit read-once resolution refutations. Thus, we focus on copy complexity with respect to unit resolution.

Additionally, we examine a variant of copy complexity which allows for copies of derived clauses in addition to copies of clauses from the original CNF formula. We also relate the copy complexity of a Horn formula to the copy complexity of the corresponding linear system. Finally, we show that the problem of determining if a 2-CNF formula has a unit read-once resolution refutation can be solved in polynomial time. This is in contrast to the result from [8] which showed that the problem of determining if a 2-CNF formula has a read-once resolution (not necessarily a unit read-once resolution) refutation is **NP-complete**.

2   Statement of Problems

In this section, we briefly discuss the terms used in this paper. We assume that the reader is familiar with elementary propositional logic.

Definition 1. *A **literal** is a variable x or its complement $\neg x$. x is called a positive literal and $\neg x$ is called a negative literal.*

Definition 2. *A **CNF clause** is a disjunction of literals. The empty clause, which is always false, is denoted as $\sqcup$.*

Definition 3. *A **CNF formula** is a conjunction of CNF clauses.*

Throughout this paper, we use m to denote the number of clauses in a CNF formula Φ and n to denote the number of variables. Note that an unsatisfiable CNF formula Φ is said to be minimal unsatisfiable if removing any clause from Φ makes Φ satisfiable.

Definition 4. *A k-**CNF clause** is a CNF clause with at most k literals.*

Definition 5. *A **Horn** clause is a CNF clause which contains at most one positive literal.*

A clause that is both a k-CNF clause and a Horn clause is called a k-Horn clause.

For a single resolution step with parent clauses $(\alpha \lor x)$ and $(\neg x \lor \beta)$ with resolvent $(\alpha \lor \beta)$, we write

$$(\alpha \lor x), (\neg x \lor \beta) \vdash_{RES}^{1} (\alpha \lor \beta).$$

The variable x is called the matching or resolution variable. If for initial clauses $\alpha_1, \ldots, \alpha_n$, a clause π can be generated by a sequence of resolution steps we write

$$\alpha_1, \ldots, \alpha_n \vdash_{RES} \pi.$$

If a resolution step involves a unit clause, a clause of the form (x) or $(\neg x)$, then it is called a unit resolution step. If a resolution refutation consists of only unit resolution steps, then it is called a unit resolution refutation.

We now formally define the types of resolution refutation discussed in this paper.

Definition 6. *A* **read-once** *resolution refutation is a refutation in which each clause, π, can be used in only one resolution step. This applies to clauses present in the original formula and those derived as a result of previous resolution steps.*

In a read-once refutation, a clause can be reused if it can be re-derived from a set of unused input clauses.

More formally, a resolution derivation $\Phi \vdash_{RES} \pi$ is a read-once resolution derivation, if for all resolution steps $\pi_1 \land \pi_2 \vdash_{RES}^{1} \pi$, we delete one instance of the clauses π_1 and π_2 from, and add a copy of the resolvent π to, the current multi-set of clauses. In other words, if U is the current multi-set of clauses, we replace U with $(U \setminus \{\pi_1, \pi_2\}) \cup \{\pi\}$.

It is important to note Read-Once resolution is an **incomplete** refutation procedure. This means that there exist infeasible CNF formulas that do not have read-once refutations.

We can similarly define unit read-once resolution.

Definition 7. *A* **unit read-once** *resolution refutation is a unit resolution refutation in which each clause, π, can be used in only one unit resolution step. This applies to clauses present in the original formula and those derived as a result of previous unit resolution steps.*

This lets us define the concept of copy complexity with respect to unit read-once resolution.

Definition 8. *A CNF formula Φ has* **copy complexity** *at most k, with respect to unit resolution, if there exists a multi-set of CNF clauses, Φ' such that:*

1. *Every clause in Φ appears at most k times in Φ'.*
2. *Every clause in Φ' appears in Φ.*
3. *Φ' has a unit read-once resolution refutation.*

In this paper, we also deal with the copy complexity of linear constraint systems that correspond to CNF formulas.

From a CNF formula Φ, we create the system of constraints $S(\Phi)$ as described in [4]. $S(\Phi)$ is constructed as follows:

1. For each boolean variable $x_j \in \Phi$ create the variable x_j.
2. For each clause $\phi_i = (x_1 \vee \ldots \vee x_p \vee \neg y_1 \vee \ldots \vee \neg y_r)$, create the constraint $\sum_{j=1}^{p} x_j - \sum_{j=1}^{r} y_j \geq 1 - r$. Let $L(\phi_i)$ denote the left-hand side of this constraint and let $R(\phi_i)$ denote the right hand side of the constraint.

For a set of clauses $\{\phi_1, \ldots, \phi_m\}$ and non-negative integers k_i we write $\sum_{i=1}^{m} k_i \cdot L(\phi_i) \geq \sum_{i=1}^{m} k_i \cdot R(\phi_i) \equiv 0 \geq 1$, if the constraint obtained by summing the constraints $k_1 \cdot S(\phi_1)$ though $k_m \cdot S(\phi_m)$ results in the constraint $0 \geq 1$.

We can now define the concept of constraint-copy complexity.

Definition 9. *A CNF formula $\Phi = \{\phi_1, \ldots, \phi_m\}$, has **constraint-copy complexity** k, if there exist non-negative integers $k_1 \ldots k_m$ such that $k_i \leq k$ for $i = 1 \ldots m$ and*

$$\sum_{i=1}^{m} k_i \cdot L(\phi_i) \geq \sum_{i=1}^{m} k_i \cdot R(\phi_i) \equiv 0 \geq 1.$$

Note that, in Definition 9, a limit is placed upon the multiplier associated with each constraint. This limits the number of times each constraint can be used in a refutation. Thus, this definition effectively extends the notion of copy complexity to systems of linear constraints.

The problem of determining if a read-once resolution refutation or unit read-once resolution refutation exists is only interesting for unsatisfiable formulas. Thus, for all of the problems studied in this paper, we assume that the formula provided is unsatisfiable. For both the formula types considered, Horn [2] and 2-CNF [12], the problems of determining satisfiability are in **P**.

3   Motivation and Related Work

In this paper, we focus on the copy complexity of Horn formulas with respect to unit resolution. This is similar to the concept of clause duplication introduced in [5]. That paper examined read-once refutations of CNF formulas under two inference rules, resolution and duplication. The added inference rule removes a clause from the formula only two add two copies of that clause. [5] classified CNF formulas based on the number of times the duplication rule needed to be used in a read-once refutation. The class $\mathbf{R}(k)$ was the set of CNF formulas with a read-once refutation using the duplication rule at most k times. Note that this differs from the concept of copy complexity in the following ways:

1. The duplication rule can be applied to derived clauses as well as clauses from the original system. In this way, the concept of clause duplication is closer to derived copy complexity than it is to regular copy complexity.

2. The class $\mathbf{R}(k)$ is interested in the total number of clauses copied instead of the number of times any particular clause is used.

[5] showed that for general CNF formulas, the class $\mathbf{R}(0)$ is **NP-complete**. Note that $\mathbf{R}(0)$ is the set of CNF formulas with a read-once resolution refutation. The paper also showed that the class $\mathbf{R}(k)\backslash\mathbf{R}(k-1)$ is $\mathbf{D^P}$-**complete** for any positive integer k.

Resolution as a proof system is widely used in the study of proof complexity. This is due to the simple nature of resolution based proofs [1]. In [15], Tseitin showed that a restricted form of resolution had an exponential lower bound. This result was improved in [3]. In that paper, it was established that CNF formulas have exponentially long refutations even without the restrictions imposed by [15]. This was accomplished by showing that any resolution based refutation of the pigeonhole principle is necessarily exponentially long in the size of the input. Additional lower bounds on the length of resolution proofs are discussed in [1,13]. The issues of copy complexity and read-once refutation in Horn clauses have also been discussed in [9]; however, the results in this paper are tangential in emphasis and scope.

4   The UROR Problem for Horn Formulas

In this section, we explore the problem of finding unit read-once resolution refutations for Horn formulas. If we restrict ourselves to unit resolution refutations then Horn formulas are no longer guaranteed to have read-once refutations.

Observation 41. For a CNF formula Φ, a unit read-once resolution refutation of Φ has at most $(m-1)$ resolution steps, where m is the number of clauses in the formula.

Proof. Recall that a read-once resolution step is equivalent to removing the two parent clauses from the formula and adding the resolvent. Thus, each read-once resolution step effectively reduces the number of clauses in the formula by 1. Since Φ initially has m clauses, there can be at most $(m-1)$ such resolution steps. $\qquad\square$

It was shown in [10] that the UROR problem for Horn formulas is **NP-complete**.

We now provide an alternative proof that the UROR problem is **NP-complete** for Horn formulas. This is done by a reduction from the set packing problem. This problem is defined as follows:

Definition 10. *The* **set packing** *problem is the following: Given a set S of size n, m subsets $S_1,\ldots,S_m$ of S, and an integer k, does $\{S_1,\ldots,S_m\}$ contain k mutually disjoint sets?*

This problem is known to be **NP-complete** [6].

Theorem 1. *The UROR problem for Horn formulas is* **NP-complete***.*

Proof. The number of resolutions in a unit read-once resolution refutation is limited by the number of clauses in the Horn formula. Thus, a unit read-once resolution can also be verified in polynomial time. Consequently, the UROR problem is in **NP**.

Let us consider an instance of the set packing problem. We construct the Horn formula Φ as follows.

1. For each $x_i \in S$, create the boolean variable x_i and the clause (x_i).
2. For $j = 1 \ldots k$, create the boolean variable v_j.
3. For each subset S_l, $l = 1 \ldots m$ create the k clauses

$$(v_j \vee \bigvee_{x_i \in S_l} \neg x_i) \qquad j = 1 \ldots k.$$

4. Finally create clause $(\neg v_1 \vee \ldots \vee \neg v_k)$.

We now show that Φ has a unit read-once resolution refutation if and only if $\{S_1, \ldots, S_m\}$ contains k mutually disjoint sets.

Suppose that $\{S_1, \ldots, S_m\}$ does contain k mutually disjoint sets. Without loss of generality, assume that these are the sets $S_1, \ldots, S_k$.

Let us consider the sets of clauses

$$\Phi_j = \{(v_j \vee \bigvee_{x_i \in S_j} \neg x_i)\} \cup \{(x_i) \mid x_i \in S_j\} \qquad j = 1 \ldots k.$$

By the construction of Φ, $\Phi_j \subseteq \Phi$ for $j = 1 \ldots k$. Since the sets $S_1, \ldots, S_k$ are mutually disjoint, so are the sets $\Phi_1, \ldots, \Phi_k$.

It is easy to see that the clause (v_j) can be derived from the set Φ_j by unit read-once resolution. Since this holds for every $j = 1 \ldots k$ and since the sets $\Phi_1, \ldots, \Phi_k$ are mutually disjoint, the set of clauses $\{(v_1), \ldots, (v_k)\}$ can be derived from Φ by unit read-once resolution.

Together with the clause $(\neg v_1 \vee \ldots \vee \neg v_k)$, this set of clauses has a unit read-once derivation of the empty clause. Thus, Φ has a unit read-once resolution refutation.

Now suppose that Φ has a unit read-once resolution refutation **R**. Note that $\Phi/\{(\neg v_1 \vee \ldots \vee \neg v_k)\}$ can be satisfied by setting every variable x_i and v_j to **true**. Thus, **R** must use the clause $(\neg v_1 \vee \ldots \vee \neg v_k)$.

To eliminate the clause $(\neg v_1 \vee \ldots \vee \neg v_k)$, **R** must either derive the clauses $(v_1), \ldots, (v_k)$, or reduce $(\neg v_1 \vee \ldots \vee \neg v_k)$ to a unit clause and then resolve it with another clause. Without loss of generality, we can assume that this unit clause is $(\neg v_1)$. In either case, **R** must derive the clauses $(v_2), \ldots, (v_k)$.

Thus, we must derive the clauses $(v_2), \ldots, (v_k)$. Let us consider the clause (v_j), $2 \le j \le k$. By the construction of Φ, this clause must be derived from one of the clauses

$$(v_j \vee \bigvee_{x_i \in S_l} \neg x_i) \qquad l = 1 \ldots m.$$

To perform this derivation, we must use the set of clauses $\Psi_{l_j} = \{(x_i) \mid x_i \in S_{l_j}\}$ for some $l_j \leq m$.

Since the refutation is read-once, the sets Ψ_{l_j} for $j = 2 \ldots k$ are mutually disjoint. Thus, the sets S_{l_j} for $j = 2 \ldots k$ are also mutually disjoint.

If **R** derives the clause (v_1), then, by the construction of Φ, this clause must be derived from one of the clauses

$$(v_1 \vee \bigvee_{x_i \in S_l} \neg x_i) \qquad l = 1 \ldots m.$$

To perform this derivation, we must use the set of clauses $\Psi_{l_1} = \{(x_i) \mid x_i \in S_{l_1}\}$ for some $l_1 \leq m$.

If **R** reduces $(\neg v_1 \vee \ldots \vee \neg v_k)$ to the clause $(\neg v_1)$, then we must resolve $(\neg v_1)$ with one of the clauses

$$(v_1 \vee \bigvee_{x_i \in S_l} \neg x_i) \qquad l = 1 \ldots m.$$

This results in the clause $(\bigvee_{x_i \in S_{l_1}} \neg x_i)$ for some $l_1 \leq m$. To eliminate this clause, we must use the set of clauses $\Psi_{l_1} = \{(x_i) \mid x_i \in S_{l_1}\}$.

Since **R** is read-once, the set Ψ_{l_1} does not share any clauses with the sets $\Psi_{l_j}, j = 2 \ldots k$. Thus, the sets S_{l_j} for $j = 1 \ldots k$ are also mutually disjoint. This means that $\{S_1, \ldots, S_m\}$ contains k mutually disjoint sets.

Thus, Φ has a unit read-once resolution refutation if and only if $\{S_1, \ldots, S_m\}$ contains k mutually disjoint sets. As a result of this, the UROR problem for Horn formulas is **NP-complete**.   □

5   Copy Complexity and Horn Constraints

In this section, we relate the copy complexity of Horn formulas to the copy complexity of the corresponding linear system.

Theorem 2. *If Φ has copy complexity k with respect to unit resolution, then Φ has constraint-copy complexity k.*

Proof. Let $\Phi = \{\phi_1, \ldots, \phi_m\}$ be a CNF formula. Assume that Φ has a copy complexity of k with respect to unit resolution. Thus, we can construct the multi-set Φ' where each clause ϕ_i appears $k_i \leq k$ times for $i = 1 \ldots m$ and Φ' has a unit read-once resolution refutation. We will show that $\sum_{i=1}^{m} k_i \cdot L(\phi_i) \geq \sum_{i=1}^{m} k_i \cdot R(\phi_i) \equiv 0 \geq 1$.

Let $(x_i) \wedge (\neg x_i \vee \beta) \vdash_{\overline{RES}} \beta$ be a resolution step in the unit read-once resolution refutation of ϕ. Observe that (x_i) corresponds to the constraint $x_i \geq 1$ and $(\neg x_i \vee \beta)$ corresponds to the constraint $L(\beta) - x_i \geq R(\beta) - 1$. Summing these constraints results in $L(\beta) \geq R(\beta)$ which is the constraint that corresponds to β. Thus, each resolution step corresponds to the summation of the corresponding constraints.

Since the last resolution step is $(x_i) \wedge (\neg x_i) \vdash_{\overline{RES}}^{\frac{1}{}} \sqcup$, it follows that the last summation is summing $x_i \geq 1$ and $-x_i \geq 0$ to get $0 \geq 1$. Thus, $0 \geq 1$ is the result of the entire summation and $\sum_{i=1}^{m} k_i \cdot L(\phi_i) \geq \sum_{i=1}^{m} k_i \cdot R(\phi_i) \equiv 0 \geq 1$ as desired.   □

Theorem 3. *If a CNF formula Φ has constraint-copy complexity k, then Φ has copy complexity k with respect to unit resolution.*

Proof. We will show a stronger result: If $\sum_{i=1}^{m} k_i \cdot L(\phi_i) \geq \sum_{i=1}^{m} k_i \cdot R(\phi_i) \equiv 0 \geq 1$ for some coefficients k_i, $i = 1 \ldots m$, then Φ has a unit resolution refutation using k_i copies of the clause ϕ_i. The theorem follows immediately from this.

Let $K = \sum_{i=1}^{m} k_i$, we will obtain the desired result by induction on K.

If $K = 2$, then Φ must have two clauses with $k_1 = k_2 = 1$. Thus, $\Phi = (x) \wedge (\neg x)$. This formula obviously has a unit read-once resolution refutation.

Now assume that the desired result holds for all formulas Φ for which there exists a set of coefficients such that $K = h$.

Let Φ be a formula with coefficients k_i such that $K = h + 1$. Let $Pos(\phi_i)$ be the number of positive literals in clause ϕ_i and let $Neg(\phi_i)$ be the number of negative literals.

Since $R(\phi_i) = 1 - Neg(\phi_i)$, we have that

$$\sum_{i=1}^{m} k_i \cdot L(\phi_i) \geq \sum_{i=1}^{m} k_i \cdot (1 - Neg(\phi_i)) \equiv 0 \geq 1.$$

Since $\sum_{i=1}^{m} k_i \cdot L(\phi_i) = 0$, we must have that the total number of negative literals is equal to the total number of positive literals. This means that $\sum_{i=1}^{m} k_i \cdot Neg(\phi_i) = \sum_{i=1}^{m} k_i \cdot Pos(\phi_i)$. Thus, we have the following:

$$2 \cdot \left(\sum_{i=1}^{m} k_i \cdot (1 - Neg(\phi_i)) \right) = 2 \cdot \sum_{i=1}^{m} k_i - \sum_{i=1}^{m} k_i \cdot Neg(\phi_i) - \sum_{i=1}^{m} k_i \cdot Pos(\phi_i)$$

$$= 2 \sum_{i=1}^{m} k_i - \sum_{i=1}^{m} k_i \cdot (Neg(\phi_i) + Pos(\phi_i))$$

$$= 2 \sum_{i=1}^{m} k_i - \sum_{i=1}^{m} k_i \cdot |\phi_i| = \sum_{i=1}^{m} k_i \cdot (2 - |\phi_i|).$$

We want this sum to be 2. The only way for this sum to be positive is if there exists a clause such that $|\phi_i| = 1$. Thus, Φ must have a unit clause. Let (λ) be that unit clause. Thus we have that Φ is of the form:

$$\Phi = (\lambda), (\neg \lambda \vee \alpha_1), \ldots, (\neg \lambda \vee \alpha_r), (\lambda \vee \beta_{r+1}), \cdots, (\lambda \vee \beta_{r+s}), \pi_{r+s+1}, \ldots, \pi_m.$$

We can resolve (λ) and $(\neg \lambda \vee \alpha_1)$ to obtain α_1. If $\alpha_1 \notin \Phi$, this results in the formula Φ' where

$$\Phi' = (\lambda), (\neg \lambda \vee \alpha_1), \ldots, (\neg \lambda \vee \alpha_r), (\lambda \vee \beta_{r+1}), \cdots, (\lambda \vee \beta_{r+s}), \pi_{r+s+1}, \ldots, \pi_m, \alpha_1.$$

Let us consider the summation $\sum_{i=1}^{m} k_i \cdot L(\phi_i) \geq \sum_{i=1}^{m} k_i \cdot R(\phi_i) \equiv 0 \geq 1$. Since addition is commutative, we can assume without loss of generality that the first addition performed in this summation is

$$(L(\lambda) \geq R(\lambda)) + (L(\neg\lambda \vee \alpha_1) \geq R(\neg\lambda \vee \alpha_1)) \equiv (L(\alpha_1) \geq R(\alpha_1)).$$

Thus, in the summation $\sum_{i=1}^{m} k_i \cdot L(\phi_i) \geq \sum_{i=1}^{m} k_i \cdot R(\phi_i) \equiv 0 \geq 1$, we can replace one instance of the constraints $L(\lambda) \geq R(\lambda)$ and $L(\neg\lambda \vee \alpha_1) \geq R(\neg\lambda \vee \alpha_1)$ with the constraint $L(\alpha_1) \geq R(\alpha_1)$.

Thus, for each clause $\phi_i' \in \Phi'$ we can create the coefficient k_i' such that $\sum_{i=1}^{m+1} k_i' \cdot L(\phi_i') \geq \sum_{i=1}^{m+1} k_i' \cdot R(\phi_i') \equiv 0 \geq 1$. This set of coefficients has the following properties:

1. $k_0' = k_0 - 1$ since we have removed an instance of the constraint $L(\lambda) \geq R(\lambda)$ from the summation.
2. $k_1' = k_1 - 1$ since we have removed an instance of the constraint $L(\neg\lambda \vee \alpha_1) \geq R(\neg\lambda \vee \alpha_1)$ from the summation.
3. $k_{m+1}' = 1$ since we have added the new constraint $L(\alpha_1) \geq R(\alpha_1)$.
4. $k_i' = k_i$ for $2 \leq i \leq m$, since we have not modified the number of times any other constraint appears in the summation.

Thus, we have

$$K' = \sum_{i=1}^{m+1} k_i' = (k_0 - 1) + (k_1 - 1) + \sum_{i=1}^{m} k_i + 1 = \sum_{i=1}^{m} k_i - 1 = K - 1 = h.$$

By the inductive hypothesis, Φ' has a unit resolution refutation in which the clause ϕ_i' is used k_i' times. We can add the resolution step $(\lambda) \wedge (\neg\lambda \vee \alpha_1) \vdash_{RES}^{1} \alpha_1$ onto the beginning of this refutation. This results in a unit resolution refutation of Φ where the clause ϕ_i is used k_i times, as desired.

If $\alpha_1 \in \Phi$, then there exists a clause $\pi_j \in \Phi$ such that $\pi_j = \alpha_1$. We can assume without loss of generality that $\pi_m = \alpha_1$. Thus, the resolution step $(\lambda) \wedge (\neg\lambda \vee \alpha_1) \vdash_{RES}^{1} \alpha_1$ results in the formula $\Phi' = \Phi$. As before, for each clause $\phi_i' \in \Phi'$ we can create the coefficient k_i' such that $\sum_{i=1}^{m} k_i' \cdot L(\phi_i') \geq \sum_{i=1}^{m} k_i' \cdot R(\phi_i') \equiv 0 \geq 1$. This set of coefficients has the following properties:

1. $k_0' = k_0 - 1$ since we have removed an instance of the constraint $L(\lambda) \geq R(\lambda)$ from the summation.
2. $k_1' = k_1 - 1$ since we have removed an instance of the constraint $L(\neg\lambda \vee \alpha_1) \geq R(\neg\lambda \vee \alpha_1)$ from the summation.
3. $k_m' = k_m + 1$ since we have added an instance of the constraint $(L(\alpha_1) \geq R(\alpha_1)) \equiv (L(\pi_m) \geq R(\pi_m))$.
4. $k_i' = k_i$ for $2 \leq i \leq m - 1$, since we have not modified the number of times any other constraint appears in the summation.

Thus, we have

$$K' = \sum_{i=1}^{m} k_i' = (k_0 - 1) + (k_1 - 1) + \sum_{i=2}^{m-1} k_i + (k_m + 1) = \sum_{i=1}^{m} k_i - 1 = K - 1 = h.$$

By the inductive hypothesis, Φ' has a unit resolution refutation in which the clause ϕ'_i is used k'_i times. We can add the resolution step $(\lambda) \wedge (\neg\lambda \vee \alpha_1) \vdash_{RES}^{1} \alpha_1$ onto the beginning of this refutation. This results in a unit resolution refutation of Φ where the clause ϕ_i is used k_i times, as desired.   □

Theorems 2 and 3 imply the following corollaries.

Corollary 1. *A CNF formula Φ has a constraint-copy complexity of k, if and only if Φ has a copy complexity of k with respect to unit resolution.*

Corollary 2. *A CNF formula Φ has a constraint-copy complexity of 1, if and only if Φ has a unit read-once resolution refutation.*

We now show that for any fixed k, determining if a Horn formula has copy complexity 2^k with respect to unit resolution is **NP-complete**.

Theorem 4. *For any fixed k, determining if a Horn formula has copy complexity 2^k with respect to unit resolution is **NP-complete**.*

Proof. This problem is in **NP**, because k is fixed. We have only to guess coefficients less than or equal to 2^k and to compute the weighted sum of the inequalities of the corresponding constraint system.

The **NP-hardness** will be shown by induction on k.

For $k = 0$, the problem is exactly the problem of deciding whether a unit read-once resolution exists. This problem has been shown to be **NP-complete** for Horn formulas [10] (See Theorem 1).

Assume that the problem of determining if a Horn formula has copy complexity 2^k with respect to unit resolution is **NP-complete**.

Let $\Phi = \phi_1 \wedge \ldots \wedge \phi_m$ be a Horn formula. In polynomial time, we can construct a Horn formula Φ' as follows:

1. For each clause $\phi_i \in \Phi$, create the variables y_i, z_i, and a_i.
2. For each clause $\phi_i \in \Phi$, add the clauses (y_i), $(\neg y_i \vee z_i)$, $(\neg y_i \vee \neg z_i \vee a_i)$, and $(\neg a_i \vee \phi_i)$ to Φ'.

Thus, we have $\Phi' = \bigwedge_{1 \le i \le m} (y_i) \wedge (\neg y_i \vee z_i) \wedge (\neg y_i \vee \neg z_i \vee a_i) \wedge (\neg a_i \vee \phi_i)$.

By construction, Φ' is a Horn formula.

We now show that Φ has a copy complexity of 2^k if and only if Φ' has a copy complexity of 2^{k+1}.

Assume that Φ has a copy complexity of 2^k with respect to unit resolution. Then, by Corollary 1, there exist coefficients $k_1, \ldots, k_m \le 2^k$ such that:

$$\sum_{i=1}^{m} k_i \cdot L(\phi_i) \ge \sum_{i=1}^{m} k_i \cdot R(\phi_i) \equiv 0 \ge 1.$$

We construct a set of coefficients for Φ' as follows:

1. For each clause (y_i), create the coefficient $t_i = 2 \cdot k_i$.
2. For each clause $(\neg y_i \vee z_i)$, create the coefficient $l_i = k_i$.

3. For each clause $(\neg y_i \vee \neg z_i \vee a_i)$, create the coefficient $p_i = k_i$.
4. For each clause $(\neg a_i \vee \phi_i)$, create the coefficient $q_i = k_i$.

Then we obtain:

$$\sum_{i=1}^{m} [(2 \cdot k_i \cdot y_i) + k_i \cdot (-y_i + z_i) + k_i \cdot (-y_i - z_i + a_i) + k_i \cdot (-a_i + L(\phi_i))] = \sum_{i=1}^{m} k_i \cdot L(\phi_i) = 0$$

and

$$\sum_{i=1}^{m} [2 \cdot k_i \cdot R(y_i) + k_i \cdot R(\neg y_i \vee z_i) + k_i \cdot R(\neg y_i \vee \neg z_i \vee a_i) + k_i \cdot R(\neg a_i \vee \phi_i)]$$

$$= \sum_{i=1}^{m} [2 \cdot k_i - k_i + k_i \cdot (R(\phi_i) - 1)] = \sum_{i=1}^{m} k_i \cdot R(\alpha_i) = 1.$$

Thus, Φ' has copy complexity 2^{k+1} with respect to unit resolution.

Now assume that Φ' has copy complexity 2^{k+1} with respect to unit resolution. Thus, there are coefficients t_i, l_i, p_i and k_i less than or equal to 2^{k+1} such that:

$$\sum_{i} [t_i \cdot (y_i) + l_i \cdot (-y_i + z_i) + p_i \cdot (-y_i - z_i + a_i) + k_i \cdot (-a_i + L(\phi_i))] = 0.$$

In this summation a_i appears positively p_i times and negatively k_i times, thus $p_i = k_i$. Similarly, z_i appears positively l_i times and negatively p_i times, thus $l_i = p_i = k_i$. We also have that y_i appears positively t_i times and negatively $l_i + p_i = 2 \cdot k_i$ times, thus $t_i = 2 \cdot k_i$.

This means that $\sum_i k_i \cdot L(\phi_i) = 0$ and $k_i \leq 2^k$.

We also have that

$$1 = \sum_{i=1}^{m} [t_i \cdot R(y_i) + l_i \cdot R(\neg y_i \vee z_i) + p_i \cdot R(\neg y_i \vee \neg z_i \vee a_i) + k_i \cdot R(\neg a_i \vee \phi_i)]$$

$$= \sum_{i=1}^{m} [2 \cdot k_i \cdot R(y_i) + k_i \cdot R(\neg y_i \vee z_i) + k_i \cdot R(\neg y_i \vee \neg z_i \vee a_i) + k_i \cdot R(\neg a_i \vee \phi_i)]$$

$$= \sum_{i=1}^{m} [2 \cdot k_i - k_i + k_i \cdot (R(\phi_i) - 1)] = \sum_{i=1}^{m} k_i \cdot R(\alpha_i).$$

Thus, Φ has copy complexity 2^k with respect to unit resolution.   □

6   Derived-Copy Complexity

In this section, we study a variant of copy complexity which allows for copies of derived clauses. This is different from regular copy complexity which allows only copies of clauses in the original system. We refer to this as derived-copy complexity.

Theorem 5. *The derived-copy complexity of a Horn formula with m clauses with respect to unit resolution is at most $(m-1)$.*

Proof. Let Φ be an unsatisfiable Horn formula. Note that adding clauses to Φ cannot increase the copy complexity. Thus, we can assume without loss of generality that Φ is minimal unsatisfiable.

First, assume that Φ has $m = 2$ clauses. Thus, Φ has the form $(x) \wedge (\neg x)$ and the derived-copy complexity is 1.

Now assume that every Horn formula with k clauses has derived-copy complexity $k - 1$.

Let Φ be a Horn formula with $m = k + 1$ clauses. Thus, Φ has the form $(x) \wedge (\neg x \vee \alpha_1) \wedge \ldots \wedge (\neg x \vee \alpha_t) \wedge \sigma_{t+1} \wedge \ldots \wedge \sigma_k$ for some $t \leq k$.

Since Φ is minimal unsatisfiable, α_i and σ_j do not contain the literal x. By construction, they also do not contain the literal $\neg x$.

If we generate t copies of the clause (x), then for each $i = 1 \ldots t$, we can apply the resolution step $(x) \wedge (\neg x \vee \alpha_i) \vdash_{\overline{RES}}^{1} \alpha_i$. Applying these resolution steps results in the Horn formula $\Phi' = \alpha_1 \wedge \ldots \wedge \alpha_t \wedge \sigma_{t+1} \wedge \ldots \wedge \sigma_k$.

Φ' is minimal unsatisfiable and consists of k clauses. By the induction hypothesis, we know that the derived-copy complexity of Φ' is at most $(k-1)$.

Thus the derived-copy complexity of Φ is at most $\max\{t, k-1\} \leq k$ as desired. □

Theorem 6. *For each value of m, there exists a Horn formula with m clauses and derived-copy complexity $(m-1)$*

Proof. We inductively define Φ_m as follows: 1. $\Phi_2 = (x_1) \wedge (\neg x_1)$.
2. If $\Phi_m = \alpha_1 \wedge \ldots \wedge \alpha_m$, then $\Phi_{m+1} = (x_m) \wedge (\neg x_m \vee \alpha_1) \wedge \ldots \wedge (\neg x_m \vee \alpha_m)$.

Note that for each m, Φ_m consists of m clauses and is an unsatisfiable Horn formula.

We now show by induction, that Φ_m has derived copy complexity $(m-1)$. Φ_2 has a unit read-once resolution refutation consisting of the single resolution step $(x_1) \wedge (\neg x_1) \vdash_{\overline{RES}}^{1} \sqcup$.

Now assume that Φ_{m-1} has derived copy complexity $(m-2)$. To derive Φ_{m-1} from Φ_m by unit read-once resolution, we need to resolve the clause (x_m) with every other clause in Φ_m. This requires $(m-1)$ copies of the clause (x_n). By the inductive hypothesis, this derived formula has derived-copy complexity $(m-2)$. Thus, the formula Φ_m has derived-copy complexity $\max\{m-1, m-2\} = (m-1)$ as desired. □

From Theorem 5 and Theorem 6, it follows that the derived copy complexity for Horn formulas with m clauses is $(m-1)$.

7   The UROR Problem for 2-CNF Formulas

In this section, we show that the UROR Problem for 2-CNF formulas is in **P**. This is done by reducing the problem to minimum weight perfect matching using a similar construction to the one in [14]. Note that the problem of finding read-once resolution refutations for 2-CNF formulas is **NP-complete** [8].

Let Φ be a 2-CNF formula with m clauses over n variables. We construct a weighted undirected graph $\mathbf{G} = \langle \mathbf{V}, \mathbf{E}, \mathbf{b} \rangle$ as follows:

1. For each variable x_i in Φ, add the vertices x_i^+, $x_i'^+$, x_i^-, and $x_i'^-$ to $\mathbf{V}$. Additionally, add the edges $x_i^- \overset{0}{\text{———}} x_i^+$ and $x_i'^- \overset{0}{\text{———}} x_i'^+$ to $\mathbf{E}$.
2. Add the vertices x_0^+ and x_0^- to $\mathbf{V}$.
3. For each constraint ϕ_k of Φ, add the vertices ϕ_k and ϕ_k' to $\mathbf{V}$ and the edge $\phi_k \overset{0}{\text{———}} \phi_k'$ to $\mathbf{E}$. Additionally:
   (a) If ϕ_k is of the form $(x_i \lor x_j)$, add the edges $x_i^+ \overset{-1}{\text{———}} \phi_k$, $x_i'^+ \overset{-1}{\text{———}} \phi_k$, $x_j^+ \overset{-1}{\text{———}} \phi_k'$, and $x_j'^+ \overset{-1}{\text{———}} \phi_k'$ to $\mathbf{E}$.
   (b) If ϕ_k is of the form $(x_i \lor \neg x_j)$, add the edges $x_i^+ \overset{-1}{\text{———}} \phi_k$, $x_i'^+ \overset{-1}{\text{———}} \phi_k$, $x_j^- \overset{-1}{\text{———}} \phi_k'$, and $x_j'^- \overset{-1}{\text{———}} \phi_k'$ to $\mathbf{E}$.
   (c) If ϕ_k is of the form $(\neg x_i \lor x_j)$, add the edges $x_i^- \overset{-1}{\text{———}} \phi_k$, $x_i'^- \overset{-1}{\text{———}} \phi_k$, $x_j^+ \overset{-1}{\text{———}} \phi_k'$, and $x_j'^+ \overset{-1}{\text{———}} \phi_k'$ to $\mathbf{E}''$.
   (d) If ϕ_k is of the form $(\neg x_i \lor \neg x_j)$, add the edges $x_i^- \overset{-1}{\text{———}} \phi_k$, $x_i'^- \overset{-1}{\text{———}} \phi_k$, $x_j^- \overset{-1}{\text{———}} \phi_k'$, and $x_j'^- \overset{-1}{\text{———}} \phi_k'$ to $\mathbf{E}$.
   (e) If ϕ_k is of the form (x_i), add the edges $x_i^+ \overset{-1}{\text{———}} \phi_k$, $x_i'^+ \overset{-1}{\text{———}} \phi_k$, $x_0^+ \overset{-1}{\text{———}} \phi_k'$, and $x_0^- \overset{-1}{\text{———}} \phi_k'$ to $\mathbf{E}$.
   (f) If ϕ_k is of the form $(\neg x_i)$, add the edges $x_i^- \overset{-1}{\text{———}} \phi_k$, $x_i'^- \overset{-1}{\text{———}} \phi_k$, $x_0^+ \overset{-1}{\text{———}} \phi_k'$, and $x_0^- \overset{-1}{\text{———}} \phi_k'$ to $\mathbf{E}$.

Under this construction, each variable is represented by a pair of 0-weight edges. Thus, each variable can be used twice by a refutation. Note that each 2-CNF formula has a refutation that uses each variable at most twice [7]. However, we only have one 0-weight edge for each clause. This prevents the refutation from re-using clauses. Observe that $\mathbf{G}$ has $O(m + n)$ vertices and $O(m + n)$ edges.

Theorem 7. *Φ has a unit read-once resolution refutation if and only if $\mathbf{G}$ has a negative weight perfect matching.*

Proof. First, assume that Φ has a unit read-once resolution refutation R. Let $\Phi_R \subseteq \Phi$ be the set of clauses in Φ used by any resolution step in R. Since R is a sequence of resolution steps, we can order the clauses in Φ_R according to the order in which they are used by resolution steps in R. Thus, if a clause ϕ_j is used by an earlier resolution step in R than the clause ϕ_k, then ϕ_j is earlier in the ordering of Φ_R than ϕ_k.

We can construct a negative weight perfect matching P of $\mathbf{G}$ as follows:

1. For each variable x_i in Φ:
   (a) If no clause in Φ_R contains the literal x_i, add the edges $x_i^+ \overset{0}{\text{———}} x_i^-$ and $x_i'^+ \overset{0}{\text{———}} x_i'^-$ to P.

(b) If only one clause in Φ_R contains the literal x_i, add the edge $x_i'^+ \overset{0}{\rule{1cm}{0.4pt}} x_i'^-$ to P.

2. For each clause ϕ_k in Φ:

(a) If $\phi_k \notin \Phi_R$, add the edge $\phi_k \overset{0}{\rule{1cm}{0.4pt}} \phi_k'$ to P.

(b) If $\phi_k \in \Phi_R$ is a two variable clause:

   i. If ϕ_k is the first clause in Φ_R to contain the literal x_i, add the edge $x_i^+ \overset{-1}{\rule{1cm}{0.4pt}} \phi_k$ (or $x_i^+ \overset{-1}{\rule{1cm}{0.4pt}} \phi_k'$) to P. If it is the second, add the edge $x_i'^+ \overset{-1}{\rule{1cm}{0.4pt}} \phi_k$ (or $x_i'^+ \overset{-1}{\rule{1cm}{0.4pt}} \phi_k'$) instead.

   ii. If ϕ_k is the first clause in Φ_R to contain the literal $\neg x_i$, add the edge $x_i^- \overset{-1}{\rule{1cm}{0.4pt}} \phi_k$ (or $x_i^- \overset{-1}{\rule{1cm}{0.4pt}} \phi_k'$) to P. If it is the second, add the edge $x_i'^- \overset{-1}{\rule{1cm}{0.4pt}} \phi_k$ (or $x_i'^- \overset{-1}{\rule{1cm}{0.4pt}} \phi_k'$) instead.

(c) If $\phi_k \in R$ is a unit clause:

   i. If ϕ_k is the first clause in Φ_R to contain the literal x_i, add the edge $x_i^+ \overset{-1}{\rule{1cm}{0.4pt}} \phi_k$ to P. If it is the second, add the edge $x_i'^+ \overset{-1}{\rule{1cm}{0.4pt}} \phi_k$ instead.

   ii. If ϕ_k is the first clause in Φ_R to contain the literal $-x_i$, add the edge $x_i^- \overset{-1}{\rule{1cm}{0.4pt}} \phi_k$ to P. If it is the second, add the edge $x_i'^- \overset{-1}{\rule{1cm}{0.4pt}} \phi_k$ instead.

   iii. If ϕ_k is the first unit clause in Φ_R, add the edge $x_0^+ \overset{-1}{\rule{1cm}{0.4pt}} \phi_k'$ to P. If it is the second, add the edge $x_0^- \overset{-1}{\rule{1cm}{0.4pt}} \phi_k'$ instead.

Every vertex in **G** is an endpoint of exactly one edge in P. Thus, P is a perfect matching. Since $\sum_{\phi_k \in R} -1 < 0$, P has negative weight.

Now assume that **G** has a negative weight perfect matching P. We can construct a unit read-once resolution refutation R as follows:

1. Since there is no edge between x_0^+ and x_0^-, P must use the edge $x_0^+ \overset{-1}{\rule{1cm}{0.4pt}} \phi_k'$ for some unit clause ϕ_k. Thus, the edge $\phi_k \overset{0}{\rule{1cm}{0.4pt}} \phi_k'$ is not in P. Note that ϕ_k is the clause (x_i) or $(\neg x_i)$ for some x_i.

2. Without loss of generality, assume ϕ_k is the clause (x_i). This means that the edge $x_i^+ \overset{-1}{\rule{1cm}{0.4pt}} \phi_k$ (or $x_i'^+ \overset{-1}{\rule{1cm}{0.4pt}} \phi_k$) must be in P. Thus, the edge $x_i^+ \overset{0}{\rule{1cm}{0.4pt}} x_i^-$ (or $x_i'^+ \overset{0}{\rule{1cm}{0.4pt}} x_i'^-$) is not in P. This means that for some clause ϕ_l, the edge $x_i^- \overset{-1}{\rule{1cm}{0.4pt}} \phi_l$ (or $x_i'^- \overset{-1}{\rule{1cm}{0.4pt}} \phi_l$) is in P. If ϕ_l corresponds to the clause $(\neg x_i \vee x_j)$ or $(\neg x_i \vee \neg x_j)$ for some x_j, then we add either $(x_i), (\neg x_i \vee x_j) \vdash_{RES}^{1} (x_j)$ or $(x_i), (\neg x_i \vee \neg x_j) \vdash_{RES}^{1} (\neg x_j)$ to R. If ϕ_l corresponds to the clause $(\neg x_i)$, then we add $(x_i), (\neg x_i) \vdash_{RES}^{1} \sqcup$ to R.

3. If ϕ_l is a non-unit clause, then we can repeat step 2 from either x_j or $\neg x_j$. This continues until a second unit clause is encountered, completing the refutation. By construction, R is a unit read-once resolution refutation. $\square$

Observe that the minimum weight perfect matching of an undirected graph having n' vertices and m' edges can be found in $O(m' \cdot n' + n'^2 \cdot \log n')$ time using the algorithm in [11]. Since in our case, **G** has $O(m+n)$ vertices and $O(m+n)$ edges, it follows that we can detect the presence of a negative weight perfect matching in $O((m+n)^2 \cdot \log(m+n))$ time. Hence, using the above reduction, the UROR problem for 2-CNF formulas can be solved in $O((m+n)^2 \cdot \log(m+n))$ time.

8 Conclusion

In this paper, we established the relationship between copy complexity of Horn formulas with respect to unit read-once resolution and the copy complexity of the corresponding system of constraints. We also studied the derived copy complexity of Horn formulas and derived tight bounds for the same. Finally, we described a polynomial time algorithm to check if a 2-CNF formula has a UROR.

Acknowledgments. We would like to thank Hans Kleine Büning for his insights into the problems examined in this paper.

References

1. Beame, P., Pitassi, T.: Propositional proof complexity: past, present, future. Bull. EATCS **65**, 66–89 (1998)
2. Dowling, W.F., Gallier, J.H.: Linear-time algorithms for testing the satisfiability of propositional Horn formulae. J. Log. Program. **1**(3), 267–284 (1984)
3. Haken, A.: The intractability of resolution. Theor. Comput. Sci. **39**(2–3), 297–308 (1985)
4. Hooker, J.N.: Generalized resolution and cutting planes. Ann. Oper. Res. **12**(1–4), 217–239 (1988)
5. Iwama, K., Miyano, E.: Intractability of read-once resolution. In: Proceedings of the 10th Annual Conference on Structure in Complexity Theory (SCTC 1995), Los Alamitos, CA, USA, June 1995, pp. 29–36. IEEE Computer Society Press (1995)
6. Karp, R.M.: Reducibility among combinatorial problems. In: Miller, R.E., Thatcher, J.W. (eds.) Complexity of Computer Computations, pp. 85–103. Plenum Press, New York (1972). https://doi.org/10.1007/978-1-4684-2001-2_9
7. Büning, H.K., Wojciechowski, P., Subramani, K.: The complexity of finding read-once NAE-resolution refutations. In: Ghosh, S., Prasad, S. (eds.) ICLA 2017. LNCS, vol. 10119, pp. 64–76. Springer, Heidelberg (2017). https://doi.org/10.1007/978-3-662-54069-5_6
8. Büning, H.K., Wojciechowski, P.J., Subramani, K.: Finding read-once resolution refutations in systems of 2CNF clauses. Theor. Comput. Sci. **729**, 42–56 (2018)
9. Kleine Büning, H., Wojciechowski, P., Subramani, K.: Read-once resolutions in horn formulas. In: Chen, Y., Deng, X., Lu, M. (eds.) FAW 2019. LNCS, vol. 11458, pp. 100–110. Springer, Cham (2019). https://doi.org/10.1007/978-3-030-18126-0_9
10. Kleine Büning, H., Zhao, X.: Read-once unit resolution. In: Giunchiglia, E., Tacchella, A. (eds.) SAT 2003. LNCS, vol. 2919, pp. 356–369. Springer, Heidelberg (2004). https://doi.org/10.1007/978-3-540-24605-3_27
11. Korte, B., Vygen, J.: Combinatorial Optimization. Number 21 in Algorithms and Combinatorics. 4th edn. Springer, New York (2010)
12. Krom, M.R.: The decision problem for a class of first-order formulas in which all disjunctions are binary. Math. Logic Q. **13**(1–2), 15–20 (1967)

13. Pudlák, P.: Lower bounds for resolution and cutting plane proofs and monotone computations. J. Symb. Log. **62**(3), 981–998 (1997)
14. Subramani, K., Wojciechowki, P.: A polynomial time algorithm for read-once certification of linear infeasibility in UTVPI constraints. Algorithmica **81**(7), 2765–2794 (2019)
15. Tseitin, G.S.: On the Complexity of Derivation in Propositional Calculus, pp. 466–483. Springer, Heidelberg (1983)

Propositional Projection Temporal Logic Specification Mining

Nan Zhang, Xiaoshuai Yuan, and Zhenhua Duan[✉]

Institute of Computing Theory and Technology, and ISN Laboratory,
Xidian University, Xi'an 710071, China
nanzhang@xidian.edu.cn, yuanxiaoshuai@stu.xidian.edu.cn,
zhhduan@mail.xidian.edu.cn

Abstract. This paper proposes a dynamic approach of specification mining for Propositional Projection Temporal Logic (PPTL). To this end, a pattern library is built to collect some common temporal relation among events. Further, several algorithms of specification mining for PPTL are designed. With our approach, PPTL specifications are mined from a trace set of a target program by using patterns in the library. In addition, a specification mining tool PPTLMiner supporting this approach is developed. In practice, given a trace set and user selected patterns, PPTLMiner can capture PPTL specifications of target programs.

Keywords: Propositional projection temporal logic · Pattern · Trace · Specification mining

1 Introduction

A software system specification is a formal description of the system requirements. Formal languages are often employed to write specifications so as to prevent the ambiguity written in natural languages. The common used formal languages include Temporal Logic (TL) and Finite State Automata (FSA). Software system specification can be used to test and verify the correctness and reliability of software systems [13]. However, due to various kinds of reasons, a great number of software systems lack formal specifications. In particular, for most of legacy software systems, formal specifications are missed. This makes the maintenance of software systems difficult. To fight this problem, various kinds of specification mining approaches are proposed [10–12,14,15,17,19–21].

Walkinshaw et al. [19] present a semi-automated approach to inferring FSAs from dynamic execution traces that builds on the QSM algorithm [8]. This algorithm infers a finite state automaton by successively merging states. Lo et al.

This research is supported by National Key Research and Development Program of China under Grant No. 2018AAA0103202, National Natural Science Foundation of China under Grant Nos. 61751207 and 61732013, and Shaanxi Key Science and Technology Innovation Team Project under Grant No. 2019TD-001.

© Springer Nature Switzerland AG 2020
W. Wu and Z. Zhang (Eds.): COCOA 2020, LNCS 12577, pp. 289–303, 2020.
https://doi.org/10.1007/978-3-030-64843-5_20

propose Deep Specification Mining (DSM) approach that performs deep learning for mining FSA-based specifications [11]. FSA specifications are intuitive and easily to be used for verifying and testing programs. However, most of FSA specification mining approaches suffer from accuracy and correctness for representing properties of programs. Yang et al. [21] present an interesting work on mining two-event temporal logic rules (i.e., of the form $G(a \rightarrow XF(b))$, where G, X and F are LTL operators, which are statistically significant with respect to a user-defined "satisfaction rate". Wasylkowski et al. [20] mine temporal rules as Computational Tree Logic (CTL) properties by leveraging a model checking algorithm and using concept analysis. Lemieux et al. [12] propose an approach to mine LTL properties of arbitrary length and complexity. Similar to the above research work, most of specification mining approaches employ LTL and CTL as the property description languages. Due to the limitation of the expressiveness of LTL and CTL, some temporal properties such as periodic repetition properties cannot be characterized.

Since the expressiveness of Propositional Projection Temporal Logic (PPTL) is full regular [3,18], in this paper, we propose a dynamic approach to mining PPTL properties based on a pattern library. PPTL contains three primitive temporal operators: next ($\bigcirc$), projection (prj) and chop-plus ($+$). Apart from some common temporal properties that can be formalized in LTL and CTL, PPTL is able to describe two other kinds of properties: interval sensitive properties and periodic repetition properties. With the proposed approach, we abstract API/method calls as events. A trace is a sequence of API/method calls occurred during program execution. Daikon [1] is used to generate raw traces first, then a tool *DtraceFilter* we developed is employed to further refine the traces. Patterns are used to characterize common temporal relations among events. Two categories of patterns, Occurrence and Order, are used. These patterns are predefined in a pattern library. The proposed mining algorithms require two inputs: an instantiated pattern formula P and a refined execution trace τ. To obtain an instantiated pattern formula, we need to specify a pattern formula which can be either a user-defined one or a predefined one in the library. The pattern is instantiated by substituting atomic propositions with concrete events. After pattern instantiation, several mining algorithms based on PPTL normal form [3–7] are employed to recursively check whether τ satisfies P.

The contribution of the paper is three-fold. First, we propose a PPTL temporal rule specification mining approach so that full regular properties can be mined. Second, we develop a tool PPTLMiner which supports the proposed mining approach. Third, we build a pattern library to cover all common patterns accumulated from literatures and abstracted from the existing software systems. The library is open, user-editable and in constant expansion and growth.

This paper is organized as follows. In the next section, PPTL is briefly introduced. In Sect. 3, the trace generation and the construction of the pattern library are presented. In Sect. 4, the overall framework of PPTLMiner and key algorithms are elaborated. Finally, conclusions are drawn in Sect. 5.

2   Propositional Projection Temporal Logic

In this section, we briefly introduce our underlying logic, Propositional Projection Temporal Logic (PPTL), including its syntax and semantics. It is used to describe specifications of programs. For more detail, please refer to [3,7].

Syntax of PPTL. Let $Prop$ be a set of atomic propositions and $p \in Prop$. The syntax of PPTL is inductively defined as follows.

$$P ::= p \mid \bigcirc P \mid \neg P \mid P \vee Q \mid (P_1, ..., P_m) \; prj \; Q \mid P^+$$

where $P_1, ..., P_m, P$ and Q are well-formed PPTL formulas. Here, $\bigcirc$ (next), prj (projection) and $+$ (chop-plus) are primitive temporal operators.

Semantics of PPTL. Let $B = \{true, false\}$ and N be the set of non-negative integers. Let ω denote infinity. PPTL formulas are interpreted over intervals. An interval σ is a finite or infinite sequence of states, denoted by $\sigma = \langle s_0, s_1, ... \rangle$. A state s_i is a mapping from $Prop$ to B. An interpretation $\mathcal{I} = (\sigma, k, j)$ is a subinterval $\langle s_k, ..., s_j \rangle$ of σ with the current state being s_k. An auxiliary operator $\downarrow$ is defined as $\sigma \downarrow (r_1, ..., r_m) = \langle s_{t_1}, s_{t_2}, ..., s_{t_n} \rangle$, where $t_1, ..., t_n$ are obtained from $r_1, ..., r_m$ by deleting all duplicates. That is, $t_1, ..., t_n$ is the longest strictly increasing subsequence of $r_1, ..., r_m$. The semantics of PPTL formulas is inductively defined as a satisfaction relation below.

(1) $\mathcal{I} \models p$ iff $s_k[p] = true$.

(2) $\mathcal{I} \models \bigcirc P$ iff $(\sigma, k+1, j) \models P$.

(3) $\mathcal{I} \models \neg P$ iff $\mathcal{I} \not\models P$.

(4) $\mathcal{I} \models P \vee Q$ iff $\mathcal{I} \models P$ or $\mathcal{I} \models Q$.

(5) $\mathcal{I} \models (P_1, ..., P_m) \; prj \; Q$ iff there exist m integers $k = r_0 \leq r_1 \leq ... \leq r_m \leq j$ such that $(\sigma, r_{l-1}, r_l) \models P_l$ for all $1 \leq l \leq m$, and one of the following two cases holds:

- if $r_m = j$, there exists r_h such that $0 \leq h \leq m$ and $\sigma \downarrow (r_0, ..., r_h) \models Q$;
- if $r_m < j$, then $\sigma \downarrow (r_0, ..., r_m) \cdot \sigma_{(r_m+1..j)} \models Q$.

(6) $\mathcal{I} \models P^+$ iff there exist m integers $k = r_0 \leq r_1 \leq ... \leq r_m = j$ $(m \in N)$ such that $(\sigma, r_{l-1}, r_l) \models P$ for all $1 \leq l \leq m$.

Derived Formulas. Some derived formulas in PPTL are defined in Table 1.

Operator Priority. To avoid an excessive number of parentheses, the precedence rules shown in Table 2 are used, where $1 =$ highest and $9 =$ lowest.

Definition 1 (PPTL Normal Formal). Let Q be a PPTL formula and Q_p denote the set of atomic propositions appearing in Q. Q is in normal form if Q has been rewritten as

$$Q \equiv \bigvee_{j=1}^{n_0} (Q_{ej} \wedge \epsilon) \vee \bigvee_{i=1}^{n} (Q_{ci} \wedge \bigcirc Q_i')$$

where $Q_{ej} \equiv \bigwedge_{k=1}^{m_0} \dot{q}_{jk}$, $Q_{ci} \equiv \bigwedge_{h=1}^{m} \dot{q}_{ih}$, $l = |Q_p|$, $1 \leq n \leq 3^l$, $1 \leq n_0 \leq 3^l$, $1 \leq m \leq l$, $1 \leq m_0 \leq l$; $q_j k, q_i h \in Q_p$, for any $r \in Q_p$, $\dot{r}$ means r or $\neg r$; Q_i' is a general PPTL formula.

Table 1. Derived formulas

A1	$\varepsilon \stackrel{\text{def}}{=} \neg \bigcirc true$	A2	$more \stackrel{\text{def}}{=} \bigcirc true$
A3	$\bigcirc^0 P \stackrel{\text{def}}{=} P$	A4	$\bigcirc^n P \stackrel{\text{def}}{=} \bigcirc(\bigcirc^{n-1}P)(n > 0)$
A5	$\odot P \stackrel{\text{def}}{=} \varepsilon \vee \bigcirc P$	A6	$P; Q \stackrel{\text{def}}{=} (P, Q) \, prj \, \varepsilon$
A7	$\Diamond P \stackrel{\text{def}}{=} true; P$	A8	$\Box P \stackrel{\text{def}}{=} \neg \Diamond \neg P$
A9	$len(n) \stackrel{\text{def}}{=} \bigcirc^n \varepsilon$	A10	$skip \stackrel{\text{def}}{=} len(1)$
A11	$P^* \stackrel{\text{def}}{=} P^+ \vee \varepsilon$	A12	$P \parallel Q \stackrel{\text{def}}{=} (P; true) \wedge Q \vee P \wedge (Q; true)$
A13	$fin \stackrel{\text{def}}{=} \Diamond \varepsilon$	A14	$inf \stackrel{\text{def}}{=} \neg fin$

Table 2. Operator priority

1.	$\neg$	2.	$+, *$	3.	$\bigcirc, \odot, \Diamond, \Box$
4.	$\wedge$	5.	$;$	6.	$\vee$
7.	prj	8.	$\parallel$	9.	$\rightarrow, \leftrightarrow$

In some circumstances, for convenience, we write $Q_e \wedge \epsilon$ instead of $\bigvee_{j=1}^{n_0}(Q_{ej} \wedge \epsilon)$ and $\bigvee_{i=1}^{r}(Q_i \wedge \bigcirc Q_i')$ instead of $\bigvee_{i=1}^{n}(Q_{ci} \wedge \bigcirc Q_i')$. Thus,

$$Q \equiv (Q_e \wedge \epsilon) \vee \bigvee_{i=1}^{r}(Q_i \wedge \bigcirc Q_i')$$

where Q_e and Q_i are state formulas. The algorithm of translating a PPTL formula into its normal form can be found in [4–6].

3   Pattern Library Construction and Trace Generation

Our specification mining algorithm relies on two inputs: a pattern and a program execution trace. A pattern is a property template in which the atomic proposition symbols need to be instantiated as events (namely, API or method calls) occurred during program execution. A trace is a sequence of method calls in the execution of a program. In this section, we present how to build the pattern library and traces.

3.1   Pattern and Pattern Library

Patterns are abstracted from common software behaviors and used to describe occurrence of events or states during program execution [9]. A pattern is a logical representation of certain event relation. The *APIs* and methods in a target program are defined as events. We say that an event occurs whenever it is called in the execution of the program. In the following, we define a quadruples to represent and store patterns.

Definition 2 (Pattern). A pattern $T = < C, N, R, A >$ is a tuple where C is a pattern category indicating occurrence or order of events, N a pattern name, R a PPTL formula, and A an annotation.

Following Dwyer et al.'s SPS [9] and Autili et al.'s PSP framework [2], we also classify patterns into two categories, *Occurrence* and *Order*.

The *Occurrence* category contains 18 patterns that indicate presence or absence of certain events or states during program execution. For instance, (1) *Absence* means that an event never happens; (2) *Universality* indicates that an event always occurs during program execution; (3) *Existence* shows that an event occurs at least once during program execution; and (4) *Bounded Existence* tells us that an event has a limited number of occurrences during program execution, e.g. event $f.open()$ occurs twice.

The *Order* category contains 19 patterns that represent relative temporal orders among multiple events or states occurred during program execution. For example, (1)"*s precedes p*" indicates that if event p occurs, event s definitely occurs before p; (2) "*s responds p*" means that if event p occurs, event s definitely occurs after p; (3)*Chain* (s,t) means that a combination chain of events s and t. (s,t) *precedes p* means that if event p happens, chain events (s,t) certainly happen before p, and (s,t) *responds p* means that if event p happens, (s,t) certainly responds to p [2,9].

Pattern Library. A pattern library L is a set containing all patterns p we collected. After an in-depth investigation of the existing literature and programs specified behavior characteristics, we build a pattern library and some patterns are shown in Table 3 and Table 4.

Table 3. Pattern library - occurrence category

No.	Pattern Name	PPTL Formula	Annotation
1	Universality	$\Box p$	Event p always occurs
2	Absence	$\Box \neg p$	Event p never occur
3	Existence	$\Diamond p$	Event p occurs at least once
4	Frequency	$\Box \Diamond p$	Event p occurs frequently
5	Both Occur	$\Diamond p \wedge \Diamond q$	Events p and q both occur
6	Simultaneity	$\Diamond(p \wedge q)$	Events p and q occur at the same time
7	Prefix of Trace	$\Box \Diamond p; more$	Event p occurs frequently at a prefix of a trace
8	Suffix of Trace	$\Diamond \Box p$	Event p occurs continuously at a suffix of a trace

3.2 Trace Generation

We concern only specifications of temporal relations among the methods or API calls occurred during program execution.

Table 4. Pattern library - order category

No.	Pattern Name	PPTL Formula	Annotation
1	Precedence (1-1)	$\Diamond p \rightarrow (\Box\neg p; s)$	Event s takes precedence over event p
2	Response (1-1)	$\Box(s \rightarrow \bigcirc\Diamond p)$	Event p responds to event s
3	Until	$(\Box p; \bigcirc s) \vee s$	Event p occurs until event s occurs
4	Response Invariance	$\Box(p \rightarrow \bigcirc\Box s)$	If p has occurred, then in response s holds continually
5	Chop	$\Box p; \bigcirc\Box q$	There exists a time point t such that event p occurs continuously before t and event q continuously after t
6	Never Follow	$\Box(p \rightarrow \bigcirc\Box\neg q)$	Event p is never followed by event q

Definition 3 (Trace). A trace is a sequence of methods or API calls (namely events) with parameters.

Example 1. A trace of a program using stack structure.
trace $\tau_1 = \langle StackAr(int), isFull(), isEmpty(), top(), isEmpty(), topAndPop(), isEmpty(), isFull(), isEmpty(), top(), isEmpty(), push(java.lang.Object), isFull()\rangle$

Example 2. A trace of a program manipulating files.
trace $\tau_2 = \langle open(f1), write(f1), read(f1), close(f1), open(f2), delete(f1), read(f2), write(f2), write(f2), read(f2), close(f2), delete(f2)\rangle$

We use Daikon [1] as an auxiliary tool to generate traces. Daikon can dynamically detect program invariants. A program invariant is a property that remains unchanged at one or more positions of program execution. The common invariants are APIs, functions, global or local variables, arguments, return values and so on. Invariants can be used to analyze behavior of a program. Dynamic invariant detection refers to a process of running a program so as to check variables and assertions detected in the program execution [16].

Daikon generates a sequence containing all invariants and stores it in a *dtrace* file in which the invariants are stored line by line. The program execution traces we need are contained in this sequence. Since there exists an amount of redundant information, the *dtrace* file needs to be further refined.

The whole process of generating a trace is shown in Fig. 1.

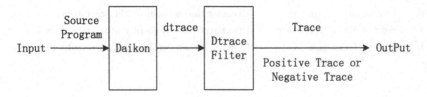

Fig. 1. The process of trace generation

Step 1. Generating sequences of program invariants

A source program and its arguments are input to Daikon so that a sequence of program invariants is generated. The sequence is written in a file $f.dtrace$ in the *dtrace* format. When the program is executed with different arguments for a desired number n of times, we obtain a set $Pool_1 = \{f_i.dtrace | i = 1, \ldots, n\}$ of program traces.

Step 2. Filtering of sequences of program invariants

A filter tool *DtraceFilter* has been developed to filter out redundant information, including parameters, variables, return values and useless spaces, in each file $f_i.dtrace$ of $Pool_1$. As a result, sequences consisting of only *APIs* and method calls constitute a new set $Pool_2 = \{f_i.trace | i = 1, \ldots, n\}$.

Step 3. Parsing traces in $Pool_2$

Each trace $f_i.trace$ in $Pool_2$ needs to be parsed so as to obtain a API/method-name list $f_i.event$. These lists constitute a set $Event = \{f_i.event | i = 1, \ldots, n\}$.

Step 4. Optimizing traces in $Pool_2$

We can specify desired API/method names from the lists in $Event$ according to the requirements. *DtraceFilter* can be used to select the events we concern from each list in $Event$ to build a positive list $f_i.pevent$ of events, and generate a set $PositiveEvent = \{f_i.pevent | i = 1, \ldots, n\}$.

Based on $PositiveEvent$, *DtraceFilter* further refines each $f_i.trace$ in $Pool_2$ to get a positive trace $f_i.ptrace$ consisting of only the events in $f_i.pevent$, and obtain a set $PositiveTrace = \{f_i.ptrace | i = 1, \ldots, n\}$.

We can also specify undesired API/method names from the lists in $Event$. In a similar way, *DtraceFilter* can be used to build a negative list $f_i.nevent$ of events and generate $NegativeEvent = \{f_i.nevent | i = 1, \ldots, n\}$. After deleting the negative events from each trace $f_i.trace$ in $Pool_2$, *DtraceFilter* builds a set $NegativeTrace = \{f_i.ntrace | i = 1, \ldots, n\}$.

4   PPTL Specification Mining

Based on the Pattern Library and set of refined traces presented in the previous section, an approach to PPTL specification mining is proposed and a specification mining tool, PPTLMiner, is developed. In this section, the framework of PPTLMiner and some key algorithms are presented in detail.

4.1   The Framework of PPTLMiner

The integrated design of PPTLMiner is shown in Fig. 2. It consists of the following six parts.

(1) Pattern Library. The Pattern Library covers all patterns we obtain after investigating literatures and programs. Our Pattern Library is open, user-editable and in constant expansion and growth. New patterns can be inserted into the library from time to time. For more details, refer to Sect. 3.1.

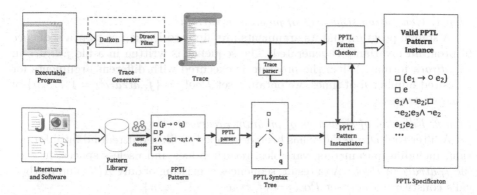

Fig. 2. The framework of PPTLMiner

(2) Trace Generator. The function of the Trace Generator is to generate traces from an executable program. To do so, an executable program and its arguments are input into Daikon to produce raw traces (dtrace files). Then *Dtracefilter* is employed to filter out redundant information in dtrace files to obtain trace files, which are further refined to obtain positive and negative traces. For more details, refer to Sect. 3.2.

(3) PPTL Parser. The input of PPTL Parser is a PPTL formula. PPTL Parser is developed by means of Flex and Bison. It can be used to generate a PPTL syntax tree for any PPTL formula.

(4) Trace Parser. The function of Trace Parser is two-fold. The first is to parse traces generated by the Trace Generator and restore them in an appropriate data structure so that the traces can conveniently be used by PPTL Pattern Checker. The second is to calculate a set $E = \{e_1, e_2, \ldots, e_n\}$ of events appeared in the traces so as to instantiate PPTL patterns.

(5) PPTL Pattern Formula Instantiator. The instantiator requires two inputs: (*a*) a PPTL pattern formula P, and (*b*) E, the set of events produced by Trace Parser. The function of the instantiator is to instantiate a pattern formula P by substituting atomic propositions in P by events in *Events*.

(6) PPTL Pattern Checker. PPTL Pattern Checker also requires two inputs: (*a*) a trace τ produced by Trace Generator, and (*b*) an instantiated pattern formula P generated by PPTL Pattern Formula Instantiator. The function of the Checker is to decide whether trace τ satisfies P.

4.2   Mining Process and Algorithms

In this subsection, we present the mining process and algorithms in detail.

(1) Syntax Tree of PPTL Formula

By syntax analysis, a PPTL Pattern Formula is parsed into a syntax tree. A syntax tree consists of a root node and two child nodes. The root node is of two attributes, NODETYPE and STRING, which indicate the type and name of the root node, respectively. All nodes having two null child nodes in the syntax tree of a PPTL pattern formula P constitute a set $S(P)$ of atomic propositions. For instance, for an atomic proposition p, its NODETYPE is "atomic proposition" while its STRING is "p". Two child nodes are all null. For formula $P_1; P_2$, its NODETYPE is "chop" while its STRING is ";". It has two non-null child nodes, $child_1$ and $child_2$, where $child_1$ is the root of P_1 while $child_2$ is the root of P_2. $S(P_1; P_2) = S(P_1) \cup S(P_2)$. More Examples are shown in Fig. 3.

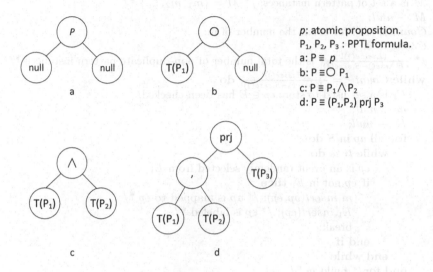

Fig. 3. PPTL syntax tree

(2) Instantiating PPTL Pattern Formulas

Based on the set $S(P)$ of atomic propositions and set E collected by Trace Parser, a PPTL Pattern Formula P is instantiated by Algorithm 1.

(3) PPTL Pattern Check

We use Algorithm 2, Algorithm 3 and Algorithm 4 to check whether τ satisfies Q, where τ is a refined trace generated in Sect. 3.2 while Q is an instantiated PPTL pattern formula obtained in part (2). These algorithms are based on PPTL Normal Form.

In particular, Algorithm 2, i.e. CheckBasedonNF(P, τ), first checks the satisfiability of P. If P is satisfiable, it is translated into its normal form P_{nf} by calling the existing external function $NF(\cdot)$ given in [3]. Then Algorithm 3 NFCheckTrace(P_{nf}, τ) is called to decide whether τ satisfies P_{nf}.

In function NFCheckTrace(P_{nf}, τ), the first disjunct $P_{nf}.child_1$ is first checked. If NFCheckTrace$(P_{nf}.child_1, \tau)$ is true, P_{nf} is already satisfied by τ. Otherwise the rest disjuncts $P_{nf}.child_2$ are further checked.

Algorithm 1. function Instantiator(E, S, P)

Input: E: a set of events;
Input: S: a set of atomic propositions appearing in P;
Input: P: a syntax tree of a PPTL pattern formula;
Output: P_s: a set of instantiated PPTL pattern formulas.
1: **begin**
2:   $P_s \leftarrow null$;
3:   m is a patttern instance;
4:   /* $m = \{(ap_i, ep_i) \mid ap_i \in S \;\&\; ep_i \in E \;\&\; 1 \leq i \leq |S| \;\&\; ap_i \neq ap_j \text{ if } i \neq j)\}$ */
5:   M is a set of pattern instances; /* $M = \{m_1, m_2, ...\}$ */
6:   $M \leftarrow null$;
7:   $Count$ is used for count the number of m;
8:   $Count \leftarrow 0$;
9:   /* $\frac{(E.size())!}{(E.size() - S.size())!}$ is the total number of non-duplicate pattern instances */
10: **while** $Count <= \frac{(E.size())!}{(E.size() - S.size())!}$ **do**
11:    E_1 is a set used to store $ep \in E$ has been checked;
12:    $m \leftarrow null$;
13:    $E_1 \leftarrow null$;
14:    **for all** ap in S **do**
15:       **while** $true$ **do**
16:          ep is an event randomly selected from E;
17:          **if** ep not in E_1 **then**
18:             $m.insert(ap, ep)$; /* ap is mapped to ep */
19:             $E_1.insert(ep)$; /* ep is labeled */
20:             break;
21:          **end if**
22:       **end while**
23:    **end for**/* build m */
24:    **if** m not in M **then**
25:       $M.insert(m)$;
26:       $count + +$;
27:    **end if**
28: **end while**/* build M */
29: **for all** m in M **do**
30:    P_{ins} is a copy of P; /* P_{ins} is used for instantiation */
31:    $P_{ins} \leftarrow P$;
32:    **for all** $node$ in P_{ins} **do**
33:       **if** $node.type == AtomicProp$ **then**
34:          **for all** m_i in m **do**
35:             **if** $m_i.ap == node.name$ **then**
36:                $node.name \leftarrow m_i.ep$;
37:             **end if**
38:          **end for**
39:       **end if**
40:    **end for**
41:    $P_s.insert(P_{ins})$; /* insert pattern instance P_{ins} into set P_s */
42: **end for**
43: **return** P_s
44: **end**

Algorithm 2. function CheckBasedonNF(P, τ)

Input: P: An instantiated PPTL pattern formula;
Input: τ: A program execution trace;
Output: *True* if τ satisfies P, *False* otherwise.
1: **begin**
2:   q is a boolean variable;
3:   q = CheckSatisfiability(P); /* check satisfiability of P [5] */
4:   **if** $\neg q$ **then**
5:     **return** *False*;
6:   **else**
7:     $P_{nf} = NF(P)$; /* transform P into its normal form [5] */
8:     **return** NFCheckTrace(P_{nf},τ);
9: **end if**
10: **end**

To check a disjunct, two cases need to be considered: (1)$P_e \wedge \varepsilon$ and (2)$P_c \wedge \bigcirc P_f$. For the first case, the function checks whether τ satisfies P_e and whether τ is empty. If both are true, $P_e \wedge \varepsilon$ is satisfied by τ. For the second case, the function checks whether τ satisfies P_c and whether $tail(\tau)$ satisfies P_f. If both are true, $P_c \wedge \bigcirc P_f$ is satisfied by τ. In checking whether τ satisfies the state formula P_e or P_c, Algorithm 4 StateFormulaCheck(P_s,τ) is called. For doing so, function StateFormulaCheck(P_s,τ) is simply to check the satisfiability of state formula P_s over τ by considering several syntax constructs of P_s.

Algorithm 3. function NFCheckTrace(P_{nf}, τ)

Input: P_{nf}: A PPTL formula in its normal form;
Input: τ: A program execution trace;
Output: *True* if τ satisfies P_{nf}, *False* otherwise.

1: **begin**
2:     $\tau_{bak} = \tau$;
3:     **switch** $P_{nf}.type$ **do**
4:         **case** *OrProp*
5:             q_1 is a boolean variable;
6:             q_1 = NFCheckTrace($P_{nf}.child_1$, τ);
7:             **if** q_1 **then** /* first disjunct is satisfied by τ */
8:                 **return** *True*;
9:             **else**/* select another disjunct */
10:                 $\tau = \tau_{bak}$;
11:                 **return** NFCheckTrace($P_{nf}.child_2$, τ);
12:             **end if**
13:         **case** *AndProp*
14:             $P_c = P_{nf}.child_1$; /* if $P_{nf}.child_2$ is ε, P_c stands for P_e */
15:             q_2 is a boolean variable;
16:             q_2 = StateFormulaCheck(P_c, τ); /* check satisfiability of P_c over τ */
17:             **if** q_2 **then**
18:                 **if** $P_{nf}.child_2.type$ is ε **then**
19:                     **if** $|\tau| == 0$ **then** /* check whether the trace is empty */
20:                         **return** *True*;
21:                     **else**
22:                         **return** *False*;
23:                     **end if**
24:                 **else**
25:                     $P_f = P_{nf}.child_2.child_1$; /* obtain next formula P_f */
26:                     **if** $|\tau| == 0$ **then**
27:                         **return** *False*;
28:                     **else**
29:                         $\tau = tail(\tau)$; /* update τ by its first proper suffix */
30:                         **return** CheckBasedOnNF(P_f, τ);
31:                     **end if**
32:                 **end if**
33:             **else**
34:                 **return** *False*;
35:             **end if**
36: **end**

Algorithm 4. function StateFormulaCheck(P_s, τ)

Input: P_s: A state PPTL formula;
Input: τ: A program execution trace;
Output: *True* if τ satisfies P_s, *False* otherwise.

```
 1: begin
 2:   switch P_s.type do
 3:     case OrProp /* P_s ≡ P_1 ∨ P_2 */
 4:       P_1 = P_s.child_1;
 5:       P_2 = P_s.child_2;
 6:       q_1 is a boolean variable;
 7:       q_1 = StateFormulaCheck(P_1, τ);
 8:       if q_1 then
 9:         return True;
10:       else
11:         return StateFormulaCheck(P_2, τ)
12:       end if
13:     case AndProp /* P_s ≡ P_1 ∧ P_2 */
14:       P_1 = P_s.child_1;
15:       P_2 = P_s.child_2;
16:       q_2 is a boolean variable;
17:       q_2 = StateFormulaCheck(P_1, τ);
18:       if q_2 then
19:         return StateFormulaCheck(P_2, τ)
20:       else
21:         return False;
22:       end if
23:     case NegationProp /* P_s ≡ ¬P_1 */
24:       P_1 = P_s.child_1;
25:       if StateFormulaCheck(P_1, τ) then
26:         return False;
27:       else
28:         return True;
29:       end if
30:     case AtomicProp /* P_s ≡ p */
31:       if head(τ) satisfies P_s then
32:         return True;
33:       else
34:         return False;
35:       end if
36:     case TrueProp /* P_s ≡ true */
37:       return True;
38:     case FalseProp /* P_s ≡ false */
39:       return False;
40: end
```

---

5   Conclusion

This paper presents an approach to mining PPTL specification from program execution traces. A tool PPTLMiner has been developed to support the proposed approach. This allows us to mine full regular temporal rules represented by PPTL formulas from traces. However, a mined PPTL formula has to be checked over all traces so as to ensure its validity. This is not a easy job since there might be error traces involved.

In the future, we will investigate how to evaluate the mined properties so that desired properties can be found. Further, we will optimize PPTLMiner to improve its mining quality and efficiency.

References

1. The Daikon Invariant Detector. http://plse.cs.washington.edu/daikon/
2. Autili, M., Grunske, L., Lumpe, M., Pelliccione, P., Tang, A.: Aligning qualitative, real-time, and probabilistic property specification patterns using a structured English grammar. IEEE Trans. Softw. Eng. 41(7), 1 (2015)
3. Duan, Z.: Temporal logic and Temporal Logic Programming. Science Press, Beijing (2005)
4. Duan, Z., Tian, C.: A practical decision procedure for propositional projection temporal logic with infinite models. Theoret. Comput. Sci. 554, 169–190 (2014)
5. Duan, Z., Tian, C., Zhang, L.: A decision procedure for propositional projection temporal logic with infinite models. Acta Informatica 45(1), 43–78 (2008)
6. Duan, Z., Tian, C., Zhang, N.: A canonical form based decision procedure and model checking approach for propositional projection temporal logic. Theor. Comput. Sci. 609, 544–560 (2016)
7. Duan, Z., Zhang, N., Koutny, M.: A complete proof system for propositional projection temporal logic. Theor. Comput. Sci. 497, 84–107 (2013)
8. Dupont, P., Lambeau, B., Damas, C., Lamsweerde, A.: The QSM algorithm and its application to software behavior model induction. Appl. Artif. Intell. 22(1&2), 77–115 (2008)
9. Dwyer, M.B., Avrunin, G.S., Corbett, J.C.: Patterns in property specifications for finite-state verification. In: Proceedings of the 1999 International Conference on Software Engineering (IEEE Cat. No.99CB37002), pp. 411–420 (1999)
10. Iegorov, O., Fischmeister, S.: Mining task precedence graphs from real-time embedded system traces. pp. 251–260 (2018)
11. Le, T.B., Lo, D.: Deep specification mining. In: Proceedings of the 27th ACM SIGSOFT International Symposium on Software Testing and Analysis, pp. 106–117 (2018)
12. Lemieux, C., Park, D., Beschastnikh, I.: General LTL specification mining (T). In: Proceedings of the 2015 IEEE/ACM International Conference on Automated Software Engineering (ASE), pp. 81–92 (2015)
13. Li, H., Shen, L.M., Ma, C., Liu, M.Y.: Role behavior detection method of privilege escalation attacks for android applications. Int. J. Perform. Eng. 15(6), 1631–1641 (2019)
14. Narayan, A., Cutulenco, G., Joshi, Y., Fischmeister, S.: Mining timed regular specifications from system traces. ACM Trans. Embed. Comput. Syst. 17(2), 1–21 (2018)

15. Pradel, M., Gross, T.R.: Automatic generation of object usage specifications from large method traces. In: Proceedings of the 2009 IEEE/ACM International Conference on Automated Software Engineering, pp. 371–382 (2009)
16. Ratcliff, S., White, D., Clark, J.: Searching for invariants using genetic programming and mutation testing. In: Proceedings of the 2011 Annual Genetic and Evolutionary Computation Conference, pp. 1907–1914 (2011)
17. Reger, G., Havelund, K.: What is a trace? A runtime verification perspective. In: Margaria, T., Steffen, B. (eds.) ISoLA 2016. LNCS, vol. 9953, pp. 339–355. Springer, Cham (2016). https://doi.org/10.1007/978-3-319-47169-3_25
18. Tian, C., Duan, Z.: Expressiveness of propositional projection temporal logic with star. Theor. Comput. Sci. **412**(18), 1729–1744 (2011)
19. Walkinshaw, N., Bogdanov, K., Holcombe, M., Salahuddin, S.: Reverse engineering state machines by interactive grammar inference. In: Proceedings of the 2007 Working Conference on Reverse Engineering, pp. 209–218 (2007)
20. Wasylkowski, A., Zeller, A.: Mining temporal specifications from object usage. In: Proceedings of the 2009 IEEE/ACM International Conference on Automated Software Engineering, pp. 295–306 (2009)
21. Yang, J., Evans, D., Bhardwaj, D., Bhat, T., Das, M.: Perracotta: mining temporal API rules from imperfect traces. In: Proceedings of the 2006 International Conference on Software Engineering, pp. 282–291 (2006)

An Improved Exact Algorithm for the Exact Satisfiability Problem

Gordon Hoi[(✉)]

School of Computing, National University of Singapore, 13 Computing Drive,
Block COM1, Singapore 117417, Republic of Singapore
e0013185@u.nus.edu

Abstract. The Exact Satisfiability problem, XSAT, is defined as the
problem of finding a satisfying assignment to a formula φ in CNF such
that exactly one literal in each clause is assigned to be "1" and the other
literals in the same clause are set to "0". Since it is an important variant
of the satisfiability problem, XSAT has also been studied heavily and has
seen numerous improvements to the development of its exact algorithms
over the years.

The fastest known exact algorithm to solve XSAT runs in $O(1.1730^n)$
time, where n is the number of variables in the formula. In this paper, we
propose a faster exact algorithm that solves the problem in $O(1.1674^n)$
time. Like many of the authors working on this problem, we give a DPLL
algorithm to solve it. The novelty of this paper lies on the design of the
nonstandard measure, to help us to tighten the analysis of the algorithm
further.

Keywords: XSAT · Measure and conquer · Exponential time
algorithms

1 Introduction

Given a propositional formula φ in conjunctive normal form (CNF), a common
question to ask would be if there is a satisfying assignment to φ. This is known
as the satisfiability problem, or SAT. SAT is seen to be a problem that is at
the center of computational complexity because it has been commonly used as
a framework to solve other combinatorial problems. In addition, SAT has found
many uses in practice as well. Some of these examples include: AI-planning,
software model checking, etc. [3].

Because of its importance, many other variants of the satisfiability problem
have also been explored. One such important variant is the Exact Satisfiability
problem, XSAT, where it asks if one can find a satisfying assignment such that
exactly one of the literal in each clause is assigned the value "1" and all other
literals in the same clause are assigned "0". All the mentioned problems, SAT
and XSAT, are both known to be NP-complete [1,2].

W. Wu and Z. Zhang (Eds.): COCOA 2020, LNCS 12577, pp. 304–319, 2020.
https://doi.org/10.1007/978-3-030-64843-5_21

In this paper[1], we will focus on the XSAT problem and in particular, exact algorithms to solve it. XSAT is a well-studied problem and has seen numerous improvements [4–7] to it, with the fastest solving it in $O(1.1730^n)$ time.

We propose an algorithm to solve XSAT in $O(1.1674^n)$ time, using polynomial space. Like most of the earlier authors, we will design a Davis-Putnam-Logemann-Loveland (DPLL) [11] style algorithm to solve this problem. We build our work upon the works of the earlier authors. While the earlier authors all used the standard measure, which is the number of variables n, the novelty here lies on the design of a nonstandard measure to help us to tighten the analysis of the algorithm further.

2 Preliminaries

In this section, we will introduce some definitions and also the techniques needed to understand the analysis of DPLL algorithm.

2.1 Branching Factor and Vector

Our algorithm is a DPLL style algorithm, or also known as a branch and bound algorithm. DPLL algorithms are recursive in nature and have two kinds of rules associated with them: Simplification and Branching rules. Simplification rules help us to simplify a problem instance or to act as a case to terminate the algorithm. Branching rules on the other hand, help us to solve a problem instance by recursively solving smaller instances of the problem. To help us to better understand the execution of a DPLL algorithm, the notion of a search tree is commonly used. We can assign the root node of the search tree to be the original problem, while subsequent child nodes are assigned to be the smaller instances of the problem whenever we invoke a branching rule. For more information of this area, one may refer to the textbook written by Fomin and Kratsch [9].

Let μ be our measure of complexity. To analyse the running time of DPLL algorithms, one just needs to bound the number of leaves generated in the search tree. This is because the complexity of such algorithm is proportional to the number of leaves, modulo polynomial factors, that is, $O(poly(|\varphi|, \mu) \times$ number of leaves in the search tree$) = O^*($number of leaves in the search tree$)$, where the function $poly(|\varphi|, \mu)$ is some polynomial dependent on $|\varphi|$ and μ, and $O^*(f(\mu))$ is the class of all function g bounded by some polynomial $p(\cdot) \times f(\mu)$.

Then we let $T(\mu)$ denote the maximum number of leaf nodes generated by the algorithm when we have μ as the parameter for the input problem. Since the search tree is only generated by applying a branching rule, it suffices to consider the number of leaf nodes generated by that rule (as simplification rules take only polynomial time). To do this, we use techniques in [10]. Suppose a branching rule has $r \geq 2$ children, with $t_1, t_2, \ldots, t_r$ decrease in measure for these children.

---

[1] Some details in this paper has been omitted due to space constraints, the full paper is available in arxiv: https://arxiv.org/pdf/2010.03850.pdf.

Then, any function $T(\mu)$ which satisfies $T(\mu) \geq T(\mu-t_1)+T(\mu-t_2)+\ldots T(\mu-t_r)$, with appropriate base cases, would satisfy the bounds for the branching rule. To solve the above linear recurrence, one can model this as $x^{-t_1}+x^{-t_2}+\ldots+x^{-t_r} = 1$. Let β be the unique positive root of this recurrence, where $\beta \geq 1$. Then any $T(\mu) \geq \beta^\mu$ would satisfy the recurrence for this branching rule. In addition, we denote the branching factor $\tau(t_1, t_2, \ldots, t_r)$ as β. If there are k branching rules in the DPLL algorithm, then the overall complexity of the algorithm is the largest branching factor among all k branching rules; i.e. $c = max\{\beta_1, \beta_2, \ldots, \beta_k\}$, and therefore the time complexity of the algorithm is bounded above by $O^*(c^\mu)$.

Next, we will introduce some known results about branching factors. If $k < k'$, then we have that $\tau(k', j) < \tau(k, j)$, for all positive k, j. In other words, comparing two branching factor, if one eliminates more weights, then this will result in a a smaller branching factor. Suppose that $i + j = 2\alpha$, for some α, then $\tau(\alpha, \alpha) \leq \tau(i, j)$. In other words, a more balanced tree will result in a smaller branching factor.

Finally, the correctness of DPLL algorithms usually follows from the fact that all cases have been covered.

2.2   Definitions

Definition 1. *A clause is a disjunction of literals. We also say that a clause is a multiset of literals. A k-literal clause is a clause C with $|C| = k$. Let C be a clause, then δ is a subclause of C if $\delta \subset C$.*

Suppose we have $C = (a \vee b \vee c \vee d)$, then C is a 4-literal clause. In addition, $\delta = (a \vee b \vee c)$ is a subclause of C. We may also write $C = (\delta \vee d)$. For now, we define a clause as a multiset of literals as the same literal may appear twice in a clause. When no simplification rules[2] can be applied, we may then think of a clause as a set of literals instead.

Definition 2. *Two clauses are called neighbours if they share at least a common variable. Two variables are called neighbours if they appear in some clause together. Let C_1 and C_2 be two clauses that are neighbours. Now if $|C_1 \cap C_2| = k \geq 2$, we say that C_1 and C_2 have k overlapping variables. In addition, the variables in $C_1 - C_2$ and $C_2 - C_1$ are known as outside variables. Let $|C_1 - C_2| = i$ and $|C_2 - C_1| = j$, $i, j \geq 1$. Then we say that there are $i + j$ outside variables, in an i-j orientation.*

Note that this definition (i-j orientation) is strictly used for the case when we have $k \geq 2$ overlapping variables between any two clauses[3]. We only consider $i, j \geq 1$ because if i or j is 0, then one of the clause must be a subclause of the other. Consider the following example.

---

[2] More details later in Sect. 3, when the algorithm is given.
[3] Mainly in Sect. 4.3.

Example 1. Let $C_1 = (a \lor b \lor c \lor d \lor e)$ and $C_2 = (d \lor e \lor f \lor g \lor h)$. Then in this case, since $C_1 \cap C_2 = \{d, e\}$, there are 2 literals in the intersection and we say that C_1 and C_2 have 2 overlapping variables. In addition, $C_1 - C_2 = \{a, b, c\}$ and $C_2 - C_1 = \{f, g, h\}$. Now, we say C_1 and C_2 have 6 outside variables in a 3-3 orientation.

Definition 3. *Let x be a literal. Now the degree of a variable, $deg(x)$, denotes the total number of times that the literal x and $\neg x$ appears in φ. If $deg(x) \geq 3$, then we say that the variable x is heavy. Further, for a heavy variable x that appears in clauses $C_1, C_2, ..., C_k$, $k \geq 3$, we say that x is in $(l_1, l_2, ..., l_k)$, where $|C_i| = l_i$, $1 \leq i \leq k$. Adding on to this,*

1. *if $\neg x$ appears in C_i, then we say x is in $(l_1, l_2, ..., \neg l_i, ..., l_k)$.*
2. *if $|C_i| \geq l_i$, then we say x is in $(l_1, l_2, ..., \geq l_i, ..., l_k)$.*

Note that if x is a heavy variable, we will only use this definition that x is in $(l_1, l_2, ..., l_k)$, whenever given any two clauses that x is in, they have at most 1 overlapping variable between them.

Example 2. Suppose we have the following clauses: $(x \lor a \lor b \lor c \lor d)$, $(\neg x \lor e \lor f \lor g)$, $(x \lor h \lor i \lor j \lor k)$. Then in this case, we have x in $(5, \neg 4, 5)$. We can also say that x is in $(\geq 4, \neg 4, 5)$ and we use "$\geq i$" whenever we just need to know that the clause length is at least i. Note that the order in which the clause length is presented here does not matter, i.e. $(5, \neg 4, 5)$ can also be written as $(\neg 4, 5, 5)$.

Definition 4. *We say that two variables, x and y, are linked when we can deduce either $x = y$ or $x = \neg y$. When this happens, we can proceed to remove one of the linked variable, either x or y, and replace by the other.*

Suppose we have a 3-literal clause $(0 \lor x \lor y)$, by definition of being exact satisfiable, we can deduce that $x = \neg y$ in this case, and proceed to remove one variable, say x, by replacing all instances of x by $\neg y$ and $\neg x$ by y respectively.

Definition 5. *Given a formula φ and δ a multiset of literals.*

1. *If $|\delta| = 1$, then let x be the only literal in δ. Now $\varphi[x = 1]$ and $\varphi[x = 0]$ denotes the new formula obtained after assigning $x = 1$ and $x = 0$ respectively.*
2. *If $|\delta| \geq 2$, then we only allow the following when $\delta \subset C$, for some clause C in φ. $\varphi[\delta = 1]$ denotes the new formula obtained after assigning all the $C - \delta$ to be 0. By definition of being exact-satisfiable, this is saying that the "1" must only appear in one of the literals in δ. Therefore, all the literals in $C - \delta$ are assigned 0. On the other hand, $\varphi[\delta = 0]$ denotes the new formula obtained after assigning all the literals in δ to be 0.*

Similarly, given two literals x and y, we say that $\varphi[x = y]$ is the new formula obtained by replacing all occurrences of x by y.

Example 3. Suppose $\varphi = (a \lor b \lor c \lor d)$ and $\delta = (a \lor b \lor c)$. Then $\varphi[\delta = 1] = (a \lor b \lor c \lor 0)$ since we are saying that the "1" appears in either a, b, or c. On the other hand, $\varphi[\delta = 0] = (0 \lor 0 \lor 0 \lor d)$.

Definition 5.1 is used whenever we are branching a variable. On the other hand, Definition 5.2 is used when we want to branch a subclause, especially when we deal with $k \geq 2$ overlapping variables between two clauses. In addition, when we have a subclause δ such that $|\delta| = 2$, then let x and y be the literals in δ. Saying that $\varphi[\delta = 1]$ is the same as saying $\varphi[x = \neg y]$, linking $x = \neg y$.

A common technique used by the earlier authors is known as resolution. If there are clauses $C_1 = (C \lor x)$ and $C_2 = (C' \lor \neg x)$, where x is a literal, C and C' are subclauses of C_1 and C_2 respectively, then we can replace every clause $(x \lor \alpha)$ by $(C' \lor \alpha)$, and every clause $(\neg x \lor \beta)$ by $(C \lor \beta)$, for some subclause α, β. In addition, every literal in $C \cap C'$ can be assigned 0. This can help us to remove literals appearing as x and $\neg x$ in different clauses.

2.3   A Nonstandard Measure

Instead of using the number of variables as our measure, we will design a nonstandard measure to help us to improve the worst case time complexity of our algorithm. Let $\{x_1, x_2, ..., x_n\}$ be the set of variables in φ. For $1 \leq i \leq n$, we define the weight w_i for x_i as:

$$w_i = \begin{cases} 0.8823, & \text{if } x_i \text{ is on a 3-literal clause such that all 3 variables in that} \\ & \text{clause do not have the same neighbour} \\ 1, & \text{otherwise} \end{cases}$$

We then define our choice of measure as $\mu = \sum_i w_i$, where $\mu \leq n$ by definition. This value of 0.8823 is chosen by a linear search program to bring down the overall runtime of the algorithm to as low as possible. Therefore, we have $O(c^\mu) \subseteq O(c^n)$, for some constant $c \geq 1$ by definition.

Example 4. Suppose we have the following clauses: $(x \lor y \lor z \lor a), (x \lor u \lor w \lor v), (x \lor r \lor s \lor t), (a \lor v \lor t)$ and the clause $(y \lor e \lor f)$. The variables x, z, u, w, r and s have weight 1. By definition, variables a, v and t are assigned the weight 1 because these variables have x as their neighbour. Variables y, e, f have weights 0.8823 because these 3 variables do not have the same neighbour.

3   Algorithm

All of our simplification rules and branching rules are designed to ensure that the overall measure does not increase after applying them. That is, the measure before applying any of the rule, μ, and the measure after applying any of the rule, μ', is always $\mu' \leq \mu$. We call our DPLL algorithm $XSAT(.)$. Note that if every variable x has $deg(x) \leq 2$, then we can solve XSAT in polynomial time [5].

With this in mind, we'll design our algorithm by branching all heavy variables. Note that each line of the algorithm has decreasing priority; Line 1 has higher priority than Line 2, Line 2 than Line 3 etc. Let α, β, δ be subclauses.

Algorithm: $XSAT$

Input: A formula φ

Output: 1 if φ is exact satisfiable, else 0

1. If there is a clause that is not exact-satisfiable, then return 0.
2. If there is a clause $C = (1 \vee \delta)$ or $C = (x \vee \neg x \vee \delta)$, for some variable x, then set all literals in δ to 0 and drop the clause C. Return $XSAT(\varphi[\delta = 0])$.
3. If there exist a clause $C = (0 \vee \delta)$, then update $C = \delta$. Update φ' as the new formula and return $XSAT(\varphi')$.
4. If there exist a 1-literal clause containing the literal l, then drop that clause. Return $XSAT(\varphi[l = 1])$.
5. If there exist a 2-literal clause containing the literal l and l', then drop that clause. Return $XSAT(\varphi[l = \neg l'])$.
6. If there exist a clause C with a literal l appearing at least twice, then return $XSAT(\varphi[l = 0])$.
7. If there exist clauses of the type $(\alpha \vee x \vee y)$ and $(\beta \vee x \vee \neg y)$, for some literal x and y, then return $XSAT(\varphi[x = 0])$.
8. If there exist clauses of the type $(\alpha \vee x \vee y)$ and $(\beta \vee \neg x \vee \neg y)$, then return $XSAT(\varphi[x = \neg y])$.
9. If there are clauses C and C' such that $C \subset C'$, then set all literals in $\delta = C' - C$ as 0, remove the clause C' and return $XSAT(\varphi[\delta = 0])$.
10. If there is a variable x appearing in at least three 3-literal clauses, then we either simplify it or branch x. If we simplify it, let φ' be the new formula after simplifying. Return $XSAT(\varphi')$. If we branch x, return $XSAT(\varphi[x = 1]) \vee XSAT(\varphi[x = 0])$.
11. If there are clauses C_1 containing x and C_2 containing $\neg x$, for some literal x. Then we apply resolution and let φ' be the new formula. Return $XSAT(\varphi')$.
12. If there are clauses C_1 and C_2 such that they have $k \geq 2$ overlapping variables, then check if the outside variables are in a 1-j orientation, $j \geq 1$. If yes, then let φ' be the new formula after applying some changes [4], then return $XSAT(\varphi')$. Else, let $\delta = C_1 \cap C_2$ and we branch the subclause δ. Return $XSAT(\varphi[\delta = 1]) \vee XSAT(\varphi[\delta = 0])$.
13. If there is a heavy variable x, then branch x. Return $XSAT(\varphi[x = 1]) \vee XSAT(\varphi[x = 0])$.
14. If all the variables x have $deg(x) \leq 2$, then solve the problem in polynomial time. Return 1 if exact-satisfiable, else return 0.

Lines 1 to 9, 11 are simplification rules, while Lines 10, 12 and 13 are branching rules. Line 14 takes only polynomial time to decide if there is an exact-satisfiable assignment to φ when $deg(x) \leq 2$ for all variable x. Line 1 says that if any clause is found not to be exact-satisfiable, then we can return 0. Line 2

---

[4] Full details given in the Sect. 4.3.

says if a clause contains a "1", then the other literals appearing in the clause must be assigned 0. Line 3 says that if we have a clause containing "0", then we can update that clause by dropping off the constant "0". Line 4 says that if we encounter a 1-literal clause, then that literal must be assigned 1. Line 5 says that if there are any 2-literal clause containing some literals x and y, then we can just link the two literals $x = \neg y$ together. After Line 5 of the algorithm, every clause in φ must be at least a 3-literal clause.

Line 6 deals with clauses containing the same literals that appear at least twice. After Line 6, every clause can only contain any literal at most once. Lines 7 and 8 deals with two clauses that have at least two variables in common, in different permutations. After Line 8, if any two clauses have at least two variables in common, then this implies that they have share at least two literals in common. After Line 9, no clause is a subclause of a larger clause in φ.

In Line 10, we deal with variables that appears in at least three 3-literal clauses. We deal with this case early on because it helps us to reduce the number of cases that we need to handle later on while branching in Sect. 4.3 and 4.4. In Line 11, we deal with clauses C_1 containing the literal x and C_2 containing $\neg x$. Line 12 deals with two clauses having $k \geq 2$ overlapping variables. First, we deal with such cases in a 1-j orientation, $j \geq 1$, followed by such cases in an i-j orientation, $i, j \geq 2$. After which, any two clauses must have only at most one variable in common. Line 13 deals with heavy variables. After that, no heavy variables exist in the formula φ and we can proceed to solve the problem in polynomial time in Line 14. We have therefore covered all cases in our algorithm.

4   Analysis of Algorithm

In this section, we will analyze the overall runtime of the algorithm given in the previous section. Note that simplification rules only take polynomial time. Therefore, we will analyse from Lines 10 to 13 of the algorithm.

Due to the way we design our measure, if a k-literal clause drops to a 3-literal clause, $k > 3$, we can factor in the change of measure of $1 - 0.8823 = 0.1177$ for each of the variables in the 3-literal clause, if there is no common neighbour. Whenever we deal with a 3-literal clause, for simplicity, we will treat all the variables in it as having a weight of 0.8823 instead of 1. This gives us an upper bound on the branching factor without the need to consider all kinds of cases.

In addition, when we are dealing with 3-literal clause, sometimes we have to increase the measure after linking. For example, suppose we have the clause $(0 \vee x \vee y)$, for some literals x and y. Now we can link $x = \neg y$ and proceed to remove one variable, say x. This means that the 3-literal clause is removed and the surviving variable y, may no longer be appearing in any other 3-literal clause. Therefore, the weight of y increases from 0.8823 to 1. This increase in weight means that we increase our measure and therefore, we have to factor in "−0.1177" whenever we are linking variables in a 3-literal clause together.

4.1 Line 10 of the Algorithm

Line 10 of the algorithm deals with a variable appearing in at least three 3-literal clauses. We can either simplify the case further, or branch x. At this point in time, Lines 11 and 12 of the algorithm has not been called. This means that we have to deal with literals appearing as x and $\neg x$, and that given any two clause, it is possible that they have $k \geq 2$ overlapping variables.

Lemma 1. *If x appears in at least three 3-literal clauses, we either simplify this case further or we branch x, incurring at most $O(1.1664^n)$ time.*

Proof. Now let x be appearing in two 3-literal clauses. We first deal with the case that for any two 3-literal clauses, there are $k \geq 2$ overlapping variables. Since simplification rules do not apply anymore, the only case we need to handle here is $(x \vee y \vee z)$ and $(x \vee y \vee w)$, for some literals w, y, z. In this case, we can link $w = z$ and drop one of these clauses.

For the remaining cases, x must appear in three 3-literal clause and there are no $k \geq 2$ overlapping variables between any two of the 3-literal clause. Therefore, for the remaining case, x must be in $(3, 3, 3)$ or $(3, 3, \neg 3)$.

For the $(3, 3, 3)$ case, let the clauses be $(x \vee v_1 \vee v_2)$, $(x \vee v_3 \vee v_4)$ and $(x \vee v_5 \vee v_6)$, where $v_1, ..., v_6$ are unique literals. We branch $x = 1$ and $x = 0$ here. When $x = 1$, we remove the variables $v_1, ..., v_6$ and x itself. This gives us a change of measure of 7×0.8823. When $x = 0$, we remove x, and link $v_1 = \neg v_2$, $v_3 = \neg v_4$ and $v_5 = \neg v_6$. This gives us a change of measure of $4 \times 0.8823 - 3 \times 0.1177$. This gives us a branching factor of $\tau(7 \times 0.8823, 4 \times 0.8823 - 3 \times 0.1177) = 1.1664$. The case for $(\neg 3, \neg 3, \neg 3)$ is symmetric.

For the $(3, 3, \neg 3)$ case, let the clauses be $(x \vee v_1 \vee v_2)$, $(x \vee v_3 \vee v_4)$ and $(\neg x \vee v_5 \vee v_6)$, where $v_1, ..., v_6$ are unique literals. Again, we branch $x = 1$ and $x = 0$. When $x = 1$, we remove x and the variables $v_1, ..., v_4$, and link the variables $v_5 = \neg v_6$. This gives us a change of measure of $6 \times 0.8823 - 0.1177$. When $x = 0$, we remove x, v_5, v_6, and link the variables $v_1 = \neg v_2$ and $v_3 = \neg v_4$. This gives us a change of measure of $5 \times 0.8823 - 2 \times 0.1177$. This gives us a branching factor of $\tau(6 \times 0.8823 - 0.1177, 5 \times 0.8823 - 2 \times 0.1177) = 1.1605$. The case for $(3, \neg 3, \neg 3)$ is symmetric. Therefore, this takes at most $O(1.1664^n)$ time.

4.2 Line 11 of the Algorithm

Line 11 of the algorithm applies resolution. One may note that our measure is designed in terms of the length of the clause. Therefore, it is possible that the measure may increase from 0.8823 to 1 after applying resolution. Applying resolution on k-literal clauses, $k \geq 4$, is fine because doing so will not increase the measure. On the other hand, applying on 3-literal clauses will increase the length of the clause and hence, increase the weights of the other variables in the clause, and finally, the overall measure. Therefore to apply resolution on such cases, we have to ensure that the removal of the variable x, is more than the increase of the weights of from 0.8823 to 1 ($1 - 0.8823 = 0.1177$). To give an upper bound, we assume that x has weight 0.8823. Taking $0.8823 \div 0.1177 = 7.5$. Therefore, if there are more than 7.5 variables increasing from 0.8823 to 1, then we refrain from doing so. This translates to x appearing in at least four 3-literal clauses. However, this has already been handled by Line 10 of the algorithm. Hence, when we come to Line 11 of the algorithm, we can safely apply resolution.

4.3   Line 12 of the Algorithm

In this section, we deal with Line 12 of the algorithm. Since simplification rules do not apply anymore when this line is reached, we may then think of clauses as sets (instead of multiset) of literals, since the same literal can no longer appear more than once in the clause. In addition, literals x and $\neg x$ do not appear in the formula. Now, we fix the following notation for the rest of this section. Let C_1 and C_2 be any clauses given such that $C_1 \cap C_2 = \delta$, with $|\delta| \geq 2$ overlapping variables, in an i-j orientation, where $|C_1 - C_2| = i$ and $|C_2 - C_1| = j$, where $i, j \geq 1$. We divide them into 3 parts, let $L = C_1 - C_2$ (left), $R = C_2 - C_1$ (right) and δ (middle). For example, in Example 1, we have $L = \{a, b, c\}$ and $R = \{f, g, h\}$. We first deal with the cases $i = 1, j \geq 1$.

Lemma 2. *The time complexity of dealing with two clauses with $k \geq 2$ overlapping variables, having 1-j orientation, $j \geq 1$, is at most $O(1.1664^n)$.*

Proof. If $j = 1$, then let $x \in L$ and $y \in R$. Then we can just link $x = y$ and this case is done. If $j \geq 2$, then let $C_1 = (x \vee \delta)$ and $C_2 = (\delta \vee R)$. From C_1, we know that $\neg x = \delta$. Therefore, C_2 can be rewritten has $(\neg x \vee R)$. With the clauses $C_1 = (x \vee \delta)$ and $C_2 = (\neg x \vee R)$, apply Line 11 (resolution) to remove the literals x and $\neg x$, or to apply branching to get a complexity of $O(1.1664^n)$.

Now, we deal with the case of having $k \geq 2$ overlapping variables in an i-j orientation, $i, j \geq 2$. Note that during the course of branching $\delta = 0$, when a longer clause drops to a 3-literal clause L (or R), then we can factor in the change of measure of $1 - 0.8823 = 0.1177$ for each of the variable in L (Normal Case). However, there are situations when we are not allowed to factor in this change. Firstly, when there is a common neighbour to the variables in L (Case 1). Secondly, when some or all variables in L already have weights 0.8823, which means the variable appears in further 3-literal clauses prior to the branching (Case 2). Details of Case 1 is in our full paper in arxiv.

For Case 2, we pay special attention to the outside variables in an i-j orientation, $i \leq 3$ or $j \leq 3$. This is because when $i, j \geq 4$, and while branching $\delta = 0$, the change in measure only comes from the removal of variables in δ. On the other hand, when $\delta = 1$, we can remove additional variables not in L, R and δ, whenever we have a variable having weight 0.8823. Let s be a variable not appearing in L, R or δ. We show all the possibilities below.

Case 2: The variables in L (or R) appear in further 3-literal clauses.

1. Case 2.1: A pair of 3-literal clauses containing s, with the neighbours of s L, R, and δ. For example, if we have $(l_1 \vee l_2 \vee \delta)$ and $(\delta \vee r_2 \vee r_1)$, and the two 3-literal clauses $(s \vee l_1 \vee r_1)$ and $(s \vee l_2 \vee r_2)$. Details of Case 2.1 is in our full paper in arxiv.

2. Case 2.2: Not Case 2.1. In other words, there is no such s that appears in two 3-literal clauses, where the neighbours of s are the variables in L and R. In this case, we have 3-literal clauses, each containing a variable from L, a variable from R, and another variable not from L, R, and δ.

By Line 10 of the algorithm, s cannot appear in a third 3-literal clause. Therefore, we must either have Case 2.1 or Case 2.2. Case 2.2 arises when it is not Case 2.1; when there is no such s, appearing in two 3-literal clauses, with the neighbours of s appearing in L and R. Case 2.2 represents the case where we can have $(l \lor s \lor r)$, where $l \in L$, $r \in R$ (s appears in exactly one 3-literal clause). In summary, Case 1, 2.1 and a few other additional cases are discussed in our full arxiv paper[5], and we will only show the Normal Case and Case 2.2.

Lemma 3. *The time complexity of dealing with two clauses with $k \geq 2$ overlapping literals, having i-j orientation, $i, j \geq 2$, is at most $O(1.1674^n)$ time.*

Proof. Let any two clauses be given with $k \geq 2$ overlapping variables and have at least 4 outside variables in a 2-2 orientation. We will show the Normal Case first, followed by Case 2.2 (only for outsides variables $i \leq 3$ or $j \leq 3$). For Case 2.2, and the appearance of each 3-literal clause, note that when branching $\delta = 1$, we can remove all the variables in the 3-literal clause, giving us 3×0.8823 per 3-literal clause that appears in this manner. Let h denote the number of further 3-literal clauses for Case 2.2 encountered below. In addition, for Case 2.2 having odd number of outside variables, we treat the variable not in any 3-literal clause as having weight 1, acting as an upper bound to our cases.

For $k = 2$, and we have 4 outside variables in a 2-2 orientation. When $\delta = 1$, we remove all 5 variables in total (4 outside variables and 1 via linking in δ). When $\delta = 0$, we remove 2 variables in δ and another 2 from linking the variables in L and R. This gives us a change of measure of 4. Therefore, we have $\tau(5,4) = 1.1674$. For Case 2.2, we can have at most two 3-literal clauses here. This gives us $\tau(h \times (3 \times 0.8823) + 2 \times (2 - h) + 1, 2 + 2 \times 0.8823)$, $h \in \{1,2\}$, which is at max of 1.1612, when $h = 1$. This completes the case for 4 outside variables.

Case: 5 outside variables in a 2-3 orientation. Branching $\delta = 1$ will remove all outside variables, and 1 of the linked variable in δ. This gives us a change of measure of 6. On the other hand, branching $\delta = 0$ will allow us to remove all the variables in δ, link the 2 variables in L, and factor in the change of measure for the remaining variables in R. This gives us $\tau(6, 3 + 3 \times 0.1177) = 1.1648$. For Case 2.2, we have at most two 3-literal clauses appearing in both L and R. Then we have $\tau(h \times (3 \times 0.8823) + 2 \times (2 - h) + 2, 2 + 0.8823 + 0.1177)$, $h \in \{1,2\}$, which is at max of 1.1636 when $h = 1$. This completes the case for 5 outside variables.

Case: 6 outside variables in a 3-3 orientation. Branching $\delta = 1$ will remove all 6 outside variables in L and R, and also remove one more variable by linking the variables in δ, a total of 7 variables. On the other hand, when $\delta = 0$, we remove all the variables in δ and also factor in the change of measure for the variables in L

---

and R, a total of $2+6\times0.1177$ for this branch. This gives us $\tau(7, 2+6\times0.1177) = 1.1664$. When Case 2.2 applies, then we can have at most three 3-literals clauses appearing. This gives us a $\tau(h\times(3\times0.8823)+2\times(3-h)+1, 2+2\times(3-h)\times0.1177)$, $h \in \{1, 2, 3\}$, which is at max of 1.1641 when $h = 1$. This completes the case for 6 outside variables.

Case: 7 outside variables in a 3-4 orientation. Branching $\delta = 1$ will allow us to remove all 7 outside variables, and 1 variable from δ via linking. This gives us a change of measure of 8. On the other hand, when $\delta = 0$, we can factor in a change of measure of 3×0.1177 from the variables. This gives us $\tau(8, 2 + 3 \times 0.1177) = 1.1630$. For Case 2.2, there are at most three 3-literal clauses between L and R. This gives us $\tau(h \times (3 \times 0.8823) + 2 \times (3 - h) + 2, 2 + (3 - h) \times 0.1177)$, which is at max of 1.1585 when $h = 1$. This completes the case for 7 outside variables.

Case: $p \geq 8$ outside variables. Branching $\delta = 1$ allows us to remove all p outside variables, and an additional variable from linking in δ, which has a change of measure of 9. For the $\delta = 0$ branch, we remove two variables. This gives us $\tau(p + 1, 2) \leq \tau(9, 2) = 1.1619$. This completes the case for $k = 2$ overlapping variables.

Now we deal with $k = 3$ overlapping variables. If there are 4 outside variables in a 2-2 orientation, then branching $\delta = 1$ will allow us to remove all 4 outside variables, which is a change of measure of 4. On the other hand, branching $\delta = 0$ will allow us to remove all the variables in δ, as well as link the two variables in L and R, removing a total of 5 variables. This gives us $\tau(4, 5) = 1.1674$. When we have Case 2.2, then we have at most two 3-literal clauses appearing. This gives us $\tau(h \times (3 \times 0.8823) + 2 \times (2 - h), 3 + 2 \times 0.8823)$, $h \in \{1, 2\}$, which is at max of 1.1588 when $h = 1$. This completes the case for 4 outside variables in a 2-2 orientation.

Case: 5 outside variables in a 2-3 orientation. Branching $\delta = 1$ will allow us to remove all 5 outside variables. On the other hand, branching $\delta = 0$ will allow us to remove all the variables in δ, as well as an additional variable from linking the two variables in L, a total of 4 variables. This gives us $\tau(5, 4) = 1.1674$. For Case 2.2, we can have at most two 3-literal clauses occurring. For simplicity, we treat the 3rd variable in R as having weight 0.8823. Then we have $\tau(h \times (3 \times 0.8823) + 2 \times (2 - h) + 1, 3 + 0.8823 + 0.1177)$, $h \in \{1, 2\}$, which is at max of 1.1563 when $h = 1$. This completes the case for 5 outside variables.

Case: 6 outside variables in a 3-3 orientation. When branching $\delta = 1$, we remove all 6 outside variables. When branching $\delta = 0$, we remove all 3 variables in δ, and we can factor in the change of measure for these of 0.1177 for these 6 variables. This gives us $\tau(6, 3 + 6 \times 0.1177) = 1.1569$. In Case 2.2, we can have at most three 3-literal appearing in L and R. Then we have $\tau(h \times (3 \times 0.8823) + 2 \times (3 - h), 3 + 2 \times (3 - h) \times 0.1177)$, $h \in \{1, 2, 3\}$, which is at max of 1.1526 when $h = 1$. This completes the case for 6 outside variables.

Case: $p \geq 7$ outside variables. Then branching $\delta = 1$ will allow us to remove at least 7 variables, and when $\delta = 0$, we remove all the variables in δ. This gives us $\tau(p, 3) \leq \tau(7, 3) = 1.1586$. For Case 2.2, we can have at most $h \leq \lfloor \frac{p}{2} \rfloor$ 3-literals clauses. Then we have $\tau(h \times (3 \times 0.8823) + 2 \times (\lfloor \frac{p}{2} \rfloor - h) + 1, 3)$, which is at max of 1.1503 when $h = 1$ and $\lfloor \frac{p}{2} \rfloor = 3$. This completes the case for $k = 3$ overlapping variables.

Now, we deal with the case of $k = 4$ overlapping variables. When we have 4 outside variables in a 2-2 orientation, then branching $\delta = 1$ will allow us to remove all 4 outside variables. On the other hand, when $\delta = 0$, we remove all the variables in δ, and link the two variables in L and R. This gives us a change of measure of 6. Therefore, we have $\tau(4, 6) = 1.1510$. For Case 2.2, we can have at most two 3-literal clauses. Then we have $\tau(h \times (3 \times 0.8823) + 2 \times (2 - h), 4 + 2 \times 0.8823)$, $h \in \{1, 2\}$, which is at max of 1.1431 when $h = 1$. This completes the case for 4 outside variables.

Case: $p \geq 5$ outside variables. Branching $\delta = 1$ will allow us to remove at least 5 variables. Branching $\delta = 0$ will remove all the variables in δ. This gives us $\tau(p, 4) \leq \tau(5, 4) = 1.1674$. For Case 2.2, we can have at most $h \leq \lfloor \frac{p}{2} \rfloor$ number of 3-literal clauses. We have $\tau(h \times (3 \times 0.8823) + 2 \times (\lfloor \frac{p}{2} \rfloor - h) + 1, 4)$, which is at max of 1.1563 when $h = 1$ and $\lfloor \frac{p}{2} \rfloor = 2$. This completes the case for 5 outside variables.

Finally for $k \geq 5$ overlapping variables and $p \geq 4$ outside variables, branching $\delta = 1$ will remove at least 4 variables, while branching $\delta = 0$ will remove at least 5 variables. This gives us $\tau(p, k) \leq \tau(4, 5) = 1.1674$. For Case 2.2, there can be at most $h \leq \lfloor \frac{p}{2} \rfloor$ number of 3-literal clauses. Then we have $\tau(h \times (3 \times 0.8823) + 2 \times (\lfloor \frac{p}{2} \rfloor) - h), k) \leq \tau(h \times (3 \times 0.8823) + 2 \times (\lfloor \frac{p}{2} \rfloor - h), 5)$, which is at max of 1.1547 when $h = 1$ and $\lfloor \frac{p}{2} \rfloor = 2$.

This completes the case for $k \geq 2$ overlapping variables and the max branching factor while executing this line of the algorithm is 1.1674.

4.4   Line 13 of the Algorithm

Now, we deal with Line 13 of the algorithm, to branch off heavy variables in the formula. After Line 12 of the algorithm, given any two clauses C_1 and C_2, there can only be at most only 1 variable appearing in them. Cases 1 and 2 in the previous section will also apply here (we deal with Case 1 in the full paper). In Sect. 4.3, we paid special attention to L and R when $|L| = 3$ or $|R| = 3$. Here, we pay special attention to x being in 4-literal clauses, because after branching $x = 0$, it will drop to a 3-literal clause. Since we have dealt with $(3, 3, 3)$ case earlier, here, we'll deal with the remaining cases; cases from $(3, 3, \geq 4)$ to $(\geq 5, \geq 5, \geq 5)$.

For Case 2, there are some changes. Here, we are dealing with 3 clauses instead of 2 in the previous section, therefore, there will be more permutation of 3-literal clauses to consider. Recall previously that we dealt with a case where s appears in two 3-literal clauses in Case 2.1 of Sect. 4.3. We shift this case to our arxiv paper.

Let s be a variable not appearing in the clauses that we are discussing about. We define Case 2.1 and Case 2.2, while keeping the Normal Case as before.

Case 2.1: If we have clauses $C_1 = (a_1 \vee a_2 \vee ... \vee x)$, $C_2 = (b_1 \vee b_2 \vee ... \vee x)$, $C_3 = (c_1 \vee c_2 \vee ... \vee x)$, $(s \vee a_1 \vee b_1)$ and $(s \vee b_2 \vee c_1)$. Note that some clause C_2

has 2 variables as neighbours of s. For the proof below, we will use this (a clause having two variables in it as neighbours of s) notation to denote the worst case.

Case 2.2: No such s occurs where we have a clause that has two variables in it as neighbours of s. Therefore, for each 3-literal clause appearing, it will only contain two variables from the clauses, and a new variable not in the three clauses. For example, if we have $(x \lor v_1 \lor v_2 \lor v_3)$, $(x \lor v_4 \lor v_5 \lor v_6)$ and $(x \lor v_7 \lor v_8 \lor v_9)$. Here, we consider 3-literal clauses appearing as $(s \lor v_1 \lor v_4)$. Similar to Case 2.2 in the previous section.

In Case 2, s cannot appear in the third 3-literal clause due to Line 10. Therefore, only Case 2.1 or 2.2 can happen. Our cases here are complete.

Lemma 4. *The time complexity of branching heavy variables is $O(1.1668^n)$.*

Proof. Let x be a heavy variable. Given (l_1, l_2, l_3), then there are $|l_1| + |l_2| + |l_3| - 2$ unique variables in these clauses. Let h denote the number of 3-literal clauses as shown in Case 2.2 above. We will give the Normal case, Case 1 (only for $(4, 4, 4)$), Case 2.1 and Case 2.2. For Case 2.1, we will treat all variables as having weight 0.8823 to lessen the number of cases we need to consider. In addition, we handle the cases in the following order: $(3, 3, \geq 4)$, then $(3, \geq 4, \geq 4)$ etc.

$(3, 3, \geq 4)$. We start with $(3, 3, 4)$. Branching $x = 1$ will allow us to remove all the variables in this case, with a change in measure of $5 \times 0.8823 + 3$. When $x = 0$, we will have a change in measure of $3 \times 0.8823 - 2 \times 0.1177$, and when the 4-literal clause drops to a 3-literal clause, another 3×0.1177. This gives us $\tau(5 \times 0.8823 + 3, 3 \times 0.8823 + 0.1177) = 1.1591$. If Case 2.1 occurs, then the worst case here would be that one of the 3-literal clauses (in $(3, 3, 4)$), contain two variables that are neighbours to s. We branch $x = 1$ and $x = 0$ to get $\tau(7 \times 0.8823 + 2, 3 \times 0.8823) = 1.1526$. If Case 2.2 occurs, then we can have at most three 3-literal clauses. We branch $x = 1$ and $x = 0$. Then we have $\tau(5 \times 0.8823 + h \times (2 \times 0.8823) + (3 - h), 3 \times 0.8823 + (3 - h) \times 0.1177 - 2 \times 0.1177)$, $h \in \{1, 2, 3\}$, which is at max of 1.1526 when $h = 1$. This completes the case for $(3, 3, 4)$. Next, we deal with $(3, 3, \geq 5)$. For such a case, when $x = 1$, we remove all variables, which gives us a change of measure of $5 \times 0.8823 + 4$. When $x = 0$, we remove x and link up the two variables in the 3-literal clauses. This gives us a change of $3 \times 0.8823 - 2 \times 0.1177$. This gives us $\tau(5 \times 0.8823 + 4, 3 \times 0.8823 - 2 \times 0.1177) = 1.1562$. If Case 2.1 or 2.2 applies here, then we give an upper bound to these cases by treating all variables as having weight 0.8823. When $x = 1$, we remove all 9 variables, this gives us 9×0.8823. On the other hand, when $x = 0$, we have $3 \times 0.8823 - 2 \times 0.1177$. This gives us at most $\tau(9 \times 0.8823, 3 \times 0.8823 - 2 \times 0.1177) = 1.1620$. This completes the case for $(3, 3, \geq 5)$ and hence $(3, 3, \geq 4)$.

$(3, \geq 4, \geq 4)$. We start with $(3, 4, 4)$. Branching $x = 1$ will allow us to remove all the variables, giving us $6 + 3 \times 0.8823$. Branching $x = 0$, gives us a change of $2 \times 0.8823 - 0.1177 + 6 \times 0.1177$ instead. This gives us $\tau(6 + 3 \times 0.8823, 2 \times 0.8823 + 5 \times 0.1177) = 1.1551$. For Case 2.1, the worst case happens when we have two variables in any of the 4-literal clauses as neighbours of s. Branching $s = 1$ will allow us to remove 7 variables, where one of which is via linking of a variable in a 3-literal clause, giving us $7 \times 0.8823 - 0.1177$. When $s = 0$, we remove x, s and 2 variables via linking in the 3-literal clause, giving us $4 \times 0.8823 - 2 \times 0.1177$. This gives $\tau(7 \times 0.8823 - 0.1177, 4 \times 0.8823 - 2 \times 0.1177) = 1.1653$. For Case 2.2, we can have at most three 3-literal clauses appearing across the two 4-literal clauses. Then branching $x = 1$ and $x = 0$ gives us $\tau(3 \times 0.8823 + h \times (3 \times 0.8823) + 2 \times (3 - h), 2 \times 0.8823 - 0.1177 + (3 - h) \times 2 \times 0.1177)$, $h \in \{1, 2, 3\}$, which is at max of 1.1571 when $h = 3$. This completes the case for $(3, 4, 4)$. For $(3, 4, \geq 5)$, branching $x = 1$ will allow us to remove all variables, giving us $3 \times 0.8823 + 7$. On the other hand, branching $x = 0$ will allow us to remove x, link a variable in the 3-literal clause and factor in the change in measure for the 4-literal clauses. This gives us $\tau(3 \times 0.8823 + 7, 2 \times 0.8823 + 2 \times 0.1177) = 1.1547$. For Case 2.1 and Case 2.2, we can find a variable s that does not appear in any of the clauses. We give an upper bound for this case by treating all variables as having weight 0.8823. When $x = 1$, we remove all 10 variables and s. This gives us 11×0.8823. When $x = 0$, we remove x and link up the other variable in the 3-literal clause, giving us $2 \times 0.8823 - 0.1177$. This gives $\tau(11 \times 0.8823, 2 \times 0.8823 - 0.1177) = 1.1666$. This completes the case for $(3, 4, \geq 5)$. Finally, for $(3, \geq 5, \geq 5)$, we give an upper bound by treating all the variables as having weight 0.8823, to deal with the Normal Case, Case 2.1 and 2.2. Branching $x = 1$ gives us 11×0.8823. When $x = 0$, this gives us $2 \times 0.8823 - 0.1177$. Therefore, we have at most $\tau(11 \times 0.8823, 2 \times 0.8823 - 0.1177) = 1.1666$ for this case. This completes the case for $(3, \geq 5, \geq 5)$ and hence $(3, \geq 4, \geq 4)$.

$(4, 4, 4)$. When $x = 1$, we remove all variables. This gives us a change of measure of 10. On the other hand, when $x = 0$, we have a change of measure of $1 + 9 \times 0.1177$. This gives us a branching factor of $\tau(10, 1 + 9 \times 0.1177) = 1.1492$. When Case 2.1 occurs, then one of the 4-literal clause must have 2 variables in it that are neighbours to s. Suppose we have $(x \vee a_1 \vee a_2 \vee a_3)$, $(x \vee b_1 \vee b_2 \vee b_3)$, $(x \vee c_1 \vee c_2 \vee c_3)$, $(s \vee a_1 \vee b_1)$ and $(s \vee b_2 \vee c_1)$. Then we branch $b_1 = 1$ and $b_1 = 0$. When $b_1 = 1$, then $s = a_1 = b_2 = b_3 = x = 0$. Since $s = b_2 = 0$, then $c_1 = 1$. Therefore, we must have $c_2 = c_3 = 0$ and we can link up $a_2 = \neg a_3$. Now, x must have weight 1, else earlier cases would have handled it. This gives a change of measure of $9 \times 0.8823 + 1$. On the other hand, when $b_1 = 0$, we link up $\neg a_1 = s$ (no increase in weights for s), giving us a change of measure of $2 \times 0.8823 + 0.1177$. This gives a branching factor of at most $\tau(9 \times 0.8823 + 1, 2 \times 0.8823 + 0.1177) = 1.1668$. When Case 2.2 arises, then the worst case happens when we have three 3-literal clauses appearing across the 4-literal clauses. This gives us $\tau(9 \times 0.8823 + 4, 1 + 3 \times 0.1177) = 1.1577$. This completes the case for $(4, 4, 4)$.

$(4, 4, \geq 5)$. When $x = 1$, we remove all 11 variables. When $x = 0$, we remove x and factor in the change of measure from the 4-literal clauses, giving us $1 + 6 \times 0.1177$. This gives us a branching factor of $\tau(11, 1 + 6 \times 0.1177) = 1.1509$. If Case 2.1 occurs, and two variables from a 4-literal clause is a neighbour to s, then choose the variable that is a neighbour to s to branch, such that we can remove all the variables in the 5-literal clause (same technique as above). The same upper bound of 1.1668 will also apply here. If two variables from a 5-literal clause is a neighbour to s, then branch any of these two variables to get the same bound of 1.1668. If Case 2.2 applies, then there are at most three 3-literal clauses between the two 4-literal clauses. Then we have $\tau(h \times (3 \times 0.8823) + 5, 1 + 2 \times (3 - h) \times 0.1177)$, $h \in \{1, 2, 3\}$, which is at max of 1.1637 when $h = 3$. This completes the case for $(4, 4, \geq 5)$.

$(4, \geq 5, \geq 5)$. When $x = 1$, we remove all 12 variables. When $x = 0$, we remove x and factor in the change of measure of $1 + 3 \times 0.1177$. Therefore, we have $\tau(12, 1 + 3 \times 0.1177) = 1.1551$. When Case 2.1 occurs, then follow the same technique as given in $(4, 4, \geq 5)$ to get the upper bound of 1.1668 here. When Case 2.2 occurs, then we can have at most three 3-literal clauses appearing across the 4-literal and the 5-literal clauses. This gives us $\tau(h \times (3 \times 0.8823) + 2 \times (3 - h) + 6, 1 + (3 - h) \times 0.1177)$, $h \in \{1, 2, 3\}$, which is at max of 1.1550 when $h = 3$. This completes the case of $(4, \geq 5, \geq 5)$.

$(\geq 5, \geq 5, \geq 5)$. When $x = 1$, we remove 13 variables and when $x = 0$, we remove only x. This gives us $\tau(13, 1) = 1.1632$. If Case 2.1 occurs, follow the same technique as given in $(4, 4, \geq 5)$ when two variables in the 5-literal clause are neighbours to s. This gives us the same upper bound of 1.1668. For Case 2.2, the worst case occurs when every variable in $(\geq 5, \geq 5, \geq 5)$ has weight 1, which gives 1.1632 (Normal Case). This is because when $x = 0$, we can only remove x and not factor in any other change in measure. On the other hand, when any of the variables have weight 0.8823, this means we can remove additional variables when $x = 1$. This completes the case for $(\geq 5, \geq 5, \geq 5)$. Hence, Line 14 of the algorithm runs in $O(1.1668^n)$ time.

Therefore, putting all the lemmas together, we have the following result:

Theorem 1. *The algorithm runs in $O(1.1674^n)$ time.*

References

1. Schaefer, T.J.: The complexity of satisfiability problems. In: Proceedings of STOC 1978, pp. 216–226. ACM (1978)
2. Cook, S.: The complexity of theorem proving procedures. In: Proceedings of 3rd Annual ACM Symposium on Theory of Computing (STOC), pp. 151–158 (1971)
3. Marques-Silva, J., Practical applications of Boolean Satisfiability. In: 9th International Workshop on Discrete Event Systems, pp. 74–80. IEEE (2008)
4. Schroeppel, R., Shamir, A.: A $T = O(2^{n/2})$, $S = O(2^{n/4})$ algorithm for certain NP-complete problems. SIAM J. Comput. **10**(3), 456–464 (1981)
5. Monien, B., Speckenmeyer, E., Vornberger, O.: Upper bounds for covering problems. Methods Oper. Res. **43**, 419–431 (1981)
6. Byskov, J.M., Madsen, B.A., Skjernaa, B.: New algorithms for exact satisfiability. Theoret. Comput. Sci. **332**(1–3), 515–541 (2005)
7. Dahllöf, V.: Exact algorithms for exact satisfiability problems. Linköping Studies in Science and Technology, Ph.D. Dissertation no 1013 (2006)
8. Fomin, F.V., Grandoni, F., Kratsch, D.: A measure and conquer approach for the analysis of exact algorithms. J. ACM (JACM) **56**(5), 25 (2009)

9. Fomin, F.V., Kratsch, D.: Exact Exponential Algorithms. Texts in Theoretical Computer Science. An EATCS Series. Springer, Heidelberg (2010). https://doi.org/10.1007/978-3-642-16533-7
10. Kullmann, O.: New methods for 3-SAT decision and worst-case analysis. Theoret. Comput. Sci. **223**(1–2), 1–72 (1999)
11. Davis, M., Logemann, G., Loveland, D.: A machine program for theorem proving. Commun. ACM **5**(7), 394–397 (1962)

Transforming Multi-matching Nested Traceable Automata to Multi-matching Nested Expressions

Jin Liu, Zhenhua Duan$^{(\boxtimes)}$, and Cong Tian

ICTT and ISN Laboratory, Xidian University,
Xi'an 710071, People's Republic of China
liujin_xd@163.com, zhhduan@mail.xidian.edu.cn, ctian@mail.xidian.edu.cn

Abstract. Multi-matching nested relation consists of a sequence of linearly ordered positions, *call*, *internal*, and *return*, augmented with one-to-one, one-to-n or n-to-one matching nested edges from *call* to *return*. After word encoding by introducing tagged letters, Multi-matching Nested Words (MNW) are obtained over a tagged alphabet. Then Multi-matching Nested Expressions (MNE) and Multi-matching Nested Traceable Automata (MNTA) are defined over MNWs. Further, a transformation method from MNTA to MNE is proposed. An extra state is introduced as the unique initial state. Three kinds of labelled arcs are created for different kinds of transitions. They are merged according to specific strategies and meanwhile the expressions are calculated. As a result, the corresponding MNE can be obtained.

Keywords: Multi-matching nested relation · One-to-n · n-to-one · Automata · Expression · Transformation

1 Introduction

Finite Automata (FA) are a useful model for many important kinds of hardware and software, such as the lexical analyzer of a typical compiler, the software for scanning large bodies of text or designing and checking the behavior of digital circuits [1]. They can describe regular languages but not languages with one-to-one matching structures.

The concept of matching relations is proposed in [2,3], consisting a sequence of linearly ordered positions (*call*, *internal* and *return*), augmented with nesting edges connecting *calls* and *returns*. The edges do not cross creating a properly nested hierarchical structure and some of the edges are allowed to be pending. The model of nested words is obtained by assigning each position with a symbol, for instance, the execution of a sequential structured program and the SAX

This research is supported by the NSFC Grant Nos. 61732013 and 61751207, the National Key Research and Development Program of China (No. 2018AAA0103202), and Shaanxi Key Science and Technology Innovation Team Project (No. 2019TD-001).

W. Wu and Z. Zhang (Eds.): COCOA 2020, LNCS 12577, pp. 320–333, 2020.
https://doi.org/10.1007/978-3-030-64843-5_22

representation of XML documentations [4–6]. Finite automata are defined over nested words, named Nested Word Automata (NWA), where the action depends on the type of positions [2,3]. At a *call*, the state propagates along both linear and nesting outgoing edges, while both linear and nesting incoming edges determine the new state for a *return*. One can easily interpret NWA as pushdown automata, named Visibly Pushdown Automata (VPA), by linear word encoding of nested words. The automaton pushes onto the stack when reading a *call* while pops the stack only at a *return*, and does not use the stack at an *internal* [3,7].

Traceable Automata (TA) are proposed in [8–10] as an extension of FA to describe the control construction of menu of user interface in integrated software engineering environment. Different tools constitutes a state set and all commands of each tool form the set of input symbols. When the environment receives a command at a state during the process, corresponding subroutine is executed. Then it turns to a specific state after the execution is completed. If the command is an exit, it will trace back to the previous state before the current one. To do this, a state stack is introduced to preserve a part of running history in order to determine the subsequent state. Two kinds of transitions are defined in TA: the current state is pushed into the stack for record in a push transition while the automaton traces to the state on the top stack which is popped then in a trace transition. Each push transition corresponds to a specific trace one. TA can describe parenthesis matching languages.

Nevertheless, the model above cannot describe one-to-n and n-to-one matching structures, which also play an important role in real-world applications. It is familiar in C programs that memory can be dynamically allocated, then referenced and manipulated by pointers, finally freed [11]. The allocation is corresponding to a specific free, which is a one-to-one matching relation. Multiple usages (read or written) and the final free constitute an n-to-one matching relation. An erroneous pointer manipulation leads to double frees. The allocation and double frees form a one-to-n matching relation. Further, an n-to-one encryption and authentication scheme is used in group-oriented cryptography for the scenario that the number of the receivers is much fewer than the number of the senders. Actually, a receiver may serve millions of senders [12,13]. Besides, one-to-n and n-to-one relations are common in mapping relations among tables in database systems [14].

With this motivation, multi-matching nested relations are studied, where nesting edges can share the same *call* or *return*. One *call* (or *return*) is corresponding to multiple *returns* (or *calls*). Its natural word encoding form is obtained by assigning each *call*, *internal* and *return* position with different tagged symbols. Accordingly, multi-matching nested expressions and multi-matching nested traceable automata are defined over multi-matching nested words. In a MNTA, the input symbols are also recorded in the stack besides states, they are used to determine the subsequent transitions together. A transformation method from MNTA to MNE is proposed. An extra state X is introduced as the unique initial state and different kinds of labeled arcs are created for transitions. The arcs are merged according to the characteristics of transitions in MNTA. Finally, the expressions for the labeled arcs from X to each final state are calculated as the one of the automaton.

The rest of paper is organized as follows. In Sect. 2, we present the definition of multi-matching nested relations and its word encoding. Section 3 defines MNTA and MNE over multi-matching nested words. A transformation method from MNTA to MNE is illustrated in Sect. 4. Finally, we conclude in Sect. 5.

2 Multi-matching Nested Words

2.1 Multi-matching Nested Relation

Given a linear sequence, the positions are divided into three types: *call*, *internal* and *return*. Matching relation is defined in [3] over a linear sequence for describing one-to-one matching relation where each *call* is matched with a specific *return*. To make the *call* or *return* shareable, we extend the definition to multi-matching nested relations including one-to-one, one-to-n and n-to-one. And the edges starting at $-\infty$ and edges ending at $+\infty$ are used to model pending edges. Assume that $-\infty < i, j < +\infty$ for integers i and j.

Definition 1 (Multi-matching Nested Relation). *A multi-matching nested relation $\rightsquigarrow$ of length m, $m \geq 0$, is a subset of $\{-\infty, 1, 2, \cdots, m\} \times \{1, 2, \cdots, m, +\infty\}$ such that*

1. *nesting edges go only forward: if $i \rightsquigarrow j$ then $i < j$;*
2. *nesting edges do not cross: if $i \rightsquigarrow j$ and $i' \rightsquigarrow j'$, then it is not the case that $i < i' \leq j < j'$;*
3. *only one end of a nesting edge can be shared with others: for $i \rightsquigarrow j$ and $i' \rightsquigarrow j'$ where $1 \leq i = i' < j' < j \leq m$ (or $1 \leq i < i' < j' = j \leq m$), there does not exist k, where $k \neq i, i', j, j'$, such that any of $k \rightsquigarrow i$, $k \rightsquigarrow j$, $j \rightsquigarrow k$, $k \rightsquigarrow j'$ and $j' \rightsquigarrow k$ (or $k \rightsquigarrow i'$, $i' \rightsquigarrow k$, $k \rightsquigarrow i$, $i \rightsquigarrow k$ and $j \rightsquigarrow k$) holds.*

In the third condition of Definition 1, consider the case that $i \rightsquigarrow j$ and $i \rightsquigarrow j'$ where $1 \leq i < j' < j \leq m$. Let $k \neq i, j, j'$. For position i, it is shareable in this case, hence only the nesting edge emanating from i is allowed, i.e. $i \rightsquigarrow k$. For position j (or j'), it cannot be shared with other nesting edges besides $i \rightsquigarrow j$ ($i \rightsquigarrow j'$), i.e. neither $k \rightsquigarrow j$ ($k \rightsquigarrow j'$) nor $j \rightsquigarrow k$ ($j' \rightsquigarrow k$) is permitted.

For n different nesting edges sharing position i: $i \rightsquigarrow j_1$, $i \rightsquigarrow j_2$, $\cdots$, $i \rightsquigarrow j_n$ where $1 \leq i < j_1 < j_2 < \cdots < j_n \leq m$ and $n \geq 1$, position i is denoted as a *call position* (*call* for short) while each position j_k, $1 \leq k \leq n$, is its *return-successor* and denoted as a *return position* (*return* for short). The *return* j_n is called *outermost-return*; while other *returns* are identified as *inner-returns*. Specifically, for the nesting edge $i \rightsquigarrow +\infty$, i is called a *pending call*; otherwise, i is a *matched call*.

Similarly, for n different nesting edges sharing position j: $i_1 \rightsquigarrow j$, $i_2 \rightsquigarrow j$, $\cdots$, $i_n \rightsquigarrow j$ where $1 \leq i_1 < i_2 < \cdots < i_n < j \leq m$ and $n \geq 1$, position j is a *return* and each *call* position i_k, $1 \leq k \leq n$, is its *call-predecessor*, where i_1 is indicated as an *outermost-call* while other *calls* are called *inner-calls*. Specifically, j is denoted as a *pending return* when there is a nesting edge $-\infty \rightsquigarrow j$; otherwise a *matched return*.

A position is an *internal position* (*internal* for short) if it is neither a *call* nor a *return*. A multi-matching nested relation is *well-matched* if there is no *pending call* or *pending return*.

Different from matching relation in [3], multi-matching nested relation requires in the third condition of Definition 1 that for n different nesting edges, they can share the same *call* (or *return*), but their *returns* (or *calls*) cannot. Specially, a multi-matching nested relation is a matching relation if there do not exist two nesting edges sharing the same *call* or *return*.

Note that for a position i: (i) i can be classified to only one type of *call*, *internal* and *return*; (ii) i is not shareable if i is a *pending call* or *pending return*.

2.2 Word Encoding

Given a multi-matching relation, a word can be obtained by assigning each position with a symbol. For distinction of the classifications of positions, different tags are introduced. Symbols decorated by tags '$<$' at the left superscript and '$>$' at the right superscript distinguish *calls* and *returns*. The dot '$\cdot$' at the top of a symbols can distinguish *calls* (or *returns*) and *inner-calls* (or *inner-returns*). For an *internal*, no tag is needed. Given an alphabet Σ, a tagged alphabet $\hat{\Sigma} = \Sigma_c \cup \dot{\Sigma}_c \cup \Sigma_i \cup \dot{\Sigma}_r \cup \Sigma_r$ is introduced, where $\Sigma_c = \{{}^{<}a | a \in \Sigma\}$, $\dot{\Sigma}_c = \{{}^{<}\dot{a} | a \in \Sigma\}$, $\Sigma_i = \Sigma$, $\dot{\Sigma}_r = \{\dot{a}^{>} | a \in \Sigma\}$ and $\Sigma_r = \{a^{>} | a \in \Sigma\}$ are the symbols of *call*, *inner-call*, *internal*, *inner-return* and *return*, respectively.

Definition 2. *For any symbols* ${}^{<}a_1 \in \Sigma_c$, ${}^{<}\dot{a}_2 \in \dot{\Sigma}_c$, $\dot{a_3}^{>} \in \dot{\Sigma}_r$ *and* $a_4^{>} \in \Sigma_r$, *(i)* ${}^{<}a_1$ *and* $a_4^{>}$; *(ii)* ${}^{<}a_1$, ${}^{<}\dot{a}_2$ *and* $a_4^{>}$; *(iii)* ${}^{<}a_1$, $\dot{a_3}^{>}$ *and* $a_4^{>}$ *are indicated to be matched if and only if (i)* $a_1 = a_4$; *(ii)* $a_1 = a_2 = a_4$; *(iii)* $a_1 = a_3 = a_4$ *respectively.*

Then, we give the definition of multi-matching nested words.

Definition 3 (Multi-matching Nested Word, MNW). *Given a multi-matching nested relation* $\rightsquigarrow$ *of length* m, *a multi-matching nested word* $w = w_1 w_2 \cdots w_m$ *over alphabet* $\hat{\Sigma}$ *can be obtained which satisfies the following conditions:*

1. *if there are* n *nesting edges* $i \rightsquigarrow j_1$, $i \rightsquigarrow j_2$, $\cdots$, $i \rightsquigarrow j_n$ *where* $1 \le i < j_1 < j_2 < \cdots < j_n \le m$ *and* $n \ge 1$, *then* $w_i \in \Sigma_c$, $w_{j_1} = w_{j_2} = \cdots = w_{j_{n-1}} \in \dot{\Sigma}_r$, $w_{j_n} \in \Sigma_r$, *and they are matched;*
2. *if there are* n *nesting edges* $i_1 \rightsquigarrow j$, $i_2 \rightsquigarrow j$, $\cdots$, $i_n \rightsquigarrow j$ *where* $1 \le i_1 < i_2 < \cdots < i_n < j \le m$ *and* $n \ge 1$, *then* $w_{i_1} \in \Sigma_c$, $w_{i_2} = w_{i_3} = \cdots = w_{i_n} \in \dot{\Sigma}_c$, $w_j \in \Sigma_r$, *and they are matched;*
3. $w_k \in \Sigma_i$ *if position* k *is an internal.*

Definition 3 requires that for n nesting edges sharing the same *call* or *return*, all symbols assigned to *calls* and *returns* of these edges must be matched. The set of all multi-matching nested words over $\hat{\Sigma}$ are denoted as $MNW(\hat{\Sigma})$. Note that due to the requirement of symbol matching, it can be obtained $MNW(\hat{\Sigma}) \subset \hat{\Sigma}^*$.

For example, given a multi-matching relation $\rightsquigarrow = \{(1,2), (1,7), (4,6), (5,6)\}$ of length 7, where 3 is an *internal*, $\{(1,2), (1,7)\}$ forms a one-to-n matching and $\{(4,6), (5,6)\}$ is an n-to-one matching. Suppose symbols a and b are assigned to the two matching relations while i indicates the *internal*, a multi-matching nested word $n = \overset{\curvearrowleft}{a} \overset{\curvearrowleft}{a} i \overset{\curvearrowright}{b} \overset{\curvearrowleft}{b} \overset{\curvearrowright}{a}$ is obtained.

3   Multi-matching Nested Languages

3.1   Multi-matching Nested Expressions

To describe multi-matching nested words over the tagged alphabet $\hat{\Sigma}$, we introduce multi-matching nested expressions. An expression e is denoted to be well-matched if there is no *pending call* or *pending return* in e.

Definition 4 (Multi-matching Nested Expression, MNE).   *A multi-matching nested expression e over $\hat{\Sigma} = \Sigma_c \cup \dot{\Sigma}_c \cup \Sigma_i \cup \dot{\Sigma}_r \cup \Sigma_r$ is inductively defined as follows:*

$$e \rightarrow \epsilon \mid \emptyset \mid x \mid (\overset{\curvearrowleft}{a}e_n \overset{\curvearrowleft}{a}e_{n-1} \cdots \overset{\curvearrowleft}{a}e_1 \overset{\curvearrowright}{a})^* \mid (\overset{\curvearrowleft}{a}e_1 \overset{\curvearrowright}{a} e_2 \overset{\curvearrowright}{a} \cdots e_n \overset{\curvearrowright}{a})^* \mid e + e \mid$$
$$e_r \cdot e_c \mid e_{rc}^* \mid (e)$$

where $x \in \Sigma_c \cup \Sigma_i \cup \Sigma_r$, $\overset{\curvearrowleft}{a} \in \Sigma_c$, $\overset{\curvearrowleft}{a} \in \dot{\Sigma}_c$, $\overset{\curvearrowright}{a} \in \dot{\Sigma}_r$, $\overset{\curvearrowright}{a} \in \Sigma_r$, $n \geq 1$.

Basically, a multi-matching nested expression e can be an empty expression ϵ or an empty set $\emptyset$. For the expression of a single symbol x, *inner-call* or *inner-return* can not exist without matched *call* and *return*.

As for expressions $(\overset{\curvearrowleft}{a}e_n \overset{\curvearrowleft}{a}e_{n-1} \cdots \overset{\curvearrowleft}{a}e_1 \overset{\curvearrowright}{a})^*$ and $(\overset{\curvearrowleft}{a}e_1 \overset{\curvearrowright}{a} e_2 \overset{\curvearrowright}{a} \cdots e_n \overset{\curvearrowright}{a})^*$, each nested expression e_i, $1 \leq i \leq n$, must be well-matched. The binary operation $+$ (called *union*) is the same as that in regular expression [1]. The expression $e_r \cdot e_c$ means the *concatenation* of e_r and e_c which satisfies that e_r containing *pending calls* and e_c containing *pending returns* are not allowed at the same time. For simplicity, the concatenation operator can be omitted. e_{rc}^* signifies the unary operation $*$ (called *closure*) augmented with the restriction that words expressed by e_{rc} cannot contain *pending calls* and *pending returns* simultaneously. Besides, the expression (e), a parenthesized e, is also a multi-matching nested expression. And the parentheses are used to group operands.

The set of words described by a multi-matching nested expression e constitutes a multi-matching nested language, denoted by $L(e)$.

The precedence order of the operators in multi-matching nested expressions is as follows. The unary operator closure $(*)$ binds stronger than the binary ones. The binary operator concatenation $(\cdot)$ takes precedence over union $(+)$. Sometimes, parentheses are used to group operands exactly as we choose.

3.2   Multi-matching Nested Traceable Automata

Traceable automata are the model that can describe one-to-one matching relations. In order to extend to multi-matching nested relations, we need to adapt the TA model by recording both input symbols and states in the stack to identify whether a one-to-n or n-to-one matching relation.

Definition 5 (Multi-matching Nested Traceable Automata, MNTA).
*A multi-matching nested traceable automaton is a tuple $M = (Q, \hat{\Sigma}, \delta, Q_0, F, \Gamma)$,
where*

1. *Q is a finite set of states;*
2. *$Q_0 \subseteq Q$ is the set of initial states;*
3. *$F \subseteq Q$ is the set of final states;*
4. *$\hat{\Sigma} = \Sigma_c \cup \dot{\Sigma}_c \cup \Sigma_i \cup \dot{\Sigma}_r \cup \Sigma_r$ is a finite set of input symbols, where Σ_c, $\dot{\Sigma}_c$, Σ_i, $\dot{\Sigma}_r$ and Σ_r denote call, inner-call, internal, inner-return and return symbols, respectively;*
5. *$\Gamma \subseteq (\Sigma_c \cup \dot{\Sigma}_c \cup \dot{\Sigma}_r) \times Q \cup \{\bot\}$ is a finite set of stack elements, where $\bot$ is a special bottom-of-stack symbol; and*
6. *δ is the transition relation consisting of three parts:*
   (a) *a push transition relation $\delta_p \subseteq Q \times \Gamma \times \Sigma_c \times Q \times (\Sigma_c \times Q \times \Gamma)$*
   (b) *an update transition relation $\delta_u \subseteq Q \times \Gamma \times (\dot{\Sigma}_c \cup \Sigma_i \cup \dot{\Sigma}_r) \times Q \times \Gamma$*
   (c) *a trace transition relation $\delta_t \subseteq Q \times \Gamma \times \Sigma_r \times \{trace\}$*

The transitions in M can be classified into three categories. When reading a *call* $\overset{\shortmid}{}a$ at state q, both q and $\overset{\shortmid}{}a$ are pushed into the stack; meanwhile, the state updates to q', denoted as $(q, \gamma, \overset{\shortmid}{}a, q', \overset{\shortmid}{}aq\gamma)$. When reading an *internal*, M takes a simple state transition, named internal update transition. For transition $(q, \overset{\shortmid}{}ap/\overset{\shortmid}{}\dot{a}p, \overset{\shortmid}{}\dot{a}, q', \overset{\shortmid}{}\dot{a}p)$, where $\overset{\shortmid}{}ap$ or $\overset{\shortmid}{}\dot{a}p$ is abbreviated as $\overset{\shortmid}{}ap/\overset{\shortmid}{}\dot{a}p$, the state changes from q to q' and the top stack element is modified from $\overset{\shortmid}{}ap/\overset{\shortmid}{}\dot{a}p$ to $\overset{\shortmid}{}\dot{a}p$, denoted as a call update transition. Similarly, a return update transition is $(q, \overset{\shortmid}{}ap/\dot{a}\hspace{-0.5mm}\overset{\shortmid}{}p, \dot{a}\hspace{-0.5mm}\overset{\shortmid}{}, q', \dot{a}\hspace{-0.5mm}\overset{\shortmid}{}p)$ where the top stack is updated to $\dot{a}\hspace{-0.5mm}\overset{\shortmid}{}p$. Upon a *return* $a\hspace{-0.5mm}\overset{\shortmid}{}$, the top stack element $\overset{\shortmid}{}ap/\overset{\shortmid}{}\dot{a}p/\dot{a}\hspace{-0.5mm}\overset{\shortmid}{}p$ is popped and M traces back to p, denoted as a trace transition. Note that if the top of stack is $\bot$, the input symbol is read only and the stack remains unchanged. Note that the input symbol must be matched with the one on the top of stack in a trace or call/return update transition transition.

Formally, a stack σ is a finite word over Γ. The set of all stacks is denoted as $St = (\Gamma \setminus \{\bot\})^* \cdot \{\bot\}$ where $\bot$ signifies the empty stack. Notation $|\sigma|$ stands for the length of the stack. Especially, $|\sigma| = 0$ if $\sigma = \bot$.

A *configuration* of M is a pair (q, σ) where $q \in Q$, and $\sigma \in St$. A *run* of M on a word $w = w_1 w_2 \cdots w_n$ is a finite non-empty sequence of configurations $\rho = (q_0, \sigma_0) \overset{w_1}{\rightarrow} (q_1, \sigma_1) \overset{w_2}{\rightarrow} \cdots \overset{w_n}{\rightarrow} (q_n, \sigma_n)$ where $q_0 \in Q_0$ and $\sigma_0 = \bot$. A run ρ is an accepting run if in the last configuration, $q_n \in F$ and $\sigma_n \in (\Sigma_c \times Q)^* \cdot \{\bot\}$ which means the stack does not contain *inner-call* or *inner-return* symbols. Hence, a word w is accepted by M if there is an accepting run of M on w. The language of M, denoted by $L(M)$, is the set of words accepted by M.

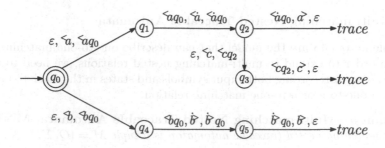

Fig. 1. An example of MNTA

For the graphical presentation of MNTA, a transition $(q, \gamma, x, q', \gamma')$ is depicted as an edge from q to q' labelled with a triple (γ, x, γ') recording the input symbol and the change of the top stack. Especially, for a push transition, γ and γ' can be denoted as ϵ and xq respectively. Note that if $x \in \Sigma_i$, γ and γ' can be omitted since the stack does not change. For a trace transition $(q, \gamma, x, trace)$, there is an edge from q to $trace$ labelled with (γ, x, ϵ), where $trace$ can appear more than once in a transition graph. For clarity, push transitions are depicted in green, trace ones in red, call and return update ones in orange and internal update ones in blue. Figure 1 presents an example MNTA accepting $\{((\overset{\curvearrowleft}{a}\overset{\curvearrowright}{a}(\overset{\curvearrowleft}{c}\overset{\curvearrowright}{c})^*\overset{\curvearrowright}{a})^* + (\overset{\curvearrowleft}{b}\overset{\curvearrowright}{b})^*)^*\}$.

4   A Transformation Method from MNTA to MNE

Now we illustrate in details how to transform MNTA to MNE. At first, different kinds of labelled arcs are introduced for transitions in MNTA.

4.1   Labelled Arcs

Given a MNTA $M = (Q, \hat{\Sigma}, \delta, Q_0, F, \Gamma)$, for any two states q and p in Q, a transition δ from q to p can be denoted as $\delta_{q \to p}$. For a word w, let $w' = w_1 w_2 \cdots w_n$ be a well-matched subword of w. A run-segment $\rho = (q, \sigma) \overset{w_1}{\to} (q_1, \sigma_1) \overset{w_2}{\to} \cdots \overset{w_n}{\to} (q_n, \sigma_n)$ over w' is a tracing run-segment if $\delta_{q \to q_1}$ is a push transition and $\delta_{q_{n-1} \to q_n}$ is its corresponding trace transition where $q = q_n$, $\sigma = \sigma_n$, $w_1 \in \Sigma_c$ and $w_n \in \Sigma_r$ are matched.

For any two states q and p, a labeled arc from q to p is denoted as $Arc_{q \to p}$. According to the characteristics of transitions and accepting conditions of MNTA, the labeled arcs are classified into three types. Let $\gamma, \gamma_1, \gamma_2$ be ϵ or in Γ and β a string over $\hat{\Sigma}$.

✓ An arc of $Acc\ Arc_{q \to p}$ represents that the run-segment from q to p is a segment of an accepting run including trace transitions with an empty stack, internal update transitions and push transitions. The label is a triple (Acc, γ, β), where only the information of the updated top stack is necessary. When $\gamma = \bot$, it can be obtained no push transition is met from q to p; otherwise, $\gamma \in \Sigma_c \times Q$.

✓ An arc of $Actr\ Arc_{q \to p}$ indicates the run-segment from q to p may be a segment of an accepting run including push transitions and internal update transitions. The label is a quadruple $(Actr, \gamma_1, \gamma_2, \beta)$ since the information about both the push transition and the updated stack top should be recorded.

✓ An arc of $Tra\ Arc_{q \to p}$ denotes that the run-segment from q to p may be a segment of a tracing run-segment in which push transitions, call/internal/return update transitions and trace transitions are involved. Similarly, the label is a quadruple $(Tra, \gamma_1, \gamma_2, \beta)$. Specifically, if the run-segment ends with a trace transition, $\gamma_2 = \epsilon$ and the arc is denoted as $Arc_{q \to trace}$.

4.2 Transformation

Given a MNTA $M = (Q, \hat{\Sigma}, \delta, Q_0, F, \Gamma)$, the transformation is performed by 4 steps.

Step 1: Preprocessing

– Initially, an extra state X is created first which will act as the unique initial state throughout the transformation. For each initial state $q_i \in Q_0$, an arc of Acc from state X to q_i labeled with $Arc_{X \to q_i} = (Acc, \bot, \epsilon)$ is created.
– Eliminating all the states and transitions which are unreachable from X.
– For each transition in δ, a labeled arc is created as follows:
  - For each transition $(q, \gamma_1, x, p, \gamma_2)$ (possibly $p = q$), where $q, p, t \in Q$,
    i. if $\gamma_1 = \gamma_2 = \bot$ and $x \in \Sigma_i \cup \Sigma_r$, an arc $Arc_{q \to p} = (Acc, \bot, x)$ is created;
    ii. if $\gamma_1 = \gamma_2 = yt$, $y \in \Sigma_c$, $x \in \Sigma_i$, or $\gamma_1 = \epsilon$, $\gamma_2 = xq$, $x \in \Sigma_c$, two arcs $Arc_{q \to p} = (Actr, \gamma_1, \gamma_2, x)$ and $Arc_{q \to p} = (Tra, \gamma_1, \gamma_2, x)$ are created, where a push transition is denoted as (q, ϵ, x, p, xq) for convenience;
    iii. if $\gamma_2 = yt$ and $y \in \dot{\Sigma}_c \cup \dot{\Sigma}_r$, an arc $Arc_{q \to p} = (Tra, \gamma_1, \gamma_2, x)$ is created.
  - For each trace transition $\delta(q, \gamma, x) = trace$ where $\gamma = yt$, $x \in \Sigma_r$ and $y \in \dot{\Sigma}_c \cup \dot{\Sigma}_c \cup \dot{\Sigma}_r$ is matched with x, an arc $Arc_{q \to trace} = (Tra, \gamma, \epsilon, x)$ is created.
– In order to reduce the number of arcs as much as possible, all created arcs can be simplified according to the following rules during the entire transformation. Suppose there are two arcs Arc_1 and Arc_2 from q to p.
  - If $Arc_1 = (Acc, \gamma_1, \beta_1)$ and $Arc_2 = (Acc, \gamma_2, \beta_2)$ where $\gamma_1 = \gamma_2$, Arc_1 and Arc_2 are eliminated and substituted by a new arc $Arc = (Acc, \gamma_1, \beta_1 + \beta_2)$;
  - If $Arc_1 = (Type, \gamma_{11}, \gamma_{12}, \beta_1)$ and $Arc_2 = (Type, \gamma_{21}, \gamma_{22}, \beta_2)$ where $\gamma_{11} = \gamma_{21}$ and $\gamma_{12} = \gamma_{22}$, Arc_1 and Arc_2 are eliminated and substituted by a new arc $Arc = (Type, \gamma_{11}, \gamma_{22}, \beta_1 + \beta_2)$, where $Type$ is $Actr$ or Tra.

Step 2: Processing all push transitions

For each arc $Arc_{q \to p}$, all push transitions (if exist) at state p are handled first. Suppose there are n push transitions and each transition is $\delta(p, \gamma, x_i) = (t_i, x_i p \gamma)$ where $1 \le i \le n$, $x_i \in \Sigma_c$ and $t_i \in Q$. Let $\beta_i = \beta_{x_i p}$ and $\beta_{new} = \sum_{i=1}^{n} \beta_i^*$. Then, for each arc $Arc_{q \to p}$,

- $Arc_{q\to p} = (Acc, \gamma, \beta)$ of Acc, it is updated as $(Acc, \gamma, \beta\beta^*_{new})$.
- $Arc_{q\to p} = (Actr/Tra, \gamma_1, \gamma_2, \beta)$ of $Actr$ or Tra, a new arc $(Actr/Tra, \gamma_1, \gamma_2, \beta\beta^*_{new})$ is replaced.

Step 3: Updating labeled arcs

In this step, all arcs relevant to each state p are modified according to the following strategies. For any two arcs $Arc_{q\to p}$ to p and $Arc_{p\to t}$ or $Arc_{p\to trace}$ from p, they are eliminated and a new arc $Arc_{q\to t}$ or $Arc_{q\to trace}$ is created. Let $\beta'_1 = \beta_1$, $\beta'_2 = \beta_2$, and $m, n \in Q$.

- For two arcs of Acc $Arc_{q\to p} = (Acc, \gamma_1, \beta_1)$ and $Arc_{p\to t} = (Acc, \gamma_2, \beta_2)$ where $\gamma_1 \neq \perp$ and $\gamma_2 = \perp$ occurring at the same time is not permitted, if $q = p \neq X$, $\beta'_1 = \beta^*_1$; similarly, if $p = t$, $\beta'_2 = \beta^*_2$. A new arc $Arc_{q\to t} = (Acc, \gamma_2, \beta'_1\beta'_2)$ is created.
- Consider two arcs of $Actr$ $Arc_{q\to p} = (Actr, \gamma_{11}, \gamma_{12}, \beta_1)$ and $Arc_{p\to t} = (Actr, \gamma_{21}, \gamma_{22}, \beta_2)$, where either $\gamma_{21} = \epsilon$ or $\gamma_{12} = \gamma_{21} = \gamma_{22}$ holds. When $\gamma_{i1} = \epsilon$, $\gamma_{i2} = x_i m_i$, $x_i \in \Sigma_c$, $m_i \in Q$ and $i \in \{1, 2\}$, that means there is a push transition at state q $(i = 1)$ or p $(i = 2)$. Then it can be inferred that $\beta'_i = (\beta_i\beta^*_{x_i m_i})^*$ if $q = p$ $(i = 1)$ or $p = t$ $(i = 2)$. Similarly, when $\gamma_{i1} = \gamma_{i2} = x_i m_i$, $\beta'_i = \beta^*_i$ if $q = p$ $(i = 1)$ or $p = t$ $(i = 2)$. Hence, the new arc $Arc_{q\to t} = (Actr, \gamma_{11}, \gamma_{22}, \beta'_1\beta'_2)$ is created.
- As to $Arc_{q\to p} = (Acc, \gamma, \beta_1)$ and $Arc_{p\to t} = (Actr, \gamma_1, \gamma_2, \beta_2)$ where $\gamma = \gamma_1 = \perp/xm$ and $x \in \Sigma_c$,
  - $\gamma_1 = \gamma_2 = \perp/xm$. When $q = p \neq X$, $\beta'_1 = \beta^*_1$. Let $\beta'_2 = \beta^*_2$ if $p = t$. Then $Arc_{q\to t} = (Acc, \gamma_2, \beta'_1\beta'_2)$.
  - $\gamma_1 = \epsilon$, $\gamma_2 = yn$ and $y \in \Sigma_c$. When $p \neq t$, we have $Arc_{q\to t} = (Acc, \gamma_2, \beta_1\beta_2)$; otherwise, $Arc_{q\to t} = (Acc, \gamma_2, \beta_1(\beta_2 + \beta_{xm})^*)$.
- Consider $Arc_{q\to p} = (Tra, \gamma_{11}, \gamma_{12}, \beta_1)$ and $Arc_{p\to t} = (Tra, \gamma_{21}, \gamma_{22}, \beta_2)$ where $\gamma_{12} = \gamma_{21} = xm$ and $x \in \Sigma_c \cup \dot{\Sigma}_c \cup \dot{\Sigma}_r$. Let $y \in \Sigma_c$ be the symbol matched with x.
  - $\gamma_{11} = \epsilon$.
    i. When $q \neq p$, a new arc $Arc_{q\to t} = (Tra, \gamma_{11}, \gamma_{22}, \beta_1\beta'_2)$ is created where $\beta'_2 = \beta^*_2$ if $p = t$ and $\gamma_{21} = \gamma_{22}$.
    ii. For the case that $q = p$, let $\beta_{ym} = \beta_1\bar{\beta}_1$ and $\beta_n = (\beta_1\beta^*_{ym})^n(\bar{\beta}_1\beta^*_{ym})^n$, $n \geq 0$. The new arc is $Arc_{q\to t} = (Tra, \gamma_{11}, \gamma_{22}, \beta_{new})$ where β_{new} is calculated as:
    $$\beta_{new} = \begin{cases} \beta_1\beta^*_n\beta_2 & p \neq t, \\ \beta_1(\beta_n + \beta_2)^* & p = t \ \& \ \gamma_{21} = \gamma_{22}, \\ \beta_1\beta^*_n\beta_2\beta^*_n & p = t \ \& \ \gamma_{21} \neq \gamma_{22}. \end{cases}$$
  - $\gamma_{11} \neq \epsilon$. We have $\beta'_1 = \beta^*_1$ if $q = p$ and $\gamma_{11} = \gamma_{12}$ while $\beta'_2 = \beta^*_2$ if $p = t$ and $\gamma_{21} = \gamma_{22}$. The new arc is $Arc_{q\to t} = (Tra, \gamma_{11}, \gamma_{22}, \beta'_1\beta'_2)$.
- Consider $Arc_{q\to p} = (Tra, \gamma_1, \gamma_2, \beta_1)$ and $Arc_{p\to trace} = (Tra, \gamma, \epsilon, \beta_2)$ where $\gamma_2 = \gamma = xm$ and $x \in \Sigma_c \cup \dot{\Sigma}_c \cup \dot{\Sigma}_r$. $y \in \Sigma_c$ is matched with x.

- When $\gamma_1 = \epsilon$, there is a push transition at state q, i.e. $q = m$. If $q = p$, $\beta_1' = \beta_1 \beta_{yq}^*$, then $Arc_{p \to trace} = (Tra, yq, \epsilon, \beta_1' \beta_2)$ is created.
- $\gamma_1 \neq \epsilon$ holds. There is $\beta_1' = \beta_1^*$ if $q = p$. $Arc_{q \to trace} = (Tra, \gamma_1, \epsilon, \beta_1' \beta_2)$.

Step 4: Merging arcs and computing MNE

- Eliminating each arc of the form $Arc_{X \to q} = (Acc, \gamma, \beta)$ $(q \notin F)$ and all arcs of Tra not in the form of $Arc_{q \to trace} = (Tra, \gamma, \epsilon, \beta)$ where $\gamma = xq$ and there is a push transition at state q by reading $call$ x.
- Calculating each β_{xq} where $x \in \Sigma_c$: check whether there exists an arc $Tra_{q \to trace} = (Tra, \gamma_{i1}, \gamma_{i2}, \beta_i)$ at q which satisfies $\gamma_{i1} = xq$ and $\gamma_{i2} = \epsilon$. If there are n such arcs, β_{xq} is calculated as $\sum_{i=1}^{n} \beta_i$; otherwise, $\beta_{xq} = \epsilon$.
- Suppose each arc of Acc is $Arc_{X \to q_{f_i}} = (Acc, \gamma_i, \beta_i)$ where $q_{f_i} \in F$. We have $E_M = \sum_{i=1}^{n} \beta_i$. In particular, if there are no Acc arcs left eventually, the multi-matching nested expression E_M of M is $\emptyset$.

Initially, for each transition in δ, according to the accepting condition of MNTA, a trace or internal update transition with the stack empty can only be in an accepting run rather than a tracing run-segment, an arc of Acc is created. As for a push transition or an internal update transition with a non-empty stack, they can not only in an accepting run but also a tracing run-segment, hence two arcs of $Actr$ and Tra are created. However, since an $inner\text{-}call$ or $inner\text{-}return$ symbol cannot exist in the stack in the final configuration of an accepting run, only the arc of Tra is created.

Note that in Step 2, push transitions are handled first since a tracing run-segment produced from state p to $trace$ by a push transition at p may occur for a few times before a non-push transition is taken at p. Within the tracing run-segment, the state traces back to p and the stack stays unchanged eventually. Hence, each arc $Arc_{q \to p}$ is updated by concatenating all expressions obtained from each tracing run-segment produced from p to $trace$ by a push at p.

The first case in Step 3 shows that two arcs of Acc can be merged directly based on the restriction that $pending\ returns$ can be followed by $pending\ calls$, but not vice versa. Besides, $pending\ calls$ can be followed by $pending\ calls$ which is dealt with in case two by the merging of two arcs of $Actr$. The case that an arc of Acc is followed by an arc of $Actr$ signifies that there has already been at least one push transition, i.e. an $pending\ call$, in an accepting run represented by the arc of Acc. Regarding to the fourth case of two arcs of Tra, two different ways of concatenations are given depending on whether the first arc starts from a push transition. Note that if there is a self-loop arc of Tra at state q starting from a push transition, the tracing run-segment from q to $trace$ can repeat for multiple times in an accepting run on the condition that a corresponding $trace$ can be met. The last case discusses that when a $trace$ is met, if the information of the stack top between two arcs of Tra matches, they are merged.

4.3   Example

Now we take the example automaton M_{ex} in Fig. 2 to intuitively illustrate the transformation.

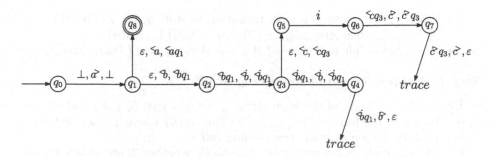

Fig. 2. A MNTA M_{ex}

Figure 3 presents the automaton after being preprocessed (Step 1). For clarity, arcs of *Actr* are depicted as black lines while arcs of *Acc* and *Tra* as red and blue dashed lines, respectively. Compared with the original automaton in Fig. 2, state X and arc $Arc_{X \to q_0} = (Acc, \perp, \epsilon)$ are created. Besides, labeled arcs of *Acc*, *Actr* or *Tra* are created for each of the transitions in M_{ex}, shown as follows.

$$Arc_{X \to q_0} = (Acc, \perp, \epsilon) \qquad Arc_{q_0 \to q_1} = (Acc, \perp, \vec{a})$$
$$Arc_{q_1 \to q_8} = (Tra, \epsilon, \overleftarrow{a}q_1, \overleftarrow{a}) \qquad Arc_{q_1 \to q_8} = (Actr, \epsilon, \overleftarrow{a}q_1, \overleftarrow{a})$$
$$Arc_{q_1 \to q_2} = (Tra, \epsilon, \overleftarrow{b}q_1, \overleftarrow{b}) \qquad Arc_{q_1 \to q_2} = (Actr, \epsilon, \overleftarrow{b}q_1, \overleftarrow{b})$$
$$Arc_{q_2 \to q_3} = (Tra, \overleftarrow{b}q_1, \overrightarrow{b}q_1, \overleftarrow{b}) \qquad Arc_{q_3 \to q_4} = (Tra, \overleftarrow{b}q_1, \overrightarrow{b}q_1, \overleftarrow{b})$$
$$Arc_{q_3 \to q_5} = (Tra, \epsilon, \overleftarrow{c}q_3, \overleftarrow{c}) \qquad Arc_{q_3 \to q_5} = (Actr, \epsilon, \overleftarrow{c}q_3, \overleftarrow{c})$$
$$Arc_{q_5 \to q_6} = (Tra, \overleftarrow{c}q_3, \overleftarrow{c}q_3, i) \qquad Arc_{q_5 \to q_6} = (Actr, \overleftarrow{c}q_3, \overleftarrow{c}q_3, i)$$
$$Arc_{q_4 \to trace} = (Tra, \overrightarrow{b}q_1, \epsilon, \overrightarrow{b}) \qquad Arc_{q_6 \to q_7} = (Tra, \overleftarrow{c}q_3, \overrightarrow{c}q_3, \overrightarrow{c})$$
$$Arc_{q_7 \to trace} = (Tra, \overrightarrow{c}q_3, \epsilon, \overrightarrow{c})$$

For instance, arc $Arc_{q_0 \to q_1} = (Acc, \perp, \vec{a})$ is created for transition $(q_0, \perp, \vec{a}, q_1, \perp)$ and arc $Arc_{q_4 \to trace} = (Tra, \overrightarrow{b}q_1, \epsilon, \overrightarrow{b})$ is created for $(q_4, \overrightarrow{b}q_1, \overrightarrow{b}, trace)$. In Fig. 3, there are 2 arcs, $Arc_{X \to q_0}$ and $Arc_{q_0 \to q_1}$, of *Acc*, 4

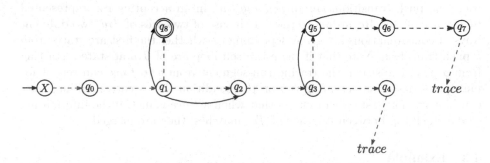

Fig. 3. Transformation from MNTA to MNE-1

arcs, $Arc_{q_1 \to q_8}$, $Arc_{q_1 \to q_2}$, $Arc_{q_3 \to q_5}$ and $Arc_{q_5 \to q_6}$, of $Actr$, and 9 arcs, $Arc_{q_1 \to q_8}$, $Arc_{q_1 \to q_2}$, $Arc_{q_2 \to q_3}$, $Arc_{q_3 \to q_4}$, $Arc_{q_4 \to trace}$, $Arc_{q_3 \to q_5}$, $Arc_{q_5 \to q_6}$, $Arc_{q_6 \to q_7}$ and $Arc_{q_7 \to trace}$ of Tra.

According to Step 2, all the push transitions are processed first. In M_{ex}, there are two push transitions $(q_1, \gamma, {}^\langle a, q_8, {}^\langle aq_1 \gamma)$ and $(q_1, \gamma, {}^\langle b, q_2, {}^\langle bq_1 \gamma)$ at state q_1, while one push $(q_3, \gamma, {}^\langle c, q_5, {}^\langle cq_3 \gamma)$ at state q_3. Hence, arcs $Arc_{q_0 \to q_1}$ of Acc and $Arc_{q_2 \to q_3}$ of Tra are updated as:

$$Arc_{q_0 \to q_1} = (Acc, \bot, \vec{a}\,(\beta_{<aq_1} + \beta_{\triangleleft bq_1})^*) \quad Arc_{q_2 \to q_3} = (Tra, {}^\triangleleft bq_1, {}^{\dot{\triangleleft}}bq_1, {}^\triangleleft b\beta^*_{<cq_3})$$

Figure 4 depicts the automaton updated according to Step 3 by merging the labeled arcs with the same type. First, for the arcs of Acc, a new arcs, $Arc_{X \to q_1}$ of Acc is created for the arc pair, $Arc_{X \to q_0}$ and $Arc_{q_0 \to q_1}$, which are eliminated then. Then the new arc $Arc_{q_3 \to q_6}$ of $Actr$ is created to substitute for $Arc_{q_3 \to q_5}$ and $Arc_{q_5 \to q_6}$. As for the arcs of Tra, in the run-segment from q_1 to $trace$ or from q_3 to $trace$, any two connected arcs satisfy the conditions of the fourth or fifth case in Step 3. Thus, $Arc_{q_1 \to trace}$ is created and reserved eventually, while the original arcs, $Arc_{q_1 \to q_2}$, $Arc_{q_2 \to q_3}$, $Arc_{q_3 \to q_4}$ and $Arc_{q_4 \to trace}$, and intermediate created arcs $Arc_{q_1 \to q_3}$ and $Arc_{q_1 \to q_4}$ are eliminated. Similarly, $Arc_{q_3 \to trace}$ is created while $Arc_{q_3 \to q_5}$, $Arc_{q_5 \to q_6}$, $Arc_{q_6 \to q_7}$ and $Arc_{q_7 \to trace}$, and the intermediate created arcs $Arc_{q_3 \to q_6}$ and $Arc_{q_3 \to q_7}$ are eliminated. Each label of new created arcs is presented below.

$$Arc_{X \to q_1} = (Acc, \bot, \vec{a}\,(\beta_{<aq_1} + \beta_{\triangleleft bq_1})^*) \quad Arc_{q_3 \to q_6} = (Actr, \epsilon, {}^\langle cq_3, {}^\langle ci)$$
$$Arc_{q_1 \to trace} = (Tra, {}^\triangleleft bq_1, \epsilon, {}^\triangleleft b \, {}^{\dot{\triangleleft}}b\beta^*_{<cq_3} \, {}^\triangleleft b\vec{b}) \quad Arc_{q_3 \to trace} = (Tra, {}^\langle cq_3, \epsilon, {}^\langle ci\vec{c}\,\vec{c})$$

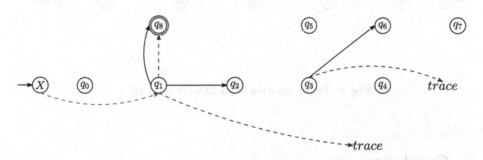

Fig. 4. Transformation from MNTA to MNE-2

For the arcs of different types, i.e. Acc and $Actr$, they are merged according to the third case of Step 3. As Fig. 5 illustrates, $Arc_{X \to q_1}$ and $Arc_{q_1 \to q_2}$ are replaced by $Arc_{X \to q_2}$ while $Arc_{X \to q_1}$ and $Arc_{q_1 \to q_8}$ by $Arc_{X \to q_8}$.

$$Arc_{X \to q_2} = (Acc, \bot, \vec{a}\,(\beta_{<aq_1} + \beta_{\triangleleft bq_1})^* {}^\triangleleft b)$$
$$Arc_{X \to q_8} = (Acc, \bot, \vec{a}\,(\beta_{<aq_1} + \beta_{\triangleleft bq_1})^* {}^\langle a)$$

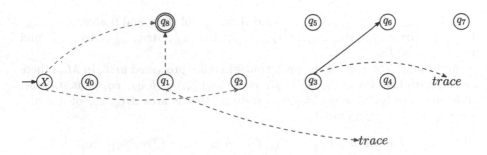

Fig. 5. Transformation from MNTA to MNE-3

Eventually, $Arc_{X \to q_2}$, $Arc_{q_1 \to q_8}$ and $Arc_{q_3 \to q_6}$ are eliminated as shown in Fig. 6. Hence, we can obtain the subexpression of each tracing run-segment:

$$\beta_{<aq_1} = \epsilon \qquad \beta_{<cq_3} = \overset{\frown}{c}\overset{\frown}{c}\,i\overset{\frown}{c} \qquad \beta_{<bq_1} = \overset{\frown}{b}\overset{\frown}{b}(\overset{\frown}{c}\overset{\frown}{c}\,i\overset{\frown}{c})^*\overset{\frown}{b}\overset{\frown}{b}.$$

And the final obtained multi-matching nested expression of M_{ex} is

$$E_{M_{ex}} = \overset{\frown}{a}(\overset{\frown}{b}\overset{\frown}{b}(\overset{\frown}{c}\overset{\frown}{c}\,i\overset{\frown}{c})^*\overset{\frown}{b}\overset{\frown}{b})^*\overset{\frown}{a}.$$

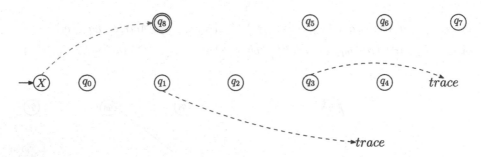

Fig. 6. Transformation from MNTA to MNE-4

5   Conclusion

In this paper, the nested words with one-to-one matching relations are extended to multi-matching nested words with one-to-n and n-to-one matching relations. MNE and MNTA are defined over multi-matching nested words. To acquire MNE from MNTA, the main idea is the construction of labeled arcs for different transitions and the merging strategies. The corresponding expression is calculated by the sum of each labeled arc from the initial state to each final state.

References

1. Hopcroft, J.E., Motwani, R., Ullman, J.D.: Introduction to Automata Theory, Languages, and Computation, 2nd edn. Pearson Education, Boston (2000). ISBN 0-201-44124-1

2. Alur, R., Madhusudan, P.: Adding nesting structure to words. In: Ibarra, O.H., Dang, Z. (eds.) DLT 2006. LNCS, vol. 4036, pp. 1–13. Springer, Heidelberg (2006). https://doi.org/10.1007/11779148_1

3. Alur, R., Madhusudan, P.: Adding nesting structure to words. J. ACM 56(3), 1–43 (2009)

4. Neven, F.: Automata, logic, and XML. In: Bradfield, J. (ed.) CSL 2002. LNCS, vol. 2471, pp. 2–26. Springer, Heidelberg (2002). https://doi.org/10.1007/3-540-45793-3_2

5. Hunter, D., Rafter, J., Fawcett, J.: Beginning XML. John Wiley & Sons, New York (2007)

6. Kumar, V., Madhusudan, P., Viswanathan, M.: Visibly pushdown automata for streaming XML. In: Proceedings of the 16th International Conference on World Wide Web, pp. 1053–1062 (2007)

7. Alur, R., Madhusudan, P.: Visibly pushdown languages. In: Proceedings of the Thirty-Sixth Annual ACM Symposium on Theory of Computing, pp. 202–211 (2004)

8. Kegang, H., Zhenhua, D., Xin, L.: On traceable automata. Chin. J. Comput. 5, 340–348 (1990)

9. Kegang, H., Zhenhua, D.: The relationship between DTA and DMTA. Microelectron. Comput. 4, 6–10 (1990)

10. Kegang, H., Zhenhua, D.: Two fundamental theorems on traceable automata. J. Northwest Univ. Nat. Sci. Ed. 20(1), 11–17 (1990)

11. Kernighan, B.W., Ritchie, D.M.: The C Programming Language. Prentice Hall, New Jersey (2006)

12. Lin, X.J., Wu, C.K., Liu, F.: Many-to-one encryption and authentication scheme and its application. J. Commun. Netw. 10(1), 18–27 (2008)

13. Zhong, H., Cui, J., Shi, R., Xia, C.: Many-to-one homomorphic encryption scheme. Secur. Commun. Netw. 9(10), 1007–1015 (2016)

14. Silberschatz, A., Korth, H.F., Sudarshan, S.: Database System Concepts. McGraw-Hill, New York (1997)

On the Complexity of Some
Facet-Defining Inequalities
of the QAP-Polytope

Pawan Aurora[1] and Hans Raj Tiwary[2(⊠)]

[1] IISER Bhopal, Bhopal, India
paurora@iiserb.ac.in
[2] Department of Applied Mathematics, Charles University, Prague, Czech Republic
hansraj@kam.mff.cuni.cz

Abstract. The Quadratic Assignment Problem (QAP) is a well-known
NP-hard problem that is equivalent to optimizing a linear objective function over the QAP polytope. The QAP polytope with parameter n –
QAP_n – is defined as the convex hull of rank-1 matrices xx^T with x as
the vectorized $n \times n$ permutation matrices.

In this paper we consider all the known exponential-sized families of
facet-defining inequalities of the QAP-polytope. We describe a new family of valid inequalities that we show to be facet-defining. We also show
that membership testing (and hence optimizing) over some of the known
classes of inequalities is coNP-complete. We complement our hardness
results by showing a lower bound of $2^{\Omega(n)}$ on the extension complexity of all relaxations of QAP_n for which any of the known classes of
inequalities are valid.

1 Introduction

The Quadratic Assignment Problem (QAP) is a fundamental combinatorial optimization problem from the category of facility location problems [23,24]. QAP
is defined as the following problem: given n facilities and n locations, distances
d_{ij} between all pairs of locations $i, j \in [n]$, flows f_{ij} between all pairs of facilities
$i, j \in [n]$ and costs c_{ij} of opening facility i at location j, for all pairs $i, j \in [n]$, find
an assignment σ of the n facilities to the n locations so that the total cost given
by the function $\sum_{i,j} f_{ij} d_{\sigma(i)\sigma(j)} + \sum_i c_{i\sigma(i)}$ is minimized. The problem is known
as QAP since it can be modeled as optimizing a quadratic function over linear
and binary constraints. However, several linearizations of the problem have been
proposed. For details refer to the book [10] and the citations therein.

Given an instance of QAP, it is NP-hard to approximate the optimum within
any constant factor [31]. What makes QAP one of the "hardest" problems in
combinatorial optimization is the fact that unlike most NP-hard combinatorial
optimization problems, it is practically intractable. It is generally considered
impossible to solve to optimality QAP instances of size larger than 20 within
reasonable time limits [10].

© Springer Nature Switzerland AG 2020
W. Wu and Z. Zhang (Eds.): COCOA 2020, LNCS 12577, pp. 334–349, 2020.
https://doi.org/10.1007/978-3-030-64843-5_23

As is common with combinatorial optimization problems, QAP can be viewed as the problem of optimizing a linear objective function over the convex hull of all feasible solutions. To this end, the QAP polytope is defined as $\mathrm{QAP}_n = conv\left(\{yy^T|y = vec(P_\sigma), \sigma \in S_n\}\right)$, where P_σ is the $n \times n$ permutation matrix corresponding to the premutation σ and $y = vec(P_\sigma)$ is its vectorization. Following the notation of [1], we denote a vertex yy^T as $P_\sigma^{[2]}$. Note that $P_\sigma^{[2]}(ij, kl) = P_\sigma(i, j) \cdot P_\sigma(k, l)$. Clearly, $\mathrm{QAP}_n \subset \mathbb{R}^{n^2 \times n^2}$. In fact QAP_n can be embedded in $\mathbb{R}^{(n^4+n^2)/2}$ since each point in the polytope is a symmetric $n^2 \times n^2$ matrix and we could only store its upper (or lower) triangular part. However, in this paper we would conveniently denote a point in QAP_n by a $n^2 \times n^2$ matrix.

One of the methods for solving hard combinatorial optimization problems is the method of branch-and-cut [27]. For this method to be effective for the QAP, it is important to identify new valid and possibly facet-defining inequalities for the QAP-polytope and to develop the corresponding separation algorithms. Given that it is NP-hard to optimize over the QAP-polytope, it is probably impossible to characterize all its facets [28]. In [18,19,26], the authors obtain early results on the combinatorial structure of the QAP-polytope and some of its facet-defining inequalities. In [1] the authors list all the known facets of the QAP-polytope besides the equations that define its affine hull. In this paper we add another exponential sized family to the list of known facets of the QAP-polytope. Optimizing the QAP objective function over any of the relaxations given by these families can provide an approximate solution to the QAP, provided the optimization problem can be efficiently solved. In this paper we also show that optimizing over the relaxations given by some of these exponential sized family of facet-defining inequalities is NP-hard. We do it by proving that the corresponding membership testing problem is coNP-complete for the appropriate classes of inequalities.

Furthermore, we prove a lower bound of $2^{\Omega(n)}$ on the extension complexity of bounded relaxations of QAP_n obtained by each of these families of inequalities.

To summarize, our main contributions are as follows.

- We identify a new family of valid inequalities for QAP_n (Sect. 2) and prove that they are facet-defining (Sect. 3),
- We prove that membership testing for three out of the five known families of valid inequalities for the QAP-polytope(including the new one we introduce) is coNP-complete (Sect. 4), and
- We prove a lower bound of $2^{\Omega(n)}$ for the extension complexity of any bounded[1] relaxation of QAP_n that has any of the known families as valid inequalities (Sect. 5).

---

[1] In fact, boundedness is not required for the results in Sect. 5. However, since we will rely on existing results, such as Theorem 1, that are published with the boundedness assumption, we will include this assumption.

1.1   Extension Complexity

Let $P \subset \mathbb{R}^n$ be a polytope. A polytope $Q \subset \mathbb{R}^{n+r}$ is called an *extension* or an *extended formulation* of P if

$$P = \{x \in \mathbb{R}^n \mid \exists y \in \mathbb{R}^r, (x, y) \in Q\}.$$

Let size(P) denote the number of facets of polytope P and let $Q \downarrow P$ denote that Q is an extended formulation of P. Then, the extension complexity of a polytope P – denoted by xc(P) – is defined to be $\min_{Q \downarrow P} \text{size}(Q)$.

Extended formulations are a very useful tool in combinatorial optimization as they allow the possibility of drastically reducing the size of a Linear Program by introducing new variables (See [12,22,32,33] for surveys). In the past decade lower bounds on the extension complexity of various polytopes have been studied [3,15,29,30] and the notion generalized and studied in various settings such as general conic extensions [17], semidefinite extensions [8,25], approximation [4,5,7,11], parameterization [9,16], generalized probabilistic theories in Physics [14], and information theoretic perspective [6,7].

Superpolynomial lower bounds on extension complexity are known for polytopes related to many NP-hard problems [3,15] as well as for the Matching polytope: the convex hull of characteristic vectors of all matchings in K_n [30]. High extension complexity of the Matching polytope highlights the fact that a superpolynomial lower bound on the extension complexity cannot be taken to mean that the underlying optimization problem is not solvable in polynomial time. However, these lowers bounds are unconditional and do not require standard complexity theoretic assumption such as $P \neq NP$. Moreover, apart from the exception of Matching polytope, linear optimization over all known polytopes with superpolynomial lower bound is infeasible. Either because the linear optimization over the polytope is NP-hard [3,15] or the polytope is not explicitly given, as is the case for some matroid polytopes [29].

For the purposes of this paper the most relevant characterization of extension complexity is given by Faenza et al. [13] where the authors prove the equivalence between existence of an extended formulation of size r with the existence of a certain two-party communication game requiring an exchange of $\Theta(\log r)$ bits. We will describe this connection here and use it as a black box in our proofs of lower bounds on the extension complexity of the polytopes considered here.

EF-Protocols: Computing a Matrix in Expectation. Let M be an $m \times n$ matrix with non-negative entries. Consider a communication game between two players: Alice and Bob. Alice and Bob both know the matrix M and can agree upon any strategy prior to the start of the game. In each round of the game, Alice receives a row index $i \in [m]$ and Bob a column index $j \in [n]$. Both Alice and Bob have no restriction on the computations that they perform and can also use (private) random bits. They can also exchange information by sending some bits to the other player. At some point one of them outputs a non-negative number and the round finishes.

Since they are allowed the use of random bits, the output X_{ij} when Alice and Bob receive inputs i and j respectively, is a random variable. We says that their strategy is an *EF-protocol* for M if $\mathbb{E}[X_{ij}] = M_{ij}$ for all $(i,j) \in [m] \times [n]$, where $\mathbb{E}[X_{ij}]$ is the expected value of the random variable X_{ij}. The complexity of an EF-protocol is defined to be the maximum number of bits exchanged between Alice and Bob for any input i,j to Alice and Bob respectively.

Let $P = \{x \mid Ax \leqslant b\} = \text{conv}(\{v_1, \ldots, v_n\})$ be a polytope where A is an $m \times d$ real matrix, $b \in \mathbb{R}^m$, $v_i \in \mathbb{R}^d$, and $\text{conv}(S)$ denotes the convex hull of the points in a set S. The slack matrix of P with respect to this representation - denoted by $S(P)$ - is the $m \times n$ (non-negative) matrix whose entry at i-th row and j-th column is $b_i - A_i v_j$, where A_i denotes the i-th row of the matrix A. Note that a polytope is not defined uniquely this way: one can always embed the polytope in higher dimensional space, and add redundant inequalities and points to the descriptions. However, in what follows, none of that makes any difference and one can choose any description that they like. This justifies the notation $S(P)$ for any slack matrix of P even though the particular description of P that defines this matrix is completely ignored in the notation.

The following connection – which we will use in Sect. 5 – was shown by Faenza et al. [13] between existence of an EF-protocol computing a slack matrix of polytope P and that of an extended formulation of P.

Theorem 1 *[13]. Let M be a non-negative matrix such that any EF-protocol for M has complexity at least c. Further, let P be a polytope such that M is a submatrix of some slack matrix $S(P)$ of P. Then, $\text{xc}(P) \geqslant 2^c$.*

2   Relaxations of QAP_n

In the following, Y is a $n^2 \times n^2$ variable matrix that is used to denote an arbitrary point in QAP_n. Further, $Y_{ij,kl}$ refers to $Y(n * (i-1) + j, n * (k-1) + l)$.

The most general family of valid inequalities that includes all known families as special cases is the following:

QAP1:

$$\sum_{ijkl} n_{ij} n_{kl} Y_{ij,kl} - (2\beta - 1) \sum_{ij} n_{ij} Y_{ij,ij} \geqslant \frac{1}{4} - (\beta - \frac{1}{2})^2 \qquad (1)$$

where $\beta \in \mathbb{Z}$ and $n_{ij} \in \mathbb{Z}$ for all $i,j \in [n]$. These inequalities were introduced in [1] as a generalization of all known facet-defining inequalities for the QAP-polytope.

QAP2:

$$\sum_{r=1}^{m} Y_{i_r j_r, kl} - Y_{kl,kl} - \sum_{r<s} Y_{i_r j_r, i_s j_s} \leqslant 0 \qquad (2)$$

where $i_1, \ldots, i_m, k$ are all distinct and $j_1, \ldots, j_m, l$ are also distinct. In addition, $n \geqslant 6, m \geqslant 3$. These inequalities were introduced in [1] and proved to be a special case of QAP1 that are facet-defining for the QAP-polytope.

QAP3:

$$(\beta - 1) \sum_{\substack{(ij) \in P \times Q}} Y_{ij,ij} - \sum_{\substack{(ij),(kl) \in P \times Q \\ i < k}} Y_{ij,kl} \leqslant \frac{\beta^2 - \beta}{2} \tag{3}$$

where $P, Q \subset [n]$. In addition (i) $\beta + 1 \leqslant |P|, |Q| \leqslant n - 3$, (ii) $|P| + |Q| \leqslant n - 3 + \beta$, (iii) $\beta \geqslant 2$. These inequalities were introduced by Jünger and Kaibel [20,21] who also proved that they are facet-defining for the QAP-polytope. They are also a special case of QAP1 [1].

QAP4:

$$- (\beta - 1) \sum_{\substack{(ij) \in P_1 \times Q}} Y_{ij,ij} + \beta \sum_{\substack{(ij) \in P_2 \times Q}} Y_{ij,ij} + \sum_{\substack{i < k \\ (ij),(kl) \in P_1 \times Q}} Y_{ij,kl}$$
$$+ \sum_{\substack{i < k \\ (ij),(kl) \in P_2 \times Q}} Y_{ij,kl} - \sum_{\substack{(ij) \in P_1 \times Q \\ (kl) \in P_2 \times Q}} Y_{ij,kl} - \frac{\beta^2 - \beta}{2} \geqslant 0 \tag{4}$$

where $P_1, P_2, Q \subset [n]$, $P_1 \cap P_2 = \emptyset$. Further, (i) $3 \leq |Q| \leq n - 3$, (ii) $|P_1| + |P_2| \leq n - 3$, (iii) $|P_1| \geqslant \min\{2, \beta + 1\}$, (iv) $|P_2| \geqslant \min\{1, -\beta + 2\}$, (v) $||P_1| - |P_2| - \beta| \leq n - |Q| - 4$, (vi) if $|P_2| = 1$: $|Q| \geqslant \min\{-\beta + 5, \beta + 2\}$; if $|P_2| \geqslant 2$: $|Q| \geqslant \min\{-\beta + 5, \beta + 3\}$ or $|Q| \geqslant \min\{-\beta + 4, \beta + 4\}$. These inequalities were also introduced by Kaibel [21] and shown to be facet-defining for the QAP-polytope. They are also a special case of QAP1.

It is known [1] that the inequalities QAP2 , QAP3 and QAP4 are special instances of the QAP1 inequalities. QAP1 inequalities are in general not facet-defining for the QAP-polytope and so it is interesting to identify conditions under which they do define facets. We identify a new special case (5) of QAP1 inequality and show in Sect. 3 that they are facet-defining. Inequality QAP5 follows from QAP1 by setting $\beta = 2$, $n_{i_1 j_1} = n_{i_2 j_2} = \cdots = n_{i_m, j_m} = 1$ for distinct $i_1, \ldots, i_m \in [n]$ and distinct $j_1, \ldots, j_m \in [n]$, and $n_{ij} = 0$ for $i \in [n] \setminus \{i_1, \ldots, i_m\}$ or $j \in [n] \setminus \{j_1, \ldots, j_m\}$.

QAP5:

$$\sum_{r=1}^{m} Y_{i_r j_r, i_r j_r} - \sum_{r < s} Y_{i_r j_r, i_s j_s} \leqslant 1 \tag{5}$$

where $i_1, \ldots, i_m$ are all distinct and $j_1, \ldots, j_m$ are also distinct. In addition, $m, n \geqslant 7$. These inequalities are new and we discuss them in the following section.

3   A New Class of Facet-Defining Inequalities

In this section we prove that the inequalities QAP5 are facet-defining. Let S_k denote the set of those vertices of QAP_n that correspond to the permutations having $i_r \mapsto j_r$ for $r \in \{r_1, r_2, \ldots, r_k\} \subseteq [m]$ and $i_r \not\mapsto j_r$ for $r \in [m] \setminus \{r_1, r_2, \ldots, r_k\}$. Here i_r, j_r, m are as in the definition of QAP5. If

V is the set of all the vertices of QAP_n then clearly $V = \cup_{i=0}^{m} S_i$, and $S_i \cap S_j = \emptyset$ for all $i \neq j \in \{0, 1, \ldots, m\}$.

Lemma 1. *The sets S_1, S_2 together constitute the vertices that satisfy the inequality (5) with equality.*

Proof. Consider a $P_{\sigma}^{[2]} \in S_k$ for $1 \leqslant k \leqslant m$. W.l.o.g. let $\sigma(i_1) = j_1, \ldots, \sigma(i_k) = j_k$ and $\sigma(i_r) \neq j_r$ for $r \in \{k+1, \ldots, m\}$. Substituting $Y = P_{\sigma}^{[2]}$ in (5) we have $\sum_{r=1}^{m} P_{\sigma}(i_r, j_r) - \sum_{r<s} P_{\sigma}(i_r, j_r) \cdot P_{\sigma}(i_s, j_s) = k - \binom{k}{2}$. For $k = 0$ we have $k - \binom{k}{2} = 0 < 1$, for $k = 1, k - \binom{k}{2} = 1$, for $k = 2, k - \binom{k}{2} = 1$ and for $3 \leqslant k \leqslant m$ we have $k - \binom{k}{2} < 1$. Hence a vertex of QAP_n satisfies the inequality (5) with equality if and only if it belongs to S_1 or S_2. $\square$

Let $S = S_1 \cup S_2$. We will show that any vertex in $V \setminus S$ can be expressed as a linear combination of the vertices in S and a fixed vertex $P_{\sigma^*}^{[2]} \in S_0$. This will establish that the dimension of the face containing S is one less than the dimension of the polytope and hence it must be a facet.

The following lemma from [1] provides a useful tool to express certain vertices as a linear combination of others.

Lemma 2 *[1, Lemma 16].* *Let $k_1, k_2, k_3, x, y \in [n]$ be distinct indices. Let $\Sigma = \{\sigma_1, \ldots, \sigma_6\}$ be a set of permutations of $[n]$ such that $\sigma_i(z) = \sigma_j(z)$ for all $z \in [n] \setminus \{k_1, k_2, k_3\}$ and for every $i, j \in \{1, \ldots, 6\}$. Further, let $\Sigma' = \{\sigma_1', \ldots, \sigma_6'\}$ where σ_i' is a transposition of σ_i on the indices x, y, for each $i = 1, \ldots, 6$. Then $\forall\, i, j, k, l \in [n], \sum_{\sigma \in \Sigma \cup \Sigma'} sign(\sigma) P_{\sigma}^{[2]}(ij, kl) = 0$.*

The next lemma shows that the vertices in the sets $S_4, \ldots, S_m$ can be expressed as a linear combination of the vertices in the sets $S_0, \ldots, S_3$.

In what follows, when it is clear from the context, we use σ to refer to a vertex $P_{\sigma}^{[2]}$.

Lemma 3. *For $k \geqslant 4$, any vertex in S_k can be expressed as a linear combination of vertices in $S_{k-1}, S_{k-2}, S_{k-3}$, and S_{k-4}.*

Proof. Let $\sigma_1 \in S_k$. Since k is at least 4, we must have indices $i_{r_1}, i_{r_2}, i_{r_3}, i_{r_4}, r_k \in [m], k = 1, \ldots, 4$, such that $\sigma_1(i_{r_k}) = j_{r_k}, k = 1, \ldots, 4$. Let $k_1 = i_{r_1}, k_2 = i_{r_2}, k_3 = i_{r_3}, x = i_{r_4}$, where k_1, k_2, k_3, x are as defined in Lemma 2. Applying Lemma 2 with y chosen as an index such that either $y \neq i_p$ for any $p \in [m]$ or when $y = i_p, p \in [m]$ then $\sigma_1(i_p) \neq j_p$, we get $\sigma_1' \in S_{k-1}, \sigma_2, \sigma_3, \sigma_6 \in S_{k-2}, \sigma_4, \sigma_5, \sigma_2', \sigma_3', \sigma_6' \in S_{k-3}, \sigma_4', \sigma_5' \in S_{k-4}$ with the property that σ_1 is a linear combination of $\sigma_2, \ldots, \sigma_6, \sigma_1', \ldots, \sigma_6'$. $\square$

Next, we show that vertices in S_3 can be expressed as linear combinations of vertices in S and S_0 as well.

Lemma 4. *Any vertex in S_3 can be expressed as a linear combination of vertices in S and S_0.*

Proof. Let $\sigma_1 \in S_3$. So we have indices $i_{r_1}, i_{r_2}, i_{r_3}, r_k \in [m], k = 1, 2, 3$, such that $\sigma_1(i_{r_k}) = j_{r_k}, k = 1, 2, 3$. Let $k_1 = i_{r_1}, k_2 = i_{r_2}, x = i_{r_3}$, where k_1, k_2, x are as defined in Lemma 2. Applying Lemma 2 with k_3, y chosen arbitrarily from $[n] \setminus \{i_{r_1}, i_{r_2}, i_{r_3}\}$, we get $\sigma_2, \ldots, \sigma_6, \sigma_1', \sigma_2', \sigma_6' \in S, \sigma_3', \sigma_4', \sigma_5' \in S_0$ with the property that σ_1 is a linear combination of $\sigma_2, \ldots, \sigma_6, \sigma_1', \ldots, \sigma_6'$. $\square$

Now that we have established that the linear hull of S_0, S_1, S_2 equals the linear hull of the QAP-polytope, the only remaining task is to show that instead of the entire set S_0, a fixed vertex in S_0 suffices to generate the entire linear hull. We will show that in fact any arbitrary vertex in S_0 is sufficient. We do this by first showing that permutations in S_0 define a connected graph if the edges connect permutations that are one transposition apart. Then, we show that any vertex $\sigma \in S_0$ is sufficient to generate all vertices in the connected component of σ by linear combination with vertices in S_1, S_2.

In the following lemma we show that it is possible to obtain the permutation corresponding to a vertex in S_0 from the permutation corresponding to any other vertex in S_0 via transpositions such that the vertices corresponding to the intermediate permutations also lie in S_0.

Lemma 5. *Consider the graph $G = (S_0, E)$ where S_0 is the set of vertices of the QAP-polytope that correspond to permutations for which $i_r \not\mapsto j_r$ for all $r \in [m]$ and $\{P_{\sigma_1}^{[2]}, P_{\sigma_2}^{[2]}\} \in E$ for some $P_{\sigma_1}^{[2]}, P_{\sigma_2}^{[2]} \in S_0$ if σ_1 and σ_2 are transpositions of each other. Then G is connected.*

Proof. Consider an arbitrary vertex $P_{\sigma}^{[2]} \in S_0$ such that $\sigma(i_r) = k_r, k_r \neq j_r$ for all $r \in [m]$. Also, consider another vertex $P_{\sigma'}^{[2]} \in S_0$ such that $\sigma'(i_r) = l_r, l_r \neq j_r$ for all $r \in [m]$. We will show that there is a path from $P_{\sigma}^{[2]}$ to $P_{\sigma'}^{[2]}$ in G. For simplicity, we will use σ to refer to $P_{\sigma}^{[2]}$. Let $\sigma, \sigma_1, \sigma_2, \ldots, \sigma_t, \sigma'$ be a path of length $t + 1$ between σ and σ'. Consider a vertex $\sigma_p, p \in [t]$ such that $\sigma_p(i_r) = l_r$ for all $r \in [s], s < m$. In the next step we will extend the path from σ_p to some vertex σ_q such that $\sigma_q(i_r) = l_r$ for all $r \in [s + 1]$. If $\sigma_p(i_x) = l_{s+1}$ such that $\sigma_p(i_{s+1}) \neq j_x$ then we can swap $\sigma_p(i_x)$ with $\sigma_p(i_{s+1})$ to get the desired vertex σ_q. Otherwise, in the first swap we can move l_{s+1} to some index $i_{x'}, x \neq x'$, such that $l_{s+1} \neq j_{x'}$ and then in the second swap get $\sigma_p(i_{s+1})$ to map to l_{s+1}. Note that both the swaps result in vertices within S_0. After the first swap we have $\sigma_{p'}(i_{x'}) = l_{s+1}$ and $\sigma_{p'}(i_x) = \sigma_p(i_{x'})$ and after the second swap we get $\sigma_q(i_{s+1}) = l_{s+1}$ and $\sigma_q(i_{x'}) = j_x$, both of which avoid a map from i_x to j_x and $i_{x'}$ to $j_{x'}$. In case it is not possible to find a suitable $i_{x'}$ to move l_{s+1}, it should be possible to move $\sigma_p(i_{s+1})$ instead. Once we have obtained a permutation σ'' such that $\sigma''(i_r) = l_r$ for all $r \in [m]$, there must exist a path from σ'' to σ' since the set of permutations having $i_r \mapsto l_r$ for all $r \in [m]$, forms a group isomorphic to the symmetric group on $n - m$ elements. $\square$

The following lemma gives a sequence of four vertices of QAP_n such that a specific linear combination of these vertices reduces the number of non-zero entries in the resulting vector to a constant independent of n. This lemma will be used crucially in Lemma 7 to express the difference of two neighboring vertices in S_0 in terms of the vertices in S.

Lemma 6. *Given a sequence of permutations over the set $[n]$, $\sigma_1, \sigma_2, \sigma_3, \sigma_4$, such that σ_2 is obtained from σ_1 by a transposition that swaps the values of $\sigma_1(i), \sigma_1(j)$; σ_3 is obtained from σ_2 by a transposition that swaps the values of $\sigma_2(i'), \sigma_2(j')$ ($i' \neq i, j' \neq j$); and σ_4 is obtained from σ_3 by a transposition that swaps the values of $\sigma_3(i), \sigma_3(j)$. Then $(P_{\sigma_1}^{[2]} - P_{\sigma_2}^{[2]}) - (P_{\sigma_4}^{[2]} - P_{\sigma_3}^{[2]})$ has a number of non-zeroes that is independent of n.*

Proof. Recall that $P_\sigma^{[2]}(ab, xy) = P_\sigma(a, b) \cdot P_\sigma(x, y)$. Since σ_1 and σ_2 differ at only i, j, we have $(P_{\sigma_1}^{[2]} - P_{\sigma_2}^{[2]})(ab, xy) = 0$ for all $a, b, x, y \in [n] \setminus \{i, j\}$. Similarly, we have $(P_{\sigma_4}^{[2]} - P_{\sigma_3}^{[2]})(ab, xy) = 0$ for all $a, b, x, y \in [n] \setminus \{i, j\}$. One can verify that $(P_{\sigma_1}^{[2]} - P_{\sigma_2}^{[2]})(ab, xy) = 1$ for all x, y such that $\sigma_1(x) = y$, when $a = i, b = \sigma_1(i)$ or when $a = j, b = \sigma_1(j)$. Symmetrically, we have $(P_{\sigma_1}^{[2]} - P_{\sigma_2}^{[2]})(ab, xy) = -1$ for all x, y such that $\sigma_2(x) = y$, when $a = i, b = \sigma_2(i)$ or when $a = j, b = \sigma_2(j)$. For the case when $a, b \notin \{i, j\}$, $(P_{\sigma_1}^{[2]} - P_{\sigma_2}^{[2]})(ab, xy) = 1$ when $x = i, y = \sigma_1(i)$ or when $x = j, y = \sigma_1(j)$ and $(P_{\sigma_1}^{[2]} - P_{\sigma_2}^{[2]})(ab, xy) = -1$ when $x = i, y = \sigma_2(i)$ or when $x = j, y = \sigma_2(j)$. Similar values follow for $P_{\sigma_4}^{[2]} - P_{\sigma_3}^{[2]}$. Note that $P_{\sigma_1}^{[2]} - P_{\sigma_2}^{[2]}$ and $P_{\sigma_4}^{[2]} - P_{\sigma_3}^{[2]}$ differ only at the indices i', j'. So subtracting the latter from the former we get, $((P_{\sigma_1}^{[2]} - P_{\sigma_2}^{[2]}) - (P_{\sigma_4}^{[2]} - P_{\sigma_3}^{[2]}))(ab, xy) = 0$ for all $a, b, x, y \notin \{i, j, i', j'\}$. The only non-zero entries that remain are the following: (i) $a = i, b = \sigma_1(i), x = i', y = \sigma_1(i')$, (ii) $a = i, b = \sigma_2(i), x = i', y = \sigma_2(i')$, (iii) $a = j, b = \sigma_1(j), x = i', y = \sigma_1(i')$, (iv) $a = j, b = \sigma_2(j), x = i', y = \sigma_2(i')$, (v) $a = i, b = \sigma_1(i), x = i', y = \sigma_3(i')$, (vi) $a = i, b = \sigma_2(i), x = i', y = \sigma_3(i')$, (vii) $a = j, b = \sigma_1(j), x = i', y = \sigma_3(i')$, (viii) $a = j, b = \sigma_2(j), x = i', y = \sigma_3(i')$. Another 8 non-zero entries correspond to the case when $x = j'$ taking the total to 16. 16 more entries follow from symmetry, by swapping a, b with x, y. Thus, we get a total of 32 non-zero entries in the resulting matrix. Half of these are $+1$ and the remaining half are -1. Note that these entries depend only on the indices where the four permutations map the indices i, j, i', j' and not on the value of n or where these permutations map the remaining indices. $\square$

Lemma 7. *For any $P_\sigma^{[2]} \in S_0$, $P_\sigma^{[2]} - P_{\sigma'}^{[2]}$ lies in the linear hull of S for every neighbor $P_{\sigma'}^{[2]} \in S_0$, provided $m \geqslant 7$.*

Proof. Let $\sigma_1, \sigma_2, \sigma_3, \sigma_4$ be as defined in Lemma 6. Let $\sigma_5, \sigma_6, \sigma_7, \sigma_8$ be four permutations different from $\sigma_1, \sigma_2, \sigma_3, \sigma_4$ but related to each other just like $\sigma_1, \sigma_2, \sigma_3, \sigma_4$ are. This means that σ_6 is obtained from σ_5 by the same transposition that is used to obtain σ_2 from σ_1, σ_7 is obtained from σ_6 by the same transposition that is used to obtain σ_3 from σ_2, and σ_8 is obtained from σ_7 by the same transposition that is used to obtain σ_4 from σ_3. So from Lemma 6, we have $P_{\sigma_1}^{[2]} - P_{\sigma_2}^{[2]} = (P_{\sigma_3}^{[2]} - P_{\sigma_4}^{[2]}) + (P_{\sigma_5}^{[2]} - P_{\sigma_6}^{[2]}) - (P_{\sigma_7}^{[2]} - P_{\sigma_8}^{[2]})$. Let $\sigma_1 = \sigma$ and $\sigma_2 = \sigma'$. If we can find $\sigma_3, \ldots, \sigma_8$ as defined above such that the corresponding vertices lie in S, then we are done. Since $\sigma, \sigma' \in S_0$, we have some index $a, a \neq i_r, r \in [m]$ such that $\sigma(a) = \sigma'(a) = j_r$. Consider the case when $a = i_p, p \in [m]$ and $\sigma(i_r) \neq j_p$. The case when $a \neq i_p$ for any $p \in [m]$ is similar. We can swap $\sigma'(a)$ with $\sigma'(i_r)$ to get $\sigma_3(a) = \sigma'(i_r), \sigma_3(i_r) = j_r$ which clearly lies in S. We obtain σ_4 from σ_3 by the transposition defined in Lemma 6 and clearly σ_4 also lies in S. Next we select a permutation $\sigma_5 \in S$ that matches with σ_4 at the four indices defined in Lemma 6 and also maps an index $i_{r'}, r \neq r'$ to $j_{r'}$. So we have $\sigma_3, \sigma_4 \in S_1, \sigma_5 \in S_2$. Obtaining $\sigma_6, \sigma_7, \sigma_8$ as outlined above, we have $\sigma_6 \in S_2, \sigma_7, \sigma_8 \in S_1$. Next, consider the case when $a = i_p, p \in [m]$ and $\sigma(i_r) = j_p$. This can happen when m is even and any transposition of σ either results in a permutation in S_0 or in S_2. It is not possible to get a permutation in S_1 by a single transposition of σ. So we obtain σ_3, σ_4 as before but this time the vertices lie in S_2 instead of S_1. Moreover, this time we select a permutation $\sigma_5 \in S$ that matches with σ_4 at the four indices defined in Lemma 6 but has a pair of indices i_x, i_y such that $\sigma_5(i_x) = j_y$ but $\sigma_5(i_y) \neq j_x$. Obtaining $\sigma_6, \sigma_7, \sigma_8$ as above, we have $\sigma_5, \sigma_6 \in S_2, \sigma_7, \sigma_8 \in S_0$. We can now repeat the above argument with $\sigma = \sigma_7, \sigma' = \sigma_8$. This time however, by the choice of $\sigma_5 \in S$, we have ensured that we are in the first case where we could express the difference vector as a combination of vertices only in S. Note that we need m to be at least 7 for the above argument to work. This is so because the transposition of σ that gives σ' can use upto four indices so that these indices are no longer available to get to S. Further, as in the second case above, two other indices swap with each other to get to S_2. So that takes up a total of six indices that are not available to get the desired σ_5. Now if m is at least 7, we are guaranteed to find a pair of indices i_x, i_y such that $\sigma_5(i_x) = j_y$ but $\sigma_5(i_y) \neq j_x$.

□

Theorem 2. *The following inequality:*

$$\sum_{r=1}^{m} Y_{i_r j_r, i_r j_r} - \sum_{r<s} Y_{i_r j_r, i_s j_s} \leq 1$$

where $i_1, \ldots, i_m$ are all distinct and $j_1, \ldots, j_m$ are also distinct, is facet-defining for the QAP-polytope when $m, n \geqslant 7$.

Proof. From Lemma 3 and Lemma 4 we can conclude that any vertex in $V \setminus (S \cup S_0)$ can be expressed as a linear combination of the vertices in $S \cup S_0$. What remains to be shown is that any vertex in S_0 can be expressed as a linear combination of the vertices in S and a fixed vertex $\sigma^* \in S_0$. From Lemma 5 we know that it is possible to go from any vertex in S_0 to any other vertex in S_0 via transpositions such that all the intermediate vertices are in S_0. Let us fix some arbitrary vertex in S_0 as σ^*. So there is a path from every other vertex in S_0 to σ^*. Consider a vertex $\sigma \in S_0$. Let $\sigma, \sigma_1, \ldots, \sigma_t, \sigma^*$ be a path from σ to σ^*. From Lemma 7 we can express the difference of any vertex $\sigma \in S_0$ with any other vertex $\sigma' \in S_0$ such that σ' is a transposition of σ, as a linear combination of the vertices in S. So we have $\sigma - \sigma_1 \in span(S), \sigma_1 - \sigma_2 \in span(S), \ldots, \sigma_t - \sigma^* \in span(S)$ which implies that $\sigma - \sigma^* \in span(S)$ or $\sigma \in span(S \cup \{\sigma^*\})$.

□

4   Membership Testing

In this section we consider the membership testing problem for each of the QAP relaxations defined in Sect. 2. That is, for each of these relaxations we wish to test whether a given point x satisfies all the constraints. Note that the separation problem where one wishes to identify a violated inequality in case the answer to membership testing is negative is a harder problem. Typically for efficient use in cutting plane methods one would like to solve the (harder) separation problem.

We show that membership testing for inequalities QAP2, QAP3, or QAP5 is coNP-complete.

Recall that QAP2 is defined by the following set of inequalities:

$$\sum_{r=1}^{m} Y_{i_r j_r, kl} - Y_{kl,kl} - \sum_{r<s} Y_{i_r j_r, i_s j_s} \leqslant 0$$

where $i_1, \ldots, i_m, k$ are all distinct and $j_1, \ldots, j_m, l$ are also distinct. In addition, $n \geqslant 6, m \geqslant 3$.

Theorem 3. *Given a point* $x \in \mathbb{R}^{n^4}$ *with* $0 \leqslant x \leqslant 1$, *it is coNP-complete to decide whether* x *satisfies all inequalities of QAP2.*

Proof. The problem is clearly in coNP since given a violated inequality it can be checked quickly that it is indeed violated.

For establishing NP-hardness we will reduce the max-clique problem to membership testing for QAP2. Let the given instance of the max-clique problem be $G = (V, E)$ where $V = [n]$. We construct a n-partite graph $G' = (V', E')$ where $V' = \{(ij)\}$ for $i, j \in [n]$ and $\{i_1 j_1, i_2 j_2\} \in E'$ if and only if $\{i_1, i_2\} \in E$. So if there is an edge $\{i, j\} \in E$ then we get a complete bi-partite graph between the partitions i and j, else there is no edge between these two partitions. Consider a clique $C = \{i_1, i_2, \ldots, i_k\}$ of size k in G. Then the set of vertices $\{i_1 j_1, i_2 j_2, \ldots, i_k j_k\}$ where j_r could be any arbitrary index in $[n]$, forms a clique of size k in G'. Conversely, given a clique $C = \{i_1 j_1, i_2 j_2, \ldots, i_k j_k\}$ of size k in G', the set of vertices $\{i_1, i_2, \ldots, i_k\}$ forms a clique of size k in G. Fix a pair of indices $k, l \in [n]$ arbitrarily and add edges $\{kl, i_r j_r\}$ for all $i_r \in \{[n] \setminus \{k\}\}$ to G' (if the edge is not already present). Now construct a point Y as follows:

$$Y_{i_1 j_1, i_2 j_2} = \begin{cases} 0, & \text{if } (kl) \notin \{(i_1 j_1), (i_2 j_2)\} \text{ and } \{i_1 j_1, i_2 j_2\} \in E' \\ n, & \text{if } (kl) \notin \{(i_1 j_1), (i_2 j_2)\} \text{ and } \{i_1 j_1, i_2 j_2\} \notin E' \\ 1, & \text{if } (i_1 \neq i_2 = k \text{ and } j_1 \neq j_2 = l) \\ & \quad \text{or } (k = i_1 \neq i_2 \text{ and } l = j_1 \neq j_2) \\ t, & \text{if } i_1 = i_2 = k \text{ and } j_1 = j_2 = l \\ n^2, & \text{if } i_1 = i_2 \text{ and } j_1 = j_2 \text{ and } (i_1 \neq k \text{ or } j_1 \neq l) \end{cases}$$

where $t \geqslant 2$ is a natural number. Notice that any point Y satisfies all the inequalities of QAP2 if and only if αY satisfies them for all $\alpha \geqslant 0$. Therefore Y can be scaled to satisfy $0 \leqslant Y \leqslant 1$. We will ignore this scale factor and continue our argument with Y as constructed above to avoid cluttered equations.

We claim that Y satisfies all the inequalities of QAP2 if and only if every clique in the subgraph induced by the neighborhood of the vertex (kl) in G', has size at most t.

Suppose that the largest clique $C = \{i_1j_1, i_2j_2, \ldots, i_{t'}j_{t'}\}$ such that $\{i_rj_r, kl\}$ $\in E'$ for $r \in [t']$, has size $t' > t$. Without loss of generality, we can assume that $i_1, \ldots, i_{t'}$ as well as $j_1, \ldots, j_{t'}$ are distinct. Consider the inequality h_C corresponding to the choice of indices $\{i_1, \ldots, i_{t'}, k\}, \{j_1, \ldots, j_{t'}, l\}$. From the above construction, $\sum_{r=1}^{t'} Y_{i_rj_r,kl} = t'$, $Y_{kl,kl} = t$ and $\sum_{r<s, r,s\in[t']} Y_{i_rj_r,i_sj_s} = 0$ giving $t' \leqslant t$ and Y violates h_C.

Now suppose that every clique C in the subgraph induced by the neighborhood of (kl), has size at most t and there exists an inequality of QAP2 defined by the sets $\{i_1, \ldots, i_m, k'\}, \{j_1, \ldots, j_m, l'\}$ that is violated by the above point Y. Notice that $k' = k$ and $l' = l$ must hold. If not then $\sum_{r=1}^m Y_{i_rj_r,k'l'} - \sum_{r<s,r} Y_{i_rj_r,i_sj_s} \leqslant nm \leqslant n^2 = Y_{k'l',k'l'}$ and Y does not violate the inequality. So any violated inequality must have $\sum_{r=1}^m Y_{i_rj_r,kl} > t + \sum_{r<s} Y_{i_rj_r,i_sj_s}$. Since $Y_{i_rj_r,kl} = 1$ for all $i_r \in \{[n] \setminus \{k\}\}$, we have $m > t + \sum_{r<s} Y_{i_rj_r,i_sj_s}$ for any violated inequality, which is not possible if $Y_{i_rj_r,i_sj_s} = n$ since $m \leqslant n$ and $t \geqslant 2$. Therefore, $Y_{i_rj_r,i_sj_s} = 0$ for all distinct $r, s \in [m]$ and so $\{i_rj_r, i_sj_s\} \in E'$. But then, $m > t$ giving a clique of size larger than t in the neighborhood of (kl) contradicting the assumption that the every such clique has size at most t.

Therefore, given a membership oracle for QAP2 we can compute the size of the largest clique in any graph except K_n by calling such an oracle for various choices of k, l and t and outputting the largest value of t for which the above constructed point satisfies all the inequalities.

□

Next, recall that QAP3 is defined by the following set of inequalities:

$$(\beta - 1) \sum_{(ij)\in P\times Q} Y_{ij,ij} - \sum_{\substack{(ij),(kl)\in P\times Q \\ i<k}} Y_{ij,kl} \leqslant \frac{\beta^2 - \beta}{2}$$

where $P, Q \subset [n]$. In addition (i) $\beta+1 \leqslant |P|, |Q| \leqslant n-3$, (ii) $|P|+|Q| \leqslant n-3+\beta$, (iii) $\beta \geqslant 2$.

Theorem 4. *Given a point $x \in \mathbb{R}^{n^4}$ with $x \geqslant 0$, it is coNP-complete to decide whether x satisfies all inequalities of QAP3.*

Proof. Again we will reduce the max-clique problem to membership testing for QAP3. Given an instance of the max-clique problem, $G = (V, E)$ with $|V| = n$, we construct a point Y as follows:

$$Y_{i_1j_1,i_2j_2} = \begin{cases} 0, & \text{if } \{i_1,i_2\} \in E \\ n^2, & \text{if } \{i_1,i_2\} \notin E \ (i_1 \neq i_2) \\ 1/t, & \text{if } i_1 = i_2 \text{ and } j_1 = j_2 = 1 \\ 0, & \text{if } i_1 = i_2 \text{ and } j_1 = j_2 \text{ and } j_1 > 1 \end{cases}$$

where $1 \leqslant t \leqslant n - 4$ is some natural number. We claim that Y satisfies all inequalities of QAP3 if and only if G doesn't contain a clique of size larger than t. This gives an algorithm to find the size of largest clique in G by increasing t gradually and computing the smallest value of t for which Y becomes feasible. If Y remains infeasible for $t = n - 4$ then the largest clique in G has size $n - 3$ or more. This can be determined by checking the $O(n^3)$ possible subsets of vertices of G.

To prove the claim, let P be a clique in G with $t < |P| \leqslant n - 3$. Define $Q = \{1\}, \beta = 2$ and consider the inequality h defined by P, Q, β. It can be checked that Y violates h.

Conversely, let h be an inequality defined by P, Q, β that is violated by Y.

We first observe that $1 \in Q$. Suppose not, then for any $(i,j) \in P \times Q$ we have $Y_{ij,ij} = 0$. Since $\beta^2 - \beta \geqslant 0$ for all natural $\beta \geqslant 2$, h cannot be violated by Y. It follows that if $i, k \in P$ and $i \neq k$, then $\{i, k\} \in E$. Again, suppose not. Then, $Y_{ij,kl} = n^2$ for all j, l but $(\beta - 1) \sum_{(ij) \in P \times Q} Y_{ij,ij} \leqslant (n-4) \cdot (n-3) < n^2$. So Y cannot violate h as $(\beta - 1) \sum_{(ij) \in P \times Q} Y_{ij,ij} - \sum_{(ij),(kl) \in P \times Q, i<k} Y_{ij,kl}$ is negative but $\beta^2 - \beta$ is nonnegative.

So we have that if Y violates h then $Y_{ij,kl} = 0$ for all $i \neq k$. Further, such an inequality must have $\{i, k\} \in E$ for all distinct $i, k \in P$. That is P must form a clique in G. Recall that only $1 \in Q$ contributes a non-zero value to the left hand side expression of h. Therefore, if Y violates h then $|P|/t = \sum_{(ij) \in P \times Q} Y_{ij,ij} > \beta/2$ and G contains a clique of size larger than $t\beta/2$, that is, larger than t.   □

Finally, recall that QAP5 is defined by the following set of inequalities:

$$\sum_{r=1}^{m} Y_{i_rj_r,i_rj_r} - \sum_{r<s} Y_{i_rj_r,i_sj_s} \leqslant 1$$

where $i_1, \ldots, i_m$ are all distinct and $j_1, \ldots, j_m$ are also distinct. In addition, $m, n \geqslant 7$.

Theorem 5. *Given a point $x \in \mathbb{R}^{n^4}$ with $x \geqslant 0$, it is coNP-complete to decide whether x satisfies all inequalities of QAP5.*

Proof. Given an instance of the max-clique problem, $G = (V, E)$ with $|V| = n$, we construct a point Y as follows:

$$Y_{i_1j_1,i_2j_2} = \begin{cases} 0, & \text{if } \{i_1,i_2\} \in E \\ n/6, & \text{if } i_1 \neq i_2 \text{ and } \{i_1,i_2\} \notin E \\ 1/t, & \text{if } i_1 = i_2 \text{ and } j_1 = j_2 \\ 0, & \text{otherwise} \end{cases}$$

where, $t \geqslant 6$ is a natural number.

We claim that Y is infeasible if and only if there exists a clique in G of size at least $t + 1$.

Suppose there exists a clique $C = \{p_1, \ldots, p_m\}$ in G with $m \geqslant t+1$. Consider the inequality h_C defined by the indices $i_1, \ldots, i_m$ and $j_1, \ldots, j_m$ with $i_k = j_k = p_k$ for all $k \in [m]$. Then, Y violates h_C because $m/t > 1$.

Conversely, suppose every clique in G has size at most t and let h be a violated inequality defined by indices $i_1, \ldots, i_m$ and $j_1, \ldots, j_m$. It must hold that $\{i_r, i_s\} \in E$ otherwise the left hand side in the inequality h with respect to Y is at most $m/t - n/6$ which is at most zero since $t \geqslant 6$ and $m \leqslant n$ and so Y cannot violate h. So for a violation, G must contain a clique of size m. But then $Y_{i_r j_r, i_s j_s} = 0$ for all distinct $r, s \in [m]$ and so $m/t > 1$ contradicting the assumption that G contains no cliques of size larger than t.

Therefore, given a graph $G = (V, E)$ if Y is feasible for all values of $t \geqslant 6$ then the size of a largest clique is at most 6 and can be computed in polynomial time. Otherwise, the largest value of t for which Y is infeasible equals the size of the largest clique in G minus one.   □

5   Extension Complexity

In this section we will prove that any relaxation of the QAP-polytope for which any of the families of inequalities defined in Sect. 2 are valid, has superpolynomial extension complexity. A set of linear inequalities is said to be a relaxation of the QAP-polytope if it contains the QAP-polytope.

We will need the following lemma whose proof we omit[2] due to space constraints.

Lemma 8. *Let N_n^k be a $2^n \times 2^n$ matrix with rows and columns indexed by binary vectors of length n and whose entries are $N_n^k(a, b) := (a^\mathsf{T} b - k) \cdot (a^\mathsf{T} b - k - 1)$. Then, any EF-protocol for N_n^k requires $\Omega(n - k)$ bits to be exchanged.*

As noted in Sect. 2, QAP1 are the most general family of valid inequalities for the QAP-polytope. Therefore any lower bound on the extension complexity of a relaxation R_n corresponding to any of the families QAP2-QAP5 also hold for QAP1. Proofs for each of QAP2-QAP5 are somewhat similar. Therefore due to space constraints we only provide partial proof of the next theorem.

Theorem 6. *Let R_n^i be any bounded relaxation of the QAP-polytope such that the inequalities of QAPi are valid for R_n^i. Then $xc(R_n^i) \geqslant 2^{\Omega(n)}$.*

Proof. (**Proof for QAP2**): Consider any relaxation of QAP_n for which the inequalities of QAP2 are valid. Let $i_1, i_2, \ldots, i_m, k$ be distinct indices in $[n]$, and $j_1, j_2, \ldots, j_m, l$ be all distinct indices in $[n]$ as well. Let $\sigma \in S_n$ be such that q indices (i_r, j_r) satisfy $P_\sigma(i_r, j_r) = 1$. Then, the slack of $\sum_{r=1}^{m} Y_{i_r j_r, kl} - Y_{kl, kl} - $

---

[2] For the omitted proofs refer to the full version of this paper [2].

$\sum_{r<s} Y_{i_r j_r, i_s j_s} \leqslant 0$ with respect to $P_\sigma^{[2]}$ is $P_\sigma(k,l) + \binom{q}{2} - q P_\sigma(k,l)$. If $P_\sigma(k,l) = 0$ then this equals $\binom{q}{2}$. If $P_\sigma(k,l) = 1$ then this equals $1 + \binom{q}{2} - q = \frac{2+q(q-1)-2q}{2} = \binom{q-1}{2}$. Therefore the slack is

$$\binom{q - P_\sigma(k,l)}{2} \tag{6}$$

Now, suppose there exists an EF-protocol for computing the slack matrix of R_n^2 that requires at most c bits to be exchanged. That is if Alice is given any valid inequality for R_n^2 and Bob any feasible point in R_n^2 they can compute the corresponding slack in expectation by exchanging at most c bits. We will show that they can modify this EF-protocol to get an EF-protocol for matrix N_n^1 with at most $O(c + \log n)$ bits exchanged. By Lemma 8 this requires $\Omega(n)$ bits to be exchanged and so $c = \Omega(n)$. Finally, applying Theorem 1 we will get that $\mathrm{xc}(R_n^2) \geqslant 2^{\Omega(n)}$.

So suppose, Alice and Bob get $a, b \in \{0,1\}^n$ respectively and wish to compute $N_n^1(a,b) = (a^\mathsf{T} b - 1)(a^\mathsf{T} b - 2)$ in expectation. We can assume that Alice receives neither the all zero nor the all one vector. If $a = (0,\ldots,0)$ then she can output zero and stop. If $a = (1,\ldots,1)$ then she can tell Bob this using one bit and Bob can output the number of nonzero entries in b. Further, we can assume that the vector b contains at least three zero entries. Otherwise Bob can tell Alice using at most $2\log n$ bits the indices where b is zero and Alice can output the correct value.

Let $p_1, \ldots, p_m$ be the indices where a is non-zero and let p be an arbitrary index such that $a_p = 0$. Alice creates the inequality corresponding to sets $i_1, \ldots, i_m, k$ and $j_1, \ldots, j_m, l$ where $i_1 = j_1 = p_1, \ldots, i_m = j_m = p_m$ and $k = l = p$. Alice then sends the index p to Bob who sets $b_p = 1$ if it is not already so. Bob then creates any permutation σ_b such that $\sigma_b(i) = i$ if $b_i = 1$ and $\sigma_b(i) \neq i$ if $b_i = 0$. This is clearly possible since b still contains at least two zeroes. Bob selects the vertex $P_{\sigma_b}^{[2]}$ of QAP_n corresponding to this permutation. Clearly, $a^\mathsf{T} b$ equals the number of index pairs (i_r, j_r) in the set created by Alice such that $P_{\sigma_b}^{[2]}(i_r, j_r) = 1$.

Using $P_\sigma(k,l) = P_{\sigma_b}(p,p) = 1$ and $q = a^\mathsf{T} b$ in equation (6) we see that the slack of Alice's inequality with respect to Bob's vertex of QAP_n is exactly $\binom{a^\mathsf{T} b - 1}{2} = \frac{1}{2} N_n^1(a,b)$ and hence they can just use the protocol for computing the slack matrix of R_n^2 for computing N_n^1 by agreeing that every time they wish to output something they would output twice as much.   □

Acknowledgement. Pawan Aurora is partially supported by grant MTR/2018/000861 of the Science and Engineering Research Board, Government of India. Hans Raj Tiwary was partially supported by grant 17-09142S of GAČR.

References

1. Aurora, P., Mehta, S.K.: The QAP-polytope and the graph isomorphism problem. J. Comb. Optim. **36**(3), 965–1006 (2018)

2. Aurora, P., Tiwary, H.R.: On the complexity of some facet-defining inequalities of the QAP-polytope. CoRR abs/2010.06401 (2020)
3. Avis, D., Tiwary, H.R.: On the extension complexity of combinatorial polytopes. Math. Program. **153**(1), 95–115 (2014). https://doi.org/10.1007/s10107-014-0764-2
4. Bazzi, A., Fiorini, S., Pokutta, S., Svensson, O.: No small linear program approximates vertex cover within a factor 2 - ε. Math. Oper. Res. **44**(1), 147–172 (2019)
5. Braun, G., Fiorini, S., Pokutta, S., Steurer, D.: Approximation limits of linear programs (beyond hierarchies). Math. Oper. Res. **40**(3), 756–772 (2015)
6. Braun, G., Jain, R., Lee, T., Pokutta, S.: Information-theoretic approximations of the nonnegative rank. Comput. Complex. **26**(1), 147–197 (2016). https://doi.org/10.1007/s00037-016-0125-z
7. Braverman, M., Moitra, A.: An information complexity approach to extended formulations. In: Proceedings of the STOC 2013, pp. 161–170 (2013)
8. Briët, J., Dadush, D., Pokutta, S.: On the existence of 0/1 polytopes with high semidefinite extension complexity. Math. Program. **153**(1), 179–199 (2014). https://doi.org/10.1007/s10107-014-0785-x
9. Buchanan, A.: Extended formulations for vertex cover. Oper. Res. Lett. **44**(3), 374–378 (2016)
10. Cela, E.: The Quadratic Assignment Problem: Theory and Algorithms. Combinatorial Optimization. Springer, Boston (1997)
11. Chan, S.O., Lee, J.R., Raghavendra, P., Steurer, D.: Approximate constraint satisfaction requires large LP relaxations. In: Proceedings of the FOCS 2013, pp. 350–359 (2013)
12. Conforti, M., Cornuéjols, G., Zambelli, G.: Extended formulations in combinatorial optimization. Ann. Oper. Res. **204**(1), 97–143 (2013)
13. Faenza, Y., Fiorini, S., Grappe, R., Tiwary, H.R.: Extended formulations, nonnegative factorizations, and randomized communication protocols. Math. Program. **153**(1), 75–94 (2014). https://doi.org/10.1007/s10107-014-0755-3
14. Fiorini, S., Massar, S., Patra, M.K., Tiwary, H.R.: Generalized probabilistic theories and conic extensions of polytopes. J. Phys. A: Math. Theor. **48**(2), 025302 (2014)
15. Fiorini, S., Massar, S., Pokutta, S., Tiwary, H.R., de Wolf, R.: Exponential lower bounds for polytopes in combinatorial optimization. J. ACM **62**(2), 17 (2015)
16. Gajarský, J., Hliněný, P., Tiwary, H.R.: Parameterized extension complexity of independent set and related problems. Discrete Appl. Math. **248**, 56–67 (2018)
17. Gouveia, J., Parrilo, P.A., Thomas, R.R.: Lifts of convex sets and cone factorizations. Math. Oper. Res. **38**(2), 248–264 (2013)
18. Jünger, M., Kaibel, V.: A basic study of the QAP-polytope. Technical report, Institut Für Informatik, Universität zu Köln, Germany (1996)
19. Jünger, M., Kaibel, V.: On the SQAP polytope. Technical report, Institut Für Informatik, Universität zu Köln, Germany (1996)
20. Jünger, M., Kaibel, V.: The QAP-polytope and the star transformation. Discrete Appl. Math. **111**(3), 283–306 (2001)
21. Kaibel, V.: Polyhedral combinatorics of the quadratic assignment problem. Ph.D. thesis, Faculty of Mathematics and Natural Sciences, University of Cologne (1997)
22. Kaibel, V.: Extended formulations in combinatorial optimization. Optima **85**, 2–7 (2011)
23. Koopmans, T., Beckmann, M.: Assignment problems and the location of economic activities. Technical report, Cowles Foundation, Yale University (1955)

24. Lawler, E.L.: The quadratic assignment problem. Manag. Sci. **9**(4), 586–599 (1963)
25. Lee, J.R., Raghavendra, P., Steurer, D.: Lower bounds on the size of semidefinite programming relaxations. In: Proceedings of the STOC 2015, pp. 567–576 (2015)
26. Padberg, M., Rijal, M.: Location, Scheduling, Design and Integer Programming. International Series in Operations Research & Management Science. Springer, Boston (1996)
27. Padberg, M., Rinaldi, G.: A branch-and-cut algorithm for the resolution of large-scale symmetric traveling salesman problems. SIAM Rev. **33**(1), 60–100 (1991)
28. Pitowsky, I.: Quantum Probability – Quantum Logic. Lecture Notes in Physics, vol. 321. Springer, Berlin (1989)
29. Rothvoß, T.: Some 0/1 polytopes need exponential size extended formulations. Math. Program. **142**(1–2), 255–268 (2013)
30. Rothvoß, T.: The matching polytope has exponential extension complexity. J. ACM **64**(6), 41:1–41:19 (2017)
31. Sahni, S., Gonzalez, T.: P-complete approximation problems. J. ACM **23**(3), 555–565 (1976)
32. Vanderbeck, F., Wolsey, L.A.: Reformulation and decomposition of integer programs. In: Jünger, M., et al. (eds.) 50 Years of Integer Programming 1958-2008, pp. 431–502. Springer, Heidelberg (2010). https://doi.org/10.1007/978-3-540-68279-0_13
33. Wolsey, L.A.: Using extended formulations in practice. Optima **85**, 7–9 (2011)

Hardness of Segment Cover, Contiguous SAT and Visibility with Uncertain Obstacles

Sharareh Alipour[1] and Salman Parsa[2(✉)]

[1] School of Computer Science, Institute for Research in Fundamental Sciences (IPM), Tehran, Iran
alipour@ipm.ir
[2] Computer Science Department, Saint Louis University, St. Louis, USA
salman.parsa@slu.edu

Abstract. We define the problem segment cover as follows. We are given a set of pairs of sub-intervals of the unit interval. The problem asks if there is a choice of a single interval from each pair such that the union of the chosen intervals covers the entire unit interval. This problem arises naturally while attempting to compute visibility between a point and a line segment in the plane in the presence of uncertain obstacles. Segment cover is equivalent to a restricted version of SAT which we call contiguous SAT. Consider a SAT with the following restrictions. An input formula is in CNF form and an ordering of the clauses is given in which clauses containing any fixed literal appear contiguously. We call this restricted problem contiguous SAT. Our main result is that the problems segment cover and contiguous SAT are NP-hard. We also discuss hardness of approximation for these problems.

1 Introduction

In this paper we consider two very related problems. One of them we call the segment cover problem and the other one contiguous SAT. These problems are encountered when trying to introduce a specific model of uncertainty into visibility problems, see Sect. 1.2 below for the connection to this uncertain visibility model which has been the origin of this work. These two problems are very natural and we expect that they would be encountered in similar situations.

1.1 Problem Statements and Results

Our first problem is called the segment cover problem. Let I be an interval of the real line, this interval is fixed once and for all and for simplicity we take $I = [0,1]$. We call a closed sub-interval of I a *segment*. An *uncertain segment*

The work of Salman Parsa is supported by the National Science Foundation under Grant CCF-1614562 and funding from the Saint Louis University Research Institute.

W. Wu and Z. Zhang (Eds.): COCOA 2020, LNCS 12577, pp. 350–363, 2020.
https://doi.org/10.1007/978-3-030-64843-5_24

is a pair $s = \{l, r\}$ of two segments. An uncertain segment models the situation where we know that the "real" segment is one of l or r but we do not know which one. Let $S = \{s_i, i = 1, \ldots, n\}$ be a set of uncertain segments. The segment cover problem asks: Is there a choice of l_i or r_i (but not both) for each i such that the union of the chosen segments is I. See Fig. 1 for an example. In other words, if the uncertain segment s_i is l_i with probability $0 \leq p_i \leq 1$ and is r_i with probability $1 - p_i$, the problem asks to decide if the probability of the entire interval I being covered is non-zero. We show in Sect. 2 that the segment cover problem is NP-hard. We remark that if we require that any two segments (of all uncertain segments) are disjoint or coincide, then the NP-hardness result does not apply. Therefore the NP-hardness is not immediate because of the presence of a choice. Also one is justified for having the first impression that this problem should not be NP-hard. We reduce 3SAT to segment cover. The method we use for the reduction is simple, however we believe it is novel.

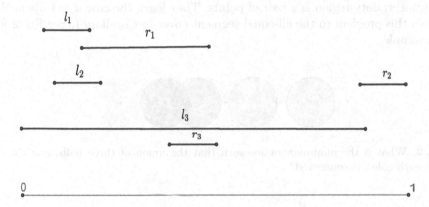

Fig. 1. An instance of segment cover with 3 uncertain segments. The segments are depicted above the interval for clarity.

Our second problem, the contiguous SAT is a restricted SAT problem. The input to contiguous SAT is i) a SAT instance in CNF form $C_1 \wedge C_2 \cdots \wedge C_m$, where the C_i's are m clauses, and ii) an ordering on the C_i. The instance must satisfy the *contiguity* condition: any literal appears in a contiguous set of clauses with respect to the given ordering. It is convenient to assume that the subscripts of the C_i respect the given ordering. Then the contiguity condition requires that for any positive literal x, the i such that $x \in C_i$ form a contiguous set of numbers, and similarly for the literal $\neg x$. For example, the following formula is an input to contiguous SAT with the left-to-right ordering of the clauses.

$$(x_1 \vee x_2 \vee x_3) \wedge (x_1 \vee x_2 \vee \neg x_3) \wedge (x_1 \vee \neg x_3 \vee \neg x_1)$$

But the following is not a valid input.

$$(x_1 \vee x_2 \vee x_3) \wedge (\neg x_1 \vee x_2 \vee \neg x_3) \wedge (x_1 \vee \neg x_3 \vee \neg x_1)$$

Lemma 1 states that the above two problems are linear-time equivalent. Therefore, by the results of this paper, the contiguous SAT problem is also NP-hard. The segment cover problem provides a geometric view to contiguous SAT. This point of view has been useful in proving our results.

We also consider a special case of the segment cover problem in which the input segments are all of equal length. We call this problem *all-equal segment cover*. We show in Theorem 2 that all-equal segment cover is NP-hard. From this result follows that a problem called Best-Case Connectivity (BCU) in [9,10] is NP-hard even in dimension 1. Therefore we strengthen a main result of [9,10] considerably.

The BCU problem is defined as follows. Let $R_1, \ldots, R_n$ be n closed regions in the d-dimensional Euclidean space. These are called *uncertainty regions*. Find the minimum value of r satisfying: there exist points $p_i \in R_i$ such that the union of the balls of radius r centered at the p_i, $\bigcup_i B(p_i, r)$, is connected. In [9,10] it is shown that this problem is NP-hard in the plane, $d = 2$, even when each uncertainty region is a pair of points. They leave the case $d = 1$ open. We reduce this problem to the all-equal segment cover in Corollary 1. See Fig. 2 for an example.

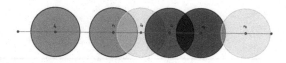

Fig. 2. What is the minimum radius such that the union of three balls, one chosen from each color, is connected?

In Sect. 4 we also prove hardness of approximation results for the segment cover problem.

1.2 Relations to Uncertain Visibility

This section describes the motivation behind our problems and can be skipped. One of the basic problems in computational geometry is computing visibility in various configurations of points and obstacles in the Euclidean plane. Visibility plays an important role in robotics and computer graphics, among other areas. In robotics, for example, the efficient exploration of an unknown environment requires computing the visibility polygon of the robot or test whether the robot sees a specific object or not.

Suppose that we are given a set S of n obstacles in the plane, say in the form of convex polygons. Two points p and q are *visible* to each other if their connecting line segment does not intersect any of the obstacles in S. A line segment $t \subset \mathbb{R}^2$ is visible to a point p if p is visible to at least one point of the line segment t.

In a version of *visibility testing problem* we are given a set S of n obstacles in $\mathbb{R}^2$ and our goal is to preprocess S so that we can quickly answer a query of the form: is a query segment $t \subset \mathbb{R}^2$ visible from a query point $q \in \mathbb{R}^2$? For instance q is a camera and the segment represents an object of interest.

Recently, there has been some attention to situations wherein there is uncertainty in the obstacle positions. Then, two points are visible to each other with a certain probability. Examples of these kind of problem can be found in [1,7]. We introduce uncertainty into the set of obstacles S as follows. Each obstacle exists in one of two possible locations with given probabilities. Consequently, the set S is replaced by a set of pairs $\{l_i, r_i\}$ of obstacles. We denote the set of pairs again by S. The *probabilistic visibility testing problem* asks to preprocess S to answer the following query: What is the probability that a given segment $I \subset \mathbb{R}^2$ be visible from a given query point q? It was shown in [1] that computing this probability exactly is #P-hard. Note that for a fixed q, each possible location of an obstacle covers a sub-segment of I. Then the probability that the interval I be visible to q is 1 if and only if, for any choices of l_i or r_i the projections of the chosen obstacles to I do not cover the interval I. This is our segment cover problem. See Fig. 3. Our results imply that deciding that a given point q can always see a given segment I is NP-hard, in the presence of uncertain obstacles.

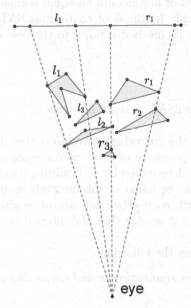

Fig. 3. 3 uncertain obstacles, each with two possibilities l_i, r_i, the projection each uncertain obstacle gives an uncertain segment, one such uncertain segment is depicted

1.3 Related Work

The problems around the visibility concept have been an active research area since the beginning of computational geometry. Point and edge visibility [4,5,11, 18], the art gallery problem [27], the watchman route problem [8,23], visibility graphs and their recognition [15,17] are among the topics of interest in this field. Uncertainty is very natural in applications and indeed has been studied from a more practical viewpoint, like robot motion planning [6,12,14,24,26].

There are many restrictions of SAT that are NP-complete. They include k-SAT, $k \geq 3$, NAE-SAT, 1-in-3 SAT [30], planar 3-SAT [22], planar 1-in-3 SAT [21], monotone planar cubic 1-in-3 SAT [25], 4-bounded planar 3-connected 3-SAT [20]. Let us denote by (r, s)-SAT the SAT problem restricted to clauses containing exactly r variables, and each variable appearing in at most s clauses. Tovey [31] has shown that $(3, 3)$-SAT is always satisfiable and $(3, 4)$-SAT is NP-Complete. In addition, it is proved in [28] that 3-SAT restricted to instances where each variable appears at most three times is NP-Complete. A stronger result, proved in Dahlhaus et al. [13], states that planar 3-SAT in which each variables appears exactly three times, and twice with one literal, a third time as the other literal, is still NP-Complete. These results will be used in Sect. 2. We make use of the hardness of SAT instances where each variable is restricted to a few clauses in our proof of hardness of all-equal segment cover. None of these special cases directly imply hardness of contiguous SAT. Indeed although our reduction is quite simple its method is novel to the best of our knowledge.

2 Reduction

In this section we reduce 3SAT to segment cover.

Convention. We make the following convention that by a *literal* of a SAT instance we mean an appearance of x or $\neg x$ for some variable x in ϕ. Therefore, a literal is uniquely determined by determining its clause and a number. In addition, if an interval is the union of sub-intervals such that the sub-intervals share only endpoints with each other and are otherwise disjoint, by abuse of notation and for simplicity, we say that the interval is a *disjoint union* of the sub-intervals.

We begin by observing the following.

Lemma 1. *The problems segment cover and contiguous SAT are equivalent with linear-time reductions.*

Proof. Given an instant of segment cover construct an instance of contiguous SAT as follows. Partition the interval I using the endpoints of all of the given segments into subintervals J_i. The J_i are closed sub-intervals that only share (one or two) endpoints with the neighboring sub-intervals and otherwise are disjoint. For each sub-interval J_i we define a clause C_i. For any uncertain segment $s_j = \{l_j, r_j\}$ define a variable x_j. For all i and j, add the literal x_j to the C_i

if J_i is covered by l_j and add the literal $\neg x_j$ to the C_i if J_i is covered by r_j. Order the C_i using the left-to-right ordering of the sub-intervals J_i. This defines an instance of contiguous SAT. See Fig. 4.

If the contiguous SAT instance we constructed is satisfiable then we choose for s_j, the segment l_j if $x_j = 1$ and r_j if $x_j = 0$. These choices cover all of I. If the segment cover is satisfiable then the contiguous SAT instance is satisfiable. Therefore segment cover reduces to contiguous SAT.

Similarly given an instance of contiguous SAT with m clauses one can construct an instance of segment cover by i) partitioning the segment I into m sub-intervals $J_i = [(i-1)/m, i/m]$, $i = 1, \cdots, m$, and associating J_i to C_i and ii) defining a segment s_j for the variable x_j and defining l_j and r_j by concatenating the sub-intervals corresponding to clauses in which x_j or $\neg x_j$ appear. We omit the details.   □

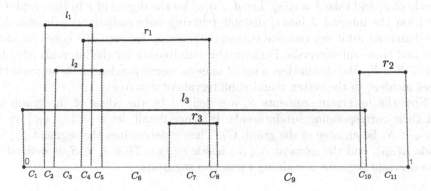

Fig. 4. Defining a SAT instance from a segment cover instance, for example $C_5 = (\neg x_1 \vee x_2 \vee x_3)$

We now start our reduction of 3SAT to segment cover. More over we assume that clauses contain exactly three literals. Let ϕ be the given 3SAT formula with s clauses, $C_1, \ldots, C_s$. For simplicity of presentation assume each variable appears at least once as a positive literal and at least once as a negative literal in ϕ. We first divide the interval I into s disjoint sub-intervals $B_j = [(j-1)/s, j/s]$, $j = 1, \ldots, s$. Next, we define an arbitrary one-one correspondence between the clauses and the intervals B_j. For simplicity we take B_j to be the sub-interval corresponding to C_j.

Clause Uncertain Segments. We partition each B_j into three equal parts, B_{j1}, B_{j2} and B_{j3}. Next we consider the set $T_j = \{\{B_{j1}, B_{j2}\}, \{B_{j2}, B_{j3}\}\}$ containing the two uncertain segments $\{B_{j1}, B_{j2}\}$ and $\{B_{j2}, B_{j3}\}$ as shown in Fig. 5. T_j has the following property.

Lemma 2. *For any choice of segments from the uncertain segments of T_j, at most two intervals among B_{j1}, B_{j2} and B_{j3} are covered.*

We define an arbitrary one-one correspondence between the literals of C_j and the sub-intervals B_{j1}, B_{j2} and B_{j3}. We denote this correspondence by $\alpha_j :$ $L_j(\phi) \rightarrow \{B_{j1}, B_{j2}, B_{j3}\}$, where $L_j(\phi)$ denotes the set of of literals of C_j. We denote by α the one-one correspondence between all appearances of literals in ϕ and all B_{jk}, $j = 1, \ldots, s$, $k = 1, 2, 3$ which is defined by the α_j. Again for simplicity we can take α to be the correspondence suggested by the subscripts, that is if $C_j = (\lambda_1 \vee \lambda_2 \vee \lambda_3)$ then $\alpha(\lambda_i) = B_{ji}$, $i = 1, 2, 3$.

Variable Uncertain Segments. Let $x_1, \ldots, x_m$ be the variables of the given formula ϕ. We shall construct a collection of uncertain intervals S_i for the variable x_i. For each variable, these uncertain intervals are defined by means of a complete bipartite graph denoted G_i, $i = 1, \ldots, m$. The vertices of G_i are the literals of ϕ. The vertices are divided into two parts denoted P_i and N_i, namely positive and negative literals. This finishes the definition of G_i.

For each i, using G_i, we define the set S_i as follows. Let $v \in P_i \cup N_i$ be a literals of x_i and take $J = \alpha(v)$. Let $d = d(v)$ be the degree of v in the graph G_i. Partition the interval J into d disjoint (sharing only endpoints) sub-intervals, and define an arbitrary one-one correspondence β_v between the edges incident on v and these sub-intervals. Perform this subdivision for the intervals $\alpha(v)$ for all $v \in P_i \cup N_i$. We obtain thus a set of one-one correspondences β_v between the edges incident to the vertex v and subintervals of $J = \alpha(v)$.

Now the uncertain segments S_i are defined by the edges of the graph G_i and their corresponding sub-intervals. In more detail, let $e = \{v_P, v_N\}, v_P \in P_i, v_N \in N_i$ be an edge of the graph G_i. Then e determines the segment $\beta_{v_P}(e)$ inside $\alpha(v_P)$, and the segment $\beta_{v_N}(e)$ inside $\alpha(v_N)$. Then $s_e \in S_i$ is defined as the uncertain segment containing these two segments.

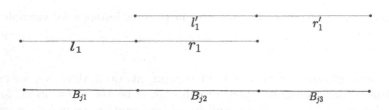

Fig. 5. The uncertain segments defining T_j, the segment are shown above the interval for visual purposes

The segment cover instance for ϕ is the set of uncertain segments $S = \bigcup_{i=1}^{m} S_i \cup \bigcup_{i=1}^{s} T_i$.

2.1 Correctness of the Reduction

In this section we show that there is a covering of the unit interval with the uncertain segments S if and only if the given sentence ϕ is satisfiable.

Assume that ϕ is satisfiable. Observe that each uncertain segment $s \in S_i \subset S$, has a *positive segment* and a *negative segment*. Namely, the positive segment is the one which corresponds to the incidence of the edge to a vertex in the positive part of G_i, i.e., P_i, and analogously for the negative segment. Hence we can write $s = \{s_p, s_n\}$ where $s_p = \beta_{v_P}(e), s_n = \beta_{v_N}(e)$, where $e = \{v_P, v_N\}$ is the edge defining s. Now assume x_i takes the value 1 (=true) in the assignment that satisfies ϕ. We choose s_p, otherwise we choose s_n, for all $s \in S_i$.

We spell out the following fact.

Lemma 3. *An interval $B_{j_0 k_0}$, for some $j_0 \in \{1, \ldots, s\}$, $k_0 \in \{1, 2, 3\}$, satisfies $\alpha(v) = B_{j_0 k_0}$ for some $v \in P_i$ if and only if x_i has a positive literal in C_{j_0}. Analogously, an interval $B_{j_0 k_0}$ satisfies $\alpha(v) = B_{j_0 k_0}$ for some $v \in N_i$ if and only if x_i has a negative literal in C_{j_0}.*

From the above lemma, whenever we choose the uncertain segments as above, since each clause is satisfied, each clause C_j has a literal v (=vertex in some graph) all of whose incident edges have chosen that vertex. Hence, $\alpha(v)$ among B_{j1}, B_{j2} and B_{j3} is covered. It remains to cover the two remaining intervals. This is easily done by a suitable choice for the uncertain segments of the set T_j. This finishes one direction of the proof.

Consider now the other direction. We have to show that if there is a choice for each uncertain segment $s \in S$, such that the unit interval is covered, then, there is an assignment of 0 and 1 to the variables x_i that satisfies the given formula ϕ. Consider a clause $C_j = (\lambda_1 \vee \lambda_2 \vee \lambda_3)$, where λ_i are literals. And let $x_{i_1}, x_{i_2}, x_{i_3}$ be the corresponding variables. The interval B_j is covered by the chosen segments. Recall that the uncertain segments correspond to the edges of the graphs G_i (other than elements of the T_j) and that a choice of an interval for an uncertain segment is equivalent to choosing one endpoint of the corresponding edge.

Lemma 4. *Assume there is a choice of uncertain segments such that I is covered. Consider the graphs $G_{i_1}, G_{i_2}, G_{i_3}$ of variables of any clause $C_j = (\lambda_1 \vee \lambda_2 \vee \lambda_3)$. There exists at least one vertex λ among λ_i, $i = 1, 2, 3$, such that, each edge incident on λ has chosen λ.*

Proof. The uncertain segments in T_j leave at least one of B_{j1}, B_{j2} and B_{j3} uncovered, wlog, let it be B_{j1}. The interval B_{j1} has to be covered using the uncertain segments of S. All the edges incident on λ_1 are required to choose λ_1, otherwise, some part of the interval $B_{j1} = \alpha(\lambda_1)$ would remain uncovered. $\square$

To construct an assignment from the choices of uncertain segments S we do as follows. Consider any clause B_j. Lemma 4 gives a $k \in \{1, 2, 3\}$ such that a vertex in G_{i_k} is chosen by all its incident edges. If the vertex is in P_{i_k}, set $x_{i_k} = 1$, otherwise set x_{i_k} to be 0.

First, we prove our assignment is well-defined. Assume for the contrary that x_i has been assigned both 0 and 1. Let $v_p \in P_i$ be the vertex based on which we have assigned 1 to x_i, and let $v_n \in N_i$ be the vertex based on which we have assigned 0 to x_i. Therefore all of the edges incident to v_p have chosen v_p and all the edges incident on v_n have chosen v_n. But this is a contradiction since G_i is a complete bi-partite graph.

Second, we show that the assignment satisfies all of the clauses. Take a clause C_j and consider its corresponding interval B_j. Let x be the vertex returned by Lemma 4 and v the vertex of the graph of x all of whose incident edges have chosen it. If v is a positive literal in C_j, v is in the positive part of G. Hence our procedure setting $x = 1$ satisfies the clause. If x has a negative literal in C_j, then v appears in the negative part of G. Hence setting $x = 0$ will satisfy C_j. This finishes the proof of the correctness of the reduction.

We now bound the run-time of the reduction procedure. Assume the given formula ϕ is an arbitrary 3SAT instance. In linear time in number of clauses we construct the sets T_j of uncertain segments. Let variable x_i appear in p_i clauses as a positive literal and in n_i clauses as a negative literal. Then the graph G_i is K_{p_i,n_i} and has $p_i n_i$ edges. Thus our reduction is of complexity $O(s + \sum_{i=1}^{m} p_i n_i)$.

Theorem 1. *The problems segment cover and contiguous SAT are NP-Complete.*

Remark. If we start by the NP-Complete problem studied by [13] in which each variable appears at most three times then the number of our uncertain segments is exactly $2s + 2m$ (this would require also dealing with clauses with two literals).

3   All-Equal Segment Cover

In this section we strengthen our result to show that the segment cover remains NP-complete even when we require that the lengths of the intervals all be equal. We call this problem all-equal segment cover. We will later deduce that the problem BCU of [9, 10] (defined in Sect. 1) is also NP-complete for $d = 1$.

We now describe the modifications to the reduction necessary to keep all the intervals the same length. Observe that we can make sure that the intervals B_{j1}, B_{j2}, B_{j3}, for all j, have equal length. It remains to make sure that the intervals in the uncertain segments from the S_i have equal length and have also length equal to the B_{ji}. For simplicity in this argument, we will start by a special 3SAT problem, namely, the one considered by [13]. They have proved that planar 3-SAT remains NP-Complete when each variable appears at most three times, once as one literal, twice as the other. If a clause contains only two literals we change the intervals T_j of Fig. 5 accordingly. When applying our reduction to this type of formulas, we will see that in the final uncertain segments intervals B_{j1}, B_{j2}, B_{j3} are divided into at most two smaller intervals. With these preliminaries in mind, we will substitute the intervals in the Fig. 6 for the corresponding intervals from the construction of Sect. 2. In this figure B_{j1}, B_{j3}, B_{j5} play the roles of B_{j1}, B_{j2} and B_{j3} of the original reduction. Note that we have assumed in the figure that the worst case happens, i.e., each three of the sub-intervals is divided. The other cases are simpler.

Theorem 2. *The problem all-equal segment cover is NP-Complete.*

Proof. Consider a clause C_j and its sub-interval B_j. To the set T_j of the original reduction we add $s - 1$ new uncertain segments, each of them consisting of two copies of the same segment. This insures that certain subsets of the interval I are always covered, see Fig. 6 (for instance, $\{s_1, s_1'\}$ is such an uncertain segment). The set S_i of uncertain segments for the variable x_i is defined just as in the original reduction, but with the modification that a vertex interval B_{jk} is not partitioned, rather the sub-intervals for the at most two incident edges are copies of the interval B_{jk}, one of them slightly moved to the right, the other slightly moved to the left.

We need to check that the new intervals have the required properties used in the reduction. As before, at most two of the intervals B_{j1}, B_{j3} and B_{j5} can be covered by the uncertain segments from (updated) T_j. It is easily checked that any interval, say B_{j1}, which is not covered, can only be covered when both of the intervals of the incident edges are present. Hence, the same correctness argument applies here as well.   □

We next show that the optimization problem called BCU and studied in [9, 10] is NP-complete on the real line. For the definition of BCU refer to Sect. 1.

Corollary 1. *The 1-dimensional BCU is NP-complete.*

Proof. Let the set $\{s_1, \ldots, s_n\}$ of uncertain segments be an instance of all-equal segment cover. For each s_i, construct an uncertain region u_i containing two points, namely, the midpoints of the two intervals in s_i. We add two more regions defined as follows. Let x_l be the smallest coordinate and x_r be the largest coordinate of any midpoint. Moreover, let r be half the length of an interval. Define $u_0 = \{x_l - 2r, x_l - 3r\}$ and $u_{n+1} = \{x_r + 2r, x_r + 3r\}$. Add these two sets to the problem instance. Then the u_i define an instance of BCU. An algorithm solving BCU returns a minimum r' such that there are $n + 2$ disks of radius r', with centers at the points of the u_i, one center from each u_i, such that the area they cover is connected. Because of u_0, u_{n+1} we have always $r' \geq r$. Moreover, $r' = r$ if and only if the answer to the original all-equal segment cover is affirmative.

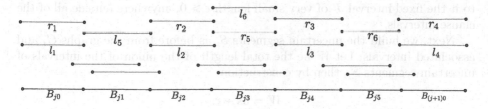

Fig. 6. The labelled intervals on the top define uncertain segments T_j, unlabelled ones in the bottom define sets S_i.

4 Approximation

In this section we consider the approximation of the segment cover problem. We can define two natural approximation problems. The first, called *max-segment cover*, or *max-SC* for short, asks to choose one interval from each uncertain segment such that the union of the resulting intervals is of maximum length possible among all the choices. Here we have extended the meaning of length of an interval to the length of a union of disjoint intervals in the obvious way. Therefor, max-SC asks for maximum coverage.

The second, called *contiguous max-SC*, requires a choice of an interval from each uncertain segment such that a maximum-length connected interval is obtained, among all connected intervals of all choices. Therefore contiguous max-SC asks for maximum connected coverage. In this section we discuss the approximation problem max-sc and leave more specialized study of contiguous max-SC as an open problem.

4.1 Hardness of Approximation for Max-SC

We first prove hardness of approximation for max-SC. Let max-E3SAT be max-3SAT restricted to formulas in which each clause contains exactly three literals.

Theorem 3. *Let c' be a ratio beyond which it is NP-hard to approximate max-E3SAT. Then it is NP-hard to approximate max-SC with ratio larger than $c = \frac{c'+2}{3}$.*

Proof. Suppose we are given an instance ϕ of max-E3SAT with n variables. We shall apply the reduction of Sect. 2 to ϕ and obtain an instance of max-SC, however, we need some modifications. Consider a graph G_i constructed in the reduction. If $|P_i| = |N_i|$ we leave the graph as it is, otherwise, let $|P_i| < |N_i|$. We add $|N_i| - |P_i|$ dummy vertices to $|P_i|$ to make the two sets equal. We do analogously in the other case. Let $\tilde{G}_i$ denote the modified graphs, $i = 1, \ldots, n$. We build the uncertain segments S_i from $\tilde{G}_i$ as follows. Let v be a vertex of $\tilde{G}_i$ and $m = |\tilde{P}_i| = |\tilde{N}_i|$. If v is not a dummy vertex, it has associated with it a sub-interval of a clause-interval. We make sure all these sub-intervals have length 1, and a clause interval has length 3. If v is a dummy vertex, associate to it the fixed interval J' of very small length $\epsilon > 0$, anywhere outside all of the clause intervals.

Next, we build the uncertain segments S_i as before from the graphs $\tilde{G}_i$ and associated intervals. Let W be the total length of the union of the intervals of uncertain segments S_i, then by construction

$$W = 3s + \epsilon.$$

Note that any two intervals of (possibly different) uncertain segments defined here are disjoint other than when both intervals are sub-intervals of J'.

We run the approximation algorithm for max-SC on our instance. The algorithm makes a choice from each uncertain segment. We modify this choice slightly. If any uncertain segment has chosen a sub-interval of J' we reverse this choice. It is clear that at the end we have at worst decreased the total approximated length by ϵ. And we have not decreased the approximated length over the original clause intervals.

Observe that the total length that the uncertain segments chosen from T_i contribute is at most $2W/3 = s$. If from any clause-interval the choice from T_i covers only $1/3$ of the interval, then the middle interval B_{j2} is covered. We change the choices so that only $1/3$ of the interval is not covered, by covering either of B_{j1} or B_{j3}. This insures that from any clause interval exactly one sub-interval is not covered by the T_i.

Now from the modified choice of uncertain segments and the graphs G_i define the graphs G'_i as follows. For each i, from the graph $\tilde{G}_i$, remove any vertex whose interval is covered in the approximation by intervals from T_i. Denote the new graph by G'_i. The total length of the intervals corresponding to the non-dummy vertices of G'_i is $W/3 = s$. Next, define an assignment as follows. We distinguish five cases from each other.

- Case 1: The graph has original vertices in positive part only, and, dummy vertices are in positive part. For any edge $e \in \tilde{G}_i$ that is not incident with a dummy vertex, we redirect the choice to the positive side. Note that since any interval we uncover is covered by T_i this does not decrease the length of the approximation. After these re-directions, any non-dummy vertex in the positive side of $\tilde{G}_i$ has all its sub-intervals chosen.
- Case 2: The graph has original vertices in positive part only, and, dummy vertices are in negative part. For any edge $e \in \tilde{G}_i$ that is not incident with a dummy vertex, we redirect the choice to the positive side. Recall that all the other edges have also chosen the positive side. Then again after this re-direction of choice all the vertices in positive part of $\tilde{G}_i$, have their intervals covered. Again this operation does not decrease the total approximated length.
- Cases 3, 4: These are analogous to the previous cases, where non-dummy vertices appear in the negative part only. We perform analogously as in those cases.
- Case 5: The graph has non-dummy vertices in both parts, or it has only dummy vertices. In this case, we can assign an arbitrary value to x_i. We choose the side which does not have dummy vertices and redirect all the edges of $\tilde{G}_i$ towards that side. Re-direction of the choice for an edge not incident on a dummy vertex does not change the approximated weight. Also we had set the choice for edges incident on dummy vertices away from them. It follows that all the intervals associated to the vertices of the chosen side are covered.

Thus we have defined an assignment. Now we compute the number of clauses satisfied by our assignment. The length not covered by the T_i and covered by the S_i in the approximation is at least $c(W + \epsilon) - \frac{2}{3}W$. After the above redirection of the choices, an interval corresponding to a clause is either all covered or covered in exactly $2/3$ of its length. Therefore, $c(W + \epsilon) - \frac{2}{3}W$ is (lower bound for) the total number of the intervals satisfied by our assignment. For any algorithm that runs in polynomial times we must have $c(W + \epsilon) - \frac{2}{3}W = 3cs + c\epsilon - 2s < c's$. This implies

$$c < \frac{c' + 2}{3 + \frac{\epsilon}{s}}.$$

The claim follows. □

Remark. By a seminal result of Håstad [19] max-E3SAT cannot be approximated by a ratio larger than 7/8. Using this result the above theorem implies that max-SC cannot be approximated beyond the ration 23/24, unless P = NP.

Approximation of Max-SC. To approximate max-SC, we can the existing algorithms for weighted max-SAT, which is a well-studied problem in the literature. We refer to the sequence of papers [2,3,16,29]. We form a SAT from our segment cover instance as follows. Any maximal sub-interval $J \subset I = [0,1]$ that does not contain an endpoint defines a clause, and in it are literals corresponding to uncertain segments covering the interval J. See the proof of Lemma 1. We assign the length of J as the weight of the corresponding clause. Given we have an algorithm for weighted max-SAT with approximation ratio $0 < c' < 1$, then clearly we have an algorithm with the same ratio for max-SC.

It is interesting to see these upper and/or lower bounds improved.

References

1. Abam, M.A., Alipour, S., Ghodsi, M., Mahdian, M.: Visibility testing and counting for uncertain segments. Theor. Comput. Sci. **779**, 1–7 (2019)
2. Asano, T.: An improved analysis of Goemans and Williamson's LP-relaxation for MAX SAT. Theor. Comput. Sci. **354**(3), 339–353 (2006)
3. Asano, T., Williamson, D.P.: Improved approximation algorithms for MAX SAT. J. Algorithms **42**(1), 173–202 (2002)
4. Avis, D., Toussaint, G.T.: An optimal algorithm for determining the visibility of a polygon from an edge. IEEE Trans. Comput. **30**(12), 910–914 (1981)
5. Ben-Moshe, B., Hall-Holt, O.A., Katz, M.J., Mitchell, J.S.B.: Computing the visibility graph of points within a polygon. In: Proceedings of the 20th ACM Symposium on Computational Geometry, Brooklyn, New York, USA, 8–11 June 2004, pp. 27–35 (2004)
6. Briggs, A.J.: An efficient algorithm for one-step planar compliant motion planning with uncertainty. Algorithmica **8**(3), 195–208 (1992)
7. Buchin, K., Kostitsyna, I., Löffler, M., Silveira, R.I.: Region-based approximation algorithms for visibility between imprecise locations. In: Proceedings of the Seventeenth Workshop on Algorithm Engineering and Experiments, ALENEX 2015, San Diego, CA, USA, 5 January 2015, pp. 94–103 (2015)
8. Carlsson, S., Jonsson, H., Nilsson, B.J.: Finding the shortest Watchman route in a simple polygon. Discrete Comput. Geom. **22**(3), 377–402 (1999)
9. Chambers, E., et al.: Connectivity graphs of uncertainty regions. In: Cheong, O., Chwa, K.-Y., Park, K. (eds.) ISAAC 2010. LNCS, vol. 6507, pp. 434–445. Springer, Heidelberg (2010). https://doi.org/10.1007/978-3-642-17514-5_37
10. Chambers, E.W., et al.: Connectivity graphs of uncertainty regions. Algorithmica **78**(3), 990–1019 (2017)

11. Chazelle, B., Guibas, L.J.: Visibility and intersection problems in plane geometry. Discrete Comput. Geom. **4**(6), 551–581 (1989). https://doi.org/10.1007/BF02187747
12. Cohen, R., Peleg, D.: Convergence of autonomous mobile robots with inaccurate sensors and movements. SIAM J. Comput. **38**(1), 276–302 (2008)
13. Dahlhaus, E., Johnson, D.S., Papadimitriou, C.H., Seymour, P.D., Yannakakis, M.: The complexity of multiterminal cuts. SIAM J. Comput. **23**(4), 864–894 (1994)
14. Donald, B.R.: The complexity of planar compliant motion planning under uncertainty. Algorithmica **5**(3), 353–382 (1990)
15. Ghosh, S.K.: Visibility Algorithms in the Plane. Cambridge University Press, New York (2007)
16. Goemans, M.X., Williamson, D.P.: Improved approximation algorithms for maximum cut and satisfiability problems using semidefinite programming. J. ACM **42**(6), 1115–1145 (1995)
17. Gudmundsson, J., Morin, P.: Planar visibility: testing and counting. In: Proceedings of the 26th ACM Symposium on Computational Geometry, Snowbird, Utah, USA, 13–16 June 2010, pp. 77–86 (2010)
18. Guibas, L.J., Hershberger, J., Leven, D., Sharir, M., Tarjan, R.E.: Linear-time algorithms for visibility and shortest path problems inside triangulated simple polygons. Algorithmica **2**, 209–233 (1987)
19. Håstad, J.: Some optimal inapproximability results. J. ACM **48**(4), 798–859 (2001)
20. Kratochvíl, J.: A special planar satisfiability problem and a consequence of its NP-completeness. Discrete Appl. Math. **52**(3), 233–252 (1994)
21. Laroche, P.: Planar 1-in-3 satisfiability is NP-complete. Comptes Rendus de L Academie des Sciences Serie I-Mathematique **316**(4), 389–392 (1993)
22. Lichtenstein, D.: Planar formulae and their uses. SIAM J. Comput. **11**(2), 329–343 (1982)
23. Mitchell, J.S.B.: Approximating Watchman routes. In: Proceedings of the Twenty-Fourth Annual ACM-SIAM Symposium on Discrete Algorithms, SODA 2013, New Orleans, Louisiana, USA, 6–8 January 2013, pp. 844–855 (2013)
24. Moon, I., Miura, J., Shirai, Y.: On-line viewpoint and motion planning for efficient visual navigation under uncertainty. Robot. Auton. Syst. **28**(2–3), 237–248 (1999)
25. Moore, C., Robson, J.M.: Hard tiling problems with simple tiles. Discrete Comput. Geom. **26**(4), 573–590 (2001)
26. Murrieta-Cid, R., González-Baños, H.H., Tovar, B.: A reactive motion planner to maintain visibility of unpredictable targets. In: Proceedings of the 2002 IEEE International Conference on Robotics and Automation, ICRA 2002, Washington, DC, USA, 11–15 May 2002, pp. 4242–4248 (2002)
27. O'rourke, J.: Art Gallery Theorems and Algorithms, vol. 57. Oxford University Press, Oxford (1987)
28. Papadimitriou, C.H., Yannakakis, M.: Optimization, approximation, and complexity classes. J. Comput. Syst. Sci. **43**(3), 425–440 (1991)
29. Poloczek, M., Schnitger, G.: Randomized variants of Johnson's algorithm for MAX SAT. In: Proceedings of the Twenty-Second Annual ACM-SIAM Symposium on Discrete Algorithms, SODA 2011, San Francisco, California, USA, 23–25 January 2011, pp. 656–663 (2011)
30. Schaefer, T.J.: The complexity of satisfiability problems. In: Proceedings of the 10th Annual ACM Symposium on Theory of Computing, San Diego, California, USA, 1–3 May 1978, pp. 216–226 (1978)
31. Tovey, C.A.: A simplified NP-complete satisfiability problem. Discrete Appl. Math. **8**(1), 85–89 (1984)

On the Complexity of Minimum Maximal Uniquely Restricted Matching

Juhi Chaudhary and B. S. Panda[✉]

Computer Science and Application Group, Department of Mathematics,
Indian Institute of Technology Delhi, Hauz Khas, New Delhi 110016, India
{maz168189,bspanda}@maths.iitd.ac.in

Abstract. A subset $M \subseteq E$ of edges of a graph $G = (V, E)$ is called
a *matching* if no two edges of M share a common vertex. A matching
M in a graph G is called a *uniquely restricted matching* if $G[V(M)]$, the
subgraph of G induced by the M-saturated vertices of G, contains exactly
one perfect matching. A uniquely restricted matching M is *maximal* if
M is not properly contained in any other uniquely restricted matching of
G. Given a graph G, the MIN-MAX-UR MATCHING problem asks to find
a maximal uniquely restricted matching of minimum cardinality in G. In
general, the decision version of the MIN-MAX-UR MATCHING problem
is known to be NP-complete for general graphs and remains so even
for bipartite graphs. In this paper, we strengthen this result by proving
that this problem remains NP-complete for chordal bipartite graphs and
chordal graphs. On the positive side, we prove that the MIN-MAX-UR
MATCHING problem is polynomial time solvable for bipartite permutation
graphs and proper interval graphs. Finally, we show that the MIN-MAX-
UR MATCHING problem is APX-complete for bounded degree graphs.

Keywords: Matching · Uniquely restricted matching · Minimum
maximal matching · Graph algorithms · NP-completeness ·
APX-completeness

1 Introduction

A subset $M \subseteq E$ of edges of a graph $G = (V, E)$ is called a *matching* if no two
edges of M share a common vertex. Given a matching M in G, a vertex $v \in V$ is
called M-*saturated* if there exists an edge $e \in M$ incident on v. We shall denote
$V(M)$ as the set of M-saturated vertices in G. A matching M in G is called a
uniquely restricted matching if the subgraph $G[V(M)]$ of G induced by $V(M)$
has only one perfect matching. The UNIQUELY RESTRICTED MATCHING problem
asks to find a uniquely restricted matching of maximum size in a given graph

J. Chaudhary—The author has been supported by the Department of Science and
Technology through INSPIRE Fellowship for this research.
B. S. Panda—The author wants to thank the SERB, Department of science and tech-
nology for their support vide Diary No. SERB/F/12949/2018-2019.

W. Wu and Z. Zhang (Eds.): COCOA 2020, LNCS 12577, pp. 364–376, 2020.
https://doi.org/10.1007/978-3-030-64843-5_25

G [6]. We shall denote $\mu_{urm}(G)$ as the size of a maximum uniquely restricted matching in G. Given a graph G, the MIN-MAX-UR MATCHING problem asks to find a maximal uniquely restricted matching of minimum size in G, and we shall denote $\mu'_{urm}(G)$ as the size of a minimum maximal uniquely restricted matching in G. The decision versions of the UNIQUELY RESTRICTED MATCHING problem and the MIN-MAX-UR MATCHING problem are defined as follows:

DECIDE-UR MATCHING problem
Instance: A graph $G = (V, E)$ and a positive integer $k \leq |V|$.
Question: Does there exist a uniquely restricted matching M in G such that $|M| \geq k$?

DECIDE-MIN-MAX-UR MATCHING problem
Instance: A graph $G = (V, E)$ and a positive integer $k \leq |V|$.
Question: Does there exist a **maximal** uniquely restricted matching M in G such that $|M| \leq k$?

The concept of minimum maximal uniquely restricted matching was introduced by Goddard et al. in [4], along with some other restricted variations of matching. Further, Panda et al. [14] proved that the DECIDE-MIN-MAX-UR MATCHING problem is NP-complete for bipartite graphs of maximum degree 7 and linear time solvable for chain graphs. They also proved that it is hard to approximate the MIN-MAX-UR MATCHING problem for bipartite graphs within a ratio of $n^{1-\epsilon}$ for any $\epsilon > 0$ unless P = NP.

In this paper, we extend the algorithmic study of the MIN-MAX-UR MATCHING problem in some subclasses of graphs. The main contributions of this paper are summarized below.

1. We strengthen the hardness result of the DECIDE-MIN-MAX-UR MATCHING problem for bipartite graph by showing that this problem remains NP-complete for chordal bipartite graphs, a subclass of bipartite graphs. On the positive side, we propose a polynomial time algorithm for computing a minimum maximal uniquely restricted matching in bipartite permutation graphs, a subclass of chordal bipartite graphs.
2. We prove that the DECIDE-MIN-MAX-UR MATCHING problem is NP-complete for chordal graphs and propose a polynomial time algorithm for computing a minimum maximal uniquely restricted matching in proper interval graphs, a subclass of chordal graphs.
3. We show that the MIN-MAX-UR MATCHING problem is APX-complete for bounded degree graphs.

The concept of uniquely restricted matching, motivated by a problem in Linear Algebra, was introduced by Golumbic et al. in [6]. Some more results related to uniquely restricted matching can be found in [2,4,10,14]. Given a graph G, an even cycle in G is said to be an *alternating cycle* with respect to a matching M if every second edge of the cycle belongs to M. The following theorem characterizes uniquely restricted matchings in terms of alternating cycles.

Theorem 1 [6]. *Let $G = (V, E)$ be a graph. A matching M in G is uniquely restricted if and only if there is no alternating cycle with respect to M in G.*

Observation 2 [6]. *Let $G = (X, Y, E)$ be a bipartite graph and M be a uniquely restricted matching in G. If $|X| = p$, $|Y| = q$ and $|M| = k$, then we can order the vertices of X as $(x_1, x_2, \ldots x_p)$ and the vertices of Y as $(y_1, y_2, \ldots y_q)$ in such a way that $x_i y_i \in M$, $1 \le i \le |M|$, and $x_i y_j \notin E$ for $1 \le j < i \le k$.*

2   Preliminaries

All graphs considered in this paper are simple and connected unless otherwise stated. For a graph $G = (V, E)$, let n denote the number of vertices and m denote the number of edges in G. The open and closed neighborhood of a vertex $u \in V$ is denoted by $N(u)$ and $N[u]$ respectively, where $N(u) = \{w \mid wu \in E\}$ and $N[u] = N(u) \cup \{u\}$. The degree of a vertex u is $|N(u)|$ and is denoted by $d(u)$. For a graph $G = (V, E)$, the subgraph of G induced by $U \subseteq V$ is denoted by $G[U]$, where $G[U] = (U, E_U)$ and $E_U = \{xy \in E \mid x, y \in U\}$. In a graph G, a *clique* is a subset of vertices of G such that every two distinct vertices in the clique are adjacent.

A graph $G = (V, E)$ is called a k-regular graph if $d(v) = k$ for every vertex v of G. A graph $G = (V, E)$ is called a *bipartite graph* if its vertex set V can be partitioned into two independent sets X and Y, such that every edge of G joins a vertex in X to a vertex in Y. A bipartite graph with bipartition (X, Y) of V is denoted by $G = (X, Y, E)$. A bipartite graph $G = (X, Y, E)$ is called a *chordal bipartite graph* if every cycle in G of length at least six has a *chord*, that is, an edge joining two non-consecutive vertices of the cycle. A graph $G = (V, E)$ is called a *permutation graph* if there exists a one to one correspondence between $V(G)$ and a set of line segments between two parallel lines such that two vertices are adjacent if and only if their corresponding line segments intersect. In other words, a permutation graph is a graph whose vertices represent the elements of a permutation, and whose edges represent pairs of elements that are reversed by the permutation. If $G = (V, E)$ is both a bipartite graph and a permutation graph, it is called a *bipartite permutation graph*.

A graph $G = (V, E)$ is called a *chordal graph* if every cycle in G of length at least four has a *chord*. It can be noted that a chordal bipartite graph may not be a chordal graph as a cycle on four vertices is a chordal bipartite graph but not a chordal graph. Let $\mathscr{F}$ be a family of sets. The *intersection graph* of $\mathscr{F}$ is obtained by taking each set in $\mathscr{F}$ as a vertex and joining two sets in $\mathscr{F}$ if and only if they have a non-empty intersection. A graph G is called a *proper interval graph* if it is the intersection graph of a family $\mathscr{F}$ of intervals on the real line such that no interval in $\mathscr{F}$ properly contains another interval in $\mathscr{F}$.

The class APX is the set of all optimization problems which admit a polynomial time α-approximation algorithm, where α is a constant [15]. For most of the approximation related terminologies, we refer the reader to [1].

3   Minimum Maximal Uniquely Restricted Matching in Subclasses of Bipartite Graphs

3.1   Chordal Bipartite Graphs

In this subsection, we show that the DECIDE-MIN-MAX-UR MATCHING problem is NP-complete for chordal bipartite graphs by providing a polynomial time reduction from the DECIDE-UR MATCHING problem in chordal bipartite graphs, which is already known to be NP-complete [11].

Theorem 3. *The* DECIDE-MIN-MAX-UR MATCHING *problem is* NP-*complete for chordal bipartite graphs.*

Proof. Clearly, the DECIDE-MIN-MAX-UR MATCHING problem is in NP for chordal bipartite graphs. To show the NP-completeness, consider the following reduction. Given a chordal bipartite graph $G = (V, E)$, where $V = \{v_1, v_2, \ldots v_n\}$, an instance of the DECIDE-UR MATCHING problem, we construct a chordal bipartite graph $H = (V_H, E_H)$, an instance of the DECIDE-MIN-MAX-UR MATCHING problem by adding a pendant edge $v_i a_i$ to each vertex $v_i \in V$. More formally, $V_H = V \cup \{a_i \mid 1 \leq i \leq n\}$, $E_H = E \cup \{v_i a_i \mid 1 \leq i \leq n\}$.

It is clear that the constructed graph H is a chordal bipartite graph. Also, given a chordal bipartite graph G, the graph H can be constructed in polynomial time. Now, the following claims are sufficient to complete the proof of the theorem.

Claim. If M is a maximal uniquely restricted matching in H, then for each $1 \leq i \leq n$, v_i is saturated by M.

Proof. Let M be a maximal uniquely restricted matching in H such that for some fixed k, v_k is not saturated by M. Then, $M \cup \{v_k a_k\}$ is also a uniquely restricted matching in H. Since M is maximal; it is a contradiction.   □

Claim. If M_H^* is a minimum maximal uniquely restricted matching in H and M_G^* is a maximum uniquely restricted matching in G, then $|M_H^*| = n - |M_G^*|$.

Proof. Let M_G^* be a maximum uniquely restricted matching in G. This implies that $2|M_G^*|$ vertices are saturated, and $n - 2|M_G^*|$ vertices are unsaturated by M_G^* in G. Define $M_H = M_G^* \cup \{v_i a_i \mid v_i$ is not saturated by $M_G^*\}$ in H. It is easy to note that M_H is a maximal uniquely restricted matching in H, and since $|M_H| = (|M_G^*| + n - 2|M_G^*|) = (n - |M_G^*|)$, $|M_H^*| \leq n - |M_G^*|$.

Conversely, let M_H^* be a minimum maximal uniquely restricted matching in H. Let edges of the form $\{v_i a_i \mid 1 \leq i \leq n\}$ are called Type-A edges and edges in the set E are called Type-B edges. Let $T^* \cup S^*$ be a partition of M_H^* such that T^* contains Type-A edges and, S^* contains Type-B edges. Since M_H^* is maximal, $|T^*| = (n - 2|S^*|)$. It implies that $|M_H^*| = |S^*| + (n - 2|S^*|) = n - |S^*|$. Since $S^* \subset M_H^*$, S^* is a uniquely restricted matching in G and $|S^*| \leq |M_G^*|$. As $|S^*| = n - |M_H^*|$, $n - |M_H^*| \leq |M_G^*|$. This completes the proof of the claim.   □

Hence, the theorem is proved. □

Next, we show that a maximal uniquely restricted matching of minimum size can be computed in polynomial time for bipartite permutation graphs, a subclass of chordal bipartite graphs.

3.2 Bipartite Permutation Graphs

Let $G = (X, Y, E)$, where $|X| = p$ and $|Y| = q$ be a bipartite graph and let $\gamma = (y_1, y_2, \ldots, y_q)$ be some ordering of Y. A subset of Y is called a *segment* of Y if its elements are consecutive in γ. The ordering γ is said to have the *convex property* if for each vertex $x \in X$, $N(x)$ is a segment in γ. An ordering of Y with the convex property is called a *convex ordering*. A bipartite graph $G = (X, Y, E)$ is said to be convex on Y if there exists a convex ordering of Y. The term convex on X is defined similarly.

Let $G = (X, Y, E)$, where $X = \{x_1, x_2, \ldots x_p\}$ and $Y = \{y_1, y_2, \ldots y_q\}$ be a bipartite permutation graph. If $\sigma = (x_1, x_2, \ldots, x_p, y_1, y_2, \ldots, y_q)$ is an ordering of $X \cup Y$, then let $f(x)$(or $f(y)$) denote the first neighbor of vertex x(or y) and $l(x)$(or $l(y)$) denote the last neighbor of vertex x(or y) in σ. Further, the ordering $\sigma = (x_1, x_2, \ldots, x_p, y_1, y_2, \ldots, y_q)$ of $X \cup Y$ is called a *forward-convex ordering* if the following conditions hold:

1. The orderings $(x_1, x_2, \ldots, x_p)$ and $(y_1, y_2, \ldots, y_q)$ are convex orderings of the sets X and Y respectively.
2. For any $x_i < x_j$ in σ, $f(x_i) \leq f(x_j)$ and $l(x_i) \leq l(x_j)$ in σ, where $f(x)$ denote the first neighbor of vertex x and $l(x)$ denote the last neighbor of vertex x.

A bipartite graph is said to be a forward-convex bipartite graph if there exists a forward-convex ordering of its vertices. It can be noted that the class of bipartite permutation graphs is the same as the class of forward-convex bipartite graphs [9]. Before discussing the algorithm, let us first, discuss the idea behind our algorithm. In a given bipartite permutation graph G, we first, identify an edge that belongs to some minimum maximal uniquely restricted matching M. After adding the desired edge to the matching, we remove those vertices from G that cannot be saturated by M in the later stages of the algorithm. Since bipartite permutation graphs are hereditary, the subgraph obtained will also be a bipartite permutation graph. In this graph, we can again identify the edge that belongs to some minimum maximal uniquely restricted matching. This process is repeated unless we are left with an empty graph.

Let $Bef(x_i, y_j) = \{x_k y_l \mid x_k \leq x_i \text{ and } y_l \leq y_j \text{ in } \sigma\}$.

Lemma 1. *Let $G = (X, Y, E)$ be a bipartite permutation graph with a forward-convex ordering $\sigma = (x_1, x_2, \ldots, x_p, y_1, y_2, \ldots, y_q)$. Then, there exists a minimum maximal uniquely restricted matching M in G such that $l(y_1)l(x_1) \in M$.*

Proof. Let M be a minimum maximal uniquely restricted matching in G. In order to prove the lemma, let us first prove the following claim.

Claim. There exists a minimum maximal uniquely restricted matching M in G such that $|M \cap Bef(l(y_1), l(x_1))| = 1$.

Proof. Due to space restriction, the proof is omitted. $\square$

Let M be a minimum maximal uniquely restricted matching in G. If $l(y_1)l(x_1) \in M$, then we are done. So assume that $x_a y_b (\neq l(y_1)l(x_1)) \in M \cap Bef(l(y_1), l(x_1))$. Let $M' = (M \setminus \{x_a y_b\}) \cup \{l(y_1)l(x_1)\}$. If M' is a uniquely restricted matching, then we are done. Otherwise, there must exist an edge $x_u y_v$ such that either $x_u y_v$ forms an alternating cycle with $l(y_1)l(x_1)$ or M' does not form a matching (because either $l(y_1) = x_u$ or $l(x_1) = y_v$). So, let $M' = M' \setminus \{x_u y_v\}$. Since $|M'| < |M|$, M' cannot be a maximal uniquely restricted matching. Now, we are left to show that exactly one more edge is required to make M' a maximal uniquely restricted matching. On the contrary, let us assume that $M'' = M' \cup \{x_i y_j, x_k y_l\}$ is a uniquely restricted matching. Without loss of generality, let $x_i < x_k$ and $y_j < y_l$ in σ. By the definition of a forward-convex ordering, $l(y_1)y_l, l(x_1)x_k \in E$. So, edges $x_i y_j$ and $x_k y_l$ forms an alternating cycle in $G[V(M)]$, which is a contradiction. Thus, exactly one more edge is required to make M' a maximal uniquely restricted matching. $\square$

Let $\sigma = (x_1, x_2, ..., x_p, y_1, y_2, ..., y_q)$ be a forward-convex ordering of a bipartite permutation graph G and let $\sigma_i = (x_a, x_b, ..., x_k, y_{a'}, y_{b'}, ..., y_{k'})$ be an ordering obtained by removing some vertices from σ. Then, σ_i is a forward-convex ordering of some bipartite permutation graph $G_i = (X_i, Y_i, E_i)$, where G_i is a subgraph of G. Hence, we have the following corollary to Lemma 1.

Corollary 1. *If $\sigma_i = (x_a, x_b, ..., x_k, y_{a'}, y_{b'}, ..., y_{k'})$ is a forward-convex ordering of a subgraph $G_i = (X_i, Y_i, E_i)$ of a bipartite permutation graph G, then there exists a minimum maximal uniquely restricted matching M in G_i such that $l(y_{a'})l(x_a) \in M$.*

Based on the above lemmas, we now present a polynomial time algorithm URM-BPG(G), which computes a maximal uniquely restricted matching of minimum size in a given bipartite permutation graph G. The pseudocode of the algorithm is given in Fig. 1.

Theorem 4. *Given a bipartite permutation graph $G_1 = (X_1, Y_1, E_1)$ with a forward-convex ordering $\sigma(G_1)$, the algorithm URM-BPG(G_1) correctly computes a minimum size maximal uniquely restricted matching in G_1.*

Proof. Due to space restriction, the proof is omitted. $\square$

4   Minimum Maximal Uniquely Restricted Matching in Subclasses of Chordal Graphs

4.1   Chordal Graphs

In this subsection, we show that the DECIDE-MIN-MAX-UR MATCHING problem is NP-complete for chordal graphs by providing a polynomial time reduction

---

Algorithm 1 URM-BPG(G)

Input: A bipartite permutation graph $G_1 = (X_1, Y_1, E_1)$ with a forward-convex order-
     ing $\sigma(G_1) = (x_1, x_2, \ldots, x_p, y_1, y_2, \ldots, y_q)$;
Output: A minimum maximal uniquely restricted matching M;
  $G_i = (X_i, Y_i, E_i)$ is a bipartite permutation graph for each $i \geq 1$;
  $\alpha_i \to$ first vertex of X_i in $\sigma(G_i)$ for $i \geq 1$;
  $\beta_i \to$ first vertex of Y_i in $\sigma(G_i)$ for $i \geq 1$;
  $S_i^X \to \{v \in N(l(\alpha_i)) \mid v > l(\beta_i) \text{ in } \sigma(G_i)\}$;
  $M \to \emptyset, i \to 1$;
  while $(E_i \neq \emptyset)$ **do**
    │ $M \to M \cup \{l(\beta_i)l(\alpha_i)\}$;
    │ $i = i + 1$;
    │ $G_i \to G_{i-1} \setminus \{\alpha_{i-1}, \ldots, l(\beta_{i-1}), \beta_{i-1}, \ldots, l(\alpha_{i-1})\}$;
    │ **for** (*each* $u \in S_{i-1}^X$) **do**
    │   │ **if** $(l(u) = l(l(\beta_{i-1})))$ **then**
    │   └ $\lfloor$ $G_i \to G_i \setminus \{u\}$;

return M;

Fig. 1. Algorithm to compute a minimum maximal uniquely restricted matching in a bipartite permutation graph G.

from the DECIDE-UR MATCHING problem in chordal graphs, which is already known to be NP-complete [6]. The reduction is similar to the reduction given in Subsect. 3.1.

Theorem 5. *The* DECIDE-MIN-MAX-UR MATCHING *problem is* NP-*complete for chordal graphs.*

Proof. Clearly, the DECIDE-MIN-MAX-UR MATCHING problem is in NP for chordal graphs. To show the NP-completeness, consider the following reduction. Given a chordal graph $G = (V, E)$, where $V = \{v_1, v_2, \ldots v_n\}$, an instance of the DECIDE-UR MATCHING problem in chordal graphs, we construct a chordal graph $H = (V_H, E_H)$, an instance of the DECIDE-MIN-MAX-UR MATCH-ING problem by adding a pendant edge $v_i a_i$ to each vertex $v_i \in V$. More formally, $V_H = V \cup \{a_i \mid 1 \leq i \leq n\}$, $E_H = E \cup \{v_i a_i \mid 1 \leq i \leq n\}$.

It is clear that the constructed graph H is a chordal graph. Also, given a chordal graph G, the graph H can be constructed in polynomial time. Now, the following claims are sufficient to complete the proof of the theorem.

Claim. If M is a maximal uniquely restricted matching in H, then for each $1 \leq i \leq n$, v_i is saturated by M.

Proof. Let M be a maximal uniquely restricted matching in H such that for some fixed k, v_k is not saturated by M. Then, $M \cup \{v_k a_k\}$ is also a uniquely restricted matching in H. Since M is maximal; it is a contradiction. □

Claim. If M_H^* is a minimum maximal uniquely restricted matching in H and M_G^* is a maximum uniquely restricted matching in G, then $|M_H^*| = n - |M_G^*|$.

Proof. Let M_G^* be a maximum uniquely restricted matching in G. This implies that $2|M_G^*|$ vertices are saturated, and $n - 2|M_G^*|$ vertices are unsaturated by M_G^* in G. Define $M_H = M_G^* \cup \{v_i a_i \mid v_i$ is not saturated by $M_G^*\}$ in H. It is easy to note that M_H is a maximal uniquely restricted matching in H, and since $|M_H| = (|M_G^*| + n - 2|M_G^*|) = (n - |M_G^*|)$, $|M_H^*| \leq n - |M_G^*|$.

Conversely, let M_H^* be a minimum maximal uniquely restricted matching in H. Let edges of the form $\{v_i a_i \mid 1 \leq i \leq n\}$ are called Type-A edges and edges in the edge set E are called Type-B edges. Let $T^* \cup S^*$ be a partition of M_H^* such that T^* contains Type-A edges and, S^* contains Type-B edges. Since M_H^* is maximal, $|T^*| = (n - 2|S^*|)$. It implies that $|M_H^*| = |S^*| + (n - 2|S^*|) = n - |S^*|$. Since $S^* \subset M_H^*$, S^* is a uniquely restricted matching in G and $|S^*| \leq |M_G^*|$. As $|S^*| = n - |M_H^*|$, $n - |M_H^*| \leq |M_G^*|$. This completes the proof of the claim.   □

Hence, the theorem is proved.   □

Next, we show that a maximal uniquely restricted matching of minimum size can be computed in polynomial time for proper interval graphs, a subclass of chordal graphs, where the problem is NP-complete.

4.2   Proper Interval Graphs

Let $G = (V, E)$ be a given graph. A vertex $v \in V$ is called a simplicial vertex, if $N[v]$ induces a clique in G. An ordering $\alpha = (v_1, v_2, \ldots, v_n)$ of vertices is called a perfect elimination ordering (PEO) of G if v_i is a simplicial vertex in $G_i = G[\{v_i, v_{i+1}, \ldots, v_n\}]$ for all $1 \leq i \leq n$. A PEO $\alpha = (v_1, v_2, \ldots, v_n)$ of a graph G is called a bi-compatible elimination ordering (BCO) if $\alpha^{-1} = (v_n, v_{n-1}, \ldots, v_1)$ i.e., the reverse of α, is also a PEO of G. It has been characterized in [7] that a graph is a proper interval graph if and only if it has a BCO.

Before discussing the algorithm, let us first, discuss the idea behind our algorithm. In a given proper interval graph G, we first identify an edge that belongs to some minimum maximal uniquely restricted matching M. After adding the desired edge to the matching, the algorithm remove those vertices from G that cannot be saturated by M in the later stages of the algorithm. We can find such vertices with the help of Theorem 1. Since proper interval graphs are hereditary, the subgraph thus obtained will also be a proper interval graph, in which an edge belonging to some minimum maximal uniquely restricted matching will be readily available. This process is repeated unless we are left with an empty graph.

In a given BCO σ of G, let $v^- (v^+)$ denote the vertex just before (after) a vertex v in σ. For $1 \leq i \leq n$ and $k \geq 2$, let $L(v_i)$ or $L^1(v_i)$ denote the maximum indexed neighbor of vertex v_i in σ and let $L^k(v_i)$ denote the maximum indexed neighbor of vertex $L^{k-1}(v_i)$ in σ.

Observation 6 [13]. *Let $\sigma = (v_1, v_2, \ldots, v_n)$ be a BCO of a proper interval graph $G = (V, E)$. If $v_i v_j \in E$, then $v_k v_j \in E$ for all $k, i \leq k \leq j - 1$.*

Observation 7 [12]. *Let $\sigma = (v_1, v_2, \ldots, v_n)$ be a BCO of a proper interval graph $G = (V, E)$ and let $L(v_i)$ denote the maximum indexed neighbor of vertex v_i in σ. If $v_i < v_j$ in σ, then $L(v_i) \leq L(v_j)$ in σ.*

Lemma 2 [3,6]. *Let* $\sigma = (v_1, v_2, \ldots, v_n)$ *be a BCO of a proper interval graph* $G = (V, E)$ *and let* M *be a uniquely restricted matching in* G. *If the edges* $u_1 w_1, u_2 w_2 \in M$ *such that* $u_1 < w_1$ *and* $u_2 < w_2$ *in* σ, *then either* $w_1 < u_2$ *or* $w_2 < u_1$.

This lemma immediately gives the following corollary.

Corollary 2. *Let* $\sigma = (v_1, v_2, \ldots, v_n)$ *be a BCO of a proper interval graph* $G = (V, E)$ *and let* $M_k = \{e_1, e_2, \ldots, e_k\}$ *be a uniquely restricted matching in* G. *If* $l(e)$ *and* $r(e)$ *denote the left and right endpoints of an edge* e, *respectively. Then, we can rename the edges in* M_k *such that* $l(e_1) < r(e_1) < \ldots < l(e_k) < r(e_k)$ *in* $\sigma(G)$.

Lemma 3 [3]. *Let* $\sigma = (v_1, v_2, \ldots, v_n)$ *be a BCO of a proper interval graph* $G = (V, E)$. *Let* e_1, e_2, e_3 *be distinct edges of* G *such that* $l(e_1) \leq l(e_2) \leq l(e_3)$ *and* $r(e_1) \leq r(e_2) \leq r(e_3)$ *in* σ. *If* $\{e_1, e_3\}$ *is not a uniquely restricted matching in* G, *then neither* $\{e_1, e_2\}$ *nor* $\{e_2, e_3\}$ *is a uniquely restricted matching in* G.

Lemma 4. *Let* $G = (V, E)$ *be a proper interval graph with a BCO* $\sigma = (v_1, v_2, \ldots, v_n)$. *Then, there exists a minimum maximal uniquely restricted matching* M *in* G *such that*

$$(i) \quad L(v_1)L(v_2) \in M \text{ if } L(v_1) \neq L(v_2),$$
$$(ii) \quad L(v_2)^- L(v_2) \in M \text{ if } L(v_1) = L(v_2).$$

Proof. Due to space restriction, the proof is omitted. □

It can be noted that in Lemma 4, the edge $L(v_1)L(v_2)$ (or $L(v_2)^- L(v_2)$) will be the first edge with respect to σ that belongs to M.

Let $\sigma = (v_1, v_2, \ldots, v_n)$ be a BCO of a proper interval graph G and let $\sigma' = (v_a, v_b, \ldots, v_k)$ be an ordering obtained from σ by removing some vertices from σ. Then, σ' is also a BCO of some proper interval graph G', where G' is a subgraph of G. Hence, we have the following corollary to Lemma 4.

Corollary 3. *If* $\sigma' = (v_a, v_b, \ldots, v_k)$ *is a BCO of a subgraph* $G' = (V', E')$ *of a proper interval graph* $G = (V, E)$, *then there exists a minimum maximal uniquely restricted matching* M *in* G *such that*

$$L(v_a)L(v_b) \in M \text{ if } L(v_a) \neq L(v_b),$$
$$L(v_b)^- L(v_b) \in M \text{ if } L(v_a) = L(v_b).$$

Based on the above lemmas, we now present a polynomial time algorithm URM-PIG(G), which computes a maximal uniquely restricted matching of minimum size in a given proper interval graph G. The pseudocode of the algorithm is given in Fig. 2.

Theorem 8. *Given a proper interval graph* $G_1 = (V_1, E_1)$ *with a bi-compatible elimination ordering* $\sigma(G_1)$, *URM-PIG(G_1) correctly computes a minimum size maximal uniquely restricted matching in* G_1.

Proof. Due to space restriction, the proof is omitted. □

Algorithm 2 URM-PIG(G)

Input: A proper interval graph $G_1 = (V_1, E_1)$ with a bi-compatible elimination ordering $\sigma(G_1) = (v_1, v_2, \ldots, v_n) = (\alpha_1, \beta_1, \ldots, v_n)$;

Output: A minimum maximal uniquely restricted matching M;

$G_i = (V_i, E_i)$ is a proper interval graph for each $i \geq 1$;

$M \to \emptyset$, $i \to 1$, counter;

$L(v) \to$ maximum indexed neighbor of vertex v in $\sigma(G_i)$ for each $i \geq 1$;

$\alpha_i \to$ first vertex in $\sigma(G_i)$ for each $i \geq 1$;

$\beta_i \to$ second vertex in $\sigma(G_i)$ for each $i \geq 1$;

$first_i \to (L(\beta_i))^+$ for each $i \geq 1$;

$second_i \to (first_i)^+$ for each $i \geq 1$;

while $(E_i \neq \emptyset)$ **do**

   if $(L(\alpha_i) \neq L(\beta_i))$ **then**

      $M \to M \cup \{L(\alpha_i)L(\beta_i)\}$;

      counter $\to 1$;

   else

      $M = M \cup \{L(\beta_i)^- L(\beta_i)\}$;

      counter $\to 1$;

   $i \to i + 1$;

   $G_i \to G_{i-1} \setminus \{\alpha_{i-1}, \beta_{i-1}, \ldots, L(\beta_{i-1})\}$;

   if $(E_i = \emptyset)$ **then**

      break;

   while (counter $= 1$) **do**

      if $(L(first_{i-1}) \in N(L(\alpha_{i-1}))$ **and** $L(second_{i-1}) \in N(L(\beta_{i-1})))$ **then**

         $G_i \to G_i \setminus \{first_{i-1}\}$;

         $first_{i-1} \to second_{i-1}$;

         $second_{i-1} \to first_{i-1}^+$;

         if $(E_i = \emptyset)$ **then**

            break;

      else

         counter $\to 0$;

return M;

Fig. 2. Algorithm to compute a minimum maximal uniquely restricted matching in a proper interval graph G.

5 APX-completeness for Bounded Degree Graphs

In this section, we show that the MIN-MAX-UR MATCHING problem is APX-complete, even if the degree of the graph is bounded by 4. We first show that the MIN-MAX-UR MATCHING problem belongs to the class APX for r-regular graphs, where $r \geq 3$ is a constant.

Observation 9. *For a graph G with n vertices, $\mu_{urm}(G) \leq \frac{n}{2}$.*

Lemma 5. *Let $r \geq 3$ be a positive integer. If G is a r-regular graph then*
$\mu'_{urm}(G) \geq \frac{rn}{2[2r(r-1)+1]}$.

Proof. Choose an edge $uv \in M$ and remove the edges incident on the vertices in $N(u) \cup N(v)$. For $uv \in M$, at most $(2r(r-1)+1)$ edges are incident on $N(u) \cup N(v)$ (since G is a r-regular graph). Therefore, $|M| \geq \frac{rn}{2[2r(r-1)+1]}$.   □

Based on these bounds, we have the following theorem.

Theorem 10. *The* Min-Max-UR Matching *problem can be approximated within a ratio of* $\frac{(2r(r-1)+1)}{r}$ *in r-regular graphs, where* $r \geq 3$ *is a constant.*

For a bounded degree graph G, $\Delta(G) \leq k$ for some integer constant k, the following corollary follows from Theorem 10.

Corollary 4. *The* Min-Max-UR Matching *problem for bounded degree graphs is in* APX.

Now, we are ready to show the APX-completeness of the Min-Max-UR Matching problem for bounded degree graphs. For this purpose, we recall the concept of L-reduction. Given two NP optimization problems π_1 and π_2 and a polynomial time transformation f from instances of π_1 to instances of π_2, we say that f is an L-reduction if there are positive constants α and β such that for every instance x of π_1:

1. $opt_{\pi_2}(f(x)) \leq \alpha.opt_{\pi_1}(x)$;
2. for every feasible solution y of $f(x)$ with objective value $m_{\pi_2}(f(x), y) = c_2$, we can find a solution y' of x in polynomial time with $m_{\pi_1}(x, y') = c_1$ such that $|opt_{\pi_1}(x) - c_1| \leq \beta.|opt_{\pi_2}(f(x)) - c_2|$.

Next, we give a L-reduction from the Min-Ind-Dom-Set problem for 3-regular graphs, which is already known to be APX-complete [8].

Theorem 11. *The* Min-Max-UR Matching *problem is* APX-*complete for graphs with maximum degree 4.*

Proof. By Corollary 4, the Min-Max-UR Matching problem is in APX for bounded degree graphs. Given a 3-regular graph $G = (V, E)$, an instance of the Min-Ind-Dom-Set problem, we construct a graph $H = (V_H, E_H)$ with maximum degree 4, an instance of the Min-Max-UR Matching problem as follows:

For each vertex $v_i \in V$, we take two copies of v_i, namely v_i^1 and v_i^2 in V_H, and join them by the edge $v_i^1 v_i^2$. For each edge $v_i v_j \in E$, we connect the vertices v_i^1 and v_j^1 using the gadget H_{ij}^1 (see Fig. 3), and connect the vertices v_i^2 and v_j^2 using the gadget H_{ij}^2 (see Fig. 3). Also, Fig. 3 illustrates the subgraph H_{ij} of H corresponding to the edge $v_i v_j$ in G.

Since G is a 3-regular graph, it is clear from the construction that graph H has maximum degree 4. Also, H can be constructed from G in polynomial time as $|V_H| = 20|V|$.

Claim. G has an independent dominating set of size k if and only if H has a maximal uniquely restricted matching of size $k + 2m$.

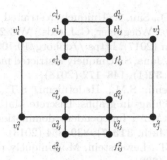

Fig. 3. Subgraph H_{ij} of H corresponding to edge $v_i v_j$ in G.

Proof. Due to space restriction, the proof is omitted. □

It is easy to observe that $2m = 3n$ as G is a 3-regular graph. From Claim 5, we have the following result.

Claim. If M^* is a minimum size maximal uniquely restricted matching in H and I^* is a minimum size independent dominating set in G, then $|M^*| = |I^*| + 3n$.

We now return to the proof of Theorem 11. It is known that any independent dominating set I in a 3-regular graph G satisfies the inequality $\frac{n}{4} \leq |I| \leq \frac{n}{2}$ [5]. Therefore, we have $|M^*| = |I^*| + 3n \leq |I^*| + 12|I^*| = 13|I^*|$. Further, by Claim 5, from any maximal uniquely restricted matching M' in H, we can obtain an independent dominating set I in G such that $|I| \leq |M'| - 3n$. Now, $|M'| - |M^*| = |M'| - |M^*| - 3n + 3n = (|M'| - 3n) - (|M^*| - 3n) \geq (|I| - |I^*|)$. From these two inequalities it follows that it is an L-reduction with $\alpha = 13$ and $\beta = 1$. □

Therefore, the MIN-MAX-UR MATCHING problem is APX-complete for graphs with maximum degree 4. □

6   Conclusions

In this paper, we have shown that the DECIDE-MIN-MAX-UR MATCHING problem is NP-complete for chordal bipartite graphs and chordal graphs. Also, the MIN-MAX-UR MATCHING problem is shown to be polynomial time solvable for bipartite permutation graphs and proper interval graphs. We also show that the MIN-MAX-UR MATCHING problem is APX-complete for bounded degree graphs. It will be interesting to study the complexity status of the MIN-MAX-UR MATCHING problem in graph classes like split graphs and permutation graphs.

References

1. Ausiello, G., Crescenzi, P., Gambosi, G., Kann, V., Spaccamela, A.M., Protasi, M.: Complexity and Approximation: Combinatorial Optimization Problems and Their Approximability Properties. Springer, Heidelberg (2012)

2. Baste, J., Rautenbach, D., Sau, I.: Uniquely restricted matchings and edge colorings. In: Bodlaender, H.L., Woeginger, G.J. (eds.) WG 2017. LNCS, vol. 10520, pp. 100–112. Springer, Cham (2017). https://doi.org/10.1007/978-3-319-68705-6_8
3. Francis, M.C., Jacob, D., Jana, S.: Uniquely restricted matchings in interval graphs. SIAM J. Discrete Math. **32**(1), 148–172 (2018)
4. Goddard, W., Hedetniemi, S.M., Hedetniemi, S.T., Laskar, R.: Generalized subgraph-restricted matchings in graphs. Discrete Math. **293**(1), 129–138 (2005)
5. Goddard, W., Henning, M.A.: Independent domination in graphs: a survey and recent results. Discrete Math. **313**(7), 839–854 (2013)
6. Golumbic, M.C., Hirst, T., Lewenstein, M.: Uniquely restricted matchings. Algorithmica **31**(2), 139–154 (2001)
7. Jamison, R.E., Laskar, R.: Elimination orderings of chordal graphs. Combin. Appl. 192–200 (1982)
8. Kann, V.: On the approximability of NP-complete optimization problems. Ph.D. thesis, Royal Institute of Technology Stockholm (1992)
9. Lai, T.H., Wei, S.S.: Bipartite permutation graphs with application to the minimum buffer size problem. Discrete Appl. Math. **74**(1), 33–55 (1997)
10. Mishra, S.: On the maximum uniquely restricted matching for bipartite graphs. Electron. Notes Discrete Math. **37**, 345–350 (2011)
11. Müller, H.: Alternating cycle-free matchings. Order **7**(1), 11–21 (1990)
12. Panda, B.S., Chaudhary, J.: Acyclic matching in some subclasses of graphs. In: Gąsieniec, L., Klasing, R., Radzik, T. (eds.) IWOCA 2020. LNCS, vol. 12126, pp. 409–421. Springer, Cham (2020). https://doi.org/10.1007/978-3-030-48966-3_31
13. Panda, B.S., Das, S.K.: A linear time recognition algorithm for proper interval graphs. Inf. Process. Lett. **87**(3), 153–161 (2003)
14. Panda, B.S., Pandey, A.: On the complexity of minimum cardinality maximal uniquely restricted matching in graphs. In: Arumugam, S., Bagga, J., Beineke, L.W., Panda, B.S. (eds.) ICTCSDM 2016. LNCS, vol. 10398, pp. 218–227. Springer, Cham (2017). https://doi.org/10.1007/978-3-319-64419-6_29
15. Papadimitriou, C.H., Yannakakis, M.: Optimization, approximation, and complexity classes. J. Comput. Syst. Sci. **43**(3), 425–440 (1991)

Search, Facility and Graphs

A Two-Layers Heuristic Search Algorithm for Milk Run with a New PDPTW Model

Xuhong Cai, Li Jiang, Songhu Guo, Hejiao Huang[✉], and Hongwei Du

Harbin Institute of Technology, Shenzhen, China
huanghejiao@hit.edu.cn

Abstract. With the growth of market competition on manufacture, milk-run becomes a popular just-in-time (JIT) logistic strategy to ensure vehicle pickups and delivers goods on multiple round trips with fixed time window. Reasonable milk-run vehicle routing planning is able to improve the utilization of vehicle, so that logistic cost can be reduced. In order to better capture the real-world scenes, we build a novel milk-run model called MOPDPTW (Multiple-Orders Pickup and Delivery Problem with Time-bound Window) based on PDPTW (Pickup and Delivery Problem with Time Window). Aiming at minimizing the number of used vehicles and total travel distance in this model, a two-layers heuristic search algorithm is proposed to solve this problem. The inner layer of proposed algorithm searches possible solutions in global and sends them to the outer layer to find local optimal solution. We validate our algorithm against an improved large neighborhood algorithm on standard Li and Lim's benchmark and instances modified for MOPDPTW. The experiment results show that our algorithm performs better in reducing the logistic cost of milk-run.

Keywords: Milk run · MOPDPTW · Large neighborhood search · Multiple orders

1 Introduction

With the increasing market competition and business globalization, manufacturing enterprises are facing huge challenge in reducing logistic costs. Moreover, the enterprise's warehouses are insufficient to meet production and storage needs. Under this setting, a logistic mode of JIT that picks up goods in small batch and high frequency is adopted in enterprises' logistic. In JIT mode, the direct shipment logistic strategy leads to a great increase in transportation cost. Milk-run is a good logistic strategy that can achieve JIT supply with minimum logistic cost, which is widely applied in automakers and gradually applied in electronic product manufacturing, electric appliance manufacturing, as well as e-tailing industry. In milk-run transport network, all members share vehicles, vehicle departs

This work is financially supported by National Key R&D Program of China under Grant No. 2017YFB0803002 and No. 2016YFB0800804, National Natural Science Foundation of China under Grant No. 61672195 and No. 61732022.

W. Wu and Z. Zhang (Eds.): COCOA 2020, LNCS 12577, pp. 379–392, 2020.
https://doi.org/10.1007/978-3-030-64843-5_26

from the distribution center, then travel to the vendors to pick up goods on multiple round trips with fixed time window among the designed route plan, finally return back to the distribution center. Though automakers have applied milk-run for many years, the application of milk-run in other manufacturing is immature and there are many problems in practice. Low vehicle loading rate, poor vehicle management efficiency, high transportation cost and idealized practical constraints lead to high logistic cost of milk-run system. Most milk-run models in automakers determine a route on which a vehicle either delivers parts from a single vendor to multiple customers (outbound logistic) or travels from multiple vendors to a single customer location (inbound logistic). Both logistics simplify the delivery scenario. However, in the real manufacturing scenario, the manufacturing enterprise needs to delivery its raw materials to its production plants. Thus, the limited vehicles need to serve multiple-vendors and multiple-customers in practice. Reasonable milk-run vehicle routing plan is a primary task in many enterprises.

Technically, various approaches have been studied and proposed to solve milk-run problems. Specifically for milk-run system of the automakers, Miao et al. [4] used ant colony optimization (ACO) method to solve the problem with minimizing manufacturing cost, delivery cost and inventory cost. Based on the same milk-run problem, Nguyen et al. [7] proposed a hybrid ACO to improve the solution. Instead of taking all logistic costs into account, researchers focus more on vehicle routing optimization which directly influences logistic costs and inventory levels. Ma et al. [10] and Huang et al. [9] built mathematical model on milk-run with time window, then used particle swarm optimization and improved C-W algorithm to solve the problem of given example, respectively. Gyulai et al. [15] studied milk-run vehicle routing for shop floors manufacturing enviroments. In addition, Alnahhal et al. [14], Urru et al. [16] studied inbound milk-run problems. Obviously, the outbound and inbound milk-run logistic are studied separately. With considering both logistics, Solomon et al. [5] and Kong et al. [3] put forward some new model and optimization algorithms for milk-run vehicle routing problems that were reduced to VRPTW (Vehicle Routing Problem with Time Window) and VRPPD (Vehicle Routing Problem with Pickup and Delivery). Based on the fact that milk-run vehicle routing problem can be reduced to VRPPD, many researchers are committed to studying VRPPD and its variants. VRPPD is a NP-hard problem, Berbeglia et al. [11] provided a complete review for the VRPPD. Montero et al. [12] developed an integer linear programming for the VRPPD with solving the problem by local search procedure. Männel et al. [2] extended the VRPPD to a three-dimensional loading problem, and modified a large neighborhood search heuristic to solve the problem. PDPTW is a variant of VRPPD, many methods of PDPTW can be found. Few literatures applied exact algorithms to solve PDPTW because of its high time complexity, so heuristic and metaheuristic algorithms are the main methods to solve PDPTW. The heuristic algorithms have been studied to solve PDPTW include large neighborhood search algorithm [6], guided ejection search algorithm [8], memetic algorithm [17] etc., while metaheuristic algorithms consist of simulated annealing [1], genetic algorithm [18], ant colony optimization [19] etc. There are many algorithms aiming at solving milk-run vehicle routing problem, while seldom do they obtain optimization and none of them study inbound and outbound milk-run together.

In this paper, we design a novel milk-run model called MOPDPTW (Multiple-Orders Pickup and Delivery Problem with Time Window), which is a more practical model in real scenario. Compared with classical PDPTW, MOPDPTW introduces an additional practical constraint called multiple orders constraint. Since the features of small batch and high frequency in milk-run, multiple orders constraint requires that customer's demand is represented by multiple orders rather than single request. MOPDPTW can be described as scheduling vehicle routes to serve a group of customers with multiple orders. With the objectives of minimizing the number of vehicles (main objective) and the total travel distance (secondary objective), we proposed a new two-layers heuristic search algorithm to solve the problem. To enhance the quality of solution, our algorithm explores solution space in two layers. The inner layer searches possible solutions in global and the outer layer searches local optimal solution so as to converge to best solution. In addition, the two-layers algorithm is equipped with a packing scheme, which effectively reduces the waste of vehicle resources. Based on the experiments on Li and Lim's benchmark, we validate that the proposed algorithm is able to retrieve high-quality solutions. We also present an in-depth analysis of the results on MOPDPTW instances to investigate the algorithm behavior and solution quality. Experiments results show that our algorithm gets better solution than 1D_LNS [2] on the real milk-run instances.

The rest of this paper is organized as follows. Section 2 gives problem description and formulation. In Sect. 3, we describe the algorithm to solve the problem. Simulations and results are discussed in Sect. 4. Finally, we draw our conclusions in Sect. 4.1.

2 Problem Description and Model

In our proposed MOPDPTW model, we consider a practical off-line vehicle routing problem with hard time window, capacity, precedence, coupling and multiple orders constraints, while relaxing the constraint that one customer (location) can be only visited once. The MOPDPTW problem is described as scheduling routes for a limited number of vehicles to serve a group of customers (vendors and plants) in the specified time windows. The fleet of vehicles is assumed to be homogeneous, all the vehicles are located at a single depot, where they start and end their route, and each vehicle has a limited capacity. The total demand on the route does not exceed the vehicles capacity. Moreover, any paired pickup and delivery locations must be serviced by the same vehicle and the pickup location must be scheduled before the corresponding delivery location in the route. the customer with multiple orders is allowed to be visited more than once and its orders can be packed together. The problem setting studied here includes customers, orders, central depot, vehicle speed, time windows and a network connecting the depot and customers. The objective of our problem is to find routes for vehicles that serve all orders at a minimum total costs (in terms of number of vehicles, travel distance, etc.). The fewer the number of vehicles, the higher the load rate of vehicle and the less the logistic costs. We defined the number of vehicles, the travel distance as the main and secondary objective respectively.

Since customers may be visited repeatedly, we treat the orders instead of customers as virtual nodes. The node set consists of all orders, $V = \{0, 1, 2, \ldots, N\}$, and the arc set is defined between each pair of nodes, $E = \{<i, j>|i, j \in V, i \neq j\}$. The problem description can be defined on a complete graph $G = \{V, E\}$. The notations and definitions of the variables of our problem are listed in Table 1.

Table 1. Notations and definitions

Notation	Definition
0	The central depot
C_k	Maximal load of vehicle k
V_0	Set of orders, $V_0 = \{1, 2, \ldots, N\}$
V	Nodes set in the transport network, $V = V_0 \cup \{0\}$
M	Maximum number of vehicles used
K	A fleet of identical vehicles, $K = \{1, 2, \ldots, M\}$
$G = (V, E)$	The undirected graph of milk-run transport network
P^+	Set of pickup orders
P^-	Set of delivery orders, $P^+ \cup P^- = V_0$
q_i	Demand at order i, $q_i > 0$, i is a pickup order, $q_i < 0$, i is a delivery order
$[e_i, l_i]$	Time window, order i must be served within $[e_i, l_i]$
i_d	Delivery order of pickup order $i \in P^+$, $i_d \in P^-$
s_i	Time needed to service at order i
$speed$	The vehicle speed is 1
d_{ij}	The distance between locations of order i and order j
t_{ij}	The time needed for vehicle to travel from order i to order j
x_{ijk}	Decision variable, equals to 1 if vehicle k traverses from order i to order j, otherwise equals to 0; $i, j \in V, k \in K$
T_{ik}	Decision variable, time for vehicle k to begin service at order i
L_{ik}	Decision variable, load for vehicle k finishing order i
α, β	Weight of the number of vehicles and total travel distances, $\alpha + \beta = 1$

Based on the notations and variable definitions, the proposed MOPDPTW model is established as follows.

The objective function is formulated in (1). It seeks to minimize the number of vehicles and total travel distance. Constraints (2) and (4) ensure that each order is only visited once. Constraint (3) is limiting the vehicle number. Constraints (5) and (6) ensure that each vehicle departs from depot to pickup order location, and return to depot from delivery order location. Constraint (7) ensure the pickup order and its coupling delivery order to be finished by the same vehicle. Constraint (8) ensures that the next order can be served only after finishing the current order. Constraint (9) ensures that the pickup order should be served before its coupling delivery order. Constraint (10) ensures that the time vehicle begin to serve at an order must be within the order's time windows. Constraints (11) and (12) are the capacity constraints.

$$\min \quad \alpha \sum_{k \in K} \sum_{j \in V} x_{0jk} + \beta \sum_{k \in K} \sum_{i \in V} \sum_{j \in V, i \neq j} x_{ijk} d_{ij} \tag{1}$$

$$\text{s.t.} \quad \sum_{i \in V} \sum_{k \in K} x_{ijk} = 1, \forall j \in V_0 \tag{2}$$

$$\sum_{k \in K} \sum_{j \in V_0} x_{0jk} \leq |K| \tag{3}$$

$$\sum_{i \in V_0} x_{ijk} - \sum_{i \in V_0} x_{jik} = 0, \forall j \in V_0, k \in K \tag{4}$$

$$\sum_{j \in P^+} x_{0jk} = 1, \forall k \in K \tag{5}$$

$$\sum_{i \in P^-} x_{i0k} = 1, \forall k \in K \tag{6}$$

$$\sum_{j \in V} x_{ijk} - \sum_{j \in V} x_{ji_d k} = 0, \forall i \in P^+, k \in K \tag{7}$$

$$(T_{ik} + s_i + t_{ij}) x_{ijk} \leq T_{jk}, \forall i, j \in V \tag{8}$$

$$T_{ik} + s_i + t_{i,i_d} \leq T_{i_d,k}, \forall i \in P^+ \tag{9}$$

$$e_i \leq T_{ik} \leq l_i, \forall i \in V, k \in K \tag{10}$$

$$(L_{ik} + q_j) x_{ikj} \leq L_{jk}, \forall i \in V, k \in K \tag{11}$$

$$0 \leq L_{ik} \leq C_k, \forall i \in V, k \in K \tag{12}$$

3   Algorithm Design

The proposed two-layers heuristic search algorithm consists of two stages. In the first stage, we adopt Solomon insertion heuristic [22] to construct an initial solution due to its success in generating feasible solution on the PDPTW and the simplicity of its implementation. The second stage can be viewed as an optimizer, an two-layers heuristic search mechanism is used to further improve the initial solution from the first stage. In the two-layers heuristic search procedure, the inner layer runs global search methods in parallel, which offers the possibility to speed up the search, then the outer layer further searches local optimal to make solution converge to best solution. In addition, we use packing scheme in each iteration to reduce the waste of vehicle resources. The framework procedure of the algorithm is presented in Algorithm 1.

Algorithm 1: Framework procedure

Input: Order information
Output: Best solution S_{best}
1 Initialization: use Insertion Heuristic to construct the initial solution S_{ini},
   $S_{best} \leftarrow S_{ini}$, $S_{cur} \leftarrow S_{ini}$, $S_{trans} \leftarrow \Phi$, $Iter \leftarrow 0$;
2 **while** $Iter \leq MINI$ **do**
3 | //parallelly compute neighbor solutions from three neighborhood structure;
4 | $S_{trans} \leftarrow$ InnerSearch(S_{cur});
5 | //get local optimal solution of each neighbor solution by the outer layer;
6 | $S_{trans} \leftarrow$ OuterSearch(S_{trans});
7 | **if** $S_{trans} < S_{cur}$ **then**
8 | | **if** $S_{trans} < S_{best}$ **then**
9 | | | $S_{best} \leftarrow S_{trans}$;
10 | | **end**
11 | | $S_{cur} \leftarrow S_{trans}$;
12 | | $Iter \leftarrow 0$;
13 | **end**
14 | **else**
15 | | $Iter \leftarrow Iter + 1$;
16 | **end**
17 **end**
18 **return** S_{best};

Table 2. Neighborhood structure heuristics

Heuristic	Description
Random and Greedy (RAG)	Remove iteratively PD-orders that are selected at random, then insert iteratively PD-orders into the solution such that the increase of the distance cost is minimal
Tour and Regret-2 (TAR)	Remove all PD-orders from a randomly chosen route. If less than m PD-orders are removed in this way, further PD-orders will be removed with Random removal. Then insert iteratively PD-orders into the solution such that the gap in the cost function between inserting the PD-orders into its best and its second best route is maximal
Worst and Priority (WAP)	Remove iteratively a PD-order whose removal leads to the largest cost (total travel distance) reduction, then insert the PD-order into its best position according to priority sequence

3.1   Inner Search of Algorithm

It is not possible to search the entire solution space for vehicle routing problem in reasonable time. Thus, we use different large neighborhood structures to ensure that solutions are searched in global as far as possible. The inner search procedure explores solutions in parallel by three large neighborhood structures, which are briefly summarized in Table 2. We apply several simple and effective removal and insertion heuristics from literature [2]. In addition, we propose a priority-based insertion heuristic for searching better solution. All operators have their own search bias. RAG operator aims to expand the search space, TAR operator is used to reduce number of vehicles and WAP operator is utilized to reduce the total distance cost.

Algorithm 2 shows the pseudo-code of WAP procedure in detail. The WAP procedure starts with parameters initialization (line 1). For each remove rate (lines 2–16), remove the number of largest cost orders (lines 3–4), then insert the removed orders to current solution according to their cost priorities (line 5). If insert successful, then updates the current solution and best solution (lines 6–15).

Algorithm 2: Worst and Priority procedure

Input: Solution S_{ini} maximal remove rate p_{max}, remove rate increasement Δp
Output: Solution S_{best}
1 Initialization: $S_{best} \leftarrow S_{ini}$, $S_{cur} \leftarrow S_{ini}$, $p \leftarrow p_0$;
2 **while** $p \leq p_{max}$ **do**
3   remove numbers $num \leftarrow N * p$;
4     $removeOrders \leftarrow$ WorstRemoval(S_{cur}, num);
5   Insert orders: PriorityInsert($S_{cur}, removeOrders$);
6   **if** *insert successful* **then**
7   Reduce vehicle number of S_{cur};
8   **if** $S_{cur} < S_{best}$ **then**
9             $S_{best} \leftarrow S_{cur}$;
10   **end**
11   **else**
12         $p \leftarrow p + \Delta p$;
13         $S_{cur} \leftarrow S_{best}$;
14   **end**
15   **end**
16 **end**
17 **return** S_{best};

Under the multiple-orders constraint, the occurrence of serving two or more orders of one customer with different vehicles is possible. In order to reduce the unnecessary waste of resource, in each insert iteration, the packing procedure tries to merge adjacent orders that were loaded or unloaded in the same customer. The pseudo-code of packing procedure is presented in Algorithm 3. In the process of inserting PD-orders, if there were two connected orders belonging to same customer and their time windows were overlapping, then the packing procedure packs the two orders together (lines 2–3). After packing, the time window of both orders become $[e_i', l_i']$, where $e_i' = max\{e_1, e_2\}$, $l_i' = min\{l_1, l_2\}$. Finally, check whether the solution still satisfies the capacity and time window constraints (line 4). If pack successful, return the new solution (lines 5–10).

Algorithm 3: Packing procedure

Input: Solution S_{input}
Output: Solution S_{new}
1 **for** *each connected orders in* S_{input} **do**
2   **if** *two orders are served in same customer* **then**
3   Pack two orders in S_{input}, update the time window, get solution S_{temp};
4   Check capacity constraint and time window constraint;
5   **if** *pack successful* **then**
6              $S_{new} \leftarrow S_{temp}$;
7   **end**
8   **end**
9 **end**
10 **return** S_{new};

3.2 Outer Search of Algorithm

In the outer search, we propose a local search neighborhood operator called IM (Irreversible Move). The IM is defined as moving PD-orders from one route to another route at all costs. An interesting feature of IM operator is the diversity of its possible solution. First, we select and remove a pair of PD-orders say $PD1$ from $Route1$. Next, treat $PD1$ as a seed to create a new route $Route2'$ and successively insert PD-orders of $Route2$ into $Route2'$ without violating problem constraints. If all orders of $Route2$ were inserted into $Route2'$, IM operator can get a solution like PD-Shift in [1]. Otherwise, there is one of $Route2$ failed to be inserted into $Route2'$. If the remaining one was inserted to $Route1$ successfully, IM operator obtained a solution like PD-exchange in [1]. If not, IM operator moved PD-orders several times, which makes search space more widely. The outer search procedure's pseudo-code is presented in algorithm 4. We adopt the acceptance test derived from [2] here. The outer search procedure begins with initial best solution at null (line 1). For each solution from inner search (lines 2–13), adopt IM operator with given times (lines 3–12). For each move, randomly select and remove a PD-orders from one route and insert it to another route at any cost (lines 4–7), then check whether to update the best solution (lines 8–10). Finally, return the best solution (line 14).

4   Computational Experiments

The computational experiments are organized in two parts. In the first part, we examine our two-layers heuristic search algorithm (denoted by TLHSA) using

---

Algorithm 4: Outer Search Algorithm

Input: Solutions set *neighbors*, move times *times*
Output: Solution S_{best}
1 Initialization: $S_{best} \leftarrow \Phi$, $k \leftarrow 1$;
2 **for** *each solution S_i in neighbors* **do**
3   **for** $k < times$ **do**
4   Random select and remove a PD-orders $PD1$ from $Route1$;
5   Create a new route $Route2'$ for $PD1$;
6   Random select a $Route2$, insert its orders to $Route2'$ first then other routes;
7   Replace the $Route2$ with $Route2'$ in S_i, $S_{cur} \leftarrow S_i$;
8   **if** $S_{cur} < S_{best}$ **then**
9             $S_{best} \leftarrow S_{cur}$;
10   **end**
11          $k \leftarrow k + 1$;
12   **end**
13 **end**
14 **return** S_{best};

---

the well-known PDP instances pdp_100 by Li and Lim [1]. In the second part, we mainly test the superiority of our algorithm with 56 new multi-orders variant of PDPTW instances. Based on pdp_100, we kept the first 80% of the orders unchanged and change the location of the remaining orders. The latter's new location is one of the former's, then we generated 56 MOPDPTW instances mopdp_100[1]. Six classes of instances (molc1, molc2, molr1, molr2, molrc1, molrc2) reflect different real-life milk-run scenarios. The characteristics of mopdp_100 is shown in Table 3.

Table 3. MOPDPTW data's test characteristics

	molc1	molc2	molr1	molr2	molrc1	molrc2
Distribution	Clustered	Clustered	Random	Random	Mix	Mix
Vehicle capacity	200	700	200	1000	200	1000
Time window	Tight	Wide	Tight	Wide	Tight	Wide
Rate of multiple orders	20%	20%	20%	20%	20%	20%

Our experimental environment is the Intel Core i7-6700 K (3.40 GHz, 8 GB RAM). The algorithms were coded in the Java programming language under Eclipse 4.6.3. The values of the parameters used are $MINI = 10$, $p_0 = 0.1$, $p_{max} = 0.5$, $\Delta p = 0.05$, $times = 20$. Since the number of vehicles is our main objective and total travel distance is the secondary objective, their weight are set as $\alpha = 0.999$, $\beta = 0.001$.

---

[1] https://github.com/caixuhong/TLHSA.

A. Results for 56 PDP Problem Instances

We tested our algorithm on Li and Lim's instances with 100 customers, and compared the results of our algorithms with those of 1D_LNS [2] which is based on large neighborhood search and tree search and the best known solutions. The test instances and best known solutions were taken from the website Sintef [13]. Table 4 shows the results obtained as indicated in the above sources (NV is number of used vehicles, TD is total travel distance). Though [2] states that they have obtained all optimal solutions for 56 instances, they did not give the optimal solutions of lrc2 instances. With the effective of two-layers heuristic search strategy, TLHSA finds the best known solution for all the instances. The results show that our designed TLHSA can effectively tackle the PDPTW problems.

Table 4. Best solutions for pdp_100 instances

Instances	TLHSA		1D_LNS		Best known solution		Instances	TLHSA		1D_LNS		Best known solution	
	NV	TD	NV	TD	NV	TD		NV	TD	NV	TD	NV	TD
lc101	10	828.94	10	828.94	10	828.94	lc201	3	591.56	3	591.56	3	591.56
lc102	10	828.94	10	828.94	10	828.94	lc202	3	591.56	3	591.56	3	591.56
lc103	9	1035.35	9	1035.35	9	1035.35	lc203	3	591.17	3	591.17	3	591.17
lc104	9	860.01	9	860.01	9	860.01	lc204	3	590.60	3	590.60	3	590.60
lc105	10	828.94	10	828.94	10	828.94	lc205	3	588.88	3	588.88	3	588.88
lc106	10	828.94	10	828.94	10	828.94	lc206	3	588.49	3	588.49	3	588.49
lc107	10	828.94	10	828.94	10	828.94	lc207	3	588.29	3	588.29	3	588.29
lc108	10	826.44	10	826.44	10	826.44	lc208	3	588.32	3	588.32	3	588.32
lc109	9	1000.60	9	1000.60	9	1000.60							
lr101	19	1650.80	19	1650.80	19	1650.80	lr201	4	1253.23	4	1253.23	4	1253.23
lr102	17	1487.57	17	1487.57	17	1487.57	lr202	3	1197.67	3	1197.67	3	1197.67
lr103	13	1292.68	13	1292.68	13	1292.68	lr203	3	949.40	3	949.40	3	949.40
lr104	9	1013.39	9	1013.39	9	1013.39	lr204	2	849.05	2	849.05	2	849.05
lr105	14	1377.11	14	1377.11	14	1377.11	lr205	3	1054.02	3	1054.02	3	1054.02
lr106	12	1252.62	12	1252.62	12	1252.62	lr206	3	931.63	3	931.63	3	931.63
lr107	10	1111.31	10	1111.31	10	1111.31	lr207	2	903.06	2	903.06	2	903.06
lr108	9	968.97	9	968.97	9	968.97	lr208	2	734.85	2	734.85	2	734.85
lr109	11	1208.96	11	1208.96	11	1208.96	lr209	3	930.59	3	930.59	3	930.59
lr110	10	1159.35	10	1159.35	10	1159.35	lr210	3	964.22	3	964.22	3	964.22
lr111	10	1108.90	10	1108.90	10	1108.90	lr211	2	911.52	2	911.52	2	911.52
lr112	9	1003.77	9	1003.77	9	1003.77							
lrc101	14	1708.80	14	1708.80	14	1708.80	lrc201	4	1406.94	–	–	4	1406.94
lrc102	12	1558.07	12	1558.07	12	1558.07	lrc202	3	1374.27	–	–	3	1374.27
lrc103	11	1258.74	11	1258.74	11	1258.74	lrc203	3	1089.07	–	–	3	1089.07
lrc104	10	1128.40	10	1128.40	10	1128.40	lrc204	3	818.66	–	–	3	818.66
lrc105	13	1637.62	13	1637.62	13	1637.62	lrc205	4	1302.20	–	–	4	1302.20
lrc106	11	1424.73	11	1424.73	11	1424.73	lrc206	3	1159.03	–	–	3	1159.03
lrc107	11	1230.14	11	1230.14	11	1230.14	lrc207	3	1062.05	–	–	3	1062.05
lrc108	10	1147.43	10	1147.43	10	1147.43	lrc208	3	852.76	–	–	3	852.76

B. Results for Newly-Generated 56 MOPDP Problem Instances

In this part, we reimplement 1D_LNS algorithm for MOPDPTW instances and get detailed results. We choose the best result of each instance over 5 runs. To validate the difference between PDPTW and MOPDPTW, we compute the gap of the number of vehicles (NV_gaps) and the gap of total travel distance (TD_gaps) between pdp_100 and mopdp_100 for two algorithms. The NV_gaps and TD_gaps of six classes instances are presented in Fig. 1 and Fig. 2. The range of NV_gaps is [−3, 2] for TLHSA and [0, 5] for 1D_LNS, while the range of TD_gaps is [−17.29%, 42.20%] for TLHSA and [−4.26%, 194.38%] for 1D_LNS. Obviously, the number of vehicles and total travel distances decrease significantly if the MOPDPTW instances are solved by TLHSA instead of 1D_LNS.

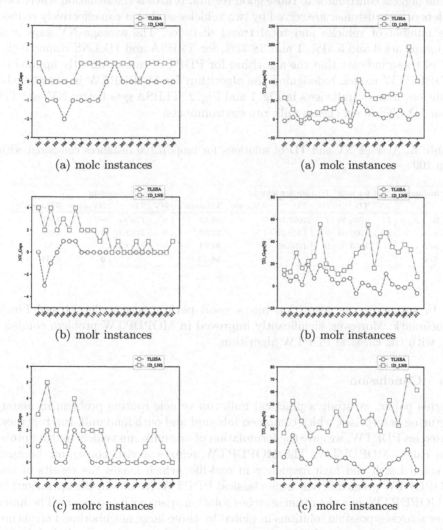

(a) molc instances

(a) molc instances

(b) molr instances

(b) molr instances

(c) molrc instances

(c) molrc instances

Fig. 1. NV_gaps between mopdp and pdp instances

Fig. 2. TD_gaps between mopdp and pdp instances

C. Simulation Result Analysis

The results clearly show that both 1D_LNS and TLHSA have a good performance on Li and Lim's benchmark. However, when the problem includes more realistic constraints like a customer owns multiple orders, the results of the two algorithms are very different. To further evaluate our algorithm, we analyze the results with algorithm behaviors. We compute the average NV, TD, NV_gaps, TD_gaps for six classes instances, which is presented in Table 5. As seen in Table 5, the results of 1D_LNS in PDPTW and MOPDPTW are very different, it indicates that the more realistic MOPDPTW model leads to a significant increase of vehicle numbers and total travel distances. Nevertheless, the results of our algorithm in MOPDPTW is very closed to PDPTW. The packing strategy of our algorithm is the biggest contributor to those good results. It avoids the situation where two orders of one customer are served by two vehicles so that it can effectively reduce the number of vehicles and total travel distance. The average NV_gaps and TD_gaps are 0 and 5.31%, 1 and 38.31%, for TLHSA and 1D_LNS respectively. All of those indicate that the algorithm for PDPTW cannot directly applied to MOPDPTW model, redesigning the algorithm for MOPDPTW is essential. In addition, from overall views in Fig. 1 and Fig. 2, TLHSA gets better NV and TD than 1D_LNS in six classes milk-run environments.

Table 5. Average NV and TD of solutions for mopdp_100 instances compared with pdp_100

Instances	Best for pdp		1D_LNS for mopdp				TLHSA for mopdp			
	NV	TD	NV	TD	NV_gaps	TD_gaps%	NV	TD	NV_gaps	TD_gaps%
lc	7	740.35	7	1097.10	0	48.19	6	776.84	−1	4.93
lr	8	1100.64	9	1392.29	1	26.50	8	1152.28	0	4.69
lrc	7	1259.93	9	1766.85	2	40.23	8	1339.38	1	6.31
Avg					1	**38.31**			0	**5.31**

In summary, our algorithm has a good performance under Li and Lim's benchmark. Moreover, significantly improved in MOPDPTW problem comparing with the classical PDPTW algorithm.

4.1   Conclusion

In this paper, we study a practical milk-run vehicle routing problem in manufacturing enterprises, which combined inbound and outbound milk-run together. Based on PDPTW, we gave the formulation of our milk-run vehicle routing problem called MOPDPTW. The MOPDPTW relaxes a visit constraint because of small batch and high frequency in real-life, which makes the results of the MOPDPTW great different from classical PDPTW. To obtain the optimization of MOPDPTW, our algorithm searches solution space with two layers. The inner layer searches possible solutions in global by three large neighborhood structure and the outer layer searches local optimal solution for converging to best solution by a small but effective local operator. In addition, we provide a packing scheme in search iteration to improve the optimization of solution for multiple-orders cases. Experiment results show that our proposed algorithm is able to retrieve high-quality solutions, better than 1D_LNS on the different milk-run instances.

References

1. Li, H., Lim, A.: A metaheuristic for the pickup and delivery problem with time windows. In: Proceedings 13th IEEE International Conference on Tools with Artificial Intelligence, ICTAI 2001, Dallas, TX, USA, pp. 160–167 (2001). https://doi.org/10.1109/ICTAI.2001.974461
2. Männel, D., Bortfeldt, A.: A hybrid algorithm for the vehicle routing problem with pickup and delivery and three-dimensional loading constraints. Eur. J. Oper. Res. **254**(3), 840–858 (2016)
3. Kong, J.-L., Jia, G.-Z., Gan, C.-Y.: A new mathematical model of vehicle routing problem based on milk-run. In: International Conference on Management Science & Engineering. IEEE (2013)
4. Miao, Z., Xu, K.L.: Modeling and simulation of lean supply chain with the consideration of delivery consolidation. Key Eng. Mater. **467–469**(467–469), 853–858 (2011)
5. Desrochers, M., Desrosiers, J., Solomon, M.: A new optimization algorithm for the vehicle routing problem with time windows. Oper. Res. **40**, 342–354 (1992)
6. Ropke, S., Pisinger, D.: An adaptive large neighborhood search heuristic for the pickup and delivery problem with time windows. Transp. Sci. **40**, 455–472 (2006)
7. Nguyen, T.H.D. Dao, T.M.: Novel approach to optimize milk-run delivery: a case study. In: 2015 IEEE International Conference on Industrial Engineering and Engineering Management (IEEM), Singapore, pp. 351–355 (2015). https://doi.org/10.1109/IEEM.2015.7385667
8. Nagata, Y., Kobayashi, S.: Guided ejection search for the pickup and delivery problem with time windows. In: Cowling, P., Merz, P. (eds.) EvoCOP 2010. LNCS, vol. 6022, pp. 202–213. Springer, Heidelberg (2010). https://doi.org/10.1007/978-3-642-12139-5_18
9. Huang, M., Yang, J., et al.: The modeling of milk-run vehicle routing problem based on improved C-W algorithm that joined time window. Transp. Res. Proc. **25**, 716–728 (2017)
10. Ma, H.J., Wei, J.: Milk-run vehicle routing optimization model and algorithm of automobile parts. Appl. Mech. Mater. 1463–1467 (2013)
11. Berbeglia, G., Cordeau, J.F., Gribkovskaia, I., et al.: Static pickup and delivery problems: aclassification scheme and survey. TOP **15**(1), 1–31 (2007)
12. Montero, A., Jose Miranda-Bront, J., Mendez-Diaz, I.: An ILP-based local search procedure for the VRP with pickups and deliveries. Ann. Oper. Res. **259**(1–2), 327–350 (2017)
13. Sintef: Li and Lim benchmark. https://www.sintef.no/projectweb/top/pdptw/li-lim-benchmark/100-customers/
14. Alnahhal, M., Ridwan, A., Noche, B.: In-plant milk run decision problems. In: International Conference on Logistics & Operations Management. IEEE (2014)
15. Gyulai, D., Pfeiffer, A., Sobottka, T., et al.: Milkrun vehicle routing approach for shop-floor logistics. Proc. Cirp **7**, 127–132 (2013)

16. Urru, A., Bonini, M., Echelmeyer, W.: Planning of a milk-run systems in high constrained industrial scenarios. In: 2018 IEEE 22nd International Conference on Intelligent Engineering Systems (INES). IEEE (2018)
17. Nalepa, J., Blocho, M.: A parallel memetic algorithm for the pickup and delivery problem with time windows. In: Euromicro International Conference on Parallel. IEEE (2017)
18. Alaia, E.B., Dridi, I.H., Bouchriha, H., et al.: Optimization of the multi-depot & multi-vehicle pickup and delivery problem with time windows using genetic algorithm. In: International Conference on Control. IEEE (2013)
19. Huang, Y., Ting, C.: Ant Colony optimization for the single vehicle pickup and delivery problem with time window. In: International Conference on Technologies & Applications of Artificial Intelligence. IEEE Computer Society (2010)
20. Li, L., Yaohua, W., Hongchun, H., et al.: A hybrid intelligent algorithm for vehicle pick-up and delivery problem with time windows. In: Control Conference. IEEE (2007)
21. Bent, R., Hentenryck, P.V.: A two-stage hybrid algorithm for pickup and delivery vehicle routing problems with time windows. Comput. Oper. Res. **33**(4), 875–893 (2006)
22. Solomon, M.M.: Algorithms for the vehicle routing and scheduling problems with time window constraints. Oper. Res. **35**(2), 254–265 (1987)

Optimal Deterministic Group Testing Algorithms to Estimate the Number of Defectives

Nader H. Bshouty and Catherine A. Haddad-Zaknoon$^{(\boxtimes)}$

Technion - Israel Institute of Technology, Haifa, Israel
{bshouty,catherine}@cs.technion.ac.il

Abstract. We study the problem of estimating the number of defective items d within a pile of n elements up to a multiplicative factor of $\Delta > 1$, using deterministic group testing algorithms. We bring lower and upper bounds on the number of tests required in both the adaptive and the non-adaptive deterministic settings given an upper bound D on the defectives number. For the adaptive deterministic settings, our results show that, any algorithm for estimating the defectives number up to a multiplicative factor of Δ must make at least $\Omega\left((D/\Delta^2)\log(n/D)\right)$ tests. This extends the same lower bound achieved in [1] for non-adaptive algorithms. Moreover, we give a polynomial time adaptive algorithm that shows that our bound is tight up to a small additive term.

For non-adaptive algorithms, an upper bound of $O((D/\Delta^2)$ $(\log(n/D) + \log\Delta))$ is achieved by means of non-constructive proof. This improves the lower bound $\Omega((\log D)/(\log \Delta))D\log n)$ from [1] and matches the lower bound up to a small additive term.

In addition, we study polynomial time constructive algorithms. We use existing polynomial time constructible *expander regular bipartite graphs*, *extractors* and *condensers* to construct two polynomial time algorithms. The first algorithm makes $O((D^{1+o(1)}/\Delta^2) \cdot \log n)$ tests, and the second makes $(D/\Delta^2) \cdot Quazipoly (\log n)$ tests. This is the first explicit construction with an almost optimal test complexity.

Keywords: Group testing · Pooling design · Deterministic group testing

1 Introduction

The problem of *group testing* is the problem of identifying or, in some cases, examining the properties of a small amount of items known as *defective* items within a pile of elements using *group tests*. Let X be a set of n items, and let $I \subseteq X$ be the set of defective items. A *group test* is a subset $Q \subseteq X$ of items. The result of the test Q with respect to I is defined by $Q(I) := 1$ if $Q \cap I \neq \emptyset$ and $Q(I) := 0$ otherwise. While the defective set I is unknown to the algorithm, in many cases we might be interested in finding the size of the defective set $|I|$, or at least an estimation of that value with a minimum number of tests.

© Springer Nature Switzerland AG 2020
W. Wu and Z. Zhang (Eds.): COCOA 2020, LNCS 12577, pp. 393–410, 2020.
https://doi.org/10.1007/978-3-030-64843-5_27

Group testing was originally proposed as a potential solution for economising mass blood testing during WWII [11]. Since then, group testing approach has been diversely applied in a wide area of practical applications including DNA library screening [20], product testing quality control [22], file searching in storage systems [17], sequential screening of experimental variables [18], efficient contention algorithms for MAC [17, 26], data compression [16], and computations in data stream model [8]. Recently, during the COVID-19 pandemic outbreak, a number of researches adopted the group testing paradigm not only to accelerate mass testing process, but also to dramatically reduce the number of kits required for testing due to severe shortages in the testing kits supply [14, 19, 27].

While an up-front knowledge of the value of d or at least an upper bound on it is required in many of the algorithms aimed at identifying the defective items, estimating or finding the number of defectives is an interesting problem on its own as well. Defectives estimation via group testing has been applied vastly in biological and medical applications [7, 13, 23–25]. In [24], for example, group testing algorithms are used to estimate aster-yellow virus transmitters proportion over the organisms in a natural population of leafhoppers. Similarly, in [25], the authors estimate the infection rate of the yellow-fever virus in mosquito population using group testing methods. On the other hand, in [13], group-testing-based estimation of rare diseases prevalence is employed not only for its effectiveness but also because it naturally preserves individual anonymity of the subjects.

Algorithms dedicated for this task might operate in *stages* or *rounds*. In each round, the tests are defined in advance and tested in a single parallel step. Tests on some round might depend on the test results of the preceding rounds. A single round algorithm is called *non-adaptive* algorithm, while a multi-round algorithm is called *adaptive algorithm*.

In recent years, there has been an increasing interest in the problem of estimating the number of defective items via group testing [2, 3, 5, 7, 9, 10, 12, 21]. The target in some of these papers is to find an estimation $\hat{d}$ within an additive factor of $\epsilon < 1$ such that $(1 - \epsilon)d \leq \hat{d} \leq (1 + \epsilon)d$. For randomized adaptive algorithms we have the following results. Falhatgar et.al. [12] give a randomised adaptive algorithm that estimates d using $2 \log \log d + O(1/\epsilon^2 \log 1/\delta)$ queries in expectation where δ is the failure probability of the algorithm. Bshouty et. al. [3] modified this result and gave an algorithm that uses $(1 - \delta) \log \log d + O((1/\epsilon^2) \log 1/\delta)$ expected number of queries. Moreover, they proved a lower bound of $(1 - \delta) \log \log d + \Omega((1/\epsilon) \log(1/\delta))$ queries.

For randomized non-adaptive algorithms with constant estimation, Damaschke and Sheikh Muhammad give in [10] a randomized non-adaptive algorithm that makes $O((\log(1/\delta)) \log n)$ tests and in [2], Bshouty gives the lower bound $\Omega(\log n / \log \log n)$.

In this paper, we are interested in *deterministic* adaptive and non-adaptive algorithms that estimate the defective items set size d up to a multiplicative factor of $\Delta > 1$. Formally, let $|I| := d$ and let $D \geq d$. We say that a deterministic algorithm $\mathcal{A}$ estimates d up to a multiplicative factor of Δ if, given D as an

Table 1. Upper and lower bounds on the number of tests required for estimating defectives in deterministic group testing.

Bounds	Adaptive/Non-adapt.	Result	Explicit/Non-expl.	Ref.
Lower B.	Non-Adapt.	$\frac{D}{\Delta^2}\log\frac{n}{D}$	–	[1]
Lower B.	Adaptive	$\frac{D}{\Delta^2}\log\frac{n}{D}$	–	Ours
Upper B.	Adaptive	$\frac{D}{\Delta^2}\left(\log\frac{n}{D}+\log\Delta\right)$	Explicit	Ours
Upper B.	Non-Adapt.	$\frac{\log D}{\log\Delta}D\log n$	Non-Expl.	[1]
Upper B.	Non-Adapt.	$\frac{D}{\Delta^2}\left(\log\frac{n}{D}+\log\Delta\right)$	Non-expl.	Ours
Upper B.	Non-Adapt.	$\frac{D^{1+o(1)}}{\Delta^2}\log n$	Explicit[a]	Ours
Upper B.	Non-Adapt.	$\frac{D}{\Delta^2}\cdot$ Quazipoly$(\log n)$	Explicit	Ours

[a]This result is true for $\Delta > C$ for some constant C. See Sect. 6.2.

input to the algorithm, it evaluates an estimation $\hat{d}$ such that $d/\Delta \leq \hat{d} \leq d\Delta$. Bshouty et al. show in [3] that, if no upper bound D is given to the algorithm, then any deterministic adaptive algorithm (and therefore also non-deterministic algorithm) for this problem must make at least $\Omega(n)$ tests. This is equivalent to testing all the items. This justifies the fact that any non-trivial efficient algorithm must have some upper bound D for d.

Agarwal et.al. [1] consider this problem. They first give the lower bound of $\Omega((D/\Delta^2)\log(n/D))$ queries for any non-adaptive deterministic algorithm. Moreover, using a non-constructive proof, they give an upper bound of $O(((\log D)/(\log\Delta))D\log n)$ queries.

We further investigate this problem. We bring new lower and upper bounds on the number of tests required both in adaptive and non-adaptive deterministic algorithms. For the adaptive deterministic settings, our results show that, any algorithm for estimating the defectives number up to a multiplicative factor of Δ must make at least $\Omega\left((D/\Delta^2)\log(n/D)\right)$ tests. This extends the same lower bound achieved in [1] for non-adaptive algorithms. Furthermore, we give a polynomial time adaptive algorithm that shows that our bound is tight up to a small additive term.

For non-adaptive algorithms, we achieve an upper bound of $O((D/\Delta^2)$ $(\log(n/D)+\log\Delta))$ by means of non-constructive proof. This improves the lower bound $O((\log D)/(\log\Delta))D\log n)$ from [1], and matches the lower bound up to a small additive term.

We then study polynomial time constructive algorithms. For this task, we use existing polynomial time constructible *expander regular bipartite graphs*, *extractors* and *condensers* to construct two polynomial time algorithms. The first algorithm makes $O((D^{1+o(1)}/\Delta^2)\cdot\log n)$ tests, and the second makes $(D/\Delta^2)\cdot quazipoly(\log n)$ tests. To the best of our knowledge, this is the first explicit construction with an almost optimal test complexity. Our results are summarised in Table 1.

2 Definitions and Preliminary Results

In this section, we give some notations and definition that will be used in this paper.

Let $X = [n] := \{1, \cdots, n\}$ be a set of *items*. Let $I \subseteq X$ be a set of *defective* items, and let d denote its size, i.e. $d = |I|$. In the group testing settings, a *test* is a subset $Q \subseteq X$ of items. An answer to a test Q with respect to the defective items set I, is denoted by $Q(I)$, such that $Q(I) := 1$ if $Q \cap I \neq \emptyset$ and 0 otherwise. We denote by $\mathcal{O}_I$ an *oracle* that for a test Q returns $Q(I)$.

Let $\mathcal{A}$ be an algorithm with an access to $\mathcal{O}_I$, and let $d = |I|$. We say that the algorithm $\mathcal{A}$ *estimates* d *up to a multiplicative factor of* Δ, if $\mathcal{A}$ gets as an input an upper bound $D \geq d$ and a parameter $\Delta > 1$, and outputs $\hat{d}$ such that $d/\Delta \leq \hat{d} \leq d\Delta$. We say that $\mathcal{A}$ is an *adaptive* algorithm, if its queries depend on the result of previous queries, and *non-adaptive* if its queries are independent of previous ones and therefore, can be executed in a single parallel step. We may assume that $D \geq \Delta^2$, otherwise, the algorithm trivially outputs $\hat{d} = D/\Delta$. We note here that $\Delta \geq 1 + \Omega(1)$, that is, it is greater than a constant that is greater than 1 and it may depend[1] on n and/or D. This is implicit in [1] and is also constrained in this paper. It is also interesting to investigate this problem when $\Delta = 1 + o(n)$ where $o()$ (small o) is with respect to D and/or n.

We will use the following

Lemma 1. Chernoff's Bound. *Let* $X_1, \ldots, X_m$ *be independent random variables taking values in* $\{0, 1\}$. *Let* $X = \sum_{i=1}^{m} X_i$ *denotes their sum and* $\mu = \mathbf{E}[X]$ *denotes the sum's expected value. Then*

$$\Pr[X > (1+\lambda)\mu] \leq \left(\frac{e^\lambda}{(1+\lambda)^{(1+\lambda)}}\right)^\mu \leq e^{-\frac{\lambda^2\mu}{2+\lambda}} \leq \begin{cases} e^{-\frac{\lambda^2\mu}{3}} & \text{if } 0 < \lambda \leq 1 \\ e^{-\frac{\lambda\mu}{3}} & \text{if } \lambda > 1 \end{cases} . (1)$$

In particular,

$$\Pr[X > \Lambda] \leq \left(\frac{e\mu}{\Lambda}\right)^\Lambda . \tag{2}$$

For $0 \leq \lambda \leq 1$ *we have*

$$\Pr[X < (1-\lambda)\mu] \leq \left(\frac{e^{-\lambda}}{(1-\lambda)^{(1-\lambda)}}\right)^\mu \leq e^{-\frac{\lambda^2\mu}{2}} . \tag{3}$$

Moreover, we will often use the inequality

$$\left(\frac{n}{k}\right)^k \leq \binom{n}{k} \leq \sum_{i=0}^{k} \binom{n}{i} \leq \left(\frac{en}{k}\right)^k . \tag{4}$$

---

[1] For example $\Delta = \log \log n + \log D$.

3   Upper Bound for Non-adaptive Deterministic Algorithms

In this section, we give the upper bound for deterministic non-adaptive algorithm that estimates d up to a multiplicative factor of Δ. We prove:

Theorem 1. *Let D be some upper bound on the number of defective items d and $\Delta > 1$. Then, there is a deterministic non-adaptive algorithm that makes*

$$O\left(\frac{D}{\Delta^2}\left(\log\frac{n}{D} + \log\Delta\right)\right)$$

tests and outputs $\hat{d}$ such that $\frac{d}{\Delta} \le \hat{d} \le d\Delta$.

To prove the Theorem we need the following:

Lemma 2. *Let $\Delta > 1$ and $\ell \ge 2\Delta^2$. There is a non-adaptive deterministic algorithm that makes*

$$t = O\left(\frac{\ell}{\Delta^2}\left(\log\frac{n}{\ell} + \log\Delta\right)\right)$$

tests such that,

1. If the number of defectives d is less than ℓ/Δ^2, it outputs 0.
2. If it is greater than ℓ/Δ, it outputs 1.

Proof. We choose a constant c such that $(1 - \Delta^2/(c\ell))^{\ell/\Delta^2} = 1/e$. Note that

$$\left(1 - \frac{\Delta^2}{2\ell}\right)^{\ell/\Delta^2} \ge 1 - \frac{\Delta^2\ell}{2\ell\Delta^2} = \frac{1}{2} > \frac{1}{e}$$

and

$$\left(1 - \frac{2\Delta^2}{\ell}\right)^{\ell/\Delta^2} = \left(\left(1 - \frac{2\Delta^2}{\ell}\right)^{\frac{\ell}{2\Delta^2}}\right)^2 \le \frac{1}{e^2} < \frac{1}{e}.$$

Therefore, such c exists and we have $1/2 \le c \le 2$.

Consider a test $Q \subseteq [n]$ chosen at random where each item $i \in [n]$ is chosen to be in Q with probability $\Delta^2/(c\ell)$. Let I be the set of defective items such that $|I| = d$, and let $Q(I)$ be the result of the test Q with respect to the set I. Then,

$$\Pr[Q(I) = 0] = \left(1 - \frac{\Delta^2}{c\ell}\right)^d. \tag{5}$$

If $d \le \ell/\Delta^2$,

$$\Pr[Q(I) = 0] \ge \left(1 - \frac{\Delta^2}{c\ell}\right)^{\ell/\Delta^2} = e^{-1}, \tag{6}$$

if $d = 2\ell/\Delta^2$,

$$\Pr[Q(I) = 0] = \left(1 - \frac{\Delta^2}{c\ell}\right)^{2\ell/\Delta^2} = e^{-2}, \tag{7}$$

and if $d = \ell/\Delta$, we get:

$$\Pr[Q(I) = 0] = \left(\left(1 - \frac{\Delta^2}{c\ell}\right)^{\frac{\ell}{\Delta^2}}\right)^{\Delta} = e^{-\Delta}. \tag{8}$$

Let $Q_1, Q_2, \ldots, Q_t$ be a sequence of t i.i.d tests such that

$$t = \frac{c'\ell}{(\Delta - 1)^2} \ln \frac{c''\Delta^2 n}{\ell}$$

where $c' = 54e^2$ and $c'' = 4e$.

Let

$$\eta = e^{-1}\left(\frac{1}{2} + \frac{1}{2\Delta}\right).$$

Consider the following two events:

1. A: There is a set of defectives I of size $|I| \leq \ell/\Delta^2$ such that the number of tests with 0 answer is less than ηt.
2. B: There is a set of defectives J of size $|J| > \ell/\Delta$ such that the number of tests with 0 answer is at least ηt.

Notice that, to prove the lemma it is enough to prove that $\Pr[A \vee B] < 1$. We will show that $\Pr[A], \Pr[B] < 1/2$.

Let $X_1, \ldots, X_t$ be random variables such that $X_i = 1$ if and only if $Q_i(I) = 0$. Let X be the number of tests that yield the result 0. Therefore, $X = \sum_{i=1}^{t} X_i$ and define $\mu := \mathbf{E}[X]$.

If $|I| = d \leq \ell/\Delta^2$, then $\mu = t \cdot \mathbf{E}[X_i] = t \cdot \Pr[X_i = 1]$. By (6) we have

$$\mu = \mathbf{E}[X] \geq t \cdot e^{-1}. \tag{9}$$

By (3) in Lemma 1, for $\lambda = 1/2 - 1/(2\Delta)$ we have

$$\Pr[X \leq \eta t] = \Pr[X \leq (1 - \lambda)te^{-1}] \leq \Pr[X \leq (1 - \lambda)\mu] \leq e^{-\frac{\lambda^2 \mu}{2}} \leq e^{-\frac{(1 - \Delta^{-1})^2 t}{8e}}.$$

Using this result, Eq. (4) and the union bound, we can conclude that

$$\Pr[A] \leq \left(\sum_{i=0}^{\ell/\Delta^2} \binom{n}{i}\right) e^{-\frac{(1 - \Delta^{-1})^2 t}{8e}} \leq \left(\frac{e\Delta^2 n}{\ell}\right)^{\frac{\ell}{\Delta^2}} e^{-\frac{(1 - \Delta^{-1})^2 t}{8e}}$$

$$= \left(\frac{e\Delta^2 n}{\ell}\right)^{\frac{\ell}{\Delta^2}} e^{-\frac{c'\ell}{8e\Delta^2} \ln \frac{c''\Delta^2 n}{\ell}} = \left(\frac{e\Delta^2 n}{\ell}\right)^{\frac{\ell}{\Delta^2}} \left(\frac{c''\Delta^2 n}{\ell}\right)^{-\frac{c'\ell}{\Delta^2}} < \frac{1}{2}.$$

On the other hand, for the event B, we have two cases.

Case I. $1 < \Delta \leq 2$.

If there is a set of defectives J of size $|J| > \ell/\Delta$ such that more than ηt of the tests yield the answer 0, then there is a set of defectives J' of size $|J'| = \ell/\Delta$ such that more than ηt of the tests answers are 0. Denote by B' the latter event. Then, by (8) we have $\mu = \mathbf{E}[X] = e^{-\Delta}t$ and for $\lambda = (e^{\Delta-1} - 1)/2 \geq (\Delta - 1)/2$, $\eta' = (e^{-1} + e^{-\Delta})/2 \leq \eta$ we get

$$\Pr[B] \leq \Pr[B'] \leq \binom{n}{\ell/\Delta} \Pr[X \geq \eta t] \leq \binom{n}{\ell/\Delta} \Pr[X \geq \eta' t]$$

$$= \binom{n}{\ell/\Delta} \Pr[X \geq (1+\lambda)\mu]$$

$$\leq \left(\frac{e\Delta n}{\ell}\right)^{\frac{\ell}{\Delta}} \Pr[X \geq (1+\lambda)\mu]$$

If $1 < \Delta \leq 2$ then $0 \leq \lambda \leq 1$ and then by (1) in Lemma 1, we have

$$\left(\frac{e\Delta n}{\ell}\right)^{\frac{\ell}{\Delta}} \Pr[X \geq (1+\lambda)\mu] \leq \left(\frac{e\Delta n}{\ell}\right)^{\frac{\ell}{\Delta}} e^{-\lambda^2\mu/3}$$

$$\leq \left(\frac{e\Delta n}{\ell}\right)^{\frac{\ell}{\Delta}} e^{-(\Delta-1)^2\mu/12} \quad \text{since } \lambda \geq (\Delta-1)/2$$

$$= \left(\frac{e\Delta n}{\ell}\right)^{\frac{\ell}{\Delta}} e^{-(\Delta-1)^2 e^{-\Delta}t/12}$$

$$= \left(\frac{e\Delta n}{\ell}\right)^{\frac{\ell}{\Delta}} \left(\frac{c''\Delta^2 n}{\ell}\right)^{-c'\ell e^{-\Delta}/12}$$

$$\leq \left(\frac{2en}{\ell}\right)^{\ell} \left(\frac{c''n}{\ell}\right)^{-(c'e^{-2}/12)\ell} < \frac{1}{2} \quad 1 \leq \Delta < 2$$

Case II. $\Delta > 2$.

In this case we have $\ell/\Delta > 2\ell/\Delta^2$. Therefore, if there is a set of defectives J of size $|J| > \ell/\Delta$ such that more than ηt of the tests yield the answer 0, then there is a set of defectives J' of size $|J'| = 2\ell/\Delta^2$ such that more than ηt of the tests answers are 0. Denote by B'' the latter event. By (7), $\mu = \mathbf{E}[X] = e^{-2}t$. Let $\lambda = 1/3 - 1/(3\Delta) < 1$. Then $\eta t > (1+\lambda)\mu$. By (1) in Lemma 1, we have

$$\Pr[X \geq \eta t] \leq \Pr[X \geq (1+\lambda)\mu] \leq e^{-\frac{\lambda^2\mu}{3}} \leq e^{-\frac{(1-\Delta^{-1})^2t}{27e^2}}.$$

Then

$$\Pr[A] \leq \Pr[B''] \leq \binom{n}{2\ell/\Delta^2} e^{-\frac{(1-\Delta^{-1})^2t}{27e^2}} \leq \left(\frac{e\Delta^2 n}{2\ell}\right)^{\frac{2\ell}{\Delta^2}} e^{-\frac{(1-\Delta^{-1})^2t}{27e^2}}$$

$$= \left(\frac{e\Delta^2 n}{2\ell}\right)^{\frac{2\ell}{\Delta^2}} e^{-\frac{c'\ell}{27e^2\Delta^2} \ln \frac{c''\Delta^2 n}{\ell}} = \left(\frac{e\Delta^2 n}{2\ell}\right)^{\frac{2\ell}{\Delta^2}} \left(\frac{c''\Delta^2 n}{\ell}\right)^{-\frac{c'\ell}{27e^2\Delta^2}} < \frac{1}{2}.$$

We are now ready to prove Theorem 1.

Let $\mathcal{A}(\ell, \Delta)$ be the algorithm from Lemma 2. Then, $\mathcal{A}(\ell, \Delta)$ makes at most

$$\frac{c\ell}{\Delta^2} \log \frac{\Delta n}{\ell} \tag{10}$$

queries for some constant c, and

1. If $\mathcal{A}(\ell, \Delta) = 1$, then $d \geq \frac{\ell}{\Delta^2}$.
2. If $\mathcal{A}(\ell, \Delta) = 0$, then $d \leq \frac{\ell}{\Delta}$.

Consider the algorithm $\mathcal{T}(n, D, \Delta)$ that runs $\mathcal{A}(D/\Delta^i, \Delta)$ for all $i = 0, \ldots, \lceil \log D / \log \Delta \rceil$. Let r be the minimum integer such that $\mathcal{A}(D/\Delta^r, \Delta) = 1$. Algorithm $\mathcal{T}(n, D, \Delta)$ then outputs $\hat{d} = D/\Delta^{r+1}$. See algorithm $\mathcal{T}$ in Fig. 1.

$\mathcal{T}(n, D, \Delta)$
1) $r \leftarrow 0$.
2) For each $i = 0, 1, \ldots, \lceil \log D / \log \Delta \rceil$ do:
    2.1) $R \leftarrow \mathcal{A}(D/\Delta^i, \Delta)$
    2.2) If $(R = 1)$ then
        $r \leftarrow i$
        $\hat{d} \leftarrow D/\Delta^{r+1}$
    Output $(\hat{d})$.

Fig. 1. Algorithm $\mathcal{T}$

We now prove:

Lemma 3. *Algorithm $\mathcal{T}(n, D, \Delta)$ is deterministic non-adaptive that makes*

$$O\left(\frac{D}{\Delta^2} \log\left(\frac{\Delta n}{D}\right)\right)$$

tests and outputs $\hat{d}$ that satisfies

$$\frac{d}{\Delta} \leq \hat{d} \leq \Delta d.$$

Proof. For $i = 0$, if $A(D/\Delta^i, \Delta) = 1$ then $d \geq D/\Delta^2$. Then $\hat{d} = D/\Delta \leq \Delta d$ and since $D \geq d$ we also have $\hat{d} = D/\Delta \geq d/\Delta$.

For $i > 0$, if $A(D/\Delta^{i-1}, \Delta) = 0$ and $A(D/\Delta^i, \Delta) = 1$ then $d \leq D/\Delta^i$ and $d \geq D/\Delta^{i+2}$. Then $\hat{d} = D/\Delta^{i+1} \leq \Delta d$ and $\hat{d} \geq d/\Delta$.

Let $q = \lceil \log D / \log \Delta \rceil$. Let t denote the number of queries performed by algorithm $\mathcal{T}(n, D, \Delta)$. By (10), the number of tests is at most

$$\sum_{i=0}^{q} \frac{cD}{\Delta^i \Delta^2} \log \frac{n\Delta^{i+1}}{D} \leq \frac{cD}{\Delta^2} \sum_{i=0}^{\infty} \frac{1}{\Delta^i} \log \frac{n\Delta^{i+1}}{D}$$

$$= \frac{cD}{\Delta^2} \left(\left(\log \frac{n}{D}\right) \sum_{i=0}^{\infty} \frac{1}{\Delta^i} + (\log \Delta) \sum_{i=0}^{\infty} \frac{i+1}{\Delta^i} \right)$$

$$\leq \frac{cD}{\Delta^2} \left(\frac{\Delta}{\Delta - 1} \log \frac{n}{D} + \frac{\Delta^2}{(\Delta - 1)^2} \log \Delta \right).$$

For the case when $\Delta = 1 + \Theta(1)$ we get

$$t = O\left(D \log \frac{n}{D}\right)$$

and for the case when $\Delta = \omega(1)$ we get

$$t = O\left(\frac{D}{\Delta^2} \left(\log \frac{n}{D} + \log \Delta\right)\right).$$

4 Lower Bound for Adaptive Deterministic Algorithm

In this section, we prove the following lower bound.

Theorem 2. *Any deterministic adaptive group testing algorithm that given $D > d$, outputs $\hat{d}$ that satisfies $d/\Delta \leq \hat{d} \leq \Delta d$ must make at least*

$$\Omega\left(\frac{D}{\Delta^2} \log \frac{n}{D}\right)$$

queries.

For the proof, we use the following from [3].

Lemma 4. *Let A be a deterministic adaptive algorithm that for a defective sets $I \subset [n]$ makes the tests $T_1^I, T_2^I \ldots, T_{w(I)}^I$ and let $s(I)$ be the sequence of answers to these tests. If $M = |\{s(I) | I \subseteq [n]\}|$ then the test complexity of A is $\max_I w(I) \geq \log M$.*

The following Lemma assists us to prove the result declared by Theorem 2.

Lemma 5. *Any deterministic adaptive algorithm such that, if the number of defectives d is less than or equal d_1 it outputs 0 and if it is greater than d_2 it outputs 1, must make*

$$\Omega\left(d_1 \log \frac{n}{d_2}\right)$$

tests.

In particular, when $d_1 = \ell/\Delta^2$ and $d_2 = \ell/\Delta$ we get

$$\Omega\left(\frac{\ell}{\Delta^2}\left(\log\frac{n}{\ell} + \log\Delta\right)\right)$$

tests.

Proof. Let A be such algorithm. Let $s(I)$ be the sequence of answers to the tests of A when the set of defective items is I. Consider a set I of size d_1 and let $\mathcal{J} = \{J \subseteq [n] : |J| = d_1, s(J) = s(I)\}$. Let $I' = \cup_{J \in \mathcal{J}} J$. We claim that $s(I') = s(I)$. Suppose for the contrary, $s(I') \neq s(I)$. Then, since $I \subseteq I'$, there is a test $Q \subseteq [n]$ that is asked by A that gives answer 0 to I and 1 to I'. Since $I' \cap Q \neq \emptyset$, there is a subset $J' \in \mathcal{J}$ such that $J' \cap Q \neq \emptyset$ and therefore Q gives answer 1 to J'. Then $s(J') \neq s(I)$ and we get a contradiction.

Since $s(I') = s(I)$ and algorithm A outputs 0 to I, it also outputs 0 to I'. Therefore, $|I'| \leq d_2$. Therefore $|\mathcal{J}| \leq N := \binom{d_2}{d_1}$. That is, for every possible sequence of answers s' of the algorithm A, there is at most N sets of size d_1 that get the same sequence of answers. Since there are $L := \binom{n}{d_1}$ such sets, the number of different sequences of answers that A might have must be at least L/N. By Lemma 4, the number of tests that the algorithm makes is at least

$$\log\frac{\binom{n}{d_1}}{\binom{d_2}{d_1}} \geq \log\left(\frac{n}{ed_2}\right)^{d_1} = \Omega\left(d_1 \log\frac{n}{d_2}\right).$$

The conclusions established by Lemma 5 show that the upper bound from Lemma 2 is tight. Moreover, using these results, we provide the following proof for Theorem 2.

Proof. Let $d_1 = D/\Delta^2 - 1$ and $d_2 = D$. For sets of size less than or equal d_1 the algorithm returns $d_1/\Delta \leq \hat{d} \leq \Delta d_1$ and for sets of equal to d_2 the algorithm returns $d_2/\Delta < \hat{d} \leq \Delta d_2$. Since $\Delta d_1 < d_2/\Delta$, the above intervals are disjoint. So, the algorithm can distinguish between sets of size less that or equal to d_1 and sets of size greater than d_2. By Lemma 5 the algorithm must make at least

$$\Omega\left(\frac{D}{\Delta^2}\log\frac{n}{D}\right)$$

tests.

5   Polynomial Time Adaptive Algorithm

In this section, we prove:

Theorem 3. *Let D be some upper bound on the number of defective items d and $\Delta > 1$. Then, there is a linear time deterministic adaptive algorithm that makes*

$$O\left(\frac{D}{\Delta^2}\left(\log\frac{n}{D} + \log\Delta\right)\right)$$

tests and outputs $\hat{d}$ such that $\frac{d}{\Delta} \leq \hat{d} \leq d\Delta$.

We first describe the algorithm. The algorithm gets as an input the set of items $X = [n]$ and splits it into two equally-sized disjoint sets Q_1 and Q_2. The algorithm asks the queries defined by Q_1 and Q_2 and proceeds in the splitting process on the sets that yielded positive answers only. We call these sets *defective sets*. As long as the algorithm gets less than D/Δ^2 distinct defective sets, it continues to split and test. Two cases can happen. Either it gets D/Δ^2 defective sets and then the algorithm outputs $\hat{d} = D/\Delta$, or the number of the defective sets is always less than D/Δ^2 and then, the algorithm finds all the defective items and returns their exact number. The algorithm is given in Fig. 2. The algorithm invokes the procedure **Split**(X) that on an input $X = \{a_1, a_2, \ldots, a_n\}$, it returns the set W where $W := \{X_1, X_2\}$ such that $X_i \subseteq X$, $X_1 = \{a_1, a_2, \ldots, a_{\lfloor n/2 \rfloor}\}$, $X_2 = \{a_{\lfloor n/2 \rfloor + 1}, \ldots, a_n\}$ if $|X| \geq 2$, and $W := \{X\}$ otherwise.

Adaptive-dEstimate $(\mathcal{O}_I, X, \Delta, D)$
1) $Q \leftarrow X, S \leftarrow \emptyset$
2) While $(|Q| \leq D/\Delta^2)$ do:
   2.1) For each $Q_i \in Q$
      $\left\{Q_i^{(1)}, Q_i^{(2)}\right\} \leftarrow$ **Split**(Q_i)
      If $(Q_i^{(1)}(I) = 1)$ then $S \leftarrow S \cup \{Q_i^{(1)}\}$
      If $(Q_i^{(2)}(I) = 1)$ then $S \leftarrow S \cup \{Q_i^{(2)}\}$
   2.2) If $\forall S_i \in S, |S_i| = 1$
      $\hat{d} \leftarrow |S|$
      Output $(\hat{d})$
   Else
      $Q \leftarrow S, S \leftarrow \emptyset$.
3) $\hat{d} \leftarrow |Q| \cdot \Delta$.
4) Output $(\hat{d})$

Fig. 2. Algorithm **Adaptive-dEstimate** to estimate the number of defective items.

Lemma 6. *Algorithm* **Adaptive-dEstimate** *is a deterministic adaptive algorithm that makes*

$$2\frac{D}{\Delta^2} \log \frac{n\Delta^2}{D} = O\left(\frac{D}{\Delta^2}\left(\log \frac{n}{D} + \log \Delta\right)\right)$$

tests and outputs an estimation $\hat{d}$ such that:

$$\frac{d}{\Delta} \leq \hat{d} \leq d\Delta.$$

Proof. If $d \leq \frac{D}{\Delta^2}$, then the splitting process in step 2 of the algorithm proceeds until each defective item belongs to a distinct set. Eventually, the condition in step 2.2 is met and the algorithm outputs the exact value of d. If $d > D/\Delta^2$, then the splitting process stops when the number of defective sets $|Q|$ exceeds D/Δ^2. The algorithm halts and outputs $\hat{d} = |Q|\Delta$. Obviously, $|Q| \le d$. Therefore, $\hat{d} = |Q|\Delta \le d\Delta$. Moreover, $|Q| > D/\Delta^2 \ge d/\Delta^2$ which implies that $\hat{d} \ge d/\Delta$.

The number of iterations cannot exceed $\log n$ iterations. In the first $\log(D/\Delta^2)$ iterations, in the worst case scenario, the algorithm splits its current set Q_i on each iteration into two sets $Q_i^{(1)}$ and $Q_i^{(2)}$ such that $Q_i^{(1)}(I) = Q_i^{(2)}(I) = 1$. Therefore, the number of tests that the algorithm asks over all the first $\log(D/\Delta^2)$ iterations is at most

$$\sum_{i=1}^{\log(D/\Delta^2)} 2^i \le 2\frac{D}{\Delta^2}.$$

Since $|Q| \le D/\Delta^2$, in the other $\log n - \log(D/\Delta^2)$ iterations, the algorithm makes at most $2D/\Delta^2$ tests each iteration. So, the total number of tests is at most

$$2\frac{D}{\Delta^2}\left(\log n - \log\frac{D}{\Delta^2}\right) + 2\frac{D}{\Delta^2} = O\left(\frac{D}{\Delta^2}\log\frac{n\Delta^2}{D}\right)$$

.

6   Polynomial Time Non-adaptive Algorithm

In this section, we show how to use expanders, condensers and extractors to construct deterministic non-adaptive algorithms for defectives number estimation. We prove:

Theorem 4. *Let D be some upper bound on the number of defective items d and $\Delta > 1$. Then, there is a polynomial time deterministic non-adaptive algorithm that makes*

$$\min\left(D^{o(1)}, 2^{O(\log^3(\log n))}\right) \cdot \frac{D}{\Delta^2}\log n$$

tests and outputs $\hat{d}$ such that $\frac{d}{\Delta} \le \hat{d} \le d\Delta$.

6.1   Algorithms Using Expanders

Let G be a bipartite graph $G = G(L, R, E)$ with left vertices $L = [n]$, right vertices $R = [m]$ and edges $E \subseteq L \times R$. For each edge $(i, j) \in E$, it holds that the endpoint $i \in L$ and $j \in R$. For a vertex $v \in L$, define $\Gamma(v)$ to be the set of the neighbours of v in G i.e. $\Gamma(v) := \{u \in R | (v, u) \in E\}$. For a subset $S \subseteq L$, we define $\Gamma(S)$ to be the set of neighbours of S, meaning $\Gamma(S) := \cup_{v \in S}\Gamma(v)$. For a vertex $v \in L$, the *degree* of v is defined as $deg(v) := |\Gamma(v)|$. We say that a bipartite graph $G = G(L, R, E)$ is a (k, a)-*expander δ-regular bipartite graph* if, the degree of every vertex in L is δ, and for every left-subset $S \subseteq L$ of size at most k, we have $|\Gamma(S)| \ge a|S|$.

Lemma 7. *Let $X = [n]$ be a set of items and $I \subseteq [n]$ is the set of defective items such that $|I| = d$ is unknown to the algorithm. Let $G = G(L, R, E)$ be a (k, a)-expander δ-regular bipartite graph with $|L| = n$ and $|R| = m$. Then, there is a deterministic non-adaptive algorithm A, such that for n items, it makes m tests and satisfies:*

1. *If $|I| < ak/\delta$, then A outputs 0.*
2. *If $|I| \geq k$, then A outputs 1.*

Proof. For every $j \in R$, we define the test $T^{(j)} = \{i | (i, j) \in E\}$. The number of tests is $|R| = m$. If $|I| \geq k$, then $|\Gamma(I)| \geq ak$. Therefore, at least ak tests will give positive answer 1. If $|I| < ak/\delta$, then, since the degree of every vertex in L is δ, we have $|\Gamma(I)| \leq \delta |I| < ak$. This shows that, for this case, at most $ak - 1$ tests give the answer 1. Hence, we can distinguish between the two cases.

Following the same proof of Lemma 3 with algorithm $\mathcal{T}$ in Fig. 1, we have:

Lemma 8. *Let $A(\ell, \Delta)$ be a deterministic non-adaptive algorithm such that, for n items, it makes $m(\ell, \Delta)$ tests and satisfies:*

1. *If $|I| < \ell/\Delta^2$, then A outputs 0.*
2. *If $|I| \geq \ell/\Delta$, then A outputs 1.*

Then, there is a deterministic non-adaptive algorithm $\mathcal{T}$ such that, given $D > d$, for n items it makes

$$\sum_{i=0}^{\lceil \log D / \log \Delta \rceil} m\left(\frac{D}{\Delta^i}, \Delta\right)$$

tests and outputs $\hat{d}$ that satisfies $d/\Delta \leq \hat{d} \leq \Delta d$.

The parameters of the explicit construction of a (k, a)-expander δ-regular bipartite graph from [4] are summarised in the following lemma.

Lemma 9. *For any $k > 0$ and $0 < \epsilon < 1$, there is an explicit construction of a (k, a)-expander δ-regular bipartite graph with*

$$m = O(k\delta/\epsilon), \quad \delta = 2^{O(\log^3(\log n/\epsilon))}, \quad a = (1 - \epsilon)\delta.$$

We now prove:

Lemma 10. *There is a polynomial time deterministic non-adaptive algorithm that makes*

$$\frac{D}{\Delta^2} \cdot 2^{O(\log^3(\log n))} = \frac{D}{\Delta^2} \cdot \text{quasipoly}(\log n)$$

tests and outputs $\hat{d}$ that satisfies

$$\frac{d}{\Delta} \leq \hat{d} \leq \Delta d.$$

Proof. We use the expander in Lemma 9. Recall that $\Delta = 1 + \Omega(1)$. Let $r = \min(\Delta, 2)$, $\epsilon = 1 - 1/r$ and $k = r\ell/\Delta^2$. Then $a = \delta/r = 2^{O(\log^3 \log n)}$ and $m = m(\ell, \Delta) = (\ell/\Delta^2)2^{O(\log^3 \log n)}$. By Lemma 7, there is a deterministic non-adaptive algorithm A such that for n items, it makes $m(\ell, \Delta)$ tests and

1. If $|I| < ak/\delta = \ell/\Delta^2$ then A outputs 0.
2. If $|I| \geq k = r\ell/\Delta^2$ then A outputs 1.

Algorithm A trivially satisfies the first condition required by Lemma 8. Consider item 2. If $\Delta < 2$ then $r = \Delta$ and then if $|I| \geq k = \ell/\Delta$ then A outputs 1. If $\Delta > 2$ then $r = 2$ and then if $|I| \geq k = 2\ell/\Delta^2$ then A outputs 1. Since $2\ell/\Delta^2 < \ell/\Delta$, if $|I| \geq \ell/\Delta$ then A outputs 1.

Now by Lemma 8, there is a deterministic non-adaptive algorithm $\mathcal{T}$ such that, given $D > d$, for n items, it makes

$$\sum_{i=0}^{\lceil \log D / \log \Delta \rceil} m\left(\frac{D}{\Delta^i}, \Delta\right) = \frac{D}{\Delta^2} \cdot 2^{O(\log^3 (\log n))}$$

tests and outputs $\hat{d}$ that satisfies $d/\Delta \leq \hat{d} \leq \Delta d$.

6.2 Algorithms Using Extractors and Condensers

Extractors are functions that convert weak random sources into almost-perfect random sources. We use these objects to construct a non-adaptive algorithm for estimating d. We start with some definitions.

Definition 1. *Let X be a random variable over a finite set S. We say that X has min-entropy at least k if $Pr[X = x] \leq 2^{-k}$ for all $x \in S$.*

Definition 2. *Let X and Y be random variables over a finite set S. We say that X and Y are $\epsilon - close$ if $\max_{P \subseteq S} |\Pr[X \in P] - \Pr[Y \in P]| \leq \epsilon$.*

We denote by U_ℓ the uniform distribution on $\{0, 1\}^\ell$. The notations $\Pr_{x \in B}$ or $\mathbf{E}_{x \in B}$ stand for the fact that the probability and the expectation are taken when x is chosen randomly uniformly from B.

Definition 3. *A function $F : \{0, 1\}^{\hat{n}} \times \{0, 1\}^{\hat{t}} \to \{0, 1\}^{\hat{m}}$ is a $k \to_\epsilon k'$ condenser if for every X with min-entropy at least k and Y uniformly distributed on $\{0, 1\}^{\hat{t}}$, the distribution of $(Y, F(X, Y))$ is ϵ-close to a distribution $(U_{\hat{t}}, Z)$ with min-entropy $\hat{t} + k'$. A condenser is called (k, ϵ)-lossless condenser if $k' = k$. A condenser is called (k, ϵ)-extractor if $\hat{m} = k'$.*

Let $\hat{N} = \{0, 1\}^{\hat{n}}, \hat{T} = \{0, 1\}^{\hat{t}}$ and $\hat{M} = \{0, 1\}^{\hat{m}}$, and let $F : \hat{N} \times \hat{T} \to \hat{M}$ be a $k \to_\epsilon k'$ condenser. Consider the $2^{\hat{t}} \times 2^{\hat{n}}$ matrix $\mathcal{M}$ induced by F. That is, for $r \in \hat{T}$ and $s \in \hat{N}$, the entry $\mathcal{M}_{r,s}$ is equal to $F(s, r)$. For $s \in \hat{N}$, let $\mathcal{M}^{(s)}$ be the sth column of $\mathcal{M}$. Then, $\mathcal{M}_r^{(s)} = \mathcal{M}_{r,s} = F(s, r)$.

Definition 4. *Let Σ be a finite set. An n-mixture over Σ is an n-tuple $\mathcal{S} := (S_1, \cdots, S_n)$ such that for all $i \in [n]$, $S_i \subseteq \Sigma$.*

Using these definitions and notations, we restate the result proved by Cheraghchi [6] (Theorem 9) in the following lemma.

Lemma 11. *Let $F : \{0,1\}^{\hat{n}} \times \{0,1\}^{\hat{t}} \to \{0,1\}^{\hat{m}}$ be a $k \to_\epsilon k'$ condenser. Let $\mathcal{M}$ be the matrix induced by F. Then, for any $2^{\hat{t}}$-mixture $\mathcal{S} = (S_1, \cdots, S_{2^{\hat{t}}})$ over $\hat{M} := \{0,1\}^{\hat{m}}$, the number of columns s in $\mathcal{M}$ that satisfies*

$$\Pr_{r \in \hat{T}}[\mathcal{M}_r^{(s)} \in S_r] > \frac{\mathbf{E}_{r \in \hat{T}}[|S_r|]}{2^{k'}} + \epsilon$$

is less than 2^k.

Equipped with Lemma 11, we prove:

Lemma 12. *If there is a $k \to_\epsilon k'$ condenser $F : \{0,1\}^{\hat{n}} \times \{0,1\}^{\hat{t}} \to \{0,1\}^{\hat{m}}$ then, there is a deterministic non-adaptive algorithm $\mathcal{A}$ for $n = 2^{\hat{n}}$ items that makes $m = 2^{\hat{t}+\hat{m}}$ tests and satisfies the following.*

1. *If the number of defectives is less than $(1 - \epsilon)2^{k'}$ then $\mathcal{A}$ outputs 0.*
2. *If the number of defectives is greater than or equal $2^k + 1$ then $\mathcal{A}$ outputs 1.*

Proof. Consider the matrix $\mathcal{M}$ induced by the condenser F as explained above. We define the test matrix $\mathcal{T}$ from $\mathcal{M}$ as follows. Let $x \in \{0,1\}^{\hat{m}}$. Define $e(x) \in \{0,1\}^{2^{\hat{m}}}$ such that $e(x)_y = 1$ if and only if $x = y$, where the bits in $e(x)$ are indexed by the elements of $\{0,1\}^{2^{\hat{m}}}$. Each row r in the matrix $\mathcal{M}$ is replaced by $2^{\hat{m}}$ rows (in $\mathcal{T}$) such that in each entry $\mathcal{M}_{r,s} \in \{0,1\}^{\hat{m}}$ is replaced by the column vector $e(\mathcal{M}_{r,s})^T \in \{0,1\}^{2^{\hat{m}}}$. The rows of the matrix $\mathcal{T}$ are indexed by $\hat{T} \times \hat{M}$. Let $\mathcal{T}^{(i)}$ denote the ith column of $\mathcal{T}$. Therefore, for $r \in \hat{T}$ and $j \in \hat{M}$, the row (r, j) in the matrix $\mathcal{T}$ is denoted by $\mathcal{T}_{(r,j)}$. Moreover, the ith entry of the row $\mathcal{T}_{(r,j)}$ is denoted by $\mathcal{T}_{(r,j),i}$ and $\mathcal{T}_{(r,j),i} = \mathcal{T}_{(r,j)}^{(i)} = 1$ if and only if $\mathcal{M}_{r,i} = j$. The size of the test matrix $\mathcal{T}$ is $m \times n$.

Let the defective elements be $s_{i_1}, \ldots, s_{i_\ell}$ and let $y \in \{0,1\}^m$ indicate the tests result. Then, y is equal to $\mathcal{T}^{(s_{i_1})} \vee \cdots \vee \mathcal{T}^{(s_{i_\ell})}$. Let $\mathcal{S} = (S_r)_{r \in \hat{T}}$ be a $2^{\hat{t}}$-mixture over $\{0,1\}^{\hat{m}}$ where for all $r \in \hat{T}$, $S_r = \{j \in \{0,1\}^{\hat{m}} | y_{(r,j)} = \mathcal{T}_{(r,j)}^{(s_{i_1})} \vee \cdots \vee \mathcal{T}_{(r,j)}^{(s_{i_\ell})} = 1\}$. It is easy to see that:

1. $|S_r| \le \ell$. This is because, by the definition of S_r, $j \in S_r$ if and only if $y_{(r,j)} = 1$. The entry $y_{(r,j)}$ gets the value 1 if at least one of the entries $\mathcal{T}_{(r,j)}^{(s_{i_1})}, \cdots, \mathcal{T}_{(r,j)}^{(s_{i_\ell})}$ is 1. Any row in $\mathcal{T}^{(s_{i_1})}, \cdots, \mathcal{T}^{(s_{i_\ell})}$ has exactly one entry that is equal to 1 in all the $2^{\hat{m}}$ rows indexed by r. Hence, each row can cause one item to be inserted to S_r.
2. For any $s_{i_j} \in \{s_{i_1}, \ldots, s_{i_\ell}\}$, we have $\Pr_{r \in \hat{T}}[\mathcal{M}_r^{(s_{i_j})} \in S_r] = 1$

.3. Given the matrix $\mathcal{M}$, its test matrix $\mathcal{T}$ and the observed result y, for any column s the probability $\Pr_{r\in\hat{T}}[\mathcal{M}_r^{(s)} \in S_r]$ can be easily computed.

If the number of defectives is less than $(1-\epsilon)2^{k'}$ then, by Lemma 11, all columns, except for at most 2^k columns, satisfy

$$\Pr_{r\in\hat{T}}[\mathcal{M}_r^{(s)} \in S_r] \leq \frac{\mathbf{E}_{r\in\hat{T}}[|S_r|]}{2^{k'}} + \epsilon < \frac{\mathbf{E}_{r\in\hat{T}}[(1-\epsilon)2^{k'}]}{2^{k'}} + \epsilon = 1.$$

So for less than $2^k + 1$ columns we have $\Pr_{r\in\hat{T}}[\mathcal{M}_r^{(s)} \in S_r] = 1$. If the number of defectives is greater than or equal 2^k+1, then for the columns of the defectives we have $\Pr_{r\in\hat{T}}[\mathcal{M}_r^{(s)} \in S_r] = 1$. So for more than 2^k columns we have $\Pr_{r\in\hat{T}}[\mathcal{M}_r^{(s)} \in S_r] = 1$.

The following Lemma summarises the state of the art result due to Guruswami et. al. [15] on explicit construction of expanders.

Lemma 13. *For all positive integers $\hat{n}, k$ such that $\hat{n} \geq k$, and all $\epsilon > 0$, there is an explicit (k, ϵ) extractor $F : \{0,1\}^{\hat{n}} \times \{0,1\}^{\hat{t}} \to \{0,1\}^{\hat{m}}$ with $\hat{t} = \log \hat{n} + O(\log k \log (k/\epsilon))$ and $\hat{m} = k' = k - 2\log 1/\epsilon - c$ for some constant c.*

We now prove:

Lemma 14. *There is a constant C such that for every $\Delta > C$, there is a polynomial deterministic non-adaptive algorithm that estimates the number of defective items in a set of n items up to a multiplicative factor of Δ and asks*

$$O\left(\frac{D^{1+o(1)}}{\Delta^2} \log n\right)$$

queries.

Proof. We use the notations from Lemma 13. Let $C = 27 \cdot 2^{c-2}$. We choose $\epsilon = 2/3$ and k' such that $(1-\epsilon)2^{k'} = \ell/\Delta^2$. Then

$$2^k = 2^{k'+2\log(1/\epsilon)+c} = 27 \cdot 2^{c-2}\frac{\ell}{\Delta^2} < \frac{\ell}{\Delta}$$

By Lemma 12, there is a deterministic non-adaptive algorithm $\mathcal{A}$ for $n = 2^{\hat{n}}$ items that makes

$$m = 2^{\hat{t}+\hat{m}} = \hat{n}2^{O(\log k \log(k/\epsilon))}\frac{\ell}{(1-\epsilon)\Delta^2} = 2^{\log^2 \log(\ell/\Delta)}\frac{\ell}{\Delta^2}\log n$$

tests that satisfies the following:

1. If the number of defectives is less than $(1-\epsilon)2^{k'} = \ell/\Delta^2$ then $\mathcal{A}$ outputs 0.
2. If the number of defectives is greater than or equal $2^k + 1$ then $\mathcal{A}$ outputs 1 and, since $2^k < \ell/\Delta$, in particular, if the number of defectives is greater than or equal ℓ/Δ then $\mathcal{A}$ outputs 1.

By Lemma 8, the result follows.

A similar work by Capalbo et al. [4] gives an explicit construction of a lossless condenser is summarised in the following lemma:

Lemma 15. *For all positive integers $\hat{n}, k$ and all $\epsilon > 0$, there is an explicit lossless condenser $F : \{0,1\}^{\hat{n}} \times \{0,1\}^{\hat{t}} \to \{0,1\}^{\hat{m}}$ with $\hat{t} = O(\log^3(\hat{n}/\epsilon))$ and $\hat{m} = k + \log(1/\epsilon) + O(1)$.*

The construction from Lemma 15 yields a result that is similar to the one established in Lemma 14.

References

1. Agarwal, A., Flodin, L., Mazumdar, A.: Estimation of sparsity via simple measurements. In: 2017 IEEE International Symposium on Information Theory (ISIT), pp. 456–460. IEEE (2017)
2. Bshouty, N.H.: Lower bound for non-adaptive estimation of the number of defective items. In: 30th International Symposium on Algorithms and Computation, ISAAC 2019, 8–11 December 2019, Shanghai University of Finance and Economics, Shanghai, China, pp. 2:1–2:9 (2019). https://doi.org/10.4230/LIPIcs.ISAAC.2019.2
3. Bshouty, N.H., Bshouty-Hurani, V.E., Haddad, G., Hashem, T., Khoury, F., Sharafy, O.: Adaptive group testing algorithms to estimate the number of defectives. CoRR abs/1712.00615 (2017). http://arxiv.org/abs/1712.00615
4. Capalbo, M., Reingold, O., Vadhan, S., Wigderson, A.: Randomness conductors and constant-degree expansion beyond the degree/2 barrier, January 2002
5. Cheng, Y., Xu, Y.F.: An efficient FPRAS type group testing procedure to approximate the number of defectives. J. Combin. Optim. **27**, 302–314 (2014)
6. Cheraghchi, M.: Noise-resilient group testing: limitations and constructions. CoRR abs/0811.2609 (2008). http://arxiv.org/abs/0811.2609
7. Chen, C.L., Swallow, W.H.: Using group testing to estimate a proportion, and to test the binomial model. Biometrics **46**(4), 1035–1046 (1990)
8. Cormode, G., Muthukrishnan, S.: What's hot and what's not: tracking most frequent items dynamically. ACM Trans. Database Syst. **30**(1), 249–278 (2005). https://doi.org/10.1145/1061318.1061325
9. Damaschke, P., Muhammad, A.S.: Bounds for nonadaptive group tests to estimate the amount of defectives. In: Wu, W., Daescu, O. (eds.) COCOA 2010. LNCS, vol. 6509, pp. 117–130. Springer, Heidelberg (2010). https://doi.org/10.1007/978-3-642-17461-2_10
10. Damaschke, P., Sheikh Muhammad, A.: Competitive group testing and learning hidden vertex covers with minimum adaptivity. In: Kutyłowski, M., Charatonik, W., Gębala, M. (eds.) FCT 2009. LNCS, vol. 5699, pp. 84–95. Springer, Heidelberg (2009). https://doi.org/10.1007/978-3-642-03409-1_9
11. Dorfman, R.: The detection of defective members of large populations. Ann. Math. Stat. **14**(4), 436–440 (1943)
12. Falahatgar, M., Jafarpour, A., Orlitsky, A., Pichapati, V., Suresh, A.: Estimating the number of defectives with group testing, pp. 1376–1380, July 2016. https://doi.org/10.1109/ISIT.2016.7541524

13. Gastwirth, J.L., Hammick, P.A.: Estimation of the prevalence of a rare disease, preserving the anonymity of the subjects by group testing: application to estimating the prevalence of aids antibodies in blood donors. J. Stat. Plann. Inference **22**(1), 15–27 (1989). https://doi.org/10.1016/0378-3758(89)90061-X. http://www.sciencedirect.com/science/article/pii/037837588990061X
14. Gollier, C., Gossner, O.: Group testing against Covid-19. Covid Econ. 32–42 (2020)
15. Guruswami, V., Umans, C., Vadhan, S.: Unbalanced expanders and randomness extractors from Parvaresh-Vardy codes. In: Twenty-Second Annual IEEE Conference on Computational Complexity (CCC 2007), pp. 96–108 (2007)
16. Hong, E.S., Ladner, R.E.: Group testing for image compression. IEEE Trans. Image Process. **11**(8), 901–911 (2002)
17. Kautz, W., Singleton, R.: Nonrandom binary superimposed codes. IEEE Trans. Inf. Theory **10**(4), 363–377 (1964)
18. Li, C.: A sequential method for screening experimental variables. J. Am. Stat. Assoc. - J AMER STATIST ASSN **57**, 455–477 (1962). https://doi.org/10.1080/01621459.1962.10480672
19. Mentus, C., Romeo, M., DiPaola, C.: Analysis and applications of adaptive group testing methods for Covid-19. medRxiv (2020). https://doi.org/10.1101/2020.04.05.20050245. https://www.medrxiv.org/content/early/2020/04/16/2020.04.05.20050245
20. Ngo, H., Du, D.Z.: A survey on combinatorial group testing algorithms with applications to DNA library screening. DIMACS Ser. Discrete Math. Theor. Comput. Sci. **55**, 171–182 (2000). https://doi.org/10.1090/dimacs/055/13
21. Ron, D., Tsur, G.: The power of an example: hidden set size approximation using group queries and conditional sampling. CoRR abs/1404.5568 (2014). http://arxiv.org/abs/1404.5568
22. Sobel, M., Groll, P.A.: Group testing to eliminate efficiently all defectives in a binomial sample. Bell Syst. Tech. J. **38**(5), 1179–1252 (1959)
23. Swallow, W.H.: Group testing for estimating infection rates and probabilities of disease transmission. Phytopathology (USA) (1985)
24. Thompson, K.H.: Estimation of the proportion of vectors in a natural population of insects. Biometrics **18**(4), 568–578 (1962)
25. Walter, S.D., Hilderth, S.W., Beaty, B.J.: Estimation of infection rates in populations of organisms using pools of variable size. Am. J. Epidemiol. **112**(1), 124–128 (1980). https://doi.org/10.1093/oxfordjournals.aje.a112961
26. Wolf, J.: Born again group testing: multiaccess communications. IEEE Trans. Inf. Theory **31**(2), 185–191 (1985)
27. Yelin, I., et al.: Evaluation of Covid-19 RT-qPCR test in multi-sample pools. medRxiv (2020). https://doi.org/10.1101/2020.03.26.20039438. https://www.medrxiv.org/content/early/2020/03/27/2020.03.26.20039438

Nearly Complete Characterization of 2-Agent Deterministic Strategyproof Mechanisms for Single Facility Location in L_p Space

Jianan Lin[✉]

School of Computer Science, Fudan University, Shanghai 201203, China
jnlin16@fudan.edu.cn

Abstract. We consider the problem of locating a single facility for 2 agents in L_p space ($1 < p < \infty$) and give a nearly complete characterization of such deterministic strategyproof mechanisms. We use the distance between an agent and the facility in L_p space to denote the cost of the agent. A mechanism is strategyproof iff no agent can reduce her cost from misreporting her private location.

We show that in L_p space ($1 < p < \infty$) with 2 agents, any location output of a deterministic, unanimous, translation-invariant strategyproof mechanism must satisfy a set of equations and mechanisms are continuous, scalable. In one-dimensional space, the output must be one agent's location, which is easy to prove in any n agents.

However, in m-dimensional space ($m \geq 2$), the situation will be much more complex, with only 2-agent case finished. We show that the output of such a mechanism must satisfy a set of equations, and when $p = 2$ the output must locate at a sphere with the segment between the two agents as the diameter. Further more, for n-agent situations, we find that the simple extension of this the 2-agent situation cannot hold when dimension $m > 2$ and prove that the well-known general median mechanism will give an counter-example.

Particularly, in L_2 (i.e., Euclidean) space with 2 agents, such a mechanism is rotation-invariant iff it is dictatorial; and such a mechanism is anonymous iff it is one of the three mechanisms in Sect. 4. And our tool implies that any such a mechanism has a tight lower bound of 2-approximation for maximum cost in any multi-dimensional space.

Keywords: Facility location · Mechanism design · L_p space.

1 Introduction

We consider the problem of locating a single facility for n (mainly in $n = 2$) agents in L_p space ($1 < p < \infty$). This facility serves these agents and every

Thanks for my advisors Pinyan Lu and Hu Fu for giving me advise on this problem.

agent has a cost which is equal to the distance to access the facility. An agent's location is private information, i.e., only she herself knows it. A strategyproof mechanism means that no agent can gain (i.e., reduce her cost) from misreporting her location. A mechanism is deterministic if the output is a specific location. Compared to randomized mechanisms, deterministic mechanisms often receive more attention because of their simplicity and ease of use.

A basic area of facility location study is the characterization of truthful mechanisms. In many situations and settings, the goal is to design a strategyproof mechanism which can minimize the objective cost function (e.g., social cost or maximum cost) as far as possible. Therefore, giving the characterization of such mechanisms will be helpful to further study. In this area, an important work is made by Moulin [10] that in any one-dimension space (they call it single-peaked preferences), every strategyproof, efficient (the selected alternative is Pareto optimal, which is different from our setting) and anonymous voting scheme (mechanism) must be a median voter scheme (to select the median agent). After that, Border and Jordan [4] extend his result to Euclidean space and show that it induces to median voter schemes in each dimension separately. Other works include Barberà et al. [2] that the result also fits in any L_1 norm, and [3] that in a compact set of the Euclidean space, which is a more restricted domain, all those mechanisms behave like generalized median voter schemes. Nearly all the relevant works focus on deterministic mechanisms and leaves the randomized ones an open question.

As for other settings, Tang et al. [13] firstly discuss the characterization of group-strategyproof (No group of agents can reduce their cost together by misreporting their location) both in deterministic and randomized mechanisms (The former characterization is complete and the latter is nearly complete). And Feigenbaum et al. [5] discuss the characterization of 2-agent randomized strategyproof mechanism in one-dimensional space. However, before our work, there has not been any discussion of the characterization of deterministic strategyproof mechanisms in any metric space.

One measurement of the facility location mechanisms is the cost they achieve. There are two common view: maximum cost (i.e., the maximum cost between the facility and some agent) and social cost (i.e., the sum of the cost between the facility and all the agents). The ratio between the cost one mechanism achieve and the minimum cost is widely used in this study. Procaccia and Tennenholtz [11] study approximately optimal strategyproof mechanisms for facility games both in maximum cost and social cost view, focusing on one-dimensional space and randomized mechanisms. They propose an interesting randomized mechanism for the maximum cost view, which achieves a ratio of $3/2$ and they proves it to be the best. Subsequently, Alon et al. [1] study the characterization of deterministic and randomized mechanisms in more general metric space (such as network and rings).

Other related works about facility location are k-facility location problems. Compared to single facility location problems, they are more complex. Agents can have preference on different facilities and their location can be public information this time. One important and classical work is made by Fotakis and Tzamos [8]. They study mechanisms that are winner-imposing, in the sense that the mechanisms allocate facilities to agents and require that each agent allocated a facility should connect to it. Also they prove an upper bound of $4k$ in the social cost view. And there are many other follow-up works (see e.g., [6,7,9,12,14]).

Our work is motivated by [5]. Their result about the characterization of 2-agent randomized strategyproof mechanisms in a line (i.e., one-dimension) leads us studying the deterministic ones. And some of our technique is motivated by [13] (e.g., the proof of continuity). We show that in any one-dimensional L_p space with n agents, the output of a deterministic unanimous translation-invariant strategyproof mechanism should locate at one agent's location and in multi-dimensional L_p space ($1 < p < \infty$) with 2 agents, the output of such a mechanism should satisfy a set of equations. Particularly, in L_2 space (i.e., the Euclidean space), let the two agents be A, B and the output be W, then they should satisfy $\overrightarrow{WA} \cdot \overrightarrow{WB} = 0$. These characterizations are nearly complete and next we give complete characterization of two more specific situations (also restricted in 2 agents). The first one is that such a mechanism is rotation-invariant if and only if it is dictatorial (i.e., the output location is always the same agent). The second one is that such a mechanism is anonymous (i.e., all the permutations of agents does not affect the output) if and only if it is one of the three mechanisms we give in Sect. 4. In the end, we show that the general median mechanism is an counter-example of the simple extension from 2-agent situation to n-agent situation in m-dimensional space ($m > 2$), which means that the characterization of n-agent situation may be very complex, unfortunately. Also, using our tool, we ensure a tight lower bound of 2-approximation for maximum cost in any multi-dimensional space.

2 Preliminaries

We consider the single facility location game with n ($n \geq 2$) agents $N = \{1, 2, ..., n\}$. All the agents are located in a m-dimensional L_p space $\mathbb{R}_m$. Obviously for $\forall x, y \in \mathbb{R}_m$, there is $\|x\| + \|y\| \geq \|x + y\|$ and the equality holds if and only if x and y have the same directions. We use $A_i \in \mathbb{R}_m$ to denote agent i's location in the space. Therefore a location profile is a vector consisting of all the agents' locations $A = (A_1, A_2, ..., A_n)$.

A deterministic mechanism is a map $f : \mathbb{R}_m^n \to \mathbb{R}_m$ from a location profile to the location of the facility. We use $W = f(A)$ to denote the output location and the cost of agent i is the distance between her and the facility, i.e., $d(A_i, W) = \|W - A_i\|$. We will mix the two representations in this paper.

Next we formally define some properties of a deterministic mechanism.

Definition 1 (Strategyproofness). *A mechanism f is strategyproof if and only if no agent can reduce her distance to the output by misreporting her location. It means that, for $\forall A \in \mathbb{R}_m^n, \forall i \in N, \forall A_i' \in \mathbb{R}_m$, there is*

$$d(f(A), A_i) \leq d(f(A_i', A_{-i}), A_i).$$

Here $\boldsymbol{A}_{-i} = (A_1, ..., A_{i-1}, A_{i+1}, ..., A_n)$, *i.e., the profile without A_i.*

Definition 2 (Unanimity). *A mechanism f is unanimous if and only if when $\forall A_i = C$, we have*

$$f(\boldsymbol{A}) = C,$$

which means that if all agents report the same location, then the mechanism must output this location.

Definition 3 (Dictatorship). *A mechanism f is dictatorial if and only if $\exists i \in N, \forall A \in \mathbb{R}_m^n$, there is*

$$f(\boldsymbol{A}) = A_i.$$

At this time we call i is the dictator.

Definition 4 (Anonymity). *A mechanism f is anonymous if and only if when any group of the agents exchange their location reports, the output is still the same, which means that any permutation of the agents' locations does not affect the output.*

Definition 5 (Translational Invariance). *A mechanism f is translation-invariant if and only if*

$$\forall \boldsymbol{A} \in \mathbb{R}_m^n, \forall t \in \mathbb{R}_m, f(\boldsymbol{A} + t) = f(\boldsymbol{A}) + t.$$

Here, $f(\boldsymbol{A} + t) = f(A_1 + t, ..., A_n + t)$. This means that if we move all the agents the same distance in one direction, then the output location will also move the same distance in this direction.

Definition 6 (Scalability). *A mechanism f is scalable if and only if*

$$\forall \boldsymbol{A} \in \mathbb{R}_m^n, \forall k \in \mathbb{R}, k > 0, f(k \cdot \boldsymbol{A}) = k \cdot f(\boldsymbol{A}).$$

Here, $f(k \cdot \boldsymbol{A}) = f(k \cdot A_1, ..., k \cdot A_n)$.

Notice that if a mechanism f satisfies translational invariance and scalability, then we will have

$$\forall \boldsymbol{A} \in \mathbb{R}_m^n, \forall k \in \mathbb{R}, k > 0, \forall t \in \mathbb{R}_m, f(k \cdot \boldsymbol{A} + t) = k \cdot f(\boldsymbol{A}) + t.$$

Definition 7 (Rotational Invariance). *A mechanism f is rotation-invariant if and only if when all the agents are rotated at the same angle around a point (not necessary to be an agent) in the same direction in some dimensions, then the output will also be rotated at this angle around the point in such direction in these dimensions.*

For convenience, the properties of rotational invariance and anonymity are only described with natural languages. Notice that the description of rotational invariance includes situations that points are rotated on an axis and so on, because of "in some dimensions".

3   Nearly Complete Characterization of Deterministic Mechanisms

We will start with the situations in one-dimensional space as a warm-up and prove that this characterization is suitable for any n agents. Then we will discuss the multi-dimensional situations in L_p space with 2 agents for $1 < p < \infty$. The reason why we abandon L_1 and L_∞ is that in these two spaces, there exists two vectors x, y with different directions that $\|x\| + \|y\| = \|x + y\|$, which is not a friendly property. We use m to denote number of dimensions.

3.1   One-Dimensional Situation

The one-dimensional situation is simple. In any L_p space, $\forall a, b, c \in \mathbb{R}$, if $a \leq b \leq c$, then we have $\|a - b\| + \|b - c\| = \|a - c\|$. For convenience, we call the negative direction in the coordinate axis "left" and call the positive direction "right".

Lemma 1. (Continuity) *If mechanism f is strategyproof, then for $\forall i \in N$ with any fixed $\mathbf{A}_{-i} \in \mathbb{R}_m^{n-1}$, we have*

$$\|u(A_i) - u(A_i')\| \leq \|A_i - A_i'\|,$$

where $u(A_i) = \|f(A_i, \mathbf{A}_{-i}) - A_i\|$. This implies that $u(A_i)$ is a continuous function.

Proof. We assume that $\exists A_i, A_i'$ such that $\|u(A_i) - u(A_i')\| > \|A_i - A_i'\|$. Also without loss of generality, we assume that $u(A_i) > u(A_i')$ which means that $u(A_i) - u(A_i') > \|A_i - A_i'\|$, then we have

$$\|f(A_i', \mathbf{A}_{-i}) - A_i\| \leq \|f(A_i', \mathbf{A}_{-i}) - A_i'\| + \|A_i - A_i'\|$$
$$= u(A_i') + \|A_i - A_i'\|$$
$$< u(A_i) = \|f(A_i, \mathbf{A}_{-i}) - A_i\|.$$

Notice this is contradict with the strategyproofness, because A_i can misreport her location as A_i' to reduce her cost. Therefore the previous inequality in lemma must hold. When $A_i' \to A_i$, we see $u(A_i)$ is a continuous function.   □

This lemma is very useful and fits any m (dimensions) and n (agents).

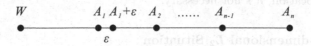

Fig. 1. Case 1 in the proof of Theorem 1

Theorem 1. *When* $m = 1$, *the output of a deterministic unanimous translation-invariant strategyproof mechanism must be one agent's location.*

Proof. Without loss of generality, we let n agents be $A_1, ..., A_n$ and assume $A_1 \leq A_2 \leq ... \leq A_n$. If all A_i are in the same location, according to unanimity, the output is A_i.

Let $W = f(A)$. We divide this into 3 different cases. We only need to prove that the output cannot locate at these three areas. Using proof by contradiction, we assume that there can be a situation that W does not locate at any agents.

Case 1, $W < A_1$: As is shown in Fig. 1, we can find a positive tiny ϵ that $\epsilon \ll d(W, A_1)$, e.g., $\epsilon < 0.01 \cdot d(W, A_1)$ and according to Lemma 1, we have $f(A_1 + \epsilon, A_{-1}) < A_1$, otherwise it will contradict with $\|u(A_1) - u(A_1 + \epsilon)\| \leq |A_1 - (A_1 + \epsilon)\| = \epsilon$.

Therefore we must have $f(A_1 + \epsilon, A_{-1}) \leq W$, otherwise agent with location A_1 gain from misreporting her location as $A_1 + \epsilon$. In the same way, we also must have $f(A_1 + \epsilon, A_{-1}) \geq W$, otherwise agent with location $A_1 + \epsilon$ can gain from misreporting her location as A_1 (fix other agents in A_{-1}). This means that $f(A_1 + \epsilon, A_{-1}) = f(A) = W$.

Let A^ϵ_{-i} denotes $(A_1 + \epsilon, ..., A_{i-1} + \epsilon, A_{i+1}, ..., A_n)$ (of course $1 \leq i \leq n$ and when $i = n$ we say it denotes $(A_1 + \epsilon, ..., A_{n-1} + \epsilon$, when $i = 1$ we say it denotes $(A_2, ..., A_n)$. In the same way, when i increases from 1 to n, we have $f(A_n + \epsilon, A^\epsilon_{-n}) = f(A_{n-1} + \epsilon, A^\epsilon_{-(n-1)}) = ... = f(A) = W$ and at last get $f(A + \epsilon) = W$. However, according to the translational-invariance, we must have $f(A) = W + \epsilon$, which leads to a contradiction. Therefore the output cannot satisfy $W < A_1$.

Case 2, $W > A_n$: This is completely symmetrical with the first case and we can use the same method by adding a tiny ϵ ($\epsilon \ll d(W, A_n)$) to all agents.

Case 3, $\exists i \in [1, n-1]$, $A_i < W < A_{i+1}$: We can still add all the agents a tiny $\epsilon \ll \min\{d(W, A_i), d(W, A_{i+1})\}$. When adding A_j with $j \leq i$, we can refer to case 1's proof and when adding A_j with $j > i$, we can refer to case 2's method.

Notice that if $A_i = A_{i+1}$, then there cannot be $A_i < W < A_{i+1}$, thus these 3 cases include all the areas except the locations of the agents. Therefore, any such strategyproof mechanism cannot output a location $W \neq A_i$ for any i. We prove this theorem. □

Therefore we can know that in one-dimensional space (including all the L_p space for any positive integer p), the output must be one agent's location. But in multi-dimensional space, the result is different. Although output still can be one agent's location, it's not necessary.

3.2 Multi-dimensional L_2 Situation

There are two reasons why we select L_2 space (i.e., Euclidean space). The first one is that it has some very friendly properties. For example, in an Euclidean space, any right triangle must satisfy that its hypotenuse is the (only) longest side. But in other space such as L_3 space, this rule may not hold. Here, the length of side is the distance between the two points in L_p space. And the second reason is that the Euclidean space is the most common and most used space. In this part, we will study the result in Euclidean space.

Lemma 2. *In any Euclidean space with n agents A_i $(i \in N)$, let the output of a deterministic unanimous translation-invariant strategyproof mechanism be W, then for $\forall A'_i$ on the segment between A_i and W (including A_i and W), we have*

$$f(A'_i, \mathbf{A}_{-i}) = W,$$

which means that if one agent move her location close to the output along the segment, the output does not change.

Proof. Obviously we only need to care the situation that $A'_i \neq A_i$. Considering the property of strategyproofness and let $W' = f(A'_i, \mathbf{A}_{-i})$, we have

$$\begin{cases} d(A_i, W) \leq d(A_i, W') \\ d(A'_i, W') \leq d(A'_i, W) \end{cases}$$

Draw the spheres (if $m = 2$ then circles and if $m > 3$ then m-spheres) O_1 and O_2 with A_i and A'_i as centers, $d(A_i, W)$ and $d(A'_i, W)$ as radius, respectively. The first inequality implies that W' cannot be inside of O_1 and the second implies that W' cannot be outside of O_2. Therefore, $W' = W$, which is the only common point between the two spheres (circles).   □

In fact, this lemma holds when $p > 2$, but at this time, what we draw is not 2 spheres any more, but 2 inscribed similar Enclosed ellipsoid on which the distance between a point and center is a constant in L_p space.

Lemma 3. *In any Euclidean space with 2 agents A and B, if the output of a deterministic unanimous translation-invariant strategyproof mechanism is on the line AB, then it can only locate at A or B.*

Proof. Similar to the proof of Theorem 1, we set the output W and divide it into three cases.

If $\exists k > 0$ that $\overrightarrow{WA} = k \cdot \overrightarrow{AB}$, then according to Lemma 2, we move A and then B towards W with a tiny distance ϵ. In this period, the output is still W, which is contradict with translational-invariance.

If $\exists k > 0$ that $\overrightarrow{WB} = k \cdot \overrightarrow{BA}$, then according to Lemma 2, we move B and then A towards W with a tiny distance ϵ. In this period, the output is still W, which is contradict with translational-invariance.

If $\exists k > 0$ that $\overrightarrow{AW} = k \cdot \overrightarrow{WB}$, then according to Lemma 2, we move A to W and have $f(W, B) = W$. Next we move B away from W to B' so that $d(B', W) = d(A, B)$. Therefore we notice that $f(W, B') \neq W + (W - A)$ because otherwise B can gain from misreporting B', and this is contradict with translational-invariance.

Therefore, the lemma is proved.   □

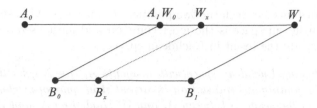

Fig. 2. Proof of Lemma 4

Lemma 4. *In any Euclidean space with 2 agents A and B, the output W can never satisfy that $0° < \angle AWB < 90°$.*

Proof. For convenience, we use term A_x and B_x ($x \in [0,1]$) and $W_x = f(A_x, B_x)$. Let $\angle A_0 W_0 B_0 \in (0°, 90°)$. Also let $\overrightarrow{B_0 B_1} = \overrightarrow{A_0 W_0}$ and $A_1 = W_0$. Therefore we have $\overrightarrow{W_0 W_1} = \overrightarrow{B_0 B_1}$. Here $\overrightarrow{A_0 A_x} = x \cdot \overrightarrow{A_0 A_1}$, $\overrightarrow{B_0 B_x} = x \cdot \overrightarrow{B_0 B_1}$ and $\overrightarrow{W_0 W_x} = x \cdot \overrightarrow{W_0 W_1}$. These are all drawn in Fig. 2.

According to Lemma 2, because $f(A_x, B_x) = W_x$, we can get $f(A_1, B_x) = W_x$. This means that if we fix A in A_1 and move B from B_0 to B_1 on a straight line, then W will move from W_0 to W_1 on a straight line. Because of strategyproofness, for $\forall x \in [0,1]$, we have $d(B_1, W_x) \geq d(B_1, W_1)$. However, since we know that $\angle A_1 W_1 B_1 = \angle A_0 W_0 B_0 < 90°$, then there must exist $x < 1$ so that $d(B_1, W_x) < d(B_1, W_1)$. Therefore the lemma is proved.   □

Lemma 5. *In any Euclidean space with 2 agents A and B, the output W can never satisfy that $90° < \angle AWB < 180°$.*

Proof. Similar to the proof of Lemma 4, we use term A_x and B_x ($x \in [0,1]$) and $W_x = f(A_x, B_x)$. Let $\angle A_0 W_0 B_0 \in (90°, 180°)$. Also let $\overrightarrow{B_0 B_1} = \overrightarrow{A_0 W_0}$ and $A_1 = W_0$. Therefore we have $\overrightarrow{W_0 W_1} = \overrightarrow{B_0 B_1}$. Here $\overrightarrow{A_0 A_x} = x \cdot \overrightarrow{A_0 A_1}$, $\overrightarrow{B_0 B_x} = x \cdot \overrightarrow{B_0 B_1}$ and $\overrightarrow{W_0 W_x} = x \cdot \overrightarrow{W_0 W_1}$.

According to Lemma 2, because $f(A_x, B_x) = W_x$, we can get $f(A_1, B_x) = W_x$. This means that if we fix A in A_1 and move B from B_0 to B_1 on a straight line, then W will move from W_0 to W_1 on a straight line. Because of strategyproofness, for $\forall x \in [0,1]$, we have $d(B_0, W_x) \geq d(B_0, W_0)$. However, since we know that $\angle W_1 A_1 B_0 < 90°$, then there must exist $x > 0$ so that $d(B_0, W_x) < d(B_0, W_0)$. Therefore the lemma is proved.   □

Theorem 2. *In any Euclidean space with 2 agents A and B, the output W of a deterministic unanimous translation-invariant strategyproof mechanism f must satisfy*

$$\overrightarrow{AW} \cdot \overrightarrow{BW} = 0,$$

which means that W must locate on a sphere with AB as the diameter.

According to Lemma 3, 4 and 5, Theorem 2 is obvious. Notice this theorem fits any dimension $m > 1$. Maybe intuitively we can guess that this can be extended to more n-agent situation, but we will show that there is an counter-example for any $m > 2$ in Sect. 5. Here is the conjecture that does not hold for $m > 2$.

Conjecture 1. In any Euclidean space with n agents $A_1, ..., A_n$, the output W of a deterministic unanimous translation-invariant strategyproof mechanism f must satisfy that

$$\exists i, j \in N, \overrightarrow{A_i W} \cdot \overrightarrow{A_j W} = 0.$$

3.3 Multi-dimensional L_p Situation

As is mentioned in the last part, we know that other L_p space has less friendly properties than L_2 space. Therefore the result is not a right angle any more because in other L_p space, a right triangle's hypotenuse may not be the longest side. We use analytical geometry to solve this problem. Obviously we only need to analyze the case that the output W does not locate at A or B.

According to translational-invariance, We can assume that $A(-x_1, ..., -x_m)$, $B(x_1, ..., x_m)$ and $W(y_1, ..., y_m)$, then the distance between two points such as A, W is $d_p(A, W) = \left(\sum_{i=1}^{m} |x_i + y_i|^p\right)^{1/p}$.

Theorem 3. *In any L_p space ($2 < p < \infty$) with 2 agents A and B, the output W of a deterministic unanimous translation-invariant strategyproof mechanism f must satisfy*

$$\begin{cases} \sum_{i=1}^{m} (x_i + y_i) \cdot (x_i - y_i) \cdot |x_i - y_i|^{p-2} = 0 \\ \sum_{i=1}^{m} (x_i - y_i) \cdot (x_i + y_i) \cdot |x_i + y_i|^{p-2} = 0 \end{cases}$$

Proof. Consider such a situation. In a L_p space, let $f(A_0, B_0) = W_0$, $A_1 = W_0$ and $\overrightarrow{B_0 B_1} = \overrightarrow{A_0 A_1}$, therefore we have $\overrightarrow{W_0 W_1} = \overrightarrow{B_0 B_1}$. We still use term A_x, B_x, W_x which mean $\overrightarrow{A_0 A_x} = \overrightarrow{A_0 A_1}$, $\overrightarrow{B_0 B_x} = \overrightarrow{B_0 B_1}$ and $\overrightarrow{W_0 W_x} = \overrightarrow{W_0 W_1}$ respectively.

Similar to the proof of Lemma 4 and Lemma 5, we can easily find that $f(A_1, B_x) = W_x$. Considering strategyproofness, because we know that when B moves from B_0 to B_1 with fixed A_1, W_x moves from W_0 to W_1, then $d(B_x, W_x)$ is the shortest distance between B_x and segment $\overline{W_0 W_1}$, otherwise agent B at B_x may misreport her location B'_x to reduce her cost.

Because of translational-invariance, let $\alpha \in [-1, 1]$. Using A, B, W instead of A_0, B_0, W_0, we set

$$g(\alpha) = \left(\sum_{i=1}^{m} |(y_i + \alpha \cdot (y_i + x_i) - x_i)|^p\right)^{1/p}.$$

Obviously, $g(\alpha)$ means the distance between B and some point in segment $\overline{AW}$. According to the last paragraph, we have $g(0) = \min_{\alpha \in [-1,1]} g(\alpha)$, and $g'(0) = 0$ which means derivative of $g(\alpha)$.

In the same way, let $h(\beta)$ $(\beta \in [-1,1])$ denotes the distance between A and some point in segment $\overline{BW}$ and we will have

$$h(\beta) = \left(\sum_{i=1}^{m} |y_i + \beta \cdot (y_i - x_i) + x_i|^p \right)^{1/p},$$

and $h'(0) = 0$.

For convenience, let $G(\alpha) = g(\alpha)^p/p$ and $H(\alpha) = h(\alpha)^p/p$, thus $G'(0) = H'(0) = 0$. We have

$$\begin{cases} G'(\alpha) = \displaystyle\sum_{i=1}^{m} (y_i + \alpha \cdot (x_i + y_i) - x_i) \cdot |y_i + \alpha \cdot (x_i + y_i) - x_i|^{p-2} \cdot (x_i + y_i) \\ H'(\beta) = \displaystyle\sum_{i=1}^{m} (y_i + \beta \cdot (y_i - x_i) + x_i) \cdot |y_i + \beta \cdot (y_i - x_i) + x_i|^{p-2} \cdot (y_i - x_i) \end{cases}$$

Considering $G'(0) = H'(0) = 0$, we have

$$\begin{cases} \displaystyle\sum_{i=1}^{m} (x_i + y_i) \cdot (x_i - y_i) \cdot |x_i - y_i|^{p-2} = 0 \\ \displaystyle\sum_{i=1}^{m} (x_i - y_i) \cdot (x_i + y_i) \cdot |x_i + y_i|^{p-2} = 0 \end{cases}$$

In summary, the equations in the theorem hold. □

We can find that the group of equations has an infinite number of solutions if and only if $p = 2$ (When $p = 2$, $|x_i \pm y_i|^{p-2}$ in the equations should be replaced with 1). And at this time the two equations are equivalent which mean the output should locate on a sphere with AB as the diameter.

Theorem 4. *In any L_p space ($1 < p < \infty$) with 2 agents A and B, a deterministic unanimous translation-invariant strategyproof mechanism f must be scalable.*

Proof. When $p > 2$, according to Theorem 3, because of finite number of valid output locations, the property scalability holds, otherwise it will contradict with continuity and translational-invariance (Let's imagine a situation: We move one agent to the other slowly, and if the output does not obey scalability, then it will "jump" in some time to another valid output location). Therefore we only need to discuss $p = 2$.

Considering translational-invariance, we can assume A locates at the origin. Then we have $f(k \cdot A, k \cdot B) = f(A, k \cdot B)$, which means we only need to move B. Notice that for any $k_1 \cdot k_2 = 1$, we find $f(k_1 \cdot A, k_1 \cdot B)$ and $f(k_2 \cdot A, k_2 \cdot B)$ are inverted to each other. Therefore we only need to analyze $0 < k < 1$ (Of course we do not need to discuss when $k = 1$). Then we divide this into 3 cases.

Case 1, $f(A, B) = A$: According to Lemma 2, $\forall k \in (0, 1)$, we have $f(A, k \cdot B) = A$.

Case 2, $f(A, B) = B$: According to Lemma 2, $\forall k \in (0, 1)$, we have $f(A + (1 - k) \cdot B, B) = B$. Considering translational-invariance, we have $f(A, k \cdot B) = f(A + (1 - k) \cdot B - (1 - k) \cdot B, B - (1 - k) \cdot B) = B - (1 - k) \cdot B = k \cdot B$.

Case 3, $f(A, B) \neq A, B$: Assume $W = f(A, B)$, $B' = k \cdot B$, and $W' = f(A, B')$. Therefore, there exists $C \in \overline{AW}$ and $D \in \overline{BW}$ that $\overrightarrow{CD} = \overrightarrow{AB'} = k \cdot \overrightarrow{AB}$. Obviously $\triangle WCD \cong \triangle W'AB'$. Thus we know $W' \in \overline{AW}$, $\overrightarrow{AW'} = k \cdot \overrightarrow{AW}$ and $\overrightarrow{W'B'} = \overrightarrow{WB}$. This means that $W' = k \cdot W$. $\qquad \square$

Also, we give our conjecture about the n-agent situation.

Conjecture 2. In any L_p $(1 < p < \infty)$ space with n agents, a deterministic unanimous translation-invariant strategyproof mechanism f is scalable.

4   Two Special Cases

Although we cannot give complete characterization of any 2-agent deterministic unanimous translation-invariant strategyproof mechanism, we finish two special cases. One is dictatorial mechanism in Euclidean spcae, and the other is anonymous mechanism in 2-dimensional Euclidean space.

4.1   Dictatorial Mechanisms

Theorem 5. *In any Euclidean space with 2 agents, f is a deterministic unanimous translation-invariant strategyproof mechanism, then f is rotation-invariant if and only if f is dictatorial.*

Proof. First of all, if f is dictatorial, then obviously it is rotation-invariant. Then we only need to analyze the case that f is rotation-invariant.

Let 2 agents be A and B, and we assume $W = f(A, B)$. If there exists A_0, B_0 that $W_0 = f(A_0, B_0) \neq A_0, B_0$, then we have $f(W_0, B_0) = W_0$. Otherwise we can imagine an agent A with location W_0 misreports her location as A_0 to reduce her cost. Besides, we know that f is scalable and translation-invariant, then if we move B_0 to some $B_1 \in \overline{A_0 B_0}$ so that $\|A_0 B_1\| = \|W_0 B_0\|$. Therefore we will find that it is contradict with rotational-invariance by observing $f(A_0, B_1)$ and $f(W_0, B_0)$. Thus the output can only locate at A or B.

Consider the following three properties: rotational-invariance, translational-invariance and scalability, and we will find that if $f(A_0, B_0) = A_0$ (or B_0, $A_0 \neq B_0$ otherwise we can solve this by unanimity), then for $\forall A, B \in \mathbb{R}_m$, we have $f(A, B) = A$ (or B), because any two points in L_p space can be transformed by A_0 and B_0 with these three properties. Therefore, f is dictatorial. $\qquad \square$

4.2 Anonymous Mechanisms

Here we give 3 anonymous mechanisms in 2-dimensional Euclidean space. Let 2 agents be A and B with different coordinates (x_A, y_A) and (x_B, y_B) respectively. For these 3 mechanisms, if $A = B$, then we select A (or B) as the facility.

Mechanism 1: (u, v)-C1 Mechanism $(u, v \in \{0, 1\})$.
If $u = 1$, then $x_W = \max\{x_A, x_B\}$; if $u = 0$, then $x_W = \min\{x_A, x_B\}$.
If $v = 1$, then $y_W = \max\{y_A, y_B\}$; if $v = 0$, then $y_W = \min\{y_A, y_B\}$.

Mechanism 2: (u)-C2 Mechanism $(u \neq 0)$. *Without loss of generality, we assume $x_A \leq x_B$. And $W = f(A, B)$ with coordinate (x_W, y_W). We divide this into three cases.*

Case 1. *When $x_A = x_B$, we let $x_W = x_A$. If $u > 0$, let $y_W = \max\{y_A, y_B\}$ and if $u < 0$, let $y_W = \min\{y_A, y_B\}$.*
Notice other 2 cases satisfy $x_A < x_B$. Let $R = (y_B - y_A)/(x_B - x_A)$.
Case 2. *$u > 0$. When $-1/u \leq R \leq u$, draw line (1) $y = u \cdot (x - x_A) + y_A$ and line (2) $y = -\frac{1}{u} \cdot (x - x_B) + y_B$. Let W be the intersection of the two lines. When $R > u$, we let $W = A$ and when $R < -1/u$, we let $W = B$.*
Case 3. *$u < 0$. When $u \leq R \leq -1/u$, draw line (1) $y = u \cdot (x - x_A) + y_A$ and line (2) $y = -\frac{1}{u} \cdot (x - x_B) + y_B$. Let W be the intersection of the two lines. When $R > -1/u$, we let $W = A$ and when $R < u$, we let $W = B$.*

Mechanism 3: (v)-C3 Mechanism $(v \neq 0)$. *Without loss of generality, we assume $y_A \leq y_B$. And $W = f(A, B)$ with coordinate (x_W, y_W). We divide this into three cases.*

Case 1. *When $y_A = y_B$, we let $y_W = y_A$. If $v > 0$, let $x_W = \max\{x_A, x_B\}$ and if $v < 0$, let $x_W = \min\{x_A, x_B\}$.*
Notice other 2 cases satisfy $x_A < x_B$. Let $S = (x_B - x_A)/(y_B - y_A)$.
Case 2. *$v > 0$. When $-1/v \leq S \leq v$, draw line (1) $x = v \cdot (y - y_A) + x_A$ and line (2) $x = -\frac{1}{v} \cdot (y - y_B) + x_B$. Let W be the intersection of the two lines. When $S > v$, we let $W = A$ and when $S < -1/v$, we let $W = B$.*
Case 3. *$v < 0$. When $v \leq S \leq -1/v$, draw line (1) $x = v \cdot (y - y_A) + x_A$ and line (2) $x = -\frac{1}{v} \cdot (y - y_B) + x_B$. Let W be the intersection of the two lines. When $S > -1/v$, we let $W = A$ and when $S < v$, we let $W = B$.*

Theorem 6. *In any 2-dimensional Euclidean space with 2 agents, a mechanism f is deterministic unanimous translation-invariant anonymous strategyproof, if and only if f is one of Mechanism 1, 2, and 3.*

Proof. Firstly we prove sufficiency. Obviously if f is one of the 3 mechanisms, it is deterministic, unanimous, translation-invariant and anonymous. So we only need to prove it is strategyproof. Considering translational-invariance, anonymity and scalability (proved in previous theorem), we only need to prove for any $A_0(0, 0)$ and $B_0(\cos\theta, \sin\theta)$, B cannot gain from misreporting her location.

Mechanism 1 is strategyproof: Out of symmetry, we only need to prove (1,1)-C1 Mechanism is strategyproof.

(1) When $\theta \in [0, \pi/2]$, B will never misreport because her cost is 0.

(2) When $\theta \in (\pi/2, \pi)$, the facility is $(0, \sin\theta)$ and B's cost is $-\cos\theta$. Because $A_0(0,0)$, then $x_W \geq 0$, therefore this is the minimum cost for B.

(3) When $\theta \in [\pi, 3\pi/2]$, the facility is $(0,0)$ and B's cost is 1. Because $A_0(0,0)$, then $x_W \geq 0$ and $y_W \geq 0$, therefore this is the minimum cost for B.

(4) When $\theta \in (3\pi/2, 2\pi)$, we can use the same way as (2) to prove.

Mechanism 2 and 3 are strategyproof: Out of symmetry, we only need to prove Mechanism 2 is strategyproof for $u > 0$ and $\theta \in [-\pi/2, \pi/2]$ (And we fill find this time $x_B \geq x_A$).

(1) When $\theta \in [\arctan u - \pi/2, \arctan u]$, then $W_0 \in$ Line 1 ($y = ux$) and $x_W \geq 0$. We assume B misreports as $B'(x', y')$. When $y' \geq 0$, if $y'/x' > u$ or ≤ 0 or $x' = 0$, then the output will locate at top left of Line 1, leading the cost larger than previous one; if $y'/x' \in [0, u]$, then the output will locate at Line 1, leading the cost never smaller than the previous one because $A_0 W_0 \perp B_0 W_0$ and in Euclidean space this is the smallest distance. When $y' < 0$, in the same way, if $-1/u < y'/x' < 0$, the output will locate at Line 1; if $y'/x' < -1/u$ or $> u$ or $x' = 0$, the output will locate at A; and if $0 < y'/x' < u$, the output will locate at top left of Line 1.

(2) When $\theta \in (\arctan u, \pi/2]$, B will never misreport because $W_0 = B_0$.

(3) When $\theta \in [-\pi/2, \arctan u - \pi/2)$, then $W_0 = A_0$ and the cost is 1. We assume B misreports as $B'(x', y')$. When $y' \leq 0$, if $y'/x' \leq -1/u$ or $\geq u$ or $x' = 0$, the output will still locate at A_0; if $-1/u < y'/x' \leq 0$, the output $W' \in$ Line 1 and $\angle B_0 A_0 W' > 90°$ so the cost will increase; if $0 < y'/x' < u$, the output $W' \in$ Line 2 and $\angle B_0 A_0 W' > 90°$ so the cost will increase. When $y' > 0$, in the same way, we have $\angle B_0 A_0 W' > 90°$.

Secondly, we prove necessity. In fact, we only need to observe the output of $f((0, -1), (0, 1))$ and $f((-1, 0), (1, 0))$ and this is enough for us to characterize the whole mechanism. According to Theorem 2, any output W must satisfy $\overrightarrow{WA} \cdot \overrightarrow{WB} = 0$. We divide this into 4 cases.

(1) $f((0, -1), (0, 1)) = (0, \pm 1)$ and $f((-1, 0), (1, 0)) = (\pm 1, 0)$. Out of symmetry, we mainly discuss the positive result $(1, 0)$ and $(0, 1)$. We claim this mechanism is the same as $(1, 1)$-C1 Mechanism. The proof is rather easy. Assume there exists a group of $W_0 = f(A_0, B_0)$ that $x_W \neq \max\{x_A, x_B\}$. Obviously $y_A \neq y_B$, $x_A \neq x_B$, and W_0, A_0, B_0 cannot locate at the same line, otherwise it will contradict the condition at once. Because $\angle A_0 W_0 B_0 = 90°$, then if we move two lines $y = 0$ and $x = 0$, we will finally find that one of the two lines will have 2 intersections with broken line $A_0 W_0 B_0$ and let the two intersections on $\overline{A_0 W_0}$ and $\overline{W_0 B_0}$ be C and D (if there are infinite intersections, we can let one of the two points be W_0 and the other be A_0 or B_0, then there must be one contradiction in these two situations. In other 3 cases, this discussion will be omitted due to the length), thus we have $f(C, D) = W_0$, which contradicts with the conditions. In the same way we can also prove $y_W \neq \max\{y_A, y_B\}$ will lead to a contradiction, too. Therefore, this is the same as C1 Mechanism. In summary, the four situations are the 4 combinations of C1 Mechanism with different parameters.

(2) $f((0,-1),(0,1)) = (0,\pm 1)$ and $f((-1,0),(1,0)) = (\cos\theta_1, \sin\theta_1)$, where $\theta_1 \neq 0, \pi$ (same in the following (4)). Out of symmetry, we only consider $f((0,-1),(0,1)) = (0,\pm 1)$ and $\theta_1 \in (0,\pi)$ (If $\theta_1 \in (\pi, 2\pi)$ then we can use the same method below to lead to a contradiction). According to Lemma 2, for $\forall C \in \overline{A_0 W_0}, \forall D \in \overline{B_0 W_0}$, we have $f(C,D) = W_0$, which is the same as case 2 in C2 Mechanism when $(y_A - y_B)/(x_A - x_B) \in [\sin\theta_1/(\cos\theta_1 - 1), \sin\theta_1/(1 + \cos\theta_1)]$. When $(y_A - y_B)/(x_A - x_B) > \sin\theta_1/(1 + \cos\theta_1)$ or $< \sin\theta_1/(\cos\theta_1 - 1)$, we can use the same method as (1) to prove W locates at agents with larger y axis, which is the same as C2 Mechanism. In summary, this case means $(\frac{\sin\theta_1}{1+\cos\theta_1})$-C2 Mechanism.

(3) $f((0,-1),(0,1)) = (\cos\theta_2, \sin\theta_2)$ and $f((-1,0),(1,0)) = (\pm 1, 0)$, where $\theta_2 \neq 0.5\pi, 1.5\pi$ (same in the following (4)). This case is the same as $(\frac{\cos\theta_2}{1+\sin\theta_2})$-C3 Mechanism. The proof is similar to (2), thus omitted.

(4) $f((0,-1),(0,1)) = (\cos\theta_2, \sin\theta_2)$ and $f((-1,0),(1,0)) = (\cos\theta_1, \sin\theta_1)$. In fact, this case cannot exist. Let $A_0(-1,0), B_0(1,0), A_1(0,-1), B_1(0,1)$. We can draw a line which can be moved in the space and let it has 2 intersections with broken lines $A_0 W_0 B_0$ and $A_1 W_1 B_1$ and the intersections are C_0, D_0 and C_1, D_1 respectively. Considering $f(C_0, D_0) = W_0$ and $f(C_1, D_1) = W_1$, we find that considering translational-invariance and scalability, the contradiction is obvious. □

5   Discussion

In this section we will discuss the general median mechanism and the lower bound of the maximum cost view.

Mechanism 4: (General Median Mechanism). *Given location of n agents, let W be the output, then in every dimension, W's coordinate is equal to agents' median coordinate in this dimension. If there are 2 median coordinates, then we select the larger one.*

In fact, when there are 2 median coordinates, it does not matter if we select the larger one or the smaller one.

Lemma 6. *The General Median Mechanism in Euclidean space fits Conjecture 1 if and only if dimension $m \leq 2$.*

Proof. Obviously this mechanism is unanimous, translation-invariant, scalable, and much literature have proved that it's strategyproof in L_p space. When $m = 1$, Conjecture 1 is also obvious.

When $m = 2$, Let's recall Theorem 2. If $\exists i \in N, W = A_i$, then for any $j \in N$, we have $\overrightarrow{WA_i} \cdot \overrightarrow{WA_j} = 0$. If for $\forall i \in N, W \neq A_i$, then assuming $W(0,...,0)$, there must exists A_s, A_t that they locate on different axes, meaning $\overrightarrow{WA_s} \cdot \overrightarrow{WA_t} = 0$.

When $m \geq 3$, we can give an counter-example with 3 agents. Let them be $A_1(0,1,-1,0,...,0)$, $A_2(-1,0,1,0,...,0)$ and $A_3(1,-1,0,0,...,0)$. Then the output is $(0,0,...,0)$. But this is contradict with Conjecture 1. □

In the end, we use our tool to prove a lower bound of 2 in the maximum cost view.

Lemma 7. *In any L_p space and maximum cost view, the lower bound of deterministic strategyproof mechanism is 2.*

Proof. Assume $f(A, B) = W$ and A, B are the only two agents, then according to the tool for the proof of Theorem 5, we have $f(W, B) = B$, which means that if there two agents located at W and B, then the maximum cost of f is always at least twice the optimal maximum cost $d(W, B)/2$. $\square$

Many works proves this result in one-dimensional space and we give a simple proof of the multi-dimensional space.

References

1. Alon, N., Feldman, M., Procaccia, A.D., Tennenholtz, M.: Strategyproof approximation of the minimax on networks. Math. Oper. Res. **35**(3), 513–526 (2010)
2. Barberà, S., Gul, F., Stacchetti, E.: Generalized median voter schemes and committees. J. Econ. Theory **61**(2), 262–289 (1993)
3. Barberà, S., Massó, J., Serizawa, S.: Strategy-proof voting on compact ranges. Games Econ. Behav. **25**(2), 272–291 (1998)
4. Border, K.C., Jordan, J.S.: Straightforward elections, unanimity and phantom voters. Rev. Econ. Stud. **50**(1), 153–170 (1983)
5. Feigenbaum, I., Sethuraman, J., Ye, C.: Approximately optimal mechanisms for strategyproof facility location: minimizing LP norm of costs. Math. Oper. Res. **42**(2), 434–447 (2017)
6. Filos-Ratsikas, A., Li, M., Zhang, J., Zhang, Q.: Facility location with double-peaked preferences. Auton. Agents Multi-Agent Syst. **31**(6), 1209–1235 (2017). https://doi.org/10.1007/s10458-017-9361-0
7. Fong, K.C., Li, M., Lu, P., Todo, T., Yokoo, M.: Facility location games with fractional preferences. In: 32nd AAAI Conference on Artificial Intelligence, AAAI 2018, pp. 1039–1046. AAAI Press (2018)
8. Fotakis, D., Tzamos, C.: Winner-imposing strategyproof mechanisms for multiple facility location games. Theor. Comput. Sci. **472**, 90–103 (2013)
9. Li, M., Lu, P., Yao, Y., Zhang, J.: Strategyproof mechanism for two heterogeneous facilities with constant approximation ratio. arXiv preprint arXiv:1907.08918 (2019)
10. Moulin, H.: On strategy-proofness and single peakedness. Public Choice **35**(4), 437–455 (1980)
11. Procaccia, A.D., Tennenholtz, M.: Approximate mechanism design without money. In: Proceedings of the 10th ACM Conference on Electronic Commerce, pp. 177–186 (2009)
12. Serafino, P., Ventre, C.: Heterogeneous facility location without money on the line. In: ECAI, pp. 807–812 (2014)
13. Tang, P., Yu, D., Zhao, S.: Characterization of group-strategyproof mechanisms for facility location in strictly convex space. In: Proceedings of the 21st ACM Conference on Economics and Computation, pp. 133–157 (2020)
14. Yuan, H., Wang, K., Fong, K.C., Zhang, Y., Li, M.: Facility location games with optional preference. In: Proceedings of the Twenty-Second European Conference on Artificial Intelligence, pp. 1520–1527 (2016)

Packing and Covering Triangles in Dense Random Graphs

Zhongzheng Tang[1] and Zhuo Diao[2(✉)]

[1] School of Sciences, Beijing University of Posts and Telecommunications,
Beijing 100876, China
tangzhongzheng@amss.ac.cn
[2] School of Statistics and Mathematics, Central University of Finance
and Economics, Beijing 100081, China
diaozhuo@amss.ac.cn

Abstract. Given a simple graph $G = (V, E)$, a subset of E is called a triangle cover if it intersects each triangle of G. Let $\nu_t(G)$ and $\tau_t(G)$ denote the maximum number of pairwise edge-disjoint triangles in G and the minimum cardinality of a triangle cover of G, respectively. Tuza [25] conjectured in 1981 that $\tau_t(G)/\nu_t(G) \leq 2$ holds for every graph G. In this paper, we consider Tuza's Conjecture on dense random graphs. We prove that under $\mathcal{G}(n, p)$ model with $p = \Omega(1)$, for any $0 < \epsilon < 1$, $\tau_t(G) \leq 1.5(1 + \epsilon)\nu_t(G)$ holds with high probability, and under $\mathcal{G}(n, m)$ model with $m = \Omega(n^2)$, for any $0 < \epsilon < 1$, $\tau_t(G) \leq 1.5(1 + \epsilon)\nu_t(G)$ holds with high probability. In some sense, on dense random graphs, these conclusions verify Tuza's Conjecture.

Keywords: Triangle cover · Triangle packing · Random graph · $\mathcal{G}(n,p)$ model · $\mathcal{G}(n,m)$ model.

1 Introduction

Graphs considered in this paper are undirected, finite and may have multiple edges. Given a graph $G = (V, E)$ with vertex set $V(G) = V$ and edge set $E(G) = E$, for convenience, we often identify a triangle in G with its edge set. A subset of E is called a *triangle cover* if it intersects each triangle of G. Let $\tau_t(G)$ denote the minimum cardinality of a triangle cover of G, referred to as the *triangle covering number* of G. A set of pairwise edge-disjoint triangles in G is called a *triangle packing* of G. Let $\nu_t(G)$ denote the maximum cardinality of a triangle packing of G, referred to as the *triangle packing number* of G. It is clear that $1 \leq \tau_t(G)/\nu_t(G) \leq 3$ holds for every graph G. Our research is motivated by the following conjecture raised by Tuza [25] in 1981, and its weighted generalization by Chapuy et al. [7] in 2014.

This research is supported part by National Natural Science Foundation of China under Grant No. 11901605, and by the disciplinary funding of Central University of Finance and Economics.

W. Wu and Z. Zhang (Eds.): COCOA 2020, LNCS 12577, pp. 426–439, 2020.
https://doi.org/10.1007/978-3-030-64843-5_29

Conjecture 1. (**Tuza's Conjecture** [25]). $\tau_t(G)/\nu_t(G) \leq 2$ holds for every simple graph G.

To the best of our knowledge, the conjecture is still unsolved in general. If it is true, then the upper bound 2 is sharp as shown by K_4 and K_5 – the complete graphs of orders 4 and 5.

Related Work. The only known universal upper bound smaller than 3 was given by Haxell [14], who shown that $\tau_t(G)/\nu_t(G) \leq 66/23 = 2.8695...$ holds for all simple graphs G. Haxell's proof [14] implies a polynomial-time algorithm for finding a triangle cover of cardinality at most 66/23 times that of a maximal triangle packing. Other results on Tuza's conjecture concern with special classes of graphs.

Tuza [26] proved his conjecture holds for planar simple graphs, K_5-free chordal simple graphs and simple graphs with n vertices and at least $7n^2/16$ edges. The proof for planar graphs [26] gives an elegant polynomial-time algorithm for finding a triangle cover in planar simple graphs with cardinality at most twice that of a maximal triangle packing. The validity of Tuza's conjecture on the class of planar graphs was later generalized by Krivelevich [18] to the class of simple graphs without $K_{3,3}$-subdivision. Haxell and Kohayakawa [15] showed that $\tau_t(G)/\nu_t(G) \leq 2-\epsilon$ for tripartite simple graphs G, where $\epsilon > 0.044$. Haxell, Kostochka and Thomasse [13] proved that every K_4-free planar simple graph G satisfies $\tau_t(G)/\nu_t(G) \leq 1.5$.

Regarding the tightness of the conjectured upper bound 2, Tuza [26] noticed that there exists infinitely many simple graphs G attaining the conjectured upper bound $\tau_t(G)/\nu_t(G) = 2$. Cui et al. [11] characterized planar simple graphs G satisfying $\tau_t(G)/\nu_t(G) = 2$; these graphs are edge-disjoint unions of K_4's plus possibly some vertices and edges that are not in triangles. Baron and Kahn [2] proved that Tuza's conjecture is asymptotically tight for dense simple graphs.

Fractional and weighted variants of Conjecture 1 were studied in literature. Krivelevich [18] proved two fractional versions of the conjecture: $\tau_t(G) \leq 2\nu_t^*(G)$ and $\tau_t^*(G) \leq 2\nu_t(G)$, where $\tau_t^*(G)$ and $\nu_t^*(G)$ are the values of an optimal fractional triangle cover and an optimal fractional triangle packing of simple graph G, respectively. [16] proved if G is a graph with n vertices, then $\nu_t^*(G) - \nu_t(G) = o(n^2)$.

We can regard the classic random graph models $\mathcal{G}(n,p)$ and $\mathcal{G}(n,m)$ as special graph classes, and we can also consider the probabilistic properties between $\tau_t(G)$ and $\nu_t(G)$. Bennett et al. [3] showed that $\tau_t(G) \leq 2\nu_t(G)$ holds with high probability in $\mathcal{G}(n,m)$ model where $m \leq 0.2403n^{1.5}$ or $m \geq 2.1243n^{1.5}$. Relevant studies in random graph models were discussed in [1,19,24]. Other extensions related to Conjecture 1 can be found in [4–6,8–10,12,17,20–23].

Our Contributions. We consider Tuza's conjecture on random graph, under two probability models $\mathcal{G}(n,p)$ and $\mathcal{G}(n,m)$.

- Given $0 \leq p \leq 1$, under $\mathcal{G}(n,p)$ model, $\mathbf{Pr}(\{v_i,v_j\} \in G) = p$ for all v_i,v_j with these probabilities mutually independent. Our main theorem is following: If

$G \in \mathcal{G}(n,p)$ and $p = \Omega(1)$, then for any $0 < \epsilon < 1$, it holds that

$$\mathbf{Pr}\,[\tau_t(G) \leq 1.5(1 + \epsilon)\nu_t(G)] = 1 - o(1).$$

- Given $0 \leq m \leq n(n-1)/2$, under $\mathcal{G}(n,m)$ model, let G be defined by randomly picking m edges from all v_i, v_j pairs. Our main theorem is following: If $G \in \mathcal{G}(n,m)$ and $m = \Omega(n^2)$, then for any $0 < \epsilon < 1$, it holds that

$$\mathbf{Pr}\,[\tau_t(G) \leq 1.5(1 + \epsilon)\nu_t(G)] = 1 - o(1).$$

The main content of the article is organized as follows: In Sect. 2, the theorem in $\mathcal{G}(n,p)$ random graph model is proved; In Sect. 3, the theorem in $\mathcal{G}(n,m)$ random graph model is proved; In Sect. 4, the conclusions are summarized and some future works are proposed. The appendix provides a list of mathematical symbols and classical theorems.

2 $\mathcal{G}(n,p)$ Random Graph Model

In this section, we discuss the probability properties of graphs in $\mathcal{G}(n,p)$. Given $0 \leq p \leq 1$, under $\mathcal{G}(n,p)$ model, $\mathbf{Pr}(\{v_i, v_j\} \in G) = p$ for all v_i, v_j with these probabilities mutually independent. Theorem 1 is our main result: If $G \in \mathcal{G}(n,p)$ and $p = \Omega(1)$, then for any $0 < \epsilon < 1$, it holds that

$$\mathbf{Pr}\,[\tau_t(G) \leq 1.5(1 + \epsilon)\nu_t(G)] = 1 - o(1).$$

The primary idea behind the theorem is as follows:

- First, in Lemma 2, Lemma 3, we prove that $\tau_t(G) \leq (1 + \epsilon)\frac{n(n-1)}{4}p$ holds with high probability by combining the Chernoff's bounds technique;
- Second, in Lemma 4, Lemma 6, we prove that $\nu_t(G) \geq (1 - \epsilon)\frac{n(n-1)}{6}p$ holds with high probability through combining the Chernoff's bounds technique and the relationship between $\nu^*(G)$ and $\nu_t(G)$[16].
- By using the previous two properties, Theorem 1 holds.

The following simple property will be used frequently in our discussions.

Lemma 1. *Let $A(n)$ and $B(n)$ be two events related to parameter n. If $\mathbf{Pr}[A(n)] = 1 - o(1)$, then $\mathbf{Pr}[B(n)] \geq \mathbf{Pr}[B(n)|A(n)] - o(1)$ where $o(1) \to 0$ as $n \to \infty$.*

Proof. This can be seen from the fact that $\mathbf{Pr}[A] \cdot \mathbf{Pr}[B] = \mathbf{Pr}[B] - o(1) \geq \mathbf{Pr}[A \cap B] - o(1)$ and $o(1)/\mathbf{Pr}[A] = o(1)$.

Denote the edge number of graph G as m. Let $b(G)$ be the maximum number of edges of sub-bipartite in G. There are four basic properties of graph parameters. The first three holds in every graph, while the last one shows the boundary condition of triangle-free in $\mathcal{G}(n,p)$.

Lemma 2.

(i) $b(G) \geq m/2$ for every graph G.
(ii) $\tau_t(G) \leq m/2$ for every graph G.
(iii) $\nu_t(G) \leq m/3$ for every graph G.
(iv) If $G \in \mathcal{G}(n,p)$ and $p = o(1/n)$, then G is triangle-free with high probability.

Proof. Suppose $b(G) < m/2$ and the corresponding sub-bipartite is $B = (V_1, V_2)$. Thus, there exists one vertex, without loss of generality, $u \in V_1$ satisfies that $d_B(u) < d_G(u)/2$. We can move vertex u from V_1 to V_2, and let $\widetilde{B} = (\widetilde{V}_1, \widetilde{V}_2)$ where $\widetilde{V}_1 = V_1 \backslash \{u\}, \widetilde{V}_2 = V_2 \cup \{u\}$. We have $|E(\widetilde{B})| > |E(B)| = b(G)$, which contradicts with the definition of $b(G)$. Therefore, statement (i) holds.
Statement (ii) follows from the definition of $b(G)$ and the result of statement (i).
Statement (iii) is trivial.
Applying Union Bound Inequality, Statement (iv) is due to

$$\mathbf{Pr}[G \text{ contains at least a triangle}] \leq \binom{n}{3} \cdot p^3 = o(1)$$

In view of Lemma 2(iv), we consider henceforth $\mathcal{G}(n,p)$ with $p = \Omega(1/n)$. Under this condition, we give the following upper bounds for $\tau_t(G)$ and $\nu_t(G)$ with high probability.

Lemma 3. *If $G \in \mathcal{G}(n,p)$ and $p = \Omega(1/n)$, for any $0 < \epsilon < 1$, it holds that*

$$\mathbf{Pr}\left[\tau_t(G) \leq (1+\epsilon)\frac{n(n-1)}{4}p\right] = 1 - o(1). \tag{1}$$

$$\mathbf{Pr}\left[\nu_t(G) \leq (1+\epsilon)\frac{n(n-1)}{6}p\right] = 1 - o(1). \tag{2}$$

Proof. For each edge e in complete graph K_n, Let X_e be the random variable defined by: $X_e = 1$ if $e \in E(G)$ and $X_e = 0$ otherwise. Then $X_e, e \in K_n$, are independent 0–1 variables, $\mathbf{E}[X_e] = p$, $m = \sum_{e \in K_n} X_e$ and $\mathbf{E}[m] = n(n-1)p/2 = \Omega(n)$. By Chernoff's Inequality, for any $0 < \epsilon < 1$ we have

$$\mathbf{Pr}[m \geq (1+\epsilon)\mathbf{E}[m]] \leq \exp\left(-\frac{\epsilon^2 \mathbf{E}[m]}{3}\right) = o(1).$$

Thus, it follows from Lemma 2(ii) and (iii) that

$$\mathbf{Pr}\left[\tau_t(G) \leq (1+\epsilon)\frac{n(n-1)}{4}p\right]$$

$$= \mathbf{Pr}\left[2\tau_t(G) \leq (1+\epsilon)\frac{n(n-1)}{2}p\right]$$

$$\geq \mathbf{Pr}\left[m \leq (1+\epsilon)\frac{n(n-1)}{2}p\right]$$

$$= \mathbf{Pr}\left[m \leq (1+\epsilon)\mathbf{E}(m)\right]$$

$$= 1 - o(1)$$

Similarly,

$$\mathbf{Pr}\left[\nu_t(G) \le (1+\epsilon)\frac{n(n-1)}{6}p\right]$$

$$= \mathbf{Pr}\left[3\nu_t(G) \le (1+\epsilon)\frac{n(n-1)}{2}p\right]$$

$$\ge \mathbf{Pr}\left[m \le (1+\epsilon)\frac{n(n-1)}{2}p\right]$$

$$= 1 - o(1)$$

proving the lemma.

Along a different line, we consider the probability result of the lower bounds of the fractional triangle packing $\nu_t^*(G)$ as follows:

Lemma 4. *If $G \in \mathcal{G}(n,p)$ and $p = \Omega(1)$, then for any $0 < \epsilon < 1$, it holds that*

$$\mathbf{Pr}\left[\nu_t^*(G) \ge (1-\epsilon)\frac{n(n-1)p}{6}\right] = 1 - o(1).$$

Proof. Consider an arbitrary edge $uv \in K_n$. For each $w \in V(G) \setminus \{u,v\}$. Let X_w be the random variable defined by: $X_w = 1$ if $uw, vw \in E(G)$ and $X_w = 0$ otherwise. Assuming $uv \in E(G)$, let T_{uv} denote the number of triangles of G that contain uv. Notice that X_w, $w \in V(G) \setminus \{u,v\}$, are independent 0–1 variables, $\mathbf{E}[X_w] = p^2$, $T_{uv} = \sum_{w \in V(G) \setminus \{u,v\}} X_w$, and $\mathbf{E}[T_{uv}] = (n-2)p^2$. By Chernoff's Inequality, we have

$$\mathbf{Pr}\left[T_{uv} \ge \left(1+\frac{\epsilon}{2}\right)(n-2)p^2\right] \le \exp\left(-\frac{\epsilon^2(n-2)p^2}{12}\right),$$

and by using Union Bound Inequality

$$\mathbf{Pr}\left[T_e \ge \left(1+\frac{\epsilon}{2}\right)(n-2)p^2 \text{ for some } e \in E(G)\right] \le n^2 \cdot \exp\left(-\frac{\epsilon^2(n-2)p^2}{12}\right) = o(1).$$

Now taking every triangle of G with an amount of $\dfrac{1}{(1+\frac{\epsilon}{2})(n-2)p^2}$, we obtain a feasible fractional triangle packing of G with high probability, giving

$$\mathbf{Pr}\left[\nu_t^*(G) \ge \sum_{T \in \mathcal{T}(G)} \frac{1}{(1+\frac{\epsilon}{2})(n-2)p^2}\right]$$

$$= \mathbf{Pr}\left[\nu_t^*(G) \ge \frac{\mathcal{T}(G)}{(1+\frac{\epsilon}{2})(n-2)p^2}\right] \tag{3}$$

$$= 1 - o(1)$$

For each triangle $T \in K_n$, let X_T be the random variable defined by: $X_T = 1$ if $T \subseteq G$ and $X_T = 0$ otherwise. Then

$$\mathbf{E}[X_T] = \mathbf{Pr}[X_T = 1] = p^3 \text{ and } \mathbf{Var}[X_T] = p^3(1-p^3).$$

For any two distinct triangles T_1, T_2 in K_n, we have

$$\mathbf{Cov}[X_{T_1}, X_{T_2}] = \mathbf{E}[X_{T_1} X_{T_2}] - \mathbf{E}[X_{T_1}]\mathbf{E}[X_{T_2}] = \begin{cases} p^5 - p^6, & \text{if } E(T_1) \cap E(T_2) \neq \emptyset \\ 0, & \text{otherwise.} \end{cases}$$

Denote $T(G) = \sum_{T \in \mathcal{T}(K_n)} X_T$. Combining $p = \Omega(1)$, we can compute

$$\mathbf{E}[T(G)] = \binom{n}{3} p^3 = \Theta(n^3).$$

$$\mathbf{Var}[T(G)] = \binom{n}{3} p^3 (1 - p^3) + 2 \binom{n}{2} \binom{n-2}{2} (p^5 - p^6) = \Theta(n^4).$$

Thus, Chebyshev's Inequality gives

$$\begin{aligned}
&\mathbf{Pr} \left[T(G) \leq \left(1 - \frac{\epsilon}{2}\right) \mathbf{E}[T(G)] \right] \\
&\leq \mathbf{Pr} \left[|T(G) - \mathbf{E}[T(G)]| \geq \frac{\epsilon}{2} \mathbf{E}[T(G)] \right] \\
&\leq \frac{4\mathbf{Var}[T(G)]}{\epsilon^2 (\mathbf{E}[T(G)])^2} \\
&= o(1)
\end{aligned} \tag{4}$$

Then, since $\dfrac{1 - \epsilon/2}{1 + \epsilon/2} > 1 - \epsilon$ when $0 < \epsilon < 1$, we obtain

$$\begin{aligned}
&\mathbf{Pr} \left[\nu_t^*(G) \geq (1 - \epsilon) \frac{n(n-1)}{6} p \right] \\
&\geq \mathbf{Pr} \left[\nu_t^*(G) \geq \frac{1 - \epsilon/2}{1 + \epsilon/2} \cdot \frac{n(n-1)}{6} p \right] \\
&\geq \mathbf{Pr} \left[\nu_t^*(G) \geq \frac{1 - \epsilon/2}{1 + \epsilon/2} \cdot \frac{n(n-1)}{6} p \,\Big|\, \nu_t^*(G) \geq \frac{T(G)}{(1 + \epsilon/2)(n - 2)p^2} \right] - o(1) \\
&\geq \mathbf{Pr} \left[\frac{T(G)}{(1 + \epsilon/2)(n - 2)p^2} \geq \frac{1 - \epsilon/2}{1 + \epsilon/2} \cdot \frac{n(n-1)}{6} p \right] - o(1) \\
&= \mathbf{Pr} \left[T(G) \geq (1 - \epsilon/2) \mathbf{E}[T(G)] \right] - o(1) \\
&= 1 - o(1),
\end{aligned}$$

where the second inequality is implied by Lemma 1 and (3), and the last equality is implied by (4). The lemma is established.

We take advantage of the following result in [16] to bridge the relationship of $\nu_t^*(G)$ and $\nu_t(G)$. This result shows that the gap between these two parameters is very small when graph G is dense.

Lemma 5. ([16]). *If G is a graph with n vertices, then $\nu_t^*(G) - \nu_t(G) = o(n^2)$.*

Combining the above lemma, we derive naturally the lower bound of $\nu_t(G)$ with high probability.

Lemma 6. *If $G \in \mathcal{G}(n, p)$ and $p = \Omega(1)$, then for any $0 < \epsilon < 1$, it holds that*

$$\mathbf{Pr}\left[\nu_t(G) \geq (1 - \epsilon) \cdot \frac{n(n-1)p}{6}\right] = 1 - o(1).$$

Proof. Using Lemma 5, when n is sufficiently large we have

$$\mathbf{Pr}\left[\nu_t(G) \geq (1 - \epsilon) \cdot \frac{n(n-1)p}{6}\right]$$

$$= \mathbf{Pr}\left[\nu_t^*(G) \geq (1 - \epsilon) \cdot \frac{n(n-1)p}{6} + o(n^2)\right]$$

$$\geq \mathbf{Pr}\left[\nu_t^*(G) \geq (1 - \epsilon) \cdot \frac{n(n-1)p}{6} + \frac{\epsilon}{2} \cdot \frac{n(n-1)p}{6}\right]$$

$$= \mathbf{Pr}\left[\nu_t^*(G) \geq \left(1 - \frac{\epsilon}{2}\right) \frac{n(n-1)p}{6}\right].$$

The result follows from Lemma 4.

Now we are ready to prove one of the two main theorems:

Theorem 1. *If $G \in \mathcal{G}(n, p)$ and $p = \Omega(1)$, then for any $0 < \epsilon < 1$, it holds that*

$$\mathbf{Pr}\left[\tau_t(G) \leq 1.5(1 + \epsilon)\nu_t(G)\right] = 1 - o(1).$$

Proof. Let A denote the event that

$$\tau_t(G) \leq \left(1 + \frac{\epsilon}{3}\right) \frac{n(n-1)}{4}p \quad \text{and} \quad \nu_t(G) \geq \left(1 - \frac{\epsilon}{3}\right) \frac{n(n-1)p}{6}.$$

Combining Lemmas 3 and 6 we have $\mathbf{Pr}[A] = 1 - o(1)$. Note that $1 + \epsilon > \dfrac{1 + \epsilon/3}{1 - \epsilon/3}$. Therefore, recalling Lemma 1, we deduce that

$$\mathbf{Pr}\left[\tau_t(G) \leq 1.5(1 + \epsilon)\nu_t(G)\right]$$

$$\geq \mathbf{Pr}\left[\tau_t(G) \leq 1.5 \cdot \frac{1 + \epsilon/3}{1 - \epsilon/3}\nu_t(G)\right]$$

$$\geq \mathbf{Pr}\left[\tau_t(G) \leq 1.5 \cdot \frac{1 + \epsilon/3}{1 - \epsilon/3}\nu_t(G) \,\Big|\, A\right] - o(1)$$

$$= 1 - o(1),$$

which establishes the theorem.

Remark 1. In $\mathcal{G}(n, p)$, $p = \Omega(1)$ implies $\mathbf{E}[m] = \binom{n}{2}p = n(n-1)p/2 = \Omega(n^2)$, thus our main theorem is a result in dense random graphs.

3 $\mathcal{G}(n, m)$ Random Graph Model

In this section, we discuss the probability properties of graphs in $\mathcal{G}(n, m)$. Given $0 \leq m \leq n(n-1)/2$, under $\mathcal{G}(n, m)$ model, let G be defined by randomly picking m edges from all v_i, v_j pairs. Theorem 2 is our main result: If $G \in \mathcal{G}(n, m)$ and $m = \Omega(n^2)$, then for any $0 < \epsilon < 1$, it holds that

$$\mathbf{Pr}\left[\tau_t(G) \leq 1.5(1 + \epsilon)\nu_t(G)\right] = 1 - o(1).$$

The primary idea behind the theorem is as follows:

– First, in Lemma 2, $\tau_t(G) \leq m/2$ holds;
– Second, in Lemma 7, Lemma 8, we prove that $\nu_t(G) \geq (1 - \epsilon)m/3$ holds with high probability through combining the Chernoff's bounds technique and the relationship between $\nu^*(G)$ and $\nu_t(G)$ [16];
– By using the previous two properties, Theorem 2 holds.

For easy of presentation, we use N to denote $\binom{n}{2}$.

Now we give the high probability result of the lower bound of $\nu_t^*(G)$ in $\mathcal{G}(n, m)$ model:

Lemma 7. *If $G \in \mathcal{G}(n, m)$ and $m = \Omega(n^2)$, then for any $0 < \epsilon < 1$, it holds that*

$$\mathbf{Pr}[\nu_t^*(G) \geq (1 - \epsilon)m/3] = 1 - o(1).$$

Proof. Consider an arbitrary edge $uv \in K_n$. For each $w \in V(G) \setminus \{u, v\}$. Let X_w be the random variable defined by; $X_w = 1$ if $uw, vw \in E(G)$ and $X_w = 0$ otherwise. Assuming $uv \in E(G)$, let T_{uv} denote the number of triangles of G that contain uv. Then we have

$$\mathbf{E}[X_w] = \frac{m(m - 1)}{N(N - 1)},$$

$$\mathbf{Var}[X_w] = \frac{m(m - 1)}{N(N - 1)}\left(1 - \frac{m(m - 1)}{N(N - 1)}\right)$$

$$\mathbf{Cov}[X_w, X_{w'}] = \frac{m(m - 1)(m - 2)(m - 3)}{N(N - 1)(N - 2)(N - 3)} - \left(\frac{m(m - 1)}{N(N - 1)}\right)^2 \leq 0$$

where $w, w' \in V(G) \setminus \{u, v\}$. It follows from $T_{uv} = \sum_{w \in V(G) \setminus \{u, v\}} X_w$ that

$$\mathbf{E}[T_{uv}] = (n - 2)\frac{m(m - 1)}{N(N - 1)} = \Theta(n).$$

Using Chernoff's Inequality, we derive

$$\mathbf{Pr}\left[T_{uv} \geq \left(1 + \frac{\epsilon}{2}\right)\frac{(n - 2)m(m - 1)}{N(N - 1)}\right] \leq \exp\left(-\frac{\epsilon^2 \mathbf{E}[T_{uv}]}{12}\right) \leq \exp\left(-\frac{\epsilon^2 \Theta(n)}{12}\right);$$

$$\mathbf{Pr}\left[T_e \geq \left(1 + \frac{\epsilon}{2}\right)\frac{(n - 2)m(m - 1)}{N(N - 1)} \; \exists \, e \in E(G)\right] \leq n^2 \exp\left(-\frac{\epsilon^2 \Theta(n)}{12}\right) = o(1).$$

So taking every triangle of G with an amount of $\left[(1 + \frac{\epsilon}{2}) \cdot \frac{(n-2)m(m-1)}{N(N-1)}\right]^{-1}$
makes a feasible fractional packing of G with high probability. Thus

$$
\begin{aligned}
\mathbf{Pr}&\left[\nu_t^*(G) \geq \sum_{\forall T} \frac{1}{(1 + \frac{\epsilon}{2}) \cdot \frac{(n-2)m(m-1)}{N(N-1)}}\right]\\
&= \mathbf{Pr}\left[\nu_t^*(G) \geq \frac{\mathcal{T}(G)}{(1 + \frac{\epsilon}{2}) \cdot \frac{(n-2)m(m-1)}{N(N-1)}}\right]\\
&= 1 - o(1).
\end{aligned}
\tag{5}
$$

For each triangle $T \in K_n$, let X_T be the random variable defined by: $X_T = 1$ if $T \subseteq G$ and $X_T = 0$ otherwise. Then

$$
\mathbf{E}[X_T] = \frac{m(m-1)(m-2)}{N(N-1)(N-2)}.
$$

$$
\mathbf{Var}[X_T] = \frac{m(m-1)(m-2)}{N(N-1)(N-2)}\left(1 - \frac{m(m-1)(m-2)}{N(N-1)(N-2)}\right).
$$

For any two distinct triangles T_1, T_2 in K_n, we have

$$
\begin{aligned}
&\mathbf{Cov}(X_{T_1}, X_{T_2})\\
&= \mathbf{E}[X_{T_1} X_{T_2}] - \mathbf{E}[X_{T_1}] \cdot \mathbf{E}[X_{T_2}]\\
&= \begin{cases}\dfrac{m(m-1)(m-2)(m-3)(m-4)}{N(N-1)(N-2)(N-3)(N-4)} - \left(\dfrac{m(m-1)(m-2)}{N(N-1)(N-2)}\right)^2,\\[2mm]
\qquad\qquad\qquad \text{if } E(T_1) \cap E(T_2) \neq \emptyset;\\[2mm]
0, \qquad\qquad\qquad \text{otherwise.}\end{cases}
\end{aligned}
$$

Notice that

$$
\mathbf{E}[\mathcal{T}(G)] = \binom{n}{3}\frac{m(m-1)(m-2)}{N(N-1)(N-2)} = \Theta(n^3)
$$

$$
\begin{aligned}
\mathbf{Var}[\mathcal{T}(G)] = {}&\binom{n}{3}\frac{m(m-1)(m-2)}{N(N-1)(N-2)}\left(1 - \frac{m(m-1)(m-2)}{N(N-1)(N-2)}\right) +\\
&2\binom{n}{2}\binom{n-2}{2}\left(\frac{m(m-1)(m-2)(m-3)(m-4)}{N(N-1)(N-2)(N-3)(N-4)} - \left(\frac{m(m-1)(m-2)}{N(N-1)(N-2)}\right)^2\right)\\
={}&\Theta(n^4).
\end{aligned}
$$

By Chebyshev's Inequality, we have:

$$
\begin{aligned}
&\mathbf{Pr}\left[\mathcal{T}(G) \leq \left(1 - \frac{\epsilon}{4}\right)\mathbf{E}[\mathcal{T}(G)]\right]\\
&\leq \mathbf{Pr}\left[|\mathcal{T}(G) - \mathbf{E}[\mathcal{T}(G)]| \geq \frac{\epsilon}{4}\mathbf{E}[\mathcal{T}(G)]\right]\\
&\leq \frac{16\mathbf{Var}[\mathcal{T}(G)]}{\epsilon^2(\mathbf{E}[\mathcal{T}(G)])^2} = o(1).
\end{aligned}
\tag{6}
$$

Since $\dfrac{1-\epsilon/2}{1+\epsilon/2} > 1-\epsilon$, we deduce from (5) and Lemma 1 that

$$\mathbf{Pr}\left[\nu_t^*(G) \geq (1-\epsilon)\frac{m}{3}\right]$$

$$\geq \mathbf{Pr}\left[\nu_t^*(G) \geq \frac{1-\epsilon/2}{1+\epsilon/2} \cdot \frac{m}{3}\right]$$

$$\geq \mathbf{Pr}\left[\nu_t^*(G) \geq \frac{1-\epsilon/2}{1+\epsilon/2} \cdot \frac{m}{3} \;\middle|\; \nu_t^*(G) \geq \frac{\mathcal{T}(G)}{(1+\frac{\epsilon}{2})\frac{(n-2)m(m-1)}{N(N-1)}}\right] - o(1)$$

$$\geq \mathbf{Pr}\left[\frac{\mathcal{T}(G)}{(1+\frac{\epsilon}{2})\frac{(n-2)m(m-1)}{N(N-1)}} \geq \frac{1-\epsilon/2}{1+\epsilon/2} \cdot \frac{m}{3}\right] - o(1)$$

$$= \mathbf{Pr}\left[\mathcal{T}(G) \geq \left(1-\frac{\epsilon}{2}\right)\binom{n}{3}\frac{m^2(m-1)}{N^2(N-1)}\right] - o(1)$$

As $(1+\frac{\epsilon}{4})\dfrac{m-2}{N-2} > \dfrac{m}{N}$ holds for sufficiently large n, we have

$$\mathbf{Pr}\left[\nu_t^*(G) \geq (1-\epsilon)\frac{m}{3}\right]$$

$$\geq \mathbf{Pr}\left[\mathcal{T}(G) \geq \left(1-\frac{\epsilon}{2}\right)\left(1+\frac{\epsilon}{4}\right)\binom{n}{3}\frac{m(m-1)(m-2)}{N(N-1)(N-2)}\right] - o(1)$$

$$\geq \mathbf{Pr}\left[\mathcal{T}(G) \geq \left(1-\frac{\epsilon}{4}\right)\mathbf{E}[\mathcal{T}(G)]\right] - o(1)$$

$$= 1 - o(1),$$

where the second inequality is implied by $(1-\epsilon/2)(1+\epsilon/4) \leq 1-\epsilon/4$, and the last equality is guaranteed by (6). This complete the proof of the lemma.

Similar to the the proof of Lemma 6, the combination of Lemma 5 and Lemma 7 gives the following Lemma 8.

Lemma 8. *If $G \in \mathcal{G}(n,m)$ and $m = \Omega(n^2)$, then for any $0 < \epsilon < 1$, it holds that*

$$\mathbf{Pr}[\nu_t(G) \geq (1-\epsilon)m/3] = 1 - o(1).$$

Proof. Using Lemma 5, when n is sufficiently large we have

$$\mathbf{Pr}\left[\nu_t(G) \geq (1-\epsilon)m/3\right]$$
$$= \mathbf{Pr}\left[\nu_t^*(G) \geq (1-\epsilon)m/3 + o(n^2)\right]$$
$$\geq \mathbf{Pr}\left[\nu_t^*(G) \geq (1-\epsilon)m/3 + \frac{\epsilon}{2} \cdot m/3\right]$$
$$= \mathbf{Pr}\left[\nu_t^*(G) \geq \left(1-\frac{\epsilon}{2}\right)m/3\right].$$

The result follows from Lemma 7.

Now, we are ready to prove the main theorem in $\mathcal{G}(n, m)$ as follows:

Theorem 2. *If $G \in \mathcal{G}(n, m)$ and $m = \Omega(n^2)$, then for any $0 < \epsilon < 1$, it holds that*

$$\mathbf{Pr}\left[\tau_t(G) \leq 1.5(1 + \epsilon)\nu_t(G)\right] = 1 - o(1).$$

Proof. Let A denote the event that

$$\tau_t(G) \leq \frac{m}{2} \quad \text{and} \quad \nu_t(G) \geq (1 - \frac{\epsilon}{2})\frac{m}{3}.$$

It follows from Lemmas 2(ii) and 8 that $\mathbf{Pr}[A] = 1 - o(1)$. Since $1 + \epsilon > (1 - \epsilon/2)^{-1}$, we deduce from Lemma 1 that

$$
\begin{aligned}
&\mathbf{Pr}[\tau_t(G) \leq 1.5(1 + \epsilon)\nu_t(G)] \\
&\geq \mathbf{Pr}\left[(1 - \epsilon/2) \cdot \tau_t(G) \leq 1.5\nu_t(G)\right] \\
&\geq \mathbf{Pr}\left[(1 - \epsilon/2) \cdot \tau_t(G) \leq 1.5\nu_t(G) \mid A\right] - o(1) \\
&= 1 - o(1)
\end{aligned}
$$

verifying the theorem.

Remark 2. In $\mathcal{G}(n, m)$, the condition $m = \Omega(n^2)$ implies that our main theorem is a result in dense random graphs.

4 Conclusion and Future Work

We consider Tuza's conjecture on random graphs, under two probability models $\mathcal{G}(n, p)$ and $\mathcal{G}(n, m)$. Two results are following:

- If $G \in \mathcal{G}(n, p)$ and $p = \Omega(1)$, then for any $0 < \epsilon < 1$, it holds that

$$\mathbf{Pr}\left[\tau_t(G) \leq 1.5(1 + \epsilon)\nu_t(G)\right] = 1 - o(1).$$

- If $G \in \mathcal{G}(n, m)$ and $m = \Omega(n^2)$, then for any $0 < \epsilon < 1$, it holds that

$$\mathbf{Pr}\left[\tau_t(G) \leq 1.5(1 + \epsilon)\nu_t(G)\right] = 1 - o(1).$$

In some sense, on dense random graph, these two inequalities verify Tuza's conjecture.

Future work: In dense random graphs, these two results nearly imply $\tau_t(G) \leq 1.5\nu_t(G)$ holds with high probability. It is interesting to consider the same problem in sparse random graphs.

Acknowledgement. The authors are very indebted to Professor Xujin Chen and Professor Xiaodong Hu for their invaluable suggestions and comments.

Appendix: A List of Mathematical Symbols

$\mathcal{G}(n,p)$	Given $0 \leq p \leq 1$, $\mathbf{Pr}(\{v_i, v_j\} \in G) = p$ for all v_i, v_j
	With these probabilities mutually independent
$\mathcal{G}(n,m)$	Given $0 \leq m \leq n(n-1)/2$, let G be defined by
	Randomly picking m edges from all v_i, v_j pairs
$\tau_t(G)$	The minimum cardinality of a triangle cover in G
$\nu_t(G)$	The maximum cardinality of a triangle packing in G
$\tau_t^*(G)$	The minimum cardinality of a fractional triangle cover in G
$\nu_t^*(G)$	The maximum cardinality of a fractional triangle packing in G
$b(G)$	The maximum number of edges of sub-bipartite in G
$\delta(G)$	The minimum degree of graph G
$f(n) = O(g(n))$	$\exists\, c > 0, n_0 \in \mathbb{N}_+, \forall n \geq n_0, 0 \leq f(n) \leq cg(n)$
$f(n) = \Omega(g(n))$	$\exists\, c > 0, n_0 \in \mathbb{N}_+, \forall n \geq n_0, 0 \leq cg(n) \leq f(n)$
$f(n) = \Theta(g(n))$	$\exists\, c_1 > 0, c_2 > 0, n_0 \in \mathbb{N}_+, \forall n \geq n_0, 0 \leq c_1 g(n) \leq f(n) \leq c_2 g(n)$
$f(n) = o(g(n))$	$\forall\, c > 0, \exists\, n_0 \in \mathbb{N}_+, \forall n \geq n_0, 0 \leq f(n) < cg(n)$
$f(n) = \omega(g(n))$	$\forall\, c > 0, \exists\, n_0 \in \mathbb{N}_+, \forall n \geq n_0, 0 \leq cg(n) < f(n)$

Union Bound Inequality:

For any finite or countably infinite sequence of events $E_1, E_2, \ldots$, then

$$\mathbf{Pr}\left[\bigcup_{i \geq 1} E_i\right] \leq \sum_{i \geq 1} \mathbf{Pr}(E_i).$$

Chernoff's Inequalities:

Let $X_1, X_2, \ldots, X_n$ be mutually independent 0–1 random variables with $\mathbf{Pr}[X_i = 1] = p_i$. Let $X = \sum_{i=1}^n X_i$ and $\mu = \mathbf{E}[X]$. For $0 < \epsilon \leq 1$, then the following bounds hold:

$$\mathbf{Pr}[X \geq (1+\epsilon)\mu] \leq e^{-\epsilon^2 \mu/3}, \quad \mathbf{Pr}[X \leq (1-\epsilon)\mu] \leq e^{-\epsilon^2 \mu/2}.$$

Chebyshev's Inequality:

For any $a > 0$,

$$\mathbf{Pr}[|X - \mathbf{E}[X]| \geq a] \leq \frac{\mathrm{Var}[X]}{a^2}.$$

References

1. Baron, J.D.: Two problems on cycles in random graphs. Ph.D. thesis, Rutgers University-Graduate School-New Brunswick (2016)

2. Baron, J.D., Kahn, J.: Tuza's conjecture is asymptotically tight for dense graphs. Comb. Probab. Comput. **25**(5), 645–667 (2016)
3. Bennett, P., Dudek, A., Zerbib, S.: Large triangle packings and Tuza's conjecture in sparse random graphs. Comb. Probab. Comput. **29**(5), 757–779 (2020)
4. Botler, F., Fernandes, C., Gutiérrez, J.: On Tuza's conjecture for triangulations and graphs with small treewidth. Electron. Notes Theor. Comput. Sci. **346**, 171–183 (2019)
5. Botler, F., Fernandes, C.G., Gutiérrez, J.: On Tuza's conjecture for graphs with treewidth at most 6. In: Anais do III Encontro de Teoria da Computação. SBC (2018)
6. Chalermsook, P., Khuller, S., Sukprasert, P., Uniyal, S.: Multi-transversals for triangles and the Tuza's conjecture. In: Proceedings of the Fourteenth Annual ACM-SIAM Symposium on Discrete Algorithms. pp. 1955–1974. SIAM (2020)
7. Chapuy, G., DeVos, M., McDonald, J., Mohar, B., Scheide, D.: Packing triangles in weighted graphs. SIAM Journal on Discrete Mathematics **28**(1), 226–239 (2014)
8. Chen, X., Diao, Z., Hu, X., Tang, Z.: Sufficient conditions for Tuza's conjecture on packing and covering triangles. In: Mäkinen, V., Puglisi, S.J., Salmela, L. (eds.) IWOCA 2016. LNCS, vol. 9843, pp. 266–277. Springer, Cham (2016). https://doi.org/10.1007/978-3-319-44543-4_21
9. Chen, X., Diao, Z., Hu, X., Tang, Z.: Total dual integrality of triangle covering. In: Chan, T.-H.H., Li, M., Wang, L. (eds.) COCOA 2016. LNCS, vol. 10043, pp. 128–143. Springer, Cham (2016). https://doi.org/10.1007/978-3-319-48749-6_10
10. Chen, X., Diao, Z., Hu, X., Tang, Z.: Covering triangles in edge-weighted graphs. Theory Comput. Syst. **62**(6), 1525–1552 (2018)
11. Cui, Q., Haxell, P., Ma, W.: Packing and covering triangles in planar graphs. Graphs and Combinatorics **25**(6), 817–824 (2009)
12. Erdös, P., Gallai, T., Tuza, Z.: Covering and independence in triangle structures. Discret. Math. **150**(1–3), 89–101 (1996)
13. Haxell, P., Kostochka, A., Thomassé, S.: Packing and covering triangles in K_4-free planar graphs. Graphs and Combinatorics **28**(5), 653–662 (2012)
14. Haxell, P.E.: Packing and covering triangles in graphs. Discret. Math. **195**(1), 251–254 (1999)
15. Haxell, P.E., Kohayakawa, Y.: Packing and covering triangles in tripartite graphs. Graphs and Combinatorics **14**(1), 1–10 (1998)
16. Haxell, P.E., Rödl, V.: Integer and fractional packings in dense graphs. Combinatorica **21**(1), 13–38 (2001)
17. Hosseinzadeh, H., Soltankhah, N.: Relations between some packing and covering parameters of graphs. In: The 46th Annual Iranian Mathematics Conference, p. 715 (2015)
18. Krivelevich, M.: On a conjecture of Tuza about packing and covering of triangles. Discret. Math. **142**(1), 281–286 (1995)
19. Krivelevich, M.: Triangle factors in random graphs. Comb. Probab. Comput. **6**(3), 337–347 (1997)
20. Lakshmanan, A., Bujtás, C., Tuza, Z.: Induced cycles in triangle graphs. Discret. Appl. Math. **209**, 264–275 (2016)
21. Munaro, A.: Triangle packings and transversals of some K_4-freegraphs. Graphs and Combinatorics **34**(4), 647–668 (2018)
22. Puleo, G.J.: Tuza's conjecture for graphs with maximum average degree less than 7. Eur. J. Comb. **49**, 134–152 (2015)
23. Puleo, G.J.: Maximal k-edge-colorable subgraphs, Vizing's Theorem, and Tuza's Conjecture. Discret. Math. **340**(7), 1573–1580 (2017)

24. Ruciński, A.: Matching and covering the vertices of a random graph by copies of a given graph. Discret. Math. **105**(1–3), 185–197 (1992)
25. Tuza, Z.: Conjecture. In: Finite and Infinite Sets, Proc. Colloq. Math. Soc. Janos Bolyai, p. 888 (1981)
26. Tuza, Z.: A conjecture on triangles of graphs. Graphs and Combinatorics **6**(4), 373–380 (1990)

Mechanism Design for Facility Location Games with Candidate Locations

Zhongzheng Tang[1], Chenhao Wang[2], Mengqi Zhang[3,4(✉)], and Yingchao Zhao[5]

[1] School of Sciences, Beijing University of Posts and Telecommunications,
Beijing 100876, China
tangzhongzheng@amss.ac.cn
[2] Department of Computer Science and Engineering, University of Nebraska-Lincoln,
Lincoln, NE, USA
wangch@amss.ac.cn
[3] Academy of Mathematics and Systems Science, Chinese Academy of Sciences,
Beijing 100190, China
mqzhang@amss.ac.cn
[4] School of Mathematical Sciences, University of Chinese Academy of Sciences,
Beijing 100049, China
[5] Caritas Institute of Higher Education, HKSAR,
Hong Kong, China
zhaoyingchao@gmail.com

Abstract. We study the facility location games with candidate locations from a mechanism design perspective. Suppose there are n agents located in a metric space whose locations are their private information, and a group of candidate locations for building facilities. The authority plans to build some homogeneous facilities among these candidates to serve the agents, who bears a cost equal to the distance to the closest facility. The goal is to design mechanisms for minimizing the total/maximum cost among the agents. For the single-facility problem under the maximum-cost objective, we give a deterministic 3-approximation group strategy-proof mechanism, and prove that no deterministic (or randomized) strategy-proof mechanism can have an approximation ratio better than 3 (or 2). For the two-facility problem on a line, we give an anonymous deterministic group strategy-proof mechanism that is $(2n-3)$-approximation for the total-cost objective, and 3-approximation for the maximum-cost objective. We also provide (asymptotically) tight lower bounds on the approximation ratio.

Keywords: Facility location · Social choice · Mechanism design.

1 Introduction

We consider a well-studied facility location problem of deciding where some public facilities should be built to serve a population of agents with their locations as private information. For example, a government needs to decide the locations

© Springer Nature Switzerland AG 2020
W. Wu and Z. Zhang (Eds.): COCOA 2020, LNCS 12577, pp. 440–452, 2020.
https://doi.org/10.1007/978-3-030-64843-5_30

of public supermarkets or hospitals. It is often modeled in a metric space or a network, where there are some agents (customers or citizens) who may benefit by misreporting their locations. This manipulation can be problematic for a decision maker to find a system optimal solution, and leads to the mechanism design problem of providing (approximately) optimal solutions while also being strategy-proof (SP), i.e., no agent can be better off by misreporting their locations, regardless of what others report.

This setup, where the agents are located in a network that is represented as a contiguous graph, is initially studied by Schummer and Vohra [18], and has many applications (e.g., traffic network). Alon et al. [1] give an example of telecommunications networks such as a local computer network or the Internet. In these cases, the agents are the network users or service providers, and the facility can be a filesharing server or a router. Interestingly, in computer networks, an agent's perceived network location can be easily manipulated, for example, by generating a false IP address or rerouting incoming and outgoing communication, etc. This explains the incentive of agents for misreporting, and thus a strategy-proof mechanism is necessary.

In the classic model [12,17], all points in the metric space or the network are feasible for building facilities. However, this is often impractical in many applications. For example, due to land use restrictions, the facilities can only be built in some feasible regions, while other lands are urban green space, residential buildings and office buildings, etc. Therefore, we assume that there is a set of candidate locations, and study the facility location games with candidate locations in this paper.

We notice that our setting somewhat coincides a metric social choice problem [8], where the voters (agents) have their preferences over the candidates, and all participants are located in a metric space, represented as a point. The voters prefer candidates that are closer to them to the ones that are further away. The goal is to choose a candidate as a winner, such that the total distance to all voters is as small as possible. When the voters are required to report their locations, this problem is the same with the facility location game with candidate locations.

Our setting is sometimes referred to as the "constrained facility location" problems [20], as the feasible locations for facilities are constrained. Sui and Boutilier [20] provide possibility and impossibility results with respect to (additive) approximate individual and group strategy-proofness, whereas do not consider the approximation ratios for system objectives.

Our Results

In this paper we study the problem of locating one or two facilities in a metric space, where there are n agents and a set of feasible locations for building the facilities. For the single-facility problem, we consider the objective of minimizing the maximum cost among the agents, while the social-cost (i.e., the total cost of agents) objective has been well studied in [8] as a voting process. We present a mechanism that deterministically selects the closest candidate location to an arbitrary dictator agent, and prove that it is group strategy-proof (GSP, no group of agents being better off by misreporting) and 3-approximation. In particular, when the space is a line, the mechanism that selects the closest candidate location to the leftmost agent is additionally anonymous, that is, the outcome is the same for all permutations of the agents' locations on the line. We provide a lower bound 3 for deterministic SP mechanisms, and 2 for randomized SP mechanisms; both lower bounds hold even on a line.

For the two-facility problem on a line, we present an anonymous GSP mechanism that deterministically selects two candidates closest to the leftmost and rightmost agents, respectively. It is $(2n - 3)$-approximation for the social-cost objective, and 3-approximation for the maximum-cost objective. On the negative side, we prove that, for the maximum-cost objective, no deterministic (resp. randomized) strategy-proof mechanism can have an approximation ratio better than 3 (resp. 2).

Our results for deterministic mechanism on a line are summarized in Table 1 in bold font, where a "$\star$" indicates that the upper bound holds for general metric spaces. All inapproximability results are obtained on a line, and thus hold for more general metric spaces.

Table 1. Results for deterministic strategyproof mechanisms on a line.

Objective	Social cost	Maximum cost
Single-facility	UB: $3^\star$ [8]	UB: $3^\star$
	LB: 3 [8]	LB: 3
Two-facility	UB: $2n - 3$	UB: 3
	LB: $n - 2$ [9]	LB: 3

Related Work

A range of works on social choice study the constrained single-facility location games for the social-cost objective, where agents can be placed anywhere, but only a subset of locations is valid for the facility. The *random dictatorship* (RD) mechanism, which selects each candidate with probability equal to the fraction of agents who vote for it, obtains an approximation ratio of $3 - 2/n$, and this is tight for all strategyproof mechanisms [8,15]. The upper bound holds for any metric spaces, whereas the lower bound requires specific constructions on the n-dimensional binary cube. Anshelevich and Postl [2] show a smooth transition of the RD approximation ratio from $2 - 2/n$ to $3 - 2/n$ as the location of the facility becomes more constrained. Meir [14] (Sect. 5.3) provides an overview of approximation results for the single-facility problem.

Approximate Mechanism Design in the Classic Setting. For the classic facility location games wherein the locations have no constraint, Procaccia and Tennenholtz [17] first consider it from the perspective of approximate mechanism design. For single-facility location on a line, they give a "median" mechanism that is GSP and optimal for minimizing the social cost. Under the maximum-cost objective, they provide a deterministic 2-approximation and a randomized 1.5 approximation GSP mechanisms; both bounds are best possible. For two-facility location, they give a 2-approximation mechanism that always places the facilities at the leftmost and the rightmost locations of agents. Fotakis and Tzamos [9] characterize deterministic mechanisms for the problem of locating two facilities on the line, and prove a lower bound of $n - 2$. Randomized mechanisms are considered in [12,13].

Characterizations. Dokow *et al.* [7] study SP mechanisms for locating a facility in a discrete graph, where the agents are located on vertices of the graph, and the possible facility locations are exactly the vertices of the graph. They give a full characterization of SP mechanisms on lines and sufficiently large cycles. For continuous lines, the set of SP and onto mechanisms has been characterized as all generalized median voting schemes [4,18].

Other Settings. There are many different settings for facility location games in recent years. Aziz *et al.* [3] study the mechanism design problem where the public facility is capacity constrained, where the capacity constraints limit the number of agents who can benefit from the facility's services. Chen *et al.* [6] study a dual-role game where each agent can allow a facility to be opened at his place and he may strategically report his opening cost. By introducing payment, they characterize truthful mechanisms and provide approximate mechanisms. After that, Li *et al.* [11] study a model with payment under a budget constraint. Kyropoulou *et al.* [10] initiate the study of constrained heterogeneous facility location problems, wherein selfish agents can either like or dislike the facility and facilities can be located in a given feasible region of the Euclidean plane. Other works on heterogeneous facilities can be found in [5,19,21].

2   Model

Let k be the number of facilities to be built. In an instance of facility location game with candidate locations, the agent set is $N = \{1, \ldots, n\}$, and each agent $i \in N$ has a private location $x_i \in S$ in a metric space (S, d), where $d : S^2 \to \mathbb{R}$ is the metric (distance function). We denote by $\mathbf{x} = (x_1, \ldots, x_n)$ the location profile of agents. The set of m candidate locations is $M \subseteq S$. A deterministic mechanism f takes the reported agents' location profile $\mathbf{x}$ as input, and outputs a facility location profile $\mathbf{y} = (y_1, \ldots, y_k) \in M^k$, that is, selecting k candidates for building facilities. A randomized mechanism outputs a probability distribution over M^k. Given an outcome $\mathbf{y}$, the *cost* of each agent $i \in N$ is the distance to the closest facility, i.e., $c_i(\mathbf{y}) = d(x_i, \mathbf{y}) := \min_{1 \leq j \leq k} d(x_i, y_j)$.

A mechanism f is *strategy-proof* (SP), if no agent $i \in N$ can decrease his cost by misreporting, regardless of the location profile $\mathbf{x}_{-i}$ of others, that is, for any $x_i' \in S$, $c_i(f(x_i, \mathbf{x}_{-i})) \leq c_i(f(x_i', \mathbf{x}_{-i}))$. Further, f is *group strategy-proof* (GSP), if no coalition $G \subseteq N$ of agent can decrease the cost of every agent in G by misreporting, regardless of the location profile $\mathbf{x}_{-G}$ of others, that is, for any $\mathbf{x}'_G$, there exists an agent $i \in G$ such that $c_i(f(\mathbf{x}_G, \mathbf{x}_{-G})) \leq c_i(f(\mathbf{x}'_G, \mathbf{x}_{-G}))$. A mechanism f is *anonymous*, if for every profile $\mathbf{x}$ and every permutation of agents $\pi : N \to N$, it holds that $f(x_1, \ldots, x_n) = f(x_{\pi(1)}, \ldots, x_{\pi(n)})$.

Denote an instance by $I(\mathbf{x}, M)$ or simply I. We consider two objective functions, minimizing the social cost and minimizing the maximum cost.

Social Cost. Given a location profile $\mathbf{x}$, the social cost of solution $\mathbf{y}$ is the total distance to all agents, that is,

$$SC(\mathbf{x}, \mathbf{y}) = \sum_{i \in N} c_i(\mathbf{y}) = \sum_{i \in N} d(x_i, \mathbf{y}).$$

Maximum Cost. Given a location profile $\mathbf{x}$, the maximum cost of solution $\mathbf{y}$ is the maximum distance to all agents, that is,

$$MC(\mathbf{x}, \mathbf{y}) = \max_{i \in N} c_i(\mathbf{y}) = \max_{i \in N} d(x_i, \mathbf{y}).$$

When evaluating a mechanism's performance, we use the standard worst-case approximation notion. Formally, given an instance $I(\mathbf{x}, M)$, let $opt(\mathbf{x}) \in \arg\min_{\mathbf{y} \in M^k} C(\mathbf{x}, \mathbf{y})$ be an optimal facility location profile, and $OPT(\mathbf{x})$ be the optimum value. We say that a mechanism f provides an α-approximation if for every instance $I(\mathbf{x}, M)$,

$$C(\mathbf{x}, f(\mathbf{x})) \leq \alpha \cdot C(\mathbf{x}, opt(\mathbf{x})) = \alpha \cdot OPT(\mathbf{x}),$$

where the objective function C can be SC or MC. The goal is to design deterministic or randomized strategy-proof mechanisms with small approximation ratios.

3   Single-Facility Location Games

In this section, we study the single-facility location games, i.e., $k = 1$. Feldman et al. [8] thoroughly study this problem for the social-cost objective. When the space is a line, they prove tight bounds on the approximations: the Median mechanism that places the facility at the nearest candidate of the median agent is SP and 3-approximation, and no deterministic SP mechanism can do better. They also propose a randomized SP and 2-approximation mechanism (called the Spike mechanism) that selects the nearest candidate of each agent with specific probabilities, and prove that this approximation ratio is the best possible for any randomized SP mechanism. When it is a general metric space, they show that random dictatorship has a best possible approximation ratio of $3 - \frac{2}{n}$.

Hence, we only consider the objective of minimizing the maximum cost among the agents. We study the problem on a line and in a general metric space, respectively.

3.1   Line Space

Suppose that the space is a line. Let x_l (resp. x_r) be the location of the leftmost (resp. rightmost) agent with respect to location profile $\mathbf{x}$. Consider the following mechanism.

Mechanism 1. *Given a location profile* $\mathbf{x}$, *select the candidate location which is closest to the leftmost agent, that is, select a candidate in location* $\arg\min_{y \in M} |y - x_l|$, *breaking ties in any deterministic way.*

Theorem 1. *For the single-facility problem on a line, Mechanism 1 is an anonymous GSP and 3-approximation mechanism, under the maximum-cost objective.*

Proof. Denoted by f Mechanism 1. It is clearly anonymous, because the outcome depends only on the agent locations, not on their identities. Namely, for any permutation of the agent locations on the line, the facility locations do not change. Let $y = f(\mathbf{x})$ be the outcome of the mechanism, and define $L = |x_r - x_l|$. We discuss three cases with respect to the location of y.

Case 1: $x_l - L/2 \le y \le x_r$. The maximum cost of y is $MC(\mathbf{x}, y) = \max\{|y - x_l|, |y - x_r|\} \le 3L/2$, and the optimal maximum cost is at least $L/2$. So we have $\frac{MC(\mathbf{x},y)}{OPT(\mathbf{x})} \le 3$.

Case 2: $y > x_r$. It is easy to see that y is an optimal solution with maximum cost $|y - x_l|$, because the closest candidate y' to the left of x_l induces a maximum cost of at least $|y - x_l| + L$, and there is no candidate between y' and y.

Case 3: $y < x_l - L/2$. The maximum cost of y is $MC(\mathbf{x}, y) = x_r - y$. The optimal candidate has a distance at least $x_l - y$ to x_l. So we have

$$\frac{MC(\mathbf{x}, y)}{OPT(\mathbf{x})} \le \frac{x_r - y}{x_l - y} = \frac{x_l + L - y}{x_l - y} = 1 + \frac{L}{x_l - y} < 1 + \frac{L}{L/2} = 3,$$

which establishes the proof for approximation ratio.

It remains to show the group strategy-proofness. For any group of agents G, we want to show at least one agent in G cannot gain by misreporting. Clearly the agent located at x_l has no incentive to join G, because he already attains the minimum possible cost. The only way for G to influence the output of mechanism f is someone reporting a location to the left of x_l. However, this cannot move the facility location to the right, and thus no agent in G can benefit by misreporting. □

Let $\epsilon > 0$ be a sufficiently small number. We prove lower bounds for the approximation ratio of (deterministic and randomized) SP mechanisms, matching the upper bound in Theorem 1.

Theorem 2. *For the single-facility problem on a line, no deterministic (resp. randomized) SP mechanism can have an approximation ratio better than 3 (resp. 2), under the maximum-cost objective.*

Proof. Suppose f is a deterministic strategy-proof mechanism with approximation ratio $3 - \delta$ for some $\delta > 0$. Consider an instance I (as shown in Fig. 1)
with agents' location profile $\mathbf{x} = (1 - \epsilon, 1 + \epsilon)$, and $M = \{0, 2\}$. By symmetry,
assume w.l.o.g. that $f(\mathbf{x}) = 0$. The cost of agent 2 is $c_2(0) = |1 + \epsilon - 0| = 1 + \epsilon$.
Now consider another instance I' with agents' location profile $\mathbf{x}' = (1 - \epsilon, 3)$, and
$M = \{0, 2\}$. The optimal solution is candidate 2, and the optimal maximum cost
is $1 + \epsilon$. The maximum cost induced by candidate 0 is 3. Since the approximation
ratio of f is $3 - \delta$ and $\epsilon \to 0$, it must select $f(\mathbf{x}') = 2$. It indicates that, under
instance I, agent 2 located at $x_2 = 1 + \epsilon$ can decrease his cost from $c_2(0)$ to
$c_2(f(\mathbf{x}')) = |1 + \epsilon - 2| = 1 - \epsilon$, by misreporting his location as $x_2' = 3$. This is a
contradiction with strategy-proofness.

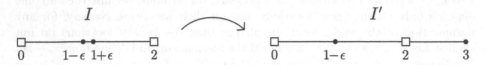

Fig. 1. Two instances I and I', where hollow squares indicate candidates, and solid
circles indicate agents.

Suppose f is a randomized strategy-proof mechanism with approximation
ratio $2 - \delta$ for some $\delta > 0$. Also consider instance I. W.l.o.g. assume that $f(\mathbf{x}) = 0$
with probability at least $\frac{1}{2}$. The cost of agent 2 is $c_2(f(\mathbf{x})) \geq \frac{1}{2}(1+\epsilon) + \frac{1}{2}(1-\epsilon) = 1$. Then consider instance I'. Let P_2' be the probability of $f(\mathbf{x}') = 2$. Since the
approximation ratio of f is $2 - \delta$, we have

$$P_2' \cdot \frac{MC(\mathbf{x}', 2)}{OPT(\mathbf{x}')} + (1 - P_2') \cdot \frac{MC(\mathbf{x}', 0)}{OPT(\mathbf{x}')} = P_2' + (1 - P_2') \cdot \frac{3}{1 + \epsilon} \leq 2 - \delta,$$

which implies that $P_2' > \frac{1}{2}$ as $\epsilon \to 0$. Hence, under instance I, agent 2 located at
$x_2 = 1 + \epsilon$ can decrease his cost to $c_2(f(\mathbf{x}')) < \frac{1}{2}(1-\epsilon) + \frac{1}{2}(1+\epsilon) = 1 \leq c_2(f(\mathbf{x}))$,
by misreporting his location as $x_2' = 3$. □

Remark 1. For randomized mechanisms, we have shown that the lower bound
is 2, and we are failed to find a matching upper bound. We are concerned with
weighted percentile voting (WPV) mechanisms (see [8]), which locate the facility
on the i-th percentile agent's closest candidate with some probability p_i, where
p_i does not depend on the location profile $\mathbf{x}$. For example, Mechanism 1 is WPV
by setting $p_0 = 1$ and $p_i = 0$ for $i > 0$. We remark that no WPV mechanism
can beat the ratio of 3. Consider an instance with agents' location profile $(1, 3)$
and candidates' location profile $(\epsilon, 2, 4 - \epsilon)$. The optimal maximum cost is 1,
attained by selecting candidate 2, while any WPV mechanism must select either
candidate ϵ or $4 - \epsilon$, inducing a maximum cost of $3 - \epsilon$. The ratio approaches 3
when ϵ tends to 0. It leaves an open question to narrow this gap.

3.2   General Metric Spaces

In this subsection, we extend the model from a line to a general metric space. In this setting, the locations of all agents and facility candidates are in a metric space (S, d). Our objective is to minimize the maximum cost of agents. We give the following dictatorial mechanism, in which the dictator can be an arbitrary agent.

Mechanism 2 (Dictatorship). *Given a location profile* $\mathbf{x}$, *for an arbitrary agent* $k \in N$, *select the closest candidate location to agent* k, *that is,* $\arg\min_{y \in M} d(x_k, y)$, *breaking ties in any deterministic way.*

Theorem 3. *For the single-facility problem in a metric space, Mechanism 2 is GSP and 3-approximation, under the maximum-cost objective.*

Proof. Denote f by Mechanism 2. Let $y = f_k(\mathbf{x})$ for a fixed k, and $y^* = opt(\mathbf{x})$ be the optimal solution. Then we have $d(x_i, y^*) \leq OPT(\mathbf{x})$ for each agent $i \in N$. As the distance function has the triangle inequality property in a metric space, we derive the following for each $i \in N$:

$$
\begin{aligned}
d(x_i, y) &\leq d(y^*, y) + d(x_i, y^*) \\
&\leq d(x_k, y) + d(x_k, y^*) + d(x_i, y^*) \\
&\leq 2d(x_k, y^*) + d(x_i, y^*) \\
&\leq 3\, OPT,
\end{aligned}
\tag{1}
$$

The group strategy-proofness is trivial, because Mechanism 2 is dictatorial.
□

One can find that, though losing the anonymity, Mechanism 2 is indeed a generalization of Mechanism 1, and the approximation ratio in Theorem 3 implies that in Theorem 1.

Recall that random dictatorship locates the facility on agent i's closest candidate with probability $1/n$ for all $i \in N$. It has an approximation ratio of $3 - \frac{2}{n}$ for the social-cost objective in any metric space [2]. However, it does not help to improve the deterministic upper bound 3 in Theorem 3, even if on the line.

4   Two-Facility Location Games

In this section, we consider the two-facility location games on a line, under both objectives of minimizing the social cost and minimizing the maximum cost. We give a linear approximation for the social-cost objective, which asymptotically tight, and a 3-approximation for the maximum-cost objective, which is best possible.

4.1   Social-Cost Objective

For the classic (unconstrained) facility location games in a continuous line under the social-cost objective, Fotakis and Tzamos [9] prove that no deterministic mechanism has an approximation ratio less than $n - 2$. Note that the lower bound $n - 2$ also holds in our setting, because when all points on the line are candidates, our problem is equivalent to the classic problem. For the same setting in [9], Procaccia *et al.* [16] give a GSP $(n-2)$-approximation mechanism, which selects the two extreme agent locations. We generalize this mechanism to our setting.

Mechanism 3. *Given a location profile* **x** *on a line, select the candidate location which is closest to the leftmost agent, (i.e.,* $\arg\min_{y \in M} |y - x_l|$), *breaking ties in favor of the candidate to the right; and select the one closest to the rightmost agent (i.e.,* $\arg\min_{y \in M} |y - x_r|$), *breaking ties in favor of the candidate to the left.*

Lemma 4. *Mechanism 3 is GSP.*

Proof. For any group G of agents, we want to show at least one agent in G cannot gain by misreporting. Clearly the agent located at x_l or x_r has no incentive to join the coalition G, because he already attains the minimum possible cost. The only way for G to influence the output of the mechanism is some member reporting a location to the left of x_l or the right of x_r. However, this can move neither of the two facility locations closer to the members. So no agent in G can benefit by misreporting.   □

Theorem 5. *For the two-facility problem on a line, Mechanism 3 is GSP, anonymous, and $(2n - 3)$-approximation under the social-cost objective.*

Proof. The group strategy-proofness is given in Lemma 4. Let $\mathbf{y}^* = (y_1^*, y_2^*)$ be an optimal solution with $y_1^* \le y_2^*$, and $\mathbf{y} = (y_1, y_2)$ with $y_1 \le y_2$ be the solution output by Mechanism 3. Let $N_1 = \{i \in N | d(x_i, y_1^*) \le d(x_i, y_2^*)\}$ be the set of agents who are closer to y_1^* in the optimal solution, and $N_2 = \{i \in N | d(x_i, y_1^*) > d(x_i, y_2^*)\}$ be the complement set. Renaming if necessary, we assume $x_l = x_1 \le \cdots \le x_n = x_r$. If $|N_1| = 0$, then y_2^* must be the closest candidate to every agent in N, including x_1 (i.e., $y_2^* \in \arg\min_{y \in M} d(x_1, y)$). By the specific way of tie-breaking, Mechanism 3 must select y_2^*, and achieve the optimality. The symmetric analysis holds for the case when $|N_2| = 0$.

So we only need to consider the case when $|N_1| \ge 1$ and $|N_2| \ge 1$. Clearly, $1 \in N_1$ and $n \in N_2$. The social cost of the outcome $\mathbf{y}$ by Mechanism 3 is

$$\sum_{i \in N} \min\{d(x_i, y_1), d(x_i, y_2)\} \le \sum_{i \in N_1} d(x_i, y_1) + \sum_{i \in N_2} d(x_i, y_2)$$

$$\le \sum_{i \in N_1 \setminus \{1\}} [\, d(x_i, y_1^*) + d(y_1^*, y_1) \,] + \sum_{i \in N_2 \setminus \{n\}} [\, d(x_i, y_2^*) + d(y_2^*, y_2) \,]$$

$$+ d(x_l, y_1^*) + d(x_r, y_2^*)$$

$$= OPT + \sum_{i \in N_1 \backslash \{1\}} d(y_1^*, y_1) + \sum_{i \in N_2 \backslash \{n\}} d(y_2^*, y_2)$$

$$\leq OPT + \sum_{i \in N_1 \backslash \{1\}} [\, d(x_l, y_1^*) + d(x_l, y_1) \,] + \sum_{i \in N_2 \backslash \{n\}} [\, d(x_r, y_2^*) + d(x_r, y_2) \,]$$

$$\leq OPT + 2(|N_1| - 1) \cdot d(x_l, y_1^*) + 2(|N_2| - 1) \cdot d(x_r, y_2^*)$$

$$\leq OPT + 2(\max\{|N_1|, |N_2|\} - 1) \cdot OPT$$

$$\leq (2n - 3) \cdot OPT,$$

where the second last inequality holds because $OPT \geq d(x_l, y_1^*) + d(x_r, y_2^*)$, and the last inequality holds because $\max\{|N_1|, |N_2|\} \leq n - 1$. $\square$

Next we give an example to show that the analysis in Theorem 5 for the approximation ratio of Mechanism 3 is tight.

Example 1. Consider an instance on a line with agents' location profile $\mathbf{x} = (1, \frac{4}{3}, \ldots, \frac{4}{3}, 2)$ and candidates' location profile $(\frac{2}{3} + \epsilon, \frac{4}{3}, 2)$. The optimal social cost is $\frac{1}{3}$, attained by solution $(\frac{4}{3}, 2)$. Mechanism 3 outputs solution $(\frac{2}{3} + \epsilon, 2)$, and the social cost is $(\frac{4}{3} - \frac{2}{3} - \epsilon) \cdot (n - 2) + \frac{1}{3} - \epsilon$. Then we have $\frac{(2/3 - \epsilon) \cdot (n-2) + 1/3 - \epsilon}{1/3} \rightarrow 2n - 3$, when ϵ tends to 0.

4.2   Maximum-Cost Objective

Next, we turn to consider the maximum-cost objective.

Theorem 6. *For the two-facility problem on a line, Mechanism 3 is GSP, anonymous, and 3-approximation under the maximum-cost objective.*

Proof. The group strategy-proofness is given in Lemma 4. For any location profile $\mathbf{x}$, let $\mathbf{y}^* = (y_1^*, y_2^*)$ be an optimal solution with $y_1^* \leq y_2^*$, and $\mathbf{y} = (y_1, y_2)$ with $y_1 \leq y_2$ be the solution output by Mechanism 3. Assume w.l.o.g. that $x_1 \leq \cdots \leq x_n$. Let $N_1 = \{i \in N | d(x_i, y_1^*) \leq d(x_i, y_2^*)\}$ be the set of agents who are closer to y_1^* in the optimal solution, and $N_2 = \{i \in N | d(x_i, y_1^*) > d(x_i, y_2^*)\}$ be the complement set. Let $n_1 = |N_1|$ and $n_2 = |N_2|$. Define $C_1 = \max_{i \in N_1} d(x_i, y_1^*)$ and $C_2 = \max_{i \in N_2} d(x_i, y_2^*)$. It is easy to see that the optimal maximum cost is $\max\{C_1, C_2\}$.

Next, we consider a restricted instance $(x_1, \ldots, x_{n_1})$ of the single-facility location problem. By the definition of Mechanism 3, candidate y_1 is the closest one to agent 1. By Theorem 1, we have $\max_{i \in N_1} d(y_1, x_i) \leq 3C_1$. Similarly, consider another restricted instance $(x_{n_1+1}, \ldots, x_n)$, we have $\max_{i \in N_2} d(y_2, x_i) \leq 3C_2$. Therefore,

$$\max_{i \in N} d(x_i, \mathbf{y}) \leq 3 \max\{C_1, C_2\},$$

which completes the proof. $\square$

In the following we give a lower bound 2 for randomized SP mechanisms, and a lower bound 3 for deterministic SP mechanisms, matching the bound

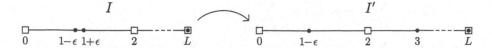

Fig. 2. There is an agent and a candidate in a very far away location L.

in Theorem 6. We use the same construction as in the proof of Theorem 2 for 2 agents, and locate an additional agent at a very far away point in all the location profiles used in the proof.

Theorem 7. *For the two-facility problem on a line, no deterministic (resp. randomized) SP mechanism can have an approximation ratio better than 3 (resp. 2), under the maximum-cost objective.*

Proof. Suppose f is a deterministic SP mechanism with approximation ratio $3 - \delta$ for some $\delta > 0$. Consider an instance I (as shown in Fig. 2) with agents' location profile $\mathbf{x} = (1 - \epsilon, 1 + \epsilon, L)$, and $M = \{0, 2, L\}$, where L is sufficiently large and $\epsilon > 0$ is sufficiently small. Note that candidate L must be selected (to serve agent 3) by any mechanism that has a good approximation ratio. We can assume w.l.o.g. that $0 \in f(\mathbf{x})$. The cost of agent 2 is $c_2(f(\mathbf{x})) = |1 + \epsilon - 0| = 1 + \epsilon$. Now consider another instance I' with agents' location profile $\mathbf{x}' = (1 - \epsilon, 3, L)$, and $M = \{0, 2, L\}$. The optimal maximum cost is $1 + \epsilon$, attained by selecting candidates 2 and L. The maximum cost induced by any solution that selects candidate 0 is at least 3. Since the approximation ratio of f is $3 - \delta$ and $\epsilon \to 0$, it must select candidate 2, i.e., $2 \in f(\mathbf{x}')$. It indicates that, under instance I, agent 2 located at $1 + \epsilon$ can decrease his cost from $c_2(f(\mathbf{x})) = 1 + \epsilon$ to $c_2(f(\mathbf{x}')) = |1 + \epsilon - 2| = 1 - \epsilon$, by misreporting his location as $x_2' = 3$. This is a contradiction with the strategy-proofness.

Suppose f is a randomized SP mechanism with approximation ratio $2 - \delta$ for some $\delta > 0$. Also consider instance I. Note that candidate L must be selected with probability 1 by any mechanisms that have a good approximation ratio, since L tends to ∞. We can assume w.l.o.g. that $0 \in f(\mathbf{x})$ with probability at least $\frac{1}{2}$. The cost of agent 2 is $c_2(f(\mathbf{x})) \geq \frac{1}{2}(1 + \epsilon) + \frac{1}{2}(1 - \epsilon) = 1$. Then consider instance I'. The optimal maximum cost is $1 + \epsilon$, which is attained by selecting candidates 2 and L. Let P_2' be the probability of f selecting candidate 2. The maximum cost induced by any solution that selects candidate 0 is at least 3. Since the approximation ratio of f is $2 - \delta$, we have

$$P_2' \cdot 1 + (1 - P_2') \cdot \frac{3}{1 + \epsilon} \leq 2 - \delta,$$

which implies that $P_2' > \frac{1}{2}$ as $\epsilon \to 0$. Hence, under instance I, agent 2 located at $1 + \epsilon$ can decrease his cost to $c_2(f(\mathbf{x}')) < \frac{1}{2}(1 - \epsilon) + \frac{1}{2}(1 + \epsilon) \leq c_2(f(\mathbf{x}))$, by misreporting his location as $x_2' = 3$. □

5   Conclusion

For the classic k-facility location games, Fotakis and Tzamos [9] show that for every $k \geq 3$, there do not exist any deterministic anonymous SP mechanisms with a bounded approximation ratio for the social-cost objective on the line, even for simple instances with $k + 1$ agents. It directly follows a corollary that there exists no such mechanism with a bounded approximation ratio for the maximum-cost objective. Therefore, in our constrained setting with candidate locations, we cannot expect to beat such lower bounds when $k \geq 3$.

In this paper we are concerned with designing truthful deterministic mechanisms for the setting with candidates. It remains an open question to find randomized mechanisms matching the lower bound 2 in Theorems 2 and 7, though we have excluded the possibility of WPV mechanisms.

Acknowledgement. The authors thank Minming Li and three anonymous referees for their invaluable suggestions and comments. Minming Li is supported by NNSF of China under Grant No. 11771365, and sponsored by Project No. CityU 11205619 from Research Grants Council of HKSAR.

References

1. Alon, N., Feldman, M., Procaccia, A.D., Tennenholtz, M.: Strategyproof approximation of the minimax on networks. Math. Oper. Res. **35**(3), 513–526 (2010)
2. Anshelevich, E., Postl, J.: Randomized social choice functions under metric preferences. J. Artif. Intell. Res. **58**, 797–827 (2017)
3. Aziz, H., Chan, H., Lee, B.E., Parkes, D.C.: The capacity constrained facility location problem. In: Proceeding of the 15th Conference on Web and Internet Economics (WINE), p. 336 (2019)
4. Border, K.C., Jordan, J.S.: Straightforward elections, unanimity and phantom voters. Rev. Econ. Stud. **50**(1), 153–170 (1983)
5. Chen, X., Hu, X., Jia, X., Li, M., Tang, Z., Wang, C.: Mechanism design for two-opposite-facility location games with penalties on distance. In: Deng, X. (ed.) SAGT 2018. LNCS, vol. 11059, pp. 256–260. Springer, Cham (2018). https://doi.org/10.1007/978-3-319-99660-8_24
6. Chen, X., Li, M., Wang, C., Wang, C., Zhao, Y.: Truthful mechanisms for location games of dual-role facilities. In: Proceedings of the 18th International Conference on Autonomous Agents and MultiAgent Systems (AAMAS), pp. 1470–1478 (2019)
7. Dokow, E., Feldman, M., Meir, R., Nehama, I.: Mechanism design on discrete lines and cycles. In: Proceedings of the 13th ACM Conference on Electronic Commerce (ACM-EC), pp. 423–440 (2012)
8. Feldman, M., Fiat, A., Golomb, I.: On voting and facility location. In Proceedings of the 17th ACM Conference on Economics and Computation (ACM-EC), pp. 269–286 (2016)
9. Fotakis, D., Tzamos, C.: On the power of deterministic mechanisms for facility location games. ACM Trans. Econ. Comput. (TEAC) **2**(4), 1–37 (2014)
10. Kyropoulou, M., Ventre, C., Zhang, X.: Mechanism design for constrained heterogeneous facility location. In: Fotakis, D., Markakis, E. (eds.) SAGT 2019. LNCS, vol. 11801, pp. 63–76. Springer, Cham (2019). https://doi.org/10.1007/978-3-030-30473-7_5

11. Li, M., Wang, C., Zhang, M.: Budgeted facility location games with strategic facilities. In: Proceedings of the 29th International Joint Conference on Artificial Intelligence (IJCAI), pp. 400–406 (2020)
12. Lu, P., Sun, X., Wang, Y., Zhu, Z.A.: Asymptotically optimal strategy-proof mechanisms for two-facility games. In: Proceedings of the 11th ACM Conference on Electronic Commerce (ACM-EC), pp. 315–324 (2010)
13. Lu, P., Wang, Y., Zhou, Y.: Tighter bounds for facility games. In: Leonardi, S. (ed.) WINE 2009. LNCS, vol. 5929, pp. 137–148. Springer, Heidelberg (2009). https://doi.org/10.1007/978-3-642-10841-9_14
14. Meir, R.: Strategic voting. Synthesis Lectures on Artificial Intelligence and Machine Learning **13**(1), 1–167 (2018)
15. Meir, R., Procaccia, A.D., Rosenschein, J.S.: Algorithms for strategyproof classification. Artif. Intell. **186**, 123–156 (2012)
16. Procaccia, A.D., Tennenholtz, M.: Approximate mechanism design without money. In: Proceedings of the 10th ACM Conference on Electronic Commerce (ACM-EC), pp. 177–186 (2009)
17. Procaccia, A.D., Tennenholtz, M.: Approximate mechanism design without money. ACM Trans. Econ. Comput. (TEAC) **1**(4), 1–26 (2013)
18. Schummer, J., Vohra, R.V.: Strategy-proof location on a network. J. Econ. Theory **104**(2), 405–428 (2002)
19. Serafino, P., Ventre, C.: Truthful mechanisms without money for non-utilitarian heterogeneous facility location. In: Proceedings of the 29th AAAI Conference on Artificial Intelligence (AAAI), pp. 1029–1035 (2015)
20. Sui, X., Boutilier, C.: Approximately strategy-proof mechanisms for (constrained) facility location. In: Proceedings of the 14th International Conference on Autonomous Agents and Multiagent Systems (AAMAS), pp. 605–613 (2015)
21. Zou, S., Li, M.: Facility location games with dual preference. In: Proceedings of the 14th International Conference on Autonomous Agents and Multiagent Systems (AAMAS), pp. 615–623 (2015)

Geometric Problem

Online Maximum k-Interval Coverage Problem

Songhua Li[1]($\boxtimes$), Minming Li[1], Lingjie Duan[2], and Victor C. S. Lee[1]

[1] City University of Hong Kong, Kowloon, Hong Kong SAR, China
songhuali3-c@my.cityu.edu.hk, {minming.li,csvlee}@cityu.edu.hk
[2] Singapore University of Technology and Design, Singapore, Singapore
lingjie_duan@sutd.edu.sg

Abstract. We study the online maximum coverage problem on a line, in which, given an online sequence of sub-intervals (which may intersect among each other) of a target large interval and an integer k, we aim to select at most k of the sub-intervals such that the total covered length of the target interval is maximized. The decision to accept or reject each sub-interval is made immediately and irrevocably (no preemption) right at the release timestamp of the sub-interval. We comprehensively study different settings of the problem, regarding the number of total released sub-intervals, we consider the unique-number (UN) setting where the total number is known in advance and the arbitrary-number (AN) setting where the total number is not known, respectively; regarding the length of a released sub-interval, we generally consider three settings: each sub-interval is of a normalized unit-length (UL), a flexible-length (FL) in a known range, or an arbitrary-length (AL). In addition, we extend the UL setting to a generalized unit-sum (US) setting, where a batch of a finite number of disjoint sub-intervals of the unit total length is released instead at each timestamp, and accordingly k batches can be accepted. We first prove in the AL setting that no online deterministic algorithm can achieve a bounded competitive ratio. Then, we present lower bounds on the competitive ratio for the other settings concerned in this paper. For the offline problem where the sequence of all the released sub-intervals is known in advance to the decision-maker, we propose a dynamic-programming-based optimal approach as the benchmark. For the online problem, we first propose a single-threshold-based deterministic algorithm SOA by adding a sub-interval if the added length exceeds a certain threshold, achieving competitive ratios close to the lower bounds, respectively. Then, we extend to a double-thresholds-based algorithm DOA, by using the first threshold for exploration and the second threshold (larger than the first one) for exploitation. With the two thresholds solved by our proposed program, we show that DOA improves SOA in the worst-case performance. Moreover, we prove that a deterministic algorithm that accepts sub-intervals by multi non-increasing thresholds cannot outperform even SOA.

Keywords: Maximum k-coverage problem · Budgeted maximum coverage problem · Interval coverage · Online algorithm

© Springer Nature Switzerland AG 2020
W. Wu and Z. Zhang (Eds.): COCOA 2020, LNCS 12577, pp. 455–470, 2020.
https://doi.org/10.1007/978-3-030-64843-5_31

1   Introduction

In the classical MAXIMUM k-COVERAGE PROBLEM, we are given a universal set of elements $U = \{U_1, \cdots, U_m\}$ in which each is associated with a weight $w : U \to \mathbb{R}$, a collection of subsets $S = \{S_1, \cdots, S_n\}$ of U and an integer k, and we aim to select k sets from S that maximize the total weight of covered elements in U. Hochbaum et al. [1] showed that this problem is NP-hard and presented a $(1 - \frac{1}{e})$-approximation algorithm that greedily selects a set that maximally increases the current overall coverage. The BUDGETED MAXIMUM COVERAGE (BMC) problem generalizes the classical coverage problem above by further associating each $S_i \in S$ with a cost $c : S \to \mathbb{R}$ and relaxing the budget k from an integer to a real number, in which the goal is replaced by selecting a sub-collection of the sets in S that maximizes the total weight of the covered elements in U while adhering to the budget k. Clearly, the BMC problem is also NP-hard and actually has a $(1 - \frac{1}{e})$-approximation algorithm [3]. In the online version of the above maximum coverage problems, where at each timestamp i a set $S_i \in S$ is released together with its elements and associated values, an algorithm must decide whether to accept or reject each set S_i at its release timestamp i and may also drop previously accepted sets (*preemption*). However, each rejected or dropped set cannot be retrieved at a later timestamp.

In this paper, we consider the online maximum k-coverage problem on a line without preemption. Given an online sequence of sub-intervals of a target interval, we aim to accept k of the sub-intervals irrevocably such that the total covered length of the target interval is maximized. We refer to this variant as the ONLINE MAXIMUM k-INTERVAL COVERAGE PROBLEM as formally defined in Sect. 2. Regarding the length of a sub-interval, we generally consider the Unit-Length (UL), the Flexible-Length (FL), and the Arbitrary-Length (AL) settings, respectively. We consider the Unique-Number (UN) and the Arbitrary-Number (AN) settings, respectively, regarding the total number of released sub-intervals. In particular, our problem under the UN setting is essentially the classical maximum k-coverage problem (or say, the BMC with unit-cost sets only and an integer budget k) without preemption, by the following *reduction* method: we partition the target interval of our problem into discrete small intervals by the boundary points of all the released sub-intervals, then, the small intervals are equivalent to the elements of a universal set U in which each element has a weight equal to the length of its corresponding small interval, and the released sub-intervals are equivalent to the sets in the collection $S = \{S_1, \cdots, S_n\}$. The objective remains the same.

Related Works. We survey relevant researches along two threads. *The first thread* is about the Online Budgeted Maximum Coverage (OBMC) problem, Saha et al. [7] presented a 4-competitive deterministic algorithm for the setting where sets have unit costs. Rawitz and Rosén [5] showed that the competitive ratio of any deterministic online algorithm for the OBMC problem must depend on the maximum ratio r between the cost of a set and the total budget, and also presented a lower bound of $\Omega(\frac{1}{\sqrt{1-r}})$ and a $\frac{4}{1-r}$-competitive deterministic algorithm. Ausiello et al. [2] studied a special variant of online maximum k-coverage problem, the maximum k-vertex coverage problem, where each element belongs to exactly two sets and the intersection of any two sets has size at most one. They presented a deterministic 2-competitive algorithm and gave a lower bound of $\frac{3}{2}$. The *second thread* is about the online k-secretary problem [12], which was introduced by Kleinberg [11] and aimed to select k out of n independent values for maximizing the expected sum of individual secretary values. Bateni et al. [4] studied a more general version called the submodular secretary problem, which aims to maximize the expectation of a submodular function that defines the efficiency of selected candidates based on their overlapping skills. Our problem is similar to theirs as the objective function of our problem is also submodular (see $Len(\cdot)$ of our model in Sect. 2). However, we focus on the adversarial release order of sub-intervals (secretaries) in the worst-case analysis of deterministic algorithms while [4] focused on a random release order of secretaries in the average-case analysis of algorithms. Other works related to this paper include the interval scheduling problem, the set cover problem, and the online knapsack problem. Interested readers may refer to [8,9,13–16].

Table 1. Main results in this paper

Settings		Lower bounds	Upper bounds
UL	UN	$\sqrt{2}$ for $k = 2$ decrease as $k \geq 3$ increase (Theorem 2)	<2 (Theorems 4 & 8)
	AN	$\sqrt{2}$ for $k = 2$ decrease as $k \geq 3$ increase (Corollary 1)	$\frac{\sqrt{9k^2-14k+9}-k-1}{2(k-1)} + 1$ (Corollary 4)
FL	UN	$\frac{2km}{2km+(1-m)\min\{k,n-k\}}(<2)$ (Theorem 3)	$<1 + \frac{k}{k-1}\sqrt{\frac{1+8m}{4}}$ (Theorem 6)
	AN	$\frac{2m}{m+1}$ (Corollary 2)	$\frac{\sqrt{(1+8m)k^2-(6+8m)k+9}-k-1}{2(k-1)} + 1$ (Corollary 5)
AL	UN or AN	$+\infty$ (Theorem 1)	-
US	UN	$\sqrt{2}$ for $k = 2$ decrease as $k \geq 3$ increase (Corollary 3)	<2 (Theorem 5)

Our Contribution. Results of this paper are three-fold. *First*, we show that no online deterministic algorithm can achieve a bounded competitive ratio in the AL setting, and present lower bounds on the competitive ratio for the other settings, respectively, in a constructive way. *Second*, we give an $O(kn + n \log n)$-time optimal solution to the offline problem where the sequence of all the released sub-intervals is known in advance to the decision-maker, by applying a dynamic programming-based approach. *Third*, for the online problems, we propose two $O(n)$-time deterministic algorithms, SOA and DOA, with their competitive ratios proved to be close to the lower bounds in the settings, respectively. We also extend our results in UL to a generalized unit-sum (US) setting, where at each timestamp, a batch of a finite number of disjoint sub-intervals is released instead and accordingly one can accept at most k released batches. In addition, we show that any deterministic algorithm, that accepts sub-intervals by non-decreasing thresholds, cannot achieve better performance even than the SOA does.

Main results of this paper are summarized in Table 1, in which, for ease of understanding, some complicated parameter-dependent results are approximated by formulations in bold. For precise results, please refer to the corresponding theorems or corollaries. *Due to space constraints, omitted proofs can be found in the full version of this paper* [6].

2   Preliminaries

Table 2. Notations in this paper.

Notations	Descriptions
$[0, a]$	The target interval
k	The maximum number of sub-intervals to accept
$V_i = [o_i, d_i]$	The ith released sub-interval
$\mathbb{V}_i = \{V_1, V_2, \cdots, V_i\}$	The sequence of the first i released sub-intervals
$\chi(\mathbb{V}_n, k)$	The optimal solution for the offline problem, given both the set $\mathbb{V}$ of offline sub-intervals and the quota k beforehand
$\Lambda(V_i, V_j)$	The length of the intersection between sub-intervals V_i and V_j, i.e., $\Lambda(V_i, V_j) = |V_i \cap V_j|$
$\Phi(\mathbb{V}_i)$	The subset of $\mathbb{V}_i$ that are accepted by our algorithm
$Len(U)$	The cumulative length of the parts of $[0, a]$ that are covered by sub-intervals in a given set U, i.e., $Len(U) = |\bigcup_{V_i \in U} V_i|$. Also, we use $Len(V_i)$ to denote the length of a sub-interval V_i, i.e., $Len(V_i) = |V_i|$

The Model. Table 2 summarizes key notations in this paper. An online sequence $\mathbb{V} = \{V_1, V_2, \cdots\}$ of sub-intervals of a large target interval $[0, a]$ are released in an adversarial order to the decision-maker, in which $V_i = [o_i, d_i] \subseteq [0, a]$ for each $V_i \in \mathbb{V}$. Upon the arrival of each $V_i \in \mathbb{V}$, the decision-maker must make a decision whether to accept or reject V_i immediately and irrevocably.

For example, when recruiting at most k employees across different domains of expertise in the target interval, each released sub-interval represents a candidate's expertise domain. The hiring decision on each sub-interval is irrevocable and must be made on candidate arrival without knowing future sub-intervals. Due to the quota limitation, the decision-maker can accept no more than k (≥ 2) sub-intervals[1]. Any two different sub-intervals $V_i, V_j \in \mathbb{V}$ may intersect (i.e., $[o_i, d_j] \cap [o_j, d_j] \neq \varnothing$) considering that the expertise of candidates may overlap in reality. Now, we formally define the settings studied in this paper: with respect to the length $(d_i - o_i)$ of each $V_i \in \mathbb{V}$, we consider three settings.

- **Unit Length (UL):** $|d_i - o_i| = 1$ is normalized with regard to a;
- **Flexible Length (FL):** $|d_i - o_i|$ varies in a known range $[1, m]$, in which $m > 1$ as $m = 1$ degenerates the case to the UL setting;
- **Arbitrary Length (AL):** $|d_i - o_i|$ varies arbitrarily in $[0, a]$;

In addition, we also consider a generalized version of the UL setting, which is the **Unit Sum (US)** setting: each $V_i \in \mathbb{V}$ is no longer restricted to contain only one sub-interval, but a batch of a finite number of disjoint sub-intervals of $[0, a]$ whose sum length is equal to 1. This tells that a candidate masters different domains of expertise. We keep the same unit-sum for all the sub-intervals to tell similar strength of all the job candidates. Accordingly, k batches of sub-intervals can be accepted in the US setting. With respect to the number $|\mathbb{V}|$ of total released sub-intervals, we consider the following two settings respectively.

- **Unique Number (UN):** $|\mathbb{V}|$ is known in advance as a constant $n \in \mathbb{N}^*$. We further restrict $n \geq k + 1$ as otherwise (when $n \leq k$) an optimal solution can be easily achieved by just accepting all sub-intervals;
- **Arbitrary Number (AN):** $|\mathbb{V}|$ is not known;

When two settings are linked by a "-", we refer to the case that the two settings hold together. For example, we use UL-UN to refer to the setting where all sub-intervals have unit length and the total number of released sub-intervals are known in advance. Whenever we specify a single setting in one dimension, we do not distinguish among settings in the other dimension. For example, when specifying the UN setting only, we actually refer to the context as any setting in {UL-UN, FL-UN, AL-UN}.

Given a sequence $\mathbb{V} = \{V_1, V_2, \cdots\}$ of online sub-intervals of $[0, a]$, the **objective** is to accept a subset $U \subseteq \mathbb{V}$ of sub-intervals such that $|U| \leq k$ and the cumulative length $Len(U)$ of the parts of $[0, a]$ that are covered by accepted sub-intervals in U is maximized. Denote $\text{ALG}(\mathbb{V})$ and $\text{OPT}(\mathbb{V})$ as the covered length by an online algorithm ALG and by an optimal offline solution with complete information of all sub-intervals known beforehand, respectively. We slightly abuse notations by rewriting $\text{ALG}(\mathbb{V})$ and $\text{OPT}(\mathbb{V})$ to ALG and OPT, respectively. For $\rho \geq 1$, a deterministic online algorithm ALG is called ρ-competitive [10] for the problem if $\text{OPT}(\mathbb{V}) \leq \rho \text{ALG}(\mathbb{V})$ for every instance $\mathbb{V}$. Alternatively,

---

[1] When $k = 1$, our problem degenerates to the classical secretary problem without expertise sub-interval overlap.

we also say the competitive ratio of ALG is ρ for the problem. Further, when a number $\gamma \geq 1$ ensures that $\gamma \leq \rho$ holds for all deterministic online algorithms, we say γ is a lower bound the on competitive ratio for the problem.

3 Lower Bounds

We construct lower bounds on the competitive ratio for the settings studied in this paper, respectively.

Theorem 1. *In the AL setting, no online deterministic algorithm can achieve a bounded competitive ratio.*

Proof. Let ε be a small positive number, i.e., $0 < \varepsilon \ll 1$. Suppose the first k sub-intervals released as $\mathbb{V}_k = \{[0, \varepsilon^{k+1-i}] | i = 1, 2, \cdots, k\}$. We discuss two cases.

Case 1. Online algorithm (ALG) rejects some sub-interval $V_j = [0, \varepsilon^{k+1-j}] \in \mathbb{V}_k$. Afterwards, the adversary only release sub-intervals as $[0, \varepsilon^{k+1-j+1}]$ instead. This way, the optimal solution (OPT) is able to achieve an overall length at least ε^{k+1-j} by accepting V_j, while ALG can achieve an overall length at most $\varepsilon^{k+1-j+1}$ by sub-intervals in $\mathbb{V}_{j-1}$, we have $\rho \leq \frac{\varepsilon^{k+1-j}}{\varepsilon^{k+1-j+1}} = \frac{1}{\varepsilon} \to +\infty$ when $\varepsilon \to 0$;

Case 2. ALG accepts all the k sub-intervals in $\mathbb{V}_k$ and hence runs out of its quota. Afterward, the adversary only release sub-intervals as $[0, 1]$. Then, OPT is able to achieve an overall length 1 by accepting some $[0, 1]$, while ALG achieves an overall length exactly equal to ε by $\mathbb{V}_k$, we have $\rho \leq \frac{1}{\varepsilon} \to +\infty$ when $\varepsilon \to 0$.

Theorem 2. *In the UL-UN setting, no online deterministic algorithm can achieve a competitive ratio better than (1), in which* $\alpha = \left\lfloor 1 - \frac{\log(k^{\frac{1}{k}}-1)}{\log(k^{\frac{1}{k}})} \right\rfloor$.

$$
\begin{cases}
\sqrt{2}, & \text{if } k = 2 \\
\min\{k^{\frac{1}{k}}, \frac{k^{\frac{\alpha}{k}}+k-\alpha-1}{k^{\frac{\alpha}{k}}+k-\alpha-2}, \frac{k}{k^{\frac{\alpha}{k}}+k-\alpha-1}\}, & \text{if } 3 \leq k \leq n-\alpha-1 \\
\min\{k^{\frac{1}{k}}, \frac{k^{\frac{\alpha}{k}}+k-\alpha-1}{k^{\frac{\alpha}{k}}+k-\alpha-2}, \frac{n-\alpha+2+k^{\frac{\alpha}{k}}-k^{\frac{n-k}{k}}}{k^{\frac{\alpha}{k}}+k-\alpha-1}\}, & \text{if } n-\alpha \leq k \leq n-1
\end{cases}
\tag{1}
$$

Corollary 1. *For UL-AN, no online deterministic algorithm can achieve a competitive ratio better than (2), in which* $\alpha = \left\lfloor 1 - \frac{\log(k^{\frac{1}{k}}-1)}{\log(k^{\frac{1}{k}})} \right\rfloor$.

$$
\begin{cases}
\sqrt{2}, & \text{if } k = 2 \\
\min\{k^{\frac{1}{k}}, \frac{k^{\frac{\alpha}{k}}+k-\alpha-1}{k^{\frac{\alpha}{k}}+k-\alpha-2}, \frac{k}{k^{\frac{\alpha}{k}}+k-\alpha-1}\}, & \text{if } 3 \leq k
\end{cases}
\tag{2}
$$

Theorem 3. *For FL-UN, no online deterministic algorithm can achieve a competitive ratio better than* $\frac{2km}{2km+(1-m)\min\{k,n-k\}}$ *which is strictly smaller than 2.*

Corollary 2. *For FL-AN, no online deterministic algorithm can achieve a competitive ratio better than* $\frac{2m}{m+1}$ *which is strictly smaller than 2.*

Corollary 3. *For US-UN, no online deterministic algorithm can achieve a competitive ratio better than (1), where $\alpha = \left\lfloor 1 - \frac{\log(k^{\frac{1}{k}} - 1)}{\log(k^{\frac{1}{k}})} \right\rfloor$.*

4 Upper Bounds

We present two online deterministic algorithms in Subsects. 4.2 and 4.3 respectively. Before that, we give an $O(kn + n \log n)$ time dynamic programming approach as a benchmark, which optimally solves the offline problem where the sequence of all the released sub-intervals are given beforehand.

4.1 Dynamic Programming Based Optimal Offline Solution

Since both the UL and the FL settings are special cases of the AL setting, we present our offline solution in the AL setting[2]. Suppose, without loss of generality, that the total number of released sub-intervals in the offline problem equals n.

First, we sort sub-intervals in $\mathbb{V}_n = \{V_1, V_2, \cdots, V_n\}$ in non-decreasing order of their end locations (i.e., the d_i of each V_i), which runs in $O(n \log n)$ *time.* We abuse notations, in this offline solution only, to denote $(V_1, V_2, \cdots, V_n)$ as the sequence of sorted sub-intervals, i.e., $d_1 \leq d_2 \leq \cdots \leq d_n$, and further $\mathbb{V}_i = \{V_1, \cdots, V_i\}$ as the first i sub-intervals in the sequence. Suppose the decision-maker accepts sub-intervals in $\mathbb{V}_n$ in decreasing order of their subscripts as well.

Definition 1. $V_{\psi(i)} = \underset{\{V_j \in \mathbb{V}_{i-1} | o_j < o_i \leq d_j\}}{\arg \max} \{o_i - o_j\}$ *indicates the sub-interval in* $\mathbb{V}_{i-1}$ *that intersects with V_i and has the left-most start location.*

Definition 2. $V_{\phi(i)} = \underset{\{V_j \in \mathbb{V}_{i-1} | d_j < o_i\}}{\arg \min} \{o_i - d_j\}$ *indicates the sub-interval in* $\mathbb{V}_{i-1}$ *that is disjoint from but is closest to V_i.*

Proposition 1. *Once an offline OPT accepts V_i, OPT accepts either $V_{\psi(i)}$ or a sub-interval in $\{V_1, V_2, \cdots, V_{\phi(i)}\}$.*

Second. Since OPT, denoted as $\chi(\mathbb{V}_n, k)$, accepts sub-intervals in $\mathbb{V}_n$ in decreasing order of their subscripts as well, we write the *Bellman Equation* in our dynamic programming as (3) and (4) by setting $i = n$ and $j = k$ initially. Specifically, we discuss the following cases when handling an arbitrary $V_i \in \mathbb{V}_i$.

1. OPT rejects V_i. Then, we have $\chi(\mathbb{V}_i, j) = Len(\mathbb{V}_i)$ if OPT has enough quota, i.e., $i \leq j$, to accept all sub-intervals in $\mathbb{V}_i$; or $\chi(\mathbb{V}_i, j) = \chi(\mathbb{V}_{i-1}, j)$ otherwise;
2. OPT accepts V_i and hence runs out of quota ($j = 0$). Then, $\chi(\mathbb{V}_i, j) = 0$;
3. OPT accepts V_i and remains quota ($j \geq 1$). By Proposition 1,

---

[2] We do not distinguish our offline solution in the other dimension since our solution performs optimally in either the UN or the AN.

(a) OPT further accepts someone in $\{V_1, V_2, \cdots, V_{\phi(i)}\}$. Since V_i is disjoint from the next accepted sub-interval, $\chi(\mathbb{V}_i, j) = Len(V_i) + \chi(\mathbb{V}_{\phi(i)}, j-1)$;

(b) OPT further accepts $V_{\psi(i)}$. To calculate $\chi(\mathbb{V}_i, j)$, we introduce an intermediate function $\kappa(\mathbb{V}_i, j)$ given in Eq. (4)[3], which always accepts the last sub-interval V_i in $\mathbb{V}_i$ and totally accepts j out of i sub-intervals in $\mathbb{V}_i$ such that the overall covered length of the interval $[0, a]$ is maximized. Then, we count the length contributed by V_i as the part without intersection with $V_{\psi(i)}$, which is $Len(V_i) - \Lambda(V_i, V_{\psi(i)})$, and transit the remaining part of OPT's overall length to $\kappa(\mathbb{V}_{\psi(i)}, j-1)$. This way, $\chi(\mathbb{V}_i, j) = Len(V_i) - \Lambda(V_i, V_{\psi(i)}) + \kappa(\mathbb{V}_{\psi(i)}, j-1)$.

$$\chi(\mathbb{V}_i, j) = \begin{cases} Len(\mathbb{V}_i), & i \leq j \\ \max\{\chi(\mathbb{V}_{i-1}, j), Len(V_i) + \chi(\mathbb{V}_{\phi(i)}, j-1), \\ \quad Len(V_i) - \Lambda(V_i, V_{\psi(i)}) + \kappa(\mathbb{V}_{\psi(i)}, j-1)\}, & 1 \leq j < i \\ 0, & j = 0 \end{cases} \quad (3)$$

$$\kappa(\mathbb{V}_i, j) = \begin{cases} \max\{ \begin{matrix} Len(V_i) + \chi(\mathbb{V}_{\phi(i)}, j-1), \\ Len(V_i) - \Lambda(V_i, V_{\psi(i)}) + \kappa(\mathbb{V}_{\psi(i)}, j-1) \end{matrix} \}, & j > 1 \\ Len(V_i), & j = 1 \\ 0, & j = 0 \end{cases} \quad (4)$$

Note that our dynamic programming solution totally generates $O(kn)$ intermediate states in which each state runs in $O(1)$ time. Together with the preliminary sorting step, our offline solution totally runs in $O(kn + n \log n)$ time.

4.2   Single-Threshold Online Algorithm

We first propose an online algorithm, named the Single-threshold based Online Algorithm (SOA), for the UN setting. Then, we extend SOA to SOA$_{AN}$ to tackle the AN setting. Note that SOA and SOA$_{AN}$ can achieve competitive ratios strictly smaller than 2 for the UN and the AN settings, respectively.

In the UN setting, SOA always accepts the first released sub-interval V_1. On the arrival of each future sub-interval $V_i \in \{V_2, ..., V_n\}$, SOA accepts V_i if and only if it meets one of the following two conditions: (i) **Quota-enough condition**, after accepting V_i, SOA has enough quota to accept all the future sub-intervals, i.e., $k - |\Phi(\mathbb{V}_{i-1})| \geq n - i + 1$; (ii) **Threshold-accepting condition**, SOA still has quota (i.e., $|\Phi(\mathbb{V}_{i-1})| \leq k - 1$) and V_i contributes an additional length of at least

$$\theta = \min\{\frac{\sqrt{1 + 2(k-1)(n-k)} - 1}{2k - 2}, \frac{\sqrt{9k^2 - 14k + 9} - k - 1}{4(k-1)}\} \quad (5)$$

to the covered length of $[0, a]$ by previously accepted sub-intervals, i.e.,

$$Len(\Phi(\mathbb{V}_{i-1}) \cup V_i) - Len(\Phi(\mathbb{V}_{i-1})) \geq \theta \quad (6)$$

We summarize SOA in Algorithm 1 and we note that

---

[3] The major difference between $\kappa(\mathbb{V}_i, j)$ and $\chi(\mathbb{V}_i, j)$ is that $\kappa(\mathbb{V}_i, j)$ always accepts the last sub-interval V_i in $\mathbb{V}_i$ while $\chi(\mathbb{V}_i, j)$ does not necessarily.

Algorithm 1. Single-threshold Online Algorithm (SOA)

Input: A sequence $\mathbb{V} = \{V_1, V_2, ..., V_n\}$ of n sub-intervals of the target interval $[0, a]$, in which $V_i = [o_i, d_i]$ for each $V_i \in \mathbb{V}$, the quota k $(2 \leq k \leq n - 1)$;
Output: A set of accepted sub-intervals, i.e., $\Phi(\mathbb{V}_n)$;

1: $\Phi(\mathbb{V}_1) = \{V_1\}$;   {always accept V_1}
2: **for** $i = 2; i + +; i \leq n$ **do**
3:   **if** $|\Phi(\mathbb{V}_{i-1})| = k$ **then**
4:     $\Phi(\mathbb{V}_n) = \Phi(\mathbb{V}_{i-1})$;
5:     **break**;   {complete accepting as SOA runs out of the quota}
6:   **else if** $k - |\Phi(\mathbb{V}_{i-1})| \geq n - i + 1$ **then**
7:     $\Phi(\mathbb{V}_i) = \Phi(\mathbb{V}_{i-1}) \cup V_i$;   {accept V_i by the quota-enough condition}
8:   **else**
9:     **if** $Len(\Phi(\mathbb{V}_{i-1}) \cup V_i) - Len(\Phi(\mathbb{V}_{i-1})) \geq \theta$ with θ given in (5) **then**
10:       $\Phi(\mathbb{V}_i) = \Phi(\mathbb{V}_{i-1}) \cup V_i$;   {accept V_i by threshold-meeting condition}
11:     **else**
12:       $\Phi(\mathbb{V}_i) = \Phi(\mathbb{V}_{i-1})$;   {reject V_i}
13:     **end if**
14:   **end if**
15: **end for**

- Once some sub-interval is accepted by the quota-enough condition, all later-released sub-intervals are accepted by SOA;
- SOA always uses up its quota to accept k sub-intervals and only breaks (in the Step 5 of Algorithm 1) when it accepts k sub-intervals according to the threshold θ from the first $(n - 1)$ released sub-intervals.

Proposition 2. *In SOA, we have* $\theta = \frac{\sqrt{1+2(k-1)(n-k)}-1}{2k-2}$ *if* $\lceil \frac{667n}{1000} \rceil \leq k \leq n-1$, *and* $\theta = \frac{\sqrt{9k^2-14k+9}-k-1}{4(k-1)}$ *if* $2 \leq k \leq \lceil \frac{667n}{1000} \rceil - 1$.

Theorem 4. *For UL-UN, SOA runs in $O(n)$ time and achieves a competitive ratio no larger than* $\min\{\frac{\sqrt{1+2(k-1)(n-k)}-1}{k-1} + 1, \frac{\sqrt{9k^2-14k+9}-k-1}{2(k-1)} + 1\}$.

Proof. SOA runs in $O(n)$ time as it runs in no more than n iterations in which each iteration runs in $O(1)$ time. To show the upper bound of SOA, we discuss in the following two cases.

Case 1. SOA accepts V_n (the last released sub-interval).

This shows that SOA triggers the quota-enough condition when accepting some $V_i \in \{V_2, V_3, ..., V_n\}$, i.e., $k - |\Phi(\mathbb{V}_{i-1})| \geq n - i + 1$. Then, the algorithm accepts all the sub-intervals $\{V_i, V_{i+1}, ..., V_n\}$ that are released later than V_i. Further, $Len(\Phi(\mathbb{V}_n)) = Len(\Phi(\mathbb{V}_{i-1}) \cup \{V_i, V_{i+1}..., V_n\})$. In the worst case, none of the accepted sub-intervals in $\{V_i, V_{i+1}, ..., V_n\}$ contributes additional length to the algorithm since these accepted sub-intervals are also available to OPT. Suppose, without loss of generality, that $\bigcup_{V \in \Phi(\mathbb{V}_{i-1})} V$ consists of a number x

of disjoint intervals, which are denoted by $\mathbb{A}_1$, $\mathbb{A}_2$,..., $\mathbb{A}_x$ respectively. Clearly, $1 \leq x \leq k$. Namely, $Len(\mathbb{A}_1)$,..., $Len(\mathbb{A}_x)$ respectively denote the length of disjoint intervals of $\Phi(\mathbb{V}_{i-1})$. Hence, $Len(\Phi(\mathbb{V}_{i-1})) = \sum_{i=1}^{x} Len(\mathbb{A}_i)$.

Note that each rejected sub-interval can contribute an additional length no more than θ to $Len(\Phi(\mathbb{V}_{i-1}))$, as otherwise it would have been accepted. On one hand, $Len(OPT) \leq Len(\Phi(\mathbb{V}_n)) + \theta(n - k)$ holds naturally since SOA totally rejects $(n - k)$ sub-intervals in $\mathbb{V}_{i-1}$. On the other hand, $Len(OPT) \leq Len(\Phi(\mathbb{V}_n)) + 2\theta x$ because there are totally x disjoint intervals formed by the sub-intervals accepted by SOA, which implies there are at most $2x$ chances that sub-interval could be missed/rejected by SOA, and each rejected sub-interval can contribute less than θ to SOA (by Step 9 of SOA). In summary, the overall length achieved by the OPT is bounded by the following Inequality (7).

$$Len(OPT) \leq Len(\Phi(\mathbb{V}_n)) + \min\{2\theta x, \theta(n - k)\} \tag{7}$$

Hence, we get the ratio

$$\rho = \frac{Len(OPT)}{Len(\Phi(\mathbb{V}_n))} \leq 1 + \frac{\min\{2\theta x, \theta(n - k)\}}{Len(\Phi(\mathbb{V}_n))} \leq 1 + \frac{2\theta x}{\sum_{i=1}^{x} Len(\mathbb{A}_i)}$$

$$\leq 1 + \min\{\frac{\sqrt{1 + 2(k - 1)(n - k)} - 1}{k - 1}, \frac{\sqrt{9k^2 - 14k + 9} - k - 1}{2(k - 1)}\}$$

in which the first inequality holds by (7), the second inequality holds by $Len(\Phi(\mathbb{V}_n)) \geq Len(\Phi(\mathbb{V}_{i-1})) = \sum_{i=1}^{x} Len(\mathbb{A}_i)$, and the last inequality holds by $\sum_{i=1}^{x} Len(\mathbb{A}_i) \geq x$ and (5).

Case 2. SOA does not accept V_n.

This means the quota-enough condition is not triggered during the execution and SOA accepts k sub-intervals by the threshold-accepting condition. This implies the following Inequality (8) because each accepted sub-interval, except V_1 (which contributes 1 to SOA), contributes at least θ to SOA, see Step 9 of SOA.

$$Len(\Phi(\mathbb{V}_n)) \geq 1 + (k - 1)\theta \tag{8}$$

Suppose that $V_i \in \{V_k, ..., V_{n-1}\}$ is the last accepted sub-interval by SOA, i.e., $|\Phi(\mathbb{V}_i)| = k$ and $Len(\Phi(\mathbb{V}_n)) = Len(\Phi(\mathbb{V}_i))$. In other words, SOA misses all sub-intervals in $\{V_{i+1}, ..., V_n\}$ which can be accepted by OPT. Since the algorithm can miss at most $n-k$ sub-intervals, OPT can get an accumulating length at most $n - k$ more than that accepted by SOA, i.e., $Len(OPT) \leq Len(\Phi(\mathbb{V}_n)) + n - k$. Also, OPT cannot get a length over its quota k in this unit-length case. In summary, the overall length accepted by the OPT is bounded by (9).

$$Len(OPT) \leq \min\{k, Len(\Phi(\mathbb{V}_n)) + n - k\} \tag{9}$$

We further discuss two sub-cases.

Case 2.1. $\lceil \frac{667n}{1000} \rceil \leq k \leq n-1$. We have $\theta = \frac{\sqrt{1+2(k-1)(n-k)}-1}{2k-2}$ by Proposition 2. Note that $\frac{\partial(\frac{\theta}{2k-n-1})}{\partial k} = \frac{8k^2-(8n+10)k+9n-3}{4(2k-n-1)^2\sqrt{1+2(k-1)(n-k)}} < 0$ for each $k \in [\lceil \frac{667n}{1000} \rceil, n] \subseteq (\frac{8n+10-\sqrt{(8n-8)^2+132}}{16}, \frac{8n+10+\sqrt{(8n-8)^2+132}}{16})$. Further, we have

Algorithm 2. SOA$_{AN}$

The SOA$_{AN}$ remains the same as the Algorithm 1 by discarding the **else if** branch of the quota-enough condition in Lines 6-7 and setting $\theta = \frac{\sqrt{9k^2-14k+9}-k-1}{4(k-1)}$;

$$\frac{\theta}{\frac{2k-n-1}{k-1}} \leq \frac{\theta}{\frac{2k-n-1}{k-1}}\Big|_{k=\lceil\frac{667n}{1000}\rceil} < 1 \tag{10}$$

Hence,

$$
\begin{aligned}
\rho &= \frac{Len(OPT)}{Len(\Phi(\mathbb{V}_n))} \\
&\leq \min\{\frac{k}{Len(\Phi(\mathbb{V}_n))}, 1+\frac{n-k}{Len(\Phi(\mathbb{V}_n))}\} && \text{by (9)} \\
&\leq \min\{\frac{k}{1+(k-1)\theta}, 1+\frac{n-k}{1+(k-1)\theta}\} && \text{by (8)} \\
&= 1+\frac{\sqrt{1+2(k-1)(n-k)}-1}{k-1} && \text{by (10) and } \theta = \frac{\sqrt{1+2(k-1)(n-k)}-1}{2k-2}
\end{aligned}
$$

Case 2.2. $2 \leq k \leq \lceil\frac{667n}{1000}\rceil - 1$. We have $\theta = \frac{\sqrt{9k^2-14k+9}-k-1}{4(k-1)} \leq \frac{\sqrt{1+2(k-1)(n-k)}-1}{2k-2}$ by Proposition 2. Hence,

$$
\begin{aligned}
\rho &= \frac{Len(OPT)}{Len(\Phi(\mathbb{V}_n))} \\
&\leq \min\{\frac{k}{Len(\Phi(\mathbb{V}_n))}, 1+\frac{n-k}{Len(\Phi(\mathbb{V}_n))}\} && \text{by (9)} \\
&\leq \min\{\frac{k}{1+(k-1)\theta}, 1+\frac{n-k}{1+(k-1)\theta}\} && \text{by (8)} \\
&\leq \frac{k}{1+(k-1)\theta} = \frac{\sqrt{9k^2-14k+9}-k-1}{2(k-1)}+1 && \text{by } \theta = \frac{\sqrt{9k^2-14k+9}-k-1}{4(k-1)}
\end{aligned}
$$

By Case 1 and Case 2, the proof completes.

Corollary 4. *For UL-AN, SOA$_{AN}$ runs in $O(n)$ time and achieves a competitive ratio no larger than $\frac{\sqrt{9k^2-14k+9}-k-1}{2(k-1)}+1$ for any limited time frame.*

The SOA algorithm can solve the the flexible-length case. Using a similar analysis idea as in Theorem 4, we have the following Theorem 5 and Theorem 6.

Theorem 5. *For US-UN, SOA runs in $O(n)$ time and achieves a competitive ratio no larger than $\min\{\frac{\sqrt{1+2(k-1)(n-k)}-1}{k-1}+1, \frac{\sqrt{9k^2-14k+9}-k-1}{2(k-1)}+1\}$.*

Theorem 6. *For FL-UN, SOA runs in $O(n)$ time and achieves a competitive ratio no larger than $\min\{\frac{\sqrt{1+2(k-1)(n-k)m-1}}{k-1}+1, \frac{\sqrt{(1+8m)k^2-(6+8m)k+9}-k-1}{2(k-1)}+1\}$ in which m indicates the maximum possible length of a sub-interval.*

Corollary 5. *For FL-AN, SOA_{AN} runs in $O(n)$ time and achieves a competitive ratio no larger than $\frac{\sqrt{(1+8m)k^2-(6+8m)k+9}-k-1}{2(k-1)}+1$ for any limited accepting time frame, in which m indicates the maximum possible length of a sub-interval.*

4.3 Double-Threshold Online Algorithm

Built upon SOA, we now present the Double-threshold Online Algorithm (**DOA**) under the UN setting, which remains the same as Algorithm 1 but extends the single threshold θ in the threshold-meeting condition to two thresholds θ_1 and θ_2 (by using θ_1 for exploration and θ_2 for exploitation). Specifically, SOA changes the threshold from θ_1 to θ_2 once accepting ω sub-intervals, in which the values of $(\omega, \theta_1, \theta_2)$ are given later by solving the non-linear program (i-viii). Before that, we first give the competitive analysis of DOA.

Denote j as the number of disjoint intervals formed by the sub-intervals accepted by DOA. When DOA accepts less than k sub-intervals by threshold, the overall length achieved by OPT is no more than $Len(\Phi(\mathbb{V}_n)) + 2j\theta$ and certainly no more than the quota k, implying Lemma 1. When DOA accepts k sub-intervals by threshold, the overall length of OPT should be no more than $(n-k+1+(\omega-1)\theta_1+(k-\omega)\theta_2)$ and also no more than k, implying Lemma 2.

Lemma 1. *In UL-UN, when DOA accepts i $(1 \leq i \leq k-1)$ sub-intervals by threshold and quota-enough accepts $k-i$ sub-intervals, OPT can achieve an overall length at most $\frac{\min\{k, j+(i-j)\theta_1+2j\theta_1\}}{j+(i-j)\theta_1}$ times of DOA length for $1 \leq i \leq \omega-1$ or at most $\frac{\min\{k, j+(\omega-1)\theta_1+(i-\omega+2j)\theta_2\}}{j+(\omega-1)\theta_1+(i-\omega)\theta_2}$ times of DOA length for $\omega \leq i \leq k-1$.*

Lemma 2. *In UL-UN, when DOA threshold-accepts k sub-intervals, OPT can achieve an overall length at most $\frac{\min\{k, n-k+1+(\omega-1)\theta_1+(k-\omega)\theta_2\}}{1+(\omega-1)\theta_1+(k-\omega)\theta_2}$ times of DOA's.*

Theorem 7. *In UL-UN, the competitive ratio of DOA is upper bounded by*

$$C = \max\{1+2\theta_1, 1+\frac{2\theta_2}{1+\frac{\omega-s}{s}\theta_1}, \frac{k}{s+1+(\omega-s-1)\theta_1}, \frac{\min\{k, n-k+q\}}{q}\}$$

$$(11)$$

where $q = 1+(\omega-1)\theta_1+(k-\omega)\theta_2$, $s = \frac{k+(1-\omega)\theta_1-2\theta_2}{1+2\theta_2-\theta_1}$.

Proof. Suppose, w.l.o.g., that DOA threshold-accepts i sub-intervals and quota-enough accepts the other $(k-i)$ sub-intervals. Let j denote the number of disjoint intervals formed by the k accepted sub-intervals. In the worst case scenario, the $k-i$ quota-enough accepted sub-intervals only contribute an additional length of zero to the covered length of the interval $[0, a]$ by previously threshold-accepted sub-intervals. We discuss in three cases,

Case 1. $1 \leq i \leq \omega - 1$. By Lemma 1, we have the ratio

$$\rho \leq \frac{\min\{k, Len(\Phi(\mathbb{V}_n)) + 2j\theta_1\}}{Len(\Phi(\mathbb{V}_n))} \leq \frac{\min\{k, j + (i-j)\theta_1 + 2j\theta_1\}}{j + (i-j)\theta_1} \leq 1 + 2\theta_1.$$

Case 2. $\omega \leq i \leq k - 1$. Denote $s+1$ as the minimum number of disjoint intervals formed by the accepted sub-intervals in an optimal solution. Suppose w.l.o.g., OPT achieves its maximum overall length k in the worst-case scenario. We have $s + 1 + (\omega - s)\theta_1 + 2(s+1)\theta_2 \geq k$ and $s + (\omega - s)\theta_1 + 2s\theta_2 < k$ in the worst case,

$$\frac{k + (1 - \omega)\theta_1 - 2\theta_2}{1 + 2\theta_2 - \theta_1} \leq s \leq \frac{k - \omega\theta_1}{1 - \theta_1 + 2\theta_2}$$

Further, we get $s = \frac{k + (1-\omega)\theta_1 - 2\theta_2}{1 + 2\theta_2 - \theta_1} \in [1, \omega - 1]$ satisfying Inequality (4.3).

Case 2.1. $j \leq s$. By Lemma 1, the ratio is upper bounded by

$$\frac{j + (\omega - j)\theta_1 + (i - \omega)\theta_2 + 2j\theta_2}{j + (\omega - j)\theta_1 + (i - \omega)\theta_2} \leq \frac{s + (\omega - s)\theta_1 + 2s\theta_2}{s + (\omega - s)\theta_1} = 1 + \frac{2\theta_2}{1 + \frac{\omega - s}{s}\theta_1}$$

Case 2.2. $j \geq s + 1$. By Lemma 1, the ratio is upper bounded by

$$\frac{k}{j + (\omega - j)\theta_1 + (i - j)\theta_2} \leq \frac{k}{s + 1 + (\omega - s - 1)\theta_1}$$

Case 3. $i = k$. By Lemma 2, the competitive ratio of DOA is upper bounded by $\frac{\min\{k, n - k + 1 + (\omega - 1)\theta_1 + (k - \omega)\theta_2\}}{1 + (\omega - 1)\theta_1 + (k - \omega)\theta_2}$.

By Cases 1, 2, 3, competitive ratio of DOA is upper bounded by (11).

To find the timing ω and the thresholds (θ_1, θ_2) that optimize the competitive ratio of DOA, we propose the following nonlinear program to minimize the maximum value (denoted by C) of the competitive ratio in Eq. (11), where constraints (ii)-(vi) are transformed from Eq. (11) respectively and constraints (vii)[4] and (viii) are naturally required.

$$\min_{(\omega, \theta_1, \theta_2)} C \tag{i}$$

$$\text{s.t.} \quad C \geq 1 + 2\theta_1, \tag{ii}$$

$$C \geq 1 + \frac{2\theta_2}{1 + \frac{\omega - s}{s}\theta_1} \tag{iii}$$

$$C \geq \frac{k}{s + 1 + (\omega - s - 1)\theta_1} \tag{iv}$$

$$C \geq \frac{\min\{k, n - k + 1 + (\omega - 1)\theta_1 + (k - \omega)\theta_2\}}{1 + (\omega - 1)\theta_1 + (k - \omega)\theta_2} \tag{v}$$

$$s = \frac{k + (1 - \omega)\theta_1 - 2\theta_2}{1 + 2\theta_2 - \theta_1} \tag{vi}$$

$$1 \leq \omega \leq k \tag{vii}$$

$$0 < \theta_1 < \theta_2 \leq 1, \tag{viii}$$

---

[4] Constraint (vii) actually can be restricted, by calculation, to $\lceil \frac{k+1}{5} \rceil \leq \omega \leq k$.

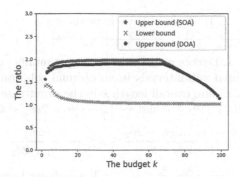

Fig. 1. Performance among DOA, SOA and the lower bound in UL-UN.

Theorem 8. *DOA with $(\omega, \theta_1, \theta_2)$ returned by program (i)–(viii) achieves the best worst-case performance of online algorithms with two thresholds.*

Since the program (i–viii) is nonlinear and is complicated when transformed into a linear programming, we search its approximated solution under the UL-UN setting by giving the precision of θ as 0.01 and $n = 100$. According to the searching result, we observe that the ω value should be set at around $0.8k$ and $\theta_1 < \theta_2$. The double-threshold algorithm DOA improves the performance of the single-threshold based algorithm (see Fig. 1 below). What is worth noting is that, when the ratio $\frac{k}{n}$ of the quota over the total number of online sub-intervals is relatively small (*resp.* large), we find that more quota induces worse (*resp.* better) performances of both SOA and DOA since OPT has more chances to gain values from those missed sub-intervals by our algorithms (*resp.* since online algorithms have fewer chances to miss values from OPT). The turning point of $\frac{k}{n}$ is around $\frac{2}{3}$ in SOA since the two items of the competitive ratio in Theorem 4 are monotone decreasing and increasing, respectively, with regard to k, and meet when $\frac{k}{n} \approx \frac{2}{3}$. Interestingly, the turning point of $\frac{k}{n}$ is also around $\frac{2}{3}$ in DOA, see the example in Fig. 1. We can also extend to more than two thresholds, yet the analysis will be more involved with only mild improvement. Particularly, when the thresholds in an algorithm are non-increasing as accepting sub-intervals, we have the following theorem.

Theorem 9. *SOA outperforms any online deterministic algorithm that accepts sub-intervals by non-increasing thresholds.*

5   Concluding Remarks

This paper studies the online maximum k-coverage problem on a line without preemption. With regard to the length of each sub-interval and the number of totally released sub-intervals, we comprehensively consider different settings in this paper. Our contribution is three-fold.

First, we present lower bounds on the competitive ratio for the settings respectively. **Second**, we propose an optimal solution for the offline problem where the sequence of offline sub-intervals is given to the decision-maker at the very beginning. **Third**, we present two online algorithms, including a single-threshold-based algorithm SOA and a double-threshold-based algorithm (DOA). DOA uses its first threshold (which is usually set below 0.5) for exploration in accepting the first $[0.8k]$ released sub-intervals and its second threshold (which is set larger than the first threshold) for exploitation in accepting the last $k - [0.8k]$ sub-intervals. We prove that SOA achieves competitive ratios close to the lower bounds, respectively, and DOA, with its parameters computed by our proposed program, improves the performance of SOA slightly. In addition, we show that any online deterministic algorithm that accepts sub-intervals by non-increasing thresholds, cannot achieve a competitive ratio better than SOA no matter how many thresholds the algorithm uses.

For the future work, we may consider the case that different sub-intervals are associated with different costs instead of unit costs in this paper, considering that candidates may call for different payments in crowding-sourcing activities.

Acknowledgements. This work was done when Songhua Li was visiting the Singapore University of Technology and Design. Minming Li is also from City University of Hong Kong Shenzhen Research Institute, Shenzhen, P.R. China. The work described in this paper was partially supported by Project 11771365 supported by NSFC. We would like to thank all the reviewers for their comments.

References

1. Hochbaum, D.S., Pathria, A.: Analysis of the greedy approach in problems of maximum k-coverage. Naval Res. Logist. (NRL) **45**(6), 615–627 (1998)
2. Ausiello, G., Boria, N., Giannakos, A., Lucarelli, G., Paschos, V.T.: Online maximum k-coverage. Discrete Appl. Math. **160**(13–14), 1901–1913 (2012)
3. Khuller, S., Moss, A., Naor, J.S.: The budgeted maximum coverage problem. Inf. Process. Lett. **70**(1), 39–45 (1999)
4. Bateni, M., Hajiaghayi, M., Zadimoghaddam, M.: Submodular secretary problem and extensions. ACM Trans. Algorithms (TALG) **9**(4), 1–23 (2013)
5. Rawitz, D., Rosén, A.: Online budgeted maximum coverage. In: the 24th Annual European Symposium on Algorithms. Schloss Dagstuhl-Leibniz-Zentrum fuer Informatik (2016)
6. Li, S., Li, M., Duan, L., Lee, V.C.S.: Online maximum k-interval coverage problem (2020). http://arxiv.org/abs/2011.10938
7. Saha, B., Getoor, L.: On maximum coverage in the streaming model and application to multi-topic blog-watch. In: Proceedings of the 2009 SIAM International Conference on Data Mining, pp. 697–708. Society for Industrial and Applied Mathematics (2009)
8. Chrobak, M., Jawor, W., Sgall, J., Tichý, T.: Online scheduling of equal-length jobs: randomization and restarts help. SIAM J. Comput. **36**(6), 1709–1728 (2007)
9. Chin, F.Y., Chrobak, M., Fung, S.P., Jawor, W., Sgall, J., Tichý, T.: Online competitive algorithms for maximizing weighted throughput of unit jobs. J. Discrete Algorithms **4**(2), 255–276 (2006)

10. Borodin, A., El-Yaniv, R.: Online Computation and Competitive Analysis. Cambridge University Press, New York (2005)
11. Kleinberg, R.D.: A multiple-choice secretary algorithm with applications to online auctions. SODA **5**, 630–631 (2005)
12. Feldman, M., Zenklusen, R.: The submodular secretary problem goes linear. SIAM J. Comput. **47**(2), 330–366 (2018)
13. Babaioff, M., Immorlica, N., Kempe, D., Kleinberg, R.: A knapsack secretary problem with applications. In: Charikar, M., Jansen, K., Reingold, O., Rolim, J.D.P. (eds.) APPROX/RANDOM -2007. LNCS, vol. 4627, pp. 16–28. Springer, Heidelberg (2007). https://doi.org/10.1007/978-3-540-74208-1_2
14. Vaze, R.: Online knapsack problem under expected capacity constraint. In: IEEE INFOCOM 2018-IEEE Conference on Computer Communications, pp. 2159–2167. IEEE (2018)
15. Alon, N., Awerbuch, B., Azar, Y., Buchbinder, N., Naor, J.: The online set cover problem. SIAM J. Comput. **39**(2), 361–370 (2009)
16. Assadi, S., Khanna, S., Li, Y.: Tight bounds for single-pass streaming complexity of the set cover problem. SIAM J. Comput. STOC 16–341 (2019)

Vertex Fault-Tolerant Spanners
for Weighted Points in Polygonal
Domains

R. Inkulu$^{(\boxtimes)}$ and Apurv Singh

Department of Computer Science and Engineering, IIT Guwahati, Guwahati, India
{rinkulu,apursingh}@iitg.ac.in

Abstract. Given a set S of n points, a weight function w to associate
a non-negative weight to each point in S, a positive integer $k \geq 1$, and
a real number $\epsilon > 0$, we devise the following algorithms to compute
a k-vertex fault-tolerant spanner network $G(S, E)$ for the metric space
induced by the weighted points in S: (1) When the points in S are
located in a simple polygon, we present an algorithm to compute G
with multiplicative stretch $\sqrt{10} + \epsilon$, and the number of edges in G is
$O(kn(\lg n)^2)$. (2) When the points in S are located in the free space of
a polygonal domain $\mathcal{P}$, we present an algorithm to compute G of size
$O(\sqrt{h}kn(\lg n)^2)$ and its multiplicative stretch is $6 + \epsilon$. Here, h is the
number of simple polygonal holes in $\mathcal{P}$.

Keywords: Computational geometry · Geometric spanners ·
Approximation algorithms

1 Introduction

In designing geometric networks on a given set of points in a metric space,
it is desirable for the network to have short paths between any pair of nodes
while being sparse with respect to the number of edges. Let $G(S, E)$ be an edge-
weighted geometric graph on a set S of n points in $\mathbb{R}^d$. The distance in G between
any two nodes p and q, denoted by $d_G(p, q)$, is the length of a shortest (that is,
a minimum-weighted) path between p and q in G. For a real number $t \geq 1$, the
graph G is called a *t-spanner* of points in S if for every two points $p, q \in S$,
$d_G(p, q)$ is at most t times the Euclidean distance between p and q. The smallest
t for which G is a t-spanner is called the *stretch factor* of G, and the number of
edges of G is called its size.

Peleg and Schäffer [31] in the context of distributed computing, and Chew [21]
in the context of geometric networks, introduced spanner networks. Althöfer
et al. [7] first attempted to study sparse spanners on edge-weighted graphs with
edge weights obeying the triangle-inequality. The text by Narasimhan and Smid

This research is supported in part by SERB MATRICS grant MTR/2017/000474.

© Springer Nature Switzerland AG 2020
W. Wu and Z. Zhang (Eds.): COCOA 2020, LNCS 12577, pp. 471–485, 2020.
https://doi.org/10.1007/978-3-030-64843-5_32

[30], Gudmundsson and Knauer [25], and the handbook chapter [24] detail various results on Euclidean spanners, including a $(1 + \epsilon)$-spanner for the set S of n points in $\mathbb{R}^d$ that has $O(\frac{n}{\epsilon^{d-1}})$ edges, for any $\epsilon > 0$.

The significant results of geometric spanner networks include spanners of low degree [8,16,20], spanners of low weight [15,23,26], spanners of low diameter [10,11], fault-tolerant spanners [2,13,22,27–29,32], and combinations of these [9,12,19]. For the case of metric space with bounded doubling metric, a few results are given in [33].

As observed in Abam et al. [3], the cost of traversing a path in a network is not only determined by the lengths of edges along the path, but also by the delays occurring at the nodes on the path. The result in [3] models these delays by associating non-negative weights to points. Let S be a set of n points in $\mathbb{R}^d$. For every $p \in S$, let $w(p)$ be the non-negative weight associated with p. The following weighted distance function d_w on S defining the metric space (S, d_w) is considered by Abam et al. in [3] and by Bhattacharjee and Inkulu in [13]: for any $p, q \in S$, $d_w(p, q)$ is equal to $w(p) + |pq| + w(q)$ if $p \neq q$; otherwise, it is equal to 0. (To remind, $|pq|$ is the Euclidean distance between p and q).

Recently, Abam et al. [4] showed that there exists a $(2 + \epsilon)$-spanner with a linear number of edges for the metric space (S, d_w) that has a bounded doubling dimension. And, [3] gives a lower bound on the stretch factor, showing that $(2 + \epsilon)$ stretch is nearly optimal. Bose et al. [17] studied the problem of computing spanners for weighted set of points. They considered points that lie on the plane to have positive weights associated with them; and defined the distance d_w between any two distinct points $p, q \in S$ as $|pq| - w(p) - w(q)$, where $|pq|$ is the Euclidean distance between p and q. Under the assumption that the distance between any pair of points is non-negative, they showed the existence of a $(1 + \epsilon)$-spanner with $O(\frac{n}{\epsilon})$ edges.

A set of $h \geq 0$ disjoint simple polygonal holes contained in a simple polygon P is the *polygonal domain* $\mathcal{P}$. (When h is 0, the polygonal domain $\mathcal{P}$ is essentially a simple polygon.) The free space $\mathcal{D}$ of the given polygonal domain $\mathcal{P}$ is defined as the closure of P, excluding the union of the interior of polygons contained in P. Any path between any two given points in $\mathcal{D}$ needs to be in the free space $\mathcal{D}$. For any two distinct points $p, q \in S$, the geodesic distance along any shortest path between p and q is denoted with $d_\pi(p, q)$. Given a set S of n points in the free space $\mathcal{D}$, computing a geodesic spanner of S is considered in Abam et al. [1]. The result in [1] showed that for the metric space (S, π), for any constant $\epsilon > 0$, there exists a $(5 + \epsilon)$-spanner of size $O(\sqrt{h}n(\lg n)^2)$. Further, for any constant $\epsilon > 0$, [1] gave a $(\sqrt{10} + \epsilon)$-spanner with $O(n(\lg n)^2)$ edges when $h = 0$, i.e., when the polygonal domain is a simple polygon. Given a set S of n points on a polyhedral terrain $\mathcal{T}$, the geodesic distance between any two points $p, q \in S$ is the distance along any shortest path between p and q on $\mathcal{T}$. The algorithm in [4] proved that for a set of unweighted points on a polyhedral terrain, for any constant $\epsilon > 0$, there exists a $(2 + \epsilon)$-geodesic spanner with $O(n \lg n)$ edges.

A graph $G(S, E)$ is a *k-vertex fault-tolerant t-spanner*, denoted by (k, t)-VFTS, for a set S of n points in $\mathbb{R}^d$ if for any subset S' of S with size at

most k, the graph $G \setminus S'$ is a t-spanner for the points in $S \setminus S'$. Algorithms in
Levcopoulos et al. [28], Lukovszki [29], and Czumaj and Zhao [22] compute a
(k, t)-VFTS for the set S of points in $\mathbb{R}^d$. These algorithms are also presented
in [30]. Levcopoulos et al. [28] devised an algorithm to compute a (k, t)-VFTS
of size $O(\frac{n}{(t-1)^{(2d-1)(k+1)}})$ in $O(\frac{n \lg n}{(t-1)^{4d-1}} + \frac{n}{(t-1)^{(2d-1)(k+1)}})$ time, and the other
algorithm to compute a (k, t)-VFTS with $O(k^2 n)$ edges in $O(\frac{kn \lg n}{(t-1)^d})$ time. The
result in [29] gives an algorithm to compute a (k, t)-VFTS of size $O(\frac{kn}{(t-1)^{d-1}})$
in $O(\frac{1}{(t-1)^d}(n \lg^{d-1} n \lg k + kn \lg \lg n))$ time. The algorithm in [22] computes a
(k, t)-VFTS having $O(\frac{kn}{(t-1)^{d-1}})$ edges in $O(\frac{1}{(t-1)^{d-1}}(kn \lg^d n + nk^2 \lg k))$ time
with total weight of edges upper bounded by an $O(\frac{k^2 \lg n}{(t-1)^d})$ multiplicative factor
of the weight of a minimum spanning tree of the given set of points.

Our Contributions

For a real number $t > 1$ and a set S of weighted points, a graph $G(S, E)$ is called
a *t-spanner for weighted points* whenever for any two points p and q in S the
distance between p and q in graph G is at most $t \cdot d_w(p, q)$. To remind, as in [3] and
in [13], the function d_w is defined on a set S of points as follows: for any $p, q \in S$,
it is equal to $w(p) + d_\pi(p, q) + w(q)$ if $p \neq q$; otherwise, $d_w(p, q)$ is 0. Here, $d_\pi(p, q)$
is the geodesic distance between p and q. Given a set S of points, a function w
to associate a non-negative weight to each point in S, an integer $k \geq 1$, and a
real number $t > 0$, a geometric graph G is called a (k, t, w)-*vertex fault-tolerant
spanner for weighted points*, denoted with (k, t, w)-VFTSWP, whenever for any
set $S' \subset S$ with cardinality at most k, the graph $G \setminus S'$ is a t-spanner for the
set $S \setminus S'$ of weighted points. Note that every edge (p, q) in G corresponds to
a shortest geodesic path between two points $p, q \in S$. In addition, the weight
associated with any edge (p, q) of G is the distance $d_\pi(p, q)$ along a geodesic
shortest path between p and q.

In [13], Bhattacharjee and Inkulu devised the following algorithms: one for
computing a $(k, 4 + \epsilon, w)$-VFTSWP when the input points are in $\mathbb{R}^d$, and the
other for computing a $(k, 4+\epsilon, w)$-VFTSWP when the given points are in a simple
polygon. Further, in [14], Bhattacharjee and Inkulu extended these algorithms to
compute a $(k, 4+\epsilon, w)$-VFTSWP when the points are in a polygonal domain and
when the points are located on a terrain. In this paper, we show the following
results for computing a (k, t, w)-VFTSWP:

* Given a simple polygon $\mathcal{P}$, a set S of n points located in $\mathcal{P}$, a weight function
  w to associate a non-negative weight to each point in S, a positive integer
  k, and a real number $0 < \epsilon \leq 1$, we present an algorithm to compute a
  $(k, \sqrt{10} + \epsilon, w)$-VFTSWP that has size $O(kn(\lg n)^2)$. (Refer to Theorem 1.)
  The stretch factor of the spanner is improved from the result in [13], and the
  number of edges is an improvement over the result in [13] when $(\lg n) < \frac{1}{\epsilon^2}$.
  Note that [13] devised an algorithm for computing a $(k, 4 + \epsilon, w)$-VFTSWP
  with size $O(\frac{kn}{\epsilon^2} \lg n)$.

* Given a polygon domain $\mathcal{P}$, a set S of n points located in the free space $\mathcal{D}$ of $\mathcal{P}$, a weight function w to associate a non-negative weight to each point in S, a positive integer k, and a real number $0 < \epsilon \leq 1$, we present an algorithm to compute a $(k, 6 + \epsilon, w)$-VFTSWP with size $O(\sqrt{h}kn(\lg n)^2)$. (Refer to Theorem 3.) Here, h is the number of simple polygonal holes in $\mathcal{P}$. Though the stretch factor of the VFTSWP given in [14] is $(4 + \epsilon)$, its size is $O(\frac{\sqrt{h}kn}{\epsilon^2}(\lg n)^2)$.

The approach in achieving these improvements is different from [13,14]. Instead of clustering (like in [13,14]), following [1], our algorithm uses the s-semi-separated pair decomposition (s-SSPD) of points projected on line segments together with the divide-and-conquer applied to input points. For a set Q of n points in $\mathbb{R}^d$, a *pair decomposition* [5,34] of Q is a set of pairs of subsets of Q, such that for every pair of points of $p, q \in Q$, there exists a pair (A, B) in the decomposition such that $p \in A$ and $q \in B$. Given a pair decomposition $\{\{A_1, B_1\}, \ldots, \{A_s, B_s\}\}$ of a point set, its weight is defined as $\sum_{i=1}^{s}(|A_i| + |B_i|)$. For a set Q of n points in $\mathbb{R}^d$, a *s-semi-separated pair decomposition (s-SSPD)* of Q is a pair decomposition of Q such that for every pair (A, B), the distance between A and B (i.e., the distance of their minimum enclosing disks) is greater than or equal to s times the minimum of the radius of A and the radius of B. (The radius of a point set X is the radius of the smallest ball enclosing all the points in X.) The SSPD has the advantage of low weight as compared to well-known well-separated pair decomposition.

The Euclidean distance between two points p and q is denoted by $|pq|$. For any point p located in the free space $\mathcal{D}$ of a polygonal domain $\mathcal{P}$, and for any line segment ℓ located in $\mathcal{D}$, let d be the geodesic distance between p and ℓ. Then any point $p_\ell \in \ell$ is called a *geodesic projection of p on ℓ* whenever the geodesic (Euclidean) distance between p and p_ℓ is d. We denote a geodesic projection of a point p on a line segment ℓ with p_ℓ. The *splitting line segment* is a line segment in free space with both of its endpoints on the boundary of the polygonal domain. (Note that the boundary of a polygonal domain is the union of boundaries of holes and the boundary of the outer polygon.) Here, π denotes a shortest path between p and q in the polygonal domain. The value of $d_w(p, q)$ is equal to $w(p) + d_\pi(p, q) + w(q)$ if $p \neq q$, and it is 0 otherwise. The length of a shortest path between p and q in a graph G is denoted by $d_G(p, q)$. The distance $d_{G \setminus S'}(p, q)$ denotes the distance along a shortest path between vertices p and q in graph G after removing set $S' \subset V'$ of vertices from G.

Section 2 presents an algorithm to compute a $(k, \sqrt{10} + \epsilon, w)$-VFTSWP when the weighted input points are in a simple polygon. When the points are located in the free space of a polygonal domain, an algorithm for computing a $(k, 6 + \epsilon, w)$-VFTSWP is detailed in Sect. 3.

2   Vertex Fault-Tolerant Spanner for Weighted Points in a Simple Polygon

Given a simple polygon $\mathcal{P}$, a set S of n points located in $\mathcal{P}$, a weight function w to associate a non-negative weight to each point in S, a positive integer k, and a real number $\epsilon > 0$, we devise an algorithm to compute a geodesic $(k, \sqrt{10}+\epsilon, w)$-VFTSWP for the set S of weighted points. That is, if for any set $S' \subset S$ with cardinality at most k, the graph $G \backslash S'$ is a $(\sqrt{10} + \epsilon)$-spanner for the set S' of weighted points. To remind, a graph $G(S', E')$ is a t-spanner for the set S' of weighted points located in a simple polygon $\mathcal{P}$ whenever the distance between any two points $p, q \in S'$ in G is upper bounded by $t \cdot (w(p) + d_\pi(p, q) + w(q))$. Here, $d_\pi(p, q)$ is the geodesic distance between points p and q in the simple polygon $\mathcal{P}$.

Algorithm 1 listed below extends the algorithm given in [1] to the case of input points associated with non-negative weights. In addition, the spanner constructed by this algorithm is vertex fault-tolerant, and it achieves $(\sqrt{10} + \epsilon)$-stretch.

Algorithm 1. VFTSWPSimplePolygon($\mathcal{P}$, S).

Input   : A simple polygon $\mathcal{P}$, a set S on n points located in $\mathcal{P}$, a weight function w to associate a non-negative weight to each point in S, an integer $k \geq 1$, and a real number $0 < \epsilon \leq 1$.
Output: A $(k, \sqrt{10} + \epsilon, w)$-VFTSWP G.

1 If $|S| \leq 1$ then return.

2 By using the polygon cutting theorem in [18], we partition $\mathcal{P}$ into two simple polygons $\mathcal{P}', \mathcal{P}''$ with a line segment ℓ joining two points on $\partial \mathcal{P}$ such that either of the sub-polygons contains at most two-thirds of the points in S. Let S' be the set of points in $\mathcal{P}'$, and let S'' be the set of points in $\mathcal{P}''$. (Without loss of generality, for any point $p \in S$, if p is located on ℓ, then we assume $p \in S'$ and $p \notin S''$.)

3 For each point $p \in S$, compute a geodesic projection p_ℓ of p on ℓ. Let S_ℓ be the set of points resulting from the geodesic projection of each point in S on ℓ.

4 Using the algorithm in [2], compute a $\frac{4}{\epsilon}$-SSPD $\mathcal{S}$ for the points in S_ℓ.

5 AddEdgesUsingSSPD($\mathcal{S}, G$). (Refer to Algorithm 2.)

6 VFTSWPSimplePolygon($\mathcal{P}', S'$).

7 VFTSWPSimplePolygon($\mathcal{P}'', S''$).

Essentially, edges added to G in Algorithm 1 help in achieving the vertex fault-tolerance, i.e., maintaining $\sqrt{10}+\epsilon$ stretch even after removing any k points in S.

We restate the following lemma from [1], which is useful in the analysis of Algorithm 1.

Algorithm 2. AddEdgesUsingSSPD($\mathcal{S}, G$)

Input : An s-SSPD $\mathcal{S}$, and a graph G.
Output: Based on $\mathcal{S}$, edges are added to G.

1 **foreach** *pair* (A, B) *in* $\mathcal{S}$ **do**
2   (In the following, without loss of generality., we assume $radius(A) \leq radius(B)$.)

3   **if** $|A| < k + 1$ **then**
4 | For every $p \in A$ and $q \in B$, add an edge (p, q) to G.
5   **else**
6   For every point p in A, associate a weight $w(p) + d_\pi(p, p_\ell)$. (Note that $d_\pi(p, p_\ell)$ is the geodesic distance between p and p_ℓ.) Let w' be the resultant restricted weight function.

7   With ties broken arbitrarily, select any $k + 1$ minimum weighted points in A, with respect to weights associated via w'; let A' be this set of points.

8   For every $p \in A \cup B$ and $q \in A'$, add an edge (p, q) to G.

9   **end**
10 **end**

Lemma 1. *(from* [1]*) Suppose ABC is a right triangle with $\angle CAB = \frac{\pi}{2}$. For some point D on line segment AC, let H be a y-monotone path between B and D such that the region bounded by AB, AD, and H is convex. Then, $3d(H) + d(D, C) \leq \sqrt{10}d(B, C)$, where $d(.)$ denotes the Euclidean length.*

We note that a y-monotone path is a path whose intersection with any line perpendicular to y-axis is connected. In the following lemma, we prove the graph G constructed in Algorithm 1 is indeed a $(k, \sqrt{10} + \epsilon, w)$-VFTSWP for the set S of points located in $\mathcal{P}$.

Lemma 2. *The spanner G computed by Algorithm 1 is a geodesic $(k, \sqrt{10}+\epsilon, w)$-VFTSWP for the set S of points located in $\mathcal{P}$.*

Proof: Consider any set $S' \subset S$ such that $|S'| \leq k$. Let p, q be any points in $S \setminus S'$. First, we note that there exists a splitting line segment ℓ at some iteration of the algorithm such that p and q lie on different sides of ℓ. Let $\pi(p, q)$ be a shortest path between p and q. Also, let r be a point at which $\pi(p, q)$ intersects ℓ. Consider a pair (A, B) in $\frac{4}{\epsilon}$-SSPD $\mathcal{S}$ such that $p_\ell \in A$ and $q_\ell \in B$ or, $q_\ell \in A$ and $p_\ell \in B$. We note that since $\mathcal{S}$ is a pair decomposition of points in S_l, such a (A, B) pair always exists in $\mathcal{S}$. Without loss of generality, assume the former holds. When $|A| < k + 1$, there exists an edge between p and q in G. Hence, $d_{G \setminus S'}(p, q) = d_w(p, q)$. Consider the other case in which $|A| \geq k + 1$. Since $|S'| \leq k$, there exists a $c_j \in A$ such that $c_j \notin S'$. Therefore, $d_{G \setminus S'}(p, q)$

$$= d_w(p, c_j) + d_w(c_j, q)$$
$$= w(p) + d_\pi(p, c_j) + w(c_j) + w(c_j) + d_\pi(c_j, q) + w(q)$$

[by the definition of A']

$$\leq w(p) + d_\pi(p, p_\ell) + |p_\ell c_{j_\ell}| + d_\pi(c_{j_\ell}, c_j) + w(c_j) + w(c_j) +$$
$$d_\pi(c_{j_\ell}, c_j) + |c_{j_\ell} q_\ell| + d_\pi(q_\ell, q) + w(q)$$

[since geodesic shortest paths follow triangle inequality]

$$\leq w(p) + d_\pi(p, p_\ell) + |p_\ell c_{j_\ell}| + w(p) + w(p) + d_\pi(p, p_\ell) + d_\pi(p, p_\ell) +$$
$$|c_{j_\ell} q_\ell| + d_\pi(q_\ell, q) + w(q)$$

[by the definition of A']

$$\leq 3[w(p) + w(q)] + 3d_\pi(p, p_\ell) + |p_\ell c_{j_\ell}| + |c_{j_\ell} q_\ell| + d_\pi(q_\ell, q)$$
$$\leq 3[w(p) + w(q)] + 3d_\pi(p, p_\ell) + |p_\ell c_{j_\ell}| + |p_\ell c_{j_\ell}| + |p_\ell r| + |r q_\ell| + d_\pi(q_\ell, q)$$

[since Euclidean distances follow triangle inequality]

$$= 3[w(p) + w(q)] + 3d_\pi(p, p_\ell) + 2|p_\ell c_{j_\ell}| + |p_\ell r| + |r q_\ell| + d_\pi(q_\ell, q)$$
$$\leq 3[w(p) + w(q)] + 3d_\pi(p, p') + 3d_\pi(p', p_\ell) + 2|p_\ell c_{j_\ell}| + |p_\ell r| +$$
$$|r q_\ell| + d_\pi(q_\ell, q') + d_\pi(q', q)$$

[by the triangle inequality of Euclidean distances; here, p'(resp.q') is the
the first point at which $\pi(p, q)$ and $\pi(p, p_\ell)$ (resp. $\pi(q, q_\ell)$) part ways]

$$\leq 3[w(p) + w(q)] + 3d_\pi(p, p') + \sqrt{10} d_\pi(p', r) + 2|p_\ell c_{j_\ell}| + \sqrt{10} d_\pi(q', r) +$$
$$d_\pi(q', q)$$

[applying Lemma 1 to triangles $q'h_{q'}r$ and $p'h_{p'}r$, where $h_{p'}$ (resp.$h_{q'}$)
is the projection on to line defined by $p_\ell(q_\ell)$and r]

$$\leq 3[w(p) + w(q)] + 3d_\pi(p, p') + \sqrt{10} d_w(p', r) + 2|p_\ell c_{j_\ell}| + \sqrt{10} d_w(q', r) +$$
$$d_\pi(q', q)$$

[since the weight associated with any point is non-negative]

$$\leq 3[w(p) + w(q)] + 3d_\pi(p, p') + \sqrt{10} d_w(p', q') + 2|p_\ell c_{j_\ell}| + d_\pi(q', q) \qquad (1)$$

[r is the point where $\pi(p, q)$ intersects ℓ; optimal substructure
property of shortest paths says $d_w(p, q) = d_w(p, r) + d_w(r, q)$]

$$\leq 3[w(p) + w(q)] + 3d_\pi(p, p') + \sqrt{10} d_w(p', q') + \epsilon d_w(p, q) + d_\pi(q', q). \qquad (2)$$

Since $\mathcal{S}$ is a $\frac{4}{\epsilon}$-SSPD for the set S_ℓ of points, for any pair (X, Y) of $\mathcal{S}$, the distance
between any two points in X is at most $\frac{\epsilon}{2}$ times of the distance between X and
Y. Hence,

$$|p_\ell c_{j_\ell}| \leq \frac{\epsilon}{2} |p_\ell q_\ell|. \qquad (3)$$

Therefore, $|p_\ell q_\ell|$

$$\leq |p_\ell r| + |r q_\ell|$$
[by the triangle inequality]
$$\leq |p_\ell p| + |pr| + |rq| + |q q_\ell|$$
[by the triangle inequality]
$$\leq d_\pi(p_\ell, p) + d_\pi(p, r) + d_\pi(r, q) + d_\pi(q, q_\ell)$$
$$\leq d_\pi(p, r) + d_\pi(p, r) + d_\pi(r, q) + d_\pi(r, q)$$

[by definition of projection of a point on l]
$$\leq d_w(p, r) + d_w(p, r) + d_w(r, q) + d_w(r, q)$$
[since the weight associated with each point is non-negative]
$$= 2 d_w(p, q) \tag{4}$$
[since r is the point where $\pi(p, q)$ intersects l].

From (3) and (4),

$$|p_\ell c_{j_\ell}| \leq \epsilon d_w(p, q). \tag{5}$$

Then, $d_{G \setminus S'}(p, q)$

$$\leq 3[w(p) + w(q)] + 3 d_\pi(p, p') + \sqrt{10} d_w(p', q') + \epsilon d_w(p, q) + d_\pi(q', q).$$
[from (2) and (5)]
$$\leq 3 d_w(p, p') + \sqrt{10} d_w(p', q') + \epsilon d_w(p, q) + d_w(q', q)$$
[since the weight associated with each point is non-negative]
$$\leq \sqrt{10} d_w(p, q) + \epsilon d_w(p, q)$$
[since $d_\pi(p, q) = d_\pi(p, p') + d_\pi(p', q') + d_\pi(q', q)$].

Hence, $d_{G \setminus S'}(p, q) \leq (\sqrt{10} + \epsilon) d_w(p, q)$. $\square$

Theorem 1. *Given a simple polygon $\mathcal{P}$, a set S of n points located in $\mathcal{P}$, a weight function w to associate a non-negative weight to each point in S, a positive integer k, and a real number $0 < \epsilon \leq 1$, Algorithm 1 computes a $(k, \sqrt{10} + \epsilon, w)$-vertex fault-tolerant geodesic spanner network G with $O(kn(\lg n)^2)$ edges, for the set S of weighted points.*

Proof: From Lemma 2, the spanner constructed is $(k, \sqrt{10} + \epsilon, w)$-VFTSWP. Let $S(n)$ be the number of edges in the spanner. Also, let n_1, n_2 be the sizes of sets obtained by partitioning the initial n points at the root node of the divide-and-conquer recursion tree. The recurrence is $S(n) = S(n_1) + S(n_2) + \sum_{(A,B) \in \frac{4}{\epsilon}\text{-SSPD}} (k(|A| + |B|))$. Since (A, B) is a pair in $\frac{4}{\epsilon}$-SSPD, $\sum(|A| + |B|) = O(n \lg n)$. Noting that $n_1, n_2 \geq n/3$, the size of the spanner is $O(kn(\lg n)^2)$. $\square$

3   Vertex Fault-Tolerant Spanner for Weighted Points in a Polygonal Domain

Given a polygonal domain $\mathcal{P}$, a set S of n points located in the free space $\mathcal{D}$ of $\mathcal{P}$, a weight function w to associate a non-negative weight to each point in S, a positive integer k, and a real number $0 < \epsilon \leq 1$, we compute a geodesic $(k, 6 + \epsilon, w)$-VFTSWP for the set S of weighted points. That is, for any set $S' \subset S$ with cardinality at most k, the graph $G \backslash S'$ is a $(6 + \epsilon)$-spanner for the set S' of weighted points.

Algorithm 3 mentioned below computes a geodesic $(6 + \epsilon)$-VFTSWP G for a set S of n points lying in the free space $\mathcal{D}$ of the polygonal domain $\mathcal{P}$, while each point in S is associated with a non-negative weight. The polygonal domain $\mathcal{P}$ consists of a simple polygon and h simple polygonal holes located interior to that simple polygon. We combine and extend the algorithms given in Sect. 2 and the algorithms in [1] to the case of input points associated with non-negative weights. In addition, the spanner constructed by this algorithm is vertex fault-tolerant and achieves a $6 + \epsilon$ multiplicative stretch.

The following theorem from [6] on computing a planar separator helps in devising a divide-and-conquer based algorithm (refer to Algorithm 3).

Theorem 2. ([6]) *Suppose $G = (V, E)$ is a planar vertex-weighted graph with $|V| = m$. Then, an $O(\sqrt{m})$-separator for G can be computed in $O(m)$ time. That is, V can be partitioned into sets P, Q, and R such that $|R| = O(\sqrt{m})$, there is no edge between P and Q, and $w(P), w(Q) \leq \frac{2}{3} w(V)$. Here, $w(X)$ is the sum of weights of all vertices in X.*

The following lemma proves that G computed by Algorithm 3 is a geodesic $(k, 6 + \epsilon, w)$-VFTSWP for the set S of points.

Lemma 3. *The spanner G is a geodesic $(k, 6 + \epsilon, w)$-VFTSWP for the set S of points located in $\mathcal{D}$.*

Proof: Consider any set $S' \subset S$ such that $|S'| \leq k$. Let p, q be any two points in $S \backslash S'$. Based on the locations of p and q, we consider the following cases: (1) The points p and q lie inside the same simple polygon and no shortest path between p and q intersects any splitting line segment from the set H. (2) The points p and q belong to two distinct simple polygons in the simple polygonal subdivision of $\mathcal{D}$, and both of these simple polygons correspond to vertices of one of P, Q, and R. (3) The points p and q belong to two distinct simple polygons in the simple polygonal subdivision of $\mathcal{D}$, but if one of these simple polygons correspond to a vertex of $P' \in \{P, Q, R\}$, then the other simple polygon correspond to a vertex of $P'' \in \{P, Q, R\}$ and $P'' \neq P'$. In Case (1), we run Algorithm 1, which implies there exists a path with $\sqrt{10} + \epsilon$ multiplicative stretch between p and q. In Cases (2) and (3), a shortest path between p and q intersects at least one of the $O(\sqrt{h})$ splitting line segments collected in the set H. Let ℓ be the splitting line segment that intersects a shortest path, say $\pi(p, q)$, between p and q. Also, let r be the point of intersection of $\pi(p, q)$ with ℓ. At this step, consider a pair (A, B) in

---

Algorithm 3. VFTSWPPolygonalDomain($\mathcal{D}, S$)

Input : The free space $\mathcal{D}$ of a polygonal domain $\mathcal{P}$, a set S on n points located in $\mathcal{D}$, a weight function w that associates a non-negative weight to each point in S, an integer $k \geq 1$, and a real number $0 < \epsilon \leq 1$.

Output: A $(k, 6 + \epsilon, w)$-VFTSWP G.

1 If $|S| \leq 1$ then return.

2 Partition the free space $\mathcal{D}$ into $O(h)$ simple polygons using $O(h)$ splitting line segments, such that no splitting line segment crosses any of the holes of $\mathcal{P}$, and each of the resultant simple polygons has at most three splitting line segments bounding it. This is done by choosing a leftmost vertex l_O (resp. a rightmost vertex r_O) along the x-axis of each obstacle $O \in \mathcal{P}$, and projecting l_O (resp. r_O) both upwards and downwards, parallel to y-axis. (If any of the resultant simple polygons have more than three splitting line segments on its boundary, then that polygon is further decomposed arbitrarily so that the resulting polygon has at most three splitting line segments on its boundary.)

3 A planar graph G_d is constructed: each vertex of G_d corresponds to a simple polygon in the above decomposition; for any two vertices v', v'' of G_d, an edge (v', v'') is introduced into G_d whenever the corresponding simple polygons of v' and v'' are adjacent to each other in the decomposition. Further, each vertex v of G_d is associated with a weight equal to the number of points that lie inside the simple polygon corresponding to v.

4 **if** *the number of vertices in G_d is 1* **then**

5   Let $\mathcal{D}'$ be the simple polygon that corresponds to the vertex of G_d, and let S' be the points in S that belong to $\mathcal{D}'$.

6   VFTSWPSimplePolygon($\mathcal{D}', S'$). (Refer to Algorithm 1.)

7   **return**

8 **end**

9 Compute an $O(\sqrt{h})$-separator R for the planar graph G_d using Theorem 2, and let P, Q, and R be the sets into which the vertices of G_d is partitioned.

10 For each vertex $r \in R$, we collect the bounding splitting line segments of the simple polygon corresponding to r into H, i.e., $O(\sqrt{h})$ splitting line segments are collected into a set H.

11 **foreach** $l \in H$ **do**

12   For each point p that lies in $\mathcal{D}$, compute a geodesic projection p_ℓ of p on ℓ. Let S_ℓ be the set of points resultant from these projections.

13   Using the algorithm from [2], compute a $\frac{8}{\epsilon}$-SSPD $\mathcal{S}$ for the points in S_ℓ.

14   AddEdgesUsingSSPD($\mathcal{S}, G$). (Refer to Algorithm 2.)

15 **end**

16 Let $\mathcal{D}'$ (resp. $\mathcal{D}''$) be the union of simple polygons that correspond to vertices of G_d in P (resp. Q). (Note that $\mathcal{D}'$ and $\mathcal{D}''$ are polygonal domains.) Also, let S' (resp. S'') be the points in S that belong to $\mathcal{D}'$ (resp. $\mathcal{D}''$).

17 VFTSWPPolygonalDomain($\mathcal{D}', S'$).

18 VFTSWPPolygonalDomain($\mathcal{D}'', S''$).

$\frac{8}{\epsilon}$-SSPD $\mathcal{S}$ such that $p_\ell \in A$ and $q_\ell \in B$ or, $q_\ell \in A$ and $p_\ell \in B$, where p_ℓ (resp. q_ℓ) is the projection of p (resp. q) on ℓ. Without loss of generality, assume the former holds. Suppose $|A| < k + 1$. Then, there exists an edge between p and q. Hence, $d_{G \backslash S'}(p, q) = d_w(p, q)$. Consider the other case in which $|A| \geq k + 1$. Since $|S'| \leq k$, there exists a $c_j \in A$ such that $c_j \notin S'$. Therefore,

$$d_{G \backslash S'}(p, q)$$
$$= d_w(p, c_j) + d_w(c_j, q)$$
$$= w(p) + d_\pi(p, c_j) + w(c_j) + w(c_j) + d_\pi(c_j, q) + w(q). \tag{6}$$
$$\leq w(p) + d_\pi(p, p_\ell) + |p_\ell c_{j_\ell}| + d_\pi(c_{j_\ell}, c_j) + w(c_j) + w(c_j) + d_\pi(c_{j_\ell}, c_j) +$$
$$|c_{j_\ell} q_\ell| + d_\pi(q_\ell, q) + w(q)$$

[since geodesic shortest paths follow triangle inequality]
$$\leq w(p) + d_w(p, q) + |p_\ell c_{j_\ell}| + d_\pi(c_{j_\ell}, c_j) + w(c_j) + w(c_j) + d_\pi(c_{j_\ell}, c_j) +$$
$$|c_{j_\ell} q_\ell| + w(q)[\text{as} d_\pi(p, p_\ell) + d_\pi(q, q_\ell) \leq d_w(p, q)]$$
$$\leq w(p) + d_w(p, q) + \frac{\epsilon}{4}|p_\ell q_\ell| + 2d_\pi(c_{j_\ell}, c_j) + 2w(c_j) +$$
$$|c_{j_\ell} q_\ell| + w(q) \quad [\text{as } \mathcal{S} \text{ is a } \frac{8}{\epsilon}\text{-SSPD, } |p_\ell c_{j_\ell}| \leq \frac{\epsilon}{4}|p_\ell q_\ell|]$$
$$\leq w(p) + d_w(p, q) + \frac{\epsilon}{4}[|p_\ell p| + |pq| + |q q_\ell|] + 2d_\pi(c_{j_\ell}, c_j) + 2w(c_j) +$$
$$|c_{j_\ell} q_\ell| + w(q)[\text{by the triangle inequality}]$$
$$\leq w(p) + d_w(p, q) + \frac{\epsilon}{4}[d_\pi(p_\ell, p) + d_\pi(p, q) + d_\pi(q, q_\ell)] + 2d_\pi(c_{j_\ell}, c_j) + 2w(c_j)$$
$$+ |c_{j_\ell} q_\ell| + w(q)[\text{as the Euclidean distance between any two points}$$
cannot be greater to the geodesic distance between them]
$$\leq w(p) + d_w(p, q) + \frac{\epsilon}{4}[d_w(p, q) + d_\pi(p, q)] + 2d_\pi(c_{j_\ell}, c_j) + 2w(c_j) +$$
$$|c_{j_\ell} q_\ell| + w(q)[\text{as } \mathcal{S} \text{ is a } \frac{8}{\epsilon}\text{-SSPD}]. \tag{7}$$

Since $\mathcal{S}$ is a $\frac{8}{\epsilon}$-SSPD for the set S_ℓ of points, for any pair (X, Y) of $\mathcal{S}$, the distance between any two points in X is at most $\frac{\epsilon}{4}$ times of the distance between X and Y. Therefore,

$$|p_\ell c_{j_\ell}| \leq \frac{\epsilon}{4}|p_\ell q_\ell|. \tag{8}$$

Since r is the point where $\pi(p, q)$ intersects ℓ, by the optimal sub-structure property of shortest paths,

$$\pi(p, q) = \pi(p, r) + \pi(r, q). \tag{9}$$

Then, $d_\pi(p, p_\ell) + d_\pi(q, q_\ell)$

$\quad \leq d_\pi(p, r) + d_\pi(q, r)$

$\qquad$ [as $d_\pi(p, p_\ell) \leq d_\pi(p, r)$ and $d_\pi(q, q_\ell) \leq d_\pi(q, r)$] $\qquad$ (10)

$\quad = d_\pi(p, q)$

$\qquad$ [since $\pi(p, q)$ intersects ℓ at r]

$\quad \leq d_w(p, q)$ $\qquad\qquad$ (11)

$\qquad$ [since the weight associated with each point is non-negative].

Moreover, $|c_{j_\ell} q_\ell|$

$\quad \leq |c_{j_\ell} p_\ell| + |p_\ell q_\ell|$

$\qquad$ [by the triangle inequality]

$\quad \leq \dfrac{\epsilon}{4} |p_\ell q_\ell| + |p_\ell q_\ell|$

$\qquad$ [from (8)]

$\quad \leq (\dfrac{\epsilon}{4} + 1)|p_\ell q_\ell|$

$\quad \leq (\dfrac{\epsilon}{4} + 1)(|p_\ell r| + |r q_\ell|)$

$\qquad$ [by the triangle inequality]

$\quad \leq (\dfrac{\epsilon}{4} + 1)(|p_\ell p| + |pq| + |q q_\ell|)$

$\qquad$ [by the triangle inequality]

$\quad \leq (\dfrac{\epsilon}{4} + 1)(d_\pi(p_\ell, p) + d_\pi(p, q) + d_\pi(q, q_\ell))$

$\qquad$ [from the definition of geodesic distance]

$\quad \leq (\dfrac{\epsilon}{4} + 1)(d_\pi(r, p) + d_\pi(p, q) + d_\pi(q, r))$

$\qquad$ [as $d_\pi(p, p_\ell) \leq d_\pi(p, r)$ and $d_\pi(q, q_\ell) \leq d_\pi(q, r)$]

$\quad = (\dfrac{\epsilon}{4} + 1)(d_w(p, q) + d_w(p, q))$

$\qquad$ [from (10)]

$\quad = 2(\dfrac{\epsilon}{4} + 1)d_w(p, q).$

Hence,

$$|p_\ell c_{j_\ell}| \leq 2(\dfrac{\epsilon}{4} + 1)d_w(p, q). \qquad (12)$$

$$d_{G \setminus S'}(p, q)$$

$$\leq w(p) + d_w(p, q) + \frac{2\epsilon}{4} d_w(p, q) + 2 d_\pi(c_{j_\ell}, c_j) + 2w(c_j) + (\frac{\epsilon}{2} + 2) d_w(p, q) +$$

$$w(q) [\text{from (7) and (11)}]$$

$$\leq w(p) + d_w(p, q) + \frac{2\epsilon}{4} d_w(p, q) + 2[w(p) + d_\pi(p, q)] + (\frac{\epsilon}{2} + 2) d_w(p, q) +$$

$$w(q) [\text{by the definition of set } A']$$

$$\leq 3[w(p) + d_\pi(p, q) + w(q)] + d_w(p, q) + \frac{2\epsilon}{4} d_w(p, q) + (\frac{\epsilon}{2} + 2) d_w(p, q)$$

$$= 3 d_w(p, q) + d_w(p, q) + \frac{2\epsilon}{4} d_w(p, q) + (\frac{\epsilon}{2} + 2) d_w(p, q).$$

Hence, $d_{G \setminus S'}(p, q) \leq (6 + \epsilon) d_w(p, q)$. □

Theorem 3. *Given a polygonal domain $\mathcal{P}$, a set S of n points located in the free space $\mathcal{D}$ of $\mathcal{P}$, a weight function w to associate a non-negative weight to each point in S, a positive integer k, and a real number $0 < \epsilon \leq 1$, Algorithm 3 computes a $(k, 6 + \epsilon, w)$-vertex fault-tolerant spanner network G with $O(kn\sqrt{h}(\lg n)^2)$ edges, for the set S of weighted points.*

Proof: From Lemma 3, the spanner constructed is $(k, 6 + \epsilon, w)$-VFTSWP. Let $S(n)$ be the number of edges in G. Since there are $O(\sqrt{h})$ splitting line segments, the number of edges included into G is $O(kn\sqrt{h} \lg n)$. At each internal node of the recursion tree of the algorithm, $S(n)$ satisfies the recurrence $S(n) = O(kn\sqrt{h} \lg n) + S(n_1) + S(n_2)$, where $n_1 + n_2 = n$ and $n_1, n_2 \geq n/3$. The number of edges included at all the leaves of the recursion tree together is $O(kn\sqrt{h}(\lg n)^2)$. Hence, the total number of edges is $O(kn\sqrt{h}(\lg n)^2)$. □

4 Conclusions

Our first algorithm computes a k-vertex fault-tolerant spanner with stretch $\sqrt{10} + \epsilon$ for weighted points located in a simple polygon. Our second algorithm computes a k-vertex fault-tolerant spanner with stretch $6 + \epsilon$ for weighted points located in a polygonal domain. It would be interesting to achieve a better bound on the stretch factor for the case of each point is unit or zero weighted. Apart from the efficient computation, it would be interesting to explore the lower bounds on the number of edges for these problems. Besides, the future work in the context of spanners for weighted points could include finding the relation between vertex-fault tolerance and edge-fault tolerance and optimizing various spanner parameters, like degree, diameter, and weight.

References

1. Abam, M.A., Adeli, M., Homapour, H., Asadollahpoor, P.Z.: Geometric spanners for points inside a polygonal domain. In: Proceedings of Symposium on Computational Geometry, pp. 186–197 (2015)

2. Abam, M.A., de Berg, M., Farshi, M., Gudmundsson, J.: Region-fault tolerant geometric spanners. Discret. Comput. Geom. **41**(4), 556–582 (2009)
3. Abam, M.A., de Berg, M., Farshi, M., Gudmundsson, J., Smid, M.H.M.: Geometric spanners for weighted point sets. Algorithmica **61**(1), 207–225 (2011)
4. Abam, M.A., de Berg, M., Seraji, M.J.R.: Geodesic spanners for points on a polyhedral terrain. In: Proceedings of Symposium on Discrete Algorithms, pp. 2434–2442 (2017)
5. Abam, M.A., Har-Peled, S.: New constructions of SSPDs and their applications. Comput. Geom. **45**(5), 200–214 (2012)
6. Alon, N., Seymour, P.D., Thomas, R.: Planar separators. SIAM J. Discret. Math. **7**(2), 184–193 (1994)
7. Althöfer, I., Das, G., Dobkin, D., Joseph, D., Soares, J.: On sparse spanners of weighted graphs. Discret. Comput. Geom. **9**(1), 81–100 (1993). https://doi.org/10.1007/BF02189308
8. Aronov, B., et al.: Sparse geometric graphs with small dilation. Comput. Geom. **40**(3), 207–219 (2008)
9. Arya, S., Das, G., Mount, D.M., Salowe, J.S., Smid, M.H.M.: Euclidean spanners: short, thin, and lanky. In: Proceedings of Annual ACM Symposium on Theory of Computing, pp. 489–498 (1995)
10. Arya, S., Mount, D.M., Smid, M.: Dynamic algorithms for geometric spanners of small diameter: randomized solutions. Comput. Geom. **13**(2), 91–107 (1999)
11. Arya, S., Mount, D.M., Smid, M.H.M.: Randomized and deterministic algorithms for geometric spanners of small diameter. In: Proceedings of Annual Symposium on Foundations of Computer Science, pp. 703–712 (1994)
12. Arya, S., Smid, M.H.M.: Efficient construction of a bounded-degree spanner with low weight. Algorithmica **17**(1), 33–54 (1997)
13. Bhattacharjee, S., Inkulu, R.: Fault-tolerant additive weighted geometric spanners. In: Pal, S.P., Vijayakumar, A. (eds.) CALDAM 2019. LNCS, vol. 11394, pp. 29–41. Springer, Cham (2019). https://doi.org/10.1007/978-3-030-11509-8_3
14. Bhattacharjee, S., Inkulu, R.: Geodesic fault-tolerant additive weighted spanners. In: Du, D.-Z., Duan, Z., Tian, C. (eds.) COCOON 2019. LNCS, vol. 11653, pp. 38–51. Springer, Cham (2019). https://doi.org/10.1007/978-3-030-26176-4_4
15. Bose, P., Carmi, P., Farshi, M., Maheshwari, A., Smid, M.: Computing the greedy spanner in near-quadratic time. Algorithmica **58**(3), 711–729 (2010)
16. Bose, P., Carmi, P., Chaitman, L., Collette, S., Katz, M.J., Langerman, S.: Stable roommates spanner. Comput. Geom. **46**(2), 120–130 (2013)
17. Bose, P., Carmi, P., Couture, M.: Spanners of additively weighted point sets. In: Gudmundsson, J. (ed.) SWAT 2008. LNCS, vol. 5124, pp. 367–377. Springer, Heidelberg (2008). https://doi.org/10.1007/978-3-540-69903-3_33
18. Bose, P., Czyzowicz, J., Kranakis, E., Krizanc, D., Maheshwari, A.: Polygon cutting: revisited. In: Akiyama, J., Kano, M., Urabe, M. (eds.) JCDCG 1998. LNCS, vol. 1763, pp. 81–92. Springer, Heidelberg (2000). https://doi.org/10.1007/978-3-540-46515-7_7
19. Bose, P., Fagerberg, R., van Renssen, A., Verdonschot, S.: On plane constrained bounded-degree spanners. Algorithmica **81**(4), 1392–1415 (2018). https://doi.org/10.1007/s00453-018-0476-8
20. Carmi, P., Chaitman, L.: Stable roommates and geometric spanners. In: Proceedings of the 22nd Annual Canadian Conference on Computational Geometry, pp. 31–34 (2010)
21. Chew, L.P.: There are planar graphs almost as good as the complete graph. J. Comput. Syst. Sci. **39**(2), 205–219 (1989)

22. Czumaj, A., Zhao, H.: Fault-tolerant geometric spanners. Discret. Comput. Geom. **32**(2), 207–230 (2004)
23. Das, G., Narasimhan, G.: A fast algorithm for constructing sparse Euclidean spanners. Int. J. Comput. Geom. Appl. **7**(4), 297–315 (1997)
24. Eppstein, D.: Spanning trees and spanners. In: Sack, J.-R., Urrutia, J. (ed.) Handbook of Computational Geometry, pp. 425–461. Elsevier (1999)
25. Gudmundsson, J., Knauer, C.: Dilation and detour in geometric networks. In: Gonzalez, T. (ed.) Handbook of Approximation Algorithms and Metaheuristics. Chapman & Hall (2007)
26. Gudmundsson, J., Levcopoulos, C., Narasimhan, G.: Fast greedy algorithms for constructing sparse geometric spanners. SIAM J. Comput. **31**(5), 1479–1500 (2002)
27. Kapoor, S., Li, X.-Y.: Efficient construction of spanners in d-dimensions. CoRR, abs/1303.7217 (2013)
28. Levcopoulos, C., Narasimhan, G., Smid, M.H.M.: Improved algorithms for constructing fault-tolerant spanners. Algorithmica **32**(1), 144–156 (2002)
29. Lukovszki, T.: New results on fault tolerant geometric spanners. In: Dehne, F., Sack, J.-R., Gupta, A., Tamassia, R. (eds.) WADS 1999. LNCS, vol. 1663, pp. 193–204. Springer, Heidelberg (1999). https://doi.org/10.1007/3-540-48447-7_20
30. Narasimhan, G., Smid, M.H.M.: Geometric Spanner Networks. Cambridge University Press, Cambridge (2007)
31. Peleg, D., Schäffer, A.: Graph spanners. J. Graph Theory **13**(1), 99–116 (1989)
32. Solomon, S.: From hierarchical partitions to hierarchical covers: optimal fault-tolerant spanners for doubling metrics. In: Proceedings of Symposium on Theory of Computing, pp. 363–372 (2014)
33. Talwar, K.: Bypassing the embedding: algorithms for low dimensional metrics. In: Proceedings of ACM Symposium on Theory of Computing, pp. 281–290 (2004)
34. Varadarajan, K.R.: A divide-and-conquer algorithm for min-cost perfect matching in the plane. In: Proceedings of Annual Symposium on Foundations of Computer Science, pp. 320–329 (1998)

Competitive Analysis for Two Variants of Online Metric Matching Problem

Toshiya Itoh[1] , Shuichi Miyazaki[2(✉)] , and Makoto Satake[3]

[1] Department of Mathematical and Computing Science,
Tokyo Institute of Technology, 2-12-1 Ookayama, Meguro-ku, Tokyo 152-8550, Japan
`titoh@c.titech.ac.jp`
[2] Academic Center for Computing and Media Studies, Kyoto University,
Yoshida-Honmachi, Sakyo-ku, Kyoto 606-8501, Japan
`shuichi@media.kyoto-u.ac.jp`
[3] Graduate School of Informatics, Kyoto University,
Yoshida-Honmachi, Sakyo-ku, Kyoto 606-8501, Japan
`satake@net.ist.i.kyoto-u.ac.jp`

Abstract. In this paper, we study two variants of the online metric matching problem. The first problem is the online metric matching problem where all the servers are placed at one of two positions in the metric space. We show that a simple greedy algorithm achieves the competitive ratio of 3 and give a matching lower bound. The second problem is the online facility assignment problem on a line, where servers have capacities, servers and requests are placed on 1-dimensional line, and the distances between any two consecutive servers are the same. We show lower bounds $1+\sqrt{6}$ (> 3.44948), $\frac{4+\sqrt{73}}{3}$ (> 4.18133) and $\frac{13}{3}$ (> 4.33333) on the competitive ratio when the numbers of servers are 3, 4 and 5, respectively.

Keywords: Online algorithm · Competitive analysis · Online matching problem

1 Introduction

The online metric matching problem was introduced independently by Kalyanasundaram and Pruhs [7] and Khuller, Mitchell and Vazirani [10]. In this problem, *n servers* are placed on a given metric space. Then *n requests*, which are points on the metric space, are given to the algorithm one-by-one in an online fashion. The task of an online algorithm is to match each request immediately to one of *n* servers. If a request is matched to a server, then it incurs a cost which is equivalent to the distance between them. The goal of the problem is to minimize the sum of the costs. The papers [7] and [10] presented a deterministic online

This work was partially supported by the joint project of Kyoto University and Toyota Motor Corporation, titled "Advanced Mathematical Science for Mobility Society" and JSPS KAKENHI Grant Numbers JP16K00017 and JP20K11677.

W. Wu and Z. Zhang (Eds.): COCOA 2020, LNCS 12577, pp. 486–498, 2020.
https://doi.org/10.1007/978-3-030-64843-5_33

algorithm (called *Permutation* in [7]) and showed that it is $(2n-1)$-competitive and optimal.

In 1998, Kalyanasundaram and Pruhs [8] posed a question whether we can have a better competitive ratio by restricting the metric space to a line, and introduced the problem called the *online matching problem on a line*. Since then, this problem has been extensively studied, but there still remains a large gap between the best known lower bound 9.001 [5] and upper bound $O(\log n)$ [16] on the competitive ratio.

In 2020, Ahmed, Rahman and Kobourov [1] proposed a problem called the *online facility assignment problem* and considered it on a line, which we denote *OFAL* for short. In this problem, all the servers (which they call *facilities*) and requests (which they call *customers*) lie on a 1-dimensional line, and the distance between every pair of adjacent servers is the same. Also, each server has a *capacity*, which is the number of requests that can be matched to the server. In their model, all the servers are assumed to have the same capacity. Let us denote OFAL(k) the OFAL problem where the number of servers is k. Ahmed et al. [1] showed that for OFAL(k) the greedy algorithm is $4k$-competitive for any k and a deterministic algorithm *Optimal-fill* is k-competitive for any $k > 2$.

1.1 Our Contributions

In this paper, we study a variant of the online metric matching problem where all the servers are placed at one of two positions in the metric space. This is equivalent to the case where there are two servers with capacities. We show that a simple greedy algorithm achieves the competitive ratio of 3 for this problem, and show that any deterministic online algorithm has competitive ratio at least 3.

We also study OFAL(k) for small k. Specifically, we show lower bounds $1 + \sqrt{6}$ (>3.44948), $\frac{4+\sqrt{73}}{3}$ (>4.18133) and $\frac{13}{3}$ (>4.33333) on the competitive ratio for OFAL(3), OFAL(4) and OFAL(5), respectively. We remark that our lower bounds $1+\sqrt{6}$ for OFAL(3) and $\frac{4+\sqrt{73}}{3}$ for OFAL(4) do not contradict the above-mentioned upper bound of Optimal-fill, since upper bounds by Ahmed et al. [1] are with respect to the *asymptotic* competitive ratio, while our lower bounds are with respect to the *strict* competitive ratio (see Sect. 2.3).

1.2 Related Work

In 1990, Karp, Vazirani and Vazirani [9] first studied an online version of the matching problem. They studied the online matching problem on unweighted bipartite graphs with $2n$ vertices that contain a perfect matching, where the goal is to maximize the size of the obtained matching. In [9], they first showed that a deterministic greedy algorithm is $\frac{1}{2}$-competitive and optimal. They also presented a randomized algorithm *Ranking* and showed that it is $(1-\frac{1}{e})$-competitive and optimal. See [12] for a survey of the online matching problem.

As mentioned before, Kalyanasundaram and Pruhs [7] studied the online metric matching problem and showed that the algorithm *Permutation* is $(2n-1)$-competitive and optimal. Probabilistic algorithms for this problem were studied in [4,13].

Kalyanasundaram and Pruhs [8] studied the online matching problem on a line. They gave two conjectures that the competitive ratio of this problem is 9 and that the *Work-Function* algorithm has a constant competitive ratio, both of which were later disproved in [11] and [5], respectively. This problem was studied in [2,3,6,14–16], and the best known deterministic algorithm is the *Robust Matching* algorithm [15], which is $\Theta(\log n)$-competitive [14,16].

Besides the problem on a line, Ahmed, Rahman and Kobourov [1] studied the online facility assignment problem on an unweighted graph $G(V, E)$. They showed that the greedy algorithm is $2|E|$-competitive and *Optimal-Fill* is $\frac{|E|k}{r}$-competitive, where $|E|$ is the number of edges of G and r is the radius of G.

2   Preliminaries

In this section, we give definitions and notations.

2.1   Online Metric Matching Problem with Two Servers

We define the online metric matching problem with two servers, denoted by OMM(2) for short. Let (X, d) be a metric space, where X is a (possibly infinite) set of points and $d(\cdot, \cdot)$ is a distance function. Let $S = \{s_1, s_2\}$ be a set of servers and $R = \{r_1, r_2, \ldots, r_n\}$ be a set of requests. A server s_i is characterized by the position $p(s_i) \in X$ and the capacity c_i that satisfies $c_1 + c_2 = n$. This means that s_i can be matched with at most c_i requests $(i = 1, 2)$. A request r_i is also characterized by the position $p(r_i) \in X$.

S is given to an online algorithm in advance, while requests are given one-by-one from r_1 to r_n. At any time of the execution of an algorithm, a server is called *free* if the number of requests matched with it is less than its capacity, and *full* otherwise. When a request r_i is revealed, an online algorithm must match r_i with one of free servers. If r_i is matched with the server s_j, the pair (r_i, s_j) is added to the current matching and the cost $d(r_i, s_j)$ is incurred for this pair. The cost of the matching is the sum of the costs of all the pairs contained in it. The goal of OMM(2) is to minimize the cost of the final matching.

2.2   Online Facility Assignment Problem on a Line

We give the definition of the online facility assignment problem on a line with k servers, denoted by OFAL(k). We state only differences from Sect. 2.1. The set of servers is $S = \{s_1, s_2, \ldots, s_k\}$ and all the servers have the same capacity ℓ, i.e., $c_i = \ell$ for all i. The number of requests must satisfy $n \leq \sum_{i=1}^{k} c_i = k\ell$. All the servers and requests are placed on a real number line, so their positions are expressed by a real, i.e., $p(s_i) \in \mathbb{R}$ and $p(r_j) \in \mathbb{R}$. Accordingly, the distance function is written as $d(r_i, s_j) = |p(r_i) - p(s_j)|$. We assume that the servers are placed in an increasing order of their indices, i.e., $p(s_1) \leq p(s_2) \leq \ldots \leq p(s_k)$. In this problem, any distance between two consecutive servers is the same, that is, $|p(s_i) - p(s_{i+1})| = d$ $(1 \leq i \leq k - 1)$ for some constant d. Without loss of generality, we let $d = 1$.

2.3 Competitive Ratio

To evaluate the performance of an online algorithm, we use the *strict competitive ratio*. (Hereafter, we omit "strict"). For an input σ, let $ALG(\sigma)$ and $OPT(\sigma)$ be the costs of the matchings obtained by an online algorithm ALG and an optimal offline algorithm OPT, respectively. Then the competitive ratio of ALG is the supremum of c that satisfies $ALG(\sigma) \leq c \cdot OPT(\sigma)$ for any input σ.

3   Online Metric Matching Problem with Two Servers

3.1   Upper Bound

In this section, we define a greedy algorithm $GREEDY$ for OMM(2) and show that it is 3-competitive.

Definition 1. *When a request is given, $GREEDY$ matches it with the closest free server. If a given request is equidistant from the two servers and both servers are free, $GREEDY$ matches this request with s_1.*

In the following discussion, we fix an optimal offline algorithm OPT. If a request r is matched with the server s_x by $GREEDY$ and with s_y by OPT, we say that r is of *type* $\langle s_x, s_y \rangle$. We then define some properties of inputs.

Definition 2. *Let σ be an input to OMM(2). If every request in σ is matched with a different server by $GREEDY$ and OPT, σ is called* anti-opt.

Definition 3. *Let σ be an input to OMM(2). Suppose that $GREEDY$ matches its first request r_1 with the server $s_x \in \{s_1, s_2\}$. If $GREEDY$ matches r_1 through r_{c_x} with s_x (note that c_x is the capacity of s_x) and r_{c_x+1} through r_n with the other server s_{3-x}, σ is called* one-sided-priority.

By the following two lemmas, we show that it suffices to consider inputs that are anti-opt and one-sided-priority. For an input σ, we define $Rate(\sigma)$ as

$$Rate(\sigma) = \begin{cases} \dfrac{GREEDY(\sigma)}{OPT(\sigma)} & \text{(if } OPT(\sigma) \neq 0) \\ 1 & \text{(if } OPT(\sigma) = GREEDY(\sigma) = 0) \\ \infty & \text{(if } OPT(\sigma) = 0 \text{ and } GREEDY(\sigma) > 0) \end{cases}$$

Lemma 1. *For any input σ, there exists an anti-opt input σ' such that $Rate(\sigma') \geq Rate(\sigma)$.*

Proof. If σ is already anti-opt, we can set $\sigma' = \sigma$. Hence, in the following, we assume that σ is not anti-opt. Then there exists a request r in σ that is matched with the same server s_x by OPT and $GREEDY$. Let σ'' be an input obtained from σ by removing r and subtracting the capacity of s_x by 1. By this modification, neither OPT nor $GREEDY$ changes a matching for the remaining requests. Therefore, $GREEDY(\sigma'') = GREEDY(\sigma) - d(r, s_x)$ and $OPT(\sigma'') = OPT(\sigma) - d(r, s_x)$, which implies $Rate(\sigma'') \geq Rate(\sigma)$.

Let σ' be the input obtained by repeating this operation until the input sequence becomes anti-opt. Then σ' satisfies the conditions of this lemma. $\quad\square$

Lemma 2. *For any anti-opt input σ, there exists an anti-opt and one-sided-priority input σ' such that $Rate(\sigma') = Rate(\sigma)$.*

Proof. If σ is already one-sided-priority, we can set $\sigma' = \sigma$. Hence, in the following, we assume that σ is not one-sided-priority.

Since σ is anti-opt, σ contains only requests of type $\langle s_1, s_2 \rangle$ or $\langle s_2, s_1 \rangle$. Without loss of generality, assume that in execution of $GREEDY$, the server s_1 becomes full before s_2, and let r_t be the request that makes s_1 full (i.e., r_t is the last request of type $\langle s_1, s_2 \rangle$).

Because σ is not one-sided-priority, σ includes at least one request r_i of type $\langle s_2, s_1 \rangle$ before r_t. Let σ'' be the input obtained from σ by moving r_i to just after r_t. Since the set of requests is unchanged in σ and σ'', an optimal matching for σ is also optimal for σ'', so $OPT(\sigma'') = OPT(\sigma)$. In the following, we show that $GREEDY$ matches each request with the same server in σ and σ''. The sequence of requests up to r_{i-1} are the same in σ'' and σ, so the claim clearly holds for r_1 through r_{i-1}. The behavior of $GREEDY$ for r_{i+1} through r_t in σ'' is also the same for those in σ, because when serving these requests, both s_1 and s_2 are free in both σ and σ''. Just after serving r_t in σ'', s_1 becomes full, so $GREEDY$ matches $r_i, r_{t+1}, \ldots, r_n$ with s_2 in σ''. Note that these requests are also matched with s_2 in σ. Hence $GREEDY(\sigma'') = GREEDY(\sigma)$ and it results that $Rate(\sigma'') = Rate(\sigma)$. Note that σ'' remains anti-opt.

Let σ' be the input obtained by repeating this operation until the input sequence becomes one-sided-priority. Then σ' satisfies the conditions of this lemma. $\quad\square$

We can now prove the upper bound.

Theorem 1. *The competitive ratio of $GREEDY$ is at most 3 for $OMM(2)$.*

Proof. By Lemma 1, it suffices to analyze only anti-opt inputs. In an anti-opt input, the number of requests of type $\langle s_1, s_2 \rangle$ and that of type $\langle s_2, s_1 \rangle$ are the same and the capacities of s_1 and s_2 are $n/2$ each. By Lemma 2, it suffices to analyze only the inputs where the first $n/2$ requests are of type $\langle s_1, s_2 \rangle$ and the remaining $n/2$ requests are of type $\langle s_2, s_1 \rangle$.

Let σ be an arbitrary such input. Then we have that

$$GREEDY(\sigma) = \sum_{i=1}^{n/2} d(r_i, s_1) + \sum_{i=n/2+1}^{n} d(r_i, s_2)$$

and

$$OPT(\sigma) = \sum_{i=1}^{n/2} d(r_i, s_2) + \sum_{i=n/2+1}^{n} d(r_i, s_1).$$

When serving $r_1, r_2, \ldots, r_{n/2}$, both servers are free but GREEDY matched them with s_1. Hence $d(r_i, s_1) \leq d(r_i, s_2)$ for $1 \leq i \leq n/2$. By the triangle inequality, we have $d(r_i, s_2) \leq d(s_1, s_2) + d(r_i, s_1)$ for $n/2 + 1 \leq i \leq n$. Again, by the triangle inequality, we have $d(s_1, s_2) \leq d(r_i, s_1) + d(r_i, s_2)$ for $1 \leq i \leq n$.

From these inequalities, we have that

$$GREEDY(\sigma) = \sum_{i=1}^{n/2} d(r_i, s_1) + \sum_{i=n/2+1}^{n} d(r_i, s_2)$$

$$\leq \sum_{i=1}^{n/2} d(r_i, s_2) + \sum_{i=n/2+1}^{n} (d(s_1, s_2) + d(r_i, s_1))$$

$$= OPT(\sigma) + \frac{n}{2} d(s_1, s_2)$$

$$= OPT(\sigma) + \frac{1}{2} \sum_{i=1}^{n} d(s_1, s_2)$$

$$\leq OPT(\sigma) + \frac{1}{2} \sum_{i=1}^{n} (d(r_i, s_1) + d(r_i, s_2))$$

$$= OPT(\sigma) + \frac{1}{2}(OPT(\sigma) + GREEDY(\sigma))$$

$$= \frac{3}{2} OPT(\sigma) + \frac{1}{2} GREEDY(\sigma).$$

Thus $GREEDY(\sigma) \leq 3OPT(\sigma)$ and the competitive ratio of $GREEDY$ is at most 3. □

3.2   Lower Bound

Theorem 2. *The competitive ratio of any deterministic online algorithm for OMM(2) is at least 3.*

Proof. We prove this lower bound on a 1-dimensional real line metric. Let $p(s_1) = -d$ and $p(s_2) = d$ for a constant d. Consider any deterministic algorithm ALG. First, our adversary gives $c_1 - 1$ requests at $p(s_1)$ and $c_2 - 1$ requests at $p(s_2)$. OPT matches the first $c_1 - 1$ requests with s_1 and the rest with s_2. If there exists a request that ALG matches differently from OPT, the adversary gives two more requests, one at $p(s_1)$ and the other at $p(s_2)$. Then, the cost of OPT is zero, while the cost of ALG is positive, so the ratio of them becomes infinity.

Next, suppose that ALG matches all these requests with the same server as OPT. Then the adversary gives the next request at the origin 0. Let s_x be the server that ALG matches this request with. Then OPT matches this request with the other server s_{3-x}. After that, the adversary gives the last request at $p(s_x)$. ALG has to match it with s_{3-x} and OPT matches it with s_x. The costs of ALG and OPT for this input are $3d$ and d, respectively. This completes the proof.   □

4   Online Facility Assignment Problem on Line

In this section, we show lower bounds on the competitive ratio of OFAL(k) for $k = 3, 4$, and 5. To simplify the proofs, we use Definitions 4 and 5 and Proposition 1, observed in [3,11], that allow us to restrict online algorithms to consider.

Definition 4. *When a request r is given, the* surrounding servers *for r are the closest free server to the left of r and the closest free server to the right of r.*

Definition 5. *If an algorithm ALG matches every request of an input σ with one of the surrounding servers, ALG is called* surrounding-oriented *for σ. If ALG is surrounding-oriented for any input, then ALG is called* surrounding-oriented.

Proposition 1. *For any algorithm ALG, there exists a surrounding-oriented algorithm ALG' such that $ALG'(\sigma) \leq ALG(\sigma)$ for any input σ.*

The proof of Proposition 1 is omitted in [3,11], so we prove it here for completeness.

Proof. Suppose that ALG is not surrounding-oriented for σ. Then ALG matches at least one request of σ with a non-surrounding server. Let r be the earliest one among such requests and s be the server matched with r by ALG. Also let s' be the surrounding server (for r) on the same side as s and r' be the request matched with s' by ALG.

We modify ALG to ALG'' so that ALG'' matches r with s' and r' with s (and behaves the same as ALG for other requests). Without loss of generality, we can assume that $p(r) < p(s)$. Then we have that $p(r) \leq p(s') < p(s)$. If $p(r') \leq p(s')$, then $ALG''(\sigma) = ALG(\sigma)$ and if $p(r') > p(s')$, then $ALG''(\sigma) < ALG(\sigma)$. In either case, we have that $ALG''(\sigma) \leq ALG(\sigma)$.

Let ALG''' be the algorithm obtained by applying this modification as long as there is a request in σ matched with a non-surrounding server. Then $ALG'''(\sigma) \leq ALG(\sigma)$ and ALG''' is surrounding-oriented for σ.

We do the above modification for all the inputs for which ALG is not surrounding-oriented, and let ALG' be the resulting algorithm. Then $ALG'(\sigma) \leq ALG(\sigma)$ and ALG' is surrounding-oriented, as required.   □

By Proposition 1, it suffices to consider only surrounding-oriented algorithms for lower bound arguments.

Theorem 3. *The competitive ratio of any deterministic online algorithm for OFAL(3) is at least $1 + \sqrt{6}$ (> 3.44948).*

Proof. Let *ALG* be any surrounding-oriented algorithm. Our adversary first gives $\ell - 1$ requests at $p(s_i)$ for each $i = 1, 2$ and 3. *OPT* matches every request r with the server at the same position $p(r)$. If *ALG* matches some request r with a server not at $p(r)$, then the adversary gives three more requests, one at each position of the server. The cost of *ALG* is positive and the cost of *OPT* is zero, so the ratio of the costs is infinity.

Next, suppose that *ALG* matches all these requests to the same server as *OPT*. Let $x = \sqrt{6} - 2$ ($\simeq 0.44949$) and $y = 3\sqrt{6} - 7$ ($\simeq 0.34847$). The adversary gives a request r_1 at $p(s_2) + x$.

Case 1. *ALG* matches r_1 with s_3

See Fig. 1. The adversary gives the next request r_2 at $p(s_3)$. *ALG* matches it with s_2. Finally, the adversary gives a request r_3 at $p(s_1)$ and *ALG* matches it with s_1. The cost of *ALG* is $2 - x = 4 - \sqrt{6}$ and the cost of *OPT* is $x = \sqrt{6} - 2$. The ratio is $\frac{4 - \sqrt{6}}{\sqrt{6} - 2} = 1 + \sqrt{6}$.

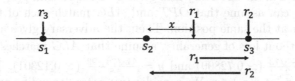

Fig. 1. Requests and *ALG*'s matching for Case 1 of Theorem 3.

Case 2. *ALG* matches r_1 with s_2

The adversary gives the next request r_2 at $p(s_2) - y$. We have two subcases.

Case 2-1. *ALG* matches r_2 with s_1

See Fig. 2. The adversary gives a request r_3 at $p(s_1)$ and *ALG* matches it with s_3. The cost of *ALG* is $3 + x - y = 8 - 2\sqrt{6}$ and the cost of *OPT* is $1 - x + y = 2\sqrt{6} - 4$. The ratio is $\frac{8 - 2\sqrt{6}}{2\sqrt{6} - 4} = 1 + \sqrt{6}$.

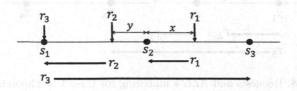

Fig. 2. Requests and *ALG*'s matching for Case 2-1 of Theorem 3.

Case 2-2. *ALG* matches r_2 with s_3

See Fig. 3. The adversary gives a request r_3 at $p(s_3)$ and *ALG* matches it with s_1. The cost of *ALG* is $3 + x + y = 4\sqrt{6} - 6$ and the cost of *OPT* is $1 + x - y = 6 - 2\sqrt{6}$. The ratio is $\frac{4\sqrt{6} - 6}{6 - 2\sqrt{6}} = 1 + \sqrt{6}$.

494   T. Itoh et al.

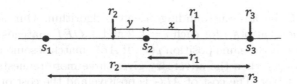

Fig. 3. Requests and ALG's matching for Case 2-2 of Theorem 3.

In any case, the ratio of ALG's cost to OPT's cost is $1 + \sqrt{6}$. This completes the proof.   □

Theorem 4. *The competitive ratio of any deterministic online algorithm for OFAL(4) is at least $\frac{4+\sqrt{73}}{3}$ (> 4.18133).*

Proof. Let ALG be any surrounding-oriented algorithm. In the same way as the proof of Theorem 3, the adversary first gives $\ell - 1$ requests at $p(s_i)$ for $i = 1, 2, 3$, and 4, and we can assume that OPT and ALG match each of these requests with the server at the same position. Then, the adversary gives a request r_1 at $\frac{p(s_2)+p(s_3)}{2}$. Without loss of generality, assume that ALG matches it with s_2.

Let $x = \frac{10-\sqrt{73}}{2}$ ($\simeq 0.72800$) and $y = \frac{11\sqrt{73}-93}{8}$ ($\simeq 0.12301$). The adversary gives a request r_2 at $p(s_1) + x$. We consider two cases depending on the behavior of ALG.

Case 1. ALG matches r_2 with s_1
See Fig. 4. The adversary gives the next request r_3 at $p(s_1)$. ALG has to match it with s_3. Finally, the adversary gives a request r_4 at $p(s_4)$ and ALG matches it with s_4. The cost of ALG is $\frac{5}{2}+x = \frac{15-\sqrt{73}}{2}$ and the cost of OPT is $\frac{3}{2}-x = \frac{\sqrt{73}-7}{2}$. The ratio is $\frac{15-\sqrt{73}}{\sqrt{73}-7} = \frac{4+\sqrt{73}}{3}$.

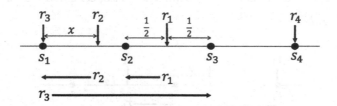

Fig. 4. Requests and ALG's matching for Case 1 of Theorem 4.

Case 2. ALG matches r_2 with s_3
The adversary gives the next request r_3 at $p(s_3) + y$. We have two subcases.

Case 2-1. ALG matches r_3 with s_4
See Fig. 5. The adversary gives a request r_4 at $p(s_4)$. ALG has to match it

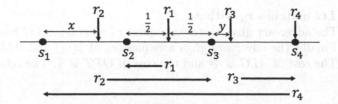

Fig. 5. Requests and ALG's matching for Case 2-1 of Theorem 4.

with s_1. The cost of ALG is $\frac{13}{2} - x - y = \frac{105-7\sqrt{73}}{8}$ and the cost of OPT is $\frac{1}{2} + x + y = \frac{7\sqrt{73}-49}{8}$. The ratio is $\frac{105-7\sqrt{73}}{7\sqrt{73}-49} = \frac{4+\sqrt{73}}{3}$.

Case 2-2. ALG matches r_3 with s_1

See Fig. 6. The adversary gives a request r_4 at $p(s_1)$ and ALG has to match it with s_4. The cost of ALG is $\frac{15}{2} - x + y = \frac{15\sqrt{73}-73}{8}$ and the cost of OPT is $\frac{5}{2} - x - y = \frac{73-7\sqrt{73}}{8}$. The ratio is $\frac{15\sqrt{73}-73}{73-7\sqrt{73}} = \frac{4+\sqrt{73}}{3}$.

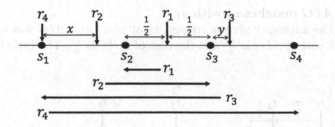

Fig. 6. Requests and ALG's matching for Case 2-2 of Theorem 4.

In any case, the ratio of ALG's cost to OPT's cost is $\frac{4+\sqrt{73}}{3}$. This completes the proof.   □

Theorem 5. *The competitive ratio of any deterministic online algorithms for OFAL(5) is at least $\frac{13}{3}$ (>4.33333).*

Proof. Let ALG be any surrounding-oriented algorithm. In the same way as the proof of Theorem 3, the adversary first gives $\ell - 1$ requests at $p(s_i)$ for $i = 1, 2, 3, 4$, and 5, and we can assume that OPT and ALG match each of these requests with the server at the same position.

Then, the adversary gives a request r_1 at $p(s_3)$. If ALG matches this with s_2 or s_4, the adversary gives the remaining requests at $p(s_1)$, $p(s_2)$, $p(s_4)$ and $p(s_5)$. OPT's cost is zero, while ALG's cost is positive, so the ratio is infinity. Therefore, assume that ALG matches r_1 with s_3. The adversary then gives a request r_2 at $p(s_3)$. Without loss of generality, assume that ALG matches it with s_2. Next, the adversary gives a request r_3 at $p(s_1) + \frac{7}{8}$. We consider two cases depending on the behavior of ALG.

Case 1. ALG matches r_3 with s_1

See Fig. 7. The adversary gives the next request r_4 at $p(s_1)$. ALG has to match it with s_4. Finally, the adversary gives a request r_5 at $p(s_5)$ and ALG matches it with s_5. The cost of ALG is $\frac{39}{8}$ and the cost of OPT is $\frac{9}{8}$. The ratio is $\frac{13}{3}$.

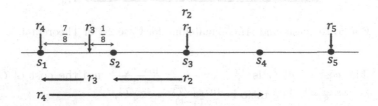

Fig. 7. Requests and ALG's matching for Case 1 of Theorem 5.

Case 2. ALG matches r_3 with s_4

The adversary gives the next request r_4 at $p(s_4)$. We have two subcases.

Case 2-1. ALG matches r_4 with s_1

See Fig. 8. The adversary gives a request r_5 at $p(s_1)$ and ALG has to match it with s_5. The cost of ALG is $\frac{81}{8}$ and the cost of OPT is $\frac{17}{8}$. The ratio is $\frac{81}{17} > \frac{13}{3}$.

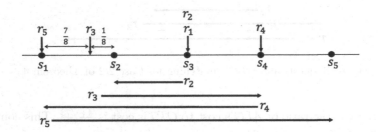

Fig. 8. Requests and ALG's matching for Case 2-1 of Theorem 5.

Case 2-2. ALG matches r_4 with s_5

See Fig. 9. The adversary gives a request r_5 at $p(s_5)$ and ALG has to match it with s_1. The cost of ALG is $\frac{65}{8}$ and the cost of OPT is $\frac{15}{8}$. The ratio is $\frac{13}{3}$.

In any case, the ratio of ALG's cost to OPT's cost is at least $\frac{13}{3}$, which completes the proof.   □

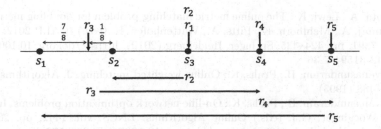

Fig. 9. Requests and *ALG*'s matching for Case 2-2 of Theorem 5.

5 Conclusion

In this paper, we studied two variants of the online metric matching problem. The first is a restriction where all the servers are placed at one of two positions in the metric space. For this problem, we presented a greedy algorithm and showed that it is 3-competitive. We also proved that any deterministic online algorithm has competitive ratio at least 3, giving a matching lower bound. The second variant is the Online Facility Assignment Problem on a line with a small number of servers. We showed lower bounds on the competitive ratio $1 + \sqrt{6}$, $\frac{4+\sqrt{73}}{3}$, and $\frac{13}{3}$ when the numbers of servers are 3, 4, and 5, respectively.

One of the future work is to analyze the online metric matching problem with three or more server positions. Another interesting direction is to consider an optimal online algorithm for the Online Facility Assignment Problem on a line when the numbers of servers are 3, 4, and 5.

Acknowledgment. The authors would like to thank the anonymous reviewers for their valuable comments.

References

1. Ahmed, A., Rahman, M., Kobourov, S.: Online facility assignment. Theoret. Comput. Sci. **806**, 455–467 (2020)
2. Antoniadis, A., Barcelo, N., Nugent, M., Pruhs, K., Scquizzato, M.: A $o(n)$-competitive deterministic algorithm for online matching on a line. In: Bampis, E., Svensson, O. (eds.) WAOA 2014. LNCS, vol. 8952, pp. 11–22. Springer, Cham (2015). https://doi.org/10.1007/978-3-319-18263-6_2
3. Antoniadis, A., Fischer, C., Tönnis, A.: A collection of lower bounds for online matching on the line. In: Bender, M.A., Farach-Colton, M., Mosteiro, M.A. (eds.) LATIN 2018. LNCS, vol. 10807, pp. 52–65. Springer, Cham (2018). https://doi.org/10.1007/978-3-319-77404-6_5
4. Bansal, N., Buchbinder, N., Gupta, A., Naor, J.S.: An $o(\log^2 k)$-competitive algorithm for metric bipartite matching. In: Arge, L., Hoffmann, M., Welzl, E. (eds.) ESA 2007. LNCS, vol. 4698, pp. 522–533. Springer, Heidelberg (2007). https://doi.org/10.1007/978-3-540-75520-3_47
5. Fuchs, B., Hochstättler, W., Kern, W.: Online matching on a line. Theoret. Comput. Sci. **332**, 251–264 (2005)

6. Gupta, A., Lewi, K.: The online metric matching problem for doubling metrics. In: Czumaj, A., Mehlhorn, K., Pitts, A., Wattenhofer, R. (eds.) ICALP 2012. LNCS, vol. 7391, pp. 424–435. Springer, Heidelberg (2012). https://doi.org/10.1007/978-3-642-31594-7_36
7. Kalyanasundaram, B., Pruhs, K.: Online weighted matching. J. Algorithms **14**(3), 478–488 (1993)
8. Kalyanasundaram, B., Pruhs, K.: On-line network optimization problems. In: Fiat, A., Woeginger, G.J. (eds.) Online Algorithms. LNCS, vol. 1442, pp. 268–280. Springer, Heidelberg (1998). https://doi.org/10.1007/BFb0029573
9. Karp, R., Vazirani, U., Vazirani, V.: An optimal algorithm for on-line bipartite matching. In: STOC 1990, pp. 352–358 (1990)
10. Khuller, S., Mitchell, S., Vazirani, V.: On-line algorithms for weighted bipartite matching and stable marriages. Theoret. Comput. Sci. **127**(2), 255–267 (1994)
11. Koutsoupias, E., Nanavati, A.: The online matching problem on a line. In: Solis-Oba, R., Jansen, K. (eds.) WAOA 2003. LNCS, vol. 2909, pp. 179–191. Springer, Heidelberg (2004). https://doi.org/10.1007/978-3-540-24592-6_14
12. Mehta, A.: Online matching and ad allocation. Theoret. Comput. Sci. **8**(4), 265–368 (2012)
13. Meyerson, A., Nanavati, A., Poplawski, L.: Randomized online algorithms for minimum metric bipartite matching. In: SODA 2006, pp. 954–959 (2006)
14. Nayyar, K., Raghvendra, S.: An input sensitive online algorithm for the metric bipartite matching problem. In: FOCS 2017, pp. 505–515 (2017)
15. Raghvendra, S.: A robust and optimal online algorithm for minimum metric bipartite matching. In: APPROX/RANDOM 2016, vol. 60, pp. 18:1–18:16 (2016)
16. Raghvendra, S.: Optimal analysis of an online algorithm for the bipartite matching problem on a line. In: SoCG 2018, pp. 67:1–67:14 (2018)

Guarding Disjoint Orthogonal Polygons
in the Plane

Ovidiu Daescu$^{(\boxtimes)}$ and Hemant Malik

University of Texas at Dallas, Richardson, TX 75080, USA
{daescu,malik}@utdallas.edu

Abstract. Given a set F of disjoint monotone orthogonal polygons, we present bounds on the number of vertex guards required to monitor the free space and the boundaries of the polygons in F. We assume that the range of vision of a guard is bounded by 180° (the region in front of the guard). For k disjoint axis-aligned monotone orthogonal polygons $H_1, H_2, \ldots, H_k$ with $m_1, m_2, \ldots, m_k$ vertices respectively, such that $\sum_{i=1}^{k} m_i = m$, we prove that $\frac{m}{2} + \lfloor \frac{k}{4} \rfloor + 4$ vertex guards are sufficient to monitor the boundaries of the polygons and the free space. When the orthogonal polygons are arbitrary oriented, we show that $\frac{m}{2} + k + 1$ vertex guards are sometimes necessary to monitor the boundaries and the free space and conjecture the bound is tight.

Keywords: Art gallery problem · Guarding · Orthogonal polygon · Monotone polygon · Visibility

1 Introduction

In 1977, Fejes Toth [22] considered the following problem: Given a set F of k disjoint compact convex sets in the plane, how many guards are sufficient to cover every point on the boundary of each set in F? In this paper, we consider a variation of this problem, specifically:

Problem 1. Given a set F of k disjoint monotone orthogonal polygons in the plane, how many vertex guards are necessary and sufficient to jointly guard the free space and the boundaries of the polygons when the range of vision of each guard is 180°?

A polygon P in the plane is called orthogonal if there exist two orthogonal directions l_1 and l_2 such that all sides of P are parallel to one of the two directions. P is monotone with respect to a direction d if its intersection with any line orthogonal to d is either empty or a single line segment. A polygon P that is orthogonal with respect to directions l_1 and l_2 is called monotone if it is monotone with respect to both l_1 and l_2. For example, an axis-aligned monotone orthogonal polygon P has each of its sides parallel to the x-axis or the y-axis, and is monotone with respect to both axes.

© Springer Nature Switzerland AG 2020
W. Wu and Z. Zhang (Eds.): COCOA 2020, LNCS 12577, pp. 499–514, 2020.
https://doi.org/10.1007/978-3-030-64843-5_34

The problem we study is related to the famous art-gallery problem introduced by Victor Klee in 1973. In almost all the studies [18, 23], the art gallery lies in the plane (2D), assuming a polygonal shape, with or without holes. In the art gallery problem, the goal is to determine the minimum number of point guards sufficient to see every point of the interior of a simple polygon. A point q is visible to a guard g if the line segment joining q and g lies completely within the polygon. When the guards are restricted to vertices of the polygon only, they are referred to as vertex guards. Sometimes the polygon is allowed to have a set of k polygonal holes in its interior. The problem we study can be modeled as the following variation of the art-gallery problem:

Problem 2. Given an axis-aligned rectangle P containing k disjoint monotone orthogonal holes, $H_1, H_2, \ldots, H_k$ with $m_1, m_2, \ldots, m_k$ vertices respectively, such that $\sum_{i=1}^{k} m_i = m$, place vertex guards on holes such that every point inside P is visible to at least one guard, where the range of vision of each guard is $180°$.

Note that Problem 1 and Problem 2 are equivalent. In all our proofs each guard is (i) placed at a vertex of an orthogonal hole and (ii) is oriented such that the seen and unseen regions of the guard are separated by a line parallel with one of the sides of the hole that are incident to the vertex where the guard is placed. From now on, we assume guards are placed as stated here, unless otherwise specified, and may omit mentioning guard orientation throughout the paper.

We address the following two versions: (A) Orthogonal holes are axis-aligned (B) Orthogonal holes are arbitrary oriented. To solve the two versions, we address the following variations of the art-gallery problem: (V1) Given an axis-aligned rectangle P with k disjoint axis-aligned monotone orthogonal holes, $H_1, H_2, \ldots, H_k$ with $m_1, m_2, \ldots, m_k$ vertices respectively, such that $\sum_{i=1}^{k} m_i = m$, place vertex guards on holes such that every point inside P is visible to at least one guard, where the range of vision of guards is $180°$. (V2) Given an axis-aligned rectangle P with k disjoint monotone orthogonal holes, $H_1, H_2, \ldots, H_k$ with $m_1, m_2, \ldots, m_k$ vertices respectively, such that $\sum_{i=1}^{k} m_i = m$, place vertex guards on holes such that every point inside P is visible to at least one guard, where the range of vision of guards is $180°$.

For the first problem (V1), we prove a sufficiency bound of $\frac{m}{2} + \lfloor \frac{k}{4} \rfloor + 4$ on the number of vertex guards. For the second problem (V2), we show that $\frac{m}{2} + k + 1$ vertex guards are sometimes necessary and conjecture the bound is tight. The bounds also hold if the holes are not monotone (see [14] for details).

2   Related Work

Simple Polygons Results: The problem of guarding polygons, with or without holes, has a long history. Given a simple polygon P in the plane, with n vertices, Chvatal [4] proved that $\lfloor n/3 \rfloor$ vertex guards are always sufficient and sometimes necessary to guard P. Chvatal's proof was later simplified by Fisk [11].

When the view of a guard is limited to $180°$, Toth [21] showed that $\lfloor \frac{n}{3} \rfloor$ point guards are always sufficient to cover the interior of P (thus, moving from 360 to 180 range of vision keeps the same sufficiency number). F. Santos conjectured that $\lfloor \frac{3n-3}{5} \rfloor \pi$ vertex guards are always sufficient and occasionally necessary to cover any polygon with n vertices. Later, in 2002, Toth [20] provided lower bounds on the number of point guards when the range of vision α is less than $180°$, specifically (i) when $\alpha < 180°$, there exist a polygon P that cannot be guarded by $\frac{2n}{3} - 3$ guards, (ii) for $\alpha < 90°$ there exist P that cannot be guarded by $\frac{3n}{4} - 1$ guards, and (iii) for $\alpha < 60°$ there exist P where the number of guards needed to guard P is at least $\lfloor \frac{60}{\alpha} \rfloor \frac{(n-1)}{2}$.

Orthogonal Polygons Results: In 1983, Kahn et al. [13] showed that if every pair of adjacent sides of the polygon form a right angle, then $\lfloor \frac{n}{4} \rfloor$ vertex guards are occasionally necessary and always sufficient to guard a polygon with n vertices. In 1983, O'Rourke [16] showed that $1 + \lfloor \frac{r}{2} \rfloor$ vertex guards are necessary and sufficient to cover the interior of an orthogonal polygon with r reflex vertices. Castro and Urrutia [9] provided a tight bound of $\lfloor \frac{3(n-1)}{8} \rfloor$ on the number of orthogonal floodlights (guards with $90°$ range of vision), placed at vertices of the polygon, sufficient to view the interior of an orthogonal polygon with n vertices.

Polygon with Holes Results: For a polygon P with n vertices and h polygonal holes, the value n denotes the sum of the number of vertices of P and the number of vertices of the holes. Let $g(n, h)$ be the minimum number of point guards and let $g^v(n, h)$ be the minimum number of vertex guards necessary to guard any polygon with n vertices and h holes. O'Rourke [17] proved that $g^v(n, h) \leq \lfloor \frac{n+2h}{3} \rfloor$. Shermer [18,19] conjectured that $g^v(n, h) \leq \lfloor \frac{n+h}{3} \rfloor$ and this is a tight bound. He was able to prove that, for $h = 1$, $g^v(n, 1) = \lfloor \frac{n+1}{3} \rfloor$. However, for $h > 1$ the conjecture remains open.

Orthogonal Polygon with Holes Results: Let $orth(n, h)$ be the minimum number of point guards and $orth^v(n, h)$ be the minimum number of vertex guards necessary to guard any orthogonal polygon with n vertices and h holes (orthogonal with respect to the coordinate axes). Note that $orth(n, h) \leq orth^v(n, h)$. O'Rourke's method [17] extends to this case and results in $orth^v(n, h) \leq \lfloor \frac{n+2h}{4} \rfloor$. Shermer [18] conjectured that $orth^v(n, h) \leq \lfloor \frac{n+h}{4} \rfloor$ which Aggarwal [2] established for $h = 1$ and $h = 2$. Zylinski [24] showed that $\lfloor \frac{n+h}{4} \rfloor$ vertex guards are always sufficient to guard any orthogonal polygon with n vertices and h holes, provided that there exists a quadrilateralization whose dual graph is a cactus. In 1996, Hoffmann and Kriegel [12] showed that $\leq \lfloor \frac{n}{3} \rfloor$ vertex guards are sufficient to guard the interior of an orthogonal polygon with holes. In 1998, Abello et al. [1] provided a first tight bound of $\lfloor \frac{3n+4(h-1))}{8} \rfloor$ for the number of orthogonal floodlights placed at the vertices of an orthogonal polygon with n vertices and h holes which are sufficient to guard the polygon, and described a simple linear-time algorithm to find the guard placement for an orthogonal polygon (with or without holes). Note that this result does not extend to our problem, since we restrict guard placement to the vertices of the holes only. In 2020, Daescu and

Malik [7] showed that $2k + \lfloor \frac{k}{4} \rfloor + 4$ vertex guards are always sufficient to guard an axis-aligned rectangle P with k disjoint axis-aligned rectangular holes, with guards placed only on vertices of the holes, when the range of vision of guards is $180°$.

Families of Convex Sets on the Plane Results: In 1977, Toth [22] considered the following problem: Given a set F of n disjoint compact convex sets in the plane, how many guards are sufficient to cover every point on the boundary of each set in F? He proved that $max\{2n, 4n - 7\}$ point guards are always sufficient. Everett and Toussaint [10] proved that n disjoint squares, for $n > 4$, can always be guarded with n point guards. For families of disjoint isothetic rectangles (rectangles with sides parallel to the coordinate axes.), Czyzowicz et al. [6] proved that $\lfloor \frac{4n+4}{3} \rfloor$ point guards suffice and conjectured that $n + c$ point guards would suffice, where c is a constant. If the rectangles have equal width, then $n + 1$ point guards suffice, and $n - 1$ point guards are occasionally necessary (see [15] for more details).

In 1994, Blanco et al. [3] considered the problem of guarding the plane with rectangular region of the plane in the presence of quadrilateral holes. Given n pairwise disjoint quadrilaterals in the plane they showed that $2n$ vertex guards are always sufficient to cover the free space and the guard locations could be found in $O(n^2)$ time. If the quadrilaterals are isothetic rectangles, all locations can be computed in $O(n)$ time. Thus, this problem is similar to ours, with the difference that the guards in [3] have $360°$ vision and the holes are quadrilaterals or isothetic rectangles. Our problem can be seen as a generalization to orthogonal holes, while also restricting the field of vision of the guards to $180°$.

Garcia-Lopez [8] proved that $\lfloor \frac{5m}{9} \rfloor$ vertex lights are always sufficient and $\lfloor \frac{m}{2} \rfloor$ are occasionally necessary to guard the free space generated by a family of disjoint polygons with a total of m vertices.

Czyzowicz et al. [5] proposed the following problem: Given a set F of n disjoint compact convex sets in the plane, how many guards are sufficient to protect each set in F? A set of F is protected by a guard g if at least one point on the boundary of that set is visible from g. They prove that $\lfloor \frac{2(n-2)}{3} \rfloor$ point guards are always sufficient and occasionally necessary to protect any family of n disjoint convex sets, $n > 2$. To protect any family of n isothetic rectangles, $\lceil \frac{n}{2} \rceil$ point guards are always sufficient, and $\lfloor \frac{n}{2} \rfloor$ point guards are sometimes necessary.

Given a set F of k disjoint isothetic rectangles in a plane, Daescu and Malik [7] showed that $2k + \lfloor \frac{k}{4} \rfloor + 4$ vertex guards are always sufficient to guard the free space, when the range of vision of guards is $180°$. They also showed that given k disjoint, arbitrary oriented, rectangular obstacles in the plane $3k + 1$ vertex guards are sometimes necessary to guard the free space and conjectured the bound is tight.

3   Axis-Aligned Orthogonal Polygons

In this section, we considered the following variation of the art-gallery problem: *Given an axis-aligned rectangle P with k disjoint axis-aligned monotone orthogonal holes, $H_1, H_2, \ldots, H_k$ with $m_1, m_2, \ldots, m_k$ vertices respectively, such that*

$\sum_{i=1}^{k} m_i = m$, place vertex guards on holes such that every point inside P is visible to at least one guard, where the range of vision of guards is $180°$.

In all our proofs each guard faces East, West, North, or South only.

Each vertex of an orthogonal polygon is either $270°$ (reflex) or $90°$ (convex), defined as follows. For each orthogonal hole H_i, pick an arbitrary vertex u and start traversing H_i in clockwise direction. For a vertex $v \in H_i$, if the turn made at v while traversing H_i is right then v is a *reflex* vertex, else v is *convex* vertex. For each hole H_i, let r_i and c_i be the total number of reflex and convex vertices, respectively. From [18], $r_i = \frac{m_i+4}{2} = \frac{m_i}{2} + 2$ and $c_i = \frac{m_i-4}{2} = \frac{m_i}{2} - 2$. All holes are orthogonal and every orthogonal polygon has an even number of vertices. Therefore, $\forall i \in [1, k]$, the value of $\frac{m_i}{2}$ is an integer. Let r be the total number of reflex vertices in P. Then,

$$r = \sum_{i=1}^{k} r_i = \sum_{i=1}^{k} (\frac{m_i}{2} + 2) = \frac{m}{2} + 2k \qquad (1)$$

Similarly, the total number of convex vertices is $\frac{m}{2} - 2k$.

For each hole H_i, let E_R^i be the rightmost edge, E_L^i be the leftmost edge, E_T^i be the topmost edge, and E_B^i be the bottom-most edge. First we perform the following two steps, shown in Fig. 1: (i) Traverse H_i from E_R^i to E_L^i counter-clockwise and extend each encountered vertical edge, including E_L^i and E_R^i, towards North (upward) direction until it hits the polygon P or any other hole. (ii) Traverse H_i from E_L^i to E_B^i counter-clockwise and extend each horizontal edge, including E_B^i, towards West (left) direction until it hits the polygon P, any other hole, or any of the extended vertical edges.

The above steps divide the polygon into shapes where each shape corresponds to a monotone staircase in x and y directions and one guard, placed at the South-East corner of each shape, is sufficient to guard the shape. The total number of guards needed to guard the polygon is equal to the number of shapes constructed while performing the two steps above.

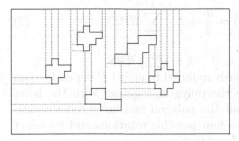
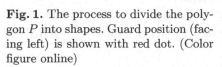
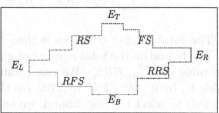

Fig. 1. The process to divide the polygon P into shapes. Guard position (facing left) is shown with red dot. (Color figure online)

Fig. 2. RS is shown in Red (dashed), FS is shown in Blue (dotted), RRS is shown in Black (dash dot dotted) and RFS is shown in Green (dash dotted). (Color figure online)

For each hole H_i, we define the following staircases (shown in Fig. 2): (i) *Rising Staircase* RS_i: Monotone staircase formed by traversing H_i in clockwise order from E_L^i to E_T^i (including E_L^i and E_T^i). (ii) *Falling Staircase* FS_i: Monotone staircase formed by traversing H_i in clockwise order from E_T^i to E_R^i (including E_T^i and E_R^i). (iii) *Reverse Rising Staircase* RRS_i: Monotone staircase formed by traversing H_i in clockwise order from E_R^i to E_B^i (including E_R^i and E_B^i). (iv) *Reverse Falling Staircase* RFS_i: Monotone staircase formed by traversing H_i in clockwise order from E_B^i to E_L^i (including E_B^i and E_L^i).

For hole H_i and staircase RS_i, let RS_i^r be the number of reflex vertices (excluding the first and last vertex of RS_i), and RS_i^c be the number of convex vertices. Similarly, we define $FS_i^r, FS_i^c, RFS_i^r, RFS_i^c, RRS_i^r$, and RRS_i^c.

Each hole contributes towards the formation of shapes. It is easy to notice that the total number of shapes contributed by a hole H_i is equal to the sum of: (i) Number of reflex vertices FS_i^r on FS_i, (ii) Number of convex vertices RS_i^c on RS_i, and (iii) Number of reflex vertices RFS_i^r on RFS_i.

The total number of shapes is then $1 + \sum_{i=1}^{k}(FS_i^r + RS_i^c + RFS_i^r)$, where one guard is placed at a corner of the bounding rectangle P.

Theorem 1. $\lceil \frac{3m}{8} + \frac{k}{2} \rceil + 1$ *vertex guards are always sufficient to guard an axis-aligned rectangle polygon with k disjoint axis-aligned monotone orthogonal holes, where m is the total number of vertices of the holes, with at most one guard placed at a corner of the bounding rectangle.*

Proof. For each hole H_i, the following conditions hold: (i) On any staircase, total reflex vertices = total convex vertices + 1 (therefore, $RS_i^r = RS_i^c + 1$, $FS_i^r = FS_i^c + 1$, $RRS_i^r = RRS_i^c + 1$ and $RFS_i^r = RFS_i^c + 1$) and (ii) Total reflex vertices = total convex vertices + 4, i.e. $r_i = c_i + 4$.

Over all holes, the relation between total reflex vertices and total convex vertices is given by $\sum_{i=1}^{k} r_i = \sum_{i=1}^{k} c_i + 4k$. From Eq. 1, $\sum_{i=1}^{k}(FS_i^r + RS_i^r + RFS_i^r + RRS_i^r) = \frac{m}{2} + 2k$. Substituting the value of RS_i^r as $RS_i^c + 1$ we have $\sum_{i=1}^{k}(FS_i^r + RS_i^c + 1 + RFS_i^r + RRS_i^r) = \frac{m}{2} + 2k$ and

$$\sum_{i=1}^{k}(FS_i^r + RS_i^c + RFS_i^r) = \frac{m}{2} + k - \sum_{i=1}^{k} RRS_i^r \qquad (2)$$

The total number of shapes is then $1 + \frac{m}{2} + k - \sum_{i=1}^{k} RRS_i^r$.

The bound on the total number of guards required to guard P depends upon the value of $\sum_{i=1}^{k} RRS_i^r$. We can rotate the polygon, together with the holes inside it, by $0°, 90°, 180°$ or $270°$ (so that the polygon and holes remain axis-aligned) to select the best bound: we have four possible rotations and we select the one which maximizes the value of $\sum_{i=1}^{k} RRS_i^r$.

From Eq. 1, the total number of reflex vertices is $\frac{m}{2} + 2k$. In the worst case, all rotations result in the same value of $\sum_{i=1}^{k} RRS_i^r$, thus $\sum_{i=1}^{k} RRS_i^r \geq \frac{1}{4} \times \lfloor \frac{m}{2} + 2k \rfloor = \lfloor \frac{m}{8} + \frac{k}{2} \rfloor$.

The total number of shapes is $1+\frac{m}{2}+k-\sum_{i=1}^{k}RRS_i^r \le 1+\frac{m}{2}+k-(\lfloor\frac{m}{8}+\frac{k}{2}\rfloor) = \lceil\frac{3m}{8}+\frac{k}{2}\rceil + 1$. ∎

If however, we do not allow a guard to be placed at a corner of the enclosing rectangle P, then the number of guards needed could increase significantly. In the rest of this section, we prove an upper bound on the number of vertex guards, placed only on vertices of the holes.

We introduce staircases RS_P, FS_P, RRS_P and RFS_P (refer to Fig. 3) defined for polygon P. RS_P is constructed as follows: $\forall i \in [1,k]$, extend the horizontal edge E_T^i of each hole towards the right (East) direction, then extend the vertical edge E_L^i towards the South direction. The closed orthogonal polygon formed by the top edge, left edge of P, and the extended edges of the holes, corresponds to a rising staircase, RS_P. Similarly, we can construct RRS_P, FS_P, and RFS_P. We can replace the guard placed on the corner of P by placing guards at the reflex vertices of RRS_P. The total number of guards required to guard such staircase is $1 + RRS_P^r$. Total number of guards required to guard $P = 1 + \sum_{i=1}^{k}(FS_i^r + RS_i^c + RFS_i^r) + RRS_P^r$. Substituting the value of $\sum_{i=1}^{k}(FS_i^r + RS_i^c + RFS_i^r)$ from Eq. 2:

$$Total\ guards = 1 + \frac{m}{2} + k - \sum_{i=1}^{k} RRS_i^r + RRS_P^r \qquad (3)$$

The bound on the total number of guards required to guard P depends upon the two terms in Eq. 3. We can always rotate the polygon to find the rotation which maximizes the value of $\sum_{i=1}^{k} RRS_i^r - RRS_P^r$.

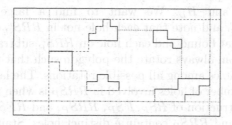

Fig. 3. The four staircases. RS_P is in Red (dashed), FS_P in Blue (dotted), RRS_P in Magenta (dash dot dotted) and RFS_P in Green (dash dotted). (Color figure online)

For hole H_i, the vertical span of H_i is the parallel strip defined by the lines supporting E_L^i and E_R^i and containing H_i. The horizontal span is defined accordingly. Let H_L be the leftmost hole, H_T be the topmost hole, H_R be the rightmost hole, and H_B be the bottom-most hole of the polygon, as defined by their leftmost, topmost, rightmost, and bottom-most edges (without loss of generality we assume here such edges are unique).

A hole H_i is called an *internal hole* if $H_i \notin \{H_L, H_T, H_R, H_B\}$. The pair of staircases (RS_P, FS_P), (FS_P, RRS_P), (RRS_P, RFS_P) and (RFS_P, RS_P) are *adjacent staircases* while the pairs (RS_P, RRS_P) and (FS_P, RFS_P) are *opposite staircases*.

Theorem 2. $\frac{m}{2} + \lfloor \frac{k}{4} \rfloor + 4$ *vertex guards are always sufficient to guard an axis-aligned rectangle polygon with k disjoint axis-aligned monotone orthogonal holes where m is the total number of vertices of the holes, with all guards placed on vertices of the holes.*

Proof. Notice it is possible that, from the set of holes (H_L, H_T), (H_T, H_R), (H_R, H_B), and (H_B, H_L), the pair in one of the sets corresponds to the same hole. In this situation (call it Situation 0), one of the staircases RS_P, FS_P, RRS_P, and RFS_P consists of only one hole. Therefore, there exists a rotation where $k - 1$ holes do not participate in the formation of RRS_P. We rotate the polygon such that the RRS_P consist of one hole. For each hole H_i not participating in the formation of RRS_P, $RRS_i^r \geq 1$ (see Fig. 3). Hence, $\sum_{i=1}^{k} RRS_i^r - RRS_P^r \geq k - 1$ and the total number of guards required to guard P is no more than $\frac{m}{2} + 2$.

Thus, from now on, we assume this is not the case. Consider the four staircases RS_P, FS_P, RFS_P, and RRS_P. We can have four possible situations (cases):

Situation 1. *No internal hole is shared by either adjacent staircases or opposite staircases.*

The staircases RS_P, FS_P, RRS_P, and RFS_P do not share any hole other than H_T, H_R, H_B, and H_L. We want to find a large lower bound for $\sum_{i=1}^{k} RRS_i^r - RRS_P^r$ and note that each hole not in RRS_P^r contributes at least 1 reflex vertex for that bound and each hole on RRS_P^r subtracts at least 1 reflex vertex from it. We can always rotate the polygon such that RRS_P contains the smallest number of holes among all possible rotations. The largest possible value for the minimum number of holes involved in RRS_P is when the number of holes involved in the construction of RS_P, FS_P, RRS_P, and RFS_P, is the same. Let the staircases RS_P and RRS_P contain δ distinct holes. Staircase FS_P contains $\delta - 2$ additional holes because hole H_T is already counted in staircase RS_P and hole H_R is counted in staircase RRS_P. Similarly, RFS_P contains $\delta - 2$ additional holes. Note that $\delta + \delta + \delta - 2 + \delta - 2 = k$ and thus $\delta = \lfloor \frac{k}{4} \rfloor + 1$. Therefore, $k - \delta = k - (\lfloor \frac{k}{4} \rfloor + 1) = \lceil \frac{3k}{4} \rceil - 1$ holes do not participate in the formation of RRS_P. Among all the holes H_i, not participating in the formation of RRS_P, $RRS_i^r \geq 1$. Therefore, $\sum_{i=1}^{k} RRS_i^r - RRS_P^r \geq \lceil \frac{3k}{4} \rceil - 1$ and the total number of guards required to guard P is no more than $\frac{m}{2} + \lfloor \frac{k}{4} \rfloor + 2$.

Situation 2. *There exists an internal hole that is shared by opposite staircases and no internal hole is shared by adjacent staircases.*

Let the staircases RS_P and RRS_P share a hole H_i and refer to Fig. 4. We define two polygons, P_1 and P_2 (turquoise and violet boundaries in Fig. 4b), included in P and such that H_i is included in both P_1 and P_2.

P_1 and P_2 are constructed as follows: Traverse H_i from E_L^i to E_R^i in clockwise direction and let u_L be the first vertex of H_i which is part of RS_P. Let e_L be the horizontal edge incident to u_L. We extend e_L to the left until it hits the left boundary of P at point a. Traverse H_i from E_R^i to E_L^i in clockwise direction and let u_R be the first vertex of H_i which is part of RRS_P. Let e_R be the horizontal edge incident to u_R. We extend e_R to the right until it hits the right boundary of P at point b.

P_1 is the polygon traced by going from u_L to a, top left vertex of P, top right vertex of P, b, u_R then counterclockwise along the boundary of H_i from u_R to u_L. P_2 is defined on the lower part of H_i in a similar way. Note that P_1, H_i, and P_2 are interior disjoint (see Fig. 4b). We then make $P_1 = P_1 \cup H_i$ and $P_2 = P_2 \cup H_i$ (see Fig. 4c).

We guard P_1 and P_2 separately. Let P_1 contain α number of holes (excluding H_i), and $m_\alpha = \sum_{i=1}^{\alpha} m_i$, and let P_2 contain β number of holes (excluding H_i), with $m_\beta = \sum_{i=1}^{\beta} m_i$. Note that $\alpha + \beta + 1 = k$.

For each hole H_j in P_1 (i) Traverse H_j from E_R^j to E_L^j in counter-clockwise direction and extend each encountered vertical edge (including E_L^j and E_R^j) towards North (upward) direction until it hits the polygon P_1 or any other hole and (ii) Traverse H_j from E_R^j to E_B^j in clockwise direction and extend each horizontal edge (including E_B^j) towards East (right) direction until it hits the polygon P_1, any other hole, or any of the extended vertical edges.

For each hole H_j in P_2 (i) Traverse H_j from E_R^j to E_L^j in clockwise direction and extend each encountered vertical edge (including E_L^j and E_R^j) towards South (downward) direction until it hits the polygon P_2 or any other hole and (ii) Traverse H_j from E_L^j to E_T^j in clockwise direction and extend each horizontal edge (including E_T^j) towards West (left) direction until it hits the polygon P_2, any other hole, or any of the extended vertical edges.

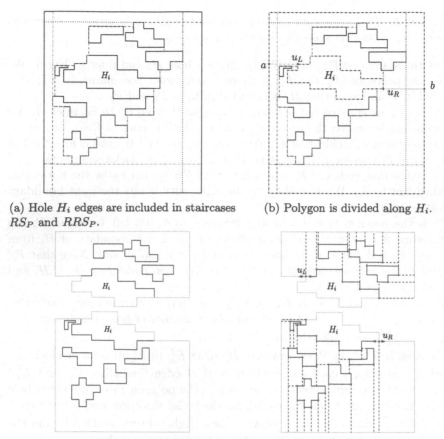

(a) Hole H_i edges are included in staircases RS_P and RRS_P.

(b) Polygon is divided along H_i.

(c) Two polygons created around hole H_i

(d) Extension of edges and placement of guards (red dots).

Fig. 4. P_1 is shown in Turquoise and P_2 is shown in violet (Color figure online)

The above steps divide the polygons P_1 and P_2 into shapes, where each shape corresponds to a monotone staircase, and one guard is sufficient to guard each shape (see Fig. 4d).

Each hole contributes towards the formation of shapes. Let S_{P_1} be the total number of shapes contributed by holes in P_1 excluding H_i. $S_{P_1} = \sum_{j=1}^{\alpha}(FS_j^c + RS_j^r + RRS_j^r)$. Let S_{P_2} be the total number of shapes contributed by holes in P_2 excluding H_i. $S_{P_2} = \sum_{j=1}^{\beta}(RFS_j^c + RS_j^r + RRS_j^r)$. Let S_{H_i} be the total number of shapes contributed by hole H_i in both P_1 and P_2, that is $S_{H_i} = \frac{m_i}{2} + 2$. We have one additional rectangular shape in each of the two polygons. In P_1, it is the leftmost shape and in P_2, it is the rightmost shape.

The total number of shapes in P_1 and P_2 is $2 + S_{P_1} + S_{P_2} + S_{H_i}$. Substituting the value of FS_i^r as $FS_i^c + 1$ in Eq. 1 we have $\sum_{i=1}^{\alpha}(RS_i^r + FS_i^c + RRS_i^r) = \frac{m_\alpha}{2} + \alpha - \sum_{i=1}^{\alpha} RFS_i^r$.

Therefore, $S_{P_1} = \frac{m_\alpha}{2} + \alpha - \sum_{i=1}^{\alpha} RFS_i^r$. For each hole, $RFS_i^r \geq 1$ and thus we have $S_{P_1} \leq \frac{m_\alpha}{2}$. Similarly, $S_{P_2} \leq \frac{m_\beta}{2}$. The total number of shapes is $S_{P_1} + S_{P_2} + S_{H_i} + 2 \leq \frac{m_\alpha}{2} + \frac{m_\beta}{2} + \frac{m_i}{2} + 4 = \frac{m}{2} + 4$.

It follows that the total number of shapes contributed by the k holes of P is no more than $\frac{m}{2} + 2$. In order to guard $\frac{m}{2} + 2$ shapes, we place one guard in each shape. Let us assume that all guards placed in P_1 are facing East (right) and all the guards placed in P_2 are facing West (left). Note that two guards are placed on each of vertex u_L and vertex u_R, one facing East and one facing West. One of the guards placed on u_L (resp. u_R) is used to cover the side rectangular shape associated with u_L (resp. u_R). Hence, the *total number of guards required to guard $P \leq \frac{m}{2} + 2$*.

Situation 3. *There exists an internal hole that is shared by adjacent staircases and no internal hole is shared by opposite staircases.*

Let H_i be a hole whose edges are included in more than one staircase. Assume the pair RS_P, FS_P shares H_i, as shown in Fig. 5. Traverse H_i from E_L^i to E_R^i in clockwise direction and let v_1^i be the first vertex encountered while traversing H_i which is part of RS_P. Similarly, let v_2^i be the last vertex encountered while traversing H_i which is part of FS_P. Let E_1^i be the vertical edge incident to v_1^i and E_2^i be the vertical edge incident to v_2^i. Extend E_1^i and E_2^i in upward direction until they hit the polygon P. Let C_1 be the set of holes which lie above H_i and inside the parallel strip defined by extended edges E_1^i and E_2^i. Let α_i be the number of holes in C_1. Note that all holes in C_1 lie within the vertical span of H_i. Let C_2 be the set containing the holes which are not in C_1 (excluding H_i) and let β_i be the number of holes in C_2. Note that $\alpha_i + \beta_i + 1 = k$. Let C be the set containing all holes H_i that fall within Situation 3. From set C, find the hole H_j that minimizes the value $|\alpha_j - \beta_j|$, such that $\alpha_j, \beta_j \geq 3$.

If such hole does not exist then each hole H_j in set C has $\alpha_j \leq 3$ or $\beta_j \leq 3$. We can use a similar argument as the one discussed in Situation 2. Recall that in the worst case all staircases should involve an equal number of holes. It is easy to notice that there exist at most three holes whose edges are included by a staircase pair (i) RS_P, FS_P (ii) FS_P, RRS_P (iii) RRS_P, RFS_P, or (iv) RFS_P, RS_P, as $\alpha_i < 3$ or $\beta_i < 3$. Let each of RS_P, RRS_P contain δ distinct holes. Staircases FS_P, RFS_P contain $\delta - 6$ additional holes, as three holes are included in each of RS_P and RRS_P. There are k holes, $2 \times \delta + 2 \times (\delta - 6) = k$, and $4 \times \delta - 12 = k$, which gives $\delta = \lfloor \frac{k}{4} \rfloor + 3$. In the worst case, $k - \delta = k - (\lfloor \frac{k}{4} \rfloor + 3) = \lceil \frac{3k}{4} \rceil - 3$ holes do not participate in the formation of RRS_P. For each hole H_i not participating in the formation of RRS_P, $RRS_i^r \geq 1$. Therefore,

$$\sum_{i=1}^{k} RRS_i^r - RRS_P^r \geq \lceil \frac{3k}{4} \rceil - 3 \qquad (4)$$

If there exists a hole H_j such that $\alpha_j, \beta_j \geq 3$, we proceed as follows. Let H_j edges be included in RS_P and FS_P. Extend the topmost edge of H_T in both directions (East and West). Let e be this extended edge. Extend E_1^j and E_2^j

North (upward) until they intersect e. Let the extension of E_1^j intersect e at a and the extension of E_2^j intersect e at b.

We introduce a new orthogonal hole N_1, such that the hole H_j and all holes in C_1 lie inside N_1 and all holes in C_2 lies outside N_1. Observe that N_1 is an orthogonal polygon traced by going from v_1^j to a, b, and v_2^j, then clockwise along the boundary of H_j from v_2^j to v_1^j (see Fig. 5).

The process above is known as *contraction*, since we have contracted $\alpha_j + 1$ holes into a new hole N_1. The reverse process of *contraction* is known as *expansion*. The updated polygon after performing the contraction consist of $(k - \alpha_j)$ holes. We keep contracting the holes if the updated polygon lies within situation 3, and there exists a hole B_d such that $\alpha_d, \beta_d \geq 3$.

Let N_i be the hole introduced during the ith contraction. The initial polygon has k holes. After repeating the process for l iterations, let the updated polygon have $k - \sum_{i=1}^{l} k_i$ holes. Assume situation 3 keeps occurring, and the process is repeated x times. Let $k^0 = k - \sum_{i=1}^{x} k_i$. At i^{th} contraction, $k_i + 1$ holes are contracted into N_i. After the last contraction (x-th one, with no further contraction possible), from Eq. 4 we have $\sum_{i=1}^{k^0} RRS_i^r - RRS_P^r \geq \lceil \frac{3k^0}{4} \rceil - 3$.

The polygon P has k^0 holes with a total of $m^0 = \sum_{i=1}^{k^0} m_i$ vertices. The number of guards required to guard $P \leq 1 + \frac{m^0}{2} + k^0 - (\lceil \frac{3k^0}{4} \rceil - 3) = \frac{m^0}{2} + \lfloor \frac{k^0}{4} \rfloor + 4$.

N_x is the last contracted hole. Assume P is rotates so that N_x be the topmost hole in P. We perform the expansion process on N_x. Remove hole N_x from P and add all the $k_x + 1$ holes contracted inside N_x. Let H_x be the hole contracted inside N_x such that the rest of k_x holes lie inside the vertical span of H_x.

After expansion, guards placed on N_x are placed on the corresponding vertices of H_x with the same orientation. If a guard is placed at a, place it on the top vertex of E_1^x and if a guard is placed b, place it on the top vertex of E_2^x, with the same orientation.

Nest, extend E_1^x and E_2^x in upward direction until they hit P. Let E_1^x intersect P at u_1^P and E_2^x intersect P at u_2^P. Let P_1 be the closed polygon traced by going from the upper vertex of E_1^x to u_1^P, u_2^P, upper vertex of E_2^x then counterclockwise along the boundary of H_x (see Fig. 5b). $P_2 = P \setminus P_1$. Note that after expansion P_2 remains guarded.

Note that there is a guard placed on the topmost edge of N_x facing North (upward). Let this guard be placed at a. After expansion, this guard is placed on the top vertex of E_1^x facing North.

Consider the polygon P. Let m^1 be the total number of vertices and let $k^1 = k - \sum_{i=1}^{x-1} k_i$ be the total number of holes inside P after expanding N_x. Let the k_x holes inside N_x (excluding H_x) have a total of m_x vertices. Let $m' = m^1 - m^0 - m_x$. We extend the edges of P_1 in a similar way as we did for situation 2. Total number of shapes formed $= 1 + \frac{m'}{2} + \sum_{i=1}^{k_x} FS_j^c + RS_j^r + RRS_j^r \leq 1 + \frac{m'}{2} + \frac{m_x}{2}$. We need one guard to guard each shape, however one of the shapes is already guarded by the guard placed on the top vertex of E_1^x.

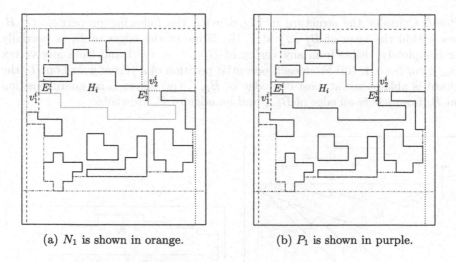

(a) N_1 is shown in orange.   (b) P_1 is shown in purple.

Fig. 5. Contraction and expansion

The total number of guards required to guard the updated P is no more than $\frac{m^1}{2} + \lfloor \frac{k^0}{4} \rfloor + 4$. After expanding all the contracted holes, the total number of guards required to guard $P = \frac{m}{2} + \lfloor \frac{k^0}{4} \rfloor + 4 \leq \frac{m}{2} + \lfloor \frac{k}{4} \rfloor + 4$. Hence, the total number guards required to guard P is no more than $\frac{m}{2} + \lfloor \frac{k}{4} \rfloor + 4$.

Situation 4. *There exists an internal hole shared by a pair of adjacent stair-cases and a pair of opposite staircases.*

Proof. We place guards according to Situation 2 and conclude that $\frac{m}{2} + 2$ guards are sufficient to cover P.   ∎

We also obtain the following Theorem:

Theorem 3. *Given a set F of k pairwise disjoint axis aligned monotone orthogonal polygons in the plane, $\frac{m}{2} + \lfloor \frac{k}{4} \rfloor + 4$ vertex guards are always sufficient to guard the free space and the boundaries of the elements in F, where m is total vertices.*

4   Arbitrary-Oriented Orthogonal Polygons

In this section, we considered the following variation of the art-gallery problem:

Given an axis-aligned rectangle P with k disjoint monotone orthogonal holes $H_1, \ldots, H_k$, with $m_1, m_2, \ldots, m_k$ vertices, respectively, such that $\sum_{i=1}^{k} m_i = m$, place vertex guards on holes such that every point inside P is visible to at least one guard, where the range of vision of guards is $180°$.

Theorem 4. *$\frac{m}{2} + k + 1$ guards are sometimes necessary to guard an axis-aligned rectangle polygon with k disjoint orthogonal holes, with all guards placed on vertices of the holes. We conjecture the bound is tight.*

Proof. Consider the structure in Fig. 6, with the following properties: (i) B_i lies within the span of B_j , $\forall j < i$. (ii) None of the edges of B_i is partially or completely visible from any vertex of B_j , $\forall j < i - 1$ and from any vertex B_m , $\forall m > i + 1$. (iii) From each potential position of a vertex guard on B_i, the guard is able to see at most one edge of B_{i+1}. (iv) There is no guard position on B_i from where an edge of B_{i-1} and an edge of B_i are visible.

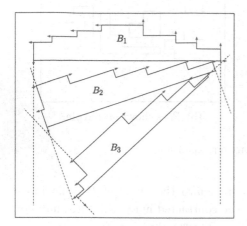

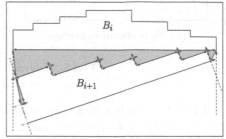

Fig. 6. City Structure with guards placement and orientation.

Fig. 7. $\wp$ is shaded in green. Potential guard positions to cover $\wp$ are shown in red and potential guard positions to cover w_L^{i+1} are shown in orange. (Color figure online)

Let E_L^i be the leftmost edge of B_i such that the interior of B_i lies towards the right of it. Similarly, we define E_T^i as the topmost edge, E_R^i as the Rightmost edge and E_B^i as the bottom-most edge. For each hole H_i, we have staircases RS_i, FS_i, RRS_i and RFS_i as defined in Sect. 3 (see Fig. 2). For a building B_i let r_i be the total number of reflex vertices of B_i and let c_i be the total number of convex vertices of B_i. Let $\sum_{i=1}^{k} r_i = r$ and $\sum_{i=1}^{k} c_i = c$.

Consider the space $\wp_i$ between two consecutive holes B_i and B_{i+1}, as shown in Fig. 7. $\wp_i$ is not visible to any guard placed on building B_j, where $j \in [1, i) \cup (i + 1, k]$ (property 2). Therefore, $\wp_i$ is only visible to the guards placed either on B_i or on B_{i+1}. Let u be the upper vertex of E_L^{i+1} and v be the bottom vertex of E_R^{i+1}. The region $\wp_i$ is visible (partially) to the following guards: (i) A guard placed on either end of E_B^i, facing along this edge, (ii) A guard placed on either end of E_B^i, facing towards B_{i+1}, (iii) Guards placed on the reflex vertices encountered while traversing B_{i+1} from u to v in clockwise direction (excluding u and v), (iv) Guards placed on the convex vertices encountered while traversing B_{i+1} from u to v (v) A guard placed on v facing towards E_T^i, (vi) A guard placed on v facing away from E_B^{i+1}, and (vii) Guards placed on u with similar orientation as of guards placed on v. Refer Fig. 7. The total number of possible guard positions on B_i and B_{i+1} from where $\wp_i$ is visible (partially) is $4(r_{i+1} - 1) + 2c_{i+1}$. These guards cover $m_{i+1} - 1$ edges: one edge of B_i, and $m_{i+1} - 2$ edges of B_{i+1}. Note that these guards do not cover any other edge (partially or entirely), and the mentioned $m_{i+1} - 1$ edges are not visible (partially or entirely) to any other potential guard. The minimum number of guards required to guard $\wp_i$ is then $r_{i+1} - 2$.

Consider the left edge w_L^i of hole B_i. Because of the structure of the hole, E_L^i is not visible to any guard placed on B_j, for $j \neq i$. Hence, E_L^i can only be guarded by a guard placed on B_i, and there are four possible guard positions from where E_L^i is visible (see Fig. 7). However, none of these guard positions covers (partially or entirely) any other edge of a hole. Therefore, we need one guard to cover the left edge of each hole.

There are $k-1$ regions $\wp_i$ in total, between consecutive holes ($i = 1, 2, \ldots, k-1$). Thus, $\sum_{i=1}^{k-1}(r_{i+1} - 2)$ guards are needed to guard their union. As argued, each left edge of a hole needs an additional guard.

For B_k, one guard is required to guard the bottom-most edge. For B_1, only two edges (left most and bottom most) are guarded by the above mentioned guards. In order to guard the rest of the edges, the minimum number of guards required is $r_1 - 2$. Therefore, in total at least $r_1 - 1 + \sum_{i=2}^{k}(r_i - 2) + k = r - k + 1$ guards are needed. Substituting the value of r from Eq. 1, the total number of guards needed is $\frac{m}{2} + k + 1$. ∎

We also obtain the following Theorem:

Theorem 5. *Given a set F of k pairwise disjoint monotone orthogonal polygons in the plane, $\frac{m}{2} + k + 1$ vertex guards are sometimes necessary to guard the free space and the boundaries of the elements in F. We conjecture the bound is tight.*

5   Conclusion

In this paper, we considered necessary and sufficient bounds on the number of vertex guards needed for guarding the free space of the plane defined by k pairwise disjoint monotone orthogonal polygons (holes) with a total of m vertices, when the range of vision of the guards is bounded by $180°$. When the holes are axis-aligned, we proved that $\frac{m}{2} + \lfloor \frac{k}{4} \rfloor + 4$ vertex guards are always sufficient to guard the free space. For k arbitrary oriented holes we showed that $\frac{m}{2} + k + 1$ vertex guards are sometimes necessary to guard the free space and conjecture that the bound is tight. Our results can be easily extended to the case when the orthogonal polygons defining the holes are not monotone (details are presented in [14]).

If the range of vision of a guard is bounded by $90°$, using the same approach presented here, it follows that $\frac{m}{2} + \lfloor \frac{k}{4} \rfloor + 5$ vertex guards are always sufficient to guard the free space. Similarly, for arbitrary oriented orthogonal polygons, it follows that $\frac{m}{2} + k + 1$ guards are sometimes necessary to guard the free space.

References

1. Abello, J., Estivill-Castro, V., Shermer, T., Urrutia, J.: Illumination of orthogonal polygons with orthogonal floodlights. Int. J. Comput. Geom. Appl. **8**(01), 25–38 (1998)
2. Aggarwal, A.: The Art Gallery Theorem: Its Variations, Applications and Algorithmic Aspects. Ph.D. thesis, AAI8501615 (1984)
3. Blanco, G., Everett, H., Lopez, J.G., Toussaint, G.: Illuminating the free space between quadrilaterals with point light sources. In: Proceedings of Computer Graphics International, World Scientific. Citeseer (1994)
4. Chvatal, V.: A combinatorial theorem in plane geometry. J. Comb. Theor. Ser. B **18**(1), 39–41 (1975)
5. Czyzowicz, J., Rivera-Campo, E., Urrutia, J., Zaks, J.: Protecting convex sets. Graphs Comb. **10**(2–4), 311–321 (1994)
6. Czyzowicz, J., Riveracampo, E., Urrutia, J.: Illuminating rectangles and triangles in the plane. J. Comb. Theor. Ser. B **57**(1), 1–17 (1993)
7. Daescu, O., Malik, H.: City guarding with limited field of view. In: Proceedings of 32nd Canadian Conference on Computational Geometry (2020). http://vga.usask.ca/cccg2020/papers/City%20Guarding%20with%20Limited%20Field%20of.pdf
8. de la Calle, J.G.L.: Problemas algorítmico-combinatorios de visibilidad. Ph.D. thesis, Universidad Politécnica de Madrid (1995)
9. Estivill-Castro, V., Urrutia, J.: Optimal floodlight illumination of orthogonal art galleries. In: Proceedings of 6th Canadian Conference on Computational Geometry, pp. 81–86 (1994)
10. Everett, H., Toussaint, G.: On illuminating isothetic rectangles in the plane (1990)
11. Fisk, S.: A short proof of chvátal's watchman theorem. J. Comb. Theor. (B) **24**, 374 (1978)
12. Hoffmann, F., Kriegel, K.: A graph-coloring result and its consequences for polygon-guarding problems. SIAM J. Discrete Math. **9**(2), 210–224 (1996)
13. Kahn, J., Klawe, M., Kleitman, D.: Traditional galleries require fewer watchmen. SIAM J. Algebraic Discrete Method **4**(2), 194–206 (1983)
14. Malik., H.: City guarding and path checking: some steps towards smart cities. Ph.D. thesis
15. Martini, H., Soltan, V.: Survey paper. Aequationes Math. **57**, 121–152 (1999)
16. O'Rourke, J.: An alternate proof of the rectilinear art gallery theorem. J. Geom. **21**(1), 118–130 (1983)
17. O'Rourke, J.: Galleries need fewer mobile guards: a variation on chvátal's theorem. Geom. Dedicata **14**(3), 273–283 (1983)
18. O'Rourke, J.: Art Gallery Theorems and Algorithms, vol. 57. Oxford University Press, Oxford (1987)
19. Shermer, T.C.: Recent results in art galleries. Proc. IEEE **80**, 1384–1384 (1992)
20. Tóth, C.D.: Art galleries with guards of uniform range of vision. Comput. Geom. **21**(3), 185–192 (2002)
21. Tóth, C.D.: Art gallery problem with guards whose range of vision is 180. Comput. Geom. Theor. Appl. **3**(17), 121–134 (2000)
22. Tóth, L.F.: Illumination of convex discs. Acta Math. Hung. **29**(3–4), 355–360 (1977)
23. Urrutia, J.: Art gallery and illumination problems (2004)
24. Żyliński, P.: Orthogonal art galleries with holes: a coloring proof of Aggarwal's theorem. Electron. J. Comb. **13**(1), 20 (2006)

Optimal Strategies in Single Round Voronoi Game on Convex Polygons with Constraints

Aritra Banik[1], Arun Kumar Das[2($\boxtimes$)], Sandip Das[2], Anil Maheshwari[3], and Swami Sarvottamananda[4]

[1] National Institute of Science Education and Research, Bhubaneswar, India
aritrabanik@gmail.com
[2] Indian Statistical Institute, Kolkata, India
arund426@gmail.com, sandipdas@isical.ac.in
[3] Carleton University, Ottawa, Canada
anil@scs.carleton.ca
[4] Ramakrishna Mission Vivekananda Educational and Research Institute, Howrah, India
sarvottamananda@rkmvu.ac.in

Abstract. We describe and study the *Voronoi games on convex polygons*. In this game, the server and then the adversary place their respective facilities by turns. We prove the lower and upper bounds of $\lceil n/3 \rceil$ and $n-1$ respectively in the single-round game for the number of clients won by the server for n clients. Both bounds are tight. Consequentially, we show that in some convex polygons the adversary wins no more than k clients in a k-round Voronoi game for any $k \leq n$. We also design $O(n \log^2 n + m \log n)$ and $O(n + m)$ time algorithms to compute the optimal locations for the server and the adversary respectively to maximize their client counts where the convex polygon has size m. Moreover, we give an $O(n \log n)$ algorithm to compute the common intersection of a set of n ellipses. This is needed in our algorithm.

1 Introduction

In the *competitive facility location* problem, the server and its adversary place one or more facilities inside an arena with a set of clients. There are different winning conditions, such as number of clients won, delivery cost, net profit, min-max risk, etc. The *Voronoi game* is a type of competitive facility location problem. In the game two players, the server and the adversary, place one or more facilities to serve a set of clients $\mathcal{C}$ in a competitive arena. Each client $c \in \mathcal{C}$ is served by the nearest facility. The goal of the players is to maximize the number of their clients through one or more rounds.

Ahn et al. [2] introduced the *Voronoi game on line segments and circles* where the server and its adversary place an equal number of facilities to serve clients located on line segments and circles respectively. Cheong et al. [5] and Fekete

© Springer Nature Switzerland AG 2020
W. Wu and Z. Zhang (Eds.): COCOA 2020, LNCS 12577, pp. 515–529, 2020.
https://doi.org/10.1007/978-3-030-64843-5_35

and Meijer [7] studied the *Voronoi game on planes.* Teramoto et al. [8] studied a discrete version of *the Voronoi game on graphs* where the clients and the players, all are located on the vertices of the graphs. They used the shortest path distance metric for their Voronoi game. Moreover, they showed that devising an optimal strategy for the adversary is **NP**-hard for a certain restricted version of the problem. Banik et al. [3] studied the discrete version of *Voronoi game in* $\Re^2$ and gave polynomial time algorithms for the optimal placement for both the server and its adversary. They were able to solve the Voronoi game on $\Re^2$ for each of the L_1, L_2 and L_∞ distance metrics. Later, Banik et al. [4] introduced the *Voronoi game on simple polygons* and devised polynomial time algorithms for the optimal placement of both the server and its adversary for the geodesic distance metric.

In this paper, we mainly study the *Voronoi games on convex polygons.* We also have some surprising results on the bounds for the *Voronoi games on convex polygons.* There are two players, the server Alice and her adversary Bob, who place *point* facilities in the planar arena, alternately. Alice is restricted to place her facilities in the closed interior of a polygon and Bob is restricted to place his facilities in the closed exterior of that polygon. The clients represented as points are on the boundary of the polygon. These restrictions can be understood as if these are their personalized areas whereas the clients are in a common area accessible to both. The objectives of Alice and Bob are to maximize their respective number of clients, denoted by $\mathcal{S}_A$ and $\mathcal{S}_B$. See Fig. 1.

In this paper, we first prove the tight bounds for the score of Alice, $\mathcal{S}_A$, which is essentially the number of her clients (shared clients are counted as half), for the single round Voronoi games on both simple polygons as well as convex polygons, if both Alice and Bob choose optimal placement for their facilities. More specifically, we show that $\lceil n/3 \rceil \leq \mathcal{S}_A \leq n - 1$ when both the server and the adversary make the optimal choices for the Voronoi game on convex polygons and $1 \leq \mathcal{S}_A \leq n - 1$ for simple polygons. These bounds imply that the adversary Bob does not always have a good strategy against the server Alice even if Alice places her facility first (since in Voronoi games the second player usually has an advantage). If Bob places his facility first, then Alice can trivially restrict Bob's score to $1/2$, if clients are distinct points, and $n/2$, if the clients are n points coincident on Bob's facility $\mathcal{B}$. For the k-round Voronoi games on simple polygons or convex polygons too, we show that there exist arenas where the adversary Bob gets no more than a $(k+1)/2$ score, if Alice is allowed to place k facilities one each in each round, and a k score, if Alice is only allowed to place a single facility in the first round. This means that the adversary Bob does not have a guaranteed good strategy to place his facilities even in a k-round Voronoi game on convex polygons. We prove that similar results hold equally true for the Voronoi games on simple polygons. Since Voronoi games are zero-sum games, the bounds on Alice's score imply the bounds on Bob's score $\mathcal{S}_B$.

We also provide algorithms to compute optimal placement for Alice for Voronoi games on convex polygons which minimizes the maximum score of Bob and for Bob after Alice has placed her facility which maximizes the score for Bob.

The algorithms respectively run in time $O(n \log^2 n + m \log n)$ and $O(n+m)$ for Alice and Bob. We present both the algorithms separately as they are significantly different. We assume that the convex polygon $\mathcal{P}$, of size m, and the n clients, on the boundary of $\mathcal{P}$, are given as input. As a part of our algorithm, as well as an independent result, we also devise a method to compute the common intersection of n ellipses, if it exists, in $O(n \log n)$ time.

The organization of the paper is as follows. We give some definitions, concepts and important characterizations of optimality criteria in Sect. 2 and Sect. 3. We present the algorithm to check and compute the common intersection of ellipses, needed later, in Sect. 4. Next, in Sect. 5, we present tight lower and upper bounds for Voronoi games on convex and simple polygons for a single round. We also prove the tight upper bound and a lower bound for the k-round game for both. In Sect. 6 we present a linear time algorithm to compute the optimal placement of the adversary Bob given any placement of Alice. Lastly, in Sect. 7, we provide an algorithm for the optimal placement of the server Alice, that minimizes the maximum possible score for the adversary Bob.

2   Definitions and Preliminary Concepts

Let the vertices of the convex polygon $\mathcal{P}$, given as the game arena, be v_0, v_1, $\ldots$, v_{m-1} in counter-clockwise order. Let there be n clients c_0, c_1, $\ldots$, c_{n-1} that are placed in counter-clockwise order on the boundary of $\mathcal{P}$. For the sake of notational brevity, the vertices v_i and $v_{(i \bmod m)}$ as well as the clients c_j and $c_{(j \bmod n)}$ are the same vertices and clients respectively.

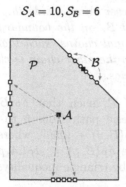
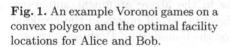

$\mathcal{S}_A = 10, \mathcal{S}_B = 6$

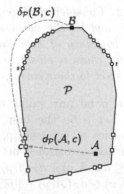

Fig. 1. An example Voronoi games on a convex polygon and the optimal facility locations for Alice and Bob.

Fig. 2. The clients of Alice and Bob are consecutive. Alice sees the distances as $d_\mathcal{P}$ whereas Bob sees the distances as $\delta_\mathcal{P}$.

In the single-round Voronoi game, Alice places her facility at $\mathcal{A}$ and Bob places his facility at $\mathcal{B}$ after Alice's turn. Let Alice's and Bob's distance from any boundary point $p \in \mathcal{P}$ be measured as internal and external geodesic distance respectively denoted by $d_\mathcal{P}(\mathcal{A}, p)$ and $\delta_\mathcal{P}(\mathcal{B}, p)$. Also, we assume that the context is always the Voronoi game on convex polygons unless mentioned otherwise. See Fig. 2. The problem studied in the paper is to find bounds on the number of clients of Alice and Bob and compute their optimal placements.

3   Characterization of Optimal Placements for Alice and Bob

We characterize Bob's optimal location first in the following lemma.

Lemma 1 (Necessary condition). *The optimal position for the adversary, $\mathcal{B}$, is always on the boundary of the convex polygon $\mathcal{P}$.*

Proof. Consider any position of $\mathcal{B}$ in the open exterior of $\mathcal{P}$. If $\mathcal{B}^*$ be the nearest point to $\mathcal{B}$ on the boundary of $\mathcal{P}$, we can show that $\delta_\mathcal{P}(\mathcal{B}, c) > \delta_\mathcal{P}(\mathcal{B}^*, c)$ holds for any client c on the boundary of the polygon using the following: (1) the external geodesic distance is the sum of Euclidean distances in some external path in the Voronoi game on polygons and (2) Euclidean distances satisfy triangle inequalities. The proof then follows.   □

3.1   A Characterization of Clients of Alice and Bob

We observe that all the clients of Alice and Bobs are partitioned into two consecutive sequences on the boundary. We prove this with the help of the following lemma.

Lemma 2. *Consider a placement of $\mathcal{A}$ of the server Alice inside the convex polygon $\mathcal{P}$ and let the adversary Bob be located at $\mathcal{B}$, on the boundary of $\mathcal{P}$. Then the boundary of $\mathcal{P}$ is divided into two polygonal chains, such that every point in one chain is closer to $\mathcal{B}$ and every point on the other chain is closer to $\mathcal{A}$ with respect to their corresponding distance metrics.*

Proof. Let p be any point on the boundary of $\mathcal{P}$, such that $\delta_\mathcal{P}(\mathcal{B}, p) \leq d_\mathcal{P}(\mathcal{A}, p)$. Let v be the first vertex on the geodesic path from p to $\mathcal{B}$. By the properties of geodesic paths, $\delta_\mathcal{P}(\mathcal{B}, p) = \delta_\mathcal{P}(\mathcal{B}, v) + ||\overline{pv}||$, where $||\overline{pv}||$ is the Euclidean distance between p and v. Since $\delta_\mathcal{P}(\mathcal{B}, p) \leq d_\mathcal{P}(\mathcal{A}, p)$, and $d_\mathcal{P}(\mathcal{A}, p) \leq d_\mathcal{P}(\mathcal{A}, v) + d_\mathcal{P}(v, p) = d_\mathcal{P}(\mathcal{A}, v) + ||\overline{pv}||$, by triangle inequality, therefore $\delta_\mathcal{P}(\mathcal{B}, v) \leq d_\mathcal{P}(\mathcal{A}, p) - ||\overline{pv}|| \leq d_\mathcal{P}(\mathcal{A}, v)$.

Next, By replacing p with the vertices in the above argument and using induction we can show that $\delta_\mathcal{P}(\mathcal{B}, v) \leq d_\mathcal{P}(\mathcal{B}, v)$ for all points in the geodesic path from p to $\mathcal{B}$. Thus the farthest points on the boundary of $\mathcal{P}$, those are closer to $\mathcal{B}$ than $\mathcal{A}$ in clockwise and anticlockwise direction, partitions the boundary as stated. Thus the lemma holds.   □

We state the following theorem that follows from Lemma 2. See Fig. 2 for an illustration.

Theorem 1. *The clients served by Bob and Alice are consecutive on the boundary of* $\mathcal{P}$.

In the discussion below, we give a necessary and sufficient condition that allows Bob to serve all the clients on a portion of the boundary of $\mathcal{P}$.

3.2   A Necessary and Sufficient Condition for Alice and Bob

Let c and c', $c \neq c'$, be two distinct clients in $\mathcal{C}$ on the boundary of the convex polygon $\mathcal{P}$. The external geodesic distance from c to c' will be $\delta_\mathcal{P}(c, c')$. We consider the locus of a point x such that $\|\overline{xc}\| + \|\overline{xc'}\| = \delta_\mathcal{P}(c, c')$ or $d_\mathcal{P}(x, c) + d_\mathcal{P}(x, c') = \delta_\mathcal{P}(c, c')$. We note that if c and c' are not on the same edge of $\mathcal{P}$ then this is an ellipse. If c and c' are on the same edge then the locus is a line segment, i.e., a degenerate ellipse. Let us denote either of these ellipses by $\mathcal{E}_{c,c'}$. We further note that, because of the properties of ellipses, in either of the two cases above, if Alice wishes to prevent Bob from serving both c and c', she must place $\mathcal{A}$ somewhere in the closed interior of $\mathcal{E}_{c,c'} \cap \mathcal{P}$. We state this fact in the following lemma.

Lemma 3. *The adversary, Bob, can serve two clients* c *and* c', $c \neq c'$, *if and only if Alice does not place* $\mathcal{A}$ *in the closed interior of* $\mathcal{E}_{c,c'} \cap \mathcal{P}$ *in the Voronoi game.*

Proof. First, let $\mathcal{A}$ be a position inside the convex polygon such that it is not in the closed interior of $\mathcal{E}_{c,c'}$, i.e., it is not in closed interior of $\mathcal{E}_{c,c'} \cap \mathcal{P}$. Then following the argument in the discussion above, $d_\mathcal{P}(\mathcal{A}, c) + d_\mathcal{P}(\mathcal{A}, c') > \delta_\mathcal{P}(c, c')$. We can find a position for $\mathcal{B}$ in the geodesic path from c to c' such that $\delta_\mathcal{P}(c, c') = \delta_\mathcal{P}(\mathcal{B}, c) + \delta_\mathcal{P}(\mathcal{B}, c')$, where $\delta_\mathcal{P}(\mathcal{B}, c) < d_\mathcal{P}(\mathcal{A}, c)$ and $\delta_\mathcal{P}(\mathcal{B}, c') < d_\mathcal{P}(\mathcal{A}, c')$. We can do this because if $x + y > z$ then we can partition z into two numbers x' and y' such that $z = x' + y'$, $x > x'$ and $y > y'$. Geometrically, this translates into finding $\mathcal{B}$ as above. Thus Bob can serve both c and c'.

Conversely, let $\mathcal{A}$ be a position inside the closed interior of $\mathcal{E}_{c,c'} \cap \mathcal{P}$. Then $d_\mathcal{P}(\mathcal{A}, c) + d_\mathcal{P}(\mathcal{A}, c') \leq \delta_\mathcal{P}(c, c')$. This will prevent Bob to serve both c and c' because (1) if Bob places $\mathcal{B}$ in the geodesic path from c to c' then either $\delta_\mathcal{P}(\mathcal{B}, c) \geq d_\mathcal{P}(\mathcal{A}, c)$ or $\delta_\mathcal{P}(\mathcal{B}, c') \geq d_\mathcal{P}(\mathcal{A}, c')$ (due to the fact that $\delta_\mathcal{P}(c, c') = \delta_\mathcal{P}(\mathcal{B}, c) + \delta_\mathcal{P}(\mathcal{B}, c')$) and (2) if Bob place $\mathcal{B}$ outside the geodesic path from c to c' then either $\delta_\mathcal{P}(\mathcal{B}, c) \geq \delta_\mathcal{P}(c, c')$ or $\delta_\mathcal{P}(\mathcal{B}, c') \geq \delta_\mathcal{P}(c, c')$. $\qquad\square$

Combining Lemma 2 and Lemma 3 we get the following theorem.

Theorem 2. *Let* s *and* t *be two points on the boundary of convex polygon* $\mathcal{P}$ *in the Voronoi game on convex polygons. The adversary Bob can serve all the clients on the geodesic path from* s *to* t *if and only if the server Alice does not place* $\mathcal{A}$ *in the intersection of the closed interior of the ellipse* $\mathcal{E}_{s,t}$ *and the polygon* $\mathcal{P}$, *where the ellipse is as described above.*

We are now in a position to present the main idea behind the paper. If Alice wishes to prevent Bob to serve more than, say r clients, in other words forcing $\mathcal{S_B} < r$, what she needs to do is to put her facility, if possible, in the intersection of all ellipses $\mathcal{E}_{c,c'}$'s for all the possible distinct pairs of clients c and c' such that the external geodesic paths from c to c' contain more than or equal to r clients. If such placement is not possible then Bob will escape the predicament and will be able to place $\mathcal{B}$ that serves more than or equal to r clients.

However, for Alice's algorithm to compute her facility's optimal placement, there is a need to check several possible positions for Bob's facility location. Correspondingly, we need to check the intersection of multiple ellipses. Therefore first we describe an algorithm to efficiently compute the intersection of a finite set of ellipses in the next section.

4   Algorithm to Compute the Common Intersection of Ellipses

Suppose E be a set of n ellipses. Let e be any ellipse in the set E. In our later algorithms, we have two very specific objectives. One is to check whether the ellipses have a common intersect at all and the second is to compute a point in the common intersection. This is a restricted goal relative to this section. However, we give a method to compute the full common intersection of the ellipses in E in this section.

We use the idea of Davenport-Schinzel sequences in our algorithm. Luckily, after some transformation, our construction satisfies the properties of Davenport-Schinzel sequences of order 2 and hence we are able to compute the common intersection of ellipses in E in $O(n \log n)$ time. We are going to use the following theorem by Agarwal et al. [1].

Theorem 3 (Agarwal et al. [1], Theorem 2.6). *The lower envelope of a set $\mathcal{F}$ of n continuous, totally defined, univariate functions, each pair of whose graphs intersect in at most s points, can be constructed, in an appropriate model of computation, in $O(\lambda_s(n) \log n)$ time. If the functions in $\mathcal{F}$ are partially defined, then $E_\mathcal{F}$ can be computed in $O(\lambda_{s+1}(n) \log n)$ time. In particular, the lower envelope of a set of n segments in the plane can be computed in optimal $O(n \log n)$ time.*

In the cited theorem above $E_\mathcal{F}$ is the lower envelope of $\mathcal{F}$. In order to use the theorem, we need to suitable modify and transform the set of ellipses E so that it satisfies the conditions of the theorem. We show in the following discussion how we do it.

Since two ellipses intersect four times, the corresponding Davenport-Schinzel sequence seems to be $\mathcal{DS}(n,4)$ which will lead to length $\lambda_4(n) = \Theta(n2^{\alpha(n)})$. This may give us a worse running time , even if we apply the theorem in our algorithm. However, if we divide the ellipses into two halves, the upper half and the lower half, the upper and lower halves will mutually intersect only two times. This gives us the Davenport-Schinzel sequence $\mathcal{DS}(n,2)$ with length $\lambda_2(n) = 2n - 1$.

Hence we need to divide all the ellipses into two halves. We describe the steps below.

We separate the boundary of every ellipse $e \in E$ into lower and upper boundary, separated at the points with the smallest and largest x-coordinate on the boundary. We call the upper part of the boundaries of the ellipses as the *upper chains* and denote them by $upper(e)$ for the ellipse e. Similarly the lower part of the boundary of an ellipse e is called the *lower chain* and denoted by $lower(e)$. It is a well known fact that the boundaries of two ellipses can intersect each other at most four times. We make similar observations for upper and lower chains.

Observation 1 *For two distinct ellipses, e and e' in the set E, the boundary of $upper(e)$ and $upper(e')$ can intersect at most 2 times.*

Observation 2 *If the set E of ellipses has a common intersection then it bounded from above by $upper(e)$'s and bounded from below by all the $lower(e)$'s, $e \in E$.*

Let l be the rightmost of the left endpoints of the ellipses in E and let r be the leftmost of the right endpoints of the ellipses in E. Let the infinite vertical strip containing points (x, y) such that $l \leq x \leq r$ be denoted by V.

Observation 3 *If the set E of ellipses has a common intersection then the infinite vertical strip V exists and all upper and lower chains for the ellipses in the set intersect every vertical line in the strip V.*

For every e, $e \in E$, $upper(e)$ and $lower(e)$ are graphs of some continuous univariate functions. The *lower envelope* and *upper envelope* of a set of functions are defined as the pointwise minimum and maximum respectively of all the functions in the set. If the set of ellipses in E has a common intersection then we can compute the lower envelope of the upper chains as well as the upper envelope of the lower chains inside the strip V. The lower envelope of the upper chains inside this strip is denoted by $\mathcal{U}_l$. Similarly the upper envelope of the lower chains of all ellipses of E inside the vertical strip V is denoted by $\mathcal{L}_u$. The following lemma states that these can be computed efficiently. We use Theorem 3 for this purpose.

Lemma 4. $\mathcal{U}_l$ *and* $\mathcal{L}_u$ *can be computed in* $O(n \log n)$ *time.*

Proof. The chains $upper(e)$ and $lower(e)$, for any e in E, can be represented by well defined, continuous, univariate, real functions in the range $l \leq x \leq r$. We number these functions from 1 to n. Each pair of their graphs intersect at most twice. Hence the two sequences of function indices occurring in $\mathcal{U}_l$ and $\mathcal{L}_u$ are Davenport-Schinzel sequences [1] of order 2, i.e., $\mathcal{DS}(n, 2)$. Then the lengths of these two sequences are bounded by $2n - 1$ and the two sequences can be computed in time $O(n \log n)$.

The main idea of the algorithm, in short, is the divide and conquer approach. We first compute the envelopes of two halves and then compute the envelope of the two resulting envelopes. Because of the limits imposed by the Davenport-Schinzel sequences, the output is always bounded by $2n - 1$ ellipse segments. $\square$

If the ellipses in the set E have a common intersection then at the intersection the lower boundaries of all the ellipses will be below the intersection and the upper boundaries of all the ellipses will be above the intersection and vice-versa.

Lemma 5. *All ellipses of E have a common intersection if and only if at least one point of $\mathcal{U}_l$ lies above $\mathcal{L}_u$.*

Proof. All the lower chains of every ellipse of E lie below $\mathcal{L}_u$ except at least one. If $\mathcal{U}_l$ goes below it at some x-coordinate. then there is at least one ellipse which has its upper chain lying below the lower chain of at least one ellipse at that x-coordinate. This means they will not form a common intersection at that x-coordinate.   □

We describe the algorithm to compute the intersection of ellipses in E in $O(n \log n)$ time. First we number the chains from 1 to n and compute all the lower chains and upper chains. Next we compute the Davenport-Schinzel sequence for both $\mathcal{U}_l$ and $\mathcal{L}_u$ using the algorithm proposed in [1]. We use these sequences to compute both $\mathcal{U}_l$ and $\mathcal{L}_u$ by Lemma 4 and check if they have intersections using Lemma 5.

Thus we have the following theorem.

Theorem 4. *The common intersection of n ellipses can be computed in $O(n \log n)$ time.*

5   Bounds for the Scores in the Voronoi Game on Polygons

In the case of the Voronoi games on simple polygons, we prove the tight upper bounds and lower bounds on both $\mathcal{S}_A$ and $\mathcal{S}_B$ for both single-round and k-round games. The k-round games are where Alice and Bob alternately place their facilities for k rounds for any fixed $k \geq 1$. Thus we have 1-round, 2-round, etc. Voronoi games. In the case of the Voronoi games on convex polygons too, we prove the tight upper bound on the score $\mathcal{S}_A$. We prove the tight lower bound for the single round Voronoi games on convex polygons and prove a lower bound for the k-rounds on the score $\mathcal{S}_A$.

5.1   Tight Lower and Upper Bounds for the Scores in the Voronoi Game on Simple Polygons

Let $\mathcal{P}$ be a simple polygon where Alice and Bob are playing the Voronoi game. For single round and k round Voronoi game we show that neither Alice nor Bob has a guaranteed strategy in the following Lemmas.

First we prove lower bounds for Alice in the following lemma.

Lemma 6. *For the single round Voronoi games on simple polygons the score of Alice $\mathcal{S}_A \geq 1/2$ for $n \geq 1$. For the k-round Voronoi games on simple polygons $\mathcal{S}_A \geq k/2$ for $n \geq k$.*

Fig. 3. Lower bound for $\mathcal{S}_A$

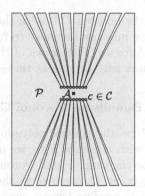

Fig. 4. Upper bound for $\mathcal{S}_A$

We can show similar bounds for Bob.

Lemma 7. *For the single round Voronoi game on simple polygons the score of Bob $\mathcal{S}_B \geq 1$ for $n \geq 2$. For the k-round Voronoi games on simple polygons $\mathcal{S}_B \geq (k+1)/2$ for $n \geq k+1$.*

Proof. We can prove this by pigeon hole principle. Except for the last round, Alice may place her facility to share clients with Bob. So Bob is guaranteed a free client only at the last turn. □

Finally, we show that the bounds of Lemma 6 and Lemma 7 are tight.

Lemma 8. *There exist k-round Voronoi games on simple polygons with n clients such that $\mathcal{S}_A = k/2$ for any n, $n \geq k$ and $\mathcal{S}_B = (k+1)/2$ for any n, $n \geq k+1$ for any k.*

Proof. We construct simple polygon similar to star shaped and reverse star shaped simple polygons where the clients are placed on the selected vertices only, as shown in Fig. 3 and Fig. 4, that satisfy the scores for optimal placements of facilities for both Alice and Bob. In fact, only one facility for Bob and Alice suffices in the example figures to restrict the scores of Alice and Bob respectively. Other placements are only necessary to deny even the full shares of clients. □

In summary, $k/2 \leq \mathcal{S}_A \leq n - (k+1)/2$ and $(k+1)/2 \leq \mathcal{S}_B \leq n - k/2$ since $\mathcal{S}_A + \mathcal{S}_B = n$ if there are at least $k+1$ clients.

If there are less clients than the rounds then trivially all clients will be shared. And moreover, as a side note, if there are multiple clients on the same edge then both Alice and Bob can perform better by placing at the median client. We state this last fact in the following lemma.

Lemma 9. *For the single round Voronoi games on simple polygons the score of the server and adversary satisfy $\mathcal{S}_A \geq r/2$ and $\mathcal{S}_B \geq r/2$ where there are at least r clients on some edge of the polygon $\mathcal{P}$.*

If the polygon P is such that there are multiple edges with multiple clients, then we may use Lemma 9 to further optimize the scores to get higher scores for both Alice and Bob. Both of them give preference to these clusters of clients on the same edge as long as these are available.

5.2 Bounds on Voronoi Games on Convex Polygons

Let $\mathcal{P}$ be the convex polygon where Alice and Bob are playing the Voronoi game with n clients. First we prove the tight lower bound on the score of Alice, $\mathcal{S}_A$, for the single round Voronoi game. In the following lemma, we use Haley's center-point Theorem [6] to show that there are always $\lfloor n/3 \rfloor$ clients for Alice. In the succeeding lemma, we construct a single round Voronoi game such that no matter where Bob places its facility it gets no more than one client, i.e., Alice serves $n - 1$ clients.

Lemma 10. *In any single-round Voronoi game on convex polygons, the server Alice can place her facility such that she serves at least $\lceil n/3 \rceil$ clients.*

Proof. Let Alice place her facility $\mathcal{A}$ at the center-point of the clients. Then every closed half plane through $\mathcal{A}$ contains at least $\lceil n/3 \rceil$ clients [3,6]. There are two cases.

Case 1: Facility $\mathcal{A}$ is strictly in the interior of $\mathcal{P}$. Let us assume that Bob is allowed to place his facility unrestrictedly in the plane with Euclidean distances to clients. Even then he will never get more than $\lfloor 2n/3 \rfloor$ clients if $\mathcal{B}$ avoids $\mathcal{A}$. Since the geodesic distances are always greater than the Euclidean distance and Bob is restricted on the exterior of the convex polygon $\mathcal{P}$, Bob will eventually serve less than $\lfloor 2n/3 \rfloor$ clients. Hence Alice will always be able to get at least $\lceil n/3 \rceil$ clients, that is, $\mathcal{S}_A \geq \lceil n/3 \rceil$.

Case 2: Facility $\mathcal{A}$ is on the boundary of $\mathcal{P}$. In this case there will be at least $\lceil n/3 \rceil$ clients on a vertex or an edge of $\mathcal{P}$. Then either (a) Bob places his facility at $\mathcal{A}$ and gets no more than $n/2$ clients, (b) he places his facility on the same edge as $\mathcal{A}$ and Alice gets at least $\lceil n/3 \rceil$ clients from the half-plane that passes through $\mathcal{A}$, avoids $\mathcal{B}$ and minimally intersects $\mathcal{P}$, or (c) he places his facility elsewhere and Alice still gets at least $\lceil n/3 \rceil$ clients.   □

We wish to prove that the above bound is tight. For this, we construct a Voronoi game on a triangle such that no matter where Alice places her facility, the adversary Bob wins at least $\lfloor 2n/3 \rfloor$ clients.

Lemma 11. *There exits a single round Voronoi game on a convex polygon such that Alice serves no more than $\lceil n/3 \rceil$ clients.*

Proof. We place n clients on an equilateral triangle as shown in Fig. 5. We can easily show that the Ellipses $\mathcal{E}_{s,t}$'s for geodesic paths from clients s to t containing $\lfloor 2n/3 \rfloor$ clients do not intersect. Note that the clients other than those near $\mathcal{A}$ are placed on the distant edge of the triangle.   □

Lemma 10 and Lemma 11 together show that the lower bound of $\lceil n/3 \rceil$ is tight for the number of clients served by Alice.

Next, we give a lower bound on the number of clients served by Bob.

Lemma 12. *For any single-round Voronoi game on convex polygons with $n \geq 2$ clients, Bob serves at least 1 client.*

We can ask a question here, whether Bob is always guaranteed a fraction of the total number of clients. Surprisingly, the answer is no. The lower bound of Lemma 12 is tight! We shall prove that the lower bound is tight by using an iterative construction with a precondition and a postcondition. We describe the construction of Fig. 6 below.

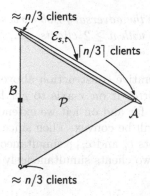

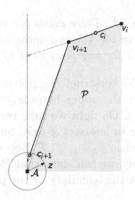

Fig. 5. Lower bound on $\mathcal{S}_{\mathcal{A}}$, Ellipses $\mathcal{E}_{s,t}$ are formed by geodesic paths of consecutive $\lfloor 2n/3 \rfloor$ clients.

Fig. 6. Inductive construction for lower bound on $\mathcal{S}_{\mathcal{B}}$.

Let $\mathcal{A}$ be the origin. Let v_i, a vertex, and c_i, a corresponding client to the vertex, be two points with positive x-coordinates, c_i is left of v_i such that the line $\overline{c_i v_i}$ has a non-zero intercept on the y-axis and a positive slope. We can place the next vertex v_{i+1} on line $\overline{c_i v_i}$ such that it has a positive x-coordinate. It is possible to shoot a ray with larger slope and smaller intercept from v_{i+1} and place c_{i+1} on it such that $||c_i v_{i+1}|| + ||v_{i+1} c_{i+1}|| = ||c_i \mathcal{A}|| + ||\mathcal{A} c_{i+1}||$. This is a consequence of triangle inequality $||c_i \mathcal{A}|| \leq ||c_i v_{i+1}|| + ||v_{i+1} \mathcal{A}||$. This implies we can have a small quantity z such that $||c_i \mathcal{A}|| + z = ||c_i v_{i+1}|| + ||v_{i+1} \mathcal{A}|| - z$. Suitably manipulating the position of c_{i+1} around distance will give us a proper candidate for c_{i+1}, i.e., $||\mathcal{A} c_{i+1}|| \approx z$ and $||v_{i+1} c_{i+1}|| \approx ||v_{i+1} \mathcal{A}|| - z$.

The construction satisfies the following lemma:

Lemma 13. *Let vertex v_i, client c_i and $\mathcal{A}$ satisfy the following preconditions:*

(1) $\mathcal{A}$ is origin,
(2) c_i and v_i has positive x-coordinates with $x(c_i) < x(v_i)$,
(3) line $\overline{c_i v_i}$ has positive slope and positive intercept on y-axis.

Then we can compute next vertex v_{i+1} on line $\overline{c_i v_i}$, left to c_i, and client position c_{i+1} such that we satisfy the following post-conditions:

(a) The ellipse $\mathcal{E}_{c_i, c_{i+1}}$ passes through $\mathcal{A}$
(b) c_{i+1} and v_{i+1} has positive x-coordinates with $x(c_{i+1}) < x(v_{i+1})$,
(c) line $\overline{c_{i+1} v_{i+1}}$ has positive slope and positive intercept on y-axis.

The construction above and the resulting post-conditions of Lemma 13 can be repeated as many times as we wish for as many clients as required. This allows us to construct a convex polygon with any number of vertices. We also note that we can slightly modify the construction so that the ellipse $\mathcal{E}_{c_i, c_{i+1}}$ properly contains $\mathcal{A}$, though this is not necessary. Thus we have the following lemma on the construction.

Lemma 14. *There exists a convex polygon where the adversary Bob serves only 1 client in the Voronoi game on convex polygons with $n \geq 2$ clients, i.e., S_B is at most 1 for any placement of B.*

Proof. We construct the polygon $\mathcal{P}$ using the iterative construction above. We construct the upper boundary first and then mirror it on x-axis to get lower boundary. On right we take vertex v_0 sufficiently far and on left we extend the last edge to intersect x-axis. Such a polygon $\mathcal{P}$ will be convex. Then since $\mathcal{A}$ is contained in $\mathcal{E}_{c_i, c_{i+1}} \cap \mathcal{P}$ Bob cannot serve clients c_i and c_{i+1} simultaneously. Iteratively, we can show that Bob cannot serve two clients simultaneously anywhere on the boundary of the polygon $\mathcal{P}$.

As a consequence of the above lemma, we also have tight lower bounds for Bob for the k-round Voronoi game on convex polygons.

Lemma 15. *For the k-round Voronoi game on convex polygons with $n \geq k$ clients and any $k \geq 1$, Bob serves at least $(k+1)/2$ client. Moreover, There exists a convex polygon where the adversary Bob serves only $(k+1)/2$ clients in the k-round Voronoi game on convex polygons with $n \geq k$ clients and any $k \geq 1$, i.e., S_B is at most $(k+1)/2$ for any placement of B.*

We combine all the results in the following theorem.

Theorem 5. *For the single round Voronoi game on convex polygons with n clients located on the boundary of the polygon $\mathcal{P}$, the score of Bob, S_B satisfies $1 \leq S_B \leq \lfloor \frac{2n}{3} \rfloor$ and the bounds are tight.*

In the successive sections, we design algorithms to compute the placements for facilities for Alice and Bob.

6   Strategy for Adversary Bob

We assume that Alice has already played her turn and she has placed her facility at $\mathcal{A}$ in $\mathcal{P}$. Bob's objective is to locate an optimal position for B. In this section, we show how Bob can compute the optimal position for $\mathcal{B}$ in linear time of the size of the input, i.e., $O(n + m)$.

As a consequence of Theorem 1, we only need to look for end to end geodesic paths between the n clients. We can simply check $\binom{n}{2}$ geodesic paths, say the geodesic path from client s to client t, and among the corresponding ellipses $\mathcal{E}_{s,t}$'s, we select the one that avoids $\mathcal{A}$ and has the maximum number of clients for $\mathcal{B}$. However, we can do this more efficiently in linear time as follows.

The main idea of the algorithm is that for each client s, $s \in \mathcal{C}$ we compute the geodesic path to t on the boundary of $\mathcal{P}$ that contains the maximum number of clients in the counter-clockwise direction such that the ellipse $\mathcal{E}_{s,t}$ avoids $\mathcal{A}$. The steps are given below.

Step 1 (Initialize): We sort and merge the lists of vertices and the clients. This is done so that we can traverse both the vertices and the clients together along the boundary.

Step 2 (Initial geodesic path): Starting from $s \leftarrow c_1(= s_1)$, we compute the maximum length geodesic path to $t \leftarrow t_1$ on boundary of $\mathcal{P}$ such that $d_{\mathcal{P}}(\mathcal{A}, s) + d_{\mathcal{P}}(\mathcal{A}, t) = \delta_P(s, t)$. This will ensure that $\mathcal{E}_{s,t}$ contains $\mathcal{A}$ on the boundary. We also compute the corresponding location for $\mathcal{B} \leftarrow \mathcal{B}_1$ that allows Bob to serve all the clients on the geodesic path.

We assume for the next step that the geodesic path from $s \leftarrow c_i(= s_i)$ to $t \leftarrow t_i$ is computed successfully and the location of $\mathcal{B} \leftarrow \mathcal{B}_i$ is stored.

Step 3 (Successive paths): We iteratively advance s to $s \leftarrow c_{i+1}$, decreasing the geodesic distance. Since Bob served the larger geodesic path earlier, he will also be able to serve the shorter path. This implies that the Geodesic path can be extended in the counter-clockwise direction. Therefore, we advance t to $t \leftarrow t_{i+1}$ such that $d_{\mathcal{P}}(\mathcal{A}, s) + d_{\mathcal{P}}(\mathcal{A}, t) = \delta_P(s, t)$. We also update $\mathcal{B} \leftarrow \mathcal{B}_{i+1}$. We repeat this step till the last client in the merged list $s \leftarrow c_n(= s_n)$.

Step 4 (Report): Once we have all the n geodesic paths containing maximal consecutive clients of Bob, we do a linear scan and report the maximum. $\square$

We show that the algorithm presented above is correct in the lemma below.

Lemma 16. *Let Alice's facility be located at $\mathcal{A}$. The algorithm described above correctly computes an optimal position of $\mathcal{B}$ for the Voronoi game on convex polygons.*

Proof. The proof follows from the necessary and sufficient condition in Theorem 2 and the fact that clients are consecutive from Theorem 1. $\square$

We can show that the algorithm above runs in linear time.

Theorem 6. *An optimal position for the adversary Bob, $\mathcal{B}$, can be computed in $O(n + m)$ time using $O(n + m)$ space, where the Voronoi game is played on a convex polygon of size m with n clients.*

We note that in the algorithm described above we do not store the geodesic paths, we only store the end points s_i's $(= c_i$'s$)$, t_i's and corresponding location of facilities $\mathcal{B}_i$'s.

7 Strategy for Server Alice

In this section, we describe an algorithm to compute the optimal facility location in the Voronoi game on convex polygons for Alice such that the maximum possible score of Bob is minimized.

The algorithm uses the computation of the common intersection of ellipses described previously in Sect. 4, to do two things: (1) To check if the common intersection is empty and (2) If the previous common intersection is non-empty, compute a point belonging to the common intersection.

In this discussion we assume that we are dealing with finite and closed ellipses, i.e., the boundary is included in the ellipses. We can use any suitable $O(1)$ representation we wish. However, the way we get these ellipses in our algorithm, is two foci and ellipses as loci of points with a fixed total distances from the two foci, so we use the representation (f_1, f_2, r), where f_1 and f_2 are the two foci points and r is the total distance of any point in the boundary to the two foci, i.e., the ellipses satisfy $\|xf_1\|_2 + \|xf_2\|_2 = r$. We proceed with the discussion of the algorithm below. The algorithm is brief as we heavily borrow most of the concepts already discussed in the previous sections.

7.1 Algorithm to Compute an Optimal Placement of $\mathcal{A}$

In the discussion below, we design an algorithm to compute the optimal placement for $\mathcal{A}$. By Theorem 5 we know that $\lceil n/3 \rceil \leq S_\mathcal{A} \leq n-1$. Also, by Theorem 2 we have a necessary and sufficient condition for the placement of the facility of Bob. Thus we can check whether it is possible to place $\mathcal{A}$ for some specific value of $S_\mathcal{A}$. We check this by computing the common intersection of all the ellipses $\mathcal{E}_{s,t}$'s such that the geodesic paths from the clients s to t contains exactly $n - S_\mathcal{A} + 1$ clients. These ellipses can be computed by a technique similar to the algorithm in Sect. 6, by advancing s and t simultaneously keeping track of geodesic distances.

Since the optimal value of the score for Alice is discrete in increments of $1/2$ and has lower and upper bounds, we can search for it using binary search. $S_\mathcal{A} = r$ is optimal if a score of r is feasible and a score of $r + 1$ is infeasible. Thus we have the following steps in the algorithm.

Step 1 (Initialize): First we set the bounds for binary search $left \leftarrow 2\lceil n/3 \rceil$, $right \leftarrow 2n - 2$. We note that the score of Alice $S_\mathcal{A}$ is discrete in multiple of $1/2$ $(2\lceil n/3 \rceil \leq 2S_\mathcal{A} \leq 2n - 2)$.

Step 2 (Recursion): We check if we need to terminate the recursion i.e. if the range is empty. Otherwise, we update mid, $mid \leftarrow (left + right)/2$. Next we need to check if $S_\mathcal{A} \leftarrow mid/2$ is feasible.

Step 3 (Feasibility check): We update $S_\mathcal{B} \leftarrow n - S_\mathcal{A} + 1$. We compute n ellipses one for each $c \in C$ such that $s \leftarrow c$ and $t \leftarrow c'$, such that there are $S_\mathcal{B}$ clients from end to end on the geodesic path from s to t. This can be done in $O(n + m)$ time.

Step 4 (Intersection of ellipses): Next, we compute the common intersection of the n ellipses mentioned in the previous step using the algorithm described in Sect. 4. This step can be completed in $O(n \log n)$ time independent of m as there are n ellipses.

Step 5 (Necessary and sufficient condition check): If the common intersection computed in the previous step is empty then the current score of $\mathcal{S}_A$ is infeasible and we update $right \leftarrow mid - 1$ otherwise it is feasible and we update $left \leftarrow mid + 1$. Then we continue the recursion. $\qquad\square$

The algorithm described above is correct as it depends on the necessary and correctness condition in the Theorem 2. However the efficiency depends on the binary search and the procedure of computing the common intersection of n ellipses. This gives us the following theorem.

Theorem 7. *An optimal position for A can be computed in $O(n \log^2 n + m \log n)$ time, where the Voronoi game is played on a convex polygon of size m with n clients.*

References

1. Agarwal, P.K., Sharir, M.: Davenport-schinzel sequences and their geometric applications. In: Sack, J.R., Urrutia, J. (eds.) Handbook of Computational Geometry, chap. 1, pp. 1–47. North-Holland, Amsterdam (2000)
2. Ahn, H.K., Cheng, S.W., Cheong, O., Golin, M., van Oostrum, R.: Competitive facility location: the voronoi game. Theoret. Comput. Sci **310**(1), 457–467 (2004)
3. Banik, A., Bhattacharya, B.B., Das, S., Mukherjee, S.: The discrete voronoi game in $\mathcal{R}^2$. Comput. Geom. **63**, 53–62 (2017)
4. Banik, A., Das, S., Maheshwari, A., Smid, M.: The discrete voronoi game in a simple polygon. Theoret. Comput. Sci. **793**, 28–35 (2019)
5. Cheong, O., Har-Peled, S., Linial, N., Matousek, J.: The one-round voronoi game. Discrete Comput. Geom. **31**(1), 125–138 (2004)
6. Edelsbrunner, H.: A Short Course in Computational Geometry and Topology. Springer Publishing Company, Incorporated, Warren (2014)
7. Fekete, S.P., Meijer, H.: The one-round voronoi game replayed. Comput. Geom. **30**(2), 81–94 (2005)
8. Teramoto, S., Demaine, E.D., Uehara, R.: The voronoi game on graphs and its complexity. J. Graph Algorithms Appl. **15**(4), 485–501 (2011)

Cutting Stock with Rotation: Packing Square Items into Square Bins

Shahin Kamali[(✉)] and Pooya Nikbakht[(✉)]

Department of Computer Science, University of Manitoba, Winnipeg, Canada
shahin.kamali@umanitoba.ca, nikbakhp@myumanitoba.ca

Abstract. In the square packing problem, the goal is to place a multi-set of square-items of various sizes into a minimum number of square-bins of equal size. Items are assumed to have various side-lengths of at most 1, and bins have uniform side-length 1. Despite being studied previously, the existing models for the problem do not allow rotation of items. In this paper, we consider the square packing problem in the presence of rotation. As expected, we can show that the problem is NP-hard. We study the problem under a resource augmented setting where an approximation algorithm can use bins of size $1 + \alpha$, for some $\alpha > 0$, while the algorithm's packing is compared to an optimal packing into bins of size 1. Under this setting, we show that the problem admits an asymptotic polynomial time scheme (APTAS) whose solutions can be encoded in a poly-logarithmic number of bits.

Keywords: Square packing with rotation · Bin packing ·
Approximation algorithms · Resource augmentation

1 Introduction

An instance of the *square packing problem* is defined with a multi-set of *squares-items* of side-lengths at most 1. The goal is to place these squares into a minimum number of unit *square-bins* in a way that two squares-items placed in the same bin do not overlap (but they can touch each other). The problem is a variant of the classical bin packing problem. As such, we refer to the square-items simply as "items" and square-bins as "bins". A square-item can be recognized by its side-length, which we refer to as the *size* of the item. Throughout, we refer to the number of bins used by an algorithm as the *cost* of the algorithm.

Square packing has many applications in practice. One application is cutting stuck, where bins represent stocks (e.g., wood boards), and items are demanded pieces of different sizes. For each item, an algorithm has to cut the stock to provide the pieces that match the requests. This cutting process is equivalent to placing items into bins. Note that the goal of cutting stock is to minimize the number of stocks, which is consistent with the objective of square packing.

There has been a rich body of research around square packing, all under the assumption that squares cannot be rotated, that is, the sides of all items

© Springer Nature Switzerland AG 2020
W. Wu and Z. Zhang (Eds.): COCOA 2020, LNCS 12577, pp. 530–544, 2020.
https://doi.org/10.1007/978-3-030-64843-5_36

Fig. 1. If all items in the input have length $\frac{\sqrt{2}}{2\sqrt{2}+1} \approx 0.35$, allowing rotation allows packing 5 items per bin instead of 4.

should be parallel to the edges of the bins. While this assumption makes the combinatorial analysis of the problem easier, it comes at a price. Consider, for example, an instance of the problem formed by n items, all of size $\frac{\sqrt{2}}{2\sqrt{2}+1} \approx 0.369$. With no rotation, any bin can include at most four items, which gives a total cost of $\lceil n/4 \rceil$. Allowing rotation, however, five items fit in each bin, and we can reduce the cost to $\lceil n/5 \rceil$ (see Fig. 1). As a result, using rotation, the number of required bins is decreased by $n/20$, which is a notable saving (e.g., for cutting stock applications).

Definition 1. *The input to the square packing problem with rotation is a multi-set of squares (items), defined with their side-lengths $\sigma = \{x_1, ..., x_n\}$, where $0 < x_i \leq 1$. The goal is to pack these squares into the minimum number of squares of unit size (bins) such that no two items overlap. When placing a square into a bin, an algorithm can translate it to any position and rotate it by any degree, as long as the item remains fully inside the bin.*

An asymptotic polynomial time scheme (APTAS) is an approximation algorithm ALG such that, for any input σ of square-items whose total area is asymptotically large, we have $Alg(\sigma) \leq (1 + \epsilon) \cdot Opt(\sigma) + c_0$ for any small parameter $\epsilon > 0$ and some constant c_0, where $Alg(\sigma)$ and $Opt(\sigma)$ denote the number of bins in the packing of ALG and an optimal packing, respectively. The running time of an APTAS needs to be polynomial in the number n of square-items, but it can be a super-polynomial function of ϵ.

Related Work. The 1-dimensional bin packing has been extensively studied (see, e.g., [9,10,20]). The problem is known to be NP-hard [20]. In a seminal paper, de la Vega and Lueker [32] provided the first APTAS for the problem. The best existing result is an algorithm that opens at most $Opt(\sigma) + O(\log Opt(\sigma))$ bins for placing any input σ [23]. There are many ways to extend bin packing to higher dimensions (see [8] for a survey). Packing square-items into square-bins (without rotation) is perhaps the most straightforward variant. In the offline setting, this problem is known to be NP-hard [27]. Bansal et al. [5] provided an APTAS for this problem (indeed, for the more general d-dimensional cube packing problem). Epstein and Levin [14] presented a "robust" approximation scheme, where robustness implies a sequential nature for the algorithm. Rectangle packing is a generalization of both bin packing and square packing, where the goal is to pack rectangles of possibly different sizes and shapes into uniform square-bins (see [8] for details). In particular, there is a variant that assumes rectangles can be rotated by exactly ninety degrees so that sides of items are still parallel to the edges of bins (see, e.g., [12]). We note that rotation by ninety degrees is not relevant for square packing.

Resource augmentation [28] is a relaxed framework for the analysis of approximation and online algorithms. Let ALG be an approximation algorithm that uses a resource when solving a problem. The resource can be, for example, the size of the cache in the online paging problem [29] or the size of the bins in the packing problems [15]. Under a resource augmented setting, the resource of ALG is increased by an *augmentation factor* $\gamma > 1$. Then, the cost of ALG is compared to an optimal solution without augmentation (with the original cache size or bin capacity). The resource augmented setting for the bin packing and related problems have been studied in several previous works [6,13,15,17,26]. In particular, the makespan scheduling problem can be considered as a resource-augmented variant of the bin packing problem (see, e.g., [2,3,18]).

Contribution. We start by reviewing a decision problem that asks whether a multi-set of squares can be placed into a single bin. We refer to this problem as the *1-Bin Square Packing* (1BSP) problem and show that it is NP-hard. It turns out that allowing rotation makes the problem much harder. It is not even clear how to answer the 1BSP problem when all square-items have uniform size, as there has been little progress in the *congruent square packing problem* [16]. The source of difficulty is that it does not appear that one can effectively discretize the problem and apply standard algorithmic techniques. A recent study [1] shows that similar packing problems are $\exists\mathbb{R}$-hard, and their verification algorithms need exponential bit-precision. The same is likely true for the 1BSP problem.

We study the square packing problem under a relaxed augmented setting, where bins of the approximation algorithm are augmented by a factor $(1 + \alpha)$ for some small $\alpha > 0$. The extra space given in the augmented bins enables free translation and rotation of squares, which ultimately allows packing items with encodable bit-precision. Under this setting, we present an APTAS for the problem. Precisely, given small, constant values for $\epsilon, \alpha > 0$, we present an algorithm that runs in polynomial time and packs any input σ in at most $(1 + \epsilon)Opt(\sigma) + 3$ augmented bins of size $1 + \alpha$. Here, $Opt(\sigma)$ denotes the number of bins in an optimal packing of σ when packed into unit bins.

Roadmap. In Sect. 2, we visit the 1BSP problem to establish the hardness of the square packing problem with rotation, and review some existing results that suggest the problem might be $\exists\mathbb{R}$-hard. In Sect. 3, we present our main result, which is an APTAS for packing any set of squares into augmented bins.

2   A Review of the 1-Bin Square Packing (1BSP) Problem

Given a multi-set S of squares, the 1-Bin Square Packing (1BSP) problem asks whether items in S can be placed into a square of size c, called a *container*, where translation to any position and rotation by any degree is allowed. The 1BSP problem is a decision variant of the square packing problem introduced in

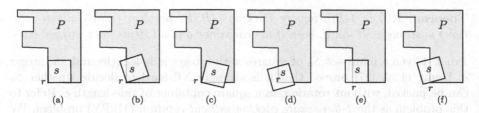

(a)   (b)   (c)   (d)   (e)   (f)

Fig. 2. To pack and cover a rectilinear polygon with a multi-set of squares, no square should be rotated.

Definition 1 (scale everything to ensure the container has size 1). In this section, we make some basic observations about the complexity of the 1BSP problem.

NP-Hardness. Many geometric packing problems are known to be NP-hard. Examples include packing squares into squares without rotation [27], packing circles into equilateral triangles [11], packing triangles into rectangles [7], and packing identical simple polygons into a larger polygon [4]. Despite similarities, none of these results establish the hardness of the 1BSP problem. As such, we provide a proof to show the problem is NP-hard, even if the container and all items have integer sizes. We start with the following lemma:

Lemma 1. *Let S be a multi-set of squares with integer side-lengths and of total area A, and let P be a rectilinear polygon of area A. It is possible to pack all squares in S into P only if no square is rotated.*

Proof. Assume S is fully packed into P, that is, P is fully covered by the squares in S. We show that no square is rotated in such packing. We start with a simple observation and then use an inductive argument.

Let r be an arbitrary convex vertex of P, that is, a vertex with an interior angle of 90°. Note that there are at least four convex vertices in any rectilinear polygon. Since P is fully covered by S, there should be a square $s \in S$ that includes or touches r. As Fig. 2 illustrates, the only case where s remains fully inside P is when it has a vertex located at r and has no rotation (Fig. 2(a)); in all other cases, a portion of s lies outside of P, which is not possible.

Given the above observation, we use an inductive argument on A to prove the lemma. For the base of induction, when $A = 1$, S is formed by one square of size 1 packed into a square P of size 1. Clearly, packing S into P involves no rotation. Next, assume we have a multi-set S of squares of total area k packed into a polygon P of area k. Consider an arbitrary convex vertex r of P. By the above observation, r is covered by a square $s \in S$ that is not rotated. Let an integer $x \geq 1$ denote the area of s. Removing the area covered by s from P results in a packing of the multi-set $S - \{s\}$ of squares into a smaller polygon P', where both $S - \{s\}$ and P' have area $k - x$. By the induction hypothesis, none of the squares in $S - \{s\}$ are rotated when packed into P'. Given that s was not rotated either, no square is rotated in the original packing of S into P. $\square$

Theorem 1. *The 1-Bin Square Packing (1BSP) problem (which allows rotation) is strongly NP-hard, even if the container and all items have integer sizes.*

Proof. Given a multi-set S_0 of squares with integer side-lengths and an integer c, Leung et al. [27] proved that it is strongly NP-hard to decide whether S_0 can be packed, without rotation, in a square container of side-length c. Refer to this problem as the *1-bin square packing without rotation* (1BSW) problem. We provide a reduction from 1BSW to 1BSP.

Assume we are given an instance of 1BSW that asks whether a multi-set S_0 of squares can be packed into a square container B of side-length c. Let u be an integer that denotes the total area of all items in S_0. Define an instance of the 1BSP problem through a multi-set S formed as the union of S_0 and $c^2 - u$ squares of unit size 1. We show that S can be packed into B with rotation if and only if S_0 can be packed into B without rotation.

Consider a $c \times c$ grid formed on B, where the grid vertices have integer coordinates. Assume S_0 can be packed into B without rotation; the total area of items in S_0 is u, and the empty area in B is $c^2 - u$. Since S_0 is packed without rotation, we can translate squares downward and towards the left such that their corners lie on the vertices of the grid (see the proof of Lemma 3.4 in [5] for details of such translation). The unused area in the resulting packing is formed by $c^2 - u$ grid cells of size 1×1. We can use the same packing and place the small squares in $S - S_0$ in the empty grid cells. This gives a valid packing of S. Next, assume S can be packed into B. Since the total area of items in S is exactly c^2, there is no empty (wasted) area in B. By Lemma 1, the packing does not involve rotation of any item. Since no rotation is involved, removing unit squares in $S - S_0$ from this packing gives a valid packing of S_0 (without rotation).   □

Congruent Square Packing. The congruent square packing problem, first studied by Erdős and Graham [16], asks for the minimum size $s(i)$ of a square that can contain i unit-sized squares. The problem is equivalent to finding the largest value of x such that i congruent squares of size x fit into a container of unit size. Clearly, an algorithm for the 1BSP problem can be used to find $s(i)$, using a binary search approach. Without rotation, it is easy to answer if a set of congruent squares fit in a square container. Allowing rotation, however, makes the problem harder. Despite being extensively studied (see, e.g., [19,21,22,31]), the value of $s(i)$ is not known for small values like $i = 11$. Figure 3 shows the packings that give the smallest known upper bounds for $s(11)$ and (18). We refer to the survey by Friedman [19] for more details on congruent square packing.

Existential Theory of the Reals. An Existential Theory of the Reals (ETR) formula can be stated as $\psi = \exists x_1, \ldots, x_n \; \Phi(x_1, \ldots, x_n)$, where Φ is a well-formed sentence over alphabet $\{0, 1, x_1, \ldots, x_n, ., =, \leq, <, \wedge, \vee, \neg\}$ [17]. A decision problem belongs to the complexity class $\exists \mathbb{R}$-complete iff it is equivalent to deciding (in polynomial time) whether an ETR formula is true or not. In particular, these problems are $\exists \mathbb{R}$-hard in that a witness, which certifies the problem is in NP, might need an exponential number of bits in any numerical representation [17]. Abrahamsen et al. [1] have recently proved that a set of problems similar to

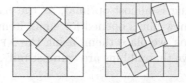

Fig. 3. The smallest known container for placing 11 (left) and 18 (right) unit squares into a larger square [19].

the 1BSP problem, such as deciding whether a set of simple polygons fit into a square container, are ∃ℝ-hard. It is not clear, however, whether packing convex objects (in particular squares) into a square container is ∃ℝ-hard. Nevertheless, Erickson et al. [17] proposed to augment a square container in order to avoid (a potential) exponential bit precision for encoding the packings. They showed that if the container's size is augmented, where a slight perturbation is applied to the augmentation parameter (as in the smoothed analysis [30]), the bit precision for encoding a solution is expected to be logarithmic to the input size. Our APTAS in the next section uses the augmentation of bin sizes, along with some of the ideas from [17], to ensure logarithmic precision when packing squares into augmented bins.

3 An APTAS for Square Packing with Rotation

In this section, we describe an APTAS for placing square-items, with rotation, into augmented square-bins. The algorithm has two constant parameters α and ϵ, which are both small, positive values. The algorithm places any input σ into at most $(1 + \epsilon)Opt(\sigma) + 3$ augmented bins of size $1 + \alpha$, where $Opt(\sigma)$ is the number of bins in an optimal packing of σ into unit squares. Throughout, we assume the total area of σ is arbitrarily large. Furthermore, we let $\varsigma = \epsilon/1081$.

Overview. The algorithm classifies items into *small*, *medium*, and *large* items. The classification is similar to the one in [5] for square packing without rotation. Medium and small items are smaller than ς (large items can also be smaller than ς). Medium items have a negligible total area. As such, we place them separately from others into a set of at most $\varsigma \cdot Opt(\sigma) + 1$ *medium bins*.

We place large items into augmented *large bins*. For that, we round up item sizes, as suggested by de la Vega and Lueker [32], to get a constant number of possible item sizes. The resulting instance is then packed, using an exhaustive approach, into a minimum number of augmented bins. The extra space in the augmented bins enables us to discretize the problem, using the ideas from [17]. We show that the number of large bins will be no more than $(1 + \varsigma)Opt(\sigma)$.

After placing large items, we partition the empty area in each large bin into a set of trapezoids and show that small items can be tightly packed into these trapezoids. For that, we partition trapezoids further into right-angled triangles and use the Next-Fit-Decreasing-Height (NFDH) algorithm of Coffman et al. [24] to pack these triangles. If all small items are placed in the trapezoids of the large bins, the resulting packing is almost-optimal, as the packing of large items is almost optimal (and no new bin is opened for small items). If some small items do not fit in the large bins, we place them, using the NFDH strategy, into *small bins*. In this case, we prove that all bins are "almost full", which eventually gives the claimed guarantee.

In what follows, we first describe how small items can be packed into trapezoids, then we explain the placement of large items into augmented bins, and finally describe the packing of arbitrary inputs.

3.1   Triangle and Trapezoid Packing

Let T be a given right-angled triangle with two legs of sides a and h (assume $h \leq a$). In what follows, we show how to pack a set of items of size at most δ into T without wasting too much area in T. In our solution, items are packed without rotation so that their sides are parallel to the legs of T. Later, we use this packing to show items of size at most δ can be packed into a trapezoid container, with rotation, so that not too much area is wasted.

In order to pack items into T, we use the Next-Fit-Decreasing-Height (NFDH) strategy [24]. We sort items in non-increasing order of their sizes and place them one by one in the following manner. Without loss of generality, assume the legs of T extend along the x- and y-axes; we refer to the two legs as 'left leg' (of length h) and 'lower leg' (of length a). We place the first item in a way that it is tangent to the two legs of T. Let h_1 denote the size of the first item. The area within a distance h_1 from the lower leg of T forms a *shelf* of height h_1. We place subsequent items on this shelf such that the left side of each item touches the right side of the previous item, while its lower side touches the lower leg of T. At some point, the next item might not fit in the shelf; at this point, we "close" the shelf and recursively pack the remaining items in the right-angled triangle formed by removing the shelf from T (see Fig. 4a). The algorithm stops when it cannot open a new shelf. Note that it is possible that no item is placed in T, which happens when the largest square in the input (the first square in the sorted order) does not fit in T (see Fig. 4b).

Lemma 2. *Assume we apply* NFDH *to pack a multi-set of items, each having a size at most* δ, *into a right-angled triangle* T *with legs of sizes* a *and* h, *where* $h \leq a$. *If the algorithm stops before packing all items (if not all items can be packed into* T*), then the wasted area in* T *is less than* $(3.5a + h)\delta$.

Proof. First, assume no item can be packed into T. That means placing the first square s_1 of the multi-set in T, in a way that its two sides are tangent with the legs of T, results in a part of s_1 lying outside of T (see Fig. 4b). In this case, a square p that is tangent to the legs of T and touches its third side is fully contained in s_1 and hence has side-length less than that of s_1 (consequently less than δ) and also less than h. So, the area of T can be partitioned into p (of area less than $h\delta$), a triangle above p (of area less than $h\delta/2$), and a triangle on the right of p (of area less than $a\delta/2$). Consequently, the total (wasted) area of T is less than $(3h/2 + a/2)\delta < (3.5a + h)\delta$.

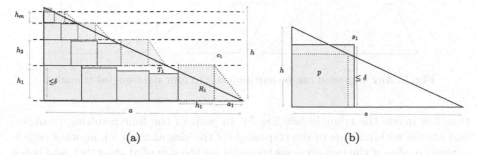

Fig. 4. (a) Applying the NFDH algorithm for packing a right-angled triangle T with squares of size at most δ. The algorithm places items in non-increasing order of their sizes in shelves that are packed from left to right. The pink squares and the green triangles are respectively the covering squares and triangles. (b) When no item can be packed into T, the area of T is less than $(3h/2 + a/2)\delta$.

Next, assume there are $m \geq 1$ shelves in the final packing (see Fig. 4a). The wasted area in the i'th shelf can be partitioned into two areas: the area on the right of the last item placed on the shelf (call it R_i), and the area on top of the squares placed on the shelf (call it T_i). R_i can be covered by two components: a *covering square* of area h_i^2 and (if required) a right-angled *covering triangle* of area $h_i a_i/2$, where a_i is the base of the covering triangle (see Fig. 4a). The bases of covering triangles of different shelves do not intersect when projected into the base of T. To see that, consider the top-right corner c_i of the covering square of shelf i. Since c_i appears outside T and just below shelf $i+1$, the covering triangle of shelf $i+1$ appears on the left of c_i, while the covering triangle of shelf i appears on its right. Consequently, we can write, $R_i < h_i^2 + h_i a_i/2 < \delta(h_i + a_i/2)$, and summing over all values of i, we get $\sum_{i=1}^{m} R_i < \delta \sum_{i}^{m}(h_i + a_i/2) < \delta(h + a/2)$. For the wasted area on top of the i'th shelf, we can write $T_i \leq a(h_i - h_{i+1})$; this is because all items placed on the shelf have a height more than h_{i+1}. Summing over all but the very last shelf, we get $\sum_{i=1}^{m-1} T_i \leq a \sum_{i=1}^{m-1}(h_i - h_{i+1}) = a(h_1 - h_m) < a\delta$. The wasted area on top of the last shelf is no more than $a\delta$, which is an upper bound for the size of the shelf. So, we have $\sum_{i=1}^{m} T_i < 2a\delta$. Finally, the unused area U on top of the whole packing (the dark area in Fig. 4a) has a size of no more than $a\delta$. The total wasted area is thus $\sum_{i=1}^{m}(R_i+T_i)+U < (a/2+h)\delta+2a\delta+a\delta = (3.5a + h)\delta$. □

Lemma 3. *Let Z be a trapezoid in which every side has length at most x, and S be a multi-set of squares of size at most δ. There is an algorithm that packs items from S into Z such that either all items are packed into Z or a subset of S is packed while the wasted area in Z is at most $54x\delta$.*

Proof. First, we partition Z into four right-angled triangles. This can be done by partitioning Z into two triangles by drawing a diagonal. By the triangle inequality, any side of these two triangles has side-lengths less than $2x$. Then, we partition each triangle into two right-angled triangles by drawing an altitude

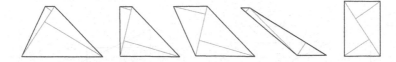

Fig. 5. Any trapezoid can be partitioned into four right-angled triangles.

that lies inside the triangle (see Fig. 5). In each of the four resulting triangles, one edge is within a side of the trapezoid (of the size at most x), another edge is within an edge of the two previous triangles (of the size of at most $2x$), and hence the last side is shorter than the sum of the other two, i.e., $3x$. Overall, each of the four triangles has sides of side-length no more than $3x$. So, we can apply Lemma 2 to pack any of the resulting right-angled triangles with items in S. The wasted area in each of these triangles is less than $(3.5 \times 3x + 3x)\delta = 13.5x\delta$. Consequently, the total wasted area in the trapezoid is no more than $4 \times 13.5x\delta = 54x\delta$.   □

3.2 Packing Large Items

We explain how to pack large square-items into augmented bins. The following lemma implies that augmenting bins enables us to encode the translation (position) and the rotation of items in a given packing, using logarithmic bit-precision. Erickson et al. [17] proved a more general statement about packing convex polygons. For completeness, we include their proof for square packing.

Lemma 4. *[17] Let S be a multi-set S of m squares that can be packed into a unit bin. For any $\alpha > 0$, it is possible to pack S into an augmented bin of size $1 + \alpha$ such that the translation and rotation of each item can be encoded in $O(\log(m/\alpha))$.*

Proof. [Sketch] The extra space given by an augmented bin can be used to ensure all items can be placed at a distance of at least $\frac{\alpha}{m+2}$ from each other and the boundary of the bin, which enables free translation and rotation of squares to a certain encodable degree. Consider a packing of S into a unit bin. Form a partial ordering of items along the x-coordinate in which for items $a, b \in S$, we have $a < b$ iff there is a horizontal line that passes through both a and b crosses a before b. Let π_x be a total ordering of items that respects the above partial ordering, and π_y be another ordering defined symmetrically based on a partial ordering through crossings of vertical lines. Let $a \in S$ be the i'th element in π_x and j'th element in π_y (we have $i, j \in \{1, \ldots, m\}$). In the augmented bin, we shift a towards the right by $\frac{i\alpha}{m+2}$ and upwards by $\frac{j\alpha}{m+2}$ (see Fig. 6). At this point, any two squares are separated by at least $\frac{\alpha}{m+2}$. So, it is possible to shift items towards left or right by $O(\alpha/m)$ and/or rotate them by $O(\alpha/m)$ degrees such that squares still do not intersect. Consequently, there is a positioning of squares in the augmented bin in which the position (translation) and rotation of each square can be presented with $O(\log(m/\alpha))$ bit-precision. See details in the proof of Corollary 26 of [17].   □

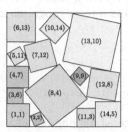   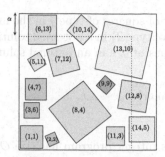

| (a) packing into a unit bin | (b) packing into an augmented bin |

Fig. 6. (a) A set of 14 square-items that are tightly packed into a unit square. The numbers (i, j) for a square indicate its respective indices in π_x and π_y. (b) The packing of squares into an augmented bin of size $1 + \alpha$, where items are separated from each other by at least $\alpha/14$. A square with indices (i, j) is shifted by $(\frac{i\alpha}{16}, \frac{j\alpha}{16})$.

Lemma 4 enables us to use an exhaustive approach for packing large items, provided that there is a constant number of item sizes.

Lemma 5. *Assume a multi-set S of n squares can be optimally packed into $Opt(S)$ unit bins. Assume all items in S have size at least $\delta > 0$, and there are only K possible sizes for them, where δ and K are both constants independent of n. There is an algorithm that packs S into $Opt(S)$ augmented bins of size $1 + \alpha$ in time $O(polylog(n))$, where $\alpha > 0$ is an arbitrary constant parameter.*

Proof. Since all items are of side-length at least δ, the number of items that fit in a bin is bounded by $m = \lceil 1/\delta^2 \rceil$. Consider multi-sets of items described by vectors of form $(x_1, x_2, \ldots, x_K)$, where $1 \leq x_i \leq m$ denotes the number of items of type (size) i in the multi-set. We say a vector is *valid* if the multi-set that it represents can be packed into a unit bin. According to Lemma 4, if a vector V is valid, the multi-set of squares associated with it can be packed into an augmented bin of size $1 + \alpha$ in a way that the exact translation and rotation of each square can be encoded in $O(\log(m/\alpha))$ bits. Since there are up to m squares in the bin and each has one of the K possible sizes, at most $C = O(m((\log K) + \log(m/\alpha)))$ bits is sufficient to encode how a multi-set associated with a valid vector should be packed into an augmented bin. Since K, α, and δ (and hence m) are constant values, C is also a constant. So, if we check all 2^C possible codes of length C, we can retrieve all valid vectors and their packing into augmented bins in a (huge) constant time. In summary, we can create a set $\{T_1, \ldots, T_Q\}$, where each T_i is associated with a unique, valid vector together with its translation and rotation and is referred to as a *bin type*. Here, Q is the number of bin types and is a constant (since $Q \leq 2^C$). From the discussion above, each bin type has an explicit description of how a multi-set of items is placed into an augmented bin of size $1 + \alpha$.

The remainder of the proof is identical to a similar proof from [5]. Let T_{ij} denote the number of squares of type j in a bin type T_i, and let n_j denote the number of squares of type j in the input. Furthermore, let y_i denote the number of bins of type T_i in a potential solution. The following integer programs indicate the values of y_i's in an optimal solution:

$$
min \sum_{i=1}^{Q} y_i, \quad s.t. \qquad \sum_{i=1}^{Q} T_{ij} y_i \geq n_j \qquad \text{for } j = 1, \ldots, K,
$$

$$
y_i \geq 0; \ y_i \in \mathbb{Z} \quad \text{for } i = 1, \ldots, Q,
$$

This integer program has size $O(\log n)$ and a constant number of variables (recall that Q is a constant). So, we can use Lenstra's algorithm [25] to find its optimal solution in $O(polylog(n))$. Such a solution indicates how many bins of each given type should be opened, and as mentioned earlier, each bin type has an explicit description of placements of items into an augmented bin.   □

Provided with Lemma 5, we can use the standard approach of de la Vega and Lueker [32] to achieve an APTAS for large items.

Lemma 6. *Assume a multi-set S of n squares, all of the size at least $\delta > 0$, where δ is a constant independent of n, can be optimally packed in $Opt(S)$ unit bins. For any constant parameters $\alpha, \varepsilon > 0$, it is possible to pack S into at most $(1 + \varepsilon)Opt(S) + 1$ augmented bins of size $1 + \alpha$ in time $O(polylog(n))$.*

Proof. We use the same notation as in [5]. Consider an arrangement of squares in S in non-increasing order of sizes. We partition S into $K = \lceil l/(\varepsilon\delta^2) \rceil$ groups, each containing at most $g = \lceil n/K \rceil$ squares. Let J be a multi-set formed from S by rounding up the size of each square $s \in S$ to the largest member of the group that it belongs to. Let J' be a multi-set defined similarly except that each square is rounded down to the smallest member of its group.

In order to pack S, we use the described algorithm in Lemma 5 to pack J into $Opt(J)$ augmented bins of size $1 + \alpha$, where $Opt(J)$ is the optimal number of unit bins that J can be packed to. Note that there are K different item sizes in J, and all have a size larger than δ, which means we can use Lemma 5 to pack J. Since squares in S are rounded up to form J, the same packing can be used to pack S. Note that it takes $O(polylog(n))$ to achieve this packing.

To analyze the packing, let $Opt(J')$ be the optimal number of unit bins that J' can be packed to. We note that $Opt(J') \leq Opt(S) \leq Opt(J)$. This is because items of S are rounded down in J' and rounded up in J. Let x denote the size of squares in group g of J', and y denote the sizes in group $g + 1$ of J. Since y appears after x in the non-increasing ordering of squares in S, we can assert any square in group g of J' has a size no smaller than a square in group $g + 1$ of J. So, if we exclude the first group, any square in J can be mapped to a square of the same or larger size in J'. Because S is formed by n squares, each of area larger than δ^2, we have $n\delta^2 < Opt(S)$. As a result, since there are $q = \lceil n/K \rceil \leq \lceil n\varepsilon\delta^2 \rceil < \lceil \varepsilon Opt(S) \rceil$ squares in the first group, we can conclude $Opt(J) \leq Opt(J') + q \leq Opt(S) + q \leq \lceil (1 + \varepsilon)Opt(S) \rceil \leq (1 + \varepsilon)Opt(S) + 1$.   □

3.3   Packing Arbitrary Input

We use the results presented in Sects. 3.1, 3.2 to describe an algorithm for packing arbitrary inputs. We start with the following lemma:

Lemma 7. *Assume a multi-set of m squares is packed into a bin. The unused area in the bin can be partitioned into at most $5m$ trapezoids.*

Proof. Create an arbitrary labelling of squares as $s_1, s_2, \ldots, s_m$. Let t, b, l and r respectively denote the topmost, bottom-most, leftmost, and rightmost points of a square s_i (break ties arbitrarily). Draw the following five horizontal line segments for s_i: 1,2) two line segments starting at l and b and extending towards the left until they touch another square or the left side of the bin. 3,4) two line segments starting at r and b and extending towards the right until they touch another square or the right side of the bin. 5) a line segment that passes through t and extends towards the left and right until it touches other squares or boundaries of the bin (see Fig. 7).

We label the line segments of s_i with the label i. When we draw the line segments for all squares, the unused area in the bin will be partitioned into trapezoids. We label each trapezoid in the partition with the label of its lower base. So, for each square s_i, there will be at most five trapezoids with the label i. Consequently, the number of trapezoids will not exceed $5m$, which completes the proof. □

In the above proof, we treated the topmost point t and bottom-most point b of squares differently. This is because the line passing b can be the lower base of two trapezoids on the two sides of the square, while the line passing t can be the lower base of one trapezoid on top of the square.

Item classification. Assume we are given an arbitrary input σ and small, constant parameters $\alpha, \epsilon > 0$. Recall that $\xi = \epsilon/1081$. Let $r = \lceil 1/\xi \rceil$, and define the following $r + 1$ classes for items smaller than ξ. Class 1 contains items with sizes in $[\xi^3, \xi)$, class 2 contains items in $[\xi^7, \xi^3)$, and generally class i $(1 \le i \le r+1)$ contains items in $[\xi^{2^{i+1}-1}, \xi^{2^i-1})$. Since there are $r + 1$ classes, the total area of squares in at least one class, say class j, is upper bounded by $area(\sigma)/(r + 1) \le \xi \cdot area(\sigma)$, where $area(\sigma)$ is the total area of squares in σ. We partition squares

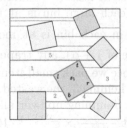

Fig. 7. The unused area in any bin with m items can be partitioned into at most $5m$ trapezoids. The five trapezoids associated with square s_1 are numbered.

in the input σ into *large*, *medium*, and *small* items as follows. Medium items are the members of class j, that is, items with size in $[\xi^{2^{j+1}-1}, \xi^{2^j-1})$. Large items are items of size at least ξ^{2^j-1}, and small items are items of size less than $\xi^{2^{j+1}-1}$.

Packing algorithm: We are now ready to explain how to pack items in σ. Medium items are placed separately from other items into unit bins. For that, we apply the NFDH algorithm of [24] to place medium items into *medium bins* without rotation. We use m_m to denote the number of resulting medium bins. We apply Lemma 6 to pack the multi-set L of large items into augmented bins of size $1 + \alpha$. We refer to these bins as *large bins* and use m_l to denote the number of large bins. It remains to pack small items. We use Lemma 7 to partition the empty area of any large bins into a set of trapezoids. We pack small items into these trapezoids. For that, we consider an arbitrary ordering of trapezoids and use the NFDH strategy (as described in Sect. 3.1) to place items into the first trapezoid (after partitioning it into four right-angled triangles). If an item does not fit, we close the trapezoids and consider the next one. The closed trapezoids are not referred to again. This process continues until either all small items are placed into trapezoids or all trapezoids of all large bins are closed. In the latter case, the remaining small items are placed using the NFDH strategy of [24] into new bins that we call *small bins*.

Analysis. When we apply NFDH to place square-items of size at most ξ into square-bins (without rotation), the wasted area in each bin is at most 2ξ [24]. Since all medium items have size at most $\xi^3 < \xi$ (for $\xi < 1$), the wasted area in each medium bin will be at most 2ξ. On the other hand, the total area of medium items is at most $\xi \cdot area(\sigma)$. So, the total number m_m of medium bins is at most $\frac{\xi \cdot area(\sigma)}{1-2\xi} + 1$, which is at most $\xi \cdot area(\sigma) + 1$ for $\xi < 1/4$. Note that $area(\sigma)$ is a lower bound for the optimal number of unit bins for packing σ. So, $m_m \le \xi Opt(\sigma) + 1$. We consider two cases for the remainder of the analysis.

- Case I: Assume the algorithm does not open a small bin. By Lemma 6, the number of large bins m_l is no more than $(1+\xi)Opt(L) + 1$. Clearly, $Opt(L) \le Opt(\sigma)$ and we can write $m_l \le (1 + \xi)Opt(\sigma) + 1$. For the total number of bins in the packing, we can write $m_m + m_l \le (1 + 2\xi)Opt(\sigma) + 2$.
- Case II: Assume the algorithm opens at least one small bin. We show that the area of all bins (except possibly the last small bin) is almost entirely used. Large items are of size at least ξ^{2^j-1} and area at least $\xi^{2^{j+1}-2}$. So, the number of large items in each large bin is at most $\frac{1}{\xi^{2^{j+1}-2}}$. By Lemma 7, the number of trapezoids in each bin is at most $\frac{5}{\xi^{2^{j+1}-2}}$. Given that bins are augmented (with a size of $1 + \alpha$), any trapezoid has side-length at most $\sqrt{2(1+\alpha)^2} < 2(1+\alpha)$. Since we pack small items of size at most $\delta = \xi^{2^{j+1}-1}$ inside these trapezoids, by Lemma 3, the wasted area in each trapezoid is less than $54\xi^{2^{j+1}-1} \times 2(1 + \alpha)$. Summing up over all trapezoids, the wasted area in each large bin is at most $\frac{5}{\xi^{2^{j+1}-2}} \times \xi^{2^{j+1}-1} \cdot 108(1 + \alpha) = 540\xi(1 + \alpha)$.

So, any large bin includes squares of total area at least $(1+\alpha)^2 - 540\xi(1+\alpha) > 1 - 540\xi$, assuming $\xi < 1/270$. Moreover, since packed by NFDH, all small bins (except potentially the last one), have a filled area of at least $1 - 2\xi > 1 - 540\xi$. In summary, with the exception of one bin, any large or small bin includes items of total area at least $1 - 540\xi$. As such, for $\xi < 1/1080$, we can write $m_l + m_s \leq \lceil area(\sigma)/(1 - 540\xi)\rceil + 1 \leq area(\sigma)(1 + 540\xi/(1 - 540\xi)) + 2 \leq (1 + 1080\xi) \cdot area(\sigma) + 2 \leq (1 + 1080\xi) \cdot Opt(\sigma) + 2$. Adding the number of medium bins, the total number of bins will be at most $(1 + 1081\xi) \cdot Opt(\sigma) + 3$.

Recall that we have $\xi = \epsilon/1081$. So, given any $\epsilon < 1$, the number of bins in the resulting packing will be at $(1 + \epsilon) \cdot Opt(\sigma) + 3$. Our algorithm's time complexity is dominated by the sorting process used for classifying items and packing small items. As such, the algorithm runs in $O(n \log n)$. We can conclude the following:

Theorem 2. *Assume a multi-set σ of n squares can be optimally packed in $Opt(S)$ unit bins. There is a polynomial-time algorithm that, for any constant $\alpha > 0$ and $\epsilon \in (0, 1)$, packs S into at most $(1 + \epsilon)Opt(\sigma) + 3$ augmented bins of size $1 + \alpha$.*

References

1. Abrahamsen, M., Miltzow, T., Seiferth, N.: Framework for *exists*r-completeness of two-dimensional packing problems. CoRR 2004.07558 (2020)
2. Albers, S.: Better bounds for online scheduling. SIAM J. Comput. **29**(2), 459–473 (1999)
3. Albers, S., Hellwig, M.: Online makespan minimization with parallel schedules. Algorithmica **78**(2), 492–520 (2017)
4. Allen, S.R., Iacono, J.: Packing identical simple polygons is NP-hard. CoRR abs/1209.5307 (2012)
5. Bansal, N., Correa, J.R., Kenyon, C., Sviridenko, M.: Bin packing in multiple dimensions: inapproximability results and approximation schemes. Math. Oper. Res. **31**(1), 31–49 (2006)
6. Boyar, J., Epstein, L., Levin, A.: Tight results for Next Fit and Worst Fit with resource augmentation. Theor. Comput. Sci. **411**(26–28), 2572–2580 (2010)
7. Chou, A.: NP-hard triangle packing problems. manuscript (2016)
8. Christensen, H.I., Khan, A., Pokutta, S., Tetali, P.: Approximation and online algorithms for multidimensional bin packing: a survey. Comput. Sci. Rev. **24**, 63–79 (2017)
9. Coffman, E.G., Garey, M.R., Johnson, D.S.: Approximation algorithms for bin packing: A survey. In: Hochbaum, D. (ed.) Approximation algorithms for NP-hard Problems. PWS Publishing Co. (1997)
10. Coffman Jr., E.G., Csirik, J., Galambos, G., Martello, S., Vigo, D.: Bin packing approximation algorithms: survey and classification. In: Pardalos, P.M., Du, D.Z., Graham, R.L. (eds.) Handbook of Combinatorial Optimization, pp. 455–531. Springer, New York (2013)
11. Demaine, E.D., Fekete, S.P., Lang, R.J.: Circle packing for origami design is hard. CoRR abs/1008.1224 (2010)
12. Epstein, L.: Two-dimensional online bin packing with rotation. Theor. Comput. Sci. **411**(31–33), 2899–2911 (2010)

13. Epstein, L., Ganot, A.: Optimal on-line algorithms to minimize makespan on two machines with resource augmentation. Theory Comput. Syst. **42**(4), 431–449 (2008)
14. Epstein, L., Levin, A.: Robust approximation schemes for cube packing. SIAM J. Optim. **23**(2), 1310–1343 (2013)
15. Epstein, L., van Stee, R.: Online bin packing with resource augmentation. Discret. Optim. **4**(3–4), 322–333 (2007)
16. Erdős, P., Graham, R.L.: On packing squares with equal squares. J. Combin. Theory Ser. A **19**, 119–123 (1975)
17. Erickson, J., van der Hoog, I., Miltzow, T.: A framework for robust realistic geometric computations. CoRR abs/1912.02278 (2019)
18. Fleischer, R., Wahl, M.: Online scheduling revisited. In: Paterson, M.S. (ed.) ESA 2000. LNCS, vol. 1879, pp. 202–210. Springer, Heidelberg (2000). https://doi.org/10.1007/3-540-45253-2_19
19. Friedman, E.: Packing unit squares in squares: a survey and new results. Elec. J. Comb. **1000**, DS7-Aug (2009)
20. Garey, M.R., Johnson, D.S.: Approximation algorithms for bin packing problems - a survey. In: Ausiello, G., Lucertini, M. (eds.) Analysis and Design of Algorithms in Combinatorial Optimization, pp. 147–172. Springer, New York (1981)
21. Gensane, T., Ryckelynck, P.: Improved dense packings of congruent squares in a square. Discret. Comput. Geom. **34**(1), 97–109 (2005)
22. Göbel, F.: Geometrical packing and covering problems. Math Centrum Tracts **106**, 179–199 (1979)
23. Hoberg, R., Rothvoss, T.: A logarithmic additive integrality gap for bin packing. In: Proceedings the 28th Annual ACM-SIAM Symposium on Discrete Algorithms (SODA), pp. 2616–2625. SIAM (2017)
24. Coffman Jr., E.G., Garey, M.R., Johnson, D.S., Tarjan, R.E.: Performance bounds for level-oriented two-dimensional packing algorithms. SIAM J. Comput. **9**(4), 808–826 (1980)
25. Lenstra Jr., H.W.: Integer programming with a fixed number of variables. Math. Oper. Res. **8**(4), 538–548 (1983)
26. Kowalski, D.R., Wong, P.W.H., Zavou, E.: Fault tolerant scheduling of tasks of two sizes under resource augmentation. J. Sched. **20**(6), 695–711 (2017). https://doi.org/10.1007/s10951-017-0541-1
27. Leung, J.Y., Tam, T.W., Wong, C.S., Young, G.H., Chin, F.Y.L.: Packing squares into a square. J. Parallel Distrib. Comput. **10**(3), 271–275 (1990)
28. Sleator, D., Tarjan, R.E.: Amortized efficiency of list update and paging rules. Commun. ACM **28**, 202–208 (1985)
29. Sleator, D.D., Tarjan, R.E.: Self-adjusting binary search trees. J. ACM **32**, 652–686 (1985)
30. Spielman, D.A., Teng, S.: Smoothed analysis of algorithms: why the simplex algorithm usually takes polynomial time. J. ACM **51**(3), 385–463 (2004)
31. Stromquist, W.: Packing 10 or 11 unit squares in a square. Elec. J. Comb. **10**, R8 (2003)
32. de la Vega, W.F., Lueker, G.S.: Bin packing can be solved within 1+ε in linear time. Combinatorica **1**(4), 349–355 (1981)

Miscellaneous

Remotely Useful Greedy Algorithms

Moritz Beck[✉] [iD]

University of Konstanz, Konstanz, Germany
beck@inf.uni-konstanz.de

Abstract. We study a class of parameterized max-min problems, called
REMOTE-$\mathcal{P}$: Given a minimization graph problem $\mathcal{P}$, find k vertices such
that the optimum value of $\mathcal{P}$ is the highest amongst all k-node subsets.
One simple example for REMOTE-$\mathcal{P}$ is computing the graph diameter
where $\mathcal{P}$ is the shortest path problem and $k = 2$. In this paper we focus
on variants of the minimum spanning tree problem for $\mathcal{P}$. In previous
literature $\mathcal{P}$ had to be defined on complete graphs. For many practically
relevant problems it is natural to define $\mathcal{P}$ on sparse graphs, such as street
networks. However, for large networks first computing the complete version of the network is impractical. Therefore, we describe greedy algorithms for REMOTE-$\mathcal{P}$ that perform well while computing only a small
amount of shortest paths. On the theoretical side we proof a constant factor approximation. Furthermore, we implement and test the algorithms
on a variety of graphs. We observe that the resulting running times are
practical and that the quality is partially even better than the theoretical
approximation guarantee, as shown via instance-based upper bounds.

Keywords: Graph optimization problem · Diversity maximization ·
Greedy algorithm

1   Introduction

Let $\mathcal{P}$ be a minimization problem on a weighted graph with an objective function
which assigns every vertex subset a solution cost. Additionaly, let $k \in \mathbb{N}$ be a
parameter that specifies the size of the vertex subset. Then the corresponding
REMOTE-$\mathcal{P}$ problem is to identify the set of k vertices that is assigned the largest
objective function value.

This class of problems, also called diversity maximization or dispersion problems, are fundamental problems in location theory [14]. Alongside theoretical interest there are also many practical applications that can be phrased as
REMOTE problems.

Application 1 (Facility Dispersion). The problem of dispersing facilities is to
choose locations such that the pairwise distance is maximized. Here k is the
number of facilities and $\mathcal{P}$ is to determine the smallest distance between two
facilities. Then the REMOTE-$\mathcal{P}$ problem seeks to choose k vertices such that the
minimum distance between any of those is maximized. This problem is known

© Springer Nature Switzerland AG 2020
W. Wu and Z. Zhang (Eds.): COCOA 2020, LNCS 12577, pp. 547–561, 2020.
https://doi.org/10.1007/978-3-030-64843-5_37

under several names and many variations have been studied for example for avoiding competition (sites of a restaurant chain, interference of radio transmitters) [11], locating undesirable facilities (e.g. power plants) [4], and protecting strategic facilities from simultaneous attacks (missile silos, ammunition dumps) [4,14].

Application 2 (Vehicle Routing). In a vehicle routing problem [3] every vehicle serves k customers at a specified drop-off zone. Here $\mathcal{P}$ is the problem of finding a minimum weight cycle that starts and ends at the depot and visits the k customers. Then REMOTE-$\mathcal{P}$ consists of finding customer locations within that zone such that the minimum delivery time is maximized. This is useful for getting location-independent upper bounds on the delivery time which for example can be used in dynamic programming and branch-and-bound approaches [5,12].

Application 3 (Local Search). One of the most popular heuristics for many NP-hard optimization problems (e.g. TSP or VRP) is local search. Thereby, the quality depends crucially on a good initial solution. As observed in [7], exact or approximate REMOTE-MST and REMOTE-TSP provide high-quality starting solutions.

The scope of this paper is to analyze variants of the minimum spanning tree problem for $\mathcal{P}$. Previous approaches for these REMOTE-$\mathcal{P}$ problems are only applicable if $\mathcal{P}$ is defined on complete graphs. Many application scenarios naturally occur on street networks, though. As computing the complete version of the graphs is impractical for large street networks, we seek for algorithms that provide satisfying results without precomputing all pairwise distances.

1.1   Related Work

The focus of previous work has been on studying REMOTE problems on metric spaces or complete, metric graphs. All these REMOTE problems are NP-hard [2]. For most of them APX-hardness has also been shown. An overview of the hardness results is given in Table 1. On the positive side there are approximation algorithms for most of these problems on metric graphs. The respective approximation factors are summarized in Table 1 as well. On non-metric graphs, though, it is not possible to approximate REMOTE-MST to a constant factor (unless P = NP) [7].

Many of the approximation algorithms on metric graphs are greedy algorithms which incrementally construct a solution vertex set by repeatedly choosing the next best vertex. In [1] a local search algorithm was proposed that exchanges one vertex as long as the objective value is rising. Ravi et al. [15] developed an approximation algorithm for REMOTE-clique in 2-dimensional euclidian space, based on an exact dynamic programming approach for the 1D case.

We will focus on designing and analyzing greedy algorithms.

Table 1. Lower and upper bound on the approximability of various REMOTE problems. MM means maximum matching.

Problem	Objective	l.b	u.b	APX-algorithm
REMOTE-MST	$w(mst(P))$	2 [7]	4 [7]	greedy
REMOTE-TSP	$w(tsp(P))$	2 [7]	3 [7]	greedy
REMOTE-ST	$w(st(P))$	$4/3$ [7]	3 [7]	greedy
REMOTE-edge	$\min_{v,u \in P} d(u,v)$	2 [15]	2 [15]	greedy
REMOTE-clique	$\sum_{v,u \in P} d(u,v)$	–	2 [9]	greedy-MM
REMOTE-star	$\min_{v \in P} \sum_{u \in P} d(u,v)$	–	2 [2]	MM
REMOTE-pseudoforest	$\sum_{v \in P} \min_{u \in P} d(u,v)$	2 [7]	$O(\log n)$ [2]	greedy+

1.2 Contribution

We consider REMOTE-$\mathcal{P}$ problems on non-complete graphs. We introduce new objective functions for $\mathcal{P}$ that are variations of the minimum spanning tree problem. Given a set of k vertices, these functions assume a value between the weight of the MST on the complete graph and the weight of the minimum Steiner tree on the given k nodes. These objective functions represent different levels of abstraction of the costs that occur in the application scenario. We show that a greedy algorithm provides a 6-approximation for all these REMOTE problems while only computing k shortest path trees. This algorithm is significantly faster than any algorithm that first has to construct the complete graph via n shortest path tree computations, where n is the number of vertices in the graph. Furthermore, we provide an algorithm that computes instance-based upper bounds to judge the practicality of the REMOTE greedy algorithms. We evaluate our algorithms on a rich set of benchmark graphs and observe that they provide practically useful results. In particular for graphs extracted from OpenStreetMap we get results that are at most a factor of 3 away from the optimum.

2   Formal Problem Statement

We are given a weighted graph $G(V, E)$ with n vertices and m edges as well as edge costs $w: E \rightarrow \mathbb{R}^+$. We assume that the given graph G is connected. The distance $d(v, w)$ between two nodes in the graph is defined as their shortest path cost in G. We are additionally given a minimization graph problem $\mathcal{P}$ with an objective function $f_{\mathcal{P}}: 2^V \rightarrow \mathbb{R}^+$ which assigns every vertex subset $W \subseteq V$ the respective minimum solution cost. Thereby, the solution cost typically is the sum of shortest path distances in the graph (e.g. induced by a path, a round tour, or a tree). The corresponding REMOTE-$\mathcal{P}$ problem seeks to find for a given parameter $k \in \{1, \ldots, n\}$ the set $W^* = \arg \max_{W \subseteq V, |W|=k} f_{\mathcal{P}}(W)$. Additionally we are interested in the maximum value $f_{\mathcal{P}}(W^*)$.

Let G_{M} be the complete, metric version of G, i.e. the graph $(V, \binom{V}{2})$ with weights $w(e_{vw}) = d(v, w)$ for every pair of vertices v, w. We look at the following problems $\mathcal{P}$ given a weighted graph G and a set of vertices W.

- **MST**: Find the minimum spanning tree (MST) of the induced subgraph $G_M[W]$.
- **UMST**: Let S be the set of shortest path edges between vertices in W, where the shortest paths correspond to the MST edges in $G_M[W]$. This includes edges in G that are part of multiple shortest paths only once. We call this problem UMST (unique edges MST).
- **MSPS**: Let S be the set of shortest paths between vertices in W, where the shortest paths are sets of edges. Find a subset $S' \subseteq S$ such that $\bigcup S'$ is a graph connecting the vertices in W and where $\sum_{e \in \bigcup S} w(e)$ is minimized. We call this problem MSPS (minimum shortest path spanning tree).
- **ST**: Find the subset of edges with the lowest weight that connects all the points in W. This is the well known Steiner tree (ST) problem.

These objective functions are illustrated in Fig. 1. These problems can be considered as a tradeoff between abstraction level of the solution and complexity: While considering the MST problems makes sense in complete graphs it might not be fitting for the application to sum up the cost of "virtual" edges. Instead the real cost might more resemble the sum of weights of the connecting edges in the underlying, non-complete graph. This gives rise to the problem UMST. Now, considering the costs of unions of shortest paths, we might ask not for the ones that, in a sense, form a minimum spanning tree, but the shortest paths minimizing the overall weight in G (instead of G_M). This is the MSPS problem. In the extreme one might only care about connecting the chosen vertices as cheaply as possible, leading to the classic ST problem. The first two problems are solvable in polynomial time while the latter two are NP-complete.

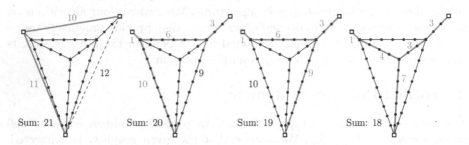

Fig. 1. The four objectives considered in this paper. From left to right: MST, UMST, MSPS, ST. The square vertices are the ones in W. Note that in this graph the function values are different for each of the objectives.

We can derive REMOTE problems from these, e.g. REMOTE-ST, where we want to find k vertices maximizing the weight of the Steiner tree on them. We consider these problems under theoretical and practical aspects.

3   Hardness and Cost Hierarchy

We first take a look at the complexity of the problems $\mathcal{P}$ themselves. As mentioned already, REMOTE problems are hard for non-fixed k. But when one wants to compute the actual value $f_{\mathcal{P}}(W)$ of a set of solution vertices W, it is also important to consider the complexity for evaluating $\mathcal{P}$.

Lemma 1. MST *and* UMST *can be computed efficiently.*

Proof. Determine the shortest paths between vertices in W and use their costs to compute the MST of $G_M[W]$. For UMST count edges of G that are part of multiple shortest paths in this MST only once.

Lemma 2. MSPS *and* ST *are NP-hard.*

Proof. By Reduction from *Exact Cover by Three Sets*. The proof works the same as the proof of NP-hardness of Steiner Tree presented in [16] as all paths of interest are shortest paths.

Next we describe the relations of the values of the different objective functions. Note that a cost hierarchy on the objectives $\mathcal{P}$ implies a cost hierarchy on the respective REMOTE problems (consider the set W^* maximizing an objective lower in the hierarchy).

Lemma 3. *For any graph $G = (V, E)$ and set $W \subseteq V$ of vertices we have*

$$f_{\text{MST}}(W) \geq f_{\text{UMST}}(W) \geq f_{\text{MSPS}}(W) \geq f_{\text{ST}}(W).$$

Proof. – $f_{\text{MST}}(W) \geq f_{\text{UMST}}(W)$: $f_{\text{MST}}(W)$ sums up the weight of the edges of an MST of $G_M[W]$, which is the sum over all path of the summed edge weight of these paths. $f_{\text{MST}}(W)$ has the same value but with the weight of every edges of G only added at most once, even if the edge is part of several shortest paths.
  – $f_{\text{UMST}}(W) \geq f_{\text{MSPS}}(W)$: Both sum up the weight of shortest paths with duplicated edges only counted once, but while $f_{\text{UMST}}(W)$ chooses the shortest paths that make up the MST on $G_M[W]$, $f_{\text{MSPS}}(W)$ chooses the paths minimizing the value of the sum.
  – $f_{\text{MSPS}}(W) \geq f_{\text{ST}}(W)$: Both $f_{\text{MSPS}}(W)$ and $f_{\text{ST}}(W)$ are the sum of edges such that the vertices in W are connected. The minimum Steiner tree, by definition, has the smallest weight sum satisfying this condition.

There are cases when the inequalities are strict, as can be seen in the example graph in Fig. 1. We follow with two further oberservations about these objective functions:

Observation 1. *If $k = n$, i.e. $W = V$, then these four functions have the same value, namely the sum of all weights in the graph.*

Observation 2. *The values of the objective functions for UMST, MSPS and ST coincide in trees because there is only one way to connect a set of chosen vertices.*

Algorithm 1: greedy_anticover(Vertex v_0, int k)

---

1 $W \leftarrow \{v_0\}$;
2 **for** $i = 1$ **to** k **do**
3 $\quad$ $v_{i+1} \leftarrow$ furthest node from W;
4 $\quad$ $W \leftarrow W \cup \{v_{i+1}\}$;
5 **return** W;

---

4   Greedy Algorithms and Approximation Guarantees

In this section we present greedy algorithms for REMOTE problems on non-complete graphs and show properties of them, in particular that one of them is a 6-approximation algorithm for REMOTE-$\mathcal{P}$ with $\mathcal{P}$ one of the above.

4.1   Algorithms

For $k = 2$ all of the mentioned REMOTE problems become the graph diameter problem. Finding the diameter of a graph takes prohibitively long on large graphs as the best known algorithm takes cubic time. A natural heuristic is to find the furthest vertex v from an arbitrary vertex u and return them, or to find the furthest vertex w from v afterwards and returning v and w. We can generalize this approach to higher values of k.

First, we define the notion of an anticover. This concept is later used to prove the approximation bound.

Definition 1 (Anticover). *An* anticover *of a graph* $G = (V, E)$ *is a set of nodes* C *such that there is a value* $r \in \mathbb{R}$ *with all vertices in* C *having pairwise distance at least* r *and all vertices* $v \in V \setminus C$ *are within distance* r *from a vertex in* C.

We consider the following two greedy algorithms (Algorithm 1+2) that aim at finding a subset $W \subseteq V$ with size k where the objective values $f_{\mathcal{P}}(W)$ are as high as possible for $\mathcal{P} \in \{\text{MST}, \text{UMST}, \text{MSPS}, \text{ST}\}$. Algorithm 1 is widely used to get approximations to REMOTE problems, marked as "greedy" in Table 1. Algorithm 1 computes an anticover (with r being the distance to the last inserted node).

The algorithms are illustraded in Fig. 2. Obviously they both do k shortest path tree computations. Both are adaptions of the greedy algorithm of previous literature, which is only defined on complete graphs, to non-complete graphs. While Algorithm 1 computes an anticover, Algorithm 2 does not but on trees it gives the optimum answer. An *extremal vertex* is any vertex that can be found by computing the furthest vertex from some starting vertex. In trees there are exactly two extremal vertices, namely the ones defining the diameter [10] (assuming the diameter is unique).

Algorithm 2: spread_points(Vertex v_0, int k)

1 $W \leftarrow \{v_0\}$;
2 $P \leftarrow \{v_0\}$;
3 **for** $i = 1$ **to** k **do**
4      $v_{i+1} \leftarrow$ furthest node from P;
5      $W \leftarrow W \cup \{v_{i+1}\}$;
    // shortest_path($\cdot, \cdot$) contains the nodes
6      $P \leftarrow P \cup$ shortest_path(P, v_{i+1});
7 **return** W;

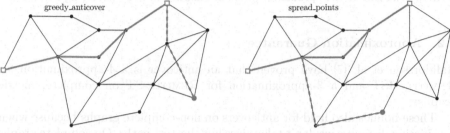

Fig. 2. The greedy algorithms for selecting a set of vertices. While Algorithm 1 select vertices that are far from the already selected vertices, Algorithm 2 selects vertices that are far from every vertex on the connecting shortest paths.

Theorem 1. *Algorithm 2 (spread_points) computes* REMOTE-*ST exactly on trees when starting on an extremal vertex.*

Proof. Sketch: We use induction over the number of already chosen vertices, when choosing a total of k vertices. By an exchange argument we show that (i) the diameter is part of any optimum solution, and (ii) the algorithm always extends the set of chosen vertices such that there still is an optimum solution of k vertices containing them.

Because, as remarked above in Observation 2, the value of the optimum solution on a tree is the same for REMOTE-ST, REMOTE-UMST and REMOTE-MSPS, this also works for the latter two problems.

For REMOTE-MST, however, neither Algorithm 2 nor Algorithm 1 computes the exact optimum set of vertices on a tree. The optimum set also doesn't even have to be only leaf vertices when k is smaller than the number of leaves, in contrast to the other REMOTE problems. For examples see Fig. 3.

In the following we look how well those greedy algorithms perform from a theoretical as well as from a practical perspective.

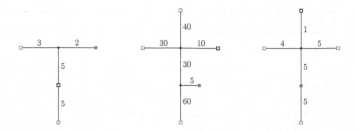

Fig. 3. Counterexamples for strategies for MST on trees. From left to right: Algorithms 1 and 2, Choose leaves first. The square vertices are the selected vertices. The vertices belonging to the optimum solution are marked in blue. (Color figure online)

4.2   Approximation Guarantee

Halldórsson et al. [7] have proven that an anticover is a 4-approximation for REMOTE-MST and a 3-approximation for REMOTE-ST on complete, metric graphs.

These bounds also hold for anticovers on non-complete graphs because when considering distances we don't take edges but shortest paths. On complete graphs only these two extreme cases (MST and ST) exist but not the intermediate problems UMST and MSPS. We show that an anticover is an approximation for their REMOTE problems, too.

Theorem 2. *Let $\mathcal{P}$ be a problem whose objective value is between $f_{ST}(W)$ and $f_{MST}(W)$ for all subsets $W \subseteq V$. Then an anticover is a 6-APX for REMOTE-$\mathcal{P}$ on metric graphs.*

Proof. The proof is similar to the one in [7]. We keep in mind that distance are not defined by direct edges but by shortest paths between vertices.

Let $G = (V, E)$ be a graph. Let W be an anticover of G with radius r and Q be an arbitrary set of vertices, both of size k. The Steiner ratio of a graph is the maximum ratio of the weights of $F_{MST}(W)$ and $f_{ST}(W)$ over all subsets $W \subset V$ and we know that the following holds [6]:

$$\frac{f_{MST}(W)}{f_{ST}(W)} \leq 2\frac{k-1}{k} \tag{1}$$

As every vertex in Q has distance at most r from a vertex of W, we have

$$f_{ST}(Q) \leq f_{ST}(W \cup Q) \leq f_{ST}(W) + kr \leq f_{\mathcal{P}}(W) + kr. \tag{2}$$

Every edge of the MST of $G_M(W)$ has length at least r.

$$f_{MST}(W) \geq (k-1)r \tag{3}$$

Now we can use these to prove the theorem.

$$
\begin{aligned}
\frac{f_P(Q)}{f_P(W)} &\overset{}{\leq} \frac{f_{\mathrm{MST}}(Q)}{f_P(W)} \overset{(1)}{\leq} 2\frac{k-1}{k}\frac{f_{\mathrm{ST}}(Q)}{f_P(W)} \overset{(2)}{\leq} 2\frac{k-1}{k}\frac{f_P(W)+kr}{f_P(W)} \\
&= 2\frac{k-1}{k}\left(1+\frac{kr}{f_P(W)}\right) \leq 2\frac{k-1}{k}\left(1+\frac{kr}{f_{\mathrm{ST}}(W)}\right) \\
&\overset{(1)}{\leq} 2\frac{k-1}{k}\left(1+2\frac{k-1}{k}\frac{kr}{f_{\mathrm{MST}}(W)}\right) \overset{(3)}{\leq} 2\frac{k-1}{k}\left(1+2\frac{k-1}{k}\frac{kr}{(k-1)r}\right) \\
&= 2\frac{k-1}{k}\left(1+2\right) = 6 - \frac{6}{k}
\end{aligned}
$$

$\square$

Corollary 1. *Algorithm 1 provides a 6-approximation algorithm for* REMOTE-*UMST and* REMOTE-*MSPS.*

Algorithm 2 does not provide an approximation guarantee but we still can measure the quality of both algorithms via instance-based upper bounds.

5   Instance-Based Upper Bounds

We want to look at instance-based upper bounds of the costs. These can be used as a baseline for the output cost of the greedy algorithms to be compared against. Our upper bounds algorithm is based on ideas that are used by Magnien et al. [13] to compute upper bounds on the diameter of a graph. They use the upper bounds (along with lower bounds) to estimate the diameter fast. It is still an open problem if there is a faster way than to compute all pairwise shortest path distances which is slow on large graphs.

They combined the following two observations about the diameter of a graph:

1. The diameter of a spanning tree is not smaller than the diameter of the base graph.
2. The diameter of a tree can be computed efficiently. [8]

The first point holds because for every edge $e \in E$ we have $\mathrm{diam}(G - e) \geq \mathrm{diam}(G)$ (as long as $G - e$ is still connected). So we can compute instance-based upper bounds on the diameter of a graph by computing any spanning tree (e.g. the shortest path tree of a single node) and then the its exact diameter.

Because both points also hold for REMOTE-ST we can use the same principle to calculate an upper bound on its optimum solution value.

Lemma 4. *The optimum solution value of an* REMOTE-*ST instance on a graph G is less than or equal to the optimal solution value of the* REMOTE-*ST instance on any spanning tree of G.*

Proof. The proof idea is that removing an edge of the graph cannot make the maximum minimum Steiner tree lighter.

After taking a spanning tree of the graph, we apply Theorem 1 to compute an upper bound. So we can efficiently compute an upper bound on REMOTE-ST as follows:

1. Compute an arbitrary spanning tree T of G.
2. Choose k terminals by running Algorithm 2 (starting from an extremal vertex) on T.
3. Compute the weight of the minimum Steiner tree of T on these k terminals.

We double this value to get an upper bound for REMOTE-MST (and thus for REMOTE-UMST and REMOTE-MSPS).

Lemma 5. *Let u be an upper bound for the optimum value of an REMOTE-ST instance with graph G and parameter k. Then $2 \cdot u$ is an upper bound for the optimum value of the REMOTE-MST instance with the same parameter and on the same graph.*

Proof. The Steiner ratio on metric graphs is 2. This means that for any metric graph and any set $W \subseteq V$ we have $2 \cdot f_{ST}(W) \geq f_{MST}$. Let $W^* \in V$ be a k-subset maximizing the value of f_{ST} and W' be a k-subset maximizing the value of f_{MST}. Then the following holds:

$$2u \geq 2 \cdot f_{ST}(W^*) \geq 2 \cdot f_{ST}(W') \geq f_{MST}(W')$$

6   Experiments

To evaluate the algorithms we have given we have implemented the algorithms in Rust, compiling with rustc 1.45.0, and tested them on a rich set of benchmark graphs. Experiments were conducted on a single core of an AMD Ryzen 7 3700X CPU (clocked at 2.2 GHz) with 128 GB main memory.

The evaluation has been run with combinations of the following parameters:

– Number of selected vertices $k \in \{3, 10, 31, 100, 316, 1000\}$
– Minimizing Problem $\mathcal{P}$: MST, UMST
– Selecting algorithm: Random, Algorithms 1 and 2

We computed sets of vertices W and the respective objective value $f_{\mathcal{P}}(W)$ ten times for all combinations of the above parameters on all graphs described in the next section (except only one for the largest street network). We first describe the graphs we used for testing and then look at the results.

6.1   Benchmark Graphs

We test the algorithms on the following set of graphs:

– Graphs derived from the national TSP data
– Street network graph from OSM data
– Generated grid graphs

National TSP Graphs. These graphs are derived from Euclidian point sets[1]. These provide, for several countries, location coordinates of cities within that country. We produced graphs out of them in the following way: Each city is one vertex. Let $r \in \mathbb{N}$ be a radius. We insert an edge between two cities, with the edge weight being their Euclidian distance rounded to the nearest whole number, if the edge weight is at most r. If r is too small to connect all cities, i.e. there is a city with minimum distance to other cities larger than r, then we pick an arbitrary vertex of each connected component and add the edges of a MST on these respresentatives. As r grows, this adds direct connections between cities and produces more complete graphs. The radii we picked can be seen in Table 2 and contain 0, ∞ (a value larger than the diameter), the smallest radius that connects the graph, and for Qatar and Oman half the diameter.

Table 2. TSP graphs: Number of cities, radius r, number of edges m

	Djibuti (38 cities)						
r	0	50	100	200	437	500	∞
m	74	80	82	132	360	430	1406
	Qatar (194 cities)						
r	0	50	100	329	729	∞	
m	386	1140	2490	12258	30394	37442	
	Oman (1979 cities)						
r	0	50	100	500	1037	5082	∞
m	3956	14842	38758	412344	1065338	3142058	3914462
	Ireland (8246 cities)						
r	0	50	100	201	500		
m	16490	60550	222800	803304	4036244		

Street Network Graphs. We tested on three street networks obtained from OpenStreetMap (OSM) data[2]. These cover the German national state of Baden-Württemberg, Germany and Europe. We've also taken three street graphs from the 9th DIMACS implementation challenge[3] covering New York City, California and mainland USA, respectively. A overview of the graphs can be seen in Table 3. The OSM graphs have a higher percentage of degree 2 nodes.

Generated Grid Graphs. We used three sets of grid graphs. They were generated as follows: Given an integer k we place vertices on a regular $k \times k$-grid and connect horizontally or vertically adjacent vertices with edges. We choose random edge

---

[1] http://www.math.uwaterloo.ca/tsp/world/countries.html.
[2] https://i11www.iti.kit.edu/resources/roadgraphs.php.
[3] http://www.dis.uniroma1.it/~challenge9.

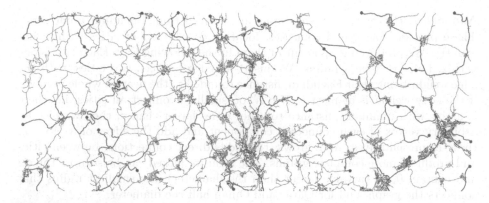

Fig. 4. Example Plot Cutout: UMST with $k = 50$ nodes

Table 3. Benchmark graphs: Street networks and grid graphs. There are 3 instance of each grid graph size.

Graph	#vertices	#edges
Grid 100 × 100	≈ 9300	17820
Grid 316 × 316	≈ 91000	179172
Grid 1000 × 1000	≈ 920000	1798200

Graph	#vertices	#edges
New York City	264346	366923
California	1890815	2328871
USA	23947347	29166672
Baden-Württemberg	3064263	3234197
Germany	20690320	21895866
Europe	173789185	182620014

weights between 1 and 20 for each row and column and then randomly remove 10% of the edges and vertices that became isolated. We generated three graphs with k equal to 100, 316 and 1000, respectively.

6.2   Results

In this section we present notable results on the conducted experiments. We consider the quality and running times of the algorithms as well as differences within those algorithms and between different types of graphs. Fig. 4 shows an example output of 50 vertices selected on a graph where the edges that constitute the UMST are highlighted.

Quality. Figure 5 shows the quality of the (non-random) greedy algorithms, comparing the cost for the points found by these algorithms with an instance-based upper bound. We see that the upper bound is mostly only a small factor larger than the computed cost. The notable exceptions are the TSP graphs, when the radius is large, i.e. the graph has significantly more edges than vertices, and k is high. But especially on grid graphs the factor is always lower than 7. On one hand this means that the greedy algorithms indeed provide good approximations, many times even better than the theoretical bound of 6. On the other hand this testifies that the computed upper bounds are not too far from the optimum values. Because of the theoretical bound for anticovers (and both algorithms performing about equally, as we well see) the higher gaps for the TSP graph don't necessarily imply that the greedy algorithms perform worse, but that the upper bound is not as tight for those graphs.

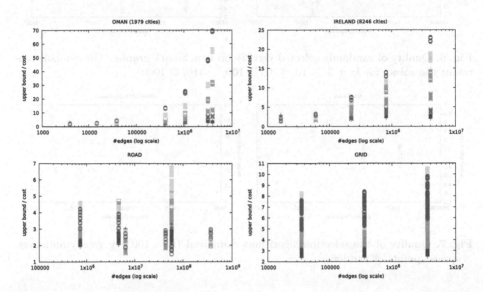

Fig. 5. Quality of the greedy algorithms on benchmark graphs. The symbols represent the values for k: $+$ 3, $\times$ 10, $*$ 31, $\square$ 100, $\blacksquare$ 316, $\bigcirc$ 1000

Furthermore, the greedy selection algorithms are a substantial improvement over random selection of vertices, especially when k is small (see Fig. 6). This seems logical because, when many vertices are selected, the choice of bad vertices can be smoothed out by other vertices. Also with small k every chosen vertex has a higher impact on the overall objective value.

Selection Algorithms. We compare the quality of the two greedy algorithms. Additionaly we tested if it is worthwhile to not select a random vertex as input/start vertex but an extremal vertex. An extremal vertex is one that is the furthest from any other vertex. Selecting an extremal vertex as the first one has very similar costs but is slightly better in most cases. Interestingly, selecting vertices with `greedy_anticover` (Algorithm 1) performs better than with `spread_points` (Algorithm 2) for all graphs but the OpenStreetMap graphs. Figure 7 shows the comparison of the selecton algorithms on two graph classes; we left out picturing random selection as the quality is too bad and the difference betwwen the other greedy algorithms could not be made out.

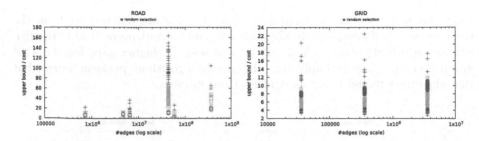

Fig. 6. Quality of randomly selected vertices on benchmark graphs. The symbols represent the values for k: $+$ 3, $\times$ 10, $*$ 31, $\square$ 100, $\blacksquare$ 316, O 1000

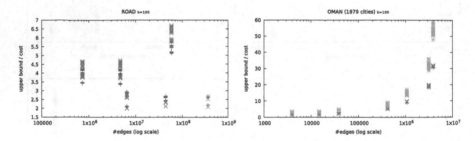

Fig. 7. Quality of the selection algorithms compared ($k = 100$): $+$ greedy_anticover, $\times$ spread_points, $*$ random

Running Time. Unsurprisingly, selecting random vertices takes close to no time (less than 2 seconds at most) when comparing to the time it takes to evaluate the cost function. While `greedy_anticover` selects slightly better vertices for most graphs, `spread_points` is a little faster. The overall running time is very much practical with a run of an algorithm taking less than 17 min on the Europe graph when selecting 1000 vertices. The maximum running time on all the graphs except for the Europe graph is hardly above one minute. The growth of the running time is plotted in Fig. 8. This figure also shows the percentage that the

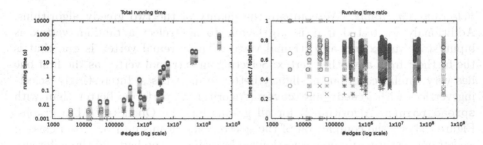

Fig. 8. Running times for the greedy algorithms. The symbols represent the values for k: $+$ 3, $\times$ 10, $*$ 31, $\square$ 100, $\blacksquare$ 316, O 1000

selection phase takes of the total running time (the rest is spent computing the objective value $f_{\mathcal{P}}$). We see that the selection phase takes up more of the running time as k grows.

7 Conclusions and Future Work

We introduced two new objectives $\mathcal{P}$ for the Remote-$\mathcal{P}$ maximization problem class on graphs that are not metric and complete, and presented greedy algorithms that aim at solving them. We proved that one of the algorithms is a 6-approximation algorithm for these REMOTE problems while the other one is also used to provide instance-based upper bounds. These bounds are then used to evaluate these algorithms in practice on a variety of graphs where we see that they perform well in regards to solution quality and running time.

For future work one could look at finding better upper bounds for dense graphs. Another direction is to improve the approximation bound for the greedy anticover algorithm on the newly introduced objectives for REMOTE problems.

References

1. Aghamolaei, S., Farhadi, M., Zarrabi-Zadeh, H.: Diversity maximization via composable coresets. In: Proceedings of the 27th Canadian Conference on Computational Geometry, CCCG 2015 (2015)
2. Chandra, B., Halldórsson, M.M.: Approximation algorithms for dispersion problems. J. Algorithms (2001)
3. Dantzig, G.B., Ramser, J.H.: The truck dispatching problem. Manag. Sci. (1959)
4. Erkut, E., Neuman, S.: Analytical models for locating undesirable facilities. Eur. J. Oper. Res. (1989)
5. Feitsch, F., Storandt, S.: The clustered dial-a-ride problem. In: Proceedings of the Twenty-Ninth International Conference on Automated Planning and Scheduling, ICAPS (2019)
6. Gilbert, E., Pollak, H.: Steiner minimal trees. SIAM J. Appl. Math. (1968)
7. Halldórsson, M.M., Iwano, K., Katoh, N., Tokuyama, T.: Finding subsets maximizing minimum structures. SIAM J. Discrete Math. (1999)
8. Handler, G.Y.: Minimax location of a facility in an undirected tree graph. Transp. Sci. (1973)
9. Hassin, R., Rubinstein, S., Tamir, A.: Approximation algorithms for maximum dispersion. Oper. Res. Lett.(1997)
10. Jordan, C.: Sur les assemblages de lignes. J. für die reine und angewandte Mathematik (1869)
11. Kuby, M.J.: Programming models for facility dispersion: the p-dispersion and maxisum dispersion problems. Geogr. Anal. (1987)
12. Lau, H.C., Sim, M., Teo, K.M.: Vehicle routing problem with time windows and a limited number of vehicles. Eur. J. Oper. Res. (2003)
13. Magnien, C., Latapy, M., Habib, M.: Fast computation of empirically tight bounds for the diameter of massive graphs. ACM J. Exp. Algorithmics (2009)
14. Moon, I.D., Chaudhry, S.S.: An analysis of network location problems with distance constraints. Manag. Sci.(1984)
15. Ravi, S.S., Rosenkrantz, D.J., Tayi, G.K.: Heuristic and special case algorithms for dispersion problems. Oper. Res. (1994)
16. Santuari, A.: Steiner tree NP-completeness proof. Technical report (2003)

Parameterized Algorithms for Fixed-Order Book Drawing with Bounded Number of Crossings per Edge

Yunlong Liu[iD], Jie Chen[iD], and Jingui Huang[(✉)][iD]

Hunan Provincial Key Laboratory of Intelligent Computing and Language
Information Processing, Hunan Normal University,
Changsha 410081, People's Republic of China
{ylliu,jie,hjg}@hunnu.edu.cn

Abstract. Given a graph $G = (V, E)$ and a fixed linear order of V, the problem FIXED-ORDER BOOK DRAWING with bounded number of crossings per edge asks whether there is a k-page book drawing of G such that the maximum number of crossings per edge is upper-bounded by an integer b. This problem was posed by Bhore et al. (GD 2019; J. Graph Algorithms Appl. 2020) and thought to be interesting for further investigation. In this paper, we study the fixed-parameter tractable algorithms for this problem. More precisely, we show that this problem parameterized by both the bounded number b of crossings per edge and the vertex cover number τ of the graph admits an algorithm running in time $(b + 2)^{O(\tau^3)} \cdot |V|$, and this problem parameterized by both the bounded number b of crossings per edge and the pathwidth κ of the vertex ordering admits an algorithm running in time $(b + 2)^{O(\kappa^2)} \cdot |V|$. Our results provide a specifical answer to Bhore et al.'s question.

1 Introduction

Book drawing is a fundamental topic in graph drawing. Combinatorially, a *k-page book drawing* $\langle \prec, \sigma \rangle$ of a graph $G = (V, E)$ consists of a linear ordering $\prec$ of its vertices along a spine and an assignment σ of each edge to one of the k pages, which are half-planes bounded by the spine [1]. The spine and the k pages construct a book. Specially, a *book drawing* of G is also called a *fixed-order book drawing* if the ordering of vertices in $V(G)$ along the spine is predetermined and fixed; see Fig. 1 for an illustration.

Fixed-order book drawing, as a specific subject on book drawings, has drawn much attention in the area of graph drawing. In particular, the FIXED LINEAR CROSSING NUMBER problem, where the number k of pages is given and the target is to minimize the number of edge crossings, has been well-studied [2–6].

This research was supported in part by the National Natural Science Foundation of China under Grants (No. 61572190, 61972423), and Hunan Provincial Science and Technology Program Foundations (No. 2018TP1018, 2018RS3065).

W. Wu and Z. Zhang (Eds.): COCOA 2020, LNCS 12577, pp. 562–576, 2020.
https://doi.org/10.1007/978-3-030-64843-5_38

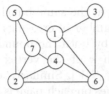 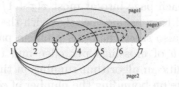

Fig. 1. A graph (left) and a 3-page fixed-order book drawing with at most 1 crossings per edge (right).

Studies on parameterized algorithms for FIXED-ORDER BOOK DRAWING, where the optimization goal is to minimize the number of pages, arise from the problem FIXED-ORDER BOOK EMBEDDING. A fixed-order book drawing is called a fixed-order book embedding if it is crossing-free. The problem FIXED-ORDER BOOK EMBEDDING decides whether a graph G admits a k-page fixed-order book embedding. The minimum k such that G admits a k-page fixed-order book embedding is the fixed-order book thickness of G, denoted by fo-bt$(G, \prec)$. Deciding whether fo-bt$(G, \prec) \leq 2$ can be solved in linear time. However, deciding if fo-bt$(G, \prec) \leq 4$ is equivalent to finding a 4-coloring of a circle graph and is **NP**-complete [7,8]. Considering the hardness of this problem even when the number of pages is fixed, Bhore et al. [9] studied fixed-parameter tractable algorithms parameterized by the vertex cover number of the graph and the pathwidth of the vertex ordering, respectively. Furthermore, Bhore et al. [9,10] posed a generalized problem, i.e., FIXED-ORDER BOOK DRAWING with bounded number of crossings *per edge* (rather than per page). Note that a k-page book drawing with bounded number of crossings per edge may contain unbounded number of crossings on some page, which makes this generalized problem more interesting.

In this paper, we focus on the problem FIXED-ORDER BOOK DRAWING with bounded number of crossings per edge. Formally, this problem asks, given a graph $G = (V, E)$ and a fixed linear order $\prec$ of V, whether there is a k-page book drawing $\langle \prec, \sigma \rangle$ of G such that the maximum number of crossings per edge is bounded by an integer b. We denote by fo-bd$(G, \prec, b)$ the minimum k such that $(G, \prec, b)$ is a YES instance. Our aim is to investigate fixed-parameter tractable algorithms for it, answering to Bhore et al.'s question. The problem FIXED-ORDER BOOK EMBEDDING (that is, $b = 0$) is **NP**-complete for small fixed values k on general graphs. Therefore, unless $\mathbf{P} = \mathbf{NP}$, there can be no fixed-parameter tractable algorithm parameterized by the natural parameters (i.e. the page number and the crossing number). We consider the problem parameterized by the sum of the natural parameter and some structural parameter of G.

Our results include two parts as follows.

(1) We show that this problem parameterized by both the bounded number b of crossings per edge and the vertex cover number τ of the graph admits an algorithm running in time $(b+2)^{O(\tau^3)} \cdot |V|$. Our main technique on bounding the number of considered edges on each page is to divide all assigned edges

on each page into at most $\tau(\tau + 1)$ bunches and in the same bunch we only need to consider one edge with the maximum number of crossings.

(2) We also show that this problem parameterized by both the bounded number b of crossings per edge and the pathwidth κ of the vertex ordering admits an algorithm running in time $(b + 2)^{O(\kappa^2)} \cdot |V|$. Similarly, our technique on bounding the number of considered edges on each page is to divide all assigned edges on each page into at most κ bunches and in the same bunch we only need to consider one edge with the maximum number of crossings.

Our results employ the dynamic programming framework in [9] and some Lemmas in [11,12]. More importantly, we give a novel approach to bound the number of considered edges on each page by a function of parameters, which will enrich the techniques on designing parameterized algorithms for graph drawing problems [13,14]. Our results are also related with a frontier research area of graph drawing where few edge crossings are allowed per edge. Recent results about this topic include [15–18].

2   Preliminaries

We use standard terminology from parameterized computation, graph theory and some notations defined in [9].

A parameterized problem consists of an input (I, k) where I is the problem instance and k is the parameter. An algorithm is called a *fixed-parameter tractable algorithm* if it solves a parameterized problem in time $f(k)n^{O(1)}$, where f is a computable function. A parameterized problem is *fixed-parameter tractable* if it admits a fixed-parameter tractable algorithm [19].

The graphs we consider in this paper are undirected and loopless. The vertex set of a graph G is denoted by $V(G)$, with $|V(G)| = n$. The edge between two vertices u and v of G is denoted by uv. For $r \in \mathbb{N}$, we use $[1, r]$ to denote the set $\{1, \ldots, r\}$.

A *vertex cover* C of a graph $G = (V, E)$ is a subset $C \subseteq V$ such that each edge in E has at least one end-point in C. A vertex $v \in V$ is a *cover vertex* if $v \in C$. The *vertex cover number* of G, denoted by $\tau(G)$, is the size of a minimum vertex cover of G. Note that a vertex cover C with size τ can be computed in time $O(1.2738^\tau + \tau \cdot n)$ [20]. In the rest of this paper, we will use C to denote a minimum vertex cover of G, and use U to denote the set $V(G) \setminus C$.

Given an n-vertex graph $G = (V, E)$ with a fixed linear order $\prec$ of V such that $v_1 \prec v_2 \prec \ldots \prec v_n$, we assume that $V(G) = \{v_1, v_2, \ldots, v_n\}$ is indexed such that $i < j$ if and only if $v_i \prec v_j$. Moreover, we use $X = \{x \in [1, n - \tau] \mid \exists c \in C : u_x$ is the immediate successor of c in $\prec\}$ to denote the set of indices of vertices in U which occur immediately after a cover vertex, and we assume that the integers in X, denoted as $x_1, x_2, \ldots, x_z$, are listed in ascending order.

Given an n-vertex graph $G = (V, E)$ with a fixed linear order $\prec$ of V such that $v_1 \prec v_2 \prec \ldots \prec v_n$, the *pathwidth* of $(G, \prec)$ is the minimum number κ such that for each vertex v_i ($i \in [1, n]$), there are at most κ vertices left of v_i that are adjacent to v_i or a vertex right of v_i. Formally, for each v_i we call the set

$P_i = \{v_j \mid j < i, \exists\, q \geq i$ such that $v_j v_q \in E\}$ the *guard set* for v_i, and the pathwidth of $(G, \prec)$ is simply $\max_{i \in [1,n]} |P_i|$ [21]. The elements of the guard sets are called the *guards* for v_i. In this paper, we will use the expanded notion of guard set in [9]. Let v_0 be a vertex with degree 0 and let v_0 be placed to the left of v_1 in $\prec$. For a vertex v_i, the expanded guard set of v_i was defined as follows: $P_{v_i}^* = \{g_1^i, g_2^i, \cdots, g_m^i\}$ where for each $j \in [1, m-1]$, g_j^i is the j-th guard of v_i in reverse order of $\prec$, and $g_m^i = v_0$.

3   Parameterization by Both the Maximum Number of Crossings per Edge and the Vertex Cover Number

The FIXED-ORDER BOOK DRAWING parameterized by both the maximum number b of crossings per edge and the vertex cover number τ, abbreviated by FDVC, is formally defined as follows.

Input: a tuple $(G, \prec)$, a non-negative integer b.
Parameters: b, τ;
Question: does there exist a k-page book drawing $(\prec, \sigma)$ of G such that the maximum number of crossings per edge on each page is no more than b ?

For the problem FDVC, we begin with introducing some special notations. Given a graph $G = (V, E)$ and a minimum vertex cover C, we denote by E_C the set of all edges whose both endpoints lie in C. For each $i \in [1, n - \tau + 1]$, we denote by $E_i = \{u_j c \in E \mid j < i, c \in C\}$ the set of all edges with one endpoint outside of C that lies to the left of u_i. Correspondingly, we denote by $U_i = \{u_x \mid x < i, u_x \in U\}$ the set of all vertices that lie to the left of u_i. An assignment $s : E_C \to [1, k]$ is called a *valid page assignment* if s assigns edges in E_C to k pages such that the maximum number of crossings per edge on each page is at most b. Given a valid page assignment s of E_C and a minimum vertex cover C in G, we call $\alpha : E_i \to [1, k]$ a *valid partial page assignment* if $\alpha \cup s$ assigns edges to k pages such that the maximum number of crossings per edge on each page is at most b. Let $u \in U \setminus U_i$ and $c \in C$. Given a valid partial page assignment α of E_i, we draw an edge uc between u and c on page p. The added edge uc is called a *potential edge* with respect to the edges assigned to page p by α.

We employ the technique on mapping assignments to matrices and the dynamic programming framework in algorithm for FIX-ORDER BOOK THICKNESS parameterized by the vertex cover number [9]. However, the specific implementation tactics in our algorithm for the problem FDVC are quite distinct from those in [9]. Most importantly of all, we adopt a more refined standard to classify all valid partial page assignments of E_i (for $i \in [1, n - \tau + 1]$). To arrive at this target, we introduce three types of crossing number matrices in the record set.

3.1   Three Types of Crossing Number Matrices in the Record Set

Let $i \in [1, n - \tau + 1]$ and $u_a \in U \setminus U_i$. We define the crossing number matrices for all potential edges incident to u_a, for some edges in E_i, and for all edges in E_C, respectively.

3.1.1   The Crossing Number Matrices for Potential Edges Incident to Vertices in $U \setminus U_i$

The notion of crossing number matrix for potential edges incident to a vertex in $U \setminus U_i$ is extended from that of visibility matrix in [9] and formally introduced in [11].

Although a valid partial page assignment in this paper refers to the assignment where the number of crossings per edge is at most b, the crossing number matrix can be defined in the same way.

Given a valid partial page assignment $\alpha \cup s$ of edges in $E_i \cup E_C$ and a vertex $u_a \in U \setminus U_i$, there are τ potential edges incident to u_a on page p ($p \in [1, k]$). The crossing number matrix $M_i^1(a, \alpha, s)$ is a $k \times \tau$ matrix, in which the entry (p, q) records the number of edge-crossings generated by the potential edge $c_q u_a$ on page p with some edge in $E_i \cup E_C$ mapped to page p by $\alpha \cup s$. In particular, when this number exceeds $b + 1$, it only record the number $b + 1$.

Similarly, the crossing number matrix admits the same property as that in [11], i.e., for some consecutive vertices in $\in U \setminus U_i$, their corresponding crossing number matrices are exactly the same one.

Lemma 1. *([11]) Let α be a valid partial page assignment of E_i, $x_j \in X$, and let x_l be the immediate successor of x_j in X. If $u_i \prec u_{x_j} \prec u_{x_h} \prec u_{x_l}$ in $\prec$ and $u_{x_h} \notin C$, then $M_i^1(x_h, \alpha, s) = M_i^1(x_j, \alpha, s)$.*

Based on Lemma 1, we only need to keep the crossing number matrices corresponding to the vertices in $\{u_i\} \cup X$, respectively. Hence, we define a crossing number *matrix queue* $\mathbb{M}_i^1(\alpha, s)$ as follows: $\mathbb{M}_i^1(\alpha, s) = (M_i^1(i, \alpha, s), M_i^1(x_1, \alpha, s), M_i^1(x_2, \alpha, s), \ldots, M_i^1(x_z, \alpha, s))$.

3.1.2   The Crossing Number Matrices for Some Edges Assigned from E_i

The number of edges in E_i assigned to page p is not bounded by a function of τ, however, we only need to consider at most $\tau(\tau + 1)$ edges for the problem FDVC. Then, we design τ crossing number matrices to capture the information about the number of crossings generated by these considered edges.

Given a valid partial assignment α of E_i and a minimum vertex cover $C = \{c_1, c_2, \ldots, c_\tau\}$, we first describe how we choose at most $\tau + 1$ edges incident to one vertex in C on page p.

The vertices in $V(G) \setminus C$ can be divided into $\tau + 1$ subsets along the order $\prec$. In other words, assume that the vertices in $V(G)$ lie on a straight line L in a left-to-right fashion with the fixed-order $\prec$, then the line L can be divided into $\tau + 1$ intervals by the τ vertices in C. For ease of expression, we denote by c_0

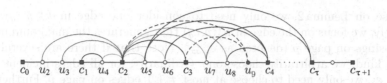

$$c_0 \quad u_2 \quad u_3 \quad c_1 \quad u_4 \quad c_2 \quad u_5 \quad u_6 \quad c_3 \quad u_7 \quad u_8 \quad u_9 \quad c_4 \quad \cdots \cdots \quad c_\tau \quad c_{\tau+1}$$

Fig. 2. An example on dividing the edges incident to the vertex c_2 on the first page into at most $\tau + 1$ bunches (note that only a part of edges are drawn for brevity).

the vertex u_1 when the first vertex in $\prec$ is not c_1. We also denote by $c_{\tau+1}$ the additional vertex of degree 0. Then, the $\tau + 1$ intervals can be uniformly denoted by (c_t, c_{t+1}) for $0 \leq t \leq \tau$. See Fig. 2 for an example.

Given a vertex c_q for $c_q \in C$, we denote by $E_i(p, c_q)$ the set of edges in E_i incident to c_q and assigned to page p by α. Corresponding to the $\tau + 1$ intervals on L, the edges in $E_i(p, c_q)$ on page p can be divided into $\tau + 1$ bunches such that for edges in the same bunch, all endpoints except the common one (i.e., c_q) lie in the same interval. We also uniformly denote these bunches as $E_i^0(p, c_q)$, $E_i^1(p, c_q)$, ..., $E_i^\tau(p, c_q)$. For example, it holds that $E_i^0(1, c_2) = \{c_2u_2, c_2u_3\}$, $E_i^1(1, c_2) = \{c_2u_4\}$, $E_i^2(1, c_2) = \{c_2u_5, c_2u_6\}$, and $E_i^3(1, c_2) = \{c_2u_7, c_2u_8\}$ in Fig. 2.

Assume that $|E_i^t(p, c_q)| \geq 2$ ($t \in [0, \tau]$). An important property can be discovered for edges in $E_i^t(p, c_q)$.

Lemma 2. *Let α be a valid partial assignment of E_i, $u_j c_r$ be a potential edge on page p, in which $i \leq j$, $c_r \in C$, and $r \neq q$. Then, either all edges in $E_i^t(p, c_q)$ are crossed by $u_j c_r$, or none edge in $E_i^t(p, c_q)$ is crossed by $u_j c_r$.*

Proof. By the definition of $E_i^t(p, c_q)$, all endpoints except c_q of edges in $E_i^t(p, c_q)$ lie in the interval (c_t, c_{t+1}). We first show that either all edges in $E_i^t(p, c_q)$ enclose c_r/u_j or none of them does. Assume towards a contradiction that one edge $c_q u_a$ in $E_i^t(p, c_q)$ encloses c_r, but another edge $c_q u_b$ in $E_i^t(p, c_q)$ does not. Then, it follows that $u_a \prec c_r \prec u_b$, contradicting the assumption that c_{t+1} is the immediate successor of c_t in $\{c_0\} \cup C \cup \{c_{\tau+1}\}$ along the given order $\prec$. Analogously, assume that $c_q u_a$ in $E_i^t(p, c_q)$ encloses u_j but $c_q u_b$ in $E_i^t(p, c_q)$ does not. It follows that $u_a \prec u_j \prec u_b$, which means that $j < b$. On the other hand, since $c_q u_b \in E_i$, i.e., $b < i$, and $i \leq j$ (see assumption), it holds that $b < j$. Hence, a contradiction is derived.

By the above discussion, we distinguish three cases for the vertices in $\{c_r, u_j\}$ based on the number of vertices uniformly enclosed by the edges in $E_i^t(p, c_q)$. Case (1): only c_r (or only u_j) in $\{c_r, u_j\}$ is uniformly enclosed by the edges in $E_i^t(p, c_q)$. Case (2): both c_r and u_j are enclosed by the edges in $E_i^t(p, c_q)$. Case (3): none of the vertices in $\{c_r, u_j\}$ is enclosed by the edges in $E_i^t(p, c_q)$. Correspondingly, in case (1), all edges in $E_i^t(p, c_q)$ are crossed by $u_j c_r$. However, in cases (2) and (3), none edge in $E_i^t(p, c_q)$ is crossed by $u_j c_r$. $\square$

Base on Lemma 2, we only need to consider one edge in $E_i^t(p, c_q)$. More precisely, we focus on the edge in $E_i^t(p, c_q)$ that generates the maximum number of crossings on page p (denoted as e_{max}). Note that if there are several edges of this kind, we arbitrarily choose one. Ordinarily, for all edges incident to the vertex c_q, we only need to choose at most $\tau + 1$ edges on page p. Furthermore, we define a $k \times (\tau + 1)$ *crossing number matrix* $M_i^2(c_q, \alpha, s)$, where an entry (p, t) is h if the edge e_{max} in $E_i^t(p, c_q)$ generates h crossings on page p. Otherwise, $E_i^t(p, c_q) = \emptyset$ and the entry (p, t) is filled with a special string "null" (see Fig. 3 for an example).

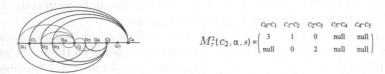

$$M_7^2(c_2, \alpha, s) = \begin{pmatrix} c_0\text{-}c_1 & c_1\text{-}c_2 & c_2\text{-}c_3 & c_3\text{-}c_4 & c_4\text{-}c_5 \\ 3 & 1 & 0 & \text{null} & \text{null} \\ \text{null} & 0 & 2 & \text{null} & \text{null} \end{pmatrix}$$

Fig. 3. A partial 2-page assignment of the edges in E_7 (left) and the corresponding crossing number matrix $M_7^2(c_2, \alpha, s)$ (right).

Since there are τ vertices in C, we define a crossing number *matrix queue* $\mathbb{M}_i^2(\alpha, s)$ for all vertices in C as follows: $\mathbb{M}_i^2(\alpha, s) = (M_i^2(c_1, \alpha, s), M_i^2(c_2, \alpha, s), \ldots, M_i^2(c_\tau, \alpha, s))$.

3.1.3   The Crossing Number Matrix for Edges in E_C

Since the number of edges in E_C is bounded by $\tau(\tau - 1)/2$, we directly design a crossing number matrix for all edges in E_C.

Assume that the edges in E_C are denoted as $e_1, e_2, \ldots, e_m$ ($m \leq \tau(\tau-1)/2$). Given a valid assignment $\alpha \cup s$ of edges in $E_i \cup E_C$, we define a $k \times \tau(\tau - 1)/2$ matrix $\mathbb{M}_i^3(\alpha, s)$, where an entry (p, q) is r if e_q ($e_q \in E_C$) satisfies two conditions: ① e_q is assigned to page p by s; ② the number of crossings generated by e_q is equal to r. Otherwise, (p, q) is filled with a special string "null".

3.2   A Parameterized Algorithm for the Problem FDVC

Based on the three types of crossing number matrices, we define the record set for a vertex $u_i \in U$ as follows: $\mathcal{R}_i(s) = \{(\mathbb{M}_i^1(\alpha, s), \mathbb{M}_i^2(\alpha, s), \mathbb{M}_i^3(\alpha, s)) \mid \exists$ valid partial page assignment $\alpha : E_i \rightarrow [1, k]\}$. By this way, all valid page assignments of edges in $E_i \cup E_C$ are divided into $|\mathcal{R}_i(s)|$ families. For ease of presentation, we will use $\mathcal{M}_i(\alpha, s)$ to denote the record $(\mathbb{M}_i^1(\alpha, s), \mathbb{M}_i^2(\alpha, s), \mathbb{M}_i^3(\alpha, s))$ in the rest of this paper. Along with $\mathcal{R}_i(s)$, we also store a mapping Λ_i^s from $\mathcal{R}_i(s)$ to valid partial page assignments of E_i which maps each record $\omega \in \mathcal{R}_i(s)$ to some α such that $\omega = \mathcal{M}_i(\alpha, s)$.

By employing the framework of dynamic programming in [9], we can obtain an algorithm for solving FDVC, denoted by ALVC.

The first step in algorithm ALVC is to branch over each case s of edge assignments on E_C. **If** s is invalid **then** discard it. Otherwise, the algorithm ALVC is to dynamically generate some partial page assignment, in which the maximum number of crossings per edge is at most b, in a left-to-right fashion. Assume the record set $\mathcal{R}_{i-1}(s)$ has been computed. Each page assignment β of edges incident to vertex u_{i-1} and each record $\rho \in \mathcal{R}_{i-1}(s)$ are branched. For each such β and $\gamma = \Lambda_{i-1}^s(\rho)$, $\gamma \cup \beta \cup s$ forms a new assignment of edges in $E_i \cup E_C$. If the maximum number of crossings per edge in $\gamma \cup \beta \cup s$ is at most b, then $\gamma \cup \beta \cup s$ is valid, the record $\mathcal{M}_i(\gamma \cup \beta, s)$ is computed and the mapping Λ_i^s is set to map to $\gamma \cup \beta$. Otherwise, $\gamma \cup \beta \cup s$ is invalid and discarded.

To prove the correctness of ALVC, we mainly argue that any two assignments in the same family defined by the record set are "interchangeable".

Let γ_1 and γ_2 be two valid page assignments of E_{i-1}, let β be a page assignment of the edges incident to the vertex u_{i-1}, and let $\gamma_1 \cup \beta$ and $\gamma_2 \cup \beta$ be the corresponding assignments of E_i.

Lemma 3. *If* $\mathcal{M}_{i-1}(\gamma_1, s) = \mathcal{M}_{i-1}(\gamma_2, s)$, *then* $\mathcal{M}_i(\gamma_1 \cup \beta, s) = \mathcal{M}_i(\gamma_2 \cup \beta, s)$.

Proof. By the definition of the record in set $\mathcal{R}_i(s)$, it is sufficient to prove that the crossing number matrix in $\mathcal{M}_i(\gamma_1 \cup \beta, s)$ is equal to that in $\mathcal{M}_i(\gamma_2 \cup \beta, s)$, respectively.

(1) we show that $\mathbb{M}_i^1(\gamma_1 \cup \beta, s) = \mathbb{M}_i^1(\gamma_2 \cup \beta, s)$, i.e., $M_i^1(x, \gamma_1 \cup \beta, s) = M_i^1(x, \gamma_2 \cup \beta, s)$ (for $x \in \{i\} \cup X$). Assume that an entry (p, q) in $M_{i-1}^1(x, \gamma_1, s)$ (for $x \in \{i-1\} \cup X$) is equal to h_0 but that in $M_i^1(x, \gamma_1 \cup \beta, s)$ (for $x \in \{i\} \cup X$) is equal to $h_0 + r$ ($r \leq \tau$). Then, in the assignment β of the (at most) τ edges incident to u_{i-1}, there must be at least r edges assigned to page p and there are exactly r edges among them cross $u_x c_q$ on page p. By the assumption that $M_{i-1}^1(x, \gamma_1, s) = M_{i-1}^1(x, \gamma_2, s)$ (for $x \in \{i-1\} \cup X$), the entry (p, q) in $M_{i-1}^1(x, \gamma_2, s)$ is equal to h_0. Correspondingly, the entry (p, q) in $M_i^1(x, \gamma_2 \cup \beta, s)$ is equal to $h_0 + r$.

(2) we show that $\mathbb{M}_i^2(\gamma_1 \cup \beta, s) = \mathbb{M}_i^2(\gamma_2 \cup \beta, s)$, i.e., $M_i^2(c, \gamma_1 \cup \beta, s) = M_i^2(c, \gamma_2 \cup \beta, s)$ (for $c \in C$). Let (p, q) be an entry in $M_{i-1}^2(c, \gamma_1, s)$. We distinguish two cases based on whether $(p, q) = $ "null" or not.

Case (2.1): $(p, q) \neq$ "null". Assume that (p, q) in $M_{i-1}^2(c, \gamma_1, s)$ is equal to h_0 but that in $M_i^2(c, \gamma_1 \cup \beta, s)$ is equal to $h_0 + r$. Let e be one of the edges in $E_i^q(p, c)$ that generate the maximum number of crossings in $\gamma_1 \cup s$. After executing the i-th step in dynamic programming, we further distinguish two subcases based on whether e still generates the maximum number of crossings in $\gamma_1 \cup \beta \cup s$ or not.

Subcase (2.1.1): the edge e does. By Lemma 2, in the assignment β of the (at most) τ edges incident to u_{i-1}, there must be at least r edges assigned to page p and there are exactly r edges among them cross all edges in $E_i^q(p, c)$ in $\gamma_1 \cup \beta \cup s$. The r edges in β will also cross the edges between c and vertices in the q-th interval on page p in $\gamma_2 \cup \beta \cup s$. Hence, the entry (p, q) in $M_i^2(c, \gamma_2 \cup \beta, s)$ is also equal to $h_0 + r$.

Subcase (2.1.2): the edge e does not. Then there must be one edge cu_{i-1} in β such that cu_{i-1} is added to $E_i^q(p,c)$ and cu_{i-1} generates h_0+r crossings on page p (see the edge c_2u_9 in Fig. 2 for an example). As shown in point (1), $M_i^1(i, \gamma_2 \cup \beta, s) = M_i^1(i, \gamma_1 \cup \beta, s)$. Hence, the number of crossings generated by cu_{i-1} in $\gamma_2 \cup \beta \cup s$ is equal to that in $\gamma_1 \cup \beta \cup s$. Therefore, the entry (p, q) in $M_i^2(c, \gamma_2 \cup \beta, s)$ is also equal to h_0+r.

Case (2.2): the entry (p, q) in $M_{i-1}^2(c, \gamma_1, s)$ is "null" but that in $M_i^2(c, \gamma_1 \cup \beta, s)$ is equal to r. Then there must be one edge cu_{i-1} in β such that cu_{i-1} is added to $E_i^q(p,c)$ and cu_{i-1} generates r crossings on page p (see the edge c_3u_9 in Fig. 2 for an example). As shown in point (1), $M_i^1(i, \gamma_2 \cup \beta, s) = M_i^1(i, \gamma_1 \cup \beta, s)$. Hence, the number of crossings generated by cu_{i-1} in $\gamma_2 \cup \beta \cup s$ is equal to that in $\gamma_1 \cup \beta \cup s$. Therefore, the entry (p, q) in $M_i^2(c, \gamma_2 \cup \beta, s)$ is also equal to r.

(3) we show that $\mathbb{M}_i^3(\gamma_1 \cup \beta, s) = \mathbb{M}_i^3(\gamma_2 \cup \beta, s)$. Since the assignment s of E_C is fixed, we need not to consider the entry that filled with "null". Assume that an entry (p, q) in $\mathbb{M}_i^3(\gamma_1 \cup \beta, s)$ is increased by r compared with that in $\mathbb{M}_{i-1}^3(\gamma_1, s)$. Along the same lines in point (1), the entry (p, q) in $\mathbb{M}_i^3(\gamma_1 \cup \beta, s)$ is also increased by r compared with that in $\mathbb{M}_{i-1}^3(\gamma_2, s)$. □

Based on Lemma 3, the algorithm $ALVC((G, \prec), \tau, b)$ correctly computes $\mathcal{R}_i(s)$ from $\mathcal{R}_{i-1}(s)$. Therefore, we obtain the following conclusion.

Theorem 1. *If $(G, \prec)$ contains at least one valid assignment, then the algorithm $ALVC((G, \prec), \tau, b)$ returns a valid page assignment.*

Lemma 4. *The algorithm ALVC for the problem FDVC runs in time $(b + 2)^{3k\tau^2 + 3k\tau + (\tau^2 + \tau + 3) \log_{(b+2)} 6k\tau} \cdot |V|$.*

Proof. Since the number of matrices with form $\mathbb{M}_i^1(\alpha, s)$ (resp. $\mathbb{M}_i^2(\alpha, s)$, $\mathbb{M}_i^3(\alpha, s)$) can be bounded by $(b+2)^{k\tau(\tau+1)}$ (resp. $(b+2)^{k\tau(\tau+1)}$, $(b+2)^{k\tau(\tau-1)/2}$), the size of $\mathcal{R}_i(s)$ can be bounded by $(b + 2)^{3k\tau^2 + 3k\tau}$. Moreover, for a given tuple $(i, \gamma \cup \beta, s)$, each matrix queue can be dynamically computed from the matrix queue corresponding to the tuple $(i-1, \gamma, s)$ in time $k\tau^2 \cdot (\tau+1)$. Additionally, the number of assignments of E_C is at most τ^{τ^2} and the number of assignments of edges incident to u_{i-1} is at most τ^τ. The total time of ALVC can be bounded by $3k\tau^2 \cdot (\tau+1) \cdot \tau^{\tau^2} \cdot \tau^\tau \cdot (b+2)^{3k\tau^2 + 3k\tau} \cdot |V| = (b+2)^{3k\tau^2 + 3k\tau + (\tau^2 + \tau + 3) \log_{(b+2)} 6k\tau} \cdot |V|$.
□

Since fo-bt$(G, \prec) < \tau$ [9] and fo-bd$(G, \prec, b) \leq$ fo-bt$(G, \prec)$, it follows that fo-bd$(G, \prec, b) < \tau$. A (fo-bd$(G, \prec, b)$)-page book drawing can be computed by executing the algorithm ALVC at most τ times. Now, we arrive at our first result.

Theorem 2. *There is an algorithm which takes as input a graph $G = (V, E)$ with a fixed vertex order $\prec$, and an integer b, computes a page assignment σ in time $(b + 2)^{O(\tau^3)} \cdot |V|$ such that $(\prec, \sigma)$ is a (fo-bd$(G, \prec, b)$)-page book drawing of G, where τ is the vertex cover number of G.*

4   Parameterization by both the Maximum Number of Crossings per Edge and the Pathwidth of the Vertex Ordering

The FIXED-ORDER BOOK DRAWING parameterized by both the maximum number b of crossings per edge and the pathwidth κ of the vertex ordering, abbreviated by FDPW, is formally defined as follows. For the problem FDPW, we also

---

Input: a tuple $(G, \prec)$, a non-negative integer b.
Parameters: b, κ;
Question: does there exist a k-page book drawing $(\prec, \sigma)$ of G such that the maximum number of crossings per edge on each page is no more than b ?

---

begin with introducing some special notations. Given a graph $G = (V, E)$ and a vertex ordering $\prec$, we use $E_i = \{v_a v_b \mid v_a v_b \in E, b > i\}$ to denote the set of all edges with at least one endpoint to the right of v_i. Correspondingly, we use $V_i = \{v_x \mid v_x \in V, x \leq i\}$ to denote the set of all vertices that lie to the left of v_{i+1}. A partial assignment $\alpha : E_i \to [1, k]$ is a *valid partial page assignment* if α maps edges to k pages such that the maximum number of crossings per edge is at most b. For ease of expression, we introduce the notion of *potential edge*. Let $i \in [1, n]$. Given a valid partial page assignment α of E_i and two vertices v_x, v_a with $x < a \leq i$, we draw an edge $v_x v_a$ on page p between v_x and v_a. The added edge $v_x v_a$ is called a potential edge with respect to the edges in E_i assigned on page p.

By extending some techniques in [9], we design an algorithm for the problem FDPW along the same lines in the above algorithm ALVC.

4.1   Two Types of Crossing Data Matrices in the Record Set

Let $i \in [1, n]$ and $v_a \in V_i$. We define the crossing data matrices for potential edges incident to v_a, and for some edges assigned from E_i, respectively.

4.1.1   The Crossing Data Matrices for Potential Edges Incident to Vertices in V_i

The crossing data matrix corresponding to a vertex $v_a \in V_i$ is originated from the notion of visibility vector in [9] and formally introduced in [12].

Although a valid partial page assignment in this paper refers to the assignment where the number of crossings per edge is at most b, the crossing data matrix can be defined in the same way. Given a valid partial page assignment $\alpha : E_i \to [1, k]$ and a vertex $v_a \in V_i$, we define a $k \times (b + 1)$ matrix $M_i^1(v_a, \alpha)$, in which the entry (p, q) is set by the following rule [12].

If there exists a guard $v_z \in P_{v_i}^*$ $(z < a)$ such that the potential edge $v_z v_a$ on page p exactly crosses q edges in E_i assigned to page p by α, **then** the entry

$(p, q) = v_z$. In case there are at least two guards that satisfy this condition, we only choose the utmost guard to the left of v_a. **Otherwise**, $(p, q) =$ "null".

Similarly, the crossing data matrix admits the same property as that in [12], i.e., for some consecutive vertices in V_i, their corresponding crossing data matrices are exactly the same one.

Lemma 5. *([12]) Let α be a valid partial assignment of E_i, $v_a \prec v_i$, and $v_a \notin P_{v_i}^*$. Assume that $v_d \in P_{v_i}^* \cup \{v_i\}$ such that $d > a$ and $d - a$ is minimized. Then $M_i^1(v_a, \alpha) = M_i^1(v_d, \alpha)$.*

Based on Lemma 5, we only need to keep the crossing data matrices corresponding to the vertices in $P_{v_i}^* \cup \{v_i\}$, respectively. Hence, we define a crossing data *matrix queue* $\mathbb{M}_i^1(\alpha)$ as follows: $\mathbb{M}_i^1(\alpha) = (M_i^1(v_i, \alpha), M_i^1(g_1^i, \alpha), M_i^1(g_2^i, \alpha), \ldots, M_i^1(g_{m-1}^i, \alpha))$.

4.1.2   The Crossing Data Matrix for Some Edges Assigned from E_i

The number of edges in E_i assigned on page p is not bounded by a function of κ, however, we only need to consider at most κ edges for the problem FDPW. We design a crossing data matrix to capture the information about the number of crossings generated by these considered edges.

How to choose at most κ edges on page p ? First of all, we exclude the edges whose both endpoints lie in $V \setminus V_{i-1}$ since the potential edge will not cross any one of them. Hence, we only focus on the edges that incident to the guards of v_i. For each vertex $g \in P_{v_i}^*$, we use $E_i(p, g)$ to denote the set of edges in E_i that incident to g and assigned to page p by α. When $|E_i(p, g)| \geq 2$, the following observation will be crucial for choosing the edge to be considered.

Lemma 6. *Let α be a valid partial assignment of E_i, $v_x v_a$ be a potential edge on page p, and $g \in P_{v_i}^*$. If $v_x \prec g \prec v_a$, then all edges in $E_i(p, g)$ are crossed by $v_x v_a$. Otherwise, none edge in $E_i(p, g)$ is crossed by $v_x v_a$.*

Proof. By the definition of potential edge, there are only three cases about the relative positions among v_x, v_a and g. More precisely, the vertex orderings include case (1): $v_x \prec g \prec v_a$; case (2): $g \prec v_x \prec v_a$; and case (3): $v_x \prec v_a \prec g$. At the same time, for the edges in $E_i(p, g)$, their common endpoint is g and other endpoints are all on the right of v_i. Note that $a \leq i$. Hence, the edge $v_x v_a$ crosses each edge in $E_i(p, g)$ in case (1). On the contrary, the edge $v_x v_a$ does not cross any edge in $E_i(p, g)$ in cases (2) and (3). $\square$

Based on Lemma 6, for edges in $E_i(p, g)$, we only need to choose the edge that generates the maximum number of crossings on page p. Ordinarily, for all edges assigned on page p, we choose at most κ edges to be considered. Since there are exactly k pages in α, we can design a $k \times \kappa$ matrix $\mathbb{M}_i^2(\alpha)$ to capture the information about the number of crossings generated by edges assigned from E_i. More precisely, an entry (p, q) in $\mathbb{M}_i^2(\alpha)$ is r if the maximum number of crossings per edge in $E_i(p, q)$ is equal to r. Otherwise, $E_i(p, q) = \emptyset$ and (p, q) is filled with a special string "null" (see Fig. 4 for an example).

$$M_6^2(\alpha) = \begin{pmatrix} v_1 & v_2 & v_4 & v_5 \\ 1 & 2 & \text{null} & 2 \\ \text{null} & 0 & 1 & \text{null} \end{pmatrix}$$

Fig. 4. A partial 2-page assignment of the edges in E_6 (left) and the corresponding crossing data matrix (right).

4.2 A Parameterized Algorithm for the Problem FDPW

Based on two types of matrices, we define the record set for a vertex v_i ($i \in [1, n]$) as follows: $\mathcal{Q}_i = \{(\mathbb{M}_i^1(\alpha), \mathbb{M}_i^2(\alpha)) \mid \exists \text{ valid partial page assignment } \alpha: E_i \to [1, k]\}$. By this way, all valid assignments of E_i are divided into at most $|\mathcal{Q}_i|$ families. For ease of presentation, we will use $\mathcal{M}_i(\alpha)$ to denote the record $(\mathbb{M}_i^1(\alpha), \mathbb{M}_i^2(\alpha))$ in the rest of this paper. Along with $\mathcal{Q}_i$, we also store a mapping Λ_i from $\mathcal{R}_i$ to valid partial page assignments of E_i which maps each record $\omega_i(\alpha) \in \mathcal{R}_i$ to some α such that $\omega_i(\alpha) = (\mathbb{M}_i^1(\alpha), \mathbb{M}_i^2(\alpha))$.

Employing the framework of dynamic programming in [9], we can obtain an algorithm ALPW for solving the problem FDPW, in which the main steps can be sketched as follows.

The basic strategy is to dynamically generate some valid partial page assignment, in which the maximum number of crossings per edge is at most b, in a right-to-left fashion. Assume the record set $\mathcal{Q}_i$ has been computed. Let $F_{i-1} = E_{i-1} \setminus E_i$. Each page assignment β of edges in F_{i-1} and each record $\rho \in \mathcal{Q}_i$ are branched. For each such β and $\alpha = \Lambda_i(\rho)$, the assignment $\alpha \cup \beta$ is checked. If the maximum number of crossings per edge in $\alpha \cup \beta$ is still at most b, then $\alpha \cup \beta$ is valid, the record $\mathcal{M}_{i-1}(\alpha \cup \beta)$ is computed and the mapping Λ_{i-1} is set to map this record to $\alpha \cup \beta$. Otherwise, $\alpha \cup \beta$ is invalid and discarded.

To show the correctness of algorithm ALPW, we mainly show that any two assignments with the same crossing data matrix in our record set are "interchangeable".

Let α_1 and α_2 be two valid page assignments of E_i, let β be a page assignment of the edges in F_{i-1}, and let $\alpha_1 \cup \beta$ and $\alpha_2 \cup \beta$ be the corresponding assignments of E_{i-1}.

Lemma 7. *If $\mathcal{M}_i(\alpha_1) = \mathcal{M}_i(\alpha_2)$, then $\mathcal{M}_{i-1}(\alpha_1 \cup \beta) = \mathcal{M}_{i-1}(\alpha_2 \cup \beta)$.*

Proof. By the definition of the record in set $\mathcal{Q}_{i-1}$, it is sufficient to prove that the crossing data matrix in $\mathcal{M}_{i-1}(\alpha_1 \cup \beta)$ is equal to that in $\mathcal{M}_{i-1}(\alpha_2 \cup \beta)$, respectively.

(1) we show that $\mathbb{M}_{i-1}^1(\alpha_1 \cup \beta) = \mathbb{M}_{i-1}^1(\alpha_2 \cup \beta)$, i.e., for each $g_j^{i-1} \in \{v_{i-1}\} \cup P_{v_{i-1}}^*$, $\mathbb{M}_{i-1}^1(g_j^{i-1}, \alpha_1 \cup \beta) = \mathbb{M}_{i-1}^1(g_j^{i-1}, \alpha_2 \cup \beta)$. Let (p, q) (for $q \leq b$) be an entry in $\mathbb{M}_{i-1}^1(g_j^{i-1}, \alpha_1 \cup \beta)$. Assume that (p, q) in $\mathbb{M}_{i-1}^1(g_j^{i-1}, \alpha_1 \cup \beta)$ is equal to v_t, but the vertex v_t in $\mathbb{M}_i^1(g_j^i, \alpha_1)$ locates in the entry (p, h).

Then, for the assignment β of edges in $E_{i-1} \setminus E_i$, there must be at least $q - h$ edges assigned to page p and there are exactly $q - h$ edges among them cross the potential edge $v_t g_j^{i-1}$ on page p. At the same time, by assumption that $M_i^1(g_j^i, \alpha_1) = M_i^1(g_j^i, \alpha_2)$, it follows that the entry (p, h) in $M_i^1(g_j^i, \alpha_2)$ is also equal to v_t. Hence, the vertex v_t also locates in the entry (p, q) in $M_{i-1}^1(g_j^{i-1}, \alpha_2 \cup \beta)$.

(2) we show that $\mathbb{M}_{i-1}^2(\alpha_1 \cup \beta) = \mathbb{M}_{i-1}^2(\alpha_2 \cup \beta)$. Let e_1^q (resp. e_2^q) be the edge that incident to v_q and generating the maximum number of crossings on page p in the assignment α_1 (resp. α_2). Without loss of generality, assume that an entry (p, q) in $\mathbb{M}_i^2(\alpha_1)$ is h_0 but that in $\mathbb{M}_{i-1}^2(\alpha_1 \cup \beta)$ is $h_0 + r$. We distinguish two cases resulting in the change of entry (p, q). Case (2.1): the edge e_1^q crosses r edges assigned by β. By Lemma 6, the edge e_2^q also crosses r edges assigned by β in $\alpha_2 \cup \beta$. Hence, the corresponding entry (p, q) in $\mathbb{M}_{i-1}^2(\alpha_2 \cup \beta)$ is also increased by r compared with that in $\mathbb{M}_i^2(\alpha_2)$. Case (2.2): e_1^q is replaced by one edge $v_q v_i$ from F_{i-1}. This means the edge $v_q v_i$ generates the maximum number of crossings among all edges incident to v_q on page p. As shown in point (1), $\mathbb{M}_{i-1}^1(\alpha_1 \cup \beta) = \mathbb{M}_{i-1}^1(\alpha_2 \cup \beta)$. Hence, $v_q v_i$ also generates the maximum number of crossings on page p in $\alpha_2 \cup \beta$ and e_2^q is also replaced by the edge $v_q v_i$.

For the special case that an entry (p, q) in $\mathbb{M}_i^2(\alpha_1)$ is equal to "null" but that in $\mathbb{M}_{i-1}^2(\alpha_1 \cup \beta)$ is equal to r, the proof is along the same line as in case (2.2). □

Based on Lemma 7, the algorithm ALPW$((G, \prec), \kappa, b)$ correctly computes $\mathcal{Q}_{i-1}$ from $\mathcal{Q}_i$. Therefore, we can obtain the following conclusion.

Theorem 3. *If $((G, \prec), \kappa, b)$ contains at least one valid assignment, then the algorithm ALPW$((G, \prec), \kappa, b)$ returns a valid page assignment.*

Lemma 8. *The algorithm ALPW for the problem FDPW runs in time $(b + 2)^{2k\kappa + 1 + (\kappa + 2) \log_{(b+2)} 3k\kappa} \cdot |V|$.*

Proof. Since the number of matrices with form $\mathbb{M}_i^1(\alpha)$ (resp. $\mathbb{M}_i^2(\alpha)$) can be bounded by $(b + 2)^{k\kappa}$ [12] (resp. $(b + 2)^{k\kappa}$), the size of $\mathcal{Q}_i$ can be bounded by $(b + 2)^{2k\kappa}$. Moreover, for a given tuple $(i, \gamma \cup \beta')$, the matrix queue with form $\mathbb{M}_i^1(\gamma \cup \beta')$ (resp. $\mathbb{M}_i^2(\gamma \cup \beta')$) can be dynamically computed from the matrix queue with form $\mathbb{M}_{i+1}^1(\gamma)$ (resp. $\mathbb{M}_{i+1}^2(\gamma)$) in time $k\kappa(\kappa + 1)(b + 1)$ (resp. $k\kappa^2$), where β' is an assignment of edges in $E_i \setminus E_{i+1}$. Furthermore, the number of assignments of edges in $E_i \setminus E_{i+1}$ is at most κ^κ. Therefore, the total time of ALPW can be bounded by $(k\kappa^2 + k\kappa(\kappa + 1)(b + 1)) \cdot \kappa^\kappa \cdot (b + 2)^{2k\kappa} \cdot |V| \leq (b + 2)^{2k\kappa + 1 + (\kappa + 2) \log_{(b+2)} 3k\kappa} \cdot |V|$. □

Since fo-bt$(G, \prec) < \kappa$ [9] and fo-bd$(G, \prec, b) \leq$ fo-bt$(G, \prec)$, it follows that fo-bt$(G, \prec, c) < \kappa$. A (fo-bd$(G, \prec, b)$)-page book drawing can be computed by executing the algorithm ALPW at most κ times. Now, we arrive at our second result.

Theorem 4. *There is an algorithm which takes as input a graph $G = (V, E)$ with a fixed vertex order $\prec$, and an integer b, computes a page assignment σ in time $(b + 2)^{O(\kappa^2)} \cdot |V|$ such that $(\prec, \sigma)$ is a (fo-bd($G, \prec, b$))-page book drawing of G, where κ is the pathwidth of $(G, \prec)$.*

5 Conclusions

In this work, we investigate fixed-parameter tractable algorithms for the problem FIXED-ORDER BOOK DRAWING with bounded number of crossings per edge. We show that this problem parameterized by both the bounded number b of crossings per edge and the vertex cover number τ of the graph admits an algorithm running in time $(b + 2)^{O(\tau^3)} \cdot |V|$, and this problem parameterized by both the bounded number b of crossings per edge and the pathwidth κ of the vertex ordering admits an algorithm running in time $(b + 2)^{O(\kappa^2)} \cdot |V|$. Our results provide a specifical answer to Bhore et al.'s question in [9, 10].

The problem FIXED-ORDER BOOK DRAWING with bounded number of crossings per edge can be parameterized by both the bounded number b of crossings per edge and other structural parameter, such as treewidth or treedepth. It would be interesting to investigate the parameterized computational complexity for them.

Acknowledgements. The authors thank the anonymous referees for their valuable comments and suggestions.

References

1. Klawitter, J., Mchedlidze, T., Nöllenburg, M.: Experimental evaluation of book drawing algorithms. In: Frati, F., Ma, K.-L. (eds.) GD 2017. LNCS, vol. 10692, pp. 224–238. Springer, Cham (2018). https://doi.org/10.1007/978-3-319-73915-1_19
2. Masuda, S., Nakajima, K., Kashiwabara, T., Fujisawa, T.: Crossing minimization in linear embeddings of graphs. IEEE Trans. Comput. **39**(1), 124–127 (1990). https://doi.org/10.1109/12.46286
3. Cimikowski, R.: Algorithms for the fixed linear crossing number problem. Discrete Appl. Math. **122**(1), 93–115 (2002). https://doi.org/10.1016/S0166-218X(01)00314-6
4. Cimikowski, R.: An analysis of some linear graph layout heuristics. J. Heuristics **12**(3), 143–153 (2006). https://doi.org/10.1007/s10732-006-4294-9
5. Buchheim, C., Zheng, L.: Fixed linear crossing minimization by reduction to the maximum cut problem. In: Chen, D.Z., Lee, D.T. (eds.) COCOON 2006. LNCS, vol. 4112, pp. 507–516. Springer, Heidelberg (2006). https://doi.org/10.1007/11809678_53
6. Cimikowski, R., Mumey, B.: Approximating the fixed linear crossing number. Discrete Appl. Math. **155**(17), 2202–2210 (2007). https://doi.org/10.1016/j.dam.2007.05.009
7. Garey, M.R., Johnson, D.S., Miller, G.L., Papadimitriou, C.H.: The complexity of coloring circular arcs and chords. SIAM J. Algebr. Discrete Methods **1**(2), 216–227 (1980). https://doi.org/10.1137/0601025

8. Unger, W.: The complexity of colouring circle graphs. In: Finkel, A., Jantzen, M. (eds.) STACS 1992. LNCS, vol. 577, pp. 389–400. Springer, Heidelberg (1992). https://doi.org/10.1007/3-540-55210-3_199

9. Bhore, S., Ganian, R., Montecchiani, F., Nöllenburg, M.: Parameterized algorithms for book embedding problems. In: Archambault, D., Tóth, C.D. (eds.) GD 2019. LNCS, vol. 11904, pp. 365–378. Springer, Cham (2019). https://doi.org/10.1007/978-3-030-35802-0_28

10. Bhore, S., Ganian, R., Montecchiani, F., Nöllenburg, M.: Parameterized algorithms for book embedding problems. J. Graph Algorithms Appl. (2020). https://doi.org/10.7155/jgaa.00526

11. Liu, Y., Chen, J., Huang, J.: Fixed-order book thickness with respect to the vertex-cover number: new observations and further analysis. In: Chen, J., Feng, Q., Xu, J. (eds.) TAMC 2020. LNCS, vol. 12337, pp. 414–425. Springer, Cham (2020). https://doi.org/10.1007/978-3-030-59267-7_35

12. Liu, Y., Chen, J., Huang, J., Wang, J.: On fixed-order book thickness parameterized by the pathwidth of the vertex ordering. In: Zhang, Z., Li, W., Du, D.-Z. (eds.) AAIM 2020. LNCS, vol. 12290, pp. 225–237. Springer, Cham (2020). https://doi.org/10.1007/978-3-030-57602-8_21

13. Bannister, M.J., Eppstein, D., Simons, J.A.: Fixed parameter tractability of crossing minimization of almost-trees. In: Wismath, S., Wolff, A. (eds.) GD 2013. LNCS, vol. 8242, pp. 340–351. Springer, Cham (2013). https://doi.org/10.1007/978-3-319-03841-4_30

14. Bannister, M.J., Eppstein, D.: Crossing minimization for 1-page and 2-page drawings of graphs with bounded treewidth. In: Duncan, C., Symvonis, A. (eds.) GD 2014. LNCS, vol. 8871, pp. 210–221. Springer, Heidelberg (2014). https://doi.org/10.1007/978-3-662-45803-7_18

15. Grigoriev, A., Bodlaender, H.L.: Algorithms for graphs embeddable with few crossings per edge. Algorithmica 49(1), 1–11 (2007). https://doi.org/10.1007/s00453-007-0010-x

16. Di Giacomo, E., Didimo, W., Liotta, G., Montecchiani, F.: Area requirement of graph drawings with few crossing per edge. Comput. Geometry 46(8), 909–916 (2013). https://doi.org/10.1016/j.comgeo.2013.03.001

17. Binucci, C., Di Giacomoa, E., Hossainb, M.I., Liotta, G.: 1-page and 2-page drawings with bounded number of crossings per edge. Eur. J. Comb. 68, 24–37 (2018). https://doi.org/10.1007/978-3-319-29516-9_4

18. Angelini, P., Bekos, M.A., Kaufmann, M., Montecchianib, F.: On 3D visibility representations of graphs with few crossings per edge. Theor. Comput. Sci. (2019). https://doi.org/10.1016/j.tcs.2019.03.029

19. Downey, R.G., Fellows, M.R.: Fundamentals of Parameterized Complexity. TCS. Springer, London (2013). https://doi.org/10.1007/978-1-4471-5559-1

20. Chen, J., Kanj, I.A., Xia, G.: Improved upper bounds for vertex cover. Theor. Comput. Sci. 411(40–42), 3736–3756 (2010). https://doi.org/10.1016/j.tcs.2010.06.026

21. Kinnersley, N.G.: The vertex separation number of a graph equals its pathwidth. Inf. Process. Lett. 42(6), 345–350 (1992). https://doi.org/10.1016/0020-0190(92)90234-M

Fractional Maker-Breaker Resolving Game

Eunjeong Yi[✉]

Texas A&M University at Galveston, Galveston, TX 77553, USA
yie@tamug.edu

Abstract. Let G be a graph with vertex set $V(G)$, and let $d(u, w)$ denote the length of a $u - w$ geodesic in G. For two distinct $x, y \in V(G)$, let $R\{x, y\} = \{z \in V(G) : d(x, z) \neq d(y, z)\}$. For a function g defined on $V(G)$ and for $U \subseteq V(G)$, let $g(U) = \sum_{s \in U} g(s)$. A real-valued function $g : V(G) \to [0, 1]$ is a resolving function of G if $g(R\{x, y\}) \geq 1$ for any distinct $x, y \in V(G)$. In this paper, we introduce the fractional Maker-Breaker resolving game (FMBRG). The game is played on a graph G by Resolver and Spoiler (denoted by R^* and S^*, respectively) who alternately assigns non-negative real values on $V(G)$ such that its sum is at most one on each turn. Moreover, the total value assigned, by R^* and S^*, on each vertex over time cannot exceed one. R^* wins if the total values assigned on $V(G)$ by R^*, after finitely many turns, form a resolving function of G, whereas S^* wins if R^* fails to assign values on $V(G)$ to form a resolving function of G. We obtain some general results on the outcome of the FMBRG and determine the outcome of the FMBRG for some graph classes.

Keywords: (fractional) Maker-Breaker resolving game · Resolving function · (fractional) metric dimension · Twin equivalence class

1 Introduction

Let G be a finite, simple, undirected, and connected graph with vertex set $V(G)$ and edge set $E(G)$. For two distinct $x, y \in V(G)$, let $R\{x, y\} = \{z \in V(G) : d(x, z) \neq d(y, z)\}$, where $d(u, w)$ denotes the length of a shortest $u - w$ path in G. A subset $W \subseteq V(G)$ is a *resolving set* of G if $|W \cap R\{x, y\}| \geq 1$ for any pair of distinct $x, y \in V(G)$, and the *metric dimension*, $\dim(G)$, of G is the minimum cardinality over all resolving sets of G. For a function g defined on $V(G)$ and for $U \subseteq V(G)$, let $g(U) = \sum_{s \in U} g(s)$. A real valued-function $g : V(G) \to [0, 1]$ is a *resolving function* of G if $g(R\{x, y\}) \geq 1$ for any distinct $x, y \in V(G)$. The *fractional metric dimension*, $\dim_f(G)$, of G is $\min\{g(V(G)) : g \text{ is a resolving function of } G\}$. Note that $\dim_f(G)$ reduces to $\dim(G)$ if the codomain of resolving functions is restricted to $\{0, 1\}$. The concept of metric dimension was introduced by Slater [20] and by Harary and Melter [10].

© Springer Nature Switzerland AG 2020
W. Wu and Z. Zhang (Eds.): COCOA 2020, LNCS 12577, pp. 577–593, 2020.
https://doi.org/10.1007/978-3-030-64843-5_39

The concept of fractional metric dimension was introduced by Currie and Oeller-mann [6], and Arumugam and Matthew [1] officially initiated the study of fractional metric dimension. Applications of metric dimension can be found in robot navigation [16], network discovery and verification [4], sonar [20], combinatorial optimization [19] and chemistry [17]. It was noted in [9] that determining the metric dimension of a general graph is an NP-hard problem.

Games played on graphs have drawn interests recently. Examples of two player games are cop and robber game [18], domination game [5], Maker-Breaker domination game [7] and Maker-Breaker resolving game [14], to name a few. Erdös and Selfridge [8] introduced the Maker-Breaker game that is played on an arbitrary hypergraph $H = (V, E)$ by two players, Maker and Breaker, who alternately select a vertex from V that was not yet chosen in the course of the game. Maker wins the game if he can select all the vertices of one of the hyperedges from E, wheras Breaker wins if she is able to prevent Maker from doing so. For further information on these games, we refer to [3] and [11].

The Maker-Breaker resolving game (MBRG for short) was introduced in [14]. Following [14], the MBRG is played on a graph G by two players, Resolver and Spoiler, denoted by R^* and S^*, respectively. R^* and S^* alternately select (without missing their turn) a vertex of G that was not yet chosen in the course of the game. R-game (S-game, respectively) denotes the game for which R^* (S^*, respectively) plays first. R^* wins the MBRG if the vertices selected by him, after finitely many turns, form a resolving set of G; if R^* fails to form a resolving set of G over the course of the MBRG, then S^* wins. The outcome of the MBRG on a graph G is denoted by $o(G)$, and it was shown in [14] that there are three realizable outcomes as follows: (i) $o(G) = \mathcal{R}$ if R^* has a winning strategy in the R-game and the S-game; (ii) $o(G) = \mathcal{S}$ if S^* has a winning strategy in the R-game and the S-game; (iii) $o(G) = \mathcal{N}$ if the first player has a winning strategy. The authors of [14] also studied the minimum number of moves needed for R^* (S^*, respectively) to win the MBRG provided R^* (S^*, respectively) has a winning strategy.

In this paper, we introduce the fractional Maker-Breaker resolving game (FMBRG) played on a graph by two players, Resolver and Spoiler. Following [14], let R^* (S^*, respectively) denote the Resolver (the Spoiler, respectively), and let R-game (S-game, respectively) denote the FMBRG for which R^* (S^*, respectively) plays first. Suppose $k \in \mathbb{Z}^+$ ($k' \in \mathbb{Z}^+$, respectively) is the maximum integer for which R^* (S^*, respectively) assigns a positive real value on at least one vertex of G on his k-th move (her k'-th move, respectively). We denote by $h_{R,i}$ ($h_{S,i}$, respectively) a function defined on $V(G)$ by R^* (S^*, respectively) on his/her i-th turn. We define the following terminology and notation.

- For $i \in \{1, 2, \ldots, k\}$, a function $h_{R,i} : V(G) \rightarrow [0,1]$ satisfying $0 < h_{R,i}(V(G)) \leq 1$ is called the *Resolver's i-th distribution function*, and a function $\Gamma_{R,i} : V(G) \rightarrow [0,1]$ defined by $\Gamma_{R,i}(u) = \sum_{\alpha=1}^{i} h_{R,\alpha}(u)$, for each $u \in V(G)$, is called the *Resolver's i-th cumulative distribution function*.
- For $j \in \{1, 2, \ldots, k'\}$, a function $h_{S,j} : V(G) \rightarrow [0,1]$ satisfying $0 < h_{S,j}(V(G)) \leq 1$ is called the *Spoiler's j-th distribution function*, and a func-

tion $\Gamma_{S,j} : V(G) \rightarrow [0,1]$ defined by $\Gamma_{S,j}(u) = \sum_{\beta=1}^{j} h_{S,\beta}(u)$, for each $u \in V(G)$, is called the *Spoiler's j-th cumulative distribution function*.

Note that, for each $u \in V(G)$, $0 \leq \Gamma_{R,k}(u) + \Gamma_{S,k'}(u) \leq 1$. In the FMBRG, R^* wins if $\Gamma_{R,t}$ forms a resolving function of G for some $t \in \mathbb{Z}^+$, and S^* wins if $\Gamma_{R,t'}$ fails to form a resolving function of G for any $t' \in \mathbb{Z}^+$. The *outcome* of the FMBRG on a graph G is denoted by $o_f(G)$, and there are three possible outcomes as follows: (i) $o_f(G) = \mathcal{R}$ if R^* has a winning strategy in the R-game and the S-game; (ii) $o_f(G) = \mathcal{S}$ if S^* has a winning strategy in the R-game and the S-game; (iii) $o_f(G) = \mathcal{N}$ if the first player has a winning strategy. We note that it is impossible for the second player to win the FMBRG given each player has an optimal strategy (see [14] on "No-Skip Lemma" for the MBRG).

This paper is organized as follows. In Sect. 2, we recall some known results on the outcome of the MBRG. In Sect. 3, we obtain some general results on the outcome of the FMBRG. We show how twin equivalence classes of a graph play a role on the outcome of the FMBRG. We also examine how the conditions of resolving functions of a graph affect the outcome of the FMBRG. In Sect. 4, we determine $o_f(G)$ for some graph classes G.

2   Some Known Results on the MBRG

In this section, we recall some known results on the MBRG. We first recall two important concepts, twin equivalence class and pairing resolving set, that are useful in determining the outcome of the MBRG.

For $v \in V(G)$, $N(v) = \{u \in V(G) : uv \in E(G)\}$. Two vertices u and w are *twins* if $N(u) - \{w\} = N(w) - \{u\}$; notice that a vertex is its own twin. Hernando et al. [12] observed that the twin relation is an equivalence relation and that an equivalence under it, called a *twin equivalence class*, induces a clique or an independent set.

Observation 1 [12]. *Let W be a resolving set of G. If x and y are distinct twin vertices in G, then $W \cap \{x, y\} \neq \emptyset$.*

Proposition 1 [14]. *Let G be a connected graph of order at least 4.*

(a) *If G has a twin equivalence class of cardinality at least 4, then $o(G) = \mathcal{S}$.*
(b) *If G has two distinct twin equivalence classes of cardinality at least 3, then $o(G) = \mathcal{S}$.*

Next, we recall the relation between a pairing resolving set and the outcome of the MBRG.

Definition 1 [14]. *Let $A = \{\{u_1, w_1\}, \ldots, \{u_k, w_k\}\}$ be a set of 2-subsets of $V(G)$ such that $|\cup_{i=1}^{k} \{u_i, w_i\}| = 2k$. If every set $\{x_1, \ldots, x_k\}$, where $x_i \in \{u_i, w_i\}$ for each $i \in \{1, 2, \ldots, k\}$, is a resolving set of G, then A is called a pairing resolving set of G.*

Proposition 2 [14]. *If a graph G admits a pairing resolving set, then $o(G) = \mathcal{R}$.*

Note that Proposition 2 implies the following.

Corollary 1. *If a connected graph G has $k \geq 1$ twin equivalence classes of cardinality 2 with $\dim(G) = k$, then $o(G) = \mathcal{R}$.*

We state the following result where the first player wins on the MBRG.

Proposition 3. *If a connected graph G has $k \geq 0$ twin equivalence class(es) of cardinality 2 and exactly one twin equivalence class of cardinality 3 with $\dim(G) = k + 2$, then $o(G) = \mathcal{N}$.*

Proof. Let W be any minimum resolving set of G, and let $E = \{x, y, z\}$ be a twin equivalence class with $|E| = 3$; then $|W \cap E| = 2$. If $k \geq 1$, let $Q_i = \{u_i, v_i\}$ be a twin equivalence class with $|Q_i| = 2$ for each $i \in \{1, 2, \ldots, k\}$; then $|W \cap Q_i| = 1$ for each $i \in \{1, 2, \ldots, k\}$. In the S-game, S^* can select two vertices of E after her second move, and R^* can select at most one vertex of E; thus, S^* wins the S-game. So, we consider the R-game. If $k = 0$, then R^* can select two vertices of E after his second move. If $k \geq 1$, then R^* can select two vertices of E after his second move, and one vertex in each Q_i thereafter. In each case, R^* wins the R-game. Thus, $o(G) = \mathcal{N}$. □

3   General Results on the Outcome of the FMBRG

In this section, we obtain some general results on the FMBRG. We show how twin equivalence classes of a graph affect the outcome of the FMBRG. We also examine how the conditions of resolving functions of a graph affect the outcome of the FMBRG. Moreover, we show the existence of graphs G satisfying the following: (i) $o(G) = \mathcal{R}$ and $o_f(G) \in \{\mathcal{R}, \mathcal{N}, \mathcal{S}\}$; (ii) $o(G) = \mathcal{N}$ and $o_f(G) \in \{\mathcal{N}, \mathcal{S}\}$; (iii) $o(G) = \mathcal{S} = o_f(G)$.

Observation 2 [21]. *Let $g : V(G) \to [0, 1]$ be a resolving function of G. If u and w are distinct twin vertices of G, then $g(u) + g(w) \geq 1$.*

Proposition 4. *If a connected graph G has a twin equivalence class of cardinality $k \geq 2$ with $\dim_f(G) = \frac{k}{2}$, then*

$$o_f(G) = \begin{cases} \mathcal{R} & \text{if } k = 2, \\ \mathcal{N} & \text{if } k = 3, \\ \mathcal{S} & \text{if } k \geq 4. \end{cases} \tag{1}$$

Proof. Let $Q = \{u_1, u_2, \ldots, u_k\} \subseteq V(G)$ be a twin equivalence class of G with $|Q| = k$, where $k \geq 2$. By Observation 2, for any resolving function g of G, $g(u_i) + g(u_j) \geq 1$ for any distinct $i, j \in \{1, 2, \ldots, k\}$. We consider three cases.

Case 1: $k = 2$. In the R-game, R^* can assign values on $V(G)$ such that $h_{R,1}(u_1) = h_{R,1}(u_2) = \frac{1}{2}$. In the S-game, suppose that $h_{S,1}(u_1) = x_1$ and $h_{S,1}(u_2) = x_2$, where $0 \leq x_1, x_2 \leq 1$ and $x_1 + x_2 \leq 1$; then R^* can assign values on $V(G)$ such that $h_{R,1}(u_1) = 1 - x_1$ and $h_{R,1}(u_2) = x_1$. In each case, $\Gamma_{R,1}$ forms a resolving function of G; thus $o_f(G) = \mathcal{R}$.

Case 2: $k = 3$. First, we consider the R-game. Suppose R^* assigns values on $V(G)$ such that $h_{R,1}(u_1) = 1$. Suppose S^* assigns values on $V(G)$ such that $h_{S,1}(u_2) = y_1$, $h_{S,1}(u_3) = y_2$, where $0 \le y_1, y_2 \le 1$ and $y_1 + y_2 \le 1$. Then R^* can assign values on $V(G)$ such that $h_{R,2}(u_2) = 1 - y_1$ and $h_{R,2}(u_3) = y_1$; note that $\Gamma_{R,2}(u_1) = 1$, $\Gamma_{R,2}(u_2) = 1 - y_1$ and $\Gamma_{R,2}(u_3) = y_1$. Since $\Gamma_{R,2}$ forms a resolving function of G, R^* wins the R-game.

Second, we consider the S-game. Suppose $h_{S,1}(u_1) = 1$. In order for R^* to win, Observation 2 implies that R^* must satisfy $\Gamma_{R,t}(u_2) = \Gamma_{R,t}(u_3) = 1$ for some t. If R^* assigns values on $V(G)$ such that $h_{R,1}(u_2) = z_1$ and $h_{R,1}(u_3) = z_2$ with $z_1 + z_2 \le 1$, then S^* can assign values on $V(G)$ such that $h_{S,2}(u_2) = 1 - z_1$ and $h_{S,2}(u_3) = z_1$. Since $\Gamma_{S,2}(u_1) + \Gamma_{S,2}(u_2) + \Gamma_{S,2}(u_3) = 2$, $\Gamma_{R,t}$ fails to form a resolving function of G for any t. So, S^* wins the S-game.

Therefore, $o_f(G) = \mathcal{N}$.

Case 3: $k \ge 4$. First, we consider the R-game. Suppose R^* assigns values on $V(G)$ such that $h_{R,1}(u_i) = x_i$ for each $i \in \{1, 2, \ldots, k\}$, where $0 \le x_i \le 1$ and $\sum_{i=1}^{k} x_i \le 1$. By relabeling the vertices of G if necessary, let $x_1 \le x_2 \le \ldots \le x_k$; then $x_1 \le \frac{1}{k} \le \frac{1}{4}$. Then S^* can assign values on $V(G)$ such that $h_{S,1}(u_1) = 1 - x_1 \ge \frac{3}{4}$ and $h_{S,1}(u_2) = x_1$. We note that, in order for R^* to win, Observation 2 implies that R^* must satisfy $\Gamma_{R,t}(u_i) \ge 1 - x_1 \ge \frac{3}{4}$ for each $i \in \{2, 3 \ldots, k\}$. Suppose R^* assigns values on $V(G)$ such that $h_{R,2}(u_2) = 1 - x_1 - x_2$ and $h_{R,2}(u_j) = y_j$ for each $j \in \{3, 4, \ldots, k\}$ with $\sum_{j=3}^{k} y_j = x_1 + x_2$; then $\Gamma_{R,2}(u_1) = x_1$, $\Gamma_{R,2}(u_2) = 1 - x_1$ and $\sum_{j=3}^{k} \Gamma_{R,2}(u_j) \le 1$, and thus $\Gamma_{R,2}(u_\alpha) \le \frac{1}{2}$ for some $\alpha \in \{3, 4, \ldots, k\}$. Then S^* can assign values on $V(G)$ such that $h_{S,2}(u_\alpha) = \frac{1}{2}$ for some $\alpha \in \{3, 4, \ldots, k\}$, which implies $\Gamma_{R,t}(u_\alpha) \le \frac{1}{2}$ for any t. Since $\Gamma_{R,t}$ fails to form a resolving function of G for any t, S^* wins the R-game.

Second, we consider the S-game. Suppose $h_{S,1}(u_1) = 1$. In order for R^* to win, Observation 2 implies that R^* must satisfy $\Gamma_{R,t}(u_i) = 1$ for each $i \in \{2, 3, \ldots, k\}$. Suppose R^* assigns values on $V(G)$ such that $h_{R,1}(u_i) = z_i$ for each $i \in \{2, 3, \ldots, k\}$, where $0 \le z_i \le 1$ and $\sum_{i=2}^{k} z_i \le 1$; further, we may assume that $z_2 \le z_3 \le \ldots \le z_k$. Then S^* can assign values on $V(G)$ such that $h_{S,2}(u_2) = 1 - z_2 > 0$; thus $\Gamma_{S,2}(u_1) + \Gamma_{S,2}(u_2) = 2 - z_2 > 1$. Since $\Gamma_{R,t}(u_1) + \Gamma_{R,t}(u_2) < 1$ for any t, $\Gamma_{R,t}$ fails to be a resolving function of G for any t; thus, S^* wins the S-game. $\square$

Proposition 5. *If a connected graph G has $k \ge 1$ twin equivalence class(es) of cardinality 3 with $\dim_f(G) = \frac{3}{2}k$, then*

$$o_f(G) = \begin{cases} \mathcal{N} & \text{if } k = 1, \\ \mathcal{S} & \text{if } k \ge 2. \end{cases} \tag{2}$$

Proof. For each $i \in \{1, 2, \ldots, k\}$, let $Q_i = \{u_{i,1}, u_{i,2}, u_{i,3}\}$ be a twin equivalence class of G with $|Q_i| = 3$, where $k \ge 1$; then $|\cup_{i=1}^{k} Q_i| = 3k$ and $Q_i \cap Q_j = \emptyset$ for distinct $i, j \in \{1, 2, \ldots, k\}$. Note that, for each $i \in \{1, 2, \ldots, k\}$ and for any distinct $j_1, j_2 \in \{1, 2, 3\}$, $g(u_{i,j_1}) + g(u_{i,j_2}) \ge 1$ for any resolving function g of G by Observation 2. If $k = 1$, then $o_f(G) = \mathcal{N}$ by Proposition 4. So, let $k \ge 2$.

First, we consider the R-game. Suppose R^* assigns values on $V(G)$ such that $h_{R,1}(V(Q_i)) = x_i$, for each $i \in \{1, 2, \ldots, k\}$, with $\sum_{i=1}^{k} x_i \leq 1$. By relabeling the vertices of G if necessary, let $x_1 \leq x_2 \leq \ldots \leq x_k$; further, let $h_{R,1}(u_{1,j}) = z_j$ for each $j \in \{1, 2, 3\}$, such that $z_1 \leq z_2 \leq z_3$. Note that $x_1 = z_1 + z_2 + z_3 \leq \frac{1}{k} \leq \frac{1}{2}$ and $z_1 \leq \frac{1}{3k} \leq \frac{1}{6}$. Then S^* can assign values on $V(G)$ such that $h_{S,1}(u_{1,1}) = 1 - z_1 \geq \frac{5}{6}$ and $h_{S,1}(u_{1,2}) = z_1$. We note that, in order for R^* to win the game, Observation 2 implies that R^* must assign values on $V(G)$ such that $\Gamma_{R,t}(u_{1,2}) \geq 1 - z_1 \geq \frac{5}{6}$ and $\Gamma_{R,t}(u_{1,3}) \geq 1 - z_1 \geq \frac{5}{6}$ for some t. Suppose R^* assigns values on $V(G)$ such that $h_{R,2}(u_{1,2}) = 1 - (z_1 + z_2)$ and $h_{R,2}(u_{1,3}) = z_1 + z_2$; then $\Gamma_{R,2}(u_{1,2}) = 1 - z_1$ and $\Gamma_{R,2}(u_{1,3}) = z_1 + z_2 + z_3 = x_1 \leq \frac{1}{2}$. So, S^* can assign values on $V(G)$ such that $h_{S,2}(u_{1,3}) = 1 - x_1 \geq \frac{1}{2}$, and thus $\Gamma_{S,2}(u_{1,3}) = 1 - x_1 \geq \frac{1}{2}$. Since $\Gamma_{R,t}(u_{1,3}) \leq \frac{1}{2} < 1 - z_1$ for any t, S^* wins the R-game.

Second, we consider the S-game. Suppose $h_{S,1}(u_{1,1}) = 1$. We note that, in order for R^* to win, Observation 2 implies that R^* must assign values on $V(G)$ such that $\Gamma_{R,t}(u_{1,2}) = \Gamma_{R,t}(u_{1,3}) = 1$ for some t. Suppose R^* assigns values on $V(G)$ such that $h_{R,1}(u_{1,2}) = y_1$ and $h_{R,1}(u_{1,3}) = y_2$, where $0 \leq y_1, y_2 \leq 1$ and $y_1 + y_2 \leq 1$. Then S^* can assign values on $V(G)$ such that $h_{S,2}(u_{1,2}) = 1 - y_1$ and $h_{S,2}(u_{1,3}) = y_1$; note that $\Gamma_{S,2}(u_{1,2}) > 0$ or $\Gamma_{S,2}(u_{1,3}) > 0$. Since $\Gamma_{R,t}$ fails to form a resolving function of G for any t, S^* wins the S-game. $\square$

Proposition 6. *If a connected graph G has $k \geq 1$ twin equivalence class(es) of cardinality 2 with $\dim_f(G) = k$, then*

$$o_f(G) = \begin{cases} \mathcal{R} & \text{if } k = 1, \\ \mathcal{N} & \text{if } k = 2, \\ \mathcal{S} & \text{if } k \geq 3. \end{cases} \tag{3}$$

Proof. For each $i \in \{1, 2, \ldots, k\}$, let $Q_i = \{u_i, w_i\}$ be a twin equivalence class of G with $|Q_i| = 2$, where $k \geq 1$; then $|\cup_{i=1}^{k} Q_i| = 2k$ and $Q_i \cap Q_j = \emptyset$ for distinct $i, j \in \{1, 2, \ldots, k\}$. Note that, for each $i \in \{1, 2, \ldots, k\}$ and for any resolving function g of G, $g(u_i) + g(w_i) \geq 1$ by Observation 2. We consider three cases.

Case 1: $k = 1$. In this case, $o_f(G) = \mathcal{R}$ by Proposition 4.

Case 2: $k = 2$. First, we consider the R-game. Suppose R^* assigns values on $V(G)$ such that $h_{R,1}(u_1) = h_{R,1}(w_1) = \frac{1}{2}$. Suppose S^* assigns values on $V(G)$ such that $h_{S,1}(u_2) = x_1$ and $h_{S,1}(w_2) = x_2$, where $0 \leq x_1, x_2 \leq 1$ and $x_1 + x_2 \leq 1$. Then R^* can assign values on $V(G)$ such that $h_{R,2}(u_2) = 1 - x_1$ and $h_{R,2}(w_2) = x_1$. Since $\Gamma_{R,2}$ forms a resolving function of G, R^* wins the R-game.

Second, we consider the S-game. Suppose $h_{S,1}(u_1) = h_{S,1}(u_2) = \frac{1}{2}$. Note that, after R^* assigns values on $V(G)$ on his first move, we have either $h_{R,1}(u_1) + h_{R,1}(w_1) \leq \frac{1}{2}$ or $h_{R,1}(u_2) + h_{R,1}(w_2) \leq \frac{1}{2}$, say the former. Then S^* can assign values on $V(G)$ such that $h_{S,2}(u_1) + h_{S,2}(w_1) = 1$; thus $\Gamma_{S,2}(u_1) + \Gamma_{S,2}(w_1) = \frac{3}{2}$, which implies $\Gamma_{R,t}(u_1) + \Gamma_{R,t}(w_1) \leq \frac{1}{2}$ for any t. Since $\Gamma_{R,t}$ fails to be a resolving function of G for any t, S^* wins the S-game.

Therefore, $o_f(G) = \mathcal{N}$.

Case 3: $k \geq 3$. In the S-game, it's easy to see that S^* wins using the argument used for Case 2. So, we consider the R-game. Suppose

$h_{R,1}(u_i) + h_{R,1}(w_i) = z_i$, for each $i \in \{1, 2, \ldots, k\}$, such that $\sum_{i=1}^{k} z_i \leq 1$. By relabeling the vertices of G if necessary, assume that $z_1 \leq z_2 \leq \ldots \leq z_k$; then $z_1 \leq \frac{1}{k} \leq \frac{1}{3}$. So, S^* can assign values on $V(G)$ such that $h_{S,1}(u_1) + h_{S,1}(w_1) = \frac{1}{2} = h_{S,1}(u_2) + h_{S,1}(w_2)$. After R^* assigns values on $V(G)$ on his second move, we have $h_{R,2}(u_1) + h_{R,2}(w_1) \leq \frac{1}{2}$ or $h_{R,2}(u_2) + h_{R,2}(w_2) \leq \frac{1}{2}$. First, suppose $h_{R,2}(u_1) + h_{R,2}(w_1) \leq \frac{1}{2}$; note that $\Gamma_{R,2}(u_1) + \Gamma_{R,2}(w_1) \leq z_1 + \frac{1}{2} \leq \frac{1}{3} + \frac{1}{2} = \frac{5}{6}$. Then S^* can assign values on $V(G)$ such that $h_{S,2}(u_1) + h_{S,2}(w_1) = \frac{2}{3}$; thus, $\Gamma_{S,2}(u_1) + \Gamma_{S,2}(w_1) = \frac{1}{2} + \frac{2}{3} = \frac{7}{6} > 1$, which implies that $\Gamma_{R,t}$ fails to form a resolving function of G for any t. Second, suppose $h_{R,2}(u_2) + h_{R,2}(w_2) \leq \frac{1}{2}$. If $z_1 \neq 0$, then $z_2 < \frac{1}{k-1} \leq \frac{1}{2}$ (i.e., $\Gamma_{R,2}(u_2) + \Gamma_{R,2}(w_2) \leq z_2 + \frac{1}{2} < 1$) and S^* can assign values on $V(G)$, on her second move, such that $h_{S,2}(u_2) + h_{S,2}(w_2) > \frac{1}{2}$, which implies $\Gamma_{S,2}(u_2) + \Gamma_{S,2}(w_2) > 1$. If $z_1 = 0$, then $z_2 \leq \frac{1}{k-1} \leq \frac{1}{2}$ and S^* can assign values on $V(G)$, on her second move, such that $\Gamma_{S,2}(u_1) + \Gamma_{S,2}(w_1) > 1$ (if $h_{R,2}(u_2) + h_{R,2}(w_2) \neq 0$) or $\Gamma_{S,2}(u_2) + \Gamma_{S,2}(w_2) > 1$ (if $h_{R,2}(u_2) + h_{R,2}(w_2) = 0$). In each case, S^* wins the R-game. So, $o_f(G) = \mathcal{S}$. $\qquad\square$

Propositions 4, 5 and 6 imply the following.

Corollary 2. *Let G be a connected graph.*

(a) *If G has a twin equivalence class of cardinality at least 4, then $o_f(G) = \mathcal{S}$.*
(b) *If G has at least two distinct twin equivalence classes of cardinality 3, then $o_f(G) = \mathcal{S}$.*
(c) *If G has at least three distinct twin equivalence classes of cardinality 2, then $o_f(G) = \mathcal{S}$.*

Next, we consider the outcome of the FMBRG on a graph that has twin equivalence classes of cardinalities 2 and 3 simultaneously.

Proposition 7. *If a connected graph G has $k_1 \geq 1$ twin equivalence class(es) of cardinality 3 and $k_2 \geq 1$ twin equivalence class(es) of cardinality 2 with $\dim_f(G) = \frac{3}{2}k_1 + k_2$, then $o_f(G) = \mathcal{S}$.*

Proof. We consider two cases.

Case 1: $k_1 \geq 2$ and $k_2 \geq 1$. By Corollary 2(b), $o_f(G) = \mathcal{S}$.

Case 2: $k_1 = 1$ and $k_2 \geq 1$. Let $U = \{u_1, u_2, u_3\}$ be a twin equivalence class of G with $|U| = 3$, and let $W_i = \{w_{i,1}, w_{i,2}\}$ be a twin equivalence class of G with $|W_i| = 2$ for each $i \in \{1, 2, \ldots, k_2\}$. For each $i \in \{1, 2, \ldots, k_2\}$, $g(W_i) \geq 1$ for any resolving function g of G by Observation 2.

First, we consider the S-game. Suppose $h_{S,1}(u_1) = 1$. In order for R^* to win, Observation 2 implies that R^* must satisfy $\Gamma_{R,t}(u_2) = \Gamma_{R,t}(u_3) = 1$ for some t. Suppose R^* assigns values on $V(G)$ such that $h_{R,1}(u_2) = x_1$ and $h_{R,1}(u_3) = x_2$, where $0 \leq x_1, x_2 \leq 1$ and $x_1 + x_2 \leq 1$. Then S^* can assign values on $V(G)$ such that $h_{S,2}(u_2) = 1 - x_1$ and $h_{S,2}(u_3) = x_1$; then $\Gamma_{S,2}(u_2) > 0$ or $\Gamma_{S,2}(u_3) > 0$, and thus $\Gamma_{R,t}(u_2) < 1$ or $\Gamma_{R,t}(u_3) < 1$ for any t. Thus, S^* wins the S-game.

Second, we consider the R-game; we show that S^* wins the R-game. We consider three subcases.

Subcase 2.1: $k_2 = 1$. Suppose R^* assigns values on $V(G)$ such that $h_{R,1}(U) = a$ and $h_{R,1}(W_1) = b$ with $a + b = 1$; further, let $h_{R,1}(u_i) = \alpha_i$ for each $i \in \{1, 2, 3\}$ with $\alpha_1 \leq \alpha_2 \leq \alpha_3$. First, suppose $a = \alpha_1 + \alpha_2 + \alpha_3 < \frac{3}{4}$; then $\alpha_1 < \frac{1}{4}$. Then S^* can assign values on $V(G)$ such that $h_{S,1}(u_1) = 1 - \alpha_1 > \frac{3}{4}$ and $h_{S,1}(u_2) = \alpha_1$. In order for R^* to win the game, Observation 2 implies that R^* must satisfy $\Gamma_{R,t}(u_2) \geq 1 - \alpha_1 > \frac{3}{4}$ and $\Gamma_{R,t}(u_3) \geq 1 - \alpha_1 > \frac{3}{4}$ for some t. Suppose R^* assigns values on $V(G)$, on his second turn, such that $h_{R,2}(u_2) = 1 - (\alpha_1 + \alpha_2)$ and $h_{R,2}(u_3) = \alpha_1 + \alpha_2$; then $\Gamma_{R,2}(u_2) = 1 - \alpha_1$ and $\Gamma_{R,2}(u_3) = \alpha_1 + \alpha_2 + \alpha_3 = a < \frac{3}{4}$. So, S^* can assign values on $V(G)$, on her second turn, such that $h_{S,2}(u_3) = 1 - a > \frac{1}{4}$, which implies $\Gamma_{R,t}(u_3) < \frac{3}{4}$ for any t. Thus, S^* wins the R-game when $a < \frac{3}{4}$. Second, suppose $a \geq \frac{3}{4}$; then $b \leq \frac{1}{4}$. Suppose S^* assigns values on $V(G)$ such that $h_{S,1}(u_1) = \min\{1 - \alpha_1, \frac{9}{10}\}$ and $h_{S,1}(w_{1,1}) = \max\{\alpha_1, \frac{1}{10}\}$; then $0 < h_{S,1}(u_1) < 1$, $0 < h_{S,1}(w_{1,1}) < 1$, and $h_{S,1}(u_1) + h_{S,1}(w_{1,1}) = 1$. In order for R^* to win the game, Observation 2 implies that R^* must satisfy $\Gamma_{R,t}(u_2) \geq h_{S,1}(u_1)$, $\Gamma_{R,t}(u_3) \geq h_{S,1}(u_1)$, and $\Gamma_{R,t}(W_1) \geq 1$ for some t. We note that R^*, on his second turn, must assign values on $V(G)$ such that $h_{R,2}(W_1) \geq 1 - b$, which implies $\Gamma_{R,2}(W_1) \geq b + (1-b) = 1$; otherwise, S^* can assign values on $V(G)$ such that $h_{S,2}(W_1) > 1 - h_{S,1}(w_{1,1})$, and thus $\Gamma_{S,2}(W_1) > 1$, which implies $\Gamma_{R,t}(W_1) < 1$ for any t. We also note that $h_{R,2}(W_1) \geq 1 - b$ implies $h_{R,2}(u_2) \leq b$; thus $\Gamma_{R,2}(u_2) \leq \alpha_2 + b < h_{S,1}(u_1) = \min\{1 - \alpha_1, \frac{9}{10}\}$ since $\alpha_1 + \alpha_2 + b < \alpha_1 + \alpha_2 + \alpha_3 + b = 1$ and $\alpha_2 + b \leq \frac{1}{2} + \frac{1}{4} = \frac{3}{4} < \frac{9}{10}$. So, S^* can assign values on $V(G)$, on her second turn, such that $h_{S,2}(u_2) = 1 - (\alpha_2 + b)$; then $\Gamma_{R,t}(u_2) \leq \alpha_2 + b < h_{S,1}(u_1)$ for any t. Thus, S^* wins the R-game when $a \geq \frac{3}{4}$.

Subcase 2.2: $k_2 = 2$. Suppose R^* assigns values on $V(G)$ such that $h_{R,1}(U) = a$ and $h_{R,1}(W_i) = b_i$ for each $i \in \{1, 2\}$, where $a + b_1 + b_2 = 1$; further, let $h_{R,1}(u_i) = \alpha_i$ for each $i \in \{1, 2, 3\}$ with $\alpha_1 \leq \alpha_2 \leq \alpha_3$. If $a < \frac{3}{4}$, then S^* wins the R-game by the argument used in the proof for Subcase 2.1 of the current proposition. So, suppose $a \geq \frac{3}{4}$; then $b_1 + b_2 \leq \frac{1}{4}$. Then S^* can assign values on $V(G)$, on her first turn, such that $h_{S,1}(w_{1,1}) = h_{S,1}(w_{2,1}) = \frac{1}{2}$. Since $\Gamma_{R,2}(W_1) + \Gamma_{R,2}(W_2) \leq b_1 + b_2 + 1 \leq \frac{1}{4} + 1 = \frac{5}{4}$, we have either $\Gamma_{R,2}(W_1) \leq \frac{5}{8}$ or $\Gamma_{R,2}(W_2) \leq \frac{5}{8}$, say the former. Then S^* can assign values on $V(G)$, on her second turn, such that $h_{S,2}(W_1) = \frac{7}{8}$; then $\Gamma_{S,2}(W_1) = \frac{1}{2} + \frac{7}{8} = \frac{11}{8} > 1$, which implies $\Gamma_{R,t}(W_1) < 1$ for any t. Since $\Gamma_{R,t}$ fails to form a resolving function of G for any t, S^* wins the R-game when $a \geq \frac{3}{4}$.

Subcase 2.3: $k_2 \geq 3$. By Corollary 2(c), S^* wins the R-game. $\square$

We note that a graph G having a pairing resolving set is guaranteed to have $o(G) = \mathcal{R}$ (see Proposition 2), whereas such a graph G has three possibilities for $o_f(G)$ (see Proposition 6). So, it is worth to explore the conditions of resolving functions of G that guarantee $o_f(G) = \mathcal{R}$.

Proposition 8. *Let a and k be positive integers, and let $b \in (0, 1]$ be a real number. Let G be a connected graph with $V_1, V_2, \ldots, V_k \subseteq V(G)$ such that $|V_i| \geq a$ for each $i \in \{1, 2, \ldots, k\}$ and $V_i \cap V_j = \emptyset$ for any distinct $i, j \in \{1, 2, \ldots, k\}$. Let $g : V(G) \to [0, 1]$ be a function defined by $g(V_i) = b$, for each $i \in \{1, 2, \ldots, k\}$,* with $g(V(G)) = bk$. If $a \geq 2b+1$ and g is a resolving function of G, then $o_f(G) = \mathcal{R}$.

Proof. We provide an algorithmic proof to show that $o_f(G) = \mathcal{R}$. We note that the optimal strategy for R^* is to assign values on $V(G)$ such that $\Gamma_{R,t}(V_i) = b$ for each $i \in \{1, 2, \ldots, k\}$ and for some t.

First, we consider the R-game. Suppose R^* assigns values on $V(G)$ such that $h_{R,1}(V_i) = \frac{b}{k}$, for each $i \in \{1, 2, \ldots, k\}$, with $\Gamma_{R,1}(V(G)) = \sum_{i=1}^{k} h_{R,1}(V_i) = \sum_{i=1}^{k} \frac{b}{k} = b \leq 1$. Let S^* assign values on $V(G)$ such that $h_{S,1}(V_i) = x_{1,i}$ for each $i \in \{1, 2, \ldots, k\}$, where $0 \leq x_{1,i} \leq 1$ and $\sum_{i=1}^{k} x_{1,i} \leq 1$. Then R^* can assign values on $V(G)$, on his second turn, such that $h_{R,2}(V_i) = \min\{x_{1,i}, b - \frac{b}{k}\}$ for each $i \in \{1, 2, \ldots, k\}$. We note the following: (i) $h_{R,2}(V(G)) \leq h_{S,1}(V(G)) \leq 1$; (ii) if $x_{1,i} < b - \frac{b}{k}$, then $\Gamma_{S,1}(V_i) < \Gamma_{R,2}(V_i) < b$; (iii) if $x_{1,i} \geq b - \frac{b}{k}$, then $\Gamma_{R,2}(V_i) = b$. We also note that, if $\Gamma_{R,2}(V_j) = b$ for some $j \in \{1, 2, \ldots, k\}$, then it is in the interest of S^* to assign the value 0 for V_j on all her future turns. Suppose S^*, on her second turn, assigns values on $V(G)$ such that $h_{S,2}(V_i) = x_{2,i}$ for each $i \in \{1, 2, \ldots, k\}$, where $0 \leq x_{2,i} \leq 1$ and $\sum_{i=1}^{k} x_{2,i} \leq 1$; here, $x_{2,j} = 0$ if $\Gamma_{R,2}(V_j) = b$. Then R^* can assign values on $V(G)$, on his third turn, as follows: (i) if $\Gamma_{R,2}(V_i) = b$, then let $h_{R,3}(V_i) = 0$; (ii) if $\Gamma_{R,2}(V_i) < b$ (i.e., $x_{1,i} < b - \frac{b}{k}$), then let $h_{R,3}(V_i) = \min\{x_{2,i}, b - \frac{b}{k} - x_{1,i}\}$. We note the following: (i) $h_{R,3}(V(G)) \leq h_{S,2}(V(G)) \leq 1$; (ii) if $x_{2,i} < b - \frac{b}{k} - x_{1,i}$, then $\Gamma_{S,2}(V_i) < \Gamma_{R,3}(V_i) < b$ and thus $\Gamma_{S,2}(V_i) + \Gamma_{R,3}(V_i) < 2b$; (iii) if $x_{2,i} \geq b - \frac{b}{k} - x_{1,i}$, then $\Gamma_{R,3}(V_i) = b$. For $r \geq 3$, let $h_{S,r}(V_i) = x_{r,i}$, where $0 \leq x_{r,i} \leq 1$ and $\sum_{i=1}^{k} x_{r,i} \leq 1$, and let $h_{R,r+1}(V_i) = \min\{x_{r,i}, b - \frac{b}{k} - \sum_{j=1}^{r-1} x_{j,i}\}$ whenever $\Gamma_{R,r}(V_i) < b$. We continue the process until $\Gamma_{R,t}(V_i) = b$ for each $i \in \{1, 2, \ldots, k\}$ and for some t; this is guaranteed since $\Gamma_{R,\alpha}(V_i) < b$ implies that $\Gamma_{S,\alpha-1}(V_i) < \Gamma_{R,\alpha}(V_i) < b$ and $\Gamma_{R,\alpha}(V_i) + \Gamma_{S,\alpha-1}(V_i) < 2b$. Thus, R^* wins the R-game.

Second, we consider the S-game. Suppose S^* assigns values on $V(G)$ such that $h_{S,1}(V_i) = z_{1,i}$ for each $i \in \{1, 2, \ldots, k\}$, where $0 \leq z_{1,i} \leq 1$ and $\sum_{i=1}^{k} z_{1,i} \leq 1$. Let R^* assign values on $V(G)$ such that $h_{R,1}(V_i) = \min\{z_{1,i}, b\}$ for each $i \in \{1, 2, \ldots, k\}$. Note that $h_{R,1}(V_i) \leq h_{S,1}(V_i)$ and $h_{R,1}(V_i) + h_{S,1}(V_i) \leq b+1$ for each $i \in \{1, 2, \ldots, k\}$. Now, suppose S^* assigns values on $V(G)$, on her second turn, such that $h_{S,2}(V_i) = z_{2,i}$ for each $i \in \{1, 2, \ldots, k\}$, where $0 \leq z_{2,i} \leq 1$ and $\sum_{i=1}^{k} z_{2,i} \leq 1$; here, $z_{2,j} = 0$ if $h_{R,1}(V_j) = b$. Let R^* assign values on $V(G)$, on his second turn, as follows: (i) if $h_{R,1}(V_i) = b$, then let $h_{R,2}(V_i) = 0$; (ii) if $h_{R,1}(V_i) < b$ (i.e., $z_{1,i} < b$), then let $h_{R,2}(V_i) = \min\{z_{2,i}, b - z_{1,i}\}$. We note the following: (i) $h_{R,2}(V(G)) \leq h_{S,2}(V(G)) \leq 1$; (ii) if $z_{2,i} < b - z_{1,i}$, then $\Gamma_{R,2}(V_i) = \Gamma_{S,2}(V_i) < b$ and thus $\Gamma_{R,2}(V_i) + \Gamma_{S,2}(V_i) < 2b$; (iii) if $z_{2,i} \geq b - z_{1,i}$, then $\Gamma_{R,2}(V_i) = b$. For $r \geq 3$, let $h_{S,r}(V_i) = z_{r,i}$, where $0 \leq z_{r,i} \leq 1$ and $\sum_{i=1}^{k} z_{r,i} \leq 1$, and let $h_{R,r}(V_i) = \min\{z_{r,i}, b - \sum_{j=1}^{r-1} z_{j,i}\}$ whenever $\Gamma_{R,r-1}(V_i) < b$. We continue the process until $\Gamma_{R,t}(V_i) = b$ for each $i \in \{1, 2, \ldots, k\}$ and for some t; this is guaranteed since $\Gamma_{R,\beta}(V_i) < b$ implies that $\Gamma_{R,\beta}(V_i) = \Gamma_{S,\beta}(V_i) < b$ and $\Gamma_{R,\beta}(V_i) + \Gamma_{S,\beta}(V_i) < 2b$. Thus, R^* wins the S-game. Therefore, $o_f(G) = \mathcal{R}$. $\square$

As an immediate consequence of Proposition 8, we have the following

Corollary 3. *Let G be a connected graph with $V_1, V_2, \ldots, V_k \subseteq V(G)$ such that $V_i \cap V_j = \emptyset$ for any distinct $i, j \in \{1, 2, \ldots, k\}$.*

(a) Let $g : V(G) \to [0,1]$ be a function defined by $g(V_i) = \frac{1}{2}$ for each $i \in \{1, 2, \ldots, k\}$ such that $g(V(G)) = \frac{k}{2}$, where $k \geq 2$. If $|V_i| \geq 2$ for each $i \in \{1, 2, \ldots, k\}$ and g is a resolving function of G, then $o_f(G) = \mathcal{R}$.

(b) Let $g' : V(G) \to [0,1]$ be a function defined by $g'(V_i) = 1$ for each $i \in \{1, 2, \ldots, k\}$ such that $g'(V(G)) = k$, where $k \geq 1$. If $|V_i| \geq 3$ for each $i \in \{1, 2, \ldots, k\}$ and g' is a resolving function of G, then $o_f(G) = \mathcal{R}$.

Next, we show the existence of graphs G satisfying the following: (i) $o(G) = \mathcal{R}$ and $o_f(G) \in \{\mathcal{R}, \mathcal{N}, \mathcal{S}\}$; (ii) $o(G) = \mathcal{N}$ and $o_f(G) \in \{\mathcal{N}, \mathcal{S}\}$; (iii) $o(G) = \mathcal{S}$ and $o_f(G) = \mathcal{S}$.

Remark 1. There exist graphs G satisfying the following outcomes: (1) $o(G) = \mathcal{R}$ and $o_f(G) = \mathcal{R}$; (2) $o(G) = \mathcal{R}$ and $o_f(G) = \mathcal{N}$; (3) $o(G) = \mathcal{R}$ and $o_f(G) = \mathcal{S}$.

Proof.(1) If G is isomorphic to Fig. 1(a), then $\dim(G) = 1 = \dim_f(G)$, $\{\{\ell, \ell'\}\}$ is a paring resolving set of G, and G has exactly one twin equivalence class of cardinality 2; thus, $o(G) = \mathcal{R}$ by Proposition 2 and $o_f(G) = \mathcal{R}$ by Proposition 6.

(2) If G is isomorphic to Fig. 1(b), then $\dim(G) = 2 = \dim_f(G)$, $\{\{\ell_1, \ell_1'\}, \{\ell_2, \ell_2'\}\}$ is a paring resolving set of G, and G has exactly two distinct twin equivalence classes of cardinality 2; thus, $o(G) = \mathcal{R}$ by Proposition 2 and $o_f(G) = \mathcal{N}$ by Proposition 6.

(3) If G is isomorphic to Fig. 1(c), then $\dim(G) = 3 = \dim_f(G)$, $\{\{\ell_1, \ell_1'\}, \{\ell_2, \ell_2'\}, \{\ell_3, \ell_3'\}\}$ is a paring resolving set of G, and G has exactly three distinct twin equivalence classes of cardinality 2; thus, $o(G) = \mathcal{R}$ by Proposition 2 and $o_f(G) = \mathcal{S}$ by Proposition 6.   □

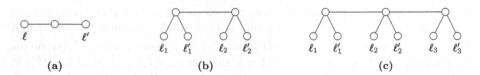

(a)   (b)   (c)

Fig. 1. Graphs G with $o(G) = \mathcal{R}$ and three different outcomes for $o_f(G)$.

Remark 2. There exist graphs G satisfying the following outcomes: (1) $o(G) = \mathcal{N}$ and $o_f(G) = \mathcal{N}$; (2) $o(G) = \mathcal{N}$ and $o_f(G) = \mathcal{S}$.

Proof.(1) If G is the 3-cycle, then $o(G) = \mathcal{N}$ by Proposition 3 and $o_f(G) = \mathcal{N}$ by Proposition 5.

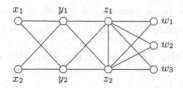

Fig. 2. A graph G with $o(G) = \mathcal{N}$ and $o_f(G) = \mathcal{S}$.

(2) If G is isomorphic to Fig. 2, then $\dim(G) = 5$, $\dim_f(G) = \frac{9}{2}$, and G has the following twin equivalence classes: $\{x_1, x_2\}$, $\{y_1, y_2\}$, $\{z_1, z_2\}$ and $\{w_1, w_2, w_3\}$. So, $o(G) = \mathcal{N}$ by Proposition 3 and $o_f(G) = \mathcal{S}$ by Proposition 7. □

Remark 3. There exists a graph G satisfying $o(G) = \mathcal{S} = o_f(G)$.

Proof. If G is a complete graph of order at least 4, then $o(G) = \mathcal{S}$ by Proposition 1(a), and $o_f(G) = \mathcal{S}$ by Corollary 2(a). □

We conclude this section with some open problems.

Question 1. (a) Is there a graph G such that $o(G) = \mathcal{N}$ and $o_f(G) = \mathcal{R}$?
(b) Is there a graph G such that $o(G) = \mathcal{S}$ and $o_f(G) = \mathcal{R}$?
(c) Is there a graph G such that $o(G) = \mathcal{S}$ and $o_f(G) = \mathcal{N}$?

4   The Outcome of the FMBRG on Some Classes of Graphs

In this section, we determine the outcome of the FMBRG for some classes of graphs. We denote by P_n, C_n, K_n and $\overline{K}_n$, respectively, the path, the cycle, the complete graph, and the edgeless graph on n vertices. We begin by recalling some known results on the fractional metric dimension of graphs. It was shown in [1] that, for any connected graph G of order at least two, $1 \leq \dim_f(G) \leq \frac{|V(G)|}{2}$. For graphs G with $\dim_f(G) = \frac{|V(G)|}{2}$, we recall the following results.

Theorem 3 [1,13]. *Let G be a connected graph of order at least two. Then $\dim_f(G) = \frac{|V(G)|}{2}$ if and only if there exists a bijection $\alpha : V(G) \to V(G)$ such that $\alpha(v) \neq v$ and $|R\{v, \alpha(v)\}| = 2$ for all $v \in V(G)$.*

An explicit characterization of graphs G satisfying $\dim_f(G) = \frac{|V(G)|}{2}$ was given in [2]. Following [2], let $\mathcal{K} = \{K_n : n \geq 2\}$ and $\overline{\mathcal{K}} = \{\overline{K}_n : n \geq 2\}$. Let $H[\mathcal{K} \cup \overline{\mathcal{K}}]$ be the family of graphs obtained from a connected graph H by (i) replacing each vertex $u_i \in V(H)$ by a graph $H_i \in \mathcal{K} \cup \overline{\mathcal{K}}$, and (ii) each vertex in H_i is adjacent to each vertex in H_j if and only if $u_i u_j \in E(H)$.

Theorem 4 [2]. *Let G be a connected graph of order at least two. Then* $\dim_f(G) = \frac{|V(G)|}{2}$ *if and only if* $G \in H[\mathcal{K} \cup \overline{\mathcal{K}}]$ *for some connected graph H.*

Next, we determine the outcome of the FMBRG on graphs G for which $\dim_f(G) = \frac{|V(G)|}{2}$ hold.

Proposition 9. *Let G be a connected graph of order $n \geq 2$ with $\dim_f(G) = \frac{n}{2}$. Then*

$$o_f(G) = \begin{cases} \mathcal{R} & \text{if } G = K_2, \\ \mathcal{N} & \text{if } G \in \{C_3, C_4, K_4 - e\}, \text{where } e \in E(K_4), \\ \mathcal{S} & \text{otherwise.} \end{cases} \tag{4}$$

Proof. Let G be a connected graph of order $n \geq 2$ with $\dim_f(G) = \frac{n}{2}$. For each $i \in \{1, 2, \ldots, k\}$ and for $k \geq 1$, let Q_i be a twin equivalence class of G with $|Q_i| \geq 2$ such that $V(G) = \cup_{i=1}^k Q_i$; notice that $Q_i \cap Q_j = \emptyset$ for any distinct $i, j \in \{1, 2, \ldots, k\}$.

First, suppose $|Q_i| \geq 4$ for some $i \in \{1, 2, \ldots, k\}$. Then $o_f(G) = \mathcal{S}$ by Corollary 2(a).

Second, suppose $|Q_i| \leq 3$ for each $i \in \{1, 2, \ldots, k\}$, and assume that $|Q_j| = 3$ for some $j \in \{1, 2, \ldots, k\}$. If $k = 1$, then $|Q_1| = 3$ and $o_f(G) = \mathcal{N}$ by Proposition 5; note that $G = C_3$ in this case. So, suppose $k \geq 2$. If $|Q_x| = 3$ and $|Q_i| = 2$ for each $i \in \{1, 2, \ldots, k\} - \{x\}$, then $o_f(G) = \mathcal{S}$ by Proposition 7. If $|Q_x| = |Q_y| = 3$ for distinct $x, y \in \{1, 2, \ldots, k\}$, then $o_f(G) = \mathcal{S}$ by Corollary 2(b).

Third, suppose $|Q_i| = 2$ for each $i \in \{1, 2, \ldots, k\}$. By Proposition 6, we have the following: (i) if $k = 1$, then $o_f(G) = \mathcal{R}$; (ii) if $k = 2$, then $o_f(G) = \mathcal{N}$; (iii) if $k \geq 3$, then $o_f(G) = \mathcal{S}$. We note that if G satisfies (i), then $G = K_2$; if G satisfies (ii), then $G \in \{C_4, K_4 - e\}$, where $e \in E(K_4)$. $\square$

Next, we determine the outcome of the FMBRG on cycles.

Theorem 5 [1]. *For $n \geq 3$,* $\dim_f(C_n) = \begin{cases} \frac{n}{n-2} & \text{if } n \text{ is even,} \\ \frac{n}{n-1} & \text{if } n \text{ is odd.} \end{cases}$

Proposition 10. *For $n \geq 3$,* $o_f(C_n) = \begin{cases} \mathcal{N} & \text{if } n \in \{3, 4\}, \\ \mathcal{R} & \text{if } n \geq 5. \end{cases}$

Proof. For $n \geq 3$, let C_n be given by $u_1, u_2, \ldots, u_n, u_1$. Note that $o_f(C_3) = \mathcal{N}$ by Theorem 5 and Proposition 4, and $o_f(C_4) = \mathcal{N}$ by Theorem 5 and Proposition 6 since C_4 has exactly two distinct twin equivalence classes of cardinality 2 with $\dim_f(C_4) = 2$. So, let $n \geq 5$, and we consider two cases.

Case 1: $n \geq 5$ is odd. Note that $|R\{u_i, u_j\}| \geq n - 1$ for any distinct $i, j \in \{1, 2, \ldots, n\}$.

First, we consider the R-game. Let $h_{R,1}(u_i) = \frac{1}{n-1}$ for each $i \in \{1, 2, \ldots, n-1\}$ with $h_{R,1}(V(C_n)) = 1$. If $h_{S,1}(u_n) \leq \frac{n-2}{n-1}$, then let $h_{R,2}(u_n) = \frac{1}{n-1}$ with $h_{R,2}(V(C_n)) = \frac{1}{n-1}$; then $\Gamma_{R,2}(u_i) = \frac{1}{n-1}$ for each $i \in \{1, 2, \ldots, n\}$, and $\Gamma_{R,2}$ is a resolving function of C_n since $\Gamma_{R,2}(R\{u_i, u_j\}) \geq (n-1)(\frac{1}{n-1}) = 1$ for any distinct $i, j \in \{1, 2, \ldots, n\}$. If $h_{S,1}(u_n) > \frac{n-2}{n-1}$, then let $h_{R,2}(u_i) = \frac{1}{n-1}$ for each $i \in \{1, 2, \ldots, n-1\}$ with $h_{R,2}(V(C_n)) = 1$; then $\Gamma_{R,2}(u_i) = \frac{2}{n-1}$ for each $i \in \{1, 2, \ldots, n-1\}$, and $\Gamma_{R,2}$ is a resolving function of C_n since $\Gamma_{R,2}(R\{u_i, u_j\}) \geq (n-2)(\frac{2}{n-1}) = \frac{2n-4}{n-1} \geq 1$ for $n \geq 5$ and for any distinct $i, j \in \{1, 2, \ldots, n\}$. In each case, R^* wins the R-game.

Second, we consider the S-game. By relabeling the vertices of C_n if necessary, let $h_{S,1}(u_1) = \max\{h_{S,1}(u_i) : 1 \leq i \leq n\}$; then $h_{S,1}(u_j) \leq \frac{1}{2}$ for each $j \in \{2, 3, \ldots, n\}$. We consider the cases for $n = 5$ and $n \geq 7$ separately. Suppose $n = 5$. Let $h_{R,1}(u_i) = \frac{1}{4}$ for each $i \in \{2, 3, 4, 5\}$ with $h_{R,1}(V(C_5)) = 1$. If $\Gamma_{S,2}(u_1) \leq \frac{3}{4}$, let $h_{R,2}(u_1) = \frac{1}{4}$ with $h_{R,2}(V(C_5)) = \frac{1}{4}$; then $\Gamma_{R,2}(u_i) = \frac{1}{4}$ for each $i \in \{1, 2, 3, 4, 5\}$, and $\Gamma_{R,2}$ is a resolving function of C_5 since $\Gamma_{R,2}(R\{u_i, u_j\}) \geq (4)(\frac{1}{4}) = 1$ for any distinct $i, j \in \{1, 2, 3, 4, 5\}$. If $\Gamma_{S,2}(u_i) \leq \frac{2}{3}$ for each $i \in \{2, 3, 4, 5\}$, then let $h_{R,2}(u_i) = \frac{1}{12}$ for each $i \in \{2, 3, 4, 5\}$ with $h_{R,2}(V(C_5)) = \frac{1}{3}$; then $\Gamma_{R,2}(u_i) = \frac{1}{4} + \frac{1}{12} = \frac{1}{3}$ for each $i \in \{2, 3, 4, 5\}$, and $\Gamma_{R,2}$ is a resolving function of C_5 since $\Gamma_{R,2}(R\{u_i, u_j\}) \geq (3)(\frac{1}{3}) = 1$ for any distinct $i, j \in \{1, 2, 3, 4, 5\}$. If $\Gamma_{S,2}(u_1) > \frac{3}{4}$ and $\Gamma_{S,2}(u_\alpha) > \frac{2}{3}$ for some $\alpha \in \{2, 3, 4, 5\}$, then $\Gamma_{S,2}(u_j) < \frac{7}{12}$ for each $j \in \{2, 3, 4, 5\} - \{\alpha\}$. In this case, let $h_{R,2}(u_j) = \frac{1}{6}$ for each $j \in \{2, 3, 4, 5\} - \{\alpha\}$ with $h_{R,2}(V(C_5)) = \frac{1}{2}$; then $\Gamma_{R,2}(u_\alpha) = \frac{1}{4}$ and $\Gamma_{R,2}(u_j) = \frac{1}{4} + \frac{1}{6} = \frac{5}{12}$ for each $j \in \{2, 3, 4, 5\} - \{\alpha\}$, and thus $\Gamma_{R,2}(R\{u_i, u_j\}) \geq \min\{(3)(\frac{5}{12}), \frac{1}{4} + 2(\frac{5}{12})\} > 1$ for any distinct $i, j \in \{1, 2, 3, 4, 5\}$, which implies $\Gamma_{R,2}$ is a resolving function of C_5. Now, suppose $n \geq 7$. Let $h_{R,1}(u_i) = \frac{1}{n-1}$ for each $i \in \{2, 3, \ldots, n\}$ with $h_{R,1}(V(C_n)) = 1$. If $\Gamma_{S,2}(u_1) \leq \frac{n-2}{n-1}$, let $h_{R,2}(u_1) = \frac{1}{n-1}$ with $h_{R,2}(V(C_n)) = \frac{1}{n-1}$; then $\Gamma_{R,2}$ is a resolving function of C_n since $\Gamma_{R,2}(R\{u_i, u_j\}) \geq (n-1)(\frac{1}{n-1}) = 1$ for any distinct $i, j \in \{1, 2, \ldots, n\}$. If $\Gamma_{S,2}(u_i) \leq \frac{n-3}{n-1}$ for each $i \in \{2, 3, \ldots, n\}$, let $h_{R,2}(u_i) = \frac{1}{n-1}$ for each $i \in \{2, 3, \ldots, n\}$ with $h_{R,2}(V(C_n)) = 1$; then $\Gamma_{R,2}$ is a resolving function of C_n since $\Gamma_{R,2}(R\{u_i, u_j\}) \geq (n-2)(\frac{2}{n-1}) \geq 1$ for any distinct $i, j \in \{1, 2, \ldots, n\}$. If $\Gamma_{S,2}(u_1) > \frac{n-2}{n-1}$ and $\Gamma_{S,2}(u_\beta) > \frac{n-3}{n-1}$ for some $\beta \in \{2, 3, \ldots, n\}$, then $\Gamma_{S,2}(u_j) < \frac{3}{n-1}$ for each $j \in \{2, 3, \ldots, n\} - \{\beta\}$. In this case, let $h_{R,2}(u_j) = \frac{1}{n-1}$ for each $j \in \{2, 3, \ldots, n\} - \{\beta\}$; then $\Gamma_{R,2}(u_\beta) = \frac{1}{n-1}$, $\Gamma_{R,2}(u_j) = \frac{2}{n-1}$ and $\Gamma_{R,2}(u_j) + \Gamma_{S,2}(u_j) < \frac{2}{n-1} + \frac{3}{n-1} = \frac{5}{n-1} < 1$ for each $j \in \{2, 3, \ldots, n\} - \{\beta\}$, and thus $\Gamma_{R,2}(R\{u_i, u_j\}) \geq \min\{(n-2)(\frac{2}{n-1}), \frac{1}{n-1} + (n-3)(\frac{2}{n-1})\} > 1$ for $n \geq 7$ and for any distinct $i, j \in \{1, 2, \ldots, n\}$, which implies $\Gamma_{R,2}$ is a resolving function of C_n for $n \geq 7$. So, R^* wins the S-game for any odd $n \geq 5$.

Therefore, if $n \geq 5$ is odd, then $o_f(C_n) = \mathcal{R}$.

Case 2: $n \geq 6$ **is even.** Note that $|R\{u_i, u_j\}| \geq n - 2$ for any distinct $i, j \in \{1, 2, \ldots, n\}$, where equality holds when $d(u_i, u_j)$ is even for $i \neq j$, and $|R\{u_i, u_j\} \cap (\cup_{a=1}^{\frac{n}{2}}\{u_a\})| = |R\{u_i, u_j\} \cap (\cup_{b=\frac{n}{2}+1}^{n}\{u_b\})| \geq \frac{n}{2} - 1$.

First, we consider the R-game. Let $h_{R,1}(u_i) = \frac{1}{n-2}$ for each $i \in \{1, 2, \ldots, n-2\}$ with $h_{R,1}(V(C_n)) = 1$. If $h_{S,1}(u_{n-1}) + h_{S,1}(u_n) \leq \frac{n-3}{n-2}$, then let $h_{R,2}(u_{n-1}) = h_{R,2}(u_n) = \frac{1}{n-2}$ with $h_{R,2}(V(C_n)) = \frac{2}{n-2}$; then $\Gamma_{R,2}(u_i) = \frac{1}{n-2}$ for each $i \in \{1, 2, \ldots, n\}$, and $\Gamma_{R,2}$ is a resolving function of C_n since $\Gamma_{R,2}(R\{u_i, u_j\}) \geq (n-2)(\frac{1}{n-2}) = 1$ for any distinct $i, j \in \{1, 2, \ldots, n\}$. If $h_{S,1}(u_{n-1}) + h_{S,1}(u_n) > \frac{n-3}{n-2}$, let $h_{R,2}(u_i) = \frac{1}{n-2}$ for each $i \in \{1, 2, \ldots, n-2\}$ with $h_{R,2}(V(C_n)) = 1$; then $\Gamma_{R,2}(u_i) = \frac{2}{n-2}$ for each $i \in \{1, 2, \ldots, n-2\}$, and $\Gamma_{R,2}$ is a resolving function of C_n since $\Gamma_{R,2}(R\{u_i, u_j\}) \geq (n-4)(\frac{2}{n-2}) = \frac{2n-8}{n-2} \geq 1$ for $n \geq 6$ and for any distinct $i, j \in \{1, 2, \ldots, n\}$. So, R^* wins the R-game.

Second, we consider the S-game. Let $x = \lceil \frac{3n}{4} \rceil$, and we assume that $h_{S,1}(u_x) = \max\{h_{S,1}(u_i) : 1 \leq i \leq n\}$ by relabeling the vertices of C_n if necessary; then $h_{S,1}(u_j) \leq \frac{1}{2}$ for each $j \in \{1, 2, \ldots, n\} - \{x\}$. Let $h_{R,1}(u_i) = \frac{2}{n}$ for each $i \in \{1, 2, \ldots, \frac{n}{2}\}$ with $h_{R,1}(V(C_n)) = 1$. If $\Gamma_{S,2}(u_i) \leq \frac{n-4}{n-2}$ for each $i \in \{1, 2, \ldots, \frac{n}{2}\}$, let $h_{R,2}(u_i) = \frac{4}{n(n-2)}$ for each $i \in \{1, 2, \ldots, \frac{n}{2}\}$ with $h_{R,2}(V(C_n)) = \frac{2}{n-2}$; then $\Gamma_{R,2}(u_i) = \frac{2}{n} + \frac{4}{n(n-2)} = \frac{2}{n-2}$ for each $i \in \{1, \ldots, \frac{n}{2}\}$, $\Gamma_{R,2}(u_j) = 0$ for each $j \in \{\frac{n}{2} + 1, \ldots, n\}$, and $\Gamma_{R,2}$ is a resolving function of C_n since $\Gamma_{R,2}(R\{u_i, u_j\}) \geq (\frac{n}{2} - 1)(\frac{2}{n-2}) = 1$ for any distinct $i, j \in \{1, 2, \ldots, n\}$. If $\Gamma_{S,2}(u_j) \leq \frac{n-1}{n}$ for each $j \in \{\frac{n}{2} + 1, \ldots, n\}$, let $h_{R,2}(u_j) = \frac{1}{n}$ for each $j \in \{\frac{n}{2} + 1, \ldots, n\}$ with $h_{R,2}(V(C_n)) = \frac{1}{2}$; then $\Gamma_{R,2}(u_i) = \frac{2}{n}$ for each $i \in \{1, 2, \ldots, \frac{n}{2}\}$, $\Gamma_{R,2}(u_j) = \frac{1}{n}$ for each $j \in \{\frac{n}{2} + 1, \ldots, n\}$, and $\Gamma_{R,2}$ is a resolving function of C_n since $\Gamma_{R,2}(R\{u_i, u_j\}) \geq (\frac{n}{2} - 1)(\frac{2}{n} + \frac{1}{n}) = \frac{3n-6}{2n} \geq 1$ for $n \geq 6$ and for any distinct $i, j \in \{1, 2, \ldots, n\}$. If $\Gamma_{S,2}(u_\alpha) > \frac{n-4}{n-2}$ for some $\alpha \in \{1, 2, \ldots, \frac{n}{2}\}$ and $\Gamma_{S,2}(u_\beta) > \frac{n-1}{n}$ for some $\beta \in \{\frac{n}{2} + 1, \ldots, n\}$, then $\Gamma_{S,2}(u_i) < 2 - (\frac{n-4}{n-2} + \frac{n-1}{n}) = \frac{3n-2}{n(n-2)} = \frac{1}{n} + \frac{2}{n-2}$ for each $i \in \{1, 2, \ldots, n\} - \{\alpha, \beta\}$. In this case, let $h_{R,2}(u_j) = \frac{2}{n}$ for each $j \in \{\frac{n}{2} + 1, \ldots, n\} - \{\beta\}$ with $h_{R,2}(V(C_n)) = \frac{n-2}{n}$; here, we note that $(\frac{1}{n} + \frac{2}{n-2}) + \frac{2}{n} = \frac{5n-6}{n(n-2)} \leq 1$ since $5n - 6 \leq n^2 - 2n$, equivalently, $n^2 - 7n + 6 = (n-6)(n-1) \geq 0$ for $n \geq 6$. Then $\Gamma_{R,2}(u_i) = \frac{2}{n}$ for each $i \in \{1, 2, \ldots, n\} - \{\beta\}$ and $\Gamma_{R,2}(R\{u_i, u_j\}) \geq (n-3)(\frac{2}{n}) \geq 1$ for $n \geq 6$ and for any distinct $i, j \in \{1, 2, \ldots, n\}$; thus, $\Gamma_{R,2}$ is a resolving function of C_n for $n \geq 6$. So, R^* wins the S-game for any even $n \geq 6$.

Therefore, if $n \geq 6$ is even, then $o_f(C_n) = \mathcal{R}$. □

Next, we determine the outcome of the FMBRG on the Petersen graph.

Theorem 6 [1]. *For the Petersen graph $\mathcal{P}$, $\dim_f(\mathcal{P}) = \frac{5}{3}$.*

Proposition 11. *For the Petersen graph $\mathcal{P}$, $o_f(\mathcal{P}) = \mathcal{R}$.*

Proof. Let $\mathcal{P}$ be the Petersen graph with $V(\mathcal{P}) = \cup_{i=1}^{10}\{u_i\}$. Note that $|R\{u_i, u_j\}| \geq 6$ for any distinct $i, j \in \{1, 2, \ldots, 10\}$.

First, we consider the R-game. Suppose R^* assigns values on $V(\mathcal{P})$ such that $h_{R,1}(u_i) = \frac{1}{10}$ for each $i \in \{1, 2, \ldots, 10\}$. Then S^* can assign values on $V(\mathcal{P})$ such that $h_{S,1}(u_i) = x_i$ for each $i \in \{1, 2, \ldots, 10\}$, where $0 \leq x_i \leq \frac{9}{10}$ and $\sum_{i=1}^{10} x_i \leq 1$; we may assume, without loss of generality, that $x_1 \geq x_2 \geq \ldots \geq x_{10}$. If $x_1 \leq \frac{4}{5}$, then R^* can assign values on $V(\mathcal{P})$ such that $h_{R,2}(u_i) = \frac{1}{10}$ for each $i \in \{1, 2, \ldots, 10\}$; then $\Gamma_{R,2}(u_i) = \frac{1}{5}$ for each $i \in \{1, 2, \ldots, 10\}$, and $\Gamma_{R,2}$ is a resolving function of $\mathcal{P}$ since $\Gamma_{R,2}(R\{u_i, u_j\}) \geq 6(\frac{1}{5}) > 1$ for any distinct $i, j \in \{1, 2, \ldots, 10\}$. If $x_1 > \frac{4}{5}$, then $x_i < \frac{1}{5}$ for each $i \in \{2, 3, \ldots, 10\}$ and R^* can assign values on $V(\mathcal{P})$ such that $h_{R,2}(u_i) = \frac{1}{9}$ for each $i \in \{2, 3, \ldots, 10\}$; then $\Gamma_{R,2}(u_1) = \frac{1}{10}$, $\Gamma_{R,2}(u_i) = \frac{1}{10} + \frac{1}{9} = \frac{19}{90}$ for each $i \in \{2, 3, \ldots, 10\}$, and $\Gamma_{R,2}$ is a resolving function of $\mathcal{P}$ since $\Gamma_{R,2}(R\{u_i, u_j\}) \geq \min\{\frac{1}{10} + 5(\frac{19}{90}), 6(\frac{19}{90})\} > 1$ for any distinct $i, j \in \{1, 2, \ldots, 10\}$. In each case, R^* wins the R-game.

Second, we consider the S-game. Let $h_{S,1}(u_1) = \max\{h_{S,1}(u_i) : 1 \leq i \leq 10\}$ by relabeling the vertices of $\mathcal{P}$ if necessary; then $h_{S,1}(u_i) \leq \frac{1}{2}$ for each $i \in \{2, 3, \ldots, 10\}$. Then R^* can assign values on $V(\mathcal{P})$ such that $h_{R,1}(u_1) = 0$ and $h_{R,1}(u_i) = \frac{1}{9}$ for each $i \in \{2, 3, \ldots, 10\}$. If $\Gamma_{S,2}(u_i) \leq \frac{4}{5}$ for each $i \in \{2, 3, \ldots, 10\}$, then R^* can assign values on $V(\mathcal{P})$ such that $h_{R,2}(u_i) = \frac{4}{45}$ for each $i \in \{2, 3, \ldots, 10\}$ with $h_{R,2}(V(\mathcal{P})) = \frac{4}{5}$; then $\Gamma_{R,2}(u_1) = 0$, $\Gamma_{R,2}(u_i) = \frac{1}{9} + \frac{4}{45} = \frac{1}{5}$ for each $i \in \{2, 3, \ldots, 10\}$, and $\Gamma_{R,2}$ is a resolving function of $\mathcal{P}$ since $\Gamma_{R,2}(R\{u_i, u_j\}) \geq 5(\frac{1}{5}) = 1$ for any distinct $i, j \in \{1, 2, \ldots, 10\}$. So, suppose $\Gamma_{S,2}(u_\alpha) > \frac{4}{5}$ for some $\alpha \in \{2, 3, \ldots, 10\}$. If $\Gamma_{S,2}(u_1) \leq \frac{4}{5}$ and $\Gamma_{S,2}(u_\alpha) > \frac{4}{5}$ for exactly one $\alpha \in \{2, 3, \ldots, 10\}$, then R^* can assign values on $V(\mathcal{P})$ such that $h_{R,2}(u_1) = \frac{1}{5}$ and $h_{R,2}(u_j) = \frac{4}{45}$ for each $j \in \{2, 3, \ldots, 10\} - \{\alpha\}$; then $\Gamma_{R,2}(u_\alpha) = \frac{1}{9}$, $\Gamma_{R,2}(u_i) = \frac{1}{5}$ for each $i \in \{1, 2, \ldots, 10\} - \{\alpha\}$, and $\Gamma_{R,2}$ is a resolving function of $\mathcal{P}$ since $\Gamma_{R,2}(R\{u_i, u_j\}) \geq \min\{\frac{1}{9} + 5(\frac{1}{5}), 6(\frac{1}{5})\} > 1$ for any distinct $i, j \in \{1, 2, \ldots, 10\}$. If $\Gamma_{S,2}(u_a) > \frac{4}{5}$ and $\Gamma_{S,2}(u_b) > \frac{4}{5}$ for distinct $a, b \in \{2, 3, \ldots, 10\}$, then $\Gamma_{S,2}(u_1) < \frac{2}{5}$ and $\Gamma_{S,2}(u_j) < \frac{2}{5}$ for each $j \in \{2, 3, \ldots, 10\} - \{a, b\}$, and R^* can assign values on $V(\mathcal{P})$ such that $h_{R,2}(u_1) = \frac{1}{5}$, $h_{R,2}(u_a) = h_{R,2}(u_b) = 0$ and $h_{R,2}(u_j) = \frac{4}{45}$ for each $j \in \{2, 3, \ldots, 10\} - \{a, b\}$; then $\Gamma_{R,2}(u_a) = \Gamma_{R,2}(u_b) = \frac{1}{9}$, $\Gamma_{R,2}(u_i) = \frac{1}{5}$ for each $i \in \{1, 2, \ldots, 10\} - \{a, b\}$, and $\Gamma_{R,2}$ is a resolving function of $\mathcal{P}$ since $\Gamma_{R,2}(R\{u_i, u_j\}) \geq 2(\frac{1}{9}) + 4(\frac{1}{5}) = \frac{46}{45} > 1$ for any distinct $i, j \in \{1, 2, \ldots, 10\}$. If $\Gamma_{S,2}(u_1) > \frac{4}{5}$ and $\Gamma_{S,2}(u_\beta) > \frac{4}{5}$ for some $\beta \in \{2, 3, \ldots, 10\}$, then $\Gamma_{S,2}(u_j) < \frac{2}{5}$ for each $j \in \{2, 3, \ldots, 10\} - \{\beta\}$, and R^* can assign values on $V(\mathcal{P})$ such that $h_{R,2}(u_j) = \frac{1}{8}$ for each $j \in \{2, 3, \ldots, 10\} - \{\beta\}$ with $h_{R,2}(V(\mathcal{P})) = 1$; then $\Gamma_{R,2}(u_1) = 0$, $\Gamma_{R,2}(u_\beta) = \frac{1}{9}$, $\Gamma_{R,2}(u_j) = \frac{1}{9} + \frac{1}{8} = \frac{17}{72}$ for each $j \in \{2, 3, \ldots, 10\} - \{\beta\}$, and $\Gamma_{R,2}$ forms a resolving function of $\mathcal{P}$ since $\Gamma_{R,2}(R\{u_i, u_j\}) \geq \min\{\frac{1}{9} + 4(\frac{17}{72}), 5(\frac{17}{72})\} > 1$ for any distinct $i, j \in \{1, 2, \ldots, 10\}$. In each case, R^* wins the S-game.

Therefore, $o_f(\mathcal{P}) = \mathcal{R}$. $\square$

Next, we consider the outcome of the FMBRG on trees T. Fix a tree T. A *major vertex* is a vertex of degree at least three. A vertex ℓ of degree 1 is called a *terminal vertex* of a major vertex v if $d(\ell, v) < d(\ell, w)$ for every other major vertex w in T. The *terminal degree*, $ter(v)$, of a major vertex v is the number of terminal vertices of v in T, and an *exterior major vertex* is a major vertex that has positive terminal degree. We denote by $ex(T)$ the number of exterior major vertices of T, and $\sigma(T)$ the number of leaves of T. Let $M(T)$ be the set of exterior major vertices of T. Let $M_1(T) = \{w \in M(T) : ter(w) = 1\}$ and let $M_2(T) = \{w \in M(T) : ter(w) \geq 2\}$; note that $M(T) = M_1(T) \cup M_2(T)$. For each $v \in M(T)$, let T_v be the subtree of T induced by v and all vertices belonging to the paths joining v with its terminal vertices, and let L_v be the set of terminal vertices of v in T.

Theorem 7. (a) [15] *For any graph G of order $n \geq 2$, $\dim_f(G) = 1$ if and only if $G = P_n$.*

(b) [21] *For any non-trivial tree T, $\dim_f(T) = \frac{1}{2}(\sigma(T) - ex_1(T))$, where $ex_1(T)$ denotes the number of exterior major vertices of T with terminal degree one.*

We first consider paths.

Proposition 12. *For $n \geq 2$, $o_f(P_n) = \mathcal{R}$.*

Proof. Let P_n be given by $u_1, u_2, \ldots, u_n$, where $n \geq 2$. If $g : V(P_n) \to [0, 1]$ is a function defined by $g(u_1) + g(u_n) = 1$, then g is a resolving function of P_n since $R\{u_i, u_j\} \supseteq \{u_1, u_n\}$ for any distinct $i, j \in \{1, 2, \ldots, n\}$. In the R-game, suppose $h_{R,1}(u_1) = \frac{1}{2} = h_{R,1}(u_n)$. In the S-game, if S^* assigns values on $V(P_n)$ such that $h_{S,1}(u_1) = x_1$, $h_{S,1}(u_n) = x_2$, where $0 \leq x_1, x_2 \leq 1$ and $x_1 + x_2 \leq 1$, then R^* can assign values on $V(P_n)$ such that $h_{R,1}(u_1) = 1 - x_1$ and $h_{R,1}(u_n) = x_1$. In each case, $h_{R,1}(u_1) + h_{R,1}(u_n) = 1$; thus $\Gamma_{R,1}$ is a resolving function of P_n. So, $o_f(P_n) = \mathcal{R}$.   □

Next, we consider the outcome of the FMBRG on trees T with $ex(T) \geq 1$. We recall the following result that is implied in [21].

Lemma 1. *Let T be a tree with $ex(T) \geq 1$, and let $M_2(T) = \{v_1, v_2, \ldots, v_k\}$. For each $i \in \{1, 2, \ldots, k\}$, let $\ell_{i,1}, \ell_{i,2}, \ldots, \ell_{i,\sigma_i}$ be the terminal vertices of v_i in T with $ter(v_i) = \sigma_i \geq 2$, and let $P^{i,j}$ be the $v_i - \ell_{i,j}$ path, excluding v_i, in T, where $j \in \{1, 2, \ldots, \sigma_i\}$. Let $g : V(T) \to [0, 1]$ be a function defined by $g(V(P^{i,j})) = \frac{1}{2}$ for each $i \in \{1, 2, \ldots, k\}$ and for each $j \in \{1, 2, \ldots, \sigma_i\}$. Then g is a resolving function of T.*

Proposition 13. *Let T be a tree with $ex(T) \geq 1$.*

(a) *If $|N(v) \cap L_v| \geq 4$ for some $v \in M_2(T)$, or $|N(u) \cap L_u| = |N(w) \cap L_w| = 3$ for two distinct $u, w \in M_2(T)$, or $|N(u) \cap L_u| = |N(v) \cap L_v| = |N(w) \cap L_w| = 2$ for three distinct $u, v, w \in M_2(T)$, then $o_f(T) = \mathcal{S}$.*
(b) *If $|N(v) \cap L_v| = 0$ for each $v \in M_2(T)$, then $o_f(T) = \mathcal{R}$.*

Proof. Let T be a tree that is not a path.

(a) If $|N(v) \cap L_v| \geq 4$ for some $v \in M_2(T)$, then T has a twin equivalence class of cardinality at least 4; thus $o_f(T) = \mathcal{S}$ by Corollary 2(a). If $|N(u) \cap L_u| = |N(w) \cap L_w| = 3$ for two distinct $u, w \in M_2(T)$, then T has two distinct twin equivalence classes of cardinality 3; thus $o_f(T) = \mathcal{S}$ by Corollary 2(b). If $|N(u) \cap L_u| = |N(v) \cap L_v| = |N(w) \cap L_w| = 2$ for three distinct $u, v, w \in M_2(T)$, then T has three distinct twin equivalence classes of cardinality 2; thus $o_f(T) = \mathcal{S}$ by Corollary 2(c).

(b) Let $M_2(T) = \{v_1, v_2, \ldots, v_k\}$, where $k \geq 1$. For each $i \in \{1, 2, \ldots, k\}$ with $ter(v_i) = \sigma_i \geq 2$, let $\ell_{i,1}, \ell_{i,2}, \ldots, \ell_{i,\sigma_i}$ be the terminal vertices of v_i in T, and let $P^{i,j}$ denote the $v_i - \ell_{i,j}$ path, excluding v_i, in T, where $j \in \{1, 2, \ldots, \sigma_i\}$. For each $i \in \{1, 2, \ldots, k\}$, let $g_i : V(T_{v_i}) \to [0, 1]$ be a function defined by $g(V(P^{i,j})) = \frac{1}{2}$ for each $j \in \{1, 2, \ldots, \sigma_i\}$. Let $g : V(T) \to [0, 1]$ be a function such that $g|_{T_{v_i}} = g_i$ (i.e., the function g restricted to T_{v_i} yields the function g_i) with $g(V(T)) = \dim_f(T)$. Then g is a resolving function of T by Lemma 1. Since the condition $|N(v) \cap L_v| = 0$ for each $v \in M_2(T)$ implies that $|V(P^{i,j})| \geq 2$ for each $i \in \{1, 2, \ldots, k\}$ and for each $j \in \{1, 2, \ldots, \sigma_i\}$, $o_f(T) = \mathcal{R}$ by Corollary 3(a).   □

Question 2. For an arbitrary tree T, can we determine $o_f(T)$?

Acknowledgement. The author thanks the anonymous referees for some helpful comments and suggestions.

References

1. Arumugam, S., Mathew, V.: The fractional metric dimension of graphs. Discret. Math. **312**, 1584–1590 (2012)
2. Arumugam, S., Mathew, V., Shen, J.: On fractional metric dimension of graphs. Discrete Math. Algorithms Appl. **5**, 1350037 (2013)
3. Beck, J.: Combinatorial Games: Tic-Tac-Toe Theory. Cambridge University Press, Cambridge (2008)
4. Beerliova, Z., et al.: Network discovery and verification. IEEE J. Sel. Areas Commun. **24**, 2168–2181 (2006)
5. Brešar, B., Klavžar, S., Rall, D.F.: Domination game and an imagination strategy. SIAM J. Discret. Math. **24**, 979–991 (2010)
6. Currie, J., Oellermann, O.R.: The metric dimension and metric independence of a graph. J. Comb. Math. Comb. Comput. **39**, 157–167 (2001)
7. Duchêne, E., Gledel, V., Parreau, A., Renault, G.: Maker-Breaker domination game. Discret. Math. 343(9) (2020). #111955
8. Erdös, P., Selfridge, J.L.: On a combinatorial game. J. Comb. Theory Ser. A **14**, 298–301 (1973)
9. Garey, M.R., Johnson, D.S.: Computers and Intractability: A Guide to the Theory of NP-Completeness. Freeman, New York (1979)
10. Harary, F., Melter, R.A.: On the metric dimension of a graph. Ars Comb. **2**, 191–195 (1976)
11. Hefetz, D., Krivelevich, M., Stojaković, M., Szabó, T.: The neighborhood conjecture. Positional Games. OS, vol. 44, pp. 113–139. Springer, Basel (2014). https://doi.org/10.1007/978-3-0348-0825-5_9
12. Hernando, C., Mora, M., Pelayo, I.M., Seara, C., Wood, D.R.: Extremal graph theory for metric dimension and diameter. Electron. J. Comb. (2010). #R30
13. Kang, C.X.: On the fractional strong metric dimension of graphs. Discret. Appl. Math. **213**, 153–161 (2016)
14. Kang, C.X., Klavžar, S., Yero, I.G., Yi, E.: Maker-Breaker resolving game. arXiv:2005.13242 (2020)
15. Kang, C.X., Yi, E.: The fractional strong metric dimension of graphs. In: Widmayer, P., Xu, Y., Zhu, B. (eds.) COCOA 2013. LNCS, vol. 8287, pp. 84–95. Springer, Cham (2013). https://doi.org/10.1007/978-3-319-03780-6_8
16. Khuller, S., Raghavachari, B., Rosenfeld, A.: Landmarks in graphs. Discret. Appl. Math. **70**, 217–229 (1996)
17. Klein, D.J., Yi, E.: A comparison on metric dimension of graphs, line graphs, and line graphs of the subdivision graphs. Eur. J. Pure Appl. Math. 5(3), 302–316 (2012)
18. Nowakawski, R., Winkler, P.: Vertex-to-vertex pursuit in a graph. Discret. Math. **43**, 235–239 (1983)
19. Sebö, A., Tannier, E.: On metric generators of graphs. Math. Oper. Res. **29**, 383–393 (2004)
20. Slater, P.J.: Leaves of trees. Congr. Numer. **14**, 549–559 (1975)
21. Yi, E.: The fractional metric dimension of permutation graphs. Acta Math. Sin. Engl. Ser. **31**, 367–382 (2015). https://doi.org/10.1007/s10114-015-4160-5

Price of Fairness in Budget Division
for Egalitarian Social Welfare

Zhongzheng Tang[1], Chenhao Wang[2], and Mengqi Zhang[3,4(✉)]

[1] School of Sciences, Beijing University of Posts and Telecommunications,
Beijing 100876, China
`tangzhongzheng@amss.ac.cn`
[2] Department of Computer Science and Engineering, University of Nebraska-Lincoln,
Lincoln, NE, USA
`wangch@amss.ac.cn`
[3] Academy of Mathematics and Systems Science, Chinese Academy of Sciences,
Beijing 100190, China
`mqzhang@amss.ac.cn`
[4] School of Mathematical Sciences, University of Chinese Academy of Sciences,
Beijing 100049, China

Abstract. We study a participatory budgeting problem of aggregating the preferences of agents and dividing a budget over the projects. A budget division solution is a probability distribution over the projects. The main purpose of our study concerns the comparison between the system optimum solution and a fair solution. We are interested in assessing the quality of fair solutions, i.e., in measuring the system efficiency loss under a fair allocation compared to the one that maximizes (egalitarian) social welfare. This indicator is called the price of fairness. We are also interested in the performance of several aggregation rules. Asymptotically tight bounds are provided both for the price of fairness and the efficiency guarantee of aggregation rules.

Keywords: Participatory budgeting · Fairness · Probabilistic voting

1 Introduction

Suppose there is a list of possible projects that require funding, and some self-interested agents (citizens, parties or players) have their preferences over the projects. Participatory budgeting is a process of aggregating the preferences of agents, and allocating a fixed budget over projects [11,12,18]. It allows citizens to identify, discuss, and prioritize public spending projects, and gives them the power to make real decisions about how to allocate part of a municipal or public budget, and how money is spent. These problems consist in sharing resources so that the agents have high satisfaction, and at the same time the budget should be utilized in an efficient way from a central point of view.

We consider participatory budgeting as a probabilistic voting process [17], which takes as input agents' preferences and returns a probability distribution

© Springer Nature Switzerland AG 2020
W. Wu and Z. Zhang (Eds.): COCOA 2020, LNCS 12577, pp. 594–607, 2020.
https://doi.org/10.1007/978-3-030-64843-5_40

over projects. That is, a budget division outcome is a vector of non-negative numbers, one for each project, summing up to 1. We focus on an important special case of *dichotomous preferences*: each agent either likes or dislikes each project, and her utility is equal to the fraction of the budget spent on projects she likes. Dichotomous preferences are of practical interest because they are easy to elicit. This process is also referred to as *approval voting*, as each voter (agent) specifies a subset of the projects that she "approves".

The decision-maker is confronted with a system objective of social welfare maximization, and looks for a budget division solution that performs well under the objective. A system optimum is any solution maximizing the social welfare. The utilitarian social welfare is defined as the sum of utilities over all agents, and the egalitarian social welfare is the minimum among the utilities of agents. On the other hand, it is desirable for a budget division solution to achieve the fairness among agents. Fairness usually concerns comparing the utility gained by one agent to the others' utilities. The concept of fairness is not uniquely defined since it strongly depends on the specific problem setting and also on the agents perception of what a fair solution is. For example, the *Individual Fair Share* requires that each one of the n agents receives a $1/n$ share of decision power, so she can ensure an outcome she likes at least $1/n$ of the time (or with probability at least $1/n$).

The system optimum may be highly unbalanced, and thus violate the fairness axioms. For instance, it could assign all budget to a single project and this may have severe negative effects in many application scenarios. Thus, it would be beneficial to reach a certain degree of agents' satisfaction by implementing some criterion of fairness. Clearly, the maximum utility of fair solutions in general deviates from the system optimum and thus incurs a loss of system efficiency. In this paper, we want to analyze such a loss implied by a fair solution from a worst-case point of view.

We are interested in assessing the quality of fair solutions, i.e., in measuring the system efficiency loss under a fair allocation compared to the one that maximizes the system objective. This indicator is called the *price of fairness*. Michorzewski *et al.* [20] study the price of fairness in participatory budgeting under the objective of maximizing the utilitarian social welfare. We consider this problem from an egalitarian perspective.

Fairness Axioms. Given a budget equal to 1 and n agents, there are some well-studied fairness criteria for a budget division solution (or simply, a distribution) [2,7,15,16]. Individual Fair Share (IFS) requires that the utility of each agent is at least $1/n$. Stronger fairness properties require that groups are fair in a sense. *Unanimous Fair Share (UFS)* gives to every group of agents with the same approval set an influence proportional to its size, that is, each agent in this kind of group will obtain a utility at least the group's related size (its size divided by n). *Group Fair Share (GFS)* requires that for any group of agents, the total fraction of the projects approved by the agents of this group is at least its relative size. *Core Fair Share (CFS)* reflects the incentive effect in the voting process. It says that for any group, each agent of the group cannot obtain a higher utility under another mixture with a probability proportional to the group size. *Average Fair Share (AFS)* requires that the average utility of any group with a common approved outcome is at least its relative size. A distribution satisfies *implementability (IMP)* if it can be decomposed into individual distributions such that no agent is asked to spend budgets on projects she considers as valueless. We remark that all other axioms mentioned above are stronger than IFS. Besides, CFS, AFS and IMP implies GFS, which implies UFS [2].

Voting Rules. The input of voting rules, also called participatory budgeting rules, includes a list of possible projects, the total available budget, and the preferences of agents. The output is a partition of budget among the projects - determining how much budget to allocate to each project- which can be seen as a distribution. We say a voting rule satisfies a certain fairness axiom, if any distribution induced by this rule satisfies that. We study the following voting rules that have been proposed for this setting.

Utilitarian (UTIL) rule selects a distribution maximizing the utilitarian social welfare, which focus only on efficiency. *Conditional Utilitarian (CUT)* rule is its simple variant, maximizing utilitarian welfare subject to the constraint that the returned distribution is implementable. *Random Priority (RP)* rule averages the outcomes of all deterministic priority rules. *Nash Max Product (NASH)* rule balances well the efficiency and fairness, which selects the distribution maximizing the product of agents' utilities. *Egalitarian (EGAL)* rule selects a distribution maximizing the minimum utility of agents. Though it is fair to individuals, it does not attempt to be fair to groups, and the egalitarian objective even treats different voters with the same approval set as if they were a single voter. *Point Voting (PV)* rule assigns each project a fraction of budget proportional to its approval score, which does not satisfy any of our fairness properties. *Uncoordinated Equal Shares (ES)* rule allocates each agent a $1/n$ share of the budget to spend equally on her approval projects.

1.1   Our Results

In this paper, we study the participatory budgeting problem under the objective of maximizing the egalitarian social welfare, i.e., the minimum utility among all agents. Two questions are considered in a worst-case analysis framework: how well can a distribution perform on the system efficiency, subject to a fairness constraint, and how much social welfare can be achieved by a certain voting rule. Suppose there are n agents and m projects, and the total budget is 1.

For the former question, we measure the system efficiency loss under a fair distribution by the price of fairness, defined as the ratio of the social welfare of the best fair distribution to the social welfare of the system optimum, under the worst-case instance. We study six fairness axioms concerning the price of fairness, and provided asymptotically tight bounds in Sect. 3. Because every system optimum satisfies IFS, the price of IFS is trivially 1. By constructing an example, we show that no distribution satisfying UFS (or GFS, IMP, AFS, CFS) can do better than $\frac{2}{n}$ for this example, and prove (almost) tight lower bounds. Our results are summarized in Table 1.

Table 1. The price of fairness for 6 axioms.

Fairness axioms	IFS	UFS	GFS	IMP	AFS	CFS
Lower bounds	1	$\frac{2}{n}$	$\frac{2}{n}$	$\frac{2}{n} - \frac{1}{n^2}$	$\frac{2}{n} - \frac{1}{n^2}$	$\frac{2}{n} - \frac{1}{n^2}$
Upper bounds	1	$\frac{2}{n}$	$\frac{2}{n}$	$\frac{2}{n}$	$\frac{2}{n}$	$\frac{2}{n}$

For the latter question, we study seven voting rules in Sect. 4. The *efficiency guarantee* [19] of a voting rule is the worst-case ratio between the social welfare induced by the rule and the system optimum. We provide asymptotically tight bounds for their efficiency guarantees, as shown in Table 2. Obviously EGAL is optimal, but it is not fair enough. CUT, NASH, ES and RP have a guarantee of $\Theta(\frac{1}{n})$, and in particular, NASH is very fair in the sense that it satisfies all axioms mentioned above.

Table 2. Efficiency guarantees for 7 voting rules

Voting rules	UTIL	CUT	NASH	EGAL	PV	ES	RP
Lower bounds	0	$\frac{1}{n}$	$\frac{2}{n} - \frac{1}{n^2}$	1	0	$\frac{1}{n}$	$\frac{2}{n}$
Upper bounds	0	$\frac{1}{n-3}$	$\frac{2}{n}$	1	$O(\frac{1}{mn})$	$\frac{1}{n} + O(\frac{1}{n^k})$	$\frac{2}{n}$

1.2 Related Work

Participatory budgeting (PB), introduced by Cabannes [12], is a process of democratic deliberation and decision-making, in which an authority allocates a fixed budget to projects, according to the preferences of multiple agents over the projects. Goel *et al.* [18] and Benade *et al.* [4] mainly focus on aggregation rules of PB for social welfare maximization. In the setting where the budget is perfectly divisible, it can be regarded as a probabilistic voting process [11,17], where a voting rule takes as input agents' (aka voters') preferences and returns a probability distribution over projects.

An important consideration on PB is what input format to use for preference elicitation - how each voter should express her preferences over the projects. While arbitrary preferences can be complicated and difficult to elicit, dichotomous preferences are simple and practical [6,7], where agents either approve or disapprove a project. For the dichotomous preference, there have been works both for divisible projects (e.g., [2,7]) and indivisible projects (e.g., [3,23]). This divisible approval-based setup is a popular input format in many settings, for which many fairness notions and voting rules have been proposed [2,10,15]. The fair share guarantee principles (e.g., IFS, GFS and AFS) are central to the fair division literature [8,21]. IMP is discussed in [10]. Brandl *et al.* [9] give a formal study of strict participation in probabilistic voting. Recently, Aziz *et al.* [2] give a detailed discussion of the above fairness notions.

For the voting rules (sometimes referred to as PB algorithms), EGAL rule maximizes the egalitarian social welfare, and is used as the lead rule in related assignment model with dichotomous preferences in [6]. NASH rule maximizes a classic collective utility function, and has featured prominently in researches [2,13,14,16]. CUT rule was first implicitly used in [15] and studied in more detail by Aziz *et al.* [2]. RP rule is discussed in [7].

Our work takes direct inspiration from Michorzewski *et al.* [20], who study the price of fairness in the divisible approval-based setting for maximizing utilitarian social welfare (while we consider the egalitarian one). Price of fairness quantifies the trade-off between fairness properties and maximization of egalitarian social welfare, and is widely studied [1,5,22].

2   Preliminaries

An instance is a triple $I = (N, P, A)$, where $N = \{1, \ldots, n\}$ is a set of agents and $P = \{p_1, \ldots, p_m\}$ is a set of projects. Each agent $i \in N$ has an *approval set* $A_i \subseteq P$ over the projects, and $A = \{A_1, \ldots, A_n\}$ is the profile of approval sets. Let $\mathcal{I}_n$ be the set of all instances with n agents. For each project $p_j \in P$, let $N(p_j) = \{i \in N : p_j \in A_i\}$ be the set of agents who approve p_j, and $|N(p_j)|$ be the *approval score* of p_j.

A budget division solution is a distribution $\mathbf{x} \in [0,1]^m$ over the projects set P, where x_j indicates the budget assigned to project p_j, and $\sum_{j=1}^m x_j = 1$. Let $\Delta(P)$ be the set of such distributions. The *utility* of agent $i \in N$ under distribution $\mathbf{x}$ is the amount of budget assigned to her approved projects, that is, $u_i(\mathbf{x}) = \sum_{p_j \in A_i} x_j$. The *(egalitarian) social welfare* of $\mathbf{x}$ is

$$sw(I, \mathbf{x}) = \min_{i \in N} u_i(\mathbf{x}).$$

Define the *normalized social welfare* of $\mathbf{x}$ as

$$\hat{sw}(I, \mathbf{x}) = \frac{sw(I, \mathbf{x})}{sw^*(I)},$$

where $sw^*(I)$ is the optimal social welfare of instance I. Clearly, $\hat{sw}(I, \mathbf{x}) \in [0,1]$. Though the system optimum (that maximizes the minimum utility of agents) is fair in some sense, it is not fair enough. We consider six fairness axioms. Given an instance $I = (N, P, A)$, a distribution $\mathbf{x}$ satisfies

- Individual Fair Share (IFS) if $u_i(\mathbf{x}) \geq 1/n$ for all agent $i \in N$;
- Unanimous Fair Share (UFS) if for every $S \subseteq N$ such that $A_i = A_j$ for all $i, j \in S$, we have $u_i(\mathbf{x}) \geq |S|/n$ for any $i \in S$.

- Group Fair Share (GFS) if for every $S \subseteq N$, we have $\sum_{p_j \in \cup_{i \in S} A_i} x_j \geq |S|/n$;
- Implementability (IMP) if we can write $\mathbf{x} = \frac{1}{n} \sum_{i \in N} \mathbf{x_i}$ for some distribution $\mathbf{x_i}$ such that $x_{i,j} > 0$ only if $p_j \in A_i$;
- Average Fair Share (AFS) if for every $S \subseteq N$ such that $\cap_{i \in S} A_i \neq \emptyset$, we have $\frac{1}{|S|} \sum_{i \in S} u_i(\mathbf{x}) \geq |S|/n$;
- Core Fair Share (CFS) if for every $S \subseteq N$, there is no vector $\mathbf{z} \in [0,1]^m$ with $\sum_{j=1}^{m} z_j = |S|/n$ such that $u_i(\mathbf{z}) > u_i(\mathbf{x})$ for all $i \in S$.

IFS is the weakest one among the above axioms. Besides, each of CFS, AFS and IMP implies GFS, which implies UFS.

A *voting rule* f is a function that maps an instance I to a distribution $f(I) \in \Delta(P)$. We consider the following voting rules:

- Utilitarian (UTIL) rule selects $\mathbf{x}$ maximizing $\sum_{i \in N} u_i(\mathbf{x})$.
- Conditional Utilitarian (CUT) rule selects the distribution $\frac{1}{n} \sum_{i \in N} \mathbf{x_i}$, where $\mathbf{x_i}$ is the uniform distribution over the projects in A_i with the highest approval score.
- Nash Max Product (NASH) rule selects $\mathbf{x}$ maximizing $\prod_{i \in N} u_i(\mathbf{x})$.
- Egalitarian (EGAL) rule selects $\mathbf{x}$ maximizing $\min_{i \in N} u_i(\mathbf{x})$.
- Point Voting (PV) rule selects $\mathbf{x}$, where $x_j = \frac{|N(p_j)|}{\sum_{p \in P} |N(p)|}$ for $p_j \in P$.
- Uncoordinated Equal Shares (ES) rule selects distribution $\frac{1}{n} \sum_{i \in N} \mathbf{x_i}$, where $\mathbf{x_i}$ is the uniform distribution over A_i, for any $i \in N$.
- Random Priority (RP) rule selects $\frac{1}{n!} \sum_{\sigma \in \Theta(N)} f^\sigma(I)$, where $\Theta(N)$ is the set of all strict orderings of N, and $f^\sigma(I) \in \arg\max_{\mathbf{x} \in \Delta(P)} \succ_{lexico}^{\sigma}$ is a distribution maximizing the utilities of agents lexicographically with ordering σ.

A voting rule f satisfies a fairness axiom if distribution $f(I)$ satisfies it for all instances I. Table 3 shows the fairness axioms satisfied by the above voting rules.

Table 3. Fairness axioms satisfied by voting rules

	UTIL	CUT	NASH	EGAL	PV	ES	RP
IFS		+	+	+		+	+
UFS		+	+			+	+
GFS		+	+			+	+
IMP		+	+			+	
AFS			+				
CFS			+				

As a warm-up, we give some properties on the optimal social welfare.

Proposition 1. *Let m^* be the minimum possible number such that there is an optimal distribution giving positive budget to exactly m^* projects. If $m^* > 1$, the optimal social welfare is at most $\frac{m^*-1}{m^*}$. If $m^* = 1$, the optimal social welfare is 1.*

Proof. Consider an optimal distribution $\mathbf{x}$ that gives positive budget to $m^* > 1$ projects. For each project $p_j \in P$, there must exist an agent (say $a_j \in N$) who does not approve p_j; otherwise, a distribution allocating budget 1 to this project is optimal, and thus $m^* = 1$, a contradiction. Further, because $\mathbf{x}$ distributes budget 1 among the m^* projects, there is a project p_k receiving a budget at least $\frac{1}{m^*}$, and agent a_k has a utility at most $1 - \frac{1}{m^*}$, establishing the proof. $\square$

Proposition 2. *Let $m' = \min_{S \subseteq P: \cup_{p \in S} N(p) = N} |S|$ be the minimum possible number of projects that cover all agents. Then the optimal social welfare is at least $\frac{1}{m'}$.*

Proof. Consider m' projects that cover all agents, i.e., each agent approves at least one of the m' projects. A distribution that allocates $\frac{1}{m'}$ to each of the m' projects induces a utility of at least $\frac{1}{m'}$ for every agent, implying the optimal social welfare at least $\frac{1}{m'}$. $\square$

Proposition 3. *For an instance (N, P, A), if the optimal social welfare is $\frac{k}{n}$ for some $k \leq n$, then there exists a project $p_j \in P$ such that at least $\lfloor k \rfloor$ agents approve it, i.e., $N(p_j) \geq \lfloor k \rfloor$.*

Proof. Suppose for contradiction that for every $p_j \in P$, $N(p_j) \leq \lfloor k \rfloor - 1$. Let $\mathbf{x}$ be an optimal distribution. Each project p_j can provide totally at most $(\lfloor k \rfloor - 1)x_j$ utility for the n agents. As $\sum_{p_j \in P} x_j = 1$, the total utility that the m projects can provide for the n agents is at most $\lfloor k \rfloor - 1$. Hence, there exists at least one agent whose utility is at most $\frac{\lfloor k \rfloor - 1}{n} < \frac{k}{n}$, a contradiction with the optimal social welfare. $\square$

3   Guarantees for Fairness Axioms

Given an instance I, the *price of fairness (POF)* of IFS with respect to I is defined as the ratio of the social welfare of the best IFS distribution to the optimal social welfare, that is,

$$\text{POF}_{IFS}(I) = \sup_{\mathbf{x} \in \Delta_{IFS}} \frac{sw(I, \mathbf{x})}{sw^*(I)} = \sup_{\mathbf{x} \in \Delta_{IFS}} \hat{sw}(I, \mathbf{x}),$$

where Δ_{IFS} is the set of distributions satisfying IFS.

The POF of IFS is the infimum over all instances, that is,

$$\text{POF}_{IFS} = \inf_{I \in \mathcal{I}_n} \text{POF}_{IFS}(I).$$

The POFs of other fairness axioms are similarly defined.

By the definition of IFS (that every agent receives a utility at least $1/n$), it is easy to see that every instance admits an IFS distribution, and thus an optimal distribution must satisfy IFS. We immediately have the following theorem.

Theorem 1. *For any instance I, there exists an IFS distribution $\mathbf{x}$ such that $\hat{sw}(I, \mathbf{x}) = 1$. That is, $POF_{IFS} = 1$.*

Also, we can give a tight lower bound for the normalized social welfare of IFS distributions. Recall that GFS implies IFS.

Theorem 2. *For any instance I and any IFS (or GFS) distribution $\mathbf{x}$, we have $\hat{sw}(I, \mathbf{x}) \geq \frac{1}{n}$. Further, there exists an instance I and a GFS distribution $\mathbf{x}$ such that $\hat{sw}(I, \mathbf{x}) = \frac{1}{n}$.*

Proof. The first claim is straightforward from the definition. For the second claim, we consider an instance I with n agents and $m = 2n + 1$ projects. For any $i \in N \setminus \{n\}$, the approval set of agent i is $A_i = \{p_{2i-1}, p_{2i}, p_{2i+1}, p_{2n+1}\}$, and $A_n = \{p_{2n-1}, p_{2n}, p_1, p_{2n+1}\}$. That is, all agents have a common approval project p_{2n+1}, and each agent i has an approval project p_{2i}, which is not approved by other agents. The optimal social welfare is $sw^*(I) = 1$, attained by placing all budget to common project p_{2n+1}. Consider a distribution $\mathbf{x}$ where $x_{2i} = \frac{1}{n}$ for each $i \in N$, and $x_j = 0$ for any other project p_j. The utility of every agent is $\frac{1}{n}$, and it is easy to check that $\mathbf{x}$ satisfies GFS, because for any group $S \subseteq N$ we have $\sum_{p_j \in \cup_{i \in S} A_i} x_j = |S|/n$. Then the social welfare induced by distribution $\mathbf{x}$ is $sw(I, \mathbf{x}) = \frac{1}{n}$, which implies $\hat{sw}(I, x) = \frac{sw(I,x)}{sw^*(I)} = \frac{1}{n}$. This completes the proof. $\square$

In the following we give a universal upper bound $\frac{2}{n}$ on the POFs of all other fairness axioms.

Theorem 3. *There exists an instance I such that for every distribution $\mathbf{x}$ satisfying UFS (or GFS, IMP, AFS, CFS), we have $\hat{sw}(I, \mathbf{x}) \leq \frac{2}{n}$. That is, the POF of UFS (or GFS, IMP, AFS, CFS) is at most $\frac{2}{n}$.*

Proof. Consider an instance I with n agents and 2 projects. Agents $1, 2, \ldots, n-1$ approve project p_1, and agent n approves p_2. That is, $N(p_1) = \{1, 2, \ldots, n-1\}$ and $N(p_2) = \{n\}$. The optimal social welfare is $sw^*(I) = 1/2$, attained by giving each project half of the budget. For any distribution $\mathbf{x}$ satisfying UFS for instance I, let the utility of each agent in $N(p_1)$ be x_1, and the utility of agent n be x_2. Applying UFS to coalition $N(p_1)$ and $N(p_2)$, respectively, we have $x_1 \geq \frac{n-1}{n}$ and $x_2 \geq \frac{1}{n}$. Since $x_1 + x_2 = 1$, it must be $x_1 = \frac{n-1}{n}, x_2 = \frac{1}{n}$. Then $sw(I, \mathbf{x}) = \min\{x_1, x_2\} = \frac{1}{n}$, and $\hat{sw}(I, x) = \frac{sw(I,x)}{sw^*(I)} = \frac{2}{n}$. This completes the proof for UFS. Since each of GFS, IMP, AFS and CFS implies UFS, this conclusion also holds for GFS, IMP, AFS and CFS. $\square$

4   Guarantees for Voting Rules

In this section, we consider seven voting rules (UTIL, CUT, NASH, EGAL, PV, ES and RP), and analyze their performance on the system objective in the worst case. Further, these analysis turn back to provide POF results for fairness axioms, as the voting rules satisfy some certain fairness axioms (see Table 3). Define the *efficiency guarantee* (or simply, *guarantee*) of voting rule f as the worst-case normalized social welfare:

$$k_{eff}(f) = \min_{I \in \mathcal{I}_n} \hat{sw}(I, f(I)).$$

Theorem 4. *The efficiency guarantee of UTIL is 0, and that of EGAL is 1. The efficiency guarantees of CUT, NASH, ES, and RP are all in $[\frac{1}{n}, \frac{2}{n}]$.*

Proof. The efficiency guarantee of EGAL is trivial. Consider the instance constructed in the proof of Theorem 3, where $N(p_1) = \{1, \ldots, n-1\}$ and $N(p_2) = \{n\}$. The optimal social welfare is $\frac{1}{2}$, attained by allocating $x_1 = x_2 = \frac{1}{2}$. Rule UTIL returns $x_1 = 1$ and $x_2 = 0$, inducing a utility 0 for agent n. So the guarantee of UTIL is 0. Rules CUT, NASH, PV, ES all return $x_1 = 1 - \frac{1}{n}$ and $x_2 = \frac{1}{n}$, inducing a utility $\frac{1}{n}$ for agent n. So the guarantee of these four rules is at most $\frac{1/n}{1/2} = \frac{2}{n}$. (Indeed, this claim simply follows from Theorem 3, since all the four rules satisfy UFS.)

On the other hand, for any instance I and the distribution $\mathbf{x}$ returned by CUT (resp., NASH, ES, RP), we have $u_i(I) \geq \frac{1}{n}$ for any $i \in N$, since the rule satisfies IFS. Then $sw(I, \mathbf{x}) \geq \frac{1}{n}$, which implies $\hat{sw}(I, \mathbf{x}) \geq \frac{1}{n}$. Therefore, the efficiency guarantees of CUT (resp., NASH, ES, RP) is at least $\frac{1}{n}$.   □

Theorem 5. *The efficiency guarantee of ES is no better than $\frac{1}{n} + O(\frac{1}{n^k})$, for all $k \in \mathbb{N}^+$.*

Proof. Consider an instance with n agents and $m = n^{k+1} + 1$ projects. Each agent approves $n^k + 1$ projects. The intersection of every two approval sets is project p_m, implying that p_m is approved by all agents, and the approval score of every other project is 1. It is easy to see that (by Proposition 1), the optimal social welfare is 1, attained by allocating all budget to p_m. The outcome of ES is $x_m = \frac{1}{n^k+1}$, and $x_j = \frac{1}{n} \cdot \frac{1}{n^k+1}$ for any $p_j \neq p_m$. Thus, ES gives each agent a utility of

$$\frac{1}{n(n^k+1)} \cdot n^k + \frac{1}{n^k+1} = \frac{n^{k-1}+1}{n^k+1} = \frac{1}{n} + O(\frac{1}{n^k}),$$

which completes the proof.   □

Theorem 6. *The efficiency guarantee of PV is $O(\frac{1}{mn})$.*

Proof. Consider an instance I with n agents and m projects. Each agent in $N \setminus \{n\}$ approves $p_1, \ldots, p_{m-1}$, and agent n approves p_m. That is, the first $m-1$ projects are approved by the first $n-1$ agents, and the last project is approved by the remaining agent. The optimal social welfare is $sw^*(I) = \frac{1}{2}$, attained by a distribution with $x_1 = x_m = \frac{1}{2}$. However, PV allocates each project in $P \setminus \{p_m\}$ a budget of $\frac{n-1}{(m-1)(n-1)+1}$, and project p_m a budget of $\frac{1}{(m-1)(n-1)+1}$. Then the social welfare induced by PV is $\frac{1}{(m-1)(n-1)+1}$, which implies the guarantee of PV is at most $\frac{2}{(m-1)(n-1)+1} = O(\frac{1}{mn})$.   □

Theorem 7. *The efficiency guarantee of CUT is no better than $\frac{1}{n-3}$.*

Proof. Consider an instance with $m = \binom{n-1}{2} + 1$. Each of the first $\binom{n-1}{2}$ projects corresponds to a unique pair of the first $n-1$ agents who disapprove it, and all other agents approve it; the last project p_m is approved by the first $n-1$ agents.

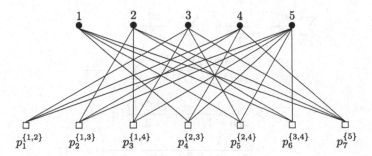

Fig. 1. An example with $n = 5$ agents and $m = 7$ projects, where p_j^S indicates that project p_j is disapproved by every agent in set $S \subseteq N$. Each of the first 6 projects corresponds to a pair of agents.

That is, each agent in $N \setminus \{n\}$ disapproves $n - 2$ projects in $P \setminus \{p_m\}$, and approves all other projects; agent n approves all $m - 1$ projects in $P \setminus \{p_m\}$, and disapproves project p_m. Figure 1 shows a 5-agent example.

Then we have $|N(p_m)| = n - 1$, and $|N(p_j)| = n - 2$ for each $j \leq m - 1$. The optimal social welfare is at least $1 - \frac{n-2}{m} > \frac{n-3}{n-1}$, achieved by allocating uniform budget to each project. However, CUT rule allocates each project in $P \setminus \{p_m\}$ a budget of $\frac{1}{n(m-1)}$, and project p_m a budget of $\frac{n-1}{n}$. Then the social welfare induced by CUT is $1/n$, (i.e., the utility of agent n) which implies the efficiency guarantee of CUT is at most $\frac{n-1}{n(n-3)} < \frac{1}{n-3}$. $\qquad\square$

Theorem 8. *The efficiency guarantee of NASH is in* $[\frac{2}{n} - \frac{1}{n^2}, \frac{2}{n}]$.

Proof. Let I be an arbitrary instance with n agents and m projects, and $f_{NS}(I)$ be the distribution returned by NASH rule. Since NASH satisfies IFS, we have $sw(I, f_{NS}(I)) \geq \frac{1}{n}$. By Theorem 3 and the fact that NASH satisfies UFS, the efficiency guarantee is at most $\frac{2}{n}$. If the social welfare induced by NASH rule is $sw(I, f_{NS}(I)) \geq \frac{2}{n}$, the proof is done. So we only need to consider the case $sw(I, f_{NS}(I)) \in [\frac{1}{n}, \frac{2}{n})$. Suppose for contradiction that

$$\frac{sw(I, f_{NS}(I))}{sw^*(I)} < \frac{2}{n} - \frac{1}{n^2}. \tag{1}$$

Let $\mathbf{u}^*$ and $\mathbf{u}^{NS}$ be the utility profiles induced by an optimal distribution and the solution output by NASH, respectively. Let $\bar{i} \in N$ be the agent with the minimum utility in the NASH solution, i.e., $u_{\bar{i}}^{NS} = sw(I, f_{NS}(I))$. Then we have

$$\sum_{i \in N} \frac{u_i^*}{u_i^{NS}} = \frac{u_{\bar{i}}^*}{u_{\bar{i}}^{NS}} + \sum_{i \in N: i \neq \bar{i}} \frac{u_i^*}{u_i^{NS}}$$

$$\geq \frac{sw^*(I)}{sw(I, f_{NS}(I))} + \sum_{i \in N: i \neq \bar{i}} \frac{u_i^*}{u_i^{NS}} \tag{2}$$

$$> \frac{1}{2/n - 1/n^2} + \frac{sw^*(I)}{1}(n-1)$$

$$> \frac{1}{2/n - 1/n^2} + \frac{1/n}{2/n - 1/n^2}(n-1) \tag{3}$$

$$= n, \tag{4}$$

where (2) comes from (1), and (3) comes from (1) and $sw(I, f_{NS}(I)) \geq \frac{1}{n}$.

The outcome of NASH rule is an optimal solution of the following convex optimization problem:

$$\max \quad h(\mathbf{u}) = \sum_{i \in N} \log(u_i)$$

$$\text{s.t.} \quad \sum_{p_j \in A_i} x_j = u_i, \quad \text{for all } i \in N$$

$$\sum_{j=1}^{m} x_j = 1,$$

$$x_j \geq 0, \quad \text{for } j = 1, \ldots, m$$

Let $\mathcal{D}$ be the feasible domain of this problem. Since all constraints are linear, $\mathcal{D}$ is a convex set. Let $\mathcal{D}_{\mathbf{u}} = \{\mathbf{u} | \exists \mathbf{x} \text{ s.t. } (\mathbf{u}, \mathbf{x}) \in \mathcal{D}\}$ be a restriction on $\mathbf{u}$. Then for any $0 \leq \alpha \leq 1$ and any utility profile $\mathbf{u}' \in \mathcal{D}_{\mathbf{u}}$, we have $\mathbf{u}^{NS} + \alpha(\mathbf{u}' - \mathbf{u}^{NS}) \in \mathcal{D}_{\mathbf{u}}$. Then, we can derive

$$\lim_{\alpha \to 0^+} \frac{h(\mathbf{u}^{NS} + \alpha(\mathbf{u}' - \mathbf{u}^{NS})) - h(\mathbf{u}^{NS})}{\alpha} \leq 0$$

$$\implies \nabla h(\mathbf{u}^{NS})^{\mathrm{T}}(\mathbf{u}' - \mathbf{u}^{NS}) \leq 0$$

$$\implies \sum_{i \in N} \frac{u_i'}{u_i^{NS}} \leq n,$$

which gives a contradiction to Eq. (4). $\qquad \square$

Because NASH rule satisfies the properties IMP, AFS and CFS, combining with Theorem 3, we have the following corollary.

Corollary 1. *The POFs of IMP, AFS and CFS are all in $[\frac{2}{n} - \frac{1}{n^2}, \frac{2}{n}]$.*

Theorem 9. *The efficiency guarantee of RP is $\frac{2}{n}$.*

Proof. Let I be an arbitrary instance, and $f_{RP}(I)$ be the distribution returned by RP rule. Since RP satisfies IFS, the social welfare of $f_{RP}(I)$ is at least $\frac{1}{n}$. If $sw^*(I) \leq \frac{1}{2}$, the normalized social welfare is $\hat{sw}(I, f_{RP}(I)) \geq \frac{2}{n}$. So it suffices to consider the case $sw^*(I) > \frac{1}{2}$.

When $sw^*(I) > \frac{1}{2}$, if there are two agents $i, j \in N$ such that $A_i \cap A_j = \emptyset$, then no distribution can give both agents a utility larger than $\frac{1}{2}$, a contradiction. So for any two agents, the intersection of their approval sets is non-empty. For each agent $i \in N$, under RP rule, the probability of ranking the first among the $n!$ permutations is $\frac{1}{n}$, where she receives a utility 1. Suppose $i = \sigma(2)$ and $j = \sigma(1)$ for a permutation $\sigma \in \Theta(N)$. Since RP maximizes the utility of agents lexicographically with respect to σ, it must allocate all budget to their intersection $A_i \cap A_j$, and the utilities of agents i and j both are 1. Note that the probability of ranking the second among the $n!$ permutations for agent i is $\frac{1}{n}$. The utility of agent i under RP rule is at least

$$\mathbf{Pr}\{i = \sigma(1)\} \cdot 1 + \mathbf{Pr}\{i = \sigma(2)\} \cdot 1 = \frac{2}{n},$$

and the normalized social welfare is also at least $\frac{2}{n}$. Combining with the upper bound in Theorem 4, the efficiency guarantee of RP is $\frac{2}{n}$. $\qquad\square$

We remark that it is still open whether RP rule can be implemented in polynomial time. Because RP rule satisfies GFS, combining with Theorem 3, we have the following corollary.

Corollary 2. *The POF of GFS is* $\frac{2}{n}$.

5   Conclusion

We quantify the trade-off between the fairness criteria and the maximization of egalitarian social welfare in a participatory budgeting problem with dichotomous preferences. Compared with the work of Michorzewski *et al.* [20], which considers this approval-based setting under the utilitarian social welfare, we additionally study a fairness axiom Unanimous Fair Share (UFS) and a voting rule Random Priority (RP). We present (asymptotically) tight bounds on the price of fairness for six fairness axioms and the efficiency guarantees for seven voting rules. In particular, both NASH and RP rules are guaranteed to provide a roughly $\frac{2}{n}$ fraction of the optimum egalitarian social welfare. The NASH solution can be computed by solving a convex program, while RP is unknown to be computed efficiently.

Both the work of [20] and this paper assume that all agents have dichotomous preferences, and an immediate future research direction would be to study the effect of fairness constraint when agents are allowed to have a more general preference. Another avenue for future research is considering the fairness in participatory budgeting from the projects' perspective. For example, the projects (e.g., facility managers and location owners) have their own thoughts, and may have a payoff from the budget division, for which a good solution should balance the system efficiency and the satisfaction of the agents and projects. So it would be interesting to study the trade-off between system efficiency and this new class of fairness concepts.

Acknowledgement. The authors thank Xiaodong Hu and Xujin Chen for their help-ful discussions, and anonymous referees for their valuable feedback.

References

1. Aumann, Y., Dombb, Y.: The efficiency of fair division with connected pieces. ACM Trans. Econ. Comput. **3**(4), 1–16 (2015)
2. Aziz, H., Bogomolnaia, A., Moulin, H.: Fair mixing: the case of dichotomous pref-erences. In: Proceedings of the 20th ACM Conference on Economics and Compu-tation (ACM-EC), pp. 753–781 (2019)
3. Aziz, H., Lee, B., Talmon, N.: Proportionally representative participatory budget-ing: axioms and algorithms. arXiv:1711.08226 (2017)
4. Benade, G., Nath, S., Procaccia, A.D., Shah, N.: Preference elicitation for par-ticipatory budgeting. In: Proceedings of the 31st AAAI Conference on Artificial Intelligence (AAAI), pp. 376–382 (2017)
5. Bertsimas, D., Farias, V.F., Trichakis, N.: The price of fairness. Oper. Res. **59**(1), 17–31 (2011)
6. Bogomolnaia, A., Moulin, H.: Random matching under dichotomous preferences. Econometrica **72**(1), 257–279 (2004)
7. Bogomolnaia, A., Moulin, H., Stong, R.: Collective choice under dichotomous pref-erences. J. Econ. Theory **122**(2), 165–184 (2005)
8. Bouveret, S., Chevaleyre, Y., Maudet, N.: Fair allocation of indivisible goods. In: Handbook of Computational Social Choice. Cambridge University Press (2016)
9. Brandl, F., Brandt, F., Hofbauer, J.: Incentives for participation and abstention in probabilistic social choice. In: Proceedings of the 14th International Conference on Autonomous Agents and Multi-Agent Systems (AAMAS), pp. 1411–1419 (2015)
10. Brandl, F., Brandt, F., Peters, D., Stricker, C., Suksompong, W.: Donor coordina-tion: collective distribution of individual contributions. Technical report (2019)
11. Brandt, F.: Rolling the dice: recent results in probabilistic social choice. Trends Comput. Soc. Choice 3–26 (2017)
12. Cabannes, Y.: Participatory budgeting: a significant contribution to participatory democracy. Environ. Urbanization **16**(1), 27–46 (2004)
13. Caragiannis, I., Kurokawa, D., Moulin, H., Procaccia, A.D., Shah, N., Wang, J.: The unreasonable fairness of maximum nash welfare. In: Proceedings of the 17th ACM Conference on Economics and Computation (ACM-EC), pp. 305–322 (2016)
14. Conitzer, V., Freeman, R., Shah, N.: Fair public decision making. In: Proceedings of the 18th ACM Conference on Economics and Computation (ACM-EC), pp. 629–646 (2017)
15. Duddy, C.: Fair sharing under dichotomous preferences. Math. Soc. Sci. **73**, 1–5 (2015)
16. Fain, B., Goel, A., Munagala, K.: The core of the participatory budgeting problem. In: Cai, Y., Vetta, A. (eds.) WINE 2016. LNCS, vol. 10123, pp. 384–399. Springer, Heidelberg (2016). https://doi.org/10.1007/978-3-662-54110-4_27

17. Gibbard, A.: Manipulation of schemes that mix voting with chance. Econometrica: J. Econom. Soc. 665–681 (1977)
18. Goel, A., Krishnaswamy, A.K., Sakshuwong, S., Aitamurto, T.: Knapsack voting for participatory budgeting. ACM Trans. Econ. Comput. **7**(2), 1–27 (2019)
19. Lackner, M., Skowron, P.: A quantitative analysis of multi-winner rules. In: Proceedings of the 28th International Joint Conference on Artificial Intelligence (IJCAI) (2019)
20. Michorzewski, M., Peters, D., Skowron, P.: Price of fairness in budget division and probabilistic social choice. In: Proceedings of the 34th AAAI Conference on Artificial Intelligence (AAAI), pp. 2184–2191 (2020)
21. Moulin, H.: Fair Division and Collective Welfare. MIT Press, Cambridge (2004)
22. Suksompong, W.: Fairly allocating contiguous blocks of indivisible items. Discret. Appl. Math. **260**, 227–236 (2019)
23. Talmon, N., Faliszewski, P.: A framework for approval-based budgeting methods. In: Proceedings of the 33th AAAI Conference on Artificial Intelligence (AAAI), pp. 2181–2188 (2019)

Inspection Strategy for On-board Fuel Sampling Within Emission Control Areas

Lingyue Li, Suixiang Gao, and Wenguo Yang[✉]

School of Mathematical Sciences, University of Chinese Academy of Sciences,
Beijing 100049, China
lilingyue17@mails.ucas.ac.cn, {sxgao,yangwg}@ucas.ac.cn

Abstract. This paper quantitatively analyzes the inspection strategy (which arriving ships are selected for inspection) for on-board fuel sampling considering limited inspection capacity and ships' violation behaviors. By establishing a semi-random input model and proposing the corresponding algorithm, the optimal inspection strategy is obtained. Furthermore, the impacts of related factors on the optimal inspection strategy are analyzed. The results show that compared to randomly select ships, the method proposed in our study can determine a more reasonable inspection strategy.

Keywords: Maritime transportation · Emission control areas · On-board fuel sampling · Semi-random input model

1 Introduction

In recent years, emission control areas (ECAs) that are proposed by the International Maritime Organization (IMO) [1] have become an important measure to reduce and control pollutant emissions from ships [2, 3]. As a basis and guarantee for the effective enforcement of the ECA regulations, governmental supervision for violations by ships is not only crucial for violation identification, but also has an important role in the achievement of expected environmental benefits. At present, port state is responsible for the supervision of ships arriving at the port and there are three commonly employed supervision methods. Among them, on-board fuel sampling that verifies the sulfur content of fuels being utilized on board is a commonly employed and reliable supervision method. However, this method has two challenges. First, the inspection capacity of a government is limited. Many ships visit one port per day, and the government can only inspect a small number of ships. Second, the violation behaviors of ships are unknown to governments. After the establishment of ECAs, there are two strategies for a ship, violation and compliance. Violation behavior is a strategic choice, indicating that the ship violates the ECA regulation. Considering violation behavior is private information of ship operators, thus governments cannot know ships' final decisions in advance. Due to the above challenges, there is no relevant research or policy to show the quantitative method of selecting inspected ships. To address this problem, this paper presents an optimization model to quantitatively analyze the inspection strategy (which arriving ships

W. Wu and Z. Zhang (Eds.): COCOA 2020, LNCS 12577, pp. 608–623, 2020.
https://doi.org/10.1007/978-3-030-64843-5_41

are selected for inspection) for on-board fuel sampling. Our aim is to develop a feasible inspection strategy that is adapted to a government's limited inspection capability and ships' violation behaviors.

Although previous studies of behavior analysis of ships and governments have achieved many results, some progress is urgently needed. First, an accurate description of the behavior of each ship is lacking. Based on the evolutionary game model in related studies [4, 5], ships are assumed to be a homogeneous population, and the individual differences of ships and the relationship among ships are disregarded. Second, few studies about inspection strategy have considered the inspection capacity. In 2018, the European Union (EU) issued regulatory guidance on sulfur emissions that employs THETIS-EU, including historical and generic information from third party alerts [6]. Nevertheless, more factors may affect the inspection strategy and need to be considered, such as the penalty policy, economy, port and ship characteristics. In addition, analysis of methods to improve the inspection strategy is lacking. Therefore, combining influencing factors to accurately describe ships' violation behaviors and considering a government's limited inspection capacity to propose a reasonable inspection strategy remains to be performed and would be extremely valuable.

Inspection strategy analysis from the perspective of mathematical description and optimization has not been addressed. Therefore, this paper is based on an optimization model to quantitatively analyze the inspection strategy with a government's limited inspection capacity and ships' violation behaviors. Moreover, the impacts of related factors on the inspection strategy are discussed. The results of this paper will provide methodological support for the effective enforcement of the ECA regulations, which is crucial for pollutant emissions reduction from ships.

The main contributions of this paper are summarized as follows: first, we describe the behavior of a ship with three properties— economic benefit, social responsibility and randomness—which provides a reasonable simulation. Second, we construct a semi-random input model to obtain the optimal inspection strategy, which can provide governments with feasible guidance. Third, we analyze the impact of related factors in terms of government and ship characteristics, which provides support for the dynamic adjustment of an inspection strategy.

2 Description of Ships' Violation Behaviors

A ship has two strategies (violation and compliance), and its final behavior is affected by many factors, such as economic benefit, social responsibility, inspection capacity and competition between ships. Thus, the challenge is how to quantify the impact of these factors and combine them.

Since the ultimate goal of ships is to make a profit, it is obvious that economic benefit is a main factor. We use fuel cost saving and penalty fine to describe the economic benefit, and environmental cost to describe social responsibility.

First, the fuel cost saving for a non-compliant ship is calculated by multiplying the fuel price difference by fuel consumption. Among them, fuel consumption is affected by many factors such as speed, load, ship type, ocean current and weather. Among these factors, speed has the greatest impact on fuel consumption, and many studies have determined that fuel consumption is proportional to the sailing speed [7–9]. Therefore, this paper also applies this relation to calculate the fuel consumption.

$$R = f \times v^a \tag{1}$$

where R is the fuel consumption per unit time, f is a conversion factor between speed and fuel consumption, v is the sailing speed, and a is the proportionality constant. We adopt the most commonly employed parameter assumptions $f = 0.012$ and $a = 3$ in the literature [10–12].

Thus, the fuel cost saving for a non-compliant ship is calculated as follows:

$$Q = (P^e - P^w)R^e = 0.012(P^e - P^w)v^2D \tag{2}$$

where Q represents the fuel cost saving; P^e and P^w represent the fuel prices per ton of compliant fuel and non-compliant fuel, respectively; R^e represents the fuel consumption when the ship sailing within the ECA. D represents the ship's sailing distance within the ECA.

Second, the penalty fine for a non-compliant ship is calculated based on the Environmental Protection Agency (EPA) [13]:

$$A = \lambda\left((P^e - P^w) \times \frac{AFC}{24} \times \frac{D}{v} + \alpha(s \times \frac{AFC}{24} \times \frac{D}{v} + G_R)\right) \tag{3}$$

where A represents the penalty fine; λ and α represent correction factors; AFC represents the average daily fuel consumption; s represents the sulfur content of non-compliant fuels; G_R represents a record-keeping violation.

Third, considering the incremental emissions caused by violations require a certain amount of pollution control cost to be removed, we assume the environmental cost of a non-compliant ship is equal to the incremental emissions multiplied by the corresponding pollution control cost:

$$C = \sum_k p^k \Delta E^k = \sum_k p^k (EF_k^e - EF_k^w)R^e \tag{4}$$

where C represents the environmental cost; k represents the type of pollution emissions; p^k represents the pollution control cost for type k emission. ΔE^k represents incremental emissions of type k emission. EF_k^w and EF_k^e represent the emission factors of type k emission for compliant fuel and non-compliant fuel, respectively. Since the motivation of ECAs is to cut SOx, NOx and PM emissions, we assume that k includes SOx, NOx and PM.

Based on the discussion above, the initial violation tendency for the i-th ship can be determined as follows:

$$c_i = \frac{Q_i}{A_i + C_i} \tag{5}$$

where c_i represents the initial violation tendency for the i-th ship. Q_i, A_i and C_i represent the fuel cost saving, penalty fine and environmental cost for the i-th ship, respectively.

Considering the government may have inspection records. That is, information about which ship was inspected and whether the ship complies with the regulations. Making this historical information available would be helpful. When the historical behaviors of violations and inspections are known, the initial violation tendency for the i-th ship can be determined as follows:

$$c_i = \frac{Q_i + \sum_{l_i}^{w_i} Q_{l_i}}{\left(A_i + \sum_{l_i}^{w_i} A_{l_i}\right) + \left(C_i + \sum_{l_i}^{w_i} C_{l_i}\right)} \tag{6}$$

where w_i represents the number of previous arrivals of the i-th ship. Q_{l_i}, A_{l_i} and C_{l_i} represent the fuel cost saving, penalty fine and environmental cost for the l-th arrival of the i-th ship, respectively.

In addition, other factors will also affect a ship's violation behavior, but they are difficult to quantify accurately. This paper assumes that the impact of these factors on ships' violation behaviors comes from the same distribution, which is characterized by a Gaussian distribution with mean 0 and standard deviation σ, denoted by $G(0, \sigma^2)$. Specifically, the initial violation tendency c_i is randomly perturbed by adding a random variable generated by $G(0, \sigma^2)$ to each c_i. The smaller σ is, the smaller the influence of other factors on initial violation tendency. The standard deviation is determined as follows:

$$\sigma = \sqrt{\frac{1}{N-1} \sum_{i=1}^{N} |c_i - \mu|^2} \tag{7}$$

where N represents the total number of ships. μ represents the mean of c, that is, $\mu = \frac{1}{N} \sum_{i=1}^{N} c_i$. Thus, the violation tendency for the i-th ship is $b_i = c_i + G(0, \sigma^2)$[1].

---

[1] According to the simulation, the numerator and denominator of c have the same order of magnitude. The range of c and σ are [0.6, 1.8] and 0.16, respectively. In addition, the determination of non-compliant ships is mainly affected by the relative value of b and has little relationship with the absolute value b. Therefore, the establishment of b is relatively reasonable.

Finally, we use the roulette selection method to generate ship behaviors (x_i) based on non-compliance rate (number of inspected ships as a percentage of the total number of ships). Based on the discussion above, the following procedure is proposed to describe ships' violation behaviors:

Step 1: Determine the violation tendency c_i by considering the historical and current economic benefit of the i-th ship.
Step 2: Determine the standard deviation σ by considering other factors of the i-th ship.
Step 3: Determine the violation tendency b_i for the i-th ship.
Step 4: Let I_i be the sum of the violation tendency of the first i ships and $I_0 = 0$. Thus, we get an interval $[I_0, \ldots I_N]$, where $I_i = I_{i-1} + b_i$.
Step 5: Randomly select a point in the interval. If the point falls within the k-th sub-interval $[I_k, I_{k+1}]$, then assume that the k-th ship is a non-compliant ship.
Step 6: Randomly select points until the non-compliance rate is reached. A ship can only be selected once. If the ship is selected again, discard the result; a new random selection is needed until a different ship is selected.

3 Model Formulation and Algorithm

Port state is responsible to conduct the inspection, and the inspection strategy is that the government where the port is located needs to determine which ship is to be inspected under consideration of ships' violation behaviors and its inspection capability. The goal is to reduce the pollution emissions of ships through government inspections. This section uses a semi-random input model to obtain inspection strategy of the government. The symbols are defined as follows:

Indices and parameters:

N Total number of ships that visit a port in one day

m Number of inspection groups, where an inspection group is composed of several inspectors who will inspect one ship at a time

r Non-compliance rate, the number of violation ships as a percentage of total ships

β Inspection rate, the number of inspected ships as a percentage of total ships

i i-th ship

j j-th inspection group

k The type of pollution emissions

w_i Number of previous arrivals of the i-th ship

l_i l-th previous arrival of the i-th ship

t_{ji} Total inspection time for the j-th group to inspect the i-th ship, including the arrival and departure times of the inspected ship

T_j Daily working hours of the j-th group

C_i Environmental cost of the i-th ship

x_i Binary variable, equals one if the i-th ship violates regulations; otherwise, it equals zero

Decision variables:

y_{ji} Binary variable, equals one if the government inspects the i-th ship; otherwise, it equals zero

---

For a government, the final goal of on-board inspection is to prevent ships from violating ECA regulations, and then reducing pollutant emissions from ships. The government's optimal inspection strategy is to inspect ships that bring greater environmental costs. Thus, the objective of inspection is to maximize the environmental cost of inspected ships.

$$\max_{y_{ji} \in \{0,1\}} \sum_{j=1}^{m} \sum_{i=1}^{N} C_i x_i y_{ji} \tag{8}$$

For each inspection group, the total inspected time is less than or equal to the working hours per day, that is,

$$\sum_{i=1}^{N} t_{ji} y_{ji} \leq T_j, \quad j = 1, \ldots, m \tag{9}$$

Each ship is at most inspected once, thus

$$\sum_{j=1}^{m} y_{ji} \leq 1, \quad i = 1, \ldots, N \tag{10}$$

We assume that the non-compliance rate of ships is known and it is reasonable to use the value obtained by monitoring or theoretical analysis in previous studies. Thus, the number of non-compliant ships is equal to the non-compliance rate multiplied by the total number of ships.

$$\sum_{i=1}^{N} x_i = rN \tag{11}$$

The proportion of ships that are inspected is equal to the policy requirement.

$$\sum_{j=1}^{m} \sum_{i=1}^{N} y_{ji} = \beta N \tag{12}$$

The semi-random input model [P1] is constructed as follows:

$$\max_{y_{ji} \in \{0,1\}} \sum_{j=1}^{m} \sum_{i=1}^{N} C_i x_i y_{ji} \tag{13}$$

$$s.t. \quad \sum_{i=1}^{N} t_{ji} y_{ji} \leq T_j, \quad j = 1, \ldots, m \tag{14}$$

$$\sum_{j=1}^{m} y_{ji} \leq 1, \quad i = 1, \ldots, N \tag{15}$$

$$\sum_{i=1}^{N} x_i = rN \tag{16}$$

$$\sum_{j=1}^{m} \sum_{i=1}^{N} y_{ji} = \beta N \tag{17}$$

$$x_i \in \{0, 1\}, \quad i = 1, \ldots, N \tag{18}$$

$$y_{ji} \in \{0, 1\}, \quad i = 1, \ldots, N, \quad j = 1, \ldots, m \tag{19}$$

The objective function (13) is to maximize the environmental cost of inspected ships. Constraint (14) is the time constraint. Constraint (15) means each ship is at most inspected once. Constraint (16) is the non-compliance rate constraint. Constraint (17) is the inspection rate constraint. Constraints (18) is a semi-random input. Constraint (19) is the 0–1 constraints.

The difficulty of solving the model [P1] is that the objective function cannot be computed exactly, but it can be estimated through simulation. Thus, the Sample Average Approximation (SAA) method [14] is applicable, which can transform the model [P1] into multiple deterministic programming models. The corresponding algorithm is designed as follows:

Algorithm 1

Input: ship $\{n_1,...,n_N\}$, penalty fine $\{A_1,...,A_N\}$, profit $\{Q_1,...,Q_N\}$, historical penalty fines $\left\{\sum_{l_i=1}^{w_1} A_{l_i},...,\sum_{l_i=1}^{w_N} A_{l_i}\right\}$, historical fuel cost saving $\left\{\sum_{l_i=1}^{w_1} Q_{l_i},...,\sum_{l_i=1}^{w_N} Q_{l_i}\right\}$, historical environmental cost $\left\{\sum_{l_i=1}^{w_1} C_{l_i},...,\sum_{l_i=1}^{w_N} C_{l_i}\right\}$, daily working time $\{T_1,...,T_M\}$, non-compliance rate γ, inspection rate β and sampling times M.

Output: optimal inspection strategy y^*

Initialize $I_0 = 0$.

h represents the h-th sampling, initialize $h = 1$.

W represents the set of violation ships for each sampling, initialize $W = \varnothing$.

The average inspection probability V^h is a matrix of dimension m by N, where

$$V_{ji}^h = \left. \sum_{\theta=1}^{h} y_{ji}^{\theta} \middle/ \sum_{j=1}^{m}\sum_{i=1}^{n}\sum_{\theta=1}^{h} y_{ji}^{\theta} \right.$$

is the percentage of the number of the i-th ship inspected by the j-th group and the total number of inspections in the h sampling.

While $h <= M$ **do**

    for $i \leftarrow 1$ to N **do**

$$I_i = I_{i-1} + \frac{Q_i + \sum_{l_i}^{w_i} Q_{l_i}}{\left(A_i + \sum_{l_i}^{w_i} A_{l_i}\right) + \left(C_i + \sum_{l_i}^{w_i} C_{l_i}\right)}$$

    end for

    while $|W| <= rN$ **do**

        randomly select a value v from the interval $[0, I_N]$, then

        $u = \arg\max\{i \mid v \geq I_i\}$

        if $u \notin W$ **then**

            $W = W \cup \{u\}$

        end if

    end while

$$x_i^h = \begin{cases} 1 & i \in W \\ 0 & i \notin W \end{cases}$$

$$x^h = \{x_1^h,...,x_N^h\}$$

For given parameters γ, β and $\{T_1,...,T_M\}$, solve the deterministic programming model [P1] for fixed x^h.

    if successfully observes solution $y^h = \begin{pmatrix} y_{11}^h & \cdots & y_{1N}^h \\ \vdots & & \vdots \\ y_{m1}^h & \cdots & y_{mN}^h \end{pmatrix}$ **then**

        continue

end if
h++
end while

$$V^M = \begin{pmatrix} \sum_{\theta=1}^{M} y_{11}^{\theta} \Big/ \sum_{j=1}^{m} \sum_{i=1}^{N} \sum_{\theta=1}^{M} y_{ji}^{\theta} & \cdots & \sum_{\theta=1}^{M} y_{1N}^{\theta} \Big/ \sum_{j=1}^{m} \sum_{i=1}^{N} \sum_{\theta=1}^{M} y_{ji}^{\theta} \\ \vdots & & \vdots \\ \sum_{\theta=1}^{M} y_{m1}^{\theta} \Big/ \sum_{j=1}^{m} \sum_{i=1}^{N} \sum_{\theta=1}^{M} y_{ji}^{\theta} & \cdots & \sum_{\theta=1}^{M} y_{mN}^{\theta} \Big/ \sum_{j=1}^{m} \sum_{i=1}^{N} \sum_{\theta=1}^{M} y_{ji}^{\theta} \end{pmatrix}$$

Return $y^* = rounding(V^M \cdot \beta N)$, which is 0-1 matrix obtained by rounding technology based on V^M.

4   Results and Discussion

This section includes three parts. Section 4.1 describes the data. Section 4.2 employs an optimization model to analyze the optimal inspection strategy, and Sect. 4.3 analyzes the impacts of different influence factors. All the employed models are solved with MATLAB 2019b software.

4.1   Data

The Atlantic coast and Gulf of Mexico coast are part of the North American ECA, which has the most complex ECA boundary shape and corresponding penalty policy, and thus, will be conducive for the quantitative study. Therefore, we consider this area as the research object and select 27 ports based on the coastal shape and port size (see Fig. 1). The fuel prices mainly come from MABUX [15] in January 2019, and missing data are replaced by the average value of the remaining ports, which are IFO380 = 420, MGO

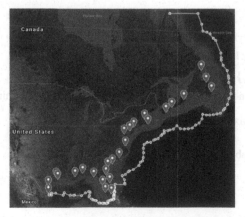

Fig. 1. ECA boundary and ports on the Atlantic coast and Gulf of Mexico coast.

= 650. Ship information comes from FleetMon [16] and sailing distances come from SeaRates [17]. The information of emissions factors and emission control costs comes from IMO [18] and CEC [19], respectively.

4.2 Optimal Inspection Strategy

This section focuses on a single target port and analyzes the optimal inspection strategy. To clarify the discussion, we define the inspection proportion of ship types (IPS), which means the number of inspected ships for one type as a percentage of the total number of ships. Since ship types are subdivided by size, to clarify expressions and avoid confusion, we use "class (subclass)" to describe a specific ship type. According to the location of the origin port and ECA, ships can be divided into two cases: origin port within the ECA (OPW) and origin port outside the ECA (OPO).

By considering the Halifax port as an example, we assume that the total number of visiting ships is 1456 in one day, which is equal to 28 × 2 × 26. Among them, 28 represents the total number of ship types, 2 represents the origin port location (within or outside the ECA) and 26 represents the number of origin ports. According to the inspection rate and non-compliance rate in Table 1, the number of inspected ships is 59. Considering that the arrival time of ships is random (either in the daytime or at night), we assume that the working time of one group for inspection is equal to 24 h. In this case, the government needs at least 10 groups to achieve the required inspection rate. Since the historical information is official data, obtaining this information in practice is difficult. Thus, in the following section, this paper considers the cases without historical information. From another point of view, if it is the first time a ship is inspected, it is obvious that there is no corresponding historical inspection information. Thus, we can regard the situation without historical information as all ships have not been inspected before.

Table 1. Assumptions of related characteristics.

Characteristics		Assumption
Ship	Ship type	28 subclass
	Origin port location	26 ports within the ECA and 26 ports outside the ECA
	Ship proportion	Each ship type has the same proportion
	Non-compliance rate	6% [20]
Government	Inspection time	4 h [21]
	Inspection group	10
	Inspection rate	4% [22]

To determine the appropriate value of M, we calculate the changes of objective value with sampling times (see Fig. 2). The changes of relative difference of objective value (the difference between two consecutive objective values divided by the previous objective value) with sampling times show that the relative difference is smaller than 0.01% after the number of sampling is 7000, which indicates that the optimal solution is stable at 7000 sampling times (see Fig. 3). In addition, the rounding method in the last step of the algorithm does cause some gaps between objective values before and after rounding. But the gap will become smaller as the number of sampling increases. The changes of relative gap of objective value (the gap between the objective value before and after rounding divided by the objective value before rounding) with sampling times show that relative gap is smaller than 0.23% after the number of sampling is 7000 (see Fig. 4). Therefore, the sampling time M is equal to 7000 in the following discussion.

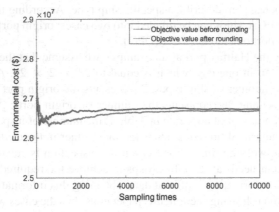

Fig. 2. The changes of objective value with sampling times.

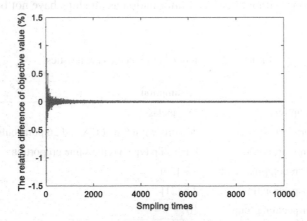

Fig. 3. The changes of relative difference of objective value with sampling times.

For the government, a naive, blind or not optimized situation is to randomly select ships arriving at the port. In this case, the average environmental benefit of inspecting the same number of ships is 1.9196×10^7\$, which is less than the environmental benefit corresponding to the optimal inspection strategy (2.6684×10^7 \$). Thus, comparing to

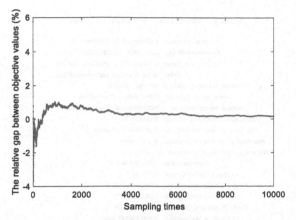

Fig. 4. The changes of relative gaps between objective value with sampling times.

randomly select ships, the method proposed in our study can determine a more reasonable inspection strategy. It can be seen in Fig. 5, the optimal inspection proportion has different IPS for different ship types and locations of origin ports. Specifically: (1) The optimal inspection proportion mainly focuses on five classes of ships, including Container, Dry bulk carrier, Crude oil tanker, Petroleum product tanker and Natural gas carrier. This is mainly because these ship classes have more subclasses, thus they account for a large proportion of ships. To ensure comparability, we divided the inspection number by the number of ships to obtain the unit inspection proportion of 11 ship classes. Their unit inspection proportions have the following relationship: Petroleum product tanker < Crude oil tanker < Chemical tanker < Dry bulk carrier < Natural gas carrier < Cruise line vessel < Container < Car carrier < Roll on-roll off ship < Other. From left to right, the government's unit inspection benefit gradually increases. (2) The location of origin port has a certain influence on the optimal inspection strategy. The IPSs of inspected ships for the OPW case and OPO case are approximately 66.54% and 33.46%, respectively. This is mainly because OPW case has a longer sailing distance within the ECA, thus the corresponding fuel cost saving and environmental cost are increased. However, the penalty policy assumes that the sailing distance within the ECA is always equal to 200 nautical miles (nmi), resulting in a lower penalty fine when the actual sailing distance is greater than 200 nmi. According to the definition of violation tendency, ships in the OPW case have higher environmental cost and violation tendency than that in the OPO case. (3) The IPSs of the same ship type with different origin port locations also change. For instance, Container and Natural gas carrier (VLCC) have similar IPSs in the OPO and OPW cases, while Dry bulk carrier, Crude oil tanker and Natural gas carrier (LGC, midsize) have smaller IPSs in the OPO case than those in the OPW case.

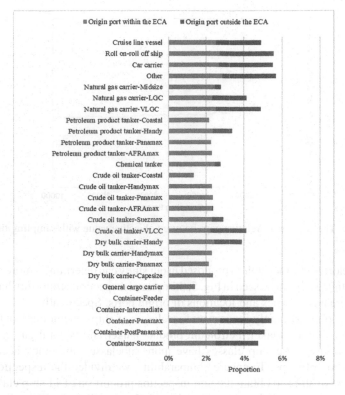

Fig. 5. Optimal inspection proportion at the Halifax port.

4.3 Analysis of Influencing Factors

This section analyzes the impacts of non-compliance and inspection rates on the optimal inspection strategy in terms of government and ship characteristics.

4.3.1 Effects of Non-compliance Rate

According to the describing process of ships' violation behaviors, the non-compliance rate plays an important role in non-compliant ships determination, which will inevitably have an impact on the optimal inspection strategy. Thus, this subsection analyzes the impact of non-compliance rate changes. To ensure comparability, we assume the government's inspection capability (inspection rate) remain unchanged and select nine values (4%, 6%, 8%, 10%, 12%, 14%, 16%, 18% and 20%) to reflect the changes of the non-compliance rate.

The results show that the changes in non-compliance rate have a significant effect on the optimal inspection strategy. As the non-compliance rate increases, the optimal inspection strategy is gradually concentrated on a part of ships, the IPSs vary for different ship types and the IPSs in the OPW case gradually increase. Specifically: (1) Container, Other, Car carrier, Roll on-roll off ship and Cruise line vessel always have higher IPSs. This indicates that these ship types have a larger violation tendency and a larger environmental cost than other ship types. (2) For General cargo carrier and Crude oil tanker (Coastal), the corresponding IPSs gradually decrease. For Container (suezmax, postpanamax, panama, intermediate), Natural gas carrier (VLGC), Other, Car carrier, Roll on-roll off ship sand Cruise line vessel, their IPSs first increase, then decrease and finally increase. The IPSs of other ship types have the opposite trends, that is, it decreases first, then increases and finally decreases. These imply the relationship of violation tendency and environmental cost between different ship types. This is mainly because ships with a greater violation tendency are more likely to become non-compliance ships, and the government tends to select ships that have larger environmental cots in the set of non-compliant ships. When the non-compliance rate is small, non-compliant ships are mainly those with relatively large violation tendencies. As the non-compliance rate continues to increase, the number of non-compliant ships also increases, and the additional non-compliant ships mainly include those with relatively small violation tendencies. Thus, the increased IPSs indicate that among the expanded non-compliant ships, these ship types have relatively large environmental costs. While the decreased IPSs indicate these ship types have relatively small environmental costs. (3) The IPSs in the OPW case have increased from 51.59% under 4% non-compliance rate to 100% under 20% non-compliance rate. This is mainly because the OPW case corresponds to a larger environmental cost. Therefore, among non-compliance ships, due to limited inspection capacity, the government tends to inspect ships with a more environmental cost.

4.3.2 Effects of Inspection Rate

The government's inspection capability also plays an important role in inspection strategy determination, this subsection mainly analyzes the impact of inspection rate changes. To ensure comparability, we assume the non-compliance rates is 15%, and the inspection rate are 2%, 4%, 6%, 8%, 10%, 12% and 14%, respectively.

The results show that the changes in inspection rate have a large effect on the optimal inspection strategy. As the inspection rate increases, the optimal inspection proportion is gradually averaged, the IPSs vary for different ship types and the IPSs in the OPW case gradually decrease. Specifically, For General cargo carrier, Dry bulk carrier (panama, handymax), Crude oil tanker (panamax, handymax, coastal) and Petroleum product tanker (AFRAmax, panama, coastal), their IPSs first increase, then decrease and finally increase. The IPSs of other ship types have the opposite trends. These imply the relationship of environmental costs between different ship types. Since the non-compliance rate is fixed, the set of non-compliant ships is also fixed. As the inspection rate continues to increase, the number of inspected ships also increases, and the additional inspected ships mainly include those with a relatively large environmental cost. Thus, the increased IPSs indicate that compared with other non-compliant ship types, these ship types have relatively large environmental costs, while the decreased IPSs indicate that these ship types have relatively small environmental costs. In addition, the IPSs in the OPW case have decreased from 100% under 2% inspection rate to 51.85% under 14% inspection rate. This is mainly because the OPW case corresponds to a larger environmental cost, so it is more likely to be inspected when the inspection rate is small.

5 Conclusions

Since the establishment of ECAs, corresponding supervision issues have introduced many challenges to the effective enforcement of regulations. Compared with a large number of ships, a government's limited inspection capacity and ships' violation behaviors are the main challenges to an inspection strategy for on-board fuel sampling that can effectively identify violations. Therefore, this paper first describes ships' violation behaviors by considering economic benefit, social responsibility and other factors. Based on this, a semi-random input model with limited inspection capacity, and the corresponding algorithm is proposed to determine the optimal inspection strategy. In addition, the impacts of related factors on the inspection strategy are discussed. The results show that compared to randomly select ships, the optimal inspection strategy corresponds to a larger objective value. This indicates the method proposed in our study can determine a more reasonable inspection strategy, which provides methodological support for on-board fuel sampling and effective enforcement of the ECA regulations. A promising direction for future research is to explore new ways to further improve the inspection strategy.

Funding. This work was supported by the National Natural Science Foundation of China (11991022).

References

1. Tichavska, M., Tovar, B., Gritsenko, D., Johansson, L., Jalkanen, J.P.: Air emissions from ships in port: does regulation make a difference? Transp. Policy 75, 128–140 (2017)
2. Perera, L.P., Mo, B.: Emission control based energy efficiency measures in ship operations. Appl. Ocean Res. **60**, 29–46 (2016)
3. Li, L., Gao, S., Yang, W., Xiong, X.: Ship's response strategy to emission control areas: from the perspective of sailing pattern optimization and evasion strategy selection. Transp. Res. Part E: Logist. Transp. Rev. **133**, 101835 (2020)
4. Liu, Y., Wang, Z.: Game simulation of policy regulation evolution in emission control area based on system dynamics. In: 2019 3rd Scientific Conference on Mechatronics Engineering and Computer Science, pp. 384–388. Francis Academic Press (2019)
5. Jiang, B., Xue, H., Li, J.: Study on regulation strategies of China's ship emission control area (ECA) based on evolutionary game (in Chinese). Logistic Sci-Tech **7**, 70–74 (2018)
6. EMSA: Sulphur Inspection Guidance. European Maritime Safety Agency (2018)
7. Ronen, D.: The effect of oil price on the optimal speed of ships. J. Oper. Res. Soc. **33**, 1035–1040 (2017)
8. Du, Y., Chen, Q., Quan, X., Long, L., Fung, R.Y.K.: Berth allocation considering fuel consumption and vessel emissions. Transp. Res. Part E: Logist. Transp. Rev. **47**, 1021–1037 (2011)
9. IMO: Reduction of GHG emissions from ships, third IMO GHG Study 2014 (2014)
10. Wang, S., Meng, Q.: Sailing speed optimization for container ships in a liner shipping network. Transp. Res. Part E: Logist. Transp. Rev. **48**, 701–714 (2012)
11. Psaraftis, H.N., Kontovas, C.A.: Speed models for energy-efficient maritime transportation: a taxonomy and survey. Transp. Res. Part C: Emerg. Technol. **26**, 331–351 (2013)
12. Dulebenets, M.A.: Advantages and disadvantages from enforcing emission restrictions within emission control areas. Marit. Bus. Rev. **1**, 107–132 (2016)

13. Phillip, A.B.: EPA penalty policy for violations by ships of the sulfur in fuel standard and related provisions. Environmental Protection Agency, United Sates (2015)
14. Kim, S., Pasupathy, R., Henderson, S.G.: A guide to sample average approximation. In: Fu, M.C. (ed.) Handbook of Simulation Optimization. ISORMS, vol. 216, pp. 207–243. Springer, New York (2015). https://doi.org/10.1007/978-1-4939-1384-8_8
15. Marine bunker exchange. http://www.mabux.com
16. https://www.fleetmon.com
17. https://www.searates.com/cn/services/distances-time/
18. IMO: Third IMO GHG Study 2014 – Final Report. International Maritime Organization (2015)
19. CEC: Best Available Technology for Air Pollution Control: Analysis Guidance and Case Studies for North America. Commission for Environmental Cooperation of North America (2005)
20. Nazha, N.: North American emission control area, canada's compliance and enforcement program. In: 5th PPCAC Conference, Transport Canada (2018)
21. Olaniyi, E.O., Prause, G.: Seca regulatory impact assessment: administrative burden costs in the baltic sea region. Transp. Telecommun. J. **20**, 62–73 (2019)
22. European Union. https://eur-lex.europa.eu/eli/dec_impl/2015/253/oj

Novel Algorithms for Maximum DS Decomposition

Shengminjie Chen[1], Wenguo Yang[1(✉)], Suixiang Gao[1], and Rong Jin[2]

[1] School of Mathematical Sciences, University of Chinese Academy of Sciences,
Beijing 100049, China
chenshengminjie19@mails.ucas.edu.cn, {yangwg,sxgao}@ucas.edu.cn
[2] Department of Computer Science, University of Texas at Dallas,
Richardson, TX 75080, USA
Rong.Jin@utdallas.edu

Abstract. DS decomposition is an important set function optimization problem. Because DS decomposition is true for any set function, how to solve DS decomposition efficiently and effectively is a heated problem to be solved. In this paper, we focus maximum DS decomposition problem and propose Deterministic Conditioned Greedy algorithm and Random Conditioned algorithm by using the difference with parameter decomposition function and combining non-negative condition. Besides, we get some novel approximation under different parameters. Also, we choose two special case to show our deterministic algorithm gets $f(S_k) - (e^{-1} - c_g) g(S_k) \geq (1 - e^{-1})[f(OPT) - g(OPT)]$ and $f(S_k) - (1 - c_g) g(S_k) \geq (1 - e^{-1}) f(OPT) - g(OPT)$ respectively for cardinality constrained problem and our random algorithm gets $E[f(S_k) - (e^{-1} - c_g) g(S_k)] \geq (1 - e^{-1})[f(OPT) - g(OPT)]$ and $E[f(S_k) - (1 - c_g)g(S_k)] \geq (1 - e^{-1}) f(OPT) - g(OPT)$ respectively for unconstrained problem, where c_g is the curvature of monotone submodular set function. Because the Conditioned Algorithm is the general framework, different users can choose the parameters that fit their problem to get a better approximation.

Keywords: Non-submodularity · DS decomposition · Conditioned Greedy

1 Introduction

Submodular Set Function Optimization, which is a hot issue in discrete optimization, attracts many researchers. In the past few decades, there have been many important results about maximum submodular set function. Nemhauser et al. (1978) [20] proposed a greedy algorithm, which adds a maximal marginal

This work was supported by the National Natural Science Foundation of China under Grant 11991022 and 12071459.

W. Wu and Z. Zhang (Eds.): COCOA 2020, LNCS 12577, pp. 624–638, 2020.
https://doi.org/10.1007/978-3-030-64843-5_42

gains element to current solution in each iteration, can get $1 - e^{-1}$ approximation for optimal solution under cardinality constrained. Meanwhile, Fisher et al. (1978) [6] proved that the greedy strategy can get $(1 + p)^{-1}$ approximation for optimal solution under intersection of p matroid. Furthermore, Conforti et al. (1984) [19] constructed a metric called curvature. Using curvature, authors proved that the greedy algorithm can get more tight approximation ($\frac{1-e^{-c}}{c}$ and $\frac{1}{1+c}$) for above two constrained. That is why greedy strategy is excellent for monotone submodular maximization. Unfortunately, they assume the set function is monotone submodular function and $f(\emptyset) = 0$. But these groundbreaking works have inspired a great deal of submodular optimization.

Relaxing the restriction of monotony, Feldman et al. (2011) [5] proposed a novel greedy which has a e^{-1} approximation for non-monotone submodular maximization under matroids constrained. Furthermore, Buchbinder et al. (2014) [3] proved that non-monotone submodular maximization under cardinality constrained also has a e^{-1} approximation polynomial algorithm using double greedy. What's more, Buchbinder et al. (2012) [2] also constructed an algorithm which can get $1/2$ approximation for non-monotone submodular maximization without constrains.

Based on these efficient algorithms, submodular optimization has played an important role in data mining, machine learning, economics and operation research such as influence maximization (Kempe et al., (2003) [13]), active learning (Golovin et al., (2011) [7]), document summarization (Lin et al., (2011) [16]), image segmentation (Jegelka et al., (2011) [12]). Unfortunately, more and more objective functions are not submodular in practical problems. Thus, how to optimize a general set function is the most important problem and has puzzled many scholars. Some researchers proposed lots of definition about approximation of sub-modularity. Krause et al. (2008) [14] constructed ϵ-Diminishing returns to evaluate what is the difference of violation marginal gains decreasing. And authors proved the standard greedy algorithm can get a $f(X) \geq (1-e^{-1})(OPT - k\epsilon)$ approximation for size constrained problem. Das and Kempe (2011) [4] used to measure violation about submodularity by submodular ratio. Besides, they proved the standard greedy strategy can get a $f(X) \geq 1 - e^{-\gamma}OPT$ approximation under cardinality constrained problem. Horel and Singer (2016) [10] proposed ϵ-approximation submodularity to calculate the difference of submodularity. According this metric, authors proved that the standard greedy algorithm can return a $f(X) \geq \frac{1}{1+\frac{4k\epsilon}{(1-\epsilon)^2}} \left(1 - e^{-1} \left(\frac{1-\epsilon}{1+\epsilon} \right)^{2k} \right) \cdot OPT$ approximation under size constrained. It is no surprise that computing these metrics is also a hard problem. It makes these results only theoretical meaning, but lack of practical application.

From the application perspective to solve set function optimization, Lu et al. (2015) [17] utilized a submodular upper bound and a submodular lower bound to constrain a set function and solved these three problems respectively. Their algorithm chose the best one of them to return called Sandwich Approach. Because this approach can get a parametric approximation, some researchers also applied

Sandwich Approach to solve their problems (Yang et al. (2020) [22], Yang et al. (2020) [23], Zhu et al. (2019) [25], Wang et al. (2017) [21]). In addition, Iyer and Bilmes (2012) [11] proved that any set function can decompose the difference of two submodular set functions called DS decomposition. Specially, the two submodular set function are monotone and non-decreasing.

Lemma 1. *(Iyer and Bilmes 2012) Any set function* $h : 2^{\Omega} \to R$ *can decompose the difference of two monotone non-decreasing submodular set functions* f *and* g, *i.e.* $h = f - g$.

According to this lemma, Iyer and Bilmes proposed SubSup, SupSub, Mod-Mod algorithms to solve minimum this problem. This excellent result has been applied in many ways (Han et al. (2018) [8], Maehara et al. (2015) [18] and Yu et al. (2016) [24]). Although DS decomposition has a bright application prospect, about a general set function, there are problems worth studying how to find the decomposition quickly and how to solve DS decomposition efficiently and effectively. These problems needed to be solved also inspire us to think how to solve maximum DS decomposition.

In this paper, we focus the maximum DS decomposition about set functions. Since DS decomposition is the difference between two submodular functions, we proposed Deterministic and Random Conditioned Greedy algorithm by using the difference with parameter decomposition function and combining non-negative condition. There are general frameworks and users can choose rational parameters according the property of problem. The major contribution of our work are as follows:

- Deterministic Conditioned Greedy is a deterministic framework which introduces some parameters about iteration rounds for cardinality constrained problem. In each iteration, the algorithm chooses the element of maximal non-negative marginal parametric gains from ground set. Under some rational assumption, the Deterministic Conditioned Greedy can get a polynomial approximation for maximum DS decomposition.
- Two special cases. We choose special parameters in Deterministic Conditioned Greedy. And the algorithm can get two novel approximations $f(S_k) - (e^{-1} - c_g)g(S_k) \geq (1 - e^{-1})[f(OPT) - g(OPT)]$ and $f(S_k) - (1 - c_g)g(S_k) \geq (1 - e^{-1})f(OPT) - g(OPT)$ respectively for cardinality constrained problem.
- Random Conditioned Greedy is a random framework which introduces some parameters about iteration rounds for unconstrained problem. In each iteration, the algorithm chooses an element from ground set uniformly. Under some rational assumption, the Random Conditioned Greedy can get a polynomial approximation for maximum DS decomposition.
- Two special cases. In Random Conditioned Greedy, we choose the same parameters as Deterministic Conditioned Greedy. And the algorithm can get two novel approximations $E[f(S_k) - (e^{-1} - c_g)g(S_k)] \geq (1 - e^{-1})[f(OPT) - g(OPT)]$ and $E[f(S_k) - (1 - c_g)g(S_k)] \geq (1 - e^{-1})f(OPT) - g(OPT)$ respectively for unconstrained problem.

The rest of this paper is organized as follow. Some related works about greedy strategy, we put them in Sect. 2. In Sect. 3, we propose Deterministic Conditioned Greedy Algorithm and prove approximation. We get two special results for Deterministic Conditioned Greedy in Sect. 4. Random Conditioned Greedy Algorithm, we introduce in Sect. 5. Also, Sect. 6 is special cases for Random Conditioned Greedy. Conclusion and future works are in Sect. 7.

2   Related Works

In this section, we introduce some related works about set function decomposition and algorithm.

Iyer and Bilmes (2012) [11] first tried to optimize the set function from the perspective of decomposition. They proved that DS decomposition exists for any set functions. What's more, they proposed three greedy strategy to solve the DS decomposition called SupSub, SubSup, ModMod respectively. But in their work, these three greedy strategies are used to solve the minimization problem. As for maximum DS decomposition, this is an urgent problem to be solved. From the decomposition perspective, Li et al. (2020) [15] proved a variation of DS decomposition, any set function can be expressed as the difference of two monotone nondecreasing supermodular functions. Similarly, they also proposed greedy strategies like Iyer and Bilmes called ModMod and SupMod.

Bai and Bilmes (2018) [1] found that some set function can be expressed as the sum of a submodular and supermodular function (BP decomposition), both of which are non-negative monotone non-decreasing. But they didn't show what circumstances exists a BP decomposition. Interestingly, they proved that greedy strategy can get a $\frac{1}{k_f} \left[1 - e^{-(1-k_g)k_f} \right]$ and $\frac{1-k_g}{(1-k_g)k_f+p}$ approximation under cardinality and matroid constrained, where k_f, k_g are curvature about submodular and supermodular functions.

Harshaw et al. (2019) [9] focused on the difference of a monotone non-negative submodular set function and a monotone non-negative modular set function. Besides, they proposed Distorted Greedy to solve this problem and get a $f(S) - g(S) \geq \left(1 - e^{-1}\right) f(OPT) - g(OPT)$ approximation. This greedy strategy inspires our idea to solve maximization of DS decomposition.

3   Deterministic Conditioned Greedy Algorithm

Greedy strategy is the useful algorithm to solve discrete optimization problem, because it is simple and efficient. In the submodular optimization problem, since greedy algorithm can get an excellent constant approximation, many researchers use greedy strategy. Although, some practical problems are not submodular, fortunately, they have DS decomposition (Lemma 1). Therefore, we proposed a deterministic conditioned greedy algorithm to solve the following problem, where the two submodular set function f, g are monotone and non-decreasing.

$$\max_{X \subseteq \Omega, |X| \leq k} h(X) = f(X) - g(X)$$

In Algorithm 1, the $A(i)$ and $B(i)$ are chosen by algorithm designers. From practical perspectives, we assume $f(\emptyset) = g(\emptyset) = 0$, $A(i) \geq 0$ and $B(i) \geq 0$. From the proof, we assume $\left(1 - \frac{1}{k}\right) A(i+1) - A(i) \geq 0$ and $B(i+1) - B(i) \geq 0$ specially. When designers choose well-defined parameters that are related to iteration round, it can get a wonderful approximation about maximum DS decomposition. Firstly, we introduce two auxiliary functions and curvature which are useful in process of approximation proof.

Algorithm 1: Deterministic Conditioned Greedy

Input: cardinality k, parameters $A(i), B(i)$

1.Initialize $S_0 \leftarrow \emptyset$

2.For $i = 0$ to $k - 1$

3. $e_i \leftarrow \arg\max_{e \in \Omega} \{A(i+1) f(e \mid S_i) - B(i+1) g(e \mid \Omega \setminus e)\}$

4.   If $A(i+1) f(e_i \mid S_i) - B(i+1) g(e_i \mid \Omega \setminus e_i) > 0$ then

5. $S_{i+1} \leftarrow S_i \cup \{e_i\}$

6.   Else

7. $S_{i+1} \leftarrow S_i$

8.End for

9.Return S_k

Definition 1. *Define two auxiliary functions*

$$\phi_i(T) = A(i) f(T) - B(i) \sum_{e \in T} g(e | \Omega \setminus e)$$

$$\psi_i(T, e) = \max\{0, A(i+1) f(e \mid T) - B(i+1) g(e \mid \Omega \setminus e)\}$$

Definition 2 (Conforti (1984)). *Given a monotone submodular set function* $f : 2^\Omega \to R$, *the curvature of* f *is*

$$c_f = 1 - min_{e \in \Omega} \frac{f(e | \Omega \setminus e)}{f(e)}$$

Look at the definition of two functions, $\psi_i(T, e)$ is the condition of the Algorithm 1. And $\phi_i(T)$ is the surrogate objective function. Next, we prove an important property about surrogate objective function.

Property 1. In each iteration

$$\phi_{i+1}(S_{i+1}) - \phi_i(S_i)$$
$$= \psi_i(S_i, e_i) + [A(i+1) - A(i)] f(S_i) - [B(i+1) - B(i)] \sum_{e \in S_i} g(e | \Omega \setminus e)$$

$$(1)$$

Proof.

$$\phi_{i+1}(S_{i+1}) - \phi_i(S_i)$$

$$= A(i+1) f(S_{i+1}) - B(i+1) \sum_{e \in S_{i+1}} g(e|\Omega \setminus e) - A(i) f(S_i) + B(i) \sum_{e \in S_i} g(e|\Omega \setminus e)$$

$$= A(i+1) [f(S_{i+1} - f(S_i))] - B(i+1) g(e_i|\Omega \setminus e_i)$$

$$+ [A(i+1) - A(i)] f(S_i) - [B(i+1) - B(i)] \sum_{e \in S_i} g(e|\Omega \setminus e)$$

$$= \psi_i(S_i, e_i) + [A(i+1) - A(i)] f(S_i) - [B(i+1) - B(i)] \sum_{e \in S_i} g(e|\Omega \setminus e)$$

$$\square$$

Using the Property 1, we construct a relationship from Deterministic Conditioned Greedy to surrogate objective function. Interestingly, the condition of Algorithm 1 has a lower bound. Therefore, in each iteration of Algorithm 1, the marginal gain of surrogate objective function has some guarantees.

Theorem 1. $\psi_i(S_i, e_i) \geq \frac{1}{k} A(i+1) [f(OPT) - f(S_i)] - \frac{1}{k} B(i+1) g(OPT)$

Proof.

$$k \cdot \psi_i(S_i, e_i) = k \cdot \max_{e \in \Omega} \{0, A(i+1) f(e \mid S_i) - B(i+1) g(e \mid \Omega \setminus e)\}$$

$$\geq |OPT| \cdot \max_{e \in \Omega} \{0, A(i+1) f(e \mid S_i) - B(i+1) g(e \mid \Omega \setminus e)\}$$

$$\geq |OPT| \cdot \max_{e \in OPT} \{A(i+1) f(e \mid S_i) - B(i+1) g(e \mid \Omega \setminus e)\}$$

$$\geq \sum_{e \in OPT} [A(i+1) f(e \mid S_i) - B(i+1) g(e \mid \Omega \setminus e)]$$

$$= A(i+1) \sum_{e \in OPT} f(e|S_i) - B(i+1) \sum_{e \in OPT} g(e|\Omega \setminus e)$$

$$\geq A(i+1) [f(OPT \cup S_i) - f(S_i)] - B(i+1) g(OPT)$$

$$\geq A(i+1) [f(OPT) - f(S_i)] - B(i+1) g(OPT)$$

The first inequality is $k \geq |OPT|$. The second inequality is $OPT \subseteq \Omega$. The third inequality is the maximum. The fourth inequality is sub-modularity, i.e. $f(OPT \cup S_i) - f(S_i) \leq \sum_{e \in OPT} f(e|S_i)$ and $\sum_{e \in OPT} g(e|\Omega \setminus e) \leq g(OPT)$. The last inequality is monotony. $\square$

Combined Property 1 and Theorem 1, the following corollary is obvious.

Corollary 1.

$$\phi_{i+1}(S_{i+1}) - \phi_i(S_i) \geq \frac{1}{k} A(i+1) f(OPT) - \frac{1}{k} B(i+1) g(OPT)$$

$$+ \left[\left(1 - \frac{1}{k}\right) A(i+1) - A(i) \right] f(S_i) - [B(i+1) - B(i)] \sum_{e \in S_i} g(e|\Omega \setminus e) \tag{2}$$

According the above assumption and Lemma 1, we have $\sum_{e \in S_i} g(e|\Omega \setminus e) \geq 0$ and $f(S_i) \geq 0$. That is to say that the proper selection of $A(i)$ and $B(i)$ is the key to the final performance of the algorithm. In our work, we assume that they should satisfy $\left(1 - \frac{1}{k}\right) A(i+1) - A(i) \geq 0$ and $B(i+1) - B(i) \geq 0$. If $\left[\left(1 - \frac{1}{k}\right) A(i+1) - A(i)\right] f(S_i) - [B(i+1) - B(i)] \sum_{e \in S_i} g(e|\Omega \setminus e) \geq 0$, the result is trivial cause we can ignore them. Therefore, we can get an approximation guarantee about maximum DS decomposition under cardinality constrained using Deterministic Conditioned Greedy.

Theorem 2. S_k *is the solution of Algorithm 1 after k iteration*

$$A(k) f(S_k) - (B(0) - c_g) g(S_k) \geq \frac{1}{k} \sum_{i=0}^{k-1} [A(i+1) f(OPT) - B(i+1) g(OPT)]$$

where c_g is the curvature of function g.

Proof. $\phi_0(S_0) = A(0)f(\emptyset) - B(0)g(\emptyset) = 0$ According the curvature $c_g = 1 - min \frac{g(e\Omega \setminus e)}{g(e)}$, we have $\frac{g(e\Omega \setminus e)}{g(e)} \geq 1 - c_g$,

$$\phi_k(S_k) = A(k)f(S_k) - B(k) \sum_{e \in S_k} g(e|\Omega \setminus e) \leq A(k)f(S_k) - B(k)(1 - c_g) \sum_{e \in S_k} g(e)$$

$$\leq A(k)f(S_k) - B(k)(1 - c_g)g(S_k)$$

Thus, we can rewrite an accumulation statement

$$A(k)f(S_k) - B(k)(1 - c_g)g(S_k) \geq \phi_k(S_k) - \phi_0(S_0) = \sum_{i=0}^{k-1} \phi_{i+1}(S_{i+1}) - \phi_i(S_i)$$

$$\geq \sum_{i=0}^{k-1} [\frac{1}{k} A(i+1)f(OPT) - \frac{1}{k} B(i+1)g(OPT) + [(1 - \frac{1}{k})A(i+1) - A(i)]f(S_i)]$$

$$- \sum_{i=0}^{k-1} [[B(i+1) - B(i)] \sum_{e \in S_i} g(e|\Omega \setminus e)]$$

$$\geq \frac{1}{k} \sum_{i=0}^{k-1} [A(i+1)f(OPT) - B(i+1)g(OPT)] - [B(k) - B(0)] \sum_{e \in S_k} g(e|\Omega \setminus e)$$

$$\geq \frac{1}{k} \sum_{i=0}^{k-1} [A(i+1)f(OPT) - B(i+1)g(OPT)] - [B(k) - B(0)]g(S_k)$$

Therefore, we can conclude

$$A(k) f(S_k) - (B(0) - c_g) g(S_k) \geq \frac{1}{k} \sum_{i=0}^{k-1} [A(i+1) f(OPT) - B(i+1) g(OPT)]$$

□

4 Two Special Cases for Deterministic Conditioned Greedy

In this section, we choose two special case to show Deterministic Conditioned Greedy strategy can get $f(S_k) - (e^{-1} - c_g)g(S_k) \geq (1 - e^{-1})[f(OPT) - g(OPT)]$ approximation and $f(S_k) - (1 - c_g)g(S_k) \geq (1 - e^{-1})f(OPT) - g(OPT)$ approximation for cadinality constrained problem.

4.1 Case 1

We set $A(i) = B(i) = \left(1 - \frac{1}{k}\right)^{(k-i)}$. Therefore, the Definition 1 and Algorithm 1 become the following. Obviously, these settings satisfy all conditions and assumptions in Sect. 3. Hence, the following results are clearly. Because the proofs are similar to Sect. 3, we omit them here.

$$\phi_i(T) = \left(1 - \frac{1}{k}\right)^{k-i} \left[f(T) - \sum_{e \in T} g(e \mid \Omega \setminus e)\right]$$

$$\psi_i(T, e) = \max \left\{0, \left(1 - \frac{1}{k}\right)^{k-(i+1)} [f(e \mid S_i) - g(e \mid \Omega \setminus e)]\right\}$$

Algorithm 2: Deterministic Conditioned Greedy I

Input: cardinality k

1. Initialize $S_0 \leftarrow \emptyset$

2. For $i = 0$ to $k - 1$

3. $e_i \leftarrow \arg\max_{e \in \Omega} \left\{\left(1 - \frac{1}{k}\right)^{k-(i+1)} [f(e \mid S_i) - g(e \mid \Omega \setminus e)]\right\}$

4.   If $\left(1 - \frac{1}{k}\right)^{k-(i+1)} [f(e_i \mid S_i) - g(e_i \mid \Omega \setminus e_i)] > 0$ then

5. $S_{i+1} \leftarrow S_i \cup \{e_i\}$

6.   Else

7. $S_{i+1} \leftarrow S_i$

8. End for

9. Return S_k

Property 2. In each iteration

$$\phi_{i+1}(S_{i+1}) - \phi_i(S_i) = \psi_i(S_i, e_i) + \frac{1}{k}\left(1 - \frac{1}{k}\right)^{-1}\phi_i(S_i)$$

Theorem 3. $\psi_i(S_i, e_i) \geq \frac{1}{k}\left(1 - \frac{1}{k}\right)^{k-(i+1)} [f(OPT) - f(S_i) - g(OPT)]$

Theorem 4. $f(S_k) - \left(e^{-1} - c_g\right) g(S_k) \geq \left(1 - e^{-1}\right)[f(OPT) - g(OPT)]$, where c_g is the curvature of function g.

From Theorem 4, we find an interesting result and get the following corollary, when $c_g = 0$, i.e. the submodular function g is a modular function.

Corollary 2. *If $c_g = 0$, i.e. g is modular. Then we have*

$$f(S_k) - e^{-1} g(S_k) \ge (1 - e^{-1}) [f(OPT) - g(OPT)]$$

4.2   Case 2

We set $A(i) = \left(1 - \frac{1}{k}\right)^{(k-i)}$, $B(i) = 1$. Therefore, the Definition 1 and Algorithm 1 become the following. Obviously, these settings satisfy all condition in Sect. 3. And the proofs are similar with Sect. 3 and Sect. 4.1. In this subsection, we have omitted all proofs.

$$\phi_i(T) = \left(1 - \frac{1}{k}\right)^{k-i} f(T) - \sum_{e \in T} g(e \mid \Omega \setminus e)$$

$$\psi_i(T, e) = \max\left\{0, \left(1 - \frac{1}{k}\right)^{k-(i+1)} f(e \mid S_i) - g(e \mid \Omega \setminus e)\right\}$$

Algorithm 3: Deterministic Conditioned Greedy II

Input: cardinality k
1. Initialize $S_0 \leftarrow \emptyset$
2. For $i = 0$ to $k - 1$
3. $e_i \leftarrow \arg\max_{e \in \Omega} \left\{ \left(1 - \frac{1}{k}\right)^{k-(i+1)} f(e \mid S_i) - g(e \mid \Omega \setminus e) \right\}$
4.   If $\left(1 - \frac{1}{k}\right)^{k-(i+1)} f(e_i \mid S_i) - g(e_i \mid \Omega \setminus e_i) > 0$ then
5. $S_{i+1} \leftarrow S_i \cup \{e_i\}$
6.   Else
7. $S_{i+1} \leftarrow S_i$
8. End for
9. Return S_k

Property 3. In each iteration

$$\phi_{i+1}(S_{i+1}) - \phi_i(S_i) = \psi_i(S_i, e_i) + \frac{1}{k}\left(1 - \frac{1}{k}\right)^{k-(i+1)} f(S_i)$$

Theorem 5. $\psi_i(S_i, e_i) \ge \frac{1}{k}\left(1 - \frac{1}{k}\right)^{k-(i+1)} [f(OPT) - f(S_i)] - \frac{1}{k} g(OPT)$

Theorem 6. $f(S_k) - (1 - c_g) g(S_k) \ge (1 - e^{-1}) f(OPT) - g(OPT)$, *where c_g is the curvature of function g.*

Corollary 3. *If $c_g = 0$, i.e. g is modular. Then we have*

$$f(S_k) - g(S_k) \geq (1 - e^{-1}) f(OPT) - g(OPT)$$

Remark: Clearly, if g is modular, then $h = f - g$ is submodular. The above approximations are different with submodular maximization under cardinality constrained problem $(1 - e^{-1})$. And also, the different parameters can also cause different approximations. We think first gap is caused by non-monotony, because h is not always monotonous. The second gap give us a clue that we can choose the appropriate parameters $A(i)$ and $B(i)$ according to the characteristics of the problem to get a better approximate ratio.

5 Random Conditioned Greedy Algorithm

Random algorithms are important algorithms for discrete optimization problem. Since randomness can bring a lot of uncertain factors and information, in some cases, random algorithms can get better approximation than deterministic algorithms. Therefore, we propose a random conditioned greedy in this section. Interestingly, our random conditioned greedy can also be used in unconstrained problems and cardinality constrained problem. The following statement is the unconstrained problems for DS decomposition.

$$\max_{X \subseteq \Omega} h(X) = f(X) - g(X)$$

According Definition 1, the Property 1 is also true for random conditioned greedy. We just have to modify the assumptions a little bit $\left(1 - \frac{1}{n}\right) A(i+1) - A(i) \geq 0$ to get the following theorems and corollary directly.

Algorithm 4: Random Conditioned Greedy

Input: ground set Ω, parameters $A(i), B(i)$
1. Initialize $S_0 \leftarrow \emptyset$
2. For $i = 0$ to $n - 1$
3. $e_i \leftarrow$ be chosen uniformly from Ω
4.   If $A(i+1) f(e_i \mid S_i) - B(i+1) g(e_i \mid \Omega \setminus e_i) > 0$ then
5. $S_{i+1} \leftarrow S_i \cup \{e_i\}$
6.   Else
7. $S_{i+1} \leftarrow S_i$
8. End for
9. Return S_n

Theorem 7. $E[\psi_i(S_i, e)] \geq \frac{1}{n} A(i+1)[f(OPT) - f(S_i)] - \frac{1}{n} B(i+1) g(OPT)$

Proof.

$$E[\psi_i(S_i, e_i)] = \frac{1}{n} \cdot \sum_{e_i \in \Omega} \psi_i(S_i, e_i)$$

$$\geq \frac{1}{n} \cdot \sum_{e_i \in OPT} [A(i+1)f(e_i|S_i) - B(i+1)g(e_i|\Omega \setminus e_i)]$$

$$= \frac{1}{n}A(i+1) \sum_{e \in OPT} f(e_i|S_i) - \frac{1}{n}B(i+1) \sum_{e \in OPT} g(e_i|\Omega \setminus e_i)$$

$$\geq \frac{1}{n}A(i+1)[f(OPT \cup S_i) - f(S_i)] - \frac{1}{n}B(i+1)g(OPT)$$

$$\geq \frac{1}{n}A(i+1)[f(OPT) - f(S_i)] - \frac{1}{n}B(i+1)g(OPT)$$

□

Corollary 4. $E[\phi_{i+1}(S_{i+1}) - \phi_i(S_i)] \geq \frac{1}{n}A(i+1)f(OPT) - \frac{1}{n}B(i+1)g(OPT)$
$+ [(1 - \frac{1}{n})A(i+1) - A(i)]f(S_i) - [B(i+1) - B(i)]\sum_{e \in S_i} g(e|\Omega \setminus e)$

Combined with Theorem 7 and Corollary 4, using the same method as Sect. 3, we can prove the Random Conditioned Greedy can get the following approximation.

Theorem 8. S_n *is the solution of Algorithm 4 after n iteration*

$$E[A(n)f(S_n) - (B(0) - c_g)g(S_n)] \geq \frac{1}{n}\sum_{i=0}^{n-1}[A(i+1)f(OPT) - B(i+1)g(OPT)]$$

where c_g is the curvature of function g.

Proof.

$$E[A(n)f(S_n) - B(n)(1 - c_g)g(S_n)]$$

$$\geq E[\phi_n(S_n)] - E[\phi_0(S_0)] = \sum_{i=0}^{n-1} E[\phi_{i+1}(S_{i+1})] - E[\phi_i(S_i)]$$

$$\geq \sum_{i=0}^{n-1}[\frac{1}{n}A(i+1)f(OPT) - \frac{1}{n}B(i+1)g(OPT) + [(1 - \frac{1}{n})A(i+1) - A(i)]f(S_i)]$$

$$- \sum_{i=0}^{n-1}[B(i+1) - B(i)] \sum_{e \in S_i} g(e|\Omega \setminus e)$$

$$\geq \frac{1}{n}\sum_{i=0}^{n-1}[A(i+1)f(OPT) - B(i+1)g(OPT)] - [B(n) - B(0)] \sum_{e \in S_n} g(e|\Omega \setminus e)$$

$$\geq \frac{1}{n}\sum_{i=0}^{n-1}[A(i+1)f(OPT) - B(i+1)g(OPT)] - [B(n) - B(0)]g(S_n)$$

Therefore, we can conclude

$$E[A(n)f(S_n) - (B(0) - c_g)g(S_n)] \geq \frac{1}{n}\sum_{i=0}^{n-1}[A(i+1)f(OPT) - B(i+1)g(OPT)]$$

$\square$

From the Theorem 7 and Theorem 8, we find an interesting phenomenon. If we decrease the number of iterations to k, this Random Conditioned Greedy can also be used in problems with cardinality constrained. Since this proof is similar with Theorem 8, we just give the statement without proof.

Theorem 9. S_k *is the solution of Algorithm 4 after k iteration*

$$E[A(k)f(S_k) - (B(0) - c_g)g(S_k)] \geq \frac{1}{n}\sum_{i=0}^{k-1}[A(i+1)f(OPT) - B(i+1)g(OPT)]$$

where c_g is the curvature of function g.

6   Two Special Cases for Random Conditioned Greedy

In this section, we choose the same parameters as Deterministic Conditioned Greedy to show Random Conditioned Greedy strategy can get $E[f(S_n) - (e^{-1} - c_g)g(S_n)] \geq (1 - e^{-1})[f(OPT) - g(OPT)]$ approximation and $E[f(S_n) - (1 - c_g)g(S_n)] \geq (1 - e^{-1})f(OPT) - g(OPT)$ approximation for unconstrained problem. Since the proof is similar with Sect. 5, we only give the statement without proofs.

6.1   Case 1

We set $A(i) = B(i) = \left(1 - \frac{1}{n}\right)^{(n-i)}$. Therefore, the Definition 1 and Algorithm 4 become the following. Obviously, these settings satisfy all assumptions in Sect. 5.

$$\phi_i(T) = \left(1 - \frac{1}{n}\right)^{n-i}\left[f(T) - \sum_{e \in T} g(e \mid \Omega \setminus e)\right]$$

$$\psi_i(T, e) = \max\left\{0, \left(1 - \frac{1}{n}\right)^{n-(i+1)}[f(e \mid S_i) - g(e \mid \Omega \setminus e)]\right\}$$

Theorem 10. $E[\psi_i(S_i, e_i)] \geq \frac{1}{n}(1 - \frac{1}{n})^{n-(i+1)}[f(OPT) - f(S_i) - g(OPT)]$

Theorem 11. $E\left[f(S_n) - (e^{-1} - c_g)g(S_n)\right] \geq (1 - e^{-1})[f(OPT) - g(OPT)]$, *where c_g is the curvature of function g*

From Theorem 11, we can draw the following corollary when g is modular function.

Corollary 5. *If $c_g = 0$, i.e. g is modular. Then we have*

$$E\left[f(S_n) - e^{-1}g(S_n)\right] \geq (1 - e^{-1})[f(OPT) - g(OPT)]$$

Algorithm 5: Random Conditioned Greedy I

Input: ground set Ω

1.Initialize $S_0 \leftarrow \emptyset$

2.For $i = 0$ to $n - 1$

3. $e_i \leftarrow$ be chosen uniformly from Ω

4.   If $\left(1 - \frac{1}{k}\right)^{k-(i+1)} \left[f\left(e_i \mid S_i\right) - g\left(e_i \mid \Omega \setminus e_i\right)\right] > 0$ then

5. $S_{i+1} \leftarrow S_i \cup \{e_i\}$

6.   Else

7. $S_{i+1} \leftarrow S_i$

8.End for

9.Return S_n

6.2   Case 2

We set $A(i) = \left(1 - \frac{1}{n}\right)^{(n-i)}$, $B(i) = 1$. Therefore, the Definition 1 and Algorithm 4 become the following. Obviously, these settings satisfy all assumptions in Sect. 5.

$$\phi_i(T) = \left(1 - \frac{1}{n}\right)^{n-i} f(T) - \sum_{e \in T} g(e \mid \Omega \setminus e)$$

$$\psi_i(T, e) = \max\left\{0, \left(1 - \frac{1}{n}\right)^{n-(i+1)} f(e \mid S_i) - g(e \mid \Omega \setminus e)\right\}$$

Algorithm 6: Random Conditioned Greedy II

Input: ground set Ω

1.Initialize $S_0 \leftarrow \emptyset$

2.For $i = 0$ to $n - 1$

3. $e_i \leftarrow$ be chosen uniformly from Ω

4.   If $\left(1 - \frac{1}{k}\right)^{k-(i+1)} f\left(e_i \mid S_i\right) - g\left(e_i \mid \Omega \setminus e_i\right) > 0$ then

5. $S_{i+1} \leftarrow S_i \cup \{e_i\}$

6.   Else

7. $S_{i+1} \leftarrow S_i$

8.End for

9.Return S_n

Theorem 12. $E[\psi_i(S_i, e_i)] \geq \frac{1}{n}\left(1 - \frac{1}{n}\right)^{n-(i+1)}[f(OPT) - f(S_i)] - \frac{1}{n}g(OPT)$

Theorem 13. $E\left[f\left(S_n\right) - \left(1 - c_g\right) g\left(S_n\right)\right] \geq \left(1 - e^{-1}\right) f\left(OPT\right) - g\left(OPT\right)$, where c_g is the curvature of function g

From Theorem 13, we can draw the following corollary when g is modular function.

Corollary 6. If $c_g = 0$, i.e. g is modular. Then we have

$$E\left[f\left(S_n\right) - g\left(S_n\right)\right] \geq \left(1 - e^{-1}\right) f\left(OPT\right) - g\left(OPT\right)$$

Remark: Obviously, if g is modular, then $h = f - g$ is submodular but non-monotone. The above approximations are different with non-monotone submodular maximization under unconstrained problem $1/2$. But we cannot measure which one is better than others. In some cases, our approximations may be better than $1/2$. This give us a clue that we can choose the appropriate parameters $A(i)$ and $B(i)$ according to the characteristics of the problem to get a better approximate ratio.

7 Conclusions

In this paper, we propose Conditioned Greedy strategy with deterministic and random which are general frameworks for maximum DS decomposition under cardinality constrained and unconstrained respectively. Users can choose some rational parameters to fit special practical problems and get a wonderful approximation about problem. Also, we choose two special cases show our strategy can get some novel approximation. In some situations, these novel approximations are better than the best approximation at the state of art.

In the future works, how to remove the curvature parameter in approximation ratio is important, because it can make the approximation much tight. What's more, how to select $A(i)$ and $B(i)$ so that the algorithm can achieve the optimal approximation ratio is also urgent problem to be solved.

References

1. Bai, W., Bilmes, J.A.: Greed is still good: maximizing monotone submodular+supermodular functions (2018)
2. Buchbinder, N., Feldman, M., Naor, J.S., Schwartz, R.: A Tight Linear Time (1/2)-Approximation for Unconstrained Submodular Maximization. IEEE Computer Society, USA (2012). https://doi.org/10.1109/FOCS.2012.73
3. Buchbinder, N., Feldman, M., Naor, J.S., Schwartz, R.: Submodular Maximization with Cardinality Constraints. Society for Industrial and Applied Mathematics, USA (2014)
4. Das, A., Kempe, D.: Submodular meets spectral: Greedy algorithms for subset selection, sparse approximation and dictionary selection. Computer Science (2011)
5. Feldman, M., Naor, J.S., Schwartz, R.: A Unified Continuous Greedy Algorithm for Submodular Maximization. IEEE Computer Society, USA (2011). https://doi.org/10.1109/FOCS.2011.46
6. Fisher, M.L., Nemhauser, G.L., Wolsey, L.A.: An analysis of approximations for maximizing submodular set functions-II. Math. Program. **8**(1), 73–87 (1978)
7. Golovin, D., Krause, A.: Adaptive submodularity: theory and applications in active learning and stochastic optimization. J. Artif. Intell. Res. **42**(1), 427–486 (2012)
8. Han, K., Xu, C., Gui, F., Tang, S., Huang, H., Luo, J.: Discount allocation for revenue maximization in online social networks, pp. 121–130, June 2018. https://doi.org/10.1145/3209582.3209595
9. Harshaw, C., Feldman, M., Ward, J., Karbasi, A.: Submodular maximization beyond non-negativity: guarantees, fast algorithms, and applications, vol. 2019-June, pp. 4684–4705. Long Beach, CA, United states (2019)

10. Horel, T., Singer, Y.: Maximization of approximately submodular functions. In: Advances in Neural Information Processing Systems (2016)
11. Iyer, R., Bilmes, J.: Algorithms for Approximate Minimization of the Difference Between Submodular Functions, with Applications. AUAI Press, Arlington, Virginia (2012)
12. Jegelka, S., Bilmes, J.: Submodularity Beyond Submodular Energies: Coupling Edges in Graph Cuts. IEEE Computer Society (2011)
13. Kempe, D.: Maximizing the spread of influence through a social network. In: Proceedings of ACM SIGKDD International Conference on Knowledge Discovery and Data Mining (2003)
14. Krause, A., Singh, A., Guestrin, C.: Near-optimal sensor placements in Gaussian processes: theory, efficient algorithms and empirical studies. J. Mach. Learn. Res. 9(3), 235–284 (2008)
15. Li, X., Du, H.G., Pardalos, P.M.: A variation of ds decomposition in set function optimization. J. Comb. Optim. (2020)
16. Lin, H., Bilmes, J.: A Class of Submodular Functions for Document Summarization. Association for Computational Linguistics, USA (2011)
17. Lu, W., Chen, W., Lakshmanan, L.V.S.: From competition to complementarity: comparative influence diffusion and maximization. Proc. VLDB Endow. 9(2), 60–71 (2015)
18. Maehara, T., Murota, K.: A framework of discrete dc programming by discrete convex analysis. Math. Program. 152, 435–466 (2015)
19. Conforti, M., Cornuéjols, G.: Submodular set functions, matroids and the greedy algorithm: tight worst-case bounds and some generalizations of the Rado-Edmonds theorem. Discret. Appl. Math. 7, 251–274 (1984)
20. Nemhauser, G.L., Wolsey, L.A., Fisher, M.L.: An analysis of approximations for maximizing submodular set functions-I. Math. Program. 14(1), 265–294 (1978)
21. Wang, Z., Yang, Y., Pei, J., Chu, L., Chen, E.: Activity maximization by effective information diffusion in social networks. IEEE Trans. Knowl. Data Eng. 29(11), 2374–2387 (2017)
22. Yang, W., Chen, S., Gao, S., Yan, R.: Boosting node activity by recommendations in social networks. J. Comb. Optim. 40(3), 825–847 (2020). https://doi.org/10.1007/s10878-020-00629-6
23. Yang, W., Zhang, Y., Du, D.-Z.: Influence maximization problem: properties and algorithms. J. Comb. Optim. 40(4), 907–928 (2020). https://doi.org/10.1007/s10878-020-00638-5
24. Yu, J., Blaschko, M.: A convex surrogate operator for general non-modular loss functions, April 2016
25. Zhu, J., Ghosh, S., Zhu, J., Wu, W.: Near-optimal convergent approach for composed influence maximization problem in social networks. IEEE Access PP(99), 1 (2019)

Reading Articles Online

Andreas Karrenbauer[1] and Elizaveta Kovalevskaya[1,2(✉)]

[1] Max Planck Institute for Informatics, Saarland Informatics Campus,
Saarbrücken, Germany
{andreas.karrenbauer,elizaveta.kovalevskaya}@mpi-inf.mpg.de
[2] Goethe University Frankfurt, Frankfurt am Main, Germany
lisa@ae.cs.uni-frankfurt.de

Abstract. We study the online problem of reading articles that are
listed in an aggregated form in a dynamic stream, e.g., in news feeds, as
abbreviated social media posts, or in the daily update of new articles on
arXiv. In such a context, the brief information on an article in the list-
ing only hints at its content. We consider readers who want to maximize
their information gain within a limited time budget, hence either discarding an article right away based on the hint or accessing it for reading.
The reader can decide at any point whether to continue with the current
article or skip the remaining part irrevocably. In this regard, Reading
Articles Online, RAO, does differ substantially from the Online Knapsack Problem, but also has its similarities. Under mild assumptions, we
show that any α-competitive algorithm for the Online Knapsack Problem in the random order model can be used as a black box to obtain an
$(e + \alpha)C$-competitive algorithm for RAO, where C measures the accuracy of the hints with respect to the information profiles of the articles.
Specifically, with the current best algorithm for Online Knapsack, which
is $6.65 < 2.45e$-competitive, we obtain an upper bound of $3.45eC$ on the
competitive ratio of RAO. Furthermore, we study a natural algorithm
that decides whether or not to read an article based on a single threshold value, which can serve as a model of human readers. We show that
this algorithmic technique is $O(C)$-competitive. Hence, our algorithms
are constant-competitive whenever the accuracy C is a constant.

1 Introduction

There are many news aggregators available on the Internet these days. However,
it is impossible to read all news items within a reasonable time budget. Hence,
millions of people face the problem of selecting the most interesting articles
out of a news stream. They typically browse a list of news items and make a
selection by clicking into an article based on brief information that is quickly
gathered, e.g., headline, photo, short abstract. They then read an article as long
as it is found interesting enough to stick to it, i.e., the information gain is still
sufficiently high compared to what is expected from the remaining items on the
list. If not, then the reader goes back to browsing the list, and the previous article
is discarded – often irrevocably due to the sheer amount of available items and a

© Springer Nature Switzerland AG 2020
W. Wu and Z. Zhang (Eds.): COCOA 2020, LNCS 12577, pp. 639–654, 2020.
https://doi.org/10.1007/978-3-030-64843-5_43

limited time budget. This problem is inspired by research in Human-Computer Interaction [8].

In this paper, we address this problem from a theoretical point of view. To this end, we formally model the Reading Articles Online Problem, RAO, show lower and upper bounds on its competitive ratio, and analyze a natural threshold algorithm, which can serve as a model of a human reader.

There are obvious parallels to the famous Secretary Problem, i.e., if we could only afford to read one article, we had a similar problem that we had to make an irrevocable decision without knowing the remaining options. However, in the classical Secretary Problem, it is assumed that we obtain a truthful valuation of each candidate upon arrival. But in our setting, we only get a hint at the content, e.g., by reading the headline, which might be a click bait. However, if there is still time left from our budget after discovering the click bait, we can dismiss that article and start browsing again, which makes the problem fundamentally more general. Moreover, a typical time budget allows for reading more than one article, or at least a bit of several articles, perhaps of different lengths. Thus, our problem is also related to Online Knapsack but with uncertainty about the true values of the items. Nevertheless, we assume that the reader obtains a hint of the information content before selecting and starting to read the actual article. This is justified because such a hint can be acquired from the headline or a teaser photo, which is negligible compared to the time it takes to read an entire article. In contrast, the actual information gain is only realized while reading and only to the extent of the portion that has already been read. For the sake of simplicity, one can assume a sequential reading strategy where the articles are read word for word and the information gain might fluctuate strongly, especially in languages like German where a predicate/verb can appear at the end of a long clause. However, in contrast to spatial information profiles, one can also consider temporal information profiles where the information gain depends on the reading strategy, e.g., cross reading. It is clear that the quality of the hint in relation to the actual information content of the article is a decisive factor for the design and analysis of corresponding online algorithms. We argue formally that the hint should be an upper bound on the information rate, i.e., the information gain per time unit. Moreover, we confirm that the hint should not be too far off the average information rate to achieve decent results compared to the offline optimum where all articles with the corresponding information profiles are known in advance. In this paper, we assume that the length of an article, i.e., the time it takes to read it to the end, is revealed together with the hint. This is a mild assumption because this attribute can be retrieved quickly by taking a quick glance at the article, e.g., at the number of pages or at the size of the scroll bar.

1.1 Related Work

To the best of our knowledge, the problem of RAO has not been studied in our suggested setting yet. The closest related problem known is the Online Knapsack Problem [2,4,10] in which an algorithm has to fill a knapsack with restricted capacity while trying to maximize the sum of the items' values. Since the input is not known at the beginning and an item is not selectable later than its arrival, optimal algorithms do not exist. In the adversarial model where an adversary chooses the order of the items to arrive, it has been shown in [12] that the competitive ratio is unbounded. Therefore, we consider the random order model where a permutation of the input is chosen uniformly at random.

A special case of the Online Knapsack Problem is the well-studied Secretary Problem, solved by [7] among others. The goal is to choose the best secretary of a sequence without knowing what values the remaining candidates will have. The presented e-competitive algorithm is optimal for this problem. The k-Secretary Problem aims to hire at most $k \geq 1$ secretaries while maximizing the sum of their values. In [11], an algorithm with a competitive ratio of $1/(1 - 5/\sqrt{k})$, for large enough k, with a matching lower bound of $\Omega(1/(1-1/\sqrt{k}))$ is presented. Furthermore, [4] contains an algorithm that is e-competitive for any k. Some progress for the case of small k was made in [3]. The Knapsack Secretary Problem introduced by [4] is equivalent to the Online Knapsack Problem in the random order model. They present a 10e-competitive algorithm. An 8.06-competitive algorithm ($8.06 < 2.97e$) is shown in [10] for the Generalized Assignment Problem, which is the Online Knapsack Problem generalized to a setting with multiple knapsacks with different capacities. The current best algorithm from [2] achieves a competitive ratio of $6.65 < 2.45e$.

There have been different approaches to studying the Knapsack Problem with uncertainty besides the random order model. One approach is the Stochastic Knapsack Problem where values or weights are drawn from a known distribution. This problem has been studied in both online [12] and offline [6] settings. In [9], an offline setting with unknown weights is considered: algorithms are allowed to query a fixed number of items to find their exact weight.

A model with resource augmentation is considered for the fractional version of the Online Knapsack Problem in [15]. There, the knapsack of the online algorithm has $1 \leq R \leq 2$ times more capacity than the knapsack of the offline optimum. Moreover, they allow items to be rejected after being accepted. In our model, this would mean that the reader gets time returned after having already read an article. Thus, their algorithms are not applicable on RAO.

1.2   Our Contribution

We introduce RAO and prove lower and upper bounds on competitive ratios under various assumptions. We present relations to the Online Knapsack problem and show how ideas from that area can be adapted to RAO. Our emphasis lies on the initiation of the study of this problem by the theory community.

We first show lower bounds that grow with the number of articles unless restrictions apply that forbid the corresponding bad instances. That is, whenever information rates may be arbitrarily larger than the hint of the corresponding article, any algorithm underestimates the possible information gain of a good article. Hence, the reader must adjust the hints such that they upper bound the information rate to allow for bounded competitive ratios. While we may assume w.l.o.g. for the Online Knapsack Problem that no item is larger than the capacity since an optimal solution cannot contain such items, we show that RAO without this or similar restrictions suffers from a lower bound of $\Omega(n)$, i.e., any algorithm is arbitrarily bad in the setting where articles are longer than the time budget. Moreover, we prove that the accuracy of the hints provides a further lower bound for the competitive ratio. We measure this accuracy as the maximum ratio C of hints and respective average information rates. Hence, a constant-competitive upper bound for RAO is only possible when C is bounded.

Under these restrictions, we present the first constant-competitive algorithm for RAO. To this end, we introduce a framework for wrapping any black box algorithm for the Online Knapsack Problem to work for RAO. Given an α-competitive algorithm for the Online Knapsack Problem as a black box, we obtain a $(e + \alpha)C$-competitive algorithm for RAO. This algorithm is $3.45eC$-competitive when using the current best algorithm for the Online Knapsack Problem from [2]. This is the current best upper bound that we can show for RAO, which is constant provided that the hints admit a constant accuracy.

However, the algorithm generated by the framework above inherits its complexity from the black box algorithm for the Online Knapsack Problem, which may yield good competitive ratios from a theoretical point of view but might be too complex to serve as a strategy for a human reader. Nevertheless, the existence of constant-competitive ratios (modulo accuracy of the hints) motivates us to strive for simple $O(C)$-algorithms. To this end, we investigate an algorithm that bases its decisions on a single threshold. The Threshold Algorithm can be seen as a formalization of human behavior. While reading, humans decide intuitively whether an article is interesting or not. This intuition is modeled by the single threshold. In case of diminishing information gain, we show that this simplistic approach suffices to obtain an upper bound of $246C < 90.5eC$ on the competitive ratio with the current analysis, which might leave room for improvement but nevertheless achieves a constant competitive ratio. Diminishing information gain means non-increasing information rates, a reasonable assumption particularly in the light of efficient reading strategies, where an article is not read word for word. In such a context, one would consider a temporal information profile that relates the information gain to the reading time. When smoothed to coarse time scale, the information rates can be considered non-increasing and lead to a saturation of the total information obtained from an article over time.

2   Preliminaries

Definition 1 (Reading Articles Online (RAO)). *There are n articles that are revealed one by one in a round-wise fashion. The reader has a time budget $T \in \mathbb{N}_{>0}$ for reading. In round i, the reader sees article i with its hint $h_i \in \mathbb{N}_{>0}$ and time length $t_i \in \mathbb{N}_{>0}$. The actual information rate $c_i : [t_i] \to [h_i]$ is an unknown function. The reader has to decide whether to start reading the article or to skip to the next article. After reading time step $j \leq t_i$, the reader obtains $c_i(j)$ information units and can decide to read the next time step of the article or to discard it irrevocably. After discarding or finishing the current article, the next*

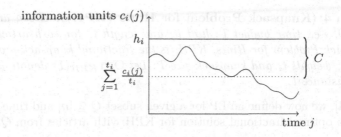

Fig. 1. Relation of hint h_i to average information gain $\sum_{j=1}^{t_i} c_i(j)/t_i$ as in Definition 3.

round begins. The objective is to maximize $\sum_{i \in [n]} \sum_{j=1}^{\tau_i} c_i(j)$ where $0 \leq \tau_i \leq t_i$ is the number of time steps read by the algorithm and $\sum_{i \in [n]} \tau_i \leq T$.

For the sake of simplicity, we have chosen a discrete formulation in Definition 1, which is justified by considering words or even characters as atomic information units. Since such tiny units might be too fine-grained compared to the length of the articles, we can also extend this formulation with a slight abuse of notation and allow that the τ_i are fractional, i.e., $\sum_{j=1}^{\tau_i} c_i(j) = \sum_{j=1}^{\lfloor \tau_i \rfloor} c_i(j) + c_i(\lceil \tau_i \rceil) \cdot \{\tau_i\}$, where $\{\tau_i\}$ denotes its fractional part. However, one could also consider a continuous formulation using integrals, i.e., the objective becomes $\sum_{i \in [n]} \int_0^{\tau_i} c_i(t) dt$. The lower bounds presented in this section hold for these models as well.

We use the random order model where input order corresponds to a permutation π chosen uniformly at random.

Definition 2 (Competitive Ratio). *We say that an algorithm* ALG *is α-competitive, if, for any instance I, the expected value of* ALG *on instance I, with respect to permutation of the input and random choices of* ALG, *is at least $1/\alpha$ of the optimal offline value* OPT(I), *i.e.,* $\mathbb{E}[\text{ALG}(I)] \geq \frac{1}{\alpha} \cdot \text{OPT}(I)$.

We measure the accuracy of hints with parameter C from Definition 3. This relation is illustrated in Fig. 1. Lemma 4 provides a lower bound in dependence on this accuracy.

Definition 3 (Accuracy of Hints). *The accuracy $C \geq 1$ is the smallest number s.t.*

$$h_i \leq C \cdot \sum_{j=1}^{t_i} \frac{c_i(j)}{t_i} \quad \forall i \in [n].$$

The hint is a single number giving a cue about a function. Therefore, no matter which measure of accuracy we consider, if the hint is perfectly accurate, the function has to be constant. In Sect. 6, we discuss other ideas on the measure of accuracy of hints such as a multi-dimensional feature vector or the hint being a random variable drawn from the information rate.

We introduce an auxiliary problem to bound the algorithm's expected value.

Definition 4 (Knapsack Problem for Hints (KPH)). *Given an instance I for RAO, i.e., time budget T, hint h_i and length t_i for each article $i \in [n]$, the Knapsack Problem for Hints, KPH, is the fractional knapsack problem with values $t_i h_i$, weights t_i and knapsack size T. Let $\text{OPT}_{KPH}(I)$ denote its optimal value on instance I.*

As in [4], we now define an LP for a given subset $Q \subseteq [n]$ and time budget x. It finds the optimal fractional solution for KPH with articles from Q and time budget x.

$$\max \sum_{i=1}^{n} t_i \cdot h_i \cdot y(i)$$
$$\text{s.t.} \sum_{i=1}^{n} \quad t_i \cdot y(i) \leq x$$
$$y(i) = 0 \qquad \forall i \notin Q$$
$$y(i) \in [0,1] \quad \forall i \in [n]$$

The variable $y_Q^{(x)}(i)$ refers to the setting of $y(i)$ in the optimal solution on articles from Q with time budget x. The optimal solution has a clear structure: There exists a threshold hint $\rho_Q^{(x)}$ such that any article $i \in Q$ with $h_i > \rho_Q^{(x)}$ has $y_Q^{(x)}(i) = 1$ and any article $i \in Q$ with $h_i < \rho_Q^{(x)}$ has $y_Q^{(x)}(i) = 0$. As in [4], we use the following notation for a subset $R \subseteq [n]$:

$$v_Q^{(x)}(R) = \sum_{i \in R} t_i \cdot h_i \cdot y_Q^{(x)}(i) \quad \text{and} \quad w_Q^{(x)}(R) = \sum_{i \in R} t_i \cdot y_Q^{(x)}(i).$$

The value $v_Q^{(x)}(R)$ and weight $w_Q^{(x)}(R)$ refer to the value and weight that set R contributes to the optimal solution. We use KPH's solution as an upper bound on the optimal solution, as shown in the following lemma:

Lemma 1. *Given instance I for RAO, let $\text{OPT}(I)$ and $\text{OPT}_{KPH}(I)$ be the respective optimal values. Then, $\text{OPT}_{KPH}(I) \geq \text{OPT}(I)$.*

Proof. Since the codomain of c_i is $[h_i]$ by Definition 1, any algorithm for RAO cannot obtain more information units than h_i in any time step. Thus, the optimal solution of KPH obtains at least the same amount of information units as the optimal solution of RAO by reading the same parts.   □

3   Lower Bounds

The proofs in this section are constructed in a way such that the instances are not producing a lower bound for the other settings. Note that in the proofs of Lemma 2 and Lemma 3, we have $C \leq 2$. Moreover, we construct the family of instances such that the lower bounds hold in a fractional setting. The key idea is to have the first time step(s) small in every information rate such that the algorithms are forced to spend a minimal amount of time on reading an article before obtaining eventually more than one information unit per time step.

Although Definition 1 already states that the codomain of c_i is $[h_i]$, we show a lower bound as a justification for this constraint on c_i.

Lemma 2. *If the functions c_i are allowed to take values larger than hint h_i, then the competitive ratio of any deterministic or randomized algorithm is $\Omega(\sqrt{n})$.*

Proof. We construct a family of instances for all $\ell \in \mathbb{N}$. Let $n := \ell^2$. For any fixed ℓ, we construct instance I as follows. Set $T := n$, $t_i := T = n$ and $h_i := n$ for all $i \in [n]$. We define two types of articles. There are $\sqrt{n}$ articles of type A where $c_i(j) := 1$ for $j \in [T] \setminus \{\sqrt{n}\}$ and $c_i(\sqrt{n}) := n^2$. The articles of type B have $c_i(j) := 1$ for $j \in [T-1]$ and $c_i(T) := n^2$.

An optimal offline algorithm reads all articles of type A up to time step $\sqrt{n}$. Therefore, $\text{OPT}(I) = \Theta(n^{2.5})$. Any online algorithm cannot distinguish between articles of type A and B until time step $\sqrt{n}$. The value of any online algorithm cannot be better than the value of algorithm ALG that reads the first $\sqrt{n}$ time steps of the first $\sqrt{n}$ articles. Since the input order is chosen uniformly at random, the expected arrival of the first type A article is at round $\sqrt{n}$. Thus, $\mathbb{E}\left[\text{ALG}(I)\right] = \Theta(n^2)$. Therefore, $\mathbb{E}\left[\text{ALG}(I)\right] \leq \text{OPT}(I)/\sqrt{n}$. □

In RAO, solutions admit reading articles fractionally. Therefore, we show a lower bound whenever articles are longer than the time budget.

Lemma 3. *If the lengths t_i are allowed to take values larger than time budget T, then the competitive ratio of any deterministic or randomized algorithm is $\Omega(n)$.*

Proof. We construct a family of instances with $T := 2$, $t_i := 3$ and $h_i := M$ for all $i \in [n]$, where $M \geq 1$ is set later. We define $c_k(1) := 1$, $c_k(2) := M$ and $c_k(3) := 1$. Any other article $i \in [n] \setminus \{k\}$ has $c_i(1) := 1$, $c_i(2) := 1$ and $c_i(3) := M$.

As the permutation is chosen uniformly at random, an online algorithm does not know which article is the one with M information units in the second time step. No algorithm can do better than the algorithm ALG that reads the first article completely while $\text{OPT}(I) = M + 1$. Its expected value is $\mathbb{E}\left[\text{ALG}(I)\right] = (1/n) \cdot (M+1) + (1 - 1/n) \cdot 2 \leq (2/n + 2/M) \cdot \text{OPT}(I)$. When setting $M = n$, we obtain the desired bound. □

A consequence of the next lemma is that if the accuracy C from Definition 3 is not a constant, then no constant-competitive algorithms can exist.

Lemma 4. *Any deterministic or randomized algorithm is $\Omega(\min\{C, n\})$-competitive.*

Proof. Consider the following family of instances I in dependence on accuracy $C \geq 1$. Let $T := 2$ and $t_i := 2$ for all $i \in [n]$. Set $c_k(1) := 1$ and $c_k(2) := C$. We define $c_i(1) := 1$ and $c_i(2) := 1$ for all $i \in [n] \setminus \{k\}$. The hints are $h_i := C$ for all $i \in [n]$, thus, they are C-accurate according to Definition 3.

Any algorithm cannot distinguish the information rate of the articles, as the hints and the first time steps are all equal. Therefore, no algorithm is better than ALG, which chooses to read the article arriving first completely. The optimal choice is to read article k; we obtain the desired bound: $\mathbb{E}\left[\text{ALG}(I)\right] = (1/n) \cdot (C+1) + (1 - 1/n) \cdot 2 \leq (2/n + 2/C) \cdot \text{OPT}(I)$. $\qquad\square$

Assumption 1. *For any article $i \in [n]$, we assume that $t_i \leq T$ and that the hints h_i and upper bounds $t_i h_i$ on the information gain in the articles are distinct.*[1]

4   Exploitation of Online Knapsack Algorithms

In this section, we develop a technique for applying any algorithm for the Online Knapsack Problem on an instance of RAO. The presented algorithm uses the classic Secretary Algorithm that is e-competitive for all positive n as shown in [5]. The Secretary Algorithm rejects the first $\lfloor n/e \rfloor$ items. Then it selects the first item that has a better value than the best so far. Note that the Secretary Algorithm selects exactly one item.

We use KPH for the analysis as an upper bound on the actual optimal solution with respect to information rates c_i. There is exactly one fractional item in the optimal solution of KPH. The idea is to make the algorithm robust against two types of instances: the ones with a fractional article of high information amount and the ones with many articles in the optimal solution.

Theorem 2. *Given an α-competitive algorithm* ALG *for the Online Knapsack Problem, the Reduction Algorithm is* $(e + \alpha)C$*-competitive.*

Proof. We fix an instance I and use Lemma 1. We split the optimal solution of KPH into the fractional article i_f that is read $x_{i_f} \cdot t_{i_f}$ time steps and the set H_{max} of articles that are read completely. Since H_{max} is a feasible solution to the integral version of KPH, the value of the articles in H_{max} is not larger than the optimal value $\text{OPT}_{KPH}^{int}(I)$ of the integral version. We denote the optimal integral solution by set H^*. Using Definition 3, we obtain:

---

[1] Disjointness is obtained by random, consistent tie-breaking as described in [4].

$$\text{OPT}(I) \le \text{OPT}_{KPH}(I) = \sum_{i \in H_{max}} t_i h_i + x_{i_f} t_{i_f} h_{i_f} \le \text{OPT}_{KPH}^{int}(I) + t_{i_f} h_{i_f}$$

$$\le \sum_{i \in H^*} t_i h_i + \max_{i \in [n]} t_i h_i \le C \cdot \left(\sum_{i \in H^*} \sum_{j=1}^{t_i} c_i(j) + \max_{i \in [n]} \sum_{j=1}^{t_i} c_i(j) \right)$$

$$\le C \cdot \left(\frac{\alpha}{\delta} \cdot \mathbb{P}\left[b = 1\right] \mathbb{E}\left[\sum_{i \in S} \sum_{j=1}^{t_i} c_i(j) \middle| b = 1 \right] \right.$$

$$\left. + \frac{e}{1 - \delta} \cdot \mathbb{P}\left[b = 0\right] \mathbb{E}\left[\sum_{i \in S} \sum_{j=1}^{t_i} c_i(j) \middle| b = 0 \right] \right)$$

$$\le C \cdot \max\left\{ \frac{\alpha}{\delta}, \frac{e}{1 - \delta} \right\} \cdot \mathbb{E}\left[\text{Reduction Algorithm}(I)\right].$$

The optimal choice of δ to minimize $\max\left\{ \frac{\alpha}{\delta}, \frac{e}{1-\delta} \right\}$ is $\delta = \frac{\alpha}{e+\alpha}$. This is exactly how the Reduction Algorithm sets the probability $\delta \in (0,1)$ in line 1, which yields a competitive ratio of $(e + \alpha) \cdot C$. $\qquad \square$

Algorithm 1: Reduction Algorithm

Input: Number of articles n, time budget T, an α-competitive algorithm ALG
    for the Online Knapsack Problem.
Output: Set S of chosen articles.

1 Set $\delta = \frac{\alpha}{e+\alpha}$ and choose $b \in \{0,1\}$ randomly with $\mathbb{P}\left[b = 1\right] = \delta$;
2 **if** $b = 1$ **then**
3 | Apply ALG with respect to values $t_i h_i$ and weights t_i;
4 **else**
5 | Apply the Secretary Algorithm with respect to values $t_i h_i$;

When using the current best algorithm for the Online Knapsack Problem presented in [2], the Reduction Algorithm has a competitive ratio of $(e + 6.65) \cdot C \le 3.45eC$. Assuming that the accuracy of the hints $C \ge 1$ from Definition 3 is constant, the RAO admits a constant upper bound on the competitive ratio.

Remark 1. (i) The Reduction Algorithm can be used to obtain an $(\alpha+e)$-competitive algorithm for the fractional version of the Online Knapsack Problem given an α-competitive algorithm for the integral version. The proof is analogous to the proof of Theorem 2. (ii) Running an α-competitive algorithm for Online Knapsack on an instance of RAO, we obtain a $2\alpha C$-competitive algorithm for RAO by a similar proof. Since the current best algorithm for Online Knapsack has $\alpha = 6.65 > e$, using the Reduction Algorithm provides better bounds. This holds for the fractional Online Knapsack Problem respectively.

5   Threshold Algorithm

While the Online Knapsack Problem has to take items completely, RAO does not require the reader to finish an article. Exploiting this possibility, we present the Threshold Algorithm, which bases its decisions on a single threshold. We adjust the algorithm and its analysis from [4]. From now on, we assume that the information rates c_i are non-increasing and that we can stop to read an article at any time, thus allowing fractional time steps. For the sake of presentation, we stick to the discrete notation (avoiding integrals). In practice, the information gain diminishes the longer an article is read, and the inherent discretization by words or characters is so fine-grained compared to the lengths of the articles that it is justified to consider the continuum limit.

Before starting to read, one has to decide at which length to stop reading any article in dependence on T. First, we show that cutting all articles of an instance after gT time steps costs at most a factor of $1/g$ in the competitive ratio.

Lemma 5. *Given an instance I with time budget T, $g \in (0, 1]$, lengths t_i, hints h_i and non-increasing information rates $c_i : [t_i] \to [h_i]$, we define the cut instance I'_g with time budget $T' = T$, lengths $t'_i = \min\{t_i, gT\}$, hints $h'_i = h_i$ and non-increasing information rates $c'_i : [t'_i] \to [h'_i]$, where $c'_i(j) = c_i(j)$ for $1 \le j \le t'_i$. Then, $\mathrm{OPT}_{KPH}(I) \le \mathrm{OPT}_{KPH}(I'_g)/g$.*

Proof. Since $gt_i \le gT$ and $gt_i \le t_i$ we have $gt_i \le \min\{gT, t_i\} = t'_i$ and obtain:
$\mathrm{OPT}_{KPH}(I) = \frac{1}{g} \sum_{i \in [n]} h_i \cdot gt_i \cdot y_{[n]}^{(T)}(i) \le \frac{1}{g} \sum_{i \in [n]} h'_i t'_i \cdot y_{[n]}^{(T)}(i) \le \frac{1}{g} \cdot \mathrm{OPT}_{KPH}(I'_g)$.
The last inequality follows as no feasible solution is better than the optimum. The time budget is respected since $\sum_{i \in [n]} t_i \cdot y_{[n]}^{(T)}(i) \le T = T'$ and $t_i \ge t'_i$. $\quad\square$

We need the following lemma that can be proven by a combination of normalization, Exercise 4.7 on page 84 and Exercise 4.19 on page 87 in [14].

Lemma 6. *Let $z_1, ..., z_n$ be mutually independent random variables from a finite subset of $[0, z_{max}]$ and $Z = \sum_{i=1}^{n} z_i$ with $\mu = \mathbb{E}[Z]$. For all $\mu_H \ge \mu$ and all $\delta > 0$,*

$$\mathbb{P}[Z \ge (1 + \delta)\mu_H] < \exp\left(-\frac{\mu_H}{z_{max}} \cdot [(1 + \delta)\ln(1 + \delta) - \delta]\right).$$

We assume that $\sum_{i \in [n]} \min\{t_i, gT\} = \sum_{i \in [n]} t'_i \ge 3T/2$ for the purpose of the analysis. The same assumption is made in [4] implicitly. If there are not enough articles, the algorithm can only improve since there are fewer articles that are not part of the optimal solution. We can now state the main theorem.

Theorem 3. *For $g = 0.0215$, the Threshold Algorithm's competitive ratio is upper bounded by $246C < 90.5eC$.*

The proof is similar to the proof of Lemma 4 in [4]. However, we introduce parameters over which we optimize the analysis of the competitive ratio. This way, we make the upper bound on the competitive ratio as tight as possible for the proof technique used here. Recall that we may assume w.l.o.g. by Assumption 1 that the hints h_i and upper bounds $t_i h_i$ are disjoint throughout the proof.

Algorithm 2: Threshold Algorithm

Input: Number of articles n, time budget T, a fraction $g \in (0, 1]$.
Article i appears in round $\pi(i)$ and reveals h_i and t_i.

Output: Number of time steps $0 \leq s_i \leq t_i$ that are read from article $i \in [n]$.

1 Sample $r \in \{1, ..., n\}$ from binomial distribution $\text{Bin}(n, 1/2)$;

2 Let $X = \{1, ..., r\}$ and $Y = \{r + 1, ..., n\}$;

3 for *round* $\pi(i) \in X$ **do**

4   Observe h_i and t_i;

5   Set $s_i = 0$;

6 Solve KPH on X with budget $T/2$, lengths $t_i' = \min\{gT, t_i\}$, and values $t_i' h_i$.

7 Let $\rho_X^{(T/2)}$ be the threshold hint;

8 for *round* $\pi(i) \in Y$ **do**

9   **if** $h_i \geq \rho_X^{(T/2)}$ **then**

10   Set $s_i = \min\left\{ t_i, gT, T - \sum_{1 \leq j < i} s_j \right\}$;

11   Read the first s_i time steps of article i;

12   **else**

13   Set $s_i = 0$;

Proof. Fix an instance I. We refer to the order by permutation π chosen uniformly at random. For simplicity, we scale the instance such that $T = 1$ and all t_i are multiplied with $1/T$, which does not affect the hints and the threshold. We use KPH to show the bound as $\text{OPT}_{KPH}(I) \geq \text{OPT}(I)$ holds by Lemma 1. As the reader always reads at most the first gT time steps, we use the bound from Lemma 5 for cutting instance I to I_g'. For better readability, we do not rename the parameters of I_g' and refer to the variables without adding a prime ′. We proceed with showing the bound on the expected value of the algorithm on instance I_g' since the algorithm reads at most g time steps of each article.

Now, we proceed as in the proof of Lemma 4 in [4]. We use two auxiliary knapsacks to bound the algorithm's expected value. Their optimal, fractional solution is computed offline on instance I. In contrast to [4], we parameterize the size of the auxiliary knapsacks to find the best possible sizes. We use a knapsack of size β and one of size γ where $0 < \beta \leq 1$ and $1 \leq \gamma \leq \sum_{i \in [n]} t_i$. Recall that we assumed that $\sum_{i \in [n]} t_i \geq 3/2 = 3T/2$. We show in the following that for all i where $y_{[n]}^{(\beta)}(i) > 0$, there is a $p \in (0, 1)$ such that $\mathbb{P}\left[s_i = t_i\right] > p$. As a consequence of p's existence, we obtain the following inequalities:

$$\text{OPT}_{KPH}(I) \leq \frac{1}{g} \cdot \text{OPT}_{KPH}(I'_g) \leq \frac{1}{g} \cdot \frac{1}{\beta} \cdot v_{[n]}^{(\beta)}([n]) = \frac{1}{g\beta} \cdot \sum_{i \in [n]} t_i h_i y_{[n]}^{(\beta)}(i)$$

$$\leq \frac{1}{g\beta p} \cdot \sum_{i \in [n]} \mathbb{P}\left[s_i = t_i\right] t_i h_i \leq \frac{1}{g\beta p} \cdot \sum_{i \in [n]} \mathbb{P}\left[s_i = t_i\right] \cdot C \cdot \sum_{j=1}^{s_i} c_i(j)$$

$$\leq \frac{C}{g\beta p} \cdot \mathbb{E}\left[\text{Threshold Algorithm}(I)\right].$$

$$(1)$$

We lose the factor C as we use the inequality from Definition 3. The best possible value for the competitive ratio is the minimum of $C/(g\beta p)$. We find it by maximizing $g\beta p$. As g and β are settable variables, we determine p first. We define random variables ζ_i for all $i \in [n]$, where

$$\zeta_i = \begin{cases} 1 \text{ if } \pi(i) \in X \\ 0 \text{ otherwise.} \end{cases}$$

As discussed in [4], conditioned on the value of r, $\pi^{-1}(X)$ is a uniformly chosen subset of $[n]$ from any subset of $[n]$ containing exactly r articles. Since r is chosen from $\text{Bin}(n, 1/2)$, it has the same distribution as the size of a uniformly at random chosen subset of $[n]$. Therefore, $\pi^{-1}(X)$ is a uniformly chosen subset of all subsets of $[n]$. The variables ζ_i are mutually independent Bernoulli random variables with $\mathbb{P}\left[\zeta_i = 1\right] = 1/2$. Now, we fix $j \in [n]$ with $y_{[n]}^{(\beta)}(j) > 0$ and define two random variables:

$$Z_1 := w_{\pi([n])}^{(\beta)}(X \setminus \{\pi(j)\}) = \sum_{i \in [n] \setminus \{j\}} t_i \cdot y_{[n]}^{(\beta)}(i) \cdot \zeta_i,$$

$$Z_2 := w_{\pi([n])}^{(\gamma)}(Y \setminus \{\pi(j)\}) = \sum_{i \in [n] \setminus \{j\}} t_i \cdot y_{[n]}^{(\gamma)}(i) \cdot (1 - \zeta_i).$$

Note that the event $\pi(j) \in Y$, i.e., $\zeta_j = 0$, is independent of Z_1 and Z_2 since they are defined without $\pi(j)$. The weights $t_i \cdot y_{[n]}^{(\beta)}(i) \cdot \zeta_i$ and $t_i \cdot y_{[n]}^{(\gamma)}(i) \cdot (1 - \zeta_i)$ within the sum are random variables taking values in $[0, g]$ since the instance is cut. Now, we reason that when article j is revealed at position $\pi(j)$ to the Threshold Algorithm, it has enough time to read j with positive probability. The next claim is only effective for $g < 0.5$ since Z_1 and Z_2 are non-negative.

Claim. Conditioned on $Z_1 < \frac{1}{2} - g$ and $Z_2 < 1 - 2g$, the Threshold Algorithm sets $s_j = t_j$ if $\pi(j) \in Y$ because $h_j \geq \rho_X^{(1/2)}$ and it has enough time left.

Proof. Since $t_j \leq gT = g$, article j can only add at most g weight to X or Y. Therefore, $w_{\pi([n])}^{(\beta)}(X) < 1/2$ and $w_{\pi([n])}^{(\gamma)}(Y) < 1 - g$. Recall that knapsacks of size β and γ are packed optimally and fractionally. Thus, a knapsack of size γ would be full, i.e., $w_{\pi([n])}^{(\gamma)}(\pi([n])) = \min\left\{\gamma, \sum_{i \in [n]} t_i\right\}$. Since we assumed in the beginning of the proof that $\gamma \leq \sum_{i \in [n]} t_i$, we have $w_{\pi([n])}^{(\gamma)}(\pi([n])) = \gamma$. We can bound the weight of X in this solution by $w_{\pi([n])}^{(\gamma)}(X) = \gamma - w_{\pi([n])}^{(\gamma)}(Y) > \gamma - (1 - g)$. We choose γ such that $\gamma - (1 - g) > 1/2$.[2] Thus, the articles in X add weights $w_{\pi([n])}^{(\beta)}(X) < 1/2$ and $w_{\pi([n])}^{(\gamma)}(X) > 1/2$ to their respective optimal solution. Since $w_{\pi([n])}^{(\gamma)}(X) = \sum_{i \in X} t_i y_n^{(\gamma)}(i) > 1/2$, it means that there are elements in X with combined t_i of at least $1/2$. Therefore, the optimal solution of KPH on X with time budget $1/2$ is satisfying the capacity constraint with equality, i.e., $w_X^{(1/2)}(X) = 1/2$. We obtain: $w_{\pi([n])}^{(\gamma)}(X) > w_X^{(1/2)}(X) > w_{\pi([n])}^{(\beta)}(X)$.

When knapsack A has a higher capacity than knapsack B on the same instance, then the threshold density of knapsack A cannot be higher than the threshold density of knapsack B. For any X, we see that the knapsack of size β uses less capacity than $1/2$ and the knapsack of size γ uses more capacity than $1/2$ on the same X respectively. Since both knapsacks are packed by an optimal, fractional solution that is computed offline on the whole instance, the respective threshold hint for articles from X and articles from $\pi([n])$ is the same. Therefore, we get the following ordering of the threshold hints: $\rho_{\pi([n])}^{(\gamma)} \leq \rho_X^{(1/2)} \leq \rho_{\pi([n])}^{(\beta)}$.

Now, we show that when the algorithm sees $\pi(j)$, it has enough time left. Let $S^+ = \left\{ \pi(i) \in Y \setminus \{\pi(j)\} : h_i \geq \rho_X^{(1/2)} \right\}$ be the set of articles that the algorithm can choose from. Thus, the algorithm reads every article from S^+ (and maybe $\pi(j)$) if it has enough time left at the point when an article from S^+ arrives. By transitivity, every article $\pi(i) \in S^+$ has $h_i \geq \rho_{\pi([n])}^{(\gamma)}$. Therefore, for all but at most one[3] $\pi(i) \in S^+$, the equation $y_{[n]}^{(\gamma)}(i) = 1$ holds. Since the only article $i \in S^+$ that could have $y_{[n]}^{(\gamma)}(i) < 1$ is not longer than g, the total length of articles in S^+ can be bounded from above by $\sum_{i \in S^+} t_i \leq g + w_{\pi([n])}^{(\gamma)}(Y \setminus \{\pi(j)\}) = g + Z_2 < 1 - g$. As $t_j \leq g$, the algorithm has enough time left to read article j completely, when it arrives at position $\pi(j)$. Moreover, if $y_{[n]}^{(\beta)}(j) > 0$, then $h_j \geq \rho_{\pi([n])}^{(\beta)} \geq \rho_X^{(1/2)}$. $\blacksquare$

We now proceed with the main proof by showing a lower bound on $p' = \mathbb{P}[Z_1 < 1/2 - g$ and $Z_2 < 1 - 2g]$. Recall that the event $\pi(j) \in Y$ is independent of Z_1 and Z_2. With the preceded claim we obtain:

$$\mathbb{P}[s_j = t_j] = \mathbb{P}\left[s_j = t_j | Z_1 < 1/2 - g \text{ and } Z_2 < 1 - 2g\right] \cdot p'$$
$$= \mathbb{P}\left[\pi(j) \in Y | Z_1 < 1/2 - g \text{ and } Z_2 < 1 - 2g\right] \cdot p' = \frac{1}{2} \cdot p' \tag{2}$$

As we are searching for the lower bound $p < \mathbb{P}[s_j = t_j]$, we use the lower bound for p' multiplied with $1/2$ as the value for p. Moreover,

$$p' = \mathbb{P}[Z_1 < 1/2 - g \text{ and } Z_2 < 1 - 2g] \geq 1 - \mathbb{P}[Z_1 \geq 1/2 - g] - \mathbb{P}[Z_2 \geq 1 - 2g]. \tag{3}$$

---

[2] In [4], it is implicitly assumed that $\sum_{i \in [n]} t_i \geq 3/2$ as their choice of γ is $3/2$.

[3] As the hints are distinct by Assumption 1.

We can bound the probabilities $\mathbb{P}\left[Z_1 \geq 1/2 - g\right]$ and $\mathbb{P}\left[Z_2 \geq 1 - 2g\right]$ by the Chernoff Bound from Lemma 6.[4] The expected values of Z_1, Z_2 are bounded by

$$\mathbb{E}\left[Z_1\right] = \frac{1}{2} \cdot \left(\beta - t_j \cdot y_{[n]}^{(\beta)}(j)\right) \leq \frac{\beta}{2} \quad \text{and} \quad \mathbb{E}\left[Z_2\right] = \frac{1}{2} \cdot \left(\gamma - t_j \cdot y_{[n]}^{(\gamma)}(j)\right) \leq \frac{\gamma}{2}.$$

We use $z_{max} = g$, $\delta_1 = (1 - 2g)/\beta - 1 > 0$ and $\delta_2 = (2 - 4g)/\gamma - 1 > 0$ to obtain:

$$\mathbb{P}\left[Z_1 \geq 1/2 - g\right] < \exp\left(\left(1 - \frac{1}{2g}\right) \cdot \ln\left(\frac{1 - 2g}{\beta}\right) - 1 + \frac{1 - \beta}{2g}\right),$$

$$\mathbb{P}\left[Z_2 \geq 1 - 2g\right] < \exp\left(\left(2 - \frac{1}{g}\right) \cdot \ln\left(\frac{2 - 4g}{\gamma}\right) - 2 + \frac{1 - \gamma/2}{g}\right). \tag{4}$$

To conclude the proof, the final step is numerically maximizing the lower bound on $g\beta p'/2$ obtained by combining Eqs. (2), (3) and (4):

$$\max_{g,\beta,\gamma} \quad \frac{\beta g}{2} \cdot \left(1 - \exp\left(\left(1 - \frac{1}{2g}\right) \cdot \ln\left(\frac{1 - 2g}{\beta}\right) - 1 + \frac{1 - \beta}{2g}\right)\right.$$
$$\left. - \exp\left(\left(2 - \frac{1}{g}\right) \cdot \ln\left(\frac{2 - 4g}{\gamma}\right) - 2 + \frac{1 - \gamma/2}{g}\right)\right) \tag{5}$$

$$\text{s.t.} \quad \gamma + g > 1.5 \quad \quad 2g + \beta < 1 \quad \quad 4g + \gamma < 2$$
$$0 < g < 0.5 \quad \quad 0 < \beta < 1 \quad \quad \gamma > 1$$

We do not use $\gamma \leq \sum_{i \in [n]} t_i$ as a constraint because the other constraints already imply that $\gamma < 2$, and we assume that the combined length of all articles is huge compared to the time budget. Numerical maximization of (5) using [13] yields $g = 0.021425$, $\beta = 0.565728$, $\gamma = 1.478575$. As we set p to the lower bound on $p'/2$, we can plug these values in Eq. (1), so $\text{OPT}(I) < 246C \cdot \mathbb{E}\left[\text{Threshold Algorithm}(I)\right]$. $\qquad \square$

It is interesting to note that our proof suggests to limit the time for each article to about 2% of the time budget in order to maximize the expected total information gain. The analysis from Lemma 4 in [4] uses $\beta = 3/4$, $\gamma = 3/2$ and $t_i \leq 1/81$ for all $i \in [n]$. Applying their analysis on the Threshold Algorithm for $g = 1/81$ yields an upper bound of $162eC$ on its competitive ratio. For these parameters, the optimized analysis in Theorem 3 gives an upper bound of $125.77eC$.

6   Open Questions

An open question is whether the analysis of the Threshold Algorithm is tight. Although we optimize to find the best possible g, the used Chernoff bound is

---

[4] Note that the Chernoff Bound is indeed applicable since the random variables $t_i y_{[n]}^{(\beta)}(i)\zeta_i$ and $t_i y_{[n]}^{(\gamma)}(i)(1 - \zeta_i)$ are discrete and ζ_i are mutually independent.

not applicable for $g \geq 1/6$. Moreover, the combined articles' lengths, cut with respect to g, have to be at least 1.48 times greater than the time budget. For the sake of improving the analysis, a different approach has to be investigated.

We informally related diminishing information gain over time to efficient reading strategies. It would be interesting to formalize it w.r.t. spatial information profiles. Further directions involve the exploration of new settings and extensions. There are different measures of the accuracy of the hints worth investigating. An example would be to interpret the information rate as a distribution of information and the hint to be a random variable drawn from this distribution. Then, the algorithm's performance is dependent on the information rate's expectation and standard deviation. An interesting task is to develop an algorithmic strategy for the setting where the length t_i of an article is not revealed when it arrives. We believe that the studied techniques in this paper can be used for this setting if the information gain diminishes, e.g., logarithmically as a function of time, while reading any article. Another reasonable setting is the one where articles appear in a non-uniform, but still random order. This is suitable for reading articles since many websites present articles in a categorized order or using recommender systems, where articles are sorted based on the user's preferences. In that light, it would also make sense to extend the scalar hint to a multi-dimensional feature vector. The investigation of related *learning-augmented* online algorithms would be a further interesting development. The idea of a threshold can be considered in that direction: Instead of the learning phase in the Threshold Algorithm, an external threshold can be considered, e.g., from past experience, gut feeling, or rating by a recommender system.

In our opinion, the most interesting extension is to allow the reader to mark a restricted number of articles and return to these articles at any point in time. For secretary problems with submodular objective functions, the setting where an algorithm is allowed to remember items and select the output after seeing the whole instance has recently been discussed by [1]. Here, they achieve a competitive ratio that is arbitrarily close to the offline version's lower bound on the approximation factor. This extension combined with a cross reading strategy, unknown reading lengths, and articles sorted by categories or preferences is the closest setting to real-life web surfing.

References

1. Agrawal, S., Shadravan, M., Stein, C.: Submodular secretary problem with short-lists. In: Blum, A. (ed.) 10th Innovations in Theoretical Computer Science Conference, ITCS 2019, San Diego, California, USA, LIPIcs, vol. 124, pp. 1:1–1:19. Schloss Dagstuhl - Leibniz-Zentrum für Informatik (2019)
2. Albers, S., Khan, A., Ladewig, L.: Improved online algorithms for knapsack and GAP in the random order model. In: Approximation, Randomization, and Combinatorial Optimization. Algorithms and Techniques, APPROX/RANDOM 2019, Massachusetts Institute of Technology, Cambridge, MA, USA, pp. 22:1–22:23 (2019)

3. Albers, S., Ladewig, L.: New results for the k-Secretary problem. In: Lu, P., Zhang, G. (eds.) 30th International Symposium on Algorithms and Computation (ISAAC 2019), Leibniz International Proceedings in Informatics (LIPIcs), vol. 149, pp. 18:1–18:19. Schloss Dagstuhl-Leibniz-Zentrum fuer Informatik, Dagstuhl (2019)
4. Babaioff, M., Immorlica, N., Kempe, D., Kleinberg, R.: A knapsack secretary problem with applications. In: Charikar, M., Jansen, K., Reingold, O., Rolim, J.D.P. (eds.) APPROX/RANDOM -2007. LNCS, vol. 4627, pp. 16–28. Springer, Heidelberg (2007). https://doi.org/10.1007/978-3-540-74208-1_2
5. Babaioff, M., Immorlica, N., Kempe, D., Kleinberg, R.: Matroid secretary problems. J. ACM **65**(6), 35:1–35:26 (2018)
6. Dean, B.C., Goemans, M.X., Vondrák, J.: Approximating the stochastic knapsack problem: the benefit of adaptivity. Math. Oper. Res. **33**(4), 945–964 (2008)
7. Dynkin, E.B.: Optimal choice of the stopping moment of a Markov process. Sov. Math. **4**, 627–629 (1963)
8. Freire, M.L.M., Potts, D., Dayama, N.R., Oulasvirta, A., Di Francesco, M.: Foraging-based optimization of pervasive displays. Pervasive Mob. Comput. **55**, 45–58 (2019)
9. Goerigk, M., Gupta, M., Ide, J., Schöbel, A., Sen, S.: The robust knapsack problem with queries. Comput.& OR **55**, 12–22 (2015)
10. Kesselheim, T., Radke, K., Tönnis, A., Vöcking, B.: Primal beats dual on online packing LPs in the random-order model. SIAM J. Comput. **47**(5), 1939–1964 (2018)
11. Kleinberg, R.D.: A multiple-choice secretary algorithm with applications to online auctions. In: Proceedings of the Sixteenth Annual ACM-SIAM Symposium on Discrete Algorithms, SODA 2005, Vancouver, British Columbia, Canada, pp. 630–631 (2005)
12. Marchetti-Spaccamela, A., Vercellis, C.: Stochastic on-line knapsack problems. Math. Program. **68**, 73–104 (1995). https://doi.org/10.1007/BF01585758
13. Maxima: Maxima, a Computer Algebra System. Version 5.43.2 (2019). http://maxima.sourceforge.net/
14. Mitzenmacher, M., Upfal, E.: Probability and Computing: Randomization and Probabilistic Techniques in Algorithms and Data Analysis. Cambridge University Press, Cambridge (2017)
15. Noga, J., Sarbua, V.: An online partially fractional knapsack problem. In: 8th International Symposium on Parallel Architectures, Algorithms, and Networks, ISPAN 2005, Las Vegas, Nevada, USA, pp. 108–112 (2005)

Sensors, Vehicles and Graphs

An Efficient Mechanism for Resource Allocation in Mobile Edge Computing

Guotai Zeng[1], Chen Zhang[2], and Hongwei Du[1(✉)]

[1] Department of Computer Science and Technology,
Harbin Institute of Technology (Shenzhen), Shenzhen, China
19S151132@stu.hit.edu.cn, hongwei.du@ieee.org
[2] Department of Computer Science, City University of Hong Kong,
Kowloon, Hong Kong, China
c.zhang@my.cityu.edu.hk

Abstract. Mobile edge computing (MEC) caches data and services from remote cloud to the edge of network. In this way, MEC lets user equipment (UE) more closer to data and services than traditional cloud computing. Service providers (SPs) deploy their own base stations (BSs) to provide high quality services to their subscribers in MEC networks. SPs get their total revenue from their subscribers, but face the cost of energy and acquiring resources. In this paper, we attempt to maximize the final profit of SPs base on a novel resource allocation method to cut down the cost of energy and acquiring resources. The simulation results indicate that our scheme increases the final profit of SPs, compared to the existing methods.

Keywords: Mobile edge computing · Resource allocation · Profit maximization

1 Introduction

With the development of wireless sensors [1] and other Internet Of Things (IOT) [2], more and more mobile devices are connected to the Internet. But limited by battery power, storage and computational ability of mobile devices, a number of mobile devices can not complete all computing tasks alone or will meet great challenges on delay or energy cost[3]. In order to deal with those problems, mobile edge computing was proposed and become a significant technology for 5G networks[4]. MEC caches services and data from the remote cloud and deploys them on the servers of base stations. Hence, service providers can serve their users with BSs equipped with MEC server which have computing and radio resources. Because the BSs are more closer to user equipment than remote cloud[5], the applications deployed on the MEC servers will have lower latency and higher quality of service than traditional cloud computing.

Resource allocation in MEC has received great attention at academic. Unlike the cloud center in cloud computing, a BS in MEC hardly extends its resource, unless SP updates the BS's hardware. Because BSs have finite resource, resource

© Springer Nature Switzerland AG 2020
W. Wu and Z. Zhang (Eds.): COCOA 2020, LNCS 12577, pp. 657–668, 2020.
https://doi.org/10.1007/978-3-030-64843-5_44

allocation is vital in MEC. A lot of previous works have done on how to allocate resource in MEC to deal with delay problems or energy saving problems, but the profit of SPs didn't draw much attention. The existing profit models of MEC are inadequate, because they only consider the different cost of acquiring resources from BSs deployed by different SPs [5], but ignore the energy expenditure that SPs need to pay. To provide high quality services to users, different SPs densely deploy their own BSs to serve their users. But the BSs are deployed for peak hours, so BSs will meet great energy waste in idle time, which reduces the final profit of SPs.

In this paper, we use a resource allocation scheme to maximize the profit of SPs in multi-SP MEC networks. In order to calculate the energy expenditure, we consider this problem in a continuous time period which can be divided into small time slots [6]. The profit maximization problem is influenced by three main aspects. First, BSs on the edge have limited resources, so a BS can only serve finite number of UEs in each time slot. Second, energy expenditure of BSs can not be ignored. Third, an SP prefers its subscribers to use its own BSs to reduce the cost of acquiring resources. The contributions of this paper are summarized as follows.

- We attempt to maximize the final profit of SPs in Multi-SP MEC network, by considering the expenditure of BSs' energy consumption. In the real world, the final profit of SPs is not only affected by cost of acquiring resources, but also affected by the total energy consumption of BSs.
- We present a novel algorithm for SPs to allocate resource to UEs appropriately, and the switch off the unprofitable BSs to reduce the energy cost. We transform the maximum profit problem (MPP) of SPs to a matching problem between BSs and UEs. We optimize the decisions of BSs switching and BS-UE association, based on local matching interactions between BSs and UEs in our algorithm.

The rest of this paper is organized as follows. Section 2 discusses the related work. Section 3 describes the system model, and Sect. 4 formulates the maximum profit problem. Section 5 introduces the details of our proposed algorithm. Simulation results presented in Sect. 6. In final, Sect. 7 concludes the paper.

2 Related Work

How to allocate resource in order to maximize SPs' final profit in MEC has been studied in previous work. In [7], Quyuan et al. proposed a incentive mechanism to allocate the resource to UEs, so as to maximize the profit of SPs. In [8], Huaqing et al. discussed how SPs select BSs, and how UEs select BSs in the fog computing scenario, so as to maximize the revenue of SPs and BSs. In [5], Chen et al. considered SPs' maximum profit problem in cellular networks; however, the resource allocation method they proposed only focus on the different cost of acquiring resources from BSs deployed by different SPs, but ignore the energy expenditure of BSs.

Energy expenditure compose of a large part of SP's cost. A main way to reduce energy expenditure in MEC is to allocate resource appropriately according to the traffic load of the network, and then switch off idle BSs dynamically [9]. In [10], the authors model the switching of MEC servers as a Minority Game; the strategy they proposed guarantees the quality of services and reduces the number of active servers. In [11], the authors take the server's operational energy into consideration and use predict mechanism to switching off the BSs. However, those works only focus on resource allocation to reduce energy consumption, and the concept of SP was not mentioned in those work. To the best of our knowledge, no existing works have studied how to maximize the final profit of SPs with the consideration of energy expenditure in Multi-SP MEC networks.

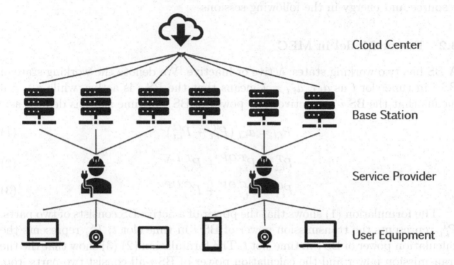

Fig. 1. System architecture

3 System Model

3.1 System Overview

The Fig. 1 shows the system architecture of Multi-SP MEC network, which consists four layers, including the could center layer, BS layer, SP layer and UE layer. Every BS is deployed by one SP to provide MEC services to the UEs within its communication range. Each UE subscribes to an SP and offloads its tasks to its subscribed SP for further process. We denote the set of BSs as B, the set of SPs as Γ, the set of UEs as U, the set of services as S.

Although a UE may be cover by many active BSs, we assume each UE can only requests one service to one BS in each time. Since the traffic load of MEC network is fluctuating in both time and space, the number of UEs and the resources that they request change over time.

An SP get its revenue from its subscribers, when its subscribers use the computing resource provided by the SP. But the computing resource is on the BSs deployed by the SP or its competitors. So the SP has to pay, when its subscribers acquire resource from BSs. Because SPs will deploy their own BSs to reduce the cost of acquiring resource, SPs have to pay for the energy expenditure of theirs own BSs. In order to calculate the energy expenditure, we divide a continuous period of time T into time slots t with an equal size. After subtracting the cost of acquiring resource and energy, we get the final profit of an SP.

To get the final profit of SPs, we will discuss power model and resource model in MEC, and then model the final profit of SPs base on the cost of acquiring resource and energy in the following sessions.

3.2   Power Model in MEC

A BS has two working states: active or inactive. We denote the working state of BS i in time slot t as $a_{i,t}$. $a_{i,t} = 1$ means that the BS i is active, while $a_{i,t} = 0$ means that the BS i is inactive. The power of BS i in time slot t is defined as:

$$P_{i,t} = a_{i,t}\left(P_{i,t}^T + P_{i,t}^C\right) \tag{1}$$

$$P_{i,t}^T = P_{i,t}^{T,OP} + P_{i,t}^{T,TX} \tag{2}$$

$$P_{i,t}^C = P_{i,t}^{C,OP} + P_{i,t}^{C,CP} \tag{3}$$

The formulation (1) shows that the power of a active BS consists of two parts. $P_{i,t}^T$ represents the transmission power of BS i in time slot t; $P_{i,t}^C$ represents the calculation power of BS i in time slot t. The formulation (2) (3) show that the the transmission power and the calculation power of BS i all consist two parts ,too. $P_{i,t}^{T,OP}$ and $P_{i,t}^{C,OP}$ are operational power of the BS's transmission system and calculation system, which are inevitable when a BS is active; $P_{i,t}^{T,TX}$ represents the BS i's effective transmission power which depends on the traffic load, and $P_{i,t}^{C,CP}$ represents the power that use to process the uploaded computing tasks.

3.3   Resource Model in MEC

To process the offloaded tasks, computing resource like CPU, memory and network I/O are needed [12]. For simplicity, we use computing resource block (CRB) to describe the computing resource. The total number of CRBs on BS i is denoted as c_i^{max}. We use $U_{i,t}$ to describe the sub set of UEs that served by BS i in the time slot t, each of UE in $U_{i,t}$ will need $c_{u,j,t}$ CRB to process its offloaded task in time slot t. The CRBs that BS i allocates to service j in time slot t is denoted as $c_{i,j,t}$, and we have:

$$c_{i,j,t} = \sum_{u \in U_{i,t}} c_{u,j,t} \ \forall i \in B \ \forall j \in S \tag{4}$$

We assume that the maximum computing power of BS i is $P_{i,t}^{C,max}$. So we can calculate the $P_{i,t}^{C,CP}$ of BS i in this way:

$$P_{i,t}^{C,CP} = \left(P_{i,t}^{C,max} - P_{i,t}^{C,OP} \right) \frac{\sum_{j \in S} c_{i,j,t}}{c_i^{max}} \tag{5}$$

To transmit data between BS and UE, radio resource is also required. Since the uploaded data is much larger than the results from BSs, we only consider the uplink radio resource in the paper. We use orthogonal frequency division multiple access (OFDMA) as network system [13]. The radio resource of a BS is organized in the form of radio resource block (RRB), and each RRB has the bandwidth W_{sub}. According to Shannon's theorem, the received data rate per RRB of BS i from UE u is:

$$e_{i,u} = W_{sub} \log_2 \left(1 + SNIR_{i,u} \right) \tag{6}$$

We denote signal-to-interference-plus-noise ratio (SINR) per RRB from UE u to BS i as $SNIR_{i,u}$. BS i has total N_i RRBs. The total required data rate of UE u to upload its data in time slot t is described as $\lambda_{u,t}$. The RRBs that BS i allocates to UEs in time slot t is denoted as $n_{i,t}$, and we have:

$$n_{i,t} = \sum_{u \in U_{i,t}} \lceil \frac{\lambda_{u,t}}{e_{i,u}} \rceil \tag{7}$$

P_{tx} denotes the transmission power per RRB, and η denotes the power amplifier efficiency. Thus, we have the transmission power $P_{i,t}^{T,TX}$ of BS i:

$$P_{i,t}^{T,TX} = \frac{P_{tx} n_{i,t}}{\eta} \tag{8}$$

3.4   Profit Model in MEC

Each UE in the network subscribes to an SP, and pay for the resource when it uses SP's service. While an SP also needs to pay the BSs or remote cloud for allocating their resources to its subscribers. The cost of acquiring resources from an SP's own BSs is lower than from other SPs' BSs, because once the SP use other SPs' BSs to provide resources to its subscribers, the SP has to pay extra money to other SPs. Thus, SPs prefer to use their own BSs to serve their subscribers.

We denote the total profit $W_{k,t}$ of SP $k \in \Gamma$ in time slot t as:

$$W_{k,t} = W_{k,t}^R - W_{k,t}^B - W_{k,t}^E \tag{9}$$

$$W_{k,t}^R = \sum_{i \in B} \sum_{u \in U_i^k} \sum_{j \in S} c_{u,j,t} m_k \tag{10}$$

$$W_{k,t}^E = \sum_{i \in B_k} P_{i,t} ty \tag{11}$$

$$W_{k,t}^B = \sum_{i \in B} \sum_{u \in U_i^k} \sum_{j \in S} x_{u,i,t} c_{u,j,t} q_{u,i} \tag{12}$$

$W_{k,t}^R$ is the total payment that SP k receives from its users; U_i^k is the sub set of UEs that subscribe to SP k on BS i; m_k is the unit price that a UE need to pay when getting a CRB from its SP. $W_{k,t}^E$ is energy expenditure of SP; B_k is the sub set of BSs deployed by SP k; y denotes the unit cost per kilowatt hour; $W_{k,t}^B$ is the total payment that SP k need to pay when getting the CRBs from BSs. The unit cost $q_{u,i}$ per CRB of BSs may be different because BSs may belongs to different SP. $q_{u,i}$ can be described as follow:

$$q_{u,i} = \begin{cases} b\left(1 + d_{u,i}^\sigma\right) & u \text{ and } i \text{ are from same SP} \tag{13} \\ b\left(\iota + d_{u,i}^\sigma\right) & u \text{ and } i \text{ are from different SP} \tag{14} \end{cases}$$

The formulation (13) (14) describe the cost from both computing resources and transmission resource. b is the base price of CRB. σ and ι are two weighted parameters. $\iota > 1$ indicates that it's more expensive for an SP to use a CRB from a BS deployed by its competitors than by its own. $d_{u,i}$ denotes the distance between UE u and BS i, and $d_{u,i}^\sigma$ accounts for the transmission cost which increase when the distance between UE and BS increase.

4   Problem Formulation

Our objective is to maximize the final profit of SPs during a continuous time T. To decrease the cost of acquiring resource, SPs prefer their subscribers to acquire resource from their own BSs. To reduce the energy expenditure, SPs will power off the BSs which are unprofitable.

To solve the maximum profit problem, we need to answer the two following questions in each time slot:

1) which BS to be powered off;
2) how to associate the remaining active BSs with UEs.

We define two variable $a_{i,t} \in \{0,1\}$ and $x_{u,i,t} \in \{0,1\}$ to answer the two questions. $a_{i,t} = 1$ indicates the BS i is active during time slot t, otherwise BS i is inactive. $x_{u,i,t} = 1$ indicates the UE u is associated with BS i in time slot t, otherwise they are disconnected. Thus, we can model the maximum profit problem as follow:

$$\max_{a_{i,t}, x_{u,i,t}, \forall u \in U, \forall j \in S, \forall i \in B} \sum_{k \in \Gamma} \sum_{t \in T} W_{k,t} \tag{15}$$

$$s.t. \sum_{u \in U, i \in B} x_{u,i,t} \leqslant 1, \forall t \in T \tag{16}$$

$$c_i^{\max} \geqslant \sum_{j \in S} c_{i,j,t}, \forall i \in B, \forall t \in T \tag{17}$$

$$N_i \geqslant n_{i,t}, \forall i \in B, \forall t \in T \tag{18}$$

The formulation (15) is our goal: to maximize the the final profit of all SPs in a period of time T. The constraint (16) indicates that a UE can only be served by a BS in each time slot. The constraint (17) (18) indicates that a BS can only process the uploaded tasks within its computing capacity and radio capacity.

5   Algorithm Design

The key to maximize the profit of SPs while considering the energy expenditure is to connect most of UEs to a few BSs and power off the unprofitable BSs. Switching off unprofitable BSs will make the active BS more busy, while cut down the unprofitable BSs' operational energy expenditure. In [6], authors has proven that the switching of a set of BSs is a weighted set cover problem, which is NP-hard. However, our problem is more complex than a set cover problem, because RRBs and CRBs on BS are limited, which means that each BS can only serve finite number of UEs; and the weight cost between a UE and a BS may vary when they belong to different SPs. Thus, we need to design a new scheme to switch off unprofitable BS, while associate the UE requests with remaining active BSs.

The connection process of BSs and UEs is a matching process [14]. Therefore, we use a matching algorithm to associate UEs and BSs and switch off the unprofitable BSs. Inspired by the improved stable marriage algorithm proposed in [5], we also use two dynamic preference lists to match UEs and BSs. However, there are three differences between our algorithm and the stable marriage algorithm: 1) UE's preference list only stores the BS within its communication range; 2) BS's preference list only stores the UEs who request to it; 3) BS will select several qualified UEs instead of one in each iteration. Inspired by the best fit algorithm in memory management of operating systems, the preference list of each UE is sorted by BSs' remain resources, and each UE will choose the BS with Minimum resources in each iteration. Thus, most UE requests will be gathered on a few BSs, and we can switch off more unprofitable BS to reduce operational energy expenditure.

Algorithm 1 illustrates the details of our dynamic match (DM) algorithm. In this algorithm, we use multiple iterations to do the matching jobs between UEs and BSs or remote cloud. In each iteration, UE u will construct a preference list from B_u which consists of BSs within UE u's communication range. The preference list of UE u is sorted by the remaining resources of BS $i \in B_u$. UE u will choose a BS i which has smallest but enough resources $N_i^r + c_i^r$ to process its upload task in each iteration. N_i^r is the remaining radio resource of BS i, and c_i^r is the remaining computing resource of BS i. UE u's upload task has to redirect to remote cloud, when its preference is empty. While UE u's request is processed on remote cloud, SP can not ensure the quality of service, and the

Algorithm 1. Dynamic Match (DM) Algorithm

Input: $B, U, S, \Gamma, c_{i,j}, N_i, c_{u,j,t}, \lambda_{u,t}, P_{i,t}^{C,max}, P_{tx}, SINR_{i,u}, d_{i,u}$
   $\forall i \in B, \forall u \in U, \forall j \in S, \forall k \in \Gamma$
Output: $a_{i,t}, x_{u,i,t}, \forall i \in B, \forall u \in U, \forall t \in T$

1 For all t in T Initialize $a_{i,t} = 1, x_{u,i,t} = 0, \forall i \in B, \forall u \in U, \forall t \in T$;
2 **foreach** $u \in U$ **do**
3 ⌊ Calculate the B_u that each u can reach;

4 **while** *have not served UE* **do**
5 **while** *have not served UE* **do**
6   **foreach** $u \in U$ **do**
7   **while** $B_u \neq \Phi$ *and* u *isn't served* **do**
8   Select BS $i^* = \mathrm{argmin}\ N_i^r + c_i^r$;
9   **if** $c_i^r \geqslant c_{u,j,t}$ *and* $N_i^r \geqslant \lceil \frac{\lambda_{u,t}}{e_{i,u}} \rceil$ **then**
10   ⌊ send request to BS i
11   **else**
12   ⌊ $B_u = B_u - \{i^*\}$

13 **foreach** $i \in B$ **do**
14   **if** *there exists UEs to request BS i* **then**
15   create two list to store UE's request for each service: U_1 to store
     UEs from SP, U_2 to store UEs from different SP ;
16   **foreach** $j \in S$ **do**
17   **if** $U_1 \neq \Phi$ **then**
18   BS i select a sub set $U_1^* \subseteq U_1$ to add to U_i^*, the
      ⌊ resources that U_1^* request can not exceed the total
       resources that BS i has;
19   **else**
20   BS i select a sub set $U_2^* \subseteq U_2$ to add to U_i^*, the
      ⌊ resources that U_2^* request can not exceed the total
       resources that BS i has;
21   BS i broadcast the remaining resources N_i^r and c_i^r to the UEs that
    it can reach;
22   ⌊ For all $u \in U_i^*$, set $x_{u,i,t} = 1$;

23 **foreach** $i \in B$ **do**
24 Calculate the final revenue $W_{i,t}$ of BS i;
25 **if** $W_{i,t} < 0$ **then**
26   Switch off the BS i, and release UE $u \in U_i^*$;
27   ⌊ Set $a_{i,t} = 0$;

cost of using resources on remote cloud is more expensive. BS i will receive a lot of UE requests in each iteration, and BS i will construct two preference lists from those UEs for each service $j \in S$: one to store the UEs from a same SP, the other to store the UEs from different SP. BS i will sort the two preference lists with the resources they request in order to serve more UEs in each time slot t. Because BS i prefer to serve the UEs from same SP, so it will ignore the second list when the first list is not empty. Only when the first list for a service is empty, will the BS i serves the UEs from second list. In each iteration, BS i will choose several UEs from the chosen list, until it uses up all its computing resource or radio resource.

In the beginning of the multiple iterations, we assume that all BSs are active in the first time, but after the matching procedure, all uploaded tasks will be served by some BSs or remote cloud. Thus, we can calculate the total profit $W_{i,t}$ of BS i, and try to power off some unprofitable BSs to reduce the loss. Since the energy expenditure may be very large comparing to the revenue received from subscribers, $W_{i,t}$ could be negative. After those unprofitable BSs are power off, we will reallocate the UEs which is once served by those unprofitable BSs to the active BSs or remote cloud, until all the UEs have been served appropriately. We will continue to switch off the unprofitable BS until all the BSs on the MEC network is profitable for SPs, and all UEs on the network is served appropriately.

6   Simulation Result

In this section, we present the configuration of our experiment and evaluate the performance of our DM algorithm for the maximum profit problem by simulation.

6.1   Simulation Settings

Consider a continuous period T of 24 h, which is divided into 24 time slots with 1 h per t. There are 5 SPs, and each of them deploy 5 BSs in the MEC network. BSs owned by different SPs are deployed randomly in 1200 m × 1200 m square with an inter-site distance of 300 m. The UEs distribution is random and changes over time. There are 6 kinds of MEC services, and each UE only requests a random service in each time slot t. The required CRBs per UE request change from 1 to 3. The required data rate per UE request varies from 1 Mpbs to 5 Mpbs. The communication range for each BS is set to be 450 m. The uplink channel model follows the free space path loss model:

$$32.44 + 20 \log_{10} d_{u,i} \, (km) + 20 \log_{10} F \, (MHz) \tag{19}$$

$d_{u,i}$ is the distance between UE u and BS i, and F is the carrier frequency. The experimental parameters are summarized in Table 1.

Table 1. Experimental parameters

Parameter	Value
Inter-site distance	300 m
Number of CRBs per BS	250
Required CRBs per UE	1–3
Number of RRBs per BS	100
Amplifier efficiency (η)	50%
Carrier frequency (F)	2 GHz
Required data rate per UE request	1 Mbps–5 Mbps
Electricity price per kilowatt hour	1.1
Service price per CRB	0.02
Maximum computing power per BS	200 W
Operational power per BS	160 W [13]
Weight parameter ι	1.5
Weight parameter σ	0.01 [5]
Power spectral density of noise	−174 dBm/Hz [6]

6.2   Performance Evaluation

In this section, we will compare our DM algorithm with other two resource allocation algorithms proposed in previous works. One is efficiency first strategy (EFS) in [6], the other is decentralized multi-SP resource allocation (DMRA) in [5]. The principle of EFS is to switch on the BSs with the high energy efficiency and turn off the idle BSs, but EFS ignores the preference of UE to specific SP. DMRA uses improve stable marriage algorithm to associate UEs and BSs to let BSs serve more UEs from a same SP, but DMRA also ignores the energy expenditure of SP. However, in our algorithm, we took both factors into consideration.

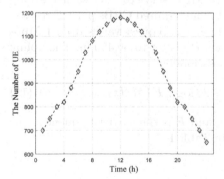

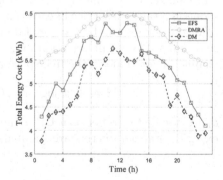

Fig. 2. The changing number of UEs

Fig. 3. The energy cost of different methods

Figure 2 illustrates the variation number of UEs in MEC network within 24 h.

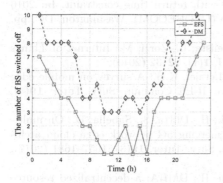

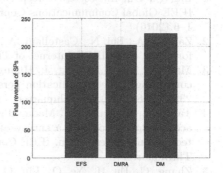

Fig. 4. The number of inactive BSs by using different methods.

Fig. 5. The final revenue by using different methods.

Figure 3 shows the energy consumption of different methods in each time slot. Since DMRA method switches on all BSs in all time, it consumes more energy than the other two methods. Figure 4 describes the number of BSs switched off in each time slot. DM switches off more BSs than EFS, because EFS will not switch off a BS if that BS have any UE requests to process, but DM will switch off any BS that can't make a profit. So DM also consumes less energy than EFS.

Figure 5 shows the final profit of different methods in the day. Since DMRA method only considers the different cost of acquiring resource from different BSs, but ignores the energy expenditure of idle BSs, DMRA has a huge expenditure of energy. Because EFS ignore the cost of acquiring resource, it doesn't make too much profit. However, DM considers the cost of both acquiring resource and energy, so DM makes more final profit than the other two.

7   Conclusion

In this paper, we study the maximum profit problem in Multi-SP MEC network, while considering the cost of energy and acquiring resources. This problem aims to maximize the final profit of SPs in MEC. Because the allocation of resource on MEC is a matching procedure, we use an improved matching algorithm to associate UEs and BSs, and switch off those unprofitable BSs to reduce operational power. By this way, we increase the final profit of SPs.

Acknowledgment. This work is supported by the Shenzhen Basic Research Program (Project No. JCYJ20190806143011274) and National Natural Science Foundation of China (No. 61772154).

References

1. Liu, C., Du, H., Ye, Q.: Sweep coverage with return time constraint. In: 2016 IEEE Global Communications Conference (GLOBECOM), Washington, DC, pp. 1–6 (2016)
2. Zanella, A., Bui, N., Castellani, A., Vangelista, L., Zorzi, M.: Internet of Things for smart cities. IEEE Internet of Things J. $1(1)$, 22–32 (2014)
3. Mao, Y., You, C., Zhang, J., Huang, K., Letaief, K.B.: A survey on mobile edge computing: the communication perspective. IEEE Commun. Surv. Tutor. $19(4)$, 2322–2358 (2017). Fourthquarter
4. Taleb, T., Samdanis, K., Mada, B., Flinck, H., Dutta, S., Sabella, D.: On multi-access edge computing: a survey of the emerging 5G network edge cloud architecture and orchestration. IEEE Commun. Surv. Tutor. $19(3)$, 1657–1681 (2017). Thirdquarter
5. Zhang, C., Du, H., Ye, Q., Liu, C., Yuan, H.: DMRA: a decentralized resource allocation scheme for Multi-SP mobile edge computing. In: 2019 IEEE 39th International Conference on Distributed Computing Systems (ICDCS), Dallas, TX, USA, pp. 390–398 (2019)
6. Yu, N., Miao, Y., Mu, L., Du, H., Huang, H., Jia, X.: Minimizing energy cost by dynamic switching ON/OFF base stations in cellular networks. IEEE Trans. Wireless Commun. $15(11)$, 7457–7469 (2016)
7. Wang, Q., Guo, S., Liu, J., Pan, C., Yang, L.: Profit maximization incentive mechanism for resource providers in mobile edge computing. IEEE Trans. Serv. Comput. (2019)
8. Zhang, H., Xiao, Y., Bu, S., Niyato, D., Yu, F.R., Han, Z.: Computing resource allocation in three-tier IoT fog networks: a joint optimization approach combining stackelberg game and matching. IEEE Internet of Things J. $4(5)$, 1204–1215 (2017)
9. Ajmone Marsan, M., Chiaraviglio, L., Ciullo, D., Meo, M.: Optimal energy savings in cellular access networks. In: 2009 IEEE International Conference on Communications Workshops, Dresden, pp. 1–5 (2009)
10. Ranadheera, S., Maghsudi, S., Hossain, E.: Computation offloading and activation of mobile edge computing servers: a minority game. IEEE Wirel. Commun. Lett. $7(5)$, 688–691 (2018)
11. Wang, Q., Xie, Q., Yu, N., Huang, H., Jia, X.: Dynamic server switching for energy efficient mobile edge networks. In: 2019 IEEE International Conference on Communications (ICC), ICC 2019, China, Shanghai, pp. 1–6 (2019)
12. Yu, N., Song, Z., Du, H., Huang, H., Jia, X.: Multi-resource allocation in cloud radio access networks. In: 2017 IEEE International Conference on Communications (ICC), Paris, pp. 1–6 (2017)
13. Yaacoub, E., Dawy, Z.: A survey on uplink resource allocation in OFDMA wireless networks. IEEE Commun. Surv. Tutor. $14(2)$, 322–337 (2012)
14. Gu, Y., Saad, W., Bennis, M., Debbah, M., Han, Z.: Matching theory for future wireless networks: fundamentals and applications. IEEE Commun. Mag. $53(5)$, 52–59 (2015)

Data Sensing with Limited Mobile Sensors in Sweep Coverage

Zixiong Nie, Chuang Liu, and Hongwei Du[✉]

Department of Computer Science and Technology,
Harbin Institute of Technology (Shenzhen),
Key Laboratory of Internet Information Collaboration, Shenzhen, China
19S051046@stu.hit.edu.cn, chuangliuhit@gmail.com, hongwei.du@ieee.org

Abstract. Sweep coverage has received great attention with the development of wireless sensor networks in the past few decades. Sweep coverage requires mobile sensors to cover and sense environmental information from Points Of Interests (POIs) in every sweep period. In some scenarios, due to the heterogeneity of POIs and a lack of mobile sensors, mobile sensors sense the different amounts of data from different POIs, and only part of POIs can be covered by mobile sensors. Therefore, how to schedule the mobile sensors to improve coverage efficiency is important. In this paper, we propose the optimization problem (MSDSC) to maximize sensed data with a limited number of mobile sensors in sweep coverage and prove it to be NP-hard. We then devise two algorithms named GD-MSDSC and MST-MSDSC for the problem. Our simulation results show that, with a limited number of mobile sensors, GD-MSDSC and MST-MSDSC are able to sense more data from POIs than algorithms from previous work. In addition, MST-MSDSC can sense more data while the time complexity of GD-MSDSC is better.

Keywords: Data sensing · Limited mobile sensors · Sweep coverage

1 Introduction

With the rapid development of Internet of Things (IoT), researches on wireless sensor networks (WSN) are gaining lots of attention lately. Coverage problem is an important part of WSNs. Generally speaking, the solution to coverage problem often includes a scheme to schedule wireless sensors to gather information of the given targets. There are mainly 3 kinds of coverage problems: full coverage, barrier coverage and sweep coverage. In full coverage and barrier coverage scenarios, the monitored targets should be covered all the time [1,3,6,7,14,16,17,19,23–25]. However, sweep coverage just requires the monitored targets to be covered periodically, it uses mobile sensors instead of static sensors. Compared with full coverage and barrier coverage, with the same number of sensors, sweep coverage can monitor more targets which are also called Points Of Interests (POIs). Sweep coverage can be applied in scenarios including forest fire prevention, patrol inspection, climate prediction, *etc.*

© Springer Nature Switzerland AG 2020
W. Wu and Z. Zhang (Eds.): COCOA 2020, LNCS 12577, pp. 669–680, 2020.
https://doi.org/10.1007/978-3-030-64843-5_45

Most studies in sweep coverage focus on minimizing the number of mobile sensors or minimizing the sweep period of POIs. Among these studies, some studies take data sensing or data gathering into consideration [2,15,26,28,30], data from a POI reflects a series of information about the POI. In some scenarios, there is only a limited number of mobile sensors and these mobile sensors may sense different amount of data from different POIs due to the heterogeneity of POIs. In this case, mobile sensors are unable to sense information from all POIs. Since gaining more information from POIs will improve the quality of monitoring, we focus on the problem to maximize sensed data from POIs with limited number of mobile sensors. And our contribution are as follows:

1) We introduce a new problem called Maximizing sensed Data in Sweep Coverage (MSDSC) and prove it to be NP-hard. In this problem, POIs are distributed in a 2D plane. Due to a limited number of mobile sensors and the heterogeneity of POIs, the main goal of the problem is to maximize sensed data from POIs in a sweep period.
2) We devise two heuristic algorithms, GD-MSDSC and MST-MSDSC, to solve the problem. The time complexity of GD-MSDSC and MST-MSDSC are $O(N^2)$ and $O(N^2 log_2 N)$ respectively.
3) We conduct simulations to compare our methods with MinExpand and CoCycle from previous literature. The results indicate that our algorithms especially MST-MSDSC have better performance than previous work in terms of maximizing sensed data from POIs.

The rest of the paper is organized as follows. Section 2 reviews some related work. In Sect. 3, formulation of the MSDSC problem is presented. Section 4 and Sect. 5 introduce GD-MSDSC and MST-MSDSC algorithms respectively. And simulation results are shown in Sect. 6. Lastly, Sect. 7 makes a conclusion.

2   Related Work

Since sweep coverage problem was first proposed in [5], there have been so many literatures studying sweep coverage. Many variations of sweep coverage problems are studied, most of these studies can be classified into two categories according to their optimization goals, one is to minimize the number of mobile sensors, the other one is to minimize the sweep period. These studies are as follows:

Some studies aim at using the least number of mobile sensors to cover all the POIs periodically. In [5] and [18], Cheng *et al.* proved the problem of minimizing the number of mobile sensors to cover all the POIs to be NP-hard. Then they present a centralized algorithm called CSWEEP and a distributed algorithm called DSWEEP to solve this problem for the first time. In [8], Du *et al.* proposed a heuristic algorithm named MinExpand under the condition that mobile sensors need to follow the same trajectory in different sweep periods. MinExpand keeps expanding its sweep route by adding the POI with the minimal path increment from the existing candidate POIs until all the POIs are included in its route, OSweep algorithm was also proposed when there is no trajectory constraint, the performance of which is far better than CSWEEP. In [20], Liu *et al.* present a heuristic algorithm called PDBA which had better performance than GSWEEP. Subsequently, more factors were taken into consideration. Li *et al.* considered data delivery in sweep coverage called Vehicle Routing Problem based Sweep Coverage (VRPSC) [27]. They also employed Simulated Annealing to solve VRPSC. In [21], Liu *et al.* designed G-MSCR and MinD-Expand based on heuristic method to determine the minimum number of mobile sensors with a time constraint in which the collected data should be delivered to base station within a preset time window. Gorain *et al.* considered the fact that the detection targets could be continuous region instead of discrete points, they investigated area and line sweep coverage problems in [12,13].

Meanwhile, some researches focus on minimizing the sweep period for given number of mobile sensors. In [9], Feng *et al.* considered the relationship between minimal spanning tree and TSP cycle, then they introduced a problem to minimize the makespan of mobile sweep routes (M^3SR) which will in turn reduce the sweep period for sweep coverage. To solve the problem, they present a greedy algorithm called GD-Sweep and an approximate algorithm called BS-Sweep with approximation ratio 6. In [10], Gao *et al.* proposed three constant-factor approximations, CycleSplit, HeteroCycleSplit and PathSplit, to minimize the longest trajectory length of mobile sensors under different scenarios.

In addition, other variations of sweep coverage problems had also been investigated. [4] and [22] utilized sensing range to optimize the sweep path of mobile sensors. In [29], Wu *et al.* considered a novel optimization problem in sweep coverage. With the help of crowdsensing technology, they converted the sweep coverage problem into a task assignment problem, the goal of which is to maximize the quality of sweep coverage.

However, most of existing studies do not consider that mobile sensors may sense different amount of data from different POIs due to the heterogeneity of POIs, and only part of POIs can be visited as a result of limited mobile sensors. In this case, it's important to maximize sensed data from POIs in order to improve the quality of sweep coverage.

3   Problem Formulation

Before introducing our problem, some basic definitions are given in Table 1. N POIs $P = \{p_1, p_2, ..., p_N\}$ with the same sweep period t are distributed in the two-dimensional target area, $d(p_i, p_j)$ is the distance between i_{th} POI and j_{th} POI. We deploy n mobile sensors to sense data from POIs in the target area periodically. In sweep coverage, a mobile sensor can sense data from a POI only when the mobile sensor reaches the position of this POI. As a result of the heterogeneity of POIs, mobile sensors can sense data at most q_i bytes from i_{th} POI per sweep period. F is the sweep coverage scheme to arrange sweep paths for each mobile sensor. Generally, the time for the sensor to move among POIs is usually much longer than the time spent on sensing data from POIs. Hence,

Table 1. Basic definitions

Symbol	Definition
t	Sweep period
v	Velocity of the mobile sensor
p_i	The i_{th} POI
q_i	Maximum amount of data which can be sensed by mobile sensors from the i_{th} POI per sweep period
s_j	The j_{th} mobile sensor
n	Total number of mobile sensors
N	Total number of POIs
$d(p_i, p_j)$	Distance between POI i and j
F	Sweep coverage scheme

the time for the sensor to sense data from POIs will not be considered in our problem. Finally, Maximizing Sensed Data in Sweep Coverage (MSDSC) will formally be defined as follows:

Definition 1 (Maximizing Sensed Data in Sweep Coverage (MSDSC)).
Given a set of POIs $P = \{p_1, p_2, ..., p_N\}$ having the same sweep period t and their corresponding maximum data amount set $Q = \{q_1, q_2, ..., q_N\}$. n identical mobile sensors with velocity v are provided. The goal of MSDSC problem is to find a sweep coverage scheme F for n mobile sensors to maximize sensed data from POIs in a sweep period t.

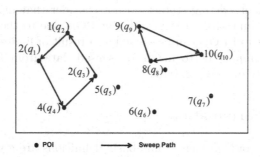

Fig. 1. An example of MSDSC

Figure 1 is an example of the MSDSC problem. In the square target region, mobile sensors can sense different amount of data from these POIs. Due to a lack of mobile sensors, POI 5, 6 and 7 are not in the sweep path of mobile sensors.

In the research, we prove the MSDSC problem is NP-hard by Theorem 1, and the details are as follows:

Theorem 1. *The MSDSC problem is NP-hard.*

Proof. The decision problem of the MSDSC problem is that whether we can find a sweep route to sense data at least m bytes per sweep period t when given n mobile sensors with velocity v. If we set $n = 1$ and $m = \sum_{i=1}^{N} q_i$, the decision problem becomes whether one mobile sensor can sense data from all POIs while ensuring that its trajectory length is no longer than vt, which implies that the mobile sensor is required to visit all the POIs and its trajectory length is no more than vt. Apparently, this problem under this special case is equivalent to TSP decision problem which is a well-known NP-hard problem. Therefore, the MSDSC problem is also NP-hard.

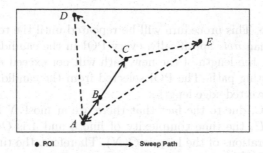

Fig. 2. An example of GD-MSDSC

4 GD-MSDSC

In this section, we present Greedy-MSDSC (GD-MSDSC) algorithm based on greedy strategy. In this algorithm, a cycle sweep path will be generated on which those n mobile sensors will be placed at equal intervals. In the process, we expand the path by selecting proper POIs one by one until the total length of the path exceeds nvt. To maximize sensed data from POIs per period, we should select those POIs which are close to the existing path and have greater value of q_i. The converted edge length takes both of the factors into consideration, and the corresponding definition is as follows:

Definition 2 (Converted Edge Length). *Suppose the current path starts from the POI p_s and the i_{th} POI p_i has just been added to the sweep path. Assume there are k candidate POIs $C = \{p_{c_1}, p_{c_2}, ..., p_{c_k}\}$, then the converted edge length of the j_{th} candidate POI is $d(p_i, p_{c_j})/q_{c_j}$.*

The GD-MSDSC is shown in Algorithm 1. Generating the cycle sweep path is an iterative process. Initially, the sweep path begins with a randomly selected POI. Then the sweep path is extended by selecting a POI from a set of candidate

---

Algorithm 1. GD-MSDSC

Input: Set $P = \{p_1, p_2, ..., p_N\}$, set $Q = \{q_1, q_2, ..., q_N\}$, n mobile sensors with velocity v, sweep period t

Output: Cycle sweep path C

1: Randomly select a POI p_s as the starting POI of C

2: **while** The Euclidean length of C is no more than nvt and $(P - C)$ is not empty **do**

3:     Find the candidate POIs set U from $(P - C)$.

4:     Find the POI p_k with the minimal converted edge length from candidate POIs set U.

5:     Add p_k to the existing path C.

6: **end while**

7: **return** C

POIs at each step. This procedure will be repeated until the total length of the path is greater than nvt. Specifically, every POI in the candidate POIs set has the property that the length of the new path will not exceed nvt if this POI is added to the existing path. The POI selected from the candidate POIs set has the minimum converted edge length.

In GD-MSDSC, due to the fact that there are at most N POIs in the candidate POIs set U, the time complexity of line 3 and 4 is $O(N)$. Meanwhile, the number of iterations of the loop is $O(N)$. Therefore, the time complexity of GD-MSDSC is $O(N^2)$

Figure 2 gives an instance of the expanding process of the sweep path, the existing sweep path is $A \rightarrow B \rightarrow C$, we assume POI D and POI E are candidate POIs. Hence, the mobile sensors are required to decide which POI should be added into the existing sweep path. By using GD-MSDSC algorithm, the converted edge lengths of D and E are calculated and we denote them as l_D and l_E. If $l_D \leq l_E$, then D is the POI to be added into the existing sweep path.

5   MST-MSDSC

To further improve the efficiency of data sensing, in this section, we present MinimalSpanningTree-MSDSC (MST-MSDSC) algorithm utilizing the relationship between minimal spanning tree and the Hamitonian cycle. MST-MSDSC attempts to generate a cycle sweep path from a spanning tree. Before introducing details of MST-MSDSC algorithm, the definition of Transformed Distance are given as follows:

Definition 3 (Transformed Distance). *The transformed distance between i_{th} and j_{th} POI is $d(p_i, p_j)/(q_i + q_j)$ denoted as $e(p_i, p_j)$.*

Algorithm 2 shows the details of MST-MSDSC. Prim algorithm will first be applied to generate the minimal spanning tree for all N POIs and we record the order in which each POI is added, denote the order sequence as $\{p_{i_1}, p_{i_2}, ..., p_{i_N}\}$, note that we adopt transformed distance $e(p_i, p_j)$ instead of $d(p_i, p_j)$ as the

---

Algorithm 2. MST-MSDSC

Input: Set $P = \{p_1, p_2, ..., p_N\}$, set $Q = \{q_1, q_2, ..., q_N\}$, n mobile sensors with velocity v, sweep period t

Output: Cycle sweep path C

1: $low \leftarrow 0, high \leftarrow N$
2: $mid \leftarrow (low + high)/2$
3: Calculate the transformed distance matrix $E = \{e(p_i, p_j)\}$, $e(p_i, p_j) = d(p_i, p_j)/(q_i + q_j)$
4: Use prim algorithm to generate minimal spanning tree MST based on transformed distance matrix E and record the addition order of each POI $seq = \{p_{i_1}, p_{i_2}, ..., p_{i_N}\}$
5: **while** $low < high$ **do**
6:     The spanning tree $T \leftarrow MST$
7:     The first mid POIs in seq will be left in T and the rest POIs will be deleted from T.
8:     Generate TSP cycle O from the spanning tree T.
9:     **if** The Euclidean length of O is no more than nvt **then**
10:         $low = mid$
11:     **else**
12:         $high = mid$
13:     **end if**
14: **end while**
15: $C \leftarrow O$
16: **return** C

---

weight of the minimal spanning tree. Due to the fact that the Euclidean length of the cycle sweep route should be no more than nvt, hence, the first k POIs in the sequence will be kept in the spanning tree and the remaining POIs in the sequence will be deleted from the spanning tree, then we get a trimmed spanning tree. After these operations, we generate the TSP cycle from the trimmed spanning tree. Specifically, binary search is applied to find the maximum k denoted as k_{max}.

In MST-MSDSC, the time complexity of line 8 is $O(N^2)$ if we apply greedy algorithm to generate TSP cycle. Meanwhile, the number of iterations from line 5 to 14 is $O(log_2 N)$. Therefore, the time complexity of MST-MSDSC is $O(N^2 log_2 N)$.

Figure 3 exhibits an instance of the MST-MSDSC algorithm. First the algorithm randomly choose the starting POI O, then the minimal spanning tree is generated from O by applying Prim algorithm and the addition order sequence is $\{O, A, B, C, D, E, F, G, H\}$. To ensure the length of the sweep path of mobile sensors does not exceed nvt, the minimal spanning tree keeps POIs O, A, B, C, D in the spanning tree and the rest POIs are deleted from the spanning tree. The TSP cycle $O \rightarrow A \rightarrow B \rightarrow C \rightarrow D \rightarrow O$ is generated from the trimmed spanning tree.

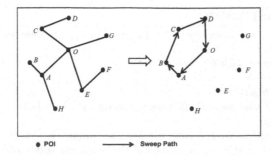

Fig. 3. An example of MST-MSDSC

6   Experimental Results

To evaluate the performance of the proposed algorithms, we use Python to conduct simulations for our algorithms. The experimental details and results will be presented in this section.

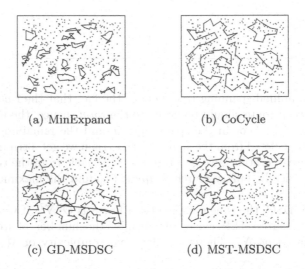

Fig. 4. The number of POIs $N = 500$ and the number of mobile sensors $n = 20$ (Color figure online)

In the simulation, a certain number of POIs with sweep period $t = 50\,\text{s}$ are randomly distributed in a $100\,\text{m}$ by $100\,\text{m}$ 2D plane, the value of all q_i randomly ranges from 1 bytes to 100 bytes. We consider two cases where the number of POIs N are 300 and 500 respectively. Meanwhile, a certain number of mobile sensors with the velocity v of $1\,\text{m/s}$ are deployed in the plane to sense data from POIs, and we increase the number of mobile sensors by 5 at each time. In addition, we run the simulation for 50 times and take the average value as the experiment results.

To validate the efficiency of our algorithms, we also implement contrast algorithms MinExpand [8] and CoCycle [11]. Note that CoCycle algorithm is used for covering all the POIs. However, in our problem, only part of the POIs will be covered when just a few mobile sensors are provided, we need to select part of the POIs while applying CoCycle algorithm. Under this condition, we use binary search to find the maximum number of POIs for CoCycle algorithm to cover, and those POIs with greater q_i will first be selected.

Figure 4 shows the coverage pattern of the four algorithms. The red dots in the figure are POIs. The black edges among the POIs illustrate the scan path of mobile sensors. In MinExpand, every ring path corresponds to the sweep path of a mobile sensor. In CoCycle, different number of mobile sensors are allocated on several ring paths. In GD-MSDSC and MST-MSDSC, there is only one ring path, on which mobile sensors are placed at the same distance and moving in the same direction. There exists cross lines in the sweep path of GD-MSDSC while MST-MSDSC does not.

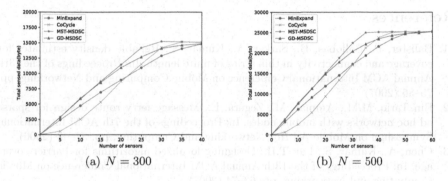

(a) $N = 300$ (b) $N = 500$

Fig. 5. Comparison of the total sensed data.

Figure 5(a) and (b) show the experimental comparison results when the number of POIs are 300 and 500 respectively. The horizontal axis represents the given number of mobile sensors and the vertical axis represents the total collected data of the mobile sensors in a sweep period. The results indicate that, for all the investigated algorithms, initially the total amount of sensed data increases as much more mobile sensors are provided. When the number of given sensors reaches a certain value, the total amount of sensed data does not grow and remain unchanged since there are sufficient number of sensors to sense data from all POIs. Compared with MinExpand and CoCycle, MST-MSDSC and GD-MSDSC are able to sense more data when insufficient sensors are provided. Furthermore, in Fig. 5(a), there are 300 POIs in the region, GD-MSDSC and MST-MSDSC have similar performance, but as the distribution of POIs gets denser when there are 500 POIs in Fig. 5(b), MST-MSDSC is more efficient in sensing data from POIs than GD-MSDSC.

7 Conclusions and Future Work

In this paper, we study the MSDSC problem, the goal of which is to maximize sensed data with limited number of mobile sensors. Then we prove the MSDSC problem to be NP-hard. Considering this, we propose two heuristic algorithms, GD-MSDSC and MST-MSDSC, to solve the problem. Compared with MinExpand and CoCycle algorithms from previous literature, the simulation results validates the efficiency of our proposed algorithms. With the same number of mobile sensors, both GD-MSDSC and MST-MSDSC are able to sense more data than MinExpand and CoCycle. Specifically, MST-MSDSC senses more data from POIs while GD-MSDSC has lower time complexity.

In the future, we plan to consider more factors like energy consumption or data delay in our problem.

Acknowledgment. This work is supported by National Natural Science Foundation of China (No. 61772154), the Shenzhen Basic Research Program (Project No. JCYJ20190806143011274).

References

1. Balister, P., Bollobas, B., Sarkar, A., Kumar, S.: Reliable density estimates for coverage and connectivity in thin strips of finite length. In: Proceedings of the 13th Annual ACM International Conference on Mobile Computing and Networking, pp. 75–86 (2007)
2. Bin Tariq, M.M., Ammar, M., Zegura, E.: Message ferry route design for sparse ad hoc networks with mobile nodes. In: Proceedings of the 7th ACM International Symposium on Mobile Ad Hoc Networking and Computing, pp. 37–48 (2006)
3. Chen, A., Kumar, S., Lai, T.H.: Designing localized algorithms for barrier coverage. In: Proceedings of the 13th Annual ACM International Conference on Mobile Computing and Networking, pp. 63–74 (2007)
4. Chen, Z., Zhu, X., Gao, X., Wu, F., Gu, J., Chen, G.: Efficient scheduling strategies for mobile sensors in sweep coverage problem. In: 2016 13th Annual IEEE International Conference on Sensing, Communication, and Networking (SECON), pp. 1–4. IEEE (2016)
5. Cheng, W., Li, M., Liu, K., Liu, Y., Li, X., Liao, X.: Sweep coverage with mobile sensors. In: 2008 IEEE International Symposium on Parallel and Distributed Processing, pp. 1–9. IEEE (2008)
6. Du, H., Luo, H.: Routing-cost constrained connected dominating set (2016)
7. Du, H., Luo, H., Zhang, J., Zhu, R., Ye, Q.: Interference-free k-barrier coverage in wireless sensor networks. In: Zhang, Z., Wu, L., Xu, W., Du, D.-Z. (eds.) COCOA 2014. LNCS, vol. 8881, pp. 173–183. Springer, Cham (2014). https://doi.org/10.1007/978-3-319-12691-3_14
8. Du, J., Li, Y., Liu, H., Sha, K.: On sweep coverage with minimum mobile sensors. In: 2010 IEEE 16th International Conference on Parallel and Distributed Systems, pp. 283–290. IEEE (2010)
9. Feng, Y., Gao, X., Wu, F., Chen, G.: Shorten the trajectory of mobile sensors in sweep coverage problem. In: 2015 IEEE Global Communications Conference (GLOBECOM), pp. 1–6. IEEE (2015)

10. Gao, X., Fan, J., Wu, F., Chen, G.: Approximation algorithms for sweep coverage problem with multiple mobile sensors. IEEE/ACM Trans. Netw. **26**(2), 990–1003 (2018)
11. Gao, X., Zhu, X., Feng, Y., Wu, F., Chen, G.: Data ferry trajectory planning for sweep coverage problem with multiple mobile sensors. In: 2016 13th Annual IEEE International Conference on Sensing, Communication, and Networking (SECON), pp. 1–9. IEEE (2016)
12. Gorain, B., Mandal, P.S.: Point and area sweep coverage in wireless sensor networks. In: 2013 11th International Symposium and Workshops on Modeling and Optimization in Mobile, Ad Hoc and Wireless Networks (WiOpt), pp. 140–145. IEEE (2013)
13. Gorain, B., Mandal, P.S.: Line sweep coverage in wireless sensor networks. In: 2014 Sixth International Conference on Communication Systems and Networks (COMSNETS), pp. 1–6. IEEE (2014)
14. Huang, C.F., Tseng, Y.C.: The coverage problem in a wireless sensor network. Mob. Netw. Appl. **10**(4), 519–528 (2005). https://doi.org/10.1007/s11036-005-1564-y
15. Kim, D., Wang, W., Li, D., Lee, J.L., Wu, W., Tokuta, A.O.: A joint optimization of data ferry trajectories and communication powers of ground sensors for long-term environmental monitoring. J. Comb. Optim. **31**(4), 1550–1568 (2016). https://doi.org/10.1007/s10878-015-9840-7
16. Kumar, S., Lai, T.H., Arora, A.: Barrier coverage with wireless sensors. In: Proceedings of the 11th Annual International Conference on Mobile Computing and Networking, pp. 284–298 (2005)
17. Kumar, S., Lai, T.H., Balogh, J.: On k-coverage in a mostly sleeping sensor network. In: Proceedings of the 10th Annual International Conference on Mobile Computing and Networking, pp. 144–158 (2004)
18. Li, M., Cheng, W., Liu, K., He, Y., Li, X., Liao, X.: Sweep coverage with mobile sensors. IEEE Trans. Mob. Comput. **10**(11), 1534–1545 (2011)
19. Lin, L., Lee, H.: Distributed algorithms for dynamic coverage in sensor networks. In: Proceedings of the Twenty-Sixth Annual ACM Symposium on Principles of Distributed Computing, pp. 392–393 (2007)
20. Liu, B.H., Nguyen, N.T., et al.: An efficient method for sweep coverage with minimum mobile sensor. In: 2014 Tenth International Conference on Intelligent Information Hiding and Multimedia Signal Processing, pp. 289–292. IEEE (2014)
21. Liu, C., Du, H., Ye, Q.: Sweep coverage with return time constraint. In: 2016 IEEE Global Communications Conference (GLOBECOM), pp. 1–6. IEEE (2016)
22. Liu, C., Du, H., Ye, Q.: Utilizing communication range to shorten the route of sweep coverage. In: 2017 IEEE International Conference on Communications (ICC), pp. 1–6. IEEE (2017)
23. Liu, C., Huang, H., Du, H., Jia, X.: Performance-guaranteed strongly connected dominating sets in heterogeneous wireless sensor networks. In: IEEE INFOCOM 2016-The 35th Annual IEEE International Conference on Computer Communications, pp. 1–9. IEEE (2016)
24. Luo, H., Du, H., Kim, D., Ye, Q., Zhu, R., Jia, J.: Imperfection better than perfection: Beyond optimal lifetime barrier coverage in wireless sensor networks. In: 2014 10th International Conference on Mobile Ad-hoc and Sensor Networks, pp. 24–29. IEEE (2014)

25. Meguerdichian, S., Koushanfar, F., Potkonjak, M., Srivastava, M.B.: Coverage problems in wireless ad-hoc sensor networks. In: Proceedings IEEE INFOCOM 2001. Conference on Computer Communications. Twentieth Annual Joint Conference of the IEEE Computer and Communications Society (Cat. No. 01CH37213), vol. 3, pp. 1380–1387. IEEE (2001)
26. Moazzez-Estanjini, R., Paschalidis, I.C.: On delay-minimized data harvesting with mobile elements in wireless sensor networks. Ad Hoc Netw. **10**(7), 1191–1203 (2012)
27. Shu, L., Wang, W., Lin, F., Liu, Z., Zhou, J.: A sweep coverage scheme based on vehicle routing problem. Telkomnika **11**(4), 2029–2036 (2013)
28. Wang, S., Gasparri, A., Krishnamachari, B.: Robotic message ferrying for wireless networks using coarse-grained backpressure control. IEEE Trans. Mob. Comput. **16**(2), 498–510 (2016)
29. Wu, L., Xiong, Y., Wu, M., He, Y., She, J.: A task assignment method for sweep coverage optimization based on crowdsensing. IEEE Internet Things J. **6**(6), 10686–10699 (2019)
30. Zhao, W., Ammar, M., Zegura, E.: Controlling the mobility of multiple data transport ferries in a delay-tolerant network. In: Proceedings IEEE 24th Annual Joint Conference of the IEEE Computer and Communications Societies, vol. 2, pp. 1407–1418. IEEE (2005)

Trip-Vehicle Assignment Algorithms for Ride-Sharing

Songhua Li[✉], Minming Li, and Victor C. S. Lee

City University of Hong Kong, Kowloon, Hong Kong SAR, China
songhuali3-c@my.cityu.edu.hk, {minming.li,csvlee}@cityu.edu.hk

Abstract. The trip-vehicle assignment problem is a central issue in most peer-to-peer ride-sharing systems. Given a set of n available vehicles with respective locations and a set of m trip requests with respective origins and destinations, the objective is to assign requests to vehicles with the minimum overall cost (which is the sum of the moving distances of the vehicles). Since the assignment constraints are well captured by edges matched in graphs, we investigate the problem from a matching algorithm point of view. Suppose there are at most two requests sharing a vehicle at any time, we answer an open question by Bei and Zhang (AAAI, 2018), that asks for a constant-approximation algorithm for the setting where the number (m) of requests is no more than twice the number (n) of vehicles, i.e., $m \leq 2n$. We propose an $O(n^4)$-time 2.5-approximation algorithm, which is built upon a solution of the Minimum-weight Fixed-size Matching problem with unmatched vertex Penalty (MFMP), in which the cost is the sum of the weights of both matched edges and unmatched vertices. Then, we study a more general setting that also allows $m > 2n$. We propose a dynamic assignment algorithm that is built upon a solution of the Minimum Weight Matching problem with unmatched vertex Penalty (MWMP). Further, we extend the dynamic assignment algorithm to an online setting where on-demand trip requests appear over time. Experiments are conducted on a real-world data set of trip records showing that our algorithms actually achieve good performances.

Keywords: Ride-sharing system · Trip-vehicle assignment · Matching algorithm

1 Introduction

The shared mobility has enjoyed rapid growth over the past few years due to the fast growth in mobile technologies, and it is proved to be effective in cost-saving, efficient in relieving traffic congestion and sustainable in increasingly crowded urban cities. Representative systems of shared mobility include bike-sharing systems (such as the *Mobike* and *Divvy*), ride-sharing systems (such as the *Uber* and the *Didi Chuxing*), and car-sharing systems (such as the *EVCard* and the *Car2go*). Among them, the ride-sharing system (RSS) seems to be the

© Springer Nature Switzerland AG 2020
W. Wu and Z. Zhang (Eds.): COCOA 2020, LNCS 12577, pp. 681–696, 2020.
https://doi.org/10.1007/978-3-030-64843-5_46

most popular shared mobility system in urban cities since some cities do not fit in the bike-sharing system well due to terrain characteristics or city regulations. Recently, optimization problems in RSS attract lots of attention. Some researchers focus on the ride-sharing routing problem to design a path for the driver to pick up and drop off passenger(s) in a ride-sharing group. Lin et al. [6] considered the demand-aware approach by leveraging travel demand statistics and proposed a pseudo-polynomial-time algorithm for the problem. Santi et al. [1] and Alonso-Mora et al. [7] studied the shortest-path-like routing algorithms.

In this paper, we focus on a fundamental RSS problem, which is to assign known trip requests (passenger orders) to available vehicles (drivers) with the sum of the vehicles' moving distances (i.e., the overall service cost) minimized. Considering that sharing a trip by two requests can reduce vehicles' cumulative trip length by 40% [1] and that users are only willing to share their trips with one other group due to the low seating capacity of RSS vehicles, we assume in this paper that each vehicle can serve at most two requests at any time. The offline problem, in which both a number m of trip requests and a number n of vehicles are given, is proved to be NP-hard [5] even for a special setting where $m = 2n$ (i.e., the number of trip requests is exactly twice the number of vehicles). Usually, there are two issues in the trip-vehicle assignment problem, including the one in grouping the requests and the other one in matching the formed groups with drivers [6]. Naturally, an algorithm with a two-stage perfect matching architecture can handle the above two issues well, and particularly perfects for the setting $m = 2n$ in which a 2.5-approximation algorithm is guaranteed by Bei et al. [5]. However, when facing with a more general setting with $m \leq 2n$, we note that *two hurdles* hinder a two-stage perfect-matching approach from achieving a provable constant approximation ratio: (a) every vehicle/driver in the setting $m = 2n$ can be assigned with exactly two trip requests [5], but this is no more the case in the setting $m \leq 2n$ since some drivers may not necessarily be assigned with trips in an optimal solution, making it hard to bound the overall moving distance of drivers in our solution; (b) a perfect matching no longer exists in grouping requests in the setting $m \leq 2n$ since some requests may not be matched for achieving the minimum overall service cost.

Related Works. We survey relevant researches along two threads. The *first thread* is on maximizing the profit earned by the system. Lowalekar et al. [11] proposes a heuristic algorithm to solve the real-time problem with the objective to maximize the number of served requests, which takes advantage of a zone (here, each zone denotes an abstraction of locations) path construction. By regarding the requests and drivers as right-side vertices and left-side vertices of a bipartite graph, respectively, the assignment problem in RSS can generally be modeled as the bipartite matching problem. Dickerson et al. [4] proposed a new model of online matching with reusable resources under known adversarial distributions, where matched resources can become available again in the future. They also presented an LP-based adaptive algorithm with a competitive ratio of $(0.5 - \epsilon)$ for any given $\epsilon > 0$. Zhao et al. [3] further considered the preferences of drivers and passengers (e.g., drivers prefer to take high-rewarding passengers while passengers prefer to select nearby drivers). They used the online stable matching to maximize the expected total profit. Recently, Huang et al. [13] introduced a fully online model of maximum cardinality matching (i.e., maximizing the number of successful assignments) in which all vertices (both drivers and passengers) arrive online, they showed that the Ranking Algorithm by Karp et al. [14] is 0.5211-competitive for general graphs. The *second thread* is on minimizing the overall cost (the distance or the time). Tong et al. [8] aimed to find a maximum-cardinality matching with minimum total distance, which is under the assumption that a new user (trip request) must be matched with an unmatched server (driver) before the next user, they showed that the greedy algorithm has a competitive ratio of 3.195 in the average-case analysis. More works on RSS can be found in [9,10,12,15].

Our contribution is three-fold. *First*, we answer an open question by [5], that asks for an approximation algorithm to assign a number m of given requests to a number n ($\geq \frac{m}{2}$) of given vehicles with two requests sharing one vehicle. To tackle the two hurdles introduced earlier in this section, we first propose a solution to find a fixed-size matching in a weighted graph (in both edges and vertices) that minimizes the total weight of matched edges and unmatched vertices. Built upon that, we develop an $O(n^4)$-time 2.5-approximation algorithm. *Second*, we consider a more general setting of the assignment problem, in which $m > 2n$ is also allowed. In the setting, each vehicle can be assigned to more than two requests but can handle at most two requests at any time. To this end, we propose a solution to the minimum-weight matching problem with unmatched vertex penalty, built upon which we propose a dynamic assignment algorithm. Experiments are conducted showing that our algorithm achieves good performances. *Third*, we extend the dynamic assignment algorithm to the online setting where on-demand requests appear over time. The effectiveness of our online algorithms is demonstrated via extensive experiments on real-world trip records.

2   The Model

In RSS, the traffic network is normally modeled as a metric space $\mathcal{N} = (\Omega, d)$, in which Ω denotes the set of distinct locations in the network, and $d : \Omega \times \Omega \rightarrow \mathbb{R}$ is the distance function of roads in the network satisfying non-negativity ($d(x, y) \geq 0$), symmetry ($d(x, y) = d(y, x)$) and triangle inequality ($d(x, y) + d(y, z) \geq d(x, z)$).[1] Further, we denote the weight of a path as $d(a_1, \ldots, a_{k-1}, a_k) = \sum_{i=1}^{k-1} d(a_i, a_{i+1})$. Given a set $D = \{1, \ldots, n\}$ of n available vehicles (drivers) with each vehicle $k \in D$ at location $d_k \in \Omega$, and a set $R = \{r_1, \ldots, r_m\}$ of m trip requests in which each request $r_i = (s_i, t_i) \in R$ contains a pick-up location $s_i \in \Omega$ and a drop-off location $t_i \in \Omega$, we aim to assign all requests in R to vehicles in D. An instance of the trip-vehicle assignment problem in RSS can be written as a tuple $I = (\mathcal{N}, D, R)$. In a trip-vehicle assignment

---

[1] Note that our implementation data in Sect. 5 does not necessarily follow these properties.

$\Pi = \{(k, R_k) | k \in D, R_k \subset R\}$, where R_k denotes the subset of requests assigned to vehicle k by Π, we define the cost function $cost(k, R_k)$ for each $(k, R_k) \in \Pi$ as the weight of the shortest path for vehicle k to serve its assigned requests in R_k. For example, $cost(k, R_k) = 0$ when $R_k = \emptyset$; $cost(k, R_k) = d(d_k, s_i) + d(s_i, t_i)$ when $R_k = \{r_i\}$; when $R_k = \{r_i, r_j\}$,

$$cost(k, R_k) = \min\{ \; d(d_k, s_i, s_j, t_i, t_j), d(d_k, s_i, s_j, t_j, t_i), d(d_k, s_j, s_i, t_i, t_j), \\ d(d_k, s_j, s_i, t_j, t_i), d(d_k, s_i, t_i, s_j, t_j), d(d_k, s_j, t_j, s_i, t_i)\} \quad (1)$$

The cost of an assignment Π is defined as $cost(\Pi) = \sum\limits_{(k, R_k) \in \Pi} cost(k, R_k)$. Given an instance $I = (\mathcal{N}, D, R)$, our model[2] is presented below.

$$\min_{\Pi} \quad cost(\Pi)$$
$$\text{s.t.} \quad R_i \cap R_j = \varnothing, \forall i, j \in D \quad (i)$$
$$\bigcup_{k \in D} R_k = R \quad (ii)$$
$$|R_k| \leq 2, \qquad \forall k \in D \quad (iii)$$

in which constraint (i) indicates that each request is assigned to only one vehicle, constraint (ii) indicates all requests are assigned successfully, and constraint (iii) indicates each vehicle can serve at most two requests (following an assumption in [5]).

An assignment Π is called *feasible* if and only if it satisfies the constraints (i)-(iii). Let D_l^Π denote the subset of vehicles that serve exactly l requests by Π (i.e., $D_l^\Pi = \{k | (k, R_k) \in \Pi, |R_k| = l\}$), where $l \in \{0, 1, 2\}$. Further, $|D_2^\Pi| + |D_1^\Pi|$ is the number of *vehicles-in-service* (i.e., the vehicles that serve at least one request) in Π.

Observation 1. $|D_1^\Pi| + 2 \cdot |D_2^\Pi| = m$ and $\lceil \frac{m}{2} \rceil \leq |D_1^\Pi| + |D_2^\Pi| \leq \min\{m, n\}$.

3   Approximation Algorithm

To answer the open question in [5] which asks for a constant-approximation algorithm for $m \leq 2n$, we propose an $O(n^4)$-time 2.5-approximation algorithm DPAA (<u>D</u>river <u>P</u>assenger <u>A</u>ssignment <u>A</u>lgorithm) in Algorithm 1. Before going to the details of DPAA, we first consider a basic matching problem of DPAA.

3.1   Minimum-Weight Fixed-Size Matching with Unmatched Vertex Penalty

DPAA is built upon a solution of the following <u>M</u>inimum-weight <u>F</u>ixed-size <u>M</u>atching with unmatched vertex <u>P</u>enalty problem (MFMP), in which we fix the

---

[2] For ease of analysis, our model admits that every pair of requests can be grouped together to share a vehicle, which is also an assumption in [5].

size restriction in matching to handle the hurdles as aforementioned in Sect. 1 (we will state this more clearly later in Sect. 3.2).

MFMP Definition. Given a complete graph $G = (V, E, w)$ in which the weight function w applies to both edges in E and vertices in V, and a fixed number $x \in \left\{0, 1, ..., \left\lfloor \frac{|V|}{2} \right\rfloor \right\}$, the goal is to find a matching M (with the matching size equal to x, i.e., $|M| = x$) in G such that the sum of weights of matched edges and unmatched vertices is minimized. Let z_e denote the indicator variable of an edge $e \in E$, and $\delta(v)$ denote the edges in E that incident to vertex $v \in V$. If an edge e is in the matching, set $z_e = 1$, otherwise set $z_e = 0$. The MFMP problem is formulated as follows.

$$\min \quad \sum_{e \in E} w(e) \cdot z_e + \sum_{v \in V \backslash V(\bigcup_{e \in E,\, z_e = 1} \{e\})} w(v)$$

$$\text{s.t.} \quad \sum_{e \in \delta(v)} z_e \leq 1, \qquad\qquad \forall v \in V \qquad\qquad \text{(iv)}$$

$$z_e \in \{0, 1\}, \qquad\qquad \forall e \in E \qquad\qquad \text{(v)}$$

$$\sum_{e \in E} z_e = x, \qquad\qquad\qquad\qquad\qquad \text{(vi)}$$

MFMPA Solution. Given the MFMP problem with graph $G = (V, E, w)$, we propose a solution named MFMPA by the following *three* steps.

Step 1. Transform the graph G to a complete graph $G_x = (V_x, E_x, w_x)$, in which the vertex set V_x is constructed by adding $(|V| - 2x)$ new vertices $V' = \left\{v_1', v_2', ..., v_{|V|-2x}'\right\}$ to V, i.e., $V_x = V \cup V'$. The weight function w_x applies only to the edges in E_x, for each $(u, v) \in E_x$,

$$w_x(u, v) = \begin{cases} w(u, v), & \text{if } u, v \in V \\ +\infty, & \text{if } u, v \in V' \\ w(u), & \text{if } u \in V, v \in V' \\ w(v), & \text{if } u \in V', v \in V \end{cases}$$

An instance of the graph transformation from G to G_x is shown from Fig. 1 to Fig. 2. In the given graph G, the vertices $\{v_1, v_2, v_3, v_4\}$ in black and the edge (in solid line) are labelled with their respective weights. Suppose the target matching has a size of $x = 1 \in \{0, 1, ..., \left\lfloor \frac{|V|}{2} \right\rfloor \}$ in the instance, it adds $(|V| - 2 \cdot x) = 2$ new vertices $\{v_1', v_2'\}$ and eight new edges (in dashed line) to the transformed graph G_x, in which vertices have zero weight and edges are labelled with their weights respectively.

Step 2. Find a minimum-weight perfect matching M in $G_x = (V_x, E_x, w_x)$. Note that such a matching M always exists because G_x is a complete graph with an even number of vertices;

Step 3. Output edges in M that also belong to the given graph G of MFMP, which is denoted as $M_T = \{e | e \in M \text{ and } e \in E\}$.

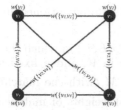

Fig. 1. The given graph G of MFMP.

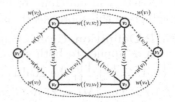

Fig. 2. The transformed graph G_x of G. *Note: the edge (v_i', v_2') with infinity weight is not shown in Fig. 2.*

Lemma 1. *MFMPA solves MFMP.*

Proof. Given a weighted complete graph $G = (V, E, w)$ of MFMP, we suppose that the target matching has a size of $x \in \left\{0, 1, ...,, \left\lfloor \frac{|V|}{2} \right\rfloor \right\}$, MFMPA first transforms the given graph G into a new complete graph $G_x = (V_x, E_x, w_x)$ with vertex set $V_x = V \cup V'$, where vertices in $V' = \left\{ v_1', v_2', ..., v_{|V|-2x}' \right\}$ are newly added.

As graph G_x is a complete graph and $|V_x| = 2|V| - 2x$ is an even number, the perfect matching exists in G_x. Then, the MFMPA can find a minimum-weight perfect matching M in G_x, and outputs all edges in $M \cap E$ to form M_T. Note that each vertex in V' is matched with a distinct vertex in V in a perfect matching, implying $|M_T| = x$. Let $w(M)$ be the weight of a matching M. According to w_x and the correlation between the M and the M_T, we know $w(M) = \sum_{e \in M_T} w(e) + \sum_{v \in V \setminus V(M_T)} w(v)$. Suppose

$$M_T \neq \underset{M' \text{ in } G \text{ with size } x}{\arg\min} \sum_{e \in M'} w(e) + \sum_{v \in V \setminus V(M')} w(v)$$

there is another matching $\overline{M_T}$ with size x in G such that

$$\sum_{e \in \overline{M_T}} w(e) + \sum_{v \in V \setminus V(\overline{M_T})} w(v) < \sum_{e \in M_T} w(e) + \sum_{v \in V \setminus V(M_T)} w(v)$$

This implies the existence of another perfect matching $\overline{M}$ in G_x that consists of x matched edges in $\overline{M_T}$ and another $(|V| - 2x)$ edges in which each one matches a distinct vertex in $V \setminus V(\overline{M_T})$ with a distinct vertex in $V_x \setminus V$. Note that

$$w(\overline{M}) = \sum_{e \in \overline{M}} w(e) + \sum_{v \in V \setminus V(\overline{M})} w(v) < w(M)$$

This contradicts with the fact that M is a minimum-weight perfect matching in G_x.

3.2   Algorithm DPAA

At the high level, DPAA attempts to fix the number $(|D_2^\Pi| + |D_1^\Pi|)^3$ of vehicles-in-service and generates a feasible assignment in which the number of drivers in-service equals to the fixed number. Since $|D_2^\Pi| + |D_1^\Pi| \in \left\{ \lceil \frac{m}{2} \rceil, \lceil \frac{m}{2} \rceil + 1, ..., \min\{m, n\} \right\}$, see Observation 1, DPAA totally generates $\min\left\{ \lfloor \frac{m}{2} \rfloor + 1, n - \lceil \frac{m}{2} \rceil + 1 \right\}$ feasible assignments and finally outputs the one with the minimum cost[4].

Given requests in R and vehicles in D, DPAA runs $\min\left\{ \lfloor \frac{m}{2} \rfloor + 1, n - \lceil \frac{m}{2} \rceil + 1 \right\}$ iterations. In each iteration $j \in \left\{ 1, 2, ..., \min\left\{ \lfloor \frac{m}{2} \rfloor, n - \lceil \frac{m}{2} \rceil \right\} + 1 \right\}$ of the matching loop[5], DPAA fixes $i = \lceil \frac{m}{2} \rceil + j - 1$ and takes a two-phase minimum-weight perfect matching approach to generate a feasible assignment.

In the first phase, DPAA constructs a complete graph G_α weighted in both edges and vertices, which is with regard to requests in R. Specifically, each request $r_i \in R$ is modeled as a vertex (in G_α) with its weight defined as the distance between the pick-up location and the drop-off location of the request, i.e., $\alpha(r_i) = d(t_i, s_i)$. Each two different requests $r_i, r_j \in R$ are connected by an edge (r_i, r_j) of G_α with its weight defined as the shortest distance (which is from the first pick-up location to the last drop-off location) to serve the two requests, i.e.,

$$\alpha(r_i, r_j) = \min\{ d(s_i, s_j, t_i, t_j), d(s_i, s_j, t_j, t_i), d(s_j, s_i, t_i, t_j),$$
$$d(s_j, s_i, t_j, t_i), d(s_i, t_i, s_j, t_j), d(s_j, t_j, s_i, t_i) \}$$

This way, we have $G_\alpha = (R, \alpha(R))$. Then, DPAA applies MFMPA in G_α to get the matching M_{1j} with the size of $(m - i)$. Further, DPAA gets i request groups including $m - i$ request pairs in M_{1j} and $2i - m$ unmatched requests, which further tackles the *hurdle* (b) presented in Sect. 1.

In the second phase, DPAA constructs a complete bipartite graph G_θ weighted only in edges. Specifically, vehicles in D are modeled as the left-side vertices while the i request groups are modeled as the right-side vertices, and the weight of each edge connecting a vehicle $k \in D$ and a request group $\{r_p, r_q\}$ (or single r_p) is defined as $cost(k, \{r_p, r_q\})$ (or $cost(k, r_p)$). Then, DPAA finds a right-side minimum-weight perfect matching M_{2j} in G_θ to assign the request groups to i distinct vehicles.

Totally, DPAA gets $\min\left\{ \lfloor \frac{m}{2} \rfloor + 1, n - \lceil \frac{m}{2} \rceil + 1 \right\}$ feasible assignments and finally outputs the one with the minimum cost.

3.3   Approximation Ratio

Before the analysis of DPAA, we first analyze a degenerated algorithm $DPAA_D$ (of DPAA) in Algorithm 2, whose approximation ratio is an upper bound of that

---

[3] This number is represented by i in Algorithm 1.

[4] This tackles the *hurdle* (a) as presented in Sect. 1.

[5] In either DPAA or $DPAA_D$, the matching loop refers to lines 7–11 of the pseudo code.

Algorithm 1. DPAA

Input: A metric space $\mathcal{N} = (\Omega, d)$ of the traffic network, a set $R = \{r_i(s_i, t_i) | 1 \leq i \leq m\}$ of m requests with pick-up locations s_i and drop-off locations t_i, a set $\{d_k | k \in D\}$ of n vehicles' locations.

Output: $\Pi_{A1} = \{(k, R_k) | k \in D, R_k \subset R \text{ with } |R_k| \leq 2\}$.

1: **for** each $r_p, r_q \in R$ and $d_k \in D$ **do**

2:    $\theta(d_k, \{r_p, r_q\}) = cost(k, \{r_p, r_q\})$ // θ is the weight function of graph G_θ, see formula (1) for the $cost$ function.

3:    $\theta(d_k, r_p) = cost(k, r_p)$

4:    $\alpha(r_p, r_q) = \min\{\ d(s_p, t_p, s_q, t_q), d(s_q, t_q, s_p, t_p), d(s_p, s_q, t_p, t_q),$
$d(s_p, s_q, t_q, t_p), d(s_q, s_p, t_q, t_p), d(s_q, s_p, t_p, t_q)\}$

5:    $\alpha(r_p) = d(s_p, t_p)$ // α is the weight function of graph G_α.

6: **end for**

7: **for** $(i = \lceil \frac{m}{2} \rceil; i \leq \min\{m, n\}; i++)$ **do**

8:    apply MFMPA in $G_\alpha := (R, \alpha(R))$ to get a matching $M_{1j} = M_T(G_\alpha, m - i)$ with the size equal to $(m - i)$;

9:    $N = R - \cup_{\{r_p, r_q\} \in M_{1j}} \{r_p, r_q\}, j = i - \lceil \frac{m}{2} \rceil + 1$;

10:    find a minimum-weight perfect matching M_{2j} on the right-side vertex set $M_{1j} \cup N$ in the weighted bipartite graph $G_\theta := (D, M_{1j} \cup N, \theta)$.
   // Please refer to constructions of G_α and G_θ in the "Details of DPAA" in Section 3.1.

11: **end for**

12: **Output** $\Pi_{A1} = \arg \min\limits_{j \in \{1, 2, \ldots, \min\{\lfloor \frac{m}{2} \rfloor, n - \lceil \frac{m}{2} \rceil\} + 1\}} cost(M_{2j})$

---

of DPAA. Note that the only difference between DPAA$_D$ and DPAA lies in the edge weight function of the bipartite graph in the second matching phase. In DPAA, the edge weight function θ is defined as the shortest path for a vehicle to serve the single request or the two requests in a pair. In DPAA$_D$, the edge weight function is replaced by λ which is defined as the distance from the vehicle's location to the closest pick-up location of the request(s) in the assigned group. The output assignment of DPAA is denoted as Π_{A1} while the output assignment of DPAA$_D$ is denoted as Π_{A2}. To show the ratio of our algorithm, we separate the cost of our output solution Π_{A2} and the cost of an optimal solution respectively into two parts, in which the first part is the sum of distances moved by the vehicles from their locations to the (first) pick-up location of the assigned request(s), and the second part is the sum of distances moved by the vehicles from their (first) pick-up locations to their (last) drop-off locations of the assigned request(s).

Lemma 2. *In each iteration*

$$j \in \left\{ 1, 2, \ldots, \min\left\{ \left\lfloor \frac{m}{2} \right\rfloor, n - \left\lceil \frac{m}{2} \right\rceil \right\} + 1 \right\}$$

of the matching loop, both DPAA and DPAA$_D$ generates a feasible assignment M_{2j} such that $D_1^{M_{2j}} = 2j - 2 + \lceil \frac{m}{2} \rceil - \lfloor \frac{m}{2} \rfloor$ and $D_2^{M_{2j}} = \lfloor \frac{m}{2} \rfloor - j + 1$.

Algorithm 2. DPAA$_D$

Remain lines (1, 4-9, 11) of DPAA the same, replace the θ function in lines (2, 3, 10, 12) of DPAA by the λ function.

2: $\lambda(d_k, r_p) = d(d_k, s_p)$;

3: $\lambda(d_k, \{r_p, r_q\}) = \min\{d(d_k, s_p), d(d_k, s_q)\}$

10: find a minimum-weight perfect matching M_{2j} over the right-side vertex set $M_{1j} \cup N$ in the weighted bipartite graph $G_\lambda := (D, M_{1j} \cup N, \lambda)$.

12: $\Pi_{A2} = \arg \min\limits_{j \in \{1,2,...,\min\{\lfloor \frac{m}{2} \rfloor, n - \lceil \frac{m}{2} \rceil\}+1\}} cost(M_{2j})$.

Proof. Recall that $\Pi = \{(k, R_k) | k \in D, R_k \subset R\}$ is called a feasible assignment if and only if (i) $R = \cup_{k \in D} R_k$, (ii) elements in $\{R_k | (k, R_k) \in \Pi\}$ are mutually disjoint, and (iii) $|R_k| \leq 2$ holds for each $k \in D$ in Π.

In each iteration $j \in \{1, 2, ..., \min\{\lfloor \frac{m}{2} \rfloor, n - \lceil \frac{m}{2} \rceil\}+1\}$ of the matching loop, both DPAA and DPAA$_D$ first fix a same number $i = \lceil \frac{m}{2} \rceil + j - 1$ of vehicles-in-service and then output a same number i of request groups (in which each contains at most two requests, i.e., (iii)) by respectively applying the MFMPA. As each request $r \in R$ only lies in one request group R_k by MFMPA, we have (ii) elements in $\{R_k | (k, R_k) \in M_{2j}\}$ are mutually disjoint. The perfect matching in MFMPA guarantees (i) $R = \cup_{R_k \in \Pi_{1j}} R_k$. When assigning the i request groups among n vehicles in the second phase, the minimum-weight right-side (of request groups) perfect matching M_{2j} can always be found because $i \leq n$. Clearly, $|D_2^{M_{2j}}| = m - i = \lfloor \frac{m}{2} \rfloor - j + 1$ which further implies $D_1^{M_{2j}} = i - |D_2^{M_{2j}}| = 2j - 2 + \lceil \frac{m}{2} \rceil - \lfloor \frac{m}{2} \rfloor$.

Given a feasible assignment $\Pi = \{(k, R_k) | k \in D, R_k \subset R\}$, let $cost_1(k, R_k, \Pi)$ denote the distance from vehicle k's location to the first pickup location (in R_k) in an assignment Π, and let $cost_2(k, R_k, \Pi)$ denote the distance from the first pickup location (in R_k) to the last drop-off location (in R_k) in Π. Further, we separate the overall cost of an assignment into two parts. The first part is the sum of distances moved by vehicles from their locations to the (first) pick-up locations of the assigned request(s), i.e., $cost_1(\Pi) = \sum_{(k,R_k) \in \Pi} cost_1(k, R_k, \Pi)$, and the second part is the sum of distances moved by vehicles from the (first) pick-up locations to the (last) drop-off locations of the assigned request(s), i.e., $cost_2(\Pi) = \sum_{(k,R_k) \in \Pi} cost_2(k, R_k, \Pi)$. Therefore, $cost(\Pi) = cost_1(\Pi) + cost_2(\Pi)$.

Given an instance $I = (\mathcal{N}, D, R)$ of the trip-vehicle assignment problem, let D_1^* and D_2^* denote the set of vehicles in an optimal assignment Π^* that serve one request and two requests respectively. Clearly, $\lceil \frac{m}{2} \rceil \leq |D_1^*| + |D_2^*| \leq \min\{m, n\}$. In the j^*-th iteration (where $j^* = i^* - \lceil \frac{m}{2} \rceil + 1$ and $i^* = |D_1^*| + |D_2^*|$) of the matching loop, DPAA$_D$ first generates a number $|D_1^*| + |D_2^*|$ of request groups by MFMPA and then generates a feasible assignment M_{2j*} that satisfies

$\left|D_1^{M_{2j*}}\right| = |D_1^*|$ and $\left|D_2^{M_{2j*}}\right| = |D_2^*|$. For ease of expression, we denote $|D_1^*| = a$ and $|D_2^*| = b$.

Lemma 3. $cost_1(M_{2j*}) \leq cost(\Pi^*) + \frac{1}{2} \cdot cost_2(\Pi^*)$.

Proof. In the j^*-th iteration of the matching loop, M_{2j*} is a minimum-weight perfect matching over the right-side vertex set $M_{1j*} \cup N$ (in which each request group in M_{1j*} contains a pair of requests and each request group in N contains a single request) in the bipartite graph $G_\lambda = (D, M_{1j*} \cup N, \lambda)$. Let us first consider the minimum-weight right-side perfect matching (denoted as $\overline{M_{2j*}}$) in a new bipartite graph $G_\lambda = (D_1^* \cup D_2^*, M_{1j*} \cup N, \lambda)$. Although we do not know M_{2j*} as we do not know the subset $D_1^* \cup D_2^*$ of vehicles, we know $w(M_{2j*}) \leq w(\overline{M_{2j*}})$ since $(D_1^* \cup D_2^*) \subseteq D$ and we can further show that $w(\overline{M_{2j*}}) \leq cost_1(\Pi^*) + cost(\Pi^*)$. To this end, we suppose in Π^* that $D_1^* = \{1, 2, \ldots, a\}, D_2^* = \{a + 1, \ldots, a+b\}$, and that $(k, r_k) \in \Pi^*$ for each $k \in D_1^*$. In $\overline{M_{2j*}}$, we sort vehicles by subscripts such that $D_1^{\overline{M_{2j*}}} = \{\delta_1, \delta_2, ..., \delta_a\}$ and $D_2^{\overline{M_{2j*}}} = \{\delta_{a+1}, \delta_{a+2}, ..., \delta_{a+b}\}$. Denote R_{δ_k} as the group assigned to vehicle δ_k in $\overline{M_{2j*}}$, i.e., $(\delta_k, R_{\delta_k}) \in \overline{M_{2j*}}$.

We construct a number a $(=|D_1^*|)$ of virtual requests $R' = \{r_1', r_2', \ldots, r_a'\}$, in which each $r_i' = (s_i', t_i') \in R'$ contains a pick-up location s_i' and a drop-off location t_i'. For each vehicle $k \in D_1^* \cup D_2^*$, we denote $\delta(k)$ as the counterpart of vehicle k in assignment $\overline{M_{2j*}}$, and $R_{\delta(k)}$ as the request group assigned to vehicle $\delta(k)$ in $\overline{M_{2j*}}$. For each $r_p' = (s_p', t_p') \in R'$ and $r_q = (s_q, t_q) \in R' \cup R$, we define $d(s_p', t_p') = 0$, and

$$
d(s_q, s_p') = \begin{cases} \min_{k \in D_1^* \cup D_2^*, r=(s,t) \in R} d(d_k, s), & \text{if } r_q = r_p \\ \max_{k \in D_1^* \cup D_2^*, r=(s,t) \in R} d(d_k, s), & \text{if } r_q \neq r_p \end{cases}
$$

in which r_p is the request assigned to vehicle p in Π^*. By replacing s_q by t_q in the above definition of $d(s_q, s_p')$, we get $d(t_q, s_p')$. For each vehicle $k \in D_1^* \cup D_2^*$,

$$
d(k, s_p') = \begin{cases} \min\{d(p, r_p), \lambda(\delta(p), R_{\delta(p)})\}, & \text{if } k = p \\ \max\{\lambda(\delta(p), R_{\delta(p)}), \lambda(\delta_p, R_{\delta_p})\}, & \text{if } k = \delta_p \\ \max_{i \in D_1^* \cup D_2^*, r=(s,t) \in R} d(d_i, s), & \text{if } other \ k \end{cases}
$$

in which λ is the weight function defined in DPAA$_D$. Now, let us consider the following problem (P*) which assigns requests in $R' \cup R$ to vehicles in $D_1^* \cup D_2^*$. By assigning each virtual request $r_k' \in R'$ respectively to vehicle k in Π^* and vehicle δ_k in $\overline{M_{2*}}$, we can get new assignments $\Pi_{P*} = \{(k, \{r_k, r_k'\}) | k \in \{1, 2, ..., a\}\} \cup \Pi^*(D_2^*)$ and $M_{P*} = \{\delta_k, \{r_{\delta_k}, r_k'\} | k \in \{1, ..., a\}\} \cup \overline{M_{2*}}(D_2^{\overline{M_{2j*}}})$ respectively, which are shown respectively in Fig. 3 and Fig. 4. According to the definition of virtual requests, we know that such assignment Π_{P*} is an optimal assignment of P*, and M_{P*} remains a perfect matching in graph $G_\lambda = (D_1^* \cup D_2^*, R \cup R', \lambda)$. Note that P* is actually an assignment problem reduced to the setting of $m = 2n$, hence we can take the idea in ([5], Lemma 4) to get the following

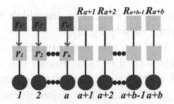

Fig. 3. The assignment Π_{P*}

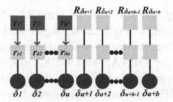

Fig. 4. The assignment M_{P*}.

Lemma 4. According to the weight function λ and the weights related to the virtual requests, we have $w\left(\overline{M_{1j*}}\right) \leq w(M_{P*})$. Hence,

$$w\left(M_{2j*}\right) \leq w\left(\overline{M_{2j*}}\right) \leq w\left(M_{P*}\right)$$

$$\leq 0.5 \sum_{(d_k, \{r_{k1}, r_{k2}\}) \in \Pi_{p*}} [d(s_{k1}, d_k) + d(s_{k2}, d_k)]$$

$$\leq 0.5 \cdot [2 \sum_{k \in D_1^*} cost_1(k, R_k, \Pi^*) + 2 \sum_{k \in D_2^*} cost_1(k, R_k, \Pi^*) + \sum_{(k, \{r_i, r_j\}) \in \Pi^*, k \in D_2^*} d(s_i, s_j)]$$

$$\leq cost_1(\Pi^*) + 0.5 \cdot cost_2(\Pi^*)$$

in which the third inequation holds by Lemma 4, and the fourth inequation holds by the triangle inequality $d(d_k, s_i) \leq d(d_k, s_j) + d(s_i, s_j)$. Further,

$$cost_1(M_{2j*}) \leq w\left(M_{2j*}\right) + \sum_{(d_k, \{r_{k1}, r_{k2}\}) \in \overline{M}_{2j*}} d(s_{k1}, s_{k2})$$

$$\leq cost(\Pi^*) + 0.5 \cdot cost_2(\Pi^*)$$

This completes the proof.

Lemma 4. *(see [5], lemma 4) For P^*,*

$$w(M_{P*}) \leq \frac{1}{2} \sum_{(d_k, \{r_{k1}, r_{k2}\}) \in \Pi_{P*}} [d(d_k, s_{k1}) + d(d_k, s_{k2})].$$

Lemma 5. $cost\left(\Pi_{A2}\right) \leq cost\left(\Pi^*\right) + \frac{3}{2} \cdot cost_2\left(\Pi^*\right)$.

Proof. Note that M_{2j*} is obtained from the j^*th iteration of the matching loop in DPAA$_D$, where $j^* = i^* - \lceil \frac{m}{2} \rceil + 1$ and $i^* = |D_1^*| + |D_2^*|$, and Π_{A2} is the one with the minimum weight among all the assignments generated by DPAA$_D$. Thus, $cost\left(\Pi_{A2}\right) \leq cost\left(M_{2j*}\right)$. As M_{2j*} is the minimum-weight perfect matching with a size of $|D_1^*| + |D_2^*|$, we have $cost_2\left(M_{2j*}\right) \leq cost_2\left(\Pi^*\right)$. Hence,

$$cost\left(\Pi_{A2}\right) \leq cost\left(M_{2j*}\right) \leq cost\left(\Pi^*\right) + \frac{3}{2} \cdot cost_2\left(\Pi^*\right)$$

Theorem 1. *For the trip-vehicle assignment problem with $m \leq 2n$, DPAA runs in time $O(n^4)$ and achieves an approximation ratio of at most 2.5.*

Proof. DPAA runs in time $O(n^4)$ since it runs $O(n)$ iterations in which the minimum-weight matchings could be found in time $O(n^3)$ by [2].

Denote, for ease of expression, M_{2j}^{A2} and M_{2j}^{A1} as the assignment generated by DPAA$_D$ and DPAA respectively in their j-th iteration. Recall the difference between the weight functions λ and θ of the second matching phase in DPAA$_D$ and DPAA, we have $cost(M_{2j}^{A1}) \leq cost(M_{2j}^{A2})$ in each iteration j, implying $cost(M_{A1}) \leq cost(M_{AT})$. By Lemma 5, we have $cost(M_{A1}) \leq cost(\Pi^*) + 1.5 \cdot cost_2(\Pi^*) \leq 2.5 \cdot cost(\Pi^*)$.

4   Dynamic Assignment Algorithm

Since vehicles may be assigned with more than two requests when $m > 2n$ is allowed, the size of the matching in grouping trip requests is not restricted anymore. Further, we replace the constraint (iii) of the model in Sect. 2 by "each vehicle can be assigned with more than two requests but can handle at most two requests at any time".

4.1   Minimum Weight Matching with Unmatched Vertex Penalty

We first consider a basic matching problem in the assignment, the Minimum Weight Matching problem with unmatched vertex Penalty (MWMP), which is defined as the MFMP problem in Sect. 3.1 with constraint (vi) excluded. Given the MWMP problem with $V = \{v_1, v_2, ..., v_{|V|}\}$ and $G = (V, E, w)$, we propose a solution named the MWMPA with the following three steps.

Step 1. Transform the given graph G to $G_y = (V_y, E_y, w_y)$ by adding a number $|V|$ of new vertices $V' = \{v_1', v_2', ..., v_{|V|}'\}$ to G, i.e., $V_y = V \cup V'$. The weight function w_y, which applies to edges in G_y only, is defined as follows: for any $v_i, v_j, v_i', v_j' \in V_y$,

$$\begin{cases} w_y(v_i, v_j) = w(v_i, v_j) \\ w_y(v_i, v_i') = w(v_i) \\ w_y(v_i, v_j') = +\infty \\ w_y(v_i', v_j') = 0 \end{cases} \tag{2}$$

Step 2. Find a minimum-weight perfect matching, denoted as $M(G)$, in G_y.
Step 3. Output those edges in $M(G)$ that also belong to the given graph G of MWMP, which is denoted as $M_T(G) = \{e | e \in M(G) \text{ and } e \in E\}$.

4.2   Driver Passenger Greedy Assignment Algorithm (DPGA)

At the high level, DPGA executes several rounds of MWMPA, dynamically either assigning requests to vehicles or grouping requests in one matching pool until all requests are assigned to vehicles. Specifically,

First. DPGA constructs a complete graph $G_\beta = (R \cup D, \beta)$ based on given requests and vehicles. The G_β is constructed as follows: to ensure constraint

(i) that each request is only assigned to one specific vehicle, we define the weight of an edge connecting a pair of vehicles $d_k, d_l \in D$ as $\beta(d_k, d_l) = +\infty$ (in practice, one can use a large enough number); the weight of an edge connecting a vehicle $d_k \in D$ and a request $r_i \in R$ is defined as $\beta(d_k, r_i) = cost(d_k, s_i)$, in which the $cost(\cdot)$ function is defined beforehand in Sect. 1; the weight of an edge connecting a pair of requests $r_i, r_j \in R$ is defined as $\beta(r_i, r_j) = \alpha(s_i, s_j)$, in which the $\alpha(\cdot)$ function is shown beforehand in DPAA algorithm in Sect. 3.2; to allow some vehicles not to be assigned with requests, we further define the weight of each vehicle vertex $d_k \in D$ as $\beta(d_k) = 0$; to guarantee that each request is either assigned to some driver or grouped with some other request in one round of MWMP matching, we define the weight of each request vertex $r_i \in R$ as $\beta(r_i) = +\infty$ (in practice, one can use a large enough number).

Then. DPGA applies MWMPA to get the MWMP matching $M_T(G_\beta)$ in G_β, attaining a set Q of matched edges that connect a vehicle and a request, and a set P of matched edges that connect a pair of requests. Next, DPGA constructs the graph G_μ as follows: the vertex weight $\mu(q_k)$ (resp. $\mu(p_i)$) for each $q_k \in Q$ (resp. $p_i \in P$) is defined as the edge weight of q_k (resp. p_i) by the previous round of matching; the weight of an edge connecting a pair of nodes $q_k, q_l \in Q$ is defined as $\mu(q_k, q_l) = +\infty$ by constraint (i) in Sect. 1; for a node $q_k \in Q$ and a node $p_i \in P$, the vehicle in q_k should first serve requests in q_k and then requests in p_i, i.e., $\mu(q_k, p_l) = \mu(q_k) + d(q_k, p_l) + \mu(p_l)$ in which $d(q_k, p_l)$ is the distance from the last drop-off location of requests in q_k to the first pick-up location of requests in p_i.

Later. DPGA applies MWMPA to get the MWMP matching $M_T(G_\mu)$ in graph $G_\mu = (P \cup Q, \mu)$ for several rounds. From the second round of MWMP matching on, the set P (resp. Q) consists of matched edges in the previous round that only contains requests (resp. contains a vehicle and some requests). Finally, DPGA terminates until every request is assigned to some vehicle, i.e., $P = \emptyset$.

Online Setting[6] **Solution.** By replacing the line 5 of the pseudo code of DPGA by "update $P = P(M_T(G_\mu)) \cup R_{new}(t)$, $Q = Q(M_T(G_\mu))$ in each time slot t", we extend DPGA to DPGA$_o$ to handle the online setting where a set $R_{new}(t)$ of on-demand requests are released to the system in each time slot t.

4.3   Experiments

To evaluate the performance of our algorithms, we conduct extensive experiments using real-world taxi records. In the implementation, we use a large number of 10^9 to represent $+\infty$ in corresponding weights of G_β and G_μ in DPGA.

Data Set Description. The original data set contains pick-up and drop-off locations of 29730 trip records generated by taxis in Chengdu, China, on August 3rd, 2014. For simplification, we first choose a representative time horizon (09:00–09:01). Then we randomly extract, from the original data set, a number (for

---

[6] In the online setting where trip requests are released over time, we do not consider the service deadline of the requests and only aim to service all the released requests with the overall moving distance of the vehicles minimized.

Algorithm 3. DPGA

Input: A metric space $\mathcal{N} = (\Omega, d)$ of the traffic network, a set $R = \{r_i(s_i, t_i)|1 \leq i \leq m\}$ of m requests with pick-up locations s_i and drop-off locations t_i, a set $\{d_k|k \in D\}$ of n vehicles' locations.

Output: $\Pi_{B1} = \{(k, R_k)|k \in D, R_k \subset R\}$ meeting (i,ii,iii).

1: Apply MWMPA in $G_\beta = (D \cup R, \beta)$ to get matching $M_T(G_\beta)$;
  //see the construction of G_β in the detailed description of DPGA in this subsection.

2: Update $Q = Q(M_T(G_\beta))$, $P = P(M_T(G_\beta))$;
3: **while** $P \neq \varnothing$ **do**
4:    apply MWMPA in $G_\mu = (P \cup Q, \mu)$ to get matching $M_T(G_\mu)$;
      //see the construction of G_μ in the detailed description of DPGA in this subsection.
5:    update $Q = Q(M_T(G_\mu))$, $P = P(M_T(G_\mu))$;
6: **end while**

---

example, $m = 400$) of trip records within the time horizon to be our input trip requests. Finally, we take the drop-off locations of a number (for example, $n = 240$) of trip requests, which are placed right before our input trip requests in the original data set, as our input drivers' locations.

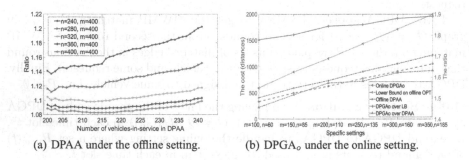

(a) DPAA under the offline setting.   (b) DPGA$_o$ under the online setting.

Fig. 5. Experimental results on DPAA and DPGA$_o$ algorithms under $m \leq 2n$.

Evaluation Set-up. Finding an optimal solution (OPT) for the problem is NP-hard [5], hence we compare the cost of our solution with a lower bound (LB) of OPT. When $m \leq 2n$, $\text{LB}_1 = \min\limits_{j \in \{1,2,...,\min\{\lfloor \frac{m}{2} \rfloor, n - \lceil \frac{m}{2} \rceil\}+1\}} \text{cost}(M_{1j})$, which is smaller than $\text{cost}(M_{1j*})$, is a natural lower bound (see Lemma 3). In either the $m \geq 2n$ setting or the online setting, we set LB_2 as the part of the cost in grouping offline requests in $R \cup \bigcup_t R_{new}(t)$. We take several groups of input, to test algorithms DPAA, DPGA, DPGA$_o$ in settings $m \leq 2n$, $m > 2n$ and the online setting respectively.

Table 1. Experimental results on DPGA under offline setting of $m > 2n$.

m	100	150	200	250	300
LB_2(km)	334	499	627	776	912
DPGA (km)	612	1216	1324	1564	2002

Results Analysis. For $m \leq 2n$, DPAA performs surprisingly excellently in practice, achieving an approximation ratio of less than 1.2, see Fig. 5(a). Moreover, the larger the ratio $\frac{\text{the number of vehicles}}{\text{the number of requests}}$ is, the better DPAA performs. For $m > 2n$, DPGA practically generates a solution with a cost at most 2.5 times of the OPT's cost, see Table 1. For the online setting with $m \leq 2n$, we compare the cost of $DPGA_o$ first with cost of LB_2 and then with the cost of DPAA in solving the offline problem, see Fig. 5(b), the cost of $DPGA_o$ is at most 1.9 times that of LB_2 (and hence of OPT).

5   Conclusions

This paper investigates the trip-vehicle assignment problem in ride-sharing systems, which is to assign trip requests to available vehicles. Suppose each vehicle serves at most two requests at any time, our contribution is three-fold.

First, we answer an open question proposed by [5], which asks for an approximation algorithm to assign a number m of given requests to a number n ($\geq \frac{m}{2}$) of given vehicles with two requests sharing one vehicle. As the new setting $m \leq 2n$ allows some vehicles to serve only one or even zero requests, the perfect matching no longer works in grouping requests. To tackle this hurdle, we propose a feasible solution to find a fixed-size matching in a weighted graph (in both edges and vertices) that minimizes the total weight of matched edges and unmatched vertices. Built upon the solution, we develop a $O(n^4)$-time 2.5-approximation algorithm for the setting $m \leq 2n$. We also show that our proposed algorithm can be easily adapted to the generalized setting, where each request contains a variable number of passengers, achieving the same approximation ratio 2.5. *Second*, we consider a more general setting of the assignment problem, in which the $m > 2n$ is also allowed. In this setting, each vehicle can be assigned to more than two requests but can handle at most two requests at any time. To this end, we propose a dynamic assignment algorithm, which is built upon our solution to the minimum-weight matching problem with an unmatched vertex penalty. Experiments are conducted showing that our dynamic assignment algorithm achieves good performances. *Third*, we extend the above algorithm to the online setting where on-demand trip requests appear over time. The effectiveness of our online algorithm is demonstrated via extensive experiments on real-world trip records.

Acknowledgements. Part of this work was done when Songhua Li was visiting the Singapore University of Technology and Design. Minming Li is also from City University of Hong Kong Shenzhen Research Institute, Shenzhen, P.R. China. The work described in this paper was partially supported by Project 11771365 supported by NSFC. We would like to thank Kaiyi Liao for his help in the implementation of our algorithms and we also thank all the anonymous reviewers for their comments.

References

1. Santi, P., Resta, G., Szell, M., Sobolevsky, S., Strogatz, S.H., Ratti, C.: Quantifying the benefits of vehicle pooling with shareability networks. Proc. Nat. Acad. Sci. **111**(37), 13290–13294 (2014)
2. Gabow, H.N.: Data structures for weighted matching and nearest common ancestors with linking. In: Proceedings of the First Annual ACM-SIAM Symposium on Discrete Algorithms, pp. 434–443 (1990)
3. Zhao, B., Xu, P., Shi, Y., Tong, Y., Zhou, Z., Zeng, Y.: Preference-aware task assignment in on-demand taxi dispatching: an online stable matching approach. In: Proceedings of the AAAI Conference on Artificial Intelligence, vol. 33, pp. 2245–2252 (2019)
4. Dickerson, J.P., Sankararaman, K.A., Srinivasan, A., Xu, P.: Allocation problems in ride-sharing platforms: online matching with reusable offline resources. In: Thirty-Second AAAI Conference on Artificial Intelligence (2018)
5. Bei, X., Zhang, S.: Algorithms for trip-vehicle assignment in ride-sharing. In: Thirty-Second AAAI Conference on Artificial Intelligence (2018)
6. Lin, Q., Dengt, L., Sun, J., Chen, M.: Optimal demand-aware ride-sharing routing. In: IEEE INFOCOM 2018-IEEE Conference on Computer Communications, pp. 2699–2707. IEEE (2018)
7. Alonso-Mora, J., Samaranayake, S., Wallar, A., Frazzoli, E., Rus, D.: On-demand high-capacity ride-sharing via dynamic trip-vehicle assignment. Proc. Nat. Acad. Sci. **114**(3), 462–467 (2017)
8. Tong, Y., She, J., Ding, B., Chen, L., Wo, T., Xu, K.: Online minimum matching in real-time spatial data: experiments and analysis. Proc. VLDB Endow. **9**(12), 1053–1064 (2016)
9. Huang, T., Fang, B., Bei, X., Fang, F.: Dynamic trip-vehicle dispatch with scheduled and on-demand requests. In: Conference on Uncertainty in Artificial Intelligence (2019)
10. Lesmana, N. S., Zhang, X., Bei, X.: Balancing efficiency and fairness in on-demand ride-sourcing. In: Advances in Neural Information Processing Systems, pp. 5309–5319 (2019)
11. Lowalekar, M., Varakantham, P., Jaillet, P.: ZAC: a zone path construction approach for effective real-time ridesharing. In: Proceedings of the International Conference on Automated Planning and Scheduling, vol. 29, no. 1, pp. 528–538 (2019)
12. Curry, M., Dickerson, J.P., Sankararaman, K.A., Srinivasan, A., Wan, Y., Xu, P.: Mix and match: Markov chains and mixing times for matching in rideshare. In: Caragiannis, I., Mirrokni, V., Nikolova, E. (eds.) WINE 2019. LNCS, vol. 11920, pp. 129–141. Springer, Cham (2019). https://doi.org/10.1007/978-3-030-35389-6_10
13. Huang, Z., Kang, N., Tang, Z.G., Wu, X., Zhang, Y., Zhu, X.: Fully online matching. J. ACM (JACM) **67**(3), 1–25 (2020)
14. Karp, R.M., Vazirani, U.V., Vazirani, V.V.: An optimal algorithm for on-line bipartite matching. In: Proceedings of the Twenty-Second Annual ACM Symposium on Theory of Computing, pp. 352–358 (1990)
15. Lowalekar, M., Varakantham, P., Jaillet, P.: Competitive ratios for online multi-capacity ridesharing. In: Proceedings of the 19th International Conference on Autonomous Agents and MultiAgent Systems, pp. 771–779 (2020)

Minimum Wireless Charger Placement
with Individual Energy Requirement

Xingjian Ding[1], Jianxiong Guo[2], Deying Li[1(✉)], and Ding-Zhu Du[2]

[1] School of Information, Renmin University of China, Beijing 100872, China
{dxj,deyingli}@ruc.edu.cn
[2] Department of Computer Science, University of Texas at Dallas,
Richardson, TX 75080, USA
{jianxiong.guo,dzdu}@utdallas.edu

Abstract. Supply energy to battery-powered sensor devices by deploying wireless chargers is a promising way to prolong the operation time of wireless sensor networks, and has attracted much attention recently. Existing works focus on maximizing the total received charging power of the network. However, this may face the unbalanced energy allocation problem, which is not beneficial to prolong the operation time of wireless sensor networks. In this paper, we consider the individual energy requirement of each sensor node, and study the problem of minimum charger placement. That is, we focus on finding a strategy for placing wireless chargers from a given candidate location set, such that each sensor node's energy requirement can be met, meanwhile the total number of used chargers can be minimized. We show that the problem to be solved is NP-hard, and present two approximation algorithms which are based on the greedy scheme and relax rounding scheme, respectively. We prove that both of the two algorithms have performance guarantees. Finally, we validate the performance of our algorithms by performing extensive numerical simulations. Simulation results show the effectiveness of our proposed algorithms.

Keywords: Wireless charger placement · Wireless sensor network · Individual energy requirement

1 Introduction

Over the past ten years, there is a growing interesting of using Wireless Sensor Networks (WSNs) to collect data from the real world. A WSN system mainly consists of lots of sensor nodes that are powered by on-board batteries. Due to the inherent constraints on the battery technology, these on-board batteries can only provide limited energy capacity, and thus it limits the operating time of the wireless sensor networks. To achieve perpetual operation of the network

This work is supported by National Natural Science Foundation of China (Grant NO. 11671400, 12071478), and partially by NSF 1907472.

W. Wu and Z. Zhang (Eds.): COCOA 2020, LNCS 12577, pp. 697–710, 2020.
https://doi.org/10.1007/978-3-030-64843-5_47

system, prolonging the operation time of these battery-powered sensor nodes has been an important task. The great progress in Wireless Power Transfer (WPT) based on magnetic resonant coupling [8] bring a novel way to replenish the batteries of wireless sensor networks. Prolonging the operation time of sensor nodes by using WPT has many advantages. Such as, WPT is insensitive to external environments and can provide relatively stable energy supply for each sensor node; the charging power to sensor nodes is controllable and thus can be flexibly adjusted according to the energy requirement of the wireless sensor network; WPT provides an efficient way to charge sensor nodes without any interconnecting conductors. As a promising way to deliver energy to wireless sensor networks, the study of deploying wireless chargers has attracted significant attention in a few years.

In previous studies, researchers mainly concentrate on charging utility maximization problem [3,4,15,18]. That is, their object is to get a maximum total charging power of a given sensor network under some certain constraints, such as the number of used chargers or the overall working power of deployed chargers. These studies may face the unbalanced energy allocation problem. For example, with the object of maximizing the total charging power of the network, there might be some sensor nodes receive lots of wireless power, but others receive rare or even no wireless power, which is not beneficial to prolong the lifetime of the wireless network. Moreover, there are some studies focus on studying how to efficient place wireless chargers so that a sensor device deployed anywhere in the network area always can receive enough energy [6]. However, these works didn't consider the individual energy requirement of sensor nodes. In a real wireless sensor network application, the energy consumption rates of different sensor nodes are significantly different [16]. On the one hand, sensor nodes may perform different sensing tasks that require different energy support. On the other hand, those sensor nodes that are around the base station need to forward data for remote nodes and thus have much higher energy consumption rates than others.

In contrast to existing works, we consider a more practice scenario that sensor nodes have different energy requirement. Our object is to get a charger placement strategy with minimum number of used chargers so that every sensor node in a network area can receive enough wireless power to meet its energy requirement. The main contributions of this paper are as follows.

- In this paper, we consider the individual energy requirement of each sensor node in a WSN, and study the problem of minimum charger placement with individual energy requirement (problems PIO). We prove the problem to be solved is NP-hard.
- We propose two approximation algorithms for the PIO problem, which are based on greedy and relax rounding, respectively. Moreover, we give detail theoretical performance analysis of the two algorithms.
- We validate the performance of the proposed algorithms by performing lots of numerical simulations. The results show the effectiveness of our designs.

The rest of this paper is organized as follows. Section 2 introduces the state-of-art work of this paper. Section 3 introduces the definition the problem to be solved. Sections 4 describes the two proposed algorithms for the PIO problem. Section 5 validates our algorithms through numerical simulations. Finally, Section 6 draws the conclusion of this paper.

2   Related Works

In the past few years, replenish energy to sensor networks by placing wireless chargers has been widely studied. There are lots of existing works that related to ours. Some works consider the charging utility maximization problem from different aspects. Dai et al. [4] focus on maximizing charging utility under the directional charging model, they aim to get a charger placement strategy for a certain number of directional chargers so that the overall charging utility of the network can be maximized. Yu et al. [18] consider that wireless chargers could communicate with each other, and they address the connected wireless charger placement problem, that is, they aim to place a certain number of wireless chargers into a WSN area to maximize the total charging utility under the constraint that the deployed wireless chargers are connected. Wang et al. [15] first deal with the problem of heterogeneous charger placement under directional charging model with obstacles. They aim to efficiently deploy a set of heterogeneous wireless chargers such that the total charging utility can be maximized while considering the effect of obstacles in the network area.

Different from the above works, some researchers focus on making sure that each sensor node could get sufficient power to achieve perpetual operation. Li et al. [9] investigate how to efficiently deploy wireless chargers to charge wearable devices, and they aim to place wireless chargers in a 2-D area with minimum cardinality to guarantee that the power non-outage probability of the wearable device is not smaller than a given threshold. The work [10] is most similar to ours, in which their object is to get a charger placement strategy with the minimum cardinality to make sure that each sensor node can receive enough energy. The key difference between this work and ours lies on we consider the individual energy requirement of sensor nodes, other than with the assumption that the same energy requirement of each sensor node is the same. Moreover, our proposed algorithms have approximation ratios which guarantee the performance in theory.

3   System Model and Problem Formulation

3.1   System Model and Assumptions

We consider a wireless sensor network that contains m rechargeable sensor nodes denoted by $\mathcal{S} = \{s_1, s_2, \ldots, s_m\}$. These sensor nodes are deployed in a limited 2-D area randomly, and their locations are fixed and known in advance. There are n candidate locations in the network area which are chosen for placing wireless chargers. The candidate locations are chosen by end-users and at most one wireless charger can be placed at each candidate location. The set of candidate

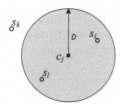

Fig. 1. Omni charging model

locations is denoted by by $\mathcal{C} = \{c_1, c_2, \ldots, c_n\}$. With a little abuse of notations, the wireless charger placed at the i-th location is also denoted by c_i.

As shown in Fig. 1, in this paper, we consider problem to be addressed under omnidirectional charging model. In the omnidirectional charging model, both rechargeable sensor nodes and wireless chargers are equipped with omni antennas. Each charger symmetrically radiates its wireless power and shape a disk charging area centering the charger. A sensor node can receive the wireless power from any direction, it can be charged by a wireless charger as long as it located within the charging area of the charger. In practice, the wireless power decays with distance increases, and thus each wireless charger has a bounded charging area. We consider the scenario that all the wireless chargers of end-users are homogeneous, and assume that each charger can only charge sensor nodes within the range of D. Next we describe the energy transfer model and explain the way to calculate the receiving power of a sensor node from wireless chargers.

Based on the Friis's free space equation [2], the receiving radio frequency (RF) power P_r of a receiver from a transmitter can be calculated as

$$P_r = G_t G_r \left(\frac{\lambda}{4\pi d}\right)^2 P_t, \tag{1}$$

where G_t and G_r are antenna gains of transmitter and receiver, respectively, d is the line-of-sight distance between transmitter and receiver, λ is the electromagnetic wavelength, and P_t is the transmitting RF power of the transmitter.

The Friis's free space function is used for far-field wireless power transmission such as satellite communication. For wireless rechargeable sensor networks, the polarization loss should be considered. Based on this, He et al. [6] use a more empirical model to formulate the wireless charging process in wireless rechargeable sensor systems:

$$P_r = \frac{G_t G_r \eta}{L_p} \left(\frac{\lambda}{4\pi(d + \beta)}\right)^2 P_t, \tag{2}$$

where η is the rectifier efficiency, L_p is the polarization loss, and β is a parameter to make Friis's free space equation suitable for short distance wireless power transmission.

As mentioned before, the wireless power decays with distance increases, it's difficult for a receiver which is very far from the transmitter to capture the wireless power as the RF signal is very weak. The symbol D is used to represent the bound distance, that is, if $d > D$, $P_r = 0$. Therefore, in our omnidirectional charging model, we use the following function to evaluate the RF power at sensor node s_i receiving from charger c_j:

$$P_{rx}(s_i, c_j) = \begin{cases} \dfrac{\alpha P_{tx}(c_j)}{(\|s_i - c_j\| + \beta)^2}, & \|s_i - c_j\| \leq D \\ 0 & \text{otherwise,} \end{cases} \quad (3)$$

where $\alpha = \frac{G_t G_r \eta}{L_p} \left(\frac{\lambda}{4\pi}\right)^2$, $P_{tx}(c_j)$ is the antenna power of the charger c_j, $\|s_i - c_j\|$ is the line-of-sight distance between s_i and c_j, and β is an empirically-determined constant determined by the hardware parameters of chargers and the surroundings.

To replenish energy to the batteries, receivers need to convert the RF energy to electric energy. In practical applications, the conversion efficiency from RF to electricity is non-linear [11]. We denote the electric power got by s_i from c_j as $P_{in}(s_i, c_j)$, and use ξ to denote the conversion efficiency, where ξ is related to the receiving RF power $P_{rx}(s_i, c_j)$, and is calculated as $\xi = f(P_{rx}(s_i, c_j))$. Then $P_{in}(s_i, c_j) = \xi P_{rx}(s_i, c_j) = f(P_{rx}(s_i, c_j)) P_{rx}(s_i, c_j)$. This function also can be expressed as $P_{in}(s_i, c_j) = g(P_{rx}(s_i, c_j))$, where $g(\cdot)$ is a non-linear function. In this work, we use the 2nd order polynomial model proposed by [17], we have

$$P_{in}(s_i, c_j) = \mu_1 (P_{rx}(s_i, c_j))^2 + \mu_2 P_{rx}(s_i, c_j) + \mu_3, \quad (4)$$

where $\mu_1, \mu_2, \mu_3 \in \mathbb{R}$ are the empirically-determined parameters.

We assume that all of the used chargers are homogeneous, and the transmitting RF power of each charger is P_{tx}. According to the above charging model, the minimum RF power that a sensor node receives from a charger can be calculated by $P_{rx}^{min} = \frac{\alpha P_{tx}}{(D+\beta)^2}$. Correspondingly, the minimum electric energy that a sensor node got from a charger is estimated as $P_{in}^{min} = g(P_{rx}^{min})$. In energy-harvesting sensor systems, each sensor node needs to manage its electric energy for achieving a long-term operation [7]. In order to make decision efficiently, the sensor node use integers rather than reals to evaluate its energy. In this paper, therefore, we use the *charging levels* present in [5] to evaluate the actual charging power of a sensor node in a discretized way. The charging levels of a sensor node s_i received from a charger c_j can be calculated as follows.

$$L(s_i, c_j) = \left\lfloor \frac{P_{in}(s_i, c_j)}{P_{rx}^{min}} \right\rfloor. \quad (5)$$

According to [4,14], we can use multiple chargers to charge a sensor node simultaneously, and the charging power of the sensor node got from these chargers is accumulative. Thus we measure the charging levels of a sensor node from multiple chargers as the summation of the charging levels provided by each charger. Limited by the hardware of the sensor nodes, the charging levels of a sensor node is bounded. We use $\mathcal{L}_{th}$ to denote the bounded charging levels of a sensor node. Then the charging levels of a sensor node s_i obtained from a given charger set C is formulated as

$$\mathcal{L}(s_i, C) = \begin{cases} \sum_{c_j \in C} L(s_i, c_j), & \text{if } \sum_{c_j \in C} L(s_i, c_j) \leq \mathcal{L}_{th} \\ \mathcal{L}_{th}, & \text{otherwise.} \end{cases} \tag{6}$$

3.2 Problem Formulation

In real wireless rechargeable sensor networks, sensor nodes may have different energy requirement, as these sensor nodes may execute different tasks, besides, the consumed energy for forwarding data is also variant. Therefore, in this work, we study the strategy to place wireless chargers to meet the energy requirement of every sensor node while the total number of used chargers can be minimized. The problem to be addressed under omnidirectional charging model is formulated as follows.

Problem 1. minimum charger Placement with Individual energy requirement under Omnidirectional charging model (PIO). *Given m rechargeable sensor nodes* $S = \{s_1, s_2, \ldots, s_m\}$, *and n pre-determined candidate locations* $C = \{c_1, c_2, \ldots, c_n\}$. *The charging levels requirement for each sensor node* $s_i \in S$ *is* $\alpha_i \leq \mathcal{L}_{th}$. *Our object is to find a subset* $C \subseteq C$ *with minimum cardinality to place wireless chargers, such that every sensor node meets its charging levels requirement.*

Formally, the PIO problem can be present as

$$\mathbf{min} \quad |C|$$
$$\mathbf{s.t.} \quad \mathcal{L}(s_i, C) \geq \alpha_i, \forall s_i \in S$$

In our study, we assume that the PIO problem always has feasible solutions, as end users will determine sufficient candidate locations to provide enough wireless power to wireless networks.

Next, we will prove that problem PIO is NP-hard through a theorem. The following introduced problems are helpful for our proof.

The Set Cover Problem (SC): Given a set S' and a collection C' of the subset of S', assume that $\cup_{c'_j \in C'} c'_j = S'$, the SC problem is to find a sub-collection $C' \subseteq C'$ with minimum cardinality such that $\cup_{c'_j \in C'} c'_j = S'$.

The Decision Version of the SC Problem (d-SC): For a given integer k, whether there is a sub-collection $C' \subseteq C'$ so that $\cup_{c'_j \in C'} c'_j = S'$ and $|C'| \leq k$?

The Decision Version of the PIO Problem (d-PIO): For a given integer l, whether there is a location subset $C \subseteq C$, such that every sensor's charging levels requirement can be met and $|C| \leq l$ if we place wireless chargers on C?

Theorem 1. *The PIO problem is NP-hard.*

Proof. We prove the theorem by reduction, where we reduce the well-known SC problem to PIO. Consider such an instance of d-SC: given an integer k, a set $\mathcal{S}' = \{s'_1, s'_2, \ldots, s'_m\}$, and a collection $\mathcal{C}' = \{c'_1, c'_2, \ldots, c'_n\}$, where $c'_j \subseteq \mathcal{S}'$ for any $c'_j \in \mathcal{C}'$. Next, we construct an instance d-PIO as follows. For each $s'_i \in \mathcal{S}'$, we generate a rechargeable sensor node s_i. We also generate a virtual candidate location c_j for each $c'_j \in \mathcal{C}'$, where the distance between c_j and s_i is less than the charging range D only when $s'_i \in c'_j$. Besides, we set the charging levels requirement α_i for each sensor s_i to be 1, and let $l = k$.

Obviously, this reduction will terminated in polynomial time, and we can get a "yes" answer from the generated instance of problem d-PIO if and only if the given instance of problem d-SC has a "yes" answer. As the SC problem is a well-known NP-complete problem [13], we know that the PIO problem is at least NP-hard.

4   Algorithms for the PIO Problem

In this section, we design two algorithms with performance guarantees for problem PIO: one is a greedy algorithm, named gPIO; another one is based on relax and rounding, named rPIO. In the following, we will describe our algorithms in detail, and given theoretical performance analysis of the two algorithms, respectively.

4.1   The Greedy Based Algorithm

Algorithm Description. We first introduce some useful concepts for making the description of algorithm gPIO more clearly. As each sensor node only needs to meet its charging levels requirement, given a set of placed wireless chargers C, we define the *useful charging levels* of a sensor s_i as $\mathcal{L}^U(s_i, C) = \min\{\mathcal{L}(s_i, C), \alpha_i\}$. The overall useful charging levels provided by charger set C for the whole network is calculated as $\mathcal{L}^U(C) = \sum_{s_i \in \mathcal{S}} \mathcal{L}^U(s_i, C)$. Clearly, $\mathcal{L}^U(\emptyset) = 0$. Consider a location set C which has been deployed with wireless chargers, the *marginal increment* about total useful charging levels is the difference between $\mathcal{L}^U(C \cup \{c_i\})$ and $\mathcal{L}^U(C)$, when a candidate location c_i is selected for placing a wireless charger.

The basic idea of algorithm gPIO is as follows. In each step, the candidate location which brings maximum marginal increment of overall useful charging levels will be selected to place wireless charger. After a candidate location is selected to be placed a charger, algorithm gPIO will update the overall useful charging levels. gPIO terminates after every sensor's charging levels requirement is achieved. Algorithm 1 shows the details of algorithm gPIO.

In the following, we give the analysis of the time complexity of gPIO. The calculation of $\mathcal{L}^U(C \cup \{c_i\})$ costs $\mathcal{O}(mn)$ time, where m and n are the number of sensor nodes and candidate locations, respectively. In each iteration of the while loop, every candidate location in $\mathcal{C} \setminus C$ needs to be checked to find the "best" one. Therefore, it costs $\mathcal{O}(mn^2)$ time for each iteration of the while loop. It's easy

---

Algorithm 1. The greedy algorithm for PIO (gPIO)

Input: $\mathcal{S}, \mathcal{C}$, and α_i for each sensor $s_i \in \mathcal{S}$
Output: a subset of candidate locations C
 1: $C \leftarrow \emptyset$
 2: **while** $\mathcal{L}^U(s_i, C) < \alpha_i, \exists s_i \in \mathcal{S}$ **do**
 3:      choose $c_i \in \mathcal{C} \setminus C$ that maximizes $\mathcal{L}^U(C \cup \{c_i\}) - \mathcal{L}^U(C)$, and break tie arbitrarily;
 4:      $C \leftarrow C \cup \{c_i\}$;
 5: **end while**
 6: **return** C

---

to know that if the feasible solution exists, the gPIO algorithm must terminate within n iterations after it scanned all the candidate locations. In summary, the time complexity of algorithm gPIO is $\mathcal{O}(mn^3)$.

Performance Analysis. We analyze the approximation ratio of algorithm gPIO through the following theorem.

Theorem 2. gPIO *is a* $(1 + \ln \gamma)$-*approximation algorithm for the PIO problem, where* $\gamma = \max_{c_i \in \mathcal{C}} L^U(\{c_i\})$.

Proof. We assume the solution found by algorithm gPIO contains g candidate locations, and we renumber the locations in the order of their selection into the solution, i.e., $C = \{c_1, c_2, \ldots, c_g\}$. We use C_i to denote the location set get by gPIO after the i-th iteration, where $i = 0, 1, \ldots, g$, i.e., $C_i = \{c_1, c_2, \ldots, c_i\}$ and $C_0 = \emptyset$. Denote the optimal solution by C^*, and assume there are t number of candidate locations in C^*. The PIO problem is to get a minimum set of candidate locations to place wireless chargers such that every sensor node meets its charging levels requirement, according to the definition of *useful charging levels*, the PIO problem also can be described as to get a minimum set of candidate locations so that the overall useful charging levels equals to $\sum_{s_i \in \mathcal{S}} \alpha_i$. Obviously, we have $\mathcal{L}^U(C) = \mathcal{L}^U(C^*) = \sum_{s_i \in \mathcal{S}} \alpha_i$.

We use $\mathcal{L}^{-U}(C_i)$ to represent the difference of useful charging levels between $\mathcal{L}^U(C_i)$ and $\mathcal{L}^U(C^*)$, i.e., $\mathcal{L}^{-U}(C_i) = \mathcal{L}^U(C^*) - \mathcal{L}^U(C_i)$. In other words, after the i-th iteration of gPIO, there still need $\mathcal{L}^{-U}(C_i)$ useful charging levels to meet the charging levels requirement of every sensor node.

As the optimal solution contains t candidate locations, it's easy to know that given a location set C_i, there exists a set with no more than t locations in $\mathcal{C} \setminus C_i$ that can provide $\mathcal{L}^{-U}(C_i)$ useful charging levels for the network. By the pigeonhole principle, there must exist a location $c_j \in \mathcal{C} \setminus C_i$ that provides at least $\frac{\mathcal{L}^{-U}(C_i)}{t}$ marginal increment of useful charging levels. According to the greedy criterion of algorithm gPIO, in each step, we select the location with maximum marginal increase of overall useful charging levels. Therefore, we have

$$\mathcal{L}^U(C_{i+1}) - \mathcal{L}^U(C_i) \geq \frac{\mathcal{L}^{-U}(C_i)}{t} = \frac{\mathcal{L}^U(C^*) - \mathcal{L}^U(C_i)}{t}. \tag{7}$$

Equivalently, we get

$$\mathcal{L}^U(C^*) - \mathcal{L}^U(C_{i+1}) \leq (\mathcal{L}^U(C^*) - \mathcal{L}^U(C_i)) \cdot \left(1 - \frac{1}{t}\right). \tag{8}$$

By induction, we have

$$\begin{aligned} \mathcal{L}^{-U}(C_i) &= \mathcal{L}^U(C^*) - \mathcal{L}^U(C_i) \\ &\leq \mathcal{L}^U(C^*) \cdot \left(1 - \frac{1}{t}\right)^i \leq \mathcal{L}^U(C^*) \cdot e^{-\frac{i}{t}}. \end{aligned} \tag{9}$$

In each iteration, $\mathcal{L}^{-U}(C_i)$ decreases from $\mathcal{L}^U(C^*)$ to 0, so we can always find an positive integer $i \leq g$ such that $\mathcal{L}^{-U}(C_{i+1}) < t \leq \mathcal{L}^{-U}(C_i)$. Each location selected by algorithm gPIO provides at least 1 useful charging levels. Thus we can conclude that after $(i+1)$-th iterations, the gPIO algorithm will terminate after at most $t - 1$ more iterations (i.e., selects at most $t - 1$ more candidate locations). Therefore, we get $g \leq i + t$. As $t \leq \mathcal{L}^{-U}(C_i) \leq \mathcal{L}^U(C^*) \cdot e^{-\frac{i}{t}}$, we have $i \leq t \cdot \ln\left(\frac{\mathcal{L}^U(C^*)}{t}\right) \leq t \cdot \ln\gamma$. Thus we have

$$g \leq i + t \leq t(1 + \ln\gamma). \tag{10}$$

Thus the theorem holds.

4.2 The Relax Rounding Algorithm

Algorithm Description. We first rewrite the PIO problem as an integer linear program problem. We use n variables $x_1, x_2, \ldots, x_n$ to be indicators to denote whether the candidate locations are selected to be placed with wireless chargers. If a location is selected, then $x_j = 1$ and $x_j = 0$ otherwise, for $1 \leq j \leq n$. Then problem PIO can be rewritten as

$$\min \quad \sum_{c_j \in C} x_j$$

$$s.t. \quad \sum_{c_j \in C} x_j \cdot L(s_i, c_j) \geq \alpha_i, \quad \forall s_i \in S \tag{11}$$

$$x_j \in \{0, 1\}, \quad 1 \leq j \leq n.$$

By relaxing the constraints of $x_j \in \{0, 1\}$ to the constraints of $0 \leq x_j \leq 1$, for $1 \leq j \leq n$, the integer linear program is transformed into a linear program:

$$\min \quad \sum_{c_j \in C} x_j$$

$$s.t. \quad \sum_{c_j \in C} x_j \cdot L(s_i, c_j) \geq \alpha_i, \quad \forall s_i \in S \tag{12}$$

$$0 \leq x_j \leq 1, \quad 1 \leq j \leq n.$$

Solving the linear program problem (12) and we get its optimal solution. To get a a feasible solution of (11), i.e., a feasible solution of PIO, we need to rounding the optimal solution of (12) to integers. Next, we will show the details of the rounding process.

Denote the optimal solution of problem (12) by $X^* = \{x_1^*, x_2^*, \ldots, x_n^*\}$, then we sort the elements in X^* in descending order of the value of x_j^* and renumber them, that is, let $x_1^* \geq x_2^* \geq \cdots \geq x_n^*$. Note that we also renumber the candidate location set according to X^* such that x_j^* indicates whether location c_j is selected to be placed with a charger. We denote a solution of problem (11) as $X_A = \{x_1^A, x_2^A, \ldots, x_n^A\}$. In the beginning, we let $x_j^A = 0$ for $1 \leq j \leq n$, and then we make X_A feasible for problem (11) through iterative operations. In the j-th operation, we let $x_j^A = 1$. The iteration terminates until X_A be a feasible solution for the integer linear program problem (11), that is, every sensor node meets its charging levels requirement. We show the details of rPIO in Algorithm 2.

Algorithm 2. The relax rounding algorithm for PIO (rPIO)

Input: $\mathcal{S}, \mathcal{C}$, and α_i for each sensor $s_i \in \mathcal{S}$
Output: a feasible solution X_A for problem (11)
1: Calculate $L(s_i, c_j)$ for each $s_i \in \mathcal{S}$ and $c_j \in \mathcal{C}$, and then convert the PIO problem to an integer linear program as shown in (11);
2: Relax problem (11) and construct a corresponding linear program (12);
3: Solve the linear program (12) and get an optimal solution X^*;
4: Sort the elements in X^* in descending order of the value of x_j^*, and renumber them such that $x_1^* \geq x_2^* \geq \cdots \geq x_n^*$;
5: Renumber candidate locations such that x_j^* indicates whether location c_j is selected;
6: $X_A = \{x_1^A, x_2^A, \ldots, x_n^A\} = \{0, 0, \ldots, 0\}$;
7: $k = 1$;
8: **while** $\sum_{c_j \in \mathcal{C}} x_j \cdot L(s_i, c_j) < \alpha_i, \exists s_i \in \mathcal{S}$ **do**
9:     $x_k^A = 1$;
10:     $k = k + 1$;
11: **end while**
12: **return** X_A

Next, we give the analysis of the time complexity of algorithm rPIO. It costs $\mathcal{O}(mn)$ time to get $L(s_i, c_j)$ for each $s_i \in \mathcal{S}$ and $c_j \in \mathcal{C}$ in the first line in algorithm rPIO. Convert problem PIO to an integer linear program and then relax it to a corresponding linear program takes $\mathcal{O}(1)$ time. Solving the linear program costs $\mathcal{O}(n^{2.5}L)$ according to [12], where L is the number of bits in the input. Sort the elements in X^* takes $\mathcal{O}(n \log n)$ time by using the Quicksort method. $L(s_i, c_j)$ has been calculated for each $s_i \in \mathcal{S}$ and $c_j \in \mathcal{C}$ in the first line, so the judgement of the while loop costs $\mathcal{O}(m)$ time. The while loop contains at most n iterations, and thus the while loop costs $\mathcal{O}(mn)$ time. To sum up, the time complexity of algorithm rPIO is $\mathcal{O}(mn) + \mathcal{O}(n^{2.5}L) + \mathcal{O}(n \log n) + \mathcal{O}(mn)$, that is $\mathcal{O}(mn + n^{2.5}L)$ time.

Performance Analysis. We use N_i to represent the summation of charging levels of sensor s_i provided by every candidate location, that is, $N_i = \sum_{c_j \in \mathcal{C}} L(s_i, c_j)$.
Then we have the following theorem.

Theorem 3. rPIO *is a δ-approximation algorithm for the PIO problem, where*
$\delta = \max_{s_i \in S}\{N_i - \alpha_i + 1\}$.

Proof. To prove the theorem, we need first prove that for any $x_j^A \in X_A$, if $x_j^A = 1$, then $x_j^* \geq \frac{1}{\delta}$. We prove this condition by contradiction. We first divide the candidate location set into two parts according to the values of elements in X^*. For any $c_j \in \mathcal{C}$, we put location c_j into C^+ if $x_j^* \geq \frac{1}{\delta}$, otherwise, we put location c_j into C^-, that is, $C^+ = \{c_j | x_j^* \geq \frac{1}{\delta}\}$ and $C^- = \{c_j | x_j^* < \frac{1}{\delta}\}$.

Assume that there exists an indicator $x_k^A \in X_A$ where $x_k^A = 1$ but $x_k^* < \frac{1}{\delta}$. According to condition of the while loop in algorithm rPIO, there must exist a sensor $s_i \in S$ such that $\sum_{j=1}^{k-1} x_j \cdot L(s_i, c_j) < \alpha_i$, otherwise, algorithm rPIO will terminates in $k-1$ iterations, and then $x_k^A = 0$. We can easy know that $|C^+| \leq k - 1$, as $x_k^* < \frac{1}{\delta}$ and the elements in X^* have been sorted in the descending order of the value of each element. In other words, the candidate location set C^+ cannot provide enough charging levels for sensor s_i. We use L_i^+ to denote the summation of the charging levels of s_i provided by each location in C^+, i.e., $L_i^+ = \sum_{c_j \in C^+} L(s_i, c_j) < \alpha_i$. Then the summation of the charging levels of s_i provided by each location in C^- can be calculated by $N_i - L_i^+$.

$$\sum_{c_j \in \mathcal{C}} x_j^* \cdot L(s_i, c_j) = \sum_{c_j \in C^+} x_j^* \cdot L(s_i, c_j) + \sum_{c_j \in C^-} x_j^* \cdot L(s_i, c_j)$$

$$< \sum_{c_j \in C^+} 1 \cdot L(s_i, c_j) + \sum_{c_j \in C^-} \frac{1}{\delta} \cdot L(s_i, c_j) \qquad (13)$$

$$= L_i^+ + (N_i - L_i^+) \cdot \frac{1}{\delta}$$

As L_i^+ is an positive integer, and $L_i^+ < \alpha_i$, so we know that $L_i^+ \leq \alpha_i - 1$. We only consider the case that problem PIO has feasible solutions, it's easy to know that $\delta \geq 1$, and then $L_i^+ + (N_i - L_i^+) \cdot \frac{1}{\delta}$ hits its maximum value when $L_i^+ = \alpha_i - 1$. Therefore, we have

$$\sum_{c_j \in \mathcal{C}} x_j^* \cdot L(s_i, c_j) < L_i^+ + (N_i - L_i^+) \cdot \frac{1}{\delta}$$

$$\qquad (14)$$

$$\leq \alpha_i - 1 + (N_i - \alpha_i + 1) \cdot \frac{1}{\delta} \leq \alpha_i.$$

Inequation (14) contradicts the fact that X^* is a feasible solution for linear program (12), i.e., it violates the condition that $\sum_{c_j \in \mathcal{C}} x_j^* \cdot L(s_i, c_j) \geq \alpha_i, \forall s_i \in S$.

Hence, we now have prove that for any $x_j^A \in X_A$, if $x_j^A = 1$, then $x_j^* \geq \frac{1}{\delta}$. Then we have the following inequation,

$$\sum_{c_j \in \mathcal{C}} x_j^A \leq \delta \cdot \sum_{c_j \in \mathcal{C}} x_j^* \tag{15}$$

We complete the proof here.

5   Performance Evaluation

We assume that there is a wireless sensor network involves 200 rechargeable sensor nodes that are randomly deployed in a 400 m × 400 m square area, and each site of a sensor node is selected as a candidate location. We set the working RF power P_{tx} of each wireless charger to be 10^6 μW. The parameters α and β are set to be 2.5 and 15, respectively. The charging distance D is set to be 70 m. For the non-linear energy conversion, according to the data measured in [1], we set $\mu_1 = -0.00001$, $\mu_2 = 0.57$ and $\mu_3 = 10$. The required charging levels of each sensor node is randomly selected in $[10, 20]$. The data points plotted in this section under different settings are the average of 100 runs.

5.1   Performance Comparison

We implement a random algorithm, named random, as the baseline for problem PIO. Specifically, algorithm random repeatedly selects a candidate location in a random way to place an omnidirectional wireless charger until every sensor node's charging levels requirement is met. Next, we compare our algorithms with the baseline with different parameters.

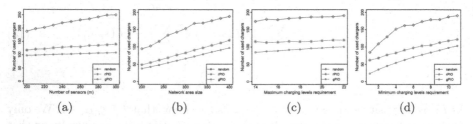

Fig. 2. Performance comparisons between our algorithms (gPIO and rPIO) and random in omnidirectional charging.

(1) Effect of the number of sensor nodes (m): Figure 2(a) shows the effect of the number of sensor nodes on the performance of our algorithms and the baseline. We can see that with the number of sensor nodes increases, all of the three algorithms will require more wireless chargers. However, our algorithms gPIO and gPIO always outperforms the random algorithm. More specifically, the growth rates of our algorithms are lower than that of the baseline algorithm. We can also see that algorithm gPIO is better than algorithm rPIO, which implies that the greedy algorithm is a simple but effective method to deal with the PIO problem.

(2) Effect of the number of the network area size: Figure 2(b) shows the effect of the network area size on the number of used wireless chargers. We keep the number of sensor nodes to be 200, and set the side length of the square area from 200 m to 400 m. We can see that, with the network area becomes larger, all of the algorithms will need more wireless chargers to meet the sensor nodes' charging levels requirements, and our algorithms always outperform the baselines.

(3) Effect of the charging levels requirements of sensor nodes: To evaluate the effect of the charging levels requirements of sensor nodes on the number of used wireless chargers, we design two different experiments. One is set the lower bound of the charging levels requirements to be 10, and range the upper bound of the charging levels requirements from 14 to 22, as shown in Fig. 2(c). It can be seen that with the upper bound of the charging levels requirements increases, the number of used wireless chargers slightly increases for all algorithms. In another set of experiments, we keep the upper bound of the charging levels requirements always be two times of the lower bound, and range the lower bound from 1 to 11, as shown in Fig. 2(d). We can see that the performance of our algorithms is always batter than the baseline, especially when the lower bound of charging levels requirement is small, for example, when the lower bound is et to be 1, the number of wireless chargers required by algorithm gPIO is only 27.6% of algorithm random, and 37.6% of algorithm rPIO.

6   Conclusions

In this study, we investigate the minimum wireless charger placement problem by considering individual energy requirement. We consider the problem under the omnidirectional charging model. We present two algorithms with performance guarantees for problem PIO. In addition, we give detail theoretical performance analysis of the two proposed algorithms. We perform lots of numerical simulations to validate the performance of our algorithms, simulation results show that our designs perform better than the baseline. The study of this problem under directional charging model will be our future work.

References

1. Powercast Corporation: P2110b Module Datasheet (2016). https://www.powercastco.com/documentation/p2110b-module-datasheet/. Accessed 20 Jan 2020
2. Balanis, C.A.: Antenna Theory: Analysis and Design. Wiley, Hoboken (2016)
3. Dai, H., et al.: Scape: safe charging with adjustable power. IEEE/ACM Trans. Netw. **26**(1), 520–533 (2018)
4. Dai, H., Wang, X., Liu, A.X., Ma, H., Chen, G.: Optimizing wireless charger placement for directional charging. In: IEEE INFOCOM 2017-IEEE Conference on Computer Communications, pp. 1–9. IEEE (2017)
5. Ding, X., et al.: Optimal charger placement for wireless power transfer. Comput. Netw. **170**, 107123 (2020)

6. He, S., Chen, J., Jiang, F., Yau, D.K., Xing, G., Sun, Y.: Energy provisioning in wireless rechargeable sensor networks. IEEE Trans. Mob. Comput. **12**(10), 1931–1942 (2012)
7. Ku, M.L., Li, W., Chen, Y., Liu, K.R.: Advances in energy harvesting communications: past, present, and future challenges. IEEE Commun. Surv. Tutor. **18**(2), 1384–1412 (2015)
8. Kurs, A., Karalis, A., Moffatt, R., Joannopoulos, J.D., Fisher, P., Soljačić, M.: Wireless power transfer via strongly coupled magnetic resonances. Science **317**(5834), 83–86 (2007)
9. Li, Y., Chen, Y., Chen, C.S., Wang, Z., Zhu, Y.: Charging while moving: deploying wireless chargers for powering wearable devices. IEEE Trans. Veh. Technol. **67**(12), 11575–11586 (2018)
10. Li, Y., Fu, L., Chen, M., Chi, K., Zhu, Y.: RF-based charger placement for duty cycle guarantee in battery-free sensor networks. IEEE Commun. Lett. **19**(10), 1802–1805 (2015)
11. Ozçelikkale, A., Koseoglu, M., Srivastava, M.: Optimization vs. reinforcement learning for wirelessly powered sensor networks. In: 2018 IEEE 19th International Workshop on Signal Processing Advances in Wireless Communications (SPAWC), pp. 1–5. IEEE (2018)
12. Vaidya, P.M.: Speeding-up linear programming using fast matrix multiplication. In: 30th Annual Symposium on Foundations of Computer Science, pp. 332–337. IEEE (1989)
13. Vazirani, V.V.: Approximation Algorithms. Springer, Heidelberg (2013). https://doi.org/10.1007/978-3-662-04565-7
14. Wang, X., Dai, H., Huang, H., Liu, Y., Chen, G., Dou, W.: Robust scheduling for wireless charger networks. In: IEEE INFOCOM 2019-IEEE Conference on Computer Communications, pp. 2323–2331. IEEE (2019)
15. Wang, X., et al.: Practical heterogeneous wireless charger placement with obstacles. IEEE Trans. Mobile Comput. (2019)
16. Xu, W., Liang, W., Jia, X., Xu, Z.: Maximizing sensor lifetime in a rechargeable sensor network via partial energy charging on sensors. In: 2016 13th Annual IEEE International Conference on Sensing, Communication, and Networking (SECON), pp. 1–9. IEEE (2016)
17. Xu, X., Özçelikkale, A., McKelvey, T., Viberg, M.: Simultaneous information and power transfer under a non-linear RF energy harvesting model. In: 2017 IEEE International Conference on Communications Workshops (ICC Workshops), pp. 179–184. IEEE (2017)
18. Yu, N., Dai, H., Chen, G., Liu, A.X., Tian, B., He, T.: Connectivity-constrained placement of wireless chargers. IEEE Trans. Mobile Comput. (2019)

An Efficient Algorithm for Routing and Recharging of Electric Vehicles

Tayebeh Bahreini, Nathan Fisher, and Daniel Grosu(✉)

Department of Computer Science, Wayne State University, Detroit, USA
{tayebeh.bahreini,fishern,dgrosu}@wayne.edu

Abstract. In this paper, we address the routing and recharging problem for electric vehicles, where charging nodes have heterogeneous prices and waiting times, and the objective is to minimize the total recharging cost. We prove that the problem is NP-hard and propose two algorithms to solve it. The first, is an algorithm which obtains the optimal solution in pseudo-polynomial time. The second, is a polynomial time algorithm that obtains a solution with the total cost of recharging not greater than the optimal cost for a more constrained instance of the problem with the maximum waiting time of $(1-\epsilon) \cdot W$, where W is the maximum allowable waiting time.

Keywords: Routing · Electric vehicles · Recharging · Rounding

1 Introduction

In the last decade, the increased awareness of the global warming brought attention toward transportation, as this sector accounts for a large amount of air pollution. In 2018, the share of transportation in greenhouse gas emissions was 28.2% [11]. The policy of replacing conventional gasoline vehicles with All-Electric Vehicles (EVs) has been followed by many countries as an effective approach toward a greener transportation system. EVs, in addition to being environmental friendly, are more energy efficient. In these vehicles, over than 77% of the electrical energy from the grid is transmitted to the wheels, while in conventional gasoline vehicles only about 12%–30% of the energy stored in gasoline is converted to power at the wheels [3]. Despite the developments in the EVs technologies over the last decade, these vehicles represent a very small fraction of the overall vehicle market, even in the countries with the largest emission of carbon dioxide. In 2018, the penetration rate of EV in the light-vehicle market of the US and China was only 2.1 and 3.9%, respectively [5]. The low public interest in EVs is partially attributed to the driving range anxiety and to the lack of extensive charging infrastructure. This challenge has motivated researchers to devote their efforts to developing efficient optimization methods for recharging and routing policies for EVs.

In fact, recharging policy optimization for EVs is analogous to refueling policy optimization for gasoline vehicles. There are factors such as overcharging

© Springer Nature Switzerland AG 2020
W. Wu and Z. Zhang (Eds.): COCOA 2020, LNCS 12577, pp. 711–726, 2020.
https://doi.org/10.1007/978-3-030-64843-5_48

cost and charging waiting time that do not apply to refueling policy optimization for gasoline vehicles, but need to be taken into account while optimizing the recharging policies for EVs. The first effort on investigating the problem of refueling policy optimization dates back to 1980s, where the aim was to find the shortest path between two nodes while the vehicle has to visit some intermediate nodes for refueling [6]. Since then, researchers have attempted to study the properties of the problem and take multiple factors into account when designing the refueling policies. Lin [8] studied the properties of the refueling policy optimization problem, and based on these properties showed that the problem of finding the optimal refueling policy can be reduced to the classical shortest path problem. Khuller et al. [7] studied refueling and routing optimization problems for conventional gasoline vehicles, assuming that each gas station has a certain price for gasoline. They considered the problem with the objective of minimizing the cost of a fixed route and showed that it can be solved in polynomial time. The authors showed that the problem of finding the cheapest tour while a given set of locations are visited is NP-complete and developed approximation algorithms for this problem. Arslan et al. [1] formulated the refueling/recharging policy optimization for plug-in hybrid EVs where the vehicle has to visit both refueling and recharging stations, and the objective is to minimize the total cost which includes fuel and energy costs, stopping costs, depreciation costs, and battery degradation costs. Nejad et al. [9] developed one approximation and two exact algorithms for the routing problem of plug-in hybrid EVs. In addition to the optimal route, their proposed algorithms identify the predominant operating mode for each segment of the path in order to minimize the fuel consumption. In a recent work, Sweda et al. [10] considered the availability of a charging station at any point in time as a probabilistic parameter and developed two heuristic methods to obtain an a priori routing and recharging policy. In real world, recharging stations might have heterogeneous prices and waiting times. To the best of our knowledge, no research has been done on the routing and recharging problem for EVs with heterogeneous prices and waiting times. In this paper, we address the routing and recharging problem for electric vehicles with the objective of minimizing the total recharging cost, where charging nodes have heterogeneous prices and waiting times. We prove that the problem is NP-hard. We propose a pseudo-polynomial algorithm to obtain the optimal solution. We also propose a polynomial time algorithm and prove that it obtains a solution with the total cost of recharging not greater than the optimal cost for a more constrained instance of the problem with the maximum waiting time of $(1 - \epsilon) \cdot W$, where W is the maximum allowable waiting time.

2   Problem Definition

We formulate the Electric Vehicle Routing and Recharging Problem (EVRRP). We consider an EV which is initially fully charged that is going to travel through a road network (i.e., a directed graph) having n charging nodes $v_1, \ldots, v_n$. We do not consider any restriction such as acyclicity and predetermined order of the nodes. The EV travels from the start node v_1 to the destination node v_n. During the trip from the source to the destination, the EV may need to be recharged at the charging nodes. The goal is to find a path from the start node to the destination node as well as a recharging policy for the EV such that the total cost of recharging is minimized, while the total waiting time for recharging does not exceed a given value, W.

As the driver selects a path from the start node to the destination node, she/he must decide whether to stop and recharge at each node, and how much to recharge at each stop. We assume that the maximum capacity of the battery is F units. We denote by h_{ij} the amount of charge consumed from node i to node j, if they are adjacent. We assume that the braking energy recuperation is negligible and thus, the amount of charge consumption of the EV to pass a road segment is non-negative (i.e., $h_{ij} \geq 0$). To have a feasible solution, we assume that for adjacent nodes i and j, h_{ij} is less than the maximum capacity of the battery, F. Also, we assume that $h_{ij} = \infty$ if there is no road segment from node i to node j. Note that in the paper, we use v_i and i alternatively when we refer to a charging node v_i.

Consider a path p that contains two consecutive nodes v_i and v_j. The charge level of the EV's battery at node j is denoted by q_j, and is recursively defined based upon: q_i, the charge level of the battery in the previously visited node i; h_{ij}, the amount of charge consumption to reach node j from node i; and r_j, the amount of recharging at node j, as follows,

$$q_j = r_j + q_i - h_{ij}. \tag{1}$$

The EV must have enough charge to travel from node i to node j,

$$h_{ij} \leq q_i. \tag{2}$$

The charge level of the EV at a node i cannot exceed the maximum capacity F,

$$q_i \leq F. \tag{3}$$

The waiting time to access node i is denoted by ω_i. The total waiting time for recharging cannot exceed a given value W,

$$\sum_{i \in p, r_i > 0} \omega_i \leq W. \tag{4}$$

The charging nodes are heterogeneous in terms of their charging price. At node i, the EV is charged at a fixed price per unit of charge, μ_i. The objective is to find a path p over all possible paths P from node 1 to node n, and a recharging policy r over all feasible recharging polices on path p, R_p, such that the total recharging cost is minimized,

$$\min_{p \in P, r \in R_p} \sum_{i \in p} \mu_i \cdot r_i. \tag{5}$$

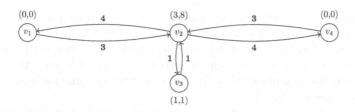

Fig. 1. An illustrative example: a road network with four charging nodes, where the start node is v_1 and the destination node is v_4. The weight on each edge indicates the amount of charge consumption, while the pair on the vertices indicates the waiting time and the price per unit of charge, respectively.

Note that in the objective function, we include the cost of recharging at the source node and the destination. In fact, since the EV is full at node v_1 at the beginning of the trip, without loss of optimality, we can assume that $\mu_1 = 0$. Therefore, the cost of recharging at this node is zero and the EV is fully charged at this node. Similarly, we do not need to recharge the EV at the destination node, we can assume that $\mu_n = 0$. Therefore, the cost of recharging at this node is zero and will not affect the value of the objective function. It is straightforward to extend the results in this paper to settings where the EV starts with a non-full charge.

EVRRP can be represented as a directed graph $G(V, E)$, where V is the set of vertices representing charging nodes and E is the set of edges representing the road segments between the nodes. Figure 1 shows an example of such a road network. In this example, there are four charging nodes. The vehicle must travel from the start node v_1 to the destination node v_4. The weight on each edge shows the amount of charge consumption required to travel the corresponding road. The pair on the vertices shows the waiting time and the price per unit of charge, respectively.

2.1 Complexity of EVRRP

In this section, we prove that EVRRP is NP-hard by showing that: (i) the decision version (EVRRP-D) of EVRRP belongs to NP, and, (ii) a well known NP-complete problem is reduced to EVRRP-D in polynomial time.

For the first condition, we can easily show that EVRRP-D is in NP. We only need to guess a solution and a value C, compute the total value of the objective function (Eq. (5)), and verify if the solution is feasible and the associated objective value is at most C. Obviously, this can be done in polynomial time. For the second condition, we show that the Shortest Weight-Constrained Path problem (SWCP), a well-known NP-complete problem (problem ND 30 in [4]), is reduced to EVRRP-D in polynomial time.

An instance of EVRRP-D is represented by a graph $G(V, E)$, where $V = \{v_1, \ldots, v_n\}$ is the set of vertices representing charging nodes, and E is the set

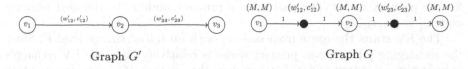

Graph G' Graph G

Fig. 2. Transforming graph G', an arbitrary instance of SWCP, to graph G, an instance of EVRRP-D.

of edges representing the road segments between the nodes. Each node is characterized by the charging price per unit of charge, μ_i, and the waiting time, ω_i. The weight of each edge (i, j) is the amount of charge consumed when traveling road segment (i, j) (i.e., h_{ij}). The decision question is whether there is a path in G from the source node v_1 to the destination node v_n with a feasible recharging policy over the path so that the total cost of recharging (Eq. (5)) does not exceed C and the total waiting time for recharging does not exceed W.

An instance of SWCP consists of a graph $G'(V', E')$, where $V' = \{v'_1, \ldots, v'_m\}$ is the set of vertices and E' is the set of edges with cost c'_{ij} and weight w'_{ij} for each $(i, j) \in E'$. The decision question is whether there is a path in G' from v'_1 to v'_m such that the total cost does not exceed C', while the total weight is not greater than W'.

Theorem 1. *EVRRP-D is NP-complete.*

Proof. We show that an arbitrary instance of SWCP is transformed into an instance of EVRRP-D. Let $F = 2$, $C = 2C'$, $W = W'$, and $n = m$. First, we build graph G with the same set of vertices as graph G' (i.e., $V = V'$), where $v_1 = v'_1$ and $v_n = v'_m$. We call these nodes the primary nodes of G. Then, we add some other nodes to the graph as the secondary nodes.

For every edge (i, j) in G', we add a secondary node in G. We denote this node by v_{ij}. Then, we add one edge from node v_i to node v_{ij}, another edge from node v_{ij} to node v_j. We set the amount of charge consumption of these edges to one. Therefore, the amount of charge consumption on path $\{v_i, v_{ij}, v_j\}$ is two.

We assume that the waiting time at the secondary node v_{ij} is w'_{ij}, while at each primary node it is $M \gg W$, a very large value. Furthermore, the charging price per unit of charge of the secondary node v_{ij} is c'_{ij}. The charging price rate at each primary node is $M \gg C$, a very large value. Therefore, it is more preferred to recharge the EV at the secondary nodes.

Figure 2 shows how the graph G is built based on graph G'. Figure 2a shows graph G' that has three nodes v_1, v_2, and v_3. The label on each edge represents the weight and the cost of that edge, respectively. Figure 2b shows graph G with three primary nodes v_1, v_2, and v_3. We add one secondary node between nodes v_1 and v_2, and one secondary node between nodes v_2 and v_3. The secondary nodes are represented by black filled circles.

Now, we show that the solution for EVRRP-D can be constructed based on the solution for SWCP. Let us assume that U' is the routing path obtained for SWCP in G'. To obtain the corresponding path U in G, we choose the same path for the primary nodes. The path from a primary node v_i to the next primary node v_j is $\{v_i, v_{ij}, v_j\}$.

The EV starts the route from node v_1 with an initial charge level F. Since the recharging price rate at primary nodes is relatively high, the EV recharges only at the secondary nodes. Furthermore, the amount of charge consumption between every two adjacent nodes on the path (primary/secondary) is $\frac{F}{2}$. Thus, an optimal recharging policy of the EV is to stop at the first secondary node after v_1 and recharge the EV to level F. The amount of recharging at this node is $\frac{F}{2}$. For the remaining path, the policy is to pass the next primary (without recharging) and stop at the next secondary node and recharge the battery fully. For the last secondary node, the node immediately before the destination node, we recharge the EV to level $\frac{F}{2}$, which is enough to reach the destination node. Thus, the amount of recharging at this node is $\frac{F}{2}$. The total waiting time of the path is equivalent to the total waiting time of the secondary nodes on the path, $\sum_{v_{ij} \in U} w'_{ij} = W'$. Since $W = W'$, Constraint (4) is satisfied.

The price per unit of charge at each secondary node v_{ij} is c'_{ij}. Since we recharge the EV for at most two units, the total cost of recharging to reach the destination is at most $\sum_{v_{ij} \in U} 2 \cdot c'_{ij} = 2C'$. Thus, we obtain a solution for EVRRP with objective value less than $2C'$. Since $C = 2C'$, the total recharging cost for EVRRP does not exceed C.

Conversely, suppose that U is the routing path in G obtained for EVRRP. To obtain path U' in G', we choose the sequence of primary nodes of path U. Since the total waiting time of path U does not exceed W, the total weight of the corresponding edges in G' does not exceed W, too. Since $W' = W$, the total weight on path U' does not exceed W'.

Furthermore, in path U, the amount of recharging at the first secondary node and the last secondary node is $\frac{F}{2}$. The amount of recharging at other secondary nodes is $F = 2$. Thus, the cost of recharging at the first secondary node and the last node is equivalent to the cost of the corresponding edge in G'. The cost of recharging at any other secondary node, is two times greater than the cost of the corresponding edge in G. Since the total cost of recharging in G does not exceed C, the total cost of the corresponding edges in graph G' does not exceed $\frac{C}{2}$. Since $C = 2C'$, the total cost of edges of path U' does not exceed C'.

3   Optimal Solution for EVRRP

Here, we present an algorithm that obtains the optimal solution for EVRRP in pseudo-polynomial time. We transform the original directed graph $G(V, E)$ into a directed graph $\tilde{G}(\tilde{V}, \tilde{E})$. In the transformed graph, we consider all possible sequences of stops for recharging. We denote by $H(i, j)$, the minimum amount of charge consumed from stop i to stop j. The value of $H(i, j)$ is obtained based on the shortest path (in terms of the amount of charge consumption) from node i to node j in G. We show that finding the optimal routing and recharging in G is equivalent to finding the shortest weighted constrained path in $\tilde{G}$. Then, we provide an algorithm to solve the problem.

In the following, we describe the Transform-Graph procedure which obtains the transformed graph $\tilde{G}(\tilde{V}, \tilde{E})$ from the original graph G. This procedure is based on the recharging rules described in the following lemmas which are extensions of the gas filling policy for the gas station problem [7].

Lemma 1. *Let node i and node j be two consecutive stops (for recharging) in the optimal solution. The path from node i to node j is the shortest path with the minimum amount of charge consumed from node i to node j in G. The following rules provide the optimal recharging policy at node i,*

(i) if $\mu_i < \mu_j$, then recharge the battery fully.

(ii) if $\mu_i \geq \mu_j$, then recharge the battery just enough to reach node j.

Proof. We can prove this by contradiction. If in the optimal solution, the path from node i to node j is not the shortest path, we can replace this path with the shortest path. Since the shortest path has the minimum amount of charge consumed from node i to node j, the level of the battery upon arriving to node j is higher than that in the optimal solution. Thus, the amount of recharging needed at node j is less than that in the optimal solution. This means that we improve the cost of recharging which is a contradiction with the optimality assumption.

Furthermore, if $\mu_i < \mu_j$ and the optimal solution does not recharge the battery fully at node i, then we can improve the cost of recharging by increasing the amount of recharging at node i and decreasing the amount of recharging at node j, which is a contradiction with the optimality assumption. Similarly, in the second case, if the optimal solution recharges the battery more than the charge amount needed to reach node j, then, we can improve the cost of recharging by decreasing the amount of recharging at node i and increasing the amount of recharging at node j.

Lemma 2. *Let nodes i, j, and k be three consecutive stops (for recharging) in the optimal solution. If $\mu_i < \mu_j$ and $\mu_j \geq \mu_k$, then $H(i, j) + H(j, k) > F$.*

Proof. According to Lemma 1, the level of the battery when the EV leaves node i is F. By contradiction, if we assume that $H(i, j) + H(j, k) \leq F$, then, the EV can reach node k without stopping at node j. In other words, the EV can improve the cost of recharging by decreasing the amount of recharging at node j (to level zero) and increasing the amount of recharging at node k. This is a contradiction with the optimality assumption.

Transform-Graph Procedure. In this procedure, we transform the original graph $G(V, E)$ into a new graph $\tilde{G}(\tilde{V}, \tilde{E})$. In the transformed graph $\tilde{G}$, each vertex represents two possible consecutive stops of the EV, and each edge represents three consecutive recharging stops. For every node i and node j in G, we add a node $<i, j>$ in $\tilde{G}$, if node j is reachable from node i (i.e., $H(i, j) \leq F$).

We also add a dummy source node $<0, 1>$ and a dummy destination node $<n, n + 1>$ to $\tilde{G}$. Since the EV is full at node v_1 at the beginning of the trip, it will not go back to this node during the trip. Therefore, we do not need to add any node $<i,1>$ (where $i > 1$) to $\tilde{G}$. Similarly, since the goal is to reach node v_n, the EV will not go back from this node to any other node. Therefore, we do not add any node $<n,i>$ to $\tilde{G}$.

For every pair of nodes $<i,j>$ and $<j,k>$, we add an edge from node $<i,j>$ to node $<j,k>$ based on a set of conditions. Each edge has a label (w_{ijk}, c_{ijk}), where w_{ijk} is the waiting time for recharging at node j, and c_{ijk} is the cost of recharging at node j.

We add an edge with label $(0,0)$ from node $<0,1>$ to every adjacent node $<1,j>$. In fact, since the cost of recharging at node v_1 is zero (i.e., $\mu_1 = 0$), the battery will be fully charged at node v_1 with cost and waiting time equal to zero. We also add an edge from every node $<i,n>$ to the destination node $<n, n+1>$ with label $(0,0)$.

For every node $<i,j>$ and node $<j,k>$, where $i > 0$ and $k \leq n$, we consider all possible cases for the values of μ_i, μ_j, and μ_k. Based on these values, we add an edge from node $<i,j>$ to node $<j,k>$, as follows:

Case I $(\mu_i < \mu_j < \mu_k)$: By Lemma 1, we should fully fill the battery at node i when node j is the next stop. Therefore, the level of the battery when arriving at node j is $F - H(i,j)$. Given that, $\mu_j < \mu_k$, we should again fill up the battery fully at node j. Thus, the cost of edge $(< i,j >, < j,k >)$ is $c_{ijk} = \mu_j \cdot H(i,j)$, and the waiting time of this edge is $w_{ijk} = \omega_j$.

Case II $(\mu_i < \mu_j$ and $\mu_j \geq \mu_k)$: According to Lemma 2, node k is the next stop after node j only if $H(i,j) + H(j,k) > F$. Therefore, we add an edge from node $<i,j>$ to node $<j,k>$ if $H(i,j) + H(j,k) > F$. By Lemma 1, the level of the battery upon arriving at node j from node i should be $F - H(i,j)$. Given that $\mu_j \geq \mu_k$, the battery should only be filled up just enough to reach node k from node j. Thus, the cost of the edge is $c_{ijk} = \mu_j \cdot (H(j,k) + H(i,j) - F)$, and the waiting time is $w_{ijk} = \omega_j$.

Case III $(\mu_i \geq \mu_j$ and $\mu_j < \mu_k)$: In this case, we have an empty battery when reaching node j from node i. Also, we want to recharge the battery fully at node j since $\mu_j < \mu_k$. Thus, we add an edge with cost $c_{ijk} = \mu_j \cdot F$, and waiting time $w_{ijk} = \omega_j$.

Case IV $(\mu_i \geq \mu_j$ and $\mu_j \geq \mu_k)$: In this case, we have an empty battery when reaching node j; however, we only want to recharge enough to reach node k. Thus, the cost of the edge is $c_{ijk} = \mu_j \cdot H(j,k)$, and the waiting time is $w_{ijk} = \omega_j$.

Theorem 2. *The optimal solution for EVRRP in graph G is equivalent to the optimal solution for EVRRP in graph $\tilde{G}$.*

Proof. We need to show that: (i) for any feasible sequence of recharging stops in G, the corresponding sequence in $\tilde{G}$ is a feasible sequence and the amount of recharging at each node is the same as in G; (ii) for any feasible sequence of recharging stops in $\tilde{G}$, the corresponding sequence in G is a feasible sequence and the EV has the same recharging policy as in $\tilde{G}$.

Let $p = \{p_1, \ldots, p_s\}$ be the sequence of stops of a feasible path in G, where s is the number of nodes in the sequence. We need to show that (1) p is a feasible sequence of stops in $\tilde{G}$; and (2) the level of the battery when the EV arrives at node p_i in both graphs is the same. We prove this by induction.

According to our assumption, the EV reaches node v_1 with an empty battery. Then, it will be recharged to level F with zero cost/waiting time ($\mu_1 = \omega_1 = 0$). Thus, node $p_1 = v_1$ is the first stop for recharging. In both G and $\tilde{G}$, node v_1 is reachable and the level of the battery when the EV reaches this node is zero.

Let us assume that $\{p_1, \ldots, p_i\}$ is a feasible sequence of stops in both graphs; and for every node $j \leq i$, the level of the battery is the same in both graphs. Now, we need to show that node $<p_i, p_{i+1}>$ in $\tilde{G}$ is reachable from $<p_{i-1}, p_i>$ via edge $(< p_{i-1}, p_i, < p_i, p_{i+1} >)$ and the level of the battery when the EV arrives at node p_{i+1} is the same in both graphs.

Since p_{i+1} is the next stop in the sequence p in G, $H(p_i, p_{i+1}) \leq F$. Thus, according to transformation rules, there is an edge from node $<p_{i-1}, p_i>$ to node $<p_i, p_{i+1}>$ in $\tilde{G}$. This implies that node $<p_i, p_{i+1}>$ is reachable from node $<p_{i-1}, p_i>$. For the second condition, we consider the possible values of μ_i and μ_{i+1},

Case 1 ($\mu_i < \mu_{i+1}$): According to Lemma 1, the level of the battery when the EV leaves node p_i in G is F. Thus, upon arriving at node p_{i+1}, the level of the battery is $F - H(p_i, p_{i+1})$. On the other hand, according to transformation rules (I) and (III), the level of the battery corresponding to edge $(< p_{i-1}, p_i >, <p_i, p_{i+1} >)$ is F. Thus, the level of the battery upon arriving at node p_{i+1} in both graphs is the same and equal to $F - H(p_i, p_{i+1})$.

Case 2 ($\mu_i \geq \mu_{i+1}$): According to Lemma 1, the level of the battery when the EV leaves node p_i in G is $H(p_i, p_{i+1})$. Thus, upon arriving at node p_{i+1}, the level of the battery is zero. On the other hand, according to transformation rules (II) and (IV), the level of the battery corresponding to edge $(< p_{i-1}, p_i >, <p_i, p_{i+1} >)$ is $H(p_i, p_{i+1})$. Thus, the level of battery upon arriving at node p_{i+1} in both graphs is the same and equal to zero.

Similarly, we can show that a feasible sequence of recharging in $\tilde{G}$ is a feasible sequence of stops in G and the level of the battery upon arriving at each node of the sequence in both graphs is the same.

Now, the problem is to find a path from the source node $<0, 1>$ to the destination node $<n, n+1>$ in the transformed graph $\tilde{G}$ such that the total cost of the path is minimized, while the total waiting time does not exceed W. Therefore, EVRRP can be viewed as an SWCP problem. In order to obtain the optimal solution, we use a dynamic programming algorithm [2], called DP-SWCP, introduced for the SWCP problem.

The general idea of DP-SWCP is to use a set of labels for each node of the graph. Each label (W_{ijl}, C_{ijl}) corresponds to a path l from the source node $<0, 1>$ to node $<i, j>$ and is composed of two elements: W_{ijl}, the total waiting time of the path when the EV leaves node j, and C_{ijl}, the total cost of that path. DP-SWCP finds all non-dominated labels on every node. The dominance relation is defined based on the total waiting time and the total cost on each label. For a given node $<i, j>$, let us assume we find two labels (W_{ijl}, C_{ijl}) and $(W_{ijl'}, C_{ijl'})$ such that $W_{ijl} \leq W_{ijl'}$, and $C_{ijl} < C_{ijl'}$. Then, path l' cannot be a part of the optimal solution, because we could replace it with path l which has a lower cost and a lower weight. Therefore, we can disregard this path. In this

---
Algorithm 1. OPT-EVRRP Algorithm

Input: $G(V, E)$: Graph representing the road network
    W: Maximum allowable waiting time
Output: $p = \{p_i\}$: Routing vector
    $r = \{r_i\}$: Recharging vector
    $cost$: Total cost
1: $\tilde{G} \leftarrow$ Transform-Graph(G)
2: $p \leftarrow$ DP-SWCP$(\tilde{G}, W)$
3: $q_1 \leftarrow F$
4: $i \leftarrow 1$
5: **for each** $u = 2, \ldots, |p| - 1$ **do**
6:     $j \leftarrow p_u$
7:     $k \leftarrow p_{u+1}$
8:     **if** $\mu_j \leq \mu_k$ **then**
9:         $q_j \leftarrow F$
10:    **else**
11:        $q_j \leftarrow H(j, k)$
12:    $r_j \leftarrow q_j - q_i + H(i, j)$
13: $cost \leftarrow \sum_{i \in p} r_i \cdot \mu_i$

---

case, we say that label (W_{ijl}, C_{ijl}) *dominates* label $(W_{ijl'}, C_{ijl'})$ and denote it by $(W_{ijl}, C_{ijl}) \triangleright (W_{ijl'}, C_{ijl'})$.

DP-SWCP starts with the source node $<0, 1>$ and assigns a label $(0, 0)$. The algorithm extends the set of labels by treating a label with the minimum cost. In the treatment of a label l, the algorithm extends the path corresponding to the label l along all outgoing edges. In fact, the treatment of a label (W_{ijl}, C_{ijl}) on node $<i, j>$ considers each adjacent node $<j, k>$ such that $W_{ijl} + w_{ijk} \leq W$: if $(W_{ijl} + w_{ijk}, C_{ijl} + c_{ijk})$ is not dominated by any label on node $<j, k>$, adds it to the set of labels on node $<j, k>$. DP-SWCP continues the procedure until all non-dominated labels are treated. Finally, it picks the path that corresponds to the label with minimum cost at the destination node $<n, n+1>$ as the optimal solution.

The algorithm for solving EVRRP, called OPT-EVRRP, is given in Algorithm 1. The input of the algorithm is the graph $G(V, E)$, while the output is the sequence of stops $p = \{p_i\}$, the recharging vector $r = \{r_i\}$, and the total cost of recharging. The algorithm calls the Transform-Graph procedure to obtain the transformed graph $\tilde{G}$ (Line 1). Then, it calls DP-SWCP to obtain the optimal sequence of recharging p in $\tilde{G}$ (Line 2). Based on Lemma 1, the OPT-EVRRP obtains the optimal amount of recharging at each stop (Lines 3–13). In Sect. 4, we provide an example on how OPT-EVRRP works on the EVRRP instance given in Figure 1.

Theorem 3. *OPT-EVRRP obtains the optimal solution for EVRRP and its time complexity is* $O(n^3 + n^2 \cdot W)$.

Proof. According to Theorem 2, the optimal solution for EVRRP in graph G is equivalent to the optimal solution in graph $\tilde{G}$. Since DP-SWCP obtains the optimal solution in $\tilde{G}$, this solution is also optimal for EVRRP in G.

To determine the time complexity of OPT-EVRRP, we need to determine the time complexity of Transform-Graph and DP-SWCP procedures. The time complexity of Transform-Graph is proportional to the number of edges in $\tilde{G}$ and is $O(n^3)$. The time complexity of DP-SWCP depends on the number of treated labels in $\tilde{G}$. DP-SWCP does not treat two labels with the same total waiting time (because one of them has a cost not less than the other one, and therefore, it dominates it). The maximum waiting time of a label is bounded by W (W is integer). Thus, there are at most $W + 1$ labels on node $<i, j>$. On the other hand, there are at most n^2 nodes in $\tilde{G}$. Thus, the total number of treated labels is $O(n^2 \cdot W)$. Thus, the time complexity of OPT-EVRRP is $O(n^3 + n^2 \cdot W)$.

4   An Illustrative Example

We provide a numerical example to illustrate how OPT-EVRRP works. In this example, we consider the road network given in Fig. 1 as the original graph G. The maximum capacity of the battery is $F = 4$, and the maximum allowable waiting time is $W = 8$. We can easily see that the optimal solution for this example is to start the trip from node v_1, visit node v_2 without recharging, stop at node v_3 and recharge for 4 units, visit node v_2 again and recharge for one unit, and finally, visit node v_4. The total cost of this recharging policy is 12, while the total waiting time is 4.

Now, we show how OPT-EVRRP obtains the optimal solution for this example. To transform graph G into $\tilde{G}$, we determine the value of $H(i, j)$, the minimum amount of charge consumed from node i to node j in G. Table 1 shows the value of $H(i, j)$ for every node i and node j in G.

Figure 3 shows the transformed graph $\tilde{G}$. The source node is node $<0, 1>$ and the destination node is $<4, 5>$. For every nodes i and j in G, we add a node $<i, j>$ to $\tilde{G}$, if node j is reachable from node i (i.e., $H(i, j) \leq F$). For example, we add node $<1, 2>$ to $\tilde{G}$ because $H(1, 2) = 3$; but we do not add node $<3, 4>$ to $\tilde{G}$ because $H(3, 4) > F$.

Table 1. Example: The values of $H(i, j)$

i/j	1	2	3	4
1	0	3	4	7
2	4	0	1	4
3	5	1	0	5
4	7	3	4	0

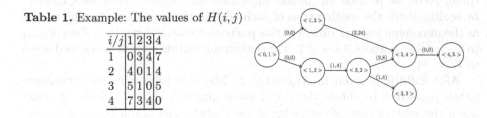

Fig. 3. Example: The transformed graph $\tilde{G}$.

For every pair of nodes $<i,j>$, and $<j,k>$, we add an edge from node $<i,j>$ to node $<j,k>$ based on the transformation rules in Transform-Graph procedure. In Figure 3, the pair on the edge $(<i,j>,<j,k>)$ shows the waiting time and the recharging cost of the edge.

We add an edge from node $<0,1>$ to adjacent node $<1,2>$ with label $(0,0)$. Similarly, we add an edge from node $<0,1>$ to node $<1,3>$ with label $(0,0)$. For nodes $<1,2>$ and $<2,4>$, since $\mu_1 < \mu_2$, and $\mu_2 > \mu_4$, we follow the transformation rule (II). In this case, since $H(1,2)+H(2,4) > F$, we add an edge from node $<1,2>$ to node $<2,4>$. The cost of this edge is $\mu_2 \cdot (H(1,2)+H(2,4)-F) = 24$ and the waiting time is $\omega_2 = 3$. Similarly, for nodes $<1,2>$ and $<2,3>$, we follow the transformation rule (II). However, since $H(1,2)+H(2,3) = 4$ which is not greater than F, we do not add any edge from node $<1,2>$ to node $<2,3>$.

For nodes $<1,3>$ and $<3,2>$, we follow the transformation rule (I). We add an edge from node $<1,3>$ to node $<3,2>$ with cost $\mu_3 \cdot H(1,3) = 4$ and the waiting time $\omega_3 = 1$. We follow the transformation rule (II) to add an edge from node $<3,2>$ to node $<2,4>$. The cost of this edge is $\mu_2 \cdot (H(3,2)+H(2,4)-F) = 8$ and the waiting time is $\omega_2 = 3$. We also follow the transformation rule (III) and add an edge from node $<2,3>$ to node $<3,2>$. The cost of this edge is $\mu_3 \cdot F = 4$ and the waiting time is $\omega_3 = 1$. For nodes $<3,2>$ and $<2,3>$, we follow the transformation rule (II). Since $H(3,2)+H(2,3) = 2$ and is not greater than F, we do not add an edge from node $<3,2>$ to node $<2,3>$. Finally, we add an edge from node $<2,4>$ to destination node $<4,5>$ with label $(0,0)$.

Now, we use DP-SWCP to obtain the optimal solution in $\tilde{G}$. Due to the limited space, we do not illustrate the procedure of DP-SWCP. There are two possible paths from source node $<0,1>$ to the destination node $<4,5>$. DP-SWCP determines path $\{<0,1>, <1,3>, <3,2>, <2,4>, <4,5>\}$ as the optimal path in $\tilde{G}$. Thus, the optimal sequence of stops is v_1, v_3, v_2, v_4, the total waiting time is 4, and the total cost of recharging is 12, which corresponds to the optimal solution in the original graph G.

5   An Efficient Algorithm for EVRRP

In the previous section, we showed that OPT-EVRRP provides the optimal solution for EVRRP and is pseudo-polynomial (in terms of the maximum total waiting time). Here, we provide an efficient algorithm for EVRRP, called APX-EVRRP, by scaling down the waiting time of each edge in the transformed graph as well as the maximum waiting time. For this purpose, we scale down the values of w_{ijk} to $\bar{\omega}_j = \lceil \frac{n \cdot \omega_j}{\epsilon \cdot W} \rceil$, where $0 < \epsilon < 1$. The maximum waiting time is also scaled down to $\bar{W} = \frac{n}{\epsilon}$.

APX-EVRRP is given in Algorithm 2. The algorithm calls the Transform-Graph procedure to obtain the transformed graph $\tilde{G}$ (Line 1). Then, it scales down the waiting time of each edge of the transformed graph (Lines 2–3), and calls DP-SWCP to solve the rounded problem with maximum allowable waiting time $\frac{n}{\epsilon}$ (Line 4). Finally, given the optimal solution p for the rounded problem, where p is the sequence of stops, it determines the recharging amount at each

---

Algorithm 2. APX-EVRRP Algorithm

Input: $G(V, E)$: Graph representing the road network
   W: Maximum allowable waiting time
Output: $p = \{p_i\}$: Routing vector
   $r = \{r_i\}$: Recharging vector
   $cost$: Total cost
1: $\tilde{G} \leftarrow$ Transform-Graph(G)
2: **for each** $(<i, j><j, k>) \in \tilde{E}$ **do**
3:    $w_{ijk} \leftarrow \lceil \frac{n \cdot w_{ijk}}{\epsilon \cdot W} \rceil$
4: $p \leftarrow$ DP-SWCP($\tilde{G}, \frac{n}{\epsilon}$)
5: $q_1 \leftarrow F$
6: $i \leftarrow 1$
7: **for each** $u = 2, \ldots, |p| - 1$ **do**
8:    $j \leftarrow p_u$
9:    $k \leftarrow p_{u+1}$
10:   **if** $\mu_j \leq \mu_k$ **then**
11:      $q_j \leftarrow F$
12:   **else**
13:      $q_j \leftarrow H(j, k)$
14:   $r_j \leftarrow q_j - q_i + H(i, j)$
15: $cost \leftarrow \sum_{i \in p} r_i \cdot \mu_i$

---

stop (Lines 5–15). We choose the sequence p with recharging policy r as the solution for EVRRP. In the next section, we show that this solution is feasible and the total cost obtained by this solution is bounded by the optimal cost for EVRRP with maximum waiting time $(1 - \epsilon) \cdot W$.

5.1   Properties of APX-EVRRP

In this section, we analyze the properties of the proposed algorithm. First, we prove the correctness of APX-EVRRP by showing that the algorithm obtains a feasible solution for EVRRP in polynomial time. Then, we show that the total recharging cost of the solution is not greater than the optimal cost for EVRRP with the maximum allowable waiting time $(1 - \epsilon) \cdot W$.

Let us denote the solution obtained by APX-EVRRP by (p, r), where p is the sequence of stops and r gives the recharging amount at each stop. We also denote by $\text{EVRRP}_{(1-\epsilon)}$, the EVRRP problem with the maximum allowable waiting time $(1 - \epsilon) \cdot W$, and by (p^*, r^*), the optimal solution for $\text{EVRRP}_{(1-\epsilon)}$.

Theorem 4. *APX-EVRRP obtains a feasible solution for EVRRP and its time complexity is* $O(\frac{n^3}{\epsilon})$.

Proof. Since the amount of charge consumption between every pair of nodes in the rounded problem is the same as in the original problem, the recharging policy r is feasible for the original problem (to reach the destination). Thus, we only need to show that the total waiting time of the solution does not exceed W.

In the rounded solution, the total rounded waiting time is not greater than $\frac{n}{\epsilon}$ (i.e., $\sum_{i \in p} \bar{\omega}_i \leq \frac{n}{\epsilon}$). On the other hand $\bar{\omega}_i = \lceil \frac{n \cdot \omega_i}{\epsilon \cdot W} \rceil$. Thus,

$$\sum_{i \in p} \frac{n \cdot \omega_i}{\epsilon \cdot W} \leq \sum_{i \in p} \bar{\omega}_i \leq \frac{n}{\epsilon}.$$

Thus,

$$\sum_{i \in p} \omega_i \leq W.$$

Therefore, (p, r) is a feasible solution for EVRRP with the total waiting time less than or equal W.

The time complexity of APX-EVRRP comes mainly from DP-SWCP. In the rounded problem, the possible value of the maximum waiting time is reduced to $\frac{n}{\epsilon}$. Therefore, the time complexity of DP-SWCP for the rounded problem is $O(\frac{n^3}{\epsilon})$. Thus, the time complexity of APX-EVRRP is $O(\frac{n^3}{\epsilon})$.

Now, we show that the total cost obtained by APX-EVRRP is not greater than the optimal cost for EVRRP$_{(1-\epsilon)}$. For this purpose, we show that (p^*, r^*) is a feasible solution for the rounded problem and the recharging cost of this solution is not less than the cost obtained from solution (p, r).

Lemma 3. *The optimal solution (p^*, r^*) for EVRRP$_{(1-\epsilon)}$ is a feasible solution for the rounded problem.*

Proof. Since the amount of charge consumption between every pair of nodes is the same in both the rounded problem and the original problem, the recharging policy r^* is feasible for the rounded problem. Thus, we only need to show that the total rounded waiting time obtained from solution (p^*, r^*) is not greater than $\frac{n}{\epsilon}$. The rounded waiting time at stop i is $\bar{\omega}_i = \lceil \frac{n \cdot \omega_i}{\epsilon \cdot W} \rceil$. Thus,

$$\sum_{i \in p^*} \bar{\omega}_i \leq \sum_{i \in p^*} (\frac{n \cdot \omega_i}{\epsilon \cdot W} + 1).$$

On the other hand, the total waiting time of $<p^*, r^*>$ is not greater that $(1-\epsilon) \cdot W$ (i.e., $\sum_{i \in p^*} \omega_i \leq (1 - \epsilon) \cdot W$). Therefore,

$$\sum_{i \in p^*} \bar{\omega}_i \leq \sum_{i \in p^*} (\frac{n \cdot \omega_i}{\epsilon \cdot W} + 1) \leq \frac{n \cdot (1 - \epsilon) \cdot W}{\epsilon \cdot W} + n \leq n \cdot (1 + \frac{1 - \epsilon}{\epsilon}) \leq \frac{n}{\epsilon}.$$

Therefore, (p^*, r^*) is a feasible solution for the rounded problem.

Theorem 5. *The total cost of the solution obtained by APX-EVRRP is not greater than the optimal cost for EVRRP$_{(1-\epsilon)}$.*

Proof. According to Lemma 3, the optimal solution for $EVRRP_{(1-\epsilon)}$ is a feasible solution for the rounded problem. On the other hand, OPT-EVRRP obtains the optimal solution $<p, r>$ for the rounded problem. Therefore,

$$\sum_{i \in p} r_i \cdot \mu_i \leq \sum_{i \in p^*} r_i^* \cdot \mu_i.$$

Therefore, the total cost of the solution obtained by APX-EVRRP is not greater than the total cost of the optimal solution for $EVRRP_{(1-\epsilon)}$.

6   Conclusion

We studied the routing and recharging optimization problem for electric vehicles, where the aim is to find the routing path from a starting point to destination such that the total recharging cost is minimized. We considered that charging nodes have heterogeneous prices and waiting times. We studied the properties of the problem and showed that the problem is NP-hard. We proposed a pseudo-polynomial algorithm for the optimal solution and a polynomial time algorithm that obtains a solution with the total recharging cost not greater than the optimal cost for the same problem but with the maximum waiting time of $(1 - \epsilon) \cdot W$, where W is the maximum allowable waiting time. As a future research, we plan on considering the heterogeneity of charging stations in terms of charging speed. Another direction for future study is to take the uncertainty of waiting times at charging stations into account.

Acknowledgments. This research was supported in part by the US National Science Foundation under grant no. IIS-1724227.

References

1. Arslan, O., Yıldız, B., Karaşan, O.E.: Minimum cost path problem for plug-in hybrid electric vehicles. Transp. Res. Part E: Logistics Transp. Rev. **80**, 123–141 (2015)
2. Desrochers, M., Soumis, F.: A generalized permanent labelling algorithm for the shortest path problem with time windows. INFOR: Inf. Syst. Oper. Res. **26**(3), 191–212 (1988)
3. DoE: All Electric Vehicles. https://www.fueleconomy.gov/feg/evtech.shtml. Accessed 27 June 2020
4. Garey, M.R., Johnson, D.S.: Computers and Intractability, vol. 174. Freeman San Francisco (1979)
5. Hertzke, P., Müller, N., Schaufuss, P., Schenk, S., Wu, T.: Expanding Electric-Vehicle Adoption Despite Early Growing Pains. McKinsey & Company Insights (2019)
6. Ichimori, T., Ishii, H., Nishida, T.: Routing a vehicle with the limitation of fuel. J. Oper. Res. Soc. Japan **24**(3), 277–281 (1981)
7. Khuller, S., Malekian, A., Mestre, J.: To fill or not to fill: the gas station problem. ACM Trans. Algorithms (TALG) **7**(3), 1–16 (2011)

8. Lin, S.-H.: Finding optimal refueling policies in transportation networks. In: Fleischer, R., Xu, J. (eds.) AAIM 2008. LNCS, vol. 5034, pp. 280–291. Springer, Heidelberg (2008). https://doi.org/10.1007/978-3-540-68880-8_27
9. Nejad, M.M., Mashayekhy, L., Grosu, D., Chinnam, R.B.: Optimal routing for plug-in hybrid electric vehicles. Transp. Sci. **51**(4), 1304–1325 (2017)
10. Sweda, T.M., Dolinskaya, I.S., Klabjan, D.: Adaptive routing and recharging policies for electric vehicles. Transp. Sci. **51**(4), 1326–1348 (2017)
11. US-EPA: Global Greenhouse Gas Emissions Data. https://www.epa.gov/ghgemissions/global-greenhouse-gas-emissions-data. Accessed 27 June 2020

The Optimization of Self-interference in Wideband Full-Duplex Phased Array with Joint Transmit and Receive Beamforming

XiaoXin Wang, Zhipeng Jiang(✉), Wenguo Yang, and Suixiang Gao

School of Mathematical Sciences, University of Chinese Academy of Sciences,
Beijing, China
jiangzhipeng@ucas.ac.cn

Abstract. Full-duplex (FD) wireless and phased arrays are both promising techniques that can significantly improve data rates in future wireless networks. However, integrating FD with transmit (Tx) and receive (Rx) phased arrays is extremely challenging, due to the large number of self-interference (SI) channels. In this paper, a model is proposed to minimize the self-interference without large TxBF and RxBF gain losses through optimizing the TxBF and RxBF weights. As the model involves complex numbers and is non-convex, an iterative algorithm by solving series convex model is given to obtain an approximate solution. Meanwhile, in each step of the iterative algorithm, the sub model is translated into real number model. Then a penalty term is added into the sub model of each step of the iterative algorithm. Simulation results show that, the iterative algorithm is effective and can get good results. The penalty term can improve the performance of the algorithm.

Keywords: Full-duplex wireless · Phased array optimization · Iterative algorithm · Non-convex optimization · Complex number optimization

1 Introduction

The conception of full-duplex (FD) wireless – simultaneous transmission and reception on the same frequency at same time – has been proposed for many years. But because of the limitation of technical level, it has not been well developed [1]. In recent years, withing the increasing number and demand of users in 5G wireless access network, the demand for FD wireless is more and more urgent.

In 4G and earlier wireless communication system, the time division duplex (TDD) and frequency division duplex (FDD) are currently widely used. Unlike FD, TDD is a kind of half-duplex (HD) wireless, it cuts the time into pieces and

W. Wu and Z. Zhang (Eds.): COCOA 2020, LNCS 12577, pp. 727–740, 2020.
https://doi.org/10.1007/978-3-030-64843-5_49

use some of them to transmit and others to reception; FDD deals with transmission and reception at same time but uses two different frequency. Therefore, the FD wireless can double the throughput theoretically [2].

The basic challenge before using full-duplex wireless is the vast amount of self-interference (SI) leaking from the transmitter (Tx) into the receiver (Rx), the self-interference could be even billions of times stronger than the desired receive signal [3]. How to cancel this interference successfully and recover the desired signal is the current important research content. In recent work the SI cancellation (SIC) has been designed and used at at the antenna interface, radio frequency(RF)/analog, and digital baseband [4,5].

Another important technology in communication is Tx and Rx phased arrays which has been widely used in radar. Phased arrays is a technology for directional signal transmission and reception utilizing spatial selectivity [6], it can enhance the communication range through analog Tx beamforming (TxBF) and Rx beamforming (RxBF). TxBF/RxBF can increase the Tx/Rx signal power significantly at the same link distance, or can the enhance link distance at the same signal-to-noise ratio (SNR).

Although application prospects of combining FD with phased arrays are very attractive, there are still many challenges we have to face. First of all, in an N-element FD phase array, a SI channel exists between every pair of Tx and Rx elements and there are a total of N^2, these SI channels need to be canceled in the RF domain. Some techniques in common phased arrays such as using circuits or alternating antenna placements to achieve wideband RF SIC [7], can not be directly applied to an FD phased array. This is because these techniques are designed for single antenna or small array system. For large phased arrays, adding extra RF cancellers is expensive, and the cancellation via antenna placements usually requires at least twice the antennas. Therefore, innovative solutions are needed to achieve FD operation in phased arrays.

In this paper, we show that by adjusting Tx and Rx *analog beamforming weights* (or called *beamformers*) carefully, FD phase arrays can simultaneously achieve wideband RF SIC with minimal TxBF and RxBF gain losses, and improved FD rate gains. In another word, this method can be seen as a RF SIC based on TxBF and RxBF. This method has many advantages: (i) this RF SIC is wideband because the SI channels in every pair of Tx and Rx elements experiences similar delays; (ii) we just adjust the beamforming weights so a specialized RF SIC circuitry is not required, thus can save energy; (iii) we achieve the RF SIC before the digital domain, which could largely reduce the analog-to-digital converter (ADC) dynamic range and power consumption; (iv) the large-scale antenna arrays are not a burden but provide a large number of modifiable TxBF and RxBF weights.

We consider a network system that BS is an N-element FD phassed array, and its user is a single-antenna user which is HD- or FD-capable. In some related works [8], the main target is to maximize the FD rate gain with SI constrains. However, in many actual application scenarios, the gain only needs to be larger than a specified threshold, while SI should be as small as possible. Thus in our case, the target is to minimize the SI while ensuring that the FD rate gain is big enough. Base on our target we formulate an optimization problem to determine the optimal Tx and Rx beamformers jointly. Because of the non-convexity and intractability of this problem, we present an approximate alternative algorithm and split the original problem into convex sub-problems, then solve these sub-problems by changing them from complex models into real models and then using logarithmic barrier method. This alternative algorithm can solve the optimization problem efficiently.

To summarize, the main contributions of this paper are as fellow: (i) building a model for the problem of minimizing SI; (ii) giving out an iterative algorithms and split the non-convex original problem into convex sub-problems; (iii) changing these sub-problems from complex into real and solving them by the logarithmic barrier method.

2 Models and Algorithms

We now consider a full-duplex antenna arrays phase problem that has a same background as the [8]. A N-element rectangular antenna array with N_x rows and N_y columns ($N = N_x \times N_y$), arranged same as in [8]. Its transmit power is P_t. The spaces between adjacent antennas are half-wavelength. Assuming the horizontal angle and the pitch angle are noted as ϕ and θ. Then the relevant phase delay $s_n(\phi, \theta)$ of the nth element at location (n_x, n_y) is

$$s_n(\phi, \theta) = e^{j\pi[(n_x-1)\cos\theta\cos\phi+(n_y-1)\cos\theta\sin\phi]}, \forall n_x, n_y. \tag{1}$$

where $n = (n_x - 1)N_y + n_y$ [9]. The steering vector of the antenna array in the direction of (ϕ, θ) is $s(\phi, \theta) = [s_n(\phi, \theta)] \in \mathbb{C}$. So for the transmitter and receiver, their relevant phase delay are noted as $s_t = s(\phi_t, \theta_t)$ and $s_r = s(\phi_r, \theta_r)$.

The variables that can be adjusted are called analog beamformers. An analog beamformer is the set of complex-valued weights applied to each element relative to that of the first element. We note the transmit (Tx) and receive (Rx) beamformers as the vectors $w = [w_n] \in \mathbb{C}^N$ and $v = [v_n] \in \mathbb{C}^N$ respectively, and they must satisfy $|w_n| \leq 1$ and $|v_n| \leq 1$. The Tx and Rx array factors in the desired beampointing directions are denoted respectively by a_t and a_r, and given by

$$a_t = s_t^T w, a_r = s_r^T v.$$

According to [9], let $w = s_t^*$ and $v = s_r^*$, array factors can achieve their theoretical maximum value N, i.e.,

$$a_t^{max} = s_t^T s_t^* = N, a_r^{max} = s_r^T s_r^* = N. \tag{2}$$

Besides, The phase of the signals should be preserved, i.e., the array factors should only contain real parts:

$$a_t, a_r \in \mathbb{R} \Leftrightarrow Im[a_t] = Im[a_r] = 0, \tag{3}$$

Naturally, we respectively set $\boldsymbol{w}_{conv} = \boldsymbol{s}_t^*$ and $\boldsymbol{v}_{conv} = \boldsymbol{s}_r^*$ as the conventional HD Tx and Rx beamformers. Therefore, the TxBF and RxBF gains in the desired Tx and Rx beampointing directions, denoted by g_t and g_r, are given by

$$g_t = |a_t|^2/N, g_r = |a_r|^2/N. \tag{4}$$

Then we consider the form of self-interference. The SI channel matrix should be defined first. Same as in [10], we denote H_{nn} as the frequency response of the SI channel from the nth Tx element to the nth Rx element for every antenna $n(n = 1, 2, ..., N)$, and H_{mn} represent the frequency response of the cross-talk SI (CTSI) channel from the nth Tx element to the mth Rx element for every two different antenna $m \neq n(m, n = 1, 2, ..., N)$. We assume $|H_{mn}| \leq 1$ for all the $m, n = 1, 2, ..., N$ because of the propagation loss of the Tx signal. The SI channel matrix is denoted as $\boldsymbol{H} = [H_{mn}] \in \mathbb{C}^{N \times N}$, notice that for every pair (m, n), $H_{mn} = H_{nm}$, so $\boldsymbol{H}$ is a symmetric matrix. The SI power under TxBF and RxBF are:

$$P_{SI}^{bf} = |\boldsymbol{v}^T \boldsymbol{H} \boldsymbol{w}|^2 \cdot \frac{P_t}{N}. \tag{5}$$

The most commonly used measure of SI is the Self-Interference-to-Noise Ratio (XINR) under TxBF and RxBF. XINR is defined as the ratio between the residual SI power after analog and digital SIC and the noise floor, and denoted as γ_{bb} for the BS and γ_{uu} in for users. In our case, we only focus on the SIC at the BS with TxBF and RxBF, so we assume that a user can always cancel its SI to below the noise floor, i.e., $\gamma_{uu} \leq 1$ [11,12]. For the BS, it has an Rx array noise floor of NP_{nf}, and an amount of achievable digital SIC denoted by SIC_{dig}. Then we have:

$$\gamma_{bb} = \frac{P_{SI}^{bf}}{SIC_{dig} \cdot NP_{nf}}.$$

$$= \frac{|\boldsymbol{v}^T \boldsymbol{H} \boldsymbol{w}|^2}{SIC_{dig} NP_{nf}} \cdot \frac{P_t}{N}. \tag{6}$$

As we said in chapter 1, the target is to minimize the SI. Let $\beta = SIC_{dig} P_{nf}$, our aim is as:

$$\min_{\boldsymbol{w}, \boldsymbol{v}} \frac{\left|\boldsymbol{v}^T \boldsymbol{H} \boldsymbol{w}\right|^2}{N\beta} \cdot \frac{Pt}{N}. \tag{7}$$

In a given FD antenna array system, N, P_t, β are positive constants, so we can use an equivalent target as:

$$\min_{\boldsymbol{w}, \boldsymbol{v}} \left|\boldsymbol{v}^T \boldsymbol{H} \boldsymbol{w}\right|^2. \tag{8}$$

Base on the objective and constrains defined above, now we can present the model:

$$\min_{w,v} \quad |v^T H w|^2, \tag{9}$$

$$\text{s.t.} \quad Re\left[s_t^T w\right] \geq a, \tag{10}$$

$$Re\left[s_r^T v\right] \geq a, \tag{11}$$

$$Im\left[s_t^T w\right] = Im\left[s_r^T v\right] = 0, \tag{12}$$

$$w_n^2 \leq 1, v_n^2 \leq 1, \forall n = 1, 2, ..., N. \tag{13}$$

in which a is an acceptable gain level, and smaller than the theoretical maximum gain value, i.e., $0 \leq a \leq N$, for example, may be $0.8N$ or $0.9N$.

In this model, w and v are both variables, so the target function (9) is not convex, this model is not a convex optimization and hard to be solved directly. But it's obvious that all the constrains are liner or convex, so we can now use an alternative optimization algorithm.

The target function can be written as:

$$\begin{aligned}
&|v^T H w|^2 \\
&= (v^T H w)^\dagger \cdot (v^T H w) \\
&= w^\dagger (H^\dagger v^* v^T H) w \\
&= w^\dagger H_v w,
\end{aligned} \tag{14}$$

in which $H_v = H^\dagger v^* v^T H$. $H_v^\dagger = H_v$ holds for a fixed Rx beamformer v, this means H_v is a Hermitian matrix. And because the SI power cannot be negative for any non-zero Tx beamformer w, i.e.,

$$w^\dagger H_v w \geq 0, \forall w \in \mathbb{C}^N, w \neq 0, \tag{15}$$

this means H_v is a positive semidefinite matrix. H_w can be defined in the same way.

Now for a fixed Rx beamformer v, we can get the optimal Tx beamformer w by solving sub-problem (P1):

$$(P1) \min_w \quad w^\dagger H_v w, \tag{16}$$

$$\text{s.t.} \quad Re\left[s_t^T w\right] \geq a, \tag{17}$$

$$Im\left[s_t^T w\right] = 0, \tag{18}$$

$$w_n^2 \leq 1, \forall n = 1, 2, ..., N \tag{19}$$

Now the objective function of (P1) is a quadratic form, and the constrains are all convex, so this is a quadratically constrained convex program.

Algorithm 1. Iterative Algorithm 1

Input: the number of antennas N, the total Tx power of P_t, the SI matrix H, steering vectors $s_t = s(\phi_t, \theta_t)$ and $s_r = s(\phi_r, \theta_r)$, the gain level a, the iterate terminal condition parameter δ and M.

Initialization: Initial values of the Tx beamformer $w^{(0)}$ and the Rx beamformer $v^{(0)}$.

For $k = 0, 1, \ldots$ do:

1: Obtain $w^{(k+1)}$ with $v^{(k)}$ by solving:

$$(P1^{(k+1)}) \min_{w} \quad w^\dagger H_{v^{(k)}} w,$$

$$\text{s.t.} \quad Re\left[s_t^T w\right] \geq a,$$

$$Im\left[s_t^T w\right] = 0,$$

$$w_n^2 \leq 1, \forall n = 1, 2, \ldots, N$$

2: Obtain $v^{(k+1)}$ with $w^{(k+1)}$ by solving:

$$(P2^{(k+1)}) \min_{v} \quad v^\dagger H_{w^{(k+1)}} v,$$

$$\text{s.t.} \quad Re\left[s_r^T v\right] \geq a,$$

$$Im\left[s_r^T v\right] = 0,$$

$$v_n^2 \leq 1, \forall n = 1, 2, \ldots, N$$

3: Keep iterating until the Tx and Rx array factor improvements are small enough:

$$||v^{(k)\dagger} H w^{(k)}|^2 - |v^{(k+1)\dagger} H w^{(k+1)}|^2| \leq \delta \cdot M.$$

Symmetrically, for a fixed Tx beamformer w, the optimal Rx beamformer v can be gotten by solving (P2):

$$(P2) \min_{v} \quad v^\dagger H_w v, \tag{20}$$

$$\text{s.t.} \quad R\acute{e}\left[s_r^T v\right] \geq a, \tag{21}$$

$$Im\left[s_r^T v\right] = 0, \tag{22}$$

$$v_n^2 \leq 1, \forall n = 1, 2, \ldots, N \tag{23}$$

Naturally, we can think of updating w and v alternately by solving (P1) and (P2) in turn, or rather, solving v for a fixed w, then updating w by using this new v.

Now we can present the iterative algorithm [8] as Algorithm 1.

In this algorithm, each iteration can be proved to have improvements, this conclusion is provided as Theorem 1.

Theorem 1. *In each iteration the target function will not decrease.*

Proof. Assume in the k step, $\boldsymbol{v}^{(k)}$ has been got, then $\boldsymbol{w}^{(k+1)}$ can be got by solving model $P1^{(k+1)}$. So $\boldsymbol{H}_{w^{(k+1)}}$ can be computed. Then $\boldsymbol{v}^{(k+1)}$ can be got by solving model $P2^{(k+1)}$.

As model $P2^{(k+1)}$ is a convex model, $\boldsymbol{v}^\dagger \boldsymbol{H}_{w^{(k+1)}} \boldsymbol{v}$ obtain the minimum for all the vector $\boldsymbol{v}$ satisfied the constrained condition of model $P2^{(k+1)}$. Meanwhile,

$$\boldsymbol{w}^{(k+1)\dagger} \boldsymbol{H}_{v^{(k)}} \boldsymbol{w}^{(k+1)} = \boldsymbol{v}^{(k)\dagger} \boldsymbol{H}_{w^{(k+1)}} \boldsymbol{v}^{(k)}, \tag{24}$$

and $\boldsymbol{v}^{(k)}$ satisfied the constrained condition of model $P2^{(k+1)}$. So

$$\boldsymbol{v}^{(k+1)\dagger} \boldsymbol{H}_{w^{(k+1)}} \boldsymbol{v}^{(k+1)} \leq \boldsymbol{v}^{(k)\dagger} \boldsymbol{H}_{w^{(k+1)}} \boldsymbol{v}^{(k)} = \boldsymbol{w}^{(k+1)\dagger} \boldsymbol{H}_{v^{(k)}} \boldsymbol{w}^{(k+1)}. \tag{25}$$

Similarly,

$$\boldsymbol{w}^{(k+2)\dagger} \boldsymbol{H}_{v^{(k+1)}} \boldsymbol{w}^{(k+2)} \leq \boldsymbol{v}^{(k+1)\dagger} \boldsymbol{H}_{w^{(k+1)}} \boldsymbol{v}^{(k+1)}. \tag{26}$$

So

$$\boldsymbol{w}^{(k+2)\dagger} \boldsymbol{H}_{v^{(k+1)}} \boldsymbol{w}^{(k+2)} \leq \boldsymbol{v}^{(k+1)\dagger} \boldsymbol{H}_{w^{(k+1)}} \boldsymbol{v}^{(k+1)} \leq \boldsymbol{w}^{(k+1)\dagger} \boldsymbol{H}_{v^{(k)}} \boldsymbol{w}^{(k+1)}. \tag{27}$$

Then Theorem 1 is proven.

Now the problem is changed into *how to solve the sub-problems*. The *Logarithmic barrier method* is a good way to solve the inequality constrained convex optimization. The details of this method is described in [13].

Next question we have to solve is that this method is just designed for real numbers and cannot be directly used for our complex-value problem. So we have to convert sub-problems P1 and P2 from complex to real. Now taking P1 as an example, P2 is same. Reviewing the form of P1:

$$(P1) \min_{\boldsymbol{w}} \quad \boldsymbol{w}^\dagger \boldsymbol{H}_v \boldsymbol{w},$$
$$\text{s.t.} \quad Re\left[\boldsymbol{s}_t^T \boldsymbol{w}\right] \geq a,$$
$$Im\left[\boldsymbol{s}_t^T \boldsymbol{w}\right] = 0,$$
$$w_n^2 \leq 1, \forall n = 1, 2, ..., N.$$

First step, we divide these vectors into real and image parts, and do not change their inner products. So we introduce a symbol: for a complex vector $\boldsymbol{c} = (c_1, c_2, ...c_n)$, the real part vector and the image part vector are denoted respectively as

$$Re[\boldsymbol{c}] = (Re[c_1], Re[c_2], ...Re[c_n]), \tag{28}$$

$$Im[\boldsymbol{c}] = (Im[c_1], Im[c_2], ...Im[c_n]). \tag{29}$$

Now set

$$\boldsymbol{w}_e = [Re[\boldsymbol{w}^T], Im[\boldsymbol{w}^T]]^T, \tag{30}$$

$$\boldsymbol{s}_{tR} = [Re[\boldsymbol{s}_t^T], -Im[\boldsymbol{s}_t^T]]^T, \boldsymbol{s}_{tI} = [Im[\boldsymbol{s}_t^T], Re[\boldsymbol{s}_t^T]]^T. \tag{31}$$

It's easily to check that

$$Re[\boldsymbol{s}_t^T \boldsymbol{w}] = \boldsymbol{s}_{tR}^T \boldsymbol{w}_e, Im[\boldsymbol{s}_t^T \boldsymbol{w}] = \boldsymbol{s}_{tI}^T \boldsymbol{w}_e. \tag{32}$$

Next set

$$H_{vR} = \begin{bmatrix} Re[H_v] & -Im[H_v] \\ Im[H_v] & Re[H_v] \end{bmatrix}, \tag{33}$$

and get

$$w^\dagger H_v w = w_e^\dagger H_{vR} w_e. \tag{34}$$

The last piece of new problem is a little complex, w_n^2, or $Re[w_n]^2 + Im[w_n]^2$ is not really an intuitive function about w_e, so we define a group of matrices:

$$U_n = [u_{ij}^n]_{2N \times 2N}, \; in \; which$$

$$u_{ij}^n = \begin{cases} 1, & i = j = n \; or \; i = j = n + N, \\ 0, & else. \end{cases} \tag{35}$$

Then we can check that $Re[w_n]^2 + Im[w_n]^2 = w_e^T U_n w_e$. Now the complex problem P1 have been converted into a real problem P1':

$$(P1') \min_{w_e} \quad w_e^\dagger H_{vR} w_e, \tag{36}$$

$$s.t. \quad -s_{tR}^T w_e + a \le 0, \tag{37}$$

$$s_{tI}^T w_e = 0, \tag{38}$$

$$w_e^T U_n w_e - 1 \le 0, \forall n = 1, 2, ..., N. \tag{39}$$

In the same way, we can define v_e, S_{rR}, S_{rI} and H_{wR}, and then get a real problem P2':

$$(P2') \min_{v_e} \quad v_e^\dagger H_{wR} v_e, \tag{40}$$

$$s.t. \quad -s_{rR}^T v_e + a \le 0, \tag{41}$$

$$s_{rI}^T v_e = 0, \tag{42}$$

$$v_e^T U_n v_e - 1 \le 0, \forall n = 1, 2, ..., N. \tag{43}$$

Now we can use the barrier method for P1' and P2'. According to [13], the Newton's method is a base step in the barrier method and it requires to calculate the gradient and second gradient of the target function and all constrains. We also don't describe the details here, but we present all these derivative here. Using P1' as an example, let

$$F(w_e) = w_e^T H_{vR} w_e, \tag{44}$$

$$gl(w_e) = -s_{tR}^T w_e + a, \tag{45}$$

$$ge(w_e) = s_{tI}^T w_e, \tag{46}$$

$$u_n(w_e) = w_e^T U_n w_e - 1, \forall n = 1, 2, ..., N. \tag{47}$$

then the results are as follow:

$$\nabla F(\boldsymbol{w}_e) = 2\boldsymbol{H}_{vR}\boldsymbol{w}_e, \nabla^2 F(\boldsymbol{w}_e) = 2\boldsymbol{H}_{vR}, \tag{48}$$

$$\nabla gl(\boldsymbol{w}_e) = -\boldsymbol{s}_{tR}^T, \qquad \nabla^2 gl(\boldsymbol{w}_e) = 0, \tag{49}$$

$$\nabla ge(\boldsymbol{w}_e) = \boldsymbol{s}_{tI}^T, \qquad \nabla^2 ge(\boldsymbol{w}_e) = 0, \tag{50}$$

$$\nabla u_n(\boldsymbol{w}_e) = 2\boldsymbol{U}_n\boldsymbol{w}_e, \nabla^2 u_n(\boldsymbol{w}_e) = 2\boldsymbol{U}_n. \tag{51}$$

After all these steps, now Algorithm 1 can finally be used and get the solution pair $\boldsymbol{w}$ and $\boldsymbol{v}$. We will check its performance in the next chapter.

In the previous model only the value of gains and the XINR are considered, so sometimes the TxBF and RxBF gains have a significant gap between each other. But in some actual application scenarios, the TxBF and RxBF gains should be balanced in the far-field, i.e., g_t and g_r (or equivalently, a_t and a_r) should not be too different, but also don't need to be exactly same. To achieve this goal, we just need to add a penalty term which is depend on the difference between those two gains.

Intuitively we use $(a_t - a_r)^2$ as the penalty term with a sequence of step size $\{\alpha_k\}$ which satisfies $1 = \alpha_1 \geq \alpha_2 \geq ... > 0$. Attention that this is a minimization problem, so the sign of penalty term is positive. For fixed $\boldsymbol{w}$, $\boldsymbol{v}$, and α, now the sub-problem are changed into P3 and P4 as follow, with noticing that $a_t = \boldsymbol{s}_t^T\boldsymbol{w}$ and $a_r = \boldsymbol{s}_r^T\boldsymbol{v}$.

$$(P3) \min_{\boldsymbol{w}} \quad \boldsymbol{w}^\dagger \boldsymbol{H}_v\boldsymbol{w} + \alpha(\boldsymbol{s}_t^T\boldsymbol{w} - \boldsymbol{s}_r^T\boldsymbol{v})^2, \tag{52}$$

$$\text{s.t.} \quad Re\left[\boldsymbol{s}_t^T\boldsymbol{w}\right] \geq a, \tag{53}$$

$$Im\left[\boldsymbol{s}_t^T\boldsymbol{w}\right] = 0, \tag{54}$$

$$w_n^2 \leq 1, \forall n = 1, 2, ..., N \tag{55}$$

$$(P4) \min_{\boldsymbol{v}} \quad \boldsymbol{v}^\dagger \boldsymbol{H}_w\boldsymbol{v} + \alpha(\boldsymbol{s}_t^T\boldsymbol{w} - \boldsymbol{s}_r^T\boldsymbol{v})^2, \tag{56}$$

$$\text{s.t.} \quad Re\left[\boldsymbol{s}_r^T\boldsymbol{v}\right] \geq a, \tag{57}$$

$$Im\left[\boldsymbol{s}_r^T\boldsymbol{v}\right] = 0, \tag{58}$$

$$v_n^2 \leq 1, \forall n = 1, 2, ..., N \tag{59}$$

The new algorithm are given as Iterative Algorithm 2.

We also use the barrier method to solve this problem, here gives the form of converted real problem P3' and P4':

$$(P3') \min_{\boldsymbol{w}_e} \quad \boldsymbol{w}_e^\dagger \boldsymbol{H}_{vR}\boldsymbol{w}_e + \alpha(\boldsymbol{s}_{tR}^T\boldsymbol{w}_e - \boldsymbol{s}_{rR}^T\boldsymbol{v}_e)^2, \tag{60}$$

$$\text{s.t.} \quad -\boldsymbol{s}_{tR}^T\boldsymbol{w}_e + a \leq 0, \tag{61}$$

$$\boldsymbol{s}_{tI}^T\boldsymbol{w}_e = 0, \tag{62}$$

$$\boldsymbol{w}_e^T\boldsymbol{U}_n\boldsymbol{w}_e - 1 \leq 0, \forall n = 1, 2, ..., N. \tag{63}$$

Algorithm 2. Iterative Algorithm 2

Input: the number of antennas N, the total Tx power of P_t, the SI matrix $\boldsymbol{H}$, steering vectors $\boldsymbol{s}_t = \boldsymbol{s}(\phi_t, \theta_t)$ and $\boldsymbol{s}_r = \boldsymbol{s}(\phi_r, \theta_r)$, the gain level a, the sequence of step size $\{\alpha_k\}$, the iterate terminal condition parameter δ and M.

Initialization: Initial values of the Tx beamformer $\boldsymbol{w}^{(0)}$ and the Rx beamformer $\boldsymbol{v}^{(0)}$.

For $k = 0, 1, ...$ do:

1: Obtain $\boldsymbol{w}^{(k+1)}$ with $\boldsymbol{v}^{(k)}$ by solving:

$$(P3^{(k+1)}) \min_{\boldsymbol{w}} \quad \boldsymbol{w}^\dagger \boldsymbol{H}_{\boldsymbol{v}^{(k)}} \boldsymbol{w} + \alpha_k (\boldsymbol{s}_t^T \boldsymbol{w} - \boldsymbol{s}_r^T \boldsymbol{v}^{(k)})^2,$$

$$\text{s.t.} \quad Re\left[\boldsymbol{s}_t^T \boldsymbol{w}\right] \geq a,$$

$$Im\left[\boldsymbol{s}_t^T \boldsymbol{w}\right] = 0,$$

$$w_n^2 \leq 1, \forall n = 1, 2, ..., N$$

2: Obtain $\boldsymbol{v}^{(k+1)}$ with $\boldsymbol{w}^{(k+1)}$ by solving:

$$(P4^{(k+1)}) \min_{\boldsymbol{v}} \quad \boldsymbol{v}^\dagger \boldsymbol{H}_{\boldsymbol{w}^{(k+1)}} \boldsymbol{v} + \alpha_k (\boldsymbol{s}_t^T \boldsymbol{w}^{(k+1)} - \boldsymbol{s}_r^T \boldsymbol{v})^2,$$

$$\text{s.t.} \quad Re\left[\boldsymbol{s}_r^T \boldsymbol{v}\right] \geq a,$$

$$Im\left[\boldsymbol{s}_r^T \boldsymbol{v}\right] = 0,$$

$$v_n^2 \leq 1, \forall n = 1, 2, ..., N$$

3: Keep iterating until the Tx and Rx array factor improvements are small enough:

$$||\boldsymbol{v}^{(k)\dagger} \boldsymbol{H} \boldsymbol{w}^{(k)}|^2 - |\boldsymbol{v}^{(k+1)\dagger} \boldsymbol{H} \boldsymbol{w}^{(k+1)}|^2| \leq \delta \cdot M.$$

$$(P4') \min_{\boldsymbol{v}_e} \quad \boldsymbol{v}_e^\dagger \boldsymbol{H}_{wR} \boldsymbol{v}_e + \alpha(\boldsymbol{s}_{rR}^T \boldsymbol{v}_e - \boldsymbol{s}_{tR}^T \boldsymbol{w}_e)^2, \tag{64}$$

$$\text{s.t.} \quad -\boldsymbol{s}_{rR}^T \boldsymbol{v}_e + a \leq 0, \tag{65}$$

$$\boldsymbol{s}_{rI}^T \boldsymbol{v}_e = 0, \tag{66}$$

$$\boldsymbol{v}_e^T \boldsymbol{U}_n \boldsymbol{v}_e - 1 \leq 0, \forall n = 1, 2, ..., N. \tag{67}$$

Because the constrains are same as above, we only show the gradient and second gradient of the new target function. Using P3' as an example, let

$$F(\boldsymbol{w}_e) = \boldsymbol{w}_e^T \boldsymbol{H}_{vR} \boldsymbol{w}_e, \tag{68}$$

then the results are as follow:

$$\nabla F(\boldsymbol{w}_e) = 2\boldsymbol{H}_{vR} \boldsymbol{w}_e + 2\alpha(\boldsymbol{s}_{tR} \boldsymbol{s}_{tR}^T \boldsymbol{w}_e - \boldsymbol{s}_{rR}^T \boldsymbol{v}_e \boldsymbol{s}_{tR}), \tag{69}$$

$$\nabla^2 F(\boldsymbol{w}_e) = 2\boldsymbol{H}_{vR} + 2\alpha \boldsymbol{s}_{tR} \boldsymbol{s}_{tR}^T, \tag{70}$$

Same as above, the performance of Algorithm 2 will be showed in the next chapter.

3 Simulations and Results

Before checking performances of these methods, we should first define some variables as performance standards. We have already defined the XINR (denoted as γ_{bb}) to show the interference level, and TxBF and RxBF gains (denoted as g_t and g_r) to show the cover ability of FD. Then what we care about are gain losses of TxBF and RxBF, which are defined as the ratio of the TxBF (resp. RxBF) gains to their theoretical maximum values, as follow:

$$GainLoss(TX) = (\boldsymbol{s}_t^T * \boldsymbol{w})/(\boldsymbol{s}_t^T * \boldsymbol{w}_{conv}), \tag{71}$$

$$GainLoss(RX) = (\boldsymbol{s}_r^T * \boldsymbol{v})/(\boldsymbol{s}_r^T * \boldsymbol{v}_{conv}). \tag{72}$$

Another important thing is that *how much better is FD than HD*, or more specifically called *FD rate gain*. Denote the FD link rate as r_{BU}^{FD} and the HD link rate as r_{BU}^{HD}, the FD rate gain is defined as r_{BU}^{FD}/r_{BU}^{HD}. Because link rates need to be calculated by Shannon's formula, we should give the signal-to-noise ratio (SNR) first. For the BS-User case, we denote $\gamma_{bu}(u$ to $b)$ and $\gamma_{ub}(b$ to $u)$ as the UL and DL SNRs respectively. Then HD link rate of BS-User is defined as follow,

$$r_{BU}^{HD} = \frac{B}{2}log2(1 + N\gamma_{bu}) + \frac{B}{2}log2(1 + N\gamma_{ub}). \tag{73}$$

Correspondingly, FD link rate is defined as follow,

$$r_{BU}^{FD} = B[log_2(1 + \frac{g_r^* \gamma_{bu}}{1 + \gamma_{bb}}) + log_2(1 + \frac{g_t^* \gamma_{ub}}{1 + \gamma_{uu}})]. \tag{74}$$

For convenience we approximate the FD sum rate in (74) by setting $\gamma_{uu} = 1$, and we use this approximate FD sum rates in our simulations:

$$\tilde{r}_{BU}^{FD} = B[log_2(1 + \frac{g_r^* \gamma_{bu}}{1 + \gamma_{bb}}) + log_2(1 + \frac{g_t^* \gamma_{ub}}{2})]. \tag{75}$$

B represents the bandwidth in these formulas, however, the FD rate gain is not influenced by B because of its definition.

Now we consider a system with $N_x = 4$ and $N_y = 8$, thus $N = 32$. Its TxBF and RxBF in the front side of the antenna array. In actual application scenarios, the elevation angle of the antenna array is not usually changed, while the azimuth angle is often adjusted. So we chose $\phi \in [-90°, 90°]$ and a fixed $\theta = 10°$. We consider $P_t = 30$ dBm, and $P_{nf} = -90$ dBm and $SIC_{dig} = 40$ dB [original paper]. Data of the SI channel matrix $\boldsymbol{H}$ are provided from the Huawei Technologies Co., Ltd which we cooperate with. We denote the acceptable gain level in (2, 3) as $a = rate \cdot N$, and chose $rate = \{0.7, 0.8, 0.9\}$. At last, let $\gamma_{bu} = \gamma_{ub} = 25$ dB, bandwidth $B = 50$ MHz.

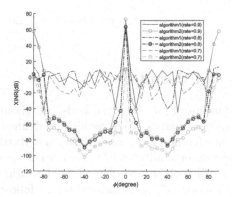

Fig. 1. The XINR in different algorithms.

Put these data into Algorithm 1 and Algorithm 2 with initial value $w(0) = 0.99 * s_t^*$ and $v(0) = 0.99 * s_r^*$. The mainly results are shown as follow. Especially in Algorithm 2, we let $\alpha_k = 1/(k+1)^2 (\forall k = 0, 1, ...)$.

Figure 1 shows the performances of XINR. It means the SI level after we adjust the beamformers. In actual application scenarios, under 0dB is small enough.

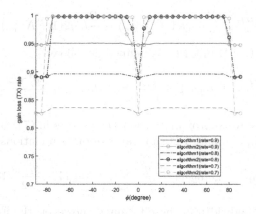

Fig. 2. The Tx gain loss in different algorithms.

Figure 2 and Fig. 3 show the performances of Tx and Rx gain loss in different algorithms and different azimuth angles. These means how much percentage of Tx and Rx gain are retained after we adjust the beamformers. Their theoretical maximums are 1.

Figure 4 shows the performances of FD gain ratio. It means how much better FD is than HD. Its theoretical maximums is 2.

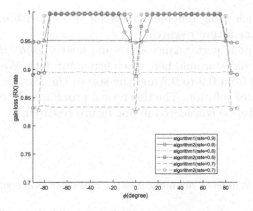

Fig. 3. The Rx gain loss in different algorithms.

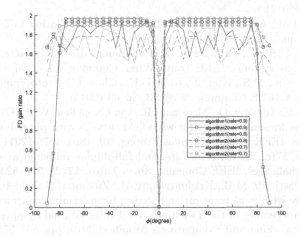

Fig. 4. The FD gain ratio in different algorithms.

We can see the performances of these two algorithms are satisfactory, especially the Algorithm 2, most of the results are not far from their theoretical maximum values, except near $0°$ $and \pm 90°$. As these simulations show, Algorithm 2 has better and more stable results than Algorithm 1. For every algorithm, the rate determine the size of the feasible domain, the lower rate leads to a bigger domain, so our target XINR should decrease with the rate. The rate decreases also means the minimum gain level decreases, so the gain loss and FD gain ratio also decrease. The results of Algorithm 2 obey this rule but Algorithm 1 are very unstable.

The Algorithm 2 has a particularly good performance, but in fact we don't proof yet that every iteration can have an improvement. We think this might be achieve by choosing a proper $\{\alpha_k\}$. And we also wander why a small penalty term can cause such a big difference between two algorithm. These issues will be considered in the future research.

We think the poor performance of results near $0°$ $and \pm 90°$, is because the constrains are too strict around here when using our data. We can see that when the rate decreases from 0.9 to 0.7 and the size of the feasible domain increases, it gets a much better solution. The theoretical proof and the existence condition of solutions will also be considered in our future research.

References

1. Goldsmith, A.: Wireless Communications. Cambridge University Press, Cambridge (2005)
2. Amjad, M., Akhtar, F., Rehmani, M.H., Reisslein, M., Umer, T.: Full-duplex communication in cognitive radio networks: a survey. IEEE Commun. Surv. Tutor. **19**, 2158–2191 (2017)
3. Hong, S., et al.: Applications of self-interference cancellation in 5G and beyond. IEEE Commun. Mag. **52**, 114–121 (2014)
4. Duarte, M., Dick, C., Sabharwal, A.: Experiment-driven characterization of full-duplex wireless systems. IEEE Trans. Wirel. Commun. **11**, 4296–4307 (2012)
5. Chung, M., Sim, M.S., Kim, J., Kim, D.K., Chae, C.: Prototyping real-time full duplex radios. IEEE Commun. Mag. **53**, 56–63 (2015)
6. Ivashina, M.V., Iupikov, O., Maaskant, R., van Cappellen, W.A., Oosterloo, T.: An optimal beamforming strategy for wide-field surveys with phased-array-fed reflector antennas. IEEE Trans. Antennas Propag. **59**, 1864–1875 (2011)
7. Liu, G., Yu, F.R., Ji, H., et al.: In-band full-duplex relaying: a survey, research issues and challenges. IEEE Commun. Surv. Tutor. **17**(2), 500–524 (2015)
8. Chen, T., Dastjerdi, M.B., Krishnaswamy, H., Zussman, G.: wideband full-duplex phased array with joint transmit and receive beamforming: optimization and rate gains. In: Proceedings of the Twentieth ACM International Symposium on Mobile Ad Hoc Networking and Computing - Mobihoc 2019, pp. 361–370. ACM Press, Catania (2019). https://doi.org/10.1145/3323679.3326534
9. Balanis, C.A.: Antenna Theory: Analysis and Design. Wiley, Hoboken (2016)
10. Everett, E., Shepard, C., Zhong, L., Sabharwal, A.: SoftNull: many-antenna full-duplex wireless via digital beamforming. IEEE Trans. Wirel. Commun. **15**, 8077–8092 (2016)
11. Chen, T., Baraani Dastjerdi, M., Zhou, J., Krishnaswamy, H., Zussman, G.: Wideband full-duplex wireless via frequency-domain equalization: design and experimentation. In: The 25th Annual International Conference on Mobile Computing and Networking, pp. 1–16. Association for Computing Machinery, New York (2019). https://doi.org/10.1145/3300061.3300138
12. Zhou, J., et al.: Integrated full duplex radios. IEEE Commun. Mag. **55**, 142–151 (2017)
13. Boyd, S., Boyd, S.P., Vandenberghe, L.: Convex Optimization. Cambridge University Press, Cambridge (2004)

Graph Problems

Oriented Coloring of msp-Digraphs
and Oriented Co-graphs
(Extended Abstract)

Frank Gurski[✉], Dominique Komander, and Marvin Lindemann

Institute of Computer Science, Algorithmics for Hard Problems Group,
Heinrich-Heine-University Düsseldorf, 40225 Düsseldorf, Germany
frank.gurski@hhu.de

Abstract. Graph coloring is an assignment of labels, so-called colors, to the objects of a graph subject to certain constraints. The coloring problem on undirected graphs has been well studied, as it is one of the most fundamental problems. Meanwhile, there are very few results for coloring problems on directed graphs. An oriented k-coloring of an oriented graph G is a partition of the vertices into k independent sets such that all the arcs, linking two of these subsets, have the same direction. The oriented chromatic number of an oriented graph G is the smallest k such that G allows an oriented k-coloring. Even deciding whether an acyclic digraph allows an oriented 4-coloring is NP-hard. This motivates to consider the problem on special graph classes.

In this paper we consider two different recursively defined classes of oriented graphs, namely msp-digraphs (short for minimal series-parallel digraphs) and oriented co-graphs (short for oriented complement reducible graphs).

We show that every msp-digraph has oriented chromatic number at most 7 and give an example that this is best possible. We use this bound together with the recursive structure of msp-digraphs to give a linear time solution for computing the oriented chromatic number of msp-digraphs.

Further, we use the concept of perfect orderable graphs in order to show that for acyclic transitive digraphs every greedy coloring along a topological ordering leads to an optimal oriented coloring, which generalizes a known solution for the oriented coloring problem on oriented co-graphs.

Keywords: Oriented coloring · msp-digraphs · Oriented co-graphs · Linear time algorithms

1 Introduction

Coloring problems are among the most famous problems in graph theory since they allow to model many real-life problems under a graph-theoretical formalism. Graph coloring is an assignment of labels, so-called colors, to the objects of a graph subject to certain constraints. Usually, vertices or edges are considered as objects. Such problems have multiple applications [4, 20, 22, 30].

© Springer Nature Switzerland AG 2020
W. Wu and Z. Zhang (Eds.): COCOA 2020, LNCS 12577, pp. 743–758, 2020.
https://doi.org/10.1007/978-3-030-64843-5_50

A k-coloring for an undirected graph G is a k-labeling of the vertices of G such that no two adjacent vertices have the same label. The smallest k such that a graph G has a k-coloring is named the chromatic number of G. As even the problem whether a graph has a 3-coloring, is NP-complete, finding the chromatic number of an undirected graph is an NP-hard problem. However, there are many efficient solutions for the coloring problem on special graph classes, like chordal graphs [17], comparability graphs [21], and co-graphs [6].

Oriented coloring, which also considers the directions of the arcs, has been introduced by Courcelle [7]. An oriented k-coloring of an oriented graph $G = (V, E)$ is a partition of the vertex set V into k independent sets, such that all the arcs linking two of these subsets have the same direction. The *oriented chromatic number* of an oriented graph G, denoted by $\chi_o(G)$, is the smallest k such that G allows an oriented k-coloring. In the oriented chromatic number problem (OCN for short) there is given an oriented graph G and an integer k and one has to decide whether there is an oriented k-coloring for G. Even the restricted problem, in which k is constant and does not belong to the input (OCN$_k$ for short), is hard. While the undirected coloring problem on graphs without cycles is easy to solve, OCN$_4$ is NP-complete even for DAGs [10]. This makes research in this field particularly interesting. Oriented coloring arises in scheduling models where incompatibilities are oriented [10].

Right now, the definition of oriented coloring is mostly considered for undirected graphs. In this case, the maximum value $\chi_o(G')$ of all possible orientations G' of an undirected graph G is considered. For several special undirected graph classes the oriented chromatic number has been bounded. Among these are outerplanar graphs [27], planar graphs [26], and Halin graphs [14]. In [15], Ganian has shown an FPT-algorithm for OCN w.r.t. the parameter tree-width (of the underlying undirected graph). Furthermore, he has shown that OCN is DET-hard[1] for classes of oriented graphs for which hold that the underlying undirected class has bounded rank-width.

So far, oriented coloring of special classes of oriented graphs seems to be nearly uninvestigated. In this paper, we consider the oriented coloring problem restricted to msp-digraphs and oriented co-graphs.

Msp-digraphs, i.e. minimal series-parallel digraphs, can be defined from the single vertex graph by applying the parallel composition and series composition. By [2, Sect. 11.1] these graphs are useful for modeling flow diagrams and dependency charts and they are used in applications for scheduling under constraints. We show that the oriented chromatic number for msp-digraphs is at most 7 and show by an example that this is best possible. With this bound and the recursive structure of msp-digraphs we achieve a linear time solution for computing the oriented chromatic number of msp-digraphs.

Oriented co-graphs, i.e. oriented complement reducible graphs, are precisely digraphs which can be defined from the single vertex graph by applying the disjoint union and order composition. Oriented co-graphs were already analyzed

---

[1] DET is the class of decision problems which are reducible in logarithmic space to the problem of computing the determinant of an integer valued $n \times n$-matrix.

by Lawler in [25] and [6, Sect. 5] using the notation of transitive series parallel (TSP) digraphs. Their recursive structure can be used to compute an optimal oriented coloring and the oriented chromatic number in linear time [18]. In this paper we use the concept of perfect orderable graphs in order to show that for acyclic transitive digraphs every greedy coloring along a topological ordering leads to an optimal oriented coloring, which generalizes the known solution for the oriented coloring problem on oriented co-graphs given in [18].

2   Preliminaries

2.1   Graphs and Digraphs

We use the notations of Bang-Jensen and Gutin [1] for graphs and digraphs. A *graph* is a pair $G = (V, E)$, where V is a finite set of *vertices* and $E \subseteq \{\{u, v\} \mid u, v \in V, u \neq v\}$ is a finite set of *edges*. A *directed graph* or *digraph* is a pair $G = (V, E)$, where V is a finite set of *vertices* and $E \subseteq \{(u, v) \mid u, v \in V, u \neq v\}$ is a finite set of ordered pairs of distinct vertices called *arcs* or *directed edges*. For a vertex $v \in V$, the sets $N^+(v) = \{u \in V \mid (v, u) \in E\}$ and $N^-(v) = \{u \in V \mid (u, v) \in E\}$ are called the *set of all successors* and the *set of all predecessors* of v. The *outdegree* of v, outdegree(v) for short, is the number of successors of v and the *indegree* of v, indegree(v) for short, is the number of predecessors of v. The *maximum (vertex) degree* is $\Delta(G) = \max_{v \in V}(\text{outdegree}(v) + \text{indegree}(v))$.

For some given digraph $G = (V, E)$, we define its underlying undirected graph by ignoring the directions of the arcs, i.e. $un(G) = (V, \{\{u, v\} \mid (u, v) \in E, u, v \in V\})$. For some (di)graph class F we define Free(F) as the set of all (di)graphs G such that no induced sub(di)graph of G is isomorphic to a member of F.

A digraph $G' = (V', E')$ is a *subdigraph* of digraph $G = (V, E)$ if $V' \subseteq V$ and $E' \subseteq E$. If every arc of E with both end vertices in V' is in E', we say that G' is an *induced subdigraph* of G and we write $G' = G[V']$.

An *oriented graph* is a digraph with no loops and no opposite arcs. We will use the following special oriented graphs. By

$$\overrightarrow{P_n} = (\{v_1, \dots, v_n\}, \{(v_1, v_2), \dots, (v_{n-1}, v_n)\}),$$

$n \geq 2$, we denote the oriented path on n vertices, and by

$$\overrightarrow{C_n} = (\{v_1, \dots, v_n\}, \{(v_1, v_2), \dots, (v_{n-1}, v_n), (v_n, v_1)\}),$$

$n \geq 2$, we denote the oriented cycle on n vertices. By

$$\overrightarrow{K_{n,m}} = (\{v_1, \dots, v_n, w_1, \dots, w_m\}, \{(v_i, w_j) \mid 1 \leq i \leq n, 1 \leq j \leq m\}),$$

$n, m \geq 1$ we denote an oriented complete bipartite digraph with $n + m$ vertices. An *oriented forest (tree)* is an orientation of a forest (tree). An *out-rooted-tree* (*in-rooted-tree*) is an orientation of a tree with a distinguished root such that all arcs are directed away from (directed to) the root. A *directed acyclic graph*

(DAG for short) is a digraph without any $\overrightarrow{C_n}$, for $n \geq 2$, as subdigraph. A *tournament* is a digraph in which there is exactly one edge between every two distinct vertices.

A vertex v is *reachable* from vertex u in G if G contains a $\overrightarrow{P_n}$ as a subdigraph having start vertex u and end vertex v. A *topological ordering* of a directed graph is a linear ordering of its vertices such that for every directed edge (u, v), vertex u is before vertex v in the ordering. A digraph is *odd cycle free* if it does not contain any $\overrightarrow{C_n}$, for odd $n \geq 3$, as subdigraph. A digraph G is bipartite if $un(G)$ is bipartite and a digraph G is planar if $un(G)$ is planar.

We call a digraph $G = (V, E)$ *transitive* if for every pair $(u, v) \in E$ and $(v, w) \in E$ of arcs with $u \neq w$ the arc (u, w) also belongs to E. The *transitive closure* $tc(G)$ of a digraph G has the same vertex set as G and for two distinct vertices u, v there is an arc (u, v) in $tc(G)$ if and only if v is reachable from u in G.

2.2 Coloring Undirected Graphs

Definition 1 (Graph coloring). *A k-coloring of a graph $G = (V, E)$ is a mapping $c : V \rightarrow \{1, \dots, k\}$ such that:*

- *$c(u) \neq c(v)$ for every $\{u, v\} \in E$.*

The chromatic number of G, denoted by $\chi(G)$, is the smallest k such that G has a k-coloring.

Name Chromatic Number (CN)
Instance A graph $G = (V, E)$ and a positive integer $k \leq |V|$.
Question Is there a k-coloring for G?

If k is a constant and not part of the input, the corresponding problem is denoted by k-Chromatic Number (CN_k). Even on 4-regular planar graphs CN_3 is NP-complete [11].

It is well known that bipartite graphs are exactly the graphs which allow a 2-coloring and that planar graphs are graphs that allow a 4-coloring. On undirected co-graphs, the graph coloring problem can be solved in linear time using two formulas from [6], which are generalized in Lemma 3 for the oriented coloring problem.

Coloring a graph $G = (V, E)$ can be done by a greedy algorithm. For some given ordering π of V, the vertices are ordered as a sequence in which each vertex is assigned to the minimum possible value that is not forbidden by the colors of its neighbors, see Algorithm 1. Obviously, different orders can lead to different numbers of colors. But there is always an ordering yielding to the minimum number of colors, which is hard to find in general.

The set of perfectly orderable graphs are those graphs for which the greedy algorithm leads to an ordering with an optimal coloring, not only for the graph itself but also for all of its induced subgraphs.

Algorithm 1. GREEDY COLORING

Data: A graph G and an ordering $\pi : v_1 < \ldots < v_n$ of its vertices.
Result: An admitted vertex coloring $c : \{v_1, \ldots, v_n\} \mapsto \mathbb{N}$ of G.
for $(i = 1$ to $n)$ do
    $\lfloor\ c(v_i) = \infty$
$c(v_1) = 1;$
for $(i = 2$ to $n)$ do
    $\lfloor\ c(v_i) = \min\{\mathbb{N} - \{c(v) \mid v \in N(v_i)\}\}$

Definition 2 (Perfectly orderable graph [5]). *Let $G = (V, E)$ be a graph. A linear ordering on V is* perfect *if a greedy coloring algorithm with that ordering optimally colors every induced subgraph of G. A graph G is* perfectly orderable *if it admits a perfect order.*

Theorem 1 ([5]). *A linear ordering π of a graph G is perfect if and only if there is no induced $P_4 = (\{a, b, c, d\}, \{\{a, b\}, \{b, c\}, \{c, d\}\})$ in G such that $\pi(a) < \pi(b)$, $\pi(b) < \pi(c)$, and $\pi(d) < \pi(c)$.*

Since they are exactly the graphs with no induced P_4, the set of all co-graphs is perfectly orderable.

2.3 Coloring Oriented Graphs

Oriented graph coloring has been introduced by Courcelle [7] in 1994. We consider oriented graph coloring on oriented graphs, i.e. digraphs with no loops and no opposite arcs.

Definition 3 (Oriented graph coloring [7]). *An* oriented k-coloring *of an oriented graph $G = (V, E)$ is a mapping $c : V \to \{1, \ldots, k\}$ such that:*

- *$c(u) \neq c(v)$ for every $(u, v) \in E$,*
- *$c(u) \neq c(y)$ for every two arcs $(u, v) \in E$ and $(x, y) \in E$ with $c(v) = c(x)$.*

The oriented chromatic number *of G, denoted by $\chi_o(G)$, is the smallest k such that G has an oriented k-coloring. The vertex sets $V_i = \{v \in V \mid c(v) = i\}$, with $1 \leq i \leq k$, divide V into a partition of so called* color classes.

For two oriented graphs $G_1 = (V_1, E_1)$ and $G_2 = (V_2, E_2)$ a *homomorphism* from G_1 to G_2, $G_1 \to G_2$ for short, is a mapping $h : V_1 \to V_2$ such that $(u, v) \in E_1$ implies $(h(u), h(v)) \in E_2$. A homomorphism from G_1 to G_2 can be regarded as an oriented coloring of G_1 that uses the vertices of G_2 as colors classes. Therefore, digraph G_2 is called the *color graph* of G_1. This leads to equivalent definitions for the oriented coloring and the oriented chromatic number. There is an oriented k-coloring of an oriented graph G_1 if and only if there is a homomorphism from G_1 to some oriented graph G_2 with k vertices. That is, the oriented chromatic number of G_1 is the minimum number of vertices in an oriented graph G_2 such that there is a homomorphism from G_1 to G_2. Obviously, it is advisable to choose G_2 as a tournament.

Observation 1. *There is an oriented k-coloring of an oriented graph G_1 if and only if there is a homomorphism from G_1 to some tournament G_2 with k vertices. Further, the oriented chromatic number of G_1 is the minimum number of vertices in a tournament G_2 such that there is a homomorphism from G_1 to G_2.*

Observation 2. *For every oriented graph G it holds that $\chi(un(G)) \leq \chi_o(G)$.*

Lemma 1. *Let G be an oriented graph and H be a subdigraph of G. Then, it holds that $\chi_o(H) \leq \chi_o(G)$.*

Name Oriented Chromatic Number (OCN)
Instance An oriented graph $G = (V, E)$ and a positive integer $k \leq |V|$.
Question Is there an oriented k-coloring for G?

If k is constant and not part of the input, the corresponding problem is denoted by k-Oriented Chromatic Number (OCN$_k$). If $k \leq 3$, then OCN$_k$ can be decided in polynomial time, while OCN$_4$ is NP-complete [23]. OCN$_4$ is even known to be NP-complete for several restricted classes of digraphs, e.g. for DAGs [10].

Up to now, the definition of oriented coloring was frequently used for undirected graphs. For an undirected graph G the maximum value $\chi_o(G')$ of all possible orientations G' of G is considered. In this sense, every tree has oriented chromatic number at most 3. For several further graph classes there exist bounds on the oriented chromatic number. Among these are outerplanar graphs [27], planar graphs [26], and Halin graphs [14].

To advance research in this field, we consider oriented graph coloring on recursively defined oriented graph classes.

3   Coloring msp-Digraphs

We recall the definitions from [2] which are based on [29]. First, we introduce two operations for two vertex-disjoint digraphs $G_1 = (V_1, E_1)$ and $G_2 = (V_2, E_2)$. Let O_1 be the set of vertices of outdegree 0 (set of sinks) in G_1 and I_2 be the set of vertices of indegree 0 (set of sources) in G_2.

- The *parallel composition* of G_1 and G_2, denoted by $G_1 \cup G_2$, is the digraph with vertex set $V_1 \cup V_2$ and arc set $E_1 \cup E_2$.
- The *series composition* of G_1 and G_2, denoted by $G_1 \times G_2$ is the digraph with vertex set $V_1 \cup V_2$ and arc set $E_1 \cup E_2 \cup \{(v, w) \mid v \in O_1, w \in I_2\}$.

Definition 4 (msp-digraphs). *The class of* minimal series-parallel digraphs, msp-digraphs *for short, is recursively defined as follows.*

1. *Every digraph on a single vertex $(\{v\}, \emptyset)$, denoted by v, is a minimal series-parallel digraph.*
2. *If G_1 and G_2 are vertex-disjoint minimal series-parallel digraphs then,*
   (a) the parallel composition $G_1 \cup G_2$ and
   (b) then series composition $G_1 \times G_2$ are minimal series-parallel digraphs.

The class of minimal series-parallel digraphs is denoted as MSP.

Every expression X using these three operations is called an *msp-expression*. The digraph defined by the expression X is denoted by digraph(X).

Example 1. The following msp-expression X defines the oriented graph shown in Fig. 1.

$$X = (v_1 \times (((v_2 \times v_3) \times v_4) \cup v_5)) \times v_6$$

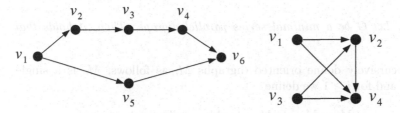

Fig. 1. Digraph in Example 1.   **Fig. 2.** Digraph in Example 4.

By removing vertex v_3 from digraph(X) in Example 1, we obtain a digraph which is no msp-digraph. This implies that the set of all msp-digraphs is not closed under taking induced subdigraphs.

A further remarkable property of msp-digraphs is that the oriented chromatic number of the disjoint union of two msp-digraphs can be larger than the maximum oriented chromatic number of the involved digraphs. This follows by the digraphs defined by expressions X_1 and X_2 in Example 2, which both have oriented chromatic number 4 but their disjoint union leads to a digraph with oriented chromatic number 5.

Example 2. In the following two msp-expressions we assume that the $\times$ operation binds more strongly than the $\cup$ operation.

$$X_1 = v_1 \times (v_2 \cup v_3 \times v_4) \times v_5 \times v_6$$

$$X_2 = w_1 \times (w_2 \cup w_3 \times (w_4 \cup w_5 \times w_6)) \times w_7$$

Obviously, for every msp-digraph we can define a tree structure T, which is denoted as *msp-tree*.[2] The leaves of an msp-tree represent the vertices of the digraph and the inner vertices of the msp-tree correspond to the operations applied on the subexpressions defined by the subtrees. For every msp-digraph one can construct an msp-tree in linear time, see [29].

---

[2] In [29], the tree-structure for an msp-digraphs is denoted as binary decomposition tree.

Several classes of digraphs are included in the set of all msp-digraphs. Among these are in- and out-rooted trees as well as oriented bipartite graphs $\overrightarrow{K_{n,m}}$.

In order to give an algorithm to compute the oriented chromatic number of msp-digraphs, we first show that this value can be bounded by a constant.

The class of undirected series-parallel graphs was considered in [27] by showing that every orientation of a series-parallel graph has oriented chromatic number at most 7. This bound can not be applied to minimal series parallel digraphs, since the set of all $\overrightarrow{K_{n,m}}$ is a subset of minimal series parallel digraphs and the underlying graphs are even of unbounded tree-width and thus, no series-parallel graphs.

Theorem 2. *Let G be a minimal series parallel digraph. Then, it holds that $\chi_o(G) \leq 7$.*

Proof. We recursively define oriented digraphs M_i as follows. M_0 is a single vertex graph and for $i \geq 1$ we define

$$M_i = M_{i-1} \cup M_{i-1} \cup (M_{i-1} \times M_{i-1}).$$

Claim. Every msp-digraph G is a (-n induced) subdigraph of some M_i such that every source in G is a source in M_i and every sink in G is a sink in M_i.

The claim can be shown by induction over the number of vertices in some msp-digraph G. If G has exactly one vertex, the claim holds true by choosing M_0.

Next, assume that G has $k > 1$ vertices. Then, it holds that $G = G_1 \square G_2$ for some $\square \in \{\cup, \times\}$ and G_1 and G_2 are msp-digraphs with less than k vertices. By the induction hypothesis we conclude that there are two integers i_1 and i_2 such that $G_1 \square G_2$ is a subdigraph of $M_{i_1} \square M_{i_2}$. (If $\square = \times$, it is important that every sink in G_1 is a sink in M_{i_1} and every source in G_2 is a source in M_{i_2}.) Thus, G is a subdigraph of $M_{i_1} \cup M_{i_2} \cup (M_{i_1} \times M_{i_2})$. W.l.o.g. we assume that $i_1 \leq i_2$. By construction it follows that M_{i_1} is a subdigraph of M_{i_2}. Consequently, G is a subdigraph of $M_{i_2} \cup M_{i_2} \cup (M_{i_2} \times M_{i_2}) = M_{i_2+1}$ and every source in G is a source in M_{i_2+1} and every sink in G is a sink in M_{i_2+1}. This completes the proof of the claim.

Since an oriented coloring of an oriented graph is also an oriented coloring for every subdigraph, we can show the theorem by coloring the digraphs M_i. Further, the first two occurrences of M_{i-1} in M_i can be colored in the same way. Thus, we can restrict to oriented graphs M_i' which are defined as follows. M_0' is a single vertex graph and for $i \geq 1$ we define

$$M_i' = M_{i-1}' \cup (M_{i-1}' \times M_{i-1}').$$

We define an oriented 7-coloring c for M_i' as follows. For some v of M_i' we define by $c(v, i)$ the color of v in M_i'. First, we color M_0' by assigning color 0 to the single vertex in M_0'.[3] For $i \geq 1$ we define the colors for the vertices v in $M_i' = M_{i-1}' \cup (M_{i-1}' \times M_{i-1}')$ according to the three copies of M_{i-1}' in M_i' (numbered from left to right). Therefore, we use the two functions $p(x) = (4 \cdot x) \bmod 7$ and $q(x) = (4 \cdot x + 1) \bmod 7$.

$$c(v, i) = \begin{cases} c(v, i-1) & \text{if } v \text{ is from the first copy} \\ p(c(v, i-1)) & \text{if } v \text{ is from the second copy} \\ q(c(v, i-1)) & \text{if } v \text{ is from the third copy} \end{cases}$$

It remains to show that c leads to an oriented coloring for M_i. Let $C_i = (W_i, F_i)$ with $W_i = \{0, 1, 2, 3, 4, 5, 6\}$ and $F_i = \{(c(u, i), c(v, i)) \mid (u, v) \in E_i\}$ be the color graph of $M_i' = (V_i, E_i)$. By the definition of M_i' we follow that

$$F_i = F_{i-1} \cup \{(p(x), p(y)) \mid (x, y) \in F_{i-1}\}$$
$$\cup \{(q(x), q(y)) \mid (x, y) \in F_{i-1}\}$$
$$\cup \{(p(c(v, i-1)), q(c(w, i-1))) \mid v \text{ sink of } M_{i-1}', w \text{ source of } M_{i-1}'\}.$$

In order to ensure an oriented coloring of M_i', we verify that C_i is an oriented graph. In Fig. 3 the color graph C_i for $i \geq 5$ is given.

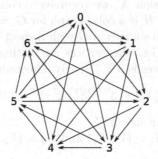

Fig. 3. Color graph C_i for $i \geq 5$ used in the proof of Theorem 2.

Every source in M_i' is colored by 0 since $p(0) = 0$. Every sink in M_i' is colored by 0, 1, or 5 since $q(0) = 1$, $q(1) = 5$, and $q(5) = 0$.

Consequently, the arcs of

$$\{(p(c(v, i-1)), q(c(w, i-1))) \mid v \text{ sink of } M_{i-1}', w \text{ source of } M_{i-1}'\}$$

belong to the set $\{(p(0), q(0)), (p(1), q(0)), (p(5), q(0))\} = \{(0, 1), (4, 1), (6, 1)\}$. For every $(u, v) \in \{(0, 1), (4, 1), (6, 1)\}$ we know that

$$(v - u) \bmod 7 \in \{1, 2, 4\} \tag{1}$$

---

[3] Please note that using colors starting at value 0 instead of 1 does not contradict Definition 3.

which implies $(u-v) \bmod 7 \notin \{1,2,4\}$ and thus, (1) does not hold for the reverse arcs of $\{(0,1),(4,1),(6,1)\}$. It remains to show that (1) remains true for all arcs (u,v) when applying p and q to M'_{i-1}:

$$
\begin{aligned}
(q(v) - q(u)) \bmod 7 &= (((4 \cdot v + 1) \bmod 7) - ((4 \cdot u + 1) \bmod 7)) \bmod 7 \\
&= (((4 \cdot v) \bmod 7) - ((4 \cdot u) \bmod 7)) \bmod 7 \\
&= (p(v) - p(u)) \bmod 7 \\
&= (4(v - u)) \bmod 7
\end{aligned}
$$

Since $(v - u) \bmod 7 \in \{1,2,4\}$ leads to $(4(v - u)) \bmod 7 \in \{1,2,4\}$, the result follows.   □

Digraph G on 27 vertices defined in Example 3 satisfies $\chi_o(G) = 7$, which was found by a computer program.[4] This implies that the bound of Theorem 2 is best possible.

Example 3. In the following msp-expression we assume that the $\times$ operation binds more strongly than the $\cup$ operation.

$$
\begin{aligned}
X = v_1 \times (v_2 \cup v_3 \times (v_4 \cup v_5 \times v_6)) \times (v_7 \cup (v_8 \cup v_9 \times v_{10}) \times (v_{11} \cup v_{12} \times v_{13})) \times \\
(v_{14} \cup (v_{15} \cup (v_{16} \cup v_{17} \times v_{18}) \times (v_{19} \cup v_{20} \times v_{21})) \times (v_{22} \cup (v_{23} \cup v_{24} \times v_{25}) \times v_{26})) \times v_{27}
\end{aligned}
$$

In oder to compute the oriented chromatic number of an msp-digraph G defined by an msp-expression X, we recursively compute the set $F(X)$ of all triples (H, L, R) such that H is a color graph for G, where L and R are the sets of colors of all sinks and all sources in G with respect to the coloring by H. The number of vertex labeled, i.e., the vertices are distinguishable from each other, oriented graphs on n vertices is $3^{n(n-1)/2}$. By Theorem 2 we can conclude that

$$
|F(X)| \le 3^{7(7-1)/2} \cdot 2^7 \cdot 2^7 \in \mathcal{O}(1)
$$

which is independent of the size of G.

For two color graphs $H_1 = (V_1, E_1)$ and $H_2 = (V_2, E_2)$ we define $H_1 + H_2 = (V_1 \cup V_2, E_1 \cup E_2)$.

Lemma 2 ($\bigstar^5$).

1. *For every $v \in V$ it holds $F(v) = \{((\{i\}, \emptyset), \{i\}, \{i\}) \mid 0 \le i \le 6\}$.*
2. *For every two msp-expressions X_1 and X_2 we obtain $F(X_1 \cup X_2)$ from $F(X_1)$ and $F(X_2)$ as follows. For every $(H_1, L_1, R_1) \in F(X_1)$ and every $(H_2, L_2, R_2) \in F(X_2)$ such that graph $H_1 + H_2$ is oriented, we put $(H_1 + H_2, L_1 \cup L_2, R_1 \cup R_2)$ into $F(X_1 \cup X_2)$.*
3. *For every two msp-expressions X_1 and X_2 we obtain $F(X_1 \times X_2)$ from $F(X_1)$ and $F(X_2)$ as follows. For every $(H_1, L_1, R_1) \in F(X_1)$ and every $(H_2, L_2, R_2) \in F(X_2)$ such that graph $H_1 + H_2$ together with the arcs in $R_1 \times L_2$ is oriented, we put $((V_1 \cup V_2, E_1 \cup E_2 \cup R_1 \times L_2), L_1, R_2)$ into $F(X_1 \times X_2)$.*

---

[4] We implemented an algorithm which takes an oriented graph G and an integer k as an input and which decides whether $\chi_o(G) \le k$.

[5] The proofs of the results marked with a $\bigstar$ are omitted due to space restrictions.

Since every possible coloring of G is part of the set $F(X)$, where X is an msp-expression for G, it is easy to find a minimum coloring for G.

Corollary 1. *There is an oriented k-coloring for some msp-digraph G which is given by some msp-expression X if and only if there is some $(H, L, R) \in F(X)$ such that color graph H has k vertices. Therefore, $\chi_o(G) = \min\{|V| \mid ((V, E), L, R) \in F(X)\}$.*

Theorem 3. *Let G be a minimal series parallel digraph. Then, the oriented chromatic number of G can be computed in linear time.*

Proof. Let G be an msp-digraph on n vertices and m edges. Further, let T be an msp-tree for G with root r. For some vertex u of T we denote by T_u the subtree rooted at u and X_u the msp-expression defined by T_u.

In order to solve OCN for some msp-digraph G, we traverse msp-tree T in a bottom-up order. For every vertex u of T we compute $F(X_u)$ following the rules given in Lemma 2. By Corollary 1 we can solve our problem by $F(X_r) = F(X)$.

An msp-tree T can be computed in $\mathcal{O}(n + m)$ time from a minimal series-parallel digraph with n vertices and m arcs, see [29]. Our rules given in Lemma 2 show the following running times.

- For every $v \in V$ set $F(v)$ is computable in $\mathcal{O}(1)$ time.
- Every set $F(X_1 \cup X_2)$ can be computed in $\mathcal{O}(1)$ time from $F(X_1)$ and $F(X_2)$.
- Every set $F(X_1 \times X_2)$ can be computed in $\mathcal{O}(1)$ time from $F(X_1)$ and $F(X_2)$.

Since we have n leaves and $n - 1$ inner vertices in msp-tree T, the running time is in $\mathcal{O}(n + m)$. □

By Corollary 2 we know that for every oriented co-graph G it holds that $\chi_o(G) = \chi(un(G))$. This does not hold for minimal series parallel digraphs by Example 3 and the next result, which can be shown by giving a 3-coloring for $un(M_i')$ for the oriented graphs M_i' used in the proof of Theorem 2.

Proposition 1 (★). *Let G be a minimal series parallel digraph. Then, it holds that $\chi(un(G)) \leq 3$.*

4 Coloring Transitive Acyclic Digraphs

Next, we will apply the concept of perfectly orderable graphs and Theorem 1 in order to color transitive acyclic digraphs.

Theorem 4. *Let G be a transitive acyclic digraph. Then, every greedy coloring along a topological ordering of G leads to an optimal oriented coloring of G and $\chi_o(G)$ can be computed in linear time.*

Proof. Since $G = (V, E)$ is an acyclic digraph there is a topological ordering t for G. Since G is transitive, it does not contain the following orientation of a P_4 as an induced subdigraph.

$$\bullet \rightarrow \bullet \rightarrow \bullet \leftarrow \bullet$$

Theorem 1 implies that every linear ordering and thus, also t is perfect on $un(G) = (V, E_u)$. Let $c : V \rightarrow \{1, \ldots, k\}$ be a coloring for $un(G)$ obtained by the greedy algorithm (Algorithm 1) for t on V. It remains to show that c is an oriented coloring for G.

- $c(u) \neq c(v)$ holds for every $(u, v) \in E$ since $c(u) \neq c(v)$ holds for every $\{u, v\} \in E_u$.
- $c(u) \neq c(y)$ for every two arcs $(u, v) \in E$ and $(x, y) \in E$ with $c(v) = c(x)$ holds by the following argumentation.
  Assume there is an arc $(v_i, v_j) \in E$ with $v_i < v_j$ in t but $c(v_i) > c(v_j)$. Then, when coloring v_i we would have taken $c(v_j)$ if possible, as we always take the minimum possible color value. Since this was not possible there must have been an other vertex $v_k < v_i$ which was colored before v_i with $c(v_k) = c(v_j)$ and $(v_k, v_i) \in E$. But if $(v_k, v_i) \in E$ and $(v_i, v_j) \in E$, due to transitivity it must also hold that $(v_k, v_j) \in E$ and consequently, $c(v_k) = c(v_j)$ is not possible. Thus, the assumption was wrong and for every arc $(v_i, v_j) \in E$ with $v_i < v_j$ in t it must hold that $c(v_i) < (c_j)$.

The optimality of oriented coloring c holds since the lower bound of Observation 2 is achieved. $\qquad\square$

In order to state the next result, let $\omega(H)$ be the number of vertices in a largest clique in the (undirected) graph H.

Corollary 2. *Let G be a transitive acyclic digraph. Then, it holds that $\chi_o(G) = \chi(un(G)) = \omega(un(G))$ and all values can be computed in linear time.*

For some oriented graph G we denote by $\ell(G)$ the length of a longest oriented path in G.

Proposition 2. *Let G be a transitive acyclic digraph. Then, it holds that $\chi_o(G) = \ell(G) + 1$.*

Proof. The proof of Theorem 4 leads to an optimal oriented coloring using $\ell(G) + 1$ colors. $\qquad\square$

Next, we consider oriented colorings of oriented graphs with bounded vertex degree. For every oriented graph its oriented chromatic number can be bounded (exponentially) by its maximum degree Δ according to [24]. For small values $\Delta \leq 7$ there are better bounds in [12] and [13].

Corollary 3. *Let G be a transitive acyclic digraph. Then, it holds that $\chi_o(G) \leq \Delta(G) + 1$.*

Proof. By Proposition 2 and the fact that the first vertex of a longest path within a transitive digraph G has outdegree at least $\ell(G)$, it follows that the oriented chromatic number of G can be estimated by $\chi_o(G) = \ell(G) + 1 \leq \Delta(G) + 1$. □

Proposition 3. *Let G be an acyclic digraph. Then, it holds that $\chi_o(G) \leq \ell(G) + 1$.*

Proof. Let G be an acyclic digraph and G' its transitive closure. Then, by Lemma 1 and Proposition 2 we know that $\chi_o(G) \leq \chi_o(G') = \ell(G') + 1 = \ell(G) + 1$. □

5 Coloring Oriented Co-graphs

Let $G_1 = (V_1, E_1)$ and $G_2 = (V_2, E_2)$ be two vertex-disjoint digraphs. The following operations have been considered by Bechet et al. in [3].

- The *disjoint union* of G_1 and G_2, denoted by $G_1 \oplus G_2$, is the digraph with vertex set $V_1 \cup V_2$ and arc set $E_1 \cup E_2$.
- The *order composition* of G_1 and G_2, denoted by $G_1 \oslash G_2$, is defined by their disjoint union plus all possible arcs from vertices of G_1 to vertices of G_2.

By omitting the series composition within the definition of directed co-graphs in [9], we obtain the class of all *oriented co-graphs*.

Definition 5 (Oriented co-graphs). *The class of oriented complement reducible graphs, oriented co-graphs for short, is recursively defined as follows.*

1. *Every digraph with a single vertex $(\{v\}, \emptyset)$, denoted by v, is an oriented co-graph.*
2. *If G_1 and G_2 are two vertex-disjoint oriented co-graphs, then*
   (a) the disjoint union $G_1 \oplus G_2$, and
   (b) the order composition $G_1 \oslash G_2$ are oriented co-graphs.

The class of directed co-graphs is denoted by OC.

Every expression X using the operations of Definition 5 is called a *di-co-expression*, see Example 4 for an illustration.

Example 4. The following di-co-expression X defines the oriented graph shown in Fig. 2.

$$X = ((v_1 \oplus v_3) \oslash (v_2 \oslash v_4))$$

The set of all oriented co-graphs is closed under taking induced subdigraphs. Following the notations of [29] we denote the following orientation of a P_4 as the N graph.

$$N = \bullet \rightarrow \bullet \leftarrow \bullet \rightarrow \bullet$$

The class of oriented co-graphs can be characterized by excluding the four forbidden induced subdigraphs $\overleftrightarrow{P_2} = (\{u, v\}, \{(u, v), (v, u)\})$, $\overrightarrow{P_3}$, $\overrightarrow{C_3}$, and N, see [18]. The class of oriented co-graphs has already been analyzed by Lawler in [25] and [6, Section 5] using the notation of *transitive series-parallel (TSP) digraphs*.

For every oriented co-graph we can define a tree structure, denoted as *di-co-tree*. The leaves of the di-co-tree represent the vertices of the digraph and the inner vertices of the di-co-tree correspond to the operations applied on the subexpressions defined by the subtrees. For every oriented co-graph one can construct a di-co-tree in linear time, see [9].

Lemma 3 ([18]). *Let G_1 and G_2 be two vertex-disjoint oriented co-graphs. Then, the following equations hold.*

1. $\chi_o(v) = 1$
2. $\chi_o(G_1 \oplus G_2) = \max(\chi_o(G_1), \chi_o(G_2))$
3. $\chi_o(G_1 \oslash G_2) = \chi_o(G_1) + \chi_o(G_2)$

Theorem 5 ([18]). *Let G be an oriented co-graph. Then, an optimal oriented coloring for G and $\chi_o(G)$ can be computed in linear time.*

The result in [18] concerning the oriented coloring on oriented co-graphs is based on a dynamic programming along a di-co-tree for the given oriented co-graph as input. Since every oriented co-graph is transitive and acyclic, Theorem 4 leads to the next result, which re-proves Theorem 5.

Corollary 4. *Let G be an oriented co-graph. Then, every greedy coloring along a topological ordering of G leads to an optimal oriented coloring of G and $\chi_o(G)$ can be computed in linear time.*

Please note that Theorem 4 is more general than Corollary 4 since it does not exclude N which is a forbidden induced subdigraph for oriented co-graphs. It holds that

$$\text{OC} = \text{Free}\{\overleftrightarrow{P_2}, \overrightarrow{P_3}, \overrightarrow{C_3}, N\} \subseteq \text{Free}\{\overleftrightarrow{P_2}, \overrightarrow{P_3}, \overrightarrow{C_3}\}$$

and $\text{Free}\{\overleftrightarrow{P_2}, \overrightarrow{P_3}, \overrightarrow{C_3}\}$ is equivalent to the set of all acyclic transitive digraphs.

Since every oriented co-graph is transitive and acyclic, Corollary 3 leads to the following bound.

Corollary 5. *Let G be an oriented co-digraph. Then, it holds that $\chi_o(G) \leq \Delta(G) + 1$.*

There are classes of oriented co-graphs, e.g., the class of all $\overrightarrow{K_{1,n}}$, for which the oriented chromatic number is even bounded by a constant and thus smaller than the shown bound. Considering transitive tournaments we conclude that the bound given in Corollary 5 is best possible.

6   Conclusions and Outlook

In this paper we considered oriented colorings of msp-digraphs and oriented co-graphs. We showed that every msp-digraph has oriented chromatic number at most 7, which is best possible. We used this bound together with the recursive structure of msp-digraphs to give a linear time solution for computing the oriented chromatic number of msp-digraphs. Further, we use the concept of perfect orderable graphs in order to show that for acyclic transitive digraphs every greedy coloring along a topological ordering leads to an optimal oriented coloring, which generalizes a known solution for the oriented coloring problem on oriented co-graphs in [18].

Comparing the solutions we conclude that a greedy coloring of $un(G)$ along a topological ordering of G does not work for computing the oriented chromatic number of an msp-digraph G. An oriented path would be colored by only two colors which is not an admitted oriented coloring. Further, a dynamic programming solution using similar formulas to Lemma 3 is not possible for computing the oriented chromatic number of msp-digraphs. Example 2 implies that the oriented chromatic number of the disjoint union of two msp-digraphs can be larger than the maximum oriented chromatic number of the involved digraphs.

For future work it could be interesting to extend our solutions to superclasses such as series-parallel digraphs [29] and graph powers [2] of minimal series-parallel digraphs. Furthermore, the parameterized complexity of OCN and OCN_k w.r.t. structural parameters has only been considered in [16] and [15]. For the parameters directed clique-width [8,19] and directed modular-width [28] the parameterized complexity of OCN remains open.

Acknowledgments. This work was funded by the Deutsche Forschungsgemeinschaft (DFG, German Research Foundation) – 388221852.

References

1. Bang-Jensen, J., Gutin, G.: Digraphs: Theory Algorithms and Applications. Springer, Berlin (2009)
2. Bang-Jensen, J., Gutin, G. (eds.): Classes of Directed Graphs. Springer, Berlin (2018)
3. Bechet, D., de Groote, P., Retoré, C.: A complete axiomatisation for the inclusion of series-parallel partial orders. In: Comon, H. (ed.) RTA 1997. LNCS, vol. 1232, pp. 230–240. Springer, Heidelberg (1997). https://doi.org/10.1007/3-540-62950-5_74
4. Byskov, J.M.: Enumerating maximal independent sets with applications to graph colouring. Oper. Res. Lett. **32**(6), 547–556 (2004)
5. Chvátal, V.: Perfectly ordered graphs. In: Berge, C., Chvátal, V. (eds.) Topics on Perfect Graphs, North-Holland Mathematics Studies, vol. 88, pp. 63–65. North-Holland (1984)
6. Corneil, D., Lerchs, H., Stewart-Burlingham, L.: Complement reducible graphs. Discrete Appl. Math. **3**, 163–174 (1981)
7. Courcelle, B.: The monadic second-order logic of graphs VI: on several representations of graphs by relational structures. Discrete Appl. Math. **54**, 117–149 (1994)
8. Courcelle, B., Olariu, S.: Upper bounds to the clique width of graphs. Discrete Appl. Math. **101**, 77–114 (2000)
9. Crespelle, C., Paul, C.: Fully dynamic recognition algorithm and certificate for directed cographs. Discrete Appl. Math. **154**(12), 1722–1741 (2006)

10. Culus, J.-F., Demange, M.: Oriented coloring: complexity and approximation. In: Wiedermann, J., Tel, G., Pokorný, J., Bieliková, M., Štuller, J. (eds.) SOFSEM 2006. LNCS, vol. 3831, pp. 226–236. Springer, Heidelberg (2006). https://doi.org/10.1007/11611257_20

11. Dailey, D.: Uniqueness of colorability and colorability of planar 4-regular graphs are NP-complete. Discrete Math. **30**(3), 289–293 (1980)

12. Duffy, C.: A note on colourings of connected oriented cubic graphs. ACM Computing Research Repository (CoRR) abs/1908.02883, 8 p (2019)

13. Dybizbański, J., Ochem, P., Pinlou, A., Szepietowski, A.: Oriented cliques and colorings of graphs with low maximum degree. Discrete Math. **343**(5), 111829 (2020)

14. Dybizbański, J., Szepietowski, A.: The oriented chromatic number of Halin graphs. Inf. Process. Lett. **114**(1–2), 45–49 (2014)

15. Ganian, R.: The parameterized complexity of oriented colouring. In: Proceedings of Doctoral Workshop on Mathematical and Engineering Methods in Computer Science, MEMICS. OASICS, vol. 13. Schloss Dagstuhl - Leibniz-Zentrum fuer Informatik, Germany (2009)

16. Ganian, R., Hlinený, P., Kneis, J., Langer, A., Obdrzálek, J., Rossmanith, P.: Digraph width measures in parameterized algorithmics. Discrete Appl. Math. **168**, 88–107 (2014)

17. Golumbic, M.: Algorithmic Graph Theory and Perfect Graphs. Academic Press (1980)

18. Gurski, F., Komander, D., Rehs, C.: Oriented coloring on recursively defined digraphs. Algorithms **12**(4), 87 (2019)

19. Gurski, F., Wanke, E., Yilmaz, E.: Directed NLC-width. Theor. Comput. Sci. **616**, 1–17 (2016)

20. Hansen, P., Kuplinsky, J., de Werra, D.: Mixed graph coloring. Math. Methods Oper. Res. **45**, 145–160 (1997)

21. Hoàng, C.: Efficient algorithms for minimum weighted colouring of some classes of perfect graphs. Discrete Appl. Math. **55**, 133–143 (1994)

22. Jansen, K., Porkolab, L.: Preemptive scheduling with dedicated processors: applications of fractional graph coloring. J. Sched. **7**, 35–48 (2004)

23. Klostermeyer, W., MacGillivray, G.: Homomorphisms and oriented colorings of equivalence classes of oriented graphs. Discrete Math. **274**, 161–172 (2004)

24. Kostochka, A., Sopena, E., Zhu, X.: Acyclic and oriented chromatic numbers of graphs. J. Graph Theory **24**(4), 331–340 (1997)

25. Lawler, E.: Graphical algorithms and their complexity. Math. Centre Tracts **81**, 3–32 (1976)

26. Marshall, T.: Homomorphism bounds for oriented planar graphs of given minimum girth. Graphs Combin. **29**, 1489–1499 (2013)

27. Sopena, É.: The chromatic number of oriented graphs. J. Graph Theory **25**, 191–205 (1997)

28. Steiner, R., Wiederrecht, S.: Parameterized algorithms for directed modular width. In: Changat, M., Das, S. (eds.) CALDAM 2020. LNCS, vol. 12016, pp. 415–426. Springer, Cham (2020). https://doi.org/10.1007/978-3-030-39219-2_33

29. Valdes, J., Tarjan, R., Lawler, E.: The recognition of series-parallel digraphs. SIAM J. Comput. **11**, 298–313 (1982)

30. de Werra, D., Eisenbeis, C., Lelait, S., Stöhr, E.: Circular-arc graph coloring: on chords and circuits in the meeting graph. Eur. J. Oper. Res. **136**(3), 483–500 (2002)

Star-Critical Ramsey Number of Large Cycle and Book

Yan Li[1], Yusheng Li[1], and Ye Wang[2]($\boxtimes$)

[1] Tongji University, Shanghai 200092, China
[2] Harbin Engineering University, Harbin 150001, China
ywang@hrbeu.edu.cn

Abstract. For graphs F, G and H, let $F \to (G, H)$ signify that any red/blue edge coloring of F contains either a red G or a blue H. Thus the Ramsey number $R(G, H)$ is defined as $\min\{r \mid K_r \to (G, H)\}$. In this note, we consider an optimization problem to define the star-critical Ramsey number $R_{\mathbb{S}}(G, H)$ as the largest s such that $K_r \setminus K_{1,s} \to (G, H)$, where $r = R(G, H)$. We shall determine $R_{\mathbb{S}}(C_n, B_n)$ for large n.

Keywords: Star-critical Ramsey number · Book · Cycle · Algorithm

1 Introduction

For vertex disjoint graphs G and H, let $G + H$ be the graph obtained from G and H by adding new edges to connect G and H completely. Let nH be the union of n disjoint copies of H, and $G \setminus H$ the graph obtained from G by deleting a copy of H from G, where we always admit that H is a subgraph of G. Slightly abusing notation, for $S \subseteq V(G)$, we denote by $G \setminus S$ the subgraph of G induced by $V(G) \setminus S$. Let $v(G)$, $e(G)$ and $\delta(G)$ be the order, the size and the minimum degree of G, respectively.

For graphs F, G and H, let $F \to (G, H)$ signify that any red/blue edge coloring of F contains either a red G or a blue H. For any graphs G and H, we can always find a graph F of large order satisfying $F \to (G, H)$. Moreover, the Ramsey number $R(G, H)$ is defined as

$$R(G, H) = \min\{r : K_r \to (G, H)\}.$$

An optimization problem is to find the largest subgraph F such that $K_r \setminus F \to (G, H)$ where $r = R(G, H)$. This problem is related to both the Ramsey number and the size Ramsey number, in which if the "largest" means the largest number of edges, then the problem is related to the size Ramsey number $\hat{r}(G, H)$ introduced by Erdős, Faudree, Rousseau and Schelp [6] as

$$\hat{r}(G, H) = \min\{e(F) : F \to (G, H)\}.$$

Supported in part by NSFC.

Most problems on the size Ramsey number are far from being solved. We find a new way to study this topic by considering the notion of the critical Ramsey number $R_{\mathbb{G}}(G, H)$ as follows.

Definition 1 [16]. *Let $k \geq 1$ be an integer and $\mathbb{G} = \{G_k, G_{k+1}, \dots\}$ be a class of graphs G_n, where each graph $G_n \in \mathbb{G}$ has minimum degree $\delta(G_n) \geq 1$. Define the critical Ramsey number $R_{\mathbb{G}}(G, H)$ of G and H with respect to $\mathbb{G}$ as*

$$R_{\mathbb{G}}(G, H) = \max \{ n \mid K_r \setminus G_n \to (G, H), \, G_n \in \mathbb{G} \},$$

where $r = R(G, H)$. Usually, we require $G_n \subset G_{n+1}$ or $e(G_n) < e(G_{n+1})$.

If $\mathbb{G}$ is a class

$$\mathbb{S} = \{K_{1,1}, K_{1,2}, \dots\}$$

of stars, we shall call $R_{\mathbb{S}}(G, H)$ the star-critical Ramsey number which was introduced by [9] in a similar way that has attracted much attention, see, e.g., [7–12,17,18].

Let C_n be a cycle of order n and B_n a book, where $B_n = K_2 + nK_1$. Bondy and Erdős [1] first studied the Ramsey number of cycles, and Rousseau and Sheehan [15] studied the Ramsey number of books, employing the Paley graphs that are the most important strongly regular graphs and quasi-random graphs, see [4].

Lemma 1 [13]. *For all sufficiently large n, it holds*

$$R(C_n, B_n) = 3n - 2.$$

For more results on cycles and books, see, e.g., [13,14].

In this note, we shall determine the star-critical Ramsey number for C_n and B_n.

Theorem 1. *For all sufficiently large n, it holds*

$$R_{\mathbb{S}}(C_n, B_n) = 2n - 6.$$

2 Proof

Before proceeding to proof, we need some notation. For a graph G, color the edges of G red and blue, and let G^R and G^B be the subgraphs of G induced by red and blue edges, respectively. Denote by $d_G(v)$ the degree of vertex v in G, and $d_G^R(v)$, $d_G^B(v)$ the number of red and blue neighbors in G, respectively. Let $d_G(v) = d_G^R(v) + d_G^B(v)$. Let $G[S]$ be the subgraph of G induced by $S \subseteq V(G)$. Note that each subgraph of G admits a red/blue edge coloring preserved from that of G.

Denote by $g(G)$ and $c(G)$ the girth and the circumference of graph G, respectively. A graph G is weakly pancyclic if it contains cycles of every length between $g(G)$ and $c(G)$. A graph G is pancyclic if it is weakly pancyclic with $g(G) = 3$

and $c(G) = v(G)$. We also need some results on the pancyclic properties and weakly pancyclic properties of graphs for the proof of our result.

A graph is 2-connected if it is connected after deleting any vertex. The following lemma is a famous result due to Dirac.

Lemma 2 [5]. *Let G be a 2-connected graph of order n with minimum degree δ. Then $c(G) \geq \min\{2\delta, n\}$.*

Dirac's result gives a fact that the $c(G)$ of a 2-connected graph G can not be too small. The following result by Bondy is an extension of the above lemma.

Lemma 3 [2]. *If a graph G of order n has minimum degree $\delta(G) \geq \frac{n}{2}$, then G is pancyclic, or $n = 2t$ and $G = K_{t,t}$.*

The following lemma is an extension of the above results by Brandt, Faudree, and Goddard.

Lemma 4 [3]. *Let G be a 2-connected nonbipartite graph of order n with minimum degree $\delta \geq n/4 + 250$. Then G is weakly pancyclic unless G has odd girth 7, in which case it has every cycle from 4 up to its circumference except C_5.*

Proof of Theorem 1. To show $R_{\mathbb{S}}(C_n, B_n) \leq 2n - 6$, consider the graph $G = K_{r-1}$ and vertex v_0 where $r = R(C_n, B_n) = 3n - 2$. Let U_1, U_2 and U_3 be red cliques of order $n - 1$. Select $u_1 \in V(U_1)$, $u_2 \in V(U_2)$ and $u_3 \in V(U_3)$. Color the edges among U_1, U_2 and U_3 blue but u_1u_3 and u_2u_3. Let v_0 be adjacent to $V(U_2)$ by blue edges completely, and to one vertex of $V(U_1) \setminus \{u_1\}$ by red. Color v_0u_1 and v_0u_3 blue. Then there is a red/blue edge coloring of $K_r \setminus K_{1,2n-5}$ containing neither a red C_n nor a blue B_n, yielding the upper bound as required.

To show $R_{\mathbb{S}}(C_n, B_n) \geq 2n - 6$, consider graph $H = K_r \setminus K_{1,s}$, where $s = 2n - 6$. Let v_0 be the central vertex of $K_{1,s}$ and $G = K_{r-1}$. We shall prove H contains either a red C_n or a blue B_n.

Assume there is a vertex $v \in V(G)$ with $d_G^B(v) \geq 2n$. Select $2n$ vertices from the blue neighbors of v in G and let the graph induced by the $2n$ vertices be G_1. If there is a vertex $u \in G_1$ such that $d_{G_1}^B(u) \geq n$, then we can find a blue B_n. Thus we may assume that $d_{G_1}^R(u) \geq n$ for any vertex $u \in G_1$. By Lemma 3, G_1^R is pancyclic, or $G_1^R = K_{n,n}$. If G_1^R is pancyclic, then we can get a red C_n. So we assume $G_1^R = K_{n,n}$ with bipartition (X, Y). Note that vertices of $V(G) \setminus (V(G_1) \cup \{v\})$ must be adjacent to X or Y by blue edges completely, otherwise we will get a red C_n. Along with vertex v, we get a blue $K_{n+2} \setminus K_2$ thus a blue B_n. Therefore we may assume that for any vertex $v \in V(G)$, we have

$$d_G^R(v) \geq r - 2 - (2n - 1) = n - 3.$$

It is easy to see that G^R is nonbipartite. Suppose G^R is 2-connected. Since

$$\delta(G^R) \geq n - 3 \geq \frac{r-1}{4} + 250,$$

by Lemma 4, G^R is weakly pancyclic unless the odd girth of G is 7, in which case G contains every cycle of length from 4 to its circumference except C_5. Therefore, by Lemma 2, $c(G^R) \geq 2(n-3) \geq n$, we get a red C_n. Then we may assume that G^R contains a cut vertex u_0. Note that the size of each component of $G^R \setminus \{u_0\}$ is at least $n-3$ since $\delta(G^R \setminus \{u_0\}) \geq n-4$. Thus $G^R \setminus \{u_0\}$ contains at most three components.

If $G^R \setminus \{u_0\}$ contains three components, then denote by U_1, U_2 and U_3 the vertex sets of three components of $G^R \setminus \{u_0\}$. Let $|U_1| \geq |U_2| \geq |U_3|$. If $|U_1| \geq n$, we have

$$n \leq |U_1| \leq r - 2 - 2(n-3) = n + 2.$$

As $\delta(G^R[U_1]) \geq n - 4 > |U_1|/2$ and by Lemma 3, if $G^R[U_1]$ is pancyclic, there is a red C_n. If $|U_1| = 2t$ and $G^R[U_1] = K_{t,t}$, we shall get a blue B_n in G. Thus there must be

$$|U_1| = |U_2| = n - 1 \text{ and } |U_3| = n - 2.$$

Moreover, U_i induces a red clique for $i = 1, 2, 3$. Otherwise if there is a blue edge in U_i, then G shall contain a blue B_{2n-3}, hence a blue B_n. Note that $d^R_{G[U_1]}(u_0) = 1$ and $d^R_{G[U_2]}(u_0) = 1$ since u_0 is the cut vertex, otherwise we get a red C_n. Thus $d^B_{G[U_3]}(u_0) = 0$, otherwise there is a blue B_n. Now we get three red cliques of order $n - 1$ with almost all edges among which being blue. Consider vertex v_0 and graph G. If $d^B_{G[U_3]}(v_0) \geq 1$, then

$$d^B_{G[U_1]}(v_0) = d^B_{G[U_2]}(v_0) = 0,$$

which will yield a red C_n. Thus $d^B_{G[U_3]}(v_0) = 0$. Furthermore, if $d^R_{G[U_3]}(v_0) = 1$, then v_0 is adjacent to at least $n + 1$ vertices in $U_1 \cup U_2$ by blue, yielding a blue B_n. Assume $d^R_{G[U_3]}(v_0) = 0$, then

$$d_{G[U_1]}(v_0) \geq 3 \text{ and } d_{G[U_2]}(v_0) \geq 3.$$

Along with $d^R_{G[U_1]}(v_0) \leq 1$ and $d^R_{G[U_2]}(v_0) \leq 1$, we get a blue B_n.

If $G^R \setminus \{u_0\}$ contains two components, then denote by U_1 and U_2 the vertex sets of two components of $G^R \setminus \{u_0\}$. Let $|U_1| \leq |U_2|$. If $n \leq |U_1| \leq (3n-4)/2$, as $\delta(G^R[U_1]) \geq n-4 > |U_1|/2$, then by Lemma 3, G contains a red C_n. Furthermore, $G^R[U_2]$ is nonbipartite. Otherwise, suppose $G^R[U_2]$ is bipartite with bipartiton (X, Y) where X and Y both induce blue cliques and all edges between U_1 and X are blue, and it follows

$$|U_1| + |X| = |U_1| + \frac{r - 2 - |U_1|}{2} \geq n + 2.$$

Then the graph induced by $U_1 \cup X$ contains a blue B_n. Thus

$$r - 2 - (2n - 1) \leq |U_1| \leq n - 1 \text{ and } r - 2 - (n - 1) \leq |U_2| \leq 2n - 1.$$

Moreover, if $G^R[U_2]$ is 2-connected, then by Lemma 2 and Lemma 4, $G^R[U_2]$ contains a red C_n as

$$\delta(G^R[U_2]) \geq n - 4 \geq \frac{|U_2|}{4} + 250.$$

Thus suppose that $G^R[U_2]$ contains a cut vertex u_1. Denote by V_1 and V_2 the vertex sets of the two components of $G^R[U_2] \setminus \{u_1\}$.

Then the order of $\{U_1, V_1, V_2\}$ must be

$$\{n-1, n-1, n-3\} \text{ or } \{n-1, n-2, n-2\}$$

since $|U_1| + |V_1| + |V_2| = 3n - 5$. Note that the graphs induced by U_1, V_1, and V_2 contain no blue edge, otherwise we will get a blue B_n.

Case 1. Suppose $|U_1| = n - 1, |V_1| = n - 1$ and $|V_2| = n - 3$. Note that u_0 and u_1 are cut vertices, so

$$d_{G[U_1]}^R(u_0) = 1, d_{G[U_1]}^R(u_1) = 0, d_{G[V_1]}^R(u_0) \leq 1 \text{ and } d_{G[V_1]}^R(u_1) = 1,$$

otherwise we will get a red C_n. Note that $d_{G[V_2]}^B(u_0) = d_{G[V_2]}^B(u_1) = 0$ or there will be a blue B_n. Thus $u_0 u_1$ must be red and $d_{G[V_1]}^R(u_0) = 0$. Hence G contains three red K_{n-1}. Consider vertex v_0 and graph G. If $d_{G[V_2]}^B(v_0) \geq 1$, then v_0 can only be adjacent to U_1 and V_1 by red edges, otherwise there will be a blue B_n. Then either $d_{G[U_1]}^R(v_0) \geq 2$ or $d_{G[V_1]}^R(v_0) \geq 2$, which will yield a red C_n. So suppose $d_{G[V_2]}^B(v_0) = 0$ and $d_{G[V_2]}^R(v_0) = 1$, then v_0 is adjacent to $n + 2$ vertices in $U_1 \cup V_1 \cup \{u_0, u_1\}$ by blue. In this case, if u_0, u_1 and v_0 have a common blue neighbor in U_1 and V_1, respectively, we will get a blue B_n. If not, $v_0 u_0$ and $v_0 u_1$ must be blue, and v_0 must be adjacent to U_1 or V_1 by blue completely, say U_1, and to the red neighbor of u_1 in V_1 by blue, which will also yield a blue B_n.

If $|U_1| = n - 3, |V_1| = n - 1$ and $|V_2| = n - 1$, G shall contain either a red C_n or a blue B_n since $d_{G[U_1]}^R(u_1) = 0$.

Case 2. Suppose $|U_1| = n - 1, |V_1| = n - 2$ and $|V_2| = n - 2$. As u_1 is the cut vertex of U_2, we have $d_{G[V_1]}^R(u_1) \geq 1$ and $d_{G[V_2]}^R(u_1) \geq 1$. Let the red neighbors of u_1 be $v_1 \in V_1$ and $v_2 \in V_2$. Then u_0 must be adjacent to at least one of $\{V_1 \setminus \{v_1\}, V_2 \setminus \{v_2\}\}$ by blue edges completely, say $V_1 \setminus \{v_1\}$, otherwise $U_2 \cup \{u_0\}$ induces a red C_n. Thus u_0 is adjacent to V_2 by red edges completely. Hence $u_0 u_1$ is blue and $d_{G[V_1]}^R(u_0) = 0$. Similarly, u_1 is adjacent to V_1 by red edges completely and to $V_2 \setminus \{v_2\}$ by blue edges completely. Now we get three red K_{n-1}. Consider vertex v_0 and graph G. If v_0 is adjacent to at least one vertex in $V_2 \setminus \{v_2\}$ by blue, then v_0 can only be adjacent to U_1 and V_1 by red, otherwise we get a blue B_n. Hence either $d_{G[U_1]}^R(v_0) \geq 2$ or $d_{G[V_1]}^R(v_0) \geq 2$, yielding a red C_n. Thus $d_{G[V_2 \setminus \{v_2\}]}^B(v_0) = 0$. So suppose $d_{G[V_2 \setminus \{v_2\}]}^R(v_0) = 1$, then clearly $v_0 v_2$ cannot be red. If $v_0 v_2$ is blue, then

$$d_{G[V_1]}^R(v_0) = d_{G[V_1]}^B(v_0) - 0$$

and v_0 is adjacent to u_0, u_1 and U_1 by blue, yielding a blue B_n. Hence v_0 is adjacent to $n + 2$ vertices in $U_1 \cup V_1 \cup \{u_0, u_1\}$ by blue, and v_0 and u_0 will have one common blue neighbor in U_1 and V_1, respectively, yielding a blue B_n.

Assume $d_{G[V_2 \setminus \{v_2\}]}^R(v_0) = 0$. If $v_0 v_2$ is red, then v_0 is adjacent to at least n vertices in $U_1 \cup V_1$ by blue, yielding a blue B_n. If $v_0 v_2$ is blue, then

$$d_{G[V_1]}^R(v_0) = 1, d_{G[V_1]}^B(v_0) = 0 \text{ and } d_{G[U_1 \cup \{u_0, u_1\}]}^B(v_0) = n + 1,$$

yielding a blue B_n.

Suppose $|U_1| = n - 2, |V_1| = n - 1$ and $|V_2| = n - 2$, then $d_{G[V_1]}^R(u_1) = 1$ and $d_{G[V_2]}^R(u_1) = n - 2$ since $d_{G[U_1]}^R(u_1) = 0$. Furthermore, $d_{G[U_1]}^R(u_0) = n - 2$ since $d_{G[V_1]}^R(u_0) \leq 1$. Then we get three red K_{n-1}. If $d_{G[U_1]}^B(v_0) \geq 1$, then $d_{G[V_2]}^B(v_0) = 0$. If $d_{G[V_1]}^B(v_0) \geq 2$, u_0 and v_0 have at least one common blue neighbor in V_1, leading to a blue B_n. So $d_{G[V_1]}^B(v_0) \leq 1$, which means $d_{G[V_1 \cup V_2]}^R(v_0) \geq 2$, yielding a red C_n. So $d_{G[U_1]}^B(v_0) = 0$. Assume $d_{G[U_1]}^R(v_0) = 1$. Note that u_0 is the cut vertex of G^R such that $d_{G[U_2]}^R(u_0) \geq 1$. So we have $d_{G[U_2]}^R(v_0) = 0$, otherwise we get a red C_n. Then either u_0 and v_0 have one common blue neighbor in V_1 and V_2, respectively, or $d_{G[V_1]}^B(v_0) = d_{G[V_1]}^B(u_0) = n - 1$ and the edges $u_0 u_1$, $u_0 v_0$, $v_0 u_1$ are all blue, yielding a blue B_n. Therefore $d_{G[U_1]}(v_0) = 0$. Since $d_{G[V_1]}^R(v_0) + d_{G[V_2]}^R(v_0) \leq 1$, v_0 is adjacent to at least n vertices in $V_1 \cup V_2$ by blue. If u_0 and v_0 fail to have one common blue neighbor in V_1 and V_2, respectively, then the edges $u_0 u_1$, $u_0 v_0$, $v_0 u_1$ are all blue and

$$d_{G[V_2]}^R(u_0) = d_{G[V_2]}^R(v_0) = 1 \text{ and } d_{G[V_1]}^B(u_0) = d_{G[V_1]}^B(v_0) = n - 1.$$

yielding a blue B_n, completing the proof.   □

References

1. Bondy, J., Erdős, P.: Ramsey numbers for cycles in graphs. J. Combin. Theory Ser. B **14**, 46–54 (1973)
2. Bondy, J.: Pancyclic graphs I. J. Combin. Theory Ser. B **11**, 80–84 (1971)
3. Brandt, S., Faudree, R., Goddard, W.: Weakly pancyclic graphs. J. Graph Theory **27**(3), 141–176 (1998)
4. Chung, F.R.K., Graham, R.L., Wilson, R.M.: Quasi-random graphs. Combinatorica **9**(4), 345–362 (1989)
5. Dirac, G.: Some theorems on abstract graphs. Proc. Lond. Math. Soc. **2**(3), 69–81 (1952)
6. Erdős, P., Faudree, R., Rousseau, C., Schelp, R.: The size Ramsey numbers. Period. Math. Hungar. **9**(1–2), 145–161 (1978)
7. Haghi, S., Maimani, H., Seify, A.: Star-critical Ramsey numbers of F_n versus K_4. Discret. Appl. Math. **217**(2), 203–209 (2017)
8. Hao, Y., Lin, Q.: Star-critical Ramsey numbers for large generalized fans and books. Discret. Math. **341**(12), 3385–3393 (2018)
9. Hook, J., Isaak, G.: Star-critical Ramsey numbers. Discret. Appl. Math. **159**(5), 328–334 (2011)
10. Hook, J.: Critical graphs for $R(P_n, P_m)$ and the star-critical Ramsey number for paths. Discuss. Math. Graph Theory **35**(4), 689–701 (2015)
11. Li, Y., Li, Y., Wang, Y.: Minimal Ramsey graphs on deleting stars for generalized fans and books. Appl. Math. Comput. **372**, 125006 (2020)
12. Li, Z., Li, Y.: Some star-critical Ramsey numbers. Discret. Appl. Math. **181**, 301–305 (2015)
13. Lin, Q., Peng, X.: Large book-cycle Ramsey numbers, arXiv preprint arXiv:1909.13533 (2019)
14. Nikiforov, V., Rousseau, C.C.: Ramsey goodness and beyond. Combinatorica **29**(2), 227–262 (2009)

15. Rousseau, C.C., Sheehan, J.: On Ramsey numbers for books. J. Graph Theory **2**(1), 77–87 (1978)
16. Wang, Y., Li, Y.: Deleting edges from Ramsey graphs. Discret. Math. **343**(3), 111743 (2020)
17. Wu, Y., Sun, Y., Radziszowski, S.: Wheel and star-critical Ramsey numbers for quadrilateral. Discret. Appl. Math. **186**, 260–271 (2015)
18. Zhang, Y., Broersma, H., Chen, Y.: On star-critical and upper size Ramsey numbers. Discret. Appl. Math. **202**, 174–180 (2016)

Computing Imbalance-Minimal Orderings for Bipartite Permutation Graphs and Threshold Graphs

Jan Gorzny[(✉)]

University of Waterloo, Waterloo, Canada
jgorzny@uwaterloo.ca

Abstract. The IMBALANCE problem is an NP-complete graph ordering problem which is only known to be polynomially solvable on some
very restricted graph classes such as proper interval graphs, trees, and
bipartite graphs of low maximum degree. In this paper, we show that
IMBALANCE can be solved in linear time for bipartite permutation graphs
and threshold graphs, resolving two open questions of Gorzny and Buss
[COCOON 2019]. The results rely on the fact that if a graph can be partitioned into a vertex cover and an independent set, there is an imbalance-
minimal ordering for which each vertex in the independent set is as balanced as possible. Furthermore, like the previous results of Gorzny and
Buss, the paper shows that optimal orderings for IMBALANCE are similar
to optimal orderings for CUTWIDTH on these graph classes. We observe
that approaches for CUTWIDTH are applicable for IMBALANCE. In particular, we observe that there is fixed-parameter tractable (FPT) algorithm
which runs in time $O(2^k n^{O(1)})$ where k is the size of a minimum vertex
cover of the input graph G and n is the number of vertices of G. This
FPT algorithm improves the best known algorithm for this problem and
parameter. Finally, we observe that IMBALANCE has no polynomial kernel
parameterized by vertex cover unless NP $\subseteq$ coNP/poly.

Keywords: Imbalance · Threshold graph · Bipartite permutation
graph · Vertex cover

1 Introduction

Given a graph G and an ordering of its vertices (also known as a linear layout of
the vertices), the imbalance of each vertex is the absolute value of the difference
in the size of its neighbourhood to its left and to its right. The imbalance of
an ordering of G is the sum of the imbalances of each vertex. IMBALANCE asks
whether or not it is possible to find an ordering with imbalance at most k for
a graph G. The IMBALANCE problem was introduced by Biedl et al. [2] in the
context of graph drawing, where such an ordering is helpful [11,12,16,18,19].

Biedl et al. [2] showed that IMBALANCE is NP-complete for bipartite graphs
with degree at most 6 and weighted trees and they provide a pseudo-polynomial

© Springer Nature Switzerland AG 2020
W. Wu and Z. Zhang (Eds.): COCOA 2020, LNCS 12577, pp. 766–779, 2020.
https://doi.org/10.1007/978-3-030-64843-5_52

time algorithm for weighted trees which runs in linear time on unweighted trees. Kára et al. [13] showed that the problem is NP-complete for arbitrary graphs of degree at most 4 and planar graphs. Gorzny and Buss [8] showed that IMBALANCE remains NP-complete on split graphs, but has a linear time algorithm on proper interval graphs, and polynomial time algorithm for superfragile graphs. Gorzny and Buss [8] incorrectly claimed that for graphs with bounded twin cover size, IMBALANCE has a fixed-parameter tractable (FPT) algorithm when the parameter is the minimum size of a twin cover of the graph; however, this requires an additional parameter (like the maximum degree of the graph). There is an FPT algorithm using the minimum size of a vertex cover for a graph as the parameter by Fellows et al. [6]. Bakken [1] showed that IMBALANCE is FPT when parameterized by the neighbourhood diversity of the graph, another generalization of vertex cover which is incomparable to the minimum twin cover size. Lokshtanov et al. [14] showed that IMBALANCE is FPT when parameterized by the solution size k by constructing an algorithm that runs in time $O(2^{O(k \log k)} n^{O(1)})$, or when parameterized by the treewidth of the graph and the maximum degree of the graph. As far as we know, there are no FPT results for IMBALANCE when the parameter is only the treewidth of the graph. Gaspers et al. [7] showed that IMBALANCE is equivalent to the GRAPH CLEANING problem, which yielded a $O(n^{\lfloor k/2 \rfloor}(n+m))$ time parameterized algorithm where k is the solution size.

In this paper, we show linear time algorithms for IMBALANCE on bipartite permutation graphs and threshold graphs, as well as an improvement over the previous best algorithms and a kernelization lower bound for IMBALANCE parameterized by vertex cover. Since bipartite permutation graphs may have arbitrarily large degree, this result shows that there are classes of bipartite graphs beyond trees for which IMBALANCE can be solved efficiently. Moreover, bipartite permutation graphs are the second class of graphs with unbounded cliquewidth for which IMBALANCE can be solved efficiently, after proper interval graphs. The result uses vertex orderings with special properties that characterizes the class of bipartite permutation graphs. Our result on threshold graphs shows that there is a non-trivial subclass of split graphs, for which IMBALANCE is NP-complete, where IMBALANCE can be solved in linear time. Threshold graphs also become the third class of unbounded cliquewidth for which IMBALANCE can be solved efficiently, after proper interval graphs and bipartite permutation graphs. Since threshold graphs are also interval graphs, this solution builds towards an understanding of IMBALANCE problem on interval graphs, for which the complexity of the problem is unknown. In the parameterized complexity setting, the fastest algorithm for IMBALANCE on graphs of bounded vertex cover runs in time $O(2^{2^{O(k)}} n^{O(1)})$ where k is the size of the vertex cover of the given graph and n is the number of vertices [6]. In this paper, we show that the approach for CUTWIDTH [5] (defined in the next section) can be applied to IMBALANCE and therefore there is an algorithm that runs in time $O(2^k n^{O(1)})$. An $O(\frac{n}{2}!)$ algorithm was known [2] for IMBALANCE on bipartite graphs; applying the approach used for CUTWIDTH [5] immediately yields an $O(2^{n/2} n^{O(1)})$ algorithm. Lastly,

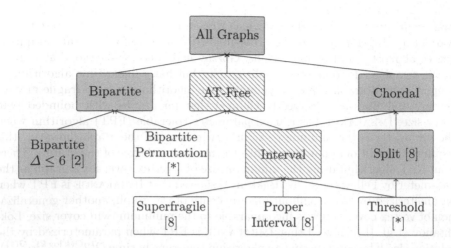

Fig. 1. The relationship between graph classes discussed in this paper. IMBALANCE is NP-complete for shaded classes, unknown for hatched classes, and polynomial for the rest. Results that appear in this work are marked with [*]. An arrow from class A to class B indicates that class A is contained within class B.

we observe that the kernelization lower bound for CUTWIDTH in [5] applies immediately to IMBALANCE.

2   Preliminaries

All graphs in this work are finite, undirected, and without multiple edges or loops. For a graph $G = (V, E)$, If $X \subseteq V$, then $G(V - X)$ refers to the graph induced by the set of vertices in $V \setminus X$, i.e. $G(V - X) = (V \setminus X, E')$ where $E' = \{(u, v) | u, v \in V \setminus X\} \cap E$. we will denote $n = |V|$ and $m = |E|$. A *complete graph* (or *clique*) is a graph whose vertices are pairwise adjacent. An *independent set* is a set $I \subseteq V$ of vertices with no edges among them (i.e., $(I \times I) \cap E = \emptyset$). A graph G is *bipartite* if V can be partitioned into to two sets X and Y such that all edges have one end-point in X and the other in Y. The *open neighbourhood* of a vertex v, denoted $N(v)$, is the set $\{u \in V \mid (v, u) \in E\}$ of vertices adjacent to v. The *closed neighbourhood* of a vertex, denoted $N[v]$, is the open neighbourhood of the vertex along with the vertex itself, i.e. $N[v] = N(v) \cup \{v\}$. A vertex v is a *pendant* if $|N(v)| = 1$. If $X \subseteq V$, then the neighbourhood of the set X, denoted $N(X)$, is $\left(\cup_{v \in X} N(v) \right) \setminus X$.

The *distance* between two vertices u and v in a graph G, denoted by $d_G(u, v)$, is the length of a shortest path between u and v in G. We will drop the subscript when it is clear from context. The *diameter* of a graph G, denoted $\mathrm{diam}(G)$, is the maximum distance between any two vertices in G, i.e. $\mathrm{diam}(G) = \max_{u,v \in V(G)} d_G(u, v)$. Two vertices u and v are *twins* if they have the same neighbours, except possibly for each other; that is, $N(u) \setminus \{v\} = N(v) \setminus \{u\}$. Two twins

u and v are *true twins* if they have an edge between them; i.e., $N[u] = N[v]$. A *vertex cover* of G is a set $C \subseteq V$ that covers every edge; i.e., $E \subseteq C \times V$.

An ordering of (a subset of) the vertices of a graph is a sequence $\langle v_1, v_2, \ldots, v_n \rangle$, with each v_i distinct. We will use set operations and notation on orderings, and also the following. For an ordering σ, $\sigma(i)$ denotes the ith vertex for $1 \leq i \leq |\sigma|$. If σ and π are disjoint orderings, $\sigma\pi$ denotes their concatenation. The relation $<_\sigma$ is defined by $u <_\sigma v$ if and only if u precedes v in σ. Relations $>_\sigma$, $\leq_\sigma$ and $\geq_\sigma$ are defined analogously. We extend these to sets of vertices: e.g., $x <_\sigma \{y, z\}$ if and only if $x <_\sigma y$ and $x <_\sigma z$. For an element x of σ, $\sigma_{<x}$ denotes the ordering induced by σ on the set $\{ y \in V \mid y <_\sigma x \}$. The orderings $\sigma_{\leq x}$, $\sigma_{>x}$, and $\sigma_{\geq x}$ are defined analogously. More generally, for a set $X \in V$, σ_X denotes the ordering induced by σ on X. Two orderings σ and τ with the same elements are said to *agree* if $u <_\sigma v$ if and only if $u <_\tau v$. If σ is an ordering, then σ^R denotes its reverse.

We now formally define IMBALANCE. Let $G = (V, E)$ be a graph and σ an ordering of V. For $v \in V$, let $\mathrm{pred}_\sigma(v)$ and $\mathrm{succ}_\sigma(v)$ respectively denote the number of neighbours of v that precede (resp. succeed) v in σ. That is, $\mathrm{pred}_\sigma(v) = |\sigma_{<v} \cap N(v)|$ and $\mathrm{succ}_\sigma(v) = |\sigma_{>v} \cap N(v)|$. The *imbalance* of v with respect to σ, denoted $\phi_\sigma(v)$, is $|\mathrm{succ}_\sigma(v) - \mathrm{pred}_\sigma(v)|$. The *imbalance* of σ is $im(\sigma) = \sum_{v \in \sigma} \phi_\sigma(v)$. The *imbalance* of G, denoted $im(G)$, is the minimum of $im(\sigma)$ over all orderings σ of V. Given a graph G and an integer k, IMBALANCE asks if $im(G) \leq k$. For convenience, given a set $X \subseteq V$ and an ordering σ of V, $\phi_\sigma(X) = \sum_{v \in X} \phi_\sigma(v)$. We say that a vertex v is *perfectly balanced* in an ordering σ if $d(v)$ is even and $\phi_\sigma(v) = 0$ or if $d(v)$ is odd and $\phi_\sigma(v) = 1$.

We also define CUTWIDTH. Let $G = (V, E)$ be a graph and σ an ordering of V. The *cutwidth after v* with respect to σ, denoted $c_\sigma(v)$, is $c_\sigma(v) = |\{(x, y) \in E | x \leq_\sigma v \text{ and } v <_\sigma y\}|$. The *cutwidth* of σ is $cw(\sigma) = \max_{v \in \sigma}\{c_\sigma(v)\}$. The *cutwidth* of G, denoted $cw(G)$, is the minimum of $cw(\sigma)$ over all orderings of V.

The next observations are helpful for understanding the imbalance of an ordering after some cliques are rearranged in an initial ordering.

Observation 1. *Let X, Y be two sets of true twins in an ordering σ such that $X <_\sigma Y$, $Y \subseteq N(X)$ and there are no vertices between X and Y in σ. Let σ' be obtained from σ by swapping the positions of X and Y in σ. Observe that*

$$\phi_{\sigma'}(X) = |\mathrm{pred}_{\sigma'}(X) - \mathrm{succ}_{\sigma'}(X)| = |(\mathrm{pred}_\sigma(X) + |Y|) - (\mathrm{succ}_\sigma(X) - |Y|)|$$
$$= |\mathrm{pred}_\sigma(X) - \mathrm{succ}_\sigma(X) + 2|Y||.$$

If $\mathrm{pred}_\sigma(X) \geq \mathrm{succ}_\sigma(X)$, $\phi_{\sigma'}(X) = \phi_\sigma(X) + 2|Y|$. Otherwise, $\mathrm{pred}_\sigma(X) < \mathrm{succ}_\sigma(X)$ and

- *if $\mathrm{succ}_\sigma(X) - |Y| < \mathrm{pred}_\sigma(X) < \mathrm{succ}_\sigma(X)$, $\phi_\sigma(X) + 2|Y| > \phi_{\sigma'}(X) > \phi_\sigma(X)$;*
- *if $\mathrm{pred}_\sigma(X) = \mathrm{succ}_\sigma(X) - |Y|$, $\phi_{\sigma'}(X) = \phi_\sigma(X)$;*
- *otherwise, $\mathrm{pred}_\sigma(X) < \mathrm{succ}_\sigma(X) - |Y|$, $\phi_{\sigma'}(X) < \phi_\sigma(X)$.*

Observation 2. *Let X, Y be two sets of true twins in an ordering σ such that $X <_\sigma Y$, $X \subseteq N(Y)$, and there are no vertices between X and Y in σ. Let σ' be obtained from σ by swapping the positions of X and Y in σ. Observe that*

$$\phi_{\sigma'}(Y) = \left| pred_{\sigma'}(Y) - succ_{\sigma'}(Y) \right| = \left| (pred_\sigma(Y) - |X|) - (succ_\sigma(Y) + |X|) \right|$$
$$= \left| pred_\sigma(Y) - succ_\sigma(Y) - 2|X| \right|.$$

If $pred_\sigma(Y) \leq succ_\sigma(Y)$, $\phi_{\sigma'}(Y) = \phi_\sigma(Y) + 2|X|$. Otherwise, $pred_\sigma(Y) > succ_\sigma(Y)$ and

- if $succ_\sigma(Y) + |X| > pred_\sigma(Y) > succ_\sigma(Y)$, $\phi_\sigma(Y) + 2|X| > \phi_{\sigma'}(Y) > \phi_\sigma(Y)$;
- if $pred_\sigma(Y) = succ_\sigma(Y) + |X|$, $\phi_{\sigma'}(Y) = \phi_\sigma(Y)$;
- otherwise, $pred_\sigma(Y) > succ_\sigma(Y) + |X|$, $\phi_{\sigma'}(Y) < \phi_\sigma(Y)$.

The next theorem shows that there is always an imbalance-minimal ordering for which a given independent set is as balanced as possible, and will be useful throughout the paper.

Theorem 1. *Let $G = (A \cup B, E)$ be a graph where B is an independent set and let σ be such that $im(\sigma) = im(G)$. There exists an optimal imbalance ordering σ^* such that for all $b \in B$, $\phi_{\sigma^*}(b) \in \{0, 1\}$; moreover, $\sigma_A = \sigma_A^*$.*

Proof. Let σ be an optimal imbalance ordering, and suppose that $b \in B$ has $\phi_\sigma(b) > 1$. Without loss of generality, suppose that b has $k \geq 2$ more neighbours to its right than its left. Let w be the vertex to the right of b in σ, and let σ' be the result of swapping b and w. If $(b, w) \notin E$, the swap does not change the total imbalance of the ordering as $\phi_{\sigma'}(b) = \phi_\sigma(b)$ and $\phi_{\sigma'}(w) = \phi_\sigma(w)$. If $(b, w) \in E$, $\phi_{\sigma'}(b) = \phi_\sigma(b) - 2$ and $\phi_{\sigma'}(w) \leq \phi_\sigma(w) + 2$. We can repeat this until $k = 1$ and $\phi_{\sigma^*}(b) \leq 1$. The case where b has more neighbours to its left than its right is analogous. Since we did not move any vertices of A past any other vertices in A, σ_A agrees with σ_A^*. $\square$

The following theorem is also helpful.

Theorem 2 ([8]). *For any graph G, there exists an imbalance-optimal ordering σ of V such that each set of true twins appears consecutively in σ.*

3   Bipartite Permutation Graphs

In this section we show that if G is a bipartite permutation graph (defined below), the imbalance of G can be computed in $O(n)$.

A graph is a *permutation* graph if it is the intersection graph of lines whose end points are on two parallel lines. A graph is a bipartite permutation graph if it is both a bipartite graph and a permutation graph. A *proper interval bigraph* is a bipartite graph that is AT-free and contains no induced cycle of size greater than four [3]; from this characterization we know that complete bipartite graphs are bipartite permutation graphs. Proper interval bipartite graphs are bipartite permutation graphs [10].

A *strong ordering* (σ^A, σ^B) of a bipartite graph $G = (A, B, E)$ consists of an ordering σ^A of A and an ordering σ^B of B such that for all $ab, a'b' \in E$, where $a, a' \in A$ and $b, b' \in B$, $a <_{\sigma^A} a'$ and $b' <_{\sigma^B} b$ implies that $ab' \in E$ and $a'b \in E$. An ordering σ^A of A has the *adjacency property* if, for every $b \in B$, $N(b)$ consists of vertices that are consecutive in σ^A. The ordering σ^A has the *enclosure property* if, for every pair b, b' of vertices of B with $N(b) \subseteq N(b')$, the vertices of $N(b') \setminus N(b)$ appear consecutively in σ^A, implying that b is adjacent to the leftmost or rightmost neighbour of b' in σ^A.

It was shown that every bipartite permutation graph has a strong ordering, and moreover, that every strong ordering of a bipartite permutation graph has both the adjacency and enclosure properties [17]. A strong ordering of a bipartite permutation graph can be computed in linear time [4].

Finally, we note that if a part A of a bipartition (A, B) is fixed as σ^A, an ordering which is imbalance-minimal among all orderings which agree with σ^A can be found in linear time.

Theorem 3 ([2]). *Given a bipartite graph $G = (A, B, E)$ and a fixed vertex-ordering σ^A of A, there is a linear time algorithm that finds an ordering of G which is imbalance-minimal with respect to all orderings that agree with σ^A.*

The main result requires handling some highly structured cases.

Lemma 1. *Let $G = (A, B, E)$ be a complete bipartite graph. Let (σ^A, σ^B) be a strong ordering of G. There is an optimal imbalance ordering σ of G such that $\sigma_A = \sigma^A$, and $\phi_\tau(A) \geq \phi_\sigma(A)$ for any ordering τ where each $b \in B$ is perfectly balanced.*

Proof. By Theorem 1, there is an optimal ordering where B is perfectly balanced; starting with σ^A (which is the same as any ordering of A, since all the vertices of A are twins), apply the algorithm of Theorem 3 to get an ordering σ with desired property; if $\sigma_B \neq \sigma^B$, re-order σ_B so they are equal: all vertices of B are twins and can be interchanged. It is clear that the desired properties hold.   □

Lemma 2. *Let $G = (A, B, E)$ be a bipartite permutation graph such that there is a vertex $b \in B$ with $N(b) = A$, and for all $b' \in B \setminus \{b\}$, $deg(b') = 1$. There is an optimal imbalance ordering σ of G such that $\sigma_A = \sigma^A$, and $\phi_\tau(A) \geq \phi_\sigma(A)$ for any ordering τ where each $b \in B$ is perfectly balanced.*

Proof. By Theorem 1, there is an optimal imbalance ordering that perfectly balances all vertices of B. Observe that the vertex $b \in B$ is perfectly balanced by placing $\lfloor \frac{|A|}{2} \rfloor$ vertices of A to its right, and $\lceil \frac{|A|}{2} \rceil$ vertices of A to its left; place the first $\lfloor \frac{|A|}{2} \rfloor$ vertices of A to the left of b and the rest of A to the right of b, and call the resulting order σ. Observe that no two vertices $a, a' \in A$ have a neighbour of $B \setminus \{b\}$ in common, as otherwise that vertex is not a pendant. Therefore, for each $a \in A$ such that $a <_\sigma b$, we can place $\lceil \frac{|N(a)|-1}{2} \rceil$ immediately to the left of a and $\lfloor \frac{|N(a)|-1}{2} \rfloor$ immediately to the right of a in σ to minimize the imbalance of a. For each $a \in A$ such that $b <_\sigma a$, we can place $\lceil \frac{|N(a)|-1}{2} \rceil$ neighbours immediately to the right of a and $\lfloor \frac{|N(a)|-1}{2} \rfloor$ neighbours immediately to the left of a in σ to minimize the imbalance of a. Each pendant of B is perfectly balanced.   □

Theorem 4. *Let* (σ^A, σ^B) *be a strong ordering of a bipartite permutation graph* $G = (A, B, E)$. *There is an ordering* σ *of* G *with* $im(\sigma) = im(G)$ *and* $\sigma_A = \sigma^A$.

Proof. We will prove the following stronger claim.

Claim. There is an optimal ordering σ of G where $\phi_\sigma(b) \in \{0, 1\}$ for all $b \in B$, $\sigma_B = \sigma^B$, $\sigma_A = \sigma^A$, and $\phi_\tau(A) \geq \phi_\sigma(A)$ for any ordering τ where each $b \in B$ is perfectly balanced.

<u>Proof of Claim:</u> The proof is by induction on n. In the base case, $n = 1$, and the result holds trivially.

Assume the result holds for $n' \geq 1$ and consider a bipartite permutation graph G where $|V| = n = n' + 1$. Let (σ^A, σ^B) be a strong ordering of G where $\sigma^A = \langle a_1, \ldots, a_s \rangle$ and $\sigma^B = \langle b_1, \ldots, b_t \rangle$ such that $d(b_1, a_s)$ is maximal (i.e., use $((\sigma^A)^R, (\sigma^B)^R)$ if $d(b_t, a_1)$ is larger than $d(b_1, a_s)$ according to the original ordering). Let $B_s \subseteq B$ be the maximal subset such that for all $b \in B_s$, $N(b) = \{a_s\}$; similarly, let $B_1 \subseteq B$ be the maximal subset such that for all $b \in B_1$, $N(b) = \{a_1\}$. Let σ_s be an optimal ordering of $G - (\{a_s\} \cup B_s)$ provided by the induction hypothesis. Let $A_1 \subseteq A$ be the maximal subset such that for all $a \in A_1$, $N(a) = \{b_1\}$. Let σ_1 be an optimal ordering of $G - (\{b_1\} \cup A_1)$ provided by the induction hypothesis.

If $d(b_1, a_s) = 1$, then $diam(G) \leq 2$ and we are done by Lemma 1; therefore we may assume $d(b_1, a_s) \geq 3$. Let $\ell(\pi, x)$ be the rightmost vertex of $N(x)$ that is to the left of x in π and let $r(\pi, x)$ be the leftmost vertex of $N(x)$ that is to the right of x in π.

Let $b' \in B$ be the rightmost vertex of B in σ_1 such that $\ell(\sigma_1, b') = \ell(\sigma_s, b')$ and $r(\sigma_1, b') = r(\sigma_s, b')$. We show that at least one such $b' \in B$ exists. Take b_i to be the vertex of B with the smallest index (with respect to σ^B) such that $a_1 \in N(b_i)$ but $N(b_i) \neq \{a_1\}$. If $a_s \notin N(b_i)$, then b_i is between the same two vertices in both orderings σ_1 and σ_s, so it is eligible. Otherwise, $a_s \in N(b_i)$. Since $a_1 \in N(b_i)$ and $a_s \in N(b_i)$, $d(a_1, a_s) = 2$. If $|A|$ is even, $\phi_{\sigma_1}(b_i) = 0$, and $\phi_{\sigma_s}(b_i) = 1$. In σ_s, b_i will be in one of the centers of $|A| - 1$ vertices; let a' be the middle vertex of A in σ_s. Pick b' to be the leftmost twin of b_i such that $a' <_{\sigma'} b'$ (if all twins of b_i are to the left of a', then $i = t$ and we are done by Lemma 2). The vertex b' has fewer neighbours to its right in σ_s than its left, and appending a_s will re-balance b_i; i.e., it will be between the same two vertices in both orderings. Otherwise, $|A|$ is odd, and $\phi_{\sigma_1}(b_i) = 1$, and $\phi_{\sigma_s}(b_i) = 0$. In σ_1, b_i will be in one of the centers of $|A|$ vertices; let a' be the middle vertex of A in σ_1. Pick b' to be the rightmost twin of b_i such that $b' <_{\sigma'} a'$ (if all twins of b_i are to the right of a', then $i = t$ and we are done by Lemma 2). The vertex b' has fewer neighbours to its left in σ_1 than its right, so removing a_s will re-balance b_i; i.e., it will be between the same two vertices in both orderings.

If $a_1 \notin N(b')$ or $a_s \notin N(b')$, let $\sigma_1' = (\sigma_1)_{\geq b'}$ and $\sigma_s' = (\sigma_s)_{<b'}$. Otherwise, let $x \in A \cup B$ be the rightmost vertex of $B_1 \cup \{a_1\}$ in σ_s, and let $\sigma_s' = (\sigma_s)_{\leq x}$, and $\sigma_1' = (\sigma_1)_{>x}$. Note that $(B \setminus B_1) \subseteq (\sigma_1)_{>x}$ since all vertices of $B \setminus B_1$ have another neighbour to the right of a_1.

Let $\sigma = \sigma'_s \cdot \sigma'_1$. Observe that each $b \in B$ is perfectly balanced in σ, as each such b has minimum of the imbalances it had in σ_1 and σ_s, at least one of which obtained the minimum of b with the entirety of $N(b)$ in the ordering. Furthermore, each $a \in A$ has minimum imbalance among all orderings which perfectly balance $N(a)$ by the same observation.

Let $A'_1 = (\sigma'_1) \cap A$ and $A'_s = (\sigma'_s) \cap A$. Let σ^* be an optimal ordering of G such that each vertex $b \in B$ has $\phi_{\sigma^*}(b) \in \{0, 1\}$, which exists by Theorem 1.

$$im(G) = im(\sigma^*) = \phi_{\sigma^*}(B) + \phi_{\sigma^*}(A'_1) + \phi_{\sigma^*}(A'_s)$$
$$\geq \phi_\sigma(B) + \phi_\sigma(A'_1) + \phi_\sigma(A'_s) = im(\sigma) \geq im(G)$$

as required.   ∎

The theorem is immediate from the claim.   □

Corollary 1. *If G is a bipartite permutation graph, $im(G)$ can be computed in linear time.*

Proof. A strong ordering of $G = (A, B, E)$ can be obtained in linear time [4]. Applying Theorem 3 using σ^A generates an optimal ordering relative to σ^A in linear time, which is optimal by Theorem 4.   □

4   Threshold Graphs

In this section we show that if G is a threshold graph (defined below), the imbalance of G can be computed in $O(n)$. Threshold graphs are a subset of split graphs. Imbalance is NP-complete on split graphs [8].

A graph is a *split graph* if its vertex set can be partitioned into a clique C and an independent set I where (C, I) is called a *split partition*. A graph is a *threshold graph* if and only if it has a split partition (C, I) such that vertices of I (and equivalently the vertices of C) can be ordered by neighbourhood inclusion. Such a split partition is called a *threshold partition*. Clearly, threshold graphs are a subset of split graphs. Determining if a graph is a threshold graph, and computing a threshold partition if it is, can be computed in linear time [15].

Let (C, I) be a split partition. We may assume that every vertex $c \in C$ has a neighbour in I, as otherwise we may take a new split partition (C', I') where $C' = C \setminus \{c\}$ and $I' = I \cup \{c\}$. We will partition I into sets $(I_0, I_1, \ldots, I_\ell)$ such that I_0 is the set of isolated vertices, and $N(I_1) \subset N(I_2) \subset \ldots N(I_\ell)$, where ℓ is largest possible. All vertices in I_j have the same neighbours (and therefore degree) for $0 \leq j \leq \ell$. The partition on I defines a partition $(C_1, C_2, \ldots, C_\ell)$ of C, where $C_1 = N(I_1)$ and $C_j = N(I_j) \setminus N(I_{j-1})$ for $2 \leq j \leq \ell$. Similarly, all vertices in C_j have the same neighbours (and therefore degree) for $1 \leq j \leq \ell$. Vertices of C_i are on the ith level of the clique, and vertices of I_i are on the ith level of the independent set. For a threshold partition (C, I), $N(C_j) \subset N(C_i)$ and therefore $|N(C_i)| > |N(C_j)|$ if $j > i$. Figure 2 shows an example of a threshold graph with a threshold partition.

The main result requires the following key lemma.

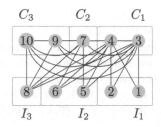

Fig. 2. A threshold graph G with levels of its threshold partition indicated. An ordering with $im(\sigma) = im(G) = 22$ is $\sigma = \langle v_1, v_2, v_5, v_3, v_4, v_6, v_7, v_8, v_9, v_{10} \rangle$.

Lemma 3. *Let G be a threshold graph on $\ell \geq 3$ levels and let σ be an ordering of G. Suppose that either $|C_1| \geq 2$, or $|C_1| = 1$ and σ is an ordering of G such that I_1 appears as the first $|I_1|$ vertices of σ. Then there is an ordering σ' such that $im(\sigma') \leq im(\sigma)$ and $c_i <_{\sigma'} c_j$ for $c_i \in C_i$ and $c_j \in C_j$ whenever $j > i$.*

Proof. We define a partial order $\prec$ on orderings of G, for which each $\prec$-minimal ordering has no inversions (c_j, c_i) where $j > i$, $c_j \in C_j$, $c_i \in C_i$, and $c_j <_{\sigma} c_i$. We define the ordering $\prec$ as follows: if $inverted(\pi) < inverted(\sigma)$, then $\pi \prec \sigma$. We then show that for any σ in which a pair (c_j, c_i) such that $j > i$, $c_j \in C_j$, $c_i \in C_i$, and $c_j <_{\sigma} c_i$ appears, there exists π with $\pi \prec \sigma$, and $im(\pi) \leq im(\sigma)$.

Given an ordering σ, we will say that a pair (c_j, c_i) is an *inverted pair* if $j > i$ and $c_j <_{\sigma} c_i$ and $c_j \in C_j$, $c_i \in C_i$; an inverted pair is a *bad pair* if it is also the case that $N(c_i) \cap \sigma_{>c_i} = N(c_j) \cap \sigma_{>c_i}$. For an ordering σ, let $inverted(\sigma)$ be the number of inverted pairs in σ.

Let σ be an imbalance ordering for G provided by the lemma hypothesis. We additionally assume $\phi_\sigma(v) \in \{0, 1\}$ for all $v \in I$ by Theorem 1. If $inverted(\sigma) = 0$, we are done; therefore we may assume that $inverted(\sigma) > 0$. Let $\sigma = L \cdot c_j \cdot M \cdot c_i \cdot R$ for a pair $c_j <_\sigma c_i$.

We will show that σ must contain a bad pair; suppose to the contrary it does not. Let (c_j, c_i) be the inverted pair that has both c_j and c_i as far right as possible: choose (c_j, c_i) so that there is no inverted pair (c'_j, c'_i) such that $c_j <_\sigma c'_j$, and given that c_j is rightmost, there is no inverted pair (c_j, c'_i) such that $c_i <_\sigma c'_i$. Let $X = (N(c_i) \cap R) \setminus (N(c_j) \cap R)$, which must be non-empty since (c_j, c_i) is an inverted pair but not a bad pair. Since X is not adjacent to both c_i and c_j, $X \subseteq I$. Since all pendants are at the beginning of the order, or there are none, and $\phi_\sigma(x) \in \{0, 1\}$ for all $x \in X \subseteq I$, there must be a c_r such that $c_i <_\sigma c_r$ and $c_r \in N(x)$ for any $x \in X$. Since $c_r \in N(x)$ but $c_j \notin N(x)$, $r < j$. However, this contradicts our choice of (c_j, c_i), as either (c_i, c_r) is an inverted pair with $c_j < c_i$, or (c_j, c_r) is an inverted pair with $c_j < c_r$. Therefore σ contains a bad pair.

Let (c_j, c_i) be the bad pair that places c_j as far right as possible and minimizes the number of vertices between c_j and c_i in σ: choose (c_j, c_i) so that there is no bad pair (c'_j, c'_i) such that $c_j <_\sigma c'_j$, and given that c_j is rightmost, there is no bad pair (c_j, c_i') such that $c_i' <_\sigma c_i$. Since (c_j, c_i) is a bad pair, $N(c_i) \cap R = N(c_i) \cap \sigma_{>c_i} = N(c_j) \cap \sigma_{>c_i} = N(c_j) \cap R$ which implies that $|N(c_i) \cap R| = |N(c_j) \cap R|$.

Suppose first that $M = \emptyset$. In this case, c_j is beside c_i. Since there is a $v \in N(c_i)$ but $v \notin N(c_j)$ and it must be the case that $v <_\sigma c_j$, c_i can be moved to the left of c_j to get $\tau \lessdot \sigma$. No new inverted pairs could have been introduced by performing such a swap: c_i remains to the right of everything in L, and c_j remains to the left of everything in R. Therefore, we may assume that $M \neq \emptyset$.

- If there is a $c_t \in M \cap C_t$ such that $t = j$, then $N(c_t) = N(c_j)$ and therefore (c_t, c_i) is a more rightmost bad pair (i.e. one with $c_j <_\sigma c_t$), contradicting our choice of (c_j, c_i).
- If there is a $c_t \in M \cap C_t$ such that $t > j$, then (c_t, c_i) is a more inverted rightmost pair. If it is bad, we have reached a contradiction; therefore it must only be inverted. Let $X = (N(c_i) \cap R) \setminus (N(c_t) \cap R)$, which must be non-empty since (c_t, c_i) is an inverted pair but not a bad pair. Since X is not adjacent to both c_i and c_t, $X \subseteq I$. Since all pendants are at the beginning of the order, or there are none, and $\phi_\sigma(x) \in \{0, 1\}$ for $x \in X$, there must be a c_r such that $c_i <_\sigma c_r$ and $c_r \in N(x)$. Since $c_r \in N(x)$ but $c_t \notin N(x)$, $r < t$; however, this contradicts our choice of (c_j, c_i), as (c_t, c_r) is a more rightmost bad pair (i.e. one with $c_j < c_t$).
- If there is a $c_t \in M \cap C_t$ such that $j > t > i$, then $N(c_j) \subseteq N(c_t)$ and therefore (c_t, c_i) is a more rightmost bad pair (i.e. one with c_j and $c_t < c_i$), contradicting our choice of (c_j, c_i).
- If there is a $c_t \in M \cap C_t$ such that $t < i$, then (c_j, c_t) is an inverted pair; if it is a bad pair, we have reached a contradiction. If (c_j, c_t) is an inverted pair, let $X = (N(c_t) \cap \sigma_{>c_t}) \setminus (N(c_j) \cap \sigma_{>c_t})$, which must be non-empty. Since X is not adjacent to both c_t and c_j, $X \subseteq I$. Since all pendants are at the beginning of the order, or there are none, and $\phi_\sigma(x) \in \{0, 1\}$ for $x \in X$, there must be a rightmost $x <_\sigma c_r$ such that $x \in N(c_r)$. Since $t < i$, $r \leq t < i$ (and in particular, $r \neq i$). If $c_r <_\sigma c_i$, then $N(c_i) \subseteq N(c_r)$ and (c_j, c_r) is a rightmost bad pair with a smaller distance between the vertices (i.e. one with c_j and $c_r <_\sigma c_i$); if $c_i <_\sigma c_r$, then (c_i, c_r) is a more rightmost bad pair (i.e. one with $c_j <_\sigma c_i$). In either case, we have a contradiction.

Therefore, $M \subseteq I \cup C_i$.

If c_j becomes strictly more balanced by moving $|M|$ vertices to the right (so that it is immediately left of c_i), since c_i has a private neighbour to its left, c_j and c_i can be swapped, resulting in $\tau \lessdot \sigma$. Since $M \subseteq I \cup C_i$, no new inverted pairs could have been introduced by performing such a swap. Therefore we may assume that this is not the case, and by Observation 1 we have

$$|N(c_j) \cap L| \geq |N(c_j) \cap (M \cup \{c_i\} \cup R)| - |M|. \qquad (1)$$

If c_i becomes strictly more balanced by moving $|M| + 1$ vertices to the left, (so that it is immediately to the *left* of c_j), then $\tau \lessdot \sigma$ as it has at least one fewer inversion: since $M \subseteq I \cup C_i$, no new inverted pairs could have been introduced by performing such a swap. Therefore we may assume that this is not the case, and by Observation 2 we have

$$|N(c_i) \cap (L \cup \{c_j\} \cup M)| \leq |N(c_i) \cap R| + (|M| + 1), \tag{2}$$

Thus, we have

$$
\begin{aligned}
|N(c_i) \cap R| &\geq |N(c_i) \cap (L \cup \{c_j\} \cup M)| - |M| - 1 && \text{by (2)} \\
&= |N(c_i) \cap L| + |\{c_j\}| + |M| - |M| - 1 \\
&= |N(c_i) \cap L| \\
&\geq |N(c_j) \cap L| \\
&\geq |N(c_j) \cap (M \cup \{c_i\} \cup R)| - |M| && \text{by (1)} \\
&\geq |N(c_j) \cap R| + |M| + |\{c_i\}| - |M| > |N(c_j) \cap R|
\end{aligned}
$$

which is a contradiction to $|N(c_i) \cap R| = |N(c_j) \cap R|$. Therefore, at least one of c_j or c_i can be moved beside the other.   □

The following lemmas enable us to prove the theorem for all threshold graphs.

Lemma 4. *Let G be a threshold graph on $\ell \geq 3$ levels such that $|C_1| = 1$. If $|I_1| \leq |G \setminus (C_1 \cup I_1)|$, then there is an ordering σ such that $im(\sigma) = im(G)$ and I_1 are the first $|I_1|$ vertices of σ.*

Proof. If there is an optimal ordering σ such that I_1 is entirely contained to the left of the unique $c_1 \in C_1$, then we can move I_1 to the beginning of the ordering; if they are contained to the right of c_1, they can be moved to the end of the ordering, at which point the reverse of the order satisfies the lemma requirement.

Suppose to the contrary that G is such that all optimal imbalance orderings σ of G have some vertices I_1^ℓ (where $I_1^\ell \neq \emptyset$) to the left of $c_1 \in C_1$, and I_1^r (where $I_1^r \neq \emptyset$) to the right of c_1 in any ordering. Let σ_0 be any such optimal ordering, and let σ be the result of simultaneously moving I_1^ℓ as left as possible in σ_0 and I_1^r as right as possible in σ_0 (since all vertices in I_1 are pendants, they can move further away from c_1 without changing the total imbalance of the ordering).

First, suppose that $|I_1^\ell| \neq |I_1^r|$; without loss of generality, assume that $|I_1^\ell| > |I_1^r|$ (as otherwise we can reverse σ). Let $d = |I_1^\ell| - |I_1^r|$ ($d > 0$). Let G' be the result of removing $|I_1^\ell| - d = |I_1^r|$ vertices from I_1^ℓ and all of I_1^r and let $\sigma' = \sigma_{G'}$. Since σ' removes the same number of vertices from both sides of c_1, $\phi_\sigma(c_1) = \phi_{\sigma'}(c_1)$. The pendants removed do not affect the imbalance of any other vertex in $G \setminus C_1$, so no other vertex's imbalance is changed. Therefore, since σ' removes $2 \cdot |I_1^r|$ pendant vertices from σ, $im(\sigma') = im(\sigma) - 2 \cdot |I_1^r|$. Since G' is still a threshold graph, and σ' places all pendants at the beginning of the ordering, Lemma 3 produces an ordering τ' with no worse imbalance than σ', and one where c_1 is to the left of $C \setminus C_1$. Applying Theorem 1 to τ' produces an ordering τ'' in which all vertices $v \in I$ have $\phi_{\tau''}(v) \in \{0, 1\}$. Since c_1 is adjacent to everything, all of the remaining vertices of I_1^ℓ are placed on the opposite side of $G' \setminus (C_1 \cup I_1)$ in τ''. Now we can pre-pend the $2 \cdot |I_1^r|$

pendants to τ'' to get an ordering σ'' of G with $im(\sigma'') = im(\tau'')$, since each pendant adds one to the total imbalance, but decreases the imbalance of c_1 by 1. However, $m(\sigma'') = im(\tau'') \leq im(\sigma') = im(\sigma) - 2 \cdot |I_1^r| < im(\sigma)$, contradicting the optimalty of σ.

Suppose instead that $|I_1^\ell| = |I_1^r|$. Let G' be the result of removing $|I_1^\ell| - 1 = x$ vertices from I_1^ℓ and all of I_1^r, and let $\sigma' = \sigma_{G'}$. Since σ' removes the same almost number of vertices from both sides of c_1, $\phi_{\sigma'}(c_1) = \phi_\sigma(c_1) + 1$. The pendants removed do not affect the imbalance of any other vertex in $G \setminus C_1$, so no other vertex's imbalance is changed. Therefore, since σ' removes $2 \cdot (x + 1)$ pendant vertices from σ, $im(\sigma') = im(\sigma) - 2 \cdot (x + 1) + 1$. Since G' is still a threshold graph, and σ' places all pendants at the beginning of the ordering, Lemma 3 produces an ordering τ' with no worse imbalance than σ', and one where c_1 is to the left of $C \setminus C_1$. Applying Theorem 1 to τ' produces an ordering τ'' in which all vertices $v \in I$ have $\phi_{\tau''} \in \{0, 1\}$. Since c_1 is adjacent to everything, all of the remaining vertices of I_1^ℓ are placed on the opposite side of $G' \setminus (C_1 \cup I_1)$ in τ''. Now we can pre-pend the $2 \cdot x + 1$ pendants to τ'' to get an ordering σ'' of G with $im(\sigma'') = im(\tau'')$, since each pendant adds one to the total imbalance, but decreases the imbalance of c_1 by 1. However, $im(\sigma'') = im(\tau'') \leq im(\sigma') = im(\sigma) - 2 \cdot (x + 1) + 1 < im(\sigma)$ (since $x \geq 1$), contradicting $im(\sigma) = im(G)$. $\square$

Lemma 5. *Let G be a threshold graph on $\ell \geq 3$ levels such that $|C_1| = 1$. If $|I_1| > |G \setminus (C_1 \cup I_1)|$, then there is an ordering σ' such that $im(\sigma') = im(G)$ and $c_i <_{\sigma'} c_j$ for $c_i \in C_i$ and $c_j \in C_j$ whenever $j > i$.*

Proof. Remove $d = |I_1| - |G \setminus (I_1 \cup C_1)|$ pendants from G to get G'. By Lemma 4, there is an optimal ordering τ' of G' that places all of the remaining pendants to the left of c_1 while minimizing the imbalance in the rest of the graph. Now add d pendant vertices to τ' on alternating sides of c_1 to get an optimal ordering of G. $\square$

Lemma 6. *If G be a threshold graph, then there is an ordering σ' such that $im(\sigma') = im(G)$ and $c_i <_{\sigma'} c_j$ for $c_i \in C_i$ and $c_j \in C_j$ whenever $j > i$.*

Proof. If G has at most two levels, then we are done by Theorem 2 (or the reverse of the ordering it provides). Suppose instead that G has at least three levels. If $|C_1| \geq 2$, apply Lemma 3 to any optimal ordering of G and we are done. Otherwise, $|C_1| = 1$. If $|I_1| \leq |G \setminus (C_1 \cup I_1)|$, apply Lemma 3 to the ordering provided by Lemma 4 and we are done. Otherwise, $|I_1| > |G \setminus (C_1 \cup I_1)|$ and we are done by Lemma 5. $\square$

Theorem 5. *Imbalance can be solved in time $O(n)$ for threshold graphs.*

Proof. Let $G = (V, E)$ with a fixed threshold partition (C, I); therefore C is partitioned into level $C_1, \ldots, C_\ell$ for some ℓ. Let $E' \subseteq E$ be the set of edges between vertices C and I, that is, $E' = \{(u, v) | u \in C, v \in I\}$. Let $G'' = (V, E')$, and construct G' by adding each edge $(u, v) \in E$ such that $u, v \in C$ and subdividing it; call the set of vertices added via subdivision V_E. The graph G' is bipartite with partition $(C, I \cup V_E)$. Let τ be an ordering of C such that $C_i <_\tau C_j$ for all

$j > i$. By Lemma 6, at least one optimal ordering σ of G is such that $\sigma_C = \tau$. Apply the algorithm of Theorem 3 to get an optimal ordering σ' of G', then move each $v_e \in V_E$ between its end points so that $\phi_{\sigma'}(v_e) = 0$, at which point replacing the each path of length two u, v_e, v by an edge (u, v) does not change the imbalance.   □

Similar to optimal CUTWIDTH orderings on threshold graphs [9], there are optimal IMBALANCE orderings where the independent set of the graph is also ordered by levels of the threshold partition.

Corollary 2 ($*^1$). *If G be a threshold graph, then there is an ordering σ' such that $im(\sigma') = im(G)$ and $c_i <_{\sigma'} c_j$ for $c_i \in C_i$ and $c_j \in C_j$ whenever $j > i$, and $v_i <_{\sigma'} v_j$ for $v_i \in I_i$ and $v_j \in I_j$ whenever $j > i$.*

5   Improved Imbalance Parameterized by Vertex Cover

In this section, we note that the approach of Cygan et al. [5] that was used to solve a similar problem, CUTWIDTH, is immediately applicable to IMBALANCE. Therefore, given a graph $G = (C \cup I, E)$ with a vertex cover of size $|C| = k$, we can compute the imbalance of G in time $O(2^k n^{O(1)})$. The solution uses dynamic programming and improves the previous algorithm by Fellows et al. [6].

Theorem 6 (*). *Let G be a graph with vertex cover of size k. There is an algorithm to solve IMBALANCE in time $O(2^k n^{O(1)})$. Therefore there is a $O(2^{n/2} n^{O(1)})$ time algorithm for IMBALANCE on bipartite graphs.*

Moreover, strengthening a lemma of [14] and using the exact same reduction of [5], we can prove the following theorem.

Theorem 7 (*). IMBALANCE *parameterized by the size of the vertex cover does not admit a polynomial kernel, unless NP $\subseteq$ coNP/poly.*

6   Conclusion

We have shown that IMBALANCE can be solved in linear time for bipartite permutation graphs and threshold graphs. The complexity of the problem remains open for some other restricted graph classes, like *cographs*, and even *trivially perfect* graphs, which form a proper subset of cographs and a superset of superfragile graphs. The complexity of IMBALANCE on *interval* graphs remains a challenging open problem. We conjecture that similar results to those presented here are possible for OPTIMAL LINEAR ARRANGEMENT.

---

[1] Statements marked with a * will be shown in the full version of the paper.

References

1. Bakken, O.R.: Arrangement problems parameterized by neighbourhood diversity. Master's thesis, The University of Bergen (2018)
2. Biedl, T., et al.: Balanced vertex-orderings of graphs. Discrete Appl. Math. **148**(1), 27–48 (2005)
3. Brandstädt, A., Spinard, J.P., Le, V.B.: Graph Classes: A Survey, volume 3. Siam (1999)
4. Chang, J.-M., Ho, C.-W., Ko, M.-T.: LexBFS-ordering in asteroidal triple-free graphs. ISAAC 1999. LNCS, vol. 1741, pp. 163–172. Springer, Heidelberg (1999). https://doi.org/10.1007/3-540-46632-0_17
5. Cygan, M., Lokshtanov, D., Pilipczuk, M., Pilipczuk, M., Saurabh, S.: On cutwidth parameterized by vertex cover. Algorithmica **68**(4), 940–953 (2014)
6. Fellows, M.R., Lokshtanov, D., Misra, N., Rosamond, F.A., Saurabh, S.: Graph layout problems parameterized by vertex cover. In: Hong, S.-H., Nagamochi, H., Fukunaga, T. (eds.) ISAAC 2008. LNCS, vol. 5369, pp. 294–305. Springer, Heidelberg (2008). https://doi.org/10.1007/978-3-540-92182-0_28
7. Gaspers, S., Messinger, M.E., Nowakowski, R.J., Pralat, P.: Clean the graph before you draw it! Inf. Process. Lett. **109**(10), 463–467 (2009)
8. Gorzny, J., Buss, J.F.: Imbalance, cutwidth, and the structure of optimal orderings. In: Du, D.-Z., Duan, Z., Tian, C. (eds.) COCOON 2019. LNCS, vol. 11653, pp. 219–231. Springer, Cham (2019). https://doi.org/10.1007/978-3-030-26176-4_18
9. Heggernes, P., Lokshtanov, D., Mihai, R., Papadopoulos, C.: Cutwidth of split graphs and threshold graphs. SIAM J. Discrete Math. **25**(3), 1418–1437 (2011). https://doi.org/10.1137/080741197
10. Hell, P., Huang, J.: Interval bigraphs and circular arc graphs. J. Graph Theory **46**(4), 313–327 (2004)
11. Kant, G.: Drawing planar graphs using the canonical ordering. Algorithmica **16**(1), 4–32 (1996)
12. Kant, G., He, X.: Regular edge labeling of 4-connected plane graphs and its applications in graph drawing problems. Theoret. Comput. Sci. **172**(1–2), 175–193 (1997)
13. Kára, J., Kratochvil, J., Wood, D.R.: On the complexity of the balanced vertex ordering problem. Discrete Math. Theoret. Comput. Sci. 9 (2007)
14. Lokshtanov, D., Misra, N., Saurabh, S.: Imbalance is fixed parameter tractable. Inf. Process. Lett. **113**(19–21), 714–718 (2013)
15. Mahadev, N.V.R., Peled, U.N.: Threshold Graphs and Related Topics, vol. 56. Elsevier (1995)
16. Papakostas, A., Tollis, I.G.: Algorithms for area-efficient orthogonal drawings. Comput. Geom. **9**(1–2), 83–110 (1998)
17. Spinrad, J., Brandstädt, A., Stewart, L.: Bipartite permutation graphs. Discrete Appl. Math. **18**(3), 279–292 (1987)
18. Wood, D.R.: Optimal three-dimensional orthogonal graph drawing in the general position model. Theoret. Comput. Sci. **299**(1–3), 151–178 (2003)
19. Wood, D.R.: Minimising the number of bends and volume in 3-dimensional orthogonal graph drawings with a diagonal vertex layout. Algorithmica **39**(3), 235–253 (2004)

Inductive Graph Invariants
and Algorithmic Applications

C. R. Subramanian[✉]

The Institute of Mathematical Sciences, HBNI,
CIT Campus, Taramani, Chennai 600113, India
crs@imsc.res.in

Abstract. We introduce and study an inductively defined analogue $f_{\mathrm{IND}}()$ of any increasing graph invariant $f()$. An invariant $f()$ is increasing if $f(H) \leq f(G)$ whenever H is an induced subgraph of G. This inductive analogue simultaneously generalizes and unifies known notions like degeneracy, inductive independence number, etc into a single generic notion. For any given increasing $f()$, this gets us several new invariants and many of which are also increasing. It is also shown that $f_{\mathrm{IND}}()$ is the minimum (over all orderings) of a value associated with each ordering.

We also explore the possibility of computing $f_{\mathrm{IND}}()$ (and a corresponding optimal vertex ordering) and identify some pairs $(\mathcal{C}, f())$ for which $f_{\mathrm{IND}}()$ can be computed efficiently for members of $\mathcal{C}$. In particular, it includes graphs of bounded $f_{\mathrm{IND}}()$ values. Some specific examples (like the class of chordal graphs) have already been studied extensively.

We further extend this new notion by (i) allowing vertex weighted graphs, (ii) allowing $f()$ take to take values from a totally ordered universe with a minimum and (iii) allowing the consideration of r-neighborhoods for arbitrary but fixed $r \geq 1$. Such a generalization is employed in designing efficient approximations of some graph optimization problems. Precisely, we obtain efficient algorithms (by generalizing the known algorithm of Ye and Borodin [30] for special cases) for approximating optimal weighted induced $\mathcal{P}$-subgraphs and optimal $\mathcal{P}$-colorings (for hereditary $\mathcal{P}$'s) within multiplicative factors of (essentially) k and $k/(m-1)$ respectively, where k denotes the inductive analogue (as defined in this work) of optimal size of an unweighted induced $\mathcal{P}$-subgraph of the input and m is the minimum size of a forbidden induced subgraph of $\mathcal{P}$. Our results generalize the previous result on efficiently approximating maximum independent sets and minimum colorings on graphs of bounded inductive independence number, to optimal $\mathcal{P}$-subgraphs and $\mathcal{P}$-colorings for arbitrary hereditary classes $\mathcal{P}$. As a corollary, it is also shown that any *maximal* $\mathcal{P}$-subgraph approximates an optimal solution within a factor of $k+1$ for unweighted graphs, where k is maximum size of any induced $\mathcal{P}$-subgraph in any local neighborhood $N_G(u)$.

1 Introduction

We consider only finite, simple and labeled graphs. We consider only classes $\mathcal{C}$ of graphs which are closed under graph isomorphism. That is, if $G \in \mathcal{C}$ and

© Springer Nature Switzerland AG 2020
W. Wu and Z. Zhang (Eds.): COCOA 2020, LNCS 12577, pp. 780–801, 2020.
https://doi.org/10.1007/978-3-030-64843-5_53

if H is isomorphic to G, then we have $H \in \mathcal{C}$. A graph invariant $f(G)$ is a nonnegative real-valued function (or more generally a $\mathcal{U}$-valued function for some arbitrary universe $\mathcal{U}$ endowed with a total order and a minimum 0.) such that $f(G) = f(H)$ if G and H are isomorphic. An invariant $f()$ is increasing if $f(H) \leq f(G)$ whenever H is an induced subgraph of G. For this version of the paper, $f()$ is assumed to be a real-valued function. For the more detailed journal version, we will make use of the more general assumption of $f()$ being $\mathcal{U}$-valued.

The minimum and maximum degrees $(\delta(G), \Delta(G))$ are two basic and widely employed graph invariants. The notion of degeneracy $d(G)$ of a graph G is defined as the minimum k such that G admits a k-degenerate ordering. A k-degenerate ordering is a linear ordering $\sigma = (u_1, \ldots, u_n)$ of V so that each u_i has at most k neighbors in $V_i^\sigma := \{u_j : j \geq i\}$. It was introduced by Szekeres and Wilf in [28] and it can be seen that $\delta(G) \leq d(G) \leq \Delta(G)$ for any G. It is known that $d(G)$ can be aribitrarily large compared to $\delta(G)$ and can also be arbitrarily small compared to $\Delta(G)$. It is also known [28] that $d(G)$ is efficiently computable for arbitrary graphs. Also, a $d(G)$-degenerate ordering of G can be computed efficiently. A k-degenerate graph is one for which $d(G) \leq k$.

Several interesting graph classes admit an universal (applicable to all members) upper bound on the value of degeneracy. For example, forests are 1-degenerate, planar graphs are 5-degenerate, graphs of treewidth $tw(G) \leq k$ are k-degenerate, any proper minor-closed family of graphs is of degeneracy at most c (for some constant c associated with the family) and so on. In general, graph classes of bounded degeneracy are of interest both from a theoretical and an algorithmic point of view. One reason is that a graph G being k-degenerate implies that every subset S of $V(G)$ induces a subgraph on at most $k|S|$ edges. Hence, for bounded values of k, G is "sparse everywhere" (as described in [16]). This property can be exploited to obtain sharp bounds on certain invariants as well as in algorithmically realising them. For example, it is well-known that $\chi(G) \leq d(G) + 1$ and $\alpha(G) \geq \frac{n}{d(G)+1}$ for any G, where $\chi(G)$ and $\alpha(G)$ denote respectively the chromatic number and independence number of G. Both of these bounds are also realisable by simple and efficient algorithms. These two bounds are respectively much better compared to better known bounds of $\chi(G) \leq \Delta(G) + 1$ and $\alpha(G) \geq \frac{n}{\Delta+1}$, when $d(G)$ is "very small". However, much of the published works on coloring and independent set are stated in terms of Δ, both for positive results (like bounds or algorithms) and also for negative results (like inapproximability results as in [7,18,31]). In particular, the inapproximability results do not hold for special classes of graphs like chordal graphs or perfect graphs where exact determination of both $\chi()$ and $\alpha()$ is efficiently realizable. *Chordal* graphs are those which admit an ordering $\sigma = (u_1, \ldots, u_n)$ of V so that for each i, $N_G(u_i) \cap V_i^\sigma$ induces a clique in G. Here, $N_G(u_i)$ denotes the set of neighbors of u_i in G. Such an ordering is referred to as a *perfect elimination ordering (PEO)*.

Jamison and Mulder [19] generalize the notion of chordal graphs further to graphs admitting a *k-simplicial elimination ordering*. Such an ordering is an ordering $(u_1, \ldots, u_n)$ which satisfies: $N_G(u_i) \cap V_i^\sigma$ can be partitioned into at most k cliques, for each i. $k = 1$ corresponds to a PEO. They did not, however, introduce an invariant defined in terms of such orderings. We do this as follows: For a graph G, define its *simplicial number* $s(G)$ to be the minimum k such that G admits a k-simplicial elimination ordering.

Recently, Ye and Borodin [30] generalize the notion of chordal graphs further in another direction to Gs admitting a k *-independence ordering*. For such an ordering σ, one requires, for each i, that $\alpha(H) \leq k$ where $H = G[V_i^\sigma \cap N_G(u_i)]$. They defined the *inductive independence number* (denoted by $\lambda(G)$) to be the minimum $k \geq 1$ such that G admits a k-independence ordering. It was also observed that each of the several well-known graph classes like chordal graphs ($\lambda = 1$), graphs of treewidth at most k ($\lambda \leq k$), etc admits a fixed bound on the value of $\lambda(G)$ of its members. It was also shown in [30] that, for every fixed $k \geq 1$, there is an efficient algorithm for recognizing if an arbitrary G satisfies $\lambda(G) \leq k$.

Each of $d(G)$, $s(G)$ and $\lambda(G)$ are defined inductively (that is, in terms of the value of an optimal ordering). In Sect. 2, we generalize and unify these notions as special cases of a more generic notion of inductive analogue of an increasing graph invariant $f()$. It is based on defining a local analogue of $f()$ and also introducing an inductive version $f_{\text{IND}}()$. $f_{\text{IND}}()$ is defined to be the minimum (over all linear orderings of $V(G)$) of the maximum of a set of inductively defined local values. It is shown that $f_{\text{IND}}()$ is sandwiched between the extremes of the local analogues just as $d(G)$ is sandwiched between $\delta(G)$ and $\Delta(G)$. A notion of f-inductive orderings is introduced and they are shown to be optimal achieving the value of $f_{\text{IND}}()$. We also obtain an alternate and equivalent definition of $f_{\text{IND}}()$. The proofs of the claims of this section are provided in Section 8 of the more detailed version [27]. In Section 9 of [27], we illustrate the generic notion of $f_{\text{IND}}()$ with several specific examples of increasing graph invariants. The examples are meant to help the reader get an appreciation of how an inductive analogue is defined and how does it compare with other invariants.

In Sect. 3, we address the computation of $f_{\text{IND}}()$ and identify some scenarios when one can efficiently compute f_{IND} and a f-inductive ordering. As a consequence, one obtains several examples of specific pairs of $(\mathcal{C}, f())$ where efficient computation of f_{IND} over members of $\mathcal{C}$ is guaranteed. Only claims are provided in this section and the proofs of these claims are provided in Section 10 of [27].

In Sect. 4, we generalize the notion of $f_{\text{IND}}()$ further in three directions. First, we allow our graphs to be vertex weighted. Second, we extend the meaning of locality to r-distance open neighborhoods (rather than just the subgraph induced by the neighbors). Thirdly, we allow $f()$ to take values from an arbitrary universe $\mathcal{U}$ endowed with a total order $\leq$ and also a minimum with respect to $\leq$. The first generalization is to extend the notion to weighted graphs, the second is to address some algorithmic applications of this notion where such an extension is called for, and the third is to provide a more general scenario where $\mathcal{U}$ could be a set of discrete structures (more general than a set of numerical values). The results of Sect. 2 have corresponding generalized analogues which are stated in this section. The complete proofs are provided in Section 11 of [27]. Again such a 3-fold generalization is needed in some applications like (i) designing efficient approximation algorithms for some optimization problems (maximum sized $\mathcal{P}$-subgraphs, minimum $\mathcal{P}$-colorings, etc.) associated with hereditary properties. A notion of f^r-inductive ordering (generalizing f-inductive orderings) is defined and is shown to be an optimal ordering. We use $f^r_{\text{IND}}(G_w)$ to denote the weighted, r-distance generalization of $f_{\text{IND}}()$ invariant. Also presented in this section is an algorithmic result (Theorem 10 based on Algorithm 1 presented in Section 11 of [27]) on obtaining an approximation to optimal inductive orderings if only an approximation to the value of $f()$ can be computed efficiently.

A graph class $\mathcal{P}$ is hereditary if $H \in \mathcal{P}$ whenever $G \in \mathcal{P}$ and H is an induced subgraph of G. It is well-known that every hereditary $\mathcal{P}$ is such that $\mathcal{P} = Free(M)$ where M is the set of vertex minimal counter examples not in $\mathcal{P}$ and $Free(M)$ is defined as the class of graphs G not containing an induced isomorphic copy of any $H \in M$. A $\mathcal{P}$-subgraph of G is an induced subgraph H of G such that $H \in \mathcal{P}$. The maximum size $(|V(H)|)$ of such a subgraph is denoted by $\alpha_{\mathcal{P}}(G)$ and is an increasing invariant. A $(k, \mathcal{P})$-coloring of G is an ordered partition $(V_1, \ldots, V_k)$ of $V(G)$ such that $G[V_i] \in \mathcal{P}$ for each i. The minimum $k \geq 1$ such that G admits a $(k, \mathcal{P})$-coloring is known as its $\mathcal{P}$-chromatic number and is denoted by $\chi_{\mathcal{P}}(G)$. When $\mathcal{P}$ is the class of edgeless graphs, a $\mathcal{P}$-subgraph is an independent in G and a $(k, \mathcal{P})$-coloring is a proper k-coloring of G. See [25], for a collection of upper bounds on $\chi_{\mathcal{P}}(G)$ and list $\mathcal{P}$-chromatic number $ch_{\mathcal{P}}(G)$. For a fixed $k \geq 1$, when we allow $G[V_i] \in \mathcal{P}_i$ for each $1 \leq i \leq k$, we obtain a $(\mathcal{P}_1, \ldots, \mathcal{P}_k)$-coloring of G. Each $\mathcal{P}_i$ is a hereditary property. See [1,4,8,13,20] for motivation and for some studies on theoretical and computational aspects of this notion.

Consider the problem **(MPS)** of computing, given a G_w, an optimal $\mathcal{P}$-subgraph. Approximating optimal $\mathcal{P}$-subgraphs have been studied before. Halldorsson [16] presents an algorithm for producing an approximation within a factor of $O\left(n(\log \log n / \log n)^2\right)$ for arbitrary graphs, *provided* $\mathcal{P}$ fails to contain some clique or some independent set. Halldorsson and Lau [17] present a $\lceil (\Delta + 1)/3 \rceil$-approximation of optimal $\mathcal{P}$-subgraphs. Nishizeki and Chiba [23] present a PTAS for planar graphs provided $\mathcal{P}$ is hereditary and is such that: for each G, $G \in \mathcal{P}$ if each connected component of G is in $\mathcal{P}$. Chen [10] extends this to PTAS for planar graphs for every hereditary $\mathcal{P}$. On the negative side, Lund and Yannakakis [22] establish that optimal $\mathcal{P}$-subgraphs are not approximable within a factor of $|V|^\epsilon$ (for some $\epsilon = \epsilon(\mathcal{P})$) for arbitrary graphs unless $P = NP$, if $\mathcal{P}$ fails to contain some clique or independent set.

In Sect. 5, we establish the existence, for every $\mathcal{P} : \mathcal{P} = Free(M)$ for some *finite* M, of an algorithm (Algorithm 2 of [27]) for computing an approximation to maximum *weighted* $\mathcal{P}$-subgraph of a given vertex weighted graph G_w. The maximum weight of a $\mathcal{P}$-subgraph is denoted by $\alpha_{\mathcal{P}}(G_w)$. We also show (see Theorem 11) that Algorithm 2 is an efficient one and obtains a $(f^d_{\text{IND}}(G) + 1)$-approximation to an optimal $\mathcal{P}$-subgraph, *provided* a f^d-inductive ordering is either efficiently computable or is available as part of the input. Here, d is the maximum diameter of any $H \in M$ and $f() = \alpha_{\mathcal{P}}()$. Here, $\alpha_{\mathcal{P}}(G)$ is with respect to the unweighted graph G. For every fixed $B \geq 0$, it is shown that there is an efficient algorithm to determine if $(\alpha_{\mathcal{P}})^d_{\text{IND}}(G) \leq B$ or not and if so, also to compute a $(\alpha_{\mathcal{P}})^d$-inductive ordering of G. *Thus, for unweighted graphs of bounded* $(\alpha_{\mathcal{P}})^d_{\text{IND}}()$ *values, we obtain an efficiently computable* $O(1)$ *-approximation for the optimal* $\mathcal{P}$ *-subgraph* of its weighted version. The algorithm is conceptually very simple and can be thought of as a generalization of a similar algorithm (based on the well-known Push-and-Pop paradigm) obtained in [3] and re-presented in [30] for maximum weighted independent sets. The algorithm of [30] obtains a $(\lambda(G) + 1)$-approximation. When $\mathcal{P}$ is such that M is not finite, we employ a stronger assumption on the availability of an efficient algorithm which, given (G, u), determines if u is part of an induced isomorphic copy (in G) of some $H \in M$. This assumption is stronger than and implies the efficient testability of $\mathcal{P}$-membership. The notion of $f^r_{\text{IND}}(G)$ plays an important role in the design and analysis of Algorithm 2. For the unweighted case, Algorithm 2 admits a simpler and natural formulation Algorithm 3 (see [27]) and we provide a simpler analysis in Theorem 12.

In Subsect. 5.1, we establish the existence of an algorithm (Algorithm 4 of [27]) which is an approximation algorithm for the problem **(MkPColPS)** of computing, given a weighted G_w and a $k \geq 1$, a maximum weighted $(k, \mathcal{P})$-colorable induced subgraph. This problem is a generalization of computing a maximum weighted induced $\mathcal{P}$-subgraph, which corresponds to the case $k = 1$. Algorithmic study of $k = 2$ has been done before. See, for example, the work of Addario-Berry, et al [2] which presents a polynomial time algorithm for computing a maximum weight induced k-colorable (k fixed but arbitrary) subgraph over i-triangulaged graphs, a subclass of perfect graphs. It is also shown by the same authors that this problem (even for $k = 2$) is NP-hard over the class of clique-separable graphs, also a subclass of perfect graphs. See also in this work a few references for practical applications of this problem like VLSI board design [11,14], genome research [21]. See also the works of [29] for exactly computing an optimal induced k-colorable subgraph over chordal graphs (for fixed values of k) and [9] for 2-approximating an optimal solution over chordal graphs (for arbitrary k).

Algorithm 4, given a G_w, k and also a $(\alpha_{\mathcal{P}})^d$-inductive ordering of G, produces a solution whose total weight approximates the optimal solution within a multiplicative factor of $(\alpha_{\mathcal{P}})^d_{\text{IND}}(G) + 1$. In particular, for every $k \geq 1$, this algorithm produces a $(k + 1)$-approximation of an optimal solution, for graphs having $(\alpha_{\mathcal{P}})^d_{\text{IND}} \leq k$. This algorithm is a generalization of Algorithm 2 with a similar analysis. This result is presented as Theorem 13. This result is a generalization of a similar result on approximating a maximum weighted induced k-colorable (standard coloring with k fixed) subgraph presented in [30].

In Subsect. 5.2, we establish the existence of an approximation algorithm (Algorithm 5 of [27]) for the problem **(MP1PkColPS)** of computing (for fixed $\mathcal{P}_1, \ldots, \mathcal{P}_k$), given a G_w and an ordering σ, a maximum weight $(\mathcal{P}_1, \ldots, \mathcal{P}_k)$-colorable induced subgraph, a problem which is a natural generalization of the MkColPS problem. This algorithm is the same as Algorithm 4 except that for each i, it makes sure that the i-th component of its output $(V_1, \ldots, V_k)$ is such that $G[V_i] \in \mathcal{P}_i$. For an appropriately defined inductive ordering (explained later), the approximation factor is $(\max\{\alpha_{\mathcal{P}_i}\}_i)^d_{\mathrm{IND}}(G_w) + 1$. Here, d is the maximum diameter of any element in $\cup_i M_i$ where $\mathcal{P}_i = Free(M_i)$ for each $i = 1, \ldots, k$. This result is presented as Theorem 14.

In Sect. 6, we study efficient approximation of minimum $\mathcal{P}$-colorings for some special classes of graphs. Achlioptas [1] established that deciding, given a G, if $\chi_{\mathcal{P}}(G) \leq k$ is NP-complete for every fixed $k \geq 2$ and for every $\mathcal{P} = Free(\{H\})$ where H is any graph on 3 or more vertices. Hence, even deciding the $(2, \mathcal{P})$-colorability of arbitrary graphs is NP-complete except for the special cases of $\mathcal{P} = Free(\{K_2\})$ or $\mathcal{P} = Free(\{\overline{K_2}\})$. For the specific case of standard coloring, it is known by the result of Zuckermann [31] that it is NP-hard to approximate $\chi(G)$ of arbitrary graphs within a factor of $n^{1-\epsilon}$, for every $\epsilon > 0$. We suspect that similar inapproximability results can be established for several other types of $\mathcal{P}$-colorings, even though we are not aware of any specific published work along these lines. However, we establish (see Theorem 15) that a simple and natural heuristic algorithm (Algorithm 6 of [27]) approximates a minimum $\mathcal{P}$-coloring in the sense that it uses a number of colors which is within a multiplicative factor of (essentially) $(\alpha_{\mathcal{P}})^d_{\mathrm{IND}}(G)/(m - 1)$ from the optimal value $\chi_{\mathcal{P}}(G)$. The meaning of d is as before and $m = m(M) = \min\{|V(H)| : H \in M\}$. As before, we assume that a $(\alpha_{\mathcal{P}})^d$-inductive ordering is either efficiently computable or is available as part of the input. Hence, for graphs of bounded $(\alpha_{\mathcal{P}})^d_{\mathrm{IND}}()$ values, Algorithm 6 computes an $O(1)$-approximation for minimum $\mathcal{P}$-coloring. This result is a generalization of a similar algorithmic result on approximating optimal (usual proper) colorings on graphs of bounded inductive independence number presented in [30].

In our opinion, this generalization highlights why the notions of $f^r_{\mathrm{IND}}()$ and f^r-inductive orderings are important and deserve to be studied further not only for their graph theoretical implications but also for their algorithmic implications. In Sect. 7, we conclude with some remarks and observations.

Outline: In Sects. 1 through 7, we present all new notions, their theoretical and algorithmic aspects including their applications to the design of efficient approximation algorithms. In Sects. 8 through 13 of the more detailed version of this paper [27], we present complete proofs, algorithms, their analyses and also examples for the benefit of the reader. Throughout, whenever we refer to one of Sects. 8 through 13, it refers to that section in [27].

A further study of inductive invariants over various graph classes may possibly lead to identification of new graph classes where one can solve exactly or approximately several NP-hard graph optimization problems. Identification of graph classes where one can improve the best-known approximation results for hard problems has always been an active area of research and our results (alongwith earlier ones) point to a new direction in this regard.

2 Inductive Versions of Graph Invariants for Unweighted Graphs

Given $G = (V, E)$ and $H = (U, F)$, we say that H is a vertex induced (shortly induced) subgraph of G if $U \subseteq V$ and $F = \{\{u, v\} \in E \mid u, v \in U\}$. We use $H = G[U]$ to denote this fact. A graph class $\mathcal{C}$ is *hereditary* if $H \in \mathcal{C}$ for every (G, H) such that $G \in \mathcal{C}$ and H is an induced subgraph of G. Let $\mathcal{G}$ denote the class of all labeled graphs and $\mathcal{R}_+$ denote the set of all nonnegative reals. A graph invariant is a function $f : \mathcal{G} \to \mathcal{R}_+$ such that $f(G) = f(H)$ whenever G and H are isomorphic. We call a graph invariant $f(G)$ *increasing* if $f(H) \le f(G)$ for every (G, H) such that H is an induced subgraph of G. For the sake of simplicity of description, we extend $f(G)$ to the graph $G = (V, E)$ where $V = E = \emptyset$ also and define $f(G) = 0$ in this case.

Notational Assumptions:

1. For reasons of notational simplicity and since all graphs (that are going to be considered in the proof arguments) are induced subgraphs of some fixed but arbitrary G, we will only use $f(U)$ to denote $f(G[U])$ for every $G = (V, E)$ and $U \subseteq V$, whenever G is clear from the context.
2. For a linear ordering $\sigma = (u_1, \ldots, u_n)$ of V and for any j, define $V_j^\sigma = \{u_j, u_{j+1}, \ldots, u_n\}$ and $G_j^\sigma = G[V_j^\sigma]$. Sometimes, we find it convenient to use u_j instead of j in the arguments to obtain a better comprehension. Hence, we also use V_u^σ to denote the set $\{v \mid u <_\sigma v\} \cup \{u\}$. Here, we use $u <_\sigma v$ to denote the fact that $\sigma^{-1}(u) < \sigma^{-1}(v)$. Define $G_u^\sigma = G[V_u^\sigma]$.

For any increasing invariant $f(G)$, we introduce three new invariants defined as follows (with $G = (V, E)$ and $|V| = n$):

Definition 1. *Given an increasing graph invaraint $f(G)$, define $f_l(G)$, $f_L(G)$ and $f_{IND}(G)$ as below:*

$$f_l(G) = \min\{f(N_G(u)) \mid u \in V\}$$
$$f_L(G) = \max\{f(N_G(u)) \mid u \in V\}$$
$$f_{IND}(G) = \min_{\sigma = (u_1, \ldots, u_n)} \max_j f(N_{G_j^\sigma}(u_j)) = \min_{\sigma : [n] \to V} M_\sigma.$$

Here, for a linear ordering σ over V, we use $M_\sigma = M_\sigma(f, G)$ to denote the quantity $\max_j f(N_{G_j^\sigma}(u_j))$.

We skip the proofs of the following propositions in this section and they are presented in Section 8 of [27].

Claim. Each of $f_l()$, $f_L()$ and $f_{IND}()$ is a graph invariant, for any invariant $f()$.

We say that an ordering $\sigma = (u_1, \ldots, u_n)$ is a *f-inductive ordering* if for each j, u_j is a vertex for which $f(N_{G_j^\sigma}(u_j)) = f_l(V_j^\sigma)$. It turns out that such orderings are optimal orderings achieving the minimum in the definition of $f_{IND}(G)$. An ordering σ is an *optimal* ordering if $M_\sigma = f_{IND}(G)$.

Lemma 1. *For any increasing $f(G)$, every f-inductive ordering is an optimal ordering.*

As a consequence, it follows that

Theorem 1. *For any increasing graph invariant $f(G)$, we have*

$$f_{IND}(G) = \max_{U \subseteq V} f_l(U).$$

Moreover, there is a unique maximal set U achieving the maximum.

Corollary 1. *For any increasing invariant $f()$, each of $f_L()$ and $f_{IND}()$ is an increasing invariant.*

Remark: $f_l()$ is not necessarily an increasing variant even if $f()$ is an increasing invariant. For example, let $f(G) = |V(G)|$, an increasing invariant. Then, $f_{IND}()$ is the same as the degeneracy of G and $f_L(G)$ is the same as $\Delta(G)$, both are increasing invariants. But $f_l(G)$ is the same as $\delta(G)$ which is not an increasing invariant.

Theorem 2. *For any increasing invariant $f(G)$ and for any graph G, we have*

$$f_l(G) \le f_{IND}(G) \le f_L(G).$$

Theorem 3. *Let f, g be increasing graph invariants such that $f(G) \le g(G)$ for every G. Then,*

$$(i) f_l(G) \le g_l(G) \; ; \; (ii) f_L(G) \le g_L(G) \; ; \; (iii) f_{IND}(G) \le g_{IND}(G)$$

for every G.

For a set of concrete examples illustrating the meanings of $f_l()$, $f_{IND}()$, $f_L()$ for various increasing $f()$'s, we refer the reader to Section 9 of [27].

3   On the Computation of $f_{IND}(G)$

A natural algorithmic question that arises in this context is: how efficiently one can achieve local computation of $f(G)$? Stated more precisely: for a fixed hereditary C and an increasing $f()$, does there exists an efficient algorithm which, given $G \in C$ and $u \in V(G)$, computes $f(N_G(u))$. The reason one wants to address this question is that an efficient computation of $f(N_G(u))$ leads to an efficient computation $f_{IND}(G)$ and also to efficiently compute a f-inductive ordering of G, as shown below.

It should be noted (also mentioned in [26]) that efficient local computation is more of a paradigm of identifying and exploiting graph classes with special structures than an algorithmic one. As for as the complexity of local computation for arbitrary graphs is concerned, it is no better than computing $f(G)$. This is because computing $f(G)$ (for an arbitrary G) can be reduced to computing

$f(N_H(u))$ where H is the graph obtained from G by adding a new universal vertex u. Hence, if there are inapproximability results like that of Hastad [18] (or Zuckermann [31]) which states that there is no efficient algorithm to approximate $\alpha(G)$ for an arbitrary G within a multiplicative factor of $n^{1-\epsilon}$ unless $NP = ZPP$ (or $P = NP$), for any $\epsilon > 0$), they carry over to the local computation over arbitrary graphs also. Hence, one needs to focus on identifying special classes of graphs where efficient local computation of f is possible.

In this section, we identify some scenarios where efficient computation of a f-inductive ordering is possible over a class $\mathcal{C}$. The following theorem identifies one such scenario. The proof is provided in Section 10 of [27].

Theorem 4. *Let $f(G)$ be an increasing invariant and let $\mathcal{C}$ be any hereditary class of graphs for which $f(G)$ is efficiently locally computable. Then, there is an efficient algorithm which, given an arbitrary $G \in \mathcal{C}$, computes $f_{IND}(G)$ and also computes a f-inductive (and hence an optimal) ordering σ (over $V(G)$) achieving $f_{IND}(G)$.*

Examples of Applications: (i) If $\mathcal{C}$ denotes the class of 3-colorable graphs, then $\lambda(G) = \alpha_{\text{IND}}(G)$ and an α-inductive ordering can be computed efficiently over $\mathcal{C}$. (ii) If $\mathcal{C}$ denotes the class of locally bounded treewidth (that is, $tw(G[N(u)]) \leq B$ for every u, for some fixed B) graphs, then each of $\lambda(G)$, $\omega_{\text{IND}}(G)$ and $\chi_{\text{IND}}(G)$ can be efficiently computed over $\mathcal{C}$.

Graphs of bounded $f_{\text{IND}}()$ value can also be recognized efficiently under assumptions. These are presented below. Before that, we note that the for any graph invariant $f()$, the set $\{f(G)\}_G$ is either finite or a countable set, since there are only countably many (upto isomorphism) finite graphs. We also make use of the following assumption that will be useful.

Assumption (X): f is increasing and for any fixed $c \geq 0$, $N_f(c) = |\{x \mid x \leq c, x = f(G) \text{ for some } G\}|$ is finite. Also, there is an efficient oracle O_f which, given c, computes not only $N_f(c)$ but also the distinct values which are at most c and which are achieved by $f(G)$ for some G.

Examples of invaraints satisfying Assumption (**X**) are nonnegative integer valued functions like chromatic number, independence number, clique number, etc. Many of the well-known graph invariants that are widely studied fall into this category. The proof of the following result is provided in Section 10 of [27].

Theorem 5. *Let $f(G)$ be any increasing invariant satisfying Assumption (**X**). Let $\mathcal{C}$ be any hereditary class of graphs. Suppose, for every fixed $B \geq 0$, there is an efficient algorithm $A_{B,f}()$ for testing, given $G \in \mathcal{C}$ and $u \in V(G)$, if $f(N_G(u)) \leq B$. Then, for every fixed B, there is an efficient algorithm $Aind_{B,f}()$ for testing, given $G \in \mathcal{C}$, if $f_{IND}(G) \leq B$ and also for computing a f-inductive ordering achieving $f_{IND}(G)$ provided $f_{IND}(G) \leq B$.*

Remarks: When $f() = \alpha()$, we obtain the algorithmic result of [30] as a special case of the above theorem. There are a number of NP-hard-to-compute parameters like $\omega()$ (maximum clique size), $\nu()$ (minimum size vertex cover) and $dom()$

(domination number), for each of which $f_{\text{IND}}()$ can be computed efficiently for an arbitrary graph, by applying the above result. However, one cannot infer efficient computation of $\chi_{\text{IND}}()$ for arbitrary graphs from the above result since $\{(G, u) : \chi(N_G(u)) \leq k\}$ is NP-complete for every fixed $k \geq 3$. For every k, a simple polynomial time computable reduction $\phi(G) = (H, u)$ (H is G plus a new universal vertex u adjacent to every $v \in V(G)$) from k-colorability instances, establishes the NP-completeness.

For $k \geq 0$, define $D_k = \{G : d(G) \leq k\}$. Since D_k (for fixed k) is a hereditary class, the following theorem follows from Theorem 5. The proof is provided in Section 10 of [27].

Theorem 6. *Let $k \geq 0$ be any fixed integer. Let $f()$ be any increasing invariant satisfying* (**X**). *Then,*

(i) $f_{IND}(G) \leq B_k$ for each $G \in D_k$, for some constant B_k.
(ii) Suppose that for every $B \leq B_k$, there exists an efficient algorithm $A_{B,f}()$ for testing, given $G \in D_k$ and $u \in V$, if $f(N_G(u)) \leq B$. Then, there is an efficient algorithm for computing, given $G \in D_k$, the value of $f_{IND}(G)$ and also a f-inductive ordering of G.

Remarks: It follows that $f_{\text{IND}}()$ is bounded for every graph class of bounded degeneracy, for every increasing f satisfying (**X**) even if it is not necessarily efficiently computable. In particular, each of $\chi_{\text{IND}}()$, $\alpha_{\text{IND}}()$, $\omega_{\text{IND}}()$, $\nu_{\text{IND}}()$ and $dom_{\text{IND}}()$ is is bounded for graphs of bounded degeneracy. This is not true for arbitrary graphs. In this context, we recall that every proper minor-closed graph class is a class of bounded degeneracy. Also, since D_4 includes 4-regular planar graphs, testing 3-colorability over D_4 is NP-complete [12,15]. This, in turn, implies that testing, given a $G \in D_5$ and $u \in V$, if $\chi(N_G(u)) \leq 3$ is NP-complete. Hence, one cannot infer efficient computation of $\chi_{\text{IND}}()$ over D_5 from Theorem 6. However, efficient computation of χ_{IND} over planar graphs ($\subseteq D_5$) is feasible as Corollary 3 below shows.

As an illustration of an application of the above theorem, we obtain the following corollaries. For any fixed $k \geq 0$, let C_k denote the class $\{G : tw(G) \leq k\}$. The next corollary is proved in Section 10 of [27].

Corollary 2. *Let $k \geq 0$ be any fixed integer and let f be an increasing invariant satisfying the assumptions of Theorem 6 with D_k replaced by C_k. Then, there is an efficient algorithm for computing, given $G \in C_k$, the value of $f_{IND}(G)$ and also a f-inductive ordering of G.*

Remarks: For each choice of $f \in \{\chi, \omega, \alpha, \nu, dom\}$, it is known that efficient computation of $f(G)$ or $f(N_G(u))$ (given $G \in C_k, u \in V(G)$) is possible for every fixed k. By Corollary 2, this implies efficient computation of $f_{\text{IND}}(G)$ for $G \in C_k$.

Corollary 3. *For $f() = \chi()$, there is an efficient algorithm for computing $f_{IND}(G)$ and also an appropriate f-inductive ordering, over planar graphs.*

Proof. Let $f() = \chi()$. It follows (from the famous Four-Color theorem of Appel and Haken [5,6]) that $\chi_{\text{IND}}(G) \leq 4$ for planar graphs. It can be seen that $N_G(u)$ induces an outerplanar graph for a planar G and $u \in V$. Also, a $\chi(G)$-coloring can be efficiently computed for outerplanar graphs as shown in [24]. Combining all of these and applying Theorem 5, we deduce that $\chi_{\text{IND}}()$ and a χ-inductive ordering are efficiently computable for planar graphs.   ∎

4 r-Distance Inductive Invariants for Weighted Graphs

In Sect. 2, $f_{\text{IND}}(G) \in \mathcal{R}_+$ was defined for unweighted graphs using only the immediate neighborhood, that is, $N(u)$. We generalize this notion further in three different directions.

1. Let $\mathcal{W}$ be *any fixed* set. We consider weighted[1] graphs G_w where $G = (V, E)$ and $w : V \to \mathcal{W}$ is an arbitrary function. G_{w_1} is isomorphic to H_{w_2} if there exists a weight-preserving isomorphism from G to H, that is an isomorphism ϕ which also satisfies $w_2(\phi(u)) = w_1(u)$ for each $u \in V(G)$. A graph invariant $f(G_w)$ is any function such that $f(G_{w_1}) = f(H_{w_2})$ whenever G_{w_1} and H_{w_2} are isomorphic. A graph G_{w_1} is an *induced subgraph* of H_{w_2} if (i) G is an induced subgraph of H, (ii) $w_1(u) = w_2(u)$ for each $u \in V(G)$.
2. We employ the more general r-distance neighborhoods to define our notions. For $r \geq 0$, define $N_G^r(u) = \{v : v \neq u, dist(u,v) \leq r\}$ is the open r-neighborhood of u in G. As usual, we omit the subscript G whenever it is clear from the context. We also omit the weight function $w()$ if it can be inferred from the context. If G is not connected, define $N^\infty(u) = \{v : v \neq u, dist(u,v) \leq \infty\} = V \setminus \{u\}$. Note that $N^0(u) = \emptyset$ for every u.
3. We allow f be a $\mathcal{U}$-valued function. Here, $\mathcal{U}$ is any universe endowed with a total order $\leq$ and also a minimum denoted by 0. For the trivial graph $G = (\emptyset, \emptyset)$, we define $f(G_w)$ to be 0. $f()$ is increasing if $f(G_{w_1}) \leq f(H_{w_2})$ whenever G_{w_1} is an induced subgraph of H_{w_2}.

Notations:

1. For a given G_w where $G = (V, E)$ and $U \subseteq V$, we define the subgraph of G_w induced by U (denoted by $G_w[U]$) to be $G[U]_{w_1}$ where $w_1 : U \to \mathcal{W}$ satisfies $w_1(u) = w(u)$ for each $u \in V$. We often omit w and simply write $G[U]$ since w_1 is uniquely inferred for induced subgraphs.
2. For a linear ordering $\sigma = (u_1, \ldots, u_n)$ of $V(G)$ and for any j, define $V_j^\sigma = \{u_j, u_{j+1}, \ldots, u_n\}$ and $G_j^\sigma = G[V_j^\sigma]$. Sometimes, we find it convenient to use u_j instead of j in the arguments to obtain better comprehension. Hence, we also use V_u^σ to denote the set $\{v \mid u <_\sigma v\} \cup \{u\}$. Here, we use $u <_\sigma v$ to denote the fact that $\sigma^{-1}(u) < \sigma^{-1}(v)$. Define $G_u^\sigma = G[V_u^\sigma]$.

---

[1] Even though we use the phrase weights for the sake of easy description of algorithmic applications where the weights are real values, we actually use "weight" to denote any labeling of vertices with elements from $\mathcal{W}$. Again, this labeling is in addition to the pairwise distinct labels that vertices already receive as parts of labeled graphs we consider.

3. For reasons of notational simplicity and since all graphs (that are going to be considered in the proof arguments) are induced subgraphs of some fixed but arbitrary graph G, we will only use $f(U)$ to denote $f(G_w[U])$ for every $U \subseteq V(G)$, whenever the graph G_w is clear from the context.

4. In particular, for an induced subgraph H of G and a $u \in V(H)$, we use $f(N_H^r(u))$ to denote $f(G_w[N_H^r(u)])$.

For every $r \in \mathcal{P}_\infty$ (where $\mathcal{P}_\infty := \mathcal{P} \cup \{\infty\}$ and $\mathcal{P}$ is the set of positive integers) and for every increasing invariant $f()$, define the r-distance inductive analogues of $f()$ as follows:

Definition 2. *Given an increasing graph invaraint* $f(G_w)$, *define* $f_l^r(G_w)$, $f_L^r(G_w)$ *and* $f_{IND}^r(G_w)$ *as below:*

$$f_l^r(G_w) = \min\{f(N_G^r(u)) \mid u \in V(G)\}$$
$$f_L^r(G_w) = \max\{f(N_G^r(u)) \mid u \in V(G)\}$$
$$f_{IND}^r(G_w) = \min_{\sigma=(u_1,\ldots,u_n)} \max_j f(N_{G_j^\sigma}^r(u_j)) = \min_{\sigma:[n]\to V} M_\sigma^r.$$

The meanings and assumptions are as before in Definition 1 except that we employ open r-distance neighborhoods instead of open immediate neighborhoods.

By specializing $r = 1$, $\mathcal{W} = \emptyset$, $\mathcal{U} = \mathcal{R}_+$, the three notions $f_l^r(), f_L^r()$ and $f_{IND}^r()$ specialize respectively to the three notions of $f_l(), f_L()$ and $f_{IND}()$ for unweighted graphs. Invariants of unweighted graphs are not the same as invariants of weighted graphs with constant weight function, since the invariant may depend on the actual value of the uniform weight. We skip the proof of the following claim.

Claim. Each of $f_l^r(), f_L^r()$ and $f_{IND}^r()$ is a graph invariant, for any invariant $f()$.

We say that an ordering $\sigma = (u_1,\ldots,u_n)$ is a f^r-*inductive ordering* if for each j, u_j is a vertex for which $f(N_{G_j^\sigma}^r(u_j)) = f_l^r((G_j^\sigma)_w)$. It will be shown that such orderings are optimal orderings achieving the minimum in the definition of $f_{IND}^r(G_w)$. We have the following generalizations of Lemma 1, Theorem 1, Corollary 1, Theorems 2 and 3 whose proofs are generalizations of those of corresponding results for the special case of $r = 1$ (outlined in Sect. 2). Full proofs are presented in Section 11 of [27].

Lemma 2. *For every $r \in \mathcal{P}_\infty$ and for any increasing f, every f^r-inductive ordering is an optimal ordering.*

As a consequence, it follows that

Theorem 7. *For any $r \in \mathcal{P}_\infty$ and for any increasing graph invariant $f(G_w)$, we have*

$$f_{IND}^r(G_w) = \max_{U \subseteq V} f_l^r(U).$$

Moreover, there is a unique maximal set U achieving the maximum.

Corollary 4. *For any* $r \in \mathcal{P}_\infty$ *and for any increasing invariant* $f()$, *each of* $f_L^r()$ *and* $f_{IND}^r()$ *is increasing.*

Theorem 8. *For any* $r \in \mathcal{P}_\infty$ *and for any increasing invariant* $f(G_w)$ *and for any graph* G_w, *we have*

$$f_l^r(G_w) \leq f_{IND}^r(G_w) \leq f_L^r(G_w).$$

Theorem 9. *Let* f, g *be increasing graph invariants such that* $f(G_w) \leq g(G_w)$ *for every* G_w. *Then, for every* G_w,

(a) *for every* $r \in \mathcal{P} \cup \{\infty\}$: (i) $f_l^r(G_w) \leq g_l^r(G_w)$; (ii) $f_L^r(G_w) \leq g_L^r(G_w)$; (iii) $f_{IND}^r(G_w) \leq g_{IND}^r(G_w)$.

(b) *for every* $r_1 \leq r_2$: (i) $f_l^{r_1}(G_w) \leq f_l^{r_2}(G_w)$; (ii) $f_L^{r_1}(G_w) \leq f_L^{r_2}(G_w)$; (iii) $f_{IND}^{r_1}(G_w) \leq f_{IND}^{r_2}(G_w)$.

An immediate consequence is the following corollary.

Corollary 5. *Let* f, g *be nonnegative real valued and increasing invariants. Then, for every* $r_1, r_2 \in \mathcal{P} \cup \{\infty\}$ *and for every* G_w, *we have:*

$$f_{IND}^{r_1}(G_w), g_{IND}^{r_2}(G_w) \leq (f + g)_{IND}^r(G_w) \; ; \; f_{IND}^{r_1}(G_w), g_{IND}^{r_2}(G_w) \leq (\max(f, g))_{IND}^r(G_w)$$

where $r = \max\{r_1, r_2\}$.

Often the function $f()$ is real-valued and happens to be NP-hard to compute exactly and only an approximation algorithm is available. In such cases, one needs to efficiently compute an approximation to f^r-inductive orderings. The following result establishes that it is indeed possible. For the rest of the paper, by a guarantee function, we mean a nonnegatively real-valued function $\rho(n) \geq 1$ which is increasing, that is, $\rho(n) \leq \rho(n+1)$. Here, $n = |V(G)|$. If there exists an efficient algorithm A which, given a G_w, computes a value $A(G_w)$ satisfying $f(G_w) \leq A(G_w) \leq f(G_w) \cdot \rho(n)$ for each G with $|V(G)| = n$, then we say that f *is efficiently approximable (from above) within a factor of* $\rho(n)$. Note that we demand approximation (from above) even for invariants (like independence number of a graph) which is a maximum-type invariant. The estimate $A(G)$ need not correspond to the value of any solution since it may exceed the maximum value $f(G)$ of any solution. The proof of this theorem is provided in Section 11 of [27].

Theorem 10. *Let* $\rho()$ *be a guarantee function. Let* f *be efficiently approximable (from above by an algorithm* A) *within a factor of* $\rho(n)$. *Then, there is an efficient algorithm (Algorithm 1 of [27]) which, given a* G_w, *computes an ordering* σ *of* V *that approximates a* f^r-*inductive ordering within a factor of* $\rho(n)$. *That is,* $M_\sigma^r \leq f_{IND}^r(G_w) \cdot \rho(n)$ *for every* G_w *on* n *vertices.*

Examples: As an illustrative application, consider the following example: (i) graphs of bounded degeneracy: As we noticed earlier, testing, given $G \in D_k$ and $u \in V$, if $\chi(N_G(u)) \leq 3$ is NP-complete, for *each* $k \geq 5$. Hence, we do not yet know if χ_{IND} can be computed efficiently over D_k. However, $\chi(N_G(u)) \leq k$ for each $G \in D_k$. Also, a k-coloring of $N_G(u)$ can be efficiently computed. Hence, any ordering σ satisfies $M_\sigma \leq k$ for each D_k.

5   Approximation of Optimal $\mathcal{P}$-Subgraphs

In this section, we show how to apply the inductive invariant notions to obtain approximation algorithms (with proven guarantees on their performance ratios) for maximum size induced $\mathcal{P}$-subgraph problems (for hereditary $\mathcal{P}$), on graphs of bounded $f_{\text{IND}}()$ values for some suitable $f()$. We study both (vertex) weighted and unweighted versions. Recall that a class $\mathcal{P}$ is hereditary if $H \in \mathcal{P}$ for every (G, H) such that $G \in \mathcal{P}$ and H is an induced subgraph of G. Recall that a class $\mathcal{P}$ is hereditary if and only if there exists a M such that $\mathcal{P} = Free(M)$ where M is the class of vertex minimal counter examples not in P. $Free(M)$ is defined as the class of graphs G such that G has no isomorphic copy of any $H \in M$ as an induced subgraph.

Assumption (Y): *Without loss of generality*, we can assume that members of M satisfy the following:

(1) For each $H \in M$ with $|V(H)| = n$, $V(H) = [n] = \{1, 2, \ldots, n\}$.
(2) For every $H_1, H_2 \in M, H_1 \neq H_2$, we have $H_1 \not\equiv H_2$. This means that M contains at most one graph (upto isomorphism) from each equivalence class of the equivalence relation defined by $\equiv$.
(3) For every $H_1, H_2 \in M, H_1 \neq H_2$, neither of them is an induced subgraph of the other.

In what follows, we focus on those properties $\mathcal{P}$ for which efficient recognition of membership in $\mathcal{P}$ is possible. Examples of such a $\mathcal{P}$ are planar graphs, bipartite graphs, $Free(M)$ for every finite M. A $\mathcal{P}$-*subgraph* of G is an induced subgraph H of G satisfying $H \in \mathcal{P}$. Its size is is $\sum_{u \in V(H)} w_u$ where $w : V(G) \to \mathcal{R}^+$ is the weight function. We use $\alpha_{\mathcal{P}}(G_w)$ to denote the maximum size of a $\mathcal{P}$-subgraph H of G_w. We often omit mentioning w for the special case of unweighted graphs: $w(u) = 1$ for each u. We refer to any such H as an *optimal* $\mathcal{P}$-subgraph of G_w. We also use the notation $\alpha_M(G_w)$ often in place of $\alpha_{\mathcal{P}}(G_w)$. For the special case of $M = \{K_2\}$, $\alpha_M(G_w)$ is the same as $\alpha(G_w)$ for any G. We refer to the problem of finding an optimal $\mathcal{P}$-subgraph as the *Maximum $\mathcal{P}$-Subgraph* (MPS) problem. Note that $\alpha_M(G_w)$ is an increasing graph invariant.

Below, we present a result on approximating the MPS problem on weighted graphs. The result (Theorem 11) presents Algorithm 2 (presented in Section 12 of [27]) for approximating an optimal $\mathcal{P}$-subgraph. The proof of the theorem is presented in Section 12 of [27].

Theorem 11. *Suppose that $\mathcal{P} = Free(M)$ is such that (i) M is finite (hence, $\mathcal{P}$-membership is efficiently testable), (ii) $diam(H) \leq d$ for each $H \in M$, for some fixed d. Then, there is an efficient algorithm (Algorithm 2) which, for a given G_w and a vertex ordering σ, produces a $\mathcal{P}$-subgraph whose total weight is within a factor of at most $M_\sigma^d + 1$ from the weight of an optimal solution.*

Note: (i) Recall that M_σ^d denotes $\max_j f(N_{G_\sigma^d}^d(u_j))$. For both weighted and unweighted cases, the same $f() = \alpha_M()$ (corresponding to the unweighted graph) is used in determining M_σ. Thus, even if the result is meant for weighted case, its approximation factor is bounded by a quantity associated with the unweighted graph. (ii) For the case of $\mathcal{P} = $ independent sets (or $M = \{K_2\}$), the approximation factor is M_σ^1. This can be noticed by a careful perusal of the analysis of the algorithm.

Remark: The Assumption (i) (M being finite) can be replaced by the following weaker assumption: there is an efficient algorithm which, given (G, u, v) ($u = v$ possibly), determines if there exists some $T \subseteq V$ such that (a) $u, v \in T$, (b) $G[T] \equiv H$ for some $H \in M$. This assumption also implies that $\mathcal{P}$-membership is efficiently testable. The assumption that M is finite implies this weaker assumption.

We have $M_\sigma^d \leq (\alpha_M)_L^d(G)$ for every σ. Also, for every fixed $B \geq 0$, an application of Theorem 5 shows that there is an efficient algorithm to determine if $(\alpha_\mathcal{P})_{\mathrm{IND}}^d(G) \leq B$ or not and if so, also to compute a $(\alpha_\mathcal{P})^d$-inductive ordering of G. Hence, the following corollary is obtained.

Corollary 6. *Let $\mathcal{P}$, M and d be as assumed in Theorem 11. The following hold:*

(1) A $(\alpha_M)_L^d(G) + 1$-approximation of a maximum weight $\mathcal{P}$-subgraph can be efficiently obtained for an arbitrary G_w.

(2) Given a $(\alpha_M)^d$-inductive ordering σ of V, Algorithm 2 produces a $(\alpha_M)_{\mathrm{IND}}^d(G) + 1$-approximation of maximum weight $\mathcal{P}$-subgraph of G_w.

(3) For every $B \geq 0$, there is an efficient algorithm to recognize graphs with $k = (\alpha_\mathcal{P})_{\mathrm{IND}}^d(G) \leq B$ and also to find a $(k+1)$-approximation of optimal $\mathcal{P}$-subgraphs of a weighted graph G_w.

To illustrate the kind of consequences one can derive further, we specialise as follows: Let $\mathcal{P} = Free(K_r)$ be the class of K_r-free graphs, for some $r \geq 3$. We have $d = 1$. When $r = 2$, a K_r-free induced subgraph is an independent set and it is known that optimal independent sets can be efficiently computed over chordal graphs. Hence we focus on $r \geq 3$ and obtain that

Corollary 7. *For each $r \geq 3$, there is an efficient algorithm which obtains a r-approximation to a maximum weight induced K_r-free subgraph in a weighted chordal graph.*

Proof. Let σ be any PEO of a chordal G. It follows that $(\alpha_\mathcal{P})_{\mathrm{IND}}(G) \leq M_\sigma \leq r - 1$. Applying Theorem 11, we deduce the claim. ∎

Remarks: For every $k \geq 1$, a chordal G is k-colorable if and only if $G \in Free(K_{k+1})$. Hence, the previous corollary leads us to a $(k+1)$-approximation of optimal k-colorable induced subgraphs of chordal graphs. However, this observation is only to illustrate Theorem 11 and it is to be noted (as explained in the following subsection) that even a 2-approximation (for even varying k) is possible for chordal graphs.

Unweighted Case: For this case, $w_u = 1$ for each u. As a result, the set S computed by the **Push Phase** of Algorithm 2 itself induces a $\mathcal{P}$-subgraph and hence the **Pop Phase** is not executed further. In other words, the algorithm gets simplified as Algorithm 3 (presented in Section 12 of [27]). Also, Assumption (i) of Theorem 11 can be replaced by the weaker assumption of: $\mathcal{P}$-membership is efficiently testable. We thus obtain the following unweighted analogue of Theorem 11. The unweighted case admits a simpler proof and is presented in Section 12 of [27] for the sake of completeness.

Theorem 12. *Suppose $\mathcal{P} = Free(M)$ is such that $diam(H) \leq d$ for each $H \in M$, for some fixed d. Suppose also that $\mathcal{P}$-membership is efficiently testable. Then, for any ordering σ of $V(G)$, the maximal $\mathcal{P}$-subgraph built (by Algorithm 3) using σ is of size within a multiplicative factor $M_\sigma^d + 1$ from the optimum.*

When $M = \{K_2\}$, the approximation factor is M_σ^1. Since $M_\sigma^d \leq (\alpha_M)_L^d(G)$ for every σ, the following interesting graph theoretical observation is obtained.

Corollary 8. *Let $\mathcal{P}$, M and d be as assumed in Theorem 12. Then, for any G, every maximal $\mathcal{P}$-subgraph of G is of size within a factor of $(\alpha_M)_L^d(G) + 1$ from an optimal solution.*

Hence, for graphs where every local neighborhood $N(u)$ can have at most $O(1)$-sized induced $\mathcal{P}$-subgraphs, one can approximate optimal $\mathcal{P}$-subgraphs within an $O(1)$ multiplicative factor by any *maximal* solution. In particular, for every $k \geq 1$, the optimal independent set is approximable by any maximal independent set within a factor of k, for graphs where every local neighborhood can only have an independent set of size at most k.

5.1   Approximation of Maximum Induced $(k, \mathcal{P})$-Colorable Subgraphs

In this subsection, we focus on a related graph optimization problem and designing approximation algorithms for it based on the inductive ordering approach outlined before. For a hereditary $\mathcal{P}$ and an integer $k \geq 1$, the class $\mathcal{P}_k = \{G : \chi_\mathcal{P}(G) \leq k\}$ is a hereditary property. Consider the problem (**MkColPS**) of computing, given a G_w, a maximum weighted induced subgraph H of G such that H is $(k, \mathcal{P})$-colorable. This is an optimization problem well-defined for every hereditary $\mathcal{P}$ and $k \geq 1$. When $k = 1$, this specializes to the maximum induced $\mathcal{P}$-subgraph problem discussed above. When $\mathcal{P} = Free(\{K_2\})$, this problem specializes to the maximum induced k-colorable (standard coloring) subgraph problem.

Since $\mathcal{P}_k$ is hereditary, it follows that MKColPS problem is the same as MPS problem discussed before (with respect to $\mathcal{P}_k$). Hence, we are essentially aiming to compute an approximation to a $\alpha_{\mathcal{P}_k}(G_w)$-sized $\mathcal{P}_k$-subgraph of G_w. From an application of Theorem 11, it follows that MkColPS is approximable within a multiplicative factor of $(\alpha_{\mathcal{P}_k})_{\text{IND}}^d(G) + 1$, *provided* a $(\alpha_{\mathcal{P}_k})_{\text{IND}}^d$-inductive ordering is available as part of the input. Here, $d = d(M_k) = \max\{diam(H) : H \in M_k\}$ where M_k is defined by $\mathcal{P}_k = Free(M_k)$.

While this approach works, it is fraught with two shortcomings compared to another approach that is presented in this subsection. The first one is: $d = d(M)$

can become large (even unbounded) when we go from $\mathcal{P}$ to $\mathcal{P}_k$. An example illustrating drawback is: $\mathcal{P} = Free(\{K_2\})$ and $k = 2$. $\mathcal{P}_2$ is the class of all 2-colorable graphs. Here, $\alpha_{\mathcal{P}}() = \alpha()$ and $\alpha_{\mathcal{P}_2}()$ is the maximum size of an induced 2-colorable subgraph. Also, $d = 1$ for $\mathcal{P}$ but $d = \infty$ for $\mathcal{P}_2$ since there are 3-color-critical graphs of arbitrarily large diameter.

The second shortcoming is: it is possible for a class of graphs that $\alpha_{\mathrm{IND}}^d()$ is bounded while $(\alpha_{\mathcal{P}_k})_{\mathrm{IND}}^d()$ is unbounded for this class. Here, d represents $d(M)$ for the appropriate family $\mathcal{P}$ or $\mathcal{P}_k$. The previous example ($M = \{K_2\}$ and $k = 2$) illustrates this again. $\alpha_{\mathrm{IND}}() \leq 5$ for planar graphs. However, $(\alpha_{\mathcal{P}_2})_{\mathrm{IND}}^\infty(G) \geq \min_{u \in V} \alpha_{\mathcal{P}_2}(G \setminus \{u\}) \geq \frac{n}{2}$ for several infinite families of planar graphs like paths ($\{P_n\}_n$), cycles ($\{C_n\}_n$) and wheels ($\{C_n + K_1\}_n$), etc. Hence, for some classes of graphs, a direct application of Theorem 11 will *not* yield a constant factor approximation algorithm for the MkColPS problem even if it yields a constant factor MPS-approximation algorithm for these classes.

Below, we present a result which establishes that it is indeed possible to ensure the same approximation guarantee (which Theorem 11 provides for MPS) for the MkColPS problem also under the same assumptions. *Even more*, we do not require k to be fixed and can allow k to grow with n. Also, we can include k as part of the input. This result is a generalization of a similar result obtained in [30] for k-colorable (standard coloring and fixed k) subgraphs, to $(k, \mathcal{P})$-subgraphs, for arbitrary hereditary properties $\mathcal{P}$.

Theorem 13. *Suppose that $\mathcal{P} = Free(M)$ is such that (i) M is finite (hence, $\mathcal{P}$-membership is efficiently testable), (ii) $diam(H) \leq d$ for each $H \in M$, for some fixed d. There is an efficient algorithm (Algorithm 4) which, for a given $G_w, k \geq 1$ and a vertex ordering σ, produces a $(k, \mathcal{P})$-colorable induced subgraph whose total weight is within a factor of at most $M_\sigma^d + 1$ from that of an optimal solution.*

As before, the assumption of M being finite can be replaced by a weaker assumption stated in the Remark that appears after the statement of Theorem 11. Algorithm 4 and its analysis are presented in Section 12 of [27]. This algorithm is essentially Algorithm 2 except that we simultaneously grow k stacks (by pushing elements) and then simultaneously build k color classes by popping from the stacks. The analysis is also along similar lines as that of Algorithm 2 and we provide it in complete details in case a reader wants to look at it. Some of the consequences of the above results are listed below.

Corollary 9. *Let $\mathcal{P}, M$ and d be as defined in Theorem 13. The following are true:*

(a) *There is an efficient algorithm which, given a G_w, k and a $(\alpha_{\mathcal{P}})^d$-inductive ordering of G, produces an induced $(k, \mathcal{P})$-colorable subgraph of G whose total weight is within a factor of $(\alpha_{\mathcal{P}})_{IND}^d(G) + 1$ from that of an optimal solution.*

(b) *For every fixed B, there is an efficient algorithm which, given arbitrary G_w, k, tests if $(\alpha_{\mathcal{P}})_{IND}^d(G) \leq B$ and if so produces an induced $(k, \mathcal{P})$-colorable subgraph within an approximation factor of $B + 1$.*

The above corollary follows immediately from Theorems 13 and 5. If we fix $\mathcal{P} = Free(\{K_2\})$ (and $d = 1$) and the input G is restricted to chordal graphs, then every PEO of G is an α-inductive ordering and $\alpha_{\text{IND}}(G) \leq 1$. Applying Claim (b) of the previous theorem, we deduce that maximum weight induced k-colorable subgraph can be approximated within a multiplicative factor of two over weighted chordal graphs even if k is *part of the input*. It can also be established (we skip the details) that one can achieve this approximation in $O(k|V| + |E|)$ time. This result was obtained by Chakravarty and Roy in [9] with an inferior running time of $O(k(|V| + |E|))$. When the input is a weighted chordal graph, the maximum weight induced k-colorable subgraph problem is polynomial time solvable for every fixed k but is NP-hard when k is part of the input as shown in [29].

5.2 Approximation of Maximum Induced $(\mathcal{P}_1, \ldots, \mathcal{P}_k)$-Colorable Subgraphs

Here, for fixed $(\mathcal{P}_1, \ldots, \mathcal{P}_k)$, we study the problem of computing an optimal $(\mathcal{P}_i)_i$-colorable induced subgraph of a given G_w. An Algorithm 5 which computes such an approximation is presented and analyzed in the proof of Theorem 14 below. The algorithm and its proof sketch are presented in Section 12 of [27].

Theorem 14. *Let $\mathcal{P}_1 = Free(M_1), \ldots, \mathcal{P}_k = Free(M_k)$ be fixed hereditary properties with $diam(H) \leq d$ for each $H \in \cup_i M_i$, for some fixed d. Suppose, for each i, there is an efficient algorithm for determining, given (G, u, v), if there exists a $T \subseteq V$ such that $u, v \in T$ and $G[T] \equiv H$ for some $H \in M_i$. Then, there is an efficient algorithm (Algorithm 5) which, for a given G_w and a vertex ordering σ, produces an induced $(\mathcal{P}_1, \ldots, \mathcal{P}_k)$-colorable subgraph whose total weight is within a factor of at most $M_\sigma^d + 1$ from that of an optimal solution. Here, the relevant increasing invariant is $\max\{\alpha_{\mathcal{P}_i} : 1 \leq i \leq k\}$.*

Let $d_c = \max\{diam(H) : H \in M_c\}$ for each $c \in [k]$ and $d = \max\{d_c : 1 \leq c \leq k\}$. Define $\beta_c = (\alpha_{\mathcal{P}_c})_{\text{IND}}^{d_c}()$ for each c. Define $\beta() = \max\{\alpha_{\mathcal{P}_c}() : 1 \leq c \leq k\}$. Then, β is an increasing invariant and by Corollary 5, we have $\beta_c() \leq \beta_{\text{IND}}^d()$ for each c. Combining, we obtain the following corollary.

Corollary 10. *Let $(\mathcal{P}_1, \ldots \mathcal{P}_k), (M_1, \ldots, M_k), \beta()$ and d be as defined in Theorem 14 and discussed as before. The following are true:*

(a) *There is an efficient algorithm which, given a G_w and a β^d-inductive ordering of G, produces an induced $(\mathcal{P}_1, \ldots, \mathcal{P}_k)$-colorable subgraph of G whose total weight is within a factor of $\beta_{IND}^d(G) + 1$ from that of an optimal solution.*

(b) *For every fixed B, there is an efficient algorithm which, given arbitrary G_w, tests if $\beta_{IND}^d(G) \leq B$ and if so produces an induced $(\mathcal{P}_c)_c$-colorable subgraph within an approximation factor of $B + 1$ from that of an optimal solution.*

6   Approximation of Minimum $\mathcal{P}$-Coloring

Below, we study the problem of designing approximation algorithms for $\mathcal{P}$-coloring a given undirected graph. For the rest of this section, we focus only on those $\mathcal{P} = Free(M)$ where $d(M) = \max\{diam(H) : H \in M\}$ exists. We also assume, without loss of generality, that $K_2 \notin M$ and hence $m_* = m_*(M) = \min\{|V(H)| : H \in M\}$ satisfies $m_* \geq 3$. When $K_2 \in M$, a $\mathcal{P}$-coloring is just a proper coloring and has been already handled in [30]. Hence, we assume that $m_* \geq 3$. We obtain the following result on the approximation factor of Algorithm 6 (a simple heuristic presented in Section 13) whose proof is presented in Section 13 of [27].

Theorem 15. *Let $\mathcal{P} = Free(M)$ be any hereditary class with $m_* \geq 3$ and $d = d(M)$ is finite. Suppose $\mathcal{P}$-membership is efficiently testable. Define $k = (\alpha_{\mathcal{P}})_{IND}^d(G)$. Assume that a $(\alpha_{\mathcal{P}})^d$-inductive ordering is part of the input or can be computed efficiently. Then,*

(a) Algorithm 6 is an efficient one and obtains a $\mathcal{P}$-coloring using at most $\left(\frac{k}{m_*-1}\right)\chi_{\mathcal{P}}(G) + 1$ colors provided $k \geq m_* - 1$.

(b) If $k < m_* - 1$, then $\chi_{\mathcal{P}}(G) = 1$ and Algorithm 6 produces an optimal coloring.

Note: The approximation factor is nearly $\frac{k}{m_*-1}$. This factor matches the bound obtained in [30] for proper colorings since $d = 1$ and $m_* = 2$ for the case of $M = \{K_2\}$.

As an illustrative application of the above theorem, restrict the input G to be a chordal graph and consider $\mathcal{P} = Free(M)$ where $M = \{K_r\}$ for some $r \geq 2$. We have $m_* = r$. $d = 1$. We call a $\mathcal{P}$-coloring as a K_r-free coloring.

Corollary 11. *For every $r \geq 3$, Algorithm 6 is an efficient one and produces, given a chordal G, a K_r-free coloring using at most $\chi_{\mathcal{P}}(G) + 1$ colors.*

Proof. For chordal graphs, $(\alpha_{\mathcal{P}})_{IND}(G) \leq r - 1$ and a $(\alpha_{\mathcal{P}})$-inductive ordering can be efficiently computed as explained in the proof of Corollary 7. Now applying Theorem 15, we obtain the desired conclusion.   ∎

As another illustrative application, let the input G be arbitrary and let $\mathcal{P} = \{H : \Delta(H) \leq r\}$ (for some $r \geq 1$). We have $\mathcal{P} = Free(M)$ where $M = \{H : |V(H)| = r + 2, \Delta(H) = r + 1\}$ and $d(M) = 2$.

Corollary 12. *For every $r \geq 1$ and every $k \geq 1$, Algorithm 6 is an efficient one and produces, given an arbitrary G with $k = (\alpha_{\mathcal{P}})_{IND}^2(G)$, a $\mathcal{P}$-coloring using at most $\left(\lfloor \frac{k}{r+1} \rfloor\right)\chi_{\mathcal{P}}(G) + 1$ colors.*

Proof. For fixed values of k, an $(\alpha_{\mathcal{P}})^2$-inductive ordering of G can be efficiently computed for those G having k as its corresponding inductive number. Now applying Theorem 15, we obtain the desired conclusion.   ∎

7 Conclusions

A further study of inductive invariants for various graph classes may possibly lead to identification of new classes where one can solve exactly or approximately several NP-hard graph optimization problems. Identification of graph classes where one can improve the best-known approximation results for hard problems has always been an active area of research and our results (along with earlier ones) point to a new direction in this regard. A few questions and directions that we believe are interesting and worth exploring further are:

1. Identify triples $(f(), g(), \mathcal{C})$ where f, g are increasing invariants and $\mathcal{C}$ is a hereditary class, satisfying: $g(G) \leq \phi(f_{\text{IND}}(G))$ for each $G \in \mathcal{C}$ and for some fixed $\phi : \mathcal{R}_+ \to \mathcal{R}_+$.
2. Can the approximation factors of k or $k/(m-1)$ where $k = (\alpha_{\mathcal{P}})_{\mathcal{P}}^d(G)$ obtained respectively for approximating an optimal $\mathcal{P}$-subgraph or an optimal $\mathcal{P}$-coloring, be improved further ?
3. Identify pairs $(f(), \mathcal{C})$ where $f()$ is an increasing invariant and $\mathcal{C}$ is hereditary, for which efficient computation of $f(N_G^r(u))$ is possible for $G \in \mathcal{C}, u \in V(G)$.
4. The notion of inductive invariants can be extended to (graph,vertex) invariants. Examples of such invariants are the maximum size of a clique containing a given vertex in a graph. A rigorous introduction to (G, u) invariants, their inductive analogues, bounds, characterizations along with some graph theoretical and algorithmic implications and concrete applications will appear in a longer journal version of this paper.

References

1. Achlioptas, D.: The complexity of G-free colourability. Discret. Math. **165–166**, 21–30 (1997). https://doi.org/10.1016/S0012-365X(97)84217-3
2. Addario-Berry, L., Kennedy, W., King, A.D., Li, Z., Reed, B.: Finding a maximum-weighted induced k-partite subgraph of an i-triangulaged graph. Discret. Appl. Math. **158**(7), 765–770 (2010). https://doi.org/10.1016/j.dam.2008.08.020
3. Akcoglu, K., Aspnes, J., Dasgupta, B., Kao, M.Y.: Opportunity-cost algorithms for combinatorial auctions. In: Proceedings of Applied Optimization 74: Computational Methods in Decision-Making, Economics and Finance, pp. 455–479. Kluwer Academic Publishers (2002)
4. Alekseev, V., Farrugia, A., Lozin, V.: New results on generalized graph coloring. Discret. Math. Theor. Comput. Sci. **6**(2) (2004). HAL:hal-00959005v1
5. Appel, K., Haken, W.: Every planar map is four colorable. Part I. Discharging. Illinois J. Math. **21**(3), 429–490 (1977). https://doi.org/10.1215/ijm/1256049011
6. Appel, K., Haken, W., Koch, J.: Every planar map is four colorable. Part II. Reducibility. Illinois J. Math. **21**(3), 491–567 (1977). https://doi.org/10.1215/ijm/1256049012
7. Austrin, P., Khot, S., Safra, M.: Inapproximability of vertex cover and independent set in bounded degree graphs. Theory Comput. **7**(1), 27–43 (2011)
8. Brown, J.: The complexity of generalized graph colorings. Discret. Appl. Math. **69**(3), 257–270 (1996). https://doi.org/10.1016/0166-218X(96)00096-0

9. Chakaravarthy, V., Roy, S.: Approximating maximum weight k-colorable subgraphs in chordal graphs. Inf. Process. Lett. **109**(7), 365–368 (2009). https://doi.org/10.1016/j.ipl.2008.12.007

10. Chen, Z.-Z.: Practical approximation schemes for maximum induced-subgraph problems on $K_{3,3}$-free or K_5-free graphs. In: Meyer, F., Monien, B. (eds.) ICALP 1996. LNCS, vol. 1099, pp. 268–279. Springer, Heidelberg (1996). https://doi.org/10.1007/3-540-61440-0_134

11. Choi, H.A., Nakajima, K., Rim, C.: Graph bipartization and via minimization. SIAM J. Discret. Math. **2**(1), 38–47 (1989). https://doi.org/10.1137/0402004

12. Dailey, D.: Uniqueness of colorability and colorability of planar 4-regular graphs are NP-complete. Discret. Math. **30**(3), 289–293 (1980). https://doi.org/10.1016/0012-365X(80)90236-8

13. Farrugia, A.: Vertex-partitioning into fixed additive induced hereditary properties is NP-hard. Electron. J. Comb. **11**(1) (2004). https://doi.org/10.37236/1799

14. Fouilhoux, P.: Graphes k-partis et conception de circuits VLSI. Universite Blaise Pascal, Clermont-Ferrand, France (2004)

15. Garey, M., Johnson, D., Stockmeyer, L.: Some simplified NP-complete graph problems. Theor. Comput. Sci. **1**(3), 237–267 (1976). https://doi.org/10.1016/0304-3975(76)90059-1

16. Halldorsson, M.M.: Approximations of weighted independent set and hereditary subset problems. J. Graph Algorithms Appl. **4**(1), 1–16 (2000)

17. Halldorsson, M., Lau, H.C.: Low-degree graph partitioning via local search with applications to constraint satisfaction, max-cut, and coloring. J. Graph Algorithms Appl. **1**(3), 1–13 (1997)

18. Håstad, J.: Clique is hard to approximate within $n^{1-\epsilon}$. Acta Mathematica **182**(1), 105–142 (1999). https://projecteuclid.org/euclid.acta/1485891205

19. Jamison, R., Mulder, H.: Tolerance intersection graphs on binary trees with constant tolerance 3. Discret. Math. **215**, 115–131 (2000)

20. Kratochvil, J., Schiermeyer, I.: On the computational complexity of (o, p)-partition problems. Discussiones Mathematicae Graph Theory **17**, 253–258 (1997)

21. Lancia, G., Bafna, V., Istrail, S., Lippert, R., Schwartz, R.: SNPs problems, complexity, and algorithms. In: auf der Heide, F.M. (ed.) ESA 2001. LNCS, vol. 2161, pp. 182–193. Springer, Heidelberg (2001). https://doi.org/10.1007/3-540-44676-1_15

22. Lund, C., Yannakakis, M.: The approximation of maximum subgraph problems. In: Lingas, A., Karlsson, R., Carlsson, S. (eds.) ICALP 1993. LNCS, vol. 700, pp. 40–51. Springer, Heidelberg (1993). https://doi.org/10.1007/3-540-56939-1_60

23. Nishizeki, T., Chiba, N.: Planar Graphs: Theory and Algorithms. Annals of Discrete Mathematics, vol. 32. Elsevier (1988)

24. Proskurowski, A., Syslo, M.: Efficient vertex- and edge-coloring of outerplanar graphs. SIAM J. Algebraic Discret. Methods **7**(1), 131–136 (2006). https://doi.org/10.1137/0607016

25. Subramanian, C.R.: List hereditary colorings of graphs. In: Proceedings of ICDM 2008. RMS Lecture Notes Series, vol. 13, pp. 191–205. Ramanujan Mathematical Society (2010)

26. Subramanian, C.R.: On approximating stochastic PIPs and independent sets (2018)

27. Subramanian, C.R.: Inductive graph invariants and algorithmic applications (2020). https://www.imsc.res.in/~crs/s19shrtL.pdf

28. Szekeres, G., Wilf, H.S.: An inequality for the chromatic number of a graph. J. Comb. Theory **4**, 1–3 (1968)

29. Yannakakis, M., Gavril, F.: The maximum k-colorable subgraph problem for chordal graphs. Inf. Process. Lett. **24**(2), 133–137 (1987). https://doi.org/10.1016/0020-0190(87)90107-4
30. Ye, Y., Borodin, A.: Elimination graphs. ACM Trans. Algorithms **8**(2), 14:1–14:23 (2012)
31. Zuckermann, D.: Linear degree extractors and the inapproximability of max clique and chromatic number. Theory Comput. **3**, 103–128 (2007). https://doi.org/10.4086/toc.2007.v003a006

Constructing Order Type Graphs Using an Axiomatic Approach

Sergey Bereg$^{(\boxtimes)}$ and Mohammadreza Haghpanah

University of Texas at Dallas, Richardson, TX 75080, USA
{besp,Mohammadreza.Haghpanah}@utdallas.edu

Abstract. A given order type in the plane can be represented by a corresponding point set. However, it might be difficult to recognize the orientations of some point triples. Recently, Aichholzer *et al.* [3] introduced exit graphs for visualizing order types in the plane. We present a new class of geometric graphs, called *OT-graphs*, using abstract order types and their axioms described in the well-known book by Knuth [15]. Each OT-graph corresponds to a unique order type. We develop efficient algorithms for recognizing OT-graphs and computing a minimal OT-graph for a given order type in the plane. We provide experimental results on all order types of up to nine points in the plane including a comparative analysis of exit graphs and OT-graphs.

Keywords: Order type · CC systems · Chirotopes · Oriented matroids

1 Introduction

The *orientation* of three noncollinear points in the plane is either clockwise CW or counterclockwise CCW. In this paper we assume that point sets are in general position the plane. Two finite point sets in the plane have the *same order type* if there is a bijection between them preserving orientation of any three distinct points. The equivalence classes defined by this equivalence relation are the *order types* [14].

Recently, Aichholzer *et al.* [3] asked "... suppose we have discovered an interesting order type, and we would like to illustrate it in a publication." This is exactly the problem that we were facing in our recent paper [5] where we found that the order type 1874 for 9 points from the database [2] provides a (tight) lower bound for Tverberg partitions with tolerance 2, see Fig. 2(a). Of course, any order type in the plane can be represented by a corresponding point set (or explicit coordinates of the points). However, it might be difficult to recognize the orientations of some point triples. Aichholzer *et al.* [3] introduced exit graphs for visualizing order types in the plane. Let S be a set n points in the plane and let $a, b, c \in S$. Then (a, b) is an *exit edge* with *witness* c if there is no $p \in S$ such that line ap separates b from c or line bp separates a from c, see Fig. 1(a). Geometrically, it means that an *hourglass* defined by a, b, c is empty. The set of exit edges form the *exit graph* of S. To verify that (a, b) is an exit edge with witness

© Springer Nature Switzerland AG 2020
W. Wu and Z. Zhang (Eds.): COCOA 2020, LNCS 12577, pp. 802–816, 2020.
https://doi.org/10.1007/978-3-030-64843-5_54

c, one can check that every point $p \in S \setminus \{a,b,c\}$ is in $A \cup B$, see Fig. 1(a). Note that $A \cap S$ and $(B \cap S) \cup \{c\}$ is a partition of $S \setminus \{a,b\}$ by line ab. We will define a new class of graphs called *OT-graphs* using such partitions.

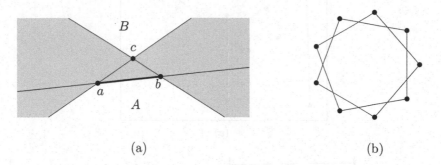

(a)   (b)

Fig. 1. (a) Exit edge (a,b) with witness c. The hourglass-shaped region (shown in gray) is empty of points. (b) An exit graph for 9 points in convex position.

We define an *OT-graph* on S using two ingredients. First, every edge (a,b) in an OT-graph is equipped with the partition of S by line ab, i.e. $S \setminus \{a,b\} = S_{ab}^+ \cup S_{ab}^-$ where S_{ab}^+ (S_{ab}^-) contains points $c \in S$ such that a,b,c has counterclockwise (clockwise) orientation. Second, we assume that an OT-graph contains a sufficient number of edges to decide the order type of points using axioms described in the well-known book by Knuth [15]. It is easy to visualize the partitions of S for the edges of an OT-graph by drawing lines through them. This may result in a dense drawing, so we omit lines in the drawing if their partitions can be easily seen. For example, the OT-graph for the order type 1874 for 9 points from the database [2] shown in Fig. 2(b) has ten edges and only two lines are sufficient. The property of this graph (since it is an OT-graph) is that the orientation of any triple abc can be decided either (a) directly from the graph if there is an edge with both endpoints in $\{a,b,c\}$, or (b) algebraically using five axioms [15].

Comparison of OT-Graphs and Exit Graphs. Both exit graphs and OT-graphs can be used for visualizing order types of points. It is not sufficient for verifying an order type to just draw such graphs. For exit graphs, one needs to see the witness and the hourglass for every exit edge. For OT-graphs, one needs to see only the lines extending the edges. The hourglasses for exit graphs and the lines for OT-graphs are needed only when some triples of points are almost collinear.

Exit graphs and OT-graphs are also different in the following sense. For a given order type (as a point set), the exit graph is unique but OT-graphs are not since OT-graphs are defined using combinatorial axioms of Knuth [15]. Therefore we have an optimization problem of computing a minimum-size OT-graph for a given order type. We believe that this optimization problem is NP-hard. For example, we believe that the OT-graph shown in Fig. 2(b) has the least number

Order type 1874

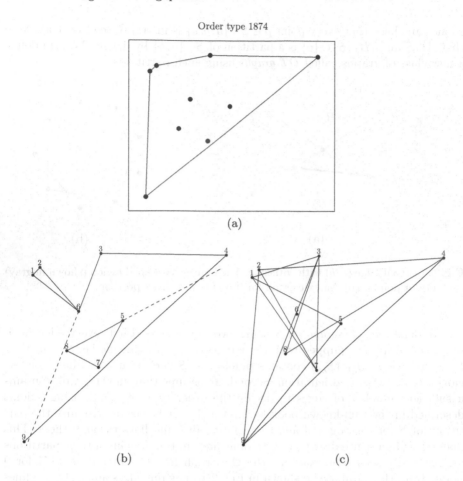

(a)

(b)   (c)

Fig. 2. (a) The order type 1874 for 9 points from the database [2]. (b) An OT-graph with 9 edges for the order type 1874 (several OT-graphs with 9 edges were computed by an extensive search). (c) The exit graph for the order type 1874.

of edges (9) for order type 1874 but we do not have a proof for it. Note that the OT-graph has 9 edges but the edge graph has 12, see Fig. 2(c) .

Identification of Order Types. Aichholzer *et al.* [3] suggested requirements for a graph representing an order type: "... we want to reduce the number of edges in the drawing as much as possible, but so that the order type remains uniquely identifiable." OT-graphs (including the set of edges and the corresponding partitions) characterize order types, i.e. each OT-graph corresponds to only one order type. Unfortunately, it does not hold for the exit graphs. As an example, Aichholzer *et al.* [3] constructed two sets each of 14 points[1] such that the exit edges are the same but the order types are different. With respect to minimiz-

---

[1] Using a pseudoline arrangement from [11].

ing the number of edges, we provide a comparative analysis of exit graphs and OT-graphs of all order types of up to 9 points in Sect. 6. Except few cases, OT-graphs have smaller number of edges. For example, Fig. 3 shows order type 1268 of 9 points where the exit graph has 15 edges but the OT-graph has only 8 edges. Furthermore, the OT-graph shown in Fig. 3(b) has non-crossing edges.

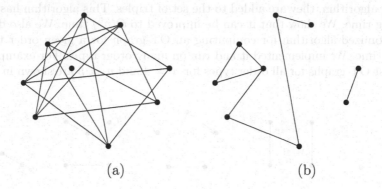

(a)   (b)

Fig. 3. The order type 1268 of 9 points represented as (a) the exit graph and (b) the OT-graph.

and we provide some algorithms for computing OT-graphs using combinatorial proofs and axioms in Sect. 5.

An interesting question is to find the smallest OT-graphs for points in convex position in the plane. Let c_n be the minimum number of edges in an OT-graph for n points in convex position.

Theorem 1. *For any $n \geq 4$, $c_n \leq \lfloor 2n/3 \rfloor$.*

It is interesting to find exact values of sequence c_n. We experimented with our randomized algorithm from Sect. 5 and conjecture that the bound in Theorem 1 is tight for all n up to 20. It is also interesting that the exit graph for n points in convex position has n edges, see Fig. 1(b) for an example.

Lower Bound. Another interesting question is to find the smallest OT-graph for an order type of n points in the plane. Based on our experiments, it is achieved for points in convex position if n is up to 9. Is it true for any n? One can argue that $\lceil n/4 \rceil$ is a lower bound for the number of edges in any OT-graph for n points in convex position. It is based on the fact that two consecutive points in the clockwise order along the boundary cannot be both isolated in an OT-graph.

Upper Bound. An obvious upper bound the smallest OT-graph for an order type of n points is $\binom{n}{2}$. We prove an upper bound in Sect. 4 which is smaller than $n^2/4$. The proof uses the idea of restricting the axioms in OT-graphs. Specifically, we prove the bound by using only Axioms 1, 2, and 3. Surprisingly, in this case, the smallest OT-graphs for any order type of n points have the same number of edges depending on n only.

Algorithms. For any set T triples with orientations, one can define its *CC-closure* $Cl(T)$ as the set of all triples that can be derived using Axioms 1–5. It is straight-forward to make an algorithm for testing in $O(n^5)$ time whether a set of triples T is the closure of itself, i.e. $Cl(T) = T$. This can be modified to an algorithm for computing the CC-closure for an OT-graph (i.e. the set of triples defined by G). The algorithm repeats the following step. If new triples are found in the testing algorithm, they are added to the set of triples. This algorithm has $O(n^8)$ running time. We show that it can be improved to $O(n^5)$ time. We also develop a randomized algorithm for computing an OT-graph for a given order type in $O(n^5)$ time. We implemented it and run on many order types. For example, the smallest OT-graphs for all order types for $n = 4$ and $n = 5$ are shown in Fig. 4.

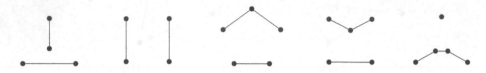

Fig. 4. Order types for $n = 4$ and $n = 5$.

Experiments. In Sect. 6 we provide experimental results using our algorithms on all order types of up to nine points in the plane. We also discuss a comparative analysis of exit graphs and OT-graphs using the size of the graphs.

Related Work. Order types are studied extensively, see for example the surveys [10,16]. Aichholzer *et al.* [4] studied representation of order types using radial orderings. Cabello [7] proved that the problem of deciding whether there is a planar straight-line embedding of a graph on a given set of points is NP-complete. Goaoc *et al.* [12] explored the application of the theory of limits of dense graphs to order types. The order types of random point sets were studied in [8,9] Goaoc and Welzl [13] studied convex hulls of random order types.

2   Preliminaries

Knuth [15] introduced and studied *CC-systems* (short for "counterclockwise systems") using properties of order types for up to five points. A *CC-system* for n points assigns true/false value for every ordered triple of points such that they satisfy the following axioms.

   Axiom 1 (cyclic symmetry). $pqr \implies qrp$.
   Axiom 2 (antisymmetry). $pqr \implies \neg prq$.
   Axiom 3 (nondegeneracy). Either pqr or prq.
   Axiom 4 (interiority). $tqr \wedge ptr \wedge pqt \implies pqr$.
   Axiom 5 (transitivity). $tsp \wedge tsq \wedge tsr \wedge tpq \wedge tqr \implies tpr$.

Any set of n points in general position in the plane induces a CC-system if we use the "counterclockwise" relation on the points. The converse is not true due to the 9-point theorem of Pappus [6,15]. When defining a graph for order types using partitions (by the lines extending the edges) one should be careful. For example, we can ask whether a given set of orientations of some triples can be extended somehow to a CC-system. If by "extended" we mean finding a CC-system such that the given orientations are preserved in the CC-system, then this problem is NP-complete. Knuth [15] proved that it is NP-complete to decide whether specified values of fewer than $\binom{n}{3}$ triples can be completed to a CC-system. We define OT-graphs using the extension of the given orientations by simply applying 5 axioms. Note that Axioms 1, 2, 4, and 5 imply some orientations. Axiom 3 also can be formulated as an implication:

Axiom 3' (nondegeneracy). $\neg pqr \implies prq$.

Definition 2. *Let G be a graph for a point set S in the plane and let T be the set of triples abc such that $ab, ac,$ or bc is an edge of G. Then G is the OT-graph if the orientation of every triple on S can be derived from T usings Axioms, 1,2,3',4, and 5.*

3   Convex Position

In this section, we explore OT-graphs for point sets in convex position and prove Theorem 1. Recall that c_n is the minimum number of edges in an OT-graph for n points in convex position. First, we prove that $c_n \leq n$ (Fig. 5).

Lemma 1. *Let S be a set of n points in convex position and let G be the graph (S, E) where E contains the edges of the convex hull of S. Then G is an OT-graph for S.*

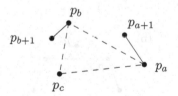

Fig. 5. Proof of Lemma 1.

Proof. Let $p_0, p_1, \ldots, p_{n-1}$ be the points of S in counterclockwise order. It suffices to prove that any triple $p_a p_b p_c$ with $0 \leq a < b < c \leq n-1$ has a CCW orientation. We prove it by induction on $m = \min\{b - a, c - b, a - c + n\}$. In the base case, $m = 1$. Then $(p_a, p_b), (p_b, p_c),$ or (p_c, p_a) is in E. Thus, p_a, p_b, p_c has a CCW orientation.

Suppose that $m > 1$ and $m = c - b$. Then $a + 1 < b$ and $b + 1 < c$. Edges (p_a, p_{a+1}) and (p_b, p_{b+1}) imply that triples $p_a p_{a+1} p_b, p_a p_{a+1} p_{b+1}, p_a p_{a+1} p_c$, and $p_a p_b p_{b+1}$ have a CCW orientation. By induction hypothesis, triple $p_a p_{b+1} p_c$ has a CCW orientation. By Axiom 5, $p_a p_b p_c$ has a CCW orientation.

Proof of Theorem 1. Let $p_0, p_1, \ldots, p_{n-1}$ be the points of S in counterclockwise order. We denote set $\{0, 1, \ldots, n-1\}$ by $[n]$.

First, suppose that $n = 3k$ for some $k \geq 2$. Consider a graph G with $2k$ edges as shown in Fig. 6. We prove that it is an OT-graph. By Lemma 1, it suffices to show that for any $i, j \in [n]$ with $j \neq i, i+1$ (modulo n), triple $p_i p_{i+1} p_j$ has a CCW orientation[2].

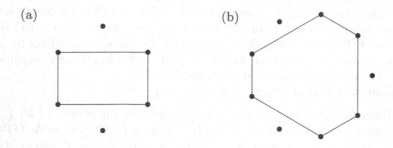

(a)   (b)

Fig. 6. OT-graphs for n points in convex position. (a) $n = 6$, (b) $n = 9$.

There are 3 cases to consider, see Fig. 7. Case (a) is clear since (p_i, p_{i+1}) is an edge of G. In Case (b), we can assume that $j \neq i+2, i-1$. Then it follows by Axiom 5 if we choose $t = p_{i+1}, s = p_{i+2}, p = p_j, q = p_{i-1}$, and $r = p_i$. In Case (c), we can assume that $j \neq i+2, i-1$. Knuth [15] proved that Axioms $1, 2, 3$, and 5 imply an axiom dual to Axiom 5.

Axiom 5' (dual transitivity). $stp \wedge stq \wedge str \wedge tpq \wedge tqr \implies tpr$.

Then Case (c) follows by Axiom 5' if we choose $t = p_i, s = p_{i-1}, p = p_{i+1}, q = p_{i+2}$, and $r = p_j$.

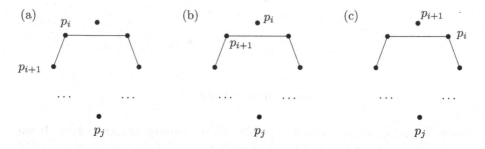

(a) p_i (b) p_i (c) p_{i+1}

p_{i+1} p_{i+1} p_i

...   .

p_j p_j p_j

Fig. 7. Proof of Theorem 1 for $n = 3k$.

Now, suppose that $n = 3k + 1$ for some $k \geq 2$. Consider a graph G with $2k$ edges as shown in Fig. 8(a). We prove that it is an OT-graph. By Lemma 1,

---

[2] This condition for a fixed i implies that (p_i, p_{i+1}) could be an edge in an OT-graph.

it suffices to show that for any $i, j \in [n]$ with $j \neq i, i+1$ (modulo n), tripe $p_i p_{i+1} p_j$ has a CCW orientation. If p_i or p_{i+1} is an isolated vertex in G then the argument is the same as in Case (b) and (c) for $n = 3k$, see Fig. 7(b) and (c). If (p_i, p_{i+1}) is an edge of G then $p_i p_{i+1} p_j$ has a CCW orientation. It remains to consider the case where (p_i, p_{i+1}) is one of two missing edges in the convex hull at the top, see Fig. 8(a). By symmetry, we assume that (p_i, p_{i+1}) is as shown in Fig. 8(b).

Suppose that vertex p_j has degree 2 in G. Let l be the length of path $p_j p_i$ in G. We show a CCW orientation of $p_i p_{i+1} p_j$ by induction on l. If $l = 1$ the orientation follows from edge $p_{i-1} p_i$ of G. If $l > 1$ then it follows by Axiom 5 if we choose $t = p_{i+1}, s = p_{i+2}, p = p_j, q = p_{j+1}$, and $r = p_i$. Note that $p_{i+1} p_j q$ has a CCW orientation since (p_j, q) is an edge of G. Also, $p_{i+1} q p_i$ has a CCW orientation by the induction hypothesis.

If vertex p_j is isolated in G then we choose p, q, r, s, t in the same way, see Fig. 8(c). Then triple tpq has a CCW orientation from the previous case (pq is an edge of convex hull). And triple tqr has a CCW orientation from the previous case (q has degree 2). By Axiom 5 tripe $p_i p_{i+1} p_j$ has a CCW orientation.

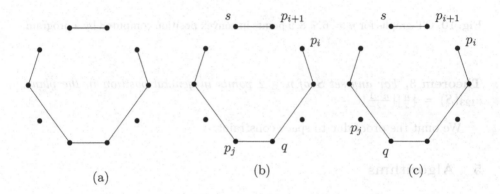

(a)   (b)   (c)

Fig. 8. Proof of Theorem 1 for $n = 3k + 1$.

Finally, suppose that $n = 3k + 2$ for some $k \geq 2$. Consider the graph shown in Fig. 9. It is an OT-graph by the same argument as for $n = 3k + 1$.   □

Remark. The OT-graphs presented in the proof of Theorem 1 are not unique. Our program finds also other graphs of the same size, see Fig. 10.

4   Axioms 1, 2, and 3 only

Let $e_{123}(S)$ be the minimum number of edges in an OT-graph for a set S of points in general position in the plane if only Axioms 1,2, and 3 are used. Surprisingly, for any set S of n points (i.e. for any order type), the smallest OT-graph *always* contains the same number of edges depending on n only.

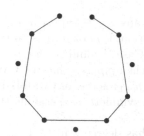

Fig. 9. OT-graph for $n = 3k + 2$.

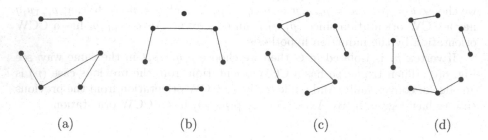

| (a) | (b) | (c) | (d) |

Fig. 10. OT-graphs for $n = 6, 7, 8, 9$ points in convex position computed by a program.

Theorem 3. *For any set S of $n \geq 2$ points in general position in the plane,* $e_{123}(S) = \lfloor \frac{n}{2} \rfloor \lfloor \frac{n-1}{2} \rfloor.$

We omit the proof due to space constraint.

5   Algorithms

Let $G = (S, E)$ be an OT-graph for a set S of n points in the plane. Let $T(G)$ be the set of triples abc such that $(a, b), (a, c)$, or (b, c) is an edge of G. Note that the orientation of abc is given by the partition of the corresponding edge. We define the *CC-closure* of G as the set all triples that can be proven by applying Axioms 1–5 from $T(G)$. Note that the *CC-closure* can be defined for any subset of triples of points with orientations.

Problem 4 (COMPUTINGCC-CLOSURE)

> **Given** *an OT-graph G.*
> **Compute** *the CC-closure of G.*

A naive approach to solve COMPUTINGCC-CLOSURE is to use an algorithm for testing CC-closure.

Problem 5 (TESTINGCC-CLOSURE)

Given *a set of triples with orientations for n points.*
Decide *whether a new triple can be derived using Axioms 1–5. If so, find a new triple using Axioms 1–5.*

By applying an algorithm for TESTINGCC-CLOSURE to $T(G)$ we can extend $T(G)$ (if possible) and solve COMPUTINGCC-CLOSURE. TESTINGCC-CLOSURE can be done in $O(n^5)$ time (since Axiom 5 requires 5 points). There are $\binom{n}{3}$ triples and, thus, the naive approach takes $O(n^8)$ time. We show that it can be done much faster.

Theorem 6. COMPUTINGCC-CLOSURE *can be solved in* $O(n^5)$ *time.*

Algorithm 1.

1. Make a list L_1 of all input triples with orientations (list L_1 stores all triples with known orientations). Copy $L_2 = L_1$.
2. While list L_2 is not empty, remove any triple abc from list L_2. Apply Axioms as follows. Find new triples using Axioms 1,2,3',4, and 5 such that triple abc is used in the condition of the axiom with the same orientation. If a new triple (i.e. not in L_1) is found, say pqr, then add it to L_1 and L_2.
3. Return list L_1.

Proof. To implement Algorithm 1 efficiently, we store triples of L_1 in a 3-dimensional array A_1. The value of $A_1[a, b, c]$ is *true/false* if abc has a CCW/CW orientation; otherwise $A_1[a, b, c] = null$. Using array A_1, we can decide in $O(1)$ time whether a triple is in list L_1 or not. Each triple abc is processed in Step 2 in $O(n^2)$ time since
(i) Axioms 1,2, and 3' can be applied at most one time,
(ii) Axiom 4 can be applied at most $n - 3$ times and
(iii) Axiom 5 can be applied at most $(n - 3)(n - 4)$ times.
Each triple is added to (and removed from) list L_2 at most one time. The number of triples removed from L_2 in Step 2 is $O(n^3)$. Therefore, the total time complexity of the algorithm is $O(n^5)$.
In our implementation of Algorithm 1, we do not maintain list L_1. Instead, we compute it in the end using array A_1.

The problem of computing the smallest OT-graph for a given order type seems complicated. Note that, if we restrict the axioms to Axioms 1,2, and 3 then a simple polynomial-time algorithm for computing the smallest OT-graph exists by Theorem 3 (by constructing two cliques). Next, we extend Algorithm 1 to a randomized algorithm for computing an OT-graph without increasing the running time. We incrementally add edges to a graph $G = (S, E)$ until G is an OT-graph for S. We store L_1, a list of triples abc such that $(a, b), (a, c)$, or (b, c) is an edge of G. Note that the orientation of abc can be computed using the coordinates of a, b, and c in $O(1)$ time. As in Algorithm 1, we have list L_2 which is useful for computing the CC-closure of G.

Algorithm 2.
Input: an order type given by a set of points S.
Output: an OT-graph G for S

1. Set $E = \emptyset$. Set $countCC = 0$, the number of triples in the CC-closure of $G = (S, E)$.
2. Compute list R of $\binom{n}{2}$ edges in the complete graph for S.
3. Initialize array $A_1[n, n, n]$ with entry values *null* and empty list L_2.
4. While $countCC < n(n-1)(n-2)$
   (a) Remove a random edge (a, b) from R.
   (b) If $A_1[a, b, c] \neq null$ for all $c \in S \setminus \{a, b\}$ then continue the "while" loop otherwise do the following steps (c) and (d).
   (c) Add (a, b) to list E. For each $c \in S \setminus \{a, b\}$ such that $A_a[a, b, c] \neq null$, add one of the triples (a, b, c) or (b, a, c) to list L_2 which has a CCW orientation.
   (d) Process list L_2 as in Algorithm 1.

Algorithm 2 (if repeated several times) can find the smallest OT-graph for a given order type, see for example Fig. 10. We also make a program that helps to verify the proof of an OT-graph. Note that a triple can be proven differently using Axioms 1–5. We develop a program for finding a human-readable proof. Once the best OT-graph for a given order type is found, the program computes a proof only for triples that require Axioms 4 and 5 (Axioms 1–3 are obvious). For example, Fig. 11 illustrates an OT-graph among all order type of 9 points and the format of the proof.

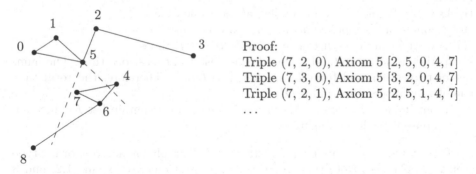

Proof:
Triple (7, 2, 0), Axiom 5 [2, 5, 0, 4, 7]
Triple (7, 3, 0), Axiom 5 [3, 2, 0, 4, 7]
Triple (7, 2, 1), Axiom 5 [2, 5, 1, 4, 7]
...

Fig. 11. An OT-graph for order type of 9 points and a part of the proof of it. The format for Axiom 4 is $[p, q, r, t]$ and the format for Axiom 5 is $[p, q, r, s, t]$.

Greedy Algorithm. Each iteration Algorithm 2 is quite fast (for relatively small n). However, it may require many runs to find small OT-graphs. Another possibility is a greedy algorithm where all possible edges for adding to the current graph are tested and the edge maximizing the size of the CC-closure is selected. Since the computation of the CC-closure takes $O(n^5)$ time, this approach is computationally expensive (it takes $O(n^7)$ time for selecting one edge and $O(n^9)$ for constructing the OT-graph). We developed a different greedy algorithm where the edge maximizing the size of the CC-closure using only Axioms 1,2,3 is selected. We found an implementation of this algorithm without increasing the running time, i.e. with running time $O(n^5)$. We add a new 2-dimensional array $C[..]$ for counting triples corresponding to the edges. Initially, $C[a, b] = n - 2$ for all pairs (a, b) of points $a \neq b$. Every time a new triple, say abc, is proven using Axioms we subtract one from $C[x, y]$ for all possible $x \neq y \in \{a, b, c\}$. Then, the greedy selection can be done by finding an edge (a, b) maximizing $C[a, b]$.

The total running time of this algorithm has two components. It is $O(n^5)$ time as in the randomized algorithm plus the total time for processing new array $C[..]$. There are $O(n^3)$ new triples and each triple requires $O(1)$ to update $C[..]$. This step takes $O(n^3)$ time in total. The computation of a new edge for G takes $O(n^2)$ time. Thus, the total time for computing the edges of $G = (S, E)$ is $O(mn^2)$ where $m = |E|$. Therefore, the total time for processing array $C[..]$ is $O(n^4)$. *Minimal OT-graphs.* When an OT-graph with m edges is computed, it can be checked for minimality. An OT-graph for some order type is *minimal* if removal of any edge results in a graph which not an OT-graph, i.e. its CC-closure does not contain all possible triples. This can be decided by applying the algorithm for COMPUTINGCC-CLOSURE m times.

6   Experiments

We implemented the randomized algorithm (Algorithm 2) and the greedy algorithm for computing OT-graphs. The programs are written in Java 8 using multi-threading and thread synchronization. We used a Linux server with 32 CPUs and 62 GB RAM to execute our program. We have computed the exit graphs and the OT-graphs on the database of order types [2] for $n = 3, 4, \ldots, 9$. To achieve current database and ensure the minimality of edges of OT-graphs, we run it around more than 3 days on the dataset. The results are shown in Table 1. Our experiments show that in many cases the greedy algorithm outperforms Algorithm 2 by the size of an OT-graph. Therefore, we iterate the greedy algorithm (with random tie-breaking) first and then iterate Algorithm 2 searching for a possible improvement. The number of iterations used for the greedy algorithm was 300,1200,10000 for $n = 7, 8, 9$ respectively. The number of iterations used for Algorithm 2 significantly larger (100000 for $n = 9$). About 70% of OT-graphs in Table 1 were computed using the greedy algorithm. The improvement achieved by Algorithm 2 was rather small: typically one edge reduction for an order type. The program implementing Algorithm 2 is still running and hopefully, new OT-graphs will be computed in a few months.

It is interesting that the smallest OT-graphs are achieved for 14 order types with $n = 6$, for 2 order types with $n = 7$, for 26 order types with $n = 8$, and for 124 order types with $n = 9$. There are only 2 order types for $n = 6$ whose

Table 1. OT-graphs for n up to 9. Column $i, i = 1, 2, \ldots, 11$ contains the number of OT-graphs with i edges.

n	1	2	3	4	5	6	7	8	9	10	11	Total
3	1											1
4		2										2
5			3									3
6				14	2							16
7			2	79	54							135
8				26	696	1,802	791					3,315
9				1	234	9,379	49,331	73,906	25,671	295		158,817

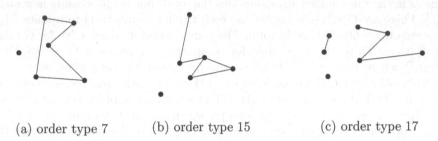

(a) order type 7   (b) order type 15   (c) order type 17

Fig. 12. Extreme OT-graphs for $n = 6$ and $n = 7$. (a),(b) Two order types for $n = 6$ maximizing the number of edges. (c) An order type for $n = 7$ (different from the convex case) minimizing the number of edges.

OT-graphs require 5 edges. They are shown in Fig. 12(a) and (b). The two order types for $n = 7$ that admit OT-graph with 4 edges are shown in Fig. 10(b) (the convex position) and in Fig. 12(c).

Let $\mu(n)$ be the minimum number of edges in an OT-graph for n points. Based on our experiments, we conjecture that $\mu(4) = 2$, $\mu(5) = 3$, $\mu(6) = \mu(7) = 4$, $\mu(8) = \mu(9) = 5$. This can be compared with exit graphs where the minimum number of edges is the same for $n = 5, 6, 7, 8$ but is larger for $n = 9$, see Fig. 13.

Figure 13(a), (c) shows the distribution of the graph sizes (OT-graph vs exit graph). Figure 13(b), (d) shows comparison of the graph sizes for each order type (the order types are sorted by the size of OT-graph and exit graph). Except one order type for $n = 8$ and 17 order types for $n = 9$, the OT-graphs are smaller that the exit graphs. For $n = 9$, the maximum size of OT-graph/exit graph is 11/16, respectively. The corresponding total number of edges is 1386819 for OT graphs and 1673757 for exit graphs which is 82.85%.

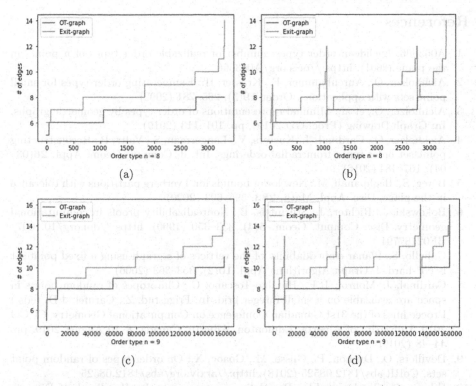

Fig. 13. The sizes of OT-graphs and exit graphs of order types for (a,b) $n = 8$ and (c,d) $n = 9$. The order types in (a) and (c) are sorted independently for OT-graphs and exit graphs. The functions in (b) and (d) use the same order type on the x-axis (the order types are sorted lexicographically).

7   Concluding Remarks

In this paper, we introduced new geometric graphs, OT-graphs, for visualizing order types in the plane. This new concept gives rise to many interesting questions. Is it true that the smallest size OT-graphs for all order types of n points are achieved for points in convex position? Is the bound in Theorem 1 tight?

In many cases there are different OT-graphs of minimum size for the same order type. One can use other criteria to optimize OT-graphs, for example, crossings. Figure 4 shows that there exist OT-graphs without crossings for all order types of 4 and 5 points. Theorem 1 shows that there are OT-graphs without crossings for points in convex position. Can it be generalized in this sense?

Finally, we plan to run our program on order types for larger values of n using the database of order types developed by Aichholzer, Aurenhammer, and Krasser [2]. It is a challenging problem since the number of order types grows as $2^{\Theta(n \log n)}$ [1,10] (there are 14,309,547 order types for $n = 10$.)

References

1. A063666: Euclidean order types: number of realizable order types of n points in the plane (2001). https://oeis.org/A063666
2. Aichholzer, O., Aurenhammer, F., Krasser, H.: Enumerating order types for small point sets with applications. Order **19**(3), 265–281 (2002)
3. Aichholzer, O., et al.: Minimal representations of order types by geometric graphs. In: Graph Drawing (Proc. GD 2019), pp. 101–113 (2019)
4. Aichholzer, O., Cardinal, J., Kusters, V., Langerman, S., Valtr, P.: Reconstructing point set order types from radial orderings. Int. J. Comput. Geom. Appl. **26**(03–04), 167–184 (2016)
5. Bereg, S., Haghpanah, M.: New lower bounds for Tverberg partitions with tolerance in the plane. Disc. Appl. Math. **283**, 596–603 (2020)
6. Bokowski, J., Richter, J., Sturmfels, B.: Nonrealizability proofs in computational geometry. Disc. Comput. Geom. **5**(4), 333–350 (1990). https://doi.org/10.1007/BF02187794
7. Cabello, S.: Planar embeddability of the vertices of a graph using a fixed point set is NP-hard. J. Graph Algorithms Appl. **10**(2), 353–363 (2006)
8. Cardinal, J., Monroy, R.F., Hidalgo-Toscano, C.: Chirotopes of random points in space are realizable on a small integer grid. In: Friggstad, Z., Carufel, J.D. (eds.) Proceedings of the 31st Canadian Conference on Computational Geometry, CCCG 2019, University of Alberta, Edmonton, Alberta, Canada, 8–10 August 2019, pp. 44–48 (2019)
9. Devillers, O., Duchon, P., Glisse, M., Goaoc, X.: On order types of random point sets. CoRR abs/1812.08525 (2018). http://arxiv.org/abs/1812.08525
10. Felsner, S., Goodman, J.E.: Pseudoline arrangements. In: Handbook of Discrete and Computational Geometry, pp. 125–157. Chapman and Hall/CRC (2017)
11. Felsner, S., Weil, H.: A theorem on higher Bruhat orders. Disc. Comput. Geom. **23**(1), 121–127 (2000)
12. Goaoc, X., Hubard, A., de Joannis de Verclos, R., Sereni, J., Volec, J.: Limits of order types. In: Arge, L., Pach, J. (eds.) 31st International Symposium on Computational Geometry, SoCG 2015, 22–25 June 2015, Eindhoven, The Netherlands. LIPIcs, vol. 34, pp. 300–314. Schloss Dagstuhl - Leibniz-Zentrum für Informatik (2015)
13. Goaoc, X., Welzl, E.: Convex hulls of random order types. In: Cabello, S., Chen, D.Z. (eds.) 36th International Symposium on Computational Geometry, SoCG 2020, LIPIcs, Zürich, Switzerland, 23–26 June 2020, vol. 164, pp. 49:1–49:15. Schloss Dagstuhl - Leibniz-Zentrum für Informatik (2020)
14. Goodman, J.E., Pollack, R.: Multidimensional sorting. SIAM J. Comput. **12**(3), 484–507 (1983)
15. Knuth, D.E. (ed.): Axioms and Hulls. LNCS, vol. 606. Springer, Heidelberg (1992). https://doi.org/10.1007/3-540-55611-7
16. Richter-Gebert, J., Ziegler, G.M.: Oriented matroids. In: Handbook of discrete and computational geometry, pp. 159–184. Chapman and Hall/CRC (2017)

FISSION: A Practical Algorithm for Computing Minimum Balanced Node Separators

Johannes Blum[✉][ORCID], Ruoying Li, and Sabine Storandt

University of Konstanz, Konstanz, Germany
johannes.blum@uni-konstanz.de

Abstract. Given an undirected graph, a balanced node separator is a set of nodes whose removal splits the graph into connected components of limited size. Balanced node separators are used for graph partitioning, for the construction of graph data structures, and for measuring network reliability. It is NP-hard to decide whether a graph has a balanced node separator of size at most k. Therefore, practical algorithms typically try to find small separators in a heuristic fashion. In this paper, we present a branching algorithm that for a given value k either outputs a balanced node separator of size at most k or certifies that k is a valid lower bound. Using this algorithm iteratively for growing values of k allows us to find a minimum balanced node separator. To make this algorithm scalable to real-world (road) networks of considerable size, we first describe pruning rules to reduce the graph size without affecting the minimum balanced separator size. In addition, we prove several structural properties of minimum balanced node separators which are then used to reduce the branching factor and search depth of our algorithm. We experimentally demonstrate the applicability of our algorithm to graphs with thousands of nodes and edges. Finally, we showcase the usefulness of having minimum balanced separators for judging the quality of existing heuristics, for improving preprocessing-based route planning techniques on road networks, and for lower bounding important graph parameters.

Keywords: Balanced node separator · Exact algorithm · Graph partitioning · Contraction hierarchies

1 Introduction

When dealing with large graphs or spatial network databases, it is often useful to partition the data into smaller components to make computations more efficient and to allow easier local data management. For many applications, as defining

© Springer Nature Switzerland AG 2020
W. Wu and Z. Zhang (Eds.): COCOA 2020, LNCS 12577, pp. 817–832, 2020.
https://doi.org/10.1007/978-3-030-64843-5_55

geographic districts [9], or for load balancing in parallel graph algorithms [21,29], it is beneficial if the resulting components are of roughly the same size or at least do not exceed some upper size bound. Moreover, it is often desired that densely connected regions end up in one component and only few nodes or edges are at the border between different components. One possibility to accomplish this is by (recursively) dividing the graph using balanced node separators.

Definition 1 (α-Balanced Node Separator). *Given a graph $G(V,E)$ and $\alpha \in [\frac{1}{2}, 1)$, a set $B \subseteq V$ is called an α-balanced node separator if after the removal of B from G all connected components have size at most $\alpha \cdot |V|$.*

The respective optimization goal is to find the smallest such B. We refer to the minimum size of an α-balanced node separator as b_α.

Balanced node separators are used in scientific computing [15], in VLSI design [2], and for network connectivity measures [19]. Furthermore, several modern route planning techniques for road networks rely on breaking down the input graph into smaller subgraphs [10,12], and for some of them the performance depends directly on the size of the chosen separators [3]. As it is NP-hard to find minimum balanced node separators in general graphs [8], practical algorithms typically compute separators in a heuristic fashion [11,18,25,26]. But without a scalable method to compute optimal solutions, their quality is difficult to judge.

We devise in this paper the first practical algorithm (called FISSION[1]) that computes minimum balanced node separators in large real-world networks. This then allows us to evaluate how close existing heuristics come to finding the best possible solution on different kinds of networks. In addition, the size of a minimum α-balanced node separator also yields a lower bound for several other important graph parameters, as e.g., the treewidth or the pathwidth [5]. There are many graph algorithms which are provably fast in case the graph exhibits small such parameter values. Hence knowing those values for a given graph allows to judge the performance of such parameterized algorithms a priori and provides structural insights. But unfortunately, the computation of the treewidth and the pathwidth is NP-hard as well, and existing exact algorithms and lower bounding techniques are limited to small and/or dense graphs. Our novel algorithm, however, is explicitly designed to perform well on sparse graphs. Figure 1 shows two examples of minimum balanced node separators computed with our approach.

---

[1] FISSION: A Framework for Improved Small Separator Identification.

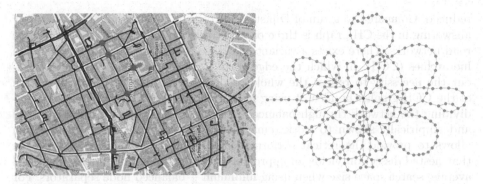

Fig. 1. Minimum balanced node separators (red nodes) computed with FISSION on a real-world road network (left) and a control-flow graph (right) using $\alpha = \frac{1}{2}$. (Color figure online)

1.1 Related Work

Theoretical Results on Balanced Separators. Finding an α-balanced node separator of size at most k is NP-hard even for $k \in \mathcal{O}(\log n)$ [13]. For constant values of k, the naive algorithm of testing all subsets of size up to k yields a polynomial time algorithm. However, a running time of $\Omega(n^k)$ where $n = |V|$ is still prohibitive for practical use. As there are no known polynomial time approximation algorithms, pseudo-approximations were investigated for this problem. In a pseudo-approximation, the balance constraint and the solution size are approximated simultaneously. In [20] it was shown that a slight relaxation of α allows to get a logarithmic approximation guarantee for b_α. For example, for $\alpha = \frac{2}{3}$ one can find a $\frac{3}{4}$-balanced node separator of size at most $\mathcal{O}(\log n) \cdot b_{2/3}$ in polynomial time. However, those techniques are not easily applicable to large real-world inputs. Hence in practice, heuristics are usually used to find small separators.

Practical Tools for Graph Partitioning. A multitude of heuristics was proposed in previous work to compute sensible graph partitions and concise (balanced) separators, as e.g. FlowCutter [18] or InertialFlow [27]. PUNCH (partitioning using natural cut heuristics) is a multi-step heuristic particularly designed to partition road networks [11] and also leverages flow computations. Software packages as KaFFPa(E) [24] and KaHIP [26] offer a wide range of general-purpose partitioning algorithms. Buffoon (featured in KaHIP) [23] is custom-tailored for road networks and uses PUNCH as a subroutine. We note that the quality of those partitioning heuristics is typically measured in comparison to other heuristics (e.g. as the percentage by which the size of the separator is smaller). This leaves the question how close their outputs are to the optimum wide open.

Separator-Based Route Planning. Contraction hierarchies (CH) [14] are currently one of the most widely used preprocessing-based route planning techniques for road networks. The preprocessing consists of the construction of a hierarchical shortcut graph, in which the hop distance of shortest paths is tremendously reduced. Compared to a run of Dijkstra's algorithm in the original graph, query answering in the CH-graph is three or more orders of magnitude faster on large road networks. There exists a variant of CH, so-called customizable contraction hierarchies (CCH), in which the edge costs can be dynamically changed without the necessity to repeat the whole preprocessing [12]. For the construction of the CCH-graph, the idea of using nested dissection, that is, recursively subdividing the network through balanced node separators, was introduced in [12] and empirically shown to work remarkably well. Moreover, this construction allows to provide theoretical performance guarantees. In [1,3], it was proven that nested dissection yields an approximation guarantee for the maximum and average search space size when using minimum $\frac{2}{3}$-balanced node separators. For practical construction, so far, only heuristics were used to identify good separators. Recently, a combination of FlowCutter and InertialFlow was proposed to enable efficient CCH preprocessing [17]. But with the use of heuristics, the theoretical performance guarantees are lost.

Bounding Techniques for Graph Parameters There exist several graph parameters which play an important role in the design of parameterized algorithms and graph analysis, as e.g. the treewidth tw, the pathwidth pw, or the treedepth td. Often, a low parameter value warrants an efficient algorithm for problems that are hard in general [4,22]. But as the computation of those parameters is NP-hard as well, upper and lower bounds have been extensively investigated. For the treewidth, lower bound techniques using e.g. contraction [7] or brambles [6] were proposed. Furthermore, the 2016 PACE challenge[2] focused on exact treewidth computation as well as algorithms for determining good upper bounds. For the treedepth, a recent paper [28] proposes an exact algorithm and an experimental evaluation on graphs with up to 78 nodes. In [5], it was proven that the following hierarchy of graph parameters holds for any $\alpha \in [\frac{1}{2}, 1)$: $b_\alpha \leq tw \leq pw \leq td$ Therefore, if we can compute b_α exactly (or determine a lower bound), we automatically derive a valid lower bound for the other parameters as well.

1.2   Contribution

The main contribution of this paper is the FISSION algorithm, which is the first method to compute minimum balanced node separators in large networks. The algorithm features a preprocessing phase in which the input graph size is significantly reduced, and a branching phase, in which for given $\alpha \in [\frac{1}{2}, 1)$ and $k \in \mathbb{N}$ either an α-balanced node separator of size at most k is computed; or it is certified that no such separator exists. We prove several interesting structural properties of balanced node separators which are then leveraged to make the preprocessing as well as the branching phase viable. In the experimental evaluation, we consider road networks of different size as well as a benchmark set of diverse graphs (with instances up to two orders of magnitude larger than the networks considered in [28]) to demonstrate the applicability of our algorithm

---

[2] https://pacechallenge.wordpress.com/pace-2016/track-a-treewidth/.

and to highlight its benefits and limits. Moreover, we evaluate the usefulness of FISSION for judging the quality of heuristics, for preprocessing-based route planning, and for lower bounding important graph parameters.

2 Properties of Minimum Balanced Node Separators

In this section, we prove several structural properties of balanced node separators in general graphs. While these might be of independent interest, we explicitly exploit them later to make our FISSION algorithm scalable to large networks.

Given an undirected, connected graph $G(V, E)$, we first investigate the relationship between structural characteristics of induced subgraphs of G and their possible share of separator nodes.

Theorem 1 *(Induced Subgraphs). Let $G'[V']$ be a subgraph of $G(V, E)$ induced by $V' \subset V$ with $|V'| \leq \alpha|V|$. Further let N be the set of nodes in $V \setminus V'$ with a neighbor in V'. Then there exists an α-balanced node separator B in G with $|B \cap V'| < |N|$.*

Proof. Assume for contradiction, that any minimum α-balanced node separator B in G contains at least $|N|$ nodes from V'. Then we can take any such separator and construct a new separator $B' := (B \setminus V') \cup N$. As $|B \setminus V'| \leq |B| - |N|$ by assumption, it follows $|B'| \leq |B|$. It remains to show that B' is also an α-balanced separator to reach a contradiction. Let $CC(B)$ be the node sets of the connected components in G after the removal of B (that by definition are all of size at most $\alpha \cdot |V|$), and $CC(B')$ the respective sets after the removal of B'. We define for each component $C \in CC(B)$ the node subsets $C_N = C \cap N$ and $C_{V'} = C \cap V'$. For each $C \in CC(B)$, there is now one new component $C' := C \setminus (V' \cup N)$ in $CC(B')$, if $C' \neq \emptyset$. We observe that $|C'| \leq |C|$ holds and hence C' adheres to the size bound. Moreover the union of all these components C' plus B' has to cover all elements in $V \setminus V'$. Therefore the only other components in $CC(B')$ have to be subsets of V'. But as for V' as a whole we have $|V'| \leq \alpha|V|$, none of those can exceed the size bound. Therefore B' is an α-balanced node separator that contains no nodes from V' but has size at most the size of B, leading to a contradiction to our initial assumption. □

Figure 2 illustrates the theorem on an example.

For the second property, we distinguish two types of nodes in a separator B: We say a node $v \in B$ is in the foam B_f of the separator, if all of its neighbors are either in B as well or end up in the same connected component after the removal of B. The other separator nodes are said to be in the separator core B_c. Note that foam nodes may be necessary to realize the α-bound on the size of the component, see Fig. 3. But they are also interchangeable with other nodes from the same (otherwise too large) component. So if the separator core is known, the foam can easily be added. Hence we now focus on the core.

Theorem 2. *Let B be a minimum α-balanced node separator in $G(V, E)$. Furthermore, let $v \in B_c$ be a separator node and $\Pi(v)$ be any subgraph of G that connects all neighbors of v. Then $B \cap \Pi(v) \neq \emptyset$.*

Proof. Assume that $B \cap \Pi(v) = \emptyset$. As v is contained in the separator core, we know v has two neighbors u and w in G, which end up in different connected components after the removal of B. However, if there exists a subgraph $\Pi(v)$ which connects all pairs of neighbors of v, but does not contain any separator node, all neighbors of v would end up in the same component, which yields a contradiction.   □

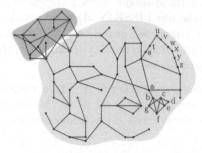

Fig. 2. The three red nodes separate the upper turquoise subgraph (which contains less than half of the graph nodes) from the remaining graph (green). Hence, there is a minimum $\frac{1}{2}$-balanced separator, which contains at most two nodes from the turquoise subgraph. (Color figure on;ine)

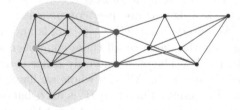

Fig. 3. Example graph with 16 nodes and a core separator of size two (red nodes). For $\alpha = \frac{1}{2}$, the left induced component (marked blue) would be too large as it contains 9 nodes. Selecting any node in that component as foam makes the resulting separator α-balanced. (Color figure on;ine)

3   The FISSION Algorithm

The FISSION algorithm consists of two phases: An initial preprocessing phase in which (i) the size of the input graph is reduced without compromising the size of an optimal α-balanced node separator, and (ii) cluster information are gained. The second phase then follows the branching paradigm to identify an α-balanced node separator of size k in case such a separator exists.

3.1   Preprocessing Phase

Based on the results from Sect. 2, we will now establish pruning rules which help to reduce the graph size before entering the branching phase of the algorithm. Furthermore, we will partition the reduced graph into different clusters using KaHIP [26] and precompute suitable cluster information which will further help to accelerate the branching phase.

Pruning. Our pruning rules either allow to delete some nodes and edges from the graph $G(V, E)$ completely without affecting the size of the minimum balanced node separator therein, or, they allow to merge certain subgraphs of G into single nodes. To account for the sizes of these subgraphs with respect to our balance constraint later on, all nodes will be assigned weights which indicate the number of nodes that were merged into it.

Pruning Rule 1 (Small Components). *Any connected component in G of size at most $\alpha \cdot |V|$ can be deleted.*

The connected components in the graph are easily detected using depth-first search (DFS), and their sizes can be computed along in time linear in the size of the graph. For any $\alpha \geq \frac{1}{2}$ at most one connected component of size $> \alpha \cdot |V|$ remains after this step.

Based on Theorem 1, the following property of minimum separators holds.

Corollary 1. *Every graph has a minimum α-balanced node separator that contains at most one node from a chain of up to $\alpha \cdot |V|$ nodes of degree two.*

This corollary allows to establish the following pruning rule.

Pruning Rule 2 (Degree-2 Chains). *Chains of degree-two nodes of length up to $\alpha \cdot |V|$ can be merged into a single node.*

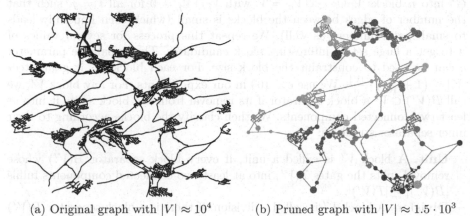

(a) Original graph with $|V| \approx 10^4$ (b) Pruned graph with $|V| \approx 1.5 \cdot 10^3$

Fig. 4. Figure (a) shows a real-world road network of an area in Germany. Figure (b) shows the pruned graph and the partition into 20 blocks. Each color represents a block. The size of a node indicates the number of nodes that were merged into it. (Color figure online)

This can be implemented by selecting a node of degree two which also has a neighboring node of degree two and to merge these nodes, assigning their summed weights to the merged node. This process can be repeated as long as such node pairs exist in the (reduced) graph. In Fig. 2, nodes $s, t, \ldots, y, z$ would be merged into a single degree-two node of weight 8.

For our third pruning rule, we consider cut nodes in the graph. A cut node in a connected graph $G(V, E)$ is a node $v \in V$ whose removal splits G into at least two connected components.

Pruning Rule 3 (Cut Components). *Let v be a cut node in $G(V, E)$ and let C be a connected component that remains after removing v. If the size of C is at most $\alpha \cdot |V|$ then C can be merged into v.*

In Fig. 2, node a is a cut node. The nodes b, c, d, e, f, g then form a component C which can be completely merged into a. Node a then receives a weight of 7. The set of cut nodes in the graph can be computed with a variant of DFS.

The pruning rules are applied in the given order. This only takes linear time in total. Nevertheless, the achieved graph size reduction might be tremendous, as shown for an example road-network in Fig. 4.

Clustering. In the reduced graph $G^*(V, E)$, we next try to detect clusters or so called blocks which exhibit a structure that can be exploited in the branching phase of the algorithm. More precisely, let $V' \subseteq V$ be a set of nodes, then a block is the subgraph of G^* induced by V'. A node $v \in V'$ is called a gate if there is an edge $(v, w) \in E$ with $w \in V \setminus V'$. With $T(V') \subset V'$, we denote the set of gates in block V'. According to Theorem 1, we know that an optimal α-balanced node separator in G^* contains at most $|T(V')|$ nodes from such a block. Hence our goal is to identify (large) blocks with few gates as this helps to test significantly fewer node combinations in our branching algorithm.

To compute such blocks, we use KaFFPa from KaHIP to partition the graph G^* into n blocks $V_i \cup \cdots \cup V_n = V$ with $V_i \cap V_j = \emptyset$ for all $i \neq j$, such that the number of edges between the blocks is small (which then hopefully leads to small gate numbers as well). We repeat this process for several values of n to get a large set of interesting block candidates. The imbalance parameter ϵ can be used to constraint the block size. For each block V_i, it guarantees $|V_i| \leq \left(1 + \frac{\epsilon}{100}\right) \cdot \frac{|V|}{n}$. We use $\epsilon = 100$ in our experiments. For any block V', we call $B(V') \subset V'$ a block separator if its removal from the block splits it into at least two connected components. We then classify the blocks according to their inner separator structure:

- **Unit.** A block V' is called a unit, if every block separator $B(V')$ whose removal splits the gates $T(V')$ into at least two connected components fulfils $|B(V')| \geq |T(V')|$.
- **Division.** A block V' is called a division, if there is a block separator $B(V')$ with $|B(V')| < |T(V')|$, whose removal splits the gates $T(V')$ into at least two connected components.

Algorithm 1: CompleteSeparator(B, v_i)

```
 1  if |B| > k then
 2  │   return false
 3  if N(v_i) is connected in G*[V \ B] then
 4  │   S ← nodes connecting N(v_i)
 5  │   for v_j ∈ S do
 6  │   │   if CompleteSeparator(B ∪ {v_j}, v_i) then
 7  │   │   │   return true
 8  else
 9  │   compute heaviest connected component X of G*[V \ B]
10  │   if X has weight at most α · |V| then
11  │   │   return true
12  │   else
13  │   │   S ← {v_j ∈ X | j > i}
14  │   │   for v_j ∈ S do
15  │   │   │   if CompleteSeparator(B ∪ {v_j}, v_j) then
16  │   │   │   │   return true
17  return false
```

3.2 Branching Phase

In general, a branching algorithm has two main ingredients: branching rules and reduction rules. Branching rules convert the problem to two or more smaller problems which are then solved recursively, while the reduction rules are used to simplify the (sub)problem, to avoid unnecessary branching, and to halt the algorithm. We will now describe how to design a branching algorithm for the problem of balanced node separator computation.

Basic Algorithm. In the branching phase we want to decide whether the pruned graph $G^*(V, E)$ has a node separator of size at most k, whose removal splits the graph into connected components of weight at most $\alpha \cdot |V|$.

In every step, our algorithm holds a node set B and attempts to complete it to a balanced separator of size at most k by adding a node from a candidate set S. Initially, we call the algorithm for every node $v_i \in V$ for $i = 1, \ldots, n$ with $B = \emptyset$. The algorithm then computes a candidate set S and branches on each node therein. Without dedicated branching and reduction rules, the candidate set S would simply be the set of all nodes not contained in B, and the algorithm would basically enumerate all possible node sets of size k to check whether they are balanced node separators, which would be impractical. Therefore, we next describe our rules that allow to significantly reduce the candidate set size in detail. Algorithm 1 depicts the high-level pseudo-code of our approach.

Branching Rules. The selection of a more concise candidate set S is performed as follows. Let v_i be the first node that was added to B and assume that we can complete $B = \{v_i\}$ to a balanced separator B^*, whose core contains v_i. Theorem 2 states that the neighborhood $N(v_i)$ of v_i cannot be connected in G^* after removing B^*. Hence if $G^*[V \setminus B]$ contains some subgraph H connecting $N(v_i)$, then we have $B^* \cap H \neq \emptyset$. After selecting v, which we also call a *root node*, we therefore perform a breadth-first search (BFS) in every subsequent step to determine whether $N(v_i)$ is connected in $G^*[V \setminus B]$. If this is the case, we backtrack the paths connecting $N(v_i)$ and use the nodes on this path as the candidate set S. This step drastically reduces the candidate set especially in sparse graphs. If $N(v_i)$ is already disconnected in $G^*[V \setminus B]$, we have cut off some part of the graph and can check whether the resulting connected components are already balanced or whether we can make them balanced by adding $k - |B|$ foam nodes to the separator. If this is not possible, we have to select a new root node v_j, whose neighborhood we try to disconnect in the following steps. To avoid generating the same separator several times, we only use v_j with $j > i$, i.e. only nodes with a higher index than the current root (using any fixed order on the node set). Hence, we get $S = \{v_j \in X \mid j > i\}$ as the candidate set where X is the remaining connected component violating the balance constraint.

Reduction Rules. Based on the blocks computed in the clustering phase, we can exclude certain sets B during the branching. Let B^* be a minimum α-balanced separator and consider some block U with gate set $T(U)$. If U is a unit, it follows from Theorem 1 that the core of B^* contains no node of $U \setminus T(U)$. Hence, we do not need to consider $U \setminus T(U)$ during the branching at all. If U is a division, Theorem 1 implies that we have either $B^* \cap U = T(U)$ or $|B^* \cap U| < |T(U)|$. This means that during the branching it suffices to consider sets B with $B \cap U = T(U)$ or $|B \cap U| < |T(U)|$. Especially in large divisions with small gate sets, this leads to a tremendous reduction of combinations to be checked.

Running Time. Under worst case assumptions, the branching phase of FISSION takes exponential time, as it might compute all $O(n^k)$ node subsets of size up to k and check for each whether it constitutes an α-balanced node separator. However, with the help of our pruning and reduction rules as well as the precomputed clusters, we can avoid the enumeration of many unnecessary node subsets. The effectiveness will be empirically justified in our experiments.

4   Experimental Evaluation

We implemented the preprocessing in Python and the branching algorithm in C++. To calculate graph partitions for cluster computation, we used KaFFPa from KaHIP 2.10. KaHIP 2.10 was compiled using Clang 11.0.0 with OpenMPI 1.4.1. Experiments were conducted on a single core of an AMDRyzen Threadripper 1950× CPU (clocked at 2.2 GHz) with 128 GB main memory.

4.1   Data and Settings

We use two benchmark data sets in our evaluation:

OSM Road Networks. The main focus of our work is to enable optimal balanced node separator computation in road networks. For validation, we extracted real-world road networks of varying size from the OpenStreetMap project[3]. The respective graph details are summarized in Table 1.

Table 1. Number of nodes and edges of our OSM benchmark data (rectangular cutouts of the German road network). Each graph consists of one connected component.

Name	$|V|$	$|E|$
OSM1	1000	2065
OSM2	2500	5209
OSM3	4676	9880
OSM4	9200	18970

PACE 2016 Graphs. For a structurally diverse set of graphs, we additionally consider the 147 public instances of the Parametrized Algorithms and Computational Experiments challenge of 2016. The challenge was concerned with exact and heuristic computation of the graph parameter treewidth. The graph set contains wireless sensor networks, instances from the DIMACS graph colouring challenge, and social networks among others. The instances are further categorized as easy, medium, hard, and random. Table 2 provides a quick overview. The graph sizes range from graphs with 9 nodes to graphs with over 30.000 nodes, with different edge densities. The time limit was set to 30 min per instance in the challenge.

Table 2. Distribution of the 147 PACE graphs.

Type	Easy	Medium	Hard	Random
Exact	50	5	5	9
Heuristic	50	16	2	10

FISSION works for any balance-factor $\alpha \in \left[\frac{1}{2}, 1\right)$. For evaluation, we focus on $\alpha = \frac{1}{2}$ and $\alpha = \frac{2}{3}$ as those are the most commonly used parameter choices in practical as well as theoretical works.

---

[3] openstreetmap.org.

4.2 Preprocessing Results

We first want to evaluate the effectiveness of our pruning rules. Subsequently, we discuss how much cluster information was gained in the preprocessing phase.

For each of our 151 input graphs (4 OSM graphs + 147 PACE graph), the reduced graph could be computed in well beyond a second. In Table 3, the pruning results for the OSM graphs are given. We see a significant reduction in the number of nodes and edges for all four graphs. For the smallest graph (OSM1), this is most pronounced but also for the larger networks we observe a reduction by 80% on average.

For the PACE graphs, we get a more diverse picture. For some of the graphs, we have no size reduction at all. For others, less than one percent of the nodes is still present in the reduced graph. On average, the number of nodes was reduced by 33% and the number of edges by 30%. For both, OSM and PACE graphs, the pruning results for $\alpha = \frac{1}{2}$ and $\alpha = \frac{2}{3}$ were almost identical. This indicates that our pruning rules were not limited by the requirement that nodes cannot be merged if the resulting weight exceeds $\alpha \cdot |V|$.

Regarding the clustering step, running times were typically beyond a second as well but peaked at around 6 seconds for some of the PACE graphs. For the OSM graphs, the number of identified unit and division blocks can also be found in Table 3, and a visualization of a partitioning of OSM4 in Fig. 4.

4.3 Branching Results

Our branching algorithm decides for a given value of $k \in \mathbb{N}$ whether an α-balanced node separator of size k exists or not. Hence to find the minimum α-balanced node separator, we need to feed the algorithm the correct value $k = b_\alpha$ and also certify that no smaller value exists by testing $k = b_\alpha - 1$. As we do not know the value of b_α a priori, we simply test values $k = 1, 2, \ldots$ until we detect the smallest value k for which we get a "true" answer from our branching algorithm. The respective k value is then automatically the minimum balanced node separator size for the chosen α, and we of course can also return a separator node set of that size.

Table 3. Preprocessing results for the OSM graphs and $\alpha = \frac{1}{2}$. The first two columns describe the percentage of nodes and edges of the original graph that remain after pruning. The last two columns indicate the number of interesting blocks that were identified.

Name	Pruning %V	%E	Partitioning # units	# divisions
OSM1	0.8	0.4	2	0
OSM2	22.4	13.5	18	21
OSM3	22.0	13.1	43	44
OSM4	16.2	9.9	48	56

For all four OSM graphs as well as the 69 PACE exact graphs, we could compute the optimum solution quickly. For the PACE heuristic graphs, we set a time out of 30 min per instance. As the category indicates, exact computation of the treewidth was not necessarily expected to be possible in that time frame. We could compute the minimum $\frac{1}{2}$-balanced node separator for 28 easy instances, one medium instance, two random instances and both hard instances, hence in total 33 out of 78 instances. For $\alpha = \frac{2}{3}$, we could solve three more of the easy instances. The medium instances that could not be solved all have a very high edge density. This is disadvantageous for our algorithm, as large node degrees lead to large candidate set sizes.

For sparse graphs with small separators, though, our branching algorithm returns solutions often within a few seconds. Figure 5 shows the running time and the resulting separator size for the 106 instances for which we computed the minimum $\frac{1}{2}$-balanced node separator. For the remaining 45 instances, we at least get valid lower bounds for b_α by considering the largest value of k for which our algorithm returned "false".

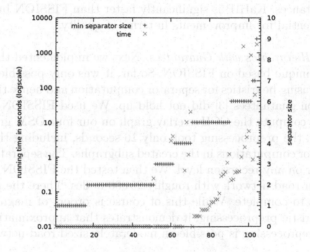

Fig. 5. Running times and separator sizes for the benchmark graphs on which FIS-SION computed the minimum balanced node separator for $\alpha = \frac{1}{2}$ within 30 min. The instances are sorted by the resulting separator size to better illustrate the correlation of separator size and running time.

5   Showcases

Finally, we want to briefly discuss some applications where being able to compute minimum α-balanced node separators with FISSION is beneficial.

Quality Assessment of Heuristics. Heuristics for α-balanced node separator computation are usually evaluated by comparing them to other heuristic approaches in terms of computed separator size and running time. But the gap between the computed separator size and minimum balanced separator size b_α cannot be judged this way. If b_α is known, however, the quality of heuristics can be assessed more distinctly. As there exist many different heuristics (that usually also come with a bunch of tuning parameters), it is beyond the scope of this paper to comprehensively analyze the quality of them all. But we used "node_separator" from KaHIP with 0-imbalance to compute upper bounds for $b_{1/2}$ on the PACE graphs and then compared them to the results of FISSION. For the 102 PACE instances for which we computed the minimum $\frac{1}{2}$-balanced separator size, the heuristic returned an optimal solution in 53 cases. For the remaining 49 instances, the minimum separator contained on average 2.6 fewer nodes than the solution returned by the heuristic, with a maximum difference of 20 on an instance where we certified $b_\alpha = 6$. On the exact instances, the running time of FISSION and KaHIP was very similar. Hence for the 25 exact instances where FISSION computed a smaller separator than KaHIP, it was the overall better algorithm. For the other instances, KaHIP is significantly faster than FISSION but our results show the potential for improvements in terms of quality.

Contraction Hierarchies with Guarantees. Next we implemented the CCH route planning technique based on FISSION. So far, it was only possible to construct CCH-graphs using heuristics for separator computation and hence the theoretical approximation guarantees [3] did not hold up. We used FISSION on all recursion levels to compute the CCH-overlay graph on our four OSM graphs. On the OSM4 graph, the preprocessing took only 16 seconds, including the 1347 minimum separator computations in the created subgraphs. The separator sizes never exceeded four on any recursion level. We then tested the FISSION algorithm on an even larger road network with roughly 25 000 nodes. There the preprocessing took 6 hours to complete. While this of course is orders of magnitudes slower than the heuristic preprocessing, it demonstrates that approximation guarantee preserving preprocessing is possible on moderately sized road networks.

Lower Bounds for Graph Parameters. As a last application, we use the separator sizes computed with FISSION to lower bound the treewidth tw, the pathwidth pw and the treedepth td of the graph by exploiting $b_\alpha \leq tw \leq pw \leq td$. In our evaluation, we focus on the comparison to another lower bounding technique for tw that computes the so called minor-min-width of the graph (see [16] for details). For the exact instances, we got a larger lower bound for 24 instances and a matching lower bound for 19 instances. For the heuristic instances, we got a better bound for 20 instances and a matching bound for 8 instances. This demonstrates that FISSION is able to produce meaningful lower bounds. Furthermore, as far as we are aware, it is also the first approach to give non-trivial lower bounds for b_α itself.

6   Conclusions and Future Work

We introduced FISSION to decide the NP-hard problem whether a graph exhibits an α-balanced node separator of a given size k in practice. For sparse graphs and small values of k our approach produces the answer quickly based on our carefully designed pruning and branching rules. Future work should explore how to improve the scalability of our algorithm further. For example, the existence of almost tight upper bounds could help to reduce the number of k-values that have to be tested. Furthermore, the development of lower bounding techniques would permit to turn our branching algorithm into a branch-and-bound algorithm that could abort the exploration of branches if the current node set size plus the lower bound on the number of nodes needed to complete it to a balanced separator exceeds k.

References

1. Bauer, R., Columbus, T., Rutter, I., Wagner, D.: Search-space size in contraction hierarchies. Theor. Comput. Sci. **645**, 112–127 (2016)
2. Bhatt, S.N., Leighton, F.T.: A framework for solving VLSI graph layout problems. J. Comput. Syst. Sci. **28**(2), 300–343 (1984)
3. Blum, J., Storandt, S.: Lower bounds and approximation algorithms for search space sizes in contraction hierarchies. In: European Symposium on Algorithms. LIPIcs, vol. 173, pp. 20:1–20:14. Schloss Dagstuhl - Leibniz-Zentrum für Informatik (2020). https://doi.org/10.4230/LIPIcs.ESA.2020.20
4. Bodlaender, H.L.: Dynamic programming on graphs with bounded treewidth. In: Lepistö, T., Salomaa, A. (eds.) ICALP 1988. LNCS, vol. 317, pp. 105–118. Springer, Heidelberg (1988). https://doi.org/10.1007/3-540-19488-6_110
5. Bodlaender, H.L., Gilbert, J.R., Hafsteinsson, H., Kloks, T.: Approximating treewidth, pathwidth, frontsize, and shortest elimination tree. J. Algorithms **18**(2), 238–255 (1995)
6. Bodlaender, H.L., Grigoriev, A., Koster, A.M.: Treewidth lower bounds with brambles. Algorithmica **51**(1), 81–98 (2008)
7. Bodlaender, H.L., Koster, A.M.C.A., Wolle, T.: Contraction and treewidth lower bounds. In: Albers, S., Radzik, T. (eds.) ESA 2004. LNCS, vol. 3221, pp. 628–639. Springer, Heidelberg (2004). https://doi.org/10.1007/978-3-540-30140-0_56
8. Bui, T.N., Jones, C.: Finding good approximate vertex and edge partitions is NP-hard. Inf. Process. Lett. **42**(3), 153–159 (1992)
9. Cohen-Addad, V., Klein, P.N., Young, N.E.: Balanced centroidal power diagrams for redistricting. In: Proceedings of the 26th ACM SIGSPATIAL International Conference on Advances in Geographic Information Systems, pp. 389–396 (2018)
10. Delling, D., Goldberg, A.V., Pajor, T., Werneck, R.F.: Customizable route planning. In: Pardalos, P.M., Rebennack, S. (eds.) SEA 2011. LNCS, vol. 6630, pp. 376–387. Springer, Heidelberg (2011). https://doi.org/10.1007/978-3-642-20662-7_32
11. Delling, D., Goldberg, A.V., Razenshteyn, I., Werneck, R.F.: Graph partitioning with natural cuts. In: 2011 IEEE International Parallel & Distributed Processing Symposium, pp. 1135–1146. IEEE (2011)

12. Dibbelt, J., Strasser, B., Wagner, D.: Customizable contraction hierarchies. J. Exp. Algorithmics (JEA) **21**, 1–49 (2016)
13. Feige, U., Mahdian, M.: Finding small balanced separators. In: Proceedings of the Thirty-Eighth Annual ACM Symposium on Theory of Computing, pp. 375–384 (2006)
14. Geisberger, R., Sanders, P., Schultes, D., Vetter, C.: Exact routing in large road networks using contraction hierarchies. Transp. Sci. **46**(3), 388–404 (2012)
15. George, A.: Nested dissection of a regular finite element mesh. SIAM J. Numer. Anal. **10**(2), 345–363 (1973)
16. Gogate, V., Dechter, R.: A complete anytime algorithm for treewidth. arXiv preprint arXiv:1207.4109 (2012)
17. Gottesbüren, L., Hamann, M., Uhl, T.N., Wagner, D.: Faster and better nested dissection orders for customizable contraction hierarchies. Algorithms **12**(9), 196 (2019)
18. Hamann, M., Strasser, B.: Graph bisection with pareto optimization. J. Exp. Algorithmics (JEA) **23**, 1–34 (2018)
19. Kratsch, D., Kloks, T., Muller, H.: Measuring the vulnerability for classes of intersection graphs. Disc. Appl. Math. **77**(3), 259–270 (1997)
20. Leighton, F.T., Rao, S.: An approximate max-flow min-cut theorem for uniform multicommodity flow problems with applications to approximation algorithms. In: Symposium on Foundations of Computer Science, pp. 422–431. IEEE Computer Society (1988). https://doi.org/10.1109/SFCS.1988.21958
21. Li, L., et al.: A simple yet effective balanced edge partition model for parallel computing. Proc. ACM Meas. Anal. Comput. Syst. **1**(1), 1–21 (2017)
22. Planken, L.R., de Weerdt, M.M., van der Krogt, R.P.: Computing all-pairs shortest paths by leveraging low treewidth. J. Artif. Intell. Res. **43**, 353–388 (2012)
23. Sanders, P., Schulz, C.: Distributed evolutionary graph partitioning. In: 2012 Proceedings of the Fourteenth Workshop on Algorithm Engineering and Experiments (ALENEX), pp. 16–29. SIAM (2012)
24. Sanders, P., Schulz, C.: High quality graph partitioning. Graph Partitioning Graph Clustering **588**(1), 1–17 (2012)
25. Sanders, P., Schulz, C.: Think locally, act globally: highly balanced graph partitioning. In: International Symposium on Experimental Algorithms. pp. 164–175. Springer (2013). https://doi.org/10.1007/978-3-642-38527-8_16
26. Sanders, P., Schulz, C.: Kahip-karlsruhe high quality partitioning (2019)
27. Schild, A., Sommer, C.: On balanced separators in road networks. In: Bampis, E. (ed.) SEA 2015. LNCS, vol. 9125, pp. 286–297. Springer, Cham (2015). https://doi.org/10.1007/978-3-319-20086-6_22
28. Trimble, J.: An algorithm for the exact treedepth problem. In: 18th International Symposium on Experimental Algorithms (2020)
29. Zeng, J., Yu, H.: A study of graph partitioning schemes for parallel graph community detection. Parallel Comput. **58**, 131–139 (2016)

Author Index

Printed in the United States
By Bookmasters